Boolean Models and Methods in Mathematics, Computer Science, and Engineering

This collection of papers presents a series of in-depth examinations of a variety of advanced topics related to Boolean functions and expressions. The chapters are written by some of the most prominent experts in their respective fields and cover topics ranging from algebra and propositional logic to learning theory, cryptography, computational complexity, electrical engineering, and reliability theory. Beyond the diversity of the questions raised and investigated in different chapters, a remarkable feature of the collection is the common thread created by the fundamental language, concepts, models, and tools provided by Boolean theory. Many readers will be surprised to discover the countless links between seemingly remote topics discussed in various chapters of the book. This text will help them draw on such connections to further their understanding of their own scientific discipline and to explore new avenues for research.

Dr. Yves Crama is Professor of Operations Research and Production Management and former Dean of the HEC Management School of the University of Liège, Belgium. He is widely recognized as a prominent expert in the field of Boolean functions, combinatorial optimization, and operations research, and he has coauthored more than 70 papers and 3 books on these subjects. Dr. Crama is a member of the editorial board of *Discrete Optimization, Journal of Scheduling*, and *4OR – The Quarterly Journal of the Belgian, French and Italian Operations Research Societies.*

The late Peter L. Hammer (1936–2006) was a Professor of Operations Research, Mathematics, Computer Science, Management Science, and Information Systems at Rutgers University and the Director of the Rutgers University Center for Operations Research (RUTCOR). He was the founder and editor-in-chief of the journals *Annals of Operations Research, Discrete Mathematics, Discrete Applied Mathematics, Discrete Optimization*, and *Electronic Notes in Discrete Mathematics*. Dr. Hammer was the initiator of numerous pioneering investigations of the use of Boolean functions in operations research and related areas, of the theory of pseudo-Boolean functions, and of the logical analysis of data. He published more than 240 papers and 19 books on these topics.

ENCYCLOPEDIA OF MATHEMATICS AND ITS APPLICATIONS

The titles below, and earlier volumes in the series, are available from booksellers or from Cambridge University Press at www.cambridge.org.

100 E. Olivieri and M. Eulália Vares *Large Deviations and Metastability*
101 A. Kushner, V. Lychagin and V. Rubtsov *Contact Geometry and Nonlinear Differential Equations*
102 L. W. Beineke and R. J. Wilson (eds.) with P. J. Cameron *Topics in Algebraic Graph Theory*
103 O. J. Staffans *Well-Posed Linear Systems*
104 J. M. Lewis, S. Lakshmivarahan and S. K. Dhall *Dynamic Data Assimilation*
105 M. Lothaire *Applied Combinatorics on Words*
106 A. Markoe *Analytic Tomography*
107 P. A. Martin *Multiple Scattering*
108 R. A. Brualdi *Combinatorial Matrix Classes*
109 J. M. Borwein and J. D. Vanderwerff *Convex Functions*
110 M.-J. Lai and L. L. Schumaker *Spline Functions on Triangulations*
111 R. T. Curtis *Symmetric Generation of Groups*
112 H. Salzmann et al. *The Classical Fields*
113 S. Peszat and J. Zabczyk *Stochastic Partial Differential Equations with Lévy Noise*
114 J. Beck *Combinatorial Games*
115 L. Barreira and Y. Pesin *Nonuniform Hyperbolicity*
116 D. Z. Arov and H. Dym *J-Contractive Matrix Valued Functions and Related Topics*
117 R. Glowinski, J.-L. Lions and J. He *Exact and Approximate Controllability for Distributed Parameter Systems*
118 A. A. Borovkov and K. A. Borovkov *Asymptotic Analysis of Random Walks*
119 M. Deza and M. Dutour Sikirić *Geometry of Chemical Graphs*
120 T. Nishiura *Absolute Measurable Spaces*
121 M. Prest Purity *Spectra and Localisation*
122 S. Khrushchev *Orthogonal Polynomials and Continued Fractions*
123 H. Nagamochi and T. Ibaraki *Algorithmic Aspects of Graph Connectivity*
124 F. W. King Hilbert *Transforms I*
125 F. W. King Hilbert *Transforms II*
126 O. Calin and D.-C. Chang *Sub-Riemannian Geometry*
127 M. Grabisch et al. *Aggregation Functions*
128 L. W. Beineke and R. J. Wilson (eds.) with J. L. Gross and T. W. Tucker *Topics in Topological Graph Theory*
129 J. Berstel, D. Perrin and C. Reutenauer *Codes and Automata*
130 T. G. Faticoni *Modules over Endomorphism Rings*
131 H. Morimoto *Stochastic Control and Mathematical Modeling*
132 G. Schmidt *Relational Mathematics*
133 P. Kornerup and D. W. Matula *Finite Precision Numbers Systems and Arithmetic*

ENCYCLOPEDIA OF MATHEMATICS AND ITS APPLICATIONS

Boolean Models and Methods in Mathematics, Computer Science, and Engineering

Edited by

YVES CRAMA

Université de Liège

PETER L. HAMMER

CAMBRIDGE
UNIVERSITY PRESS

CAMBRIDGE UNIVERSITY PRESS
Cambridge, New York, Melbourne, Madrid, Cape Town,
Singapore, São Paulo, Delhi, Mexico City

Cambridge University Press
32 Avenue of the Americas, New York, NY 10013-2473, USA

www.cambridge.org
Information on this title: www.cambridge.org/9780521847520

First published 2010
Reprinted 2012 (twice)

A catalog record for this publication is available from the British Library.

Library of Congress Cataloging in Publication Data

Boolean models and methods in mathematics, computer science, and engineering /
edited by Yves Crama, Peter L. Hammer.
p. cm. – (Encyclopedia of mathematics and its applications ; 134)
Includes bibliographical references and index.
ISBN 978-0-521-84752-0
1. Algebra, Boolean. 2. Probabilities. I. Crama, Yves, 1958–
II. Hammer, P. L., 1936– III. Title. IV. Series.
QA10.3.B658 2010
511.3′24–dc22 2010017816

ISBN 978-0-521-84752-0 Hardback

Contents

Preface

Boolean models and methods play a fundamental role in the analysis of a broad diversity of situations encountered in various branches of science.

The objective of this collection of papers is to highlight the role of Boolean theory in a number of such areas, ranging from algebra and propositional logic to learning theory, cryptography, computational complexity, electrical engineering, and reliability theory.

The chapters are written by some of the most prominent experts in their fields and are intended for advanced undergraduate or graduate students, as well as for researchers or engineers. Each chapter provides an introduction to the main questions investigated in a particular field of science, as well as an in-depth discussion of selected issues and a survey of numerous important or representative results. As such, the collection can be used in a variety of ways: some readers may simply skim some of the chapters in order to get the flavor of unfamiliar areas, whereas others may rely on them as authoritative references or as extensive surveys of fundamental results.

Beyond the diversity of the questions raised and investigated in different chapters, a remarkable feature of the collection is the presence of an "Ariane's thread" created by the common language, concepts, models, and tools of Boolean theory. Many readers will certainly be surprised to discover countless links between seemingly remote topics discussed in various chapters of the book. It is hoped that they will be able to draw on such connections to further their understanding of their own scientific disciplines and to explore new avenues for research.

The collection intends to be a useful companion and complement to the monograph by Yves Crama and Peter L. Hammer, *Boolean Functions: Theory, Algorithms, and Applications*. Cambridge University Press, Cambridge, U.K., 2010, which provides the basic concepts and theoretical background for much of the material handled here.

Introduction

The first part of the book, "Algebraic Structures," deals with compositions and decompositions of Boolean functions.

A set F of Boolean functions is called *complete* if every Boolean function is a composition of functions from F; it is a *clone* if it is composition-closed and contains all projections. In 1921, E. L. Post found a completeness criterion, that is, a necessary and sufficient condition for a set F of Boolean functions to be complete. Twenty years later, he gave a full description of the lattice of Boolean clones. Chapter 1, by Reinhard Pöschel and Ivo Rosenberg, provides an accessible and self-contained discussion of "Compositions and Clones of Boolean Functions" and of the classical results of Post.

Functional decomposition of Boolean functions was introduced in switching theory in the late 1950s. In Chapter 2, "Decomposition of Boolean Functions," Jan C. Bioch proposes a unified treatment of this topic. The chapter contains both a presentation of the main structural properties of modular decompositions and a discussion of the algorithmic aspects of decomposition.

Part II of the collection covers topics in logic, where Boolean models find their historical roots.

In Chapter 3, "Proof Theory," Alasdair Urquhart briefly describes the more important proof systems for propositional logic, including a discussion of equational calculus, of axiomatic proof systems, and of sequent calculus and resolution proofs. The author compares the relative computational efficiency of these different systems and concludes with a presentation of Haken's classical result that resolution proofs have exponential length for certain families of formulas.

The issue of the complexity of proof systems is further investigated by John Franco in Chapter 4, "Probabilistic Analysis of Satisfiability Algorithms." Central questions addressed in this chapter are: How efficient is a particular algorithm when applied to a random satisfiability instance? And what distinguishes "hard" from "easy" instances? Franco provides a thorough analysis of these questions, starting with a presentation of the basic probabilistic tools and models and covering advanced results based on a broad range of approaches.

In Chapter 5, "Optimization Methods in Logic," John Hooker shows how mathematical programming methods can be applied to the solution of Boolean inference and satisfiability problems. This line of research relies on the interpretation of the logical symbols 0 and 1 as numbers, rather than meaningless symbols. It leads both to fruitful algorithmic approaches and to the identification of tractable classes of problems.

The remainder of the book is devoted to applications of Boolean models in various fields of computer science and engineering, starting with "Learning Theory and Cryptography" in Part III.

In Chapter 6, "Probabilistic Learning and Boolean Functions," Martin Anthony explains how an unknown Boolean function can be "correctly approximated," in a probabilistic sense, when the only available information is the value of the function on a random sample of points. Questions investigated here relate to the quality of the approximation that can be attained as a function of the sample size, and to the algorithmic complexity of computing the approximating function.

A different learning model is presented by Robert H. Sloan, Balázs Szörényi, and György Turán in Chapter 7, "Learning Boolean Functions with Queries." Here, the objective is to identify the unknown function *exactly* by asking questions about it. The efficiency of learning algorithms, in this context, depends on prior information available about the properties of the target function, about the type of representation that should be computed, about the nature of the queries that can be formulated, and so forth. Also, the notion of "efficiency" can be measured either by the number of queries required by the learning algorithm (*information complexity*) or by the total of amount of computational steps required by the algorithm (*computational complexity*). The chapter provides an introduction and surveys a large variety of results along these lines.

In Chapter 8, Claude Carlet provides a very complete overview of the use of "Boolean Functions for Cryptography and Error-Correcting Codes." Both cryptography and coding theory are fundamentally concerned with the transformation of binary strings into binary strings. It is only natural, therefore, that Boolean functions constitute a basic tool and object of study in these fields. Carlet discusses quality criteria that must be satisfied by error-correcting codes and by cryptographic functions (high algebraic degree, nonlinearity, balancedness, resiliency, immunity, etc.) and explains how these criteria relate to characteristics of Boolean functions and of their representations. He introduces several remarkable classes of functions such as bent functions, resilient functions, algebraically immune functions, and symmetric functions, and he explores the properties of these classes of functions with respect to the aforementioned criteria.

In Chapter 9, "Vectorial Boolean Functions for Cryptography," Carlet extends the discussion to functions with multiple outputs. Many of the notions introduced in Chapter 8 can be naturally generalized in this extended framework: families of representations, quality criteria, and special classes of functions are introduced and analyzed in a similar fashion.

Part IV concentrates on "Graph Representations and Efficient Computation Models" for Boolean functions.

Beate Bollig, Martin Sauerhoff, Detlef Sieling, and the late Ingo Wegener discuss "Binary Decision Diagrams" (BDDs) in Chapter 10. A BDD for function f is a directed acyclic graph representation of f that allows efficient computation of the value of $f(x)$ at any point x. Different types of BDDs can be defined by placing restrictions on the underlying digraph, by allowing probabilistic choices, and so forth. Questions surveyed in Chapter 10 are, among others: What is the size of a smallest BDD representation of a given function? How can a BDD be efficiently generated? How difficult is it to solve certain problems on Boolean functions (satisfiability, minimization, etc.) when the input is represented as a BDD?

Matthias Krause and Ingo Wegener discuss a different type of graph representations in Chapter 11, "Circuit Complexity." Boolean circuits provide a convenient model for the hardware realization of Boolean functions. Krause and Wegener describe efficient circuits for simple arithmetic operations, such as addition and multiplication. Further, they investigate the possibility of realizing arbitrary functions by circuits with small size or small depth. Although lower bounds or upper bounds on these complexity measures can be derived under various assumptions on the structure of the circuit or on the properties of the function to be represented, the authors also underline the existence of many fundamental open questions on this challenging topic.

Fourier transforms are a powerful tool of classical analysis. More recently, they have also proved useful for the investigation of complex problems in discrete mathematics. In Chapter 12, "Fourier Transforms and Threshold Circuit Complexity," Jehoshua Bruck provides an introduction to the basic techniques of Fourier analysis as they apply to the investigation of Boolean functions and neural networks. He explains, in particular, how they can be used to derive bounds on the size of the weights and on the depth of Boolean circuits consisting of threshold units.

The topic of "Neural Networks and Boolean Functions" is taken up again by Martin Anthony in Chapter 13. The author focuses first on the number and on the properties of individual threshold units, which can be viewed as linear, as nonlinear, or as "delayed" (*spiking*) threshold Boolean functions. He next discusses the expressive power of feed-forward artificial neural networks made up of threshold units.

Martin Anthony considers yet another class of graph representations in Chapter 14, "Decision Lists and Related Classes of Boolean Functions." A decision list for function f can be seen as a sequence of Boolean tests, the outcome of which determines the value of the function on a given point x. Every Boolean function can be represented as a decision list. However, when the type or the number of tests involved in the list is restricted, interesting subclasses of Boolean functions arise. Anthony investigates several such restrictions. He also considers the algorithmic complexity of problems on decision lists (recognition, learning, equivalence), and he discusses various connections between threshold functions and decision lists.

The last part of the book focuses on "Applications in Engineering."

Since the 1950s, electrical engineering has provided a main impetus for the development of Boolean logic. In Chapter 15, J.-H. Roland Jiang and Tiziano Villa survey the use of Boolean methods for "Hardware Equivalence and Property Verification." A main objective, in this area of system design, is to verify that a synthesized digital circuit conforms to its intended design. The chapter introduces the reader to the problem of formal verification, examines the complexity of different versions of equivalence checking ("given two Boolean circuits, decide whether they are equivalent"), and describes approaches to this problem. For the solution of these engineering problems, the authors frequently refer to models and methods covered in earlier chapters of the book, such as satisfiability problems or binary decision diagrams.

In Chapter 16, Tiziano Villa, Robert K. Brayton, and Alberto L. Sangiovanni-Vincentelli discuss the "Synthesis of Multilevel Boolean Networks." A multilevel representation of a Boolean function is a circuit representation, similar to those considered in Chapter 11 or in Chapter 13. From the engineering viewpoint, the objective of multilevel implementations is to minimize the physical area occupied by the circuit, to reduce its depth, to improve its testability, and so on. Villa, Brayton, and Sangiovanni-Vincentelli survey efficient heuristic approaches for the solution of these hard computational problems. They describe, in particular, factoring and division procedures that can be implemented in "divide-and-conquer" algorithms for multilevel synthesis.

The combinatorial structure of operating or failed states of a complex system can be reflected through a Boolean function, called the *structure function* of the system. The probability that the system operates is then simply the probability that the structure function takes value 1. In Chapter 17, Charles J. Colbourn explores in great detail the "Boolean Aspects of Network Reliability." He reviews several exact methods for reliability computations, based either on "orthogonalization" or decomposition, or on inclusion-exclusion and domination. He also explains the intimate, though insufficiently explored, connections between Boolean models and combinatorial simplicial complexes, as they arise in deriving bounds on system reliability.

Acknowledgments

The making of this book has been a long process, and it has benefited over the years from the help and advice provided by several individuals. The editors gratefully acknowledge the contribution of these colleagues to the success of the endeavor.

First and foremost, all chapter contributors are to be thanked for the quality of the material that they have delivered, as well as for their patience and understanding during the editorial process.

Several authors have contributed to the reviewing process by cross-reading each other's work. Additional reviews, suggestions, and comments on early versions of the chapters have been kindly provided by Endre Boros, Nadia Creignou, Tibor Hegedűs, Lisa Hellerstein, Toshi Ibaraki, Jörg Keller, Michel Minoux, Rolf Möhring, Vera Pless, Gabor Rudolf, Mike Saks, Winfrid Schneeweiss, and Ewald Speckenmeyer.

Special thanks are due to Endre Boros, who provided constant encouragement and tireless advice to the editors over the gestation period of the volume. Marty Golumbic gave a decisive push to the process by bringing most contributors together in Haifa, in January 2008, on the occasion of the first meeting on "Boolean Functions: Theory, Algorithms, and Applications." Terry Hart provided the efficient administrative assistance that allowed the editors to keep track of countless mail exchanges.

Finally, I must thank my mentor, colleague, and friend, Peter L. Hammer, for helping me launch this ambitious editorial project, many years ago. Unfortunately, Peter did not live to see the outcome of our joint efforts. I am sure that he would have loved it, and that he would have been very proud of this contribution to the dissemination of Boolean models and methods.

Yves Crama
Liège, Belgium, January 2010

Contributors

Martin Anthony
Department of Mathematics
London School of Economics and Political Science, UK

Jan C. Bioch
Department of Econometrics
Erasmus University Rotterdam, The Netherlands

Beate Bollig
Department of Computer Science
Technische Universität Dortmund, Germany

Robert K. Brayton
Department of Electrical Engineering & Computer Sciences
University of California at Berkeley, USA

Jehoshua Bruck
Computation and Neural Systems and Electrical Engineering
California Institute of Technology, USA

Claude Carlet
Department of Mathematics
University of Paris 8, France

Charles J. Colbourn
Computer Science and Engineering
Arizona State University, USA

John Franco
Department of Computer Science
University of Cincinnati, USA

John Hooker
Tepper School of Business
Carnegie Mellon University, USA

J.-H. Roland Jiang
Department of Electrical Engineering
National Taiwan University, Taiwan

Matthias Krause
Theoretical Computer Science
Mannheim University, Germany

Reinhard Pöschel
Institut für Algebra
Technische Universität Dresden, Germany

Ivo Rosenberg
Département de Mathématiques et de Statistique
Université de Montréal, Canada

Alberto L. Sangiovanni-Vincentelli
Department of Electrical Engineering & Computer Sciences
University of California at Berkeley, USA

Martin Sauerhoff
Department of Computer Science
Technische Universität Dortmund, Germany

Detlef Sieling
Department of Computer Science
Technische Universität Dortmund, Germany

Robert H. Sloan
Department of Computer Science
University of Illinois at Chicago, USA

Balázs Szörényi
Hungarian Academy of Sciences
University of Szeged, Hungary

György Turán
Department of Mathematics, Statistics, and Computer Science
University of Illinois at Chicago, USA

Alasdair Urquhart
Department of Philosophy
University of Toronto, Canada

Tiziano Villa
Dipartimento d'Informatica
University of Verona, Italy

Ingo Wegener[†]
Department of Computer Science
Technische Universität Dortmund, Germany

[†]Professor Wegener passed away in November 2008.

Acronyms and Abbreviations

AB	almost bent
AIG	AND-Inverter graph
ANF	algebraic normal form
APN	almost perfect nonlinear
ATPG	Automatic Test Pattern Generation (p. 698)
BDD	binary decision diagram
BED	Boolean Expression Diagram
BMC	bounded model checking
BP	branching program
C-1-D	complete-1-distinguishability
CDMA	code division multiple access
CEC	combinational equivalence checking
CNF	conjunctive normal form
CQ	complete quadratic
CTL	computation tree logic
DD	decision diagram
DNF	disjunctive normal form
DPLL	Davis-Putnam-Logemann-Loveland
EDA	electronic design automation
FBDD	free binary decision diagram
FCSR	feedback with carry shift register
FFT	fast Fourier transform
FRAIG	Functionally Reduced AIG
FSM	finite-state machine
FSR	feedback shift register
GPS	generalized partial spread
HDL	hardware description language
HFSM	hardware finite-state machine
HSTG	hardware state transition graph
IBQ	incomplete boundary query

LFSR	linear feedback shift register
LP	linear programming
LTL	linear temporal logic
MTBDD	multiterminal binary decision diagram
NNF	numerical normal form
OBDD	ordered binary decision diagram
PAC	probably approximately correct
PBDD	partitioned binary decision diagram
PC	propagation criterion
QBF	quantified Boolean formula
ROBDD	reduced ordered binary decision diagram
RTL	register-transfer level
SAC	strict avalanche criterion
SAT	satisfiability [not an acronym]
SBS	stochastic binary system
SCC	strongly connected component
SEC	sequential equivalence checking
SEM	sample error minimization
SOP	sum-of-product
SQ	statistical query
STG	state transition graph
UBQ	unreliable boundary query
UMC	unbounded model checking
VC	Vapnik-Chervonenkis
XBDD	extended binary decision diagram

Boolean Models and Methods in Mathematics, Computer Science, and Engineering

Part I

Algebraic Structures

1

Compositions and Clones of Boolean Functions

Reinhard Pöschel and Ivo Rosenberg

1.1 Boolean Polynomials

The representations of Boolean functions are frequently based on the fundamental operations $\{\vee, \wedge, '\}$, where the disjunction $x \vee y$ represents the logical OR, the conjunction $x \wedge y$ represents the logical AND and is often denoted by $x \cdot y$ or simply by the juxtaposition xy, and x' stands for the negation, or complement, of x and is often denoted by $\overline{x}$. This system naturally appeals to logicians and, for some reasons, also to electrical engineers, as illustrated by many chapters of this volume and by the monograph [7]. Its popularity may be explained by the validity of many identities or laws: for example, the associativity, commutativity, idempotence, distributive, and De Morgan laws making $\mathcal{B} := \langle B; \vee, \wedge, ', 0, 1\rangle$ a Boolean algebra, where $B = \{0, 1\}$; in fact, $\mathcal{B}$ is the least nontrivial Boolean algebra.

It is natural to ask whether there is a system of basic Boolean functions other than $\{\vee, \wedge, '\}$, but equally powerful in the sense that each Boolean function may be represented over this system. To get such a system, we introduce the following binary (i.e., two-variable) Boolean function $\dotplus$ defined by setting $x \dotplus y = 0$ if $x = y$ and $x \dotplus y = 1$ if $x \neq y$; its truth table is

x	y	$x \dotplus y$
0	0	0
0	1	1
1	0	1
1	1	0

Clearly $x \dotplus y = 1$ if and only if the arithmetical sum $x + y$ is odd, and for this reason $\dotplus$ is also referred to as the sum *mod* 2. It corresponds to the "exclusive or" of logic, whereby the "exclusive or" of two statements P and Q is true if "either P or Q" is true. Notice that some natural languages, such as, French, distinguish "or" from "exclusive or" ("ou" et "soit"), whereas most natural languages are less precise. The function $\dot{+}$ is also denoted $\oplus$ or $+$ and, in more recent engineering literature, by EXOR.

Consider the system $\{\dot{+}, \cdot, 0, 1\}$, where 0 and 1 are constants. A reader familiar with groups will notice that $\langle B; \dot{+}\rangle$ is an abelian group with neutral element 0 satisfying $x\dot{+}x \approx 0$ (also called an elementary 2-group), where $\approx$ stands for an identity on B. Moreover, $GF(2) := \langle B; \dot{+}, \cdot, 0, 1\rangle$ is a field, called a Galois field and denoted $\mathbf{Z}_2$ or F_2. Thus, in $GF(2)$ we may use all the arithmetic properties valid in familiar fields (such as the fields $\mathbf{Q}$, $\mathbf{R}$, and $\mathbf{C}$ of all rational, real, and complex numbers) but not their order or topological properties. In addition, $GF(2)$ also satisfies $x\dot{+}x \approx 0$ and $x^2 \approx x$. Clearly $GF(2)$ is the field of the least possible size, and so one may be inclined to dismiss it as a trivial and unimportant object. Surprisingly, it has serious applications. A practical one is in cryptography and coding theory (for secret or secure data transmission, for example, for governments, banks, or from satellites; see Chapters 8 and 9 in this volume).

Denote $x_1\dot{+}\cdots\dot{+}x_n$ by $\dot{\sum}_{i=1}^n x_i$. Let f be an n-ary Boolean function distinct from the constant c_0^n (which is the n-variable Boolean function with constant value 0). In its complete disjunctive normal form (DNF), replace the disjunction $\vee$ (of complete elementary conjunctions) by their sum mod 2 $\dot{\sum}$. This is still a representation of f because for every $(a_1, \ldots, a_n) \in \mathbf{B}^n$, at most one of the elementary conjunctions takes value 1 and $1\dot{+}0\dot{+}\cdots\dot{+}0 = 1$ and $0\dot{+}0\dot{+}\cdots\dot{+}0 = 0$. Using $x^0 \approx x' \approx 1\dot{+}x$ and $x^1 \approx x$ throughout, we obtain a representation of f over $\{\dot{+}, \cdot, 0, 1\}$. The following proposition makes this more precise. Here the symbol $\prod$ denotes the usual arithmetical product. We make the usual convention that in expressions involving $\dot{+}, \cdot$ and 1 products are calculated before sums, for example, $xy\dot{+}z$ stands for $(xy)\dot{+}z$, and that $\prod$ and $\dot{\sum}$ over the empty set are 1 and 0, respectively.

Proposition 1.1. [28] *For every n-ary Boolean function f, there exists a unique family F of subsets of $N = \{1, \ldots, n\}$ such that*

$$f(x_1, \ldots, x_n) \approx \dot{\sum_{I\in F}} \prod_{i\in I} x_i. \tag{1.1}$$

For example, $x_1 \wedge x_2 \approx x_1\dot{+}x_2\dot{+}x_1x_2$ *with* $F = \{\{1\}, \{2\}, \{1, 2\}\}$ *(direct verification). Call the right-hand side of (1.1) a* Boolean polynomial.

Proof. Let f be an n-ary Boolean function. If $f = c_0^n$, take $F = \emptyset$. Thus, let $f \neq c_0^n$. In the discussion leading to the proposition, we saw that f may be represented over $\{\dot{+}, \cdot, 0, 1\}$. Multiplying out the parentheses, we obtain a representation of f as a polynomial in variables $x_1, \ldots, x_n$ over $GF(2)$. In view of $x^2 \approx x$, it may be reduced to a sum of square-free monomials. From $x\dot{+}x \approx 0$, it follows that it may be further reduced to such a sum in which every monomial appears at most once, proving the representability of f by a Boolean polynomial. It remains to prove the uniqueness. For every $F \subseteq \mathcal{P}(N)$, that is, a family of subsets of N, denote by $\varphi(F)$ the corresponding Boolean polynomial. Clearly φ is a map from the set $\mathcal{P}(\mathcal{P}(N))$ of families of subsets of N into the set $O^{(n)}$ of n-ary Boolean functions.

Claim. *The map φ is injective.*

Indeed, by the way of contradiction, suppose $\varphi(F) = \varphi(G)$ for some $F, G \subseteq \mathcal{P}(N)$ with $F \neq G$. Choose $I \in (F\backslash G) \cup (G\backslash F)$ of the least possible cardinality, say, $I \in F\backslash G$. Put $a_i = 1$ for $i \in I$ and $a_i = 0$ otherwise. Then for $\underline{a} = (a_1, \ldots, a_n)$, it is easy to see that $\varphi(F)(\underline{a}) = \varphi(G)(\underline{a}) \dot{+} 1$ (as every subset of I is either in both families F and G or in neither). This contradiction shows $G = F$. Now $|\mathcal{P}(\mathcal{P}(N))| = 2^{2^n} = |O^{(n)}|$, and hence φ is a bijection from $\mathcal{P}(\mathcal{P}(N))$ onto $O^{(n)}$, proving the uniqueness. ■

Remark 1.1. *The representation from Proposition 1.1 is sometimes referred to as the Reed-Muller expression or the algebraic normal form of f (see, e.g., Chapter 8). So far, Boolean polynomials have been less frequently used than the disjunctive and conjunctive normal forms, but they have proved indispensable in certain theoretical studies, such as enumeration or coding theory. More recently, electrical engineers have also become interested in Boolean polynomials.*

Remark 1.2. *Boolean polynomials may be manipulated in a conceptually simple way. For example, using the representations $x_1 \vee x_2 \approx x_1 \dot{+} x_2 \dot{+} x_1 x_2$ and $x_1 \rightarrow x_2 \approx 1 \dot{+} x_1 \dot{+} x_1 x_2$, we can compute*

$$\begin{aligned}(x_1 \wedge x_2)(x_1 \rightarrow x_2) &= (x_1 \dot{+} x_2 \dot{+} x_1 x_2)(1 \dot{+} x_1 \dot{+} x_1 x_2) \\ &= x_1 \dot{+} x_2 \dot{+} x_1 x_2 \dot{+} x_1^2 \dot{+} x_1 x_2 \dot{+} x_1^2 x_2 \dot{+} x_1^2 x_2 \dot{+} x_1 x_2^2 \dot{+} x_1^2 x_2^2 \\ &= x_1 \dot{+} x_2 \dot{+} x_1 x_2 \dot{+} x_1 \dot{+} x_1 x_2 \dot{+} x_1 x_2 \dot{+} x_1 x_2 \dot{+} x_1 x_2 \dot{+} x_1 x_2 \\ &= x_2.\end{aligned}$$

This can be easily performed by a computer program, but we may face an explosion in the number of terms.

Remark 1.3. *Suppose that an n-ary Boolean function is given by a table. How do we find its Boolean polynomial or, equivalently, the corresponding family F? We could proceed via the complete DNF (as indicated earlier), but this again may produce a large number of monomials at the intermediate stages. A direct algorithm is as follows.*

Let $f \in O^{(n)}$. On the hypercube B^n we have the standard (partial) order: $(a_1, \ldots, a_n) \leq (b_1, \ldots, b_n)$ if $a_1 \leq b_1, \ldots, a_n \leq b_n$; for example; for $n = 2$ we have that $(1,0) \leq (1,1)$, but $(1,0) \leq (0,1)$ does not hold. As usual, we write $\underline{a} < \underline{b}$ if $\underline{a} \leq \underline{b}$ and $\underline{a} \neq \underline{b}$. For $i = 1, \ldots, n$, set $S_i = \{(a_1, \ldots, a_n) \in B^n : a_1 + \cdots + a_n = i\}$; hence, S_i consists of all $\underline{a} \in B^n$ having exactly i coordinates equal to 1. Recursively, we construct $H_i \subseteq S_i$ $(i = 0, \ldots, n)$. Set $H_0 = S_0$ if $f(0, \ldots, 0) = 1$, and $H_0 = \emptyset$ otherwise. Suppose $0 \leq i < n$ and $H_0, \ldots, H_i$ have been constructed. For $\underline{a} \in S_{i+1}$, set

$$T_{\underline{a}} = \{\underline{b} \in H_0 \cup \cdots \cup H_i \ : \underline{b} \leq \underline{a}\}$$

and

$$H_{i+1} = \{\underline{a} \in S_{i+1} : f(\underline{a}) + |T_{\underline{a}}| \text{ is odd}\}.$$

For $(a_1, \ldots, a_n) \in B^n$, *set*

$$\chi(\underline{a}) = \{1 \leq i \leq n : a_i = 1\}$$

and set $F = \chi(H_0 \cup \cdots \cup H_n)$. *A straightforward proof shows that*

$$f(\underline{x}) \approx \dot{\sum_{I \in F}} \prod_{i \in I} x_i .$$

For example, if $n = 2$ *and* f *is the implication* $\to$, *we get* $H_0 = \{(0,0)\}$, $H_1 = \{(1,0)\}$, $H_2 = \{(1,1)\}$, $F = \{\emptyset, \{1\}, \{1,2\}\}$, *and* $x_1 \to x_2 \approx 1 \dot{+} x_1 \dot{+} x_1 x_2$.

Remark 1.4. *Contrary to disjunctive normal forms, there is no minimization problem for representations of the form (1.1). However, the Boolean polynomials of some Boolean functions are long. The worst one is the Boolean polynomial of* $f(x_1, \ldots, x_n) \approx x_1' \ldots x_n'$ *whose family* F *is the whole* $\mathcal{P}(N)$*; for example, for* $n = 2$ *we have* $x_1' x_2' \approx 1 \dot{+} x_1 \dot{+} x_2 \dot{+} x_1 x_2$. *If, along with* $\dot{+}, \cdot, 1$, *we also allow the negation, certain Boolean polynomials may be shortened. Here we can use some additional rules such as* $x'y \approx y \dot{+} xy$ *and* $xx' \approx 0$. *For these more general polynomials, we face minimization problems. It seems that these and the systematic use of the more general polynomials have not been investigated in depth.*

Remark 1.5. *As the reader may suspect, Proposition 1.1 can be extended to any finite field* $\mathbf{F}$ *(e.g.,* $\mathbf{Z}_3$*) and maps* $f : \mathbf{F}^n \to \mathbf{F}$.

Remark 1.6. *A long list of practical problems (mostly from operations research) leads to the following problem. Let* f *be a map from* B^n *into the set* $\mathbf{Z}$ *of integers and assume that we want to find the minimum value of* f *on* B^n. *The potential of Boolean polynomials for this problem was realized quite early in [6]. A more direct variant and a related duality are in [23, 24], but generally this approach seems to be dormant.*

1.2 Completeness and Maximal Clones

It is well known that every Boolean function may be represented over $\{\wedge, \cdot, '\}$ (e.g., through a DNF or a conjunctive normal form, CNF). In Section 1.1, we saw a representation of Boolean functions over $\{\dot{+}, \cdot, 1\}$ (through Boolean polynomials). In general, call a set F of Boolean functions *complete* if every Boolean function is a composition of functions from F. In some older and East European literature, a complete set is often referred to as *functionally complete*. In universal algebra, the corresponding algebra $\langle B; F\rangle$ is termed *primal*.

Naturally we may ask about other complete sets and their size. The two examples we have seen so far consist of three functions each. However, the set $\{\wedge, \cdot, '\}$ is actually redundant. Indeed, $\{\wedge, {}'\}$ is also complete as $x_1x_2 \approx (x_1' \wedge x_2')'$ (this identity is known as one of the De Morgan laws).

A Boolean function f is *Sheffer* if the singleton set $\{f\}$ is complete. Clearly a Sheffer function is at least binary. The following functions NAND and NOR are Sheffer:

$$x \text{ NAND } y :\approx x' \vee y', \quad x \text{ NOR } y :\approx x' \wedge y'.$$

The suggestive symbols NAND and NOR were adopted by electrical engineers to describe the functioning of certain transistor gates after these were invented in the 1950s. However, the two functions were introduced in logic long ago by Sheffer [25] and Nicod [19]; they are known as Sheffer strokes or Nicod connectives and are variously denoted by $|$, $\perp$, $\top$, $\uparrow$, $\downarrow$, and so forth. The fact that NAND is Sheffer follows from

$$x' \approx x \text{ NAND } x, \quad x \wedge y \approx x' \text{ NAND } y';$$

consequently, the complete set $\{\wedge, {}'\}$ can be constructed from NAND alone, and hence NAND is Sheffer. The proof for NOR is similar. The fact that NAND and NOR are the only binary Sheffer functions will follow from Corollary 1.5, which provides a complete characterization of Sheffer functions.

Having seen a few complete sets of Boolean functions, we may ask about other complete sets or – even better – for a completeness criterion, that is, for a necessary and sufficient condition for a set F of Boolean functions to be complete. E. L. Post found such a criterion in [20]. To formulate the criterion in a modern way, we need two crucial concepts. For $1 \leq i \leq n$, the ith *n-ary projection* is the Boolean n-ary function e_i^n satisfying $e_i^n(x_1, \ldots, x_n) \approx x_i$. Notice that e_i^n just replicates its ith argument and ignores all other arguments. The *composition* of an m-ary Boolean function f with the n-ary Boolean functions $g_1, \ldots, g_m$ is the n-ary Boolean function h, defined by setting $h(x_1, \ldots, x_n) = f(g_1(x_1, \ldots, x_n), \ldots, g_m(x_1, \ldots, x_n))$ for all n-tuples $(x_1, \ldots, x_n)$ in B^n.

Now, a set of Boolean functions is a clone if it is composition-closed and contains all projections. Thus, a clone is rich enough in the sense that we cannot exit from it via any composition of its members (each possibly used several times). The concept is similar to the concept of a transformation monoid (whereby the set of all projections plays a role analogous to that of the identity selfmap in monoids).

It is not difficult to check that the intersection of an arbitrary set of clones is again a clone. Thus, for every set F of Boolean functions, there exists a unique least clone containing F. This clone, denoted by $\langle F \rangle$, is the clone generated by F.

The clone $\langle F \rangle$ can be alternatively interpreted in terms of combinatorial switching circuits. Suppose that for each n-ary $f \in F$ there is a gate (switching device, typically a transistor) with n inputs and a single output realizing $f(a_1, \ldots, a_n)$ on the output whenever, for $i = 1, \ldots, n$, the zero-one input a_i is applied to the ith input. An *F-based combinatorial circuit* is obtained from gates realizing functions from F by attaching the output of each gate (with a single exception) to an input of another gate so that (1) the circuit has a single external output; (2) to each input of

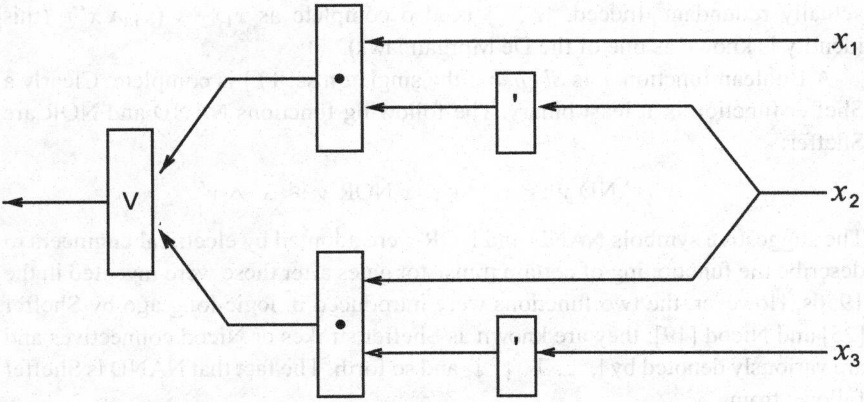

Figure 1.1. Combinatorial circuit realizing $f(x_1, x_2, x_3) \approx x_1x_2' \vee x_2x_3'$.

a gate is attached a single external input or a single output of another gate but not both; and (3) the circuit is feedback-free (i.e., following the arcs from any external input, we always arrive at the external output without ever making a loop). An example is in Figure 1.1; see also Chapters 11 and 16.

Now the clone $\langle F \rangle$ generated by F consists of all Boolean functions realizable by F-based combinatorial circuits. This interpretation also allows us to present an alternative definition of the concept of completeness: in this terminology, a set F of Boolean functions is complete exactly if $\langle F \rangle$ is the set O of all Boolean functions. Clearly O is the greatest clone (with respect to inclusion $\subseteq$). A clone C distinct from O is maximal (also precomplete or preprimal) if $C \subset D \subset O$ holds for no clone D: in other words, if O covers C in the containment relation $\subseteq$. Thus, C is maximal exactly if C is incomplete, but $C \cup \{f\}$ is complete for each $f \in O \setminus C$. The concept is a direct analog of a maximal subgroup, a maximal subring, and so on.

We prove in Theorem 1.2 that each clone distinct from O is contained in at least one maximal clone. This leads to the following almost immediate but still basic fact.

Fact 1. *A set F of Boolean functions is complete if and only if F is contained in no maximal clone.*

Proof. ($\Rightarrow$) By contradiction, if F is contained in some maximal clone M, then the clone $\langle F \rangle$ generated by F is included in M, and so F is not complete.
($\Leftarrow$) If F is incomplete, then $\langle F \rangle$ extends to a maximal clone M, and so $F \subseteq \langle F \rangle \subseteq M$. ■

Of course, Fact 1 is fully applicable only if we know the entire list of maximal clones. Then a set F of Boolean functions is complete if and only if for each maximal clone M (from the list) we can find some $f \in F \setminus M$.

Most clones and all maximal clones may be described by relations in the following way. Recall that for a positive integer h, an *h-ary relation* on B is a subset ρ of B^h (i.e., a set of zero-one h-tuples; some logicians prefer to view it as a map φ from B^h into $\{+, -\}$, whereby $\rho = \varphi^{-1}(+)$, or to call it a predicate). For $h = 1$, the relation is called *unary* and is just a subset of B, whereas for $h = 2$, the relation is *binary*.

An n-ary Boolean function f preserves an h-ary relation ρ if for every $h \times n$ matrix A whose column vectors all belong to ρ, the h-tuple of the values of f on the rows of A belongs to ρ. In symbols, if $A = [a_{ij}]$ with $(a_{1j}, \ldots, a_{hj}) \in \rho$ for all $j = 1, \ldots, n$, then

$$(f(a_{11}, \ldots, a_{1n}), \ldots, f(a_{h1}, \ldots, a_{hn})) \in \rho.$$

Notice the "rows versus columns" type of the definition. Several examples are given next. In the universal algebra terminology, f preserves ρ if ρ is a subuniverse of $\langle B; f\rangle^h$. The first impression may be that the definition is too artificial or covers only special cases. It turns out that it is the right one not only for clones of Boolean functions but also for clones on any finite universe. It seems that it was first explicitly formulated in [15], and it has been reinvented under various names. We illustrate this crucial concept using a few examples needed for the promised completeness criterion.

Example 1.1. *Let $h = 1$ and $\rho = \{0\}$. Then $A = [0 \cdots 0]$ (a $1 \times n$ matrix), and f preserves $\{0\}$ if and only if $f(0, \ldots, 0) = 0$ (in universal algebra terminology, if and only if $\{0\}$ is a subuniverse of $\langle B; f\rangle$).*

Example 1.2. *Similarly, a Boolean function f preserves the unary relation $\{1\}$ if and only if $f(1, \ldots, 1) = 1$.*

Example 1.3. *Consider the binary relation $\sigma := \{(0, 1), (1, 0)\}$ on B. Note that σ is the diagram (or graph) π^0 of the permutation $\pi : x \mapsto x'$. It can also be viewed as a graph on B with the single edge $\{0, 1\}$. A $2 \times n$ matrix $A = [a_{ij}]$ has all columns in σ if and only if $a_{2j} = a'_{1j}$ holds for all $j = 1, \ldots, n$. Writing a_j for a_{1j}, we obtain that f preserves σ if and only if*

$$(f(a_1, \ldots, a_n), f(a'_1, \ldots, a'_n)) \in \sigma$$

holds for all $a_1, \ldots, a_n \in B$. This is equivalent to the identity

$$f(x'_1, \ldots, x'_n) \approx f(x_1, \ldots, x_n)'.$$

The standard name for a Boolean function satisfying this identity is selfdual *(see [7] and Section 1.3). Expressed algebraically, f is selfdual if and only if the negation $'$ is an automorphism of the algebra $\langle B; f\rangle$. A selfdual $f \in O^{(n)}$ is fully determined by its values $f(0, a_2, \ldots, a_n)$ with $a_2, \ldots, a_n \in B$ (because $f(1, b'_2, \ldots, b'_n) = f'(0, b'_2, \ldots, b'_n)$). Thus, there are exactly $2^{2^{n-1}}$ selfdual n-ary Boolean functions, and the probability that a randomly chosen $f \in O^{(n)}$ is selfdual is the low value $2^{-2^{n-1}}$: for example, it is approximately* 0.0000152 *for $n = 5$.*

Example 1.4. *Next consider* $\rho = \{(0,0),(0,1),(1,1)\}$. *Notice that* $(x,y) \in \rho$ *if and only if* $x \le y$ *(where* $\le$ *is the natural order on* B*), and so we write* $x \le y$ *instead of* $(x,y) \in \rho$. *A* $2 \times n$ *matrix* $A = [a_{ij}]$ *has all columns in* $\le$ *whenever* $a_{1j} \le a_{2j}$ *holds for all* $j = 1, \ldots, n$, *and therefore* f *preserves* $\le$ *if and only if*

$$a_1 \le b_1, \ldots, a_n \le b_n \Rightarrow f(a_1, \ldots, a_n) \le f(b_1, \ldots, b_n);$$

that is, if every argument is kept the same or increased, then the value is the same or increases. This is the standard definition of a monotone (also isotone, order-respecting, or order-compatible) Boolean function; see [7].

Example 1.5. *As a final example, consider the 4-ary (or quaternary) relation*

$$\lambda = \{(x_1, x_2, x_3, x_1 \dot{+} x_2 \dot{+} x_3) : x_1, x_2, x_3 \in B\},$$

where $\dot{+}$ *is the sum mod 2 introduced in Section 1.1. Expressed differently, the last coordinate in a 4-tuple from* λ *is exactly the parity check (making the coordinate sum even), a basic error check used in computer hardware and other digital devices; equivalently, a 4-tuple belongs to* λ *if and only if it contains an even number of 1s. The description of the functions preserving* λ *is not as transparent as in the preceding examples. In order to establish it, let us say that a Boolean function* $f \in O^{(n)}$ *is linear (or affine) if there are* $c \in B$ *and* $1 \le i_1 < \ldots < i_k \le n$ *such that* $f(x_1, \ldots, x_n) \approx c \dot{+} x_{i_1} \dot{+} \cdots \dot{+} x_{i_k}$.

Fact 2. *A Boolean function preserves* λ *if and only if it is linear.*

Proof. ($\Rightarrow$) Let $f \in O^{(n)}$ preserve λ. Then

$$f(x_1, \ldots, x_n) \approx f(x_1, 0, \ldots, 0) \dot{+} f(0, x_2, \ldots, x_n) \dot{+} f(0, \ldots, 0). \qquad (1.2)$$

Here $f(0, x_2, \ldots, x_n) \in O^{(n-1)}$ also preserves λ, and so we can apply (1.2) to it. Continuing in this fashion, we obtain

$$f(x_1, \ldots, x_n) \approx \dot{\sum}_{i=1}^{n} f(0, \ldots, 0, x_i, 0, \ldots, 0) \dot{+} d \qquad (1.3)$$

where $d = f(0, \ldots, 0) \dot{+} \ldots \dot{+} f(0, \ldots, 0)$ (n times). Each unary Boolean function $f(0, \ldots, 0, x_i, 0, \ldots, 0)$ is of the form $a_i x \dot{+} b_i$ for some $a_i, b_i \in B$, and thus the right-hand side of (1.3) is $a_1 x_1 \dot{+} \cdots \dot{+} a_n x_n \dot{+} c$ where $c = b_1 \dot{+} \cdots \dot{+} b_n \dot{+} d$. This proves that f is linear.
($\Leftarrow$) It can be easily verified that every linear function preserves λ. ∎

The set of Boolean functions preserving a given h-ary relation ρ on B is denoted by Pol ρ. It is easy to verify that Pol ρ is a clone by showing that it contains all projections and that it is composition closed.

Now we are ready for the promised completeness criterion due to Post [20].

Theorem 1.2. (Completeness Criterion.)

(1) *A set F of Boolean functions is complete if and only if F is contained in none of the clones*

$$Pol\{0\}, Pol\{1\}, Pol\sigma, Pol \leq, Pol\lambda. \tag{1.4}$$

(2) *Each clone distinct from O extends to a maximal one, and the foregoing five clones are exactly all maximal clones.*

Remark 1.7. *Part (2) is just a rephrasing of part (1). The criterion is often given in the following equivalent form: a set F of Boolean functions is complete if and only if there exist $f_1, \ldots, f_5 \in F$ such that*

$$f_1(0, \ldots, 0) = 1, f_2(1, \ldots, 1) = 0,$$

f_3 is not selfdual, f_4 is not monotone, and f_5 is not linear (here $f_1, \ldots, f_5$ need not be pairwise distinct; e.g., they are all equal for $F = \{f\}$, where f is Sheffer).

The proof given next follows A. V. Kuznetsov's proof (see [10], pp. 18–20). The relatively direct proof is of some interest due to a Slupecki-type criterion (Lemma 1.4 later), but it does not reveal how the completeness criterion was discovered.

Proof. The necessity is obvious, as all five clones listed are distinct from O. As for sufficiency, in the following lemma we start by characterizing clones included in neither of the first two clones in (1.4) and end up with the same for all clones except the last. Remarkably, all this is done through the unary operations of the clone. Recall that $O^{(n)}$ denotes the set of n-ary Boolean functions.

Lemma 1.3. *If C is a clone and $f \in O^{(n)}$, then:*

(i) *$C \not\subseteq Pol\{0\}$ if and only if C contains the unary constant c_1 or C contains the negation.*
(ii) *$C \not\subseteq Pol\{1\}$ if and only if C contains the unary constant c_0 or C contains the negation.*
(iii) *$f \notin Pol \leq$ if and only if*

$$f(a_1, \ldots, a_{i-1}, 0, a_{i+1}, \ldots, a_n) > f(a_1, \ldots, a_{i-1}, 1, a_{i+1}, \ldots, a_n) \tag{1.5}$$

for some $1 \leq i \leq n$ and some $a_1, \ldots, a_n \in B$.
(iv) *$C \not\subseteq Pol\{0\}$, $C \not\subseteq Pol\{1\}$, and $C \not\subseteq Pol \leq$ if and only if C contains the negation.*
(v) *C is a subclone of none of $Pol\{0\}$, $Pol\{1\}$, $Pol \leq$, and $Pol\sigma$ if and only if $C \supseteq O^{(1)}$, that is, C contains all four unary Boolean functions.*

Proof.

(i) Let us first show that if $C \not\subseteq \text{Pol}\{0\}$, then C contains the unary constant $c_1 \approx 1$ or the negation. Because $C \not\subseteq \text{Pol}\{0\}$, clearly $h(0, \ldots, 0) = 1$ for some $h \in C$.

(a) Suppose $h(1, \ldots, 1) = 0$. Then $g(x) \approx h(x, \ldots, x) \approx x'$. Here $g \in C$ because $g(x, \ldots, x) \approx h(e_1^1(x), \ldots, e_1^1(x))$ where the clone C contains both h and the projection e_1^1 and is composition closed. Thus, C contains the negation.

(b) Suppose next that $h(1, \ldots, 1) = 1$. By the same token as in (a), we get $c_1 \in C$. This proves the statement.

(ii) A similar argument as above applied to $\text{Pol}\{1\}$ shows that C contains the unary constant $c_0 \approx 0$ or the negation.

(iii) Let $\leq$ be the componentwise order on B^n introduced in Remark 1.3 of Section 1.1. From $f \notin \text{Pol} \leq$, we obtain $f(\underline{a}) \not\leq f(\underline{b})$ for some $\underline{a} \leq \underline{b}$. Here already $f(\underline{a}) = 1$ and $f(\underline{b}) = 0$. For notational simplicity, we assume that $\underline{a} = (0, \ldots, 0, a_{k+1}, \ldots, a_n)$, $\underline{b} = (1, \ldots, 1, a_{k+1}, \ldots, a_n)$ for some $0 < k < n$. For $i = 0, \ldots, k$, set $\underline{a}_i = (1, \ldots, 1, 0, \ldots, 0, a_{k+1}, \ldots, a_n)$, where the first i coordinates are 1 and the next $k - i$ coordinates are 0. In view of $f(\underline{a}_0) = f(\underline{a}) = 1$ and $f(\underline{a}_k) = f(\underline{b}) = 0$, there exists $0 \leq i < k$ with $f(\underline{a}_i) = 1$ and $f(\underline{a}_{i+1}) = 0$, proving (iii).

(iv) By (i) and (ii), it suffices to consider the case of C containing both constants c_0 and c_1. Because $C \not\subseteq \text{Pol} \leq$, by (iii) there exists $f \in C$, satisfying (1.5). Define a unary Boolean function h by setting

$$h(x) \approx f(g_1(x), \ldots, g_n(x))$$

where $g_i(x) \approx e_1^1(x) \approx x$ and g_j is the unary constant function with value a_j for all $j \neq i$. Clearly $h \in C$ and $h(x) \approx x'$.

(v) By (iv), the clone C contains the negation. As $C \not\subseteq \text{Pol}\sigma$, the clone C also contains a nonselfdual n-ary function f, and hence $f(a_1, \ldots, a_n) = f(a_1', \ldots, a_n')$ for some $a_1, \ldots, a_n \in B$. Set $x^0 \approx x'$ and $x^1 \approx x$ and define a unary g by $g(x) \approx f(x^{a_1}, \ldots, x^{a_n})$. Clearly g is constant and $g \in C$ because e_1^1, the negation, and f are in the clone C. The other constant is $g' \in C$. Thus, $C \supseteq O^{(1)}$. ■

This lemma leads to the question: what are the maximal clones containing all four unary operations? The following lemma asserts that the clone $\text{Pol}\lambda$ (of all linear functions, see Fact 2) is the unique maximal clone containing $O^{(1)}$.

Lemma 1.4. *Let C be a clone such that $C \supseteq O^{(1)}$. Then $C = O$ if and only if C contains a nonlinear function.*

Proof. The necessity is obvious. For sufficiency, let f be a nonlinear function from C. In view of the De Morgan law: $x \wedge y \approx (x'y')'$, the set $\langle \cdot, ' \rangle$ is complete.

Now C already contains the negation, and so it suffices to show that the conjunction belongs to the clone $\langle O^{(1)} \cup \{f\}\rangle$ generated by the four unary operations and f. By Proposition 1.1, the function f can be represented by a Boolean polynomial

$$\dot{\sum_{I \in F}} \prod_{i \in I} x_i,$$

which obviously is nonlinear, and so there exists $J \in F$ with $|J| > 1$. Choose such J of the least possible size, and for notational convenience let $J = \{1, \ldots, j\}$. Define $h \in O^{(2)}$ by

$$h(x_1, x_2) \approx f(x_1, x_2, 1, \ldots, 1, 0, \ldots, 0), \tag{1.6}$$

where 1 appears $j - 2$ times. Clearly, $h \in C$. We obtain that

$$h(x_1, x_2) \approx x_1x_2 \dot{+} ax_1 \dot{+} bx_2 \dot{+} c \tag{1.7}$$

for some $a, b, c \in B$. Indeed, by the minimality of J, every $I \in F \setminus \{J\}$ with $|I| > 1$ meets the set $\{j + 1, \ldots, n\}$, and so the corresponding product in (1.6) vanishes.

Now $h(x_1 \dot{+} b, x_2 \dot{+} a) \dot{+} ab \dot{+} c$ belongs to C because h, e_1^1, and the negation belong to C. It can be verified that this Boolean function is actually the conjunction x_1x_2. ■

To complete the proof of Theorem 1.2, simply combine Lemma 1.3(v) and Lemma 1.4. ■

Remark 1.8. *Post's criterion was rediscovered at least ten times, but it would serve no purpose to list all the references. Its beauty lies in the simplicity of the five conditions, which may be verified directly on the given set F of Boolean functions; in particular, no construction or reference to another structure is necessary. Indeed, the first two conditions may be checked by inspecting the values* $f(0, \ldots, 0)$ *and* $f(1, \ldots, 1)$ *for* $f \in F$. *To find out whether a given n-ary Boolean function is selfdual, we need at most* 2^{n-1} *checks (see Example 1.3). Next, by Lemma 1.3(iii), the function f is monotone if and only if* $f(a_1, \ldots, a_n) \leq f(b_1, \ldots, b_n)$ *whenever* $a_i = 0$, $b_i = 1$ *for a single i and* $a_j = b_j$ *otherwise. Thus, we need to check at most* $n2^{n-1}$ *pairs* $\underline{a}, \underline{b}$. *To check whether f is linear, put* $a_0 = f(0, \ldots, 0)$ *and* $a_i = (0, \ldots, 0, 1, 0, \ldots, 0)$, *where 1 is at the ith place and* $i = 1, \ldots, n$. *Clearly, f is linear exactly if*

$$f(x_1, \ldots, x_n) \approx a_0 \dot{+} (a_0 \dot{+} a_1)x_1 \dot{+} \cdots \dot{+} (a_0 \dot{+} a_n)x_n,$$

requiring at most $2^n - n - 1$ *checks.*

Using these observations, one could write a computer program that could decide whether an arbitrary finite set F of Boolean function is complete. This would be executed in a priori bounded time where the (upper) bound depends on the sum of the arities of functions from F. In other words, there is an effective algorithm (in the foregoing sense) for the completeness problem.

Remark 1.9. *Although it may take a little while to discover that a given set F of Boolean functions is complete, afterward a few values of at most five function from F are enough to convince anybody of its completeness. For an incomplete set F of Boolean functions, we can find all the maximal clones containing F. This information yields the necessary and sufficient conditions for the choice of an additional set G of Boolean functions capable of making $F \cup G$ complete (this may play a role in switching theory when F is the set of Boolean functions describing the functioning of new types of gates).*

Post's completeness criterion yields an elegant characterization of Sheffer functions (i.e., functions f such that the singleton $\{f\}$ is complete; see Section 1.2).

Corollary 1.5. *A Boolean function $f \in O^{(n)}$ is Sheffer if and only if*

(i) $f(x, \ldots, x) \approx x'$, *and*
(ii) $f(a_1, \ldots, a_n) = f(a_1', \ldots, a_n')$ *for some* $a_1, \ldots, a_n \in B$.

Proof. The condition (i) is equivalent to $f \notin \mathrm{Pol}\{0\} \cup \mathrm{Pol}\{1\}$, whereas the condition (ii) says that f is not selfdual.
($\Rightarrow$) Both conditions are necessary by Post's completeness criterion.
($\Leftarrow$) Let f satisfy (i) and (ii). Then $f \notin \mathrm{Pol}\{0\} \cup \mathrm{Pol}\{1\} \cup \mathrm{Pol}\sigma$. Next we claim that

$$\mathrm{Pol} \leq \subseteq \mathrm{Pol}\{0\} \cup \mathrm{Pol}\{1\}.$$

Indeed, by contraposition, let $h \notin \mathrm{Pol}\{0\} \cup \mathrm{Pol}\{1\}$ be arbitrary. Then

$$h(0, \ldots, 0) = 1 > 0 = h(1, \ldots, 1),$$

proving $h \notin \mathrm{Pol} \leq$. In particular, $f \notin \mathrm{Pol} \leq$. Suppose now by contraposition that f is linear. In view of (i), clearly $f(x_1, \ldots, x_n) \approx 1 \dotplus a_1x_1 \dotplus \cdots \dotplus a_nx_n$, where $a_1, \ldots, a_n \in B$ satisfy $a_1 \dotplus \cdots \dotplus a_n = 1$. Now

$$\begin{aligned} f(x_1', \ldots, x_n')' &\approx 1 \dotplus f(1 \dotplus x_1, \ldots, 1 \dotplus x_n) \\ &\approx 1 \dotplus 1 \dotplus a_1 \dotplus \cdots \dotplus a_n \dotplus a_1x_1 \dotplus \cdots \dotplus a_nx_n \\ &\approx 1 \dotplus a_1x_1 \dotplus \cdots \dotplus a_nx_n \\ &\approx f(x_1, \ldots, x_n), \end{aligned}$$

proving that f is selfdual. Thus f is nonlinear, and by Post's completeness criterion $\{f\}$ is complete. ■

Remark 1.10. *Corollary 1.5 may be rephrased. Denote by $\mathcal{M}$ the union of the five maximal clones. Then $Sh := O \setminus \mathcal{M}$ is the set of Sheffer functions. Corollary 1.5 in fact states that*

$$\mathcal{M} = Pol\{0\} \cup Pol\{1\} \cup Pol\sigma,$$

and so the maximal clones $Pol\{0\}$, $Pol\{1\}$, and $Pol\sigma$ cover $\mathcal{M}$. In fact, they also provide a unique irredundant cover of $\mathcal{M}$ (meaning that every cover of $\mathcal{M}$ by *maximal clones must include the foregoing three clones). To show this, it suffices to find in each of the three maximal clones a function belonging to no other maximal clone. The functions $x'y$ and $x \to y$ are such functions for Pol{0} and Pol{1}, whereas for Polσ, although there is no such binary function, there are four such ternary functions.*

We proceed to enumerate the n-ary Sheffer functions. For this we classify them by the "first" n-tuple $(a_1, \ldots, a_n)$ in their table satisfying the condition (ii) of Corollary 1.5. More precisely, the weight $w(\underline{a})$ of $\underline{a} = (a_1, \ldots, a_n)$ is defined by $2^{n-1}a_1 + 2^{n-2}a_2 + \cdots + a_n$, and the chain (also called a linear order or total order) on B^n induced by the weights is the lexicographic order $\preceq$ defined by: $\underline{a} \preceq \underline{b}$ if $w(\underline{a}) \leq w(\underline{b})$. For $i = 1, \ldots, 2^{n-1} - 1$, denote by S_i the set of n-ary Sheffer functions f such that $f(x, \ldots, x) \approx x'$ and such that the condition (ii) from Corollary 1.5 holds for $\underline{a} \in B^n$ with $w(\underline{a}) = i$, but not for any $\underline{b} \in B^n$ with $w(\underline{b}) < i$ (i.e., $\underline{a}$ is the least element in the lexicographic order satisfying (ii)). From Corollary 1.5, we obtain the following corollary.

Corollary 1.6. *If $n > 1$, then:*

(i) *the sets $S_1, \ldots, S_{2^{n-1}-1}$ partition the set $Sh^{(n)}$ of n-ary Sheffer functions,*
(ii) $|S_i| = 2^{2^n-i-2}$ $(i = 1, \ldots, 2^{n-1} - 1)$, *and*
(iii) *there are exactly*

$$2^{2^n-2} - 2^{2^{n-1}-1}$$

n-ary Sheffer functions.

Proof.

(i) The sets $S_1, \ldots, S_{2^{n-1}-1}$ cover $Sh^{(n)}$ by Corollary 1.5, and they are obviously pairwise disjoint.
(ii) Let $1 \leq i \leq 2^{n-1} - 1$. For $f \in S_i$ and $\underline{a} = (a_1, \ldots, a_n)$ with $w(\underline{a}) < i$, the value $f(a_1', \ldots, a_n')$ equals $f(\underline{a})'$ and so cannot be chosen freely. Moreover, $f(0, \ldots, 0) = 1$ and $f(1, \ldots, 1) = 0$, and hence we have exactly 2^{2^n-i-2} free choices, proving (ii).
(iii) According to (i) and (ii),

$$Sh^{(n)} = \sum_{i=1}^{2^{n-1}-1} 2^{2^n-i-2} = 2^{2^n-3} \sum_{j=0}^{2^{n-1}-2} 2^{-j}.$$

This is a finite geometric series with the quotient 2^{-1}, and a well-known formula yields (iii). ■

Remark 1.11. *According to Corollary 1.6(iii), there are $2^{2^2-2} - 2^{2^1-1} = 2^2 - 2 = 2$ binary Sheffer functions. In fact, these are NOR and NAND, introduced earlier in this section. There are already 56 ternary Sheffer functions, and the* *number increases rather rapidly with n. Thus we should rather ask about the proportion τ_n of n-ary Sheffer functions among all n-ary Boolean functions. From Corollary 1.6(iii), we get*

$$\tau_n = |Sh^{(n)}|/|O^{(n)}| = 2^{-2} - 2^{-2^{n-1}-1},$$

and so $\lim_{n\to\infty} \tau_n = 1/4$. The numbers τ_n grow very fast to $1/4$, for example, τ_6 already shares the first 10 decimal places with 0.25. Thus, the probability that, for n big enough, a randomly chosen n-ary Boolean function is Sheffer is practically 0.25.

In switching circuits, the constant unary functions c_0 and c_1 are usually available (as constant signals) or can be realized very cheaply. This leads to the following definition.

Definition 1.1. *A set F of Boolean functions is* complete with constants *(or functionally complete) if $F \cup \{c_0, c_1\}$ is complete. Similarly, a Boolean function f is* Sheffer with constants *if $\{f, c_0, c_1\}$ is complete.*

The two constant functions take care of the first three maximal clones in Post's completeness criterion (Theorem 1.2), and so we have the following corollary.

Corollary 1.7. *A set of Boolean functions is complete with constants if and only if it contains a nonmonotone function and a nonlinear function. In particular, a Boolean function f is Sheffer with constants if and only if f is neither monotone nor linear.*

Denote by γ_n the number of n-ary Boolean functions that are Sheffer with constants. Further, let φ_n denote the number of n-ary monotone Boolean functions. The number φ_n, called the *Dedekind number*, is of interest on its own. The relation between γ_n and φ_n is given in [8].

Corollary 1.8.

(1) $\gamma_n = 2^{2^n} - 2^{n+1} - \varphi_n + n + 2$.
(2) $\lim_{n\to\infty} \gamma_n/|O^{(n)}| = 1$.

Proof. We start with two claims:

Claims.

(i) $e_1^n, \ldots, e_n^n, c_0^n, c_1^n$ are the only n-ary monotone and linear Boolean functions.
(ii) There are 2^{n+1} linear n-ary functions.

To prove the claims, consider a linear and monotone function $f \in O^{(n)}$. Then $f(x_1, \ldots, x_n) = b \dotplus x_{i_1} \dotplus \cdots \dotplus x_{i_k}$ for some $b \in B$, $k \geq 0$ and $1 \leq i_1 < \cdots < i_k \leq n$. If $k = 0$ then clearly $f = c_0^n$ or $f = c_1^n$. Thus let $k \geq 1$. For notational simplicity, suppose that $i_j = j$ for $j = 1, \ldots, k$. First, $b = 0$, because $b = 1$ leads to

$$1 = f(0, \ldots, 0) > f(1, 0, \ldots, 0) = 0,$$

which contradicts monotonicity. Finally, $k = 1$, because for $k \geq 2$ we get a contradiction from

$$1 = f(1, 0, \ldots, 0) > f(1, 1, 0, \ldots, 0) = 0.$$

Now, for $k = 1$, clearly $f = e^n_{i_1}$. It is immediate that $e^n_1, \ldots, e^n_n, c^n_0, c^n_1$ are both monotone and linear, proving (i).

The number of linear n-ary Boolean functions follows from their general form and from the unicity of their representation (Proposition 1.1). This proves the claims.

Now (1) follows from Corollary 1.7 and the claims (by inclusion-exclusion), and (2) is a consequence of (1) and an asymptotic for φ_n from [13]. ■

Remark 1.12. *The last argument slightly cuts corners, because the asymptotic given in [13] is not in a form showing immediately that* $\varphi_n 2^{-2^n}$ *goes to* 0. *Because (2) is not our main objective, we are not providing a detailed proof. The known values of* φ_n *for* $n = 1, \ldots, 8$ *are* 3, 6, 20, 168, 7581, 7828354, 2414682040998, 56130437228687557907788 *[4, 27]. The first three values can be verified directly.*

The differences $1 - \gamma_n 2^{-2^n}$ $n = 1, \ldots, 7$ are approximately

$$1, 0.625, 0.121, 0.0029, 1.7 \times 10^{-6}, 4.243 \times 10^{-3}, 7 \times 10^{-27}.$$

This indicates that the proportion of n-ary functions that are Sheffer with constants indeed goes very fast to 1. This may be interpreted as follows. For a large n, an n-ary Boolean function picked at random is almost surely Sheffer with constants.

Consider now a set F of Boolean functions, $f \in F$, and a Boolean function g. Suppose

$$f \in M \Leftrightarrow g \in M$$

holds for M running through the five maximal clones. Then clearly F is complete if and only if $(F \setminus \{f\}) \cup \{g\}$ is complete. This leads to the following more formal definition.

Definition 1.2. *The characteristic set of a Boolean function* f *is the subset* f^* *of* $\{1, \ldots, 5\}$ *such that*

$$\begin{aligned}
1 \in f^* &\Leftrightarrow f \notin Pol\{0\},\\
2 \in f^* &\Leftrightarrow f \notin Pol\{1\},\\
3 \in f^* &\Leftrightarrow f \notin Pol\sigma,\\
4 \in f^* &\Leftrightarrow f \notin Pol \leq,\\
5 \in f^* &\Leftrightarrow f \notin Pol\lambda.
\end{aligned}$$

For example, $f^* = \{1, 3, 5\}$ exactly if $f(0, \ldots, 0) = f(1, \ldots, 1) = 1$, f is not selfdual and f is monotone but not linear. Boolean functions f and g are equivalent – in symbols, $f \sim g$ – if $f^* = g^*$: in other words, if for each maximal clone M

either both $f, g \in M$ or both $f, g \notin M$. As mentioned before, two equivalent functions are interchangeable with respect to completeness. Note that according to the Completeness Criterion, a set F of Boolean functions is complete if and only if $F^* := \{f^* : f \in F\}$ covers the set $\{1, \dots, 5\}$. Notice that the relation $\sim$ on O is the kernel of the map $f \mapsto f^*$ and, as such, is an equivalence relation on O.

Clearly the blocks (or equivalence classes) of $\sim$ are the (inclusion) minimal nonempty intersections of the five maximal clones and their complements (in O). In principle, there could be as many as 32 blocks (if the map were onto all subsets of $\{1, \dots, 5\}$), but in reality there are only 15. To derive this result, we need the following lemma, where CF stands for $O \setminus F$ (the complement of F in O).

Lemma 1.9. *For $i = 0, 1$:*

(1) $CPol\{0\} \cap CPol\{1\} \subseteq CPol \leq$,
(2) $Pol\{i\} \cap Pol\sigma \subseteq Pol\{1-i\}$,
(3) $CPol\{0\} \cap CPol\{1\} \cap Pol\lambda \subseteq Pol\sigma$,
(4) $Pol\{0\} \cap Pol\{1\} \cap Pol\lambda \subseteq Pol\sigma$,
(5) $Pol\{i\} \cap CPol\{1-i\} \cap Pol(\leq) = \langle c_i \rangle \subseteq Pol\lambda$,
(6) $Pol(\leq) \cap Pol\lambda = \langle c_0, c_1 \rangle$.

Proof. (1) and (2): Immediate. (3) and (4): See the proof of Corollary 1.5. We only prove (5) for $i = 0$. Let an n-ary f belong to the left-hand side of (5), and let $a_1, \dots, a_n \in B$. Then

$$0 = f(0, \dots, 0) \leq f(a_1, \dots, a_n) \leq f(1, \dots, 1) = 0.$$

Hence, $f = c_0^n$ (= the n-ary constant 0), and so $f \in \langle c_0 \rangle \subset \mathrm{Pol}\lambda$. (6): See the Claims in the proof of Corollary 1.8. ■

Now we are ready for the explicit list of the 15 blocks of $\sim$ or, equivalently, of the minimal nonempty intersections of the five maximal clones and their complements. We number the blocks as follows. With each subset $A \subseteq \{1, \dots, 5\}$, we associate $w(A) := \sum_{a \in A} 2^{5-a}$ and we set $\Gamma_{w(A)} := \{f \in O : f^* = A\}$.

Denote by J the clone of all projections and recall that Sh is the set of all Sheffer functions.

Proposition 1.10. *Among the sets $\Gamma_0, \dots, \Gamma_{31}$, exactly the following fifteen sets are nonempty:*

$$\Gamma_0, \Gamma_1, \Gamma_2, \Gamma_3, \Gamma_5, \Gamma_7, \Gamma_{12}, \Gamma_{14}, \Gamma_{15}, \Gamma_{20}, \Gamma_{22}, \Gamma_{23}, \Gamma_{26}, \Gamma_{27}, \Gamma_{31}.$$

Moreover,

$$\begin{aligned}
\Gamma_0 &= J,\\
\Gamma_2 &= \{a_1x_1\dot{+}\cdots\dot{+}a_nx_n : a_1+\cdots+a_n > 1 \text{ and odd}\},\\
\Gamma_{12} &= \langle c_0\rangle \setminus J,\\
\Gamma_{14} &= \{a_1x_1\dot{+}\cdots\dot{+}a_nx_n : a_1+\cdots+a_n > 0 \text{ and even}\},\\
\Gamma_{20} &= \langle c_1\rangle \setminus J,\\
\Gamma_{22} &= \{1\dot{+}a_1x_1\dot{+}\cdots\dot{+}a_nx_n : a_1+\cdots+a_n > 0 \text{ and even}\},\\
\Gamma_{26} &= \{1\dot{+}a_1x_1\dot{+}\cdots\dot{+}a_nx_n : a_1+\cdots+a_n > 1 \text{ and odd}\},\\
\Gamma_{31} &= Sh.
\end{aligned}$$

Proof. Call $f \in O$ *idempotent* if $f \in \mathrm{Pol}\{0\} \cap \mathrm{Pol}\{1\}$, that is, $f(x,\ldots,x) \approx x$.

(a) We show $\Gamma_0 = J$. First Γ_0 is the intersection of the five maximal clones, hence a clone by Section 1.2, and therefore $\Gamma_0 \supseteq J$. Suppose to the contrary that there exists $f \in \Gamma_0 \setminus J$. Then f is constant by Lemma 1.9(6), in contradiction to the idempotency of f. Thus, $\Gamma_0 = J$.
(b) The Boolean function $f(x_1,x_2,x_3) \approx x_1x_2\dot{+}x_1x_3\dot{+}x_2x_3$ belongs to Γ_1. Indeed, f is idempotent, selfdual (because $f(x_1\dot{+}1,x_2\dot{+}1,x_3\dot{+}1)\dot{+}1 \approx f(x_1,x_2,x_3)$), monotone (due to $f(x_1,x_2,x_3)=1 \Leftrightarrow x_1+x_2+x_3 \geq 2$), and clearly nonlinear.
(c) Γ_2 consists of idempotent, selfdual and nonmonotone functions, which translates into the condition stated in the second part of the proposition. The same applies to Γ_{26}.
(d) Let $f(x,y,z) \approx xy\dot{+}xz\dot{+}yz\dot{+}x\dot{+}y$. Clearly f is idempotent and nonlinear. It is also selfdual and nonmonotone because, for example; $f(1,0,0)=1 > 0 = f(1,0,1)$. Thus, $f \in \Gamma_3$.
(e) The disjunction $\vee$ belongs to Γ_5.
(f) $f(x,y,z) \approx x\dot{+}xy\dot{+}xyz$ belongs to Γ_7 because it is idempotent, nonselfdual ($f(1,1,0)=f(0,0,1)=0$), nonmonotone ($f(1,0,0)=1>0=f(1,1,0)$), and nonlinear.
(g) Γ_{12} and Γ_{20} are of the form indicated in the second part of the statement.
(h) Γ_{14} and Γ_{22} are of the form indicated in the second part of the statement.
(i) $f(x,y) \approx xy'$ belongs to Γ_{15}.
(j) Γ_{31} is the set of Sheffer functions.

The following table indicates which statement of Lemma 1.9 can be used to prove that Γ_i is void for the remaining values of i:

i	4,6	8–11	13	16–19	21	24, 25, 28, 29	30
Statement #	(4)	(2)	(5)	(2)	(5)	(1)	(3)

■

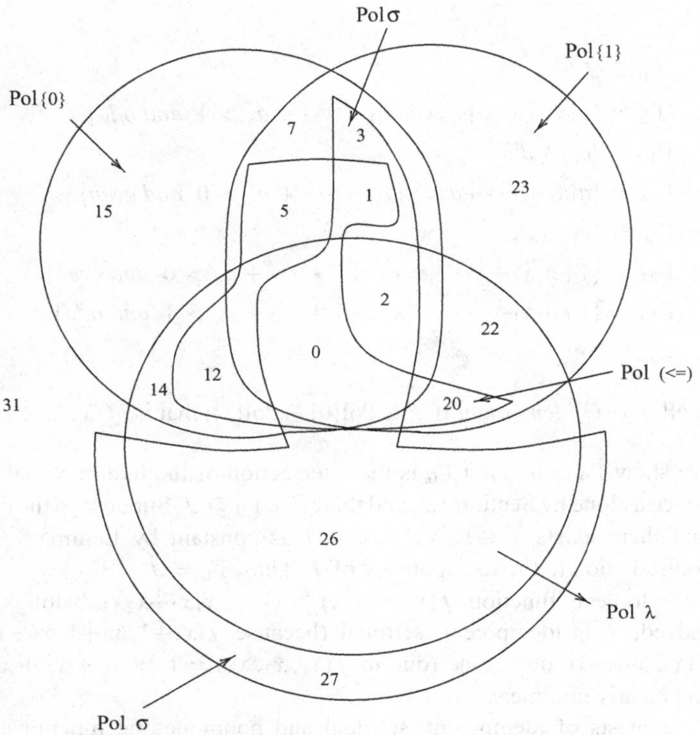

Figure 1.2. Venn diagram of the five maximal clones:

$$\begin{aligned}
\mathrm{Pol}\{0\} &= \Gamma_0 \cup \cdots \Gamma_3 \cup \Gamma_5 \cup \Gamma_7 \cup \Gamma_{12} \cup \Gamma_{14} \cup \Gamma_{15},\\
\mathrm{Pol}\{1\} &= \Gamma_0 \cup \cdots \Gamma_3 \cup \Gamma_5 \cup \Gamma_7 \cup \Gamma_{20} \cup \Gamma_{22} \cup \Gamma_{23},\\
\mathrm{Pol}\sigma &= \Gamma_0 \cup \cdots \Gamma_3 \cup \Gamma_{26} \cup \Gamma_{27},\\
\mathrm{Pol} \leq &= \Gamma_0 \cup \Gamma_1 \cup \Gamma_5 \cup \Gamma_{12} \cup \Gamma_{20},\\
\mathrm{Pol}\lambda &= \Gamma_0 \cup \Gamma_2 \cup \Gamma_{12} \cup \Gamma_{14} \cup \Gamma_{20} \cup \Gamma_{22} \cup \Gamma_{26}.
\end{aligned}$$

Remark 1.13. *In the second part of the foregoing proposition, we have characterized only some of the* Γ_i. *The description of the other* Γ_i *can be obtained from their definitions. For example,* $\Gamma_1 = (Pol\{0\} \cap Pol\{1\} \cap Pol\sigma \cap Pol \leq \cap\ CPol\lambda)$*, is the set of all selfdual, monotone, and nonlinear* f *satisfying* $f(0,\ldots,0) = 0$.

Remark 1.14. *The situation is depicted in Figure 1.2. In this Venn diagram of the five maximal clones, the ith region represents the set* Γ_i. *For example, the central region* 0 *represents the set* Γ_0 *(of all projections), which is the intersection of all maximal clones.*

Remark 1.15. *The sizes of certain* $\Gamma_i^{(n)}$ *(= the n-ary functions from* Γ_i*) were given in [14]:*

$$\begin{aligned}
|\Gamma_2^{(n)}| &= 2^{n-1} - n, \\
|\Gamma_{12}^{(n)}| &= |\Gamma_{20}^{(n)}| = 1, \\
|\Gamma_{14}^{(n)}| &= |\Gamma_{22}^{(n)}| = 2^{n-1} - 1, \\
|\Gamma_{15}^{(n)}| &= |\Gamma_{23}^{(n)}| = 2^{2^{n-2}} - 2^{n-1}, \\
|\Gamma_{26}^{(n)}| &= 2^{n-1}, \\
|\Gamma_{27}^{(n)}| &= 2^{2^{n-1}} - 2^{n-1}.
\end{aligned}$$

Because Γ_0 *is the clone of all projections, we have* $|\Gamma_0^{(n)}| = n$*. Next,* $|\Gamma_{31}^{(n)}|$ *was given in Corollary 1.6 (iii). Finally,*

$$\begin{aligned}
|\Gamma_1^{(n)}| + |\Gamma_3^{(n)}| &= 2^{2^{n-1}-1} - 2^{n-1}, \\
|\Gamma_5^{(n)}| + |\Gamma_7^{(n)}| &= 2^{2^n-2} - 2^{2^{n-1}-1} - 2^{n-1}.
\end{aligned}$$

Definition 1.3. *A complete set F of Boolean functions is a* basis *if no proper subset of F is complete, that is, if F is irredundant (or irreducible) with respect to completeness.*

The characteristic sets (see Definition 1.2) provide a tool for the description of bases. According to the Completeness Criterion, a set F of Boolean functions is a basis exactly if $F^* := \{f^* \in F\}$ is an irredundant cover of $\{1, \ldots, 5\}$ (i.e., F^* is a cover of $\{1, \ldots, 5\}$ such that each proper subfamily of F^* misses at least one element of $\{1, \ldots, 5\}$). Each Boolean function belongs to exactly one of the 15 sets listed in Proposition 1.10. The corresponding sets are $\emptyset$ and

$$\begin{aligned}
&\{5\}, \{4\}, \{4,5\}, \{3,5\}, \{3,4,5\}, \{2,3\}, \{2,3,4\}, \{2,3,4,5\}, \\
&\{1,3\}, \{1,3,4\}, \{1,3,4,5\}, \{1,2,4\}, \{1,2,4,5\}, \{1,2,3,4,5\} \qquad (1.8)
\end{aligned}$$

(for example, $\Gamma_{14}^* = \{2, 3, 4\}$ because $14 = 2^3 + 2^2 + 2^1$). The determination of all possible types of bases amounts to the problem of finding all irredundant covers of $\{1, \ldots, 5\}$ formed from the 14 sets listed in (1.8). This is a purely technical problem that can be solved by a simple computer program. It is so small that it was solved by hand [9, 12, 14]. There are exactly 1, 17, 22 and 2 irredundant covers consisting of 1, 2, 3 and 4 sets, respectively. For example, the cover by a single set is $\{1, \ldots, 5\}$, which evidently corresponds to the set Γ_{31} of all Sheffer functions. An example of an irredundant 2-set cover is $\{1, 2, 4, 5\}$, $\{2, 3, 4, 5\}$. A corresponding basis is $\{f, g\}$ with $f \in \Gamma_{27}$ and $g \in \Gamma_{15}$, that is, f is selfdual but belongs to no other maximal clone and g preserves 0 but belongs to no other maximal clone.

The total number of bases consisting of two n-ary functions is given in [14]. From this it follows that the proportion of such bases among all pairs of n-ary functions goes fast to $\frac{1}{8}$ as $n \to \infty$. Together with Corollary 1.6(iii), this shows that a randomly chosen pair of n-ary Boolean functions is complete with probability practically equal to 0.375.

Two bases F and G are of the same type if $F = \{f_1, \ldots, f_n\}$ and $G = \{g_1, \ldots, g_n\}$ where $f_1 \sim g_1, \ldots, f_n \sim g_n$. The result quoted earlier states: There are exactly 42 types of bases, and each basis consists of at most four functions. The latter statement ([10] p. 20) follows from the fact that $f(0, \ldots, 0) = 1$ for some f in the basis, and so either $f(1, \ldots, 1) = 0$ or f is not selfdual. A four-element basis is $\{c_0, c_1, \cdot, s_3\}$ where $s_3(x_1, x_2, x_3) \approx x_1 \dot{+} x_2 \dot{+} x_3$.

1.3 A Description of the Post Lattice

1.3.1 Definition of the Post Lattice

In the previous section, we saw the role of the five maximal clones of Boolean functions and their intersections. It is natural to ask about the other clones of Boolean functions and their inclusions; this leads to the following problems. Denote by $\mathcal{C}$ the set of clones. One of the first questions about $\mathcal{C}$ may be its size. As clones are subsets of the countably infinite set O (of all Boolean functions), a priori $|\mathcal{C}|$ could be any cardinal less than or equal to $2^{\aleph_0}$, the cardinality of the set of reals. Twenty years after his completeness paper, E. L. Post showed in [21] that $|\mathcal{C}| = \aleph_0$. (It turned out later that this is exceptional because almost all other infinite variants are of continuum cardinality; a fact that, e.g., distinguishes the classical two-valued logic from many-valued logics.)

Consider a nonvoid subset $\{C_i : i \in I\}$ of $\mathcal{C}$. It is easy to see that

$$C = \bigcap_{i \in I} C_i$$

is the greatest clone contained in every C_i. The clone C is called the meet (or infimum) of $\{C_i : i \in I\}$.

Consider now any subset F of O and set

$$\mathcal{C}(F) = \{X \in \mathcal{C} : X \supseteq F\}.$$

Clearly $\mathcal{C}(F)$, the family of all clones containing F, is nonvoid because it contains the greatest clone O. Thus $\bigcap_{X \in \mathcal{C}(F)} X$ is a clone, denoted by $\langle F \rangle$, and called the clone generated by F. Clearly $\langle F \rangle$ contains F; hence $\langle F \rangle$ is a member of $\mathcal{C}(F)$, and thus $\langle F \rangle$ is the least clone containing F. (The argument we just presented is the standard one for the existence of the vector subspace spanned by a set of vectors, the subgroup generated by a subset of a group, etc.) It follows that

$$\langle \bigcup_{i \in I} C_i \rangle$$

is the least clone containing all clones C_i ($i \in I$). It is called the join (or supremum) of $\{C_i : i \in I\}$.

Because of the existence of meets and joins, the ordered set $\mathcal{L} = (\mathcal{C}, \subseteq)$, where $\subseteq$ is the set inclusion (or containment), is a so-called complete lattice. In his landmark 100-page paper [21], Post gave a full description of the lattice $\mathcal{L}$ (today called the Post lattice). Post actually characterized all composition-closed sets of Boolean functions: however, there are only seven such sets that are not clones – that is, do not contain all projections – and these very small sets can be easily described.

1.3.2 Duality

We start with a helpful symmetry of the lattice $\mathcal{L}$.

The dual of an n-ary Boolean function f is the Boolean function f^∂ defined by setting

$$f^\partial(x_1, \ldots, x_n) \approx (f(x_1', \ldots, x_n'))'$$

(where x' is the negation). It can be checked that $f^{\partial\partial} = f$. The duals of $\vee$ and $+$ are $\cdot$ and $\leftrightarrow$, respectively. Also, f is selfdual (see Example 1.3) if and only if $f = f^\partial$. To every $F \subseteq O$, assign $F^\partial := \{f^\partial : f \in F\}$. If F is a clone, then F^∂ is also a clone and $F \mapsto F^\partial$ is a lattice automorphism of $\mathcal{L}$, that is, a bijection (or 1-1 and onto selfmap) of $\mathcal{C}$ onto itself respecting the lattice joins and meets (to see it, check that $f \mapsto f^\partial$ respects the composition).

A finite ordered set is usually represented by its (Hasse) diagram. In such a drawing, vertices correspond to the elements of the set and two vertices are joined by a line segment exactly if the element corresponding to the vertex drawn higher on the page covers the element corresponding to the vertex drawn lower on the page. Here *a covers b* means that $a > b$ but $a > c > b$ holds for no c. In certain cases we also can draw diagrams of infinite orders.

In view of the automorphism $F \mapsto F^\partial$ of $\mathcal{L}$, we can draw a diagram of $\mathcal{L}$ so that it is symmetric with respect to the central vertical line: in other words, so that the left-hand half is the mirror image of the right-hand half; that is, $F \mapsto F^\partial$ acts horizontally. This symmetry is tied to the relational description of clones as follows.

Fact 3. *If ρ is an h-ary relation on B and*

$$\rho' := \{(a_1', \ldots, a_n') : (a_1, \ldots, a_n) \in \rho\},$$

then $(Pol\rho)^\partial = Pol\rho'$.

For example, $(Pol\{0\})^\partial = Pol\{0'\} = Pol\{1\}$, $(Pol\{1\})^\partial = Pol\{0\}$, while $F^\partial = F$ for the remaining three maximal clones because $\leq^\partial$, σ^∂ and λ^∂ equal $\geq$, σ and λ, respectively.

Proof. First we show

$$(\text{Pol}\rho)^\partial \subseteq \text{Pol}\rho'. \tag{1.9}$$

Let $f \in (\mathrm{Pol})^\partial$ be n-ary and $M = [m_{ij}]$ be any $h \times n$ matrix whose columns are all in ρ'. Let $M' = [m'_{ij}]$ and, for $i = 1, \ldots, h$, denote by r_i and r'_i the ith row of M and M', respectively. Clearly $f^\partial \in \mathrm{Pol}\rho$ and every column of M' is in ρ and so $(f^\partial(r'_1), \ldots, f^\partial(r'_h)) \in \rho$. Here $f^\partial(r'_i) = f'(r_i)$, proving $(f(r_1), \ldots, f(r_h)) \in \rho'$ and $f \in \mathrm{Pol}\rho'$.

For the converse inclusion, apply (1.9) to ρ':

$$\mathrm{Pol}\rho' = (\mathrm{Pol}\rho')^{\partial\partial} \subseteq (\mathrm{Pol}\rho'')^\partial = (\mathrm{Pol}\rho)^\partial. \qquad \blacksquare$$

1.3.3 Breaking the Diagram into Pieces

Now we are ready to describe the Post lattice $\mathcal{L}$. Unfortunately, its diagram, given in Figure 1.3, looks complicated no matter how it is drawn. Actually, in most cases the diagram is either taken directly from [21] or it is only slightly modified. Following an idea from [18] we break the diagram into smaller pieces presented separately whereby the diagram of the parts are easy to draw. When this is done, we assemble the pieces into one diagram.

Notation also poses a problem. In most cases Post's original symbols – or their modifications – are used, but they are hard to remember. We mix three different notations: (1) $\mathrm{Pol}\rho$ where ρ is a relation; e.g., $\mathrm{Pol}\{0\}$; (2) $\langle F\rangle$ where $F \subseteq O$; for

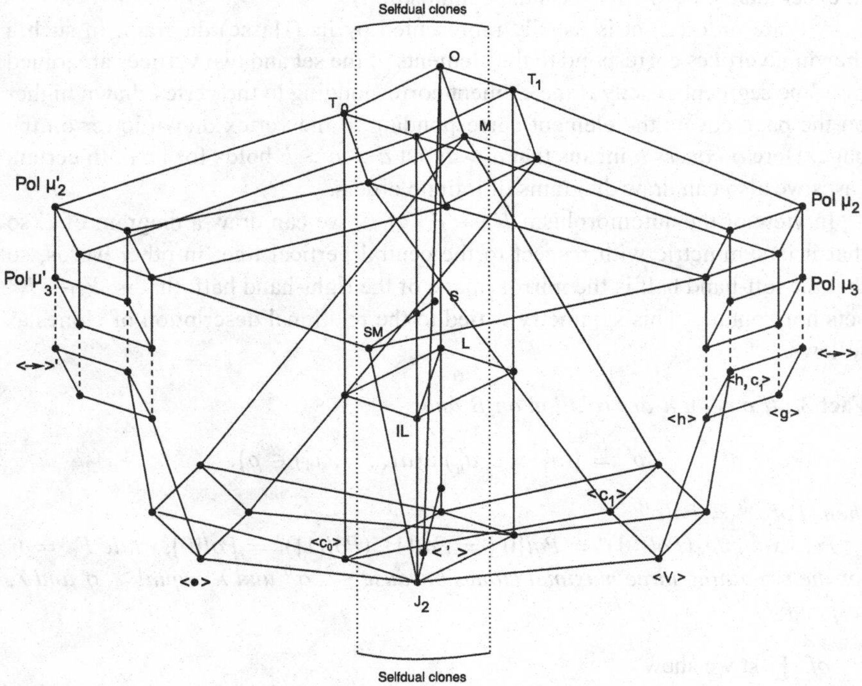

Figure 1.3. The Post lattice $\mathcal{L}$.

example, $\langle\vee\rangle$ is the clone generated by the disjunction; and (3) special notations: $T_0 = \mathrm{Pol}\{0\}$, $T_1 = \mathrm{Pol}\{1\}$, $S = \mathrm{Pol}\sigma$, $M = \mathrm{Pol} \leq$ and $L = \mathrm{Pol}\lambda$ for the maximal clones and I for $T_0 \cap T_1$. We abbreviate $S \cap M$ by SM (i.e., SM is the clone of all functions that are both selfdual and monotone), and similarly for other intersections of clones.

In order to find all clones, we start with the five maximal clones and their intersections. In principle Proposition 1.10 (see Figure 1.2) may be used to determine all the intersections of maximal clones. Nevertheless, we provide more details now and draw the diagrams of certain easily describable intervals.

We start with a trivial case. Recall that a Boolean function f is idempotent if $f(x, \ldots, x) \approx x$. Clearly $I := T_0T_1(= T_0 \cap T_1)$ is the clone of idempotent functions.

1.3.4 The Clone SM

The clone SM is the intersection of the maximal clones $S = \mathrm{Pol}\sigma$ and $M = \mathrm{Pol} \leq$. This clone consists of selfdual monotone functions. A quick check shows that among the unary and binary functions, only projections are selfdual and monotone, and this may lead to the impression that perhaps SM is the least clone J of all projections. But SM contains a nontrivial ternary function

$$m(x_1, x_2, x_3) :\approx x_1x_2 \vee x_1x_3 \vee x_2x_3,$$

called the majority or median function. It is clearly monotone. According to the De Morgan laws, its dual m^∂ satisfies

$$\begin{aligned} m^\partial(x_1, x_2, x_3) &\approx (x_1'x_2' \vee x_1'x_3' \vee x_2'x_3')' \\ &\approx (x_1'' \vee x_2'')(x_1'' \vee x_3'')(x_2'' \vee x_3'') \\ &\approx (x_1 \vee x_2)(x_1 \vee x_3)(x_2 \vee x_3). \end{aligned}$$

Applying the distributive law, it is easy to verify that

$$(x_1 \vee x_2)(x_1 \vee x_3)(x_2 \vee x_3) \approx x_1x_2 \vee x_1x_3 \vee x_2x_3,$$

proving $m^\partial = m$ and $m \in S$. (Actually, the last identity is important because it characterizes the distributive lattices among all lattices.) An easy check shows that m and the projections are the only ternary functions in SM. However, SM contains many interesting functions of higher arity. (Median algebras, see, e.g., [1], generalize the majority function and some other functions from SM and seem to have some applications in social sciences.) It is immediate that every function from SM is idempotent, and so SM is contained in all maximal clones except the clone L of linear functions. Moreover, SM is an atom of the Post lattice. Here an atom of $\mathcal{L}$, called a minimal clone, is a clone properly containing exactly the clone J. For $A, B \in \mathcal{C}$, $A \subseteq B$, the sets

$$[A) = \{X \in \mathcal{C} : X \supseteq A\}, \quad [A, B] = \{X \in \mathcal{C} : A \subseteq X \subseteq B\}$$

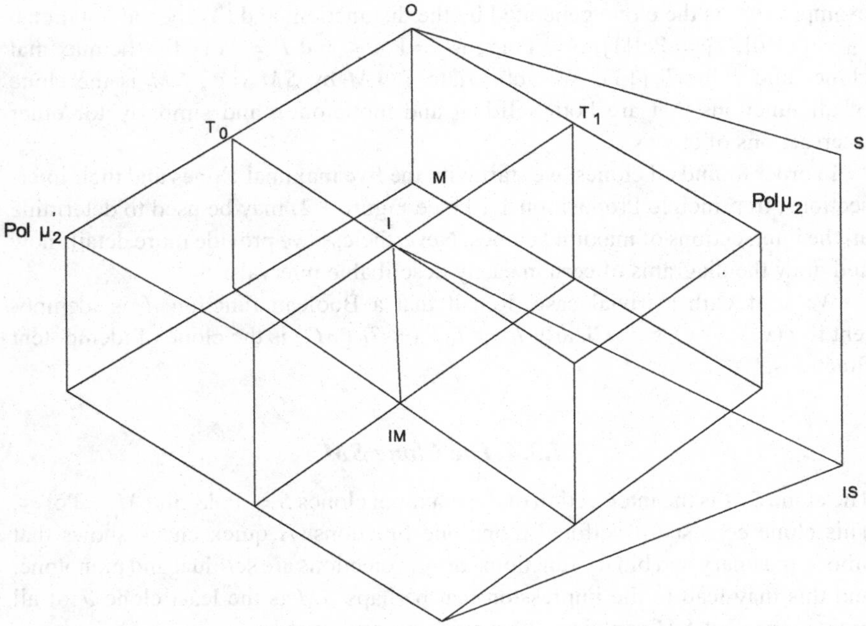

Figure 1.4. The filter $[SM)$.

are called a filter and an interval of $\mathcal{L}$, respectively. The diagram of the filter $[SM)$ (of all clones containing the clone SM) is shown in Figure 1.4. Apart from the four maximal clones T_0, T_1, S, M, the filter consists of their intersections (e.g., I, T_0M, IM, and IS) and two additional clones (which are meet-irreducible in the lattice theory terminology). One is $\mathrm{Pol}\mu_2$ where

$$\mu_2 := B^2 \setminus \{(0,0)\},$$

and the other is its dual $\mathrm{Pol}\mu_2'$ where $\mu_2' = B^2 \setminus \{(1,1)\}$.

1.3.5 The Clone IL

Another filter that can be easily described is $[IL)$. The clone

$$IL = \mathrm{Pol}\{0\}\mathrm{Pol}\{1\}\mathrm{Pol}\lambda$$

consists of linear idempotent functions. An n-ary $f \in IL$ is of the form $a \dotplus x_{i_1} \dotplus \cdots \dotplus x_{i_m}$ with $a \in B$. Here $a = 0$ due to $f(0,\ldots,0) = 0$, and m is odd due to $f(1,\ldots,1) = 1$. Consequently, IL consists of functions of the form $x_{i_1} \dotplus \cdots \dotplus x_{i_{2k+1}}$ with $k \geq 0$. It is easy to verify that IL is the intersection of the four maximal clones T_0, T_1, L, and S. The clone IL is generated by each of its nontrivial functions (i.e., $f \in IL \setminus J$ of the form $x_1 \dotplus \cdots \dotplus x_{2k+1}$ with $k > 0$).

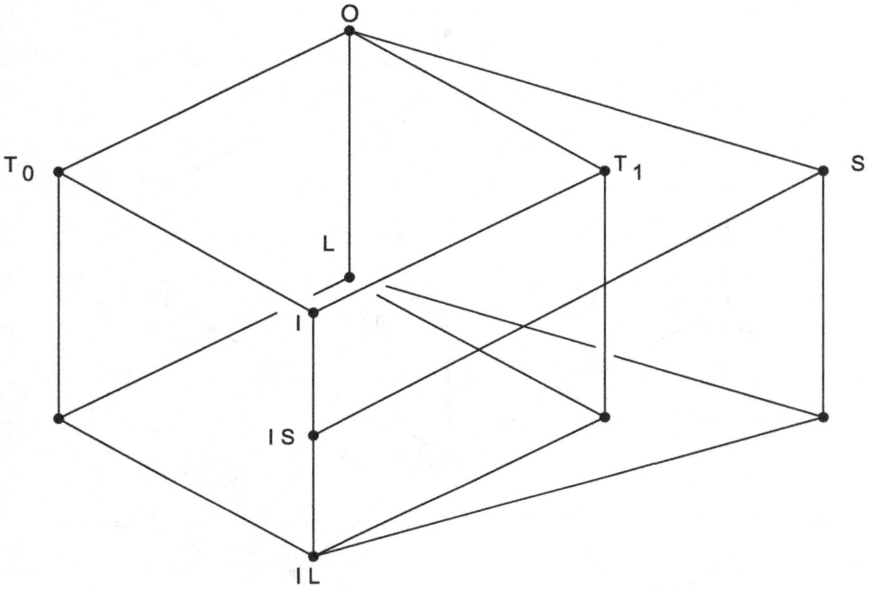

Figure 1.5. The filter $[IL)$.

Indeed, for $k > 1$,

$$x_1 \dot{+} \cdots \dot{+} x_{2k+1} \approx (\cdots((x_1 \dot{+} x_2 \dot{+} x_3) \dot{+} x_4 \dot{+} x_5) \dot{+} \cdots) \dot{+} x_{2k} \dot{+} x_{2k+1}.$$

Conversely,

$$x_1 \dot{+} x_2 \dot{+} x_3 \approx x_1 \dot{+} x_2 \dot{+} x_3 \dot{+} x_3 \dot{+} \cdots \dot{+} x_3.$$

This shows that IL is a minimal clone. The interval $[IL)$ is shown in Figure 1.5.

1.3.6 The Clone ML

The last nontrivial intersection of two maximal clones is the clone $ML := \mathrm{Pol} \leq \cap \mathrm{Pol}\lambda$ of the monotone linear functions. We have already seen in Lemma 1.9(6) that $ML = \langle c_0, c_1 \rangle$ is made up from the constants and the projections. Now ML is contained in neither of T_0, T_1 and S. Between ML and M there are exactly the two mutually dual clones $\langle \vee, c_0, c_1 \rangle$ and $\langle \cdot, c_0, c_1 \rangle$, and between ML and L there is precisely the clone $\langle\, ', c_0 \rangle$. However, ML is not minimal; in fact, ML is obviously the join of the two minimal clones $\langle c_0 \rangle$ and $\langle c_1 \rangle$, and ML covers no other clone. The interval $[J, ML]$ and the filter $[ML)$ are shown in Figure 1.6.

1.3.7 The Seven Minimal Clones

In Sections 1.3.4 through 1.3.6, we have seen all nontrivial intersections of maximal clones (the intersection of all five maximal clones is obviously the trivial clone J

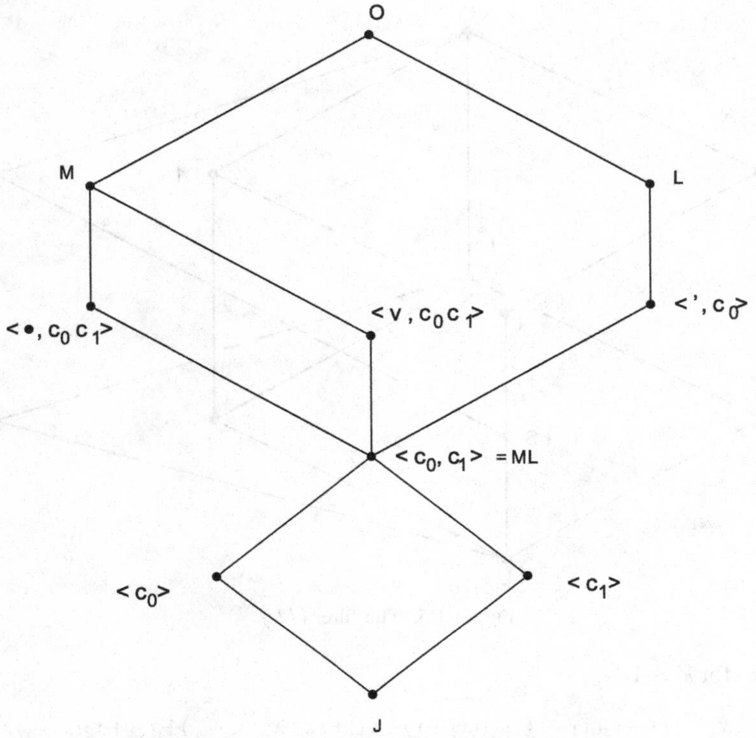

Figure 1.6. The interval $[J, ML]$ and the filter $[ML)$.

of all projections). Two of them, namely SM and IL, are even minimal clones, and the third, namely ML, covers two minimal clones $\langle c_0 \rangle$ and $\langle c_1 \rangle$. At this point the reader may be wondering about the full list of all minimal clones. Obviously there is also the minimal clone $\langle \, ' \rangle$ generated by the negation, but the existence of other minimal clones is not obvious. In fact, there are only two more minimal clones, namely $\langle \vee \rangle$ and $\langle \cdot \rangle$.

Proposition 1.11. *There are exactly seven minimal clones:*

$$\langle c_0 \rangle, \langle c_1 \rangle, \langle \, ' \rangle, \langle \vee \rangle, \langle \cdot \rangle, SM, IL$$

(where SM is generated by the majority function $x_1x_2 \vee x_1x_3 \vee x_2x_3$ and IL is generated by $x_1 \dotplus x_2 \dotplus x_3$).

The minimality of the clones $\langle \vee \rangle$ and $\langle \cdot \rangle$ is easily checked. We omit the more complex proof that there are no other minimal clones. The various joins of the seven minimal clones can be verified directly. Their diagram is drawn in Figure 1.7.

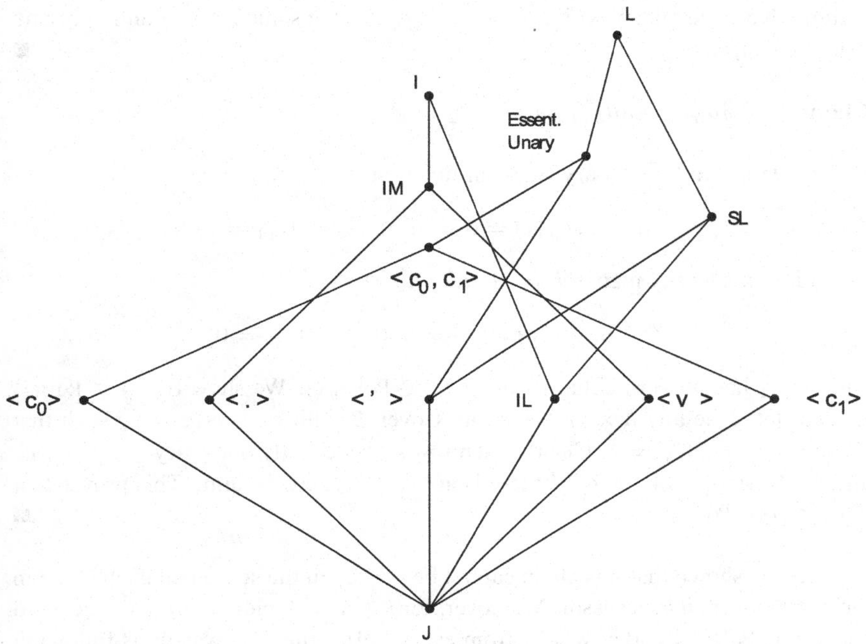

Figure 1.7. Joins of minimal clones.

1.3.8 An Infinite Descending Chain of Clones

So far, starting from the top and the bottom, we have described a few clones, but a major part of the Post lattice is still missing. The key to it is a countably infinite descending chain of clones to be described in this section.

For $h \geq 1$, set

$$\mu_h = B^h \setminus \{(0, \dots, 0)\}.$$

This very natural relation consists of all the $2^h - 1$ h-tuples having at least one coordinate 1. The fact that only $(0, \dots, 0)$ does not belong to μ_h is behind the following "backward" formulation of $f \in \mathrm{Pol}\mu_h$. An n-ary function f preserves μ_h if and only if the following holds: If the values of f on the rows of an $h \times n$ zero-one matrix A are all 0, then A has a zero column (i.e., a column $(0, \dots, 0)^T$). For example, c_1, $\vee$ and $\rightarrow$ belong to every $\mathrm{Pol}\mu_h$. Note that $\mu_1 = B \setminus \{0\} = \{1\}$. We show

$$T_1 = \mathrm{Pol}\mu_1 \supset \mathrm{Pol}\mu_2 \supset \mathrm{Pol}\mu_3 \supset \cdots.$$

Claim 1. $\mathrm{Pol}\mu_h \supseteq \mathrm{Pol}\mu_{h+1}$ *for all* $h \geq 1$.

Proof. Let $f \in \mathrm{Pol}\mu_{h+1}$ be n-ary and let A be any $h \times n$ matrix over B with rows $r_1, \dots, r_h$ such that $f(r_1) = \cdots = f(r_h) = 0$. Denote by A the $(h+1) \times n$ matrix with rows $r_1, \dots, r_h, r_h$. Because f preserves μ_{h+1}, the matrix A has a zero column. Now the matrix with rows $r_1, \dots, r_h$ has the same zero column, proving that $f \in \mathrm{Pol}\mu_h$. ■

Claim 2. $\mathrm{Pol}\mu_h \supset \mathrm{Pol}\mu_{h+1}$ *for all* $h \geq 1$.

Proof. Define an $(h+1)$-ary Boolean function f by setting

$$f(a_1, \dots, a_{h+1}) = 0 \Leftrightarrow a_1 + \dots + a_{h+1} = 1.$$

Consider the identity matrix I_{h+1} of order $h+1$. As

$$f(1, 0, \dots, 0) = \dots = f(0, \dots, 0, 1) = 0$$

while I_{h+1} has no zero column, clearly $f \notin \mathrm{Pol}\mu_{h+1}$. We show that $f \in \mathrm{Pol}\mu_h$. Indeed, let A be any $h \times (h+1)$ matrix over B with rows $r_1, \dots, r_h$ such that $f(r_1) = \dots = f(r_h) = 0$. Clearly each row r_i has exactly one entry 1, and hence among the $h+1$ column of A there is at least one zero column. This proves that $f \in \mathrm{Pol}\mu_h \setminus \mathrm{Pol}\mu_{h+1}$. ■

It can be shown that this chain cannot be refined in the sense that each clone in the chain covers its successor. Moreover, for $i \geq 1$ each clone $\mathrm{Pol}\mu_{i+1}$ is covered exactly by $\mathrm{Pol}\mu_i$; in other words, from every $\mathrm{Pol}\mu_i$, the only way up is following the chain. It is natural to ask about the intersection $\bigcap_{i=2}^{\infty} \mathrm{Pol}\mu_i$. It can be shown that it is the clone $\langle\rightarrow\rangle$. This clone consists of functions that can be built from $\rightarrow$ alone, such as, $g(x, y, z) :\approx (y \rightarrow z) \rightarrow x \approx x \vee yz'$. In fact, the clone $\langle\rightarrow\rangle$ and seven other related clones are the only clones not of the form $\mathrm{Pol}\rho$ for a (finitary) relation ρ on B. A good part of the calculus of the 2-valued propositional logic studies certain properties of the clone $\langle\rightarrow\rangle$. For example, an n-ary tautology based only on $\rightarrow$ expresses a way the n-ary constant c_1^n is composed from $\rightarrow$. The fact that $\langle\rightarrow\rangle$ is among the eight exceptional clones (that are intersections of infinite descending chains of clones) is certainly not coincidental.

1.3.9 Other Infinite Descending Chains

The intersections of the clones $\mathrm{Pol}\mu_h$ with the maximal clone $T_0 = \mathrm{Pol}\{0\}$ form another infinite descending chain

$$I = T_0\mathrm{Pol}\mu_1 \supset T_0\mathrm{Pol}\mu_2 \supset \cdots . \tag{1.10}$$

Similarly, the intersections of $\mathrm{Pol}\mu_h$ with the maximal clone $M = \mathrm{Pol} \leq$ make up the chain

$$M\mathrm{Pol}\mu_1 \supset M\mathrm{Pol}\mu_2 \supset \cdots , \tag{1.11}$$

and then we have the fourth chain

$$T_0M\mathrm{Pol}\mu_1 \supset T_0M\mathrm{Pol}\mu_2 \supset \cdots \tag{1.12}$$

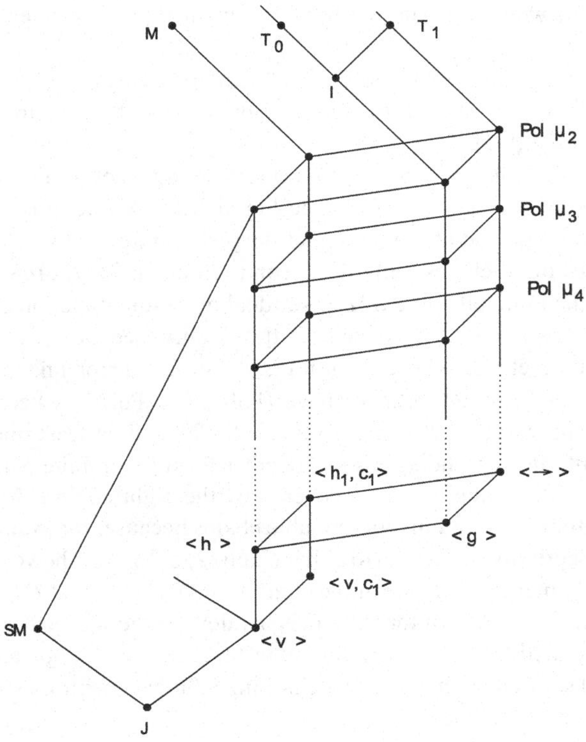

Figure 1.8. Infinite descending chains.

(see Figure 1.8). The intersection of all the clones in the chain (1.10) is the clone $T_0\langle\rightarrow\rangle$ (of all 0-preserving functions from $\langle\rightarrow\rangle$). This clone is generated by

$$g(x, y, z) \approx x \vee yz'.$$

Similarly, the intersection of all the clones of the chain (1.11) is the clone $M\langle\rightarrow\rangle$ generated by $\{h, c_1\}$, where

$$h(x, y, z) :\approx x \vee yz.$$

Finally, the intersection of the chain (1.12) is the clone $T_0M\langle\rightarrow\rangle$ generated by h.

The n-ary monotone Boolean functions are in a one-to-one correspondence with Sperner hypergraphs on $N = \{1, \ldots, n\}$: that is, the inclusion-free families of subsets of N. The correspondence is provided by the fact that such a Boolean function f has a unique negation-free disjunctive normal form

$$\bigvee_{I \in F} \prod_{i \in I} x_i,$$

where F is a Sperner hypergraph on N. The connection is important for both theories; for example, the foregoing clones $\mathrm{Pol}\mu_h$ are related to weak chromatic numbers of hypergraphs. Although there are many interesting results and problems, to present them would take us far from the main topic; we refer for instance to [2, 3].

The intersection $T_0M\langle\rightarrow\rangle$ of the chain (1.12), that is, the clone $\langle h\rangle$, covers only the minimal clone $\langle\vee\rangle$, whereas the intersection $M\langle\rightarrow\rangle$ of the chain (1.11) covers precisely the clone $\langle\vee, 1\rangle$.

The four chains and their four intersections are attached to the remainder of the Post lattice only at a few places (see Figure 1.8). At the top, this occurs at T_0, I, T_0M, and IM, and at the bottom at the just-mentioned two clones $\langle h\rangle$ and $\langle h, c_1\rangle$. Besides these clones, only the second top clone $T_0M\mathrm{Pol}\mu_2$ of the chain (1.12) covers the minimal clone SM of selfdual monotone functions.

The symmetry $C \to C^\partial$ (induced by the duality, see Section 1.3.2) yields four more infinite chains whose diagram is the exact mirror image of Fig. 1.8. As mentioned in Section 1.3.2, we have $(\mathrm{Pol}\mu_h)^\partial = \mathrm{Pol}\mu'_h$, where $\mu'_h = B^h \setminus \{(1,\ldots,1)\}$ consists of all $2^h - 1$ zero-one h tuples with at least one coordinate 0. The eight infinite descending chains together form the infinite part of the Post lattice (although this infinite part is covered by the eight chains, the Post lattice contains $\aleph_0$ infinite descending nonrefinable chains because, for example, we may switch at any depth from chain (1) to either chain (2) or (3), etc; however each such chain ultimately merges with one of the eight chains). Notice that $\mathcal{L}$ has no infinite ascending chain. It is known that this is equivalent to the fact that each clone C is finitely generated (i.e., $C = \langle F\rangle$ for some finite set F of Boolean functions). E. L. Post [21] showed much more by exhibiting a finite generating set of the least possible size for each clone.

1.3.10 Assembling the Pieces

Now we are ready to assemble the full lattice $\mathcal{L}$ from the pieces described earlier. Its full diagram is in Figure 1.3. For simplicity, only a few key symbols are listed – such as for the maximal and minimal clones – while the others can be determined from joins, meets, and the lateral symmetry $C \to C^\partial$. However, this symmetry of the left-hand and the right-hand half of the diagram is not rigidly adhered to, because otherwise all the selfdual clones would be drawn on the vertical axis of symmetry. Table 1.1 lists all clones numbered as in Figure 1.9. In this table the second column gives our notation or the description of each clone C as a join or meet of other clones. In the third column we list a basis (a generating set) of C of the least possible maximum arity. In the fourth column we give a least arity relation ρ such that $C = \mathrm{Pol}\rho$ provided it exists (i.e., for all clones except the eight intersections of the infinite descending chains). The fourth column lists the number of the dual clone C^∂ if $C^\partial \neq C$ and s if $C^\partial = C$. The second-to-last column gives Post's original notation, and finally the last column contains a description and remarks.

In the third column we list a basis B of the clone C such that the maximum arity o of its functions is as small as possible. The number o, called the order of

Table 1.1. A list of all Boolean clones

#	Not.	Basis	Relation	Dual	Post's not.	Remarks
1	O	NAND	$\emptyset$	S	C_1	Greatest clone; all functions
2	T_1	$\leftrightarrow, \vee$	(1)	3	C_2	Maximal clone; $f(1, \ldots, 1) = 1$.
3	T_0	$\dot{+}, \cdot$	(0)	2	C_3	Maximal clone; $f(0, \ldots, 0) = 0$.
4	M	$\cdot, \vee, c_0, c_1$	$\begin{pmatrix} 0 & 0 & 1 \\ 0 & 1 & 1 \end{pmatrix}$	S	A_1	Maximal clone; monotone functions.
5	S	$xy' \vee xz' \vee y'z'$	$\begin{pmatrix} 0 & 1 \\ 1 & 0 \end{pmatrix}$	S	D_3	Maximal clone; selfdual functions.
6	L	$\dot{+}, c_1$	$\begin{pmatrix} 0 & 0 & 0 & 0 & 1 & 1 & 1 & 1 \\ 0 & 0 & 1 & 1 & 0 & 0 & 1 & 1 \\ 0 & 1 & 0 & 1 & 0 & 1 & 0 & 1 \\ 0 & 1 & 1 & 0 & 1 & 0 & 0 & 1 \end{pmatrix}$	S	L_1	Maximal clone; linear (alternating) functions
7	J	$\emptyset$	$\begin{pmatrix} 0 & 0 & 1 \\ 0 & 1 & 0 \\ 1 & 0 & 0 \end{pmatrix}$	S	R_1	Least clone; Projections
8	$\langle c_1 \rangle$	c_1	$\begin{pmatrix} 0 & 0 & 0 & 1 & 1 & 1 \\ 0 & 0 & 1 & 0 & 1 & 1 \\ 0 & 1 & 0 & 1 & 0 & 1 \\ 1 & 1 & 1 & 1 & 1 & 1 \end{pmatrix}$	9	R_6	Minimal clone; essentially unary; all constants 1 MLT_1
9	$\langle c_0 \rangle$	c_0	$\begin{pmatrix} 0 & 0 & 0 & 1 & 1 & 1 \\ 0 & 0 & 1 & 0 & 1 & 1 \\ 0 & 1 & 0 & 1 & 0 & 1 \\ 0 & 0 & 0 & 0 & 0 & 0 \end{pmatrix}$	8	R_8	Minimal clone; essentially unary; all constants 0 MLT_0
10	$\langle \vee \rangle$	$\vee$	$\begin{pmatrix} 0 & 0 & 0 & 0 \\ 0 & 0 & 1 & 1 \\ 0 & 1 & 0 & 1 \\ 0 & 1 & 1 & 1 \\ 1 & 1 & 1 & 1 \end{pmatrix}$	11	S_1	Minimal clone; all disjunctions
11	$\langle \cdot \rangle$	$\cdot$	$\begin{pmatrix} 1 & 1 & 1 & 1 \\ 0 & 0 & 1 & 1 \\ 0 & 1 & 0 & 1 \\ 0 & 0 & 0 & 1 \\ 0 & 0 & 0 & 0 \end{pmatrix}$	10	P_1	Minimal clone; all conjunctions
12	$\langle ' \rangle$	$'$	$\begin{pmatrix} 0 & 0 & 0 & 1 & 1 & 1 \\ 0 & 1 & 1 & 0 & 0 & 1 \\ 1 & 0 & 1 & 0 & 1 & 0 \end{pmatrix}$	s	R_4	Minimal clone; essential unary
13	SM	m	$\begin{pmatrix} 0 & 0 & 1 \\ 0 & 1 & 0 \\ 1 & 0 & 1 \end{pmatrix}$	s	D_2	Minimal clone; All selfdual monotone
14	IL	s_3	$\begin{pmatrix} 0 & 0 & 0 & 0 \\ 0 & 0 & 1 & 1 \\ 0 & 1 & 0 & 1 \\ 1 & 0 & 0 & 1 \end{pmatrix}$	s	L_4	Minimal clone; all idempotent linear functions T_0LS

(continued)

Table 1.1. (*continued*)

#	Not.	Basis	Relation	Dual	Post's not.	Remarks
15	I	$\vee, s_3$	$\begin{pmatrix}0\\1\end{pmatrix}$	s	C_4	all idempotent functions
16	I_1M	$\cdot, \vee, c_1$	$\begin{pmatrix}0&0&1\\0&1&1\\1&1&1\end{pmatrix}$	17	A_2	all monotone except constants
17	T_0M	$\cdot, \vee, c_0$	$\begin{pmatrix}0&1&1\\0&0&1\\0&0&0\end{pmatrix}$	16	A_3	all monotone except constant 1
18	IM	$\cdot, \vee$	$\begin{pmatrix}0&0&0\\0&0&1\\0&1&1\\1&1&1\end{pmatrix}$	s	A_4	all nonconstant monotone
19	IS	m, s_3	$\begin{pmatrix}0&0\\0&1\\1&0\end{pmatrix}$	s	D_1	all 0-preserving (also 1-preserving) selfdual
20	T_1L	$\leftrightarrow$	$\begin{pmatrix}0&0&1&1\\0&1&0&1\\1&0&0&1\end{pmatrix}$	21	L_2	all 1-preserving linear
21	T_0L	$\dot{+}$	$\begin{pmatrix}0&0&1&1\\0&1&0&1\\0&1&1&0\end{pmatrix}$	20	L_3	functions of the form $x_{i_1}\dot{+}\cdots\dot{+}x_{i_k}$
22	SL	$s_3\dot{+}1$	$\begin{pmatrix}0&0&0&1&1&1&1&0\\0&0&1&0&1&1&0&1\\0&1&0&0&1&0&1&1\\1&0&0&0&0&1&1&1\end{pmatrix}$	s	L_5	functions of the form $x_{i_1}\dot{+}\cdots\dot{+}x_{i_{2k+1}}$ $\dot{+}a$
23	ML	c_0, c_1	$\begin{pmatrix}0&0&0&1\\0&0&1&1\\0&1&0&1\\0&1&1&1\end{pmatrix}$	s	R_{11}	all constants
24	$\langle O^{(1)}\rangle$	$c_0, {}'$	$\begin{pmatrix}0&0&0&1&1&1\\0&0&1&0&1&1\\0&1&0&1&0&1\end{pmatrix}$	s	R_{13}	all essentially unary
25	$\langle c_1\rangle \cup \langle\vee\rangle$	$c_1, \vee$	$\begin{pmatrix}1&1&1&1\\0&1&1&1\\0&0&1&1\\0&1&0&1\end{pmatrix}$	26	S_3	all disjunctions and constants 1
26	$\langle\cdot\rangle \cup \langle c_0\rangle$	$\cdot, c_0$	$\begin{pmatrix}0&0&0&0\\0&0&0&1\\0&0&1&1\\0&1&0&1\end{pmatrix}$	25	P_3	all conjunctions and constants 0
27	$\langle\vee\rangle \cup \langle c_0\rangle$	$\vee, c_0$	$\begin{pmatrix}0&1&1&1\\0&0&1&1\\0&1&0&1\\0&0&0&0\end{pmatrix}$	28	S_5	all disjunctions and constants 0
28	$\langle\cdot\rangle \cup \langle c_1\rangle$	$\cdot, c_1$	$\begin{pmatrix}0&0&0&1\\0&0&1&1\\0&1&0&1\\1&1&1&1\end{pmatrix}$	27	P_5	all conjunctions and constants 1

Table 1.1. *(continued)*

#	Not.	Basis	Relation	Dual	Post's not.	Remarks
29	$\langle\vee, c_0, c_1\rangle$	$\vee, c_0, c_1$	$\begin{pmatrix}0&0&1&1\\0&1&1&1\\0&1&0&1\end{pmatrix}$	30	S_6	all disjunctions and constants
30	$\langle\cdot, c_0, c_1\rangle$	$\cdot, c_0, c_1$	$\begin{pmatrix}0&0&1&1\\0&0&0&1\\0&1&0&1\end{pmatrix}$	29	P_6	all conjunctions and constants
31_h	$\mathrm{Pol}\mu_h$	$\rightarrow, f_h$	$\mu_i := B^k\backslash \{(0,\ldots,0)\}$	32_h	F_4^h	$h > 1$
32_h	$\mathrm{Pol}\mu'_h$	xy', f_h^∂	$\mu'_i := B^k\backslash \{(1,\ldots,1)\}$	31_h	F_8^h	$h > 1$
33_h	$T_0\mathrm{Pol}\mu_h$	$x \vee yz'$, f_h	$\mu_h \times \{0\}$	34_h	F_1^h	$= I\mathrm{Pol}\mu_h$
34_h	$T_1\mathrm{Pol}\mu'_h$	$x(y \vee z')$, f_h^∂	$\mu'_h \times \{1\}$	33_h	F_5^h	$= I\mathrm{Pol}\mu_h^\partial$
35_h	$M\mathrm{Pol}\mu_h$	c_1, f_h	$(\mu_h \times \{0\}) \cup\{(1,\ldots,1)\}$	36_h	F_3^h	$h > 1$
36_h	$M\mathrm{Pol}\mu'_h$	c_0, f_h^∂	$(\mu_h^\partial \times \{1\}) \cup\{(1,\ldots,1)\}$	35_h	F_7^h	$h > 1$
37_2	$IM\mathrm{Pol}\mu_2$	$xy \vee z$, f_2	$(\{(0,0)\} \times \mu_2) \cup\{(0,1,1,1)\}$	38_2	F_2^2	
37_h	$IM\mathrm{Pol}\mu_h$	f_2	$(\{(0,0)\} \times \mu_h) \cup\{(0,1,\ldots,1)\}$	38_h	F_2^h	$h > 2$
38_2	$IM\mathrm{Pol}\mu'_2$	$(x \vee y)z$, f_2^∂	$(\{(1,1)\} \times \mu'_2) \cup\{(0,0,0,0)\}$	37_2	F_6^2	
38_h	$IM\mathrm{Pol}\mu'_h$	f_h^∂	$(\{(1,1)\} \times \mu_h^\partial) \cup\{(1,0,\ldots,0)\}$	37_h	F_6^h	$h > 2$
39	$\bigcap_{h>0}\mathrm{Pol}\mu_h$	$\rightarrow$	--	40	F_4^∞	
40	$\bigcap_{h>0}\mathrm{Pol}\mu'_h$	$x'y$	--	39	F_8^∞	
41	$I\bigcap_{h>0}\mathrm{Pol}\mu_h$	$xy' \vee z$	--	42	F_1^∞	
42	$I\bigcap_{h>0}\mathrm{Pol}\mu'_h$	$(x \vee y')z$	--	41	F_5^∞	
43	$M\bigcap_{h>0}\mathrm{Pol}\mu_h$	$x \vee yz$, c_1	--	44	F_3^∞	
44	$M\bigcap_{h>0}\mathrm{Pol}\mu'_h$	$x(y \vee z)$, c_0	--	43	F_6^∞	
45	$IM\bigcap_{h>0}\mathrm{Pol}\mu_h$	$x \vee yz$	--	46	F_2^∞	
46	$IM\bigcap_{h>0}\mathrm{Pol}\mu'_h$	$x(y \vee z)$	--	45	F_6^∞	

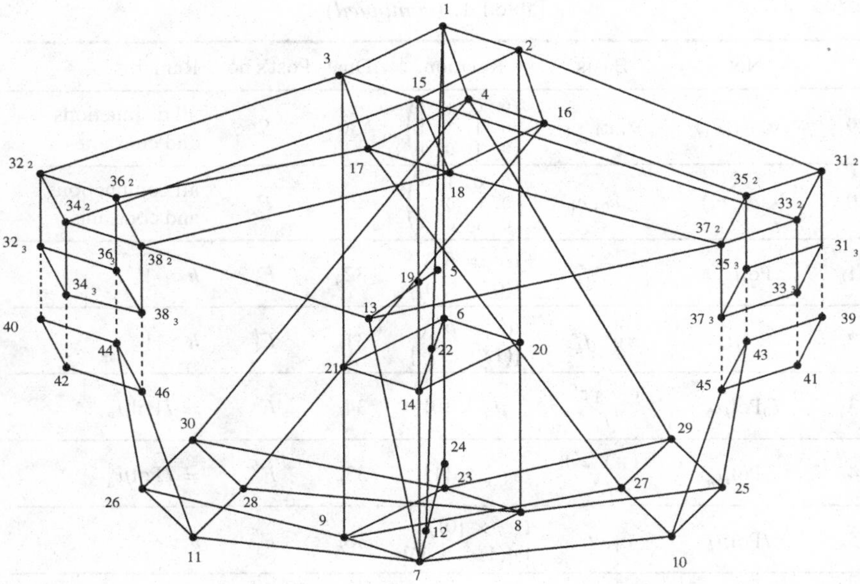

Figure 1.9.

C, satisfies

$$\langle C^{(o-1)} \rangle \subset C = \langle C^{(o)} \rangle$$

(where $C^{(n)} := C \cap O^{(n)}$). Instead of such bases, we could have listed bases of the least possible size. The latter need not be the same as those listed; for example, the clone T_1 (number 2) already has a basis $\{x \vee y'z'\}$. Relational descriptions of Boolean clones were first given in [5]; see also [16]. There are several other notations for Boolean clones in the literature [11, 17, 22], but listing them would just burden the table too much.

For $n = 1, 2, \cdots$ we use

$$f_n(x_1, \ldots, x_{n+1}) :\approx (x_2 \vee \cdots \vee x_{n+1})(x_1 \vee x_3 \vee \cdots \vee x_{n+1}) \cdots (x_1 \vee \cdots \vee x_n)$$

and its dual

$$f_n^{\partial}(x_1, \ldots, x_{n+1}) :\approx x_2 \cdots x_{n+1} \vee x_1 x_3 \cdots x_{n+1} \vee \cdots \vee x_1 \cdots x_n.$$

Notice that f_2 coincides with the ternary majority function m from Section 1.3.4. We also use $s_3(x, y, z) :\approx x \dot{+} y \dot{+} z$ introduced at the end of Section 1.2.

1.3.11 Final Comments

The original proof in Post [21] consists of a hundred-odd pages, but if one gets used to the so-called Jevon notation, it is quite readable. In fact, the first draft of the paper dating from 1921 is in existence and comprises about 1,000 pages. It is based on a simple but ingenious separation into subcases. Call two clones A and B equivalent if they share their unary Boolean functions; that is; if $A \cap O^{(1)} = B \cap O^{(1)}$. The equivalence classes (or blocks) of this equivalence relation are intervals of $\mathcal{L}$ that are in one-to-one correspondence to submonoids of $\langle O^{(1)}, \circ, e_1^1 \rangle$ (where $\circ$ denotes the usual composition $(f \circ g)(x) \approx f(g(x))$). There are only six submonoids of $O^{(1)}$, and so the separation into subcases is quite feasible.

Naturally in the past several decades there appeared shorter versions of the proof (see, e.g., [17, 18, 22, 26]) using either universal algebra results not known in 1941 or shortcuts while following the basic approach of [21].

References

[1] H.-J. Bandelt, and J. Hedlikova. Median algebras. *Discrete Mathematics* **45**, pp. 1–30, 1983.

[2] C. Benzaken. Post's closed systems and the weak chromatic number of hypergraphs. *Discrete Mathematics* 23, pp. 77–84, 1978.

[3] C. Benzaken. Critical hypergraphs for the weak chromatic number. *Journal of Combinatorial Theory B* 29, pp. 328–38, 1980.

[4] J. Berman, and P. Köhler. Cardinalities of finite distributive lattices. *Mitteilungen aus dem Mathematischen Seminar Giessen* **121**, pp. 103–24, 1976.

[5] G.N. Blochina. On the predicate description of Post classes (Russian). *Discrete Analiz* **16**, pp. 16–29, 1970.

[6] P. Camion. Une méthode de résolution par l'algèbre de Boole des problèmes combinatoires où interviennent des entiers. *Cahiers du Centre d'Etudes de Recherche Opérationnelle* **2**, pp. 234–89, 1960.

[7] Y. Crama, and P. L. Hammer. *Boolean Functions: Theory, Algorithms, and Applications*. Cambridge University Press, Cambridge, UK, 2010.

[8] S. G. Gindikin. Algebra of Logic (Russian). Nauka 1972. English translation: *Algebraic Logic*, Springer-Verlag, 1985.

[9] S. V. Iablonskiĭ. On the superposition of functions of algebra of logic (Russian). *Matematicheskii Sbornik N.S.* **30**, pp. 329–48, 1952.

[10] S. V. Iablonskiĭ. Functional constructions in many valued logics (Russian). *Trudy Mat. Inst. Steklova* **51**, pp. 5–142, 1958.

[11] S. V. Iablonskiĭ, G. P. Gavrilov, and V. B. Kudriavcev. *Functions of Algebra of Logic and Post Classes* (Russian), Nauka, Moscow, 1966. German translation: *Boolesche Funktionen und Postsche Klassen*. Akademie-Verlag, Berlin, 1970.

[12] K. Ibuki, K. Naemura, and A. Nozaki. General theory of universal logical function systems (Japanese). *Journal of the Institute of Electrical and Communication Engineers of Japan* **46**, July 1963, 28 pp.

[13] A. D. Korshunov. Solution of Dedekind's problem on the number of monotone Boolean functions (Russian). *Doklady Akademii Nauk SSSR* **233**, pp. 543–6, 1977.

[14] L. Krnić. Types of bases of algebra of logic (Russian). *Glasnik math.-fiz.i astr.* **20**(1–2), pp. 23–30, 1965.

[15] A. V. Kuznetsov. Structure with closure and criteria of functional completeness (Russian). *Uspekhi Matematicheskikh Nauk.* **16**(98), pp. 201–2, 1961.

[16] D. Lau. Über die Dimension der abgeschlossenen Teilmengen von P_2 (Ergänzungen und Berechtigungen zu Blochina G. N., On predicate description of Post's classes (Russian), (German). *Diskretny. Analiz* **16**, pp. 16–29. 1970 Preprint U. Rostock, 1988.

[17] D. Lau. On closed subsets of Boolean functions (A new proof for Post's theorem). *Journal of Information Processing and Cybernetics EIK* **27**(3), pp. 167–78, 1991.

[18] R. N. McKenzie, G. F. McNulty, and W. F. Taylor. *Algebras, Lattices, Varieties*. Manuscript of the projected vol. II, Ch. 7, 1985.

[19] J. G. P. Nicod. A reduction in the number of primitive propositions of logic. Proceedings of the Cambridge Philosophical Society **19**, pp. 32–41, 1917.

[20] E. L. Post. Introduction to a general theory of elementary propositions. *American Journal of Mathematics* **43**, pp. 163–85, 1921.

[21] E. L. Post. The Two-Valued Iterative Systems of Mathematical Logic. *Annals of Mathematics Studies*, No. 5. Princeton University Press, Princeton, NJ, 1941.

[22] M. Reschke, and K. Denecke. Ein neues Beweis für die Ergebnisse von E. L. Post über abgeschlossene Klassen Boolescher Funktionen (German). *Journal of Information Processing and Cybernetics EIK* **25**, pp. 361–80, 1989.

[23] I. G. Rosenberg. Minimization of pseudo-Boolean functions by binary developments. *Discrete Mathematics* **7**, pp. 157–65, 1974.

[24] I. G. Rosenberg. Characteristic polynomials in $GF(2)$ of zero-one inequalities and equations. *Utilitas Mathematica* **7**, pp. 323–543, 1975.

[25] H. M. Sheffer. A set of five independent postulates for Boolean algebras with application to logical constants. *Transactions of the American Mathematical Society* **14**, pp. 481–88, 1913.

[26] A. B. Ugol'nikov. On Post's closed classes (Russian). *Matematika* (Kazan) **7**, pp. 79–87, 1988.

[27] D. Wiedemann. A computation of the eighth Dedekind number. *Order* **8**, pp. 5–6, 1991.

[28] I. I. Zhegalkin. The arithmetization of symbolic logic (Russian). *Matematicheskii Sbornik* **35**, pp. 311–78, 1928; **36**, pp. 205–338, 1929.

2

Decomposition of Boolean Functions

Jan C. Bioch

2.1 Introduction

The basic step in functional decomposition of a Boolean function $f : \{0, 1\}^n \mapsto \{0, 1\}$ with input variables $N = \{x_1, x_2, \ldots x_n\}$ is essentially the partitioning of the set N into two disjoint sets $A = \{x_1, x_2, \ldots, x_p\}$ (the "modular set") and $B = \{x_{p+1}, \ldots, x_n\}$ (the "free set"), such that $f = F(g(x_A), x_B)$. The function g is called a component (subfunction) of f, and F is called a composition (quotient) function of f. The idea here is that F computes f based on the *intermediate* result computed by g and the variables in the free set. More complex (Ashenhurst) decompositions of a function f can be obtained by recursive application of the basic decomposition step to a component function or to a quotient function of f.

Functional decomposition for general Boolean functions has been introduced in switching theory in the late 1950s and early 1960s by Ashenhurst, Curtis, and Karp [1, 2, 20, 23, 24]. More or less independent from these developments, decomposition of positive functions has been initiated by Shapley [36], Billera [5, 4], and Birnbaum and Esary [12] in several contexts such as voting theory (simple games), clutters, and reliability theory. However, the results in these areas are mainly formulated in terms of set systems and set operations.

The objective of this chapter is to give a unified treatment of functional decomposition of general and positive Boolean functions. However, in the area of the design of electronic circuits, one also considers decompositions where the components are possibly multioutput functions. For these generalized decompositions, so-called *Curtis decompositions*, which are not discussed in this chapter, we refer the reader to [34].

The two most important issues in this chapter are the investigation of the properties of modular sets and the algorithmic aspects of the computation of the collection of all modular sets of a Boolean function f. To guide the reader, we briefly explain the overall structure of the following sections.

Organization of this Chapter

The two main parts of this chapter are:

(1) The theory of modular sets of general Boolean functions described in Sections 2.2 through 2.8.
(2) The refinement of this theory to the class of positive Boolean functions developed in Section 2.9. This class is interesting for theoretical reasons and for the application domains: voting theory, reliability theory, and clutters.

In Section 2.2, some basic concepts such as modular sets and congruence partitions are discussed. Moreover, we briefly discuss some application domains.

In Section 2.3 we introduce the idea of the "Shannon representation" of a decomposition as an alternative for the well-known decomposition charts [20] to analyze the decomposability of a Boolean function. This representation, although formally equivalent to a decomposition chart, appears to be very useful in the proofs of several results.

Every decomposition induces a partitioning of the set of variables. This partitioning is called a congruence partition in [28, 29]. In Section 2.4 it is shown that a congruence partition is just a partition with modular classes.

In Section 2.5 the properties of modular sets are investigated. A natural question concerning modular sets is: what is the relationship between the modular sets of a Boolean function f and the modular sets of, respectively, a component function of f and a quotient function of f? Another important question is whether the collection of all modular sets is closed under set operations such as intersection, union, and complementation. The answers to these questions are summarized in Theorem 2.17, which lists the properties of modular sets.

Because the number of modular sets of a Boolean function can be exponential in the number n of variables, it comes as a surprise that the total collection of modular sets can be represented by a tree structure of size $O(n)$ called a "(de)composition tree." Composition trees, discussed in Section 2.6, were first studied by Shapley [36] in the context of simple games. These trees are determined by using congruence partitions consisting of maximal modular sets. It appears that maximal modular sets have an interesting property: either they are all disjoint or every pair of maximal modular sets has a nonempty intersection. To show that the composition tree is a correct representation of the collection of all modular sets, all concepts and results of the preceding sections are used. Therefore, a composition tree also (implicitly) embodies the properties of modular sets and congruence partitions discussed thus far.

Because the congruence partitions of a Boolean function reflect all its possible decompositions, the collection of all congruence partitions is as important as the collection of all modular sets. In Section 2.7 it is shown that the collection of all congruence partitions itself has an interesting structure: it appears to be an upper semimodular sublattice of the lattice of all partitions of the set of variables.

Therefore, all maximal chains between two congruence partitions have the same length (the Jordan-Dedekind condition).

It appears that the determination of the smallest modular subset containing a subset of variables (called the modular closure) is a central step in the computation of the composition tree (modular tree) of a Boolean function. In Section 2.8, it is shown that a modular tree can be determined by the computation of $O(n^3)$ times the modular closure of a subset of variables. Furthermore, it can be shown that the problem of recognizing whether a subset of variables of a Boolean function given by a disjunctive normal form (DNF) expression is modular or not is coNP-complete.

In Section 2.9, several criteria for the decomposability of positive Boolean functions are discussed. A component of a positive Boolean function is uniquely determined and can be identified with the so-called contraction of a positive function on a set of variables. Several characterizations of modular sets exist, and they are used to show that the modular closure of a set and modular tree of a positive Boolean function can be computed in polynomial time.

In Section 2.10, we relate the theory developed in the preceding section to simple games (voting games). Moreover, it is shown that decisive simple games can be uniquely factorized into prime decisive simple games.

In the last section, related and future research is discussed.

2.2 Basic Concepts and Applications

2.2.1 Functional Decomposition

Let $f : \{0,1\}^n \mapsto \{0,1\}$ be a Boolean function and $N = \{1, 2, \ldots, n\}$. Identify each $i \in N$ with the variable x_i. Then f is said to be a function defined on N. Furthermore, if $N = A_1 \cup A_2 \cup \cdots A_m$ is a partition of N $(A_i \cap A_j = \emptyset,\ i \neq j)$, then we will denote this by $x_N = (x_{A_1}, \ldots, x_{A_m})$ and $f(x_N) = f(x_{A_1}, \ldots, x_{A_m})$. Let $F(y_I)$ and $g_i(x_{A_i})$ be Boolean functions defined on the mutually disjoint sets $I = \{1, \ldots, m\}$ and $A_i,\ i \in I$, and let $N = \cup_{i=1}^m A_i$. Then the Boolean function defined by

$$f(x_N) = F(g_1(x_{A_1}), \ldots, g_m(x_{A_m}))$$

is called the *composition* of the functions F and $g_i,\ i \in I$, obtained by *substitution* of the variables y_i in F by the functions $g_i,\ i \in I$. This composition is denoted by $F[g_i,\ i \in I]$. A composition is called *proper* if $|I| > 1$ and $|A_i| > 1$ for some $i \in I$. A Boolean function is said to be *decomposable* if it has a representation as a proper composition. Otherwise, the function f is called *indecomposable* or *prime*. If f is given, then $f = F[g_i,\ i \in I]$ is called a disjoint decomposition of f.[1] It appears that each (complex) decomposition of a Boolean function f can

[1] In the classical literature, this is also called a *disjunctive* (complex) decomposition. All types of decompositions discussed in this chapter are also called substitution decompositions in the classical literature or functional decompositions in the recent literature.

be obtained by a series of so-called *simple disjoint decompositions*. These are decompositions of the form

$$f(x_N) = F(g(x_A), x_B),$$

where $\{A, B\}$ is a partition of N. It is known [23, 20] that the study of nondisjoint decompositions can also be reduced to that of simple disjoint decompositions. However, a discussion of nondisjoint decompositions is outside the framework of this chapter; see [20].

Definition 2.1. *Let f be a Boolean function defined on N. Then $A \subseteq N$ is called a modular set of f if f has a simple disjoint decomposition of the form $f(x_N) = F(g(x_A), x_B)$. The function g is called a component of f.*

From this definition it follows that the base set N and its singleton subsets are modular. It appears that the components of a Boolean function are determined up to complementation, see Corollary 2.3. In particular if f is a positive function then a component g defined on A is uniquely determined and called the *contraction* of f on A.

If $F[g_i,\ i \in I]$ is a (complex) decomposition of the function f then the partition $\pi = \{A_i,\ i \in I\}$ of N is called a *congruence partition*[2] and F is called the *quotient* of f modulo π. This is denoted by $F = f/\pi$.

Recall (see e.g. [8]) that the dual of a Boolean function f is the function f^d defined by: $f^d(x_1, x_2, \ldots, x_n) = \overline{f}(\overline{x}_1, \overline{x}_2, \ldots, \overline{x}_n)$ for all Boolean vectors $(x_1, x_2, \ldots, x_n)$. From the definition of decomposition it follows easily that

$$f = F[g_i,\ i \in I] \Leftrightarrow f^d = F^d[g_i^d,\ i \in I]. \tag{2.1}$$

Therefore, we have $F = f/\pi \Leftrightarrow F^d = f^d/\pi$.

Example 2.1. *The Boolean function $f = \bar{x}_1\bar{x}_2\bar{x}_3\bar{x}_4 \vee \bar{x}_1x_2x_3 \vee x_1x_2\bar{x}_3 \vee x_1\bar{x}_2x_3 \vee x_1\bar{x}_2x_4 \vee x_1\bar{x}_3x_4$, has the decomposition $f(x_1, x_2, x_3, x_4) = F(x_1, g(x_2, x_3, x_4))$, where $F(x, y) = \bar{x}\bar{y} \vee xy$ and $g(x, y, z) = (x \oplus y) \vee \bar{x}z \vee \bar{y}z$. Therefore, $A = \{2, 3, 4\}$ is a modular set of f, and $\{\{1\}, \{2, 3, 4\}\}$ is a congruence partition of f.*

2.2.2 Applications

Functional decomposition has been studied thoroughly by researchers in many different contexts such as switching theory, game theory, reliability theory, network theory, graph theory and hypergraph theory. Not surprisingly, the concept of a modular set has been rediscovered several times under various names: bound

[2] The concept congruence partition was first used in [29] by Möhring and Radermacher, who discussed substructures and quotient structures of several discrete structures in an algebraic setting. Note, that the concept of congruence relation is well known in lattice theory.

sets, autonomous sets, closed sets, stable sets, clumps, committees, externally related sets, intervals, nonsimplifiable subnetworks, partitive sets and modules, see [15, 29] and references therein. An excellent survey for the various applications of substitution decomposition and connections with combinatorial optimization is given by Möhring and Radermacher in [28, 29]. The decomposition of monotone Boolean functions has been studied in several contexts: game theory (decomposition of n-person games [36]), reliability theory (decomposition of coherent systems [12]) and set systems (clutters [5, 4]).

In switching theory [34] decomposition of general Boolean functions is still an important tool in the design and analysis of circuits, see also Section 2.11. Some applications of decompositions of positive Boolean functions to be discussed briefly here are in the areas of reliability theory, game theory and combinatorial optimization; see also Crama and Hammer [19].

Application 2.1 (Reliability Theory). *In reliability theory a system S consisting of n components is modeled by a positive (monotone) Boolean function f_S called the structure function of f_S. This function indicates whether system S is operating or not depending on the states of the n components: operative ($x_i = 1$) or failed ($x_i = 0$) (see [19] and Chapter 17 in this volume). Modular sets play a role in the design and analysis of a complex system S because they reflect the decomposability possibilities of S in subsystems.*

Application 2.2 (Game Theory). *An n-person simple game (or voting game) G can be modeled by a positive Boolean function f_G such that the winning coalitions of G correspond to the prime implicants of f_G (see [19]). The factorization of compound simple games, as introduced by Shapley [36], is equivalent to the decomposition of the associated positive Boolean functions; see also Section 2.10.*

Application 2.3 (Clutters). *Combinatorial optimization over set systems has motivated research on the decomposition of clutters, or Sperner families (see, e.g., [5, 4, 29]). The interface between a clutter C and its associated positive function f_C is given by the correspondence between the elements of C and the prime implicants of f_C; see [19].*

Notations

We say that a set $A \subseteq N$ is *essential* for f, or that f *depends on* A, if A contains at least one essential variable (see [8]) of f. The set of all modular sets of f is denoted by $\mu_0(f)$ and the set of all modular sets of f that are essential for f is denoted by $\mu(f)$. A congruence partition π is called *essential* if all the classes of π are essential for f. Furthermore, the set of all essential congruence partitions of f is denoted by $\Pi(f)$.

If a Boolean function f is constant, then either $f \equiv 0$ (denoted by $f = \mathbf{0}$) or $f \equiv 1$ (denoted by $f = \mathbf{1}$).

2.3 Shannon Representation of a Decomposition

To analyze the decomposability of a Boolean function, Ashenhurst [1, 2] introduced the well-known Ashenhurst decomposition charts. As an alternative to these charts, we use the idea of the "Shannon representation" of a decomposition of a Boolean function. This representation, although formally equivalent to a decomposition chart, appears to be very useful in the proofs of several results.

Let f be a Boolean function on N. Then for all $j \in N$ the following decomposition holds (see [19]):

$$f = \bar{x}_j f_{|x_j=0} \vee x_j f_{|x_j=1}. \tag{2.2}$$

Equation (2.2) is known as a *Shannon expansion(decomposition)* of f. Now consider the simple disjoint decomposition

$$f(x_N) = F(g(x_A), x_B). \tag{2.3}$$

Then by applying Equation (2.2) to F, we get

$$f(x_N) = \bar{g}(x_A)F_0(x_B) \vee g(x_A)F_1(x_B), \tag{2.4}$$

where $F_0(x_B) = F(0, x_B)$ and $F_1(x_B) = F(1, x_B)$.

Conversely, let g and h_0, h_1 be arbitrary Boolean functions defined respectively on A and B such that $f = \bar{g}h_0 \vee gh_1$, and let the function F be defined by $F(y, x_B) := \bar{y}h_0 \vee yh_1$. Then $f(x_N) = F(g(x_A), x_B)$ is a simple disjoint decomposition of f, where $F_0(x_B) = h_0$ and $F_1(x_B) = h_1$. Therefore, we have proved the following fundamental lemma.

Lemma 2.1. *Let f be a Boolean function on N. Then $A \subseteq N$ is a modular set of f iff there exists a Boolean function g on A and functions h_0 and h_1 on $B = N \setminus A$ such that $f = \bar{g}h_0 \vee gh_1$.*

We call the decomposition in the previous lemma *a generalized Shannon decomposition*. In particular, we call the decomposition in Equation (2.4) a Shannon representation of the simple disjoint decomposition (2.3). If A is a modular set of the function f such that A contains at least one essential variable of f, then it follows from the decomposition

$$f = \bar{g}h_0 \vee gh_1 \tag{2.5}$$

that the function g is nonconstant and that the functions h_0 and h_1 are not identical. Therefore, there exists a binary vector b_0 such that $h_0(b_0) \neq h_1(b_0)$. Therefore, either $g(x_A) = f(x_A, b_0)$ or $\bar{g} = f(x_A, b_0)$. In general, Equation (2.5) shows that if b is a fixed vector, then the function $f(x_A, b)$ is either constant or identical to g or identical to $\bar{g}$. It is not difficult to see that the converse also holds. Therefore, the following theorem holds.

Theorem 2.2. *Let f be a Boolean function defined on N. If $A \subseteq N$ contains at least one essential variable of f, then the following statements are equivalent:*

(a) *A is modular.*
(b) *There exists a vector* b_0 *such that the function* $g(x_A) := f(x_A, b_0)$ *is nonconstant and for all fixed b, the function* $f_b := f(x_A, b)$ *is either constant or identical to either g or* $\bar{g}$.

Corollary 2.3. *Suppose* $f(x_N) = F(g(x_A), x_B) = G(h(x_A), x_B)$, *and that A is essential for f. Then either* $g = h$ *and* $F = G$ *or* $g = \bar{h}$ *and* $F(y, x_B) = G(\bar{y}, x_B)$.

Proof. We leave this as an exercise. ■

Corollary 2.4. *If A is a modular set of f, then there exists a component g of f on A that is a restriction of f on A. We denote this by* $g = f_A$.

Example 2.2. *Consider the functions f and g defined in Example 2.1. Then* $f = \bar{g}\bar{x}_1 \vee gx_1$, *implying that g is a component of f.*

2.4 Modular Sets and Congruence Partitions

Recall that a congruence partition is induced by a disjoint (complex) decomposition of the form $f(x_N) = F(g_1(x_{A_1}), \ldots, g_m(x_{A_m}))$. Now the question arises how these decompositions are related to the simple disjoint decompositions of the form $f(x_N) = F(g(x_A), x_B)$. This question obviously is equivalent to the question of how modular sets and congruence partitions are related. The following theorem shows that the modular sets of a function f are precisely the classes of the congruence partitions of f.

Theorem 2.5. *Let f be a Boolean function defined on N, and let* $\pi = \{A_i, i \in I\}$ *be a partition of N. Then,* $\pi \in \Pi(f)$ *if and only if* $A_i \in \mu(f)$ *for all* $i \in I$.

Proof. If $f(x_N) = F[g_i(x_{A_i}), i \in I]$, then obviously for all $i \in I$ there exists a Boolean function F_i such that $f(x_N) = F_i(g_i(x_{A_i}), x_{\bar{A}_i})$, implying that $A_i \in \mu(f)$. To prove the converse, we first assume that $|I| = 2$. If $A, B \in \mu(f)$ and $\pi = \{A, B\}$ is a partition of N, then $f(x_A, x_B) = F(g(x_A), x_B) = G(h(x_B), x_A)$. Because A and B are essential for f, there exist binary vectors $a(y)$ and $b(y)$ such that $y = g(a(y)) = h(b(y))$, where $y \in \{0, 1\}$. Now we define the function H by

$$H(y_1, y_2) = H(g(a(y_1)), h(b(y_2)) := f(a(y_1), b(y_2)).$$

To prove that $f(x_A, x_B) = H(g(x_A), h(x_B))$, we have to show that the function H is properly defined. We note that $g(a_1) = g(a_2)$ implies $f(a_1, x_B) = F(g(a_1), x_B) = F(g(a_2), x_B) = f(a_2, x_B)$. Furthermore, $h(b_1) = h(b_2)$ implies $f(x_A, b_1) = f(x_A, b_2)$. Therefore, we conclude $f(a_1, b_1) = f(a_2, b_1) = f(a_2, b_2)$. Conclusion: if $A, B \in \mu(f)$, then $\pi = \{A, B\} \in \Pi(f)$. The case $|I| \geq 2$ is a straightforward generalization. ■

Note that in view of the preceding theorem, a congruence partition is just a partitioning of the set of variables N such that each class is a modular set.

We will use this fact in Theorem 2.8. Other properties of congruence partitions are discussed in another section.

2.5 Properties of Modular Sets

In this section we derive a number of properties of modular sets by proving decomposition theorems as in Curtis [20]. The main tool we use here is the Shannon representation of a simple decomposition and Theorem 2.2. These properties describe:

(a) The behavior of the collection of all modular sets under set operations.
(b) The relationship between the modular sets of a Boolean function f and the modular sets of respectively a component function of f and a quotient function of f.

Lemma 2.6. *Let $f(x, y)$ be a Boolean function depending on x and y. Then $f(x, y) = y_1 \star y_2$, where $y_1 = x$ or $\bar{x}$, $y_2 = y$ or $\bar{y}$, and $\star$ denotes either $\vee$ or $\wedge$ or $\oplus$.*

Proof. Consider the decomposition $f(x, y) = \bar{y} f(x, 0) \vee y f(x, 1)$. Because x is a component of f, we must, according to Theorem 2.2, consider the following cases: $x = f(x, 0)$, $x = f(x, 1)$, $\bar{x} = f(x, 0)$ or $\bar{x} = f(x, 1)$. If $x = f(x, 0)$, then $f(x, 1) \in \{0, 1, \bar{x}\}$, implying that $f(x, y) \in \{x\bar{y}, x \vee y, x \oplus y\}$. If $x = f(x, 1)$, then $f(x, y) \in \{xy, x \vee \bar{y}, x \oplus \bar{y}\}$. Both cases together can be expressed as $f(x, y) \in \{x \star y, x \star \bar{y}\}$. Similarly, the other two cases yield $f(x, y) \in \{\bar{x} \star y, \bar{x} \star \bar{y}\}$. ∎

Corollary 2.7. *There are 10 Boolean functions of two essential variables.*

To answer the question whether the complement of a modular set is also modular, we consider a congruence partition with only two classes (bipartition).

Theorem 2.8. *Suppose $A \in \mu(f)$ and $\bar{A}$ is essential for f. Then $\bar{A} \in \mu(f)$ if and only if f has a decomposition $f(x_N) = g(x_A) \star h(x_{\bar{A}})$, where $\star$ denotes either $\vee$ or $\wedge$ or $\oplus$.*

Proof. Suppose $B = \bar{A} \in \mu(f)$. Then by Theorem 2.5, f can be written as $f(x_A, x_B) = F(g_1(x_A), h_1(x_B))$. Because A and B are essential for f, the function F has two essential variables. So, by Lemma 2.6 it follows that $f(x_N) = g(x_A) \star h(x_{\bar{A}})$, where g and h are respectively equal to g_1 and h_1 modulo complementation. The converse is obvious. ∎

So the last theorem shows that the complement of a modular set is itself modular only in special cases.

Example 2.3. *Consider the functions f and g discussed in Examples 2.1 and 2.2. Then $\{1\}$ and $\{2, 3, 4\}$ are complementary modular sets, and $f = \bar{g}\bar{x}_1 \vee g x_1 = \bar{x}_1 \oplus g$. This is in agreement with the last theorem.*

According to the following theorem, a subset of a modular set of a component of a Boolean function is a modular set of this component iff it is already a modular set of f.

Theorem 2.9. *Let $A \in \mu(f)$ and let g be a component of f defined on A. Then $\mu_0(g) = \{C \subseteq A \mid C \in \mu_0(f)\}$. In addition, if f depends on all the variables in A, then $\mu(g) = \{C \subseteq A \mid C \in \mu(f)\}$.*

Proof. According to Corollary 2.4, we may assume that $g = f_A$. So, $f(x_A, x_{\bar{A}}) = F(g(x_A), x_{\bar{A}})$, and there exists a vector b such that $g(x_A) = f(x_A, b)$. If $C \subseteq A$ and $C \in \mu_0(f)$, then we also have $f(x_C, x_{\bar{C}}) = G(h(x_C), x_{\bar{C}})$. Therefore, $g(x_A) = G(h(x_C), x_{A\setminus C}, b)$. Let the function H be defined by $H(y, y_{A\setminus C}) := G(y, x_{A\setminus C}, b)$. Then $g(x_C, x_{A\setminus C}) = H(h(x_C), x_{A\setminus C})$, so we have $C \in \mu_0(g)$. If in addition f depends on all variables in A, then $C \in \mu(g)$.

Conversely, suppose $C \in \mu_0(g)$. Then $g(x_A) = G(h(x_C), x_{A\setminus C})$. Therefore, $f(x_A, x_{\bar{A}}) = F(G(h(x_C), x_{A\setminus C}), x_{\bar{A}})$. Let the function H be defined by $H(y_C, y_{\bar{C}}) := F(G(y_C, y_{A\setminus C}), y_{\bar{A}})$. Then $f(x_C, x_{\bar{C}}) = H(h(x_C), x_{\bar{C}})$, so $C \in \mu_0(f)$. If in addition f depends on all variables in A, then we have $C \in \mu(f)$. ∎

Because by the last theorem a component of a component of a Boolean function f is itself a component of f, it is not necessary to consider *nested* congruence partitions. For example, the nested partition $\{\{\{1,2\},\{3,4\}\},\{5,6\}\}$ is equivalent to the flat partition $\{\{1,2\},\{3,4\},\{5,6\}\}$. This has an important consequence: the collection of all congruence partitions represents all possible disjoint decompositions of a Boolean function. Otherwise stated, it is sufficient to consider only "flat" decompositions of the form $f(x_N) = F(g_1(x_{A_1}), \ldots, g_m(x_{A_m}))$, even if the subfunctions g_i are further decomposed.

An important question regarding the collection of all modular sets of a Boolean function is how it behaves under set operations such as union and intersection. It is clearly not closed under set union as evidenced by the fact that singleton sets are modular. However, the following two theorems, 2.10 and 2.11, discuss the behavior of the set of modular sets under set operations. For an interpretation, we refer to Theorem 2.12.

Theorem 2.10. *Let f be a Boolean function defined on the partition $\{A, B, C\}$. Let A, B, C be essential for f. If $A \cup B$ and $B \cup C$ are modular sets of f, then $A, B, C \in \mu(f)$ and $f = f_A \star f_B \star f_C$, where $\star$ denotes either $\vee$ or $\wedge$ or $\oplus$.*

Proof. We may assume that $f(x_N) = F(g(x_A, x_B), x_C) = G(x_A, h(x_B, x_C))$, where $g = f_{A\cup B}$ and $h = f_{B\cup C}$. According to Theorem 2.2 there exists a c such that $g(x_A, x_B) = f(x_A, x_B, c) = G(x_A, h(x_B, c)) = G(x_A, k(x_B))$, where $k(x_B) = h(x_B, c)$. Therefore, B is a modular set of the component g of f, and because B is essential for f, Theorem 2.9 implies that $B \in \mu(f)$. Similarly, there exists a vector a such that $h(x_B, x_C) = f(a, x_B, x_C) = F(g(a, x_B), x_C) = F(G(a, k(x_B)), x_C)$. Furthermore, we have $G(a, k(x_B)) = f(a, x_B, c) = g(a, x_B) = h(x_B, c) = k(x_B)$. From this we conclude that $k = f_B$

and that

$$f(x_N) = F(G(x_A, k(x_B)), x_C) = G(x_A, F(k(x_B), x_C)). \tag{2.6}$$

Because C is essential for F, there exists a vector u such that $F_u(y) := F(y, u) \neq y$. Therefore, $F_u(y) \in \{0, 1, \bar{y}\}$. We now consider the following three cases:

(1) $F_u(y) = 0$. Then Equation (2.6) implies $G_0(x_A) := G(x_A, 0) = 0$. Therefore, we have $g(x_A, x_B) = \bar{k}(x_B)G_0(x_A) \vee k(x_B)G_1(x_A) = k(x_B)G_1(x_A)$, where $G_1(x_A) := G(x_A, 1)$. There exists a vector b such that $k(b) = 1$. So, we can write G_1 as $G_1(x_A) = g(x_A, b) = f(x_A, b, c)$. Therefore $G_1 = f_A$.
(2) $F_u(y) = 1$. In this case we have $G(x_A, 1) = 1$, implying that $g(x_A, x_B) = k(x_B) \vee G_0(x_A)$, and $G_0 = f_A$.
(3) $F_u(y) = \bar{y}$. In this case Equation (2.6) yields $\bar{G}(x_A, k(x_B)) = G(x_A, \bar{k}(x_B))$. In particular, because the function k is not identical to 1, we have $\bar{G}_0(x_A) = G_1(x_A)$. Therefore, $g(x_A, x_B) = k(x_B) \oplus G_0(x_A)$, and $G_0 = f_A$.

Note, that the cases $G_0 = 0$, $G_1 = 1$ and $\bar{G}_0 = G_1$ are mutually exclusive. For example, if $G_0 = 0$ and $G_1 = 1$, then $g(x_A, x_B) = k(x_B)$, contrary to our assumption that f depends on A. Conclusion: A and B are modular sets of f and $g = f_B \star f_A$, where $\star$ denotes either $\vee$ or $\wedge$ or $\oplus$. Similarly, C is a modular set of f and exactly one of the following cases occurs: $F_0 = 0$, $F_1 = 1$ and $\bar{F}_0 = F_1$. Now consider the following decompositions:

$$f = \bar{g}F_0 \vee gF_1, \quad g = \bar{k}G_0 \vee kG_1 \tag{2.7}$$

and the cases:

(a) $F_0 = G_0 = 0$. Then (2.7) implies $f = kG_1F_1 = f_B f_A f_C$,
(b) $F_1 = G_1 = 1$. Then $f = k \vee G_0 \vee F_0 = f_B \vee f_A \vee f_C$,
(c) $\bar{G}_0 = G_1$ and $\bar{F}_0 = F_1$. Then $f = k \oplus G_0 \oplus F_0 = f_B \oplus f_A \oplus f_C$.

To show that there are no other possible cases, we consider the following cases:

(d) If $F_0 = 0$ and $G_1 = 1$, then:

$$h(x_B, x_C) = k(x_B)F_1(x_C) \text{ and } g(x_A, x_B) = k(x_B) \vee G_0(x_A). \tag{2.8}$$

Because f depends on C, there exists a vector c such that $F_1(c) = 0$. Now assume that $k(b) = 1$ holds. Then by (2.8) $h(b, c) = 0$ and $g(x_A, b) = 1$, implying that $G_0 = G(x_A, h(b, c)) = F(g(x_A, b), c) = F_1(c) = 0$. This contradicts the assumption $G_1 = 1$. Therefore, $F_1(c)$ implies $k(b) = 0$ for all b: contradiction.

(e) If $F_0 = 0$ and $\bar{G}_0 = G_1$, then

$$h(x_B, x_C) = k(x_B)F_1(x_C) \text{ and } g(x_A, x_B) = k(x_B) \oplus G_0(x_A). \tag{2.9}$$

There exists a vector c such that $F_1(c) = 1$. Now assume that $k(b) = 0$ holds. Then by (2.9) $h(b, c) = 0$ and $g(x_A, b) = G_0(x_A)$, implying that

$$G_0 = G(x_A, h(b, c)) = F(g(x_A, b), c) = F(G_0(x_A), c) = 0. \qquad (2.10)$$

Because f depends on A, there exists a vector a such that $G_0(A) = 0$. Then (2.10) implies $F_0(c) = 0$, contrary to our assumption $F_1(c) = 1$. From this we conclude $k(b) = 1$ for all b: contradiction.

The cases $F_0 = 1$ and $G_0 = 0$, and $F_0 = 1$ and $F_0 = 0$ and $\bar{G}_0 = G_1$, are symmetrical with (d) and (e). Similarly, the case $\bar{G}_0 = G_1$, and $\bar{F}_0 = F_1$, also leads to a contradiction (we leave this as an exercise).

Conclusion: Cases (a), (b), and (c) are the only possible ones. Therefore, we have proved that $f = f_A \star f_B \star f_C$, where $\star$ is uniquely determined as either $\vee$ or $\wedge$ or $\oplus$. ■

Theorem 2.11. *Suppose f is a Boolean function defined on the partition $\{A, B, C, D\}$, and f depends on A, B, and C. If $A \cup B$ and $B \cup C$ are modular sets of f, then A, B, C and $A \cup C$, $A \cup B \cup C \in \mu(f)$. Moreover, $f_{A \cup B \cup C} = f_A \star f_B \star f_C$, where $\star$ denotes either $\vee$ or $\wedge$ or $\oplus$.*

Proof. Because $A \cup B$ and $B \cup C$ are modular, there exist functions F, G, and h such that $f(x_N) = F(g(x_A, x_B), x_C, x_D) = G(x_A, h(x_B, x_C), x_D)$, where $g = f_{A \cup B}$ and $h = f_{B \cup C}$. Moreover, g depends on A and B and h depends on B and C. Because f depends on A, B, and C, there exists at least one vector u such that the function $f_u = f(x_A, x_B, x_C, d) \notin \{\mathbf{1}, \mathbf{0}\}$. We first prove the following

Claim 1. If $f_u \notin \{\mathbf{1}, \mathbf{0}\}$, then f_u depends on A, B, and C.
Suppose $F_u(y, x_C)$ depends on the variable y. Then since g depends on A and B and $f_u = F_u(g(x_A, x_B), x_C)$, the sets A, B, and C are essential for f_u. Similarly, if $G_u(z, x_A, x_D)$ depends on z, then f_u depends on B and C. Now assume that $F_u(y, x_C)$ does not depend on y. Then we will derive a contradiction as follows: Because $f_u = F_u(g(x_A, x_B), x_C)$, f_u depends on C. Therefore, since $f_u = G(x_A, h(x_B, x_C))$, the function G_u depends on the variable z, implying that f_u depends on B and C. Consequently, f_u depends on y, contrary to our assumption. Conclusion: F_u and G_u depend respectively on y and z, and f_u depends on A, B and C.

Claim 2. If $f_u, f_e \notin \{\mathbf{1}, \mathbf{0}\}$, then $f_e \in \{f_u, \bar{f}_u\}$.
Suppose $f_u \notin \{\mathbf{1}, \mathbf{0}\}$. Then Theorem 2.10 implies $f_u = \phi_1 \star \phi_2 \star \phi_3$, such that $g = \phi_1 \star \phi_2$ and $h = \phi_2 \star \phi_3$, where $\star$ is uniquely determined as $\vee$, $\wedge$, and $\oplus$. Because $\phi_1 \in \mu(g)$ and $g \in \mu(f)$, Theorem 2.9 implies that $\phi_1 \in \mu(f)$. Since g and h are subfunctions of f, we have $\phi_1 = f_A$, $\phi_2 = f_B$ and $\phi_3 = f_C$. Similarly, if $f_e \notin \{\mathbf{1}, \mathbf{0}\}$, then $f_e = \psi_1 \circ \psi_2 \circ \psi_3$ where $\circ$ is uniquely determined as $\vee$, $\wedge$, and $\oplus$. Moreover, $g = \psi_1 \circ \psi_2$, $h = \psi_2 \circ \psi_3$, and $\psi_i \in \{\phi_i, \bar{\phi}_i\}$. These constraints imply that $f_e \in \{f_u, \bar{f}_u\}$. Therefore, we have proved that for all e, $f_e \in \{\mathbf{1}, \mathbf{0}, f_u, \bar{f}_u\}$. According to Theorem 2.2, this means that $A \cup B \cup C$ is a modular set of f. So by

Theorem 2.10, we have $f_{A \cup B \cup C} = f_A \star f_B \star f_C$, where $\star$ denotes either $\vee$, *or* $\wedge$, or $\oplus$, and $A \cup C \in \mu(f)$. ■

Let $A, B \subseteq N$. Then A and B are called *overlapping* iff A and B are not comparable and $A \cap B \neq \emptyset$. The following theorem is a useful reformulation of Theorems 2.10 and 2.11.

Theorem 2.12. *Let f be a Boolean function. Suppose that A and B are overlapping modular sets of f, and that f depends on $A \cap B$, $A \cap \bar{B}$, and $\bar{A} \cap B$. Then $A \cap B$, $A \cap \bar{B}$, $\bar{A} \cap B$, $(A \cap \bar{B}) \cup (\bar{A} \cap B)$, and, $A \cup B$ are modular sets of f, and $f_{A \cup B} = f_{A \cap \bar{B}} \star f_{A \cap B} \star f_{\bar{A} \cap B}$, where $\star$ is either $\wedge$ or $\vee$ or $\oplus$.*

Theorem 2.12 is a well-known and important result, sometimes called the *Three Modules Theorem* of Ashenhurst [2] (although as far as we know, it is due to Singer [37]). It shows that the union of two *overlapping* modular sets gives rise to a congruence partition of three "modules" (modular sets), such that each possible union of these "modules" is itself a modular set. For monotone Boolean functions, this theorem has been rediscovered in game theory and reliability theory [32]. This fundamental theorem is proved in the literature using Ashenhurst decomposition charts, expansions of Boolean functions, or differential calculus ([1, 2, 20, 23, 21]).

Example 2.4. *Let the function f be defined by $f = x_1x_3x_4 \vee x_2x_3x_4 \vee x_1x_3x_5 \vee x_2x_3x_5$. Let $A = \{1, 2, 3\}$, and $B = \{3, 4, 5\}$. Then A, B, $A \cap B$, $A \cap \bar{B}$, and $\bar{A} \cap B$ are modular and $f = (x_1 \vee x_2)x_3(x_4 \vee x_5)$.*

Finally, we discuss the relationship between the modular sets of a function and the modular sets of its quotients. The following definition is taken from Möhring and Radermacher [29].

Definition 2.2. *Let f be a Boolean function defined on the set N, and let $\pi = \{A_i \mid i \in I\}$ be a congruence partition of f. The set of classes of π will also denoted by N/π. The quotient $F = f/\pi$ is a function defined on the set I. By identifying I and N/π we define the natural mapping $\theta_\pi : N \mapsto N/\pi$ by: $\theta_\pi(j) = i \Leftrightarrow j \in A_i$. Furthermore, we define the* completion *of a set $C \subseteq N$ as $\pi(C) := \bigcup\{A_i \mid C \cap A_i \neq \emptyset\}$.*

Note that the preceding definition implies $\pi(C) = \bigcup\{A_i \mid i \in \theta_\pi(C)\}$.

Proposition 2.13. *Let $\pi = \{A_i, i \in I\} \in \Pi(f)$ and let $F = f/\pi$. Suppose $\emptyset \neq J \subset I$. If $B = \bigcup\{A_j, j \in J\}$, then $J \in \mu(F) \Leftrightarrow B \in \mu(f)$.*

Proof. Suppose $f(x_N) = F[g_i(x_{A_i}), i \in I]$ and $J \in \mu(F)$. Without loss of generality, we may assume $J = \{1, 2, \ldots, l\} \subset I = \{1, 2, \ldots, m\}$, where $1 < l < m$. Then $F(y_I) = G(h(y_J), y_{\bar{J}})$ and

$$f(x_N) = G(h(g_1(x_{A_1}), \ldots, g_l(x_{A_l})), g_{l+1}(x_{A_{l+1}}), \ldots, g_m(x_{A_m})). \tag{2.11}$$

Let the functions k and H be defined by

$$k(x_B) := h[g_j(x_{A_j}), j \in J]$$

and

$$H(y, x_{\bar{B}}) := G(y, g_{l+1}(x_{A_{l+1}}), \ldots, g_m(x_{A_m})),$$

where $B = \bigcup\{A_j, j \in J\}$. Then Equation (2.11) implies $f(x_N) = H(k(x_B), x_{\bar{B}})$, so that $B \in \mu(f)$.

Conversely, let $B \in \mu(f)$, where $B = \bigcup\{A_j, j \in J\}$, and $J = \{1, 2, \ldots, l\}$. Then according to Theorem 2.5, $\{B, A_{l+1}, \ldots, A_m\} \in \Pi(f)$, so that f can be written as

$$f(x_N) = G(g(x_{A_1}, \ldots, x_{A_l}), g_{l+1}(x_{A_{l+1}}), \ldots, g(x_m A_m)). \tag{2.12}$$

Suppose f depends on $A_i, i \in I$, for all $i \in I$ and $y \in \{0, 1\}$. Then there exists a binary vector $a_i(y)$ such that $y = g_i(a_i(y))$. If h is the function defined by

$$h(y_1, \ldots, y_l) = h(g_1(a_1(y_1)), \ldots, g_l(a_l(y_l))) := g(a_1(y_1), \ldots, a_l(y_l)),$$

then Equation (2.12) implies

$$\begin{aligned} F(y_I) &= F[g_i(a_i(y_i))), i \in I] = f[a_i(y_i),\ i \in I] \\ &= G(g(a_1(y_1), \ldots, a_l(y_l)), y_{l+1}, \ldots, y_m) \\ &= G(h(y_J), y_{\bar{J}}). \end{aligned}$$

This shows that $J \in \mu(F)$. ■

By noting that $\theta_\pi^{-1}(J) = \bigcup\{A_j, j \in J\}$, it is easy to see that the following theorem is a reformulation of Proposition 2.13.

Theorem 2.14. *$J \in \mu(F)$ if and only if $\theta_\pi^{-1}(J) \in \mu(f)$.*

Proposition 2.15. *If $\pi \in \Pi(f)$ and $B \in \mu(f)$, then $\pi(B) \in \mu(f)$.*

Proof. Let $\pi = \{A_i \mid i \in I\}$. Then by definition $\pi(B) = \bigcup\{A_i \mid B \cap A_i \neq \emptyset\}$. If $B \subseteq A_i$ for some i then $\pi(B) = A_i$. Furthermore, if $\pi(B) = \bigcup\{A_j \mid j \in J \subseteq I\}$, then $\pi(B) = B$. In all other cases there exists a j such that B and A_j are overlapping. According to Theorem 2.12, we have $B \cup A_j \in \mu(f)$. Therefore, $\pi(B) = B \cup \{A_j \mid B$ and A_j are overlapping$\} \in \mu(f)$. ■

Theorem 2.16. *If $\pi \in \Pi(f)$ and $B \in \mu(f)$, then $\theta_\pi(B) \in \mu(F)$.*

Proof. Because $\theta_\pi(B) = \theta_\pi(\pi(B))$, this follows from Proposition 2.13 and Proposition 2.15. ■

We now collect a number of properties of modular sets proved sofar:

Theorem 2.17. *Let f be a Boolean function defined on N depending on all its variables, and let g be a component of f defined on $A \in \mu(f)$. Suppose $\pi \in \Pi(f)$ and let F be the quotient f/π. Then:*

M_0 : $\mu(f) = \mu(f^d)$.
M_1 : $N \in \mu(f)$ *and* $\{i\} \in \mu(f)$ *for all* $i \in N$.
M_2 : $A, \bar{A} \in \mu(f) \Leftrightarrow f = f_A \star f_{\bar{A}}$, *where $\star$ denotes either $\vee$ or $\wedge$ or $\oplus$.*

M_3 : *If* $A, B \in \mu(f)$ *are overlapping then the sets* $A \cap \bar{B}$, $A \cap B$, $B \cap \bar{A}$, $A \cap B$ *and* $A \oplus B$ *all belong to* $\mu(f)$.
M_4 : $\mu(g) = \{B \in \mu(f) \mid B \subseteq A\}$.
M_5 : $J \in \mu(F) \Leftrightarrow \theta_\pi^{-1}(J) \in \mu(f)$.
M_6 : *If* $B \in \mu(f)$ *then* $\pi(B) \in \mu(f)$.
M_7 : *If* $B \in \mu(f)$ *then* $\theta_\pi(B) \in \mu(F)$.

Proof. M_0 follows from Equation (2.1). M_1 is an immediate consequence of the definition of modular sets. The other properties are respectively proved in Theorems 2.9, 2.11, 2.15 and Theorems 2.16, 2.14. ∎

2.6 Composition Trees

In this section we consider only Boolean functions depending on all their variables. Composition trees were first studied by Shapley [36] in the context of simple games (monotone Boolean functions). These trees represent in a compact way all the information on the modular sets of a Boolean function. Although the number of modular sets may be exponential in the number n of variables, it appears that the number of nodes in a composition tree is linear in n. Let f be a Boolean function defined on N. Then $C \in N$ is called a *maximal modular set* of f if $C \in \mu(f)$, and for all B with $C \subset B \neq N$, we have $B \notin \mu(f)$. The set of all maximal modular sets is denoted by $m(f)$. A function f is of *composition type I* if every two maximal modular sets have an empty intersection; otherwise f is of *composition type II*. We will show that in the latter case, the set of complements of the maximal modular sets, denoted by $\pi_\star$, is a partition of N.

Definition 2.3. *A Boolean function* f *defined on* N *is called modular degenerated if and only if every nonempty set* $A \subseteq N$ *is a modular set of* f.

Theorem 2.18. *Let* f *be a function on* $N = \{x_1, x_2, \ldots, x_n\}$. *Then* f *is modular degenerated if and only if* $f = y_1 \star y_2 \star \cdots \star y_n$, *where* $y_i = x_i$ *or* $\bar{x}_i$, *and* $\star$ *denotes either* $\vee$ *or* $\wedge$ *or* $\oplus$.

The proof of Theorem 2.18 is left as an exercise.

Definition 2.4. *Let* f *be a Boolean function. Then* $\Delta(f)$ *is the set of all congruence partitions* π *such that the quotient* f/π *is modular degenerated.*

Corollary 2.19. *Let* $\pi \in \Delta(f)$. *Then every union of classes of* π *belongs to* $\mu(f)$. *In particular, if* $A \in \pi$ *then* $\bar{A} \in \mu(f)$.

Theorem 2.20. *If* f *is of type II, then* $|m(f)| \geq 3$ *and* $\pi_\star \in \Delta(f)$.

Proof. Suppose $A, B \in m(f)$ and $A \cap B \neq \emptyset$. Then we have respectively: A and B are overlapping, $N = A \cup B$, $\bar{A} \cap \bar{B} = \emptyset$ and $\bar{A} \cup \bar{B} = A \oplus B \in \mu(f)$. Because $A \cap B \neq \emptyset$ implies $\bar{A} \cup \bar{B} \neq N$, there exists a $C \in m(f)$ such that $C \supseteq \bar{A} \cup \bar{B}$. From this we conclude that $|m(f| \geq 3$ and that $\forall C, D \in m(f)$ the following holds: $C \cap D \neq \emptyset$, $\bar{C} \cap \bar{D} = \emptyset$, $\bar{C} \cup \bar{D} \neq N$ and $\bar{C} \cup \bar{D} \in \mu(f)$.

According to Theorem 2.12 this implies that $U := \bar{C}_1 \cup \bar{C}_2 \cup \cdots \cup \bar{C}_m \in \mu(f)$. Moreover, we claim that $U = N$, for otherwise there would exist a $C_j \in m(f)$ such that $U \le C_j$, which is clearly a contradiction. Therefore, $\pi_\star$ is a partition of f. Now let $J \subseteq I = \{1, 2, \ldots, m\}$. Then $\bigcup\{\bar{C}_j \mid j \in J\} = \overline{\bigcup\{\bar{C}_i \mid i \in \bar{J}\}} = \bigcap\{C_i \mid i \in \bar{J}\} \in \mu(f)$. This proves that $\pi_\star \in \Delta(f)$. ■

Theorem 2.21. *A Boolean function f is of type II if and only if $|m(f)| \ge 3$ and $\Delta(f) \ne \emptyset$.*

Proof. The only-if part of this theorem follows from Theorem 2.20. Conversely, assume $|m(f)| \ge 3$ and $\pi \in \Delta(f)$. Suppose f is of type I and π^* is the maximal disjoint congruence partition of f. Let A and B be two classes of π such that $A \subseteq C_i$ and $B \subseteq C_j$, where C_i and C_j are different classes of π^*. Since $A \cup B \in \mu(f)$ and $\forall\, C_k \in \pi^* : A \cup B \nsubseteq C_k$, we have a contradiction. Therefore, f is of type II. ■

Corollary 2.22. *A Boolean function f is of type I if and only if $|m(f)| = 2$ or $\Delta(f) = \emptyset$.*

Theorem 2.23. *Suppose f is of type II and let $\pi_\star = \{C_1, C_2, \ldots, C_m\}$. Furthermore, let $\mu_\star(f)$ denote the set of all unions of classes in $\pi_\star$. Then*

$$\mu(f) = \mu(f_{C_1}) \cup \mu(f_{C_2}) \cup \cdots \cup \mu(f_{C_m}) \cup \mu_\star(f).$$

Proof. Let $B \in \mu(f)$. Assume that B is not properly contained in any class of $\pi_\star$ and B is not a union of classes of $\pi_\star$. Then there exists a class $C \in \pi_\star$ such that either $B \supset \bar{C}$ or B and $\bar{C}$ are overlapping, implying $B \cup \bar{C} \in \mu(f)$. However, this contradicts the maximality of $\bar{C}$. ■

Theorem 2.24. *If f is of type II and $\pi \in \Delta(f)$, then $\pi_\star \le \pi$.*

Proof. Suppose $\pi \in \Delta(f)$ and A is a class of π. Then by Corollary 2.19 $\bar{A} \in \mu(f)$. Therefore, the class A cannot be properly contained in a class C of $\pi_\star$. For otherwise $\bar{C} \subset \bar{A}$, contrary to the maximality of $\bar{C}$. So according to Theorem 2.23 every class of π is a union of classes of $\pi_\star$, implying that $\pi_\star \le \pi$. ■

The following theorem shows that $\pi_\star$ is the finest partition such that f/π is modular degenerated.

Theorem 2.25. *If $\pi_1, \pi_2 \in \Delta(f)$, then $\pi_1 \wedge \pi_2 \in \Delta(f)$.*

Proof. Suppose $\pi_1, \pi_2 \in \Delta(f)$. Then $\pi_\star \le \pi_1 \wedge \pi_2$. Therefore, $f/\pi_1 \wedge \pi_2$ is a quotient of $f/\pi_\star$. Since $f/\pi_\star$ is modular degenerated, this implies that $\pi_1 \wedge \pi_2 \in \Delta(f)$. ■

Based on the two composition types, we can construct a *composition tree* $\mathcal{T}(f)$ for a Boolean function f defined on N:

(1) The root of $\mathcal{T}$ is the set N. Each node of $\mathcal{T}$ is a modular set of f.

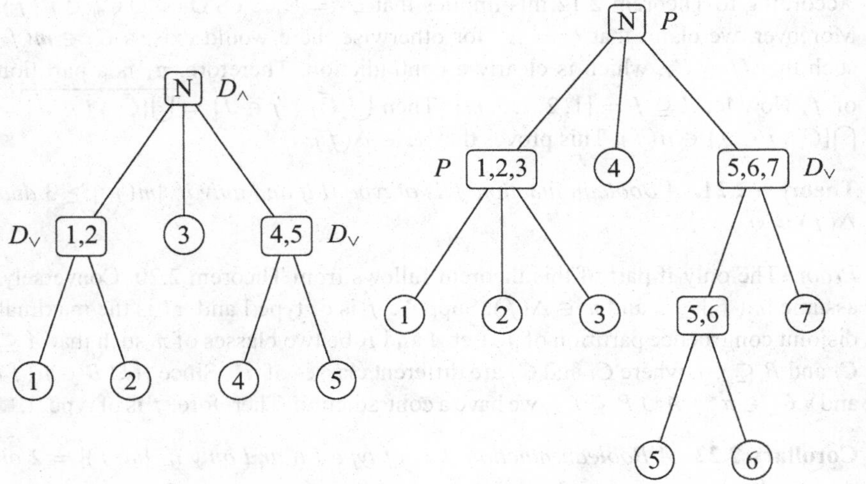

Figure 2.1. Modular decomposition of f and g.

(2) If C is a node and f_C is of type I, then f_C has a maximal disjoint decomposition $\pi^\star = \{C_1, C_2, \ldots, C_m\}$. Then $C_1, C_2, \ldots, C_m$ are the children of node C, and node C is labeled with $P(rime)$.

(3) If f_C is of type II, then $\Delta(f_C)$ has a finest partition $\pi_\star = \{C_1, C_2, \ldots, C_m\}$, with $m \geq 3$. Then $C_1, C_2, \ldots, C_m$ are the children of node C, and node C is labeled with $D(egenerated)$: $D_\wedge$, $D_\vee$ or $D_\oplus$.

(4) The leaves of $T(f)$ are the singleton sets $\{i\} \in \mu(f)$.

Example 2.5. *Let f and g be positive functions defined by*

$$f = (x_1 \vee x_2)x_3(x_4 \vee x_5) = x_1x_3x_4 \vee x_2x_3x_4 \vee x_1x_3x_5 \vee x_2x_3x_5, \textit{ and}$$
$$g = x_1x_2x_4 \vee x_1x_3x_4 \vee x_2x_3x_4 \vee x_1x_2x_5x_6 \vee x_1x_3x_5x_6 \vee x_2x_3x_5x_6 \vee x_4x_5x_6$$
$$\vee x_1x_2x_7 \vee x_1x_3x_7 \vee x_2x_3x_7 \vee x_4x_7.$$

Then $m(f) = \{\{1,2,3\}, \{3,4,5\}, \{1,2,4,5\}\}$ and $\pi_\star(f) = \{\{4,5\}, \{1,2\}, \{3\}\}$. Moreover, $m(g) = \{\{1,2,3\}, \{4\}, \{5,6,7\}\}$. The modular trees of these functions are given in Figure 2.1. Note that although a function with $|m(f)| = 2$ is prime, the corresponding node in the tree is labeled as D.

Theorem 2.26. *Let f be a Boolean function defined on N. A subset of $C \subseteq N$ is a modular set of f if and only if one of the following holds:*

(a) *C is a node of $T(f)$.*
(b) *C is the union of children of a node of type D.*

Proof. This follows from Theorem 2.17 (M_3) and Theorem 2.23. ■

Theorem 2.27. *Let f be a Boolean function defined on N, and let $|\mathcal{T}|$ denote the number of nodes of the modular tree $\mathcal{T}$. Then $|\mathcal{T}(f)| \leq 2|N| - 1$.*

Proof. We use induction on $n = |N|$. The assertion is trivial for $n = 2$. Suppose the the assertion is true for all Boolean functions with $n \leq k$. Now consider a function f on N with $|N| = k + 1$. Let $C_1, C_2, \ldots, C_m$ be the children of node N, where $m \geq 2$. Then $\sum_{i=1}^{m} |C_i| = |N|$. We then have

$$|\mathcal{T}(f)| = 1 + \sum_{i=1}^{m} |\mathcal{T}(C_i)| \leq 1 + \sum_{i=1}^{m} (2|C_i| - 1) = 1 + 2|N| - m \leq 2|N| - 1.$$

■

Computational aspects concerning composition trees are discussed in Section 2.8.

2.7 The Set of Congruence Partitions

Let f be a Boolean function defined on N. The set of partitions on N will be denoted by $P(f) = P(N)$. In the previous section we have argued that because Theorem 2.9, the collection of all congruence partitions $\Pi(f) \subseteq P(f)$ reflects all the decomposition possibilities of f. So this collection carries the same information as a composition tree. In this section we show that the congruence partitions form an upper semimodular lattice. As a consequence this lattice satisfies the Jordan-Dedekind condition. We follow here the presentation of [29].

It is known ([11]) that $P(f)$ is a finite lattice, with the ordering relation $\pi_1 \leq \pi_2$ denoting that each class of π_1 is contained in a class of π_2. In that case, π_1 is called a *refinement* of π_2, and π_2 is called a *coarsening* of π_1. The least upper bound, respectively greatest lower bound, of π_1 and π_2 is denoted by $\pi_1 \vee \pi_2$, respectively $\pi_1 \wedge \pi_2$. The partition $\pi_1 \wedge \pi_2$ consists of all nonempty intersections of a class of π_1 and of a class of π_2. The partition $\pi_1 \vee \pi_2$ is the intersection of all partitions π containing π_1 and π_2. It is easy to see that if R is a class of $\pi_1 \vee \pi_2$ and if C is either a class of π_1 or of π_2, then $C \cap R \neq \emptyset \Rightarrow C \subseteq R$. Moreover, the class R is a union of classes of π_1 as well as a union of classes of π_2. Therefore each class R of $\pi_1 \vee \pi_2$ can be written as

$$R = P_1 \cup Q_2 \cup P_3 \cup Q_4 \cup \cdots \cup P_l, \tag{2.13}$$

where respectively the P_i are (possibly *not* different) classes of π_1 and the Q_j are (possibly *not* different) classes of π_2. Furthermore, $P_i \cap Q_{i+1} \neq \emptyset$ and $Q_i \cap P_{i+1} \neq \emptyset$.

Proposition 2.28. *$\Pi(f)$ is a sublattice of $P(f)$.*

Proof. We have to prove that $\Pi(f)$ is closed under the meet and join operations of $P(f)$. Suppose $\pi_1, \pi_2 \in \Pi(f)$. Then each class of $\pi_1 \wedge \pi_2$ is the nonempty intersection of two modular sets. Therefore, it follows from

Theorem 2.12 that $\pi_1 \wedge \pi_2 \in \Pi(f)$. Similarly, Theorem 2.12 and Equation (2.13) imply that $\pi_1 \vee \pi_2 \in \Pi(f)$. ∎

$P(f)$ and $\Pi(f)$ contain a finest partition π^0 consisting of all singleton subsets of N and a coarsest partition π^1 consisting of a single class namely the set N. Let π_1, π_2 be two partitions with $\pi_1 < \pi_2$ then π_2 *covers* π_1 if for all partitions σ with $\pi_1 < \sigma < \pi_2$ we have either $\sigma = \pi_1$ or $\sigma = \pi_2$. A partition $\pi \in \Pi(f)$ is called an *atom* of $\Pi(f)$ if π covers π^0. It is easy to see that π is an atom of $\Pi(f)$ if and only if $\pi = \{A, \{i\} \mid i \in N \setminus A\}$, where $A \in \mu(f)$ and f_A is prime. Let $\mathcal{P}$ be a finite poset (partially ordered set). If $a, b \in \mathcal{P}$ and $a < b$ then a sequence $a = a_0, a_1, \ldots a_n = b$ is called a *chain* between the endpoints a and b of length n if $a_{i-1} < a_i$ for $i = 1, 2, \ldots, n$. Moreover, a chain is called *maximal* if a_i covers a_{i-1}, for $i = 1, 2, \ldots, n$. The poset $\mathcal{P}$ satisfies the *Jordan-Dedekind chain condition* if all maximal chains in $\mathcal{P}$ between two endpoints a and b have the same length.

Definition 2.5. *A finite lattice $\mathcal{L}$ is called upper semimodular if a_i covers $a_1 \wedge a_2$, i = 1,2, implies that $a_1 \vee a_2$ covers both a_1 and a_2.*

Theorem 2.29. *Let f be a Boolean function. Then $\Pi(f)$ is upper semimodular.*

Proof. Suppose $\pi_1, \pi_2 \in \Pi(f)$ cover $\pi_1 \wedge \pi_2 = \{P_j \mid j \in J\}$. Then there exists exactly one class A_i of π_i such that A_i is a union of classes of $\pi_1 \wedge \pi_2$ and $\pi_i = \{A_i, P_k \mid k \in J_i\}$, where $J_i \subset J$, $i = 1, 2$ and f_{A_i} is prime. If $J_1 \cap J_2 = \emptyset$, then $\pi_1 \vee \pi_2 = \{J_1, J_2, P_j \mid j \in J \setminus (J_1 \cup J_2)\}$. If $J_1 \cap J_2 \neq \emptyset$, then $\pi_1 \vee \pi_2 = \{J_1 \cup J_2, P_j \mid j \in J \setminus (J_1 \cup J_2)\}$. Therefore, in both cases $\pi_1 \vee \pi_2$ covers π_1 and π_2. ∎

Because every finite upper semimodular lattice satisfies the Jordan-Dedekind condition (see, e.g., [11]), we have the following corollary.

Corollary 2.30. *Let f be a Boolean function. Then $\Pi(f)$ satisfies the Jordan-Dedekind condition.*

Example 2.6. *Let f be the positive functions defined by*

$$f = (x_1 \vee x_2)x_3(x_4 \vee x_5) = x_1x_3x_4 \vee x_2x_3x_4 \vee x_1x_3x_5 \vee x_2x_3x_5;$$

see Example 2.5 for the composition tree of f. Then,

$$\{\{1,2,3,4,5\}\} > \{\{1,2,3\},\{4,5\}\} > \{\{1,2\},\{3\},\{4,5\}\} > \{\{1\},\{2\},\{3\},\{4,5\}\} > \{\{1\},\{2\},\{3\},\{4\},\{5\}\}$$

and

$$\{\{1,2,3,4,5\}\} > \{\{1,2,4,5\},\{3\}\} > \{\{1,2\},\{3\},\{4,5\}\} > \{\{1\},\{2\},\{3\},\{4,5\}\} > \{\{1\},\{2\},\{3\},\{4\},\{5\}\}$$

are two maximal composition series of length 4 in $\Pi(f)$.

Other algebraic properties of congruence partitions of Boolean functions and other discrete structures are discussed in [28, 29].

2.8 Computational Aspects

Because the nonempty intersection of two modular sets of a Boolean function is again modular (see Theorem 2.12), each subset A of variables is contained in a smallest modular set called the *modular closure* of A.

Definition 2.6. *Let f be a Boolean function defined on N. The closure of a non-empty set $A \subseteq N$ is defined by $Cl_f(A) = \cap\{B \mid A \subseteq B,\ B$ is a modular set of $f\}$.*

We show in this section that the modular tree of a Boolean function with n variables can be obtained by computing $O(n^3)$ times the modular closure of a subset of variables. Therefore, a central step in the determination of the modular tree of a Boolean function is the computation of the modular closure of a set. For the class of positive functions represented in DNF, an efficient algorithm to compute the modular closure of a set is presented in the next section. However, as we show in this section, such an efficient algorithm for general Boolean functions (represented in DNF) does not exist.

2.8.1 Computing the Modular Tree of Boolean Function

Let f denote a Boolean function defined on N with $n = |N| \geq 2$ represented by a DNF expression with m terms. We assume that f depends on all its variables. Recall that $m(f)$ denotes the set of all maximal modular sets of f. We also refer to the results in Section 2.6. Let $T(m, n)$ denote the complexity of the computation of the modular closure of a nonempty subset of variables of f.

Lemma 2.31. *A set $C \in m(f)$ can be computed in time $O(nT(m, n))$.*

Proof. Let $i \in N$; then we can construct the series of modular closures $\{i\} = C_0 \subset C_1 \cdots \subset C_k = C$, where $C_{i+1} = Cl_f(C_i \bigcup \{j\})$ by choosing some $j \in \bar{C}_i$, with $Cl_f(C_i \bigcup \{j\}) \neq N$. If such an element j does not exist, then $C = C_i \in m(f)$. Because $k \leq n$, the set C can be computed in time $O(nT(m, n))$. ∎

Proposition 2.32. *The set $m(f)$ can be computed in time $O(n^2 T(m, n))$.*

Proof. We first construct a set $C_1 \in m(f)$ using the procedure discussed in Lemma 2.31. In the same way we can construct a maximal modular set C_2 using an element $i \in \bar{C}_1$. Suppose $C_1 \bigcap C_2 = \emptyset$ and $|m(f)| = k$. If $C_1, C_2 \ldots, C_l \in m(f)$, then $l < k$ if and only if $D := C_1 \bigcup C_2 \ldots \bigcup C_l \subset N$. Let $j \in \bar{D} \neq \emptyset$. Then we determine $C_{l+1} \in m(f)$ such that $j \in C_{l+1}$. If $C_1 \bigcap C_2 \neq \emptyset$, then $l < k$ if and only if $E := \bar{C}_1 \bigcup \bar{C}_2 \ldots \bigcup \bar{C}_l \subset N$. Therefore, if $l < k$ and $C \in m(f) \setminus \{C_1, C_2 \ldots, C_l\}$, then $C \supseteq E$. Now we construct $C_{l+1} \in m(f)$ such that $C_{l+1} \supseteq E$. Since $|m(f)| \leq n$, it follows that $m(f)$ can be computed in time $O(n^2 T(m, n))$. ∎

According to Theorem 2.9 the modular sets of a component f_C of f defined on $C \subseteq N$ are just the modular sets of f contained in C. Therefore, by using the same arguments as in Proposition 2.32, we can also compute the set $m(f_C)$ in time $O(n^2 T(m, n))$, without the explicit determination of a DNF expression of f_C.

Corollary 2.33. *Let f_C be a component of f defined on C. Then the set $m(f_C)$ can be computed in time $O(n^2 T(m, n))$.*

Theorem 2.34. *The modular tree of f can be computed in time $O(n^3 T(m, n))$.*

Proof. Let $\mathcal{T}(f)$ denote the modular tree of f. We have already established in Theorem 2.27 that $|\mathcal{T}(f)| \leq 2n - 1$. Because the leaves of $\mathcal{T}(f)$ are the singleton sets of N, it follows that the number of internal nodes of the tree is less than or equal to $n - 1$. Suppose C is an internal node and $m(f_C) = \{C_1, C_2, \ldots, C_k\}$, where $k \leq n$. Then by Corollary 2.33, $m(f_C)$ can be determined in time $O(n^2 T(m, n))$. If $C_1 \bigcap C_2 = \emptyset$, then the children of C are the nodes $C_1, C_2, \ldots, C_k$. Otherwise, the children of C are $\bar{C}_1, \bar{C}_2, \ldots, \bar{C}_k$. Since there are at most $n - 1$ internal nodes, it follows that $\mathcal{T}(f)$ can be determined in time $O(n^3 T(m, n))$. ∎

It is easy to see that the determination of the modular closure can be solved by solving $O(n)$ times the following search problem:

Problem MODULAR-SET
Input: A Boolean function f represented by a DNF expression with m terms defined on N, where $|N| = n$, and $A \subseteq N$.
Output: "A is modular" if A is modular. An element $x \in Cl_f(A) \setminus A$ otherwise.

2.8.2 The Complexity of Recognizing Modular Sets

In this section we prove that for general Boolean functions the problem of recognizing modular sets (called MODULAR) is coNP-complete. In switching theory, this complexity has not been discussed. In this context, modular sets and decompositions are based on the evaluation of Ashenhurst decomposition charts or on differential calculus techniques [1, 2, 20, 21, 23]. It has been shown in [28, 29] that the algorithms for the determination of modular sets are exponential in the number of variables. However, here we study the complexity of the recognition problem of Boolean functions given in DNF form. In particular we discuss the following restricted version of the MODULAR-SET problem:

Problem MODULAR
Given: A Boolean function f in DNF defined on N and a set $A \subset N$ that contains at least one essential variable of f.
Question: Is A a modular set of f?

We relate this problem to the following recognition problem:

Problem COMPLEMENT
Given: Boolean functions g_1 and g_2 in DNF.
Question: Is $g_2 = \overline{g}_1$?

It is well known that this problem is coNP-complete, because the problem is already coNP-complete if $g_1 = \mathbf{0}$. It can also be shown that problem COMPLEMENT remains coNP-complete if we assume that $f, g \notin \{\mathbf{0}, \mathbf{1}\}$.

Theorem 2.35. *Problem MODULAR is coNP-complete.*

Proof. Suppose g_1 and g_2 are nonconstant Boolean functions given in DNF on $A = \{x_1, \cdots, x_n\}$. Let $\{x, y\} \bigcap A = \emptyset$. Define the function f on $A \cup \{x, y\}$ as

$$f = xg_1 \vee yg_2. \tag{2.14}$$

We first show that $g_2 = \overline{g}_1$ if and only if A is modular, implying that the problem MODULAR is coNP-hard. So suppose that $g_2 = \overline{g}_1$, then A is a modular set of f. Moreover, since $g_1, g_2 \notin \{\mathbf{0}, \mathbf{1}\}$, A also contains at least one essential variable. Conversely, suppose A is modular and A contains essential variables of f. Then there exists a pair of binary values (x_0, y_0) such that the function g defined by $g = f(x_0, y_0, x_A)$ is nonconstant. Furthermore, according to Theorem 2.2, for all *fixed* x and y the function $h(x_A) = f(x, y, x_A)$ is constant or identical to the function g or its complement. From Equation (2.14) it follows that $h \in \{\mathbf{0}, g_2, g_1, g_1 \vee g_2\}$. Therefore, we have $g_2 = \overline{g}_1$. Conclusion: $g_2 = \overline{g}_1 \Leftrightarrow A$ is modular.

To prove that the problem MODULAR is in coNP, we note that according to Theorem 2.2, A is a modular set of f if and only if for all binary vectors b the function $f_b := f(x_A, b) \in \{\mathbf{1}, \mathbf{0}, g, \overline{g}\}$, where g is a component of f on A. Therefore, the set A is *not* modular if and only if there exist binary vectors b_1 and b_2 such that $f_{b_1}, f_{b_2} \notin \{\mathbf{1}, \mathbf{0}\}$ and $f_{b_1} \neq \overline{f_{b_2}}$. Equivalently, the set A is not modular if and only if there exist three different binary vectors a, a_1, a_2, and two different vectors b_1, b_2 such that $f_{b_1}(a) = f_{b_2}(a) \neq f_{b_1}(a_1) = f_{b_2}(a_2)$. This shows that problem MODULAR is in coNP. ∎

2.9 Decomposition of Monotone Boolean Functions

2.9.1 Introduction

In this section we will refine the theory of decomposition of Boolean functions to the subclass of monotone Boolean functions. This subclass is interesting not only for theoretical reasons but also for several application domains. Various decomposition algorithms in these domains are known. A discussion of all algorithms (up to 1990) related to modular sets known in game theory, reliability theory and set systems (clutters) is given by Ramamurthy [32]. It is the objective of this section to give a unified Boolean treatment of the theory of decompositions of monotone Boolean functions. Because the development of decomposition algorithms of monotone Boolean functions has a long history in several domains, we start with a brief historical review.

We have already seen that the recognition problem MODULAR for general Boolean functions is coNP-complete. For positive Boolean functions, the situation is quite different. Let f be a positive function defined on the set N, where $|N| = n$, and let m be the number of prime implicants of f. Then according to Möhring and Radermacher [29] (see also Section 2.8), the modular tree can be computed in time $O(n^3 T(m, n))$, where $T(m, n)$ is the complexity of computing the modular closure of a set $A \subseteq N$. The first known algorithm to compute the modular closure due to Billera [5] was based on computing the dual of f. Although this problem is hard in general, for positive functions the complexity of the dualization problem is still not known, although this problem is unlikely to be hard: see, for example, [9] and [8]. An improvement of Billera's algorithm by Ramamurthy and Parthasarathy [33], also based on dualization, has a similar complexity. The first polynomial algorithm given by Möhring and Radermacher [29] reduced the complexity to $T(m, n) = O(m^3 n^4)$. Subsequently, the complexity was further reduced by Ramamurthy and Parthasarathy [33] and Ramamurthy [32] to respectively $T(m, n) = O(m^3 n^2)$ and $T(m, n) = O(m^2 n^2)$. A final reduction of this complexity to $T(m, n) = O(mn^2)$ is obtained by Bioch in [8].

2.9.2 Preliminaries

If f is a monotone Boolean function defined on N and $f(x_N) = F(g(x_A), x_B)$, then there exist uniquely determined monotone Boolean functions G and h such that $f(x_N) = G(h(x_A), x_B)$. This shows that the decomposability of a monotone function within the class of monotone functions is the same as in the class of general Boolean functions. Therefore, we will restrict the decomposition of a monotone Boolean function to the class of all monotone Boolean functions. This implies that if $A \in \mu(f)$, then a (monotone) component g of f defined on A is uniquely determined. The other results obtained in the previous sections also apply to monotone Boolean functions. In particular, the decomposition tree contains all the available information about the modular sets of a monotone function f. However, if f is monotone function and if a node of the composition tree is of type II, then only the cases $D_\vee$ or $D_\wedge$ can occur.

Notations: Let f be a positive function defined on N. Then a subset $A \subseteq N$ will frequently be represented by its characteristic vector $a := char(A) \in \{0, 1\}^n$, with $n = |N|$. If $A = \emptyset$, then this will be denoted by $a = \mathbf{0}$, where $\mathbf{0}$ is the all-zero vector. Furthermore, the set of all essential variables of f is called the *support set* of f. This set is denoted by $S(f)$, and the vector $char(S(f))$ is denoted by $s(f)$. If $v, w \in \{0, 1\}^n$, then $v \wedge w$ (also denoted by vw), and $v \vee w$ denote respectively the vectors obtained by applying the and-operation and the or-operation componentwise to the vectors v and w. Finally, recall that f^d denotes the dual of the function f.

The following lemma is easy to prove.

Lemma 2.36. *Let f be a monotone function. Then $S(f^d) = S(f)$.*

2.9.3 Decomposition of Positive Boolean Functions

In this section we introduce two functions associated with a positive function f and a set of variables A. The first one, denoted f^a, is obtained by deleting from f all prime implicants that have no variables in common with A. The second one, called the *contraction* from f to A and denoted f_a, is obtained by deleting from the prime implicants of f^a all variables not contained in A. Now the main result of this subsection is that A is a modular set of f if and only if the following holds true:

$$f^a = f_a h, \tag{2.15}$$

where h is the function obtained from f^a by deleting from the prime implicants of f^a all variables contained in A. This result will be proved by using the Shannon representation of the decomposition of a positive function (cf. Section 2.3). It shows that a component of a positive function is unique and equal to the contraction of f. Moreover, it also shows that the modularity of A can be checked using Equation (2.15) in time $O(m^2n^2)$, where m and n are respectively the number of prime implicants and variables of f. Let us now establish these results.

Definition 2.7. *Let f be a positive function defined on N and let $A \subseteq N$. If f depends on A (i.e $s(f) \wedge a \neq \mathbf{0}$), then the positive function f^a on A is defined by $\min T(f^a) = \{v \mid v \in \min T(f), v \wedge a \neq \mathbf{0}\}$, where $a = char(A)$. Otherwise $f^a := \mathbf{0}$.*

From this definition it follows that every positive Boolean function f can be decomposed as

$$f = f_{|x_A=\mathbf{0}} \vee f^a, \quad \text{where } A \subseteq N. \tag{2.16}$$

Furthermore, for a monotone Boolean function f, Shannon's decomposition has the form

$$f(x) = f_{|x_j=0} \vee x_j f_{|x_j=1}. \tag{2.17}$$

Definition 2.8. *Let f be a positive function defined on N, and $A \subseteq N$. Then the contraction f_a of f on A is defined by $f_a(x_A) = f^a{}_{|x_{\bar{A}}=\mathbf{1}}(x_A)$, where $a = char(A)$.*

Example 2.7. *Let f be the positive function on $\{1, 2, \ldots, 6\}$ defined by $f = x_1x_2x_4x_5 \vee x_1x_2x_6 \vee x_2x_3x_4x_5 \vee x_2x_3x_6 \vee x_4x_6$ and* let $A = \{1, 2, 3\}$. *Then $a = char(A) = 111000$, $f_{|x_A=\mathbf{0}} = x_4x_6$, $f^a = x_1x_2x_4x_5 \vee x_1x_2x_6 \vee x_2x_3x_4x_5 \vee x_2x_3x_6$, and $f_a = x_1x_2 \vee x_2x_3$.*

It is easy to verify that the following lemma holds.

Lemma 2.37. *Let f be a positive function defined on N, and let $A \subseteq N$. Then $f_a(x) = 1$ if and only if there exists $v \in \min T(f)$ such that $x \geq v \wedge a > \mathbf{0}$.*

The following characterization of the contraction is well known; see [32].

Theorem 2.38. *Let f be a positive function defined on N and let $A \subseteq N$. Suppose that $a = char(A)$ and $x \leq a$. Then $f_a(x) = 1$ if and only if there exists $y \leq \bar{a}$ such that $f(y) = 0$ and $f(x \vee y) = 1$.*

Proof. Suppose that $x \leq a$ and that $f_a(x) = 1$. Then by Lemma 2.37 there exists $v \in \min T(f)$ such that $x \geq v \wedge a > \mathbf{0}$. Let $y = v \wedge \bar{a}$. Then $y \leq \bar{a}$ and $x \vee y \geq v$. This implies that $f(x \vee y) = 1$. Moreover, because $v \wedge a > \mathbf{0}$, we have $v \wedge \bar{a} < v$. Therefore: $f(y) = 0$.

Conversely, suppose there exists $y \leq \bar{a}$ such that $f(y) = 0$ and $f(x \vee y) = 1$. Then $x \vee y \geq v$ for some $v \in \min T(f)$. From this we conclude that $x \geq v \wedge a$ and that $y = y \wedge \bar{a} \geq v \wedge \bar{a}$. From this we derive that $v \not\geq \bar{a}$, for otherwise we would have $y \geq v$, contrary to our assumption that $f(y) = 0$. Conclusion: there exists v such that $v \wedge a \neq \mathbf{0}$, and $x \geq v \wedge a$. According to Lemma 2.37, this is equivalent to $f_a(x) = 1$. ∎

The following theorem shows that if f is a positive function and if $A \in \mu(f)$, then the component $g(x_A)$ of f is just the contraction of f on A.

Theorem 2.39. *Let f be a positive Boolean function defined on N and let $A \subseteq N$. Then, A is modular if and only if $f^a = f^a{}_{|x_A=1} f_a$. Moreover, the component $g(x_A)$ is the contraction of f on A: $g = f_a$.*

Proof. If f does not depend on A, then $f^a = \mathbf{0}$, so the theorem is obviously true. If $A \in \mu(f)$, then by definition $s(f) \wedge a \neq \mathbf{0}$ and $f = F(g(x_A), x_B)$, where $\{A, B\}$ is a partition of N. Then Shannon's decomposition, $F(y, x_B) = F_{|y=0} \vee yF_{|y=1}$, implies the fundamental equation:

$$f = f_{|x_A=\mathbf{0}} \vee g f_{|x_A=\mathbf{1}}. \tag{2.18}$$

Furthermore, according to Equation (2.16), $f_{|x_A=\mathbf{1}} = f^a{}_{|x_A=\mathbf{1}} \vee f_{|x_A=\mathbf{0}}$. Therefore, Equation (2.18) implies that $f^a = g f^a{}_{|x_A=\mathbf{1}}$, and that

$$f = f_{|x_A=\mathbf{0}} \vee g f^a{}_{|x_A=\mathbf{1}}. \tag{2.19}$$

Using the fact that the functions $f_{|x_A=\mathbf{0}}$ and $f^a{}_{|x_A=\mathbf{1}}$ depend only on $B = N \setminus A$, Equation (2.19) implies

$$f_a(x_A) = f^a{}_{|x_{\bar{A}}=\mathbf{1}}(x_A) = g(x_A),$$

implying that g is the contraction of f on A. Therefore, we have the decomposition:

$$f = f_{|x_A=\mathbf{0}} \vee f^a{}_{|x_A=\mathbf{1}} f_a. \tag{2.20}$$

However, Equation (2.20) is equivalent to

$$f^a = f^a{}_{|x_A=\mathbf{1}} f_a. \tag{2.21}$$

Conversely, if Equation (2.21) holds, then A is modular. ∎

Corollary 2.40. *Let f be a positive function defined on N such that f depends on $A \subseteq N$. If $f(x_N) = F(g(x_A), x_B)$, where F and g are positive functions, then $g = f_a$, with $a = char(A)$.*

Corollary 2.41. *Let f be a positive function defined on N and $A \subseteq N$. Then the following assertions are equivalent:*

(a) *$A \in \mu(f)$.*
(b) *There exists a positive function g defined on A such that for all b : $f_b(x_A) := f(x_A, b) \in \{\mathbf{0}, \mathbf{1}, g\}$.*

Proof. This is a consequence of Theorem 2.2 and of the uniqueness of g. ∎

Remark 2.1. *Note that according to Theorem 2.39, the problem of deciding whether a set A is modular or not can be solved in time $O(m^2n^2)$ by checking whether $f^a = f^a{}_{|x_A=\mathbf{1}} f_a$ is true or not!*

Example 2.8. *Consider the function f of Example 2.7, and let $A = \{1, 2, 3\}$. Then $f^a = f^a{}_{|x_A=\mathbf{1}} f_a = (x_4x_5 \vee x_6)(x_1x_2 \vee x_2x_3)$.*

2.9.4 Characterizations of a Modular Set

The first characterization of a modular set of a positive Boolean function f, which also characterizes the associated component of f, was already given in Theorem 2.39. The following characterizations of a modular set (except possibly (e)) are well-known; see, for example, [32].

Theorem 2.42. *Suppose that f is a positive function defined on N, and $A \subseteq N$. Furthermore, let $s(f) \wedge a \neq \mathbf{0}$, where $a = char(A)$. Then the following assertions are equivalent:*

(a) *A is a modular set of f.*
(b) *f_a is a component of f.*
(c) *A is a modular set of f^a.*
(d) *$(f^d)_a = (f_a)^d$.*
(e) *There exists a positive function $g(x_A)$ such that for all b: $f_b(x_A) := f(x_A, b) \in \{\mathbf{0}, \mathbf{1}, g\}$.*
(f) *For all $v, w \in \min T(f^a) : f(va \vee w\bar{a}) = 1$.*
(g) *$\min T(f^a) = \{va \vee w\bar{a} \mid v, w \in \min T(f^a)\}$.*
(h) *$s(((f^a)^d)^a) = s(f) \wedge a$.*

Proof. (a) ⇔ (b)⇔ (c) ⇔ (d) ⇔ (e) ⇔ (f) The equivalence of the assertions (a), (b), (c), (e), and (f) follows from Theorem 2.39. The equivalence of (a) and (d) follows from (b) and the fact that g is a component of f on A if and only if g^d is a component of f^d on A.

(f) ⇔ (g) Obviously, (g) implies (f). Conversely, suppose assertion (f) holds true, and $z = xa \vee y\bar{a} \in T(f^a)$, with $x, y \in \min T(f^a)$. If $z \notin \min T(f^a)$, then $z > v$ for some $v \in \min T(f^a)$. So at least one of the following inequalities holds true: $xa > va$ or $y\bar{a} > v\bar{a}$. However, the first inequality implies that $x = xa \vee x\bar{a} > va \vee x\bar{a} \in T(f^a)$, contrary to the minimality of x. Similarly, $y\bar{a} > v\bar{a}$, implies that $y = ya \vee y\bar{a} > ya \vee v\bar{a} \in T(f^a)$, contrary to the minimality of y. From this we conclude that (f) and (g) are equivalent.

(a) $\Leftrightarrow$ (h) Finally, we note that according to Theorem 2.39, $A \in \mu(f)$ is equivalent to $f^a = gh$, where g and h are monotone functions with $s(g) = s(f) \wedge a$, and $s(h) \leq \bar{a}$. However, $f^{ad} = (f^a)^d = g^d \vee h^d$ implies that $f^{ada} = g^d$. Therefore, if $A \in \mu(f)$, then $s(f^{ada}) = s(g^d) = s(g) = s(f) \wedge a$. Conversely, if $s(f^{ada}) = s(f) \wedge a$, then there exist monotone functions g and h such that $f^{ad} = g^d \vee h^d$, with $s(g) = s(f) \wedge a$, and $s(h) \leq \bar{a}$, implying that $f^a = gh$. This establishes the equivalence of the assertions (a) and (h). ∎

Example 2.9. *Consider the function* $f = (x_1x_2 \vee x_2x_3)(x_4x_5 \vee x_6) = x_1x_2x_4x_5 \vee x_1x_2x_6 \vee x_2x_3x_4x_5 \vee x_2x_3x_6$. *If* $A = \{1, 2\}$ *or* $A = \{1, 2, 3\}$, *then* $f^{ad} = x_2 \vee x_1x_3 \vee x_4x_6 \vee x_5x_6$. *If* $A = \{1, 2, 3\}$, *then* $s(f^{ada}) = a$. *However, if* $A = \{1, 2\}$, *then* A *is not modular because* $s(f^{ada}) \neq a$.

Some early attempts to solve the recognition problem of modular sets are based on the dualization of positive Boolean functions (see Theorem 2.42(d) or Theorem 2.42(h)). However, the complexity of this problem is still not known; see [9]. Nevertheless, the support set of the function $((f^a)^d)^a$ plays a prominent role in the determination of the modular closure of A.

2.9.5 The Modular Closure

In the previous section, it was shown that the modular tree of a Boolean function f can be generated in time $O(n^3T(m, n))$, where $T(m, n)$ denotes the complexity of the problem to determine the modular closure of a set A. For positive Boolean functions, $T(m, n)$ is the complexity of solving $O(n)$ times the following search problem:

Problem PMODULAR-SET
Input: A positive Boolean function f with m prime implicants defined on N, where $|N| = n$, and $A \subseteq N$.
Output: "A is modular" if A is modular. An element $x \in Cl_f(A) \setminus A$ otherwise.

In this subsection we discuss the problem of finding efficiently an element $x \in Cl_f(A) \setminus A$, if A is not modular. The solution of this problem depends on two key ideas. The first idea is to restrict the problem by considering only elements x in the support set of the function f^{ada}; see property (h) of Theorem 2.42. The second idea is based on property (f) of the same theorem: if A is not modular, then there exist minimal vectors v and w of the function f^a such that $f(va \vee w\bar{a}) = 0$. We call the vector $va \vee w\bar{a}$ a *culprit*. It can be shown that given a culprit, an element $x \in Cl_f(A) \setminus A$ can be determined in time $O(mn)$. However, the arguments are sometimes quite subtle.

Proposition 2.43. *Let f be a positive function on N and $A \subseteq B \subseteq S(f^a)$, where $a = char(A)$. Then $B \in \mu(f^a)$ if and only if for all $v \in \min T(f^{ad})$, $b \geq v$ or $b \leq \bar{v}$, where $b = char(B)$.*

Proof. Let L and R denote respectively the left side and right side of the equivalence of the proposition. Suppose R is false; then there exists $v \in \min T(f^{ad})$ such that $v \wedge b \neq \mathbf{0}$ and $v \wedge \bar{b} \neq \mathbf{0}$. This implies that $f^{ad}(bv) = f^{ad}(b\bar{v}) = 0$. Therefore, $f^a(\overline{vb}) = f^a(\overline{v\bar{b}}) = 1$, and according to Theorem 2.42(f), there exist $x, y \in \min T(f^a)$ such that $x \leq \overline{vb}$ and $y \leq \overline{v\bar{b}}$. Let $z = xb \vee y\bar{b}$. Then it is easy to verify that $z \leq \bar{v}$. Suppose $B \in \mu(f^a)$. Then $z \in \min T(f^a)$, implying that $f^a(\bar{v}) = 1$. This contradicts the fact that $f^{ad}(v) = 1$. Conclusion: $L \Rightarrow R$.

Conversely, suppose that R is true. If $B \notin \mu(f^a)$, then there exist $x, y \in \min T(f^a)$ such that $z := xb \vee y\bar{b} \notin T(f^a)$. Therefore, $\bar{z} = \bar{x}b \vee \bar{y}\bar{b} \notin T(f^{ad})$. From this it follows that there exists $w \in \min T(f^{ad})$ such that $w \leq \bar{x}b \vee \bar{y}\bar{b}$. Since we assume that R is true, we have that either $b \geq w$ or $b \leq \bar{w}$. This means that at least one of the vectors $\bar{x}$ or $\bar{y}$ belongs to $T(f^{ad})$, contrary to the fact that $x, y \in T(f^a)$. Conclusion: $R \Rightarrow L$. ■

Definition 2.9. *Let f be a positive function on N and $\emptyset \neq A \subseteq N$. Then we define an equivalence relation θ on N by: $i\theta j \Leftrightarrow i = j$ or there exists a sequence $i = i_1, \ldots i_k = j$, with $k \geq 2$, such that i_l and i_{l+1} both occur in some $v \in \min T(f^{ad})$, $l = 1, \ldots, k-1$.*

Proposition 2.44. *Let f be a positive function on N and $\emptyset \neq A \subseteq N$. Then we have*

$$Cl_{f^a}(A) = \{i : i \in N \text{ and } i\theta j \text{ for some } j \in A\}.$$

Proof. Let $B = Cl_{f^a}(A)$ and $R = \{i : i \in N \text{ and } i\theta j \text{ for some } j \in A\}$. Then $A \subseteq B \subseteq S(f^a)$, and $B \in \mu(f^a)$. According to Proposition 2.43, we have: for all $v \in \min T(f^{ad})$, either $v \leq b$ or $v \leq \bar{b}$, where $b = char(B)$. Using Definition 2.9, we conclude that $R \subseteq B$. On the other hand this definition implies that for all $v \in \min T(f^{ad})$ either $v \leq r$ or $v \leq \bar{r}$, where $r = char(R)$. Since $A \subseteq R \subseteq S(f^{ad})$, Proposition 2.43 implies that $R \in \mu(f^a)$. Therefore, we have $R \supseteq B$. This shows that $R = B$. ■

Lemma 2.45. *Let f be a positive function defined on $N, \emptyset \neq A \subseteq B \subseteq N$. If $B \in \mu(f)$, then $B \in \mu(f^a)$, where $a = char(A)$.*

Proof. This follows from Theorem 2.42(g). ■

The following theorem [32] relates the modular closure of f^a to that of the dual of f^a.

Theorem 2.46. *Let f be a positive function on N and $\emptyset \neq A \subseteq N$, with $a = char(A)$. Then $A \subseteq S(f^{ada}) \subseteq Cl_{f^a}(A) \subseteq Cl_f(A)$.*

Proof. Because by assumption f depends on all variables in N, we have $A \subseteq S(f^{ada})$. Furthermore, it is easy to verify that Proposition 2.44 implies that $S(f^{ada}) \subseteq Cl_{f^a}(A)$. Finally, because $Cl_f(A) \in \mu(f)$, we note that according to Lemma 2.45, $A \subseteq Cl_f(A) \in \mu(f^a)$. This implies that $Cl_{f^a}(A) \subseteq Cl_f(A)$. ■

Recall that if A is not modular, then according to Theorem 2.42(f) there exist minimal vectors v and w of f^a such that $f(va \vee w\bar{a}) = 0$. We now show that the complement of the vector $va \vee w\bar{a}$ can be used to find an element $x \in Cl_f(A) \setminus A$.

Definition 2.10. *Suppose there exist $u, v \in \min T(f^a)$ such that $f(ua \vee v\bar{a}) = 0$. Then we call the vector $ua \vee v\bar{a}$ a culprit of f with respect to a.*

Theorem 2.47. *Suppose f is a positive function and $u, v \in \min T(f^a)$. If $f(ua \vee v\bar{a}) = 0$, then the vector $t := \bar{u}a \vee \bar{v}\bar{a} \in T(f^{ad})$. Furthermore, $\forall w \in \min T(f^{ad})$ such that $w \leq t$ we have $\mathbf{0} \not\leq w\bar{a} \leq s(f^{ada})$.*

Proof. It is easy to see that $\bar{t} = ua \vee v\bar{a}$, so $t \in T(f^{ad})$. Furthermore, the assumptions imply $w \leq \bar{u}a \vee \bar{v}\bar{a}$, and $\bar{u}, \bar{v} \in F(f^{ad})$. Therefore, because $w \in \min T(f^{ad})$, we conclude $w \not\leq \bar{u}a$ and $w \not\leq \bar{v}\bar{a}$, implying $w\bar{u}a \neq \mathbf{0}$ and $w\bar{v}\bar{a} \neq \mathbf{0}$. From this we conclude that $w \leq s(f^{ada})$ and that $w\bar{a} \neq \mathbf{0}$. ■

Given t, then a vector w in Theorem 2.47 can be determined in time $O(mn^2)$, since it is known that a minimal transversal w can be obtained from a transversal t in $O(n)$ steps. Therefore, if A is not modular, then the last theorem shows that we can determine an element in $Cl_f(A) \setminus A$ given t in time $O(mn^2)$. However, an interesting and more refined analysis due to Ramamurthy [32] shows that such an element can be found in time $O(mn)$. To see this we first present the following lemma of independent interest.

Lemma 2.48. *Let f be a positive Boolean function and $f^d(w) = 1$. Let $v \in \operatorname{argmin}\{|uw| \mid u \in \min T(f)\}$. Then for all unit vectors $e \leq wv$ there exists a vector $w_0 \in \min T(f^d)$ such that $e \leq w_0 \leq w$.*

Proof. Because $w\bar{v} \wedge v = \mathbf{0}$ and $v \in \min T(f)$ we conclude that $w\bar{v} \notin T(f^d)$. On the other hand, we claim that

$$w\bar{v} \vee e \in T(f^d). \tag{2.22}$$

To prove this claim, we suppose that $u \in \min T(f)$ but $(w\bar{v} \vee e) \wedge u = \mathbf{0}$. Then we have $e \not\leq u$ and $w\bar{v}u = \mathbf{0}$. However, the last equality implies $wu \leq v$, so that

$$\mathbf{0} \neq wu \leq wv. \tag{2.23}$$

By the minimality assumption we then have $wu = wv$. Because $e \not\leq u$ and $e \leq wv$, this is a contradiction. This proves our claim (2.22). Furthermore, we claim that

$$\text{For all } w_0 \in \min T(f^d) \text{ such that } w_0 \leq w\bar{v} \vee e, \text{ we have } e \leq w_0. \tag{2.24}$$

To prove claim (2.24), assume $e \not\leq w_0$. Then we would have $w_0 \leq w\bar{v}$. However, $w\bar{v} \notin T(f^d)$, so $w_0 \not\leq w\bar{v}$. Contradiction. This finishes our proof. ■

The following lemma is a reformulation of a proposition in [32].

Lemma 2.49. *Suppose f is a positive Boolean function and $f^d(w) = 1$. Let c be a vector such that $U = \{u \in \min T(f) \mid uwc = \mathbf{0}\} \neq \emptyset$. Let $v \in \text{argmin}_{u \in U}\{\ |uw|\ \}$. Then for all unit vectors $e \leq wv$ there exists a vector $w_0 \in \min T(f^d)$ such that $e \leq w_0 \leq w$.*

Proof. Note that the inequality (2.23) implies $wuc \leq wvc = \mathbf{0}$, so $u \in U$. Using this observation, the proof of this lemma is the same as the proof of Lemma 2.48. ■

The following fundamental theorem, a variation of a theorem in [32], finally solves the second part of problem PMODULAR-SET posed at the beginning of this subsection.

Theorem 2.50. *Let f be a positive function. Suppose t is the complement of a culprit of f with respect to a. Then $U = \{u \in \min T(f^a) \mid uta = \mathbf{0}\} \neq \emptyset$. Furthermore, if $u_0 \in argmin_{u \in U}\{\ |ut|\ \}$, then $\mathbf{0} \neq u_0 t = u_0 t\bar{a} \leq Cl_f(a)$.*

Proof. Since t is the complement of a culprit, we have that there exist $v, w \in \min T(f^a)$ such that $t = \bar{v}a \vee \bar{w}\bar{a}$, and $f^{ad}(t) = 1$. Furthermore, since $\bar{v}a \notin T(f^{ad})$, there must exist a vector $u_0 \in \min T(f^a)$ such that $u_0\bar{v}a = \mathbf{0}$. From $u_0 ta = u_0\bar{v}a = \mathbf{0}$, it follows that $u_0 \in U$. Now suppose $u_0 \in argmin_{u \in U}\{\ |ut|\ \}$. Then according to Lemma 2.49: for all unit vectors $e \leq u_0 t$, there exists $t_0 \in \min T(f^{ad})$ such that $e \leq t_0 \leq t$. Now Theorem 2.46 implies $\mathbf{0} \neq t_0\bar{a} \leq s(f^{ada})$. Therefore, we conclude: $\mathbf{0} \neq u_0 t = u_0 t\bar{a} \leq Cl_f(a)$. ■

Corollary 2.51. *The vector $u_0 t$ can be determined in $O(mn)$ time. Therefore, if a culprit is known, then we can determine in $O(mn)$ time an element in $Cl_f(A) \setminus A$.*

2.9.6 Computational Aspects

In this section we show that problem PMODULAR-SET can be solved in time $O(mn)$, implying that the modular closure of a set can be determined in time $T(m, n) = O(mn^2)$. It follows from the discussion in the previous subsection that to solve PMODULAR-SET, there are yet two subproblems to be solved:

(I) The recognition problem: deciding whether a set A is modular or not.
(II) Finding a culprit when A is not modular.

It was proved by Ramamurthy [32] that these subproblems can be solved in time $O(m^2 n)$. This complexity can be further reduced to $O(mn)$ as shown by Bioch in [8].

I: Recognition of Modular Sets

Let f be a positive Boolean function f on N, $\emptyset \neq A \subseteq N$, and $a = char(A)$. Then we denote $M = \min T(f^a) = \{v_1, \ldots, v_m\}$, $S = \{va \mid v \in M\}$, $T = \{v\bar{a} \mid v \in M\}$, $p = |S|$ and $q = |T|$. Furthermore, without loss of generality we may assume that $s(f) = N$, or equivalently that $M \neq \emptyset$, and that for all $v \in M = \min T(f^a)$ we have $v \not\leq a$. For each $v \in M$, we can write $N = va \vee v\bar{a}$ as a $2n$-vector: $(va|v\bar{a})$. Note that by assumption both vectors va and $v\bar{a}$ are nonzero. We now consider the list of all (column) vectors:

$$\begin{vmatrix} v_1a & v_2a & \cdots & \cdots & v_ma \\ v_1\bar{a} & v_2\bar{a} & \cdots & \cdots & v_m\bar{a} \end{vmatrix}.$$

According to [35], the set of all these $2n$-vectors can be lexicographically sorted in time $O(mn)$.

Example 2.10. *Let $f = x_1x_5 \vee x_1x_6 \vee x_2x_4x_5 \vee x_3x_5 \vee x_3x_6 \vee x_4x_6$, and $A = \{1, 2, 3, 4\}$. Then $f^a = f$ and the sorted list is given by*

$$\mathcal{S} = \begin{vmatrix} 1 & 1 & 24 & 3 & 3 & 4 \\ 5 & 6 & 5 & 5 & 6 & 6 \end{vmatrix}.$$

Note here that the 2n-vector $(va|v\bar{a})$ is denoted by a pair of subsets: for example, the third column vector (010100|000010) *is denoted by* (24|5).

Theorem 2.52. *A is modular if and only if the sorted list of all 2n-vectors has the following structure:*

$$\mathcal{S} = \begin{vmatrix} s_1 \cdots s_1 \\ t_1 \cdots t_q \end{vmatrix} \begin{matrix} s_2 \cdots s_2 \\ t_1 \cdots t_q \end{matrix} \begin{vmatrix} \cdots\cdots \\ \cdots\cdots \end{vmatrix} \begin{matrix} s_p \cdots s_p \\ t_1 \cdots t_q \end{matrix} \Bigg|, \text{ where } s_i \in S \text{ and } t_j \in T,$$

and we have $S = \min T(f_a)$ and $T = \min T(f^a{}_{|x_A=1})$. So if A is modular, then the list $\mathcal{S}$ consists of p segments of length q, and $m = pq$.

Proof. According to Theorem 2.42, we have $A \in \mu(f)$ if and only if $f^a = f_a f^a{}_{|x_A=1}$, hence $S = \min T(f_a)$ and $T = \min T(f^a{}_{|x_A=1})$. Furthermore, if $v_1, v_2, w_1, w_2 \in \min T(f^a)$, then $v_1a \vee w_1\bar{a} = v_2a \vee w_2\bar{a} \Leftrightarrow v_1a = v_2a$ and $w_1\bar{a} = w_2\bar{a}$. ∎

Example 2.11. *Let f be the function of Example 2.7, and let $A = \{1, 2, 3\}$. Then we have $f^a = x_1x_2x_6 \vee x_2x_3x_6 \vee x_1x_2x_4x_5 \vee x_2x_3x_4x_5$, and the sorted list is given by*

$$\mathcal{S} = \begin{vmatrix} 12 & 12 \\ 45 & 6 \end{vmatrix} \begin{matrix} 23 & 23 \\ 45 & 6 \end{matrix} \Bigg|.$$

Therefore, $A \in \mu(f)$ and $p = q = 2$. Similarly, it can be checked that $\{1, 3\} \in \mu(f)$.

It is easy to see that the structure $\mathcal{S}$ can be identified in time $O(mn)$, by scanning the list $\mathcal{S}$ from left to right. Therefore, it can be determined in time $O(mn)$ whether a set A is modular or not. However, the more difficult part is to detect an element $x \in Closure(A) \setminus A$ in time $O(mn)$ if A is not modular. According to Theorem 2.50 this can be done in time $O(mn)$ if we can find a culprit in time $O(mn)$.

II: Finding a Culprit in Time $O(mn)$

Recall that the vector $va \vee w\bar{a}$, with $v, w \in \min T(f^a)$ is called a culprit with respect to to A if $f(va \vee w\bar{a}) = 0$. Note that if A is not modular, the scanning process described only indicates the existence of a culprit but does not find it. In order to find a culprit, we have to scan and compare the two rows of the structure $\mathcal{S}$ separately. This is actually the reason why the algorithm of Ramamurthy [32] finds a culprit in time $O(m^2n)$. However, it can be shown that the next basic lemma can be used several times in order to find a culprit if it exists in time $O(mn)$. In this lemma the following notations are used: $v \sim w \Leftrightarrow (v < w \text{ or } v > w)$, and $v \simeq w \Leftrightarrow (v \leq w \text{ or } v > w)$.

Lemma 2.53. *Let $(s_1|t_1)$ and $(s_2|t_2)$ denote any two different columns of the list $\mathcal{S}$. Then:*

(a) $s_1 = s_2 \Rightarrow t_1 \not\simeq t_2$.
(b) $t_1 = t_2 \Rightarrow s_1 \not\simeq s_2$.
(c) *If $s_1 \sim s_2$ then either $s_1 \vee t_2$ or $s_2 \vee t_1$ is a culprit.*
(d) *If $t_1 \sim t_2$, then either $s_1 \vee t_2$ or $s_2 \vee t_1$ is a culprit.*
(e) *If the $2n$-vector $(s_1|t_2)$ does not occur in the list $\mathcal{S}$ and s_1 and t_2 are minimal, then $s_1 \vee t_2$ is a culprit.*

Proof. Let v and w be minimal vectors of f^a such that $s_1 = va$, $s_2 = wa$, $t_1 = v\bar{a}$ and $t_2 = w\bar{a}$.

(a) Suppose $s_1 = s_2$ so $va = wa$. Then obviously $v\bar{a} \neq w\bar{a}$; otherwise, we would have $v = w$. Therefore, $t_1 \neq t_2$. Now assume that $t_1 \sim t_2$, for example, $v\bar{a} < w\bar{a}$. Then $v = va \vee v\bar{a} < wa \vee w\bar{a} = w$, contrary to our assumption that N and w are minimal vectors of f^a.

(b) This is proved similarly to (a).

(c) Suppose $s_1 \sim s_2$, e.g $va > wa$. Then $v = va \vee v\bar{a} > wa \vee v\bar{a}$. Because v is a minimal vector of f^a, the vector $wa \vee v\bar{a}$ is a culprit: $f(wa \vee v\bar{a}) = 0$, see Theorem 2.42.f.

(d) This assertion is proved similarly to (c).

(e) If the vector $va \vee w\bar{a}$ is not a culprit, then $f(va \vee w\bar{a}) = 1$. Hence, there exists a vector $u \in \min T(f^a)$ such that $u \leq va \vee w\bar{a}$. This implies $ua \leq va$ and $u\bar{a} \leq w\bar{a}$. Because by assumption va and $w\bar{a}$ are minimal, we have $ua = va$ and $u\bar{a} = w\bar{a}$. Therefore, the vector $(va|w\bar{a}) = (ua|u\bar{a})$ is a column vector of $\mathcal{S}$, contrary to our assumption. So the vector $va \vee w\bar{a}$ is a culprit. ∎

Suppose that $(s_1|t_2)$ does not occur in the list $\mathcal{S}$. Then we can check in $O(mn)$ time whether the elements s_1 and t_2 are minimal. If both elements are minimal, then we can apply assertion (e) of Lemma 2.53. Otherwise, we can apply either (c) or (d). Therefore, we have the following corollary.

Corollary 2.54. *If $(s_1|t_2)$ does not occur in the list $\mathcal{S}$, then a culprit can be found in time $O(mn)$.*

Example 2.12. *Consider the sorted list in Example 2.10:*

$$\mathcal{S} = \begin{vmatrix} 1 & 1 & 24 & 3 & 3 & 4 \\ 5 & 6 & 5 & 5 & 6 & 6 \end{vmatrix}.$$

Then the first segment has length $q = 2$. Since the first element of the fourth column is not equal to 24, we detect that the column $(24|6)$ is not in $\mathcal{S}$. However, $246(= 010101)$ is not a culprit, because the element 24 is not minimal. By scanning the first row we discover that 4 is comparable with 24. Hence, by Lemma 2.53.(c) applied to the third and last column, either 246 or 45 is not a true vector of f^a. In this case 45 (= 000110) is a culprit, because $(4|5)$ is not in $\mathcal{S}$ (see Lemma 2.53.(a) and the elements 4 and 5 are minimal.

In [8] a detailed algorithm based on Lemma 2.53 is given to find a culprit in time $O(mn)$. Combined with Theorem 2.50, this proves our final result.

Theorem 2.55. *Problem PMODULAR-SET is solvable in time $O(mn)$.*

Corollary 2.56. *The modular closure of set A can be determined in time $O(mn^2)$.*

2.10 Applications in Game Theory

Recall that a positive Boolean function is the characteristic function of a simple game (or voting game; see [19]). The variables represent voters, and winning coalitions are associated with the true vectors of the positive function. Therefore, the prime implicants of a positive function represent the minimal winning coalitions of the associated game. In the sequel, we identify a simple game by its characteristic function. A game f is called *proper* if f is dual-minor ($f \leq f^d$) and f is called *strong* if f is dual-major ($f^d \leq f$). A proper strong game is called a *decisive game* (and corresponds to a positive self-dual Boolean function). Shapley [36] and Billera [4] studied the composition of simple games in "tournaments" [25]: $f = F[g_1, g_2, \ldots, g_m]$. Each of the m players in the game F represents the winning coalition in game g_i. The voters of each g_i vote independently as "committees." The group decisions of each committee g_i are inputs to the collective choice function F, which combines the decisions of the committees. So a committee of a simple game f is a component of f. The modular set on which a committee is defined is called a *committee set*. A compound simple game f is a decomposable positive Boolean function; otherwise, it is called a prime simple game. A *full game* f is one without *dummy* players, so each voter occurs in at least one minimal winning coalition. So a full simple game f is a positive function of which all the variables are essential.

Decomposition into Prime Simple Games

Let f be a simple game defined on N. In this section we only consider full simple games f. Recall that a node C in the composition tree of f is called prime if the corresponding function f_C is of type I, that is, f_C has a maximal disjoint decomposition on the partition $\{C_1, C_2, \ldots, C_m\}$ of C. So f_C has the decomposition

$$f_C = F(g_1, g_2, \ldots, g_m). \tag{2.25}$$

In this case the function F is called *prime* because F has no proper modular sets. Moreover, the decomposition (2.25) is called a *prime decomposition* of f_C. If a node in the composition tree is of type II, then it is either a $D_\vee$ node or a $D_\wedge$ node. If f is a positive function, then the functions in the nodes in the composition tree are uniquely defined. Therefore, a composition tree of a simple game is also called a *unique factorization* of f. It turns out that this factorization has a particular nice form in the important case of a decisive simple game.

Lemma 2.57. *Let f be a decisive simple game. Suppose g is a component game of f with quotient game F. Then f and F are also decisive simple games.*

Proof. Assume that f is a positive self-dual Boolean function. According to Corollary 2.41, there exists a Boolean vector b_0 such that $g(x_A) = f(x_A, b_0)$. Therefore, $g^d(x_A) = \bar{f}(\bar{x_A}, b_0) = f^d(x_A, \bar{b}_0) = f(x_A, \bar{b}_0)$. From Corollary 2.41, it also follows that $f(x_A, \bar{b}_0) \in \{\mathbf{1}, \mathbf{0}, g\}$. Because g is nontrivial, we conclude that $g^d = g$. The fact that F is self-dual follows from Equation (2.1). ∎

In order to show that all the nodes in a decomposition tree of a decisive simple game are prime (of type I), we prove the following fundamental result.

Lemma 2.58. *Let f be a decisive game defined on N. Suppose that A and B are non-comparable committee-sets of f. Then $A \bigcap B = \emptyset$.*

Proof. Suppose A and B are non-comparable modular sets of f such that $A \bigcap B \neq \emptyset$. Then by the Three Modules Theorem (see Theorem 2.12), the function $g = f_{A \cup B}$ is a component of f and

$$g = f_1 \star f_2 \star f_3, \text{ where } \star \text{ denotes either } \wedge \text{ or } \vee\,. \tag{2.26}$$

According to Lemma 2.57, the functions f_i are self-dual for all $i \in \{1, 2, 3\}$. Moreover, the function g is self-dual. Here, we only consider the case that $\star$ denotes $\vee$. The other case is similar and left to the reader. So, suppose $g = f_1 \vee f_2 \vee f_3$, $g^d = g$ and $f_i^d = f_i$, $i \in \{1, 2, 3\}$. Then $g = g^d = f_1 f_2 f_3$, so that $f_1 f_2 f_3 = f_1 \vee f_2 \vee f_3$. This implies e.g. that $f_1 f_2 f_3 = f_1$. Because the functions f_i are defined on disjoint sets of variables and f depends on all its variables, this is clearly a contradiction. ∎

So if f is a decisive game, then the maximal committee sets have empty intersection. Therefore, f has the decomposition

$$f = F(g_1, g_2, \ldots, g_m), \tag{2.27}$$

where F and g_i are decisive games and F is prime. The components g_i can be further decomposed until they become variables or prime decisive games. Therefore, we have proved the following theorem.

Theorem 2.59. *A decisive simple game f admits a unique factorization of f in prime simple games. Moreover, each component game and quotient game in the composition tree is a decisive simple game.*

Remark 2.2. *Theorem 2.59 has been proven from first principles by D. Loeb in [25]. In comparison with the unique prime factorization of integers as the fundamental theorem of arithmetic, Theorem 2.59 is also called the "fundamental theorem of voting schemes."*

The result of Lemma 2.58 can be sharpened in the following way.

Theorem 2.60. *Let f be a simple game defined on N. Suppose that A and B are noncomparable committee sets of f, and that the corresponding components defined on A and B are both decisive simple games. Then $A \bigcap B = \emptyset$.*

Proof. Let g be a component of f defined on A. Then we define the "lifting"

$$h(x, y, x_N) = xy \vee xf \vee yf^d, \tag{2.28}$$

where x, y are variables distinct from each other and from all other variables. It is easily shown that the function h is self-dual. Moreover, we show that A is also a modular set of h and that g is a component of h on A as follows. Because g is a modular component of f, we have according to Corollary 2.41: for all $c \in \{0, 1\}^{|N|-|A|}$: $f(x_A, c) \in \{\mathbf{0}, \mathbf{1}, g\}$. Therefore, for all $c \in \{0, 1\}^{|N|-|A|}$: $f^d(x_A, c) \in \{\mathbf{1}, \mathbf{0}, g^d\}$. If g is self-dual, then $f^d(x_A, c) \in \{\mathbf{1}, \mathbf{0}, g\}$. So, for all c and for all $a, b \in \{0, 1\}$, we have $h(a, b, x_A, c) \in \{\mathbf{0}, \mathbf{1}, g\}$. According to Corollary 2.41, this shows that g is a component of h on A. ■

Remark 2.3. *The construction in Equation (2.28) to "lift" an arbitrary Boolean function to a self-dual function appears to be very useful; see, for example, [9]. Later this construction was introduced by Taylor and Zwicker in [38] for positive functions (simple games). In their book this "lifting" is called the "two-point extension" of f.*

Note that a theorem similar to that of Theorem 2.59 does not hold for proper simple games, because a component of a proper game is not necessarily itself a proper simple game, and Lemma 2.58 does not hold. However, Theorem 2.59 can be combined by a result discussed in [10]: if f is a proper game on N, then f admits a nondisjoint decomposition into decisive games of the form

$$f = f_1 f_2 \ldots f_k,$$

where f_i is a decisive game defined on N, and k is at most less than or equal to the number of voters in each minimal winning coalition. So a proper game can be replaced by a limited number of decisive games defined on the same set of voters. Each of these games can be decomposed according to Theorem 2.59.

Finally, we note that the results of this section can be easily extended to the class of general Boolean functions.

2.11 Discussion, Further Research

Decompositions of Boolean functions, as well as modular sets, are very fundamental notions in many contexts and applications. In these contexts, the collection of all modular sets is efficiently represented by the so-called *composition tree* introduced by Shapley in [36]. A unified set-theoretic treatment of all algorithms related to modular sets known in game theory, reliability theory and set systems (clutters) up to 1990 is given by Ramamurthy [32]. These results and several extensions are covered in our discussion of modular sets of monotone Boolean functions.

Connections between functional decomposition of discrete structures and optimization have been discussed by Möhring and Radermacher [28, 29]. The complexity of deciding whether a set of variables is modular or not has been investigated in Bioch [7, 8]. In graph theory efficient algorithms are known to compute so-called modules in a graph [15, 27, 17]. The notion of a module in a graph has been generalized to hypergraphs in [13].

Recently, (generalized) functional decomposition has attracted renewed attention as an important tool in the design of switching circuits(FPGAs, Field Programmable Arrays), see Scholl [34] and Perkowski [30] for an overview. However, in this area the idea of functional decomposition has been generalized by allowing multiple components defined on the same set of variables (Curtis decompositions). Algorithms for functional decomposition in this sense have been developed based on OBDDs (ordered binary decision diagrams; see Chapter 10 by Bollig et al. in this volume). Note that although an OBDD is an efficient representation of a Boolean function, the order of the variables in the BDD depends on the proposed modular set.

Functional decomposition can also be extended to incomplete specified Boolean functions and partially defined discrete functions. Unfortunately, many theoretical results discussed in this chapter are not valid in all these cases. Because partially defined Boolean functions play an important role in many data mining tasks, we consider functional decomposition in data mining as an important area for further research. In this direction the paper of Zupan et al. [39] on concept hierarchies deserves further attention. The complexity of nondisjoint decompositions of several classes of partially defined Boolean functions has been determined in Boros et al. [14] and in Makino et al. [26]; see also the chapter on partially defined Boolean functions in [19].

Decompositions with components restricted to a certain class, such as matroids (Benzaken, Crama, and Hammer [3, 18]), self-dual functions (Bioch et al. [10]), regular or threshold functions (Hammer and Mahadev [22]), committees (game theory), etc., are also an interesting topic for future research. Finally, the theory of nondisjoint decompositions of (partially defined) Boolean functions deserves further study.

In contrast to the classical treatment of functional decomposition based on partition matrices, we have based our presentation on the representation of a disjoint decomposition as a generalized Shannon decomposition that appears to be an effective tool to study decompositions of Boolean functions and to unify the set theoretic approaches used in the literature.

References

[1] R. L. Ashenhurst. The decomposition of switching functions. Bell Laboratories' Report No. BL-1(II) (reprinted in [20]), 1952.

[2] R. L. Ashenhurst. The decomposition of switching functions. Proc. International Symposium on the Theory of Switching, Part I (vol. XXIX, Ann. Computation Lab. Harvard), Harvard University Press, Cambridge, MA, pp. 75–116, 1959.

[3] C. Benzaken and P. L. Hammer. Boolean techniques for matroidal decomposition of independence systems and applications to graphs, Discrete Mathematics 56, pp. 7–34, 1985.

[4] L. J. Billera. Clutter decomposition and monotonic Boolean functions. Annals of the New York Academy of Sciences 175, pp. 41–8, 1970.

[5] L. J. Billera. On the composition and decomposition of clutters. Journal of Combinatorial Theory 11, pp. 234–45, 1970.

[6] J. C. Bioch. Modular decomposition of Boolean functions. Erim Report Series ERS-2002-37-LIS, 36p. Erasmus University Rotterdam, www.erim.nl, 2002.

[7] J. C. Bioch. Complexity of Boolean function decomposition. In 5th International Workshop on Boolean Problems, B. Steinbach, (ed.) Freiberg, pp. 77–82, 2002. ISBN 3-86012-180-4 (http://dnb.ddb.de).

[8] J. C. Bioch. The complexity of modular decomposition of Boolean functions. Discrete Applied Mathematics 149, pp. 1–13, 2005.

[9] J. C. Bioch and T. Ibaraki. Complexity of identification and dualization of positive Boolean functions. Information and Computation 123, pp. 50–63, 1995.

[10] J. C. Bioch, T. Ibaraki, and K. Makino. Minimum self-dual decompositions of positive dual-minor Boolean functions. Discrete Applied Mathematics, pp. 307–26, 1999.

[11] G. Birkhoff. Lattice Theory. AMS Colloquium Publications, Vol. 25. AMS, USA, 1967.

[12] Z. W. Birnbaum and J. D. Esary. Modules of coherent systems. SIAM Journal of Applied Mathematics 13, pp. 444–62, 1965.

[13] P. Bonizzoni and G. D. Vedova. An algorithm for the modular decomposition of hypergraphs. Journal of Algorithms 32, pp. 65–86, 1999.

[14] E. Boros, V. Gurvich, P. L. Hammer, T. Ibaraki, and A. Kogan. Decomposition of partially defined Boolean functions. Discrete Applied Mathematics 62, pp. 51–75, 1995.

[15] A. Brandstädt, V. B. Le, and J. P. Spinrad. Graph Classes: A Survey. SIAM Monographs on Discrete Mathematics and Applications, 1999.

[16] R. W. Butterworth. A set theoretic treatment of coherent systems. SIAM Journal of Applied Mathematics 22, pp. 590–8, 1972.

[17] A. Cournier and M. Habib. A new linear algorithm for modular decomposition, LNCS, vol. 787, Springer-Verlag, Berlin, pp. 68–84, 1994.

[18] Y. Crama and P. L. Hammer. Bimatroidal independence systems. Zeitschrift für Operations Research 33, pp. 149–165, 1989.

[19] Y. Crama and P. L. Hammer. Boolean Functions: Theory, Algorithms, and Applications. Cambridge University Press, Cambridge, UK, 2010.

[20] H. A. Curtis. A New Approach to the Design of Switching Circuits. Van Nostrand, Princeton, NJ, 1962.
[21] M. Davio, J. P. Deschamps, and A. Thayse. Discrete and Switching Functions. McGraw-Hill, New York, 1978.
[22] P. L. Hammer and N. V. R. Mahadev. Bithreshold graphs. SIAM Journal on Applied Mathematics 6, pp. 497–506, 1985.
[23] S. T. Hu. Mathematical Theory of Switching Circuits and Automata. University of California Press, Berkeley, 1968.
[24] R. M. Karp. Functional decomposition and switching circuit design. Journal of the Society for Industrial and Applied Mathematics 11(2), 291–335, 1963.
[25] D. Loeb. The fundamental theorem of voting schemes. Journal of Combinatorial Theory (A) 73, pp. 120–9, 1996.
[26] K. Makino, K. Yano, and T. Ibaraki. Positive and Horn decomposibility of partially defined Boolean functions. Discrete Applied Mathematics. 74, 251–74, 1997.
[27] R. M. McConnell and J. P. Spinrad. Linear-time modular decomposition of undirected graphs and efficient orientation of comparability graphs. Proceedings of the 5th ACM-SIAM Symposium on Discrete Algorithms, SODA '96, pp. 536–45, 1996.
[28] R. H. Möhring. Algorithmic aspects of the substitution of decomposition in optimization over relations, set systems and Boolean functions. Annals of Operations Research 4, pp. 195–225, 1985–6.
[29] R. H. Möhring and F. J. Radermacher. Substitution decomposition of discrete structures and connections to combinatorial optimization. Annals of Discrete Mathematics 19, pp. 257–356, 1984.
[30] M. A. Perkowski. A survey of literature on functional decomposition, 1995. Available on mperkowski@ee.pdx.edu.
[31] K. G. Ramamurthy. A new algorithm to find the smallest committee containing a given set of players. Opsearch 25, pp. 49–56, 1988.
[32] K. G. Ramamurthy. Coherent Structures and Simple Games. Theory and Decision Library, Series C, Kluwer, Dordrecht, 1990.
[33] K. G. Ramamurthy and T. Parthasarathy. An algorithm to find the smallest committee containing a given set. Opsearch, 23, pp. 1–6, 1986.
[34] C. Scholl. Functional Decomposition with Application to FPGA Synthesis. Kluwer Academic Publishers, Dordrecht, 2001.
[35] R. Sedgewick. Algorithms in C. Addison-Wesley, Reading, MA, 1990.
[36] L. S. Shapley. On committees. In New Methods of Thought and Procedure, F. Zwicky and A. G. Wilson, eds. Springer-Verlag, New York, pp. 246–270.
[37] T. Singer. The Decomposition Chart as a Theoretical Aid. Bell Laboratories' Report No., III-1-III-28, 1953. (Reprinted in [20]).
[38] A. D. Taylor and W. S. Zwicker. Simple Games: Desirability Relations, Trading, Pseudoweightings; Princeton University Press, Princetor, NJ, 1999.
[39] B. Zupan, M. Bohanec, J. Demsar, and I. Bratko. Learning by discovering concept hierarchies. Artifical Intelligence, 109, pp. 211–42, 1999.

Part II

Logic

3

Proof Theory

Alasdair Urquhart

3.1 Introduction

The literature contains a wide variety of proof systems for propositional logic. In this chapter, we outline the more important of such proof systems, beginning with an equational calculus, then describing a traditional axiomatic proof system in the style of Frege and Hilbert. We also describe the systems of sequent calculus and resolution that have played an important part in proof theory and automated theorem proving. The chapter concludes with a discussion of the problem of the complexity of propositional proofs, an important area in recent logical investigations. In the last section, we give a proof that any consensus proof of the pigeonhole formulas has exponential length.

3.2 An Equational Calculus

The earliest proof systems for propositional logic belong to the tradition of algebraic logic and represent proofs as sequences of equations between Boolean expressions. The proof systems of Boole, Venn, and Schröder are all of this type. In this section, we present such a system, and prove its completeness, by showing that all valid equations between Boolean expressions can be deduced formally.

We start from the concept of Boolean expression defined in Chapter 1 of the monograph Crama and Hammer [9]. If ϕ and ψ are Boolean expressions, then we write $\phi[\psi/x_i]$ for the expression resulting from ϕ by substituting ψ for all occurrences of the variable x_i in ϕ. With this notational convention, we can state the formal rules for deduction in our equational calculus.

Definition 3.1. *The rules for equational deductions are as follows:*

(1) *For any expression ϕ, $\phi = \phi$ is an axiom;*
(2) *From $\phi = \psi$, deduce $\psi = \phi$;*
(3) *From $\phi = \psi$ and $\psi = \chi$, deduce $\phi = \chi$;*

(4) *From* $\phi = \psi$, *deduce* $\phi[\chi/x_i] = \psi[\chi/x_i]$;
(5) *From* $\phi = \psi$, *deduce* $\chi[\phi/x_i] = \chi[\psi/x_i]$.

The foregoing set of rules applies to any equational logic. The calculus of Boolean equations is obtained by postulating special axioms. We choose these from the set of identities of Theorem 1.1 in [9].

Definition 3.2. *The nonlogical axioms of the Boolean equational calculus are as follows:*

(1) $x \vee 1 = 1$ *and* $x0 = 0$;
(2) $x \vee 0 = x$ *and* $x1 = x$;
(3) $x \vee y = y \vee x$ *and* $xy = yx$ *(Commutativity);*
(4) $(x \vee y) \vee z = x \vee (y \vee z)$ *and* $(xy)z = x(yz)$ *(Associativity);*
(5) $x \vee x = x$ *and* $xx = x$ *(Idempotency);*
(6) $x \vee (yz) = (x \vee y)(x \vee z)$ *and* $x(y \vee z) = (xy) \vee (xz)$ *(Distributivity);*
(7) $x \vee \overline{x} = 1$ *and* $x\overline{x} = 0$;
(8) $\overline{(x \vee y)} = \overline{x}\,\overline{y}$ *and* $\overline{(xy)} = \overline{x} \vee \overline{y}$ *(De Morgan's laws).*

The axiom system is self-dual, in the sense that if $\phi = \psi$ is an axiom, then its dual $\phi^d = \psi^d$ is also an axiom. It follows that if an equation can be derived from the axioms, then we can conclude immediately that its dual can also be derived, simply by replacing all expressions by their duals in the derivation.

The reader may notice that we have not listed all of the identities listed in [9] Theorem 1.1. In fact, the identities we have not listed can be derived from the ones just listed by the equational rules of inference. As an example of a derivation in the equational calculus, we prove one of the last two identities of [9], Theorem 1.1:

$$\begin{aligned} x \vee (\overline{x}y) &= (x \vee \overline{x})(x \vee y) \\ &= 1(x \vee y) \\ &= (x \vee y)1 \\ &= (x \vee y). \end{aligned}$$

Here, the steps are justified successively by axioms 6, 7, 3, and 2; the rules for equational logic are applied tacitly. In fact, all valid equations can be proved in the Boolean equational calculus, as we show in the next proposition.

Proposition 3.1. *If two Boolean expressions* ϕ *and* ψ *are equivalent, the equation* $\phi = \psi$ *can be derived in the Boolean equational calculus.*

Proof. Let ϕ be a Boolean expression, expressed in terms of the variables $x_1, \ldots, x_n$. Then there is an equivalent Boolean expression ϕ', in DNF, so that the equation $\phi = \phi'$ is derivable in the Boolean equation calculus. To see this, it is sufficient to observe that the process of converting a Boolean expression to DNF, as described in Section 1.6 of [9], can be carried out in the equation calculus. More precisely, we can prove by induction on the complexity of the expression ϕ that there is a DNF expression ψ and a CNF expression χ so that both $\phi = \psi$

and $\phi = \chi$ are derivable in the Boolean equation calculus. If ϕ is a variable, then the claim is immediate. Let us assume that the claim holds for the immediate subexpressions of ϕ. If ϕ is the expression $\phi_1 \wedge \phi_2$, then by assumption there are CNF expressions χ_1 and χ_2 so that $\phi_1 = \chi_1$ and $\phi_2 = \chi_2$ are derivable. Hence, $\phi = (\chi_1 \wedge \chi_2)*$ is derivable using postulate 5 of Definition 3.2, where $(\chi_1 \wedge \chi_2)*$ is the CNF expression resulting from $(\chi_1 \wedge \chi_2)$ by deleting repeated clauses. By inductive assumption, there are also DNF expressions ψ_1 and ψ_2 so that $\phi_1 = \psi_1$ and $\phi_2 = \psi_2$ are derivable. By repeated use of the distributive laws, we can show that there is an expression θ so that $\psi_1 \wedge \psi_2 = \theta$, and θ is of the form $\theta_1 \vee \cdots \vee \theta_k$, where each θ_i is a conjunction of literals. Let θ^* be the result of deleting any conjunction in θ containing contradictory literals x and $\overline{x}$, and then removing any repeated literals in a conjunction. We can derive $\phi = (\psi_1 \wedge \psi_2) = \theta = \theta^*$, by using Postulates 1, 2, 5, 6, and 7 of Definition 3.2, and the last expression θ^* is in DNF. The remaining steps in the inductive proof are left to the reader.

We can show that there is in fact a minterm expression that is provably equivalent to ϕ. This is because if a variable x_i does not occur in a term C in a DNF expression, then we can replace C by the disjunction $Cx_i \vee C\overline{x_i}$, using axioms 2 and 7, together with distribution.

Let ϕ and ψ be two equivalent Boolean expressions, expressed in terms of the variables $x_1, \ldots, x_n$. Then there are minterm expressions ϕ' and ψ' so that $\phi = \phi'$ and $\psi = \psi'$ are provable. We claim that the expressions ϕ' and ψ' must contain exactly the same terms (possibly in a different order). For, suppose this is not so, and that there is a term C occurring in ϕ', but not in ψ'. Then the assignment of values to $x_1, \ldots, x_n$ that makes C true must make ψ' false, because every term in ψ' must differ from C in at least one literal; this contradicts our assumption that ϕ and ψ are equivalent. Hence, $\phi' = \psi'$ must be provable in the equational calculus using associativity and commutativity, and so $\phi = \psi$ is also provable. ■

The proof just constructed is essentially a "brute force" construction, resulting in an equational derivation of size exponential in the number of variables in the equation. Whether there is an efficient proof system for the tautologies is one of the most important open questions in complexity theory (see Section 3.5).

3.3 Frege Systems and Gentzen Systems

Although the Boolean equational calculus is the most natural proof system in the context of algebraic logic and switching circuit theory, it is not the system usually presented in introductory texts on logic. More commonly, such texts base their presentations on Frege-style axiomatizations, natural deduction systems, or Gentzen-style sequent systems.

Axiomatic systems in the tradition of Frege, Whitehead, Russell, and Hilbert are based on a family of tautologies adopted as axioms, together with a few rules such as modus ponens (From ϕ and $\phi \implies \psi$, infer ψ). As an example of such a system, we present the formulation of the propositional calculus that appears in the famous textbook of Kleene [13]. The primitive symbols are $\Longrightarrow$, $\wedge$, $\vee$, and negation (complementation).

Definition 3.3. *The axioms and rules of Kleene's formulation of the propositional calculus ([13, p. 82]) are as follows:*

(1) $\phi \Longrightarrow (\psi \Longrightarrow \phi)$;
(2) $(\phi \Longrightarrow \psi) \Longrightarrow ((\phi \Longrightarrow (\psi \Longrightarrow \chi)) \Longrightarrow (\phi \Longrightarrow \chi))$;
(3) $\phi \Longrightarrow (\psi \Longrightarrow (\phi \wedge \psi))$;
(4) $(\phi \wedge \psi) \Longrightarrow \phi$ *and* $(\phi \wedge \psi) \Longrightarrow \psi$;
(5) $\phi \Longrightarrow (\phi \vee \psi)$ *and* $\psi \Longrightarrow (\phi \vee \psi)$;
(6) $(\phi \Longrightarrow \chi) \Longrightarrow ((\psi \Longrightarrow \chi) \Longrightarrow ((\phi \vee \psi) \Longrightarrow \chi)$;
(7) $(\phi \Longrightarrow \psi) \Longrightarrow ((\phi \Longrightarrow \overline{\psi}) \Longrightarrow \overline{\phi})$;
(8) $\overline{\overline{\phi}} \Longrightarrow \phi$;
(9) *From* ϕ *and* $\phi \Longrightarrow \psi$ *infer* ψ.

The listed axioms are intended as axiom schemes: that is to say, in the case of the first axiom (for instance), any expression of the form $\phi \Longrightarrow (\psi \Longrightarrow \phi)$ is an axiom. The use of axiom schemes obviates the necessity for a rule of substitution such as we employed earlier in the equational calculus. In the literature of propositional proof complexity such systems are usually called "Frege systems," although this is not quite historically accurate, because Frege included in his own system a (tacitly applied) rule of uniform substitution. Frege systems with the rule of substitution added are usually called "substitution Frege systems," or "s-Frege systems."

The axiomatic system just described is complete, in the sense that any tautology is derivable. A proof of this fact is in Kleene's textbook [13], Section 29. A proof can also be obtained from Proposition 3.1 by observing that the equational calculus can be simulated in Kleene's axiomatic system, in the sense that if $\phi = \psi$ is derivable in the equational calculus, then the tautology $\phi \Leftrightarrow \psi$ is derivable in the axiomatic system. A converse simulation also holds, because if ϕ is a tautology, expressed in the language of Boolean expressions, then ϕ and 1 are equivalent, so $\phi = 1$ is provable in the equational calculus, by Proposition 3.1.

Although Frege systems often provide compact and convenient formulations of proof systems, they suffer from certain drawbacks as tools of analysis. This becomes clear if we consider the problem of searching for a proof of a tautology in Kleene's system (as part of an automated theorem-proving system, for example). If we are trying to prove a tautology ψ, then (unless it is an axiom) it must be derived using the sole rule of modus ponens. However, this is of no great help to us, because ψ could be derived from infinitely many premises of the form $\phi \Longrightarrow \psi$ and ϕ – there is no real constraint on our proof search. This defect of Frege systems was corrected by Gentzen in his sequent calculus.

The basic objects in the sequent calculus are paired sets of Boolean expressions. If X and Y are sets of expressions (possibly empty), then we write the pair $\langle X, Y \rangle$ as the *sequent* $X \vdash Y$. The intended meaning of the sequent $X \vdash Y$ is that if all the expressions in X are true, then at least one expression in Y is true – alternatively, that the expression $\bigwedge X \implies \bigvee Y$ is a tautology. An expression such as $X, \phi, \psi \vdash Y, \chi$ is to be read as $X \cup \{\phi, \psi\} \vdash Y \cup \{\chi\}$, where (for example) Y and $\{\chi\}$ may or may not be disjoint sets. We say that a sequent $X \vdash Y$ is *valid* if every Boolean assignment making all the formulas in X true makes at least one formula in Y true.

Definition 3.4. *The axioms and rules of a basic Gentzen sequent calculus for classical logic are as follows:*

$$X, \phi \vdash Y, \phi \text{ (Axiom)}$$

$$\frac{X \vdash \phi, Y \quad Z \vdash W, \psi}{X, Z \vdash W, Y, \phi \wedge \psi} (\vdash \wedge) \qquad \frac{X, \phi, \psi \vdash Y}{X, \phi \wedge \psi \vdash Y} (\wedge \vdash)$$

$$\frac{X \vdash Y, \phi, \psi}{X \vdash Y, \phi \vee \psi} (\vee \vdash) \qquad \frac{X, \phi \vdash Y \quad Z, \psi \vdash W}{X, Z, \phi \vee \psi \vdash W, Y} (\vdash \vee)$$

$$\frac{X \vdash \phi, Y}{X, \overline{\phi} \vdash Y} (\bar{\cdot} \vdash) \qquad \frac{X, \phi \vdash Y}{X \vdash Y, \overline{\phi}} (\vdash \bar{\cdot})$$

The basic Gentzen system just given has a pleasing symmetry. There are two rules for each connective, one for introducing it on the left of the sequent, one for introducing it on the right. The rules for $\wedge$ and $\vee$ make manifest the duality between the two connectives. In an application of a rule, we say that the formula introduced is the *principal formula* in the inference. For example, in the rule $(\wedge \vdash)$, the formula $\phi \wedge \psi$ is the principal formula.

Example 3.1. *As an example of a proof in the style of Gentzen, we give a proof of a distribution principle:*

$$\dfrac{\dfrac{\dfrac{x \vdash x, (x \wedge z) \quad y \vdash y, (x \wedge z)}{x, y \vdash (x \wedge y), (x \wedge z)} (\vdash \wedge)}{x, y \vdash (x \wedge y) \vee (x \wedge z)} (\vdash \vee) \quad \dfrac{\dfrac{x \vdash (x \wedge y), x \quad z \vdash (x \wedge y), z}{x, z \vdash (x \wedge y), (x \wedge z)} (\vdash \wedge)}{x, z \vdash (x \wedge y) \vee (x \wedge z)} (\vdash \vee)}{\dfrac{x, (y \vee z) \vdash (x \wedge y) \vee (x \wedge z)}{x \wedge (y \vee z) \vdash (x \wedge y) \vee (x \wedge z)} (\wedge \vdash)} (\vee \vdash)$$

The proof system has the crucial *subformula property*, that is to say, in the proof of a sequent $X \vdash Y$, only subformulas of formulas in $X \vdash Y$ occur in the proof. As a consequence of this key property, there are only a finite number of rules that allow the derivation of a given sequent. This fact makes possible easy procedures that search for proofs of sequents in this calculus.

Proposition 3.2. *A sequent $X \vdash Y$ is valid if and only if it is derivable in the basic Gentzen system.*

Proof. Starting with a given sequent $X \vdash Y$, we apply a simple proof search procedure that consists in applying one of the inference rules in reverse to a formula in the sequent to which a rule has not previously been applied. One of two things must occur. Either every branch in the proof search tree terminates in an axiom, in which case we have found a proof of $X \vdash Y$, or at least one branch fails to terminate in an axiom, in which case we can find a falsifying assignment by examining the formulas in the branch.

We now carry out the idea just sketched in more detail. At each stage in our search, we have generated a tree in which the nodes are labeled with sequents. The formulas occurring on the left or right of these sequents are marked as "dead" or "alive." We terminate our search if either (a) all the sequents labeling the leaves of the tree are axioms, so that the tree is a proof, or (b) one of the sequents labeling a leaf of the tree is not an axiom, but all of the formulas on the left or right of the sequent either are dead or are propositional variables.

At stage 0, the tree consists of the single node $X \vdash Y$, and all the formulas in X and Y are marked as alive. Now suppose that we have generated the tree at a given stage, but the search has not yet terminated. This means that there is a sequent labeling a leaf of the tree that is not an axiom, but in which there is a complex formula ϕ on the left or right of the sequent that is alive. Apply the appropriate inference rule in reverse to this formula (that is to say, extend the tree by the appropriate inference in which ϕ is the principal formula). Furthermore, apply the rule in such a way that if a formula occurs on the left or right of the conclusion of the inference, then it also occurs in any premise. The formula ϕ is marked as dead in any premise of the inference; any new formula appearing in a premise is marked as alive, while all other markings remain the same. We have now generated the proof tree for the next stage in the procedure. Notice that every application of a rule decreases the complexity of the remaining set of formulas that are alive, so that the search must eventually terminate, as defined above.

If all branches in the proof search tree end in axioms, then clearly we have found a proof of our original sequent $X \vdash Y$. However, suppose that the proof search terminates as in case (b), that is to say, at some point we have generated a proof tree in which a sequent $Z \vdash W$ labeling a leaf is not an axiom, but all complex formulas on the left or right are dead. Define an assignment of values as follows: if a variable x_i is in Z, then assign it the value 1, otherwise give it the value 0. We claim that if a formula occurs on the left of a sequent in the branch terminating in $Z \vdash W$, then it takes the value 1, whereas if it occurs on the right of a sequent in the branch, it must take the value 0 (note that the a formula cannot occur both on the left and the right because of the way we constructed the tree). This claim is easily verified by checking the basic rules. It follows that all of the formulas in X must take the value 1, whereas all the formulas in Y must take the value 0. Hence, we have shown in case (b) that the initial sequent $X \vdash Y$ is invalid. ∎

The completeness theorem for our cut-free Gentzen system has an additional very useful corollary, the admissibility of the cut rule:

$$\frac{X, \phi \vdash Y \quad X \vdash Y, \phi}{X \vdash Y} \ (Cut).$$

That is to say, a sequent that is deducible using the Cut rule is also deducible without it (although there are cases where the use of the Cut rule can result in much shorter proofs).

Another example of a theorem that is simple to prove using a cut-free proof system is Craig's interpolation theorem. We write Var(ϕ) and Var(X) for the set of variables occurring in a formula ϕ and a set of formulas X. If $X \vdash Y$ is a valid sequent, we say that a Boolean expression ϕ is an *interpolant* for $X \vdash Y$ if the sequents $X \vdash \phi$ and $\phi \vdash Y$ are valid, and in addition, $\mathrm{Var}(\phi) \subseteq \mathrm{Var}(X) \cap \mathrm{Var}(Y)$.

Proposition 3.3. *Every valid sequent has an interpolant.*

Proof. The basic idea of the proof is to construct an interpolant by induction on the length of the sequent calculus proof of a valid sequent $X \vdash Y$. In order to allow the inductive steps to go through smoothly, we prove a more general form of the proposition stated.

The form of the interpolation theorem that we prove by induction can be stated as follows: if $X, Y \vdash Z, W$ is a valid sequent, then there is a formula ϕ so that the sequents $X \vdash \phi, Z$ and $\phi, Y \vdash W$ are both valid, and in addition $\mathrm{Var}(\phi) \subseteq \mathrm{Var}(X \cup Z) \cap \mathrm{Var}(Y \cup W)$.

The inductive proof of this assertion is straightforward but has quite a number of cases. We content ourselves with a few cases as an example, the first being the case of an axiom of the form $X, Y, \phi \vdash Z, W, \phi$, and the second a case where the principal formula is a conjunction.

In the case of an axiom such as $X, Y, \phi \vdash Z, W, \phi$, four cases arise depending on how we partition the sets of formulas on the left and the right of the sequent. If we partition the sequent as $X; Y, \phi \vdash Z; W, \phi$ (where the semicolon represents the partition), then the interpolant is the constant 1. If the partition is $X, \phi; Y \vdash Z; W, \phi$, then the interpolant is ϕ itself. If the partition is $X, \phi; Y \vdash Z, \phi; W$, then the interpolant is the constant 0. Finally, if the partition is $X; Y, \phi \vdash Z, \phi; W$, then the interpolant is $\overline{\phi}$.

We now deal with the case where the sequent of interest has been inferred by the rule $\vdash \wedge$. As before, several cases arise, depending on how we partition the sets on the left and right of the sequent. Let us suppose that the sequent takes the form $X_1, Z_1; X_2, Z_2 \vdash W_1, Y_1; W_2, Y_2, \phi \wedge \psi$, and is inferred from the sequents $X_1; X_2 \vdash Y_1; Y_2, \phi$ and $Z_1; Z_2 \vdash W_1; W_2, \psi$. By induction hypothesis, there are interpolants θ and χ for these last two sequents. It is now straightforward to verify that the expression $\theta \wedge \chi$ is an interpolant for the original sequent.

The main proposition now follows by setting $Z = Y = \emptyset$ in the more general version that we stated previously. ∎

3.4 Consensus and Resolution

Although the cut-free Gentzen systems described in the previous section have been used fairly widely in automated theorem proving, an even more important role has been played by proof systems related to consensus and resolution.

The reader is referred to Chapter 2 of [9], particularly Section 2.7, for a discussion of the consensus rule. Recall that an application of the consensus rule consists in deriving the elementary conjunction CD from two elementary conjunctions xC and $\overline{x}D$. A *consensus proof* of a DNF expression ϕ is a derivation of the constant 1 from the terms of ϕ, using the consensus rule.

The resolution rule is simply the dual form of the consensus rule. Because most of the literature on automated theorem proving is written in terms of the resolution rule, we use it as the basis of our discussion in this chapter. In fact, any result on resolution proofs can be translated into a result on consensus proofs simply by replacing elementary disjunctions by elementary conjunctions; the reverse translation is equally easy.

Although resolution operates only on clauses, it can be converted into a general-purpose theorem prover for tautologies by employing an efficient method of conversion to conjunctive normal form, first used by Tseitin [17]. Tseitin's method is the dual version of the procedure used to compute an implicit DNF representation of a function, described in [9], Section 1.6. Let ϕ be a formula containing various binary connectives such as $\Longrightarrow$ and $\Leftrightarrow$; associate a literal with each subformula of ϕ so that the literal associated with a subformula $\overline{\psi}$ is the complement of the literal associated with ψ. If the subformula is a propositional variable, then the associated literal is simply the variable itself. We write l_χ for the literal associated with the subformula χ. If α is a subformula having the form $\beta \circ \gamma$, where $\circ$ is a binary connective, then $\mathrm{Cl}(\alpha)$ is the set of clauses making up the conjunctive normal form of $l_\alpha \Leftrightarrow (l_\beta \circ l_\gamma)$. For example, if α has the form $(\beta \Leftrightarrow \gamma)$, then $\mathrm{Cl}(\alpha)$ is the set of clauses

$$\{\, \overline{l_\alpha} \vee \overline{l_\beta} \vee l_\gamma,\ \overline{l_\alpha} \vee l_\beta \vee \overline{l_\gamma},\ l_\alpha \vee \overline{l_\beta} \vee \overline{l_\gamma},\ l_\alpha \vee l_\beta \vee l_\gamma \,\}.$$

The set of clauses $\mathrm{Def}(\phi)$ corresponding to ϕ is defined as the union of all $\mathrm{Cl}(\psi)$, where ψ is a compound subformula of α. With this conversion method in mind, the resolution method provides a complete proof procedure for tautologies, as proved in [9]. That is to say, given a tautology ϕ, the set of clauses $\mathrm{Def}(\phi) \cup \{\overline{l_\phi}\}$ is contradictory and hence has a resolution refutation.

Resolution proofs represented in tree form ("tree resolution") derive their importance from their close connection with the branching procedures described in [9], Section 2.5. Here we consider the branching procedure from [9] as a proof procedure, so that we consider only the case in which the procedure shows that no solution of an equation $\phi = 1$ exists, where ϕ is a CNF expression. Thus we consider a binary tree, in which the root is labeled with the empty assignment. If an internal node in the tree is labeled with the assignment σ, then the two children of that node are labeled with assignments extending σ in which a previously unset variable is set to 0 and 1, respectively. The leaves of the tree are labeled with partial assignments, each of which sets a clause in ϕ to 0. Thus, the whole labeled tree can be considered as a proof that the equation $\phi = 1$ has no solution, that is to say, ϕ itself is a contradiction.

A tree resolution proof is defined to be *regular* if on any path in the tree from a leaf to the root, no variable is resolved upon twice. It is not hard to show that irregularities can be eliminated from resolution proofs, in the sense that if there is an irregular tree resolution refutation of a given set of clauses, then there is a regular refutation that is smaller than the original irregular proof.

Any regular tree resolution proof of a CNF contradiction ϕ can be converted easily into a branching procedure proof. Such a proof consists of a binary tree, in which the leaves are labeled with clauses of ϕ, and the root with the constant 0. If an interior node is labeled with a clause A, then $A = C \vee D$ is derived by resolution from clauses $x \vee C$ and $\overline{x} \vee D$ labeling the children of the node. We now convert this into a branching procedure proof. First, we associate the empty assignment with the root. Suppose that we have associated an assignment σ with an interior node labeled with $C \vee D$, derived from the premises $x \vee C$ and $\overline{x} \vee D$ labeling the children of the node. Then we associate the assignment σ_1 extending σ by setting x to 0 with the first child, and the assignment σ_2 extending σ by setting x to 1 with the second. By construction, all of these assignments set to 0 the clauses labeling the nodes in question, and in particular all the clauses labeling the leaves are set to 0 by the corresponding assignments. Hence the newly labeled tree is a branching procedure proof.

The conversion in the opposite direction can also be performed, though here we have to be a little more careful, because not every branching procedure proof corresponds exactly to a tree resolution proof. Suppose that we are given a branching procedure proof that a CNF expression ϕ is a contradiction. We now show how to convert this into a tree resolution proof in two steps. At each stage of the conversion, we associate with nodes in the tree clauses that are set to 0 by the assignment associated with that node. We proceed from the leaves of the tree toward the root.

We begin by associating with each leaf in the tree a clause of ϕ that is set to 0 by the partial assignment at that leaf. Now assume, for the inductive construction, that we have associated clauses C and D with the children of a node in the tree, and that at these children, the assignment σ labeling the parent node was extended by setting the variable x to 1 and 0, respectively. If the literal x occurs in C and the literal $\overline{x}$ in D, then we associate with the parent node the clause derived from C and D by resolution. If either condition fails, then we associate with the parent node a clause (either C or D) not containing the variable x.

Because the empty assignment labels the root of the tree, it follows that the clause associated with the root must be the constant 0, since it is the only elementary disjunction set to 0 by the empty assignment. The resulting labeled tree may not be a tree resolution proof, since there can be some repetitions of clauses along some of its branches. However, we can eliminate these repetitions by merging them together, and the result is a regular tree resolution proof.

Refutations can be represented as trees or as sequences of clauses; the worst-case complexity differs considerably depending on the representation. We distinguish between the two by describing the first system as "tree resolution," the second simply as "resolution." The close correspondence between branching procedure proofs and tree resolution proofs shows that we can prove lower bounds on the complexity of the branching procedure by establishing lower bounds on the size of resolution proofs in tree form.

3.5 Relative Complexity of Proofs

In the decades since 1970, there has been intensive research activity in the area of propositional proof systems. The reason for this is the close connection between the complexity of propositional proofs and fundamental questions in complexity theory.

Although the problem of the complexity of propositional proofs is very natural, it has been investigated systematically only since the late 1960s. Interest in the problem arose from two fields connected with computers, automated theorem proving and computational complexity theory. The earliest paper in the subject is a ground-breaking article by Tseitin [17], the published version of a talk given in 1966 at a Leningrad seminar. In the decades since that talk, substantial progress has been made in determining the relative complexity of proof systems and in proving strong lower bounds for some restricted proof systems. However, major problems remain to challenge researchers.

The literature of mathematical logic contains a very wide variety of proof systems; the systems discussed in this chapter are only a small sample of the many such systems that have been proposed. To compare their efficiency, we need a general definition of a proof system. In this section, we give such a definition, together with another that formalizes the relation holding between two proof systems when one can simulate the other efficiently. The definitions are adapted from Cook and Reckhow [8].

Let Σ be a finite alphabet; we write Σ^* for the set of all finite strings over Σ. A *language* is defined as a subset of Σ^*, that is, a set of strings over a fixed alphabet Σ. The length of a string x is written as $|x|$.

Definition 3.5. *If Σ_1 and Σ_2 are finite alphabets, a function f from Σ_1^* into Σ_2^* is in $\mathcal{L}$ if it can be computed by a deterministic Turing machine in time bounded by a polynomial in the length of the input.*

The class $\mathcal{L}$ of polynomial-time computable functions is a way of making precise the vague notion of "feasibly computable function."

Definition 3.6. *If $L \subseteq \Sigma^*$, a* proof system *for L is a function $f : \Sigma_1^* \rightarrow L$ for some alphabet Σ_1, where $f \in \mathcal{L}$ and f is onto. A proof system f is* polynomially bounded *if there is a polynomial* $p(n)$ *such that for all* $y \in L$, *there is an* $x \in \Sigma_1^*$ *such that* $y = f(x)$ *and* $|x| \leq p(|y|)$.

The intention of this definition is that $f(x) = y$ is to hold if x is a proof of y. The crucial property of a proof system as defined earlier is that, given an alleged proof, there is a feasible method for checking whether or not it really is a proof, and if so, of what it is a proof. A standard axiomatic proof system for the tautologies, for example, can be brought under the definition by associating the following function f with the proof system $\mathcal{F}$: if a string of symbols σ is a legitimate proof in $\mathcal{F}$ of a formula ϕ, then let $f(\sigma) = \phi$; if it is not a proof in $\mathcal{F}$, then let $f(\sigma) = T$, where T is some standard tautology, say $x_1 \vee \overline{x_1}$.

Let us recall here some of the basic definitions in computational complexity theory (for details, the reader is referred to the texts [10, 12, 15]). A set of strings is in the class $\mathcal{P}$ ($\mathcal{NP}$) if it is recognized by a deterministic (nondeterministic) Turing machine in time polynomial in the length of the input. A set of strings is in the class co-$\mathcal{NP}$ if it is the complement of a language in $\mathcal{NP}$. In more logical terms, a set S of strings is in $\mathcal{P}$ if its characteristic function is in $\mathcal{L}$, whereas it is in $\mathcal{NP}$ if the condition $y \in S$ can be expressed in the form $(\exists x)(|x| \leq p(|y|) \wedge R(x, y))$, where p is a polynomial, and R is a polynomial-time computable relation. Thus, $\mathcal{P}$ is the polynomial-time analog of the recursive sets, whereas $\mathcal{NP}$ corresponds to the recursively enumerable sets, showing that the basic question $\mathcal{P} =?\mathcal{NP}$ is the polynomial-time analog of the halting problem.

The importance of our main question for theoretical computer science lies in the following result of Cook and Reckhow [8].

Theorem 3.4. $\mathcal{NP} =$ *co-*$\mathcal{NP}$ *if and only if there is a polynomially bounded proof system for the classical tautologies.*

Proof. If $\mathcal{NP} =$ co-$\mathcal{NP}$, then because the set TAUT of classical tautologies is in *co-*$\mathcal{NP}$, TAUT would be in $\mathcal{NP}$, that is to say, there would be a nondeterministic Turing machine M accepting TAUT. Let f be the function such that $f(x) = y$ if and only if x encodes a computation of M that accepts y; then f is a polynomially bounded proof system for TAUT.

Conversely, let us assume that there is a polynomially bounded proof system for TAUT. Let L be a language in $\mathcal{NP}$. By the basic $\mathcal{NP}$-completeness result of Cook [6], L is reducible to the complement of TAUT in the sense that there is a function $f \in \mathcal{L}$ so that for any string x, $x \in L$ if and only if $f(x) \notin TAUT$. Hence a nondeterministic polynomial-time procedure for accepting the complement of L is: on input x, compute $f(x)$ and accept x if $f(x)$ has a proof in the proof system. Hence, $\mathcal{NP}$ is closed under complementation, so $\mathcal{NP} =$ co-$\mathcal{NP}$. ∎

This equivalence result underlines the very far-reaching nature of the widely believed conjecture $\mathcal{NP} \neq$ co-$\mathcal{NP}$. The conjecture implies that there is an infinite sequence of tautologies ϕ_n so that the size of their proofs grows faster than any polynomial, even if we use all the resources of current mathematics, as expressed in axiomatic set theory.

Because the complexity class $\mathcal{P}$ is closed under complementation, it follows that if $\mathcal{P} = \mathcal{NP}$, then $\mathcal{NP} = \text{co-}\mathcal{NP}$. This suggests that we might attack the problem $\mathcal{P} =?\mathcal{NP}$ by trying to prove that $\mathcal{NP} \neq \text{co-}\mathcal{NP}$; by Theorem 3.4, this is the same as trying to show that there is no polynomially bounded proof system for the classical tautologies. This line of research was first suggested in papers by Cook and Reckhow [7, 8]. At the moment, the goal of settling the question $\mathcal{NP} \neq \text{co-}\mathcal{NP}$ seems rather distant. However, progress has been made in classifying the relative complexity of well-known proof systems and in proving lower bounds for restricted systems. An attractive feature of the research program is that we can hope to approach the goal step by step, developing ideas and techniques for simpler systems first. We now give a precise definition of what it means for one proof system to simulate another efficiently. This is a more precise version of the simulation notion introduced informally by Crama and Hammer in [9], Section 2.10.

Definition 3.7. *If $f_1 : \Sigma_1^* \rightarrow L$ and $f_2 : \Sigma_2^* \rightarrow L$ are proof systems for L, then* f_2 p-simulates f_1 *provided that there is a polynomial-time computable function $g : \Sigma_1^* \rightarrow \Sigma_2^*$ such that $f_2(g(x)) = f_1(x)$ for all x.*

The intention of this definition is that g is a feasible translation function that translates proofs in f_1 into proofs in f_2. It is an immediate consequence of the definition that if a proof system is polynomially bounded, then any proof system p-simulating it is also polynomially bounded.

The diagram in Figure 3.1 is a map showing the relative efficiency of various systems. The boxes in the diagram indicate equivalence classes of proof systems under the symmetric closure of the p-simulation relation. Systems below the dotted line have been shown to be not polynomially bounded, whereas no such lower bounds are known for those that lie above the line. Hence, the dotted line represents the current frontier of research on the main problem. Although systems below the line are no longer candidates for the role of a polynomially bounded proof system, there are still some interesting open problems concerning the relative complexity of such systems. Questions of this sort, although not directly related to such problems as $\mathcal{NP} =?\text{co-}\mathcal{NP}$, have some relevance to the more practical problem of constructing efficient automatic theorem provers. Although the more powerful systems above the dotted line are the current focus of interest in the complex of questions surrounding the $\mathcal{NP} =?\text{co-}\mathcal{NP}$ problem, the systems below allow simple and easily mechanized search strategies and so are still of considerable interest in automated theorem proving.

An arrow from one box to the other in the diagram indicates that any proof system in the first box can p-simulate any system in the second box. In the case of cut-free Gentzen systems, this simulation must be understood as referring to a particular language on which both systems are based. An arrow with a slash through it indicates that no p-simulation is possible between any two systems in the classes in question. If a simulation is possible in the reverse direction, then we can say that systems in one class are strictly more powerful than systems in the other (up to a polynomial).

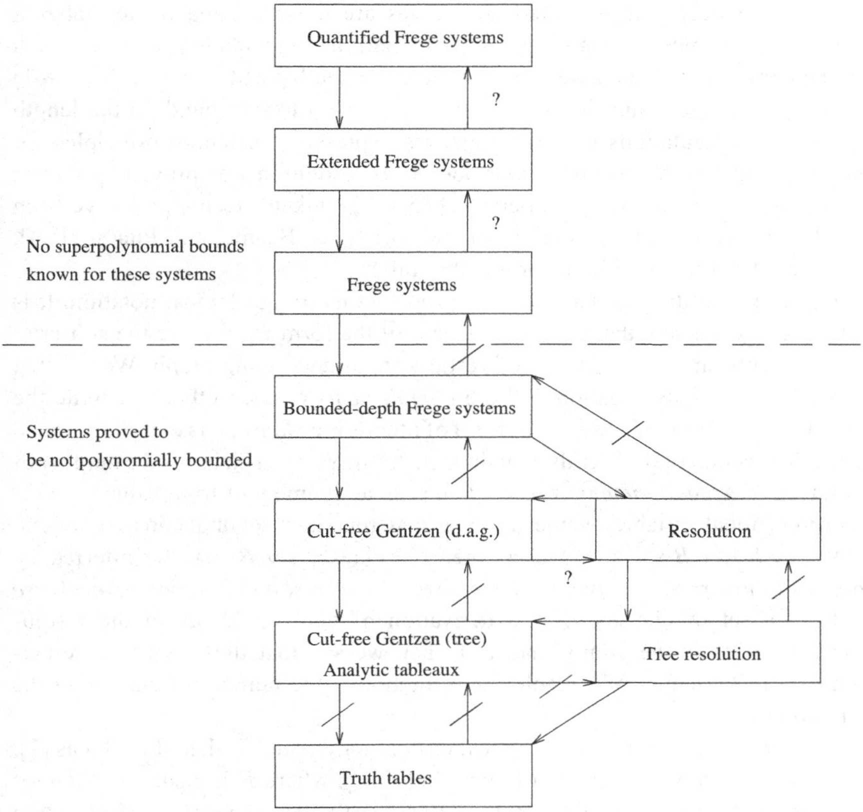

Figure 3.1. Proof system map.

This chapter does not describe fully all of the systems in the diagram; we content ourselves here with short sketches of the systems not described here in detail. Both resolution and cut-free Gentzen systems are more powerful when proofs are represented as directed acyclic graphs, rather than trees; exponential speed-up is possible in both cases, as shown in the diagram. Bounded-depth Frege systems restrict formulas to a fixed depth, where unbounded fan-in AND and OR gates are allowed. Extended Frege systems allow the possibility of abbreviative definitions (thus a single variable can stand for a more complicated formula). Quantified Frege systems include inference rules involving quantified Boolean formulas. Extended Frege systems are p-equivalent to s-Frege systems. It is easy to see that the equational calculus with which we began the chapter is p-equivalent to an s-Frege system, and hence to any extended Frege system.

3.6 Complexity of Resolution

In this section, we prove lower bounds on the complexity of proofs in the resolution system. Exactly the same lower bounds apply to the consensus system, by duality.

The most deeply investigated tautologies are those created by formalizing matching principles in graphs, and in particular, the pigeonhole principle, which can be considered as the assertion that the complete bipartite graph $K(n, n-1)$ has no perfect matching. In this section, we prove a lower bound on the length of resolution refutations of sets of clauses expressing matching principles for bipartite graphs. The original lower bound for resolution was proved by Armin Haken, in a groundbreaking paper of 1985 [11]. Haken's techniques have been simplified considerably as a result of work of Buss, Beame, and Pitassi [1, 4], whose methods are used in the proofs that follow.

Before proceeding to details of the proof, let us fix our logical notation. It is convenient to assume that our variables are of the form P_{xy}, where the subscript xy represents an unordered pair of vertices in an underlying graph. We say that a variable P and its negation FP are *complements* of each other; we write the complement of a literal l as $\bar{l}$. A finite set of literals is a *clause*; it is to be interpreted as the disjunction of the literals contained in it. The *size* of a clause is the number of literals in it. An *assignment* or *restriction* is an assignment of truth values to a set of propositional variables; some variables may remain unset under an assignment.

If $A \vee l$ and $B \vee \bar{l}$ are clauses, then the clause $A \vee B$ may be inferred by the resolution rule, *resolving on* the literal l. A *resolution proof of a clause* C from a set of clauses Σ is a derivation of C from Σ, using the resolution rule. If C is the empty clause 0, then we say that the proof is a refutation of Σ. The length of a resolution refutation is the number of clauses in the derivation.

We now define our graph-theoretical terminology, which is that of Bollobás [2]. A *graph* G is an ordered pair of finite sets (V, E), where E is a subset of the set of unordered pairs of V; the set $V = V(G)$ is the set of *vertices* of G, while $E = E(G)$ is the set of *edges* of G. If x, y are vertices, then we write xy for the edge containing x and y. We say that two edges are *adjacent* if they have exactly one vertex in common; a set of edges in a graph is *independent* if no two edges in the set are adjacent. A matching in a graph G is an independent subset of $E(G)$; we say that a vertex in G is covered by a matching if it belongs to one of the edges in the matching.

Let $G = (V, E)$ be a bipartite graph, where $V = D \cup R = D(G) \cup R(G)$, and G is a subgraph of $D \times R$. We can formulate the assertion that there is a matching of D with a subset of R as a set of clauses $\mathrm{M}(G)$. Where $x \in D$, $y \in R$, the propositional variable P_{xy} is to be read as asserting, "Vertex x is matched with vertex y." If x is a vertex in D, the disjunction $D_x = \bigvee\{P_{xy} \mid y \text{ adjacent to } x\}$ asserts that x is matched with one of its neighbors. Similarly, the disjunction $E_{xyz} = (FP_{xz} \vee FP_{yz})$ asserts that z cannot be matched with both x and y. The set of clauses $\mathrm{M}(G)$ contains all the disjunctions D_x, for $x \in D$, together with all the disjunctions E_{xyz}, where $x, y \in D$, $z \in R$, z is adjacent to x and y, and $x \neq y$. If $|D| = m$, $|R| = n$, then $\mathrm{M}(G)$ has size $O(m^2 n)$. All of the graphs discussed here are bipartite, so we shall henceforth usually omit the adjective "bipartite" when talking about graphs.

If there is no matching covering all of D, then $\mathrm{M}(G)$ is contradictory. Let $K(n, n-1)$ be the complete bipartite graph with $V = D \cup R$, $|D| = n$, $|R| = n-1$, and $E = \{xy \mid x \in D, y \in R\}$. The set of clauses $\mathrm{M}(K(n, n-1))$ is contradictory; the statement of this fact is a formulation of the pigeonhole principle, so we refer to these clauses as the pigeonhole clauses PHC^n_{n-1}.

If μ is a matching in a graph G, then it induces a (partial) assignment of truth values to the variables P_{xy}, where $xy \in E(G)$. If either of the two vertices x or y is covered by μ, then we give P_{xy} the value 1 if $xy \in \mu$, otherwise 0. If neither x nor y is covered by μ, then P_{xy} remains unset. We identify a matching with the assignment it determines, so that we can write $\mu(C) = 0$ or $\mu(C) = 1$, where C is a clause. If Σ is a set of clauses, then we write $\mu(\Sigma) = 1$ if μ makes all the clauses in Σ true.

If G is a graph, then we refer to the set of variables P_{xy} associated with G as the language of G, $L(G)$; we also say that clauses built from variables in $L(G)$ belong to $L(G)$. Let $\Sigma \cup \{C\}$ be a set of clauses in $L(G)$. We say that C is a consequence of Σ with respect to the family of matching assignments in G, or an m-consequence of Σ, if for every matching μ in G, whenever $\mu(\Sigma) = 1$, then $\mu(C) = 1$. It is clear that this concept of consequence differs from the ordinary concept; for example, if P_{ik} is a variable in $L(G)$, and $i \neq j$, then FP_{jk} is an m-consequence of P_{ik}. If Σ is a set of monotone clauses in $L(G)$, then Σ is m-contradictory if the empty clause is an m-consequence of Σ. It is possible for Σ to be m-contradictory, but not contradictory in the sense of ordinary consequence; we can take $\Sigma = \{P_{ik}, P_{jk}\}$, for $i \neq j$, as an example. However, for the proofs that follow, we only need the fact that if $\Sigma = M(G)$ is contradictory in the ordinary sense, then it is also m-contradictory.

In proving lower bounds for resolution proofs of matching principles, it is convenient to transform these proofs in such a way that negation no longer occurs. The idea of this transformation is due to Buss and Pitassi [4]. Let G be a graph, and C a clause in $L(G)$. Then the monotone transform of C is the clause C^m in which every negative variable FP_{ij} is replaced by the disjunction of the set of variables

$$\{P_{ik} \mid k \neq j \,\&\, ik \in E(G)\}.$$

If Σ is a set of clauses in $L(G)$, then we write Σ^m for the set $\{C^m \mid C \in \Sigma\}$. If we replace clauses in a resolution proof with their monotone transforms, then the result is no longer a resolution proof, since there are no negated literals. However, the proof remains sound, provided we restrict our attention to the assignments determined by matchings.

Lemma 3.5. *If $C \vee P_{ij}$ and $D \vee FP_{ij}$ are clauses in $L(G)$, then $(C \vee D)^m$ is an m-consequence of $(C \vee P_{ij})^m$ and $(D \vee FP_{ij})^m$.*

Proof. The monotone transforms of $C \vee P_{ij}$ and $D \vee FP_{ij}$ are $C^m \vee P_{ij}$ and $D^m \vee \bigvee\{P_{ik} \mid k \neq j \,\&\, ik \in E(G)\}$. Let us suppose that there is a matching μ making both of these monotone transforms true, but not $(C \vee D)^m$. Because μ is a matching, it cannot make both P_{ij} true and one of the variables in $\{P_{ik} \mid k \neq j \,\&\, ik \in E(G)\}$ true. Hence, it must make at least one of the variables in C^m or D^m true, contrary to assumption. ■

If $C_1, \ldots, C_k$ is a resolution proof from a set of clauses in $L(G)$, then we define the monotone transform of $C_1, \ldots, C_k$ to be the sequence of monotone clauses $C_1^m, \ldots, C_k^m$. Let us define a sequence of monotone clauses in $L(G)$ to be a monotone proof from a set of clauses Σ if every clause in the sequence is either a clause in Σ or a monotone consequence of at most two earlier clauses in the sequence. A monotone proof is a refutation if it is a proof of the empty clause. The lemma just proved shows that if $C_1, \ldots, C_k$ is a resolution proof from Σ, then its monotone transform is a monotone proof from Σ^m.

It should be noted that the notion of monotone consequence is relative to a graph G, and hence the concept of a monotone proof is also defined relative to a given graph. In what follows, the graph in question is usually clear from the context, but in cases where we wish to make it explicit, we shall refer to a "monotone G-proof" or a "monotone G-refutation" where the graph G serves to define the concept of m-consequence.

If μ is a matching in a bipartite graph G, and C is a clause in $L(G)$, then we write $C \restriction \mu$ for the result of applying μ to the clause C. That is to say, if there is a literal l in C so that $\mu(l) = 1$, then $C \restriction \mu = 1$; otherwise $C \restriction \mu$ is the clause resulting from C by removing all the literals in C that are set to 0 by μ. If G is a graph, and μ a matching in G, then we define the graph $G \restriction \mu$ to be the result of removing from G all of the vertices in G covered by μ, together with all the edges attached to them.

Lemma 3.6. *Let G be a graph, and μ a matching in G. If the sequence $C_1, \ldots, C_k$ is a monotone proof from Σ, where all the clauses C_i belong to $L(G)$, then the sequence $C_1 \restriction \mu, \ldots, C_k \restriction \mu$ is a monotone proof from $\Sigma \restriction \mu$.*

Proof. We need to show that if $\Gamma \cup \{C\}$ is a set of monotone clauses in $L(G)$, and C is an m-consequence of Γ (relative to the set of matchings in G), then $C \restriction \mu$ is an m-consequence of $\Gamma \restriction \mu$ (relative to the set of matchings in $G \restriction \mu$).

Thus, let ν be a matching in $G \restriction \mu$ so that $\nu(\Gamma \restriction \mu) = 1$. Extend ν to a matching ν' in G by setting $\nu' = \nu \cup \mu$. Then $\nu'(\Gamma) = 1$, so $\nu'(C) = 1$. If $\mu(C) = 1$, then $C \restriction \mu = 1$, whereas if $\nu(C) = 1$, then $\nu(C \restriction \mu) = 1$, proving the lemma. ■

If G is a bipartite graph, and xy an edge in G, then we define $G - xy$ to be the graph resulting from G by removing the edge xy. If C is a monotone clause in $L(G)$, then $C - xy$ is the clause resulting from C by removing the variable P_{xy}; similarly, if Σ is a set of clauses in $L(G)$, $\Sigma - xy$ is the set $\{C - xy \mid C \in \Sigma\}$.

Lemma 3.7. *Let G be a bipartite graph, $\Sigma \cup \{C\}$ a set of monotone clauses in $L(G)$, and xy an edge in G. If the clause C is an m-consequence of Σ (relative to the family of matchings in G), then $C - xy$ is an m-consequence of $\Sigma - xy$ (relative to the family of matchings in $G - xy$).*

Proof. The lemma follows immediately from the fact that any matching in $G - xy$ is also a matching in G. ■

Let G be a graph, and C a monotone clause in $L(G)$ that is an m-consequence of $M(G)^m$. We define the complexity of C (relative to G) to be the size of the smallest subset X of $D(G)$ so that C is an m-consequence of $\{D_x \mid x \in X\}$. Relative to the graph $K(m, n)$, the clauses D_x have complexity 1, while the clauses $(E_{xyz})^m$ have complexity 2.

Lemma 3.8. *Let G be a graph with $|D(G)| = m$, $|R(G)| = n$, and C an m-consequence of $M(G)^m$ of complexity $k \leq m$. Then there is a subset X of $D(G)$, and a subset Y of $R(G)$ so that $|X| = k$, $|Y| = n - k + 1$, and C contains all the variables P_{xy} where $x \in X$, $y \in Y$, and $xy \in E(G)$.*

Proof. By definition, there is a subset X of $D(G)$ where $|X| = k$, and C is an m-consequence of $\{D_x \mid x \in X\}$, but not of any proper subset. Thus for $Y \subseteq X$ with $k - 1$ elements, there is a matching μ_Y so that μ_Y makes all the clauses in $\{D_y \mid y \in Y\}$ true, but not C. Let G' be the subgraph of G containing all the edges in all such matchings μ_Y. Then $|D(G')| = k$, $|R(G')| \geq k - 1$, and none of the variables P_{ij}, where ij is an edge in G', can appear in C.

By the construction of the bipartite graph G', every subset $Y \subseteq D(G')$ with at most $k - 1$ elements is connected with a subset $Z \subseteq R(G')$ where $Z \geq k - 1$. If $|R(G')| \geq k$, then the conditions for Hall's theorem ([3], p. 6) would be satisfied, so that G' would have a perfect matching. This would contradict the assumption that C is an m-consequence of $\{D_x \mid x \in X\}$, because any such matching would give rise to a matching assignment making all of the D_x, for $x \in X$, true, but C false. Hence, $|R(G')| = k - 1$.

Define Y to be $R(G) \setminus R(G')$. Let xy be an edge in G, where $x \in X$ and $y \in Y$. By assumption, there is a matching μ in G' so that μ makes all the clauses in $\{D_z \mid z \in X \setminus \{x\}\}$ true, but $\mu(C) \neq 1$. If P_{xy} does not occur in C, then we can extend μ to a matching $\mu' = \mu \cup \{xy\}$ so that μ' makes $\{D_z \mid z \in X\}$ true, but $\mu'(C) \neq 1$, contrary to assumption. It follows that C must contain all such variables P_{xy}. ■

Lemma 3.8 allows us to show that any monotone refutation of certain sets must contain a large clause.

Lemma 3.9. *Let Σ be an m-contradictory set of monotone clauses expressed in the language $L(K(n, n-1))$ such that every clause in Σ has complexity $\leq n/3$ relative to $K(n, n-1)$. Then any monotone $K(n, n-1)$-refutation of Σ contains a clause with at least $n^2/9$ variables.*

Proof. Because the initial clauses have complexity $\leq n/3$, and the empty clause has complexity n, relative to $K(n, n-1)$, it follows that there must be a clause C in the refutation having complexity k, where $n/3 \leq k < 2n/3$. By Lemma 3.8, there is a subset $X \subseteq D$ and $Y \subseteq R$, where $|X| = k$, $|Y| = n - k$, and C contains all of the variables P_{xy}, where $x \in X$, $y \in Y$. Because $|X|, |Y| \geq n/3$, it follows that C contains at least $n^2/9$ variables. ■

Lower bounds on the size of resolution proofs are based on a simple idea. If we subject a set of clauses to a random restriction, then there is a high probability that all large clauses are set to 1. The next lemma (due to Beame and Pitassi [1]) makes this idea precise.

Lemma 3.10. *Let G be a bipartite graph, and Σ a set of M monotone clauses in $L(G)$. Let v be the number of edges in G, and for $m > 1$, let $p = m/v$. Then if $k > \ln M/p$, there is a matching μ in G of size k so that $\Sigma \upharpoonright \mu$ contains no clauses of size $\geq m$.*

Proof. We prove by induction on k that we can find a matching μ in G of size k so that $\Sigma \upharpoonright \mu$ contains at most $M_k = M(1-p)^k$ clauses of size $\geq m$. Assume that this condition holds for k, and let N be the number of clauses of size $\geq m$ in $\Sigma \upharpoonright \mu$. If we choose an edge xy at random in $G \upharpoonright \mu$, then the expected number of clauses of size $\geq m$ set to 1 by the restriction $\{xy\}$ is at least Np; this follows by writing the random variable counting the number of such clauses set to 1 as a sum of indicator variables, and then using the linearity of expectation. Thus we can choose an edge xy in $G \upharpoonright \mu$ so that at least Np such clauses are set to 1 by setting P_{xy} to 1. Extend μ to a new restriction μ' by setting $\mu' = \mu \cup \{xy\}$. Then the remaining number of clauses of size $\geq m$ in $\Sigma \upharpoonright \mu$ is at most

$$N - Np \leq M_k(1-p) = M(1-p)^{k+1},$$

completing the induction step. The proof of the lemma is completed by observing that if $k > \ln M/p$, then

$$M(1-p)^k < Me^{-pk} < 1,$$

so that $\Sigma \upharpoonright \mu$ contains no clauses of size $\geq m$. ■

Lemmas 3.9 and 3.10 are sufficient to prove Haken's exponential lower bound [11] for resolution proofs of the pigeonhole principle. The proof given here is essentially that of Beame and Pitassi [1].

Theorem 3.11. *Any resolution refutation of PHC^n_{n-1} contains at least $2^{n/45}$ clauses.*

Proof. The monotone transform of a resolution refutation of PHC^n_{n-1} has the same length as the original refutation, so it is sufficient to consider a monotone refutation of PHC^n_{n-1}. Define a large clause to be one containing at least 1/16th of the $n(n-1)$ variables in $L(K(n, n-1))$. Let Σ be the set of clauses in a monotone refutation of PHC^n_{n-1}, and let us assume that $|\Sigma| < 2^{n/45}$. Setting $m = n(n-1)/16$, $v = n(n-1)$, we have $p = m/v = 1/16$. Thus $n/4 > \ln(2^{n/45})/p$, so by Lemma 3.10, there is a matching μ of size $n/4$ so that $\Sigma \upharpoonright \mu$ contains no large clauses.

By Lemma 3.6, the result of applying the restriction μ to the refutation is a monotone $K(n', n'-1)$-refutation of $\mathrm{PHC}^n_{n-1} \restriction \mu$, where $\mathrm{PHC}^n_{n-1} \restriction \mu$ is a set of clauses in $L(K(n', n'-1))$, $n' = 3n/4$. Hence, by Lemma 3.9, the restricted refutation contains a clause with at least $(n')^2/9 = n^2/16$ variables. This contradicts the fact that $\Sigma \restriction \mu$ contains no large clauses. ■

The preceding theorem is quite powerful, as it shows that any proof procedure based solely on resolution must take an exponentially long time to show that certain CNF formulas are contradictory, no matter what heuristics are used in the search process. In particular, any algorithm based on the branching procedure of [9], Section 2.4, requires an exponential number of steps on the pigeonhole formulas (see also [9], Section 2.10). This lower bound holds independently of any choice of branching heuristics.

In this chapter, we have covered only a small part of what is currently known about the complexity of propositional proofs. The reader who wishes to learn more should consult the excellent texts by Krajíček and by Clote and Kranakis [5, 14].

3.7 Exercises

(i) Prove that the absorption laws $x \vee (xy) = x$ and $x(x \vee y) = x$ can be derived in the Boolean equation calculus.

(ii) Prove that the involution law $\overline{\overline{x}} = x$ can be derived in the Boolean equation calclulus.

(iii) Complete the remaining parts of the inductive step in Proposition 3.1.

(iv) Prove that if an expression ϕ is derivable in Kleene's formulation of the propositional calculus, then $\vdash \phi$ is derivable in the basic Gentzen sequent calculus.

(v) Complete the proof of the interpolation theorem (Proposition 3.3) by proving the remaining cases of the induction step.

(vi) Show that the equational calculus of Definition 3.2 is p-equivalent to the proof system of Definition 3.3, with the rule of substitution added.

(vii) Show that if there is an irregular tree resolution refutation of a set of clauses, then there is a smaller regular tree resolution refutation of the same set.

(viii) Show that resolution p-simulates the basic Gentzen sequent calculus.

References

[1] Paul Beame and Toniann Pitassi. Simplified and improved resolution lower bounds. *Proceedings of the 37th Annual IEEE Symposium on the Foundations of Computer Science*, pp. 274–82, 1996.

[2] Béla Bollobás. *Graph Theory*. Springer-Verlag, Berlin, 1979.

[3] Béla Bollobás. *Combinatorics*. Cambridge University Press, Cambridge, UK, 1986.

[4] Samuel R. Buss and Toniann Pitassi. Resolution and the weak pigeonhole principle. *Proceedings, Computer Science Logic (CSL '97)*, Lecture Notes in Computer Science 1414, pp. 149–56, 1998.

[5] Peter Clote and Evangelos Kranakis. *Boolean Functions and Computation Models*. Springer-Verlag, Berlin, 2002.

[6] Stephen A. Cook. The complexity of theorem-proving procedures. *Proceedings of the Third Annual ACM Symposium on Theory of Computing*, pp. 151–8, 1971.

[7] Stephen A. Cook and Robert A. Reckhow. On the lengths of proofs in the propositional calculus. *Proceedings of the Sixth Annual ACM Symposium on Theory of Computing*, 1974. See also corrections in *SIGACT News*, 6, pp. 15–22, 1974.

[8] Stephen A. Cook and Robert A. Reckhow. The relative efficiency of propositional proof systems. *Journal of Symbolic Logic*, 44, pp. 36–50, 1979.

[9] Yves Crama and Peter L. Hammer. *Boolean Functions: Theory, Algorithms, and Applications*. Cambridge University Press, Cambridge, UK, 2010.

[10] Michael R. Garey and David S. Johnson. *Computers and Intractability. A Guide to the Theory of NP-completeness*. W. H. Freeman, New York, 1979.

[11] Armin Haken. The intractability of resolution. *Theoretical Computer Science* 39, pp. 297–308, 1985.

[12] John E. Hopcroft and Jeffrey D. Ullman. *Introduction to Automata Theory, Languages and Computation*. Addison-Wesley, Reading, MA, 1979.

[13] S. C. Kleene. *Introduction to Metamathematics*. Van Nostrand, Princeton, NJ, 1952.

[14] Jan Krajíček. *Bounded Arithmetic, Propositional Logic and Complexity Theory*. Cambridge University Press, Cambridge, UK, 1996.

[15] Christos H. Papadimitriou. *Computational Complexity*. Addison-Wesley, Reading, MA, 1994.

[16] Jörg Siekmann and Graham Wrightson, eds. *Automation of Reasoning*. Springer-Verlag, New York, 1983.

[17] G. S. Tseitin. On the complexity of derivation in propositional calculus. In Studies in Constructive Mathematics and Mathematical Logic, A. O. Slisenko, ed., Part 2, pp. 115–25. Consultants Bureau, New York, 1970. Reprinted in [16], Vol. 2, pp. 466–83.

4

Probabilistic Analysis of Satisfiability Algorithms

John Franco

4.1 Introduction

Probabilistic and average-case analysis can give useful insight into the question of what algorithms for testing satisfiability might be effective and why. Under certain circumstances, one or more structural properties shared by each of a family or class of expressions may be exploited to solve such expressions efficiently; or structural properties might force a class of algorithms to require superpolynomial time. Such properties may be identified and then, using probabilistic analysis, one may argue that these properties are so common that the performance of an algorithm or class of algorithms can be predicted for most of a class of expressions. Perhaps most important, sometimes an analysis provides the intutition needed to suggest improved algorithms.

A classic example is the work on resolution for random (n, m, k)-CNF expressions. By a random (n, m, k)-CNF expression we mean a CNF expression of m clauses, each chosen uniformly at random and with replacement from among the set of $2^k\binom{n}{k}$ elementary disjunctions of k literals on a set of n Boolean variables and their complements. Simple variants of the well known Davis-Putnam-Logemann-Loveland algorithm (DPLL) [32], which never reassign a value to a variable once it is set, can solve large random (n, m, k)-CNF expressions with probability tending to 1 if they are generated with $m/n < c_k 2^k/k$, where c_k is a constant plus a term of complexity $o(k)$. Because variables are assigned at most one time, those variants run in polynomial time. The probabilistic analyses leading to these results help explain what is actually happening as the algorithms operate and actually suggest improvements to those algorithms. This has led to remarkable progress in understanding this class of algorithms [63]. The prospect of further progress along these lines is reasonably good.

On the other hand, resolution is remarkably poor in trying to prove that no solution exists on random (n, m, k)-CNF expressions that have no solution. In particular, the probability that the length of a shortest resolution proof is bounded by a polynomial in n when $m/n > c'_k 2^k$, c'_k some constant, and $\lim_{m,n\to\infty} m/n =$

$O(1)$ tends to 0 [24, 13]. Again, the analysis illuminates the reason. But, if the number of clauses in a random (n, m, k)-CNF expression is great enough, that expression can, with high probability, efficiently be shown to have no solution. For example, if $m/n = \Theta(n^{k-1})$, the probability that there exist 2^k clauses containing the same variables spanning all different literal complementation patterns tends to 1, and because such a pattern can be identified in polynomial time and certifies there is no solution, random expressions can be proven not to have a solution with probability tending to 1 in this case. Investigation of resolution proof size to find the relationship between m and n that defines the crossover from hard to easy expressions, although brilliant, has not progressed considerably. The best bound on the crossover is not much different from $m/n = \Theta(n^{k-1})$ [12]. This lack of success has motivated consideration of alternatives to resolution. One such alternative is to cast SAT as a HITTING SET problem and sum the number of variables that are *forced* to be set to 1 and the number that are forced to be set to 0; if this sum can be shown to be greater than n in polynomial time, a "short" proof of unsatisfiability follows. This idea has been applied to random (n, m, k)-CNF expressions and yields a hard-easy crossover no worse than $m = n^{k/2+o(1)}$, a considerable improvement over resolution results [52]. This is an example of how probabilistic analysis has driven the search for improved, alternative algorithms.

But it has been difficult for some to take probabilistic results seriously. The main drawbacks of relying on random (n, m, k)-CNF results for practical problems are (1) some distribution of input expressions must be assumed and a chosen distribution may not represent reality very well; (2) results are usually sensitive to the choice of distribution; (3) the state of analytical tools is such that distributions yielding to analysis are typically symmetric with independent components; and (4) few algorithms have yielded to analysis.

Despite these drawbacks, probabilistic results can be a useful supplement to worst-case results, which can be overly pessimistic for NP-complete problems, in understanding algorithmic behavior. One reason for this is that CNF expressions that are presented to a SAT solver are typically translated from another logic form that may even be a binarization of integer variable expressions. Such translations tend to smear or garble original domain-specific structural relationships and make the resulting CNF translation look somewhat like a random expression. We have noticed that this happens on some problems related to circuit verification, for example, which are very difficult for advanced SAT solvers such as Chaff.

Hence, in order to better understand the nature of expressions that are hard for current SAT solvers, it seems reasonable to study the relationship between hardness and random expressions. There has been much work on this subject in recent years (see, for example, [1] and [40] for a bibliography), most focusing on random (n, m, k)-CNF expressions. As m and n tend to infinity with limiting ratio $m/n \to \alpha$, probabilistic analysis and experimental results have provided evidence for the existence of a phase transition at some $\alpha = r_k$. That is, when in the limit $m/n < r_k$, the probability that a random (n, m, k)-CNF expression is satisfiable tends to 1, and when $m/n > r_k$, that probability tends to 0. Results show, for each fixed k, that there is a sharp threshold [47], with associated critical ratio $r_k \geq 2^{k-2}/k$ [23] (it is known that $\lim_{k\to\infty} r_k \to c \cdot 2^k$, c constant [4, 5], but the result of those papers is not close to the actual threshold when k is around 3). It has also been observed that random expressions become harder for SAT solvers when generated with values of m and n where the ratio m/n is close to r_k and easier when m/n is distant from r_k: the more distant being easier.

These results and observations have suggested a relationship between hardness and threshold. Further investigation has identified long "backbones," or chains of inferences, to be a good candidate for the underlying cause of the sharp thresholds and poor algorithm performance near the thresholds, because it appears to be the high density of well-separated "almost solutions" induced by the backbones that leads to thrashing in search algorithms [22]. In [77] and other articles it has been suggested that there is a strong connection between the "order" of threshold sharpness, that is, whether the transition is smooth or discontinuous, and hardness.

Recent advances [28, 30] have revealed the importance of minimal monotonic structures to the existence of sharp transitions. Those results have been used, for example, to show how limited most succinctly defined polynomial-time classes of SAT are. Notable examples of such classes are Horn [33, 58], renameable Horn [73], q-Horn [16, 15], extended Horn [18], SLUR [89], balanced [25], and matched [46], to name a few. These classes have been studied partly in the belief that they will yield some distinction between hard and easy problems. For example, in [16] a satisfiability index is presented such that a class with index greater than $1 + \varepsilon$, for any positive constant ε, is NP-complete, but the q-Horn class has satisfiability index 1. Thus, it seems that q-Horn is situated right at the point delineating hard and easy satisfiability problems. This hypothesis has been tested somewhat using m/n as a scale for determining the boundaries, in a probabilistic sense, of q-Horn and other classes: it has been found that a random (n, m, k)-CNF expression is q-Horn with probability tending to 0 if $m/n > 2/k(k-1)$ and that the probability that a random (n, m, k)-CNF expression is q-Horn is bounded away from 0 if $m/n < 1/k(k-1)$ [46]. Similar results have been obtained for other polynomial-time solvable classes. They illuminate the fact that most instances of such classes are satisfiable because their extent on the m/n scale is far below the r_k satisfiability threshold. Because their boundaries, in a probabilistic sense, are so distant from the threshold, all the polynomial-time classes mentioned above may be considered *extremely* easy, especially when compared to the good probabilistic performance shown for polynomial-time incomplete algorithms in the range $m/n < c_k \cdot 2^k/k$ [23].

But the limitations of these classes seem to be related to thresholds. Except for the matched class, the foregoing classes, including q-Horn, are "vulnerable" to cyclic clause structures, any one of which prevents an expression containing such a structure from being a member of the class. These structures have the recently discovered minimality and monotonic properties which are necessary for sharp thresholds and are defined in [28]. So, it seems that, to find challenging polynomial-time solvable classes, it is advisable to look for classes that are not so vulnerable: that is, those for which expressions cannot be excluded by adding certain minimal monotonic structural components. This is a case where probabilistic tools might prove useful in the development of broad succinctly definable polynomial-time SAT classes.

At this point, it is to be hoped that the reader is convinced that probabilistic analysis can be an important tool in understanding and coping with SAT expressions. In what follows we review some notable probabilistic results, the mathematical tools they are based on, and suggest how these results add to our intuition about SAT.

4.2 CNF Expressions

All results in this chapter apply to Boolean expressions in conjunctive normal form (CNF). Important elements of CNF expressions as well as the definition of CNF expressions are presented here for completeness.

Let v be a Boolean variable (in what follows we use *variable* to mean Boolean variable). Then v can be assigned a value from $\{0, 1\}$. A *positive literal* is a variable. A *negative literal* is the complement of a variable: that is, it takes value 1 if and only if its corresponding variable takes value 0. We use the symbol $\bar{}$ to represent the operation of complementing a literal: that is, if l is a literal, then $\bar{l}$ is its complement. Thus, if v is a positive literal, its complement is denoted by $\bar{v}$, and if $\bar{v}$ is a negative literal, its complement is $\bar{\bar{v}}$, which is also v. Whether a literal is complemented or not is referred to as its *polarity*.

A *clause* is a disjunction of literals and evaluates to 1 if and only if one of its literals has value 1. In this chapter we represent a clause as a set of literals. The *width* of a clause is the number of literals it contains. A clause of width 1 is called a *unit clause*.

A *CNF* expression is a set of clauses. An expression evaluates to 1 if and only if all its clauses evaluate to 1. We use C to represent a clause, ϕ to represent a CNF expression, and V to represent the set of variables existing in ϕ as positive or negative literals. If $\exists C \in \phi : l \in C$ and $\forall C \in \phi, \bar{l} \notin C$ then l is called a *pure literal*. An assignment of values to V is called a *truth assignment* or *assignment*. If T is an assignment, the notation $\phi(T)$ is used to mean the value of ϕ under T. If there exists an assignment M such that $\phi(M) = 1$, then M is called a *model* for ϕ and ϕ is said to have a solution. In this chapter we represent a model M as a set of variables: the value of all variables in M is considered to be 1, and the value of all other variables is considered to be 0.

We use the term k-CNF to mean any of a family of CNF expressions all of whose clauses have width k. It is well known that 2-CNF expressions are solved in polynomial time [38]. Therefore, we will only be interested in generating k-CNF expressions where $k \geq 3$.

Several generators of CNF expressions have been studied, but one has withstood the test of time and has received much more attention than the rest. Because of this, we focus attention on that generator in this chapter. This generator (from [45]) has three parameters, n, m, and k. An expression returned by this generator contains

m clauses generated uniformly, independently, and with replacement from the set of all elementary disjunctions of k literals on a set of n Boolean variables and their complements. Such an expression is referred to as a random (n, m, k)-CNF expression.

4.3 Basic Tools of Analysis

In this section we discuss some basic probabilistic tools that are commonly applied to Satisfiability algorithms.

4.3.1 First Moment Method

The first moment method is used to provide an upper bound on the probability that a specified property P holds on a class of random structures. Let X be a positive real-valued random variable. Then

$$Pr(X \geq 1) = \int_{t=1}^{\infty} p_X(t)dt \leq \int_{t=1}^{\infty} t p_X(t)dt \leq E\{X\},$$

where we have used p_X to denote the distribution density function of X and $E\{X\}$ to denote the mean of X. The foregoing is known as Markov's inequality. It says the mean of X is an upper bound on the probability that X takes value at least 1. Suppose X takes value at least 1 if and only if a random structure has property P and takes value 0 otherwise. Then the mean is an upper bound on the probability that a random structure has property P. The bound will be useful if the mean is small.

For example, let P be the property that there exists a model for a random (n, m, k)-CNF expression ϕ, and suppose we wish to find the probability that P holds in ϕ. Define X to be the *number* of models for ϕ. The probability that P holds is identical to the probability that $X \geq 1$. By Markov's inequality, this is bounded by $E\{X\}$. Index all possible assignments of variables to values (there are 2^n such assignments). Define indicator random variables $I_1, \ldots, I_{2^n}$ such that I_i has value 1 if and only if the ith assignment is a model for ϕ and has value 0 otherwise. For all i, the probability that $I_i = 1$ is $(1 - 2^{-k})^m$, since $(1 - 2^{-k})$ is the probability that a clause has value 1 under the ith assignment and clauses are constructed uniformly, independently, and with replacement. Then, since $X = I_1 + \cdots + I_{2^n}$,

$$E\{X\} = \sum_{i=1}^{2^n} E\{I_i\} = \sum_{i=1}^{2^n} Pr(I_i = 1) = 2^n(1 - 2^{-k})^m.$$

Thus, the probability that P holds is bounded from above by $2^n(1 - 2^{-k})^m$. This bound is quite useful if

$$m/n > -\ln(2)/\ln(1 - 2^{-k}), \tag{4.1}$$

for then $\lim_{n,m\to\infty}(2(1 - 2^{-k})^{m/n})^n \to 0$, and by the first moment method, a random $(n, m, 3)$-CNF expression has no model with probability tending to 1 (with increasing m and n) if $m/n > 5.19$.

4.3.2 Second Moment Method

The second moment method is used to prove that a specified property P holds on a class of random structures with high probability as the size parameter of the class tends to ∞. It is applied here in two major ways: to determine bounds, in probability, on the running time of a proposed algorithm for finding a model; and to determine a bound on the probability that a random expression is a member of a particular class of expressions and therefore exhibits certain properties that may be exploited by an algorithm for finding a model.

For a given random structure (in our case, a random (n, m, k)-CNF expression ϕ), a *witness* w is a substructure (in our case, a witness may be a set of clauses, $w \subseteq \phi$) whose presence implies that the structure has property P. Let W be the set of all possible witnesses for the class. The idea is to prove the probability that a randomly chosen structure *fails to contain any witness* tends to zero with increasing size parameter.

Let w also represent the *event* that $w \subseteq \phi$: which meaning is intended will be clear from context. Usually the set of witnesses W is chosen so its elements are *symmetric*: that is, for any pair $w, z \in W$, there is an automorphism of the probability space that maps w to z. We assume that this is the case. Thus, $p = Pr(w)$ is independent of w. Let I_w be the indicator random variable that has value 1 if event w occurs and 0 otherwise. Then,

$$E\{I_w\} = p \quad \text{and} \quad \text{var}(I_w) = E\{(I_w - p)^2\} = p(1 - p).$$

Define the random variable $X = \sum_{w \in W} I_w$, and let $\mu = E\{X\} = |W|p$ and $\sigma^2 = \text{var}(X)$. A special case of the Chebyshev inequality states that

$$Pr(X = 0) \leq \frac{\sigma^2}{\mu^2}. \tag{4.2}$$

Thus it suffices to show that this ratio tends to zero as the size parameter increases (in our case, as $n \to \infty$).

If the events w are independent, then $\sigma^2 = |W|p(1 - p) = O(\mu)$, and it becomes sufficient to show that $\mu \to \infty$ as $n \to \infty$. But the events w are usually not independent. In that case the crucial aspect of the second moment method is to show that, although the events w *are not* independent, the dependencies are weak enough that $\sigma^2 = o(\mu^2)$.

To analyze σ^2 in the case of random (n, m, k)-CNF expressions, we introduce the following notation:

$A(w)$ is the set of witnesses having some clause in common with w, other than w itself.

We can now state a basic lemma of the second moment method.

Lemma 4.1. (Alon and Spencer [10, Ch. 4.3, Cor. 3.5]) *If:*

(i) *the elements of W are symmetric,*
(ii) $\mu \to \infty$ *as* $n \to \infty$,
(iii) $\sum_{z \in A(w)} Pr(z|w) = o(\mu)$ *for an arbitrary* $w \in W$,

then $Pr(P) \to 1$ *as* $n \to \infty$. ■

Observe that bending the definition of "witness" slightly to mean a model[1] for ϕ and correspondingly letting X be the number of models for ϕ, as before, one obtains $\sigma^2 = (m^2/n)O(\mu^2)$ [19]. Because an "interesting" lower bound is no less than $m/n = 1$ from Page 120, the second moment method cannot be applied to obtain a meaningful lower bound on the probability that a random expression has a model when X is the number of models. However, recent work, presented in Section 4.6.3, shows how to achieve a very good bound using the second moment method.

4.3.3 Markovian Analysis and Approximation by Differential Equations

Probabilistic analysis was originally applied to search algorithms, most notably variants of DPLL, which is shown in Figure 4.1, to gain intuition about how an effective search heuristic should be designed. More recently, it has been used to determine lower bounds on the satisfiability threshold of random (n, m, k)-CNF expressions. The variants that we would like to analyze are completely deterministic, their heuristic variable setting choices depending only on the reduced CNF expression implied by the current assignment of values to variables. Moreover, the search process induced by any of the variants jumps between states, where each state represents a reduced CNF expression and the partial assignment of values that causes it.[2] Thus, a search process may be modeled as a Markovian process.

Given DPLL with some search heuristic, if one can calculate the probability that terminal states representing models are reached, then one can determine the probability that a random (n, m, k)-CNF expression has a model or even the probability that a model will be found in some given number of jumps. Currently, this task is too ambitious because of the large number of states in question. Hence, analysis is typically applied to some other, simpler algorithm for which the states of the corresponding search space are a tiny subset of the states of the search space for DPLL: if the probability that a model is found by the simpler algorithm tends to 1 (or even is bounded by a constant in case we are looking for thresholds – see later discussion), then the probability that DPLL finds a model also tends to 1.

[1] A model is merely a collection of unit clauses encompassing literals associated with all variables and a disjunction of all such representations of models can be added to ϕ without changing it. The resulting expression can easily be translated to a CNF expression.

[2] The reduced expression is the original expression minus the clauses containing at least one literal which has value 1 in the partial assignment and minus literals which have value 0 in the partial assignment. In Figure 4.1 the lines $\phi_{d+1} \leftarrow \{c - \{\bar{v}\} : c \in \phi_d, v \notin c\}$ create reduced expressions.

```
Procedure DPLL (φ)
Input: a CNF expression φ.
Output: either "unsatisfiable" or a model for φ.

begin
    d := 0;
    M_d := ∅;
    φ_d := φ;
    V_P := ∅.
    repeat the following until some statement outputs a value {
        if φ_d = ∅ then return M_d; // M_d is a model for φ
        if ∅ ∈ φ_d then {            // M_d falsifies a clause
            repeat the following until encountering a v that is "tagged" {
                if V_P = ∅ then return "unsatisfiable";
                pop v ← V_P;
                d := d − 1;
            }
            push V_P ← v̄;
            untag v;
            if v is a complemented literal then M_{d+1} := M_d ∪ {v̄};
            φ_{d+1} := {C − {v̄} : C ∈ φ_d, v ∉ C};
            d := d + 1;
        } else {
            if there exists a pure literal in φ_d then
                choose a pure literal v;
            else if there is a unit clause in φ_d then
                choose a literal v in a unit clause and "tag" v;
            else
                choose a literal v in φ_d and "tag" v;
            push V_P ← v;
            if v is an uncomplemented literal then M_{d+1} := M_d ∪ {v};
            φ_{d+1} := {C − {v̄} : C ∈ φ_d, v ∉ C};
            d := d + 1;
        }
    }
end;
```

Figure 4.1. DPLL algorithm for CNF expressions. If M_d is output, a model for ϕ has been found; otherwise no model exists for ϕ.

Whereas DPLL may unassign and reassign values to a variable many times, the algorithms that have best yielded to analysis on random (n, m, k)-CNF expressions iteratively choose a variable, assign a value to it, then never change that value up to termination of the algorithm. We call this class of algorithms *straight line* algorithms. All straight line algorithms are *incomplete*: that is, they may terminate without providing an answer. Although the number of states is reduced for straight line algorithms, the greatest reduction in the state space comes from coalescing into one state all states that have some attributes in common: for example, the clause counts of every possible clause width from 1 up to k, and the count of assigned variables. This particular coalescence is quite important, and later we refer to it as the *spectral coalescence of states*.

Now the state space is small enough to work with, but calculating state probabilities is trickier because, for a given search algorithm, the distribution of expressions at each coalesced state may depend on how that state was reached. Showing that such dependence does not exist, or is so weak that the results are not affected by it, is a crucial part of the probabilistic analysis of search algorithms. This requirement limits the type of algorithms that may be considered.[3] A general search algorithm typically introduces strong dependencies when reassigning values to variables. The dependence problem can be controlled to some extent by straight line algorithms, and this is why this class of algorithms is so widely considered.

Myopic Algorithms

Most performance results on random (n, m, k)-CNF expressions have been obtained for myopic algorithms. A straight line algorithm is called *myopic* [3] if, under the spectral coalescence of states, the distribution of expressions corresponding to a particular coalesced state can be expressed by its spectral components alone: that is, by the *number* of clauses of width i, for all $1 \leq i \leq k$, and the *number* of assigned variables. Thus, given random (n, m, k)-CNF expressions, the distribution of expressions for the coalesced "start" state is determined by the distribution of such expressions; and the distribution for the coalesced state corresponding to j assigned variables and $m_1, m_2, \ldots, m_k$ clauses of width $1, 2, \ldots, k$ is that of random $(m_1, n-j, 1)$-CNF, $(m_2, n-j, 2)$-CNF, $\ldots$, $(m_k, n-j, k)$-CNF expressions, respectively for each clause width.

To determine whether an algorithm is myopic, it is sufficient to show that no information about remaining clauses and literals, other than number, is revealed after assigning a value to a variable and eliminating clauses satisfied and literals falsified by that value. This is the case if variables are chosen at random and assigned values at random. In Section 4.6.1 more illuminating examples are given.

But some choices of variables will prevent an algorithm from being myopic. For example, if a pure literal is always chosen when one exists among remaining clauses, we will not have a myopic algorithm because the distribution of number of occurrences in clauses depends on whether a pure literal is chosen or not. To see this, consider the set of expressions containing 6 variables and 4 clauses, each of width 3. Label the variables $v_1, v_2, \ldots, v_6$. For each expression D, let

[3] Observe that the expression generator also has an effect on state dependence, so we are also limited in the types of generators chosen for analysis.

$p_D(e_1, e_2, e_3)$ be the probability that v_4, v_5, or v_6 is chosen first *and* the new expression resulting from the elimination of clauses (no literals are falsified since the chosen variable is a pure literal) is $\{\{v_1, v_2, v_3\}, \{\langle e_1, v_1\rangle, \langle e_2, v_2\rangle, \langle e_3, v_3\rangle\}, *\}$, where $\langle e, v\rangle$ means variable v occurs as an uncomplemented literal if $e = 1$ and as a complemented literal if $e = 0$, and $*$ means arbitrary. Let $\sigma(e_1, e_2, e_3) = \sum_{|D|=4} p_D(e_1, e_2, e_3)$. Obviously, $\sigma(e_1, e_2, e_3)$ should be independent of e_1, e_2, e_3 if choosing pure literals is myopic. More specifically, $\sigma(1, 1, 1)$ and $\sigma(0, 0, 0)$ should be equal. But, since v_1, v_2, nor v_3 can be pure in the latter case but cannot be pure in the former case, this requires that the pure literal selection heuristic ignore v_1, v_2, and v_3 and pick only from v_4, v_5, or v_6. Reversing the roles of v_1, v_2, v_3 and v_4, v_5, v_6, the pure literal selection heuristic would have to ignore v_4, v_5, and v_6. Then no literals can be chosen! Thus, any form of the pure literal heuristic[4] cannot be myopic. Another way to choose literals that leads to nonmyopic algorithms is via the well-known "Johnson heuristic," where clauses are weighted by the number of literals they contain and a literal is given a weight that is the sum of the weights of the clauses containing it.

In Section 4.6.1 some examples of myopic algorithms are discussed. What they have in common is that choosing variables can be based on the number of occurrences of literals in remaining clauses, but ties must be broken randomly.

Differential Equations to Approximate Discrete Processes

The idea of using differential equations to approximate discrete random processes goes back at least to [72], and its application to the analysis of algorithms goes back to [66]. Given an initial system of mk literal clauses taken from a set of n variables, the process we will approximate is the movement of clauses out of the system as they become satisfied and the movement of i literal clauses down to $i - 1$ literal clauses as literals are falsified during iterations of a straight line algorithm. To this end let $m_i(j)$ be the number of clauses containing i literals at the start of the jth iteration of a straight line algorithm. Initially, $m_k(1) = m$ and $m_i(1) = 0$ for $0 < i < k$. Observe that a coalesced state is represented as the vector $\langle j, m_1(j), m_2(j), \ldots, m_k(j)\rangle$.

The following theorem from [1] (based on Theorem 2 of [97]) is used to approximate clause flows by differential equations. Hypothesis (i) ensures that $m_i(j)$ does not change too quickly from iteration to iteration of an algorithm; hypothesis (ii) tells us what we expect the rate of change of $m_i(j)$ to be and involves functions that are calculated from a knowledge of what the algorithm is doing; and hypothesis (iii) ensures that this rate of change does not change too quickly. For a random variable X, it is said $X = o(f(n))$ *always* if $\max\{x : Pr(X = x) \neq 0\} = o(f(n))$. The term *uniformly* refers to the convergence implicit in the $o(.)$ terms. By function $f(u_1, \ldots, u_k)$ satisfies a Lipschitz condition on $D \subseteq \mathbb{R}^j$, it is meant

[4] For example, if some weighting scheme is applied to the set of pure literals existing in an expression. For more details on this, see [87].

that there exists a constant $L > 0$ such that $|f(u_1, \ldots, u_j) - f(v_1, \ldots, v_j)| \leq L \sum_{i=1}^{j} |u_i - v_i|$ for all $(u_1, \ldots, u_j)$ and $(v_1, \ldots, v_j)$ in D.

Theorem 4.2. *Let $m_i(j)$ be a sequence of real-valued random variables, $0 < i \leq k$ for some fixed k, such that for all i, all j, and all n, $|m_i(j)| \leq Bn$ for some constant B. Let $\mathbf{H}(j) = \langle\langle m_1(1), \ldots, m_k(1)\rangle, \ldots, \langle m_1(j), \ldots, m_k(j)\rangle\rangle$ be the state history of sequences.*

Let $I = \{\langle c_1, \ldots, c_k\rangle : Pr(\langle m_1(1), \ldots, m_k(1)\rangle = \langle c_1 n, \ldots, c_k n\rangle) \neq 0$ for some $n\}$.

Let D be some bounded connected open set containing the intersection of $\{\langle s, c_1, \ldots, c_k\rangle : s \geq 0\}$ with a neighborhood of $\{\langle 0, c_1, \ldots, c_k\rangle : \langle c_1, \ldots, c_k\rangle \in I\}$.

Let $f_i : \mathbb{R}^{k+1} \mapsto \mathbb{R}$, $0 < i \leq k$, and suppose that for some function $m = m(n)$

(i) *for all i and uniformly over all $j < m$*

$$Pr(|m_i(j+1) - m_i(j)| > n^{1/5} | \mathbf{H}(j)) = o(n^{-3}) \text{ always;}$$

(ii) *for all i and uniformly over all $j < m$*

$$E\{m_i(j+1) - m_i(j) | \mathbf{H}(j)\} = f_i(j/n, m_1(j)/n, \ldots, m_k(j)/n) + o(1) \text{ always;}$$

(iii) *for each i, the function f_i is continuous and satisfies a Lipschitz condition on D.*

Then

(a) *for $\langle 0, \hat{z}(1), \ldots, \hat{z}(k)\rangle \in D$ the system of differential equations*

$$\frac{dz_i}{ds} = f_i(s, z_1, \ldots, z_k), 0 < i \leq k$$

has a unique solution in D for $z_i : \mathbb{R} \mapsto \mathbb{R}$ passing through $z_i(0) = \hat{z}(i)$, $0 < i \leq k$, and which extends to points arbitrarily close to the boundary of D;

(b) *almost surely,*

$$m_i(j) = z_i(j/n) \cdot n + o(n),$$

uniformly for $0 \leq j \leq \min\{\sigma n, m\}$ and for each i, where $z_i(j)$ is the solution in (a) with $\hat{z}(i) = m_i(j)/n$, and $\sigma = \sigma(n)$ is the supremum of those s to which the solution can be extended.

How to Use the Solution to the Differential Equations

Theorem 4.2 is useful particularly because the differential equations are developed using *expectations* of clause counts and flows, which in many cases are relatively easy to compute. Moreover, these expectations in the discrete world are translated to actual flow and count values in the solution to corresponding differential equations. Thus, the solution found in Theorem 4.2(*b*) for $m_2(j)$, say, does not deviate significantly from $E\{m_2(j)\}$ asymptotically. This is significant because in that case the function $m_2(j)$ is often enough to predict where a unit-clause-based algorithm will be successful probabilistically, in some sense.

Theorem 4.3. (From [1].) *Let A be a myopic algorithm that always chooses to satisfy a unit clause, when one exists among nonsatisfied clauses. Let U_j be the event that on iteration j of A there are no empty or unit clauses existing among remaining nonsatisfied clauses. Suppose*

$$m_2(j) < (1-\delta)(n-j)$$

for all $1 \le j < (1-\varepsilon)n$, $0 < \varepsilon$ and $0 < \delta$ fixed, almost always. Then, there exists $\rho = \rho(\delta, \varepsilon)$, $\rho > 0$, such that $Pr(U_{(1-\varepsilon)n}) > \rho$.

Theorem 4.3 can be applied directly to the result of Theorem 4.2(b) but there are two mysteries that need some clarification first. For one thing, Theorem 4.3 only applies to $j < (1-\varepsilon)n$. This problem is disposed of on an ad-hoc basis. For example, in Section 4.6.1, the following is obtained for a particular myopic algorithm:

$$m_i(j) = \frac{1}{2^{k-i}}\binom{k}{i}\left(1-\frac{j}{n}\right)^i\left(\frac{j}{n}\right)^{k-i} m.$$

Then, using the binomial theorem,

$$\sum_{i=2}^{k} m_i(j) = \left(\left(1-\frac{j}{2n}\right)^k - \left(\frac{j}{2n}\right)^k - k\left(1-\frac{j}{n}\right)\left(\frac{j}{2n}\right)^{k-1}\right) m. \quad (4.3)$$

Let $m/n = 1/\lambda$ and suppose[5] $\lambda < 1$ and[6] $1/2^k < \lambda$. Set $\varepsilon = \lambda/\binom{k}{2}$. After substituting $1-\varepsilon$ for j/n on the right side of (4.3), expanding terms, and collecting powers of ε, it can be seen that (4.3) is bounded from above by $8\binom{k}{2}(\lambda/\binom{k}{2})^2 m/2^k = 8\lambda n/2^k\binom{k}{2} = (8/2^k)\varepsilon n \le \varepsilon n$. In other words, the number of width 2 and greater clauses remaining when $j = (1-\varepsilon)n$ is no greater than the number of unset variables, with high probability. By Theorem 4.3, there are no unit clauses or empty clauses remaining with bounded probability. But, assuming the algorithm is myopic and inputs are random (n, m, k)-CNF expressions, one may randomly remove all but two literals from each clause, resulting in a random $(n, m, 2)$-CNF expression. Such an expression is satisfiable with high probability and can be taken care of trivially.

The second mystery concerns the bound ρ, which is only guaranteed to be a constant and may not be close to 1. In the case where the analysis is used to

[5] The case $\lambda > 1$ is not interesting in this context, because the simple minded strategy of randomly removing all but two literals from every clause and applying a 2-CNF algorithm to the result succeeds with high probability in that case.

[6] Straight line algorithms for finding models will do poorly if $1/2^k \ge \lambda$, because almost all expressions have no models in that case.

determine a bound on the probability of the existence of a model, finding a constant bound is all that is needed to prove the probability tends to 1. This is discussed in Section 4.3.5.

4.3.4 Sharp Thresholds, Minimality, and Monotonicity

Let $X = \{x_1, \ldots, x_{n_e}\}$ be a set of n_e elements. Let A_X, a subset of the power set of X (denoted 2^X), be called a *property*. Call A_X a *monotone property* if for any $s \in A_X$, if $s \subset s'$, then $s' \in A_X$. Typically, A_X follows from a high-level description. For example, let $X = \mathcal{C}_{k,n}$ be the set of all nontautological, width k clauses that can be constructed from n variables. Thus $n_e = 2^k\binom{n}{k}$ and any $s \in 2^{\mathcal{C}_{k,n}}$ is a k-CNF expression. Let $\mathcal{UNSAT}_{\mathcal{C}_{k,n}}$ be the property that a k-CNF expression constructed from n variables has no model. That is, any $s \in \mathcal{UNSAT}_{\mathcal{C}_{k,n}}$ has no model and any $s \in 2^{\mathcal{C}_{k,n}} \setminus \mathcal{UNSAT}_{\mathcal{C}_{k,n}}$ has a model. If $s \in \mathcal{UNSAT}_{\mathcal{C}_{k,n}}$ and $c \in \mathcal{C}_{k,n}$ such that $c \notin s$, then $s \cup \{c\} \in \mathcal{UNSAT}_{\mathcal{C}_{k,n}}$ so the property $\mathcal{UNSAT}_{\mathcal{C}_{k,n}}$ is monotone.

An element $s \in A_X$ is said to be *minimal* if for all $s' \subset s$, $s' \in 2^X \setminus A_X$. For any $0 \le p \le 1$ and any monotone property $A_X \subset 2^X$, define

$$\mu_p(A_X) = \sum_{s \in A_X} p^{|s|}(1-p)^{n_e - |s|}$$

to be the probability that a random set has the monotone property. For the property $\mathcal{UNSAT}_{\mathcal{C}_{k,n}}$ (among others), s is a set of clauses; hence, this probability measure does not match that for what we call random k-CNF expressions but comes very close with $p = m/(2^k\binom{n}{k}) \approx m \cdot k!/(2n)^k$.

Observe that $\mu_p(A_X)$ is an increasing function of p. Let $p_c(X)$ denote that value of p for which $\mu_p(A_X) = c$. If for any small, positive ε, $\lim_{|X|\to\infty}(p_{1-\varepsilon}(X) - p_\varepsilon(X))/p_{1/2}(X) = 0$ then A_X is said to have a sharp threshold. If $\lim_{|X|\to\infty}(p_{1-\varepsilon}(X) - p_\varepsilon(X))/p_{1/2}(X) > 0$, then A_X is said to have a coarse threshold. The following criteria for sharp thresholds are found in [29] and are developed from [47].

Theorem 4.4. *Let A_X be a monotone property such that $\lim_{|X|\to\infty} p_{1/2}(X) \to 0$. If the following two conditions are satisfied, then A_X has a sharp threshold.*

(i) *For all $0 < c < 1$ and for all positive integers λ,*

$$\lim_{|X|\to\infty} \mu_{p_c(X)}(s' \subseteq s,\ s' \text{ is minimal for } A_X,\ \text{and } |s'| \le \lambda) \to 0.$$

(ii) *For all $0 < c < 1$, for all positive integers λ, and for all $s^* \in 2^X \setminus A_X$ with $|s^*| = \lambda$,*

$$\lim_{|X|\to\infty} \mu_{p_c(X)}(s \in A_X, s \setminus s^* \in 2^X \setminus A_X | s^* \subseteq s) \to 0.$$

The first condition of Theorem 4.4 says that elements of A_X of bounded size have negligible probability of appearing. The second condition says that the probability that a given s is in A_X is not affected much by conditioning on an element of bounded size that is not in A_X. But there are better conditions for sharp thresholds when one is dealing with k-CNF. Let $A_{\mathcal{C}_{k,n}}$ be a monotone property on width k CNF expressions in the sense that if s is a set of clauses verifying such a property, then so does any set s' of clauses containing s. Let E be a set of values.[7]

Theorem 4.5. (Adaptation from [30].) *If the following three conditions are satisfied, then the monotone property $A_{\mathcal{C}_{k,n}}$ has a sharp threshold.*

(i) *For all $0 < c < 1$, $p_c(\mathcal{C}_{k,n}) = O(n^{1-k})$.*
(ii) *For all s minimal for $A_{\mathcal{C}_{k,n}}$, the number of variables of s is no greater than $(k-1) \cdot |s| - 1$.*
(iii) *For all $0 < c < 1$, for all t, for all $(\delta_1, \ldots, \delta_t) \in E^t$, and all $\gamma > 0$,*

$$\mu_{p_c(\mathcal{C}_{k,n})}(s \in 2^{\mathcal{C}_{k,n}} \setminus A^{\delta}_{\mathcal{C}_{k,n}}, |C^{\delta}_s| \geq \gamma \cdot n^{k-1}) \to 0,$$

where $A^{\delta}_{\mathcal{C}_{k,n}}$ denotes the property $A_{\mathcal{C}_{k,n}}$ with the assignment $v_1 = \delta_1, \ldots, v_t = \delta_t$, and C^{δ}_s denotes the set of clauses C having at least one variable in $\{v_1, \ldots, v_t\}$ and is such that $s \cup C \in A^{\delta}_{\mathcal{C}_{k,n}}$.

Theorem 4.5 may be used in a variety of ways depending on the monotone property $A_{\mathcal{C}_{k,n}}$. For example, the property that random (n, m, k)-CNF expressions are members of some polynomial time solvable class can be shown to have a sharp threshold for a number of polynomial time solvable classes. Two such illustrations are given in [31] for the classes of hidden Horn and of q-Horn expressions.[8]

4.3.5 Lifting Constant Probability Bounds to Almost Surely Bounds

If a straight line algorithm is shown to find a model for a random expression with constant probability for a range of densities, this implies that a random expression has a model, almost surely, for that range of densities. The appropriate theorem is the following [47].

Theorem 4.6. *Let $p_k(m, n)$ denote the probability that a random (n, m, k)-CNF expression has a model. For every $k \geq 2$ there exists a sequence $r_k(n)$ such that for any $0 < \varepsilon$,*

$$\lim_{n\to\infty} p_k((r_k(n) - \varepsilon)n, n) = 1 \quad \text{and} \quad \lim_{n\to\infty} p_k((r_k(n) + \varepsilon)n, n) = 0.$$

Theorem 4.6 says that for each k there is a sharp threshold $r_k(n)$ at some density for every n such that a random expression almost surely has a model below the threshold and almost surely does not have a model above it. It follows that proving a model exists with probability bounded from below by a constant for a range of densities is sufficient to imply that a model exists almost surely for the same range of densities.

[7] In some applications $E = \{0, 1\}$, in some E has more than two elements.
[8] The monotone property $A_{\mathcal{C}_{k,n}}$ has to do with the emergence of structures that cause an expression *not* to be q-Horn or hidden Horn. See Section 4.4 for class definitions and Section 4.6.6 for some results and their meaning.

4.4 Some Efficiently Solvable Classes of CNF Expressions

All members of some easily identified and in several cases well-known, classes of CNF expressions can be solved efficiently. For these it is natural to ask how often one encounters such an expression, whether expressions of a class can be recognized efficiently, or whether they even need to be. Some of these classes are incomparable, meaning, for each of a pair of classes, there exists an expression that is a member of one but not the other. Nevertheless, probabilistic analysis can reveal some interesting properties of expressions in these classes, such as:

(i) What weakness does the class possess? That is, what critically distinguishes the class from more difficult classes?
(ii) Is one class much larger than another incomparable class?

In this section a few examples of polynomial time solvable classes are defined. In Section 4.6.6 probabilistic results on these classes will be reviewed. We omit discussion of a large number of polynomial time solvable classes such as 2-CNF, extended Horn [18], simple extended Horn [92], CC-balanced expressions [25], renameable Horn [73], and many others, because we are interested here in how to use probability to make some statement about the relative size of such classes.

Definition 4.1. *A CNF expression is* Horn *if every clause it contains has at most one uncomplemented literal. A CNF expression is* hidden Horn *if reversing the polarity of the literals associated with some subset of variables (called a* switch set*) causes it to be Horn.*

This class is widely studied, in part because of its close association with logic programming. Namely, a Horn clause $\{\bar{v}_1, \bar{v}_2, \ldots, \bar{v}_i, y\}$ is equivalent to the rule $v_1 \wedge v_2 \wedge \ldots \wedge v_i \rightarrow y$ or the implication $v_1 \rightarrow v_2 \rightarrow \cdots \rightarrow v_i \rightarrow y$ (where association is from right to left). However, the notion of causality is generally lost when translating from rules to Horn sets. Horn sets can be solved in linear time using unit resolution [33, 58, 91] (see algorithm UR, Figure 4.2). An important

Procedure UR(ϕ, M): `// Applies unit resolution until exhaustion`
Input: a CNF expression ϕ and partial assignment M.
Output: a CNF expression and an updated M.

begin
  repeat the following {
    let $\{l\} \in \phi$ be a unit clause;
    if l is an uncomplemented literal then $M := M \cup \{l\}$;
    $\phi := \{C - \{complement(l)\} : C \in \phi, l \notin C\}$;
  } while $\emptyset \notin \phi$ and there is a unit clauses in ϕ;
  return (ϕ);
end;

Figure 4.2. Unit resolution algorithm.

Procedure SLUR(ϕ):
Input: a CNF expression ϕ.
Output: a model M for ϕ, if one exists, or "unsatisfiable," or "give up."

```
begin
    M := ∅;
    φ := UR(φ, M);
    if ∅ ∈ φ then return ("unsatisfiable");
    level := 0;
    repeat the following while φ ≠ ∅ {
        choose arbitrarily a variable v ∈ φ;
        M1 := M;
        M2 := M
        φ1 := UR({C − {v} : C ∈ φ, v̄ ∉ C}, M1)
        φ2 := UR({C − {v̄} : C ∈ φ, v ∉ C}, M2)
        if ∅ ∈ φ1 and ∅ ∈ φ2 {
            if level = 0 then return ("unsatisfiable");
            else return ("give up");
        } else {
            arbitrarily choose i so that ∅ ∉ φi;
            φ := φi;
            M := Mi;
            if i = 2 then M := M ∪ {x}
        }
        level := 1;
    }
    return (M);
end;
```

Figure 4.3. SLUR algorithm.

property of Horn sets is that a model, if one exists, that is minimal in the number of variables set to 1 is a *unique* minimum model with respect to 1 (this is referred to later as the *minimal model*). This property has important implications in, among other things, efficient algorithms for some other classes of CNF expressions.

Definition 4.2. *A CNF expression is a member of the class* SLUR *if, for all possible sequences of selected variables v, the* SLUR *algorithm of Figure 4.3 does not give up on that expression.*

This class was developed in [89] as a generalization of other classes including Horn, extended Horn [18], simple extended Horn [92], and CC-balanced expressions [25]. It is peculiar in that it is defined based on an algorithm rather than on properties of expressions. The algorithm of Figure 4.3, called SLUR for Single Lookahead Unit Resolution, selects variables sequentially and arbitrarily and considers a one-level lookahead, under unit resolution, of both possible values that the selected variable can take. Observe that due to the definition of this class, the question of class recognition is avoided.

The origin of the class called q-Horn is [16, 15], but it is also a special case of work considered in [94, 95]. Represent a CNF expression $\phi = \{c_1, c_2, \ldots, c_m\}$ as a $(0, \pm 1)$ matrix A^{ϕ}, columns indexed on variables, rows indexed on clauses, and such that $A^{\phi}_{i,j} = 1$ if literal $v_j \in c_i$, $A^{\phi}_{i,j} = -1$ if literal $\bar{v}_j \in c_i$ and $A^{\phi}_{i,j} = 0$ if $v_j \notin c_i$ and $\bar{v}_j \notin c_i$.

Definition 4.3. *Suppose some of the columns of A^{ϕ} can be scaled by -1 and the rows and columns permuted so that there is a partition of the resulting matrix into quadrants*

$$\left(\frac{A^1 \mid E}{D \mid A^2} \right)$$

where submatrix A^1 has at most one $+1$ entry per row, submatrix D contains only -1 or 0 entries, submatrix A^2 has no more than two nonzero entries per row, and E has only 0 entries. Then ϕ is q-Horn.

Such a partition, if it exists, can be found in linear time. To solve a q-Horn expression, perform the partition, solve A^1 (a Horn expression), cancel rows in D and A^2 that are satisfied by the minimal model for A^1 (or return "unsatisfiable" if there is no model), solve A^2 less the canceled rows (a 2-CNF expression), and return models for A^1 and A^2 (or "unsatisfiable" if A^2 has no model).

Expressions in the class q-Horn had been thought to be close to what might be regarded as the largest easily definable class of polynomially solvable propositional expressions due to results in [16]. Let $\{v_1, v_2, \ldots, v_n\}$ be a set of variables, and P_k and N_k, $P_k \cap N_k = \emptyset$, be subsets of $\{1, 2, \ldots, n\}$ such that the kth clause in ϕ is given by $\vee_{i \in P_k} v_i \vee_{i \in N_k} \bar{v}_i$. Construct the following system of inequalities:

$$\sum_{i \in P_k} \alpha_i + \sum_{i \in N_k} (1 - \alpha_i) \leq Z, \quad (k = 1, 2, \ldots, m), \text{ and} \tag{4.4}$$

$$0 \leq \alpha_i \leq 1 \quad (i = 1, 2, \ldots, n), \tag{4.5}$$

where $Z \in R^+$. The minimum value of Z that satisfies (4.4) and (4.5) is the *satisfiability index* for ϕ. If ϕ's satisfiability index is no greater than 1, then it is q-Horn. However, the class of all expressions with a satisfiability index greater than $1 + 1/n^{\varepsilon}$, for any fixed $\varepsilon < 1$, is NP-complete.

The last class of this section is called the Matched class [46].[9] For a given expression ϕ, define G^{ϕ} to be an undirected bipartite graph with vertex sets C^{ϕ}, whose elements are the clauses of ϕ, and V^{ϕ}, whose elements are the variables of ϕ, and edge set $E^{\phi} = \{\langle v, c \rangle : v \in V^{\phi}, c \in C^{\phi},$ and variable v is in clause c$\}$. A matching in G^{ϕ} is a disjoint subset of edges $B \subset E^{\phi}$. A maximum matching

9 The observation that a total matching implies a model was credited to Adam Rosenberg by Tovey [93].

in G^ϕ is a matching in G^ϕ containing the maximum possible number of edges. A total matching in G^ϕ is a matching in G^ϕ where every $c \in C^\phi$ is in some edge $e \in B$.

Definition 4.4. *A CNF expression ϕ is a member of the Matched class if G^ϕ has a total matching.*

It is well known that, due to an augmenting path algorithm of Edmonds, a maximum matching can be found for bipartite graphs in polynomial time. This implies that a total matching can be found for G^ϕ in polynomial time, if one exists.

It is straightforward to show that the SLUR, q-Horn, and Matched classes are incomparable.

4.5 A Digest of Probabilistic Results

Goldberg was among the first to investigate the frequency with which DPLL algorithms return a model quickly or return a short proof that no model exists. He provided an average-case analysis of a variant of DPLL that does not handle pure literals or unit clauses [53]. The analysis was based on a different parameterized distribution that generates random (n, m, p)-CNF expressions.

Definition 4.5. *A random (n, m, p)-CNF expression contains m clauses, each generated independently according to a random process described as follows. Let V be a set of n variables and let $0 \leq p \leq 1/2$. For every variable $v \in V$, independently and uniformly do one of the following: add literal v to clause C with probability p, add literal $\bar{v}$ to C with probability p, and add neither v nor $\bar{v}$ to C with probability $1 - 2p$.*

Goldberg showed that, for random (n, m, p)-CNF expressions, the DPLL variant has average complexity for bounded from above by $O(m^{-1/\log(p)}n)$ for any fixed $0 < p < 1$. This includes the "unbiased" sample space when $p = 1/3$ and all expressions are equally likely. Later work [54] showed the same average-case complexity even if pure literals are handled as well. The scientific and engineering communities are often interested in refutation proofs, but Goldberg made no mention of the frequency of unsatisfiable random (n, m, p)-CNF expressions over the parameter space of that distribution.

However, Franco and Paull [45] pointed out that large sets of random (n, m, p)-CNF expressions, for fixed $0 < p < 1/2$, are dominated by easily *satisfiable* expressions: that is, a random assignment of values to the variables of a random expression is a model for that expression with high probability. This result is refined somewhat in [41], where it is shown that a random assignment is a model for a random (n, m, p)-CNF expression with high probability if $p > \ln(m)/n$ and a random (n, m, p)-CNF expression is unsatisfiable with high probability if $p < \ln(m)/2n$. In the latter case, a "proof" of unsatisfiability is trivially found with high probability because a random (n, m, k)-CNF expression for this range of p

usually contains at least one empty clause, which can easily be located, and implies unsatisfiability. The case $p = c\ln(m)/n$, $1/2 \le c \le 1$, was looked at in [44], where it was shown that a random (n, m, p)-CNF expression is satisfiable with high probability if $\lim_{n,m\to\infty} m^{1-c}/n^{1-\varepsilon} < \infty$, for any $0 < \varepsilon < 1$. These results show two things regarding random (n, m, p)-CNF expressions: (1) there seems to be some threshold, probably sharp, for the probability that a random expression is satisfiable, and (2) expressions generated on both sides of the threshold are usually trivially solved because they either contain empty clauses or can be satisfied by a random assignment. In other words, only a small region of the parameter space is capable of not being dominated by trivially solved satisfiable or unsatisfiable expressions: namely, when the average clause width is $c\ln(m)/n$, $1/2 \le c \le 1$. Because of this, random (n, m, p)-CNF generators are not considered interesting by many.

Nevertheless, there has been significant interest in Goldberg's distribution, and the results of some of this work are summarized in the two-dimensional chart of Figure 4.4, which partitions the entire parameter space of that distribution: the vertical axis measures the average clause width $(p \cdot n)$ and the horizontal axis measures the density (m/n). Each result is presented as a line through the chart with a perpendicular arrow. Each line is a boundary for the algorithm labeling the line, and the arrow indicates that the algorithm has polynomial average time performance in that region of the parameter space that is on the arrow side of the line (constant factors are ignored for simplicity). Goldberg's result appears in the upper right corner, labeled **Goldberg**, occupying only a boundary of the parameter space (hence no arrow) and shows that the algorithm analyzed by Goldberg has polynomial average time performance if p is a constant. Iwama's counting algorithm [59], labeled **Counting**, does better. If clauses are resolved only if the pivot appears in two (remaining) clauses, then a null clause verifying unsatisfiablity will be obtained in polynomial average time below the lines labeled **Limited Resolution** [43]. If pure literal reductions are added to the algorithm analyzed by Goldberg, then polynomial average time is achieved below the line labeled **Pure Literals** as well as above the **Goldberg** line [84]. But an improved result, shown by the lines labeled **Unit Clauses**, is obtained merely by repeatedly satisfying unit clauses until a null clause is generated (or the algorithm gives up) [42]. Results for Search Rearrangement Backtracking [85] are disappointing (shown bounded by the two lines labeled **Backtracking**), but a slight variant, always choosing variables that are in a positive clause (if there is no positive clause, satisfiability is determined by setting the remaining variables to 0), is fantastic, as is shown by the line labeled **Probe Order Backtracking** [83]. Observe that the combination of probe order backtracking and unit clauses yield polynomial average time almost everywhere for random (n, m, p)-CNF expressions. Now we turn our attention to the possibly more robust random (n, m, k)-CNF expressions.

Franco and Paull in [45] (see [82] for corrections) also considered the probabilistic performance of Goldberg's variant of DPLL for random (n, m, k)-CNF expressions. They showed that for all $k \ge 3$ and every fixed $m/n > 0$, with

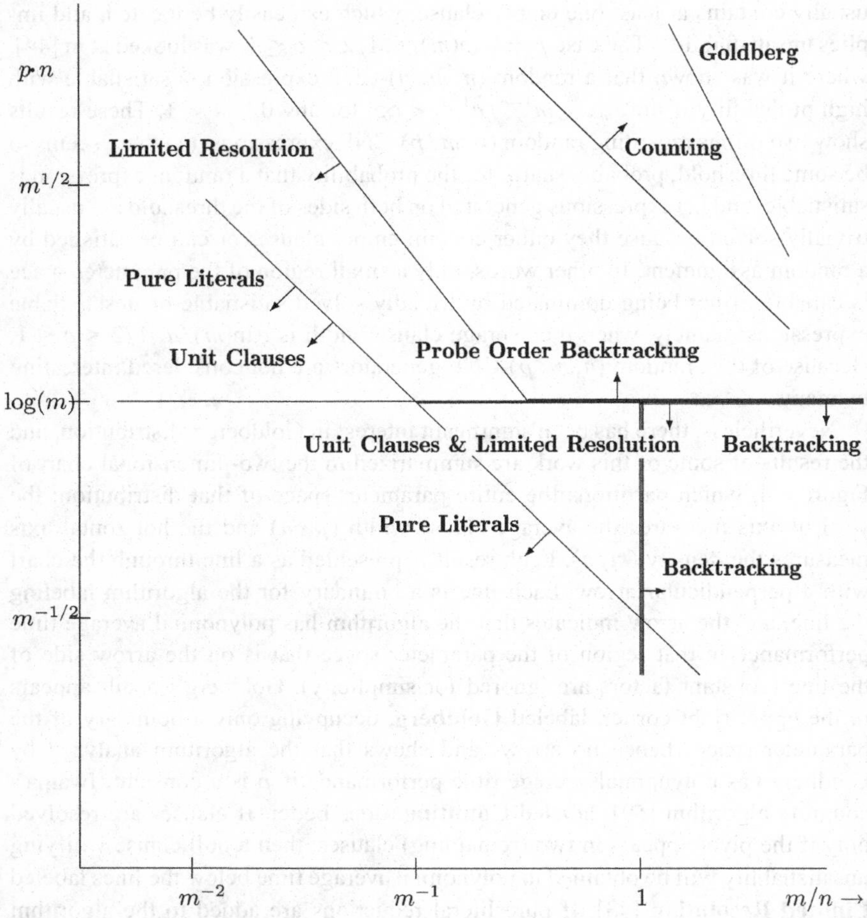

Figure 4.4. The parameter space of Goldberg's distribution partitioned by polynomial average time solvability. Pick a point in the parameter space. Locate the lines with names of algorithms and arrows on the side of the line facing the chosen point. Random formulas generated with parameters set at the specified point are solved in polynomial average time by the named algorithms.

probability $1 - o(1)$, the variant takes an *exponential* number of steps to report a result: that is, either to report all ("cylinders" of) models, or that no model exists. The first upper bound on r_k, namely the smallest value of m/n such that the expected number of (n, m, k)-CNF models tends to 0, was also presented in that paper: the probability that a random expression is unsatisfiable is $1 - o(1)$ if $m/n > -\ln(2)/\ln(1 - 2^{-k})$ (the result of Section 4.3.1, which is approximately $0.69 \cdot 2^k$).

Later, in a series of two papers [20, 21], Chao and Franco presented some useful insights that influenced the lower bound probabilistic analysis of Satisfiability and

Procedure SCA (ϕ, s):
Input: a CNF expression ϕ, integer s.
Output: either a model for ϕ or "give up".

```
begin
    M := ∅;
    repeat the following until some statement outputs a value {
        if φ = ∅ then return M;
        if ∅ ∈ φ then return "gives up";
        for i := 0 to k do C_i := {C : C ∈ φ and |C| = i};
        q := min{i : C_i ≠ ∅};
        L_1 := {l : ∃c ∈ C_q such that l ∈ c};
        L_2 := {l : ∃c ∈ φ such that l ∈ c};
        if q ≤ s then choose l randomly from L_1;
        else choose l randomly from L_2;
        if l is an uncomplemented literal then M := M ∪ {l};
        φ := {c − {l̄} : c ∈ φ and l ∉ c};
    }
end;
```

Figure 4.5. Smallest Clause Algorithm for CNF expressions.

Coloring algorithms in subsequent years (for example, [2, 6, 7, 8, 23, 48]). Unlike upper bounds, which are probabilistic counting arguments, they produced lower bounds for r_k that are algorithmic. The algorithms they considered are shown in Figure 4.5 as a single algorithm, called SCA, with one parameter s. SCA performs exactly like DPLL until either a model is found or the first backtrack is attempted, and, in the latter case, the algorithm gives up. Thus, a positive result for algorithm SCA is also a positive result for DPLL. The analysis of this algorithm was intended primarily to determine conditions under which efficient performance was likely and secondarily to determine a lower bound for r_k. It was based on considering the "flow" of clauses between levels of clause sets, where level i consists of all clauses of i literals at any iteration j of SCA. The technique of flow analysis is discussed in detail in Section 4.6.1.

Using flows, in [20] it is shown that the UNIT CLAUSE (UC) algorithm (Algorithm SCA, Figure 4.5 with $s = 1$) has positive probability of finding a model for random (n, m, k)-CNF expressions when $m/n < 8/3 = 2.66..$ and, when combined with a "majority" rule, for $m/n < 2.9$. In [21] the GENERALIZED UNIT CLAUSE (GUC) algorithm (Algorithm SCA with $s = k$) is shown to find a model with bounded probability when $m/n < (0.77)2^k((k-1)/(k-2))^{k-2}/(k+1)$ and $4 \le k \le 40$ and with probability $1 - o(1)$ when $m/n < (0.46)2^k((k-1)/(k-2))^{k-2}/(k+1)$ and $4 \le k \le 40$. This was improved by Chvátal and Reed [23], who showed that the SHORTEST CLAUSE (SC) algorithm (Algorithm SCA with $s = 2$) finds a model with probability $1 - o(1)$ when $m/n < (0.125)2^k((k-1)/(k-3))^{k-3}/k$ and $3 \le k$. Observe that, combined with previous results, this tells us random

(n, m, k)-CNF expressions are nearly always satisfiable if $m/n < c_1 2^k/k$ and nearly always unsatisfiable if $m/n > c_2 2^k$ for some positive constants c_1 and c_2 and sufficiently large n.

According to a result of Friedgut (see Theorem 4.6), there is a sharp satisfiability threshold r_k for every $k \geq 3$, and we know from the preceding results that it must be no greater than some constant times 2^k, but no less than some constant times $2^k/k$. But where, precisely, is it? The answer is interesting for various reasons, including that it may provide some insight about the performance of DPLL variants. These variants seem to do well up to m/n about equal to $2^k/k$, but after that, for fixed m/n, their performance seems to suffer. Thus, a threshold result of order 2^k suggests a rather "hard" region for DPLL algorithms. The question has been essentially solved recently. Before this, nearly all concerned researchers had believed the threshold is of order 2^k, and for this reason it became known in 1999 as the "*Why* 2^k?" problem. In 2002 Achlioptas and Moore [4] showed that nearly everyone was right, using the second moment method on a variant of Satisfiability that bounds it in some way. Details of the analysis are instructive and are presented in Section 4.6.3.

The most active research on that portion of the parameter space where expressions with models are numerous has considered the cases $k = 2$ and $k = 3$. Because 2-CNF expressions are polynomial time solvable, for $k = 2$ the issue can only be whether the threshold exists and, if so, where is it. Chvátal and Reed [23], Goerdt [51], and Fernandez de la Vega [39] independently answered these questions: they determined $r_2 = 1$. It is important to observe that 2-CNF expressions being solvable in polynomial time [26] means that there is a *simple* characterization of unsatisfiable 2-CNF expressions. Indeed, both [23] and [51] make full use of this characterization as they proceed by focusing on the emergence of the "most likely" unsatisfiable random (n, m, k)-CNF expressions. Also using this characterization, Bollobás et al. [14] completely determined the "scaling window" for random $(n, m, 2)$-CNF expressions, showing that the transition from satisfiability to unsatisfiability occurs for $m = n + \lambda n^{2/3}$ as λ goes from $-\infty$ to $+\infty$.

For $k = 3$, less progress has been made: the value of r_3 has not been established, although ever-improving bounds for it have progressively been reported. Because $r_2 = 1, 1 \leq r_3$ follows trivially. Broder, Frieze, and Upfal [17] proved that the pure literal heuristic alone, with no backtracking, almost always sets all the variables for $m/n \leq 1.63$. They used Martingales to offset distribution dependency problems as pure literals are uncovered. (Martingales are not discussed here, because more recent techniques appear to be more powerful.) Mitzenmacher [76] showed that this bound is tight for the pure literal heuristic. Frieze and Suen [48] determined the probability of success of algorithms SC and GUC. In particular, they showed that for $m/n < 3.003\ldots$, both heuristics succeed with positive probability. Moreover, they proved that a modified version of GUC, which performs a very limited form of backtracking, succeeds almost surely for such values of m/n. Years later, Achlioptas [2] showed, using a flow analysis, that for $m/n \leq 3.145$, a random (n, m, k)-CNF expression is satisfiable with probability $1 - o(1)$ by changing SC slightly to choose two literals at a time coming from a clause with two unfalsified literals remaining. In [3], Achlioptas and Sorkin take this approach to the limit with the discovery of an algorithm that almost surely finds a model if $m/n < 3.26$. By "take to the limit" we mean that no other myopic algorithm for Satisfiability can perform better probabilistically. However, Kaporis et al. [63] found a workaround to this "barrier" and analyzed a simple nonmyopic greedy algorithm for Satisfiability. A description is outlined in Section 4.6.2. The algorithm controls the dependence problem by considering a different generator of expressions such that probabilistic results also hold for random $(n, m, 3)$-CNF expressions. Their result is that there exists an algorithm for Satisfiability that almost always finds a model when $m/n < 3.42$. With this analysis machinery in place, and considering the results of numerous experiments, it will not be long before better results are reported. Because an analysis tends to direct algorithm development, it is hoped that future probabilistic results will reveal new generally useful search heuristics, or at least explain the mechanics of existing search heuristics.

Algorithmic methods, applied successfully to DPLL on the satisfiable side, do not help in finding good upper bounds on r_k. Instead, successful results, based on counting, are due to the application of the first and second moment methods. It was stated earlier that a simple bound of $r_k < -\ln(2)/\ln(1 - 2^{-k})$ is obtained from the first moment method applied to the number of models of a random expression, later referred to as N_M, and the second moment method on N_M does not give a better bound because the variance is too large. However, the first moment method can yield better results if the count being measured is changed. For example, from the foregoing, $r_3 < 5.19$ but in [62] the count is allowed to be reduced because of the observation that, with high probability, a large number of variables have no effect on models. Taking the expectation of N_M assuming an appropriately reduced number of variables leads to $r_3 < 4.762$. A weaker bound using the same idea, but on a slight variation of the generation of random 3-CNF expressions, was obtained independently in [37]. But a more clever idea, introduced independently in [67] and [34], is based on identifying, for a given expression ϕ, a set of "critical" models that is far smaller than the complete set of models of ϕ. The expected number of critical models is a tighter upper bound on the probability that an expression is satisfiable than the expected number of models, and it can be found with some cleverness as follows.

Call a truth assignment a *critical model* for CNF expression ϕ if it is a model for ϕ and flipping exactly the value of one 0-valued variable to 1 results in an assignment that is not a model for ϕ. A critical model can be obtained from any model by flipping 0 values to 1 until such flips can only result in an assignment that is not a model. Therefore, every satisfiable expression has at least one critical model, and a bound on the expected number of critical models is a bound on the probability that there exists a model. Given a critical model M and a specific flip, there are $\binom{n}{k}$ possible clauses that are excluded from ϕ, and there is at least one of $\binom{n-1}{k-1}$ possible clauses containing the flipped variable that must be included to cause ϕ to be falsified when that variable is flipped. Hence, the probability that an assignment is a model is $(1 - 2^{-k})^m$, and the probability that a clause is the one that causes ϕ to be falsified given ϕ is satisfied by M is $\binom{n-1}{k-1}/(2^k-1)\binom{n}{k}$, which is approximately $k/((2^k-1)n)$. From the latter, the probability that M is critical but a given single flip results in not satisfying ϕ is $1-(1-k/((2^k-1)n))^m$. If s is the number of possible single flips for M, then the probability that M is critical given it is a model for ϕ is bounded from above[10] by $(1-(1-k/((2^k-1)n))^m)^s$. The probability that M is critical given s flips are possible is $(1-2^{-k})^m(1-(1-k/((2^k-1)n))^m)^s$, and this is also the expected value of the indicator variable that takes value 1 if and only if the corresponding M is critical. The expected number of critical models is then

$$(1-2^{-k})^m \sum_{s=0}^{n} \binom{n}{s} (1-(1-k/((2^k-1)n))^m)^s$$
$$= (1-2^{-k})^m (2-(1-k/((2^k-1)n))^m)^n.$$

Setting this equal to 1 provides the bound (approximately)

$$r_k < \frac{\ln(2-e^{-(k/(2^k-1))r_k})}{\ln(1-2^{-k})}. \tag{4.6}$$

Comparing inequalities (4.1) and (4.6), we see that the difference is only in the exponential term. For $k=3$, the upper bound given by inequality (4.6) is $r_3 < 4.667$. However, a generalization of this method results in a bound $r_3 < 4.601$. In [36], a still further improvement to $r_3 < 4.506$ is reported. Experimental results suggest that $r_3 = 4.26$. For more information on this area of research, the reader is referred to [35].

Returning to algorithms, pessimistic results have been reported for resolution-based algorithms, including all DPLL variants, in the case of random unsatisfiable expressions. Resolution is a general procedure used primarily to certify that a given CNF expression has no model. The idea predates the often cited work reported in [86]. Let c_1 and c_2 be clauses such that there is exactly one variable v that occurs negated in one clause and unnegated in the other. Then, the *resolvent* of c_1 and c_2, denoted by $\mathcal{R}^{c_1}_{c_2}$, is a clause that contains all the literals of c_1 and all the literals of c_2 except for v and its complement. That is, $\mathcal{R}^{c_1}_{c_2} = \{l : l \in c_1 \cup c_2 \setminus \{v, \bar{v}\}\}$. The variable v is called a *pivot* variable.

The usefulness of resolvents derives from the following lemma, which is straightforward to prove.

Lemma 4.7. *Let ϕ be a CNF expression. Suppose there exists a pair $c_1, c_2 \in \phi$ of clauses such that $\mathcal{R}^{c_1}_{c_2} \notin \phi$ exists. Then the CNF expression $\phi \cup \{\mathcal{R}^{c_1}_{c_2}\}$ is functionally equivalent to ϕ.*

A resolution algorithm for CNF expressions, making use of Lemma 4.7 and the fact that a clause containing no literals cannot be satisfied by any truth assignment, is presented in Figure 4.6. If the algorithm outputs "unsatisfiable," then the set of

[10] A few details concerning dependence must be added here to verify the bound.

Procedure RESOLUTION (ϕ)
Input: a CNF expression ϕ.
Output: either "unsatisfiable" or a model for ϕ.

```
begin
  M1 := ∅;
  M2 := ∅;
  repeat the following until some statement below outputs a value {
    if ∅ ∈ φ then return "unsatisfiable";
    if there are two clauses C1, C2 ∈ φ such that R^{C1}_{C2} ∉ φ exists then
      φ := φ ∪ {R^{C1}_{C2}};
    else {
      repeat the following while there is a pure literal l ∈ φ {
        if l is an uncomplemented literal then M1 := M1 ∪ {l};
        φ := {C : C ∈ φ, l ∉ C};
      }
      repeat the following while there is an uncomplemented clause C ∈ φ {
        choose variable v ∈ C;
        reverse the polarity of all occurrences of v and v̄ in φ;
        M2 := M2 ∪ {v};
      }
      M := M1 ∪ M2;
      return M;
    }
  }
end;
```

Figure 4.6. Resolution algorithm for CNF expressions.

all resolvents generated by the algorithm is a *resolution refutation* for the input expression, certifying that it has no model. If the output is a set of variables, then the assignment obtained by setting those variables to 1 and all others to 0 is a model for the input expression.

In [27] it is shown that if there is a short DPLL refutation for a given unsatisfiable expression, then there must be a short resolution refutation for the same expression. Therefore, resolution is potentially more efficient than DPLL. However, finding a *shortest* resolution refutation is generally hard. Moreover, the more restricted nature of DPLL algorithms, namely finding all resolvents involving a particular pivot variable v and then removing clauses containing v and $\bar{v}$ from ϕ, seems to result in a better intuitive grasp of effective search heuristics, and this seems to be the reason DPLL is preferred to resolution, in general.

In any case, the pessimistic probabilistic results for resolution, which also apply to DPLL, begin with establishing the root cause: namely, expression sparseness [24]. Without getting too technical at this time, sparseness is a measure of the number of times pairs of clauses have a common literal or complementary pair of literals; any "moderately large" subset $\mathcal{C}$ of clauses taken from a sparse expression must contain a "large" number of variables that occur exactly once in $\mathcal{C}$.

Sparse expressions force a superpolynomial number of resolution steps for the following reason. Let $\mathcal{P}^{\phi}$ be the minimum set of clauses that is the result of repeated applications of the resolution rule starting from a sparse unsatisfiable CNF expression ϕ and ending with the null clause. By sparsity, any moderately sized subset $\mathcal{C}$ of clauses taken from ϕ must contain a large number of variables that occur exactly once in $\mathcal{C}$. This forces at least one clause of high width to exist in $\mathcal{P}^{\phi}$. But, almost all "short" resolution refutations contain no long clauses after eliminating all clauses satisfied by a particular small random partial assignment ρ. Moreover, resolution refutations for almost all unsatisfiable random (n, m, k)-CNF expressions with clauses satisfied by ρ removed are sparse and, therefore, must have at least one high-width clause. Consequently, almost all unsatisfiable random (n, m, k)-CNF expressions have long resolution refutations. Ideas leading up to a concise understanding of this phenomenon can be found in [11–13,24,50,56,96]. Despite considerable tweaking, the best we can say right now is that the probability that the length of a shortest resolution proof is bounded by a polynomial in n when $m/n > c'_k \cdot 2^k$, c'_k some constant, and $\lim_{m,n\to\infty} m/n^{(k+2)/(4-\varepsilon)} < 1$, where ε is some small constant, tends to 0 and when $\lim_{m,n\to\infty} m/n > (n/\log(n))^{k-2}$ tends to 1. These results are restated with a little more detail in Section 4.6.4.

The failure of resolution has led to the development of new techniques for certifying unsatisfiability. A major success with respect to probabilistic analysis is reported in [52] and described in Section 4.6.5. In essence, for a random (n, m, k)-CNF expression ϕ, one finds a bound $N^+_{m,n}$ on the number of variables needed to be set to 1 to satisfy clauses containing only uncomplemented literals and a bound $N^-_{m,n}$ on the number of variables needed to be set to 0 to satisfy clauses containing only complemented literals. If $N^+_{m,n} + N^-_{m,n} > n$, then at least one variable would have to be set to both 1 and 0 to satisfy the given expression. Because this is impossible if ϕ is satisfiable, this test can provide certification that ϕ is unsatisfiable. In [52], spectral techniques are used to obtain bounds sufficient to show that certifying unsatisfiability in polynomial time can be accomplished with high probability when $\lim_{m,n\to\infty} m/n > n^{k/2-1+o(1)}$ Results on resolution are discussed further in Section 4.6.4. Figure 4.7 summarizes probabilistic results on algorithms for random (n, m, k)-CNF expressions.

The probabilistic analysis of properties of CNF expressions can develop insights into the nature of "hard" problems as well as the potential effectiveness of preprocessing expressions before search algorithms are applied. For example, consider checking whether a given expression is a member of a polynomial time solveable class before search. If so, the search can be eliminated entirely. Such a strategy might result in a drastic reduction of overall execution time. However, this seems unlikely to succeed for any of the well-known efficiently solvable classes because of probabilistic studies of random expressions such as those described in Section 4.6.6. According to those results, the likelihood of the existence of

Algorithm	Efficient in Probability: $m/n < \cdots$ $k = 3$	$k > 3$ $(Pr > c)$	$k > 3$ $(Pr \to 1)$
Goldberg	None	None	None
Pure Literals (Broder...)	1.63	Not reported	Not reported
UC (unit clauses)	2.66	$\frac{2^k}{k}\frac{1}{2}\left(\frac{k-1}{k-2}\right)^{k-2}$	None
UC + majority	2.9	Not determined	Not determined
GUC (shortest clauses)	3.003	$\frac{3 \cdot 2^k}{4k}\left(\frac{k-1}{k-2}\right)^{k-2}\left(\frac{k}{k+1}\right)$	$0.46\frac{2^k}{k}\left(\frac{k-1}{k-2}\right)^{k-2}\left(\frac{k}{k+1}\right)$
SC (1,2 width clauses)	3.003	$O(2^k/k)$	$\frac{2^k}{8k}\left(\frac{k-1}{k-3}\right)^{k-3}\left(\frac{k-1}{k-2}\right)$
SC (2 literals at a time)	3.145	$O(2^k/k)$	$O(2^k/k)$
Best Myopic	3.26	$O(2^k/k)$	$O(2^k/k)$
Greedy (Kaporis...)	3.42	$O(2^k/k)$	$O(2^k/k)$

Algorithm	Efficient in probability: $m/n > \cdots$ $k = 3$	$k > 3$ $(Pr > c)$	$k > 3$ $(Pr \to 1)$
Restricted Width	0.66n	–	$n^{k-2}\left(\frac{2^k}{2k!}\right)$
DPLL variant	$n/\log(n)$	–	$n^{k-2}\left(\frac{1}{\log(n)}\right)^{k-2}$
Hitting Set	$n^{1/2+\varepsilon}$	–	$n^{k/2-1+o(1)}$

Figure 4.7. Probabilistic behavior of algorithms on random (n, m, k)-CNF expression as a function of density (m/n). The top table, except for Goldberg's algorithm, applies to incomplete algorithms that either find a model or give up. Such algorithms perform well in probability for densities (m/n) below a certain quantity. Table entries indicate known bounds on those quantities. Two types of results are shown for $k > 3$: probability of success bounded from below by a constant (column heading includes $Pr > c$) and probability of success tending to 1 (column heading includes $Pr \to 1$). Much work has been done for the case $k = 3$, and those results are in a separate column. In that case, the probability of success tends to 1 only for the algorithm of Broder. However, the addition of a limited form of backtracking brings GUC and SC into that category as well. Table entries $O(2^k/k)$ have not been published but either are obvious or have been communicated personally. These are interesting in light of the result that the Satisfiability threshold is $O(2^k)$. The bottom table applies to algorithms that intend to verify unsatisfiability.

instances of such classes is extremely low among reasonably hard random expressions. The analysis explains why: namely, the highly likely presence of specific cyclic structures (this is explained in Section 4.6.6).

Insights developed from a probabilistic analysis also can be quite surprising. For example, using density (m/n) as a parameter, it is found that several well studied, different, complex polynomially solvable classes are rare when $m/n > 4/(k^2 - k)$ [46]. On the other hand, some polynomial time solvable classes that are apparently so trivial that they have been all but neglected in the literature and are not vulnerable to cyclic structures are common even out to $m/n = 1$. See Section 4.6.6 for details.

Before leaving this section, we mention the recent important nonrigorous contributions of the statistical physics community leading to a better understanding of the nature of hard problems and proposals for algorithms that can deal with them. A discussion of this topic is left to Section 4.6.8.

4.6 The Probabilistic Analysis of Algorithms

In this section we present examples of analysis using the tools of Section 4.3. We choose examples that represent, without complication, ideas that have led to progressively better results. This means, in some cases, the results obtained here are not the "best" known.

4.6.1 Myopic Algorithms for Satisfiable Expressions

For densities where random expressions are satisfiable with high probability, the behavior of some DPLL variants can be understood adequately by means of performance results on corresponding straight line algorithms. A first wave of many results were obtained for myopic algorithms of the form of Algorithm SCA in Figure 4.5, known as the Shortest Clause Algorithm because each variable assignment is chosen to satisfy a nonsatisfied clause of least width. The study of these algorithms is motivated by the observation that if a straight line algorithm is to succeed, it cannot allow an accumulation of unit clauses because too many unit clauses increases the likelihood of a pair of unit clauses that have opposite polarity and that would terminate the algorithm without success. Intuitively, eliminating shortest clauses first should then be a priority. Probabilistic analysis confirms this intuition to some extent: results for the class of shortest clause first algorithms are quite good compared to other heuristics, for example the pure literal heuristic alone. In addition, the analysis is based on *clause flows* and illuminates the algorithmic mechanics that cause successful behavior. The remainder of this section is a high-level presentation of the analysis of several straight line algorithms. The main intention is to focus on motivation and intuition with a minimum of details.

A clause-flow model can be used to analyze the mechanics of many straight line algorithms. In particular, in this section we concentrate on the family identified as Algorithm SCA in Figure 4.5 and known as the *shortest clause algorithms*. Let ϕ be a random (n, m, k)-CNF expression. The shortest clause algorithms choose an unassigned literal on each iteration. Such a choice causes some clauses of ϕ to disappear because they have become satisfied, and some clauses to have a literal removed because they have become falsified by the implicit assignment. Thus, at the start of the jth iteration of SCA, some nonsatisfied, remaining clauses have k literals, some $k-1$ literals, and so on. Define $C_i^{\phi}(j)$ to be the set of clauses of ϕ that have exactly i literals at the start of the jth iteration (henceforth, the

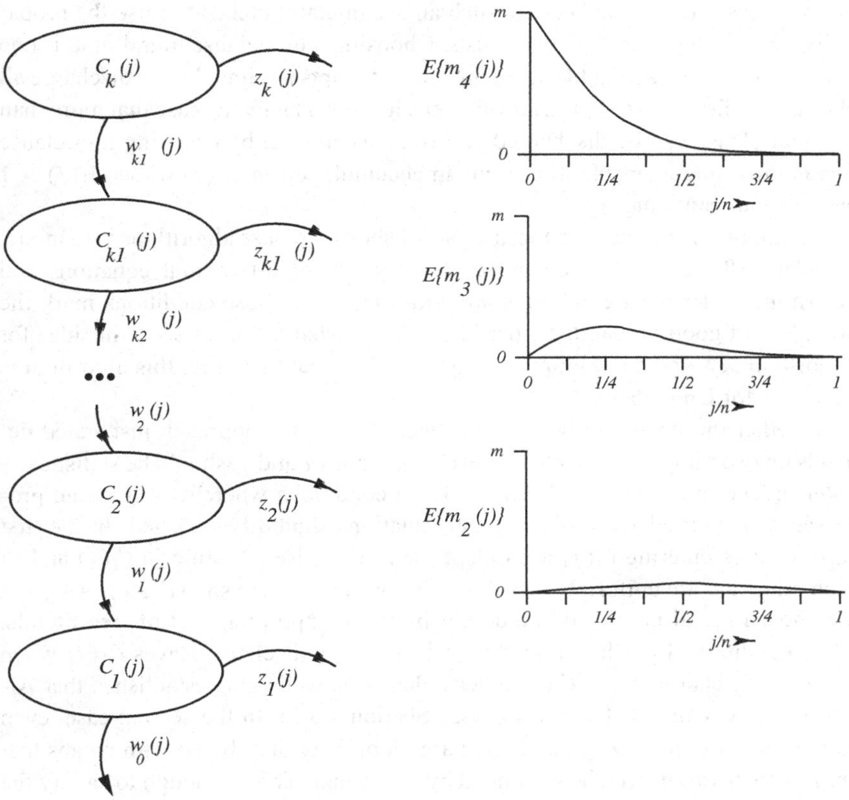

Figure 4.8. Clause sets and flows for a linear algorithm. On the left is a schematic representation. On the right are plots of the expected number of clauses in $C_i(j)$ versus j/n for the case $k = 4$ with $i = 2, 3, 4$.

superscript will be dropped for simplicity). Define $w_i(j)$ to be the number of i-literal clauses added to $C_i(j)$ as a result of choosing a literal on the jth iteration. That is, $w_i(j)$ is the flow of clauses into $C_i(j)$ due to picking a literal on the jth iteration. Define $z_i(j)$ to be the number of clauses eliminated from $C_i(j)$ (satisfied) as a result of choosing a literal on the jth iteration. That is, $z_i(j)$ is the flow out of $C_i(j)$ due to picking a literal on the jth iteration. Let $m_i(j) = |C_i(j)|$. Figure 4.8 shows these clause sets, represented by ovals, and clause flows, represented by arcs with arrows indicating flow direction.

The success of a shortest clause algorithm depends critically on what is happening to $w_0(j)$. If $w_0(j) > 0$ for any j, any shortest clause algorithm stops and gives up because some clause has just had all its literals falsified by the current partial truth assignment. In turn, $w_0(j)$ can be controlled by keeping complememtary pairs of clauses out of $C_1(j)$, for all j, because, if such a pair exists in $C_1(j)$ for some $j = j'$, then eventually $w_0(j) > 0$ for some $j = j^* > j'$. Complementary pairs may be kept out of $C_1(j)$, for all j, by preventing a significant accumulation of unit clauses over time because such an accumulation tends to raise the probability that a complementary pair exists. Choosing a unit clause literal first, if one exists, does this by acting like a "pump" that attempts to immediately discharge all clauses that flow into $C_1(j)$. Unfortunately, it is not usually the case that more than one unit clause can be discharged at a time. Therefore, by choosing unit clause literals first, one is unable to prevent an accumulation in $C_1(j)$ when $w_1(j) > 1$ over a significant range of j.

An important approach to the analysis of shortest clause algorithms is to model the clause flows and accumulations as a system of differential equations and determine under what conditions $max_j\{w_1(j)\} = 1$. Those conditions mark the boundary of good probabilistic performance. In what follows we try this idea for Algorithm SCA given in Figure 4.5 with $s = 1$. In the literature, this algorithm is called UC for Unit Clause.

Whether the flows can be modeled according to the approach just stated depends on two things: (1) the clauses in $\mathcal{C}_i(j)$, for any i and j, should be statistically independent and uniformly distributed; (2) conditions whereby Markovian processes may be modeled as differential equations should be satisfied. In the first case, clauses entering $\mathcal{C}_i(j)$ are independent of clauses existing in $\mathcal{C}_i(j)$ and of each other and are uniformly distributed because they were so in $\mathcal{C}_{i+1}(j-1)$, and the chosen literal is selected randomly from the appropriate set of free literals. Also, conditioned on the event that at least one unit clause leaves $\mathcal{C}_1(j)$ when $|\mathcal{C}_1(j)| > 0$, clauses leave $\mathcal{C}_i(j)$ independently as well. This establishes that Algorithm SCA with $s = 1$ is myopic (see Section 4.3.3). In the second case, even as n grows, the flows $w_i(j)$ and $z_i(j)$ are binomially distributed with means that are less than km/n, which is bounded by a constant. This is enough to satisfy the conditions of Theorem 4.2 so there is no loss in modeling the discrete flows and accumulations by a system of differential equations.

With this out of the way we can write, for $1 \le i \le k$, $0 < j < n$,

$$m_i(j+1) = m_i(j) + w_i(j) - z_i(j).$$

Taking expectations gives

$$E\{m_i(j+1)\} = E\{m_i(j)\} + E\{w_i(j)\} - E\{z_i(j)\},$$

which can be written

$$E\{m_i(j+1)\} - E\{m_i(j)\} = E\{w_i(j)\} - E\{z_i(j)\}. \tag{4.7}$$

But, because of the statistical independence of clauses at every j and the fact that ϕ is originally constructed from a set of n variables, we have, for all $2 \le i \le k$, $0 < j < n$,

$$\begin{aligned} E\{z_i(j)\} &= E\{E\{z_i(j)|m_i(j)\}\} \\ &= E\left\{\frac{i * m_i(j)}{n-j}\right\} = \frac{i * E\{m_i(j)\}}{n-j}. \end{aligned}$$

Also, for all $1 \le i < k, 0 < j < n$,

$$E\{w_i(j)\} = E\{E\{w_i(j)|m_{i+1}(j)\}\}$$
$$= E\left\{\frac{(i+1) * m_{i+1}(j)}{2(n-j)}\right\} = \frac{(i+1) * E\{m_{i+1}(j)\}}{2(n-j)} \quad \text{and}$$
$$E\{w_k(j)\} = 0.$$

After substituting these results into Equation 4.7 and verifying conditions in Theorem 4.2, we can write the corresponding difference equations as follows: for $2 \le i < k$,

$$E\{m_i(j+1) - m_i(j)\} = \frac{(i+1) * E\{m_{i+1}(j)\}}{2(n-j)} - \frac{i * E\{m_i(j)\}}{n-j} \quad \text{and}$$
$$E\{m_k(j)\} = -\frac{k * E\{m_k(j)\}}{n-j}.$$

The corresponding differential equations are

$$\frac{d\bar{m}_i(x)}{dx} = \frac{(i+1) * \bar{m}_{i+1}(x)}{2(1-x)n} - \frac{i * \bar{m}_i(x)}{(1-x)n} \quad \text{and}$$
$$\frac{d\bar{m}_k(x)}{dx} = -\frac{k * \bar{m}_k(x)}{(1-x)n}$$

where $\bar{m}_i(x)$ is $z_i(x)$ of Theorem 4.2.

Boundary conditions, assuming m clauses of k literals in ϕ initially, are $\bar{m}_k(0) = m/n$ and $\bar{m}_i(0) = 0$ for all $1 \le i < k$. The solution to the equations with these boundary conditions is, for all $2 \le i \le k$,

$$\bar{m}_i(x) = \frac{1}{2^{k-i}}\binom{k}{i}(1-x)^i (x)^{k-i}\, m/n.$$

Thus,

$$E\{m_i(j)\} = \frac{1}{2^{k-i}}\binom{k}{i}(1-j/n)^i (j/n)^{k-i}\, m.$$

Plots of these functions for $k = 4$ were given in Figure 4.8.

The important flow is given by

$$E\{w_1(j)\} = \frac{E\{m_2(j)\}}{n-j} = \frac{1}{2^{k-2}}\binom{k}{2}(1-j/n)(j/n)^{k-2}(m/n).$$

By Theorem 4.3 and because $|E\{m_2(j)\} - m_2(j)| < \beta$, for any $0 < \beta$, almost always, from Theorem 4.2(b), Algorithm UC succeeds with probability bounded from below by a constant if

$$E\{w_1(j)\} = \frac{E\{m_2(j)\}}{n-j} < \frac{m_2(j)+\beta}{n-j} = \frac{m_2(j)}{n-j} + o(1) < 1 - \varepsilon + o(1)$$

for any $0 < \varepsilon$. That is, the algorithm succeeds as long as the average rate of production of unit clauses is no greater than 1 per iteration at any iteration.

Taking the derivative with respect to j and setting to 0 yields a maximum for $E\{w_1(j)\}$ at $j = j^* = \frac{k-2}{k-1}n$. The value of $E\{w_1(j^*)\}$ is less than 1 if

$$\frac{m}{n} < \frac{2^{k-1}}{k}\left(\frac{k-1}{k-2}\right)^{k-2}.$$

We can now conclude the following.

Theorem 4.8. *Algorithm* UC *determines that a given random (n, m, k)-CNF expression has a model with probability bounded from below by a constant if*

$$\frac{m}{n} < \frac{2^{k-1}}{k}\left(\frac{k-1}{k-2}\right)^{k-2}.$$

Observe that for $k = 3$, UC succeeds with bounded probability when $m/n < 2.666$. By Theorem 4.6, almost all random $(n, m, 3)$-CNF expressions have at least one model if $m/n < 2.666$.

A feature of the flow analysis just outlined is that it reveals mechanisms that suggest other, improved heuristics. For example, because increasing z flows decreases w flows, the following adjustment to Algorithm UC is suggested: if there are no unit clauses, choose a literal l randomly from ϕ, then compare the number of occurrences of l with the number of occurrences of $\bar{l}$ and choose the literal that occurs most often (this tends to increase z flows at the expense of w flows). This isn't quite good enough, however, because $C_i(j)$ clauses are approximately twice as influential as $C_{i+1}(j)$ clauses. This is because, roughly, one clause is accounted for in $z_i(j)$ for every two clauses in $z_{i+1}(j)$. Thus, it is better to compare *weights* of literals where the weight of a literal is given by

$$\omega(l) = \sum_{c \in \phi : l \in c} 2^{-|c|}.$$

An analysis of such a heuristic is not known to us, but the algorithm for 3-CNF expressions (shown in Figure 4.9) called UCL, using the original idea of counting clauses, comes fairly close.

Whereas in the case of Algorithm UC applied to random $(n, m, 3)$-CNF expressions a model can be found with bounded probability when $m/n < 2.66$, in the case of Algorithm UCL we have the following:

Theorem 4.9. *Algorithm* UCL *determines that a given random $(n, m, 3)$-CNF expression has a model with probability bounded from below by a constant if $m/n < 2.9$.*

An improved analysis and improved bound to $m/n < 3.001$ is given in [1].

The reader may be curious about why the number of occurrences of literals in $C_2(j)$ was not taken into account in Algorithm UCL. Flow analysis tells us that it is unnecessary. Suppose that, in the case $C_1(j) = \emptyset$, the $(j+1)$st literal is chosen on the number of times it occurs in both $C_3(j)$ and $C_2(j)$. Assume the most optimistic case: the literal appears in more clauses of both $C_3(j)$ and $C_2(j)$ than

```
Procedure UCL(φ)
Input: a CNF expression φ.
Output: either "unsatisfiable" or a model for φ.

begin
   M := ∅;
   j := 0;
   L := {x, x̄ : ∃C ∈ φ such that either x ∈ C or x̄ ∈ C};
   repeat the following {
      if C1(j) ≠ ∅ then randomly choose literal x ∈ C1(j);
      else {
         choose literal y randomly from L;
         if # occurrences of ȳ in C3(j) > # occurrences of y in C3(j) then
            x := ȳ;
         else
            x := y;
      }
      L := L − {x, x̄};
      remove from φ all clauses containing x;
      remove from φ all occurrences of x̄;
      if x is an uncomplemented literal then M := M ∪ {x}.
      j := j + 1;
   } until φ = ∅ or there are two complementary unit clauses in φ;
   if φ = ∅ then return M;
   else return("cannot determine whether a model exists");
end;
```

Figure 4.9. A heuristic that chooses variables randomly but assigns values based on the difference in the number of occurrences of complemented and uncomplemented literals.

its complement (then the flow into $C_1(j+1)$ is minimized because the number of 2- and 3-literal clauses removed due to the $(j+1)$st chosen literal is maximized). Let $E\{w_1^*(j)\}$ denote the new average flow of clauses into $C_1(j)$. Then

$$E\{w_1^*(j)\} = E\{w_1(j)\} - h_1(j)(1 - E\{w_1^*(j)\}),$$

where $h_1(j)$ is the extra number of clauses removed from the flow into $C_1(j)$ when the chosen literal is not a unit clause, and $1 - E\{w_1^*(j)\}$ is the probability (to within $O(\frac{1}{n})$) that the chosen literal is not a unit clause. Therefore,

$$E\{w_1^*(j)\} = \frac{E\{w_1(j)\} - h_1(j)}{1 - h_1(j)}.$$

Thus, $E\{w_1^*(j)\} < 1$ is equivalent to $E\{w_1(j)\} < 1$, and no benefit is gained by considering the number of occurrences of the chosen literal in $C_2(j)$.

There is another improved heuristic suggested by flow analysis. If "pumping" clauses at the bottom level by means of the unit clause rule is effective, putting "pumps" at all levels should be more effective. This amounts to adopting the following strategy for literal selection, which is a generalization of the unit clause rule called the smallest clause rule: choose a literal from a clause of smallest size. Thus, if there is at least one unit clause, choose from one of them; otherwise, if there is at least one 2-literal clause, choose a literal from a 2-literal clauses; otherwise, if there is at least one 3-literal clause, choose a literal from a 3-literal clause, and so on. This is algorithm SCA, Figure 4.5, with $s = k$. We call this Algorithm GUC.

The effectiveness of "pumping" at all levels is revealed by a flow analysis applied to algorithm **GUC**. The results, taken from [21], are

Theorem 4.10. *Algorithm* **GUC** *determines that a random* (n, m, k)*-CNF expression has a model with probability bounded from below by a constant when*

$$\frac{m}{n} < \frac{3.09 * 2^{k-2}}{k+1}\left(\frac{k-1}{k-2}\right)^{k-2} \quad \text{and } 4 \le k \le 40.$$

Theorem 4.11. *Algorithm* **GUC** *determines that a random* (n, m, k)*-CNF expression has a model with probability tending to 1 when*

$$\frac{m}{n} < \frac{1.845 * 2^{k-2}}{k+1}\left(\frac{k-1}{k-2}\right)^{k-2} - 1 \text{ and } 4 \le k \le 40.$$

But it is not necessary to "pump" at all levels to get this kind of result. Consider algorithm SCA with $s = 2$ and call it algorithm SC. The following analysis sketch of algorithm SC is based on results by Chvátal and Reed [23]. Let p_j denote the probability that a fixed input clause shrinks to two literals after exactly j iterations of Algorithm SC. Then $p_j m$ is the average flow into $C_2(j)$. The following simple expression for p_j can be obtained straightforwardly:

$$p_j = \frac{\binom{j-1}{k-3}\binom{n-j}{2}}{\binom{n}{k}} \frac{1}{2^{k-2}}.$$

This can be bounded from above by setting j to the value that maximizes p_j (the maximum occurs when the ratio p_{j+1}/p_j crosses the value 1). We are interested in conditions that imply the bound for p_j is less than $1/m$, because that gives an average flow into $C_2(j)$ that is less than 1. Straightforward calculation reveals, for all j,

$$p_j < \frac{1-\varepsilon}{m} \text{ for any fixed } \varepsilon > 0 \text{ such that } \frac{m}{n} < \left(\frac{1-\varepsilon}{1+\varepsilon}\right)\frac{2^k}{8k}\left(\frac{k-1}{k-3}\right)^{k-2}\frac{k-1}{k-2}.$$

This yields the following.

Theorem 4.12. *Algorithm* **SC** *determines that a random* (n, m, k)*-CNF expression,* $k \ge 3$*, has a model with probability tending to 1 when*

$$\frac{m}{n} < \frac{2^k}{8k}\left(\frac{k-1}{k-3}\right)^{k-3}\frac{k-1}{k-2}.$$

The difference between the result of Theorem 4.12 and that of Theorem 4.11 is due to improved analysis facilitated by working with an easier algorithm.

By adding a limited amount of backtracking to GUC, Frieze and Suen produced an algorithm, called GUCB, for 3-CNF expressions that finds a model, with probability tending to 1, when $m/n < 3.003$ [48]. Probabilistic analysis of a backtracking algorithm given random (n, m, k)-CNF expressions ϕ can be exceedingly difficult because statistical dependences can easily show up when returning to subexpressions after a failure to locate a model. However, Frieze and Suen showed it is possible to carefully manage a limited amount of backtracking so that this does not happen. The following explains how the backtracking in GUCB is accomplished. The initial operation of GUCB is the same as that of GUC. Suppose GUCB has successfully completed t iterations and has chosen the sequence of literals $\{x_{\pi_1}, x_{\pi_2}, \ldots x_{\pi_t}\}$ and set them to 1. Suppose $|C_1(t')| = 0$, $t' < t$ and $|C_1(j)| > 0$ for all $t' < j \leq t$ so the last iteration that saw no unit clauses was iteration t'. Suppose further that choosing and setting literal $x_{\pi_{t+1}}$ to 1 results in the existence of complementary unit clauses in $C_1(t+1)$. Then GUCB backtracks by setting $x_{\pi_{t'}} = x_{\pi_{t'+1}} = \ldots = x_{\pi_t} = x_{\pi_{t+1}} = 0$. Corresponding adjustments are made to the clauses and literals of ϕ. That is, clauses now satisfied are removed, removed clauses now *not* satisfied are reinstated, literals now falsified are removed, and removed literals now *not* falsified and in nonsatisfied clauses are reinstated. After backtracking, GUCB continues choosing and setting literals to 1 as before. The algorithm succeeds if all clauses are eliminated. The algorithm fails in two ways: (1) the resetting of literals from 1 to 0 results in a null clause, and (2) a complementary pair of unit clauses is encountered before $|C_1|$ has become 0 after a backtrack. The analysis of GUCB is possible because the effect on the distribution of $C_i(j)$ is slight and because, with probability tending to 1, GUCB backtracks at most $\ln^5(n)$ times when $m/n < 3.003$.

Frieze and Suen also apply limited backtracking to algorithm SC and call the resulting algorithm SCB. The result is the following:

Theorem 4.13. *For $k \geq 4$, SCB succeeds, with probability tending to 1, when $m/n < \eta_k 2^k/k$ where $\eta_4 \approx 1.3836$, $\eta_5 \approx 1.504$, and $\lim_{k\to\infty} \eta_k \approx 1.817$.*

This may be compared to the previous result for GUC, which, at $k = 40$, has a performance bound of $m/n < (1.2376)2^{40}/40$ (probability tending to 1), and the foregoing result for SC, which has a performance bound of $\lim_{k\to\infty} m/n < (0.9236)2^k/k$ (probability tending to 1).

There is a limit to the probabilistic performance of straight line, myopic algorithms. That is, there is an optimal policy for choosing a variable and value on any iteration, depending on the iteration. For $k = 3$, the optimal policy is given by the following:

Optimal Myopic Literal Selection Policy If $m_1(j) = 0$ choose 2-literal clause $\{v, w\}$ or $\{v, \bar{w}\}$, or $\{\bar{v}, w\}$ or $\{\bar{v}, \bar{w}\}$ at random. Select v, $\bar{v}$, w or $\bar{w}$ at random and temporarily assign the value to the chosen variable that satisfies its clause. Then fix the assignment according to the following.

Let $M_3(j)$ be the number of 3-literal clauses satisfied minus the number of literals falsified in 3-literal clauses. Let $M_2(j)$ be the number of 2-literal clauses satisfied minus the number of literals falsified in 2-literal clauses. Let $d_3(j) = m_3(j)/(n - j)$. Let $d_2(j) = m_2(j)/(n - j)$. Define $\Theta(j) = (1.5 \cdot d_3(j) - 2 \cdot d_2(j) + \gamma)/(1 - d_2(j))$ where γ is some constant. Reverse the assignment if $\Theta(j) > M_3(j)/M_2(j)$.

Theorem 4.14. ([3].) *Algorithm* SCA *with* $s = 1$ *and using the optimal myopic literal selection policy instead of choosing a literal randomly from* L_2 *succeeds in finding a model given random* $(n, m, 3)$*-CNF expressions as input with probability bounded from below by a constant when* $m/n < 3.22$*. This is the best possible literal selection policy for* SCA *on random* $(n, m, 3)$*-CNF expressions.*

According to the optimal myopic literal selection policy, if there is a literal that will maximize both the number of 3-literal clauses and 2-literal clauses satisfied, one will be selected. Otherwise, a literal is selected as a compromise between reducing the number of 2-literal clauses immediately and increasing the chance of reducing the number of 2-literal clauses at some point in the future.

A better result is due to a literal selection policy that may actually look at two literals on an iteration instead of just one. The result is:

Theorem 4.15. *There exists a myopic straight line algorithm that succeeds in finding a model given random* $(n, m, 3)$*-CNF expressions as input with probability bounded from below by a constant when* $m/n < 3.26$*. This is the best performance possible by any myopic straight line algorithm.*

4.6.2 Nonmyopic Algorithms for Satisfiable Expressions

In light of experimental evidence suggesting r_3 is approximately 4.25 and the results of Theorem 4.15, there appears to be an unfortunate gap between what is achievable by myopic algorithms and what we would hope is possible from straight line algorithms. However, the tools of Section 4.3.3 may still be applied to the analysis of nonmyopic straight line algorithms, under some restrictions. For example, flow analysis is still possible if algorithmic operations include [65]:

(i) Select uniformly at random a pure literal, assign it the value 1, and remove all satisfied clauses.

(ii) Select uniformly at random a literal occuring exactly once in the expression and whose occurrence is in a 3-literal clause, assign it the value 0, remove it from the clause in which it appears, and remove all clauses containing its complementary literal (these are satisfied).

(iii) Select uniformly at random a literal occuring exactly once in the expression and whose occurrence is in a 2-literal clause, assign it the value 0, remove it from the clause in which it appears, and remove all clauses containing its complementary literal (these are satisfied). Then apply the unit clause unit to exhaustion (until no unit clauses remain).

Procedure GPL(ϕ)
Input: a CNF expression ϕ.
Output: either "gives up" or returns a model for ϕ.

```
begin
    M := ∅;
    repeat the following {
        if φ = ∅ then return M;
        if ∅ ∈ φ then return("cannot determine whether model exists");
        let l be a literal in φ occurring at least as frequently as any other;
        if l is an uncomplemented literal then M := M ∪ {l};
        φ := {C − {l̄} : C ∈ φ, l ∉ C}
        repeat the following until no unit clauses exist in φ {
            let {l} be a unit clause in φ;
            if l is an uncomplemented literal then M := M ∪ {l};
            φ := {C − {l̄} : C ∈ φ, l ∉ C};
        }
    }
end;
```

Figure 4.10. A nonmyopic straight line algorithm.

An example, which we call GPL, from [63] is shown in Figure 4.10.

Theorem 4.16. *Algorithm* GPL *succeeds in finding a model given random $(n, m, 3)$-CNF expressions as input with probability bounded from below by a constant when $m/n < 3.42$.*

Similar algorithms have been reported to have good performance out to $m/n < 3.52$ [55, 64].

Other nonmyopic straight line algorithms have shown even better performance on random (n, m, k)-CNF inputs. For example, the algorithm of Figure 4.11 seems to do well for $m/n < 3.6$ on random $(n, m, 3)$-CNF inputs. The algorithm is designed to iteratively select a variable and value that maximizes the expected number of models possessed by what is left of the given expression after satisfied clauses and falsified literals are removed, assuming somehow the clauses of the new expression are statistically independent. Of course, that is not the case. But it is conceivable that the fundamental idea of maximizing the expected number of models or maximizing the probability that a model exists can be enhanced to provide algorithms of greater performance. An analysis of this and other algorithms like it has not yet been reported.

4.6.3 Lower Bounds on the Satisfiability Threshold

The results of the previous two sections may be used to obtain lower bounds for the random (n, m, k)-CNF threshold. Thus, from algorithmic analysis we know

```
Procedure MPS(φ)
Input: a CNF expression φ.
Output: either "gives up" or returns a model for φ.

begin
    M := ∅;
    repeat the following {
        if φ = ∅ then return M;
        if ∅ ∈ φ then return("cannot determine whether model exists");
        repeat the following until no unit clauses exist in φ {
            let {l} be a unit clause in φ;
            if l is an uncomplemented literal then M := M ∪ {l};
            φ := {C − {l̄} : C ∈ φ, l ∉ C};
        }
        for all unassigned literals l do the following {
            φ_l := {C − {l̄} : C ∈ φ, l ∉ C};
            define w(l), the weight of literal l, to be ∏_{C∈φ_l} (1 − 2^{−|C|});
        }
        choose l so that w(l) is maximum over all unassigned literals;
        if l is an uncomplemented literal then M := M ∪ {l}
        φ := {C − {l̄} : C ∈ φ, l ∉ C};
    }
end;
```

Figure 4.11. Another nonmyopic straight line algorithm.

$r_3 > 3.42$. Unfortunately, the best that can be said for the general threshold, based on algorithmic analysis, is $r_k > c \cdot 2^k/k$, c a constant. But recently the second moment method has been applied ([5]) to show that

$$r_k \geq 2^k \ln(2) - (k+1)\ln 2/2 - 1 - \delta_k$$

where $\delta_k \to 0$ for increasing k. These results improve and are built upon those of [4]. We sketch the motivation leading to these results in this section.

The crucial question is to which random variable X the second moment method can be applied. As stated in Section 4.3.2, if X is the number of models, the bound provided by the second moment method is inadequate. This is because for many random (n, m, k)-CNF expressions, the number of models is far from the mean number, causing the variance of X to be too high. More specifically, $Pr(z|w)$ (see Lemma 4.1) is too high. This can be seen as follows. Let z and w be assignments to variables in random (n, m, k)-CNF expression ϕ agreeing in αn variables. Then

$$Pr(z \text{ is a model for } \phi | w \text{ is a model for } \phi) = \left(1 - \frac{1-\alpha^k}{2^k - 1}\right)^m.$$

Therefore, to apply the second moment method, we will need to control

$$\sum_{0\le\alpha\le1} \binom{n}{\alpha n}\left(1 - \frac{1-\alpha^k}{2^k-1}\right)^m .$$

Observe that the maximum of this sum does not occur at $\alpha = 1/2$ and is not $o(\mu)$ as needed (recall $\mu = 2^n(1-2^{-k})^m$).

The fix is to use a bound for a closely related problem for which the maximum of the sum occurs at $\alpha = 1/2$. The problem is called Not-All-Equal k-CNF (which we denote by NAE_k). Given random (n, m, k)-CNF expression ϕ, NAE_k is the problem of finding a model for ϕ such that every clause has at least one literal falsified as well as at least one satisfied. Because a NAE_k model for ϕ is also a model for ϕ,

$$Pr(\exists \text{ model for } \phi) > Pr(\exists\ \text{NAE}_k \text{ model for } \phi)$$

But

$$Pr(z \text{ is a NAE}_k\ \textit{model for}\ \phi | w \text{ is a NAE}_k\ \textit{model for}\ \phi)$$
$$= \left(1 - \frac{1-\alpha^k-(1-\alpha)^k}{2^{k-1}-1}\right)^m$$

and

$$\sum_{0\le\alpha\le1} \binom{n}{\alpha n}\left(1 - \frac{1-\alpha^k-(1-\alpha)^k}{2^{k-1}-1}\right)^m \approx \frac{2^n(1-2^{-k+1})^m}{\sqrt{n}} = o(\mu)$$

because all the significant contributions to the sum occur near the maximum of both terms in the summand, which is at $\alpha = 1/2$ and μ for NAE_k on random (n, m, k)-CNF inputs is $2^n(1-2^{-k+1})^m$.

4.6.4 Algorithms for Verifying Unsatisfiability: Resolution

The algorithms just discussed are useless for verifying the unsatisfiability of an expression where it is required to prove that no truth assignment is a satisfying one. Resolution is one popular method capable of doing this and can be extremely effective when certain clause patterns are present, with high probability. Recall from Page 122 that the resolvent of two clauses c_1 and c_2 such that there is exactly one variable v that occurs as a complemented literal in one clause and as an uncomplemented literal in the other, denoted by $\mathcal{R}^{c_1}_{c_2}$, is a clause that contains all the literals of c_1 and all the literals of c_2 except for v and $\bar{v}$. That is, $\mathcal{R}^{c_1}_{c_2} = \{l : l \in c_1 \cup c_2 \setminus \{v, \bar{v}\}\}$. If resolvents are generated until one is the empty set, then the original expression is unsatisfiable.

There are many resolution-based algorithms: that is, algorithms using the generation of resolvents to determine whether a given expression is satisfiable. They differ in the restrictions that are applied to help improve the total number of resolvents generated. The following restricted resolution algorithm, called RW for

Restricted Width, has polynomial time complexity because resolvents are restricted to be no greater than k in width:

Procedure RW(ϕ)
Input: a CNF expression ϕ.
Output: either "unsatisfiable" or "gives up".

begin
  repeat the following {
    if there are clauses $C_1, C_2 \in \phi$ s.t. $\mathcal{R}_{C_2}^{C_1} \notin \phi$ and $|\mathcal{R}_{C_2}^{C_1}| \le k$ then
      $\phi := \phi \cup \{\mathcal{R}_{C_2}^{C_1}\}$;
    else
      break out of the loop;
  }
  if $\emptyset \in \phi$ then return ("unsatisfiable");
  else return ("cannot determine whether ϕ is unsatisfiable");
end;

Clearly, algorithm RW cannot be mistaken if it outputs "unsatisfiable."

It can be shown that Algorithm RW is effective on random (n, m, k)-CNF expressions up to a point. Particular patterns of clauses that RW can exploit to produce a null resolvent are developed as follows:

Let $\mathcal{S}_{i,j}$ denote the family of clause sets defined as follows over a set of variables $X = \{x_{11}, x_{21}, \ldots, x_{12}, x_{22}, \ldots\}$:

$$\mathcal{S}_{i,j} = \begin{cases} \{\{x_{i1}\}\{\bar{x}_{i1}\}\} & if\ j = 1; \\ \{P \cup \{x_{ij}\} : P \in \mathcal{S}_{2i-1,j-1}\} \cup \{P \cup \{\bar{x}_{ij}\} : P \in \mathcal{S}_{2i,j-1}\} & \text{if } j > 1. \end{cases}$$

Thus,

$$\begin{aligned} \mathcal{S}_{1,1} &= \{\{x_{11}\}, \{\bar{x}_{11}\}\} \\ \mathcal{S}_{1,2} &= \{\{x_{12}, x_{11}\}, \{x_{12}, \bar{x}_{11}\}, \{\bar{x}_{12}, x_{21}\}, \{\bar{x}_{12}, \bar{x}_{21}\}\} \\ \mathcal{S}_{1,3} &= \{\{x_{13}, x_{12}, x_{11}\}, \{x_{13}, x_{12}, \bar{x}_{11}\}, \{x_{13}, \bar{x}_{12}, x_{21}\}, \{x_{13}, \bar{x}_{12}, \bar{x}_{21}\}, \\ &\quad \{\bar{x}_{13}, x_{22}, x_{31}\}, \{\bar{x}_{13}, x_{22}, \bar{x}_{31}\}, \{\bar{x}_{13}, \bar{x}_{22}, x_{41}\}, \{\bar{x}_{13}, \bar{x}_{22}, \bar{x}_{41}\}\}. \end{aligned}$$

Over variable set $X \cup \{y_0, y_1, y_2, \ldots\}$, define

$$\mathcal{R}_{k,r} = \left\{ \begin{array}{l} \{P \cup \{y_0, \bar{y}_1\} : P \in \mathcal{S}_{1,k-2}\} \cup \\ \{\cup_{i=1}^{r-2}\{P \cup \{y_i, \bar{y}_{i+1}\} : P \in \mathcal{S}_{i+1,k-2}\}\} \cup \\ \{P \cup \{y_{r-1}, y_0\} : P \in \mathcal{S}_{r,k-2}\} \cup \\ \{P \cup \{\bar{y}_0, \bar{y}_r\} : P \in \mathcal{S}_{r+1,k-2}\} \cup \\ \{\cup_{i=r+1}^{2r-2}\{P \cup \{y_{i-1}, \bar{y}_i\} : P \in \mathcal{S}_{i+1,k-2}\}\} \cup \\ \{P \cup \{y_{2r-2}, \bar{y}_0\} : P \in \mathcal{S}_{2r,k-2}\} \end{array} \right\}.$$

For example,

$$\mathcal{R}_{3,3} = \left\{ \begin{array}{l} \{y_0, \bar{y}_1, x_{11}\}, \{y_0, \bar{y}_1, \bar{x}_{11}\}, \{y_1, \bar{y}_2, x_{21}\}, \{y_1, \bar{y}_2, \bar{x}_{21}\}, \\ \{y_2, y_0, x_{31}\}, \{y_2, y_0, \bar{x}_{31}\}, \{\bar{y}_0, \bar{y}_3, x_{41}\}, \{\bar{y}_0, \bar{y}_3, \bar{x}_{41}\}, \\ \{y_3, \bar{y}_4, x_{51}\}, \{y_3, \bar{y}_4, \bar{x}_{51}\}, \{y_4, \bar{y}_0, x_{61}\}, \{y_4, \bar{y}_0, \bar{x}_{61}\} \end{array} \right\}.$$

Algorithm RW, applied to any k-CNF expression that contains a subset of clauses that can be mapped to $\mathcal{R}_{k,r}$ ($r \geq 2$) after renaming of variables, always results in a certificate of unsatisfiability for that expression. Therefore, if random (n, m, k)-CNF expression ϕ has such a subset with high probability, then RW will conclude unsatisfiability with high probability.

It is straightforward to obtain the probability that there is a subset of clauses that maps to $\mathcal{R}_{k,r}$ for some r. Expression $\mathcal{R}_{k,r}$ contains $r2^{k-1}$ clauses and $r2^{k-1} - 1$ distinct variables and is *minimally unsatisfiable*: that is, it is unsatisfiable, but removal of any clause produces a satisfiable set. This property is important to the analysis because a minimally unsatisfiable expression can be padded with additional clauses to get other unsatisfiable clause patterns with a much higher ratio of variables to clauses, but the probability that such a padded pattern exists is not greater than the probability that one of its base minimally unsatisfiable sets exist. The existence probability crosses from tending to 0 to tending to 1 at a particular ratio of m/n. This can be estimated by finding conditions that set the average number of minimally unsatisfiable sets to ∞ in the limit. The probability that a particular subset of $r2^{k-1}$ clauses matches the pattern $\mathcal{R}_{k,r}$ for a particular choice of variables is $(r2^{k-1})!/\left(2^k\binom{n}{k}\right)^{r2^{k-1}}$. There are $\binom{n}{r2^{k-1}-1}$ ways to choose the variables, and each of $2^{r2^{k-1}-1}$ ways to complement the chosen variables presents a pattern that RW can handle. Futhermore, any of the $(r2^{k-1} - 1)!$ permutations of those variables also presents a solvable pattern. Finally, there are $\binom{m}{r2^{k-1}}$ ways to choose clauses. Therefore, the average number of $\mathcal{R}_{k,r}$ patterns in ϕ is the product of the preceding terms, which is about $(k!m/(2^{k-1}n^{(k-1)+1/r2^{k-1}}))^{r2^{k-1}}$. This tends to ∞ when $m/n^{(k-1)} > (n\omega(n))^{1/r2^{k-1}}$ where $\omega(n)$, is any slowly growing function of n. Setting $r = \lfloor \ln^{1+\varepsilon}(n) \rfloor$, where ε is a small constant, is sufficient to give the following result (in this case patterns are small enough for a second moment analysis to carry through).

Theorem 4.17. *Algorithm* RW *succeeds, with probability tending to 1, on random (n, m, k)-CNF expressions if $m/n > n^{k-2}2^{k-1}/k!$.*

This analysis shows the importance of the ratio of the number of variables to the number of clauses in a minimally unsatisfiable expression. A higher ratio means a lower cutoff for m/n^{k-1}. We remark that the first results of this nature appearing in print, as far as we know, are due to Xudong Fu [49]. Fu used a minimally unsatisfiable pattern that he called a *Flower* to achieve the foregoing result. But Flowers have $2^{k-1}(r + k - 1)$ clauses and $2^{k-1}r + k - 1$ variables. Because the difference between clauses and variables in $R_{k,r}$ is 1, whereas the difference is $(2^{k-1} - 1)(k - 1)$ in the case of Flowers, we chose to present $R_{k,r}$ instead.

Improving on the preceding result is hard. An obvious initial attempt consists of finding minimally unsatisfiable clause patterns where the ratio of variables to clauses is higher than 1 and it is hoped, close to k (recall that a ratio of 1 is roughly what is achieved earlier). But, the following result, which can be proved by induction on the number of variables in a minimally unsatisfiable set and first appeared in [9] (see also [11] and [68]), is discouraging:

Theorem 4.18. *If ϕ is a minimally unsatisfiable CNF expression with m clauses, then ϕ has less than m variables.*

Notice that Theorem 4.18 does not say anything about the number of literals in a clause: it could be fixed at any k or different for each clause and the result still holds.

Theorems 4.17 and 4.18 do not say that resolution fails to prove unsatisfiability in polynomial time with high probability when $m/n^{(k-1)} \to 0$. These theorems suggest only that looking for clause patterns that cause an algorithm such as RW to succeed often for such values of m/n is likely not to bear fruit. However, a deeper and illuminating argument shows that resolution cannot be successful, with high probability, for a wide range of m/n values.

A number of papers consider the probabilistic performance of resolution for random unsatisfiable (n, m, k)-CNF expressions (for example, [11, 12, 24, 49]). The following two results are noteworthy.

Theorem 4.19. *For $k > 3$, an unsatisfiable random (n, m, k)-CNF expression has only exponential size resolution proofs, with probability tending to 1, if $m/n^{(k+2)/4-\varepsilon} < 1$ for any $\varepsilon > 0$. A random $(n, m, 3)$-CNF expression has only exponential size resolution proofs, with probability tending to 1, if $m/n^{6/5-\varepsilon} < 1$, for any $\varepsilon > 0$.*

Theorem 4.20. *There exists a DPLL variant that verifies the unsatisfiability of a random (n, m, k)-CNF expression in polynomial time with probability tending to 1 when $m/n > (n/\log(n))^{k-2}$.*

The fact that no resolution result better than that of Theorem 4.20 has been found is surprising: it seems hard to imagine that RW is so close to what is best possible for resolution. But a result of [12] shows that at least some natural DPLL variants cannot do much better. These pessimistic results motivate the results of the next section.

4.6.5 *Algorithms for Unsatisfiability: A Spectral Analysis*

The pessimistic results obtained for verifying unsatisfiability via resolution algorithms has motivated the search for alternatives. In this section a simple one is presented and an analysis outlined. A decision is made based on counting variables that must be set to 1 or 0 if a model exists.

The rough idea is as follows. Given a random (n, m, k)-CNF expression ϕ, throw out all clauses except those containing all uncomplemented literals and those containing all complemented literals. Let n_+ be the minimum number of variables that must be set to 1 to satisfy the uncomplemented clauses, and let n_- be the number of variables that must be set to 0 to satisfy the complemented clauses. If $n_+ + n_- > n$, then some variable must be set to both 1 and 0 if ϕ has a model. But this is impossible, so ϕ cannot have a model. Hence, this simple counting argument provides a means to verify unsatisfiability. The only problem is that finding n_+ and n_- is NP-complete. However, it may be sufficient merely to approximate n_+ and n_- closely enough. The following explains how based on work reported in [52] (warning: some simplifications have been made). We start with this theorem, which is obvious from the previous discussion:

Theorem 4.21. *If k-CNF expression ϕ has a model, then there is a subset V' of $n/2$ variables such that either ϕ has no all-uncomplemented clause or no all-complemented clause taken strictly from V'.*

For purposes of discussion, fix $k = 4$. Construct two graphs G_+ and G_-, corresponding to the uncomplemented and complemented clause sets of ϕ, respectively. Both G_+ and G_- have $\binom{n}{2}$ vertices, and each vertex of G_+ and G_- is uniquely labeled by a pair of variables. An edge between two vertices of G_+ exists if and only if the uncomplemented clause set of ϕ contains a clause consisting of all variables labeling both its endpoints. An edge between two vertices of G_- exists if and only if the all-complemented clause set of ϕ contains a clause consisting of all variables labeling both its endpoints. It follows from Theorem 4.21 that:

Theorem 4.22. *If ϕ has a model, either G_+ or G_- has an independent set of size greater than $\binom{n/2}{2} \approx n^2/8$.*

The following theorem connects independent sets with eigenvalues.

Theorem 4.23. ([69].) *Let $G(V, E)$ be an undirected graph with vertex set $V = \{1, \ldots, n'\}$. Let $0 \leq p \leq 1$. Define $n' \times n'$ matrix $A_{G,p}$ such that $A_{G,p}(i, j) = 1$ if edge $\langle i, j \rangle \notin E$ and $A_{G,p}(i, j) = 1 - 1/p$ otherwise ($A_{G,p}(i, i) = 1$). Let $\lambda_1(A_{G,p})$ be the largest eigenvalue of $A_{G,p}$. Let $\alpha(G)$ be the size of the largest independent set of $G(V, E)$. For any possible p, $\lambda_1(A_{G,p}) \geq \alpha(G)$.*

The problem of determining the eigenvalues of $A_{G,p}$ can be handled by the following

Theorem 4.24. (For example, [79].) *The eigenvalues of $A_{G,p}$ can be computed with relative error less than 2^{-b} in time $O((n')^3 + (n' \log^2(n')) \log(b))$.*

The G of matrix $A_{G,p}$ will be G_+ once and G_- once. The p parameter is set to represent the probability that a pair of vertices of $A_{G,p}$ is connected by an edge. That is,[11] the number of clauses represented in G is about $m' = m/16$ and $p = m'\binom{n'}{2} = (m/16)\binom{n^2}{2}$. A theorem such as the following can then be proved.

Theorem 4.25. (Adapted from [52].) *Let* $m' = (\ln^7(n')/2)n'$ *and* $p = m'\binom{n'}{2} = \ln^7(n')/(n'-1)$. *With high probability,*

$$\lambda_1(A_{G,p}) \le 2n'(1/(\ln(n'))^{7/2})(1+o(1)).$$

Theorems 4.23 and 4.25 give an adequate bound on the size of the independent sets of G_+ and G_-. These results are used as follows. Suppose $m = 16(\ln(n^2))^7 n^2 = 16(2^7)(\ln^7(n))n^2$. Because $n' = n^2$ and $m' = m/16$, $A_{G,p}$ has about $2^7(\ln^7(\sqrt{n'}))n' = (\ln^7(n'))n'$ edges chosen with probability $p = \ln^7(n')/(n'-1)$. From Theorems 4.23 and 4.25 $\alpha(G_+) \le n^2/16$ and $\alpha(G_-) \le n^2/16$ with high probability. Hence, with high probability the maximum independent set of G_+ and of G_- is less than that required in Theorem 4.22, and finding the eigenvalues is enough to prove it in polynomial time. The general result is:

Theorem 4.26. *Random* (n, m, k)*-CNF expressions can be certified as unsatisfiable in polynomial time with high probability if* $m > n^{k/2+o(1)}$, $k \ge 4$, *and if* $m > n^{3/2+\varepsilon}$, ε *an arbitrarily small constant,* $k = 3$.

4.6.6 Polynomial Time Solvable Classes

Expressions that are members of certain polynomial time solvable classes, such as those defined in Section 4.4, are not generated frequently enough, for interesting densities m/n, to assist in determining whether random (n, m, k)-CNF expressions have models. This is unlike the case for random (n, m, p)-CNF expressions (Definition 4.5 and results in Section 4.5). We illustrate with a few examples. Whenever we use ϕ, it is to be understood that ϕ is a random (n, m, k)-CNF expression.

First consider the frequency with which random (n, m, k)-CNF expressions are Horn or hidden Horn. The probability that a clause is Horn is $(k+1)/2^k$. Therefore, the probability that ϕ is Horn is $((k+1)/2^k)^m$, which tends to 0 for any fixed k. For a hidden Horn expression, regardless of switch set, there are only $k+1$ out of 2^k ways (negation patterns) that a random clause can become Horn. Therefore, the expected number of successful switch sets is $2^n((k+1)/2^k)^m$, which tends to 0 if $m/n > 1/(k - \log_2(k+1))$. Thus, ϕ is not hidden Horn, with probability tending to 1, if $m/n > 1/(k - \log_2(k+1))$. Even when $k = 3$, this is $m/n > 1$. But this bound can be improved considerably by finding complex

[11] The factor of $1/16$ is due to G_+ and G_- representing, with high probability, $1/16$ of the m input clauses in the case that $k = 4$.

structures that imply an expression cannot be hidden Horn. Such a structure is presented next.

The following result for q-Horn expressions is taken from [46]. For $p = \lfloor \ln(n) \rfloor \geq 4$, call a set of p clauses a *c-cycle* if all but two literals can be removed from each of $p-2$ clauses, all but three literals can be removed from two clauses, the variables can be renamed, and the clauses can be reordered in the following sequence:

$$\{v_1, \bar{v}_2\}, \{v_2, \bar{v}_3\} \ldots \{v_i, \bar{v}_{i+1}, v_{p+1}\} \ldots \{v_j, \bar{v}_{j+1}, \bar{v}_{p+1}\} \ldots \{v_p, \bar{v}_1\}, \quad (4.8)$$

where $v_i \neq v_j$ if $i \neq j$. We use the term cycle to signify the existence of cyclic paths through clauses that share a variable: that is, by jumping from one clause to another clause only if the two clauses share a variable, one may eventually return to the starting clause. Given a c-cycle $\mathcal{C} \subset \phi$, if no two literals removed from $\mathcal{C}$ are the same or complementary, then $\mathcal{C}$ is called a *q-blocked c-cycle*.

If ϕ has a q-blocked c-cycle, then it is not q-Horn. Let a q-blocked c-cycle in ϕ be represented as before. Develop inequalities (4.4) and (4.5) for ϕ. After rearranging terms in each, a subset of these inequalities is as follows:

$$\alpha_1 \leq Z - 1 + \alpha_2 - \cdots \quad (4.9)$$

$$\ldots$$

$$\alpha_i \leq Z - 1 + \alpha_{i+1} - \alpha_{p+1} - \cdots$$

$$\ldots$$

$$\alpha_j \leq Z - 1 + \alpha_{j+1} - (1 - \alpha_{p+1}) - \cdots$$

$$\ldots$$

$$\alpha_p \leq Z - 1 + \alpha_1 - \cdots . \quad (4.10)$$

From inequalities (4.9) to (4.10) we deduce

$$\alpha_1 \leq pZ - p + \alpha_1 - (1 - \alpha_{p+1} + \alpha_{p+1}) - \cdots$$

or

$$0 \leq pZ - p - 1 - \cdots ,$$

where all the terms in $\cdots$ are nonnegative. Thus, all solutions to (4.9) through (4.10) require $Z > (p+1)/p = 1 + 1/p = 1 + 1/\lfloor \ln^2 n \rfloor > 1 + 1/n^\beta$ for any fixed $\beta < 1$. This violates the requirement that $Z \leq 1$ in order for ϕ to be q-Horn.

The expected number of q-blocked c-cycles can be found and the second moment method applied to give the following result.

Theorem 4.27. *A random (n, m, k)-CNF expression is not q-Horn, with probability tending to 1, if $m/n > 4/(k^2 - k)$.*

For $k = 3$, this is $m/n > 2/3$.

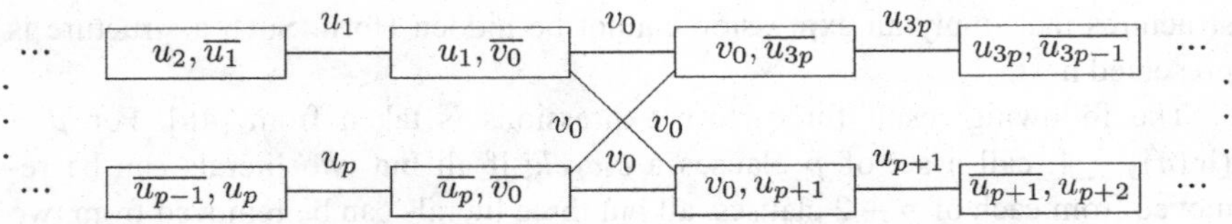

Figure 4.12. A "criss-cross loop" of $t = 3p + 2$ clause nodes. Each box or clause node represents a clause. An edge exists between two clause nodes if and only if the two corresponding clauses have a variable in common and the edge is labeled by that variable. Only "cycle literals" are shown in the nodes; "padding literals," required for $k \geq 3$, are present but are not shown. Padding literals are different from cycle variables of a criss-cross loop.

A similar analysis yields the same results for hidden Horn, SLUR, CC-balanced, or extended Horn expressions. The critical substructure that causes ϕ not to be SLUR is called a criss-cross loop. An example is shown in Figure 4.12. Looking at Figure 4.12 and expression (4.8), we see that the SLUR and q-Horn classes are vulnerable to certain types of "cyclic" structures. Most other polynomial time solvable classes are similarly vulnerable to cyclic structures of various kinds. But the generator of random expressions is relatively blind to such cyclic structures: as m/n is increased, at some point cycles begin to appear in ϕ in abundance, and when this happens, "killer" cycles also show up. Hence, cycles appear in abundance when $m/n > 1/O(k^2)$, and this is where "killer" cycles appear, too.

But the Matched class (see Definition 4.4), is not affected by cycles, and consequently probabilistic analysis tells us that there are many "more" Matched expressions than SLUR or q-Horn expressions. Let $Q \subset \phi$ be any subset of clauses of ϕ. Define the *neighborhood* of Q, denoted $\mathbf{V}(Q)$, to be the set of variables that occur in Q. Recall that the bipartite graph G^ϕ contains two vertex sets C^ϕ and V^ϕ, where the elements of C^ϕ are the clauses of ϕ, the elements of V^ϕ are the variables of ϕ, and the edges of G^ϕ connect $v \in V^\phi$ to $c \in C^\phi$ just when variable v appears (complemented or uncomplemented) in clause c. Thus $\mathbf{V}(Q)$ is also the set of vertices in V^ϕ adjacent to the vertices corresponding to Q.

Definition 4.6. *Define the* deficiency *of Q, denoted $\delta(Q)$, as $\delta(Q) = |Q| - |\mathbf{V}(Q)|$, that is, the excess of clauses over distinct variables in those clauses. A subset $Q \subseteq C^\phi$ is said to be* deficient *if $\delta(Q) > 0$.*

The following theorem is well known.

Theorem 4.28. (Hall's Theorem [57].)
Given a bipartite graph with vertex sets V^ϕ and C^ϕ, a matching that includes every vertex of C^ϕ exists if and only if no subset of C^ϕ is deficient.

Theorem 4.29. *Random (n, m, k)-CNF expressions are Matched expressions with probability tending to 1 if $m/n < r(k)$ where $r(k)$ is given by the* *following table [46]:*

k	$r(k)$
3	0.64
4	0.84
5	0.92
6	0.96
7	0.98
8	0.990
9	0.995
10	0.997

Theorem 4.29 may be proved by finding a lower bound on the probability that G^ϕ has a matching that includes every vertex of C^ϕ. By Theorem 4.28 it is sufficient to prove an upper bound on the probability that there exists a deficient subset of C^ϕ, then show that the bound tends to 0 for $m/n < r(k)$ as given in the theorem. Using the first moment method, a sufficient upper bound is given by the expected number of deficient subsets.

These results are interesting for at least two reasons. First, Theorem 4.29 says that random expressions are Matched expressions with high probability if $m/n < r(k)$ and $r(k)$ is roughly 1. But, by Theorem 4.27 and similar results, random expressions are almost never one of the well-studied classes mentioned earlier unless $m/n < 4/(k^2 - k)$. This is somewhat disappointing because all the other classes were proposed for rather profound reasons, usually reflecting cases when corresponding instances of integer programming present polytopes with some special properties. And in spite of all the theory that helped establish these classes, the Matched class, ignored in the literature because it is so trivial in nature, turns out to be, in some probabilistic sense, much bigger than all the others.

Second, the results provide insight into the nature of larger classes of polynomial time solvable expressions. Classes vulnerable to "killer" cycles appear to be handicapped relative to classes that are not. In fact, the Matched class may be generalized considerably to larger polynomial time solvable classes such as Linear Autarkies [71, 74].

4.6.7 Randomized Algorithms: Upper Bounds

Exploitable properties have been used to find $2^{n(1-\varepsilon)}$ upper bounds on the number of steps needed to solve a given k-CNF expression, ϕ. The first nontrivial upper bound is found in the classic paper of Monien and Speckenmeyer [78] and is based on the notion of autark assignments. An assignment to a set of variables is said to be *autark* if all clauses that contain at least one of those variables are satsfied by the assignment. If an autark assignment is found during search, it is sufficient to fix the autark assignment below that point in the search: that is, backtracking to consider other values for the variables of the autark assignment is unnecessary at the point the autark assignment is discovered. An example is an assignment causing a pure literal to have value 1: clearly, it is unnecessary to expand the search space with a pure literal set to 0. The search algorithm analyzed in [78] is a DPLL variant that branches on all assignments to the variables of a shortest width clause except the one falsifying the clause. Such assignments either produce a clause of shorter width or are autark. This provides an upper bound of $O(\alpha_k^n)$, where α_k is the largest real root of the equation

$$\alpha_k^k - 2\alpha_k^{k-1} + 1 = 0.$$

Thus, if $k = 3$, the upper bound is $O(1.618^n)$ steps. The bound increases with increasing k and approaches $O(2^n)$. A number of more complex variants have been studied, and the best bounds obtained in this way, so far, appear to be $O(1.505^n)$ [70] and $O(1.497^n)$ [88] for 3-SAT.

Better bounds have been obtained for probabilistic algorithms. A family of such algorithms exploit a relationship between the structure of a k-CNF expression and structure of the space of models expressed in terms of isolated models and critical clauses [80]. A *j-isolated model* is a model with exactly $n - j$ assignment neighbors differing in one variable that are also models. If j is close to n, the set of such models, called nearly isolated, for a given k-CNF expression has a relatively "short" description. Thus, a nearly isolated model, if one exists, is relatively easy to find by searching the space of relatively "short" descriptions. The important observation is that any satisfiable k-CNF expression either has a nearly isolated model or has very many models. In either case, a model can be found relatively quickly.

Short descriptions depend on the notion of critical clauses. If M is a model but reversing variable v_i in M is not a model, then there must exist a clause with exactly one literal that has value 1 under M, and that literal is either v_i if $v_i \in M$ or $\bar{v}_i$ if $v_i \notin M$. This clause is said to be *critical* for variable i at model M. A critical clause cannot be critical for two different variables at the same model. Therefore, a j-isolated model has j critical clauses.

A description of a model can be encoded in fewer than n bits as follows. Let π be a permutation of $\{1, 2, \ldots, n\}$. Let M be a model for a k-CNF expression ϕ. Define a string $x^M \in \{0, 1\}^n$ such that $x_i^M = 1$ if and only if $v_i \in M$ (that is, v_i is assigned value 1). An encoding of x^M with respect to π, written $I_\pi(x^M)$, is a permuation of its bits according to π but with the ith bit deleted for any i such that (1) there is a critical clause for the variable v_{π_i} at M, and (2) v_{π_i} or $\bar{v}_{\pi_i}$ is the last literal in the clause if literals are ordered by π.

An encoding can be efficiently decoded. That is, a model M can be recovered efficiently from its encoding as described previously. This is done by starting with the original expression ϕ, assigning values to variables one at a time, in order, according to π, and reducing ϕ accordingly. Bits of the encoding are used to decide values of assignment M except in the case that the reduced ϕ has a unit clause consisting of the current variable, in order, or its complement. In that case, the current variable is set to satisfy that clause. This leads to the following lemma.

Lemma 4.30. (The Satisfiability Coding Lemma, from [80].) *If M is a j-isolated model of a k-CNF expression ϕ, then its average (over all permutations π) description length under the encoding $I_\pi(x^M)$ is at most $n - j/k$.*

This has inspired the algorithm shown in Figure 4.13. Notice that the algorithm essentially randomly chooses a permutation π and then simultaneously (1) attempts a decoding according to π and (2) checks a potential satisfying assignment according to π. Performance of the algorithm is given by the following.

Theorem 4.31. *(From [80].) The algorithm of Figure 4.13 runs in $O(n^2|\phi|2^{n-n/k})$ time and finds a model for a given satisfiable k-CNF expression with probability tending to 1.*

Procedure RANDOMSATSOLVER (ϕ)
Input: a k-CNF expression ϕ with n variables.
Output: either "give up" or a model for ϕ.

begin
  repeat the following $n^2 2^{n-n/k}$ times {
    $M := \emptyset$;
    $V := \{v : \exists C \in \phi$ such that $v \in C$ or $\bar{v} \in C\}$;
    $\phi' := \phi$;
    while $V \neq \emptyset$ do the following {
      if $\emptyset \in \phi'$ break `// No assignment extension will be a model`
      if $\phi' = \emptyset$ return M;
      randomly choose $v \in V$;
      if $\{v\} \in \phi'$ then
        $M := M \cup \{v\}$;
        $\phi' := \{C - \{\bar{v}\} : C \in \phi', v \notin C\}$;
      else if $\{\bar{v}\} \in \phi'$ then
        $\phi' := \{C - \{v\} : C \in \phi', \bar{v} \notin C\}$;
      else with probability 1/2 do the following
        $M := M \cup \{v\}$;
        $\phi' := \{C - \{\bar{v}\} : C \in \phi', v \notin C\}$;
      else do the following
        $\phi' := \{C - \{v\} : C \in \phi', \bar{v} \notin C\}$;
      $V := V \setminus \{v\}$;
    }
  }
  return "give up";
end;

Figure 4.13. A randomized algorithm for finding a model for a satisfiable k-CNF expression.

Procedure SLS (ϕ)
Input: a k-CNF expression ϕ with n variables.
Output: either "give up" or a model for ϕ.

```
begin
  V := {v : ∃C ∈ φ such that v ∈ C or v̄ ∈ C};
  repeat the following n(2(1 − 1/k))^n times {
    M := ∅;
    for each v ∈ V set M := M ∪ {v} with probability 1/2;
    repeat the following 3n times {
      Let φ' := {C ∈ φ : ∀v ∈ C, v ∉ M and ∀v̄ ∈ C, v ∈ M};
      if φ' = ∅ return M;
      randomly choose C ∈ φ';
      randomly choose literal l ∈ C;
      if l is a positive literal v then M := M ∪ {v};
      otherwise, if l is a negative literal v̄ then M := M \ {v};
    }
  }
  return "give up";
end;
```

Figure 4.14. A local-search randomized algorithm for finding a model for a satisfiable k-CNF expression. The factor n in the `repeat` loop can be improved but is good enough for the purposes of this exposition.

The complexity is easy to see. The following outlines the reason the probability of finding a model tends to 1. Suppose ϕ has a j-isolated model x^M. Fix j critical clauses as before. By Lemma 4.30, the average number (over all π) of critical variables that appear last among the variables in their critical clauses is at least j/k. Because the maximum number of critical variables is n, for at least $1/n$ fraction of the permutations the number of such variables is at least j/k. Therefore, the probability that there are at least j/k critical clauses where the critical variables occur last for π (the variable choices for a single round of the algorithm) is at least $1/n$. Because x^M is j-isolated, if the algorithm is to output x^M, it makes at most $n - j/k$ random assignments. So the probability that the algorithm outputs x^M on a round is at least $2^{-n+j/k}/n$. A straightforward summation over all models for ϕ determines a bound on the probability that the algorithm gives up.

An improved randomized algorithm that uses RANDOMSATSOLVER of Figure 4.13 is given in [81]. It finds a model, when one exists, for a 3-CNF expression with probability tending to 1 in $O(1.362^n)$ steps. The algorithm is peculiar because its performance actually deteriorates as the number of models for a given expression increases [60]. This is likely due to the emphasis of the algorithm on finding nearly isolated models and the lack of intelligence in looking for nonisolated models.

A better randomized algorithm for 3-CNF expressions, which is based on local search, is presented in [90] and shown in Figure 4.14. Each round of this algorithm begins with an assignment that is hoped to be fairly close to a model. Then, for $3n$ iterations, single variable assignment flips are made, attempting to eliminate all falsified clauses. After $3n$ iterations, if a model is not found, it is unlikely that the assignment chosen at the beginning of the round is close to a model, so the round terminates, another assignment is picked on the next round, and the process repeats.

Theorem 4.32. (Adapted from [90].) *The algorithm of Figure 4.14 has complexity $O(n|\phi|(2(1-1/k))^n)$ and finds a model for a given satisfiable k-CNF expression with probability tending to 1.*

For 3-CNF expressions, this algorithm has $O(1.333^n)$ complexity. Contrary to the randomized algorithm of Figure 4.13, the performance of this algorithm improves as the number of models of an input increases. This observation is exploited in an algorithm presented in [60], where a round begins with a random assignment τ and then the $3n$ local search steps of SLS and the $O(n)$ Davis-Putnam steps of RANDOMSATSOLVER are run on τ. The resulting algorithm has complexity $O(1.324^n)$ for 3-CNF expressions.

4.6.8 Results Inspired by Statistical Mechanics

New, startlingly successful CNF algorithms have been developed by observing similarities in CNF structure and models of matter. We describe one here for the purpose of illustrating the impact nonrigorous methods may have on future SAT algorithm development.

We are concerned with the CNF structure that arises from the bipartite graph G^ϕ introduced at the end of Section 4.4. Recall that, given CNF expression ϕ, G^ϕ has vertex set V^ϕ corresponding to variables, vertex set C^ϕ corresponding to clauses, and edge set E such that for $v \in V^\phi$ and $c \in C^\phi$, $\langle v, c\rangle \in E$ if and only if clause c has either literal v or $\bar{v}$. It has been observed that if G^ϕ has few "short" cycles and ϕ has no model, then ϕ is a hard problem for resolution. For random (n, m, k)-CNF expressions, not only are cycles "long," but the minimum length cycle grows logarithmically with n for fixed m/n. Even when m/n is such that a random expression has a model with high probability, it is difficult for a variety of proposed algorithms to find such a model when m/n is greater than a certain threshold (about 4 in the case of random $(n, m, 3)$-CNF expressions).

At the core of the statistical physics of disordered systems is the spin glass problem. A model of a spin glass contains binary variables, $\sigma_1, \sigma_2, \sigma_3, \ldots$, called spins taking values $+1$ or -1, and energy relationships $E_1, E_2, E_3, \ldots$ among small groups of neighboring spins. This model can be represented graphically, as depicted in Figure 4.15, where circled vertices are spins, boxed vertices are energy functions, and edges show the dependence of each energy function on spins. Such a graph is bipartite: there is no edge between a pair of circle vertices and no edge between a pair of box vertices. It is a general rule in physics that systems tend to migrate toward their minimum energy (that is, maximum entropy) state. Thus,

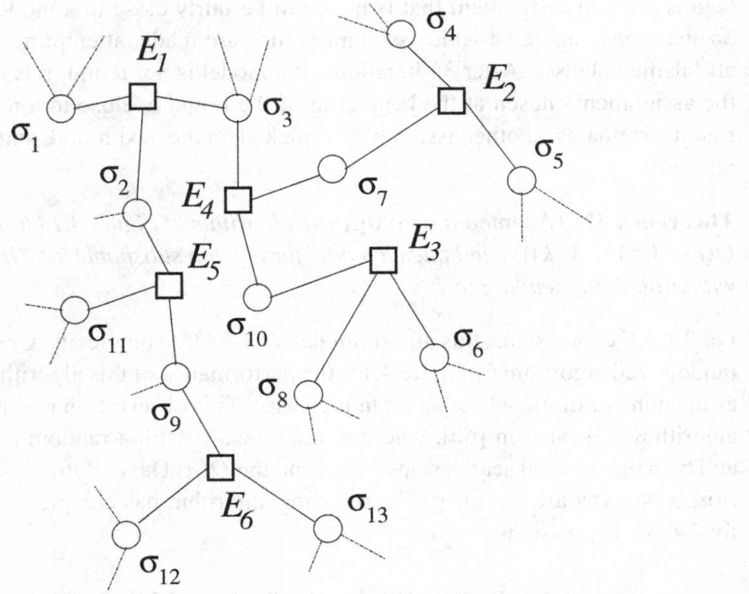

Figure 4.15. Model of a spin glass. Variables $\sigma_1, \sigma_2, \ldots,$ called spins, are binary taking values $+1$ or -1. Functions $E_1, E_2, \ldots$ are energy functions, each of which depends on a small number of neighboring spins. The spin glass problem is to find spin values that minimize the total energy of a given collection of energy functions.

the spin glass problem is to determine the minimum energy of a given spin glass model.

The spin glass problem is analogous to the problem of determining a model for a given CNF expression: spins take the role of Boolean variables, energy functions take the role of clauses, and energy corresponds to clauses satisfied. Thus, if the minimum energy of a spin glass is 0, the corresponding CNF expression has a model, and if the minimum energy is greater than 0, the expression has no model. Energy functions consistent with the so-called *Ising model* and expressing this behavior for k literal clauses composed of literals taken from variables $v_{i_1}, v_{i_2}, \ldots, v_{i_k}$ corresponding to spins $\sigma_{i_1}, \sigma_{i_2}, \ldots, \sigma_{i_k}$ are

$$E_i = 2\prod_{r=1}^{k}\frac{1 + J_i^r \sigma_{i_r}}{2},$$

where J_i^r is $+1$ if the rth literal of the ith clause is complemented and -1 if it is uncomplemented.[12] Thus, the energy of clause $\{v_1, \bar{v}_2, \bar{v}_3\}$ is $(1 - \sigma_1)(1 + \sigma_2)$

[12] The factor of 2 is explained later.

$(1 + \sigma_3)/4$. Observe that this is 0 if $\sigma_1 = +1$, $\sigma_2 = -1$, or $\sigma_3 = -1$ and is 2 otherwise. The energy of an entire system of m energy functions is

$$E = \sum_{i=1}^{m} E_i.$$

All clauses are satisfied if and only if the energy of the analogous spin glass is 0.

Where can physicists help? They can make assumptions analogous to those that apply to the physical world and thereby aim for a level of understanding that computer scientists would not have thought of. It is natural in physics to consider the probability distribution of spins:

$$Pr(\sigma_1, \sigma_2, ..., \sigma_n) = \frac{1}{Z} e^{-\frac{1}{T}E}$$

where T is temperature and Z is a normalizing constant. This distribution follows from the observation that a system in equilibrium tends to seek its highest entropy, or equivalently, lowest energy state. At $T = 0$, the lowest energy states are the only significant terms contributing to this distribution (nonrigorously) and, given a system of energy functions as defined earlier, one is interested in the distribution at the 0, or lowest possible, energy state. It can be calculated in a fairly straightforward manner for each spin separately due to the assumption of a thermodynamic limit: that is, $\lim_{n\to\infty} E/n$ is bounded. By this assumption, what is happening at any particular spin is independent of what is happening at nearly all other spins. Thus, probability distributions are assumed to be decomposed into products of probability distributions of influential variables.

Let us see how this may apply to k-CNF analogs. For simplicity of notation, and without loss of generality, let E_i be a function of spins $\sigma_1, \ldots, \sigma_k$. Let $h_{j\to i} \cdot \sigma_j$ be the energy contributed to E_i by the spin σ_j assuming that the effect of E_i on σ_j is disregarded. The h terms are called magnetic fields in physics. Let $u_{i\to j}$ be the contribution to the magnetic field of spin σ_j from E_i. Consider the marginal distribution for σ_1 of E_i. If the Ising model is assumed, one writes

$$\sum_{\sigma_2,\ldots,\sigma_k} e^{-\frac{1}{T}(E_i - h_{2\to i}\cdot\sigma_2 - \ldots - h_{k\to i}\cdot\sigma_k)} = e^{\frac{1}{T}(w_{i\to 1} + u_{i\to 1}\cdot\sigma_1)}.$$

Because one is interested in the case $T = 0$, this simplifies to

$$\min_{\sigma_2,\ldots,\sigma_k} \{E_i - h_{2\to i}\cdot\sigma_2 - \ldots - h_{k\to i}\cdot\sigma_k\} = -w_{i\to 1} - u_{i\to 1}\cdot\sigma_1.$$

For the k-CNF problem, E_i has been given earlier and

$$\begin{aligned} w_{i\to 1} &= |h_{2\to i}| + \ldots + |h_{k\to i}| - \theta(J_i^2 \cdot h_{2\to i}) \cdot \ldots \cdot \theta(J_i^k \cdot h_{k\to i}) \\ u_{i\to 1} &= -J_i^1 \cdot \theta(J_i^2 \cdot h_{2\to i}) \cdot \ldots \cdot \theta(J_i^k \cdot h_{k\to i}), \end{aligned}$$

where $\theta(x) = 1$ if $x > 0$ and $\theta(x) = 0$ if $x \le 0$. Interpreting $h_{j\to i} > 0$ as evidence that σ_j should be $+1$ and $h_{j\to i} < 0$ as evidence that σ_j should be -1 to minimize the ith energy function (satisfy the ith clause), $w_{i\to 1} + u_{i\to 1} \cdot \sigma_1 = |h_{2\to i}| + ... + |h_{k\to i}|$ if for any $\sigma_2 ... \sigma_k$ the evidence *supports* the current value of the corresponding spin, or if no support by these h variables is given and σ_1 has a value that minimizes the ith energy function. But $w_{i\to 1} + u_{i\to 1} \cdot \sigma_1 = |h_{2\to i}| + ... + |h_{k\to i}| - 2$ if σ_1 is not set to minimize E_i and no support is given for any of $\sigma_2, \ldots, \sigma_k$ (the ith clause is falsified). This explains the factor of 2 used in defining clausal energy E_i. The minimum energy of the entire system given that spin, say σ_j, has a particular value is

$$E|_{\sigma_j} = C - \sum_{i:\sigma_j \in E_i} w_{i\to j} - \sigma_j \sum_{i:\sigma_j \in E_i} u_{i\to j},$$

where C is some constant. Write

$$h_j = \sum_{i:\sigma_j \in E_i} u_{i\to j}. \tag{4.11}$$

Let $Q_{i\to j}(u)$ be the probability distribution of $u_{i\to j}$. Let $P_{j\to i}(h)$ be the probability distribution of the contribution of h_j to function E_i. Suppose energy functions $E_{\pi_1}, \ldots, E_{\pi_p}$ influence spin σ_j. By independence of distributions, which follows from the assumption of thermodynamic limit,

$$P_{j\to i}(h) = C_{j\to i} \sum_{u_1,\ldots,u_p : \sum_{x=1}^{p} u_x = h} Q_{\pi_1\to j}(u_1) \cdot \ldots \cdot Q_{\pi_p\to j}(u_p) e^{y|h|}.$$

For each spin σ_j of E_i, let $\sigma_{\pi'_1}, \ldots, \sigma_{\pi'_{p'}}$ be the remaining spins of E_i. Let $\theta(J_i^{\pi'_1} \cdot h_1) \cdot \ldots \cdot \theta(J_i^{\pi'_{p'}} \cdot h_{p'})$ be denoted by $w_{i,1\ldots p'}$. Write

$$Q_{i\to j}(u) = C_{i\to j} \sum_{\substack{h_1,\ldots,h_{p'}: \\ u = -J_i^j \cdot w_{i,1\ldots p'}}} P_{\pi'_1\to i}(h_1) \cdot \ldots \cdot P_{\pi'_{p'}\to i}(h_{p'}) e^{-y\cdot w_{i,1\ldots p'}}.$$

The terms $C_{i\to j}$ and $C_{j\to i}$ are normalizing constants, and y is a parameter that expresses the rate of change of complexity with respect to E/n, and complexity is a measure of the number of different energy states possible. Because this information is not known generally, y must be guessed.

The values for $Q_{i\to j}(u)$ and $P_{j\to i}(h)$ may be computed by the algorithm of Figure 4.16. It is then a simple matter to determine $P_j(h)$, the distribution for the field acting on spin σ_j. But the value of $P_j(h)$ suggests a setting for spin σ_j that minimizes energy: namely, $+1$ if $\sum_{h>0} P_j(h) > \sum_{h<0} P_j(h)$ and -1 if $\sum_{h>0} P_j(h) < \sum_{h<0} P_j(h)$. If $\sum_{h>0} P_j(h) = \sum_{h<0} P_j(h)$ or $P_h(0) = 1$, no bias is detected. This suggests the algorithm of Figure 4.17 for solving CNF expressions: assign values to "biased" variables first, recomputing $P_j(h)$ after each variable is assigned, then apply a standard SAT solver to complete the assignment.

Procedure SP (G^ϕ)
Input: bipartite graph G^ϕ relating energy functions to spins.
Output: set of distributions of all spin fields.

begin
for all energy function E_i to spin σ_j edges in G^ϕ define:

$$Q_{i\to j}(u) = \begin{cases} c_{i\to j} & \text{if } u = 0 \\ 1 - c_{i\to j} & \text{if } u = J_i^j \\ 0 & \text{otherwise} \end{cases}$$

and initialize $c_{i\to j}$ randomly;
repeat the following until convergence {
select an energy function E_i;
for each spin σ_j input to E_i compute $P_{j\to i}(h) =$

$$C_{j\to i} \sum_{u_1,\ldots,u_p:\sum_{x=1}^{p} u_x = h} Q_{\pi_1\to j}(u_1)\cdot \ldots \cdot Q_{\pi_p\to j}(u_p)e^{y|h|};$$

where $E_{\pi_1}, \ldots, E_{\pi_p}$ are functions depending on spin σ_j except E_i and $C_{j\to i}$ is a normalizing constant;
for each spin σ_j input to E_i let $\theta(J_i^{\pi_1}\cdot h_1)\cdot \ldots \cdot \theta(J_i^{\pi_{p'}}\cdot h_{p'})$ be denoted by $w_{i,1\ldots p'}$ (that is, from Page 151, $w_{i,1\ldots p'}$ denotes $w_{i\to j} - |h_{\pi'_1}| - \ldots - |h_{\pi'_{p'}}|$ where $\sigma_{\pi'_1}\ldots\sigma_{\pi'_{p'}}$ are spins of E_i) and compute $Q_{i\to j}(u) =$

$$C_{i\to j} \sum_{\substack{h_1,\ldots,h_{p'}:\\ u = -J_i^j\cdot w_{i,1\ldots p'}}} P_{\pi'_1\to i}(h_1)\cdot \ldots \cdot P_{\pi'_{p'}\to i}(h_{p'})e^{-y\cdot w_{i,1\ldots p'}}$$

}
Compute $P_j(h) =$

$$C_j \sum_{u_1,\ldots,u_{p+1}:\sum_{x=1}^{p+1} u_x = h} Q_{\pi_1\to j}(u_1)\cdot \ldots \cdot Q_{\pi_{p+1}\to j}(u_{p+1})e^{y|h|}.$$

where $E_{\pi_1}\ldots E_{\pi_{p+1}}$ are functions depending on spin σ_j;
return $P_j(h)$ for all j;
end;

Figure 4.16. An algorithm for computing the distributions $P_j(h)$.

Procedure SID (ϕ)
Input: a CNF expression ϕ.
Output: either "unsatisfiable" or a model for ϕ.

```
begin
    repeat the following indefinitely {
        M := ∅;
        establish G^φ and variables u_{i→j};
        randomly choose values for u_{i→j};
        repeat the following until ∅ ∈ φ {
            run SP on G^φ to evaluate P_j(h);
            if P_j(h) = 0 except at h = 0 for all j then {
                apply a SAT solver to φ;
                if it finds a model then return that model;
            }
            for all σ_j, let w_j^- = Σ_{h<0} P_j(h) and w_j^+ = Σ_{h>0} P_j(h);
            choose σ_j such that |w_j^+ − w_j^-| is a maximum;
            if w_j^+ > w_j^- then σ_j := +1; else σ_j := −1;
            if σ_j = +1 then {
                M := M ∪ {v_j};
                φ := {C \ {v̄_j} : C ∈ φ, v_j ∉ C};
            } else
                φ := {C \ {v_j} : C ∈ φ, v̄_j ∉ C};
            while there exists a unit clause {l} ∈ φ do the following {
                if l is an uncomplemented literal then M := M ∪ {l};
                φ := {C \ {l̄} : C ∈ φ, l ∉ C};
            }
            if φ = ∅ then return M;
        }
    }
end;
```

Figure 4.17. An algorithm for solving a CNF expression based on maximizing entropy.

Considerable success has been reported with this approach [75] on random (n, m, k)-CNF expressions and also some benchmarks that have been considered hard for a variety of SAT solvers. What is different about the method discussed here? First, it provides a way to choose initial variables and values, based on apparent probability distributions, that is intelligent enough to know when to stop. Traditional methods for choosing the first so many variables to branch on will choose as many values and variables that the user would like. Mistakes higher up the search process are very serious and, if made, can result in very long searches at the lower end. The method of Mézard and Zecchina seems to make few mistakes on many expressions. By maximizing entropy, variable/value choices tend to result in reduced expressions such that an additional variable choice will yield essentially the same reduced subexpressions regardless of its value. Second, more interdependence of CNF components is taken into account. We illustrate with the well-known Johnson heuristic [61], which chooses a variable and assigns a value so as to maximize the probability that a 0 energy state (a model) exists assuming clauses are statistically independent (an unlikely situation). Although successful in several cases, this heuristic cannot really see very far ahead of the current state. But the method of Mézard and Zecchina is designed to, in some sense, explore all possible energy states for a given expression or subexpression, particularly the lowest energy states, and present statistics on those states and their causes.

References

[1] D. Achlioptas. Lower bounds for random 3-SAT via differential equations. *Theoretical Computer Science* 265, pp. 159–85, 2001.

[2] D. Achlioptas. Setting 2 variables at a time yields a new lower bound for random 3-sat. *32nd ACM Symposium on Theory of Computing*, pp. 28–37. Association for Computing Machinery, New York, 2000.

[3] D. Achlioptas and G. Sorkin. Optimal myopic algorithms for random 3-SAT. *41st Annual Symposium on Foundations of Computer Science*, pp. 590–600. IEEE Computer Society Press, Los Alamitos, CA, 2000.

[4] D. Achlioptas and C. Moore. The asymptotic order of the random k-SAT thresholds. *43rd Annual Symposium on Foundations of Computer Science*, pp. 779–88. IEEE Computer Society Press, Los Alamitos, CA, 2002.

[5] D. Achlioptas and Y. Peres. The threshold for random k-SAT is $2^k \log 2 - O(k)$. *Journal of the American Mathematical Society* 17, pp. 947–73, 2004.

[6] D. Achlioptas, L. M. Kirousis, E. Kranakis, D. Krizanc, M. Molloy, and Y. Stamatiou. Random constraint satisfaction: A more accurate picture. *3rd Conference on the Principles and Practice of Constraint Programming*, Lecture Notes in Computer Science 1330, pp. 107–20, 1997.

[7] D. Achlioptas, L. M. Kirousis, E. Kranakis, and D. Krizanc. Rigorous results for random $(2 + p)$-SAT. *Theoretical Computer Science* 265, pp. 109–29, 2001.

[8] D. Achlioptas and M. Molloy. The analysis of a list-coloring algorithm on a random graph. *38th Annual Symposium on Foundations of Computer Science*, pp. 204–12. IEEE Computer Society Press, Los Alamitos, CA, 1997.

[9] R. Aharoni and N. Linial. Minimal non-two colorable hypergraphs and minimal unsatisfiable formulas. *Journal of Combinatorial Theory, Series A* 43, pp. 196–204, 1986.

[10] N. Alon and J. Spencer. *The Probabilistic Method* (2nd ed.). Wiley, New York, 2000.

[11] P. Beame and T. Pitassi. Simplified and improved resolution lower bounds. *37th Annual Symposium on Foundations of Computer Science*, pp. 274–82. IEEE Computer Society Press, Los Alamitos, CA, 1996.

[12] P. Beame, R. M. Karp, T. Pitassi, and M. Saks. On the complexity of unsatisfiability proofs for random k-CNF formulas. *30th Annual Symposium on the Theory of Computing*, Association for Computing Machinery, pp. 561–71. New York, 1998.

[13] E. Ben-Sasson and A. Wigderson. Short proofs are narrow – resolution made simple. *Journal of the Association for Computing Machinery* 48, pp. 149–69, 2001.

[14] B. Bollobás, C. Borgs, J. Chayes, J. H. Kim, and D. B. Wilson. The scaling window of the 2-SAT transition. *Random Structures and Algorithms* 18, pp. 201–56, 2001.

[15] E. Boros, P. L. Hammer, and X. Sun. Recognition of q-Horn formulae in linear time. *Discrete Applied Mathematics* 55, pp. 1–13, 1994.

[16] E. Boros, Y. Crama, P. L. Hammer, and M. Saks. A complexity index for satisfiability problems. *SIAM Journal on Computing* 23, pp. 45–9, 1994.

[17] A. Z. Broder, A. M. Frieze, and E. Upfal. On the satisfiability and maximum satisfiability of random 3-CNF formulas. *4th Annual ACM-SIAM Symposium on Discrete Algorithms*, Association for Computing Machinery, pp. 322–30. New York, 1993.

[18] V. Chandru and J. N. Hooker. Extended Horn sets in propositional logic. *Journal of the Association for Computing Machinery* 38, pp. 205–21, 1991.

[19] M.-T. Chao. Probabilistic analysis and performance measurement of algorithms for the satisfiability problem. PhD thesis, 41–45, Department of Computer Engineering and Science, Case Western Reserve University, 1985.

[20] M.-T. Chao and J. Franco. Probabilistic analysis of two heuristics for the 3-satisfiability problem. *SIAM Journal on Computing* 15, pp. 1106–18, 1986.

[21] M.-T. Chao and J. Franco. Probabilistic analysis of a generalization of the unit-clause literal selection heuristics for the k-satisfiability problem. *Information Sciences* 51, pp. 289–314, 1990.

[22] P. Cheeseman, B. Kanefsky, and W. M. Taylor. Where the really hard problems are. *12th International Joint Conference on Artificial Intelligence*, pp. 331–40. Morgan Kaufmann, 1991.

[23] V. Chvátal and Bruce Reed. Mick gets some (the odds are on his side). *33rd Annual Symposium on Foundations of Computer Science*, IEEE Computer Society Press, pp. 620–7. Los Alamitos, CA, 1992.

[24] V. Chvátal and E. Szemerédi. Many hard examples for resolution. *Journal of the Association for Computing Machinery* 35, pp. 759–68, 1988.

[25] M. Conforti, G. Cornuéjols, A. Kapoor, K. Vušković, and M. R. Rao. Balanced matrices. *Mathematical Programming: State of the Art*, J. R. Birge and K. G. Murty, eds. Braun-Brumfield, United States. Produced in association with the 15th International Symposium on Mathematical Programming, University of Michigan, 1994.

[26] S. A. Cook. The complexity of theorem-proving procedures. *3rd Annual ACM Symposium on Theory of Computing*, Association for Computing Machinery, pp. 151–8. New York, 1971.

[27] S. A. Cook and R. A. Reckhow. Corrections for "On the lengths of proofs in the propositional calculus preliminary version," *SIGACT News (ACM Special Interest Group on Automata and Computability Theory)* 6, pp. 15–22, 1974.

[28] N. Creignou and H. Daudé. Generalized satisfiability problems: Minimal elements and phase transitions. *Theoretical Computer Science* 302, pp. 417–30, 2003.

[29] N. Creignou and H. Daudé. Smooth and sharp thresholds for random k-XOR-CNF satisfiability, *Theoretical Informatics and Applications* 37, pp. 127–48, 2003.

[30] N. Creignou and H. Daudé. Combinatorial sharpness criterion and phase transition classification for random CSPs, *Information and Computation* 190, pp. 220–38, 2004.

[31] N. Creignou, H. Daudé, and J. Franco. A sharp threshold for q-Horn, *Discrete Applied Mathematics* 153, pp. 48–57, 2005.

[32] M. Davis, G. Logemann, and D. Loveland. A machine program for theorem proving, *Communications of the ACM* 5, pp. 394–7, 1962.

[33] W. F. Dowling and J. H. Gallier. Linear-time algorithms for testing the satisfiability of propositional Horn formulae. *Journal of Logic Programming* 1, pp. 267–84, 1984.

[34] O. Dubois and Y. Boufkhad. A general upper bound for the satisfiability threshold of random r-SAT formulae, *Journal of Algorithms* 24, pp. 395–420, 1997.

[35] O. Dubois. Upper bounds on the satisfiability threshold. *Theoretical Computer Science* 265, pp. 187–97, 2001.

[36] O. Dubois, Y. Boufkhad, and J. Mandler. Typical random 3-SAT formulae and the satisfiability threshold. *11th ACM-SIAM Symposium on Discrete Algorithms*, pp. 124–6. Association for Computing Machinery, New York, 2000.

[37] A. El Maftouhi and W. Fernandez de la Vega. On random 3-SAT. *Combinatorics, Probability, and Computing* 4, pp. 189–95, 1995.

[38] S. Even, A. Itai, and A. Shamir. On the complexity of timetable and multi-commodity flow problems, *SIAM Journal on Computing* 5, pp. 691–703, 1976.

[39] W. Fernandez de la Vega. On random 2-sat. unpublished manuscript, 1992.

[40] J. Franco. Results related to threshold phenomena research in Satisfiability: Lower bounds. *Theoretical Computer Science* 265, pp. 147–57, 2001.

[41] J. Franco. On the probabilistic performance of algorithms for the Satisfiability problem. *Information Processing Letters* 23, pp. 103–6, 1986.

[42] J. Franco. On the occurrence of null clauses in random instances of Satisfiability, *Discrete Applied Mathematics* 41, pp. 203–9, 1993.

[43] J. Franco. Elimination of infrequent variables improves average case performance of Satisfiability algorithms, *SIAM Journal on Computing* 20, pp. 1119–27, 1991.

[44] J. Franco and Y. C. Ho. Probabilistic performance of heuristic for the Satisfiability problem. *Discrete Applied Mathematics* 22, pp. 35–51, 1988–9.

[45] J. Franco and M. Paull. Probabilistic analysis of the Davis-Putnam procedure for solving the Satisfiability problem, *Discrete Applied Mathematics* 5, pp. 77–87, 1983.

[46] J. Franco and A. Van Gelder. A perspective on certain polynomial time solvable classes of Satisfiability. *Discrete Applied Mathematics* 125, pp. 177–214, 2003.

[47] E. Friedgut, with an appendix by J. Bourgain. Sharp thresholds of graph properties, and the k-SAT problem, *Journal of the American Mathematical Society* 12, pp. 1017–54, 1999.

[48] A. M. Frieze and S. Suen. Analysis of two simple heuristics on a random instance of k-SAT. *Journal of Algorithms* 20, pp. 312–55, 1996.

[49] X. Fu, On the complexity of proof systems. PhD thesis, University of Toronto, Canada, 1995.

[50] Z. Galil. On resolution with clauses of bounded size. *SIAM Journal on Computing* 6, pp. 444–59, 1977.

[51] A. Goerdt. A threshold for unsatisfiability, *Journal of Computer System Science* 53, pp. 469–86, 1996.

[52] A. Goerdt and M. Krivelevich. Efficient recognition of random unsatisfiable k-SAT instances by spectral methods. *Lecture Notes in Computer Science* 2010, pp. 294–304.

[53] A. Goldberg. On the complexity of the satisfiability problem. In *4th Workshop on Automated Deduction*, pp. 1–6. Academic Press, New York, 1979.

[54] A. Goldberg, P. W. Purdom, and C. Brown. Average time analysis of simplified Davis-Putnam procedures, *Information Processing Letters* 15, pp. 72–5, 1982.

[55] M. T. Hajiaghayi and G. B. Sorkin. The Satisfiability threshold of random 3-SAT is at least 3.52, 2003, Available at http://arxiv.org/pdf/math.CO/0310193.

[56] A. Haken. The intractability of resolution, *Theoretical Computer Science* 39, pp. 297–308, 1985.

[57] P. Hall. On representatives of subsets. *Journal of London Mathematical Society* 10, pp. 26–30, 1935.

[58] A. Itai and J. Makowsky. On the complexity of Herbrand's theorem, Working Paper 243, Department of Computer Science, Israel Institute of Technology, Haifa, 1982.

[59] K. Iwama. CNF Satisfiability test by counting and polynomial average time. *SIAM Journal on Computing* 18, pp. 385–91, 1989.

[60] K. Iwama. Improved upper bounds for 3-SAT, *Electronic Colloquium on Computational Complexity* Report No. 53, 2003, available at http://www.eccc.uni-trier.de/eccc-reports/2003/TR03-053/Paper.pdf.

[61] D. S. Johnson. Approximation algorithms for combinatorial problems. *Journal of Computer and Systems Sciences* 9, pp. 256–78, 1974.

[62] A. Kamath, R. Motwani, K. Palem, and P. Spirakis. Tail bounds for occupancy and the satisfiability conjecture. *Random Structures and Algorithms* 7, pp. 59–80, 1995.

[63] A. C. Kaporis, L. M. Kirousis, and E. G. Lalas. The probabilistic analysis of a greedy satisfiability algorithm. *Lecture Notes in Computer Science* 2461, pp. 574–86, 2002.

[64] A. C. Kaporis, L. M. Kirousis, and E. G. Lalas. Selecting complementary pairs of literals. *Electronic Notes in Discrete Mathematics*, Vol. 16, pp. 47–70, 2003.

[65] A. C. Kaporis, L. M. Kirousis, and Y. C. Stamatiou. How to prove conditional randomness using the principle of deferred decisions. 2002. Available at http://www.ceid.upatras.gr/faculty/kirousis/kks-pdd02.ps.

[66] R. M. Karp and M. Sipser. Maximum matchings in sparse random graphs. In *22nd Annual Symposium on the Foundations of Computer Science*, pp. 364–375. IEEE Computer Society Press, Los Alamitos, CA, 1981.

[67] L. M. Kirousis, E. Kranakis, D. Krizanc, and Y. C. Stamatiou, Approximating the unsatisfiability threshold of random formulae. *Random Structures and Algorithms* 12, pp. 253–69, 1998.

[68] H. Kleine Büning. On the minimal unsatisfiability problem for some subclasses of CNF. In *Abstracts of 16th International Symposium on Mathematical Programming*. Mathematical Programming Society, 1997.

[69] M. Krivelevich and V. H. Vu. Approximating the independence number and the chromatic number in expected polynomial time. *Lecture Notes in Computer Science* 1853, pp. 13–24, 2000.

[70] O. Kullmann. New methods for 3-SAT decision and worst-case analysis. *Theoretical Computer Science* 223, pp. 1–72, 1999.

[71] O. Kullmann. Investigations on autark assignments. *Discrete Applied Mathematics* 107, pp. 99–137, 2000.

[72] T. G. Kurtz. Solutions of ordinary differential equations as limits of pure jump Markov processes. *Journal of Applied Probability* 7, pp. 49–58, 1970.

[73] H. R. Lewis. Renaming a set of clauses as a Horn set. *Journal of the Association for Computing Machinery* 25, pp. 134–5, 1978.

[74] H. van Maaren. A short note on linear autarkies, q-Horn formulas, and the complexity index. Technical Report 99-26, Dimacs. Rutgers University, New Brunswick, NJ, 1999.

[75] M. Mézard and Riccardo Zecchina. The random k-satisfiability problem: From an analytic solution to an efficient algorithm. *Physical Review E* 66, pp. 056126–056126-27, 2002.

[76] M. Mitzenmacher. Tight thresholds for the pure literal rule. DEC/SRC Technical Note 1997-011, June 1997.

[77] R. Monasson, R. Zecchina, S. Kirkpatrick, B. Selman, and L. Troyansky. Determining computational complexity from characteristic "phase transitions." *Nature* 400, pp. 133–7, 1999.

[78] B. Monien and E. Speckenmeyer. Solving Satisfiability in less than 2^n steps. *Discrete Applied Mathematics* 10, pp. 287–95, 1985.

[79] V. Y. Pan and Z. Q. Chen. The complexity of the matrix eigenproblem. *31st ACM Symposium on Theory of Computing*, pp. 506–16. Association for Computing Machinery, New York, 1999.

[80] R. Paturi, P. Pudlák, and F. Zane. Satisfiability coding lemma. *38th Annual Symposium on the Foundations of Computer Science*, pp. 566–74. IEEE Computer Society Press, Los Alamitos, CA, 1997.

[81] R. Paturi, P. Pudlák, M. E. Saks, and F. Zane. An improved exponential-time algorithm for k-SAT. In *39th Annual Symposium on the Foundations of Computer Science*, pp. 628–37. IEEE Computer Society Press, Los Alamitos, CA, 1998.

[82] J. W. Plotkin, J. W. Rosenthal, and J. Franco. Correction to probabilistic analysis of the Davis-Putnam procedure for solving the satisfiability problem. *Discrete Applied Mathematics* 17, pp. 295–9, 1987.

[83] P. W. Purdom and G. N. Haven. Probe order backtracking, *SIAM Journal on Computing* 26, pp. 456–83, 1997.

[84] P. W. Purdom and C. A. Brown. The pure literal rule and polynomial average time, *SIAM Journal on Computing* 14, pp. 943–53, 1985.

[85] P. W. Purdom. Search rearrangement backtracking and polynomial average time. *Artificial Intelligence* 21, pp. 117–33, 1983.

[86] J. A. Robinson. A machine-oriented logic based on the resolution principle. *Journal of the ACM* 12, pp. 23–41, 1965.

[87] J. W. Rosenthal, J. W. Plotkin and J. Franco. The probability of pure literals. *Journal of Logic and Computation* 9, pp. 501–13, 1999.

[88] I. Schiermeyer. Pure literal look ahead: An $O(1.497^n)$ 3-Satisfiability algorithm. *1st Workshop on Satisfiability*, Certosa di Pontignano, Università degli Studi di Siena, Italy, 1996. Available at http://www.ececs.uc.edu/ franco/Sat-workshop/ps/Schiersat.ps.

[89] J. S. Schlipf, F. Annexstein, J. Franco, and R. Swaminathan. On finding solutions for extended Horn formulas. *Information Processing Letters* 54, pp. 133–7, 1995.

[90] U. Schöning. A probabilistic algorithm for k-SAT and constraint satisfaction problems. In *40th Annual Symposium on the Foundations of Computer Science*, pp. 410–14. IEEE Computer Society Press, Los Alamitos, CA, 1999.

[91] M. G. Scutella. A note on Dowling and Gallier's top-down algorithm for propositional Horn Satisfiability. *Journal of Logic Programming* 8, pp. 265–73, 1990.

[92] R. P. Swaminathan and D. K. Wagner. The arborescence–realization problem. *Discrete Applied Mathematics* 59, pp. 267–83, 1995.

[93] C. A. Tovey. A simplified NP-complete satisfiability problem. *Discrete Applied Mathematics* 8, pp. 85–9, 1984.

[94] K. Truemper. Monotone decomposition of matrices. Technical Report UTDCS-1-94, University of Texas at Dallas, 1994.

[95] K. Truemper. Effective Logic Computation. Wiley, New York, 1998.

[96] A. Urquhart. Hard examples for resolution. *Journal of the Association for Computing Machinery* 34, pp. 209–19, 1987.

[97] N. C. Wormald. Differential equations for random processes and random graphs. *Annals of Applied Probability* 5, pp. 1217–35, 1995.

5

Optimization Methods in Logic

John Hooker

5.1 Numerical Semantics for Logic

Optimization can make at least two contributions to Boolean logic. Its solution methods can address inference and satisfiability problems, and its style of analysis can reveal tractable classes of Boolean problems that might otherwise have gone unnoticed.

The key to linking optimization with logic is to provide logical formulas a numerical interpretation or semantics. Whereas *syntax* concerns the structure of logical expressions, *semantics* gives them meaning. Boolean semantics, for instance, focuses on truth functions that capture the meaning of logical propositions. To take an example, the function $f(x_1, x_2)$ given by $f(0, 1) = 0$ and $f(0, 0) = f(1, 0) = f(1, 1) = 1$ interprets the expression $x_1 \vee \bar{x}_2$, where 0 stands for "false" and 1 for "true."

The Boolean function f does not say a great deal about the meaning of $x_1 \vee \bar{x}_2$, but this is by design. The point of formal logic is to investigate how one can reason correctly based solely on the *form* of propositions. The meaning of the atoms x_1 and x_2 is irrelevant, aside from the fact either can be true or false. Only the "or" ($\vee$) and the "not" ($\bar{}$) require interpretation for the purposes of formal logic, and the function f indicates how they behave in the expression $x_1 \vee \bar{x}_2$. In general, interpretations of logic are chosen to be as lean as possible in order to reflect only the formal properties of logical expressions.

For purposes of solving inference and satisfiability problems, however, it may be advantageous to give logical expressions a more specific meaning. This chapter presents the idea of interpreting 0 and 1 as actual numerical values rather than simply as markers for "false" and "true." Boolean truth values signify nothing beyond the fact that there are two of them, but the numbers 0 and 1 derive additional meaning from their role in mathematics. For example, they allow Boolean expressions to be regarded as inequalities, as when $x_1 \vee \bar{x}_2$ is read as $x_1 + (1 - x_2) \geq 1$.

This maneuver makes such optimization techniques as linear and 0-1 programming available to logical inference and satisfiability problems. In addition it helps to reveal the structure of logical problems and calls attention to classes of problems that are more easily solved.

George Boole seems to give a numerical interpretation of logic in his seminal work, *The Mathematical Analysis of Logic*, because he notates disjunction and conjunction with symbols for addition and multiplication. Yet his point in doing so is to emphasize that one can calculate with propositions no less than with numbers. The notation does not indicate a numerical interpretation of logic, because Boole's main contribution is to demonstrate a *non*numeric calculus for deductive reasoning. The present chapter, however, develops the numerical interpretation suggested by Boole's original notation.

We begin by showing how Boolean inference and satisfiability problems can be solved as optimization problems. We then use the numerical interpretation of logic to identify tractable classes of satisfiability problems. We conclude with some computational considerations.

5.2 Solution Methods

The Boolean inference problem can be straightforwardly converted to an integer programming problem and solved in that form, and we begin by showing how. It is preferable, however, to take advantage of the peculiar structure of satisfiability problems rather than solving them as general integer programming problems. We explain how to generate specialized separating cuts based on the resolution method for inference and how to isolate Horn substructure for use in branching and Benders decomposition. We also consider Lagrangean approaches that exploit the characteristics of satisfiability problems.

5.2.1 The Integer Programming Formulation

Recall that a *literal* has the form x_j or $\bar{x}_j$, where x_j is an atom. A *clause* is a disjunction of literals. A clause C implies clause D if and only if C *absorbs* D, that is, if all the literals of C occur in D.

To check whether a Boolean expression P is satisfiable, we can convert P to conjunctive normal form or CNF (a conjunction of clauses) and write the resulting clauses as linear 0-1 inequalities. P is satisfiable if and only if the 0-1 inequalities have a feasible solution, as determined by integer programming.

Example 5.1. *To check the proposition* $x_1\bar{x}_2 \vee x_3$ *for satisfiability, write it as a conjunction of two clauses,* $(x_1 \vee x_3)(\bar{x}_2 \vee x_3)$, *and convert them to the inequalities*

$$\begin{aligned} &x_1 + x_3 \geq 1 \\ &(1 - x_2) + x_3 \geq 1, \end{aligned} \tag{5.1}$$

where $x_1, x_2, x_3 \in \{0, 1\}$. *An integer programming algorithm can determine that* (5.1) *has at least one feasible solution, such as* $(x_1, x_2, x_3) = (0, 1, 1)$. *The proposition* $x_1\bar{x}_2 \vee x_3$ *is therefore satisfiable.*

To state this in general, let S be the set of clauses that result when P is converted to CNF. Each clause C has the form

$$C_{AB} = \bigvee_{j \in A} x_j \vee \bigvee_{j \in B} \bar{x}_j \tag{5.2}$$

and can be converted to the inequality

$$C^{01} = \sum_{j \in A} x_j + \sum_{j \in B} (1 - x_j) \geq 1,$$

where each $x_i \in \{0, 1\}$. It is convenient to write C^{01} using the shorthand notation

$$x(A) + \bar{x}(B) \geq 1. \tag{5.3}$$

If S^{01} is the set of 0-1 inequalities corresponding to clauses in S, then P is *satisfiable* if and only if S^{01} has a feasible solution.

P *implies* a given clause C_{AB} if and only if the optimization problem

$$\begin{array}{ll} \text{minimize} & x(A) + \bar{x}(B) \\ \text{subject to} & S^{01} \\ & x_j \in \{0, 1\}, \text{ all } j \end{array}$$

has an optimal value of at least 1. Alternatively, P implies C_{AB} if and only if $P\bar{C}_{AB}$ is unsatisfiable. $P\bar{C}_{AB}$ is obviously equivalent to the clause set

$$S \cup \{\bar{x}_j \mid j \in A\} \cup \{x_j \mid j \in B\},$$

which can be checked for satisfiability by using 0-1 programming.

Example 5.2. *The proposition*

$$P = (x_1 \vee x_3 \vee x_4)(x_1 \vee x_3 \vee \bar{x}_4)(x_2 \vee \bar{x}_3 \vee x_4)(x_2 \vee \bar{x}_3 \vee \bar{x}_4)$$

implies $x_1 \vee x_2$ *if and only if the following problem has an optimal value of at least 1:*

$$\begin{array}{ll} \text{minimize} & x_1 + x_2 \\ \text{subject to} & x_1 + x_3 + x_4 \geq 1 \\ & x_1 + x_3 + (1 - x_4) \geq 1 \\ & x_2 + (1 - x_3) + x_4 \geq 1 \\ & x_2 + (1 - x_3) + (1 - x_4) \geq 1 \\ & x_1, \ldots, x_4 \in \{0, 1\}. \end{array}$$

An integer programming algorithm can determine that the optimal value is 1.

Alternatively, P *implies* $x_1 \vee x_2$ *if the clause set*

$$\begin{array}{l} x_1 \vee x_3 \vee x_4 \\ x_1 \vee x_3 \vee \bar{x}_4 \\ x_2 \vee \bar{x}_3 \vee x_4 \\ x_2 \vee \bar{x}_3 \vee \bar{x}_4 \\ \bar{x}_1 \\ \bar{x}_2 \end{array} \tag{5.4}$$

is infeasible. An integer programming algorithm can determine that the corresponding 0-1 constraint set is infeasible:

$$\begin{aligned}
&x_1 + x_3 + x_4 \geq 1 \\
&x_1 + x_3 + (1 - x_4) \geq 1 \\
&x_2 + (1 - x_3) + x_4 \geq 1 \\
&x_2 + (1 - x_3) + (1 - x_4) \geq 1 \\
&(1 - x_1) \geq 1 \\
&(1 - x_2) \geq 1 \\
&x_1, \ldots, x_4 \in \{0, 1\}.
\end{aligned} \tag{5.5}$$

It follows that (5.4) *is unsatisfiable and that* P *implies* $x_1 \vee x_2$.

A Boolean expression can be written as an inequality without first converting to CNF, but this is normally impractical for purposes of optimization. For example, one could write the proposition $x_1\bar{x}_2 \vee x_3$ as the inequality

$$x_1(1 - x_2) + x_3 \geq 1.$$

This nonlinear inequality results in a much harder 0-1 programming problem than the linear inequalities that represent clauses.

The most popular solution approach for linear 0-1 programming is *branch and cut*. In its simplest form it begins by solving the *linear programming (LP) relaxation* of the problem, which results from replacing the binary conditions $x_j \in \{0, 1\}$ with ranges $0 \leq x_j \leq 1$. If all variables happen to have integral values in the solution of the relaxation, the problem is solved. If the relaxation has a nonintegral solution, the algorithm *branches* on a variable x_j with a nonintegral value. To branch is to repeat the process just described, once using the LP relaxation with the additional constraint $x_j = 0$, and a second time with the constraint $x_j = 1$. The recursion builds a finite but possibly large binary tree of LP problems, each of which relaxes the original problem with some variables fixed to 0 or 1. The original LP relaxation lies at the root of the tree.

The leaf nodes of the search tree represent LP relaxations with an integral solution, a dominated solution, or no solution. A dominated solution is one whose value is no better than that of the best integral solution found so far (so that there is no point in branching further). The best integral solution found in the course of the search is optimal in the original problem. If no integral solution is found, the problem is infeasible. The method is obviously more efficient if it reaches leaf nodes before descending too deeply into the tree.

The branching search is often enhanced by generating *valid cuts* or *cutting planes* at some nodes of the search tree. These are inequality constraints that are added to the LP relaxation so as to "cut off" part of its feasible polyhedron without cutting off any of the feasible 0-1 points. Valid cuts tighten the relaxation and increase the probability that its solution will be integral or dominated.

The practical success of branch and bound rests on two mechanisms that may result in shallow leaf nodes.

Integrality. LP relaxations may, by luck, have integral solutions when only a few variables have been fixed to 0 or 1, perhaps because of the polyhedral structure of problems that typically arise in practice.

Bounding. LP relaxations may become tight enough to be dominated before one descends too deeply into the search tree, particularly if strong cutting planes are available.

In both cases the nature of the relaxation plays a central role. We therefore study the LP relaxation of a Boolean problem.

5.2.2 The Linear Programming Relaxation

Linear programming provides an incomplete check for Boolean satisfiability. For clause set S, let S^{LP} denote the LP relaxation of the 0-1 formulation S^{01}. If S^{LP} is infeasible, then S is unsatisfiable. If S^{LP} is feasible, however, the satisfiability question remains unresolved.

This raises the question, how much power does linear programming have to detect unsatisfiability when it exists? It has the same power as a simple inference method known as unit resolution, in the sense that the LP relaxation is infeasible precisely when unit resolution proves unsatisfiability.

Unit resolution is a linear-time inference procedure that essentially performs back substitution. Each step of unit resolution fixes a *unit clause* $U \in S$ (i.e., a single-literal clause) to true. It then eliminates U and $\bar{U}$ from the problem by removing from S all clauses that contain the literal U, and removing the literal $\bar{U}$ from all clauses in S that contain it. The procedure repeats until no unit clauses remain. It may result in an *empty clause* (a clause without literals), which is a sufficient condition for the unsatisfiability of S.

Because unit resolution is much faster than linear programming, it makes sense to apply unit resolution before solving the LP relaxation. Linear programming is therefore used only when the relaxation is known to be feasible and is not a useful check for satisfiability. It is more useful for providing bounds and finding integral solutions. Let a *unit refutation* be a unit resolution proof that obtains the empty clause.

Proposition 5.1. *A clause set S has a unit refutation if and only if S^{LP} is infeasible.*

Two examples motivate the proof of the proposition. The first illustrates why the LP relaxation is feasible when unit resolution fails to detect unsatisfiability.

Example 5.3. *Consider the clause set* (5.4). *Unit resolution fixes $x_1 = x_2 = 0$ and leaves the clause set*

$$\begin{aligned} & x_3 \vee x_4 \\ & x_3 \vee \bar{x}_4 \\ & \bar{x}_3 \vee x_4 \\ & \bar{x}_3 \vee \bar{x}_4 . \end{aligned} \tag{5.6}$$

Unit resolution therefore fails to detect the unsatisfiability of (5.4). *To see that the* LP *relaxation* (5.5) *is feasible, set the unfixed variables* x_3 *and* x_4 *to* 1/2. *This is a feasible solution (regardless of the values of the fixed variables) because every inequality in* (5.5) *has at least two unfixed terms of the form* x_j *or* $1 - x_j$.

We can also observe that unit resolution has the effect of summing inequalities.

Example 5.4. *Suppose we apply unit resolution to the first three clauses below to obtain the following.*

$$\begin{array}{lll} \bar{x}_1 & \text{(a)} & \\ x_1 \vee x_2 & \text{(b)} & \\ x_1 \vee \bar{x}_2 & \text{(c)} & \\ \hline x_2 & \text{(d)} & \textit{from } \text{(a)} + \text{(b)} \\ \bar{x}_2 & \text{(e)} & \textit{from } \text{(a)} + \text{(c)} \\ \emptyset & & \textit{from } \text{(d)} + \text{(e)} \end{array} \tag{5.7}$$

Resolvent (d) *corresponds to the sum of* $1 - x_1 \geq 1$ *and* $x_1 + x_2 \geq 1$, *and similarly for resolvent* (e). *Now* x_2 *and* $\bar{x}_2$ *resolve to produce the empty clause, which corresponds to summing* $x_2 \geq 1$ *and* $(1 - x_2) \geq 1$ *to get* $0 \geq 1$. *Thus applying unit resolution to* (5.7) *has the effect of taking a nonnegative linear combination* $2 \cdot \text{(a)} + \text{(b)} + \text{(c)}$ *to yield* $0 \geq 1$.

Proof of Proposition 5.1. If unit resolution fails to demonstrate unsatisfiability of S, then it creates no empty clause, and every remaining clause contains at least two literals. Thus every inequality in S^{LP} contains at least two unfixed terms of the form x_j or $1 - x_j$ and can be satisfied by setting the unfixed variables to 1/2.

Conversely, if unit resolution proves unsatisfiability of S, then some nonnegative linear combination of inequalities in S^{01} yields $0 \geq 1$. This means that S^{LP} is infeasible. ■

There are special classes of Boolean problems for which unit resolution always detects unsatisfiability when it exists. Section 5.3 shows how polyhedral analysis can help identify such classes.

We can now consider how the branch and cut mechanisms discussed earlier might perform in the context of Boolean methods.

Integrality. Because one can always solve a feasible LP relaxation of a satisfiability problem by setting unfixed variables to 1/2, it may appear that the integrality mechanism will not work in the Boolean case. However, the simplex method for linear programming finds a solution that lies at a vertex of the feasible polyhedron. There may be many integral vertex solutions, and the solution consisting of 1/2s may not even be a vertex. For instance, the fractional solution $(x_1, x_2, x_3) = (1/2, 1/2, 1/2)$ is feasible for the LP relaxation of (5.1), but it is not a vertex solution. In fact, all of the vertex solutions are integral (i.e., the LP relaxation defines an *integral polyhedron*). The integrality mechanism can therefore be useful in a Boolean context, albeit only empirical investigation can reveal how useful it is.

Bounding. The bounding mechanism is more effective if we can identify cutting planes that exploit the structure of Boolean problems. In fact we can, as shown in the next section.

5.2.3 Cutting Planes

A cutting plane is a particular type of logical implication. We should therefore expect to see a connection between cutting plane theory and logical inference, and such a connection exists. The most straightforward link is the fact that the well-known resolution method for inference generates cutting planes.

Resolution generalizes the unit resolution procedure discussed above (see Chapters 2 and 3 in the monograph [25]). Two clauses C, D have a *resolvent* R if exactly one variable x_j occurs as a positive literal x_j in one clause and as a negative literal $\bar{x}_j$ in the other. The resolvent R consists of all the literals in C or D except x_j and $\bar{x}_j$. C and D are the *parents* of R. For example, the clauses $x_1 \vee x_2 \vee x_3$ and $\bar{x}_1 \vee x_2 \vee \bar{x}_4$ yield the resolvent $x_2 \vee x_3 \vee \bar{x}_4$.

Given a clause set S, each step of the resolution algorithm finds a pair of clauses in S that have a resolvent R that is absorbed by no clause in S, removes from S all clauses absorbed by R, and adds R to S. The process continues until no such resolvents exist. Quine [68, 69] showed that the resolution procedure yields precisely the prime implicates of S. It yields the empty clause if and only if S is unsatisfiable.

Unlike a unit resolvent, a resolvent R of C and D need not correspond to the sum of C^{01} and D^{01}. However, it corresponds to a cutting plane. A *cutting plane* of a system $Ax \geq b$ of 0-1 inequalities is an inequality that is satisfied by all 0-1 points that satisfy $Ax \geq b$. A *rank 1* cutting plane has the form $\lceil uA \rceil x \geq \lceil ub \rceil$, where $u \geq 0$ and $\lceil \alpha \rceil$ rounds α up to the nearest integer. Thus, rank 1 cuts result from rounding up a nonnegative linear combination of inequalities.

Proposition 5.2. *The resolvent of clauses C, D is a rank 1 cutting plane for C^{01}, D^{01} and bounds $0 \leq x_j \leq 1$.*

The reasoning behind the proposition is clear in an example.

Example 5.5. *Consider again the clauses $x_1 \vee x_2 \vee x_3$ and $\bar{x}_1 \vee x_2 \vee \bar{x}_4$, which yield the resolvent $x_2 \vee x_3 \vee \bar{x}_4$. Consider a linear combination of the corresponding 0-1 inequalities and 0-1 bounds, where each inequality has a multiplier 1/2:*

$$\begin{array}{llll}
x_1 & +x_2+x_3 & & \geq 1 \\
(1-x_1) & +x_2 & +(1-x_4) & \geq 1 \\
 & \quad\;\; x_3 & & \geq 0 \\
 & & (1-x_4) & \geq 0 \\
\hline
 & x_2+x_3 & +(1-x_4) & \geq 1/2
\end{array}$$

Rounding up the right-hand side of the resulting inequality (below the line) yields a rank 1 cutting plane that corresponds to the resolvent $x_2 \vee x_3 \vee \bar{x}_4$.

Proof of Proposition 5.2. Let $C = C_{A\cup\{k\},B}$ and $D = D_{A',B'\cup\{k\}}$, so that the resolution takes place on variable x_k. The resolvent is $R = R_{A\cup A',B\cup B'}$. Consider the linear combination of C^{01}, D^{01}, $x_j \geq 0$ for $j \in A\Delta A'$, and $(1 - x_j) \geq 0$ for $j \in B\Delta B'$ in which each inequality has weight 1/2, and $A\Delta A'$ is the symmetric difference of A and A'. This linear combination is $\sum_{j\in A\cup A'} x_j + \sum_{j\in B\cup B'}(1 - x_j) \geq 1/2$. By rounding up the right-hand side, we obtain R^{01}, which is therefore a rank 1 cut. ■

Let the *input resolution* algorithm for a given clause set S be the resolution algorithm applied to S with the restriction that at least one parent of every resolvent belongs to the original set S. The following is proved in [40].

Proposition 5.3. *The input resolution algorithm applied to a clause set S generates precisely the set of clauses that correspond to rank 1 cuts of S^{01}.*

One way to obtain cutting planes for branch and cut is to generate resolvents from the current clause set S. The number of resolvents tends to grow very rapidly, however, and most of them are likely to be unhelpful. Some criterion is needed to identify useful resolvents, and the numerical interpretation of logic provides such a criterion. One can generate only *separating cuts*, which are cutting planes that are violated by the solution of the current LP relaxation. Separating cuts are so called because they cut off the solution of the LP relaxation and thereby "separate" it from the set of feasible 0-1 points.

The aim, then, is to identify resolvents R for which R^{01} is a separating cut. In principle this can be done by screening all resolvents for separating cuts, but there are better ways. It is straightforward, for example, to recognize a large class of clauses that cannot be the parent of a separating resolvent. Suppose that all clauses under discussion have variables in $\{x_1, \ldots, x_n\}$.

Proposition 5.4. *Consider any clause C_{AB} and any $x \in [0, 1]^n$. C_{AB} can be the parent of a separating resolvent for x only if $x(A) + \bar{x}(B) - x(\{j\}) < 1$ for some $j \in A$ or $x(A) + \bar{x}(B) - \bar{x}(\{j\}) < 1$ for some $j \in B$.*

Proof. The resolvent on x_j of clause C_{AB} with another clause has the form $R = C_{A'B'}$, where $A \setminus \{j\} \subset A'$ and $B \setminus \{j\} \subset B'$. R^{01} is separating when $x(A') + \bar{x}(B') < 1$. This implies that $x(A) + \bar{x}(B) - x(\{j\}) < 1$ if $j \in A$ and $x(A) + \bar{x}(B) - \bar{x}(\{j\}) < 1$ if $j \in B$. ■

Example 5.6. *Suppose $(x_1, x_2, x_3) = (1, 0.4, 0.3)$ in the solution of the current LP relaxation of an inference or satisfiability problem, and let $C_{AB} = x_1 \vee x_2 \vee \bar{x}_3$. Here $x(A) + \bar{x}(B) = 2.1$. C_{AB} cannot be the parent of a separating resolvent because $x(A) + \bar{x}(B) - x(\{1\}) = 1.1 \geq 1$, $x(A) + \bar{x}(B) - x(\{2\}) = 1.7 \geq 1$, and $x(A) + \bar{x}(B) - \bar{x}(\{3\}) = 1.4 \geq 1$.*

Proposition 5.4 suggests that one can apply the following separation algorithm at each node of the branch-and-cut tree. Let S_0 consist of the clauses in the satisfiability or inference problem at the current node, including unit clauses imposed by branching. Simplify S_0 by applying unit resolution, and solve S_0^{LP} (unless S_0 contains the empty clause). If the LP solution is nonintegral and undominated, generate cutting planes as follows. Let set S^{SR} (initially empty) collect separating resolvents. Remove from S_0 all clauses that cannot be parents of separating resolvents, using the criterion of Proposition 5.4. In each iteration k, beginning with $k = 1$, let S_k be initially empty. Generate all resolvents R that have both parents in S_{k-1} and are not dominated by any clause in S_{k-1}. If R^{01} is separating, remove from S^{SR} all clauses absorbed by R, and put R into S_k and S^{SR}; if R^{01} is not separating, put R into S_k if R meets the criterion in Proposition 5.4. If the resulting set S_k is nonempty, increment k by one and repeat. The following is proved in [47].

Proposition 5.5. *Given a clause set S and a solution x of S^{LP}, every clause corresponding to a rank 1 separating cut of S^{01} is absorbed by a clause in S^{SR}.*

S^{SR} may also contain cuts of higher rank.

Example 5.7. *Consider the problem of checking whether x_1 follows from the following clause set S:*

$$\begin{array}{lr} x_1 \vee x_2 \vee x_3 & \text{(a)} \\ x_1 \vee x_2 \vee \bar{x}_3 \vee x_4 & \text{(b)} \\ x_1 \vee \bar{x}_2 \qquad \vee x_4 & \text{(c)} \\ \qquad \bar{x}_2 \vee x_3 \vee \bar{x}_4 & \text{(d)} \end{array} \tag{5.8}$$

The 0-1 problem is to minimize x_1 subject to S^{01}. At the root node of the branch-and-cut tree, we first apply unit resolution to $S_0 = S$, which has no effect. We solve S_0^{LP} to obtain $(x_1, x_2, x_3, x_4) = (0, 1/3, 2/3, 1/3)$. Only clauses (a)–(c) satisfy Proposition 5.4, and we therefore remove (d) from S_0. In iteration $k = 1$ we generate the following resolvents from S_0:

$$\begin{array}{lr} x_1 \vee x_2 \qquad \vee x_4 & \text{(e)} \\ x_1 \qquad \vee \bar{x}_3 \vee x_4 & \text{(f)} \\ x_1 \qquad \vee x_3 \vee x_4 & \text{(g)} \end{array}$$

Resolvents (e) and (f) are separating and are added to both S_1 and S^{SR}. Resolvent (g) passes the test of Proposition 5.4 and is placed in S_1, which now contains (e), (f), and (g).

In iteration $k = 2$, we generate the resolvent $x_1 \vee x_4$ from S_1. Because it is separating and absorbs both (e) and (f), the latter two clauses are removed from S^{SR} and $x_1 \vee x_4$ is added to S^{SR}. Also, $x_1 \vee x_4$ becomes the sole element of S_2. Clearly S_3 is empty, and the process stops with one separating clause in S^{SR}, namely $x_1 \vee x_4$. It corresponds to the rank 1 cut $x_1 + x_4 \geq 1$.

At this point the cut $x_1 + x_4 \geq 1$ can be added to S_0^{LP}. If the LP is re-solved, an integral solution $(x_1, \ldots, x_4) = (0, 0, 1, 1)$ results, and the 0-1 problem is solved without branching. Because $x_1 = 0$ in this solution, x_1 does not follow from S.

5.2.4 Horn Problems

A promising approach to solving Boolean satisfiability problems is to exploit the fact that they often contain large "renamable Horn" subproblems. That is, fixing to true or false a few atoms $x_1, \ldots, x_p$ may create a renamable Horn clause set, which unit resolution can check for satisfiability in linear time. Thus when using a branching algorithm to check for satisfiability, one need only branch on variables $x_1, \ldots, x_p$. After one branches on these variables, the remaining subproblems are all renamable Horn and can be solved without branching.

Originally proposed by Chandru and Hooker [16], this approach is a special case of the more general strategy of finding a small set of variables that, when fixed, simplify the problem at hand. The idea later reemerged in the literature under the name of finding a *backdoor* [27, 53, 54, 61, 66, 67, 74, 78] for Boolean satisfiability and other problems. Complexity results have been derived for several types of backdoor detection [27, 61, 77]; see also [24] for related investigations.

A *Horn clause* is a clause with at most one positive literal, such as $\bar{x}_1 \vee \bar{x}_2 \vee x_3$. A clause set H is Horn if all of its clauses are Horn. H is *renamable Horn* if it becomes Horn when zero or more variables x_j are complemented by replacing them with $\bar{x}_j$ (and $\bar{x}_j$ with x_j). Horn problems are discussed further in Chapter 6 of the monograph [25].

Example 5.8. *The clause set*

$$\begin{array}{c} x_1 \vee x_2 \\ \bar{x}_1 \vee \bar{x}_2 \end{array}$$

is renamable Horn because complementing x_1 makes it Horn. On the other hand, the following clause set is not renamable Horn:

$$\begin{array}{c} x_1 \vee x_2 \vee x_3 \\ \bar{x}_1 \vee \bar{x}_2 \vee \bar{x}_3. \end{array} \tag{5.9}$$

Let *positive unit resolution* be a unit resolution algorithm in which every resolvent has at least one parent that is a positive unit clause.

Proposition 5.6. *A Horn set is satisfiable if and only if there is no positive unit refutation. A renamable Horn set is satisfiable if and only if there is no unit refutation.*

Proof. Suppose that positive unit resolution applied to Horn clause set S generates no empty clause. Let S' be the clause set that remains. Then every clause in S' is a negative unit cause or contains two or more literals, at least one of which is negative. The clauses in S can be satisfied by setting all variables in S' to false and all variables on which the algorithm resolved to true. Now suppose that unit resolution applied to a renamable Horn clause set S generates no empty clause. For some renaming of the variables in S, the resulting clause set S^+ is Horn. There is no positive unit refutation of S^+, because otherwise, un-renaming the variables in this proof would yield a unit refutation of S. Thus S^+ is satisfiable, which means that S is satisfiable. ■

From this and Proposition 5.1 we immediately infer the following proposition.

Proposition 5.7. *A renamable Horn set S is satisfiable if and only if S^{LP} is feasible.*

Given a clause set S, we would like to find a smallest backdoor set. In particular, we seek the shortest vector of variables $x' = (x_{j_1}, \ldots, x_{j_p})$ that, when fixed to any value, simplifies S to a renamable Horn set. Following [16], let $v = (v_1, \ldots, v_p)$ be a 0-1 vector, and define $S(x', v)$ to be the result of fixing each x_{j_ℓ} in x' to v_ℓ. Thus if $v_\ell = 1$, all clauses in S containing literal x_{j_ℓ} are removed from S, and all literals $\bar{x}_{j_\ell}$ are removed from clauses in S, and analogously if $v_\ell = 0$. Thus, we wish to find the smallest variable set x' such that $S(x', v)$ is renamable Horn for all valuations v.

Let $S(x')$ be the result of removing from clauses in S every literal that contains a variable in x'. Then since $S(x', v) \subset S(x')$ for every v, $S(x', v)$ is renamable Horn for every v if $S(x')$ is renamable Horn. To find the shortest x' for which $S(x')$ is renamable Horn, we introduce 0-1 variables $y_j, \bar{y}_j$. Let $y_j = 1$ when x_j is not renamed, $\bar{y}_j = 1$ when x_j is renamed, and $y_j = \bar{y}_j = 0$ when x_j is removed from S by including it in x'. Then we wish to solve the set packing problem

$$\begin{aligned}
\text{maximize} \quad & \sum_j (y_j + \bar{y}_j) \\
\text{subject to} \quad & y(A) + \bar{y}(B) \leq 1, \quad \text{all clauses } C_{AB} \in S \\
& y_j + \bar{y}_j \leq 1, \quad \text{all } j \\
& y_j, \bar{y}_j \in \{0, 1\}
\end{aligned} \tag{5.10}$$

The first constraint ensures that each renamed clause in S contains at most one positive literal.

A set packing problem can always be solved as a maximum clique problem. In this case, we define an undirected graph G that contains a vertex for each literal $x_j, \bar{x}_j$ and an edge between two vertices whenever the corresponding literals never occur in the same clause of S and are not complements of each other. A clique of G is a set of vertices in which every pair of vertices is connected by an edge. If W is a clique of maximum size, then we put x_j into x' when $x_j, \bar{x}_j \notin W$. The maximum clique problem is NP-hard, but there are numerous exact and approximate algorithms for it [2–5, 9, 12, 29, 30, 50, 63, 64, 72, 81]. One can also solve the maximum independent set problem on the complementary graph [11, 28, 39, 55, 65, 75]. Two graph-based algorithms specifically for finding a small backdoor set are presented in [54], and heuristic methods in [66, 67].

Example 5.9. *Let us check again whether x_1 follows from the clause set (5.8). This time we do so by checking the satisfiability of*

$$\begin{array}{l} x_1 \vee x_2 \vee x_3 \\ x_1 \vee x_2 \vee \bar{x}_3 \vee x_4 \\ x_1 \vee \bar{x}_2 \qquad \vee x_4 \\ \qquad \bar{x}_2 \vee x_3 \vee \bar{x}_4 \\ \bar{x}_1 \end{array} \tag{5.11}$$

To find the shortest vector x' of variables we must fix to obtain a Horn problem, we solve the set packing problem

$$\begin{array}{lllllllllll} \text{maximize} & y_1 + y_2 + y_3 + y_4 + \bar{y}_1 + \bar{y}_2 + \bar{y}_3 + \bar{y}_4 \\ \text{subject to} & y_1 + y_2 + y_3 & & & & \leq 1 \\ & y_1 + y_2 & + y_4 & & + \bar{y}_3 & \leq 1 \\ & y_1 & + y_4 & + \bar{y}_2 & & \leq 1 \\ & \quad y_3 & & + \bar{y}_2 & + \bar{y}_4 & \leq 1 \\ & & \bar{y}_1 & & & \leq 1 \\ & y_j + \bar{y}_j \leq 1, \quad j = 1, \ldots, 4 \\ & y_j \in \{0, 1\} \end{array}$$

An optimal solution is $y = (0, 0, 0, 0)$, $\bar{y} = (1, 1, 1, 0)$. The solution indicates that only x_4 need be fixed to obtain a renamable Horn problem, which is converted to Horn by renaming x_1, x_2, x_3. Alternatively, we can find a maximum clique in the graph of Figure 5.1. One such clique is $\{\bar{x}_1, \bar{x}_2, \bar{x}_3\}$, which again indicates that x_4 can be fixed and x_1, x_2, x_3 renamed to obtain a Horn problem.

We therefore branch only on x_4. If we set $x_4 = 0$, the resulting problem is renamable Horn:

$$\begin{array}{l} x_1 \vee x_2 \vee x_3 \\ x_1 \vee x_2 \vee \bar{x}_3 \\ x_1 \vee \bar{x}_2 \\ \bar{x}_1 \end{array} \tag{5.12}$$

and unit resolution proves unsatisfiability. Taking the $x_4 = 1$ branch, we obtain the renamable Horn problem

$$\begin{array}{l} x_1 \vee x_2 \vee x_3 \\ \qquad \bar{x}_2 \vee x_3 \\ \bar{x}_1 \end{array} \tag{5.13}$$

Unit resolution fixes $x_1 = 0$ and leaves the clause set $\{x_2 \vee x_3, \bar{x}_2 \vee x_3\}$, whose clauses we satisfy by setting its variables to 0 in the renamed problem (which is Horn). Because x_2, x_3 are renamed in this case, we set $(x_1, x_2, x_3) = (0, 1, 1)$, which with $x_4 = 1$ is a satisfying solution for (5.11).

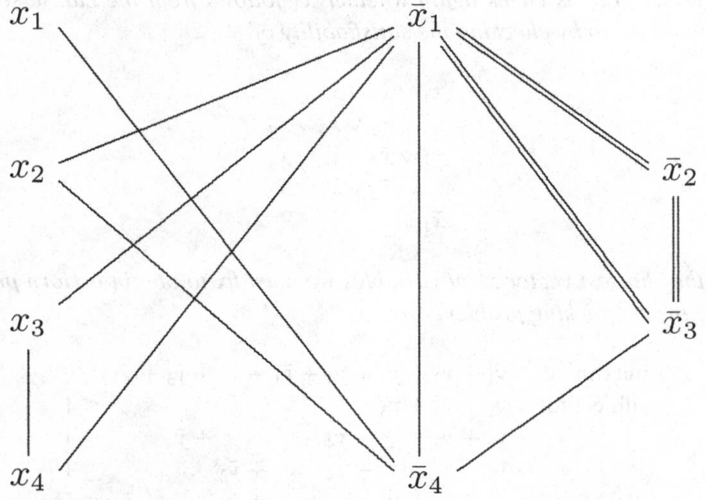

Figure 5.1. Maximum clique problem for finding renamable Horn substructure. One solution is shown in double lines.

5.2.5 Benders Decomposition

Benders decomposition [6, 31] is a well-known optimization technique that can be applied to Boolean satisfiability problems, particularly those with significant renamable Horn substructure.

Benders decomposition normally applies to optimization problems, but we focus on feasibility problems of the form

$$\begin{aligned} &Ax + g(y) \geq b \\ &x \in R^n, \; y \in Y \end{aligned} \tag{5.14}$$

where $g(y)$ is a vector of functions $g_i(y)$. The method enumerates some values of $y \in Y$ and, for each, seeks an x such that (x, y) is feasible. Thus, for each trial value $\hat{y}$, we solve the linear programming *subproblem* that results from fixing y to $\hat{y}$ in (5.14):

$$\begin{aligned} &Ax \geq b - g(\hat{y}) \\ &x \in R^n. \end{aligned} \tag{5.15}$$

If a solution $\hat{x}$ exists, then $(\hat{x}, \hat{y})$ solves (5.14). If the subproblem is infeasible, then by the well-known Farkas lemma, there is a row vector $\hat{u} \geq 0$ such that

$$\begin{aligned} &\hat{u}A = 0 \\ &\hat{u}(b - g(\hat{y})) > 0. \end{aligned}$$

Thus, to obtain a feasible subproblem, we must find a y such that

$$\hat{u}(b - g(y)) \leq 0. \tag{5.16}$$

We therefore impose (5.16) as the *Benders cut* that any future trial value of y must satisfy.

In iteration k of the Benders algorithm we obtain a trial value $\hat{y}$ of y by solving a *master problem* that contains all Benders cuts so far generated:

$$\begin{aligned} &\hat{u}^{\ell}(b - g(y)) \leq 0, \ \ell = 1, \dots, k-1 \\ &y \in Y. \end{aligned} \tag{5.17}$$

Thus $\hat{u}^1, \dots, \hat{u}^{k-1}$ are the vectors $\hat{u}$ found in the previous subproblems. We then formulate a new subproblem (5.15) and continue the process until the subproblem is feasible or until the master problem becomes infeasible. In the latter case, the original problem (5.14) is infeasible.

The classical Benders method does not apply to general Boolean satisfiability problems, because the Benders subproblem must be a continuous linear or nonlinear problem for which the multipliers u can be derived. Logic-based Benders decomposition [42, 48, 82], an extension of the method, allows solution of general Boolean problems, but here we consider a class of Boolean problems in which the logic-based method reduces to the classical method. Namely, we solve Boolean problems in which the subproblem is a renamable Horn problem and therefore equivalent to an LP problem (Proposition 5.7).

To obtain a renamable Horn subproblem, we must let the master problem contain variables that, when fixed, result in a renamable Horn structure. This can be accomplished by the methods of the previous section.

It is inefficient to solve the subproblem with an LP solver, because unit resolution is much faster. Fortunately, we can obtain u from a unit refutation, as illustrated by Example 5.4. Recall that in this example the empty clause was obtained by taking the linear combination $2 \cdot (a) + (b) + (c)$, where the coefficients represent the number of times each of the original clauses are "used" in the unit refutation. To make this precise, let C be any clause that occurs in the unit refutation, and let $n(C)$ be the number of times C is used in the refutation. Initially each $n(C) = 0$ and we execute the following recursive procedure by calling count(∅), where ∅ denotes the empty clause.

Procedure count(C)
  Increase $n(C)$ by 1.
  If C has parents P, Q in the unit refutation, then
    Perform count(P) and count(Q).

The following is a special case of a result of Jeroslow and Wang [52].

Proposition 5.8. *Suppose the feasibility problem $Ax \geq b$ represents a renamable Horn satisfiability problem, where each row i of $Ax \geq b$ corresponds to a clause C_i. If the satisfiability problem has a unit refutation, then $uA = 0$ and $ub = 1$, where each $u_i = n(C_i)$.*

Proof. When C is the resolvent of P, Q in the foregoing recursion, the inequality C^{01} is the sum of P^{01} and Q^{01}. Thus, $\emptyset^{01}$, which is the inequality $0 \geq 1$, is the sum of $n(C_i) \cdot C_i^{01}$ over all i. Because C_i^{01} is row i of $Ax \geq b$, this means that $uA = 0$ and $ub = 1$ if we set $u_i = n(C_i)$ for all i. ∎

Benders decomposition can now be applied to a Boolean satisfiability problem. First put the problem in the form of a 0-1 programming problem (5.14). The variables y are chosen so that, when fixed, the resulting subproblem (5.15) represents a renamable Horn satisfiability problem and can therefore be regarded as an LP problem. The multipliers $\hat{u}$ are obtained as in Proposition 5.8 and the Benders cut (5.16) formed accordingly.

Example 5.10. *Consider again the satisfiability problem (5.11). We found in Example 5.9 that the problem becomes renamable Horn when x_4 is fixed to any value. We therefore put x_4 in the master problem and x_1, x_2, x_3 in the subproblem. Now (5.11) becomes the 0-1 feasibility problem*

$$\begin{array}{llll}
x_1 & +x_2 & +x_3 & \geq 1 \\
x_1 & +x_2 & +(1-x_3)+x_4 & \geq 1 \\
x_1 & +(1-x_2) & +x_4 & \geq 1 \\
 & (1-x_2)+x_3 & +(1-x_4) & \geq 1 \\
(1-x_1) & & & \geq 1
\end{array}$$

where x_4 plays the role of y_1 in (5.14). This problem can be written in the form (5.14) by bringing all constants to the right-hand side:

$$\begin{array}{l}
x_1 + x_2 + x_3 \geq 1 \\
x_1 + x_2 - x_3 + x_4 \geq 0 \\
x_1 - x_2 + x_4 \geq 0 \\
- x_2 + x_3 - x_4 \geq -1 \\
-x_1 \geq 0
\end{array}$$

Initially the master problem contains no constraints, and we arbitrarily set $\hat{x}_4 = 0$. This yields the satisfiability subproblem (5.12), which has the 0-1 form

$$\begin{array}{lr}
x_1 + x_2 + x_3 \geq 1 & \text{(a)} \\
x_1 + x_2 - x_3 \geq 0 & \text{(b)} \\
x_1 - x_2 \geq 0 & \text{(c)} \\
- x_2 + x_3 \geq -1 & \text{(d)} \\
-x_1 \geq 0 & \text{(e)}
\end{array}$$

(Note that clause (d) could be dropped because it is already satisfied.) This is a renamable Horn problem with a unit refutation, as noted in Example 5.9. The refutation obtains the empty clause from the linear combination $(a) + (b) + 2 \cdot (c) + 4 \cdot (e)$, and we therefore let $\hat{u} = (1, 1, 2, 0, 4)$. The Benders

cut (5.16) becomes

$$[1\;1\;2\;0\;4]\left(\begin{bmatrix}1\\0\\0\\-1\\0\end{bmatrix}-\begin{bmatrix}0\\x_4\\x_4\\-x_4\\0\end{bmatrix}\right)\le 0$$

or $3x_4 \ge 1$. *The master problem (5.17) now consists of the Benders cut* $3x_4 \ge 1$ *and has the solution* $x_4 = 1$. *The next subproblem is therefore (5.13), for which there is no unit refutation. The subproblem is solved as in Example 5.9 by setting* $(x_1, x_2, x_3) = (0, 1, 1)$, *and the Benders algorithm terminates. These values of* x_j *along with* $x_4 = 1$ *solve the original problem.*

Rather than solve each master problem from scratch as in the classical Benders method, we can conduct a single branch-and-cut search and solve the subproblem each time a feasible solution is found. When a new Benders cut is added, the branch-and-cut search resumes where it left off with the new Benders cut in the constraint set. This approach was proposed in [42] and tested computationally on machine scheduling problems in [76], where it resulted in at least an order-of-magnitude speedup.

Although we have used Benders to take advantage of Horn substructure, it can also exploit other kinds of structure by isolating it in the subproblem. For example, a problem may decompose into separate problems when certain variables are fixed [48].

5.2.6 Lagrangean Relaxation

One way to strengthen the relaxations solved at nodes of a search tree is to replace LP relaxations with Lagrangean relaxations. A Lagrangean relaxation removes or "dualizes" some of the constraints by placing penalties for their violation in the objective function. The dualized constraints are typically chosen in such a way that the remaining constraints decouple into small problems, allowing rapid solution despite the fact that they are discrete. Thus, whereas Benders decomposition decouples the problem by removing variables, Lagrangean relaxation decouples by removing constraints.

Consider an optimization problem:

$$\begin{aligned} \text{minimize} \quad & cx \\ \text{subject to} \quad & Ax \ge b \\ & x \in X \end{aligned} \tag{5.18}$$

where $Ax \ge b$ represent the constraints to be dualized, and $x \in X$ represents constraints that are easy in some sense, perhaps because they decouple into small subsets of constraints that can be treated separately. We regard $Ax \ge b$ as consisting of rows $A_i x \ge b_i$ for $i = 1, \dots, m$. Given *Lagrange multipliers* $\lambda_1, \dots, \lambda_m \ge 0$, the following is a *Lagrangean relaxation* of (5.18):

$$\begin{aligned} \text{minimize} \quad & cx + \sum_i \lambda_i (b_i - A_i x) \\ \text{subject to} \quad & x \in X \end{aligned} \tag{5.19}$$

The constraints $Ax \geq b$ are therefore dualized by augmenting the objective function with weighted penalties $\lambda_i(b_i - A_i x)$ for their violation.

As for the choice of multipliers $\lambda = (\lambda_1, \ldots, \lambda_m)$, the simplest strategy is to set each to some convenient positive value, perhaps 1. One can also search for values of λ that yield a tighter relaxation. If $\theta(\lambda)$ is the optimal value of (5.19), the best possible relaxation is obtained by finding a λ that solves the *Lagrangean dual* problem $\max_{\lambda \geq 0} \theta(\lambda)$. Because $\theta(\lambda)$ is a concave function of λ, it suffices to find a local maximum, which is a global maximum.

Subgradient optimization is commonly used to solve the Lagrangean dual ([80], pp. 174–5). Each iteration k begins with the current estimate λ^k of λ. Problem (5.19) is solved with $\lambda = \lambda^k$ to compute $\theta(\lambda^k)$. If x^k is the solution of (5.19), then $b - Ax^k$ is a subgradient of $\theta(\lambda)$ at λ^k. The subgradient indicates a direction in which the current value of λ should move to achieve the largest possible rate of increase in $\theta(\lambda)$. Thus we set $\lambda^{k+1} = \lambda^k + \alpha_k(b - Ax^k)$, where α_k is a stepsize that decreases as k increases. Various stopping criteria are used in practice, and the aim is generally to obtain only an approximate solution of the dual.

For a Boolean satisfiability problem, $c = 0$ in (5.18), and the constraints $Ax \geq b$ are 0-1 formulations of clauses. The easy constraints $x \in X$ include binary restrictions $x_j \in \{0, 1\}$ as well as "easy" clauses that allow decoupling. When the Lagrangean relaxation has a positive optimal value $\theta(\lambda)$, the original problem is unsatisfiable, whereas $\theta(\lambda) \leq 0$ leaves the optimality question unresolved.

Example 5.11. *Consider the unsatisfiable clause set S:*

$$\begin{array}{lr} \bar{x}_1 \vee \bar{x}_2 \vee x_3 \phantom{\vee \bar{x}_4} & \text{(a)} \\ \bar{x}_1 \vee \bar{x}_2 \vee \bar{x}_4 & \text{(b)} \\ x_1 \vee x_2 & \text{(c)} \\ x_1 \vee \bar{x}_2 & \text{(d)} \\ \bar{x}_1 \vee x_2 & \text{(e)} \\ \phantom{\bar{x}_1 \vee x_2 \vee} \bar{x}_3 \vee x_4 & \text{(f)} \\ \phantom{\bar{x}_1 \vee x_2 \vee} x_3 \vee \bar{x}_4 & \text{(g)} \end{array} \tag{5.20}$$

S^{LP} is feasible and therefore does not demonstrate unsatisfiability. Thus, if we use the LP relaxation in a branch-and-cut algorithm, branching is necessary. A Lagrangean relaxation, however, can avoid branching in this case. Constraints (a) and (b) are the obvious ones to dualize, because the remainder of the problem splits into two subproblems that can be solved separately. The objective function of the Lagrangean relaxation (5.19) becomes

$$\lambda_1(-1 + x_1 + x_2 - x_3) + \lambda_2(-2 + x_1 + x_2 + x_4).$$

Collecting terms in the objective function, the relaxation (5.19) is

$$
\begin{array}{lllll}
\text{minimize} & (\lambda_1+\lambda_2)x_1+(\lambda_1+\lambda_2)x_2 & -\lambda_1x_3+\lambda_2x_4 & -\lambda_1-2\lambda_2 & \\
\text{subject to} & x_1+x_2 & & \geq 1 & \\
 & x_1-x_2 & & \geq 0 & \\
 & -x_1+x_2 & & \geq 0 & \qquad (5.21) \\
 & & -x_3+x_4 & \geq 0 & \\
 & & x_3-x_4 & \geq 0 & \\
 & x_j \in \{0,1\} & & &
\end{array}
$$

The relaxation obviously decouples into two separate subproblems, one containing x_1, x_2 *and one containing* x_3, x_4.

Starting with $\lambda^0 = (1, 1)$, *we obtain an optimal solution* $x^0 = (1, 1, 0, 0)$ *of (5.21) with* $\theta(\lambda^0) = 1$. *This demonstrates that (5.20) is unsatisfiable, without the necessity of branching.*

Bennaceur et al. [7] use a somewhat different Lagrangean relaxation to help solve satisfiability problems. They again address the problem with a branching algorithm that solves a Lagrangean relaxation at nodes of the search tree (or at least at certain nodes). This time, the dualized constraints $Ax \geq b$ are the clauses violated by the currently fixed variables, and the remaining constraints $x \in X$ are the clauses that are already satisfied. The multipliers λ_i are all set to 1, with no attempt to solve the Lagrangean dual. A local search method approximately solves the resulting relaxation (5.19). If the solution $\hat{x}$ satisfies additional clauses, the process is repeated while dualizing only the clauses that remain violated. This continues until no additional clauses can be satisfied. If all clauses are satisfied, the algorithm terminates. Otherwise, branching continues in the manner indicated by $\hat{x}$. That is, when branching on x_j, one first takes the branch corresponding to $x_j = \hat{x}_j$. This procedure is combined with intelligent backtracking to obtain a competitive satisfiability algorithm, as well as an incremental satisfiability algorithm that re-solves the problem after adding clauses. The details may be found in [7].

5.3 Tractable Problem Classes

We now turn to the task of using the quantitative analysis of logic to identify tractable classes of satisfiability problems. We focus on two classes: problems that can be solved by unit resolution, and problems whose LP relaxations define integral polyhedra.

5.3.1 Two Properties of Horn Clauses

There is no known necessary and sufficient condition for solubility by unit resolution, but some sufficient conditions are known. We have already seen that Horn and renamable Horn problems, for example, can be solved in this manner

(Proposition 5.6). Two properties of Horn sets account for this, and they are actually possessed by a much larger class of problems. This allows a generalization of Horn problems to *extended Horn* problems that can likewise be solved by unit resolution, as shown by Chandru and Hooker [15].

Unit resolution is adequate to check for satisfiability when we can always find a satisfying solution for the clauses that remain after applying unit resolution to a satisfiable clause set. We can do this in the case of Horn problems because:

- Horn problems are *closed under deletion and contraction*, which ensures that the clauses that remain after unit resolution are still Horn.
- Horn problems have a *rounding property* that allows these remaining clauses to be assigned a solution by rounding a solution of the LP relaxation in a prespecified way: in the case of Horn clauses, by always rounding down.

A class $\mathcal{C}$ of satisfiability problems is closed under *deletion* and *contraction* if, given a clause set $S \in \mathcal{C}$, S remains in $\mathcal{C}$ after (a) any clause is deleted from S and (b) any given literal is removed from every clause of S in which it occurs. Because unit resolution operates by deletion and contraction, it preserves the structure of any class that is closed under these operations. This is true of Horn sets in particular because removing literals does not increase the number of positive literals in a clause.

Horn clauses can be solved by rounding down because they have an integral least element property. An element $v \in P \subset R^n$ is a *least element* of P if $v \leq x$ for every $x \in P$. It is easy to see that if S is a satisfiable set of Horn clauses, S^{LP} defines a polyhedron that always contains an integral least element. This element is identified by fixing variables as determined by unit resolution and setting all remaining variables to zero. Thus, if S is the Horn set that remains after applying unit resolution to a satisfiable Horn set, we can obtain a satisfying solution for S by rounding down any feasible solution of S^{LP}.

Cottle and Veinott [23] state a sufficient condition under which polyhedra in general have a least element.

Proposition 5.9. *A nonempty polyhedron* $P = \{x \mid Ax \geq b,\ x \geq 0\}$ *has a least element if each row of A has at most one positive component. There is an integer least element if every positive element of A is 1.*

Proof. If $b \leq 0$, then $x = 0$ is a least element. Otherwise let b_i be the largest positive component of b. Because P is nonempty, row i of A has exactly one positive component A_{ij}. The ith inequality of $Ax \geq b$ can be written

$$x_j \geq \frac{1}{A_{ij}} \left(b_i - \sum_{k \neq j} A_{ik} x_k \right).$$

Because $A_{ik} \leq 0$ for $k \neq j$ and each $x_k \geq 0$, we have the positive lower bound $x_j \geq b_i / A_{ij}$. Thus, we can construct a lower bound $\underline{x}$ for x by setting $\underline{x}_j = b_i / A_{ij}$

and $\underline{x}_k = 0$ for $k \neq j$. If we define $\tilde{x} = x - \underline{x}$, we can translate polyhedron P to

$$\tilde{P} = \left\{\tilde{x} \mid A\tilde{x} \geq \tilde{b} = (b - A\underline{x}),\ \tilde{x} \geq 0\right\}.$$

We repeat the process and raise the lower bound $\underline{x}$ until $\tilde{b} \leq 0$. At this point $\underline{x}$ is a least element of P. Clearly $\underline{x}$ is integer if each $A_{ij} = 1$. ∎

Because the inequality set S^{01} for a Horn problem S satisfies the conditions of Proposition 5.9, Horn problems have the integral least element property.

5.3.2 Extended Horn Problems

The key to extending the concept of a Horn problem is to find a larger problem class that has the rounding property and is closed under deletion and contraction. Some sets with the rounding property can be identified through a result of Chandrasekaran [13], which relies on Cottle and Veinott's least element theorem.

Proposition 5.10. *Let $Ax \geq b$ be a linear system with integral components, where A is an $m \times n$ matrix. Let T be a nonsingular $n \times m$ matrix that satisfies the following conditions:*

(i) *T and T^{-1} are integral.*
(ii) *Each row of T^{-1} contains at most one negative entry, and any such entry is -1.*
(iii) *Each row of AT^{-1} contains at most one negative entry, and any such entry is -1.*

Then if x solves $Ax \geq b$, so does $T^{-1}\lceil Tx\rceil$.

The matrix T in effect gives instructions for how a solution of the LP can be rounded.

Proof. We rely on an immediate corollary of Proposition 5.9: a polyhedron of the form $P = \{x \mid Ax \geq b,\ x \leq a\}$ has an integral largest element if A, b and a are integral and each row of A has at most one negative entry, namely -1.

Now if $\hat{y}$ solves $AT^{-1}y \geq b$, the polyhedron $\hat{P} = \{y \mid AT^{-1}y \geq b,\ y \leq \hat{y}\}$ has an integral largest element, and $\lceil\hat{y}\rceil$ is therefore in $\hat{P}$. This shows that

$$AT^{-1}y \geq b \text{ implies } AT^{-1}\lceil y\rceil \geq b.$$

Setting $x = T^{-1}y$ we have

$$Ax \geq b \text{ implies } AT^{-1}\lceil Tx\rceil \geq b. \tag{5.22}$$

Similarly, if $\tilde{y}$ satisfies $T^{-1}y \geq 0$, the polyhedron $\tilde{P} = \{y \mid T^{-1}y \geq 0,\ y \leq \lceil\tilde{y}\rceil\}$ has an integral largest element and $\lceil\tilde{y}\rceil$ is in $\tilde{P}$. So

$$T^{-1}y \geq 0 \text{ implies } T^{-1}\lceil y\rceil \geq 0,$$

and setting $x = T^{-1}y$ we have

$$x \geq 0 \text{ implies } T^{-1}\lceil Tx \rceil \geq 0. \tag{5.23}$$

Implications (5.22) and (5.23) together prove the proposition. ■

To apply Proposition 5.10 to a clause set S, we note that S^{LP} has the form $Hx \geq h$, $0 \leq x \leq e$, where e is a vector of 1s. This is an instance of the system $Ax \geq b$, $x \geq 0$ in Proposition 5.10 when

$$A = \begin{bmatrix} H \\ -I \end{bmatrix}, b = \begin{bmatrix} h \\ -e \end{bmatrix}.$$

From condition (ii) of Proposition 5.10, each row of T^{-1} contains at most one negative entry (namely, -1), and from (iii) the same is true of $-T^{-1}$. Thus we have:

(i′) T^{-1} is a nonsingular $n \times n$ matrix.
(ii′) Each row of T^{-1} contains exactly two nonzero entries, namely 1 and -1.
(iii′) Each row of HT^{-1} contains at most one negative entry, namely -1.

Condition (ii′) implies that T^{-1} is the edge-vertex incidence matrix of a directed graph. Because T^{-1} is nonsingular, it is the edge-vertex incidence matrix of a directed tree $\mathcal{T}$ on $n+1$ vertices. For an appropriate ordering of the vertices, T^{-1} is lower triangular. The inverse T is the vertex-path incidence matrix of $\mathcal{T}$.

Example 5.12. *Figure 5.2 shows the directed tree $\mathcal{T}$ corresponding to the matrices T^{-1}, T below.*

$$T^{-1} = \begin{array}{c} \begin{array}{ccccccc|c} A & B & C & D & E & F & G & R \end{array} \\ \left[\begin{array}{ccccccc} -1 & 0 & 0 & 0 & 0 & 0 & 0 \\ 1 & -1 & 0 & 0 & 0 & 0 & 0 \\ 1 & 0 & -1 & 0 & 0 & 0 & 0 \\ 0 & 0 & 0 & -1 & 0 & 0 & 0 \\ 0 & 0 & 0 & 1 & -1 & 0 & 0 \\ 0 & 0 & 0 & 1 & 0 & -1 & 0 \\ 0 & 0 & 0 & 0 & 0 & 1 & -1 \end{array}\right] \begin{array}{c} 1 \\ 0 \\ 0 \\ 1 \\ 0 \\ 0 \\ 0 \end{array} \end{array}$$

$$T = \begin{array}{c} A \\ B \\ C \\ D \\ E \\ F \\ G \end{array} \left[\begin{array}{ccccccc} -1 & 0 & 0 & 0 & 0 & 0 & 0 \\ -1 & -1 & 0 & 0 & 0 & 0 & 0 \\ -1 & 0 & -1 & 0 & 0 & 0 & 0 \\ 0 & 0 & 0 & -1 & 0 & 0 & 0 \\ 0 & 0 & 0 & -1 & -1 & 0 & 0 \\ 0 & 0 & 0 & -1 & 0 & -1 & 0 \\ 0 & 0 & 0 & -1 & 0 & -1 & -1 \end{array}\right]$$

The column corresponding to vertex R, the root of $\mathcal{T}$, is shown to the right of T^{-1}. In this example all the arcs are directed away from the root, but this need not be so. An arc directed toward the root would result in reversed signs in the corresponding row of T^{-1} and column of T.

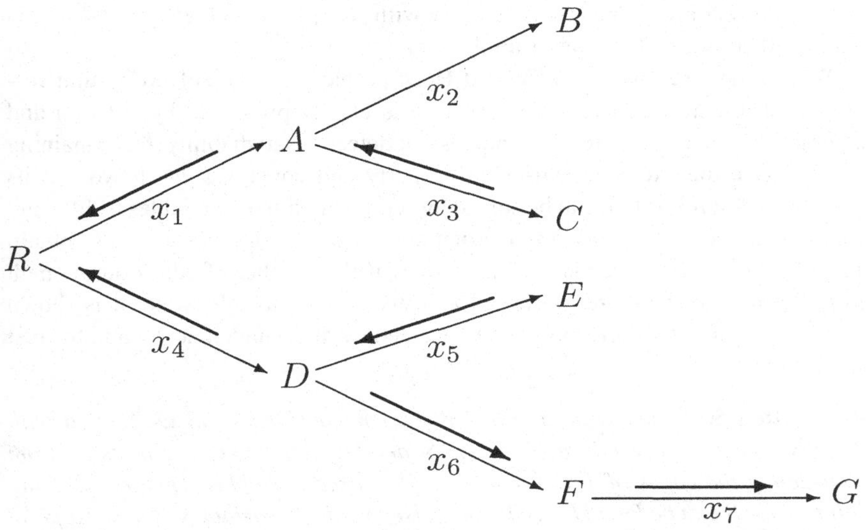

Figure 5.2. Extended star-chain flow pattern corresponding to an extended Horn clause.

The entries in each row of the propositional matrix H can be interpreted as flows on the edges of $\mathcal{T}$. Thus, each variable x_j is associated with an edge in $\mathcal{T}$. Condition (iii′) has the effect of requiring that at most one vertex (other than the root) be a net receiver of flow. To see what this means graphically, let an *extended star* be a rooted tree consisting of one or more arc-disjoint chains extending out from the root. Then (iii′) implies that any row of H has the *extended star-chain property*: it describes flows that can be partitioned into a set of unit flows into the root on some (possibly empty) extended star subtree of $\mathcal{T}$ and a unit flow on one (possibly empty) chain in $\mathcal{T}$.

Example 5.13. *Suppose* $H_i = [-1 \ 0 \ -1 \ -1 \ -1 \ 1 \ 1]$ *is a row of* H*, corresponding to the clause* $\bar{x}_1 \vee \bar{x}_3 \vee \bar{x}_4 \vee \bar{x}_5 \vee x_6 \vee x_7$*. It defines the flow depicted by arrows in Figure. 5.2. Note that a* -1 *in* H_i *indicates flow against the direction of the edge. The extended star consists of the flows* $C \to A \to R$ *and* $D \to R$*, whereas the chain consists of* $E \to D \to F \to G$*. (Flow* $E \to D$ *could also be regarded as part of the extended star.) In this case* $H_i \mathcal{T}^{-1} = [0 \ 0 \ 1 \ 1 \ 1 \ 0 \ -1]$*, which satisfies (iii′).*

A clause set S with the 0-1 formulation $Hx \geq h$ is *renamable extended Horn* if it can be associated with a directed tree $\mathcal{T}$ in which each row of H has the extended star-chain property. S is *extended Horn* if each edge of $\mathcal{T}$ is directed away from the root. If each row of H describes flows into the root on a *star* subtree of $\mathcal{T}$ (i.e., an extended star whose chains have length 1), then H is renamable Horn. Extended Horn is therefore a substantial generalization of Horn. Clause set (5.9), for example, is not renamable Horn but is renamable extended Horn. This can be seen by associating x_1, x_2, x_3 respectively with arcs (R, A), (A, B), and (B, C) on a tree with arcs directed away from the root R.

We can now see why an extended Horn problem can be solved by unit resolution. The extended star-chain structure is clearly preserved by deletion and contraction. Thus if unit resolution does not detect unsatisfiability, the remaining clauses have the extended star-chain property and contain at least two literals each. Their LP relaxation can be solved by setting all unfixed variables to $1/2$, and Chandrasekaran's theorem gives instructions for rounding this solution to obtain an 0-1 solution. Let an edge e of $\mathcal{T}$ be *even* if the number of edges on the path from the root to the closer vertex of e is even, and *odd* otherwise. It is shown in [15] that variables corresponding to even edges are rounded down, and the rest rounded up.

Proposition 5.11. *Let S be a satisfiable extended Horn set S associated with directed tree $\mathcal{T}$. Apply unit resolution to S, and let $\mathcal{T}'$ be the tree that results from contracting the edges of $\mathcal{T}$ that correspond to fixed variables. Then a satisfying solution for S can be found by assigning false to unfixed variables that correspond to even edges of $\mathcal{T}'$, and true to those corresponding to odd edges.*

The result is valid for renamable extended Horn sets if the values of renamed variables are complemented. In the special case of a Horn set, one always assigns false to unfixed variables, because all edges are adjacent to the root and therefore even. The following is a corollary.

Proposition 5.12. *A renamable extended Horn clause set S is satisfiable if and only if it has no unit refutation, and if and only if S^{LP} is feasible.*

Example 5.14. *Consider an extended Horn set S consisting of the single clause $\bar{x}_1 \vee \bar{x}_3 \vee \bar{x}_4 \vee \bar{x}_5 \vee x_6 \vee x_7$, discussed in Example 5.13. Unit resolution has no effect, and S^{LP} has the solution $\hat{x} = (1/2, 0, 1/2, 1/2, 1/2, 1/2, 1/2)$. Thus, we obtain a satisfying solution for S when we set* $x = T^{-1}\lceil T\hat{x}\rceil = T^{-1}\lceil(-1/2, -1, -1, -1/2, -1, -1, -3/2)\rceil = T^{-1}(0, -1, -1, 0, -1, -1, -1) = (0, 1, 1, 0, 1, 1, 0)$. Note that we round down on even edges and up on odd edges.

Interestingly, once a numerical interpretation has pointed the way to an extension of the Horn concept, a slightly more general extension becomes evident. Note that the extended star-chain property is preserved by contraction of edges in part because every edge in the extended star is connected by a path to the root. The same is true if the paths to the root are not disjoint. As Schlipf et al. [71] point out, we can therefore generalize the extended star-chain property as follows: a clause has the *arborescence-chain property* with respect to $\mathcal{T}$ when it describes flows that can be partitioned into a set of unit flows into the root on some (possibly empty) subtree of $\mathcal{T}$ and a unit flow on one (possibly empty) chain in $\mathcal{T}$. A clause having this structure is clearly satisfied by the same even-odd truth assignment as before. We can therefore further extend the concept of renamable extended Horn problems as those whose clauses define flows having the arborescence/chain structure on some corresponding directed tree $\mathcal{T}$.

Example 5.15. *Suppose the clause in Example 5.14 contains an additional literal* $\bar{x}_2$. *The corresponding flow pattern is that shown in Figure 5.2 with an additional flow on arc* (A, B) *directed toward the root. Thus, literals* $\bar{x}_1, \bar{x}_2, \bar{x}_3, \bar{x}_4$ *correspond to an arborescence, and* x_5, x_6, x_7 *to a chain. We conclude that the clause set* S *consisting solely of* $\bar{x}_1 \vee \bar{x}_2 \vee \bar{x}_3 \vee \bar{x}_4 \vee x_6 \vee x_7$ *is extended Horn in the expanded sense.*

From here out we understand extended Horn and renamable extended Horn problems in the expanded sense of Schlipf et al. Propositions 5.11 and 5.12 continue to hold.

5.3.3 One-Step Lookahead

Renamable Horn sets have the double advantage that (a) their satisfiability can be checked by unit resolution, and (b) a satisfying solution can be easily identified when it exists. We have (b) because there are linear-time algorithms for checking whether a clause set is renamable Horn and, if so, how to rename variables to make it Horn [1, 14, 58]. Once the renaming scheme is identified, variables that are unfixed by unit resolution can simply be set to false, or to true if they are renamed.

Renamable extended Horn sets do not have this double advantage. Although unit resolution can check for satisfiability, a satisfying solution is evident only when the associated directed tree $\mathcal{T}$ is given. There is no known polynomial-time algorithm for finding $\mathcal{T}$, even in the case of an unrenamed extended Horn problem. Swaminathan and Wagner [73] have shown how to identify $\mathcal{T}$ for a large subclass of extended Horn problems, using a graph realization algorithm that runs in slightly more than linear time. Yet their approach has not been generalized to full extended Horn sets.

Fortunately, as Schlipf et al. point out [71], a simple *one-step lookahead* algorithm can solve a renamable extended Horn problem without knowledge of $\mathcal{T}$. Let a class $\mathcal{C}$ of satisfiability problems have the *unit resolution property* if (a) a clause set in $\mathcal{C}$ is satisfiable if and only if there is no unit refutation for it, and (b) $\mathcal{C}$ is closed under deletion and contraction.

Proposition 5.13. *If a class of Boolean satisfiability problems has the unit resolution property, a one-step lookahead algorithm can check any clause set in the class for satisfiability and exhibit a satisfying solution if one exists.*

Because renamable extended Horn problems have the unit resolution property, a one-step lookahead algorithm solves their satisfiability problem.

One-step lookahead is applied to a clause set S as follows. Let S_0 be the result of applying unit resolution to S. If S_0 contains the empty clause, stop, because S

is unsatisfiable. If S_0 is empty, stop, because the variables fixed by unit resolution already satisfy S. Otherwise perform the following:

(i) Let $S_1 = S_0 \cup \{x_j\}$ for some variable x_j occurring in S_0 and apply unit resolution to S_1. If S_1 is empty, fix x_j to true and stop. If S_1 is nonempty and there is no unit refutation, fix x_j to true, let $S_0 = S_1$, and repeat this step.
(ii) Let $S_1 = S_0 \cup \{\bar{x}_j\}$ and apply unit resolution to S_1. If S_1 is empty, fix x_j to false and stop. If S_1 is nonempty and there is no unit refutation, fix x_j to false, let $S_0 = S_1$, and return to step 1.
(iii) Stop without determining whether S is satisfiable.

The algorithm runs in time proportional to nL, where n is the number of variables and L the number of literals in S.

Proof of Proposition 5.13. Because the one-step lookahead algorithm is clearly correct, it suffices to show that it cannot reach step 3 when S belongs to a problem class with the unit resolution property. Step 3 can be reached only if there is no unit refutation for S_0, but there is a unit refutation for both $S_0 \cup \{x_j\}$ and $S_0 \cup \{\bar{x}_j\}$, which means S_0 is unsatisfiable. Yet S_0 cannot be unsatisfiable because it has no unit refutation and was obtained by unit resolution from a problem S in a class with the unit resolution property. ■

Example 5.16. *Consider the clause set S below:*

$$\bar{x}_1 \vee x_2$$
$$\bar{x}_1 \vee \bar{x}_2$$
$$x_1 \vee x_2 \vee x_3$$
$$x_1 \vee x_2 \vee \bar{x}_3$$

S is not renamable Horn but is renamable extended Horn, as can be seen by renaming x_1, x_2 and associating x_1, x_2, x_3 respectively with $(A, B), (R, A), (B, C)$ in a directed tree. However, without knowledge of the tree, we can solve the problem with one-step lookahead. Unit resolution has no effect on $S_0 = S$, and we let $S_1 = S_0 \cup \{x_1\}$ in step 1. Unit resolution derives the empty clause from S_1, and we move to step 2 by setting $S_1 = S_0 \cup \{\bar{x}_1\}$. Unit resolution reduces S_1 to $\{x_2 \vee x_3, x_2 \vee \bar{x}_3\}$, and we return to step 1 with $S_0 = \{x_2 \vee x_3, x_2 \vee \bar{x}_3\}$. Setting $S_1 = S_0 \cup \{x_2\}$, unit resolution reduces S_1 to the empty set, and the algorithm terminates with $(x_1, x_2) = (0, 1)$, where x_3 can be set to either value.

At this writing no algorithms or heuristic methods have been proposed for finding a set of variables that, when fixed, leave a renamable extended Horn problem. The same deficiency exists for the integrality classes discussed next.

5.3.4 Balanced Problems and Integrality

We have identified a class of satisfiability problems that can be quickly solved by a one-step lookahead algorithm that uses unit resolution. We now consider problems that can be solved by linear programming because their LP relaxation describes an integral polytope. Such problems are doubly attractive because unit resolution can solve them even without one-step lookahead, because of the following fact [18].

Let us say that a clause set S is *ideal* if S^{LP} defines an integral polyhedron. If S^{01} is the system $Ax \geq b$, we say that A is an ideal matrix if S is ideal.

Proposition 5.14. *Let S be an ideal clause set that contains no unit clauses. Then for any variable x_j, S has a solution in which x_j is true and a solution in which x_j is false.*

Proof. Since every clause contains at least two literals, setting each $x_j = 1/2$ satisfies S^{LP}. This solution is a convex combination of the vertices of the polyhedron described by S^{LP}, which by hypothesis are integral. Thus, for every j there is a vertex v in the convex combination at which $v_j = 0$ and a vertex at which $v_j = 1$. ■

To solve an ideal problem S, first apply unit resolution to S to eliminate all unit clauses. The remaining set S is ideal. Pick any atom x_j that occurs in S and add the unit clause x_j or $\bar{x}_j$ to S, arbitrarily. By Proposition 5.14, S remains satisfiable if it was satisfiable to being with. Repeat the procedure until a unit refutation is obtained or until S is empty (and unit resolution has fixed the variables to a satisfying solution). The algorithm runs in linear time.

One known class of ideal satisfiability problems are *balanced* problems. If S^{01} is the 0-1 system $Ax \geq b$, then S is balanced when A is balanced. A 0, ± 1 matrix A is balanced when every square submatrix of A with exactly two nonzeros in each row and column has the property that its entries sum to a multiple of four. The following proposition is proved by Conforti and Cornuéjols [18].

Proposition 5.15. *Clause set S is ideal if it is balanced.*

Related results are surveyed in [20, 22]. For instance, balancedness can be checked by examining subsets of S for bicolorability. Let a $\{0, \pm 1\}$-matrix be *bicolorable* if its rows can be partitioned into blue rows and red rows such that every column with two or more nonzeros contains either two entries of the same sign in rows of different colors or two entries of different signs in rows of the same color. A clause set S is bicolorable if the coefficient matrix of S^{01} is bicolorable. Conforti and Cornuéjols [19] prove the following.

Proposition 5.16. *Clause set S is balanced if and only if every subset of S is bicolorable.*

Example 5.17. *Consider the following clause set S:*

$$\begin{array}{llll} x_1 & & \vee x_3 & \\ \bar{x}_1 & & & \vee x_4 \\ & x_2 & \vee x_3 & \\ & \bar{x}_2 \vee & & \vee x_4 \end{array} \tag{5.24}$$

By coloring the first two clauses red and the last two blue, we see from Proposition 5.16 that all subsets of S are balanced, and by Proposition 5.15, S is ideal. We can also use Proposition 5.14 to solve S. Initially, unit resolution has no effect, and we arbitrarily set $x_1 = 0$, *which yields the single clause* $\bar{x}_2 \vee x_4$ *after unit resolution. Now we arbitrarily set* $x_2 = 0$, *whereupon S becomes empty. The resulting solution is* $(x_1, x_2, x_3) = (0, 0, 1)$ *with either value for* x_4.

5.3.5 Resolution and Integrality

Because resolvents are cutting planes, one might ask whether applying the resolution algorithm to a clause set S cuts off enough of the polyhedron defined by S^{LP} to produce an integral polyhedron. The answer is that it does so if and only if the monotone subproblems of S already define integral polyhedra.

To make this precise, let a *monotone subproblem* of S be a subset $\hat{S} \subset S$ in which no variable occurs in both a positive and a negative literal. A monotone subproblem is *maximal* if it is a proper subset of no monotone subproblem. We can suppose without loss of generality that all the literals in a given monotone subproblem $\hat{S}$ are positive (by complementing variables as needed). Thus $\hat{S}^{01}$ is a set covering problem, which is a 0-1 problem of the form $Ax \geq e$ in which A is a 0-1 matrix and e a vector of 1s. $\hat{S}$ can also be viewed as a satisfiability problem with all positive literals (so that the same definition of an ideal problem applies). The following result, proved in [41], reduces the integrality question for satisfiability problems to that for set covering problems.

Proposition 5.17. *If S contains all of its prime implicates, then S is ideal if and only if every maximal monotone subproblem* $\hat{S} \subset S$ *is ideal.*

One can determine whether S contains all of its prime implicates by checking whether the resolution algorithm adds any clauses to S, that is, by checking whether any pair of clauses in S have a resolvent that is not already absorbed by a clause in S.

Guenin [34] pointed out an alternate statement of Proposition 5.17. Given a clause set S, let A be the coefficient matrix in S^{01}, and define

$$D_S = \begin{bmatrix} P & N \\ I & I \end{bmatrix}$$

where 0-1 matrices P, N are the same size as A and indicate the signs of entries in A. That is, $P_{ij} = 1$ if and only if $A_{ij} = 1$ and $N_{ij} = 1$ if and only if $A_{ij} = -1$.

Proposition 5.18. *If S contains all of its prime implicates, then S is ideal if and only if D_S is ideal.*

Example 5.18. *Consider the clause set S:*

$$\begin{array}{l} x_1 \vee x_2 \vee x_3 \\ x_1 \vee \bar{x}_2 \vee \bar{x}_3 \end{array}$$

Because S is not balanced, Proposition 5.15 does not show that S is ideal. However, Proposition 5.17 applies because S consists of its prime implicates, and its two maximal monotone subproblems (i.e., the two individual clauses) obviously define integral polyhedra. S is therefore ideal. Proposition 5.18 can also be applied if one can verify that the following matrix D_S is ideal:

$$\begin{bmatrix} 1 & 1 & 1 & 0 & 0 & 0 \\ 1 & 0 & 0 & 0 & 1 & 1 \\ 1 & 0 & 0 & 1 & 0 & 0 \\ 0 & 1 & 0 & 0 & 1 & 0 \\ 0 & 0 & 1 & 0 & 0 & 1 \end{bmatrix}$$

D_S is not balanced because there is no feasible coloring for the submatrix consisting of rows 1, 2, and 4. D_S is ideal but must be shown to be so in some fashion other than bicolorability.

Nobili and Sassano [62] strengthened Propositions 5.17 and 5.18 by pointing out that a restricted form of resolution suffices. Let a *disjoint* resolvent be the resolvent of two clauses that have no variables in common, except the one variable that appears positively in one clause and negatively in the other. The disjoint resolution algorithm is the same as the ordinary resolution algorithm except that only disjoint resolvents are generated.

Proposition 5.19. *Let S' be the result of applying the disjoint resolution algorithm to S. Then S is ideal if and only if $D_{S'}$ is ideal.*

Example 5.19. *Consider the following clause set S:*

$$\begin{array}{lll} x_1 \vee x_2 & & \\ \bar{x}_1 & & \vee x_3 \\ x_1 \vee \bar{x}_2 \vee x_3 & & \end{array}$$

Disjoint resolution generates one additional clause, so that S' consists of the foregoing and

$$x_2 \vee x_3$$

Further resolutions are possible, but they need not be carried out because the resolvents are not disjoint. The matrix $D_{S'}$ is

$$\begin{bmatrix} 1 & 1 & 0 & 0 & 0 & 0 \\ 0 & 0 & 1 & 1 & 0 & 0 \\ 1 & 0 & 1 & 0 & 1 & 0 \\ 0 & 1 & 1 & 0 & 0 & 0 \\ 1 & 0 & 0 & 1 & 0 & 0 \\ 0 & 1 & 0 & 0 & 1 & 0 \\ 0 & 0 & 1 & 0 & 0 & 1 \end{bmatrix}.$$

$D_{S'}$ is not balanced, because the submatrix consisting of rows 1, 3, and 5 is not bicolorable. In fact, $D_{S'}$ is not ideal because the corresponding set covering problem $D_{S'}y \geq e$ defines a polyhedron with the fractional vertex $y = (1/2, 1, 1/2, 1/2, 0, 1/2)$. Therefore S is not ideal, as can be verified by noting that S^{LP} defines a polyhedron with the fractional vertex $x = (1/2, 0, 1/2)$.

One can apply Proposition 5.17 by carrying out the full resolution algorithm on S, which yields $\hat{S} = \{x_1 \vee x_2, x_3\}$. $\hat{S}$ itself is its only maximum monotone subproblem, which means $\hat{S}$ is ideal because it is obviously balanced.

5.4 Computational Considerations

Computational testing of integer programming methods for satisfiability dates back at least to a 1988 paper of Blair, Jeroslow, and Lowe [8]. A 1994 experimental study [37] suggested that, at that time, an integer programming approach was roughly competitive with a pure branching strategy that uses no LP relaxations, such as a Davis-Putnam-Loveland (DPL) method [26, 57]. DPL methods have since been the subject of intense study and have improved dramatically (see [32, 33, 83] for surveys). The Chaff system of Moskewicz et al. [60], for example, solves certain industrial satisfiability problems with a million variables or more. It is well known, however, that satisfiability problems of a given size can vary enormously in difficulty (e.g., [21, 25] and Chapter 4 in this volume).

Integer programming methods for the Boolean satisfiability problem have received relatively little attention since 1994, despite substantial improvements in the LP solvers on which they rely. One might argue that the integer programming approach is impractical for large satisfiability problems, because it requires considerable memory and incurs substantial overhead in the solution of the LP relaxation. This is not the case, however, for the decomposition methods discussed earlier. They apply integer programming only to a small "core" problem and leave the easier part of the problem to be processed by unit resolution or some other fast method.

The most straightforward form of decomposition, discussed in Section 5.2.4, is to branch on a few variables that are known to yield renamable Horn problems at the leaf nodes of a shallow tree. (One might also branch so as to produce renamable extended Horn problems at the leaf nodes, but this awaits a practical heuristic method for isolating extended Horn substructure.) The branching portion of the algorithm could be either a branch-and-cut method or a DPL search. In the latter case, decomposition reduces to a DPL method with a sophisticated variable selection rule.

Benders decomposition also takes advantage of renamable Horn substructure (Section 5.2.5). A branch-and-cut method can be applied to the Benders master problem, which is a general 0-1 problem, while unit resolution solves the renamable Horn subproblems. (The subproblems might also be renamable extended Horn, which would require research into how Benders cuts can be rapidly generated.) Yan and Hooker [82] tested a specialized Benders approach on circuit verification problems, an important class of satisfiability problems. They found it to be faster than binary decision diagrams at detecting errors, although generally slower at proving correctness. A related Benders method has been applied to machine scheduling problems [42–45, 49], resulting in computational speedups of several orders of magnitude relative to state-of-the-art constraint programming and mixed integer programming solvers. However, a Benders method has apparently not been tested on satisfiability problems other than circuit verification problems.

A third form of decomposition replaces LP relaxation with Lagrangean relaxation, again in such a way as to isolate special structure (Section 5.2.6). A full Lagrangean method would solve the Lagrangean dual at some nodes of the search tree, but this strategy has not been tested computationally. The Lagrangean-based method of Bennaceur et al. [7], however, uses a fixed set of Lagrange multipliers and appears to be significantly more effective than DPL. Because such advanced methods as Chaff are highly engineered DPL algorithms, a Lagrangean approach may have the potential to outperform the best methods currently in use.

The integrality results (Sections 5.3.4, 5.3.5) remain primarily of theoretical interest, since it is hard to recognize or isolate special structure that ensures integrality. Nonetheless, they could become important in applications that involve optimization of an objective function subject to logical clauses, because they indicate when an LP solver may be used rather than a general integer solver. One such application is the maximum satisfiability problem [25, 35, 36].

The ideas presented here are part of a more general strategy of merging logical inference with optimization methods. There is growing evidence [38, 42, 46, 56, 59, 79] that logic-based methods can enhance optimization, and similar evidence that optimization methods can assist logical inference – not only in Boolean logic, but in probabilistic, nonmonotonic, and belief logics as well [17]. The guiding principle in each case is to employ concepts, whether they be logical or numerical, that best reveal the structure of the problem at hand.

5.5 Exercises

(i) Show that the resolvent $x_1 \vee x_4$ obtained in Example 5.7 corresponds to a rank 1 cut.

(ii) Formulate the problem of checking whether a clause set is extended Horn as a 2-SAT problem (i.e., a satisfiability problem in CNF with at most two literals per clause). The most straightforward 2-SAT formulation is quadratic in size, but there are linear-size formulations (e.g., [1]).

(iii) Show that disjoint resolution adds nothing to the inferential power of unit resolution. That is, unit resolution shows a clause set S to be unsatisfiable if and only if a combination of unit and disjoint resolution shows S to be unsatisfiable. Hint: note that a disjoint resolvent is the sum of its parents.

(iv) Let T be the set of satisfying solutions for a given Horn clause set. Show that if $x^1, x^2 \in T$, then $\min\{x^1, x^2\} \in T$, where the minimum is componentwise. This property of Horn sets is generalized in [51, 70].

(v) Exhibit a Horn clause set S for which S^{LP} has a nonintegral extreme point.

(vi) Suppose that S contains all of its prime implicates. Show that a one-step lookahead algorithm solves the satisfiability problem for S.

(vii) Exhibit a two-variable satisfiability problem that is not renamable extended Horn.

(viii) One can show that extended Horn sets have the unit resolution property, based solely on the arborescence-chain structure and without reference to the numerical interpretation. Construct such an argument.

(viii) Suppose that extended Horn sets were defined so that the arborescence-chain property permitted as many as two chains rather than just one. Show that extended Horn sets, under this definition, would not have the unit resolution property, by exhibiting a counterexample. Where does the argument of the previous question break down?

References

[1] B. Aspvall. Recognizing disguised NR(1) instance of the satisfiability problem. *Journal of Algorithms* 1, pp. 97–103, 1980.

[2] E. Balas and W. Niehaus. Optimized crossover-based genetic algorithms for the maximum cardinality and maximum weight clique problems. *Journal of Heuristics* **4**, pp. 107–22, 1998.

[3] E. Balas and J. Xue. Minimum weighted coloring of triangulated graphs with application to maximum weight vertex packing and clique finding in arbitrary graphs. *SIAM Journal on Computing* **20**, pp. 209–21, 1991.

[4] E. Balas and C. S. Yu. Finding a maximum clique in an arbitrary graph. *SIAM Journal on Computing* **15**, pp. 1054–68, 1986.

[5] R. Battiti and M. Protasi. Reactive local search for the maximum clique problem. *Algorithmica* **29**, pp. 610–37, 2001.

[6] J. F. Benders. Partitioning procedures for solving mixed-variables programming problems. *Numerische Mathematik* **4**, pp. 238–52, 1962.

[7] H. Bennaceur, I. Gouachi, and G. Plateau. An incremental branch-and-bound method for the satisfiability problem. *INFORMS Journal on Computing* **10**, pp. 301–8, 1998.

[8] C. Blair, R. G. Jeroslow, and J. K. Lowe. Some results and experiments in programming techniques for propositional logic. *Computers and Operations Research* **13**, pp. 633–45, 1988.

[9] I. M. Bomze, M. Budinich, M. Pellilo, and C. Rossi. Annealed replication: A new heuristic for the maximum clique problem. *Discrete Applied Mathematics* **121**, pp. 27–49, 2002.

[10] G. Boole. *The Mathematical Analysis of Logic: Being a Essay Toward a Calculus of Deductive Reasoning*. Blackwell, Oxford, UK, 1951. Original work published 1847.

[11] S. Busygin, S. Butenko, and P. M. Pardalos. A heuristic for the maximum independent set problem based on optimization of a quadratic over a sphere. *Journal of Combinatorial Optimization* **6**, pp. 287–97, 2002.

[12] R. Carraghan and P. Pardalos. An exact algorithm for the maximum clique problem. *Operations Research Letters* **9**, pp. 375–82, 1990.

[13] R. Chandrasekaran. Integer programming problems for which a simple rounding type of algorithm works. In *Progress in Combinatorial Optimization*, W. R. Pulleyblank, ed. pp. 101–6. Academic Press Canada, 1984.

[14] V. Chandru, C. R. Coullard, P. L. Hammer, M. Montañez, and X. Sun. Horn, renamable Horn and generalized Horn functions. Annals of Mathematics and Artificial Intelligence **1**, pp. 333–47, 1990.

[15] V. Chandru and J. N. Hooker. Extended Horn clauses in propositional logic. *Journal of the ACM* **38**, pp. 203–21, 1991.

[16] V. Chandru and J. N. Hooker. Detecting embedded Horn structure in propositional logic. *Information Processing Letters* **42**, pp. 109–11, 1992.

[17] V. Chandru and J. N. Hooker. *Optimization Methods for Logical Inference*. John Wiley & Sons, New York, 1999.

[18] M. Conforti and G. Cornuéjols. A class of logic programs solvable by linear programming. *Journal of the ACM* **42**, pp. 1107–13, 1995.

[19] M. Conforti and G. Cornuéjols. Balanced 0,±1 matrices, bicoloring and total dual integrality. *Mathematical Programming* **71**, pp. 249–58, 1995.

[20] M. Conforti, G. Cornuéjols, A. Kapoor, and K. Vušković. Perfect, ideal and balanced matrices. *European Journal of Operational Research* **133**, pp. 455–61, 2001.

[21] S. A. Cook and D. G. Mitchell. Finding hard instances of the satisfiability problem: A survey. In *Satisfiability Problem: Theory and Applications*, D. Du, J. Gu, and P. M. Pardalos, eds. DIMACS Series in Discrete Mathematics and Theoretical Computer Science, Vol. **35**, pp. 1–17, American Mathematical Society 1997.

[22] G. Cornuéjols and B. Guenin. Ideal clutters. *Discrete Applied Mathematics* **123**, pp. 303–38, 2002.

[23] R. W. Cottle and A. F. Veinott. Polyhedral sets having a least element. *Mathematical Programming* **3**, pp. 238–49, 1972.

[24] Y. Crama, O. Ekin, and P. L. Hammer. Variable and term removal from Boolean formulae. *Discrete Applied Mathematics* **75**, pp. 217–30, 1997.

[25] Y. Crama and P. L. Hammer. *Boolean Functions: Theory, Algorithms, and Applications*. Cambridge University Press, Cambridge, UK, 2010.

[26] M. Davis and H. Putnam. A computing procedure for quantification theory. *Journal of the ACM* **7**, pp. 201–15, 1960.

[27] B. Dilkina, C. Gomes, and A. Sabharwal. Tradeoffs in the complexity of backdoor detection. In Principles and Practice of Constraint Programming (CP 2007), C. Bessiere, ed., LNCS 4741, pp. 256–70.

[28] T. A. Feo, M. G. C. Resende, and S. H. Smith. A greedy randomized adaptive search procedure for maximum independent set. *Operations Research* **42**, pp. 860–78, 1994.

[29] N. Funabiki, Y. Takefuji, and Kuo-Chun Lee. A neural network model for finding a near-maximum clique. *Journal of Parallel and Distributed Computing* **14**, pp. 340–4, 1992.

[30] M. Gendreau, P. Soriano, and L. Salvail. Solving the maximum clique problem using a tabu search approach. *Annals of Operations Research* **41**, pp. 385–403, 1993.

[31] A. M. Geoffrion. Generalized Benders decomposition. *Journal of Optimization Theory and Applications* **10**, pp. 237–60, 1972.

[32] E. Giunchiglia, M. Maratea, A. Tacchella, and D. Zambonin. Evaluating search heuristics and optimization techniques in propositional satisfiability. In *Automated Reasoning: First International Joint Conference* (IJCAR2001), R. Gore, A. Leitsch, and T. Nipkow, eds. *Lecture Notes in Artificial Intelligence* **2083**, pp. 347–63, 2001.

[33] Jun Gu, P. W. Purdom, J. Franco, and B. W. Wah. Algorithms for the satisfiability (SAT) problem: A survey, in *Satisfiability Problem: Theory and Applications*, D. Du, J. Gu, and P. M. Pardalos, eds. DIMACS Series in Discrete Mathematics and Theoretical Computer Science Vol. **35**, pp.19-51. American Mathematical Society, 1997.

[34] B. Guenin. Perfect and ideal 0, ± 1 matrices. *Mathematics of Operations Research* **23**, pp. 322–38, 1998.

[35] P. Hansen and B. Jaumard. Algorithms for the maximum satisfiability problem. *Computing* **44**, pp. 279–303, 1990.

[36] P. Hansen, B. Jaumard, and M. P. De Aragao. Mixed-integer column generation algorithms and the probabilistic maximum satisfiability problem. *European Journal of Operational Research* **108**, pp. 671–83, 1998.

[37] F. Harche, J. N. Hooker, and G. Thompson. A computational study of satisfiability algorithms for propositional logic. *ORSA Journal on Computing* **6**, pp. 423–35, 1994.

[38] S. Heipcke. *Combined modelling and problem solving in mathematical programming and constraint programming*. PhD thesis, University of Buckingham, 1999.

[39] M. Hifi. A genetic algorithm-based heuristic for solving the weighted maximum independent set and some equivalent problems. *Journal of the Operational Research Society* **48**, pp. 612–22, 1997.

[40] J. N. Hooker. Input proofs and rank one cutting planes. *ORSA Journal on Computing* **1**, pp. 137–45, 1989.

[41] J. N. Hooker. Resolution and the integrality of satisfiability polytopes. *Mathematical Programming* **4**, pp. 1–10, 1996.

[42] J. N. Hooker. *Logic-Based Methods for Optimization: Combining Optimization and Constraint Satisfaction*. John Wiley & Sons, New York, 2000.

[43] J. N. Hooker. A hybrid method for planning and scheduling. *Constraints* **10**, pp. 385–401, 2005.

[44] J. N. Hooker. An integrated method for planning and scheduling to minimize tardiness. *Constraints* **11**, pp. 139–57, 2006.

[45] J. N. Hooker. Planning and scheduling by logic-based Benders decomposition. *Operations Research* **55**, pp. 588–602, 2007.

[46] J. N. Hooker. *Integrated Methods for Optimization*. Springer, New York, 2007.

[47] J. N. Hooker and C. Fedjki. Branch-and-cut solution of inference problems in propositional logic. *Annals of Mathematics and Artificial Intelligence* **1**, pp. 123–40,1990.

[48] J. N. Hooker and G. Ottosson. Logic-based Benders decomposition. *Mathematical Programming* **96**, pp. 33–60, 2003.

[49] V. Jain and I. E. Grossmann. Algorithms for hybrid MILP/CP models for a class of optimization problems. *INFORMS Journal on Computing* **13**, pp. 258–76, 2001.

[50] A. Jagota and L. A. Sanchis. Adaptive, restart, randomized greedy heuristics for maximum clique. *Journal of Heuristics* **7**, pp. 565–85, 2001.

[51] P. Jeavons, D. Cohen, and M. Gyssens. A test for tractability. In *Principles and Practice of Constraint Programming* (CP96). E. C. Freuder, ed. *Lecture Notes in Computer Science* **1118**, 267–81, 1996.

[52] R. E. Jeroslow and J. Wang. Solving propositional satisfiability problems. *Annals of Mathematics and Artificial Intelligence* **1**, pp. 167–88, 1990.

[53] P. Kilby, J. K. Slaney, S. Thibaux, and T. Walsh. Backbones and backdoors in satisfiability. *AAAI Proceedings*, 1368–73, 2005.

[54] S. Kottler, M. Kaufmann, and C. Sinz. Computation of renameable Horn backdoors. *Theory and Applications of Satisfiability Testing, Eleventh International Conference* (SAT 2008). *Lecture Notes in Computer Science* **4996**, pp. 154–60, 2008.

[55] H. Y. Lau and H. F. Ting. The greedier the better: An efficient algorithm for approximating maximum independent set. In *Computing and Combinatorics: 5th Annual International Conference* (COCOON'99). T. Asano, H. Imai, D. T. Lee, S. Nakano, and T. Tokuyama, eds. *Lecture Notes in Computer Science* **1627**, pp. 483–92, 1992.

[56] J. Little, and K. Darby-Dowman. The significance of constraint logic programming to operational research, in *Operational Research Tutorial Papers 1995*, M. Lawrence and C. Wilson, eds. pp. 20–45. Operational Research Society, 1995.

[57] D. W. Loveland. *Automated Theorem Proving: A Logical Basis*. North-Holland, Amsterdam, 1978.

[58] H. Mannila, and K. Mehlhorn. A fast algorithm for renaming a set of clauses as a Horn set. *Information Processing Letters* **21**, pp. 261–72, 1985.

[59] M. Milano. *Constraint and Integer Programming: Toward a Unified Methodology*. Operations Research/Computer Science Interfaces Series. Springer, New York, 2003.

[60] M. W. Moskewicz, C. F. Madigan, Y. Zhao, L. Zhang, and S. Malik. Chaff: Engineering an efficient SAT solver. In *Proceedings of the 38th Design Automation Conference*, pp. 530–5, ACM, New York, 2001.

[61] N. Nishimura, P. Ragde, and S. Szeider. Detecting backdoor sets with respect to Horn and binary clauses. *Theory and Applications of Satisfiability Testing, Seventh International Conference* (SAT 2004).

[62] P. Nobili and A. Sassano. (0, ± 1) ideal matrices. *Mathematical Programming* **80**, pp. 265–81, 1998.

[63] P. R. J. Ostergard. A fast algorithm for the maximum clique problem. *Discrete Applied Mathematics* 120, pp. 197–207, 2002.

[64] P. Pardalos and G. Rogers. A branch and bound algorithm for the maximum clique problem. *Computers and Operations Research* **19**, pp. 363–75, 1992.

[65] P. M. Pardalos and N. Desai. An algorithm for finding a maximum weighted independent set in an arbitrary graph. *International Journal of Computer Mathematics* **38**, pp. 163–75, 1991.

[66] L. Paris, R. Ostrowski, P. Siegel, and L. Sais. Computing Horn strong backdoor sets thanks to local search. *18th IEEE International Conference on Tools with Artificial Intellignece* (ICTAI 2006) pp. 139–43.

[67] L. Paris, R. Ostrowski, P. Siegel, and L. Sais. From Horn strong backdoor sets to ordered strong backdoor sets. Advances in Artificial Intelligence (MICAI 2007), *Lecture Notes in Computer Science* **4827**, pp. 105–17, 2007.

[68] W. V. Quine. The problem of simplifying truth functions. *American Mathematical Monthly* **59**, pp. 521–31, 1952.

[69] W. V. Quine. A way to simplify truth functions. *American Mathematical Monthly* **62**, pp. 627–31, 1955.

[70] T. J. Schaeffer. The complexity of satisfiability problems. *Proceedings, 10th Annual ACM Symposium on Theory of Computing* (STOC'78), pp. 216–26. ACM New York, 1978.

[71] J. S. Schlipf, F. S. Annexstein, J. V. Franco, and R. P. Swaminathan. On finding solutions for extended Horn formulas. *Information Processing Letters* **54**, pp. 133–7, 1995.

[72] M. Shindo and E. Tomita. A simple algorithm for finding a maximum clique and its worst-case time complexity. *Systems and Computers in Japan* **21**, pp. 1–13, 1990.

[73] R. P. Swaminathan and D. K. Wagner. The arborescence-realization problem. *Discrete Applied Mathematics* **59**, pp. 267–83, 1995.

[74] S. Szeider. Backdoor sets for DLL subsolvers. *Journal of Automated Reasoning* **35**, pp. 73–88, 2005.

[75] R. E. Tarjan and A. E. Trojanowski. Finding a maximum independent set. *SIAM Journal on Computing* **6**, pp. 537–46, 1977.

[76] E. S. Thorsteinsson. Branch-and-check: A hybrid framework integrating mixed integer programming and constraint logic programming. In T. Walsh. *Principles and Practice of Constraint Programming* (CP2001). T. Walsh, ed. *Lecture Notes in Computer Science* **2239**, pp. 16–30, 2001.

[77] R. Williams, C. Gomes, and B. Selman. Backdoors to typical case complexity. *Proceedings, International Joint Conference on Artificial Intelligence* (IJCAI 2003), pp. 1173–8.

[78] R. Williams, C. Gomes, and B. Selman. On the connections between heavy-tails, backdoors, and restarts in combinatorial search. *Theory and Applications of Satisfiability*

Testing, 6th International Conference (SAT 2003). *Lecture Notes in Computer Science* **2919**, pp. 222–30, 2004.

[79] H. P. Williams and J. M. Wilson. Connections between integer linear programming and constraint logic programming – An overview and introduction to the cluster of articles. *INFORMS Journal on Computing* **10**, pp. 261–4, 1998.

[80] L. A. Wolsey. *Integer Programming*. John Wiley & Sons, New York, 1998.

[81] Y. Yamada, E. Tomita, and H. Takahashi. A randomized algorithm for finding a near-maximum clique and its experimental evaluations. *Systems and Computers in Japan* **25**, pp. 1–7, 1994.

[82] H. Yan and J. N. Hooker. Logic circuit verification by Benders decomposition. In *Principles, and Practice of Constraint Programming: The Newport Papers*, V. Saraswat and P. Van Hentenryck, eds., pp. 267–88. MIT Press, Cambridge, MA, 1995.

[83] Lintao Zhang and S. Malik. The quest for efficient Boolean satisfiability solvers, in *Automated Deduction* (CADE-18), *18th International Conference on Automated Deduction*, A. Voronkov, ed. *Lecture Notes in Artificial Intelligence* **2392**, pp. 295–313, 2002.

Part III

Learning Theory and Cryptography

6

Probabilistic Learning and Boolean Functions

Martin Anthony

6.1 Introduction

This chapter explores the *learnability* of Boolean functions. Broadly speaking, the problem of interest is how to infer information about an unknown Boolean function given only information about its values on some points, together with the information that it belongs to a particular class of Boolean functions. This broad description can encompass many more precise formulations, but here we focus on probabilistic models of learning, in which the information about the function value on points is provided through its values on some randomly drawn sample, and in which the criteria for successful "learning" are defined using probability theory. Other approaches, such as "exact query learning" (see [1, 18, 20] and Chapter 7 in this volume, for instance) and "specification," "testing," or "learning with a helpful teacher" (see [12, 4, 16, 21, 26]) are possible, and particularly interesting in the context of Boolean functions. Here, however, we focus on probabilistic models and aim to give a fairly thorough account of what can be said in two such models.

In the probabilistic models discussed, there are two separate, but linked, issues of concern. First, there is the question of how much information is needed about the values of a function on points before a good approximation to the function can be found. Second, there is the question of how, algorithmically, we might find a good approximation to the function. These two issues are usually termed the *sample complexity* and *computational complexity* of learning. The chapter breaks fairly naturally into, first, an exploration of sample complexity and then a discussion of computational complexity.

6.2 Probabilistic Modeling of Learning

6.2.1 A Probabilistic Model

The primary probabilistic model of "supervised" learning we discuss here is a variant of the "probably approximately correct" (or PAC) model introduced by

Valiant [31], and further developed by a number of many others; see [32, 13, 2], for example. The probabilistic aspects of the model have their roots in work of Vapnik and Chervonenkis [33, 34], as was pointed out by [5]. Valiant's model additionally placed considerable emphasis on the computational complexity of learning.

In the model, it is assumed that we are using some class H of Boolean functions on $X = \{0, 1\}^n$ (termed the *hypothesis space*) to find a good fit to a set of data. We assume that the (labeled) data points take the form (x, b) for $x \in \{0, 1\}^n$ and $b \in \{0, 1\}$ (though most of what we discuss will apply also to the more general case in which H maps from $\mathbb{R}^n$ to $\{0, 1\}$ and the data are in $\mathbb{R}^n \times \{0, 1\}$). The learning model is probabilistic: we assume that we are presented with some randomly generated "training" data points and that we choose a hypothesis on this basis.

The simplest assumption to make about the relationship between H and the data is that the data can indeed be exactly matched by some function in H, by which we mean that each data point takes the form $(x, t(x))$ for some fixed $t \in H$ (the *target concept*). In this *realizable* case, we assume that some number m of (labeled) data points (or *labeled examples*) are generated to form a *training sample* $\mathbf{s} = ((x_1, t(x_1)), \ldots, (x_m, t(x_m))$ as follows: each x_i is chosen independently according to some fixed probability distribution μ on X. The learning problem is then, given only $\mathbf{s}$, and the knowledge that the data are labeled according to *some* target concept in H, to produce some $h \in H$ that is "close" to t (in a sense to be formalized later).

A more general framework can usefully be developed to model the case in which the data cannot necessarily be described completely by a function in H, or, indeed, when there is a stochastic, rather than deterministic, labeling of the data points. In this more general formulation, it is assumed that the data points (x, b) in the training sample are generated according to some probability distribution P on the product $X \times \{0, 1\}$. This formulation includes the realizable case just described, but also permits a given x to appear with the two different labels 0 and 1, each with certain probability. The aim of learning in this case is to find a function from H that is a good predictor of the data labels (something we will shortly make precise). It is hoped that such a function can be produced given only the training sample.

6.2.2 Definitions

We now formalize these outline descriptions of what is meant by learning. We place most emphasis on the more general framework, the realizable one being a special case of this.

A training sample is some element of Z^m, for some $m \geq 1$, where $Z = X \times \{0, 1\}$, We may therefore regard a learning algorithm as a function $L : Z^* \to H$ where $Z^* = \bigcup_{m=1}^{\infty} Z^m$ is the set of all possible training samples. (It is conceivable that we might want to define L only on part of this domain. But we could easily extend its domain to the whole of Z^* by assuming some default output in cases outside the domain of interest.) We denote by $L(\mathbf{s})$ the *output hypothesis* of the learning algorithm after being presented with training sample $\mathbf{s}$.

Because there is assumed to be some probability distribution, P, on the set $Z = X \times \{0, 1\}$ of all examples, we may define the *error*, $\text{er}_P(h)$, of a function h (with respect to P) to be the P-probability that, for a randomly chosen example, the label is not correctly predicted by h. In other words, $\text{er}_P(h) = P(\{(x, b) \in Z : h(x) \neq b\})$.

The aim is to ensure that the error of $L(\mathbf{s})$ is "usually near-optimal" provided the training sample is "large enough." Because each of the m examples in the training sample is drawn randomly and independently according to P, the sample $\mathbf{s}$ is drawn randomly from Z^m according to the product probability distribution P^m. Thus, more formally, we want it to be true that with high P^m-probability the sample $\mathbf{s}$ is such that the output function $L(\mathbf{s})$ has near-optimal error with respect to P. The smallest the error could be is $\text{opt}_P(H) = \min\{\text{er}_P(h) : h \in H\}$. (For a class of Boolean functions, because H is finite, the minimum is defined, but in general we would use the infimum.)

This leads us to the following formal definition of a version of "PAC," (probably approximately correct) learning.

Definition 6.1. (PAC learning.) *The learning algorithm L is a* PAC-learning algorithm *for the class H of Boolean functions if for any given $\delta, \varepsilon > 0$ there is a sample length $m_0(\delta, \varepsilon)$ such that for all probability distributions P on $Z = X \times \{0, 1\}$,*

$$m > m_0(\delta, \varepsilon) \Rightarrow P^m\left(\left\{\mathbf{s} \in Z^m : \text{er}_P(L(\mathbf{s})) \geq \text{opt}_P(H) + \varepsilon\right\}\right) < \delta.$$

The smallest suitable value of $m_0(\delta, \varepsilon)$, denoted $m_L(\delta, \varepsilon)$, is called the *sample complexity* of L.

The definition is fairly easy to understand in the realizable case. In this case, $\text{er}_P(h)$ is the probability that a hypothesis h disagrees with the target concept t on a randomly chosen example. So, here, informally speaking, a learning algorithm is PAC if, provided a random sample is long enough (where "long enough" is independent of P), then it is "probably" the case that after training on that sample, the output hypothesis is "approximately" correct. We often refer to ε as the *accuracy parameter* and δ as the *confidence parameter*.

Note that the probability distribution P occurs twice in the definition: first in the requirement that the P^m-probability of a sample be small and second through the fact that the error of $L(\mathbf{s})$ is measured with reference to P. The crucial feature of the definition is that we require the sample length $m_0(\delta, \varepsilon)$ to be independent of P.

6.2.3 A Learnability Result for Boolean Classes

For $h \in H$ and $\mathbf{s} = (((x_1, b_1), \ldots, (x_m, b_m))$, the *sample error* of h on $\mathbf{s}$ is

$$\hat{\text{er}}_{\mathbf{s}}(h) = \frac{1}{m} \left|\{i : h(x_i) \neq x_i\}\right|,$$

and we say that L is a SEM (sample-error minimization) algorithm if, for any $\mathbf{s}$,

$$\hat{\text{er}}_{\mathbf{s}}(L(\mathbf{s})) = \min\{\hat{\text{er}}_{\mathbf{s}}(h) : h \in H\}.$$

We now show that L is a PAC learning algorithm provided it has this fairly natural property.

Theorem 6.1. *Any SEM learning algorithm L for a set H of Boolean functions is PAC. Moreover, the sample complexity is bounded as follows:*

$$m_L(\delta, \varepsilon) \le \frac{2}{\varepsilon^2} \ln\left(\frac{2|H|}{\delta}\right).$$

Proof. By Hoeffding's inequality [14], for any particular $h \in H$,

$$P^m\left(|\hat{\text{er}}_{\mathbf{s}}(h) - \text{er}_P(h)| \ge \varepsilon/2\right) \le 2\exp(-\varepsilon^2 m/2).$$

So, for any P and ε,

$$\begin{aligned} P^m\left\{\max_{h\in H} |\hat{\text{er}}_{\mathbf{s}}(h) - \text{er}_P(h)| \ge \varepsilon/2\right\} &= P^m\left(\bigcup_{h\in H}\{\mathbf{s} \in Z^m : |\hat{\text{er}}_{\mathbf{s}}(h) - \text{er}_P(h)| \ge \varepsilon/2\}\right) \\ &\le \sum_{h\in H} P^m\left\{|\hat{\text{er}}_{\mathbf{s}}(h) - \text{er}_P(h)| \ge \varepsilon/2\right\} \\ &\le |H|\, 2\exp(-\varepsilon^2 m/2) \end{aligned}$$

as required. Now suppose $h^* \in H$ is such that $\text{er}_P(h^*) = \text{opt}_P(H)$. Then

$$P^m\left\{\max_{h\in H} |\hat{\text{er}}_{\mathbf{s}}(h) - \text{er}_P(h)| \ge \varepsilon/2\right\} \le 2|H|\exp\left(-\varepsilon^2 m/2\right),$$

and this is no more than δ if $m \ge (2/\varepsilon^2)(2|H|/\delta)$. In this case, with probability at least $1-\delta$, for *every* $h \in H$, $\text{er}_P(h) - \varepsilon/2 < \hat{\text{er}}_{\mathbf{s}}(h) < \text{er}_P(h) + \varepsilon/2$, and so,

$$\begin{aligned} \text{er}_P(L(\mathbf{s})) &\le \hat{\text{er}}_{\mathbf{s}}(L(\mathbf{s})) + \varepsilon/2 = \min_{h\in H} \hat{\text{er}}_{\mathbf{s}}(h) + \varepsilon/2 \le \hat{\text{er}}_{\mathbf{s}}(h^*) + \varepsilon/2 \\ &< (\text{er}_P(h^*) + \varepsilon/2) + \varepsilon/2 = \text{opt}_P(H) + \varepsilon. \end{aligned}$$

The result follows. ∎

We have stated the result for classes of Boolean functions, but it clearly applies also to *finite* classes of $\{0, 1\}$-valued functions defined on $\mathbb{R}^n$.

The proof of Theorem 6.1 shows that, for any $m > 0$, with probability at least $1-\delta$, L returns a function h with

$$\text{er}_P(h) < \text{opt}_P(H) + \sqrt{\frac{2}{m}\ln\left(\frac{2|H|}{\delta}\right)}.$$

Thus, $\varepsilon_0(\delta, m) = \sqrt{\frac{2}{m} \ln\left(\frac{2|H|}{\delta}\right)}$ may be thought of as a bound on the *estimation error* of the learning algorithm. The definitions and results can easily be stated in terms of estimation error rather than sample complexity, but here we mostly use sample complexity.

We state, without its proof (which is, in any case, simpler than the one just given and may be found in [5]), the following result for the realizable case. Note that, in the realizable case, the optimal error is zero, so a SEM algorithm is what is called a *consistent* algorithm. That is, the output hypothesis h is consistent with the sample, meaning that $h(x_i) = t(x_i)$ for each i, where t is the target concept.

Theorem 6.2. *Suppose that H is a set of Boolean functions. Then, for any m and δ, and any target concept $t \in H$, the following holds with probability at least $1 - \delta$: if $h \in H$ is any hypothesis consistent with a training sample $\mathbf{s}$ of length m, then with probability at least $1 - \delta$,*

$$\text{er}_P(h) < \frac{1}{m} \ln\left(\frac{|H|}{\delta}\right).$$

In particular, for realizable learning problems, any consistent learning algorithm L is PAC and has sample complexity bounded as follows: $m_L(\delta, \varepsilon) \leq (1/\varepsilon) \ln(|H|/\delta)$.

6.2.4 Learning Monomials

We give a simple example of a PAC algorithm in the realizable case. A *monomial* is a Boolean function that can be represented by a formula that is a simple conjunction of literals. There is a very simple learning algorithm for monomials, due to Valiant [31]. We begin with no information, so we assume that every one of the $2n$ literals $x_1, \bar{x}_1, \ldots, x_n, \bar{x}_n$ can occur in the target monomial. On presentation of a positive example $(x, 1)$, the algorithm deletes literals as necessary to ensure that the current hypothesis monomial is true on the example. The algorithm takes no action on negative examples: it will always be the case that the current hypothesis correctly classifies such examples as false points. The formal description is as follows. Suppose we are given a training sample $\mathbf{s}$ containing the labeled examples (x_i, b_i) $(1 \leq i \leq m)$, where each example x_i is an n–tuple of bits $(x_i)_j$. If we let h_U denote the monomial formula containing the literals in the set U, the algorithm can be expressed as follows:

```
  set U = {x_1, x̄_1, ..., x_n, x̄_n};
for i:= 1 to m do
    if b_i = 1 then
       for j:= 1 to n do
           if (x_i)_j = 1 then delete x̄_j if present in U
                              else delete x_j if present in U;
L(s):= h_U
```

It is easy to check that if **s** is a training sample corresponding to a monomial, then the algorithm outputs a monomial consistent with **s**. So the algorithm is a PAC algorithm for the realizable case. Furthermore, because the number of monomials is at most $3^n + 1$, the sample complexity of L is bounded above by

$$\frac{1}{\varepsilon} \ln\left(\frac{\ln(3^n + 1)}{\delta}\right),$$

which, ignoring constants, is of order $(n + \ln(1/\delta))/\varepsilon$. The algorithm is also computationally efficient, something we shall turn our attention to later.

6.2.5 Discussion

Theorem 6.1 and Theorem 6.2 show that the sample complexity of learning can be bounded above using the cardinality of H. But it is natural to ask if one can do better: that is, can we obtain tighter upper bounds? Furthermore, we have not yet seen any lower bounds on the sample complexity of learning. To deal with these concerns, we now look at the *VC-dimension*, which turns out to give (often better) upper bounds, and also lower bounds, on sample complexity.

6.3 The Growth Function and VC-Dimension

6.3.1 The Growth Function of a Function Class

Suppose that H is a set of Boolean functions defined on $X = \{0, 1\}^n$. Let $\mathbf{x} = (x_1, x_2, \ldots, x_m)$ be a sample (unlabeled) of length m of points of X. As in [34, 5], we define $\Pi_H(\mathbf{x})$, the *number of classifications of* $\mathbf{x}$ *by* H, to be the number of distinct vectors of the form

$$(f(x_1), f(x_2), \ldots, f(x_m)),$$

as f runs through all functions of H. (This definition works more generally if H is a set of $\{0, 1\}$-valued functions defined on some $\mathbb{R}^n$, for although in this case H may be infinite, $\Pi_H(\mathbf{x})$ will be finite.) Note that for any sample $\mathbf{x}$ of length m, $\Pi_H(\mathbf{x}) \leq 2^m$. An important quantity, and one that turns out to be crucial in PAC learning theory, is the maximum possible number of classifications by H of a sample of a given length. We define the *growth function* Π_H by

$$\Pi_H(m) = \max\left\{\Pi_H(\mathbf{x}) : \mathbf{x} \in X^m\right\}.$$

We have used the notation Π_H for both the number of classifications and the growth function, but this should cause no confusion.

6.3.2 VC-Dimension

We noted that the number of possible classifications by H of a sample of length m is at most 2^m, this being the number of binary vectors of length m. We say that a sample $\mathbf{x}$ of length m is *shattered* by H, or that H *shatters* $\mathbf{x}$, if this maximum possible value is attained: that is, if H gives all possible classifications of $\mathbf{x}$. We also find it useful to talk of a set of points, rather than a sample, being shattered. The notion is the same: the set is shattered if and only if a sample with those entries is shattered. To be shattered, $\mathbf{x}$ must clearly have m distinct examples. Then, $\mathbf{x}$ is shattered by H if and only if for each subset S of $\{x_1, x_2 \ldots, x_m\}$, there is some function f_S in H such that for $1 \leq i \leq m$, $f_S(x_i) = 1 \iff x_i \in S$.

Consistent with the intuitive notion that a set H of functions has high expressive power if it can achieve all possible classifications of a large set of examples, following [34, 5], we use as a measure of this power the *Vapnik-Chervonenkis dimension*, or *VC-dimension*, of H, which is defined to be the maximum length of a sample shattered by H. Using the notation introduced earlier, we can say that the VC-dimension of H, denoted VCdim(H), is given by

$$\text{VCdim}(H) = \max\left\{m : \Pi_H(m) = 2^m\right\}.$$

We may state this definition formally, and in a slightly different form, as follows.

Definition 6.2. (VC-dimension.) *Let H be a set of Boolean functions from a set X to $\{0, 1\}$. The* VC-dimension *of H is the maximal size of a subset E of X with the property that for each $S \subseteq E$, there is $f_S \in H$ with $f_S(x) = 1$ if $x \in S$ and $f_S(x) = 0$ if $x \in E \setminus S$.*

The VC-dimension of a set of Boolean functions can easily be bounded in terms of its cardinality.

Theorem 6.3. *For any set H of Boolean functions,* $\text{VCdim}(H) \leq \log_2 |H|$.

Proof. If d is the VC-dimension of H and $\mathbf{x} \in X^d$ is shattered by H, then $|H| \geq |H_{\mathbf{x}}| = 2^d$. (Here, $H_{\mathbf{x}}$ denotes the restriction of H to domain $E = \{x_1, x_2, \ldots, x_d\}$.) It follows that $d \leq \log_2 |H|$. ∎

It should be noted that Theorem 6.3 is sometimes loose, as we shall shortly see. However, it is reasonably tight: to see this, we need to explore further the relationship between growth function and VC-dimension.

Note: All of the definitions in this section can be made more generally for (possibly infinite) sets of functions mapping from $X = \mathbb{R}^n$ to $\{0, 1\}$. The VC-dimension can then be infinite. Theorem 6.3 applies to any finite such class.

6.4 Relating Growth Function and VC-Dimension

The growth function $\Pi_H(m)$ is a measure of how many different classifications of an m-sample into true and false points can be achieved by the functions of H, whereas the VC-dimension of H is the maximum value of m for which $\Pi_H(m) = 2^m$. Thus, the VC-dimension is defined in terms of the growth function. But there is a converse relationship: the growth function $\Pi_H(m)$ can be bounded by a polynomial function of m, and the degree of the polynomial is the VC-dimension d of H. Explicitly, we have the following theorem [24, 27], usually known as Sauer's lemma (or the Sauer-Shelah lemma).

Theorem 6.4. (Sauer's Lemma.) *Let $d \geq 0$ and $m \geq 1$ be given integers and let H be a set of $\{0, 1\}$-valued functions with* $\text{VCdim}(H) = d \geq 1$. *Then*

$$\Pi_H(m) \leq \sum_{i=0}^{d} \binom{m}{i} < \left(\frac{em}{d}\right)^d,$$

where the second inequality holds for $m \geq d$.

Proof. For $m \leq d$, the inequality is trivially true since in that case the sum is 2^m. Assume that $m > d$ and fix a set $S = \{x_1, \ldots, x_m\} \subseteq X$. We make use of the correspondence between $\{0, 1\}$-valued functions on a set and subsets of that set by defining the set system (or family of sets)

$$\mathcal{F} = \{\{x_i \in S : f(x_i) = 1\} : f \in H\}.$$

The proof proceeds, as in [29], by first creating a transformed version $\mathcal{F}^*$ of $\mathcal{F}$ that is a down-set with respect to the partial order induced by set inclusion, and which has the same cardinality as $\mathcal{F}$. (To say that $\mathcal{F}^*$ is a down-set means that if $A \in \mathcal{F}^*$ and $B \subseteq A$ then $B \in \mathcal{F}^*$.)

For an element x of S, let T_x denote the operator that, acting on a set system, removes the element x from all sets in the system, unless that would give a set that is already in the system:

$$T_x(\mathcal{F}) = \{A \setminus \{x\} : A \in \mathcal{F}\} \cup \{A \in \mathcal{F} : A \setminus \{x\} \in \mathcal{F}\}.$$

Note that $|T_x(\mathcal{F})| = |\mathcal{F}|$. Consider now $\mathcal{F}^* = T_{x_1}(T_{x_2}(\cdots T_{x_m}(\mathcal{F})\cdots))$. Clearly, $|\mathcal{F}^*| = |\mathcal{F}|$. Furthermore, for all x in S, $T_x(\mathcal{F}^*) = \mathcal{F}^*$. Clearly, $\mathcal{F}^*$ is a down-set. For, if it were not, there would be some $C \in \mathcal{F}^*$ and some $x \in C$ such that $C \setminus \{x\} \notin \mathcal{F}^*$. But we have applied the operator T_x to obtain $\mathcal{F}^*$; thus, if $C \in \mathcal{F}^*$, then this is only because $C \setminus \{x\}$ is also in $\mathcal{F}^*$.

We can define the notion of shattering for a family of subsets, in the same way as for a family of $\{0, 1\}$-valued functions. For $R \subseteq S$, we say that $\mathcal{F}$ shatters R if $\mathcal{F} \cap R = \{A \cap R : A \in \mathcal{F}\}$ is the set of all subsets of R. We next show that, whenever $\mathcal{F}^*$ shatters a set, so does $\mathcal{F}$. It suffices to show that, for any $x \in S$, if $T_x(\mathcal{F})$ shatters a set, so does $\mathcal{F}$. So suppose that x in S, $R \subseteq S$, and $T_x(\mathcal{F})$ shatters R. If x is not in R, then, trivially, $\mathcal{F}$ shatters R. If x is in R, then for all $A \subseteq R$ with $x \notin A$, because $T_x(\mathcal{F})$ shatters R we have $A \in T_x(\mathcal{F}) \cap R$ and $A \cup \{x\} \in T_x(\mathcal{F}) \cap R$. By the definition of T_x, this implies $A \in \mathcal{F} \cap R$ and $A \cup \{x\} \in \mathcal{F} \cap R$. This argument shows that $\mathcal{F}$ shatters R. It follows that $\mathcal{F}^*$ can only shatter sets of cardinality at most d. Because $\mathcal{F}^*$ is a down-set, this means that the largest set in $\mathcal{F}^*$ has cardinality no more than d. (For, if there were a set of cardinality $d + 1$ in $\mathcal{F}^*$, all its subsets would be in $\mathcal{F}^*$, too, because $\mathcal{F}^*$ is a down-set, and it would therefore be shattered.) We therefore have $|\mathcal{F}^*| \leq \sum_{i=0}^{d} \binom{m}{i}$, this expression being the number of subsets of S containing no more than d elements. The result follows, because $|\mathcal{F}| = |\mathcal{F}^*|$, and because S was chosen arbitrarily. For the second inequality, we have, as argued in [6],

$$\sum_{i=0}^{d} \binom{m}{i} \leq \left(\frac{m}{d}\right)^d \sum_{i=0}^{d} \binom{m}{i} \left(\frac{d}{m}\right)^i \leq \left(\frac{m}{d}\right)^d \sum_{i=0}^{m} \binom{m}{i} \left(\frac{d}{m}\right)^i$$
$$= \left(\frac{m}{d}\right)^d \left(1 + \frac{d}{m}\right)^m .$$

Now, for all $x > 0$, $(1 + (x/m))^m < e^x$, so this is bounded by $(m/d)^d e^d = (em/d)^d$, giving the bound. ■

The first inequality of this theorem is tight. If H corresponds to the set system $\mathcal{F}$ consisting of all subsets of $\{1, 2, \ldots, n\}$ of cardinality at most d, then $\text{VCdim}(H) = d$ and $|\mathcal{F}|$ meets the upper bound.

Now, Theorem 6.4 has the following consequence when we use the fact that $|H| = \Pi_H(2^n)$.

Theorem 6.5. *For any class H of Boolean functions defined on $\{0, 1\}^n$,*

$$\text{VCdim}(H) \geq \frac{\log_2 |H|}{n + \log_2 e},$$

and if $\text{VCdim}(H) \geq 3$, *then* $\text{VCdim}(H) \geq \log_2 |H|/n$.

Given also the earlier bound, Theorem 6.3, we see that, essentially, for a Boolean class on $\{0, 1\}^n$, $\text{VCdim}(H)$ and $\log_2 |H|$ are within a factor n of each other. This gap can be real. For example, when $H = T_n$ is the class of threshold functions, then $\text{VCdim}(T_n) = n + 1$, whereas $\log_2 |T_n| > n^2/2$. (In fact, as shown by Zuev [35], $\log_2 |T_n| \sim n^2$ as $n \to \infty$.)

6.5 VC-Dimension and PAC Learning

It turns out that the VC-dimension quantifies, in a more precise way than does the cardinality of the hypothesis space, the sample complexity of PAC learning.

6.5.1 Upper Bounds on Sample Complexity

The following results bound from above the sample complexity of PAC learning (in the general and realizable cases, respectively). It is obtained from a result of Vapnik and Chervonenkis [34]; see [2].

Theorem 6.6. *Suppose that H is a set of Boolean functions with VC-dimension $d \geq 1$, and let L be any SEM algorithm for H. Then L is a PAC learning algorithm* *for H with sample complexity bounded as follows:*

$$m_L(\delta, \varepsilon) \le m_0(\delta, \varepsilon) = \frac{64}{\varepsilon^2}\left(2d \ln\left(\frac{12}{\varepsilon}\right) + \ln\left(\frac{4}{\delta}\right)\right).$$

In fact, it is possible (using a result of Talagrand [30]; see [2]) to obtain an upper bound of order $(1/\varepsilon^2)(d + \ln(1/\delta))$. (However, the constants involved are quite large.) For the realizable case, from a result in [5], we have the following bound.

Theorem 6.7. *Suppose that H is a set of Boolean functions with VC-dimension $d \ge 1$, and let L be any consistent learning algorithm for H. Then L is a PAC learning algorithm for H in the realizable case, with sample complexity bounded as follows:*

$$m_L(\delta, \varepsilon) \le \frac{4}{\varepsilon}\left(d \ln\left(\frac{12}{\varepsilon}\right) + \ln\left(\frac{2}{\delta}\right)\right).$$

6.5.2 Lower Bounds on Sample Complexity

The following lower bounds on sample complexity are also obtainable. (These are from [2], and similar bounds can be found in [9, 28].)

Theorem 6.8. *Suppose that H is a class of $\{0, 1\}$-valued functions with VC-dimension d. For any PAC learning algorithm L for H, the sample complexity $m_L(\delta, \varepsilon)$ of L satisfies*

$$m_L(\delta, \varepsilon) \ge \frac{d}{320\varepsilon^2}$$

for all $0 < \varepsilon, \delta < 1/64$. Furthermore, if H contains at least two functions, we have

$$m_L(\delta, \varepsilon) \ge 2\left\lfloor \frac{1-\varepsilon^2}{2\varepsilon^2} \ln\left(\frac{1}{8\delta(1-2\delta)}\right)\right\rfloor$$

for all $0 < \varepsilon < 1$ and $0 < \delta < 1/4$.

The two bounds taken together imply a lower bound of order $(1/\varepsilon^2)(d + \ln(1/\delta))$.

For the realizable case, we have the following [10].

Theorem 6.9. *Suppose that H is a class of $\{0, 1\}$-valued functions of VC-dimension $d \ge 1$. For any PAC learning algorithm L for H in the realizable model, the sample complexity $m_L(\delta, \varepsilon)$ of L satisfies $m_L(\delta, \varepsilon) \ge (d-1)/(32\varepsilon)$ for all $0 < \varepsilon < 1/8$ and $0 < \delta < 1/100$. Furthermore, if H contains at least three functions, then $m_L(\delta, \varepsilon) > (1/2\varepsilon)\ln(1/\delta)$, for $0 < \varepsilon < 3/4$ and $0 < \delta < 1$.*

Thus, in the realizable case, the sample complexity of a PAC learning algorithm is at least of the order of

$$\frac{1}{\varepsilon}\left(d + \ln\left(\frac{1}{\delta}\right)\right).$$

Suppose H_n is a class of Boolean functions on $\{0, 1\}^n$. Given the connections between cardinality and VC-dimension for Boolean classes, we see that any SEM algorithm is PAC and (for fixed δ) has sample complexity at least of order $\frac{\log_2 |H_n|}{n\varepsilon^2}$ and at most of order $\frac{\log_2 |H_n|}{\varepsilon^2} \ln\left(\frac{1}{\varepsilon}\right)$. (In fact, as noted earlier, we can omit the logarithmic factor in the upper bound at the expense of worse constants.) In the realizable case, we can similarly see that any consistent algorithm is PAC and has sample complexity of order at least $\frac{\log_2 |H_n|}{n\varepsilon}$ and at most $\frac{\log_2 |H_n|}{\varepsilon} \ln\left(\frac{1}{\varepsilon}\right)$.

The cardinality therefore can be used to bound the sample complexity of learning, but the VC-dimension provides tighter bounds. (Moreoever, the bounds based on VC-dimension remain valid if we consider not Boolean classes but classes of functions mapping from $\mathbb{R}^n$ to $\{0, 1\}$: as long as such classes have finite VC-dimension – even if infinite cardinality – they are still learnable by SEM algorithms, or consistent algorithms in the realizable model.)

6.6 VC-Dimensions of Boolean Classes

6.6.1 Monomials

As an example of VC-dimension, we consider the set M_n^+ of *positive monomials*, consisting of the simple conjunctions on nonnegated literals.

Theorem 6.10. *The class M_n^+ of positive monomials on $\{0, 1\}^n$ has VC-dimension n.*

Proof. Because there are 2^n such functions, we have $\mathrm{VCdim}(M_n^+) \leq \log_2(2^n) = n$. To show that the VC-dimension is in fact exactly n, we show that there is some set $S \subseteq \{0, 1\}^n$ such that $|S| = n$ and S is shattered by M_n^+. Let S consist of all $\{0, 1\}$-vectors having exactly $n - 1$ entries equal to 1, and denote by u_i the element of s having a 0 in position i. Let R be any subset of S and let $h_R \in M_n^+$ be the conjunction of the literals x_j for all j such that $u_j \notin R$. Then $h_R(x) = 1$ for $x \in R$ and $h_R(x) = 0$ for $x \in S \setminus R$. This shows S is shattered. ■

6.6.2 Threshold Functions

It is known [7] that if $T = T_n$ is the set of threshold functions on $\{0, 1\}^n$, then

$$\Pi_T(m) \leq \psi(n, m) = 2 \sum_{i=0}^{n} \binom{m-1}{i}.$$

This result is proved by using the classical fact [7, 25] that N hyperplanes in $\mathbb{R}^n$, each passing through the origin, divide $\mathbb{R}^n$ into at most $C(N, n) = 2\sum_{i=0}^{n-1}\binom{N-1}{i}$ regions. It follows directly from this, since $\psi(n, n+1) = 2^{n+1}$ and $\psi(n, n+2) < 2^{n+2}$, that the VC-dimension of T_n is at most $n+1$. In fact, the VC-dimension is exactly $n+1$, as we now show. (In the proof, an alternative, more direct, way of seeing that the VC-dimension is at most $n+1$ is given.)

Theorem 6.11. *The class of threshold functions on* $\{0, 1\}^n$ *has VC-dimension* $n+1$.

Proof. Recall that any threshold function h is described by a weight vector $w = (w_1, w_2, \ldots, w_n)$ and a threshold θ, so that $h(x) = 1$ if and only if $\sum_{i=1}^n w_i x_i \geq \theta$. Let S be any subset of $\{0, 1\}^n$ with cardinality $n+2$. By Radon's theorem, there is a nonempty subset R of S such that $\text{conv}(R) \cap \text{conv}(S \setminus R) \neq \emptyset$, where $\text{conv}(X)$ denotes the convex hull of X. Suppose that there is a threshold function h in T_n such that R is the set of true points of h in S. We may assume that none of the points lies on the hyperplane defining h. Let H^+ be the open half-space on which h is true and H^- the open half-space on which it is false. Then $R \subseteq H^+$ and $S \setminus R \subseteq H^-$. But because half-spaces are convex subsets of $\mathbb{R}^n$, we then have

$$\text{conv}(R) \cap \text{conv}(S \setminus R) \subseteq H^+ \cap H^- = \emptyset,$$

which is a contradiction. It follows that no such t exists, and hence S is not shattered. But since S was an arbitrary subset of cardinality $n+2$, it follows that $\text{VCdim}(T_n) \leq n+1$. Now we show that $\text{VCdim}(T_n) \geq n+1$. Let $\mathbf{0}$ denote the all-0 vector and, for $1 \leq i \leq n$, let e_i be the point with a 1 in the ith coordinate and all other coordinates 0. We shall show that T_n shatters the set $S = \{\mathbf{0}, e_1, e_2, \ldots, e_n\}$. Suppose that R is any subset of S. For $i = 1, 2, \ldots, n$, let

$$w_i = \begin{cases} 1, & \text{if } e_i \in R; \\ -1, & \text{if } e_i \notin R; \end{cases}$$

and let

$$\theta = \begin{cases} -1/2, & \text{if } \mathbf{0} \in R; \\ 1/2, & \text{if } \mathbf{0} \notin R. \end{cases}$$

Then it is straightforward to verify that if h is the threshold function with weight vector w and threshold θ, then the set of true points of h in S is precisely R. Therefore S is shattered by T_n and, consequently, $\text{VCdim}(T_n) \geq n+1$. The result now follows. ■

6.6.3 k-DNF

The class of k-DNF functions on $\{0, 1\}^n$ consists of all those functions representable by a DNF in which the terms are of degree at most k. Let $D_{n,k}$ denote the set of k-DNF functions of n variables. Then, for fixed k, the VC-dimension of $D_{n,k}$ is $\Theta(n^k)$, as shown in [10].

Theorem 6.12. *Let $k \in \mathbb{N}$ be fixed and let $D_{n,k}$ be the set of k-DNF functions on $\{0,1\}^n$. Then* $\text{VCdim}(D_{n,k}) = \Theta(n^k)$.

Proof. The number of monomials or terms that are nonempty, not identically false, and of degree at most k is $\sum_{i=1}^{k} \binom{n}{i} 2^i$, which is, for fixed k, $O(n^k)$. Because any k-DNF formula is created by taking the disjunction of a set of such terms, the number of k-DNF formulas (and hence $|D_{n,k}|$) is $2^{O(n^k)}$. Therefore, $\text{VCdim}(D_{n,k}) \leq \log_2 |D_{n,k}| = O(n^k)$.

On the other hand, we can show that the VC-dimension is $\Omega(n^k)$ by proving that a sufficiently large subset is shattered. Consider the set S of examples in $\{0,1\}^n$ that have precisely k entries equal to 1. Then S can be shattered by $D_{n,k}$. Indeed, suppose R is any subset of S. For each $y = (y_1, y_2, \ldots, y_n) \in R$, form the term that is the conjunction of those literals x_i such that $y_i = 1$. Because $y \in S$, this term has k literals; further, y is the only true point in S of this term. The disjunction of these terms, one for each member of R, is therefore a function in $D_{n,k}$ whose true points in S are precisely the members of R. Hence S is shattered by $D_{n,k}$. Now, $|S| = \binom{n}{k}$, which, for a fixed k, is $\Omega(n^k)$. ∎

6.7 Efficient PAC Learning

6.7.1 Introduction

Up to now, a learning algorithm has been mainly described as a function that maps training samples into output functions (or hypotheses). We will now be more specific about the computational effectiveness of learning algorithms. If the process of PAC learning by an algorithm L is to be of practical value, it should be possible to implement the algorithm "quickly." We discuss what should be meant by an efficient PAC learning algorithm, and we highlight an important connection between the existence of efficient PAC learning algorithms and the existence of efficient procedures for producing hypotheses with small sample error. As mentioned earlier, computational efficiency was a key aspect of Valiant's learning model [31] and has been much further explored for the models of this chapter. The papers [5, 23] provided some of the important initial results, and these are further explored in the books [18, 20, 3]. The treatment here follows [2].

Consider the monomial learning algorithm (for the realizable case) described earlier. This is an efficient algorithm: its running time on a training sample of m data points in $\{0,1\}^n$ is $O(mn)$, which is linear in the size of the training sample. Furthermore, noting either that $\text{VCdim}(M_n) = O(n)$ or that $\log_2 |M_n| = O(n)$, we can see that, for given ε and δ, we can produce a hypothesis that, with probability at least $1 - \delta$, has accuracy ε, in time of order $n^2 p(1/\varepsilon, \ln(1/\delta)) = q(n, 1/\varepsilon, \ln(1/\delta))$, where p and q are (small degree) polynomials. This is an example of what we mean by an efficient learning algorithm: as n scales, the time taken to produce a PAC output hypothesis scales polynomially with n; additionally, the running time is polynomial in $1/\varepsilon$ and $\ln(1/\delta)$.

6.7.2 Graded Classes

In order to enable a more general discussion of efficient learning, we introduce the idea of a *graded* function class. Suppose H_n is a set of Boolean functions defined on $\{0, 1\}^n$. Then we say that $H = \bigcup_{n=1}^{\infty}$ is a *graded* hypothesis space. The reason for introducing this idea is that we want to analyze the running time (with respect to n) of what might be termed a "general" learning algorithm for a graded class of Boolean functions. This is an algorithm that works in essentially the same manner on each of the classes H_n. For example, the monomial learning algorithm works on M_n for any n in essentially the same way: there is no fundamental difference between its actions on, say, the monomials with 5 variables and those with 50 variables.

Denoting $\{0, 1\}^n \times \{0, 1\}$ by Z_n, a *learning algorithm* L for the graded space $H = \bigcup_{n=1}^{\infty} H_n$ is a mapping from $\bigcup_{n=1}^{\infty} Z_n^*$ to H with the property that if $\mathbf{s} \in Z_n^*$ then $L(\mathbf{s}) \in H_n$. The only difference between this definition and the basic notion of a learning algorithm for an ungraded class is that we have now encapsulated some sense of the "generality" of the algorithm in its action over all the H_n. With this, we now state formally what is meant by a PAC learning algorithm for a graded class.

Definition 6.3. *If L is a learning algorithm for $H = \bigcup H_n$, then we say that L is PAC if for all $n \in \mathbb{N}$ and $\delta, \varepsilon \in (0, 1)$, there is $m_0(n, \delta, \varepsilon)$ such that if $m \geq m_0(n, \delta, \varepsilon)$ then, for* any *probability distribution P on Z_n, if $\mathbf{s} \in Z_n^m$ is drawn randomly according to the product probability distribution P^m on Z_n^m, then with probability at least $1 - \delta$, the hypothesis $L(\mathbf{s})$ output by L satisfies* $\text{er}_P(L(\mathbf{s})) < \text{opt}_P(H_n) + \varepsilon$.

6.7.3 Definition of Efficient Learning

We now assume that learning algorithms are algorithms in the proper sense (that is, that they are computable functions). Suppose that L is a learning algorithm for a graded function class $H = \bigcup H_n$. An input to L is a training sample, which consists of m labeled binary vectors of length n. It would be possible to use $m(n + 1)$ as the measure of input size, but we will find it useful to consider dependence on m and n separately. We use the notation $R_L(m, n)$ to denote the worst-case running time of L on a training sample of m points of Z_n. Clearly, n is not the only parameter with which the running time of the learning procedure as a whole should be allowed to vary, because decreasing either the confidence parameter δ or the accuracy parameter ε makes the learning task more difficult, requiring a larger size of sample. We shall ask that the running time of a learning algorithm L be polynomial in m, and that the sample complexity $m_L(n, \delta, \varepsilon)$ depend polynomially on $1/\varepsilon$ and $\ln(1/\delta)$. If these conditions hold, then the running time required to produce a "good" output hypothesis will be polynomial in n, $\ln(1/\delta)$, and $1/\varepsilon$.

We now formally define what we mean by an *efficient learning algorithm* for a graded function class.

Definition 6.4. *Let* $H = \bigcup H_n$ *be a graded class of Boolean functions and suppose that* L *is a learning algorithm for* H*. We say that* L *is* efficient *if:*

- *The worst-case running time* $R_L(m, n)$ *of* L *on samples* $\mathbf{s} \in Z_n^m$ *is polynomial in* m *and* n*, and*
- *The sample complexity* $m_L(n, \delta, \varepsilon)$ *of* L *on* H_n *is polynomial in* n*,* $1/\varepsilon$ *and* $\ln(1/\delta)$.

We have described the outputs of learning algorithms as hypotheses. But, more precisely, they are representations of hypotheses. When discussing the complexity of learning, it is always assumed that the output lies in a representation class for the hypothesis class. All this often amounts to is that the output must be a formula of a particular type. For example, the monomial learning algorithm outputs a monomial formula (and not some other representation). This is not something we shall explore much further, but it is sometimes important.

6.7.4 Sufficient Conditions for Efficient Learning

We define a SEM algorithm for a graded Boolean class H to be an algorithm that given any sample $\mathbf{s} \in Z_n^m$, returns a function $h \in H_n$ with minimal sample error $\hat{\text{er}}_{\mathbf{s}}(h)$ on $\mathbf{s}$. The following result, which may be found in [5], follows directly from earlier results and shows that the rate of growth with n of the VC-dimension determines the sample complexity of learning algorithms.

Theorem 6.13. *Let* $H = \bigcup H_n$ *be a graded Boolean function class.*

- *If* $\text{VCdim}(H_n)$ *is polynomial in* n*, then any SEM algorithm for* H *is a PAC learning algorithm with sample complexity* $m_L(n, \delta, \varepsilon)$ *polynomial in* n*,* $1/\varepsilon$ *and* $\ln(1/\delta)$.
- *If there is an efficient PAC learning algorithm for* H*, then* $\text{VCdim}(H_n)$ *is polynomial in* n.

Note that, by Theorem 6.5, the same statement is true with $\text{VCdim}(H_n)$ replaced by $\ln |H_n|$.

We now turn our attention to the running time of SEM algorithms. Having seen that, in many circumstances, such algorithms yield PAC learning algorithms, we now investigate the *efficiency* of these derived learning algorithms. We say that a SEM algorithm for the graded Boolean function class $H = \bigcup H_n$ is efficient if, given as input $\mathbf{s} \in Z_n^m$, it returns its output in time polynomial in m and n. The following result is immediate.

Theorem 6.14. *Suppose that* $H = \bigcup H_n$ *is a graded Boolean function class and that* $\text{VCdim}(H_n)$ *is polynomial in* n*. Then, any efficient SEM algorithm for* H *is an efficient PAC learning algorithm for* H.

6.8 Randomized PAC and SEM Algorithms

There may be some advantage in allowing learning algorithms to be randomized. Furthermore, as we shall see, there are some fairly succinct characterizations of learnability provided we permit algorithms to be randomized.

For our purposes, a randomized algorithm $\mathcal{A}$ has available to it a random number generator that produces a sequence of independent, uniformly distributed bits. The randomized algorithm $\mathcal{A}$ uses these random bits as part of its input, but it is useful to think of this input as somehow "internal" to the algorithm and to think of the algorithm as defining a mapping from an "external" input to a probability distribution over outputs. The computation carried out by the algorithm is, of course, determined by its input, so that, in particular, it depends on the particular sequence produced by the random number generator, as well as on the "external" input. We may speak of the "probability" that $\mathcal{A}$ has a given outcome on an (external) input x, by which we mean the probability that the stream of random numbers gives rise to that outcome when the external input to the algorithm is x. It is useful to extend our concept of a PAC learning algorithm to allow randomization. The definition of a randomized PAC learning for a graded class is as in Definition 6.3, with the additional feature that the algorithm is randomized. (So, L can no longer be regarded as a deterministic function.)

We shall also be interested in randomized SEM algorithms.

Definition 6.5. *A randomized algorithm $\mathcal{A}$ is an* efficient randomized SEM algorithm *for the graded Boolean function class $H = \bigcup H_n$ if given any $\mathbf{s} \in Z_n^m$, $\mathcal{A}$ halts in time polynomial in n and m and outputs $h \in H_n$ that, with probability at least* $1/2$, *satisfies* $\hat{\text{er}}_{\mathbf{s}}(h) = \min_{g \in H_n} \hat{\text{er}}_{\mathbf{s}}(g)$.

Suppose we run a randomized SEM algorithm k times on a fixed sample $(\mathbf{s})$, keeping the output hypothesis $f^{(k)}$ with minimal sample error among all the k hypotheses returned. In other words, we take the *best of k iterations* of the algorithm. Then the probability that $f^{(k)}$ has sample error that is *not* minimal is at most $(1/2)^k$. This is the basis of the following result, which shows that, as far as its applications to learning are concerned, an efficient randomized SEM algorithm is as useful as its deterministic counterpart. (The key idea in establishing this result is to take the best of k iterations of $\mathcal{A}$ for a suitable k, absorbing the randomness in the action of $\mathcal{A}$ into the "δ" of learning.)

Theorem 6.15. *Suppose that $H = \bigcup H_n$ is a graded Boolean function class and that* $\text{VCdim}(H_n)$ *is polynomial in n. If there is an efficient randomized SEM algorithm $\mathcal{A}$ for H, then there is an efficient randomized PAC learning algorithm for H that uses $\mathcal{A}$ as a subroutine.*

6.9 Learning and Existence of SEM Algorithms

We have seen that efficient SEM algorithms (both deterministic and randomized) can in many cases be used to construct efficient PAC learning algorithms. The next result proves, as a converse, that if there is an efficient PAC learning algorithm for a graded class then *necessarily* there is an efficient randomized SEM algorithm. (For the realizable case, this may be found in [23, 5, 22].)

Theorem 6.16. *If there is an efficient PAC learning algorithm for the graded binary class* $H = \bigcup H_n$, *then there is an efficient randomized SEM algorithm.*

Proof. Suppose L is an efficient PAC learning algorithm for the graded class $H = \bigcup H_n$. We construct a randomized algorithm $\mathcal{A}$, which will turn out to be an efficient randomized SEM algorithm. Suppose the sample $\mathbf{s} \in Z_n^m$ is given as input to $\mathcal{A}$. Let P be the probability distribution that is uniform on the labeled examples in $\mathbf{s}$ and zero elsewhere on Z_n. (This probability is defined with multiplicity; that is, for instance, if there are two labeled examples in $\mathbf{s}$ each equal to z, we assign the labeled example z probability $2/m$ rather than $1/m$.) We use the randomization allowed in $\mathcal{A}$ to form a sample of length $m^* = m_L(n, 1/2, 1/m)$, in which each labeled example is drawn according to P. Let $\mathbf{s}^*$ denote the resulting sample. Feeding $\mathbf{s}^*$ into the learning algorithm, we receive as output $h^* = L(\mathbf{s}^*)$, and we take this to be the output of the algorithm $\mathcal{A}$; that is, $\mathcal{A}(\mathbf{s}) = h^* = L(\mathbf{s}^*)$. By the fact that L is a PAC learning algorithm, and given that $m^* = m_L(n, 1/2, 1/m)$, with probability at least $1/2$, we have $\text{er}_P(h^*) < \text{opt}_P(H) + 1/m$. But because P is discrete, with no probability mass less than $1/m$, this means $\text{er}_P(h^*) = \text{opt}_P(H)$. For any h, by the definition of P, $\text{er}_P(h) = \hat{\text{er}}_{\mathbf{s}}(h)$. So with probability at least $1/2$,

$$\hat{\text{er}}_{\mathbf{s}}(h^*) = \text{er}_P(h^*) = \text{opt}_P(H) = \min_{g \in H_n} \text{er}_P(g) = \min_{g \in H_n} \hat{\text{er}}_{\mathbf{s}}(g).$$

This means that $\mathcal{A}$ is a randomized SEM algorithm. Because L is efficient, $m^* = m_L(n, 1/2, 1/m)$ is polynomial in n and m. Since the sample $\mathbf{s}^*$ has length m^*, and since L is efficient, the time taken by L to produce h^* is polynomial in m^* and n. Hence, $\mathcal{A}$ has running time polynomial in n and m, as required. ■

We arrive at the following succinct characterization of PAC learnability (allowing randomized algorithms).

Theorem 6.17. *Suppose that* $H = \bigcup H_n$ *is a graded Boolean function class. Then there is an efficient randomized PAC learning algorithm for* H *if and only if* $\text{VCdim}(H_n)$ *is polynomial in* n and *there is an efficient randomized SEM algorithm for* H.

Given the connection of Theorem 6.5 between cardinality and VC-dimension, the same statement with $\ln |H_n|$ replacing $\text{VCdim}(H_n)$ holds. (It should be noted, however, that Theorem 6.17 holds, more generally, in the case where H_n maps from $\mathbb{R}^n$ to $\{0, 1\}$.)

6.10 Establishing Hardness of Learning

There are two quite natural decision problems associated with a graded Boolean function class $H = \bigcup H_n$:

H-FIT
Instance: $\mathbf{s} \in Z_n^m = (\{0, 1\}^n \times \{0, 1\})^m$ and an integer k between 1 and m.
Question: Is there $h \in H_n$ such that $\hat{\text{er}}_{\mathbf{s}}(h) \le k/m$?

H-CONSISTENCY
Instance: $\mathbf{s} \in Z_n^m = (\{0, 1\}^n \times \{0, 1\})^m$.
Question: Is there $h \in H_n$ such that $\hat{\text{er}}_{\mathbf{s}}(h) = 0$?

Clearly H-CONSISTENCY is a subproblem of H-FIT, obtained by setting $k = 0$. Thus, any algorithm for H-FIT can be used also to solve H-CONSISTENCY. Note that H-consistency is the decision problem associated with finding an extension in H of the partially defined Boolean function described by the sample $\mathbf{s}$ (see also Chapter 14 in Crama and Hammer [8]).

We say that a randomized algorithm $\mathcal{A}$ solves a decision problem Π if the algorithm always halts and produces an output – either "yes" or "no" – such that if the answer to Π on the given instance is "no," the output of $\mathcal{A}$ is "no," and if the answer to Π on the given instance is "yes," then, with probability at least $1/2$, the output of $\mathcal{A}$ is "yes." A randomized algorithm is *polynomial time* if its worst-case running time (over all instances) is polynomial in the size of its input. The class of decision problems Π that can be solved by a polynomial-time randomized algorithm is denoted by RP. One approach to proving that PAC learning is computationally intractable for particular classes (in the general or realizable cases) is through showing that these decision problems are hard. The reason is given in the following results. First, we have the following [19, 15].

Theorem 6.18. *Let $H = \bigcup H_n$ be a graded Boolean function class. If there is an efficient learning algorithm for H, then there is a polynomial-time randomized algorithm for H-FIT; in other words, H-FIT is in RP.*

Proof. If H is efficiently learnable, then, by Theorem 6.16, there exists an efficient randomized SEM algorithm $\mathcal{A}$ for H. Using $\mathcal{A}$, we construct a polynomial-time randomized algorithm $\mathcal{B}$ for H-FIT as follows. Suppose that $\mathbf{s} \in Z_n^m$ and k together constitute an instance of H-FIT, and hence an input to $\mathcal{B}$. The first step of the algorithm $\mathcal{B}$ is to compute $h = \mathcal{A}(\mathbf{s})$, the output of $\mathcal{A}$ on $\mathbf{s}$. This function belongs to H_n and, with probability at least $1/2$, $\hat{\text{er}}_{\mathbf{s}}(h)$ is minimal among all functions in H_n. The next step in $\mathcal{B}$ is to check whether $\hat{\text{er}}_{\mathbf{s}}(h) \le k/m$. If so, then the output of $\mathcal{B}$ is "yes" and, if not, the output is "no." It is clear that $\mathcal{B}$ is a randomized algorithm for H-FIT. Furthermore, because $\mathcal{A}$ runs in time polynomial in m and n, and because the time taken for $\mathcal{B}$ to calculate $\hat{\text{er}}_{\mathbf{s}}(h)$ is linear in the size of $\mathbf{s}$, $\mathcal{B}$ is a polynomial-time algorithm. ∎

The following result [23] applies to the realizable case.

Theorem 6.19. *Suppose that $H = \bigcup H_n$ is a graded Boolean function class. If H is efficiently learnable in the realizable model, then there is a polynomial-time randomized algorithm for H-CONSISTENCY; that is, H-CONSISTENCY is in RP.*

In particular, therefore, we have the following.

Theorem 6.20. *Suppose RP $\neq$ NP. If H-FIT is NP-hard, then there is no efficient PAC learning algorithm for H. Furthermore, if H-CONSISTENCY is NP-hard, then there is no efficient PAC learning algorithm for H in the realizable case.*

6.11 Hardness Results

We now use the theory just developed to show that PAC learnability of threshold functions is computationally intractable (although it is tractable in the realizable case). We also show the intractability of PAC learning a particular class of Boolean functions in the realizable case.

6.11.1 Threshold Functions

First, we note that it is well known that if T_n is the set of threshold Boolean functions on $\{0, 1\}^n$, then the graded class $T = \bigcup T_n$ is efficiently PAC learnable in the realizable case. Indeed, the VC-dimension of T_n is $n + 1$, which is linear, and there exist SEM algorithms based on linear programming. (See [5, 2], for instance.) However, T is *not* efficiently PAC learnable in the general case, if RP $\neq$ NP. This arises from the following result [11, 17, 15].

Theorem 6.21. *Let $T = \bigcup T_n$ be the graded class of threshold functions. Then T-FIT is NP-hard.*

We prove this by establishing that the problem it is at least as hard as the well-known NP-hard VERTEX COVER problem in graph theory.

We denote a typical graph by $G = (V, E)$, where V is the set of vertices and E the edges. We assume that the vertices are labeled with the numbers $1, 2, \ldots, n$. Then, a typical edge $\{i, j\}$ will, for convenience, be denoted by ij. A *vertex cover* of the graph is a set U of vertices such that for each edge ij of the graph, at least one of the vertices i, j belongs to U. The following decision problem is known to be NP-hard [11].

VERTEX COVER
Instance: A graph $G = (V, E)$ and an integer $k \leq |V|$.
Question: Is there a vertex cover $U \subseteq V$ such that $|U| \leq k$?

A typical instance of VERTEX COVER is a graph $G = (V, E)$ together with an integer $k \leq |V|$. We assume, for simplicity, that $V = \{1, 2, \ldots, n\}$ and we denote the number of edges, $|E|$, by r. Notice that the size of an instance of VERTEX COVER is $\Omega(r+n)$. We construct $\mathbf{s} = \mathbf{s}(G) \in (\{0,1\}^{2n} \times \{0,1\})^{2r+n}$ as follows. For any two distinct integers i, j between 1 and $2n$, let $e_{i,j}$ denote the binary vector of length $2n$ with 1s in positions i and j and 0s elsewhere. The sample $\mathbf{s}(G)$ consists of the labeled examples $(e_{i,n+i}, 1)$ for $i = 1, 2, \ldots, n$ and, for each edge $ij \in E$, the labeled examples $(e_{i,j}, 0)$ and $(e_{n+i,n+j}, 0)$. Note that the "size" of $\mathbf{s}$ is $(2r+n)(2n+1)$, which is polynomial in the size of the original instance of VERTEX COVER, and that $z(G)$ can be computed in polynomial time.

For example, if a graph G has vertex set $V = \{1, 2, 3, 4\}$ and edge set $E = \{12, 23, 14, 13\}$, then the sample $\mathbf{s}(G)$ consists of the following 12 labeled examples:

$$(10001000, 1), (01000100, 1), (00100010, 1), (00010001, 1),$$
$$(11000000, 0), (00001100, 0), (01100000, 0), (00000110, 0),$$
$$(10100000, 0), (00001010, 0), (10010000, 0), (00001001, 0).$$

Lemma 6.22. *Given any graph $G = (V, E)$ with n vertices, and any integer $k \le n$, let $\mathbf{s} = \mathbf{s}(G)$ be as defined earlier. Then, there is $h \in T_{2n}$ such that $\hat{\text{er}}_{\mathbf{s}}(h) \le k/(2n)$ if and only if there is a vertex cover of G of cardinality at most k.*

Proof. Recall that any threshold function is represented by some weight vector w and threshold θ. Suppose first that there is such an h and that this is represented by the weight vector $w = (w_1, w_2, \ldots, w_{2n})$ and threshold θ. We construct a subset U of V as follows. If $h(e_{i,n+i}) = 0$, then we include i in U; if, for $i \ne j$, $h(e_{i,j}) = 1$ or $h(e_{n+i,n+j}) = 1$, then we include *either one* of i, j in U. Because h is "wrong" on at most k of the examples in $\mathbf{s}$, the set U consists of at most k vertices. We claim that U is a vertex cover. To show this, we need to verify that given any edge $ij \in E$, at least one of i, j belongs to U. It is clear from the manner in which U is constructed that this is true if either $h(e_{i,n+i}) = 0$ or $h(e_{j,n+j}) = 0$, so suppose that neither of these holds; in other words, suppose that $h(e_{i,n+i}) = 1 = h(e_{j,n+j})$. Then we may deduce that

$$w_i + w_{n+i} \ge \theta, \quad w_j + w_{n+j} \ge \theta,$$

and so

$$w_i + w_j + w_{n+i} + w_{n+j} \ge 2\theta;$$

that is,

$$(w_i + w_j) + (w_{n+i} + w_{n+j}) \ge 2\theta.$$

From this, we see that either $w_i + w_j \ge \theta$ or $w_{n+i} + w_{n+j} \ge \theta$ (or both); thus, $h(e_{i,j}) = 1$ or $h(e_{n+i,n+j}) = 1$, or both. Because of the way in which U is constructed, it follows that at least one of the vertices i, j belongs to U. Because ij was an arbitrary edge of the graph, this shows that U is indeed a vertex cover.

We now show, conversely, that if there is a vertex cover of G consisting of at most k vertices, then there is a function in T_{2n} with sample error at most $k/(2n)$

on $\mathbf{s}(G)$. Suppose U is a vertex cover and $|U| \leq k$. Define a weight vector $w = (w_1, w_2, \ldots, w_{2n})$ and threshold θ as follows: let $\theta = 1$ and, for $i = 1, 2, \ldots, n$,

$$w_i = w_{n+i} = \begin{cases} -1 & \text{if } i \in U \\ 1 & \text{if } i \notin U. \end{cases}$$

We claim that if h is the threshold function represented by w and θ, then $\hat{\text{er}}_{\mathbf{s}}(h) \leq k/(2n)$. Observe that if $ij \in E$, then, because U is a vertex cover, at least one of i, j belongs to U, and hence the inner products $w^T e_{i,j}$ and $w^T e_{n+i,n+j}$ are both either 0 or -2, less than θ, so $h(e_{i,j}) = h(e_{n+i,n+j}) = 0$. The function h is therefore correct on all the examples in $\mathbf{s}(G)$ arising from the edges of G. We now consider the other types of labeled example in $\mathbf{s}(G)$: those of the form $(e_{i,n+i}, 1)$. Now, $w^T e_{i,n+i}$ is -2 if $i \in U$ and is 2 otherwise, so $h(e_{i,n+i}) = 0$ if $i \in U$ and $h(e_{i,n+i}) = 1$ otherwise. It follows that h is "wrong" only on the examples $e_{i,n+i}$ for $i \in U$, and hence

$$\hat{\text{er}}_{\mathbf{s}}(h) = \frac{|U|}{2n} \leq \frac{k}{2n},$$

as claimed. ∎

This result shows that the answer to T-FIT on the instance $(\mathbf{s}(G), k)$ is the same as the answer to VERTEX COVER on instance (G, k). Given that $\mathbf{s}(G)$ can be computed from G in time polynomial in the size of G, we have therefore established that T-FIT is NP-hard.

6.11.2 k-Clause CNF

Pitt and Valiant [23] were the first to give an example of a Boolean class H for which the consistency problem H-CONSISTENCY is NP-hard. Let C_n^k, the set of k-clause CNF functions, be the set of Boolean functions on $\{0, 1\}^n$ that can be represented as the conjunction of at most k clauses.

We show that, for fixed $k \geq 3$, the consistency problem for $C^k = \bigcup C_n^k$ is NP-hard. Thus, if NP$\neq$ RP, then can be no efficient PAC learning algorithm for C^k in the realizable case.

The reduction in this case is to GRAPH K-COLORABILITY. Suppose we are given a graph $G = (V, E)$, with $V = \{1, 2, \ldots, n\}$. We construct a training sample $\mathbf{s}(G)$, as follows. For each vertex $i \in V$ we take as a negative example the vector v_i that has 1 in the ith coordinate position and 0s elsewhere. For each edge $ij \in E$ we take as a positive example the vector $v_i + v_j$.

Lemma 6.23. *There is a function in C_n^k that is consistent with the training sample $\mathbf{s}(G)$ if and only if the graph G is k-colorable.*

Proof. Suppose that $h \in C_n^k$ is consistent with the training sample. By definition, h is a conjunction

$$h = h_1 \wedge h_2 \wedge \ldots \wedge h_k$$

of clauses. For each vertex i of G, $h(v_i) = 0$, and so there must be at least one clause h_f $(1 \le f \le k)$ for which $h_f(v_i) = 0$. Thus we may define a function χ from V to $\{1, 2, \ldots, k\}$ as follows:

$$\chi(i) = \min\{f : h_f(v_i) = 0\}.$$

We claim that χ is a coloring of G. Suppose that $\chi(i) = \chi(j) = f$, so that $h_f(v_i) = h_f(v_j) = 0$. Because h_f is a clause, every literal occurring in it must be 0 on v_i and on v_j. Now v_i has a 1 only in the ith position, and so $h_f(v_i) = 0$ implies that the only negated literal that can occur in h_f is $\bar{x}_i$. Because the same is true for $\bar{x}_j$, we conclude that h_f contains only some literals x_l, with $l \ne i, j$. Thus $h_f(v_i + v_j) = 0$ and $h(v_i + v_j) = 0$. Now if ij were an edge of G, then we should have $h(v_i + v_j) = 1$, because we assumed that h is consistent with $\mathbf{s}(G)$. Thus ij is not an edge of G, and χ is a coloring, as claimed.

Conversely, suppose we are given a coloring $\chi : V \to \{1, 2, \ldots, k\}$. For $1 \le f \le k$, define h_f to be the clause $\bigvee_{\chi(i) \ne f} x_i$, and define $h = h_1 \wedge h_2 \wedge \ldots \wedge h_k$. We claim that h is consistent with $\mathbf{s}(G)$.

First, given a vertex i, suppose that $\chi(i) = g$. The clause h_g is defined to contain only those (nonnegated) literals corresponding to vertices *not* colored g, and so x_i does not occur in h_g. It follows that $h_g(v_i) = 0$ and $h(v_i) = 0$. Second, let ij be any edge of G. For each color f, there is at least one of i, j that is not colored f; denote an appropriate choice by $i(f)$. Then h_f contains the literal $x_{i(f)}$, which is 1 on $v_i + v_j$. Thus every clause h_f is 1 on $v_i + v_j$, and $h(v_i + v_j) = 1$, as required. ∎

Note that when $k = 1$, we have $C_n^1 = C_n$, and there is a polynomial time learning algorithm for C_n dual to the monomial learning algorithm. The consistency problem (and hence intractability of learning) remains, however, when $k = 2$: to show this, the consistency problem can be related to the NP-complete SET-SPLITTING problem; see [23].

This hardness result is "representation dependent": part of the difficulty arises from the need to output a formula in k-clause-CNF form. Now, any k-clause-CNF formula can simply be rewritten as an equivalent k-DNF formula. So any function in C_n^k is also a k-DNF function; that is, using the notation from earler, it belongs to $D_{n,k}$. But there is a simple efficient PAC learning algorithm for $D_{n,k}$; see [31, 3]. So C_n^k is learnable if the output hypotheses are permitted to be drawn from the larger class $D_{n,k}$ (or, more precisely, if the output formula is a k-DNF).

References

[1] D. Angluin. Queries and concept learning. *Machine Learning* 2(4), pp. 319–42, 1988.

[2] M. Anthony and P. L. Bartlett. *Neural Network Learning: Theoretical Foundations*. Cambridge University Press, Cambridge, UK, 1999.

[3] Martin Anthony and Norman L. Biggs. *Computational Learning Theory: An Introduction*. Cambridge Tracts in Theoretical Computer Science, Vol. 30. Cambridge University Press, Cambridge, UK, 1992.

[4] M. Anthony, G. Brightwell, and J. Shawe-Taylor. On specifying Boolean functions by labelled examples. *Discrete Applied Mathematics* 61, pp. 1–25, 1995.

[5] A. Blumer, A. Ehrenfeucht, D. Haussler, and M. K. Warmuth. Learnability and the Vapnik-Chervonenkis dimension. *Journal of the ACM* 36(4), pp. 929–65, 1989.

[6] S. Chari, P. Rohatgi, and A. Srinivasan. Improved algorithms via approximations of probability distributions (Extended Abstract). In *Proceedings of the Twenty-Sixth Annual ACM Symposium on the Theory of Computing*, pp. 584–92. ACM Press, New York, 1994.

[7] T. M. Cover. Geometrical and Statistical properties of systems of linear inequalities with applications in pattern recognition. *IEEE Transactions on Electronic Computers*, EC-14, pp. 326–34, 1965.

[8] Y. Crama and P. L. Hammer. *Boolean Functions: Theory, Algorithms, and Applications*. Cambridge University Press, Cambridge, UK, 2010.

[9] L. Devroye and G. Lugosi. Lower bounds in pattern recognition and learning. *Pattern Recognition* 28(7), pp. 1011–18, 1995.

[10] A. Ehrenfeucht, D. Haussler, M. Kearns, and L. Valiant. A general lower bound on the number of examples needed for learning. *Information and Computation* 82, pp. 247–61, 1989.

[11] M. Garey and D. Johnson. *Computers and Intractibility: A Guide to the Theory of NP-Completeness*. Freemans, San Francisco, 1979.

[12] S. A. Goldman and M. J. Kearns. On the complexity of teaching. *Journal of Computer and System Sciences* 50(1), pp. 20–31, 1995.

[13] D. Haussler. Decision theoretic generalizations of the PAC model for neural net and other learning applications. *Information and Computation* 100(1), pp. 78–150, 1992.

[14] W. Hoeffding. Probability inequalities for sums of bounded random variables. *Journal of the American Statistical Association* 58(301), pp. 13–30, 1963.

[15] K.-U. Höffgen, H. U. Simon, and K. S. Van Horn. Robust trainability of single neurons. *Journal of Computer and System Sciences* 50(1), pp. 114–25, 1995.

[16] J. Jackson and A. Tomkins. A computational model of teaching. Proceedings of 5th Annual Workshop on Comput. Learning Theory, pp. 319–26. ACM Press, New York, 1992.

[17] D. S. Johnson and F. P. Preparata. The densest hemisphere problem. *Theoretical Computer Science* 6, pp. 93–107, 1978.

[18] M. J. Kearns. *The Computational Complexity of Machine Learning*. ACM Distinguished Dissertation Series. MIT Press, Cambridge, MA, 1989.

[19] M. J. Kearns, R. E. Schapire, and L. M. Sellie. Toward efficient agnostic learning. *Machine Learning* 17(2–3), pp. 115–42, 1994.

[20] M. J. Kearns and U. Vazirani. *Introduction to Computational Learning Theory*. MIT Press, Cambridge, MA, 1995.

[21] S. A. Goldman and H. D. Mathias. Teaching a smarter learner. *Journal of Computer and System Sciences* 52(2), pp. 255–67, 1996.

[22] B. K. Natarajan. On learning sets and functions. *Machine Learning* 4(1), pp. 67–97, 1989.

[23] L. Pitt and L. Valiant. Computational limitations on learning from examples. *Journal of the ACM* 35, pp. 965–84, 1988.

[24] N. Sauer. On the density of families of sets. *Journal of Combinatorial Theory (A)* 13, pp. 145–7, 1972.

[25] L. Schläfli. *Gesammelte Mathematische Abhandlungen I*. Birkhäuser, Basel, 1950.

[26] R. Servedio. On the limits of efficient teachability. *Information Processing Letters* 79(6), pp. 267–72, 2001.

[27] S. Shelah. A combinatorial problem: Stability and order for models and theories in infinitary languages. *Pacific Journal of Mathematics* 41, pp. 247–61, 1972.

[28] H. U. Simon. General bounds on the number of examples needed for learning probabilistic concepts. *Journal of Computer and System Sciences* 52(2), pp. 239–54, 1996.

[29] J. M. Steele. Existence of submatrices with all possible columns. *Journal of Combinatorial Theory, Series A* 24, pp. 84–8, 1978.

[30] M. Talagrand. Sharper bounds for Gaussian and empirical processes. *Annals of Probability* 22, pp. 28–76, 1994.
[31] L. G. Valiant. A theory of the learnable. *Communications of the ACM* 27(11), 1134–42, 1984.
[32] V. N. Vapnik. *Statistical Learning Theory*. Wiley, New York, 1998.
[33] V. N. Vapnik. *Estimation of Dependences Based on Empirical Data*. Springer-Verlag, New York, 1982.
[34] V. N. Vapnik and A. Y. Chervonenkis. On the uniform convergence of relative frequencies of events to their probabilities. *Theory of Probability and Its Applications* 16(2), pp. 264–80, 1971.
[35] Y. A. Zuev. Asymptotics of the logarithm of the number of threshold functions of the algebra of logic. *Soviet Mathematics Doklady* 39, pp. 512–13, 1989.

7

Learning Boolean Functions with Queries

Robert H. Sloan, Balázs Szörényi, and György Turán

7.1 Introduction

The objective of machine learning is to acquire knowledge from available information in an automated manner. As an example, one can think of obtaining rules for medical diagnosis, based on a database of patients with their test results and diagnoses. If we make the simplifying assumption that data are represented as binary vectors of a fixed length and the rule to be learned classifies these vectors into two classes, then the task is that of *learning*, or *identifying a Boolean function*. This simplifying assumption is realistic in some cases, and it provides a good intuition for more general learning problems in others.

The notion of learning is somewhat elusive, and there are a large number of approaches to defining precisely what is meant by learning. A probabilistic notion, PAC (probably approximately correct) learning, based on random sampling of examples of the function to be learned, is discussed by Anthony in Chapter 6 of this volume. Here we discuss a different approach called *learning by queries*, introduced by Angluin in the 1980s [5]. In this model, it is known in advance that the function to be learned belongs to some given class of functions, and the learner's objective is to identify this function *exactly* by asking questions, or *queries*, about it. The prespecified class is called the target class, or *concept class*, and the function to be learned is called the target function, the target concept, or simply the *target*. Thus, query learning is an *active* kind of learning, where, instead of being provided only random examples, the learner can participate in selecting the information available in the course of the learning process.

The actual learning model depends on what types of queries are allowed. A basic query type, called *membership query*, simply asks for the function value at a given vector specified by the learning algorithm. A membership query can also be thought of as an *experiment* performed by the learner. In the medical diagnosis example, this would correspond to asking an expert about the diagnosis of a patient with a hypothetical set of test results. Other query types are described later on.

The goal of the learning algorithm is to identify the target function exactly, *as efficiently as possible*.

The complexity of a learning problem, just as in the case of probabilistic learning, can be measured in two different ways. Its *information complexity* is the minimal number of queries that need to be asked in order to identify an arbitrary function from the concept class, without taking into account the amount of computation needed to generate the queries and to process the information obtained. Taking the amount of computation into account as well, one gets the notion of the *computational complexity* of the learning problem.

In this chapter we give an overview of the query learning approach to learning classes of Boolean functions. We have tried to present results that are simply stated and give a flavor of this area.

In Section 7.2 we introduce the learning problems discussed. Section 7.3 gives definitions of the learning models and the criteria of efficiency. Section 7.4 gives outlines of efficient learning algorithms, and Section 7.5 gives negative complexity results for computational and information complexity. A large number of topics in learning theory are related to query learning, such as combinatorial characterizations, attribute-efficient learning, revision models, and noise-tolerant learning. These are briefly discussed in Section 7.6. Section 7.7 contains some further remarks and pointers to the literature.

7.2 Learning Problems

A learning problem is specified by a concept class of the form $\mathcal{F} = \bigcup_{n=1}^{\infty} \mathcal{F}_n$, where $\mathcal{F}_n$ is a family of n-variable Boolean functions. (Throughout this chapter, n always denotes the number of Boolean variables.) For example, the problem of learning a read-once function corresponds to identifying an unknown function from the concept class of all read-once functions.

This formulation needs to be refined somewhat. Usually we are not learning a function, but a particular representation of it, such as a disjunctive normal form (DNF) or a decision tree. Formally, a *representation* of a Boolean function f is a string $c \in \Sigma^*$ over some alphabet Σ, such as the encoding of a DNF or a decision tree. With some abuse of notation, we write $c(x)$ for the value of the function represented by c at the vector x. Thus, as our second attempt, we can specify a learning problem by a family $\mathcal{C} = \bigcup_{n=1}^{\infty} \mathcal{C}_n$, where $\mathcal{C}_n$ is a set of representations of a set of n-variable Boolean functions.

For the formulation of some of the results it is useful to go one step further. Many important representation formalisms, such as DNF or decision trees, are universal in the sense that they can be used to represent any Boolean function. In these cases it is necessary to take the size of the representation into account as well. Intuitively, functions having "small" representations are the ones considered to be "simpler" and thus are expected to be "easier to learn". In general, it is assumed that there is a *size* $s(c)$ associated with each representation c. A function f may have several representations in a given $\mathcal{C}$, and its size with respect to $\mathcal{C}$, denoted by $sc(f)$, is defined to be the minimum of the sizes of its representations. For example, the size of DNF can be defined as the number of terms. In this case the size of a function is the minimum number of terms needed to represent it. A *learning problem*, then, is specified as a family

$$\mathcal{C} = \bigcup_{n=1}^{\infty} \bigcup_{s=1}^{\infty} \mathcal{C}_{n,s},$$

where $\mathcal{C}_{n,s}$ is the set of n-variable representations of size s, and in such cases $\mathcal{C}_n = \bigcup_{s=1}^{\infty} \mathcal{C}_{n,s}$.

The following terms are useful for the specification of particular learning problems. A *term* or *conjunction* is a conjunction of Boolean literals, and a DNF formula is a disjunction of terms. A *clause* is a disjunction of Boolean literals, and a conjunctive normal form (CNF) formula is a conjunction of clauses. A formula is *monotone* if all its literals are unnegated. For example, a term is monotone if it is a conjunction of unnegated variables, and a monotone DNF is a DNF all of whose terms are monotone. A Boolean formula is *unate* if no variable occurs in it both negated and unnegated.

Remark on terminology. Several chapters of this book use the term *positive* for what we call monotone, and *monotone* for what we call unate. The terminology we use is consistent with standard usage in learning theory and related areas, such as computational complexity theory. In particular, all papers we cite use this terminology.

A clause is *Horn* if at most one of its literals is unnegated. A *Horn sentence* or *Horn formula* is a conjunction of Horn clauses. A Boolean formula is *read-once* if no variable occurs more than once in that formula. A Boolean function is read once if it can be represented by a read-once formula. The *parity* function is true on exactly those vectors with an odd number of 1s.

7.3 Learning Models

Given a learning problem $\mathcal{C}$, a learning algorithm, or learner, is first given n, the number of variables and s, the size of the target.[1] It is assumed that there is an unknown target $c \in \mathcal{C}_{n,s}$ to be learned, representing a target function f. The learning algorithm can then make queries that are answered by the environment.[2] At the end of the learning process, the learning algorithm has to output a representation that is equivalent to the target (i.e., it represents the same function f).

A *membership query* consists of a vector $x \in \{0, 1\}^n$. The response to the query, denoted by MQ(x), is $c(x)$, the value of the target function at x. An *equivalence*

[1] Knowing the size of the target in advance is a convenient assumption that does not significantly affect the complexity of learning; see, e.g., the discussion of the "doubling trick" in [50].

[2] As we consider worst-case complexity from the point of view of the learner, i.e., we assume that the queries are answered in an adversarial manner, it is perhaps more appropriate to refer to an environment rather than a presumably helpful teacher.

query consists of a *hypothesis* h, and the learner is asking whether h represents the same function as the target c. The response, denoted by EQ(h), is *YES* if it does represent the same function. In this case the learning process terminates. Otherwise, the response is a *counterexample*, that is, a vector x such that $h(x) \neq c(x)$. If $c(x) = 1$ (resp., 0), then x is a *positive* (resp., *negative*) counterexample.

Hypotheses do not have to belong to $\mathcal{C}$ in general, but may come from another family

$$\mathcal{H} = \bigcup_{n=1}^{\infty} \bigcup_{s=1}^{\infty} \mathcal{H}_{n,s},$$

the *hypothesis class*, where $\mathcal{H}_{n,s}$ is again a set of n-variable representations of size s. In this case the learning algorithm also gets an input parameter s', which is an upper bound on the size of the hypotheses that may be asked as equivalence queries. For example, one may try to learn 3-term DNF by using 3-CNF hypotheses (the implicit upper bound on the number of clauses is $O(n^3)$). If the hypothesis h used in the query does belong to $\mathcal{C}_n$ then the equivalence query is *proper*. If it also belongs to $\mathcal{C}_{n,s'}$ for some $s' \leq s$, then the equivalence query is *strongly proper*. Thus, a proper DNF learning algorithm uses DNF as hypotheses (as opposed to, say, Boolean formulas of depth 3), and a strongly proper DNF learning algorithm uses DNF hypotheses of size not larger than the target.[3]

Proper learning can be useful in applications either where the representation of the learned concept is used by other modules of an AI system, or where it should be comprehensible to humans. If the learned concept is used only for prediction, then its representation is irrelevant.

In a particular run of a learning algorithm, we have to learn a target from $\mathcal{C}_{n,s}$ using hypotheses from $\mathcal{H}_{n,s'}$. If the parameters are clear from the context or irrelevant, we may drop the indices and simply talk about learning a target from $\mathcal{C}$ using hypotheses from $\mathcal{H}$.

In the formal model of computation for query learning algorithms, it is assumed that the queries are written on a special query tape and the answer to a query (which in our case is a single bit or a counterexample vector) is provided in a single step on an answer tape. This assumption implies that the time taken to write a query is included in the running time.

[3] The situation is tricky when size figures into the definition of the concept and hypothesis classes themselves. For example, the class n-term DNF is the union over n of all functions represented by a DNF of at most n terms. Taking the natural size measure for DNF of the number of terms, we have

$$n\text{-term DNF} = \bigcup_{n} s\text{-term DNF for } s \leq n = \bigcup_{n} \left(\bigcup_{s=1}^{n} s\text{-term DNF} \cup \bigcup_{s=n+1}^{\infty} \emptyset \right).$$

Thus, a proper learning algorithm for n-term DNF must ask hypotheses of no more than n terms, even though a proper efficient learning algorithm for (arbitrary) DNF can ask a hypothesis of poly(n) terms when the target DNF has only n terms.

A strong learning algorithm for n-term DNF is even more restricted, because the hypotheses may not contain any more terms than the target DNF, independent of n.

Although there are several other query types, most of the work on query learning considers learning with membership and equivalence queries. As the next section shows, there are many efficient learning algorithms using both types of queries, for example, for read-once formulas or Horn formulas. In these two cases and in several other ones, it can also be shown that efficient learning is not possible if only one type of query is allowed.

An example of another query type is the *subset query*. A subset query also consists of a hypothesis h, but the query now asks whether $h \subseteq c$, that is, $h(x) = 1$ implies $c(x) = 1$ for every x. The answer is *YES* if the containment holds. Otherwise it is a counterexample x such that $h(x) = 1$ and $c(x) = 0$.

A learning algorithm is *efficient* if for every n and every target function f with a representation from C_n, it identifies the target in time polynomial in n and $s_C(f)$. A problem C is *polynomial-time learnable* if it has an efficient learning algorithm. We will usually specify the kind of queries needed and say, for example, that C is polynomial-time learnable from membership and proper equivalence queries, and we will sometimes specify a hypothesis space and say, for instance, that some problem C is polynomial-time learnable with hypotheses from $\mathcal{H}$.

To complicate matters further, it is of interest to consider just the *number of queries* used by a learning algorithm, without taking into account its running time. (This is the information complexity referred to in the introduction.) We need to be somewhat careful here: even if running time is not considered, we still have to restrict the *size of the queries* to be polynomial in the size of the target, in order to get meaningful results. The *query complexity* of a particular learning algorithm is the number of queries it uses in the worst case. The *query complexity* of a problem C is the minimum number of queries required by any algorithm that learns C. It may be the case that a learning algorithm uses only a few queries but it is not computationally efficient, as it has to solve an NP-hard problem to construct its queries. We talk about *polynomial-query learnability* with the same qualifiers as in polynomial-time learnability case.

Equivalence queries and random sampling used in the PAC model are related using the following observation [5]: an equivalence query can be simulated by drawing a sufficient number of random examples. If there is a counterexample among these then we have obtained an answer to the equivalence query, and we can continue the query learning algorithm. Otherwise the equivalence query is likely to have small error, so we can output it as our final hypothesis. Thus learnability from equivalence queries implies PAC-learnability. Note that a similar simulation does not work for membership or subset queries.

7.4 Algorithms

In this section we present efficient query learning algorithms for several important classes of Boolean functions. We begin with monotone conjunctions, a simple class that illustrates some basic ideas in query learning algorithms, and then continue on to various other classes, such as monotone DNFs, Horn functions, and read-once functions. We also discuss the monotone theory, a natural framework for studying Boolean functions that has found applications outside computational learning theory as well.

It will simplify the presentation to sometimes treat vectors and monotone terms as interchangeable. Thus, if we use the vector 0110 in a place where a monotone term is expected, we mean the term x_2x_3, and if we use the term x_2x_3 in a place where a vector is expected (and n is known to be 4), we mean the vector 0110. When we speak of *flipping bit (or position) i* in vector x, we mean changing x by negating its ith bit; thus, flipping bit 3 of 0110 yields 0100.

7.4.1 Monotone Conjunctions

Let us first show that monotone conjunctions over n variables can be learned in polynomial time from both membership queries alone and equivalence queries alone.

To learn from membership queries alone, initialize a vector x to be all 1s. For each position i for $1 \leq i \leq n$ in turn, ask a membership query of x with bit i flipped, and if the membership query returns 1, update x to be x with bit i flipped. At the end, the target function is the term corresponding to the final value of x. This algorithm makes exactly n membership queries and clearly has polynomial running time.

We could instead learn monotone conjunctions from equivalence queries alone. In this case, we initialize the hypothesis φ to be the conjunction of all n Boolean variables and begin making equivalence queries.[4] Notice that the first equivalence query can return only a positive counterexample x (unless φ is the target). Any bits that are 0 in x must correspond to variables that cannot be in the target conjunction, so we remove them from hypothesis φ. We repeat asking equivalence queries until we get the target function. It must always be that we receive only positive counterexamples throughout the process. This algorithm makes at most $n + 1$ equivalence queries and again has polynomial running time.

Notice that this equivalence query algorithm can be easily modified to handle conjunctions in general. An equivalence query using an everywhere false hypothesis such as $\mathrm{EQ}(x_1\bar{x}_1)$ at the very beginning must return a positive vector x (or we successfully terminate because the target is the identically false function).[5] This vector gives us the orientation of every variable in the target formula. For instance, if the vector's first bit is 0, then we know that either the variable x_1 does not occur at all in the target, or it occurs as the negated $\bar{x}_1$; it cannot occur unnegated. We now proceed as we did for the case where negation was not allowed, except that

[4] When discussing propositional formulas, we often use Greek letters, such as φ, ψ, for representations.
[5] We assume that every monotone conjunction is satisfied by the all-1s vector, but general conjunctions may be identically false.

we start with the conjunction of all the variables with each variable having its only possible orientation. A variant of this idea is used in the monotone theory discussed later on in this section.

On the other hand, the membership query algorithm does not generalize to the case of general conjunctions. The reason for that is the lack of a start vector like the all-1s vector in the monotone case. Indeed, if all membership queries are answered negatively, then as long as we did not ask all the 2^n vectors, it is still possible that the target is satisfied by one of the unasked vectors or is identically false. Thus we get the following negative result: in order to learn an n-variable conjunction, one has to ask 2^n membership queries in the worst case. This is a lower bound to the number of queries, irrespective of the running time of the learning algorithm.

7.4.2 Monotone DNF and k-Term DNF

A monotone conjunction can be viewed as a one-term monotone DNF. It is a fundamental question of query learning, and of computational learning theory in general, whether the positive results of the previous subsection can be generalized to monotone DNF or perhaps even to DNF (in the latter case, we have seen that equivalence queries are necessary). It turns out that monotone DNF are not efficiently query learnable if we are restricted to only one of membership or equivalence queries, as discussed in the next section. On the other hand, it is one of the seminal results that monotone DNF are efficiently learnable, in fact, *strongly properly efficiently learnable*, if both membership and equivalence queries are available. More precisely, the following holds.

Theorem 7.1. (Angluin [5].) *Monotone DNF is polynomial-time learnable using $O(m)$ strongly proper equivalence queries and $O(nm)$ membership queries, where m is the number of terms in the target DNF.*

Algorithm 1 Exact learning algorithm for monotone DNF

$\varphi \leftarrow$ empty disjunction ▷ identically *false*
while EQ(φ) returns counterexample x (rather than *YES*) **do**
    for each bit b that is 1 in x **do**
        if MQ(x with b set to 0) $= 1$ **then**
            $x \leftarrow x$ with b set to 0
        end if
    end for
    $\varphi \leftarrow \varphi \vee$ (term corresponding to) x
end while

Proof. We show the algorithm as Algorithm 1. The key to its correctness is that it maintains the invariant that its hypothesis φ is always a subset of the target (i.e., implies it). Each equivalence query must produce a positive counterexample x that satisfies some term in the target but not in the current hypothesis. The **for** loop at lines 3–7 "reduces" this counterexample so that it becomes a *minterm* of the target. A *minterm* or *prime implicant* of a Boolean function f is defined to be a minimal (with respect to inclusion) term that implies f.[6] Each equivalence query except the last finds a prime implicant of the target. Whereas the number of prime implicants of a DNF may in general be exponential in the number of its terms, for monotone DNFs the number of prime implicants is at most the number of terms. Therefore, to learn a monotone DNF with m terms, the algorithm uses $m + 1$ equivalence queries. Between two successive equivalence queries, the algorithm makes at most n membership queries, and its running time is clearly polynomial in n and m. ■

Algorithm 1 is a typical example of membership and equivalence query learning algorithms in the following sense: a counterexample to an equivalence query contains new information (a new term covered by the counterexample), and several membership queries are used to extract this information (by exploring the neighborhood of the counterexample vector).

General DNF with a small number of terms are also efficiently learnable with membership and strongly proper equivalence queries. This was first shown for DNF with a constant number of terms by Angluin [4]. The complexity bounds were later improved by Blum and Rudich [27] and Berggren [22]. The following result shows that the number of terms can also be allowed to grow slowly with the number of variables.

Theorem 7.2. (Bshouty et al. [31].) *The class of $O(\sqrt{\log n})$-term DNF is polynomial-time learnable using membership and strongly proper equivalence queries.*

The class of $O(\log n)$-term DNF has received lot of attention, and its status is not resolved completely. We next describe a result that adapts the approach of Algorithm 1 to this problem, using the notion of a *universal* set of vectors. Similar ideas have been used in [31].

A set of vectors $S \subseteq \{0, 1\}^m$ is called (m, k)*-universal* if for every k variables and for every assignment to these variables, there is a vector in S that agrees with this partial truth assignment. The existence of (m, k)*-universal* sets of size $k\,2^k \log(2m)$ follows by a probabilistic argument, but there are also deterministic constructions known of size $2^k\, k^{O(\log k)} \log m$ that take time polynomial in n and 2^k (see [67]).

Theorem 7.3. (Hellerstein and Raghavan [51].) *The class of* $\log n$*-term DNF is polynomial-time learnable using membership queries and equivalence queries from n-term DNF.*

[6] The definition of *prime implicant* appears to be completely standard throughout the Boolean function literature; however, other definitions of *minterm* are sometimes used elsewhere in the literature.

Proof. [Proof sketch] Assume for now that we have a term t and would like to find out whether t implies the target $c = t_1 \vee \cdots \vee t_{\log n}$. If t does *not* imply c, then there is a vector x satisfying t but falsifying c. Then x falsifies each term of c, and hence there are literals z_i in t_i $(i = 1, \ldots, \log n)$ that are all false for x.

Consider an $(n - |t|, \log n)$-universal set of vectors S over the set of variables *not* contained in t. Let S' be the set of complete truth assignments obtained from S by assigning 1 to each *literal* in t. Then the definition of a universal set shows that t implies c if and only if c is true for every truth assignment in S'; this corresponds to $|S'|$ many membership queries. But $|S'|$ is polynomial in n, and so we get a polynomial-time procedure to decide if a term is an implicant of the target. This procedure can then be used to get a polynomial time procedure that, given a positive counterexample x, starting with the term x, repeatedly deletes literals, to find a prime implicant of the target covering x.

Now we can proceed as in Algorithm 1 and use equivalence queries on a disjunction of prime implicants to find more and more prime implicants of the target. However, unlike in the monotone case (as we noted there already), the number of prime implicants can be larger than the number of terms in the target. Nevertheless, as shown by Chandra and Markowsky [34], an ℓ-term DNF can have at most $2^{\ell} - 1$ prime implicants, and this bound is sharp (see also [78] for a characterization of extremal DNF). This implies the size bound for the queries claimed in the theorem. ■

If one is interested in polynomial query complexity and strongly proper learning, but not necessarily polynomial running time, then the following result holds. The proof uses certificates of exclusion (see Section 7.6.1.)

Theorem 7.4. (Hellerstein and Raghavan [51].) *The class of $O(\log n)$-term DNF is polynomial-query learnable using membership and strongly proper equivalence queries.*

Theorems 7.3 and 7.4 give two quite different learning algorithms for $\log n$-term DNF. Kushilevitz [57] gives a simple learning algorithm for the same class using finite automata as hypotheses. Yet another algorithm is implied by the monotone theory (see Section 7.4.4). None of these algorithms are simultaneously proper and polynomial-time. The existence of such an algorithm is open. Hellerstein and Raghavan [51] show that such an algorithm could be used to give a polynomial-time algorithm for the following computational problem: given the truth-table of an n-variable Boolean function, can the function be represented by an n-term DNF? (As the input to this algorithm is of size 2^n, polynomial time here means polynomial in 2^n.) The result of [34] implies that there is a polynomial-time algorithm if the same question is asked about $\sqrt{n}$-term DNF, but the case of n-term DNF is open (see also Allender et al. [3]). In Section 7.5 we present a negative result indicating the limitations for increasing the number of terms.

7.4.3 Horn Formulas

Horn formulas form an expressive and tractable class of Boolean expressions and play a central role in several fields of computer science and artificial intelligence (see the chapter by Boros in the monograph [35] for an extensive discussion). In the predicate logic version, Horn formulas are closely related to *logic programs*. Learning logic programs, called *inductive logic programming*, is a much studied subfield of machine learning. In our context, Horn formulas form a natural generalization of *antimonotone CNF* (the negation of monotone DNF). It turns out that this class is again polynomial-time learnable using equivalence and membership queries. The learning algorithm is proper, that is, it uses Horn formulas as hypotheses, but those may contain more clauses than the target formula.

Theorem 7.5. (Angluin, Frazier, and Pitt [11].) *Horn formulas are polynomial-time learnable using $O(nm)$ proper equivalence and $O(m^2n)$ membership queries, where m is the number of clauses in the target Horn formula.*

Proof. [Proof sketch] At the highest level the algorithm is somewhat similar to the dual of the monotone DNF algorithm, although Horn sentences are more complicated, precisely because they are not antimonotone. The counterexamples may now be both positive and negative.

The negated variables of a Horn clause C form its *body*, denoted body(C), and the unnegated variable (if it exists) is its *head*. There is one technical point: Angluin et al. achieved the best query complexity by organizing the hypothesis by distinct clause *bodies*. We call the collection of all clauses in one Horn sentence that have the same body a *clause-group*. (Angluin et al. called a clause-group a *meta-clause*.) Thus, one clause-group may contain several clauses.

We start with the hypothesis being the empty conjunction (i.e., everything is classified as true) and repeatedly, in an outer loop, make equivalence queries until the correct Horn sentence has been found. Clause-groups are maintained as an *ordered* set. Each negative counterexample is used, with the help of membership queries, to make the hypothesis more restrictive; each positive counterexample is used to make the hypothesis more general.

The key to using negative counterexamples to shrink the body of a clause-group is the statement body(C) $\leftarrow$ body(C) $\cap x$ at line 6 of Algorithm 2. The notation body(C) $\cap x$ means either the vector that is 1 in those positions i such that both body(C) contains variable x_i and vector x has a 1 in position i, or, equivalently, the conjunction of variables corresponding to the 1s in that vector.

If a negative counterexample cannot be used to shrink an existing clause, then we insert a new clause-group whose body is the conjunction of all the variables corresponding to 1s in the clause and that initially has all other variables as possible heads.

Notice that the deletion done with positive counterexamples is of individual *clauses*, not of clause-groups. That is, each positive counterexample causes at least one head to be removed from at least one clause-group. If all heads are removed

Algorithm 2 Algorithm to learn Horn formulas

```
φ ← empty conjunction (sequence) of clause-groups
while EQ(φ) receives a counterexample x do
    if x is a negative counterexample then
        for each clause-group C ∈ φ in order do
            if body(C) ∩ x ⊂ body(C) and then MQ(body(C) ∩ x) = 0 then
                body(C) ← body(C) ∩ x
                break the for loop
            end if
        end for
        if x wasn't used to shrink any clause-group in φ then
            Add new last clause-group to φ with body x and all possible heads
        end if
    else                                        ▷ x is a positive counterexample
        for all clauses C of φ such that C(x) = 0 do
            Delete C from φ
        end for
    end if
end while
```

from a clause-group in this way, the clause-group remains and now represents the single Horn clause with the obvious body and no head. ■

Concerning lower bounds, Arias et al. [16] proved that every strongly proper learning algorithm for Horn formulas uses $\Omega(nm)$ queries. They use certificates of exclusion (see Section 7.6.1 for the definition and for the connection to learning). Bshouty et al. [31] show that every learning algorithm for Horn formulas uses $\Omega\,(nm/(\log nm))$ equivalence queries, even if *arbitrary* equivalence queries are allowed, as long as it uses only polynomially many membership queries. This kind of lower bound is also of interest because when one transforms a query algorithm into the online or mistake bounded model of learning, the number of equivalence queries will become the number of mistakes.

Boros shows in [35] that *body minimization* can be solved efficiently for Horn formulas, that is, given a Horn formula, one can find in polynomial time an equivalent Horn formula with the minimal number of distinct bodies, or in the terminology of Theorem 7.5, clause-groups. Interestingly, as noted by Frazier [38], Algorithm 2 provides another proof of this result, as running it on a given Horn formula, one obtains such a representation. Arias and Balcázar [15] give a new analysis of Algorithm 2 in a more abstract lattice setting.

Learning Horn formulas has also been studied in the model of *learning from entailment* [38, 39]. Here the instances are not Boolean vectors but are Horn clauses themselves. Thus, a target concept is considered to be a theory, and the positive (resp., negative) examples of this theory are Horn clauses which are implied (resp., not implied) by the theory. Learning a theory can be thought of as modeling the process of acquiring a knowledge base. Frazier and Pitt [39] showed that Horn formulas are efficiently learnable in this model as well. One of their algorithms is obtained by a reduction to Algorithm 2, and the other one is a direct algorithm in the entailment model.

7.4.4 The Monotone Theory

Bshouty [28] developed an algorithm that can be seen as a generalization of Algorithm 1, the monotone DNF algorithm, providing an efficient query learning algorithm, for among other classes, decision trees. This algorithm is *not proper*: its hypotheses are conjunctions of DNF formulas, that is, they are Boolean expressions of depth 3. The algorithm is based on the novel *monotone theory*, a representation of Boolean functions as the conjunction of the monotone closures of the function in every possible direction. This representation is of interest in itself, and it has been found to be relevant for several problems other than learning, such as model-based representations [55] and case-based reasoning [74].

A unate formula is *a-unate* for a vector a if variable x_i occurs negated (resp., unnegated) in the formula only if $a_i = 1$ (resp., only if $a_i = 0$) – that is, if the sign of variables in the formula corresponds to the bit values in a with bit 0 corresponding to no negation. Monotone formulas are 0-unate. We can think of a as giving an alternate direction for the bottom of the hypercube $\{0, 1\}^n$.

Let us say for vectors x, y, and a that $x \leq_a y$ if $x \oplus a \leq y \oplus a$, where $\oplus$ denotes componentwise exclusive or. Thus again a may be thought of as the minimum vector for $\leq_a$. The usual partial ordering of $\{0, 1\}^n$ has minimum vector 0; that is, $x \leq y \Leftrightarrow x \leq_0 y$.

Definition 7.1. *The* a**-unate closure** *of Boolean function* f *is*

$$\mathcal{M}_a(f)(x) = \begin{cases} 1 & \text{if } \exists y \leq_a x\colon\ f(y) = 1 \\ 0 & \text{otherwise.} \end{cases}$$

Bshouty calls this the "minimal a-monotone Boolean function of f." Notice that for a monotone function f, we have that $f = \mathcal{M}_0(f)$. The following is a key observation.

Lemma 7.6. *For any Boolean function* f,

$$f = \bigwedge_{a \in \{0,1\}^n} \mathcal{M}_a(f). \tag{7.1}$$

In some cases we need not have such a large conjunction; indeed, for monotone functions we need only one conjunct rather than 2^n in (7.1). If f can be represented in the form

$$f = \bigwedge_{a \in S} \mathcal{M}_a(f),$$

then S is called a *monotone basis* for f. A set $A \subseteq \{0, 1\}^n$ is a *(monotone) basis for a class of functions* $\mathcal{F}_n$ if A is a basis for every $f \in \mathcal{F}_n$.

The following is a useful lemma for finding small monotone bases.

Lemma 7.7. *A set S of vectors is a monotone basis for f if and only if f can be written as a CNF such that every clause of this CNF is falsified by at least one vector in S.*

Thus, for example, Horn formulas have a monotone basis consisting of vectors a with at most one 0 component, and thus require at most $n + 1$ conjuncts in (7.1). Also, $O(\log n)$-term DNF have a monotone basis of polynomial size. This follows from the existence of polynomial size $(n, O(\log n))$-universal sets.

Define the *DNF size* of a Boolean function to be the minimum number of terms of a DNF for the function, and the DNF size of a set of Boolean functions on n variables to be the maximum DNF size of any function in the set. CNF size is defined similarly with respect to clauses.

After these preparations, we can state Bshouty's results. The first main theorem Bshouty proves is applicable to the case where we know a monotone basis for a class, such as Horn formulas and $\log n$-term DNF.

Theorem 7.8. (Bshouty [28].) *A class $\mathcal{C}$ of Boolean functions can be learned from membership and (improper) equivalence queries in time polynomial in n, the DNF size of $\mathcal{C}$, and the size of a monotone basis for $\mathcal{C}$.*

The formulation of the next theorem is somewhat different from the general framework presented in Sections 7.2 and 7.3. Unlike the previous cases of monotone DNF and Horn formulas, this algorithm learns an *arbitrary* Boolean function, without any previous information. The running time, and thus the query complexity, is bounded by the complexity of the target, which this time is measured by the *sum* of the DNF and CNF sizes of the function. One could think of this complexity measure as the size of the representation providing both a DNF and a CNF for the function (called a *CDNF representation*); this is slightly unnatural, however, as then deciding whether a string is a valid representation becomes presumably intractable. On the other hand, special cases of the theorem, such as for decision trees, fit our framework.

Theorem 7.9. (Bshouty [28].) *A Boolean function f can be learned using membership and equivalence queries, with running time bounded by a polynomial in n, the DNF size of f, and the CNF size of f, using conjunctions of DNF as hypotheses.*

Proof. [Proof sketch] We first generalize Algorithm 1 to an algorithm for learning an a-monotone DNF, presented as Algorithm 3. Instead of the heart of the algorithm being a loop that walks vectors toward the zero vector, we now have a loop that walks vectors toward a. Bshouty proves that if it keeps receiving positive counterexamples, then Algorithm 3 will eventually make $h = \mathcal{M}_a(f)$.

Algorithm 3 Learns $\mathcal{M}_a(f)$ given positive counterexamples

```
Require: f(a) = 0 and EQ and MQ oracles for f.
function DNFSubroutine(a)
    h ← empty disjunction                                        ▷ identically false
    while EQ(h) returns (positive) counterexample x do
        while there is i s.t. x_i ≠ a_i and MQ(x with x_i flipped) = 1 do
            x ← x with bit i flipped                              ▷ Walk x to a
        end while
        T ← conjunction of literals satisfied by x but not by a
        h ← h ∨ T
    end while
end function
```

The idea then is to run enough copies of this algorithm with different as so that intersection of all the learned $\mathcal{M}_a(f)$ will be f. In particular, based on Lemma 7.7, we want each a we use to falsify some clause of a minimal CNF for f. If f has a small CNF, then we will not use too many a's.

A potential difficulty is that we have an equivalence oracle for f, but will need counterexamples for $\mathcal{M}_a(f)$. We will use one instantiation of Algorithm 3 for every distinct a-unate closure of f that we are going to find. Each instantiation has its own value of h at all times while that instantiation is waiting for its next positive counterexample.

Now Algorithm 4 will find a conjunction of such DNFs.

Algorithm 4 Algorithm for learning function with small DNF and small CNF

```
φ ← empty conjunction                                            ▷ identically true
while EQ(φ) returns counterexample x do
    if x is a negative counterexample then
        Start a new instantiation of DNFSubroutine (x)
    else
        for all instantiations of DNFSubroutine whose hypothesis h is not
    satisfied by x do
            Pass x as positive counterexample to that instantiation
        end for
    end if
    φ ← conjunction of the h of all instantiations of DNFSubroutine
end while
```

The idea of Algorithm 4 is that each negative counterexample is a new element of a monotone basis of f and is thus used to start a new instantiation of Algorithm 3. When each DNF instance has converged, we will have precisely the target f. While the algorithm is running, each positive counterexample is used to start a new term in at least one DNF instance. ■

For any a, the size of $\mathcal{M}_a(f)$ cannot be more than polynomially larger than the DNF size of f, so the learned $\mathcal{M}_a(f)$ are not too large and do not require too many queries.

As the size of a minimal DNF and a minimal CNF for Boolean function f is polynomially bounded by the size of a minimal decision tree for f, we get the following result.

Corollary 7.10. (Bshouty [28].) *Decision trees are polynomial-time learnable from membership and improper equivalence queries.*

Also, any Boolean function is learnable in time polynomial in n and its decision-tree size from membership and improper equivalence queries.

7.4.5 Threshold Functions, Read-Once Functions, and Decision Lists

Now we discuss three more interesting classes of Boolean functions.

Zero-one threshold functions are Boolean functions defined by a weight vector $w = (w_1, \ldots, w_n) \in \{0, 1\}^n$ and a threshold $t \in \mathbf{N}$. The function is true iff

$$\sum_{i=1}^{n} w_i x_i \geq t. \tag{7.2}$$

The variables x_i with $w_i = 1$ are called *relevant* variables. Thus, the function is true if at least t relevant variables are 1. Zero-one threshold functions form a much studied concept class in computational learning theory. For example, Winnow is an attribute-efficient mistake-bounded learning algorithm for this class [60].

Theorem 7.11. (Hegedűs [48].) *Zero-one threshold functions are polynomial-time learnable using $O(n)$ membership queries.*

Proof. [Proof sketch] The algorithm is as follows: First ask a membership query on the all-1s vector to make sure it is positive. (Otherwise the target is identically *false*.)

Now initialize a vector y to be the all-1s vector. Flip the bits of y one at a time, asking a membership query after each one, and keeping the bit flipped (i.e., off) if the membership query returned 1. At the end, y will have t 1s, all in positions of relevant variables, where t is the threshold of the target.

Now for each position i that is a zero in y, ask a membership query on a vector formed by turning off one of the bits of y and turning on position i. The relevant variables are located in all the positions that are 1 in y plus all the positions that had the membership query return 1 in this last step. ∎

Hegedűs also gave a matching $\Omega(n)$ lower bound for the number of membership queries required [48].

General threshold functions are defined as in (7.2), but without any restriction on the weights and the threshold. Threshold functions have been studied for several decades (Crama and Hammer [35], Muroga [66]) and of course have gained a lot of attention with the advent of neural networks (see Chapter 13 by Anthony). We mention here a result on their efficient learnability with equivalence queries.

Theorem 7.12. (Maass and Turán [62].) *Threshold functions are polynomial-time learnable with equivalence queries.*

The proof is based on the observation that the ellipsoid method of linear programming gives an efficient learning algorithm. The progress of the algorithm is measured by the reduction of the volume of an ellipsoid containing the target. In the terminology of machine learning, this corresponds to the shrinkage of the version space (Mitchell [64]). Using an efficient implementation of the ellipsoid method (Vaidya [80]), one gets a $O(n^2 \log n)$ upper bound for the number of queries. Maass and Turán [61] also give an $\Omega(n^2)$ lower bound.

A Boolean function is *read-once* if there is a formula for it in which each variable appears at most once. In particular, for each i, one but not both of x_i or $\bar{x}_i$ may appear.

Theorem 7.13. (Angluin, Hellerstein, and Karpinski [12].) *Read-once formulas are polynomial-time learnable using $O(n^3)$ membership and $O(n)$ proper equivalence queries.*

The algorithm uses properties of read-once functions discussed by Golumbic in the monograph [35]. In general the *read number* of a Boolean formula is the maximal number of occurrences of a single variable. Thus, every variable occurs at most twice in a *read-twice* formula. The following is a positive learnability result for read-twice DNF. In the next section we show that this result cannot be generalized further to read number 3.

Theorem 7.14. (Pillaipakkamnatt and Raghavan [68].) *Read-twice DNF formulas are polynomial-time learnable from membership and equivalence queries.*

We conclude this section by mentioning a learning algorithm for learning decision lists from equivalence queries alone. To be specific, for this result, we assume the class of decision lists, sometimes called k-decision lists, having tests that are size k monomials for constant k. See Chapter 14 in this volume for more discussion of the definition of decision lists.

Theorem 7.15. (Castro and Balcázar [33], Simon [75].) *Decision lists are polynomial-time learnable from proper equivalence queries.*

The algorithm maintains its hypothesis in a form that generalizes decision lists: the tests are partitioned into groups, and tests within a group are unordered. (Nevertheless, a hypothesis can be formed using this representation by ordering tests within a group arbitrarily.) If a counterexample is received, then one or more tests responsible for the wrong classification are moved further down the list.

7.5 Negative Results

Anthony shows in Chapter 6 that the VC-dimension of a concept class provides both an upper and a lower bound to the sample complexity of learning in the PAC model, and that the NP-hardness of the fitting and consistency problems for the class implies that the class is not efficiently PAC learnable under complexity-theoretic assumptions. The first result deals with *information complexity*, whereas the second is about *computational complexity*. In this chapter we discuss related results in query models, starting with complexity-theoretic negative results and continuing with information-theoretic ones.

We have seen in the previous section that the learnability of various subclasses of DNF is an important topic in learning theory. The question whether general DNF are efficiently learnable has been a fundamental open problem for a long time. Although the problem is still not settled completely, there have been significant recent negative results. Some of these are discussed in this chapter.

7.5.1 Computational Hardness Results for Threshold Circuits and DNF

It has been noted already in the the seminal paper of Valiant [81] that there is a relationship between learnability and the security of cryptosystems: learning is akin to breaking a cryptosystem. This informal relationship can be made precise, and it can be used to prove that under assumptions on the security of cryptosystems, certain learning problems are not efficiently learnable. Boolean representation classes to which such an approach applies have to be expressive enough to represent certain cryptosystems.

A target class is *efficiently predictable with membership queries* if it is efficiently learnable using both equivalence queries from *some* hypothesis space,[7] and membership queries.

Threshold circuits are Boolean circuits composed of threshold gates. A threshold gate computes a threshold function of the form (7.2). Threshold circuits are much studied in circuit complexity theory (see Chapter 11 by Krause and Wegener in this volume).

Kharitonov [56] proved an unpredictability result for bounded depth threshold circuits in the model of PAC learning with membership queries under a standard complexity-theoretic assumption from cryptography. Using the relationship between PAC learning and query learning described at the end of Section 7.3, it implies the following.

Theorem 7.16. (Kharitonov [56].) *If factoring Blum integers is hard, then for some constant d, threshold circuits of depth d are not predictable with membership queries.*

[7] The precise formulation includes some technical conditions similar to *nice* representations, described at the beginning of Section 7.5.2. These are omitted for simplicity.

This a rather strong negative result for bounded-depth threshold circuits (for a sufficiently large constant depth bound). It means that they are not polynomial time learnable with membership and equivalence queries for *any* hypothesis space. The smallest value of d is not known, and it is also not known if a similar unpredictability result holds for DNF.

A breakthrough result on the hardness of learning DNF formulas was recently obtained by Alekhnovich et al. [2]. They showed that, under the complexity theoretic hypothesis NP $\neq$ RP, DNF formulas are not learnable with an efficient PAC algorithm using disjunctions of threshold functions as hypotheses (which is clearly an extension of DNF). This, in particular, shows that DNF are not properly learnable by an efficient PAC algorithm. Quite recently Feldman [37] extended Alekhnovich et al.'s results to include membership queries. Again, using the relationship between PAC learning and query learning described at the end of Section 7.3, this implies the following.

Theorem 7.17. (Feldman [37].) *If* NP $\neq$ RP, *then DNF formulas are not polynomial-time learnable using membership and equivalence queries, even if the equivalence queries are allowed to be disjunctions of threshold functions.*

The proof uses deep results on the inapproximability of certain NP-complete problems, which build upon the theory of probabilistically checkable proof systems (see, e.g., [63]), and we do not go into the discussion of the ideas involved.

It is tempting to ask whether Theorem 7.17 remains valid for query complexity, with equivalence queries restricted to polynomial-size DNF (or, more generally, to disjunctions of threshold functions) [51]. As far as we know, this is open.

7.5.2 Consistency and Representation Problems

In this subsection we present some earlier results that provide insights into the limitations of efficient query learnability and have implications for learning subclasses of DNF.

In order to formulate a necessary condition for polynomial-time learnability with membership and equivalence queries, we need to define some general properties of representations. Briefly, a representation is *nice* if it has the following three properties: *(i)* $c(x)$ can be evaluated in time polynomial in n and $s(c)$, *(ii)* it can be decided in time polynomial in $s(c)$ if c is a representation of a concept,[8] *(iii)* the size of every $c \in \mathcal{C}_n$ is polynomial in n. Taking the example of read-once formulas, *(i)* means that one can efficiently evaluate a formula on a given input vector, *(ii)* means that we can determine efficiently if a string encodes a read-once formula, and *(iii)* means that the length of (the encoding of) a read-once formula is polynomial in n.

Recall from Chapter 6 that given a learning problem $\mathcal{C}$, the *$\mathcal{C}$-consistency problem* is the following computational problem: given a set of labeled examples,

[8] Recall that this automatically excludes CDNF formulas unless P = NP.

that is, $(x_1, b_1), \ldots, (x_m, b_m)$, where $x_i \in \{0, 1\}^n$, $b_i \in \{0, 1\}$, is there a concept c from C_n such that $c(x_i) = b_i$ for every $i = 1, \ldots, m$? (See also Chapter 12 in the monograph [35].)

Theorem 7.18. *If a learning problem C is nice and polynomial-time learnable with equivalence queries then its consistency problem is in* P.

The idea of the proof is to use the learning algorithm to solve the consistency problem by answering the equivalence queries based on the given sample. Notice that nothing is known about the performance of the learning algorithm if the answers given to its queries are not consistent with any target. Nevertheless, the bound for the running time guarantees that in such a case, we can terminate the learning process and conclude that there is no consistent concept.

Corollary 7.19. *If* P $\neq$ NP *and the consistency problem of a nice learning problem C is* NP*-hard, then C is not polynomial-time learnable with equivalence queries.*

The following is a well-known application of this corollary. Then Theorem 7.21 serves as an example of getting around a negative result by extending the hypothesis space.

Corollary 7.20. (Pitt and Valiant [70].) *3-term DNF is not polynomial-time learnable with proper equivalence queries.*

Theorem 7.21. *3-term DNF is polynomial-time learnable with equivalence queries from 3-CNF.*

Given a learning problem C, the *C-representation problem* is the following computational problem: given a Boolean formula φ, is it equivalent to a representation from C? Looking again at the example of read-once formulas: given a Boolean formula, is it equivalent to a read-once formula? Aizenstein et al. [1] established the following connection between the complexity of this problem and polynomial-time learnability from membership and equivalence queries.

Theorem 7.22. (Aizenstein et al. [1].) *If a learning problem C is nice and polynomial-time learnable from membership and equivalence queries, then the C-representation problem is in co-*NP.

The proof idea is again to use the learning algorithm to give an algorithm for the representation problem, using the input formula φ in place of the target concept. This time the algorithm is nondeterministic, and it accepts φ if and only if the learning algorithm does not terminate successfully (i.e., when φ is not equivalent to any representation from C). Membership queries are answered using φ. Equivalence queries are answered by guessing a counterexample. If φ is not equivalent to any representation from C, then if every query is syntactically correct (which can be checked by the assumption that C is nice), such a counterexample can always be found, and after exceeding the running time bound, φ is accepted. If φ is equivalent to a representation from C, then the learning algorithm will succeed and, as no counterexample can be found for its final query, φ will never be accepted.

Corollary 7.23. (Aizenstein et al. [1].) *If* NP $\neq$ *co*-NP *and the* C*-representation problem for nice learning problem* C *is* NP*-hard, then* C *is not polynomial-time learnable with membership and equivalence queries.*

An application of Corollary 7.23 concerns DNF with a bounded number of literal occurrences. Recall from Section 7.4 that read-once formulas are polynomial-time learnable from membership and equivalence queries. If that learning algorithm is applied to learn a read-once DNF formula, then it turns out that all its equivalence queries are on read-once DNFs. Thus, it follows that read-once DNF formulas are polynomial-time learnable from membership and equivalence queries. Earlier we saw in Theorem 7.14 that read-twice DNF formulas are also polynomial-time learnable from membership and equivalence queries. As the following result (first proved by Aizenstein et al. [1] and then by Pillaipakkamnatt and Raghavan [69] with a simpler proof) shows, going to read-thrice-DNF makes the problem intractable.

Theorem 7.24. (Aizenstein et al., Pillaipakkamnatt and Raghavan [1, 69].) *The representation problem for read-thrice-DNF is* NP*-hard.*

Thus Corollary 7.23 and Theorem 7.24 imply the following.

Corollary 7.25. *If* NP $\neq$ *co*-NP *then read-thrice DNF are not polynomial time learnable with membership and equivalence queries.*

7.5.3 Information Complexity

The results presented in the previous subsections referred to computationally efficient learning algorithms. As noted earlier, another way to measure the complexity of learning algorithms is to consider only the number of queries, without taking into consideration the amount of computation necessary to generate these queries. A result showing that learning some concept class requires a superpolynomial number of queries is thus stronger than a result stating the same for running time. Now we present some such lower bound results.

Angluin [6] gave a superpolynomial lower bound for DNF if only equivalence queries are allowed.

Theorem 7.26. (Angluin [6].) *DNF is not polynomial-query learnable using only equivalence queries on DNF of size polynomial in the number of variables and the size of the target.*

In fact, the result holds if the target is restricted to monotone DNF but the hypotheses are still allowed to be arbitrary DNF. Thus, it is necessary to use membership queries in Algorithm 3.

The other result we mention is the counterpart of Theorem 7.4. It can be viewed as a partial result on the question we formulated at the end of Section 7.5.1.

Theorem 7.27. (Hellerstein and Raghavan [51].) *For every* $\varepsilon > 0$, $\Omega(\log^{3+\varepsilon} n)$-*term DNF is not polynomial-query learnable using membership and strongly proper equivalence queries.*

The proofs of Theorems 7.26 and 7.27 are based on the existence of general combinatorial parameters, which provide lower bounds to various query complexities. These are discussed in some detail in Section 7.6.1.

7.6 Related Topics

In the previous two sections we have presented efficient query learning algorithms, and computational and information negative results for query learnability. In order to complete the proofs of the latter, we owe the reader the definition of certain combinatorial parameters and a discussion of their relationship to learnability. This topic, which is interesting in its own right, is discussed in Subsection 7.6.1. The remaining subsections deal with modifications of the basic query learning model that attempt to capture features of learning that are important to make the model more realistic. These are *attribute-efficient learning*, where it is assumed that only some of the variables are relevant for defining the target; *revision*, where it is assumed that learning starts with a reasonably good approximation of the target; and various ways of modeling *noise*, that is, imperfectness and incompleteness of the information available about the target.

7.6.1 Combinatorial Characterizations

We emphasize that in this subsection we are mainly interested in query complexity and not in running time. A comprehensive survey of topics discussed in the section is given in Angluin [7].

Given a class of n-variable Boolean functions $\mathcal{F}$, and an n-variable Boolean function f not necessarily belonging to $\mathcal{F}$, a subset $A \subseteq \{0, 1\}^n$ is a specifying set of f with respect to $\mathcal{F}$, if at most one of the functions in $\mathcal{F}$ agrees with f on A.[9] Clearly such a specifying set always exists, as $\{0, 1\}^n$ is a specifying set. The specifying set size of f with respect to $\mathcal{F}$ is

$$\mathrm{spec}_{\mathcal{F}}(f) = \min \{|A| : A \text{ is a specifying set of } f \text{ w.r.t. } \mathcal{F}\}.$$

Based on this notion, we define three combinatorial parameters measuring the complexity of a class of functions. The exclusion dimension of $\mathcal{F}$ is

$$\mathrm{XD}(\mathcal{F}) = \max \left\{\mathrm{spec}_{\mathcal{F}}(f) : f \notin \mathcal{F}\right\},$$

[9] We are using $\mathcal{F}$ instead of $\mathcal{C}$ in these definitions to emphasize that these quantities depend only on the class of functions and not on the representation.

the teaching dimension of $\mathcal{F}$ is

$$\mathrm{TD}(\mathcal{F}) = \max\left\{\mathrm{spec}_{\mathcal{F}}(f) : f \in \mathcal{F}\right\},$$

and the extended teaching dimension of $\mathcal{F}$ is

$$\mathrm{XTD}(\mathcal{F}) = \max(\mathrm{XD}(\mathcal{F}), \mathrm{TD}(\mathcal{F})).$$

It is a remarkable fact that all these and other related dimensions provide both upper and lower bounds for query complexity in the different learning models.

The property we get from $\mathrm{XD}(\mathcal{F})$ is that for any $f \notin \mathcal{F}$ there is always a set of at most $\mathrm{XD}(\mathcal{F}) + 1$ examples certifying that f is not in $\mathcal{F}$. Such a set is called a certificate of exclusion. It turns out that exclusion dimension provides both a lower and an upper bound to the number of membership and proper equivalence queries. Thus, it can be considered to be a combinatorial characterization of this kind of query learning complexity, although the correspondence is not as tight as in the case of PAC learnability and VC-dimension.

Before turning to the characterization, let us note that every concept class $\mathcal{C}$ can be learned with $\log |\mathcal{C}|$ queries using the *halving algorithm* [60]. This algorithm keeps track of all the concepts that are consistent with all the counterexamples received so far. As its next query, it presents a hypothesis which classifies each instance as the *majority* of the consistent concepts. Any counterexample to this hypothesis eliminates at least half of the consistent concepts, implying the query bound. The drawback of this algorithm is that it may have highly improper and intractable hypotheses.

Theorem 7.28. (Hegedűs [49]; Hellerstein et al. [50].) *If a concept class $\mathcal{C}$ has query complexity q using membership and proper equivalence queries, then*

$$\mathrm{XD}(\mathcal{C}) \leq q \leq \mathrm{XD}(\mathcal{C}) \log |\mathcal{C}| \ .$$

Proof. [Proof sketch] The lower bound uses a simple *adversary argument* based on a function $f \notin \mathcal{F}$ having maximal specifying set size.

The upper bound is based on an implementation of the halving algorithm. If the hypothesis of the halving algorithm is proper, then we ask this hypothesis as an equivalence query. Otherwise there are $\mathrm{XD}(\mathcal{C}) + 1$ examples certifying that the hypothesis is not in $\mathcal{C}$. Asking membership queries for all these examples (except the last one), we are guaranteed to get at least one counterexample (or know the last one must be a counterexample), and then we can continue with the halving algorithm. ∎

Note that the algorithm described in Theorem 7.28's proof is not computationally efficient in general. Also, it cannot be applied directly to concept classes such as DNF, where, as we have seen in Section 7.2, the size of the representation needs to be taken into consideration as well. However, the theorem can be generalized to cover these cases as well, as follows.

The concept class $\mathcal{C} = \bigcup_{n=1}^{\infty} \bigcup_{s=1}^{\infty} \mathcal{C}_{n,s}$ has (p, q) *certificates (of exclusion)* for functions $p(n, s)$ and $q(n, s)$ if for every n-variable function $f \in \mathcal{C}$ with

$sc(f) > p(n, s)$ there is a certificate of exclusion of size at most $q(n, s)$ certifying that $f \notin C_{n,s}$. Class C has *polynomial certificates of exclusion* if it has (p, q) certificates for some polynomials p and q. Then the following variant of Theorem 7.28 holds. Recall from Section 7.3 that even if we count only queries, and not running time, the size of the queries must be polynomial in the size of the smallest representation of the target.

Theorem 7.29. (Hellerstein et al. [50].) *A concept class C is polynomial-query learnable with membership and equivalence queries if and only if it has polynomial certificates of exclusion.*

Implicit in Hellerstein et al.'s result is the stronger result stated explicitly by Arias et al. [16], who also provide some examples of specific certificate bounds.

Theorem 7.30. *Let q be a smallest function such that C has (p, q) certificates. Then*

$$\text{Query complexity of } C_{n,s} \text{ using MQ's and EQ's from } C_{n,p(n,s)} \geq q(n, s).$$

Hellerstein and Raghavan [51] used polynomial certificates to show both positive and negative results (Theorems 7.3 and 7.27).

Now we turn our attention to the cases when only one type of query is available, beginning with membership queries. Goldman and Kearns [42] introduced the notion of teaching dimension and observed that if $\mathcal{F}$ can be learned using q membership queries, then $\text{TD}(\mathcal{F}) \leq q$. For the other direction, Hellerstein et al. [50] showed that for a *projection-closed concept class* C (see Definition 7.2 in the next subsection), if the teaching dimension of $C_{n,s}$ is polynomial in n and s, then C is also polynomially learnable using membership queries only. The combinatorial characterization of learnability with membership queries is due to Hegedűs [49]. Building on the work of Moshkov [65], he showed that if only membership queries are available, then in the worst case any learning algorithm has to ask at least $\max(\text{XTD}(\mathcal{F}), \log |\mathcal{F}|)$ queries, and there is a learning algorithm asking at most $2\text{XTD}(\mathcal{F}) \log |\mathcal{F}| / \log \text{XTD}(\mathcal{F})$ queries. Consequently (noting that $\log |C_{n,s}|$ is always at most polynomial in n and s); we have the following theorem.

Theorem 7.31. (Hegedűs [49].) *A concept class C is polynomial-query learnable using membership queries alone if and only if* $\text{XTD}(C_{n,s})$ *is bounded by a polynomial of s and n.*

Proof. [Proof sketch] The "*only if*" part: Assume a learning algorithm for C that uses only membership queries. For any c (either in or not in C), the set of examples queried by that algorithm must define a specifying set of c with respect to C.

The "*if*" part: By definition, having polynomial-sized extended teaching dimension means having polynomial certificates of exclusion; thus, according to Theorem 7.29, C can be learned with polynomial query complexity using membership and equivalence queries. However, as an equivalence query for any $f \in C_{n,s}$ can always be simulated using $\text{spec}_{C_{n,s}}(f)$ many membership queries, and the extended teaching dimension always upper bounds the teaching dimension, the previous algorithm can be simulated by using polynomially many membership queries. ■

Now we turn to the case of equivalence queries alone. The first characterization of this kind actually predates the foregoing results. Angluin showed that the approximate fingerprint property (see later discussion) of a concept class implies that the class is not efficiently query learnable with equivalence queries [6]. A converse for this was proved later by Gavaldà [40].

The pair $(w, b) \in \{0,1\}^n \times \{0,1\}$ is an α-approximate fingerprint with respect to a concept class $\mathcal{F}$, if

$$|\{f \in \mathcal{F} : f(w) = b\}| < \alpha|\mathcal{F}|.$$

A concept class $\mathcal{C}$ has the approximate fingerprint property if there exists a polynomial $p(n)$ such that for any polynomial $q(n)$ there exist concept classes $\mathcal{F}_n \subseteq \mathcal{C}_{n,p(n)}$ with $|\mathcal{F}_n| \geq 2$, and infinitely many $n \in \mathbb{N}$ such that for any $c \in \mathcal{C}_{n,p(n)}$ there exists a string $w \in \{0,1\}^n$ for which $(w, c(w))$ is a $1/q(n)$-approximate fingerprint with respect to $\mathcal{F}_n$.

The idea behind the approximate fingerprint property is that when it holds, an adversary can always give an answer to any equivalence query that eliminates only a small proportion of some "hard core" set of concepts by carefully choosing its answers.

Theorem 7.32. (Angluin [6], Gavaldà [40].) *A concept class is polynomial-query learnable from equivalence queries of polynomial size if and only if it* does not have *the approximate fingerprint property.*

Theorem 7.26 is proved using this technique. Also, using this characterization Lavín and Raghavan have shown negative results in the equivalence query model for decision trees, CDNF formulas, and some other classes [59].

We should mention that although we have restricted ourselves to Boolean functions, many of the original combinatorial results in this subsection are much more general. They covered – among others – deterministic and nondeterministic finite automata, context-free grammars, and geometric concept classes. Notions like the dimensions discussed here appear to be natural in the context of predicate logic as well (see Grohe and Turán [47] and Johnson and Laskowski [52]). More recently, several new related combinatorial parameters have been studied (see, e.g., [19–21]). Balcázar et al. [19] also show that DNF can be learned in polynomial time with proper subset and superset queries.

7.6.2 Attribute-Efficient Learning

In many practical learning situations and also in learning by humans, there is a very large number of variables, but it is expected that only a few of them are *relevant*, that is, actually used in the description of the target, and the rest of them are *irrelevant*.

This suggests a modification of the definition of efficient learnability by requiring a polynomial dependence on the number of relevant variables. Such a modification, in turn, is too restrictive, as intuitively there should be some dependence on the total number of variables as well. The actual definition allows for a logarithmic dependence on the total number of variables. Notice that these considerations apply only to the number of queries, as examples are presented as n-vectors, so the running time cannot be sublinear in the total number of variables.[10]

A learning algorithm is thus called *attribute efficient* if its query complexity is polynomial in r (the number of variables of the target c), in $s(c)$ (the size of the target representation), and in $\log n$, and its running time is polynomial in n and $s(c)$. It is assumed that r, unlike n, is not known to the learning algorithm. A problem C is *attribute efficiently learnable* if it has an attribute efficient learning algorithm.

The study of attribute-efficient learning was initiated by Littlestone [60], presenting the *Winnow* algorithm, a multiplicative variant of the perceptron algorithm, that has been very influential in the development of learning theory. Winnow has the remarkable property of learning some simple concept classes, such as disjunctions or conjunctions, attribute efficiently in the mistake-bounded model of learning. In our terms, this corresponds to learning with threshold function equivalence queries. Additionally, by introducing a new variable for each term of size k, he has also shown how to turn this into an algorithm that learns the class of k-DNF formulas attribute efficiently.[11] The same sort of idea is used in the proof of Theorem 7.21.

Obviously attribute efficient learnability implies efficient learnability; the interesting direction is thus the converse. The first result of this type is due to Blum et al. [26] for the model using both types of queries. Before we state their result, some notions need to be defined. We should also mention that they originally used the mistake-bounded model (with membership queries), but the proof goes through just as well for the query model.

Definition 7.2. *A concept class C is embedding closed, if for every n, every $k < n$, and every $c \in C_k$, concept c' defined by $c'(x_1, \ldots, x_n) = c(x_1, \ldots, x_k)$ is contained in C_n. Concept class C is projection closed, if for every n, every $i < n$, and every $c \in C_n$, the $(n-1)$-variable concept obtained by fixing c's ith input to either 0 or 1 is contained in C_{n-1}.*

These notions are quite natural: the class of DNFs, k-DNFs, k-term DNFs, read-once formulas, decision lists, and CNFs all fulfill the foregoing conditions. On the other hand, there are many examples that do not: the class of majority functions[12] over some subset of the variables is not projection closed, the class

[10] Models of attribute-efficient learning have also been extended to the case of learning with *infinitely many* attributes, assuming that in every instance only finitely many variables are true, and the instance is represented by this set of variables [26].

[11] The algorithm is exponential in k, however.

[12] The majority function returns 1 if at least half of its variables are true, 0 otherwise.

of functions that have exactly one satisfying assignment is not embedding closed, and so forth.

The following theorem provides a *meta-algorithm*, which converts every polynomial-time membership and equivalence query learning algorithm into an attribute-efficient one. We have already seen such a meta-algorithm, translating from equivalence queries to the PAC model, at the end of Section 7.3, and we will see some more later on. Such algorithms are of interest in learning theory as they elucidate the relationship between different learning models and protocols.

Theorem 7.33. (Blum et al. [26].) *Let C be an embedding and projection closed concept class. If C is polynomial-time learnable from membership and equivalence queries, then C is also attribute-efficiently learnable.*

Proof. [Proof sketch] Iterate the following. Fix those variables to 1 that are not known to be relevant, and run the original learning algorithm using the rest of the variables. Each time it asks a membership query for an example, extend it by placing 1s in the missing components. If it asks an equivalence query, forward it. In case of receiving a counterexample, switch on all its "irrelevant" bits and ask a membership query for it. If the target evaluates it to the same, forward it to the learning algorithm; otherwise there is a relevant variable that needs to be uncovered. Find it via a kind of binary search using $O(\log n)$ membership queries, and restart. ■

Corollary 7.34. *Monotone DNF, Horn formulas, decision lists, threshold functions, and read-once formulas are all attribute efficiently learnable.*

Let us now turn to learnability with membership queries. It was shown in [26] that embedding and projection closed classes of *monotone* Boolean functions that are polynomial-time learnable using only membership queries, can also be learned attribute efficiently using only membership queries. Unfortunately, the monotonicity criterion cannot be dropped: as noted by Blum et al. [26], the class of parity functions cannot be learned attribute-efficiently by a deterministic algorithm (meanwhile this class is both embedding and projection closed). On the other hand, as Bshouty and Hellerstein have discovered, in a slightly strengthened model the monotonicity criterion is not needed any more.[13]

Theorem 7.35. (Bshouty and Hellerstein [29].) *Let C be an embedding and projection-closed concept class. If C is efficiently learnable using membership queries alone, then C is also attribute-efficiently learnable using membership queries alone, if the number of relevant variables is revealed to the learner in advance.*

[13] Note that the doubling trick cannot help when the learner has access only to membership queries, as there is no way to check the correctness of the current hypothesis. This is why a small amount of additional information can make such a significant difference.

7.6.3 Revision

Theory revision is concerned with the situation where the learner starts with a "good approximation" of a function. It is argued that in many machine learning applications, having an initial approximation of the target is necessary for successful learning (Wrobel [82]). We begin by discussing what is meant by a "good approximation" of a function. In theory revision the learner manipulates its current hypotheses using simple syntactic operations that are usually fixed/predefined by the given model; accordingly, the notion of closeness is based on these operations as well. These revision operators are typically fixing an occurrence of a variable to some constant (deletion model[14]), addition of a variable (addition model), or both (deletion and addition model). The last is sometimes referred to as the general model. The revision distance of an initial formula φ from the target is the minimal number of revision operators needed to apply on the initial formula to obtain a representation for the target. (Note that this distance is not necessarily symmetric.) If such a revision is not possible then the revision distance is defined to be infinite. An (efficient) revision algorithm for an initial formula φ is a learning algorithm with the extra input φ that runs in time polynomial in the revision distance, in the size of φ, and in n, and the number of queries asked is polynomial in the revision distance and in $\log n$. Note how strongly the learning task depends on the initial hypotheses.

Obviously revision implies learnability (and even attribute efficient learnability), but unfortunately – as opposed to the notions of the previous sections – there are no general results for the other direction. In fact, the following results indicate that we cannot expect results in that direction:

- When both types of queries are available to the learner, read-once formulas can be learned efficiently [12] and can also be revised efficiently in the deletion-only model [45], but for addition, it is not even clear what the right model should be.
- When both types of queries are available to the learner, Horn formulas (resp., monotone DNF formulas) can be learned efficiently [5, 11], but the revision problem of finding one deletion in an n-clause (resp., n-term) formula has query complexity $\Omega(n)$ [45, 46]. (See also Theorem 7.38.)
- Threshold functions can be learned using membership queries only, but in case of theory revision, both query types are needed for the query complexity to be polynomial in $\log n$ [44].

In the rest of this subsection we present positive results for different concept classes due to Goldsmith et al. [44–46]. The first two theorems summarize most of the known results for the query model.

[14] Note that there is an equivalence between deleting a subformula and fixing the value of a variable (or more, if necessary).

For DNF formulas an addition operation is the addition of a literal to a term, and for Horn formulas it is the addition of a literal to a clause. For the class of unate DNF formulas, the only restriction in the general model is that the complement of a deleted variable cannot be added. The first theorem contains revision algorithms for the general model of deletions and additions, and the second theorem contains results for the deletion model. The revision algorithms for Horn formulas use CNF hypotheses with each clause being a revision of a clause from the initial formula, but one initial clause can have multiple revisions in a hypothesis.

The *graph* of a Horn formula has the variables as its vertices and has an edge from every variable in the body of a rule to the variable in the head with extra vertices T and F defined in the natural way. (See Boros' chapter in [35].) A Horn formula is *depth-1 acyclic* if its graph is acyclic and has depth 1. A Horn formula is *definite with unique heads* if the heads of the clauses exist and are all different.

Theorem 7.36. (Goldsmith et al. [44–46].) *In the deletion and addition model of revisions, the following classes can be revised efficiently: monotone DNF formulas with a constant number of terms; 2-term unate DNF formulas; depth-1 acyclic Horn formulas with a constant number of terms; Horn formulas with unique head with a constant number of terms; and 0–1 threshold functions.*

Theorem 7.37. (Goldsmith et al. [45].) *In the deletion model of revisions, (general) Horn formulas with a constant number of terms and read-once formulas can be revised efficiently.*

An open problem is the revision of Horn formulas with a constant number of terms in the deletion and addition model of revisions.

We also give here one negative result for revision. Consider the variables $x_1, \ldots, x_n, y_1, \ldots, y_n$, and let

$$t_i = x_1 \wedge \cdots \wedge x_{i-1} \wedge x_{i+1} \wedge \cdots \wedge x_n \wedge y_i$$

and $\varphi_n = t_1 \vee \cdots \vee t_n$.

Theorem 7.38. (Goldsmith et al. [46].) *In the deletion model of revisions at least $n - 1$ queries are needed to revise φ_n, even if it is known that exactly one literal y_i is deleted.*

Theorems 7.36 and 7.37 show that in the deletion model a monotone DNF can be revised efficiently, if it has small terms, few terms, or few occurrences of each variable (applied to read-once DNF). Theorem 7.38, on the other hand, shows that if neither of these conditions holds, then efficient revision is not always possible.

7.6.4 Noise

There have been efforts to extend the model of query learning by incorporating a notion of *noise*. Noise has been modeled primarily by looking at alternate membership oracles that do not always return the correct value of $f(x)$ given vector x. We mostly discuss noisy membership queries, and at the end one recent result concerning noisy equivalence queries.

Sakakibara [73] proposed a model where membership queries receive the wrong answer at random, but the noise is not persistent for any one instance, so repeated queries can overcome that noise.

Angluin and Slonim introduced a model – incomplete membership queries – where membership queries receive a persistent "I don't know," response on randomly chosen instances [14]. In their model, the "I don't know" answers are determined by independent coin flips the first time each query is made. They give a polynomial-time algorithm to learn monotone DNF formulas with high probability in this setting. They also show that a variant of this algorithm can deal with one-sided errors, assuming that no negative point is classified as positive. There have been relatively few other positive results for this model; results have been given for k-term DNF [43] and regular sets [32], and for monotone DNF in a somewhat harder variation of this model [30].

Blum et al. [25] pointed out that "I don't know" answers are *not* likely to occur in a uniform random way; rather, in practice, "I don't know" answers are more likely to occur on instances that are syntactically close to the boundary between positive and negative (e.g., vectors x having small Hamming distance from some vector x' such that $f(x) \neq f(x')$, where f is the target function). They formulated models where the answers to the membership queries are all correct except in a neighborhood of a fixed radius of the boundary of the target concept. In the neighborhood of the boundary, the answers to the membership queries are assumed to be either correct or "don't know" (the *incomplete boundary query* (IBQ) model), or they are assumed to be either correct or incorrect (the *unreliable boundary query* (UBQ) model). These models are extended to allow equivalence queries by requiring that the counterexamples be chosen from outside the boundary region. They considered learning subclasses of monotone DNF in a modified version of the UBQ model with only one-sided error. Notice that every learning algorithm in the UBQ model or the one-sided UBQ model can run in the IBQ model by changing every "don't know" answer it receives to 1. Thus some subclasses of monotone DNF are efficiently learnable in the IBQ model. Sloan and Turán [79] gave an algorithm for learning threshold functions in the IBQ model, which makes use of split graphs and their generalization to split hypergraphs.

Angluin et al. [13] proposed learning from a membership oracle that could make either "I don't know" *omission* errors or *malicious* errors where the wrong answer is given. In either case, the instances in which the "I don't know" or wrong response is given are chosen by an adversary with unlimited computing power and knowledge of the learning algorithm. Many of their results are based on the notion of *finite exceptions*: the representation of a concept c from the class of functions $\mathcal{F}$ with finite exceptions consists of a representation c' of some $f \in \mathcal{F}$ together with a list of vectors on which c and c' disagree.

Angluin et al. studied the case where proper equivalence queries are used. Among other results, they showed the following theorem.

Theorem 7.39. (Angluin et al. [13].) *The class monotone DNF with finite exceptions can be learned from malicious membership and proper equivalence queries in time polynomial in n, the size of the target, and the total number of wrong answers received.*

Recently Bshouty et al. [24] obtained a general positive result for the case where we drop the requirement of proper equivalence queries. They also tolerate *malicious equivalence queries* where such a query can return a wrong counterexample (a vector that is not a counterexample).

Theorem 7.40. (Bshouty et al. [24].) *If a class is closed under projection and is learnable in polynomial time using standard membership and equivalence queries, then it is also learnable from malicious membership and malicious equivalence queries in time polynomial in n, the time of the original algorithm, and the total number of wrong answers received.*

7.7 Future Outlook and Conclusions

In this final section, we first very briefly discuss several additional models of machine learning that are all currently active areas, and all somehow related to query learning of Boolean functions, and then make a few concluding remarks.

7.7.1 Additional Related Learning Models

Value-Injection Circuit Model

Recently a new query model of learning something closely related to but distinct from traditional Boolean functions has been introduced in a series of papers by Angluin et al. [8–10]. In this model, the target function is specified by an unknown acyclic Boolean circuit, and the learner is given the number of *wires* that the circuit contains. However, from the learner's point of view, this circuit does not have distinguished inputs. The inputs, and hence every wire, has a default value.

The learner may make *value-injection queries*, which in this model play a role akin to membership queries in the traditional query model. In a value-injection query, the learner specifies a set of (wire, Boolean value) pairs, and those values override the values that would otherwise be on the wires. The learner receives back the output value of the circuit for this case. Additionally, the learner may ask an equivalence query. Now two circuits are considered equivalent if they would behave the same on all value-injection queries. This model can also be seen as asking a twenty-first-century version of the question, "Into what sort of subpieces must we break a circuit for a Boolean function so that very wide classes of Boolean functions are learnable?" which was studied much earlier by Rivest and Sloan [71]. The circuit-injection model is studied both because it is mathematically interesting in its own right, and because these circuits (suitably generalized) model gene regulatory models.

Statistical Queries

We mention Kearns's statistical query (SQ) model of learning [53] because it is an important theoretical machine learning model that has the word *query* in its name. However, it is in fact concerned with learning in the probabilistic, PAC-style setting, with important applications to learning in the presence of noise. This model is a mild restriction of the PAC model where instead of asking for labeled examples, the learner asks *directly* for various statistics over the distribution of labeled examples. The key initial result in this area is that the SQ oracle could be simulated by a PAC source of labeled examples *degraded by random classification noise* [53]. A more detailed description of the model and early results with it can be found in Kearns and Vazirani's textbook [54]. Recent results include [36, 76, 77, 83].

Active Learning

There are learning applications, for example in speech recognition and computer vision, where it is inexpensive to get unlabeled examples, but labeling an example involves a human and is therefore expensive. Selecting those examples that are important to be labeled can thus be viewed as a kind of membership query, and the objective is to minimize their number. This version of learning is called *active learning* (see, e.g., [18] and further references in that paper).

Clustering

Balcan and Blum [17] recently introduced models for interactions between a user and a *clustering* algorithm. Some of the problems studied in this framework are related to Boolean function learning problems. In particular, in the basic model, the clustering algorithm proposes a clustering, which the user may either accept or respond to with either a request to merge certain clusters or split a cluster. This can be viewed as a form of equivalence query that returns different information than the usual counterexample.

Property Testing

Property testing is a huge and very active area of theoretical computer science, which is somewhat related to PAC learning with membership queries. The goal of property testing is not to learn an unknown target, but to decide if the unknown target belongs to a specific class or is *far from it*. Being far means that many entries of the target have to be changed in order to get an object belonging to the specific class. If the target is in the class, then it has to be accepted. If it is far from the class, then it has to be rejected with some fixed probability. In the remaining case, when the target is not in the class but is close to it, there is *no* requirement for the algorithm.

Property testing has been mostly studied for graphs, but there are several results on Boolean functions as well. For example, testing the monotonicity of a

Boolean function under the uniform distribution is a well-studied problem. Consider the following testing algorithm: pick randomly a set of edges (x_i, y_i) of the n-dimensional hypercube $\{0, 1\}^n$ with $x_i < y_i$, $i = 1, \ldots, \ell$, in the componentwise ordering, and accept if $f(x_i) \leq f(y_i)$ for every i. If the target *is* monotone, then it will always be accepted. On the other hand, if the target is "far" from being monotone, then it is expected that there are many edges failing the test, and it is likely that at least one of them will be picked. Note that the testing algorithm uses both random sampling *and* (randomized) membership queries: the uniform distribution provides the x_is, but then we have to query a (random) neighbor. We refer to Glasner and Servedio [41] and to Ron [72] for surveys of property testing for Boolean functions and to Bhattacharya [23] for a short analysis of the case of monotonicity.

7.7.2 Final Remarks

In this chapter we have presented a brief overview of results on learning Boolean functions with queries. Boolean functions play a prominent role in query learning. DNF, read-once functions, and other Boolean classes provide some of the most challenging and most studied learning problems. In the other direction, query learning provides interesting notions, techniques, and problems for the theory of Boolean functions. The exclusion dimension, the monotone theory, and the consistency and representation problems we discussed are all worthy of study in their own right by the Boolean function community independent of learning theory.

Although many types of query models have been considered in the literature, it turned out that learning with membership and equivalence queries is perhaps the most important model, as it leads to many efficient algorithms that require both type of queries, and it serves as a useful framework to develop algorithms that use random examples and membership queries. Therefore, we have restricted our attention to this model.

Query learning is the first, and perhaps most studied, version of learning situations, where the learner interacts with the environment in a more active manner than just by taking a random sample. Such an interaction is not always possible, but when it is, one can try to make use of it. This may not be an easy matter: experiments show that learning algorithms may try to use membership queries for instances that are on the boundary of the target concept, and such queries may be hard to answer for a human expert [58]. These difficulties notwithstanding, the idea of using interaction in learning is a persistent one, for which query learning provides a foundational mathematical model.

References

[1] H. Aizenstein, T. Hegedűs, L. Hellerstein, and L. Pitt. Complexity theoretic results for query learning. *Computational Complexity* 7, pp. 19–53, 1998.

[2] M. Alekhnovich, M. Braverman, V. Feldman, A. R. Klivans, and T. Pitassi. The complexity of properly learning simple concept classes. *Journal of Computer and System Sciences* 74(1), pp. 16–34, 2008.
[3] E. Allender, L. Hellerstein, P. McCabe, T. Pitassi, and M. Saks. Minimizing Disjunctive Normal Form formulas and AC^0 circuits given a truth table. *SIAM Journal on Computing* 38, pp. 63–84, 2008.
[4] D. Angluin. Learning k-term DNF formulas using queries and counterexamples. Technical Report YALEU/DCS/RR-559, Department of Computer Science, Yale University, Aug. 1987.
[5] D. Angluin. Queries and concept learning. *Machine Learning* 2(4), pp. 319–42, 1988.
[6] D. Angluin. Negative results for equivalence queries. *Machine Learning* 5, pp. 121–50, 1990.
[7] D. Angluin. Queries revisited. *Theoretical Computer Science*, 313(2), pp. 175–94, 2004. Special issue for ALT 2001.
[8] D. Angluin, J. Aspnes, J. Chen, D. Eisenstat, and L. Reyzin. Learning acyclic probabilistic circuits using test paths. In *21st Annual Conference on Learning Theory – COLT 2008, Helsinki, Finland, July 9–12, 2008*, pp. 169–80. Omnipress, 2008.
[9] D. Angluin, J. Aspnes, J. Chen, and L. Reyzin. Learning large-alphabet and analog circuits with value injection queries. In *Learning Theory, 20th Annual Conference on Learning Theory, COLT 2007, San Diego, CA, June 13–15, 2007. Proceedings, Lecture Notes in Artificial Intelligence* 4539, pp. 51–65, 2007.
[10] D. Angluin, J. Aspnes, J. Chen, and Y. Wu. Learning a circuit by injecting values. *Journal of Computer and System Sciences* 75(1), pp. 60–77, 2009.
[11] D. Angluin, M. Frazier, and L. Pitt. Learning conjunctions of Horn clauses. *Machine Learning* 9, pp. 147–64, 1992.
[12] D. Angluin, L. Hellerstein, and M. Karpinski. Learning read-once formulas with queries. *Journal of the ACM* 40(1), pp. 185–210, 1993.
[13] D. Angluin, M. Kriķis, R. H. Sloan, and G. Turán. Malicious omissions and errors in answers to membership queries. *Machine Learning* 28(2–3), pp. 211–55, 1997.
[14] D. Angluin and D. K. Slonim. Randomly fallible teachers: Learning monotone DNF with an incomplete membership oracle. *Machine Learning* 14, pp. 7–26, 1994.
[15] M. Arias and J. L. Balcázar. Query learning and certificates in lattices. In *ALT, Lecture Notes in Computer Science* 5254, pp. 303–15. Springer, 2008.
[16] M. Arias, A. Feigelson, R. Khardon, and R. A. Servedio. Polynomial certificates for propositional classes. *Information and Computation* 204(5), pp. 816–34, 2006.
[17] M.-F. Balcan and A. Blum. Clustering with interactive feedback. In *Algorithmic Learning Theory, 19th International Conference, ALT 2008, Budapest, Hungary, October 2008*. Proceedings, of *Lecture Notes in Artificial Intelligence* 5254, pp. 316–28, 2008.
[18] M.-F. Balcan, S. Hanneke, and J. Wortman. The true sample complexity of active learning. In *21st Annual Conference on Learning Theory – COLT 2008, Helsinki, Finland, July 9–12, 2008*, pp. 45–56. Omnipress, 2008.
[19] J. Balcázar, J. Castro, and D. Guijarro. A new abstract combinatorial dimension for exact learning via queries. *Journal of Computer and System Sciences* 64(1), pp. 2–21, 2002. Special issue for COLT 2000.
[20] J. Balcázar, J. Castro, D. Guijarro, J. Köbler, and W. Lindner. A general dimension for query learning. *Journal of Computer and System Sciences* 73(6), pp. 924–40, 2007.
[21] J. L. Balcázar, J. Castro, D. Guijarro, and H.-U. Simon. The consistency dimension and distribution-dependendent learning from queries. *Theoretical Computer Science* 288(2), pp. 197–215, 2002. Special issue for ALT '99.
[22] U. Berggren. Linear time deterministic learning of k-term DNF. In *Proceedings of the Sixth Annual ACM Conference on Computational Learning Theory*, pp. 37–40, ACM Press, New York, 1993.
[23] A. Bhattacharyya. A note on the distance to monotonicity of Boolean functions. Technical Report TR08-012, Electronic Colloquium on Computational Complexity Report, 2008.

[24] L. Bisht, N. H. Bshouty, and L. Khoury. Learning with errors in answers to membership queries. *Journal of Computer and System Sciences* 74(1), pp. 2–15, 2008.

[25] A. Blum, P. Chalasani, S. A. Goldman, and D. K. Slonim. Learning with unreliable boundary queries. *Journal of Computer and System Sciences* 56(2), pp. 209–22, 1998. Special issue: Eighth Annual Conference on Computational Learning Theory.

[26] A. Blum, L. Hellerstein, and N. Littlestone. Learning in the presence of finitely or infinitely many irrelevant attributes. *Journal of Computer and System Sciences* 50(1), pp. 32–40, 1995. Earlier version in 4th COLT, 1991.

[27] A. Blum and S. Rudich. Fast learning of k-term DNF formulas with queries. *Journal of Computer and System Sciences* 51(3), pp. 367–73, 1995.

[28] N. Bshouty. Exact learning Boolean function via the monotone theory. *Information and Computation* 123, pp. 146–53, 1995.

[29] N. Bshouty and L. Hellerstein. Attribute-efficient learning in query and mistake-bound models. *Journal of Computer and System Sciences* 56(3), pp. 310–19, 1998.

[30] N. H. Bshouty and N. Eiron. Learning monotone DNF from a teacher that almost does not answer membership queries. *Journal of Machine Learning Research* 3, pp. 49–57, 2002.

[31] N. H. Bshouty, S. A. Goldman, T. R. Hancock, and S. Matar. Asking questions to minimize errors. *Journal of Computer and System Sciences* 52(2), pp. 268–86, 1996. Earlier version in 6th COLT, 1993.

[32] N. H. Bshouty and A. Owshanko. Learning regular sets with an incomplete membership oracle. In *14th Annual Conference on Computational Learning Theory, COLT 2001 and 5th European Conference on Computational Learning Theory, EuroCOLT 2001, Amsterdam, The Netherlands, July 2001*. Proceedings, *Lecture Notes in Artificial Intelligence* 2111, pp. 574–88, 2001.

[33] J. Castro and J. L. Balcázar. Simple PAC learning of simple decision lists. In *Algorithmic Learning Theory, 6th International Workshop, ALT '95, Fukuoka, Japan, October 18–20, 1995*. Proceedings, *Lecture Notes in Artificial Intelligence* 997, pp. 239–48, 1995.

[34] A. K. Chandra and G. Markowsky. On the number of prime implicants. *Discrete Mathematics* 24, pp. 7–11, 1978.

[35] Y. Crama and P. L. Hammer. *Boolean Functions: Theory, Algorithms, and Applications.* Cambridge University Press, Cambridge, UK, 2010.

[36] V. Feldman. Evolvability from learning algorithms. In *STOC '08: Proceedings of the 40th Annual ACM Symposium on Theory of Computing*, pp. 619–28, 2008.

[37] V. Feldman. Hardness of approximate two-level logic minimization and PAC learning with membership queries. *Journal of Computer and System Sciences* 75(1), pp. 13–26, 2009.

[38] M. Frazier. *Matters Horn and other features in the computational learning theory landscape: The notion of membership*. PhD thesis, Department of Computer Science, University of Illinois at Urbana-Champaign, 1994. Technical report UIUCDCS-R-94-1858.

[39] M. Frazier and L. Pitt. Learning from entailment: An application to propositional Horn sentences. In *Proceedings of the Tenth International Conference of Machine Learning*, pp. 120–27. Morgan Kaufmann, June 1993.

[40] R. Gavaldà. On the power of equivalence queries. In *Computational Learning Theory: EuroColt '93*, Vol. New Series 53 of *The Institute of Mathematics and Its Applications Conference Series*, pp. 193–203. Oxford University Press, Oxford, 1994.

[41] D. Glasner and R. Servedio. Distribution-free testing lower bounds for basic Boolean functions. In *Proceedings of the 10th International Workshop on Approximation and the 11th International Workshop on Randomization, and Combinatorial Optimization. Algorithms and Techniques*, pp. 494–508. Springer-Verlag, Berlin, 2007.

[42] S. A. Goldman and M. J. Kearns. On the complexity of teaching. *Journal of Computer and System Sciences* 50(1), pp. 20–31, 1995. Earlier version in 4th COLT, 1991.

[43] S. A. Goldman and H. D. Mathias. Learning k-term DNF formulas with an incomplete membership oracle. In *Proceedings of the 5th Annual ACM Workshop on Computer Learning Theory*, pp. 77–84, ACM Press, New York, 1992.

[44] J. Goldsmith, R. H. Sloan, B. Szörényi, and G. Turán. New revision algorithms. In *Algorithmic Learning Theory, 15th International Conference, ALT 2004, Padova, Italy, October 2004. Proceedings, of Lecture Notes in Artificial Intelligence* 3244, pp. 395–409, 2004.

[45] J. Goldsmith, R. H. Sloan, B. Szörényi, and G. Turán. Theory revision with queries: Horn, read-once, and parity formulas. *Artificial Intelligence* 156, pp. 139–76, 2004.

[46] J. Goldsmith, R. H. Sloan, and G. Turán. Theory revision with queries: DNF formulas. *Machine Learning* 47(2–3), pp. 257–95, 2002.

[47] M. Grohe and G. Turán. Learnability and definability in trees and similar structures. *Theory of Computing Systems* 37, pp. 193–220, 2004.

[48] T. Hegedűs. On training simple neural networks and small-weight neurons. In *Computational Learning Theory: EuroColt '93*, Vol. New Series 53 of *The Institute of Mathematics and Its Applications Conference Series*, pp. 69–82, Oxford University Press, Oxford, 1994.

[49] T. Hegedűs. Generalized teaching dimensions and the query complexity of learning. In *Proceedings of the 8th Annual Conference on Computer Learning Theory*, pp. 108–17, ACM Press, New York, 1995.

[50] L. Hellerstein, K. Pillaipakkamnatt, V. Raghavan, and D. Wilkins. How many queries are needed to learn? *Journal of the ACM* 43(5), pp. 840–62, 1996.

[51] L. Hellerstein and V. Raghavan. Exact learning of DNF formulas using DNF hypotheses. *Journal of Computer and System Sciences* 70(4), pp. 435–70, 2005. Special issue on COLT 2002.

[52] H. R. Johnson and M. C. Laskowski. Compression schemes, stable definable families, and o-minimal structures. *Discrete Computational Geometry*. To appear. Available at http://www.math.umd.edu/~mcl/.

[53] M. Kearns. Efficient noise-tolerant learning from statistical queries. In *Proceedings 25th Annual ACM Symposis Theory Computer*, pp. 392–401, ACM Press, New York, 1993.

[54] M. J. Kearns and U. V. Vazirani. *An Introduction to Computational Learning Theory*. MIT Press, Cambridge, MA, 1994.

[55] R. Khardon and D. Roth. Learning to reason. *Journal of the ACM* 44(5), pp. 697–725, 1997.

[56] M. Kharitonov. Cryptographic hardness of distribution-specific learning. In *Proceedings 25th Annual ACM Symposis Theory Computer*, pp. 372–81, ACM Press, New York, 1993.

[57] E. Kushilevitz. A simple algorithm for learning $O(\log n)$-term DNF. In *Proceedings 9th Annual Conference on Computer Learning Theory*, pp. 266–269, ACM Press, New York, 1996.

[58] K. J. Lang and E. B. Baum. Query learning can work poorly when a human oracle is used. In *International Joint Conference on Neural Networks*, Beijing, 1992.

[59] V. Lavín and V. Raghavan. Decision trees have approximate fingerprints. Technical Report NC-TR-97-016, NeuroCOLT, 1997.

[60] N. Littlestone. Learning quickly when irrelevant attributes abound: A new linear-threshold algorithm. *Machine Learning* 2(4), pp. 285–318, 1988.

[61] W. Maass and G. Turán. On the complexity of learning from counterexamples. In *Proceedings of the 30th Annual IEEE Symposium on Foundations of Computer Science*, pp. 262–7. IEEE Computer Society Press, Los Alamitos, CA, 1989.

[62] W. Maass and G. Turán. How fast can a threshold gate learn? In S. J. Hanson, G. A. Drastal, and R. L. Rivest, eds., Chapter 13, pp. 381–414. MIT Press, Cambridge, MA, 1994. Earlier versions appeared in FOCS89 and FOCS90.

[63] E. W. Mayr, H. J. Prömel, and A. Steger, eds. *Lectures on Proof Verification and Approximation Algorithms*. Vol. 1367 of *Lecture Notes in Computer Science*. Springer, Berlin, 1998.

[64] T. M. Mitchell. Version spaces: A candidate elimination approach to rule learning. In *Proc. IJCAI-77*, pp. 305–10, Cambridge, MA, Aug. 1977. International Joint Committee for Artificial Intelligence.

[65] M. Y. Moshkov. Conditional tests. *Problemy Kibernetiki* 40, 1983. (In Russian).

[66] S. Muroga. *Threshold Logic and Its Applications*. John Wiley & Sons, New York, 1971.

[67] M. Naor, L. J. Schulman, and A. Srinivasan. Splitters and near-optimal derandomization. In *Proceedings 36th Annual IEEE Symposium Foundation Computer Science*, pp. 182–93, 1995.

[68] K. Pillaipakkamnatt and V. Raghavan. Read-twice DNF formulas are properly learnable. *Information and Computation* 122(2), pp. 236–67, 1995.

[69] K. Pillaipakkamnatt and V. Raghavan. On the limits of proper learnability of subclasses of DNF formulas. *Machine Learning* 25, pp. 237–63, 1996.

[70] L. Pitt and L. G. Valiant. Computational limitations on learning from examples. *Journal of the ACM* 35(4), pp. 965–84, 1988.

[71] R. L. Rivest and R. Sloan. A formal model of hierarchical concept learning. *Information and Computation* 114, pp. 88–114, 1994.

[72] D. Ron. Property testing: A learning theory perspective. Retrieved May 13, 2009, from http://www.eng.tau.ac.il/~danar/papers.html.

[73] Y. Sakakibara. On learning from queries and counterexamples in the presence of noise. *Information Processing Letters* 37(5), pp. 279–84, 1991.

[74] K. Satoh. Analysis of case-based representability of Boolean functions by monotone theory. In *Algorithmic Learning Theory, 9th International Conference, ALT '98, Otzenhausen, Germany, October 1998, Proceedings*, *Lecture Notes in Artificial Intelligence* 1501, pp. 179–90. 1998.

[75] H. U. Simon. Learning decision lists and trees with equivalence-queries. In *Computational Learning Theory, Second European Conference, EuroCOLT '95, Barcelona, Spain, March 1995, Proceedings*, Lecture Notes in Artificial Intelligence 904, pp. 322–36, 1995.

[76] H. U. Simon. Spectral norm in learning theory: Some selected topics. In *Algorithmic Learning Theory, 17th International Conference, ALT 2006, Barcelona, Spain, October 2006, Proceedings*, Lecture Notes in Artificial Intelligence 4264, pp. 13–27, 2006.

[77] H. U. Simon. A characterization of strong learnability in the statistical query model. *STACS 2007*, pp. 393–404, 2007.

[78] R. H. Sloan, B. Szörényi, and G. Turán. On k-term DNF with the maximal number of prime implicants *SIAM Journal on Discrete Mathematics* 21(4), pp. 987–98, 2008.

[79] R. H. Sloan and G. Turán. Learning from incomplete boundary queries using split graphs and hypergraphs. In *Computational Learning Theory, Third European Conference, EuroCOLT '97, Jerusalem, Israel, March 1997, Proceedings*, Lecture Notes in Artificial Intelligence 1208, pp. 38–50, 1997.

[80] P. Vaidya. A new algorithm for minimizing convex functions over convex sets. *Mathematical Programming* 73(3), pp. 291–341, 1996.

[81] L. G. Valiant. A theory of the learnable. *Communications of the ACM* 27(11), pp. 1134–42, 1984.

[82] S. Wrobel. *Concept Formation and Knowledge Revision*. Kluwer, Dordrecht, 1994.

[83] K. Yang. New lower bounds for statistical query learning. *Journal of Computer and System Sciences* 70(4), pp. 485–509, 2005. Special issue on COLT 2002.

8

Boolean Functions for Cryptography and Error-Correcting Codes

Claude Carlet

8.1 Introduction

A fundamental objective of *cryptography* is to enable two persons to communicate over an insecure channel (a public channel such as the internet) in such a way that any other person is unable to recover their message (called the *plaintext*) from what is sent in its place over the channel (the *ciphertext*). The transformation of the plaintext into the ciphertext is called *encryption*, or enciphering. Encryption-decryption is the most ancient cryptographic activity (ciphers already existed four centuries B.C.), but its nature has deeply changed with the invention of computers, because the *cryptanalysis* (the activity of the third person, the eavesdropper, who aims at recovering the message) can use their power.

The encryption algorithm takes as input the plaintext and an encryption key K_E, and it outputs the ciphertext. If the encryption key is secret, then we speak of *conventional cryptography*, of *private key cryptography*, or of *symmetric cryptography*. In practice, the principle of conventional cryptography relies on the sharing of a private key between the sender of a message (often called Alice in cryptography) and its receiver (often called Bob). If, on the contrary, the encryption key is public, then we speak of *public key cryptography*. Public key cryptography appeared in the literature in the late 1970s.

The *decryption* (or deciphering) algorithm takes as input the ciphertext and a secret[1] decryption key K_D. It outputs the plaintext.

[1] According to principles stated in 1883 by A. Kerckhoffs [212], who cited a still more ancient manuscript by R. du Carlet [49], only the secret keys must be kept secret – the confidentiality should not rely on the secrecy of the encryption method – and a cipher cannot be considered secure if it can be decrypted by the designer himself.

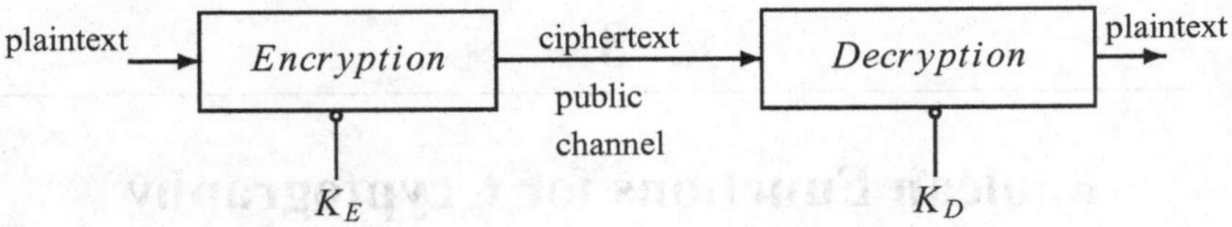

Public key cryptography is preferable to conventional cryptography, because it allows secure communication without having previously shared keys in a secure way: every person who wants to receive secret messages can keep secret a decryption key and publish an encryption key; if n persons want to secretly communicate pairwise using a public key cryptosystem, they need n encryption keys and n decryption keys, where as conventional cryptosystems will need $\binom{n}{2} = \frac{n(n-1)}{2}$ keys. But all known public key cryptosystems are much less efficient than conventional cryptosystems (they allow a much lower data throughput), and they also need much longer keys to ensure the same level of security. This is why conventional cryptography is still widely used and studied nowadays. Thanks to public key cryptosystems, the share-out of the necessary secret keys can be done without using a secure channel (the secret keys for conventional cryptosystems are strings of only a few hundred bits and can then be encrypted by public key cryptosystems). Protocols specially devoted to key exchange can also be used.

The objective of *error correcting codes* is to enable digital communication over a noisy channel in such a way that the errors in the transmission of bits can be detected[2] and corrected by the receiver. This aim is achieved by using an encoding algorithm that transforms the information before sending it over the channel. In the case of block coding,[3] the original message is treated as a list of binary words (vectors) of the same length – say k – that are encoded into *codewords* of a larger length – say n. Thanks to this extension of the length, called *redundancy*, the decoding algorithm can correct the errors of transmission (if their number is, for each sent word, less than or equal to the so-called correction capacity of the code) and recover the correct message. The set of all possible codewords is called the *code*. Sending words of length n over the channel instead of words of length k slows down the transmission of information in the ratio of $\frac{k}{n}$. This ratio, called the *transmission rate*, must be as high as possible, to allow fast communication.

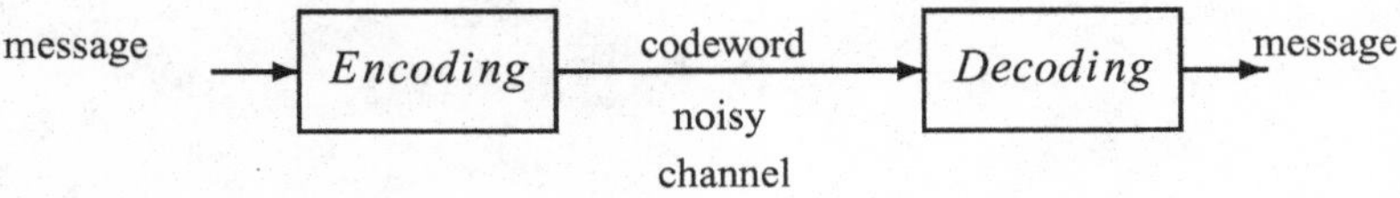

[2] If the code is used only to detect errors, then when an error is detected, the information must be requested and sent again in a so-called "automatic request" procedure.

[3] We shall not address convolutional coding here.

Table 8.1. Number of n-variable Boolean functions

n	4	5	6	7	8
$\|\mathcal{BF}_n\|$	2^{16}	2^{32}	2^{64}	2^{128}	2^{256}
$\approx$	$6 \cdot 10^4$	$4 \cdot 10^9$	10^{19}	10^{38}	10^{77}

In both cryptographic and error correcting coding activities, *Boolean functions* (that is, functions from the vectorspace $\mathbb{F}_2^n$ of all binary vectors of length n, to the finite field with two elements[4] $\mathbb{F}_2$) play roles:

- Every code of length 2^n, for some positive integer n, can be interpreted as a set of Boolean functions, because every n-variable Boolean function can be represented by its truth table (an ordering of the set of binary vectors of length n being first chosen) and thus associated with a binary word of length 2^n, and *vice versa*. Important codes (Reed-Muller, Kerdock codes) can be defined this way as sets of Boolean functions.
- The role of Boolean functions in conventional cryptography is even more important: cryptographic transformations (pseudorandom generators in stream ciphers, S-boxes in block ciphers) can be designed by appropriate composition of nonlinear Boolean functions.

In both frameworks, n is rarely large, in practice. The error correcting codes derived from n-variable Boolean functions have length 2^n; so, taking $n = 10$ already gives codes of length 1024. For reason of efficiency, the S-boxes used in most block ciphers are concatenations of sub S-boxes on at most eight variables. In the case of stream ciphers, n was in general at most equal to 10 until recently. This has changed with the algebraic attacks (see [112, 116, 150] and see the later discussion), but the number of variables is now most often limited to 20.

Despite the fact that Boolean functions are currently used in cryptography and coding with low numbers of variables, determining and studying those Boolean functions satisfying the desired conditions (see Subection 8.4.1) is not feasible through an exhaustive computer investigation: the number $|\mathcal{BF}_n| = 2^{2^n}$ of n-variable Boolean functions is too large when $n \geq 6$. In Table 8.1, we give the values of this number for n ranging between 4 and 8.

Assume that visiting an n-variable Boolean function, and determining whether it has the desired properties, requires one nanosecond (10^{-9} seconds). Then it would take millions of hours to visit all functions in six variables, and about 100 billion times the age of the universe to visit all those in seven variables. The number of eight-variable Boolean functions approximately equals the number of atoms in the whole universe! We see that trying to find functions satisfying the desired conditions by simply picking up functions at random is also impossible for these values of n, because visiting a nonnegligible part of all Boolean functions in seven or more variables is not feasible, even when parallelizing. The study

[4] Denoted by $\mathcal{B}$ in some chapters of the present collection.

of Boolean functions for constructing or studying codes or ciphers is essentially mathematical. But clever computer investigation is very useful to imagine or to test conjectures, and sometimes to generate interesting functions.

8.2 Generalities on Boolean Functions

In this chapter and in the chapter "Vectorial Boolean Functions for Cryptography," which follows, the set $\{0, 1\}$ is most often endowed with the structure of field (and denoted by $\mathbb{F}_2$), and the set $\mathbb{F}_2^n$ of all binary vectors[5] of length n is viewed as an $\mathbb{F}_2$-vectorspace. We denote simply by 0 the null vector in $\mathbb{F}_2^n$. The vectorspace $\mathbb{F}_2^n$ is sometimes also endowed with the structure of field – the field $\mathbb{F}_{2^n}$ (also denoted by $GF(2^n)$); indeed, this field being an n-dimensional vectorspace over $\mathbb{F}_2$, each of its elements can be identified with the binary vector of length n of its coordinates relative to a fixed basis. The set of all Boolean functions $f : \mathbb{F}_2^n \rightarrow \mathbb{F}_2$ is denoted as usual by $\mathcal{BF}_n$. The *Hamming weight* $w_H(x)$ of a binary vector $x \in \mathbb{F}_2^n$ being the number of its nonzero coordinates (*i.e.*, the size of $\{i \in N / \; x_i \neq 0\}$, where N denotes the set $\{1, \ldots, n\}$, called the *support of the codeword*), the Hamming weight $w_H(f)$ of a Boolean function f on $\mathbb{F}_2^n$ is (also) the size of the *support of the function*, that is, the set $\{x \in \mathbb{F}_2^n / \; f(x) \neq 0\}$. The *Hamming distance* $d_H(f, g)$ between two functions f and g is the size of the set $\{x \in \mathbb{F}_2^n / \; f(x) \neq g(x)\}$. Thus, it equals $w_H(f \oplus g)$.

Note. Some additions of bits will be considered in $\mathbb{Z}$ (in characteristic 0) and denoted then by $+$, and some will be computed modulo 2 and denoted by $\oplus$. These two different notations are necessary because some representations of Boolean functions live in characteristic 2 and some representations of the same functions live in characteristic 0. But the additions of elements of the finite field $\mathbb{F}_{2^n}$ will be denoted by $+$, as is usual in mathematics. So, for simplicity (since $\mathbb{F}_2^n$ will often be identified with $\mathbb{F}_{2^n}$) and because there will be no ambiguity, we also denote by $+$ the addition of vectors of $\mathbb{F}_2^n$ when $n > 1$.

8.2.1 Representation of Boolean Functions

Among the classical representations of Boolean functions, the one that is most usually used in cryptography and coding is the n-variable polynomial representation over $\mathbb{F}_2$, of the form

$$f(x) = \bigoplus_{I \in \mathcal{P}(N)} a_I \left(\prod_{i \in I} x_i \right) = \bigoplus_{I \in \mathcal{P}(N)} a_I \, x^I, \tag{8.1}$$

where $\mathcal{P}(N)$ denotes the power set of $N = \{1, \ldots, n\}$. Every coordinate x_i appears in this polynomial with exponents at most 1, because every bit in $\mathbb{F}_2$ equals its own

[5] Coders say "words."

square. This representation belongs to $\mathbb{F}_2[x_1, \ldots, x_n]/(x_1^2 \oplus x_1, \ldots, x_n^2 \oplus x_n)$. It is called the *algebraic normal form* (in brief, the ANF).

Example. *Let us consider the function f whose truth table is*

x_1	x_2	x_3	$f(x)$
0	0	0	0
0	0	1	1
0	1	0	0
0	1	1	0
1	0	0	0
1	0	1	1
1	1	0	0
1	1	1	1

It is the sum (modulo 2 or not, no matter) of the atomic functions *f_1, f_2 and f_3 whose truth-tables are*

x_1	x_2	x_3	$f_1(x)$	$f_2(x)$	$f_3(x)$
0	0	0	0	0	0
0	0	1	1	0	0
0	1	0	0	0	0
0	1	1	0	0	0
1	0	0	0	0	0
1	0	1	0	1	0
1	1	0	0	0	0
1	1	1	0	0	1

The function $f_1(x)$ takes value 1 if and only if $1 \oplus x_1 = 1$, $1 \oplus x_2 = 1$, and $x_3 = 1$, that is, if and only if $(1 \oplus x_1)(1 \oplus x_2)\, x_3 = 1$. Thus, the ANF of f_1 can be obtained by expanding the product $(1 \oplus x_1)(1 \oplus x_2)\, x_3$. After similar observations on f_2 and f_3, we see that the ANF of f equals $(1 \oplus x_1)(1 \oplus x_2)\, x_3 \oplus x_1(1 \oplus x_2)\, x_3 \oplus x_1x_2x_3 = x_1x_2x_3 \oplus x_2x_3 \oplus x_3$.

Another possible representation of this same ANF uses an indexation by means of vectors of $\mathbb{F}_2^n$ instead of subsets of N; if, for any such vector u, we denote by a_u what is denoted by $a_{supp(u)}$ in Relation (8.1) (where $supp(u)$ denotes the support of u), we have the equivalent representation:

$$f(x) = \bigoplus_{u \in \mathbb{F}_2^n} a_u \left(\prod_{j=1}^{n} x_j^{\,u_j} \right).$$

The monomial $\prod_{j=1}^{n} x_j^{\,u_j}$ is often denoted by x^u.

Existence and Uniqueness of the ANF. By applying the Lagrange interpolation method described in the foregoing example, it is a simple matter to show the existence of the ANF of every Boolean function. This implies that the mapping, from every polynomial $P \in \mathbb{F}_2[x_1, \ldots, x_n]/(x_1^2 \oplus x_1, \ldots, x_n^2 \oplus x_n)$ to the corresponding function $x \in \mathbb{F}_2^n \mapsto P(x)$, is onto $\mathcal{BF}_n$. Because the size of $\mathcal{BF}_n$ equals the size of $\mathbb{F}_2[x_1, \ldots, x_n]/(x_1^2 \oplus x_1, \ldots, x_n^2 \oplus x_n)$, this correspondence is one to one.[6] But more can be said.

Relationship between a Boolean Function and Its ANF. The product $x^I = \prod_{i \in I} x_i$ is nonzero if and only if x_i is nonzero (*i.e.*, equals 1) for every $i \in I$, that is, if I is included in the support of x; hence, the Boolean function $f(x) = \bigoplus_{I \in \mathcal{P}(N)} a_I\, x^I$ takes the value

$$f(x) = \bigoplus_{I \subseteq supp(x)} a_I, \tag{8.2}$$

where $supp(x)$ denotes the support of x. If we use the notation $f(x) = \bigoplus_{u \in \mathbb{F}_2^n} a_u x^u$, we obtain the relation $f(x) = \bigoplus_{u \preceq x} a_u$, where $u \preceq x$ means that $supp(u) \subseteq supp(x)$ (we say that u is *covered* by x). A Boolean function f° can be associated to the ANF of f: for every $x \in \mathbb{F}_2^n$, we set $f^\circ(x) = a_{supp(x)}$, that is, with the notation $f(x) = \bigoplus_{u \in \mathbb{F}_2^n} a_u x^u$: $f^\circ(u) = a_u$. Relation (8.2) shows that f is the image of f° by the so-called *binary Möbius transform*.

The converse is also true, as shown in the following proposition.

Proposition 8.1. *Let f be a Boolean function on $\mathbb{F}_2^n$ and let $\bigoplus_{I \in \mathcal{P}(N)} a_I\, x^I$ be its ANF. We have*

$$\forall I \in \mathcal{P}(N),\ a_I = \bigoplus_{x \in \mathbb{F}_2^n /\ supp(x) \subseteq I} f(x). \tag{8.3}$$

Proof. Let us denote $\bigoplus_{x \in \mathbb{F}_2^n /\ supp(x) \subseteq I} f(x)$ by b_I and consider the function $g(x) = \bigoplus_{I \in \mathcal{P}(N)} b_I\, x^I$. We have

$$g(x) = \bigoplus_{I \subseteq supp(x)} b_I = \bigoplus_{I \subseteq supp(x)} \left(\bigoplus_{y \in \mathbb{F}_2^n /\ supp(y) \subseteq I} f(y) \right)$$

and thus

$$g(x) = \bigoplus_{y \in \mathbb{F}_2^n} f(y) \left(\bigoplus_{I \in \mathcal{P}(N) /\ supp(y) \subseteq I \subseteq supp(x)} 1 \right).$$

The sum $\bigoplus_{I \in \mathcal{P}(N) /\ supp(y) \subseteq I \subseteq supp(x)} 1$ is null if $y \neq x$, because the set $\{I \in \mathcal{P}(N) /\ supp(y) \subseteq I \subseteq supp(x)\}$ contains $2^{w_H(x) - w_H(y)}$ elements if $supp(y) \subseteq supp(x)$, and none otherwise. Hence, $g = f$ and, by uniqueness of the ANF, $b_I = a_I$ for every I. ■

[6] Another argument is that this mapping is a linear mapping from a vectorspace over $\mathbb{F}_2$ of dimension 2^n to a vectorspace of the same dimension.

Algorithm. There exists a simple divide-and-conquer butterfly algorithm to compute the ANF from the truth table (or *vice versa*), that we can call the *fast Möbius transform*. For every $u = (u_1, \ldots, u_n) \in \mathbb{F}_2^n$, the coefficient a_u of x^u in the ANF of f equals

$$\bigoplus_{(x_1,\ldots,x_{n-1}) \preceq (u_1,\ldots,u_{n-1})} [f(x_1, \ldots, x_{n-1}, 0)] \quad \text{if } u_n = 0 \text{ and}$$

$$\bigoplus_{(x_1,\ldots,x_{n-1}) \preceq (u_1,\ldots,u_{n-1})} [f(x_1, \ldots, x_{n-1}, 0) \oplus f(x_1, \ldots, x_{n-1}, 1)] \text{ if } u_n = 1.$$

Hence if, in the truth table of f, the binary vectors are ordered in lexicographic order, with the bit of higher weight on the right, the table of the ANF equals the concatenation of the ANFs of the $(n-1)$-variable functions $f(x_1, \ldots, x_{n-1}, 0)$ and $f(x_1, \ldots, x_{n-1}, 0) \oplus f(x_1, \ldots, x_{n-1}, 1)$. We deduce the following algorithm:

(i) Write the truth table of f, in which the binary vectors of length n are in lexicographic order as decribed above;
(ii) Let f_0 and f_1 be the restrictions of f to $\mathbb{F}_2^{n-1} \times \{0\}$ and $\mathbb{F}_2^{n-1} \times \{1\}$, respectively[7]; replace the values of f_1 by those of $f_0 \oplus f_1$;
(iii) Recursively apply step 2, separately, to the functions now obtained in the places of f_0 and f_1.

When the algorithm ends (*i.e.*, when it arrives to functions in one variable each), the global table gives the values of the ANF of f. The complexity of this algorithm is of $n\,2^n$ XORs.

Remark. *The algorithm works the same if the vectors are ordered in standard lexicographic order, with the bit of higher weight on the left (indeed, this corresponds to applying it to $f(x_n, x_{n-1}, \ldots, x_1)$).*

The degree of the ANF is denoted by $d^\circ f$ and is called the *algebraic degree* of the function (this makes sense thanks to the existence and uniqueness of the ANF): $d^\circ f = \max\{|I| / a_I \neq 0\}$, where $|I|$ denotes the size of I. Some authors also call it the nonlinear order of f. According to Relation (8.3), we have the following proposition.

Proposition 8.2. *The algebraic degree $d^\circ f$ of any n-variable Boolean function f equals the maximum dimension of the subspaces $\{x \in \mathbb{F}_2^n / supp(x) \subseteq I\}$ on which f takes value 1 an odd number of times.*

[7] The truth table of f_0 (resp. f_1) corresponds to the upper (resp. lower) half of the table of f.

The algebraic degree is an *affine invariant* (it is invariant under the action of the general affine group): for every affine isomorphism

$$L : \begin{pmatrix} x_1 \\ x_2 \\ \vdots \\ x_n \end{pmatrix} \in \mathbb{F}_2^n \mapsto M \times \begin{pmatrix} x_1 \\ x_2 \\ \vdots \\ x_n \end{pmatrix} \oplus \begin{pmatrix} a_1 \\ a_2 \\ \vdots \\ a_n \end{pmatrix} \in \mathbb{F}_2^n$$

(where M is a nonsingular $n \times n$ matrix over $\mathbb{F}_2$), we have $d^\circ(f \circ L) = d^\circ f$. Indeed, the composition by L clearly cannot increase the algebraic degree, since the coordinates of $L(x)$ have degree 1. Hence, we have $d^\circ(f \circ L) \leq d^\circ f$ (this, inequality is more generally valid for every affine homomorphism). And applying this inequality to $f \circ L$ in the place of f and to L^{-1} in the place of L shows the inverse inequality.

Two functions f and $f \circ L$ where L is an $\mathbb{F}_2$-linear automorphism of $\mathbb{F}_2^n$ (in the case $a_1 = a_2 = \cdots = a_n = 0$ just given) will be called *linearly equivalent* and two functions f and $f \circ L$, where L is an affine automorphism of $\mathbb{F}_2^n$, will be called *affinely equivalent*.

The algebraic degree being an affine invariant, Proposition 8.2 implies that it also equals the maximum dimension of all the affine subspaces of $\mathbb{F}_2^n$ on which f takes value 1 an odd number of times.

It is shown in [296] that, for every nonzero n-variable Boolean function f, denoting by g the binary Möbius transform of f, we have $d^\circ f + d^\circ g \geq n$. This same paper deduces characterizations and constructions of the functions that are equal to their binary Möbius transform, called *coincident functions*.

Remarks.

(1) *Every atomic function has algebraic degree n, since its ANF equals* $(x_1 \oplus \varepsilon_1)(x_2 \oplus \varepsilon_2) \cdots (x_n \oplus \varepsilon_n)$*, where* $\varepsilon_i \in \mathbb{F}_2$*. Thus, a Boolean function* f *has algebraic degree n if and only if, in its decomposition as a sum of atomic functions, the number of these atomic functions is odd, that is, if and only if* $w_H(f)$ *is odd. This property has an important consequence on the Reed-Muller codes and it will be also useful in Section 8.3.*

(2) *If we know that the algebraic degree of an n-variable Boolean function* f *is bounded above by* $d < n$*, then the whole function can be recovered from some of its restrictions (*i.e., *a unique function corresponds to this* partially defined *Boolean function). Precisely, according to the existence and uniqueness of the ANF, the knowledge of the restriction* $f_{|E}$ *of the Boolean function* f *(of algebraic degree at most d) to a set* E *implies the knowledge of the whole function if and only if the system of the equations* $f(x) = \bigoplus_{I \in \mathcal{P}(N)/\, |I| \leq d} a_I\, x^I$*, with indeterminates* $a_I \in \mathbb{F}_2$*, and where* x *ranges over* E *(this makes* $|E|$ *equations), has a unique*

solution.[8] *This happens with the set* E_d *of all words of Hamming weights less than or equal to* d*, since Relation (8.3) gives the value of* a_I *(when* $I \in \mathcal{P}(N)$ *has size* $|I| \leq d$*). Notice that Relation (8.2) allows then to express the value of* $f(x)$ *for every* $x \in \mathbb{F}_2^n$ *by means of the values taken by* f *at all words of Hamming weights smaller than or equal to* d*. We have (using the notation* a_u *instead of* a_I*, see earlier discussion):*

$$f(x) = \bigoplus_{u \preceq x} a_u = \bigoplus_{\substack{u \preceq x \\ u \in E_d}} a_u = \bigoplus_{\substack{y \preceq x \\ y \in E_d}} f(y)\, |\{u \in E_d \,/\, y \preceq u \preceq x\}|$$

$$= \bigoplus_{\substack{y \preceq x \\ y \in E_d}} f(y) \left[\left[\sum_{i=0}^{d-w_H(y)} \binom{w_H(x) - w_H(y)}{i} \right] [\bmod 2] \right].$$

More generally, the whole function f *can be recovered from* $f_{|E}$ *for every set* E *affinely equivalent to* E_d*, according to the affine invariance of the algebraic degree. This also generalizes to "pseudo-Boolean" (that is, real-valued) functions, if we consider the numerical degree (see later discussion) instead of the the algebraic degree,* cf. *[347].*

The simplest functions, from the viewpoint of the ANF, are those Boolean functions of algebraic degrees at most 1, called *affine functions*:

$$f(x) = a_1 x_1 \oplus \cdots \oplus a_n x_n \oplus a_0.$$

They are the sums of linear and constant functions. Denoting by $a \cdot x$ the usual *inner product* $a \cdot x = a_1 x_1 \oplus \cdots \oplus a_n x_n$ in $\mathbb{F}_2^n$, or any other inner product (symmetric and such that, for every $a \neq 0$, the function $x \to a \cdot x$ is a nonzero linear form on $\mathbb{F}_2^n$), the general form of an n-variable affine function is $a \cdot x \oplus a_0$ (with $a \in \mathbb{F}_2^n$; $a_0 \in \mathbb{F}_2$).

Affine functions play an important role in coding (they are involved in the definition of the Reed-Muller code of order 1; see Subsection 8.3.1) and in cryptography (the Boolean functions used as "nonlinear functions" in cryptosystems must behave as differently as possible from affine functions; see Subsection 8.4.1).

Trace Representation(s). A second kind of representation plays an important role in sequence theory and is also used for defining and studying Boolean functions. It leads to the construction of the Kerdock codes (see Subsection 8.6.10). Recall that, for every n, there exists a (unique up to isomorphism) field $\mathbb{F}_{2^n}$ (also denoted by $GF(2^n)$) of order 2^n (see [247]). The vectorspace $\mathbb{F}_2^n$ can be endowed with the structure of this field $\mathbb{F}_{2^n}$. Indeed, we know that $\mathbb{F}_{2^n}$ has the structure of an n-dimensional $\mathbb{F}_2$-vectorspace; if we choose an $\mathbb{F}_2$-basis $(\alpha_1, \ldots, \alpha_n)$ of this

8 Taking $f_{|E}$ null leads to determining the so-called annihilators of the *indicator* of E (the function 1_E, also called characteristic function of E, defined by $1_E(x) = 1$ if $x \in E$ and $1_E(x) = 0$ otherwise); this is the core analysis of Boolean functions from the viewpoint of algebraic attacks; see Subsection 8.4.1.

vectorspace, then every element $x \in \mathbb{F}_2^n$ can be identified with $x_1\alpha_1 + \cdots + x_n\alpha_n \in \mathbb{F}_{2^n}$. We shall still denote by x this element of the field.

(1) It is shown in the chapter "Vectorial Boolean Functions for Cryptography" (see another proof later) that every mapping from $\mathbb{F}_{2^n}$ into $\mathbb{F}_{2^n}$ admits a (unique) representation as a polynomial

$$f(x) = \sum_{i=0}^{2^n-1} \delta_i x^i \tag{8.4}$$

over $\mathbb{F}_{2^n}$ in one variable and of (univariate) degree at most $2^n - 1$. Any Boolean function on $\mathbb{F}_{2^n}$ is a particular case of a vectorial function from $\mathbb{F}_{2^n}$ to $\mathbb{F}_{2^n}$ (since $\mathbb{F}_2$ is a subfield of $\mathbb{F}_{2^n}$) and therefore admits such a unique representation, which we shall call the *univariate representation* of f. For every $u, v \in \mathbb{F}_{2^n}$ we have $(u+v)^2 = u^2 + v^2$ and $u^{2^n} = u$. A univariate polynomial $\sum_{i=0}^{2^n-1} \delta_i x^i$, $\delta_i \in \mathbb{F}_{2^n}$, is then the univariate representation of a Boolean function if and only if $\left(\sum_{i=0}^{2^n-1} \delta_i x^i\right)^2 = \sum_{i=0}^{2^n-1} \delta_i^2 x^{2i} = \sum_{i=0}^{2^n-1} \delta_i x^i$ [mod $x^{2^n} + x$], that is, $\delta_0, \delta_{2^n-1} \in \mathbb{F}_2$, and, for every $i = 1, \ldots, 2^n - 2$, $\delta_{2i} = \delta_i^2$, where the index $2i$ is taken mod $2^n - 1$.

(2) The function defined on $\mathbb{F}_{2^n}$ by $tr_n(u) = u + u^2 + u^{2^2} + \cdots + u^{2^{n-1}}$ is $\mathbb{F}_2$-linear and satisfies $(tr_n(u))^2 = tr_n(u^2) = tr_n(u)$; it is therefore valued in $\mathbb{F}_2$. This function is called the *trace function* from $\mathbb{F}_{2^n}$ to its prime field $\mathbb{F}_2$ or the absolute trace function on $\mathbb{F}_{2^n}$. The function $(u, v) \mapsto tr_n(u\,v)$ is an inner product in $\mathbb{F}_{2^n}$ (that is, it is symmetric and for every $v \neq 0$, the function $u \to tr_n(u\,v)$ is a nonzero linear form on $\mathbb{F}_{2^n}$). Every Boolean function can be written in the form $f(x) = tr_n(F(x))$ where F is a mapping from $\mathbb{F}_{2^n}$ into $\mathbb{F}_{2^n}$ (an example of such mapping F is defined by $F(x) = \lambda f(x)$ where $tr_n(\lambda) = 1$ and $f(x)$ is the univariate representation). Thus, every Boolean function can be also represented in the form

$$tr_n\left(\sum_{i=0}^{2^n-1} \beta_i\, x^i\right), \tag{8.5}$$

where $\beta_i \in \mathbb{F}_{2^n}$. Such a representation is not unique. Now, thanks to the fact that $tr_n(u^2) = tr_n(u)$ for every $u \in \mathbb{F}_{2^n}$, we can restrict the exponents i with nonzero coefficients β_i so that there is at most one such exponent in each *cyclotomic class* $\{i \times 2^j\ [\bmod\ (2^n - 1)];\ j \in \mathbb{N}\}$ of 2 modulo $2^n - 1$ (but this still does not make the representation unique). We shall call this expression the *absolute trace representation* of f.

(3) We come back to the univariate representation. Let us see how it can be obtained from the truth table of the function and represented in a convenient way by using the notation tr_n. Denoting by α a *primitive element* of the field $\mathbb{F}_{2^n}$ (that is, an element such that $\mathbb{F}_{2^n} = \{0, 1, \alpha, \alpha^2, \ldots, \alpha^{2^n-2}\}$,

which always exists [247]), the *Mattson-Solomon polynomial* of the vector $(f(1), f(\alpha), f(\alpha^2), \ldots, f(\alpha^{2^n-2}))$ is the polynomial [257]

$$A(x) = \sum_{j=1}^{2^n-1} A_j x^{2^n-1-j} = \sum_{j=0}^{2^n-2} A_{-j} x^j$$

with

$$A_j = \sum_{k=0}^{2^n-2} f(\alpha^k)\alpha^{kj}.$$

Note that the Mattson-Solomon transform is a discrete Fourier transform.

We have, for every $0 \le i \le 2^n - 2$:

$$A(\alpha^i) = \sum_{j=1}^{2^n-1} A_j \alpha^{-ij} = \sum_{j=1}^{2^n-1} \sum_{k=0}^{2^n-2} f(\alpha^k)\alpha^{(k-i)j} = f(\alpha^i)$$

(since $\sum_{j=1}^{2^n-1} \alpha^{(k-i)j} = \sum_{j=0}^{2^n-2} \alpha^{(k-i)j} = \frac{\alpha^{(k-i)(2^n-1)}+1}{\alpha^{k-i}+1}$ equals 0 if $1 \le k \ne i \le 2^n - 2$), and A is therefore the univariate representation of f, if $f(0) = A_0 = \sum_{i=0}^{2^n-2} f(\alpha^i)$ (note that this works also for functions from $\mathbb{F}_{2^n}$ to $\mathbb{F}_{2^n}$), that is, if f has even weight, *meaning it* has algebraic degree strictly less than n. Otherwise, we have $f(x) = A(x) + 1 + x^{2^n-1}$, since $1 + x^{2^n-1}$ takes value 1 at 0 and 0 at every nonzero element of $\mathbb{F}_{2^n}$.

Note that $A_{2j} = A_j^2$. Denoting by $\Gamma(n)$ the set obtained by choosing one element in each cyclotomic class of 2 modulo $2^n - 1$ (the most usual choice for k is the smallest element in its cyclotomic class, called the *coset leader* of the class). This allows representing $f(x)$ in the form

$$\sum_{j\in\Gamma(n)} tr_{n_j}(A_j x^j) + \varepsilon(1 + x^{2^n-1}), \tag{8.6}$$

where $\varepsilon = w_H(f)$ [mod 2] and where n_j is the size of the cyclotomic class containing j. Note that, for every $j \in \Gamma(n)$ and every $x \in \mathbb{F}_{2^n}$, we have $A_j \in F_{2^{n_j}}$ (since $A_j^{2^{n_j}} = A_j$) and $x^j \in F_{2^{n_j}}$ as well. We shall call this expression the *trace representation* of f. Obviously, it is nothing more than an alternate expression for the univariate representation. For this reason, it is unique (if we restrict the coefficient of x^j to live in $F_{2^{n_j}}$). But it is useful to distinguish the different expressions by different names. We shall call globally "trace representations" the three expressions (8.4), (8.5), and (8.6).

Trace representations and the algebraic normal form are closely related. Let us see how the ANF can be obtained from the univariate representation: we express x in the form $\sum_{i=1}^{n} x_i\alpha_i$, where $(\alpha_1, \ldots, \alpha_n)$ is a basis of the $\mathbb{F}_2$-vectorspace $\mathbb{F}_{2^n}$. Recall that, for every $j \in \mathbb{Z}/(2^n - 1)\mathbb{Z}$, the *binary expansion* of j has the form $\sum_{s\in E} 2^s$, where $E \subseteq \{0, 1, \ldots, n - 1\}$. The size of E is often called the *2-weight* of j and written $w_2(j)$. We write more conveniently the binary expansion of j in the form $\sum_{s=0}^{n-1} j_s 2^s$, $j_s \in \{0,1\}$. We then have:

$$\begin{aligned} f(x) &= \sum_{j=0}^{2^n-1} \delta_j \left(\sum_{i=1}^{n} x_i \alpha_i\right)^j \\ &= \sum_{j=0}^{2^n-1} \delta_j \left(\sum_{i=1}^{n} x_i \alpha_i\right)^{\sum_{s=0}^{n-1} j_s 2^s} \\ &= \sum_{j=0}^{2^n-1} \delta_j \prod_{s=0}^{n-1} \left(\sum_{i=1}^{n} x_i \alpha_i^{2^s}\right)^{j_s}. \end{aligned}$$

Expanding these last products and simplifying gives the ANF of f.

Function f then has algebraic degree $\max_{j=0,\ldots,2^n-1/\,\delta_j \neq 0} w_2(j)$. Indeed, according to the foregoing equalities, its algebraic degree is clearly bounded above by this number, and it cannot be strictly smaller, because the number of Boolean n-variable functions of algebraic degrees at most d equals the number of the polynomials $\sum_{j=0}^{2^n-1} \delta_j x^j$ such that $\delta_0, \delta_{2^n-1} \in \mathbb{F}_2$, and $\delta_{2j} = \delta_j^2 \in \mathbb{F}_{2^n}$ for every $j = 1, \ldots, 2^n - 2$ and $\max_{j=0,\ldots,2^n-1/\,\delta_j \neq 0} w_2(j) \leq d$.

We also have the following proposition.

Proposition 8.3. ([50].) *Let a be any element of $\mathbb{F}_{2^n}$ and k any integer [mod $2^n - 1$]. If $f(x) = tr_n(ax^k)$ is not the null function, then it has algebraic degree $w_2(k)$.*

Proof. Let n_k be again the size of the cyclotomic class containing k. Then the univariate representation of $f(x)$ equals

$$\left(a + a^{2^{n_k}} + a^{2^{2n_k}} + \cdots + a^{2^{n-n_k}}\right) x^k + \left(a + a^{2^{n_k}} + a^{2^{2n_k}} + \cdots + a^{2^{n-n_k}}\right)^2 x^{2k}$$

$$+ \cdots + \left(a + a^{2^{n_k}} + a^{2^{2n_k}} + \cdots + a^{2^{n-n_k}}\right)^{2^{n_k-1}} x^{2^{n_k-1}k}.$$

All the exponents of x have 2-weight $w_2(k)$ and their coefficients are nonzero if and only if f is not null. ■

Remark. *Another (more complex) way of showing Proposition 8.3 is used in [50] as follows: let $r = w_2(k)$; we consider the r-linear function ϕ over the field $\mathbb{F}_{2^n}$ whose value at $(x_1, \ldots, x_r)$ equals the sum of the images by f of all the 2^r possible linear combinations of the x_j's. Then $\phi(x_1, \ldots, x_r)$ equals the sum, for all bijective mappings σ from $\{1, \ldots, r\}$ onto E (where $k = \sum_{s \in E} 2^s$) of $tr_n(a \prod_{j=1}^{r} x_j^{2^{\sigma(j)}})$. Proving that f has degree r is equivalent to proving that ϕ is not null, and it can be shown that if ϕ is null, then f is null.*

The representation over the reals has proved itself to be useful for characterizing several cryptographic criteria [62, 86, 87] (see Sections 8.6 and 8.7). It represents Boolean functions, and more generallyreal-valued functions on $\mathbb{F}_2^n$ (called

n-variable *pseudo-Boolean functions* in [117]) by elements of $\mathbb{R}[x_1, \ldots, x_n]/(x_1^2 - x_1, \ldots, x_n^2 - x_n)$ (or of $\mathbb{Z}[x_1, \ldots, x_n]/(x_1^2 - x_1, \ldots, x_n^2 - x_n)$ for integer-valued functions). We shall call it the *numerical normal form* (NNF).

The existence of this representation for every pseudo-Boolean function is easy to show with the same arguments as for the ANFs of Boolean functions (writing $1 - x_i$ instead of $1 \oplus x_i$). The linear mapping from every element of the 2^n-th dimensional $\mathbb{R}$-vectorspace $\mathbb{R}[x_1, \ldots, x_n]/(x_1^2 - x_1, \ldots, x_n^2 - x_n)$ to the corresponding pseudo-Boolean function on $\mathbb{F}_2^n$ being onto, it is therefore one to one (the $\mathbb{R}$-vectorspace of pseudo-Boolean functions on $\mathbb{F}_2^n$ having also dimension 2^n). We deduce the uniqueness of the NNF.

We call the degree of the NNF of a function its *numerical degree*. Because the ANF is the mod 2 version of the NNF, the numerical degree is always bounded below by the algebraic degree. It is shown in [285] that, if a Boolean function f has no ineffective variable (*i.e.*, if it actually depends on each of its variables), then the numerical degree of f is greater than or equal to $\log_2 n - \log_2 \log_2 n$.

The numerical degree is not an affine invariant. But the NNF leads to an affine invariant (see a proof of this fact in [87]; see also [191]) that is more discriminant than the algebraic degree:

Definition 8.1. *Let f be a Boolean function on $\mathbb{F}_2^n$. We call* generalized degree *of f the sequence $(d_i)_{i \geq 1}$ defined as follows: for every $i \geq 1$, d_i is the smallest integer $d > d_{i-1}$ (if $i > 1$) such that, for every multiindex I of size strictly greater than d, the coefficient λ_I of x^I in the NNF of f is a multiple of 2^i.*

Example. *The generalized degree of any nonzero affine function is the sequence of all positive integers.*

Similarly to the case of the ANF, a (pseudo-) Boolean function $f(x) = \sum_{I \in \mathcal{P}(N)} \lambda_I \, x^I$ takes value

$$f(x) = \sum_{I \subseteq supp(x)} \lambda_I. \tag{8.7}$$

But, contrary to what we observed for the ANF, the reverse formula is not identical to the direct formula.

Proposition 8.4. *Let f be a pseudo-Boolean function on $\mathbb{F}_2^n$ and let its NNF be $\sum_{I \in \mathcal{P}(N)} \lambda_I \, x^I$. Then*

$$\forall I \in \mathcal{P}(N), \; \lambda_I = (-1)^{|I|} \sum_{x \in \mathbb{F}_2^n \,|\, supp(x) \subseteq I} (-1)^{w_H(x)} f(x). \tag{8.8}$$

Thus, function f and its NNF are related through the *Möbius transform over integers*.

Proof. Let us denote the number $(-1)^{|I|} \sum_{x\in\mathbb{F}_2^n \,|\, supp(x)\subseteq I} (-1)^{w_H(x)} f(x)$ by μ_I and consider the function $g(x) = \sum_{I\in\mathcal{P}(N)} \mu_I\, x^I$. We have

$$g(x) = \sum_{I\subseteq supp(x)} \mu_I = \sum_{I\subseteq supp(x)} \left((-1)^{|I|} \sum_{y\in\mathbb{F}_2^n \,|\, supp(y)\subseteq I} (-1)^{w_H(y)} f(y) \right)$$

and thus

$$g(x) = \sum_{y\in\mathbb{F}_2^n} (-1)^{w_H(y)} f(y) \left(\sum_{I\in\mathcal{P}(N)/\ supp(y)\subseteq I\subseteq supp(x)} (-1)^{|I|} \right).$$

The sum $\sum_{I\in\mathcal{P}(N)/\ supp(y)\subseteq I\subseteq supp(x)} (-1)^{|I|}$ is null if $supp(y) \not\subseteq supp(x)$. It is also null if $supp(y)$ is included in $supp(x)$, but different. Indeed, denoting $|I| - w_H(y)$ by i, it equals $\pm \sum_{i=0}^{w_H(x)-w_H(y)} \binom{w_H(x)-w_H(y)}{i} (-1)^i = \pm(1-1)^{w_H(x)-w_H(y)} = 0$. Hence, $g = f$ and, by uniqueness of the NNF, we have $\mu_I = \lambda_I$ for every I. ■

We have seen that the ANF of any Boolean function can be deduced from its NNF by reducing it modulo 2. Conversely, the NNF can be deduced from the ANF because we have

$$f(x) = \bigoplus_{I\in\mathcal{P}(N)} a_I\, x^I \iff (-1)^{f(x)} = \prod_{I\in\mathcal{P}(N)} (-1)^{a_I\, x^I}$$

$$\iff 1 - 2\, f(x) = \prod_{I\in\mathcal{P}(N)} (1 - 2\, a_I\, x^I).$$

Expanding this last equality gives the NNF of $f(x)$ and we have [86]

$$\lambda_I = \sum_{k=1}^{2^n} (-2)^{k-1} \sum_{\substack{\{I_1,\ldots,I_k\}\,|\\ I_1\cup\cdots\cup I_k = I}} a_{I_1} \cdots a_{I_k}, \tag{8.9}$$

where "$\{I_1, \ldots, I_k\} \mid I_1 \cup \cdots \cup I_k = I$" means that the multiindices $I_1, \ldots, I_k$ are all distinct, in indefinite order, and that their union equals I.

A polynomial $P(x) = \sum_{J\in\mathcal{P}(N)} \lambda_J\, x^J$, with real coefficients, is the NNF of some Boolean function if and only if we have $P^2(x) = P(x)$, for every $x \in \mathbb{F}_2^n$ (which is equivalent to $P = P^2$ in $\mathbb{R}[x_1, \ldots, x_n]/(x_1^2 - x_1, \ldots, x_n^2 - x_n)$), or equivalently, denoting $supp(x)$ by I:

$$\forall I \in \mathcal{P}(N), \left(\sum_{J\subseteq I} \lambda_J \right)^2 = \sum_{J\subseteq I} \lambda_J. \tag{8.10}$$

Remark. *Imagine that we want to generate a random Boolean function through its NNF (this can be useful, because we will see later that the main cryptographic criteria, on Boolean functions, can be characterized, in simple ways, through their NNFs). Assume that we have already chosen the values λ_J for every $J \subseteq I$*

(where $I \in \mathcal{P}(N)$ is some multiindex) except for I itself. Let us denote the sum $\sum_{J\subseteq I \mid J\neq I} \lambda_J$ by μ. Relation (8.10) gives $(\lambda_I + \mu)^2 = \lambda_I + \mu$. This equation of degree 2 has two solutions (it has same discriminant as the equation ${\lambda_I}^2 = \lambda_I$, that is 1). One solution corresponds to the choice $P(x) = 0$ (where $I = supp(x)$), and the other corresponds to the choice $P(x) = 1$.

Thus, verifying that a polynomial $P(x) = \sum_{I\in\mathcal{P}(N)} \lambda_I\, x^I$ with real coefficients represents a Boolean function can be done by checking 2^n relations. But it can also be done by verifying a simple condition on P and checking a single equation.

Proposition 8.5. *Any polynomial $P \in \mathbb{R}[x_1, \ldots, x_n]/(x_1^2 - x_1, \ldots, x_n^2 - x_n)$ is the NNF of an integer-valued function if and only if all of its coefficients are integers. Assuming that this condition is satisfied, P is the NNF of a Boolean function if and only if: $\sum_{x\in\mathbb{F}_2^n} P^2(x) = \sum_{x\in\mathbb{F}_2^n} P(x)$.*

Proof. The first assertion is a direct consequence of Relations (8.7) and (8.8). If all the coefficients of P are integers, then we have $P^2(x) \geq P(x)$ for every x; this implies that the 2^n equalities, expressing that the corresponding function is Boolean, can be reduced to the single one $\sum_{x\in\mathbb{F}_2^n} P^2(x) = \sum_{x\in\mathbb{F}_2^n} P(x)$. ■

The translation of this characterization in terms of the coefficients of P is given later in Relation (8.32). We also refer to the monograph [117] for additional information regarding pseudo-Boolean functions and their representations.

8.2.2 The Discrete Fourier Transform on Pseudo-Boolean and on Boolean Functions

Almost all the characteristics needed for Boolean functions in cryptography and for sets of Boolean functions in coding can be expressed by means of the weights of some related Boolean functions (of the form $f \oplus \ell$, where ℓ is affine, or of the form $D_a f(x) = f(x) \oplus f(x + a)$). In this framework, the *discrete Fourier transform* is a very efficient tool: for a given Boolean function f, the knowledge of the discrete Fourier transform of f is equivalent with the knowledge of the weights of all the functions $f \oplus \ell$, where ℓ is linear (or affine). Also called Hadamard transform, the discrete Fourier transform is the linear mapping that maps any pseudo-Boolean function φ on $\mathbb{F}_2^n$ to the function $\widehat{\varphi}$ defined on $\mathbb{F}_2^n$ by

$$\widehat{\varphi}(u) = \sum_{x\in\mathbb{F}_2^n} \varphi(x)(-1)^{x\cdot u} \tag{8.11}$$

where $x \cdot u$ is some chosen inner product (for instance, the usual inner product $x \cdot u = x_1\, u_1 \oplus \cdots \oplus x_n\, u_n$).

Algorithm. There exists a simple divide-and-conquer butterfly algorithm to compute $\widehat{\varphi}$, called the *fast fourier transform* (FFT). For every $a = (a_1, \ldots, a_{n-1}) \in$

$\mathbb{F}_2^{n-1}$ and every $a_n \in \mathbb{F}_2$, the number $\widehat{\varphi}(a_1, \ldots, a_n)$ equals

$$\sum_{x=(x_1,\ldots,x_{n-1})\in\mathbb{F}_2^{n-1}} (-1)^{a\cdot x} \left[\varphi(x_1, \ldots, x_{n-1}, 0) + (-1)^{a_n}\varphi(x_1, \ldots, x_{n-1}, 1)\right].$$

Hence, if in the tables of values of the functions, the vectors are ordered in lexicographic order with the bit of highest weight on the right, the table of $\widehat{\varphi}$ equals the concatenation of those of the discrete Fourier transforms of the $(n-1)$-variable functions $\psi_0(x) = \varphi(x_1, \ldots, x_{n-1}, 0) + \varphi(x_1, \ldots, x_{n-1}, 1)$ and $\psi_1(x) = \varphi(x_1, \ldots, x_{n-1}, 0) - \varphi(x_1, \ldots, x_{n-1}, 1)$. We deduce the following algorithm:

(i) Write the table of the values of φ (its truth table if φ is Boolean), in which the binary vectors of length n are in lexicographic order as described earlier;
(ii) Let φ_0 be the restriction of φ to $\mathbb{F}_2^{n-1} \times \{0\}$ and φ_1 the restriction of φ to $\mathbb{F}_2^{n-1} \times \{1\}$[9]; replace the values of φ_0 by those of $\varphi_0 + \varphi_1$ and those of φ_1 by those of $\varphi_0 - \varphi_1$;
(iii) Recursively apply step 2, separately, to the functions now obtained in the places of φ_0 and φ_1.

When the algorithm ends (*i.e.,* when it arrives to functions in one variable each), the global table gives the values of $\widehat{\varphi}$. The complexity of this algorithm is of $n\,2^n$ additions/substractions.

Application to Boolean Functions. For a given Boolean function f, the discrete Fourier transform can be applied to f itself, viewed as a function valued in $\{0, 1\} \subset \mathbb{Z}$. We denote by $\widehat{f}$ the corresponding discrete Fourier transform of f. Notice that $\widehat{f}(0)$ equals the Hamming weight of f. Thus, the Hamming distance $d_H(f, g) = |\{x \in \mathbb{F}_2^n / f(x) \neq g(x)\}| = w_H(f \oplus g)$ between two functions f and g equals $\widehat{f \oplus g}(0)$.

The discrete Fourier transform can also be applied to the pseudo-Boolean function $\mathsf{f}_\chi(x) = (-1)^{f(x)}$ (often called the *sign function*[10]) instead of f itself. We have

$$\widehat{\mathsf{f}_\chi}(u) = \sum_{x\in\mathbb{F}_2^n} (-1)^{f(x)\oplus x\cdot u}.$$

[9] The table of values of φ_0 (resp. φ_1) corresponds to the upper (resp. lower) half of the table of φ.
[10] The symbol χ is used here because the sign function is the image of f by the nontrivial character over $\mathbb{F}_2$ (usually denoted by χ); to be sure that the distinction between the discrete Fourier transforms of f and of its sign function is easily done, we change the font when we deal with the sign function; many other ways of denoting the discrete Fourier transform can be found in the literature.

Table 8.2. Truth table and Walsh spectrum of $f(x) = x_1x_2x_3 \oplus x_1x_4 \oplus x_2$

x_1	x_2	x_3	x_4	$x_1x_2x_3$	x_1x_4	$f(x)$	$\mathrm{f}_\chi(x)$				$\widehat{\mathrm{f}_\chi}(x)$
0	0	0	0	0	0	0	1	2	4	0	0
1	0	0	0	0	0	0	1	0	0	0	0
0	1	0	0	0	0	1	-1	-2	-4	8	8
1	1	0	0	0	0	1	-1	0	0	0	8
0	0	1	0	0	0	0	1	2	0	0	0
1	0	1	0	0	0	0	1	0	0	0	0
0	1	1	0	0	0	1	-1	-2	0	0	0
1	1	1	0	1	0	0	1	0	0	0	0
0	0	0	1	0	0	0	1	0	0	0	4
1	0	0	1	0	1	1	-1	2	4	4	-4
0	1	0	1	0	0	1	-1	0	0	0	4
1	1	0	1	0	1	0	1	-2	0	4	-4
0	0	1	1	0	0	0	1	0	0	0	-4
1	0	1	1	0	1	1	-1	2	0	-4	4
0	1	1	1	0	0	1	-1	0	0	0	4
1	1	1	1	1	1	1	-1	2	-4	4	-4

We shall call *Walsh transform*[11] of f the Fourier transform of the sign function f_χ. In Table 8.2 we give an example of the computation of the Walsh transform, using the algorithm recalled previously.

Notice that f_χ being equal to $1 - 2f$, we have

$$\widehat{\mathrm{f}_\chi} = 2^n\,\delta_0 - 2\widehat{f} \tag{8.12}$$

where δ_0 denotes the *Dirac symbol, that is,* the indicator of the singleton $\{0\}$, defined by $\delta_0(u) = 1$ if u is the null vector and $\delta_0(u) = 0$ otherwise; see Proposition 8.8 for a proof of the relation $\widehat{1} = 2^n\,\delta_0$. Relation (8.12) gives conversely $\widehat{f} = 2^{n-1}\delta_0 - \frac{\widehat{\mathrm{f}_\chi}}{2}$ and, in particular,

$$w_H(f) = 2^{n-1} - \frac{\widehat{\mathrm{f}_\chi}(0)}{2}. \tag{8.13}$$

Relation (8.13) applied to $f \oplus \ell_a$, where $\ell_a(x) = a \cdot x$, gives

$$d_H(f, \ell_a) = w_H(f \oplus \ell_a) = 2^{n-1} - \frac{\widehat{\mathrm{f}_\chi}(a)}{2}. \tag{8.14}$$

[11] The terminology is not much more settled in the literature than is the notation; we take advantage here of the fact that many authors, when working on Boolean functions, use the term of Walsh transform instead of discrete Fourier transform: we call Fourier transform the discrete Fourier transform of the Boolean function itself and Walsh transform (some authors write "Walsh-Hadamard transform") the discrete Fourier transform of its sign function.

The mapping $f \mapsto \widehat{f_\chi}(0)$ playing an important role, and being applied in the sequel to various functions deduced from f, we shall also use the specific notation

$$\mathcal{F}(f) = \widehat{f_\chi}(0) = \sum_{x \in \mathbb{F}_2^n} (-1)^{f(x)}. \tag{8.15}$$

Properties of the Fourier Transform. The discrete Fourier transform, like any other Fourier transform, has very nice and useful properties. The number of these properties and the richness of their mutual relationship are impressive. All of these properties are very useful in practice for studying Boolean functions (we shall often refer to the following relations in the rest of the chapter; see also Chapter 12 by Bruck). Almost all properties can be deduced from the next lemma and from the next two propositions.

Lemma 8.6. *Let E be any vectorspace over $\mathbb{F}_2$ and ℓ any nonzero linear form on E. Then $\sum_{x \in E} (-1)^{\ell(x)}$ is null.*

Proof. The linear form ℓ being not null, its support is an affine hyperplane of E and has $2^{dimE-1} = \frac{|E|}{2}$ elements.[12] Thus, $\sum_{x \in E} (-1)^{\ell(x)}$ being the sum of 1s and -1s in equal numbers, it is null. ∎

Proposition 8.7. *For every pseudo-Boolean function φ on $\mathbb{F}_2^n$ and every elements a, b and u of $\mathbb{F}_2^n$, the value at u of the Fourier transform of the function $(-1)^{a \cdot x}\, \varphi(x + b)$ equals $(-1)^{b \cdot (a+u)}\, \widehat{\varphi}(a + u)$.*

Proof. The value at u of the Fourier transform of the function $(-1)^{a \cdot x}\, \varphi(x + b)$ equals $\sum_{x \in \mathbb{F}_2^n} (-1)^{(a+u) \cdot x} \varphi(x + b) = \sum_{x \in \mathbb{F}_2^n} (-1)^{(a+u) \cdot (x+b)} \varphi(x)$ and thus equals $(-1)^{b \cdot (a+u)}\, \widehat{\varphi}(a + u)$. ∎

Proposition 8.8. *Let E be any vector subspace of $\mathbb{F}_2^n$. Denote by 1_E its indicator (recall that it is the Boolean function defined by $1_E(x) = 1$ if $x \in E$ and $1_E(x) = 0$ otherwise). Then*

$$\widehat{1_E} = |E|\, 1_{E^\perp}, \tag{8.16}$$

where $E^\perp = \{x \in \mathbb{F}_2^n / \ \forall y \in E,\ x \cdot y = 0\}$ is the orthogonal *of E.*

In particular, for $E = \mathbb{F}_2^n$, we have $\widehat{1} = 2^n\, \delta_0$.

Proof. For every $u \in \mathbb{F}_2^n$, we have $\widehat{1_E}(u) = \sum_{x \in E} (-1)^{u \cdot x}$. If the linear form $x \in E \mapsto u \cdot x$ is not null on E (i.e., if $u \notin E^\perp$), then $\widehat{1_E}(u)$ is null, according to Lemma 8.6. And if $u \in E^\perp$, then it clearly equals $|E|$. ∎

We deduce from Proposition 8.8 the *Poisson summation formula*, which has been used to prove many cryptographic properties in [242, 252, 53], and later in [40, 41] and whose most general statement is the following corollary.

[12] Another way of seeing this is as follows: choose $a \in E$ such that $\ell(a) = 1$; then the mapping $x \mapsto x + a$ is one to one between $\ell^{-1}(0)$ and $\ell^{-1}(1)$.

Corollary 8.9. *For every pseudo-Boolean function φ on $\mathbb{F}_2^n$, for every vector subspace E of $\mathbb{F}_2^n$, and for every element a and b of $\mathbb{F}_2^n$, we have:*

$$\sum_{u\in a+E} (-1)^{b\cdot u}\,\widehat{\varphi}(u) = |E|\,(-1)^{a\cdot b} \sum_{x\in b+E^\perp} (-1)^{a\cdot x}\,\varphi(x). \tag{8.17}$$

Proof. Let us first assume that $a = b = 0$. The sum $\sum_{u\in E}\widehat{\varphi}(u)$, by definition, equals $\sum_{u\in E}\sum_{x\in\mathbb{F}_2^n}\varphi(x)(-1)^{u\cdot x} = \sum_{x\in\mathbb{F}_2^n}\varphi(x)\,\widehat{1_E}(x)$. Hence, according to Proposition 8.8:

$$\sum_{u\in E}\widehat{\varphi}(u) = |E| \sum_{x\in E^\perp}\varphi(x). \tag{8.18}$$

We apply this last equality to the function $(-1)^{a\cdot x}\,\varphi(x+b)$, whose Fourier transform's value at u equals $(-1)^{b\cdot(a+u)}\,\widehat{\varphi}(a+u)$, according to Proposition 8.7. We deduce $\sum_{u\in E}(-1)^{b\cdot(a+u)}\,\widehat{\varphi}(a+u) = |E|\sum_{x\in E^\perp}(-1)^{a\cdot x}\,\varphi(x+b)$, which is equivalent to Equality (8.17). ■

Relation (8.17) with $a = 0$ and $E = \mathbb{F}_2^n$ gives the following corollary.

Corollary 8.10. *For every pseudo-Boolean function φ on $\mathbb{F}_2^n$,*

$$\widehat{\widehat{\varphi}} = 2^n\,\varphi. \tag{8.19}$$

Thus, the Fourier transform is a permutation on the set of pseudo-Boolean functions on $\mathbb{F}_2^n$ and is its own inverse, up to division by a constant. In order to avoid this division, the Fourier transform is often normalized, that is, divided by $\sqrt{2^n} = 2^{n/2}$ so that it becomes its own inverse. We do not use this normalized transform here because the functions we consider are integer-valued, and we want their Fourier transforms to be also integer-valued.

Corollary 8.10 allows us to show easily that some properties, valid for the Fourier transform of any function φ having some specificities, are in fact necessary and sufficient conditions for φ having these specificities. For instance, according to Proposition 8.8, the Fourier transform of any constant function φ takes null value at every nonzero vector; because the Fourier transform of a function null at every nonzero vector is constant, Corollary 8.10 implies that a function is constant if and only if its Fourier transform is null at every nonzero vector. Similarly, φ is constant on $\mathbb{F}_2^n\setminus\{0\}$ if and only if $\widehat{\varphi}$ is constant on $\mathbb{F}_2^n\setminus\{0\}$.

A classical property of the Fourier transform is to be an isomorphism from the set of pseudo-Boolean functions on $\mathbb{F}_2^n$, endowed with the so-called convolutional product (denoted by $\otimes$), into this same set, endowed with the usual (Hadamard) product (denoted by $\times$). We recall the definition of the convolutional product between two functions φ and ψ:

$$(\varphi\otimes\psi)(x) = \sum_{y\in\mathbb{F}_2^n}\varphi(y)\psi(x+y)$$

(adding here is equivalent to substracting because the operations take place in $\mathbb{F}_2^n$).

Proposition 8.11. *Let φ and ψ be any pseudo-Boolean functions on $\mathbb{F}_2^n$. We have*

$$\widehat{\varphi \otimes \psi} = \widehat{\varphi} \times \widehat{\psi}. \tag{8.20}$$

Consequently,

$$\widehat{\varphi} \otimes \widehat{\psi} = 2^n \, \widehat{\varphi \times \psi}. \tag{8.21}$$

Proof. We have

$$\widehat{\varphi \otimes \psi}(u) = \sum_{x \in \mathbb{F}_2^n} (\varphi \otimes \psi)(x)\,(-1)^{u \cdot x} = \sum_{x \in \mathbb{F}_2^n} \sum_{y \in \mathbb{F}_2^n} \varphi(y)\psi(x+y)\,(-1)^{u \cdot x}$$

$$= \sum_{x \in \mathbb{F}_2^n} \sum_{y \in \mathbb{F}_2^n} \varphi(y)\psi(x+y)\,(-1)^{u \cdot y \oplus u \cdot (x+y)}.$$

Thus,

$$\widehat{\varphi \otimes \psi}(u) = \sum_{y \in \mathbb{F}_2^n} \varphi(y)(-1)^{u \cdot y} \left(\sum_{x \in \mathbb{F}_2^n} \psi(x+y)\,(-1)^{u \cdot (x+y)} \right)$$

$$= \left(\sum_{y \in \mathbb{F}_2^n} \varphi(y)(-1)^{u \cdot y} \right) \left(\sum_{x \in \mathbb{F}_2^n} \psi(x)\,(-1)^{u \cdot x} \right) = \widehat{\varphi}(u)\,\widehat{\psi}(u).$$

This proves the first equality. Applying it to $\widehat{\varphi}$ and $\widehat{\psi}$ in the places of φ and ψ, we obtain $\widehat{\widehat{\varphi} \otimes \widehat{\psi}} = 2^{2n} \, \varphi \times \psi$, according to Corollary 8.10. Again using this same corollary, we deduce Relation (8.21). ■

Relation (8.21) applied at 0 gives

$$\widehat{\varphi} \otimes \widehat{\psi}(0) = 2^n \, \widehat{\varphi \times \psi}(0) = 2^n \sum_{x \in \mathbb{F}_2^n} \varphi(x)\psi(x) = 2^n \, \varphi \otimes \psi(0). \tag{8.22}$$

Taking $\psi = \varphi$ in (8.22), we obtain *Parseval's relation.*

Corollary 8.12. *For every pseudo-Boolean function φ, we have:*

$$\sum_{u \in \mathbb{F}_2^n} \widehat{\varphi}^2(u) = 2^n \sum_{x \in \mathbb{F}_2^n} \varphi^2(x).$$

If φ takes values ± 1 only, this becomes

$$\sum_{u \in \mathbb{F}_2^n} \widehat{\varphi}^2(u) = 2^{2n}. \tag{8.23}$$

This is why, when dealing with Boolean functions, we most often prefer using the Walsh transform of f (that is, the Fourier transform of the function $\mathrm{f}_\chi = (-1)^{f(x)}$) instead of the Fourier transform of f.

Relation (8.20) leads to another relation involving the derivatives of a Boolean function.

Definition 8.2. *Let f be an n-variable Boolean function and let b be any vector in $\mathbb{F}_2^n$. We call* derivative *of f in the direction of b the Boolean function $D_b f(x) = f(x) \oplus f(x+b)$.*

For instance, the derivative of a function of the form $g(x_1, \ldots, x_{n-1}) \oplus x_n\, h(x_1, \ldots, x_{n-1})$ in the direction of $(0, \ldots, 0, 1)$ equals $h(x_1, \ldots, x_{n-1})$.

Relation (8.20) applied with $\psi = \varphi = \mathrm{f}_\chi$ implies the so-called *Wiener-Khintchine theorem*:

$$\widehat{\mathrm{f}_\chi \otimes \mathrm{f}_\chi} = \widehat{\mathrm{f}_\chi}^{\,2}. \tag{8.24}$$

We have $(\mathrm{f}_\chi \otimes \mathrm{f}_\chi)(b) = \sum_{x \in \mathbb{F}_2^n} (-1)^{D_b f(x)} = \mathcal{F}(D_b f)$ (the notation $\mathcal{F}$ was defined at Relation (8.15)). Thus, Relation (8.24) shows that $\widehat{\mathrm{f}_\chi}^{\,2}$ is the Fourier transform of the so-called *autocorrelation function* $b \mapsto \Delta_f(b) = \mathcal{F}(D_b f)$ (this property was first used in the domain of cryptography in [52]):

$$\forall u \in \mathbb{F}_2^n, \ \sum_{b \in \mathbb{F}_2^n} \mathcal{F}(D_b f)(-1)^{u \cdot b} = \widehat{\mathrm{f}_\chi}^{2}(u). \tag{8.25}$$

Applied at vector 0, this gives

$$\sum_{b \in \mathbb{F}_2^n} \mathcal{F}(D_b f) = \mathcal{F}^2(f). \tag{8.26}$$

Corollary 8.9 and Relation (8.25) imply that, for every vector subspace E of $\mathbb{F}_2^n$ and every vectors a and b (cf. [41]):

$$\sum_{u \in a+E} (-1)^{b \cdot u}\, \widehat{\mathrm{f}_\chi}^{2}(u) = |E|(-1)^{a \cdot b} \sum_{e \in b+E^\perp} (-1)^{a \cdot e} \mathcal{F}(D_e f)\,. \tag{8.27}$$

Another interesting relation was also shown in [41] (see also [249]).

Proposition 8.13. *Let E and E' be subspaces of $\mathbb{F}_2^n$ such that $E \cap E' = \{0\}$ and whose direct sum equals $\mathbb{F}_2^n$. For every $a \in E'$, let h_a be the restriction of f to the coset $a + E$ (h_a can be identified with a function on $\mathbb{F}_2^k$ where k is the dimension of E), Then*

$$\sum_{u \in E^\perp} \widehat{\mathrm{f}_\chi}^{2}(u) = |E^\perp| \sum_{a \in E'} \mathcal{F}^2(h_a)\,. \tag{8.28}$$

Proof. Every element of $\mathbb{F}_2^n$ can be written in a unique way in the form $x + a$ where $x \in E$ and $a \in E'$. For every $e \in E$, we have

$$\mathcal{F}(D_e f) = \sum_{x \in E; a \in E'} (-1)^{f(x+a) \oplus f(x+e+a)} = \sum_{a \in E'} \mathcal{F}(D_e h_a).$$

We deduce from Relation (8.27), applied with $E^\perp$ instead of E, and with $a = b = 0$, that

$$\sum_{u\in E^\perp} \widehat{f_\chi}^2(u) = |E^\perp| \sum_{e\in E} \mathcal{F}(D_e f) = |E^\perp| \sum_{e\in E} \left(\sum_{a\in E'} \mathcal{F}(D_e h_a) \right)$$
$$= |E^\perp| \sum_{a\in E'} \left(\sum_{e\in E} \mathcal{F}(D_e h_a) \right).$$

Thus, according to Relation (8.26) applied with E in the place of $\mathbb{F}_2^n$ (recall that E can be identified with $\mathbb{F}_2^k$ where k is the dimension of E), $\sum_{u\in E^\perp} \widehat{f_\chi}^2(u) = |E^\perp| \sum_{a\in E'} \mathcal{F}^2(h_a)$. ■

Fourier Transform and Linear Isomorphisms. A last relation that must be mentioned shows what the composition with a linear isomorphism implies on the Fourier transform of a pseudo-Boolean function.

Proposition 8.14. *Let φ be any pseudo-Boolean function on $\mathbb{F}_2^n$. Let M be a nonsingular $n \times n$ binary matrix and L the linear isomorphism*

$$L : \begin{pmatrix} x_1 \\ x_2 \\ \vdots \\ x_n \end{pmatrix} \mapsto M \times \begin{pmatrix} x_1 \\ x_2 \\ \vdots \\ x_n \end{pmatrix}.$$

Let us denote by M' the transpose of M^{-1} and by L' the linear isomorphism

$$L' : \begin{pmatrix} x_1 \\ x_2 \\ \vdots \\ x_n \end{pmatrix} \mapsto M' \times \begin{pmatrix} x_1 \\ x_2 \\ \vdots \\ x_n \end{pmatrix}$$

(note that L' is the adjoint operator *of L^{-1}, that is, satisfies $u \cdot L^{-1}(x) = L'(u) \cdot x$ for every x and u).*

Then

$$\widehat{\varphi \circ L} = \widehat{\varphi} \circ L'. \tag{8.29}$$

Proof. For every $u \in \mathbb{F}_2^n$, we have $\widehat{\varphi \circ L}(u) = \sum_{x\in\mathbb{F}_2^n} \varphi(L(x))(-1)^{u\cdot x} = \sum_{x\in\mathbb{F}_2^n} \varphi(x)(-1)^{u\cdot L^{-1}(x)} = \sum_{x\in\mathbb{F}_2^n} \varphi(x)(-1)^{L'(u)\cdot x}$. ■

A relationship between algebraic degree and Walsh transform was shown in [229] (see also [53]).

Proposition 8.15. *Let f be an n-variable Boolean function ($n \geq 2$), and let $1 \leq k \leq n$. Assume that the Walsh transform of f takes values divisible by 2^k (i.e., according to Relation (8.12), that its Fourier transform takes values divisible by 2^{k-1}, or equivalently, according to Relation (8.14), that all the Hamming* *distances between f and affine functions are divisible by 2^{k-1}). Then f has algebraic degree at most $n-k+1$.*

Proof. Let us suppose that f has algebraic degree $d > n-k+1$ and consider a term x^I of degree d in its algebraic normal form. The Poisson summation formula (8.18) applied to $\varphi = \mathrm{f}_\chi$ and to the vectorspace $E = \{u \in \mathbb{F}_2^n / \, \forall i \in I, \, u_i = 0\}$ gives $\sum_{u\in E} \widehat{\mathrm{f}_\chi}(u) = 2^{n-d} \sum_{x\in E^\perp} \mathrm{f}_\chi(x)$. The orthogonal $E^\perp$ of E equals $\{u \in \mathbb{F}_2^n / \, \forall i \notin I, \, u_i = 0\} = \{u \in \mathbb{F}_2^n / \, supp(u) \subseteq I\}$. According to Proposition 8.2, we have that $\sum_{x\in E^\perp} f(x)$ is not even and therefore $\sum_{x\in E^\perp} \mathrm{f}_\chi(x)$ is not divisible by 4. Hence, $\sum_{u\in E} \widehat{\mathrm{f}_\chi}(u)$ is not divisible by 2^{n-d+2} and it is therefore not divisible by 2^k. A contradiction. ■

The converse of Proposition 8.15 is obviously valid if $k=1$. It is also valid if $k=2$, because the n-variable Boolean functions of degrees at most $n-1$ are those Boolean functions of even Hamming weights, and $f(x) \oplus u \cdot x$ has degree at most $n-1$ too for every u, since $n \geq 2$. It is finally also valid for $k=n$, because the affine functions are characterized by the fact that their Walsh transforms take values $\pm 2^n$ and 0 only (more precisely, their Walsh transforms take value $\pm 2^n$ once, and all their other values are null, because of Parseval's relation). The converse is false for any other value of k. Indeed, we shall see that it is false for $k = n-1$ ($n \geq 4$), because there exist quadratic functions f whose Walsh transforms take values $\pm 2^{n/2}$ for n even, ≥ 4, and $\pm 2^{(n+1)/2}$ for n odd, ≥ 5. It is then an easy task to deduce that the converse of Proposition 8.15 is also false for any value of k such that $3 \leq k \leq n-1$: we choose a quadratic function g in four variables, whose Walsh transform value at 0 equals 2^2, that is, whose weight equals $2^3 - 2 = 6$, and we take $f(x) = g(x_1, x_2, x_3, x_4)x_5 \cdots x_l$ ($5 \leq l \leq n$). Such a function has algebraic degree $l-2$ and its weight equals 6; hence, its Walsh transform value at 0 equals $2^n - 12$ and is therefore not divisible by 2^k with $k = n-(l-2)+1 = n-l+3 \geq 3$.

It is possible to characterize the functions whose Walsh transform values are divisible by 2^{n-1}: they are the affine functions and the sums of an indicator of a *flat* – an affine space – of codimension 2 and of an affine function (they have degree at most 2 according to Proposition 8.15, and the characterization follows from the results of Subsection 8.5.2). Determining those Boolean functions whose Walsh transform is divisible by 2^k is an open problem for $3 \leq k \leq n-2$.

Note that it is possible to characterize the fact that a Boolean function has degree at most d by means of its Fourier or Walsh transform: because a Boolean function has algebraic degree at most d if and only if its restriction to any $(d+1)$-dimensional flat has an even weight, we can apply Poisson summation formula (8.17).

Characterizing the Fourier Transforms of Integer-Valued Pseudo-Boolean Functions and of Boolean Functions. Obviously, according to the inverse Fourier transform property (8.19), the Fourier transforms of integer-valued functions (resp. the Walsh transforms of Boolean functions) are those integer-valued functions over $\mathbb{F}_2^n$ whose Fourier transforms take values divisible by 2^n (resp. equal to $\pm 2^n$). Also, the Walsh transforms of Boolean functions being those integer-valued functions φ over $\mathbb{F}_2^n$ such that $\widehat{\varphi}^2$ equals the constant function 2^{2n}, they are those integer-valued functions φ such that $\widehat{\varphi \otimes \varphi} = 2^{2n}$ (according to Relation (8.20) applied with $\psi = \varphi$), that is, $\varphi \otimes \varphi = 2^{2n}\, \delta_0$. But these characterizations are not easy to use mathematically, and neither they are easily computable: they need to check 2^n divisibilities by 2^n for the Fourier transforms of integer-valued functions, and 2^n equalities for the Walsh transforms of Boolean functions.

Because the main cryptographic criteria on Boolean functions will be characterized later as properties of their Walsh transforms, it is important to have characterizations that are as simple as possible. We have seen that characterizing the NNFs of integer-valued (resp. Boolean) functions is easy (resp. easier than with Fourier transform). So it is useful to clarify the relationship between these two representations.

8.2.2.1 Fourier Transform and NNF

There is a similarity between the Fourier transform and the NNF:

- The functions $(-1)^{u\cdot x}$, $u \in \mathbb{F}_2^n$, constitute an orthogonal basis of the space of pseudo-Boolean functions, and the Fourier transform corresponds, up to normalization, to a decomposition over this basis.
- The NNF is defined similarly with respect to the (nonorthogonal) basis of monomials.

Let us see now how each representation can be expressed by means of the other.

Let $\varphi(x)$ be any pseudo-Boolean function, and let $\sum_{I\in\mathcal{P}(N)} \lambda_I x^I$ be its NNF. For every word $x \in \mathbb{F}_2^n$, we have $\varphi(x) = \sum_{I\subseteq supp(x)} \lambda_I$. Setting $b = (1, \ldots, 1)$, we have

$$\varphi(x+b) = \sum_{I\in\mathcal{P}(N)/\, supp(x)\cap I=\emptyset} \lambda_I$$

(since the support of $x + b$ equals $\mathbb{F}_2^n \setminus supp(x)$).

For every $I \in \mathcal{P}(N)$, the set $\{x \in \mathbb{F}_2^n / \, supp(x) \cap I = \emptyset\}$ is an $(n - |I|)$-dimensional vector subspace of $\mathbb{F}_2^n$. Let us denote it by E_I. Its orthogonal equals $\{u \in \mathbb{F}_2^n / \, supp(u) \subseteq I\}$. We have $\varphi(x+b) = \sum_{I\in\mathcal{P}(N)} \lambda_I\, 1_{E_I}$. Applying Propositions 8.7 (with $a = 0$) and 8.8, we deduce

$$\widehat{\varphi}(u) = (-1)^{w_H(u)} \sum_{I\in\mathcal{P}(N)\,|\, supp(u)\subseteq I} 2^{n-|I|}\lambda_I. \tag{8.30}$$

Using the same method as for computing λ_I by means of the values of f, it is an easy task to deduce

$$\lambda_I = 2^{-n}(-2)^{|I|} \sum_{u\in\mathbb{F}_2^n\,|\, I\subseteq supp(u)} \widehat{\varphi}(u). \tag{8.31}$$

Note that if φ has numerical degree at most D, then, according to Relation (8.30), we have $\widehat{\varphi}(u) = 0$ for every vector u of weight strictly greater than D and that the converse is true, according to Relation (8.31).

Applying Relation (8.30) to $\varphi(x) = P(x) = \sum_{I \in \mathcal{P}(N)} \lambda_I \, x^I$ and to $\varphi(x) = P^2(x) = \sum_{I \in \mathcal{P}(N)} \left(\sum_{J,J' \in \mathcal{P}(N) \,|\, I = J \cup J'} \lambda_J \, \lambda_{J'} \right) x^I$, with $u = 0$, we deduce from Proposition 8.5 that a polynomial $P(x) = \sum_{I \in \mathcal{P}(N)} \lambda_I \, x^I$, with integer coefficients, is the NNF of a Boolean function if and only if

$$\sum_{I \in \mathcal{P}(N)} 2^{n-|I|} \sum_{J,J' \in \mathcal{P}(N) \,|\, I = J \cup J'} \lambda_J \, \lambda_{J'} = \sum_{I \in \mathcal{P}(N)} 2^{n-|I|} \lambda_I . \qquad (8.32)$$

We observe that this relation can be verified in time polynomial in the size of $P(x)$.

Remark. *The NNF presents the interest of being a polynomial representation, but it can also be viewed as the transform that maps any pseudo-Boolean function* $f(x) = \sum_{I \in \mathcal{P}(N)} \lambda_I \, x^I$ *to the pseudo-Boolean function g defined by* $g(x) = \lambda_{supp(x)}$. *Let us denote this mapping by* Φ. *Three other transforms have also been used for studying Boolean functions:*

- *The mapping* Φ^{-1} *(the formulas relating this mapping and the Walsh transform are slightly simpler than for* Φ*; see [303]);*
- *A mapping defined by a formula similar to Relation (8.8), but in which* $supp(x) \subseteq I$ *is replaced by* $I \subseteq supp(x)$*; see [171];*
- *The inverse of this mapping.*

8.2.2.2 The Size of the Support of the Fourier Transform and Its Relationship with Cayley Graphs

Let f be a Boolean function and let G_f be the *Cayley graph* associated to f: the vertices of this graph are the elements of $\mathbb{F}_2^n$, and there is an edge between two vertices u and v if and only if the vector $u + v$ belongs to the support of f. Then (see [18]), if we multiply by 2^n the values $\widehat{f}(a)$, $a \in \mathbb{F}_2^n$, of the Fourier spectrum of f, we obtain the eigenvalues of the graph G_f (that is, by definition, the eigenvalues of the adjency matrix $(M_{u,v})_{u,v \in \mathbb{F}_2^n}$ of G_f, whose term $M_{u,v}$ equals 1 if $u + v$ belongs to the support of f and equals 0 otherwise).

As a consequence, the cardinality $N_{\widehat{f}}$ of the support $\{a \in \mathbb{F}_2^n / \; \widehat{f}(a) \neq 0\}$ of the Fourier transform of any n-variable Boolean function f is greater than or equal to the cardinality $N_{\widehat{g}}$ of the support of the Fourier transform of any restriction g of f, obtained by keeping constant some of its input bits. Indeed, the adjency matrix M_g of the Cayley graph G_g is a submatrix of the adjency matrix M_f of the Cayley graph G_f; the number $N_{\widehat{g}}$ equals the rank of M_g and is then less than or equal to the rank $N_{\widehat{f}}$ of M_f.

This property can be generalized to any pseudo-Boolean function φ. Moreover, a simpler proof is obtained by using the Poisson summation formula (8.17): let I

be any subset of $N = \{1, \dots, n\}$; let E be the vector subspace of $\mathbb{F}_2^n$ equal to $\{x \in \mathbb{F}_2^n / x_i = 0, \forall i \in I\}$; we have $E^\perp = \{x \in \mathbb{F}_2^n / x_i = 0, \forall i \in N \setminus I\}$ and the sum of E and of $E^\perp$ is direct; then, for every $a \in E^\perp$ and every $b \in E$, the equality $\sum_{u \in a+E} (-1)^{b \cdot u} \widehat{\varphi}(u) = |E| (-1)^{a \cdot b} \widehat{\psi}(a)$, where ψ is the restriction of φ to $b + E^\perp$, implies that, if $N_{\widehat{f}} = k$, that is, if $\widehat{\varphi}(u)$ is nonzero for exactly k vectors $u \in \mathbb{F}_2^n$, then clearly $\widehat{\psi}(a)$ is nonzero for at most k vectors $a \in E^\perp$.

If φ is chosen to be a Boolean function of algebraic degree d and if we choose for I a multiindex of size d such that x^I is part of the ANF of φ, then the restriction ψ has odd weight and its Fourier transform therefore takes nonzero values only. We deduce (as proved in [18]) that $N_{\widehat{\varphi}} \geq 2^d$. Notice that $N_{\widehat{\varphi}}$ equals 2^d if and only if at most one element (that is, exactly one) satisfying $\widehat{\varphi}(u) \neq 0$ exists in each coset of E, that is, in each set obtained by keeping constant the coordinates x_i such that $i \in I$.

The number $N_{\widehat{\varphi}}$ is also bounded above by $\sum_{i=0}^{D} \binom{n}{i}$, where D is the numerical degree of φ. This is a direct consequence of Relation (8.30) and of the observation that follows Relation (8.31).

The graph viewpoint also gives insight on the Boolean functions whose Fourier spectra have at most three values (see [18]).

A hypergraph can also be related to the ANF of a Boolean function f. A related (weak) upper bound on the nonlinearity of Boolean functions (see definition in Subsection 8.4.1) has been pointed out in [361].

8.3 Boolean Functions and Coding

We explained in the introduction how, in error correcting coding, the message is divided into vectors of the same length k, which are transformed into codewords of length $N > k$ before being sent over a noisy channel, in order to enable the correction of the errors of transmission (or of storage, in the case of CD, CD-ROM, and DVD) at their reception. A choice of the set of all possible codewords (called the code – let us denote it by C) allows us to correct up to t errors (in the transmission of each codeword) if and only if the Hamming distance between any two different codewords is greater than or equal to $2t + 1$ (so, if d is the minimum distance between two codewords, the code can enable to correct up to $\lfloor \frac{d-1}{2} \rfloor$ errors, where "$\lfloor \, \rfloor$" denotes the integer part). Indeed, the only information the receiver has, concerning the sent word, is that it belongs to C. In order to be always able to recover the correct codeword, he needs that, for every word y at distance at most t from a codeword x, there does not exist another codeword x' at distance at most t from y, and this is equivalent to saying that the Hamming distance between any two different codewords is greater than or equal to $2t + 1$. This necessary condition is also sufficient.[13] Thus, the problem of generating

[13] In practice, we still need to have an efficient decoding algorithm to recover the sent codeword; the naive method consisting in visiting all codewords and keeping the nearest one from the received word is inefficient because the number 2^k of codewords is too large, in general.

a good code consists in finding a set C of binary words of the same length whose *minimum distance* $\min_{a \neq b \in C} d_H(a, b)$ (where $d_H(a, b) = |\{i / a_i \neq b_i\}|$) is high.[14]

A code is called a *linear code* if it has the structure of a linear subspace of $\mathbb{F}_2^N$ where N is its length. The minimum distance of a linear code equals the minimum Hamming weight of all nonzero codewords, because the Hamming distance between two vectors equals the Hamming weight of their difference (*i.e.*, their sum, since we reduce ourselves here to binary vectors). We write that a linear code is an $[N, k, d]$-*code* if it has length N, dimension k, and minimum distance d. It can then be described by a *generator matrix* G, obtained by choosing a basis of this vectorspace and writing its elements as the rows of this matrix. The code equals the set of all the vectors of the form $u \times G$, where u ranges over $\mathbb{F}_2^k$ (and $\times$ is the matrix product), and a possible encoding algorithm is therefore the mapping $u \in \mathbb{F}_2^k \mapsto u \times G \in \mathbb{F}_2^N$. The generator matrix is well suited for generating the codewords, but it is not for checking whether a received word of length N is a codeword or not. A characterization of the codewords is obtained thanks to the generator matrix H of the *dual code* $C^\perp = \{x \in \mathbb{F}_2^N / \forall y \in C,\ x \cdot y = \bigoplus_{i=1}^N x_i\, y_i = 0\}$ (such a matrix is called a *parity-check matrix*): we have $x \in C$ if and only if $x \times H^t$ is the null vector. It is a simple matter to prove that the minimum distance of the code equals the minimum number of linearly dependent columns of H. For instance, the *Hamming code*, which has by definition for parity-check matrix the $n \times (2^n - 1)$ matrix whose columns are all the nonzero vectors of $\mathbb{F}_2^n$ in some order, has minimum distance 3. This code depends, *stricto sensu*, on the choice of the order, but we say that two binary codes are *equivalent codes* if they are equal, up to some permutation of the coordinates of their codewords.

We shall use in the sequel the notion of *covering radius* of a code: it is the smallest integer ρ such that the spheres of radius ρ centered at the codewords cover the whole space, *t*hat is, the minimum integer t such that every binary word of length N lies at Hamming distance at most t from at least one codeword, *that is*, the maximum multiplicity of errors that have to be corrected when maximum likelihood decoding is used on a binary symmetric channel. The covering radius of a code is an important parameter [110], which can be used for analyzing and improving the decoding algorithms devoted to this code.

A linear code C is a *cyclic code* if it is invariant under cyclic shifts of the coordinates (see [257]). Cyclic codes have been extensively studied in coding theory. They have useful properties that we briefly recall: representing each codeword $(c_0, \ldots, c_{N-1})$ by the polynomial $c_0 + c_1 X + \cdots + c_{N-1} X^{N-1}$, we obtain an ideal of the quotient algebra $\mathbb{F}_2[X]/(X^N + 1)$ (viewed as a set of polynomials of degrees at most $N - 1$, each element of the algebra being identified to its minimum degree representent). This algebra is a principal domain, and any (linear) cyclic code has a unique element having minimal degree, called its

[14] High with respect to some known bounds giving the necessary trade-offs between the length of the code, the minimum distance between codewords; and the number of codewords, see [257, 297].

generator polynomial. To simplify the presentation, we assume now that $N = 2^n - 1$ (which will be the case in the sequel). The generator polynomial being (as easily shown) a divisor of $X^{2^n-1} + 1$, its roots all belong to $\mathbb{F}_{2^n}^*$. The code equals the set of all those polynomials that include the roots of the generator polynomial among their own roots. The generator polynomial having all its coefficients in $\mathbb{F}_2$, its roots are of the form $\{\alpha^i,\ i \in I\}$ where $I \subseteq \mathbb{Z}/(2^n - 1)\mathbb{Z}$ is a union of cyclotomic classes of 2 modulo $2^n - 1$. The set I is called the *defining set* of the code. The elements $\alpha^i, i \in I$ are called the zeroes of the code, which has dimension $N - |I|$. The generator polynomial of $C^\perp$ is the reciprocal of the quotient of $X^{2^n-1} + 1$ by the generator polynomial of C, and its defining set therefore equals $\{2^n - 1 - i;\ i \in \mathbb{Z}/(2^n - 1)\mathbb{Z} \setminus I\}$.

A very efficient bound on the minimum distance of cyclic codes is the *BCH bound* [257]: if I contains a string $\{l + 1, \ldots, l + k\}$ of length k in $\mathbb{Z}/(2^n - 1)\mathbb{Z}$, then the cyclic code has minimum distance greater than or equal to $k + 1$. A proof of this bound (in the framework of Boolean functions) is given in the proof of Theorem 8.61. This bound is valid for cyclic codes over any finite field as well. When the length of such a cyclic code equals the order of the underlying field less 1, the set of zeros can be any set of nonzero elements of the field; when it is constituded of consecutive powers of a primitive element, the code is called a *Reed-Solomon code*.

A cyclic code C of length N being given, the extended code of C is the set of vectors $(c_{-\infty}, c_0, \ldots, c_{N-1})$, where $c_{-\infty} = c_0 \oplus \cdots \oplus c_{N-1}$. It is a linear code of length $N + 1$ and of the same dimension as C.

Cyclic codes over $\mathbb{F}_2$ can also be considered in terms of the trace function and therefore viewed as sets of Boolean functions (when their length is $2^n - 1$, recall that we assume this). Any codeword of a cyclic code with nonzeros α^i for i in the cyclotomic classes containing $u_1, \ldots, u_l$ can be represented as $\sum_{i=1}^{l} tr_n(a_i x^{-u_i})$, $a_i \in \mathbb{F}_{2^n}$.

8.3.1 Reed-Muller Codes

As explained in the introduction, every code whose length equals 2^n, for some positive integer n, can be interpreted as a set of Boolean functions. The existence of *Reed-Muller codes* comes from the following observation.

Theorem 8.16. *Any two distinct n-variable functions f and g of algebraic degrees at most r have mutual distances at least 2^{n-r}.*

Proof. In order to prove this property, it is necessary and sufficient to show that any nonzero Boolean function f of algebraic degree $d \leq r$ has weight at least 2^{n-r} (because the difference between two Boolean functions of algebraic degrees at most r has algebraic degree at most r). This can be proved by a double induction over r and n (see [257]), but there exists a simpler proof.

Let $\prod_{i \in I} x_i$ be a monomial of degree d in the ANF of f; consider the 2^{n-d} restrictions of f obtained by keeping constant the $n - d$ coordinates of x whose indices lie outside I. Each of these restrictions, viewed as a function on $\mathbb{F}_2^d$, has an ANF of degree d because, when fixing these $n-d$ coordinates, the monomial $\prod_{i\in I} x_i$ is unchanged and all the monomials different from $\prod_{i\in I} x_i$ in the ANF of f give monomials of degrees strictly less than d. Thus any such restriction has an odd (and hence a nonzero) weight (see Subsection 8.2.1). The weight of f being equal to the sum of the weights of its restrictions, f has weight at least 2^{n-d}, which completes the proof. ∎

The functions of Hamming weight 2^{n-r} and degree r have been characterized; see a proof in [257]. We now give a proof that brings a little more insight into the reasons for this characterization.

Proposition 8.17. *The Boolean functions of algebraic degree r and of Hamming weight 2^{n-r} are the indicators of $(n-r)$-dimensional flats (*i.e., *the functions whose supports are $(n-r)$-dimensional affine subspaces of $\mathbb{F}_2^n$).*

Proof. The indicators of $(n-r)$-dimensional flats have clearly Hamming weight 2^{n-r} and they have degree r, because every $(n-r)$-dimensional flat equals $\{x \in \mathbb{F}_2^n \,/\, \ell_i(x) = 1,\ \forall i = 1, \ldots, r\}$ where the ℓ_is are affine and have linearly independent linear parts, and the ANF of its indicator equals $\prod_{i=1}^r \ell_i(x)$. Conversely, let f be a function of algebraic degree r and of Hamming weight 2^{n-r}. Let $\prod_{i\in I} x_i$ be a monomial of degree r in the ANF of f, and let $J = \{1, \ldots, n\} \setminus I$. For every vector $\alpha \in \mathbb{F}_2^J$, let us denote by f_α the restriction of f to the flat $\{x \in \mathbb{F}_2^n;\ \forall j \in J,\ x_j = \alpha_j\}$. According to the proof of Theorem 8.16, and because f has Hamming weight 2^{n-r}, each function f_α is the indicator of a singleton $\{a_\alpha\}$. Let us prove that the mapping $a : \alpha \to a_\alpha$ is affine, *that is*, for every $\alpha, \beta, \gamma \in F_2^J$, we have $a_{\alpha+\beta+\gamma} = a_\alpha + a_\beta + a_\gamma$ (this will complete the proof of the proposition since, denoting by x_J the vector of $\mathbb{F}_2^J$ whose coordinates match the corresponding coordinates of x, the support of f equals the set $\{x \in \mathbb{F}_2^n \,/\, x_I = a_{x_J}\}$ and that the equality $x_I = a_{x_J}$ is equivalent to r linearly independent linear equations). Proving this is equivalent to proving that the function of Hamming weight at most 4 equal to $f_{\alpha+\beta+\gamma} \oplus f_\alpha \oplus f_\beta \oplus f_\gamma$ has algebraic degree at most $r-2$. But more generally, for every k-dimensional flat A of F_2^J, the function $\bigoplus_{\alpha\in A} f_\alpha$ has degree at most $r-k$ (this can be easily proved by induction on k, using that f has degree r). ∎

Remark.

(1) *The proof of Theorem 8.16 shows in fact that, if a monomial $\prod_{i\in I} x_i$ has coefficient 1 in the ANF of f, and if every other monomial $\prod_{i\in J} x_i$ such that $I \subset J$ has coefficient 0, then the function has weight at least $2^{n-|I|}$. Applying this observation to the Möbius transform f° of f – whose definition has been given after Relation (8.2) – shows that, if there exists a vector $x \in \mathbb{F}_2^n$ such that $f(x) = 1$ and $f(y) = 0$ for every vector $y \neq x$ whose support contains supp(x), then the ANF of f has at least $2^{n-w_H(x)}$*

terms (this was first observed in [361]). Indeed, the Möbius transform of f° is f.

(2) *The d-dimensional subspace $E = \{x \in \mathbb{F}_2^n \,/\, x_i = 0, \forall i \notin I\}$, in the proof of Theorem 8.16, is a* maximal odd weighting *subspace: the restriction of f to E has odd weight, and the restriction of f to any of its proper superspaces has even weight (i.e., the restriction of f to any coset of E has odd weight). Similarly as before, it can be proved (see [361]), that any Boolean function admitting a d-dimensional maximal odd weighting subspace E has weight at least 2^{n-d}.*

The Reed-Muller code of order r is by definition the set of all Boolean functions of algebraic degrees at most r (or, more precisely, the set of the binary words of length 2^n corresponding to (last columns of) the truth tables of these functions). Denoted by $R(r, n)$, it is an $\mathbb{F}_2$-vectorspace of dimension $1 + n + \binom{n}{2} + \cdots + \binom{n}{r}$ (since this is the number of monomials of degrees at most r, which constitute a basis of $R(r, n)$), and thus, it has $2^{1+n+\binom{n}{2}+\cdots+\binom{n}{r}}$ elements.

For $r = 1$, it equals the set of all affine functions. Notice that the weight of any nonconstant affine function being equal to the size of an affine hyperplane, it equals 2^{n-1}.

Historic Note. The Reed-Muller code $R(1, 5)$ was used in 1972 for transmitting the first black-and-white photographs of Mars. It has $2^6 = 64$ words of length $2^5 = 32$, with mutual distances at least $2^4 = 16$. Each codeword corresponded to a level of darkness (this made 64 different levels). Up to $\lfloor \frac{16-1}{2} \rfloor = 7$ errors could be corrected in the transmission of each codeword.

$R(r, n)$ is a linear code, *t*hat is, an $\mathbb{F}_2$-vectorspace. Thus, it can be described by a generator matrix G. For instance, a generator matrix of the Reed-Muller code $R(1, 4)$ is

$$G = \begin{bmatrix} 1&1&1&1&1&1&1&1&1&1&1&1&1&1&1&1\\ 0&1&0&0&0&1&1&1&0&0&0&1&1&1&0&1\\ 0&0&1&0&0&1&0&0&1&1&0&1&1&0&1&1\\ 0&0&0&1&0&0&1&0&1&0&1&1&0&1&1&1\\ 0&0&0&0&1&0&0&1&0&1&1&0&1&1&1&1 \end{bmatrix}$$

(the first row corresponds to the constant function 1, and the other rows correspond to the coordinate functions $x_1, \ldots, x_4$).[15]

The duals of Reed-Muller codes are Reed-Muller codes, as shown in the following theorem.

[15] We have chosen to order the words of length 4 by increasing weights. We could have chosen other orderings; this would have led to other codes, but equivalent ones, having the same parameters (a binary code C of length N is said to be equivalent to another binary code C' of the same length if there exists a permutation σ on $\{1, \ldots, N\}$ such that $C = \{(x_{\sigma(1)}, \ldots, x_{\sigma(N)}) / x \in C'\}$).

Theorem 8.18. *The dual*

$$R(r,n)^{\perp} = \{f \in \mathcal{BF}_n / \forall g \in R(r,n),\ f \cdot g = \bigoplus_{x \in \mathbb{F}_2^n} f(x)\,g(x) = 0\}$$

equals $R(n-r-1,n)$.

Proof. We have seen in Subsection 8.2.1 that the n-variable Boolean functions of even weights are the elements of $R(n-1,n)$. Thus, $R(r,n)^{\perp}$ is the set of those functions f such that, for every function g of algebraic degree at most r, the product function fg (whose value at any $x \in \mathbb{F}_2^n$ equals $f(x)g(x)$) has algebraic degree at most $n-1$. This is clearly equivalent to the fact that f has algebraic degree at most $n-r-1$. ■

The Reed-Muller codes are invariant under the action of the general affine group. The sets $R(r,n)$ or $R(r,n)/R(r',n)$ have been classified under this action for some values of r, of $r' < r$ and of n; see [29, 183, 185, 258, 338, 339].

The Reed-Muller code $R(r,n)$ is an extended cyclic code for every $r < n$ (see [257]): the zeros of the corresponding cyclic code ($R^*(r,n)$, the *punctured Reed-Muller code of order* r) are the elements α^j such that $1 \le j \le 2^n - 2$ and such that the 2-weight of j is at most equal to $n-r-1$.

The Problem of Determining the Weight Distributions of the Reed-Muller Codes, the MacWilliams Identity, and the Notion of Dual Distance. What are the possible distances between the words of $R(r,n)$, or equivalently the possible weights in $R(r,n)$ (or better, the weight distribution of $R(r,n)$)? The answer, which is useful for improving the efficiency of the decoding algorithms and for evaluating their complexities, is known for every n if $r \le 2$: see Subsection 8.5.2. For $r \ge n-3$, it can also be deduced from the very nice relationship, due to F. J. MacWilliams, existing between every linear code and its dual: let C be any binary linear code of length N; consider the polynomial $W_C(X,Y) = \sum_{i=0}^{N} A_i X^{N-i} Y^i$ where A_i is the number of codewords of weight i. This polynomial is called the *weight enumerator* of C and describes[16] the *weight distribution* $(A_i)_{0 \le i \le N}$ of C. Then (see [257, 297])

$$W_C(X+Y, X-Y) = |C|\, W_C(X,Y). \tag{8.33}$$

We give a sketch of the proof of this *MacWilliams identity*: we observe first that $W_C(X,Y) = \sum_{x\in C} \prod_{i=1}^{N} X^{1-x_i} Y^{x_i}$; we deduce $W_C(X+Y, X-Y) = \sum_{x\in C} \prod_{i=1}^{N} (X + (-1)^{x_i} Y)$; applying a classical method of expansion, we derive $W_C(X+Y, X-Y) = \sum_{x\in C} \sum_{b\in\mathbb{F}_2^N} \prod_{i=1}^{N} \left(X^{1-b_i}((-1)^{x_i} Y)^{b_i}\right)$ (choosing X in the ith factor $X + (-1)^{x_i} Y$ for $b_i = 0$ and $(-1)^{x_i} Y$ for $b_i = 1$; all the different possible choices are taken into account by considering all binary words b of length N). We obtain then $W_C(X+Y, X-Y) = \sum_{b\in\mathbb{F}_2^N} \left(X^{N-w_H(b)} Y^{w_H(b)} \sum_{x\in C} (-1)^{b\cdot x}\right)$, and we conclude by using Relation (8.16) with $E = C$.

[16] W_C is a homogeneous version of the classical generating series for the weight distribution of C.

The MacWilliams identity allows us, theoretically, to deduce the weight distribution of $R(n-r-1,n)$ from the weight distribution of $R(r,n)$ (in fact, to actually determine this weight distribution, it is necessary to be able to explicitely expand the factors $(X+Y)^{N-i}(X-Y)^i$ and to simplify the obtained expression for $W_C(X+Y, X-Y)$; this is possible by running a computer up to some value of n). But this gives no information for the cases $3 \leq r \leq n-4$, which remain unsolved (except for small values of n, see [17], and for $n = 2r$, because the code is then self-dual; see [257, 297]). *McEliece's theorem* [271] (or *Ax's theorem* [12]; see also the *Stickelberger theorem*, e.g., in [232, 236]) shows that the weights (and thus the distances) in $R(r,n)$ are all divisible by $2^{\lceil \frac{n}{r} \rceil -1} = 2^{\lfloor \frac{n-1}{r} \rfloor}$, where $\lceil u \rceil$ denotes the ceiling – the smallest integer greater than or equal to u – and $\lfloor u \rfloor$ denotes the integer part (this can also be shown by using the properties of the NNF; see [86]). Moreover, if f has degree d and g has degree $d' \leq d$, then $d_H(f,g) \equiv w_H(f) \left[\bmod 2^{\left\lceil \frac{n-d'}{d} \right\rceil} \right]$ [209] (see also [195]). In [35], A. Canteaut gives further properties of the weights in $f \oplus R(1,n)$. Kasami and Tokura [207] have shown that the only weights in $R(r,n)$ occurring in the range $[2^{n-r}; 2^{n-r+1}[$ are of the form $2^{n-r+1} - 2^i$ for some i, and they have completely characterized the codewords with these weights (and computed their number). The functions whose weights are between the minimum distance 2^{n-r} and 2.5 times the minimum distance have also been characterized, in [208].

The principle of MacWilliams' identity can also be applied to nonlinear codes. When C is not linear, the weight distribution of C has no great relevance. The distance distribution has more interest. We consider the *distance enumerator* of C: $D_C(X,Y) = \frac{1}{|C|}\sum_{i=0}^{N} B_i X^{N-i} Y^i$, where B_i is the size of the set $\{(x,y) \in C^2 / d_H(x,y) = i\}$. Note that, if C is linear, then $D_C = W_C$. Similarly as above, we see that $D_C(X,Y) = \frac{1}{|C|}\sum_{(x,y)\in C^2} \prod_{i=1}^{N} X^{1-(x_i \oplus y_i)} Y^{x_i \oplus y_i}$; we deduce that the polynomial $D_C(X+Y, X-Y)$ equals $\frac{1}{|C|}\sum_{(x,y)\in C^2} \prod_{i=1}^{N} (X + (-1)^{x_i \oplus y_i} Y)$. Expanding these products, we obtain $\frac{1}{|C|}\sum_{(x,y)\in C^2} \sum_{b \in \mathbb{F}_2^N} \prod_{i=1}^{N} \left(X^{1-b_i}((-1)^{x_i \oplus y_i} Y)^{b_i}\right)$, that is,

$$D_C(X+Y, X-Y) = \frac{1}{|C|}\sum_{b \in \mathbb{F}_2^N} X^{N-w_H(b)} Y^{w_H(b)} \left(\sum_{x \in C} (-1)^{b \cdot x}\right)^2. \tag{8.34}$$

Hence, $D_C(X+Y, X-Y)$ has nonnegative coefficients.

The minimum exponent of Y with nonzero coefficient in the polynomial $D_C(X+Y, X-Y)$, that is, the number $\min\{w_H(b);\, b \neq 0, \sum_{x \in C}(-1)^{b \cdot x} \neq 0\}$, is usually denoted by $d^\perp$ and is called the *dual distance* of C. Note that the maximum number j such that the sum $\sum_{x \in C}(-1)^{b \cdot x}$ is null, for every nonzero vector b of weight at most j, equals $d^\perp - 1$ (see more in [129, 130]). This property will be useful in Subsection 8.4.1.

It is shown in [51] (see the remark of Subsection 8.5.2 in the present chapter) that for every Boolean function f on $\mathbb{F}_2^n$, there exists an integer m and a Boolean

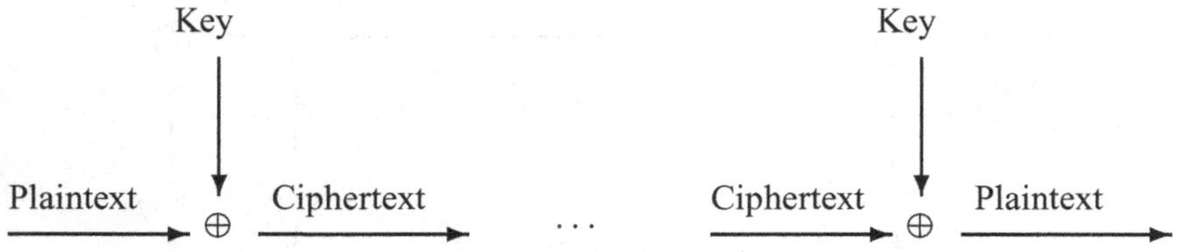

Figure 8.1. Vernam Cipher.

function g of algebraic degree at most 3 on $\mathbb{F}_2^{n+2m}$ whose Walsh transform satisfies $\widehat{g_x}(0) = 2^m \widehat{f_\chi}(0)$. This means that the weight of f is related to the weight of a function of degree at most 3 (but in a number of variables which can be exponentially larger) in a simple way. This shows that the distances in $R(3, n)$ can be very diverse, contrary to those in $R(2, n)$.

8.4 Boolean Functions and Cryptography

Stream ciphers are based on the so-called *Vernam cipher* (see Figure 8.1) in which the plaintext (a binary string of some length) is bitwise added to a (binary) secret key of the same length, in order to produce the ciphertext. The Vernam cipher is also called the *one-time pad* because a new random secret key must be used for every encryption. Indeed, the bitwise addition of two ciphertexts corresponding to the same key equals the addition of the corresponding plaintexts, which gives much information on these plaintexts (it is often enough to recover both plaintexts; some secret services and spies learned this at their own expense!).

The Vernam cipher, which is the only known cipher offering unconditional security (see [329]) if the key is truly random and if it is changed for every new encryption, was used for communication between the heads of the United States and the USSR during the cold war (the keys being carried by diplomats) and by some secret services.

In practice, because the length of the private key must be equal to the length of the plaintext, pseudorandom generators are most often used in order to minimize the size of the private key (but the unconditional security is then no longer ensured): a method is chosen for producing long *pseudorandom sequences* from short random secret keys (only the latter are actually shared; the method is supposed to be public; according to the Kerckhoff principle, only the parameters that can be used by the attacker to break the system must be kept secret). The pseudorandom sequence is used in the place of the key in a Vernam cipher. For this reason, it is also called the *keystream*. If the keystream depends only on the key (and not on the plaintext), the cipher is called *synchronous*.[17] Stream ciphers, because they operate on data units as small as a bit or a few bits, are suitable

[17] There also exist self-synchronous stream ciphers, in which each keystream bit depends on the n preceding ciphertext bits, which allows re-synchronising after n bits if an error of transmission occurs between Alice and Bob.

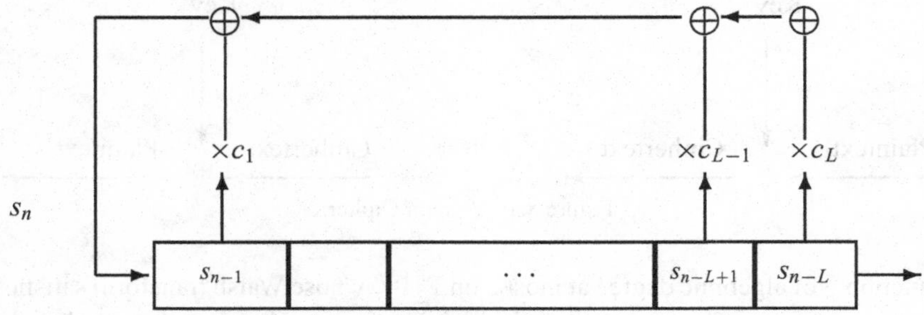

Figure 8.2. LFSR.

for fast telecommunication applications. Having also a very simple construction, they are easily implemented in both hardware and software.

The first method for generating a pseudorandom sequence from a secret key used *linear feedback shift registers* (*LFSRs*). In such an LFSR (see Figure 8.2, where $\times$ means multiplication), at every clock cycle, the bits $s_{n-1}, \ldots, s_{n-L}$ contained in the flip-flops of the LFSR move to the right. The rightmost bit is the current output and the leftmost flip-flop is fed with the linear combination $\bigoplus_{i=1}^{L} c_i s_{n-i}$, where the c_is are bits. Thus, such an LFSR outputs a recurrent sequence satisfying the relation

$$s_n = \bigoplus_{i=1}^{L} c_i s_{n-i}.$$

Such a sequence is always ultimately periodic[18] (if $c_L = 1$, then it is periodic; we shall assume that $c_L = 1$ in the sequel, because otherwise, the same sequence can be output by an LFSR of a shorter length, except for its first bit, and this can be exploited in attacks) with period at most $2^L - 1$. The generating series $s(X) = \bigoplus_{i \geq 0} s_i X^i$ of the sequence can be expressed in a nice way (see the chapter by Helleseth and Kumar in [297]): $s(X) = \frac{G(X)}{F(X)}$, where $G(X) = \bigoplus_{i=0}^{L-1} X^i \left(\bigoplus_{j=0}^{i} c_{i-j} s_j \right)$ is a polynomial of degree smaller than L and $F(X) = 1 \oplus c_1 X \oplus \cdots \oplus c_L X^L$ is the *feedback polynomial*.

The short secret key contains the initialization $s_0, \ldots, s_{L-1}$ of the LFSR and the values of the *feedback coefficients* c_i (these must be kept secret; otherwise, the observation of L consecutive bits of the key would allow recovering all the subsequent sequence).

But these LFSRs are cryptographically weak because of the *Berlekamp-Massey algorithm* [268]: let $\mathcal{L}$ be the length of a minimum length LFSR producing the same sequence (this length, called the *linear complexity* of the sequence, is assumed to be unknown from the attacker; note that it equals L if and only if the polynomials F and G above are coprime), then if we know at least $2\mathcal{L}$ consecutive bits,

[18] Conversely, every ultimately periodic sequence can be generated by an LFSR.

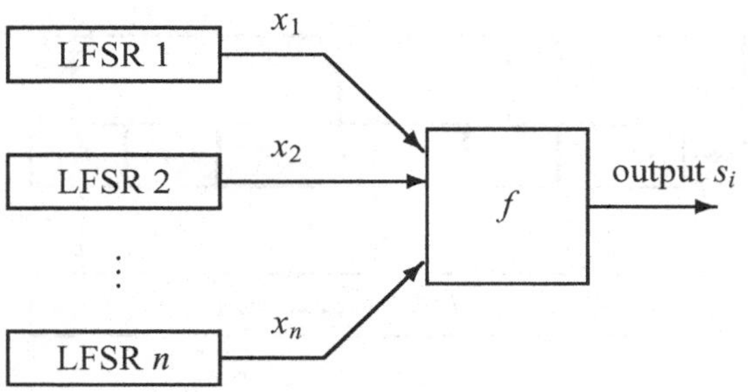

Figure 8.3. Combiner Model.

the Berlekamp-Massey algorithm recovers the values of $\mathcal{L}$ and of the feedback coefficients of an LFSR of length $\mathcal{L}$ generating the sequence, and the initialization of this LFSR in $O(\mathcal{L}^2)$ elementary operations. A modern way of avoiding this attack is by using Boolean functions. The first model that appeared in the literature for using Boolean functions is the *combiner model* (see Figure 8.3).

Notice that the feedback coefficients of the n LFSRs used in such a generator can be public. The Boolean function is also public, in general, and the short secret key gives only the initialization of the n LFSRs: if we want to use, for instance, a 128-bit-long secret key, this allows using n LFSRs of lengths $L_1, \ldots, L_n$ such that $L_1 + \cdots + L_n = 128$.

Such a system clearly outputs a periodic sequence whose period is at most the LCM of the periods of the sequences output by the n LFSRs (assuming that $c_L = 1$ in each LFSR; otherwise, the sequence is ultimately periodic). So, this sequence is also recurrent and can therefore be produced by a single LFSR. However, as we shall see, well-chosen Boolean functions allow the linear complexity of the sequence to be much larger than the sum of the lengths of the n LFSRs. Nevertheless, choosing LFSRs producing sequences of large periods,[19] choosing these periods pairwise coprime so as to have the largest possible global period and choosing f such that the linear complexity is large enough, too, are not sufficient. As we shall see, the combining function should also not leak information about the individual LFSRs and behave as differently as possible from affine functions, in several different ways.

The combiner model is only a model, useful for studying attacks and related criteria. In practice, the systems are more complex (see, for instance, at http://www.ecrypt.eu.org/stream/ how the stream ciphers of the *eSTREAM Project* are designed).

[19] *E.g.*, *m-sequences*, also called *maximum length sequences*, that is, sequences of period $2^{\mathcal{L}} - 1$ where $\mathcal{L}$ is the linear complexity – assuming that $L = \mathcal{L}$, this corresponds to taking a primitive feedback polynomial – which can be represented in the form $s_i = tr_n(a\alpha^i)$, where α is a primitive element of $\mathbb{F}_{2^n}$, and which have very strong properties; see the chapter by Helleseth and Kumar in [297].

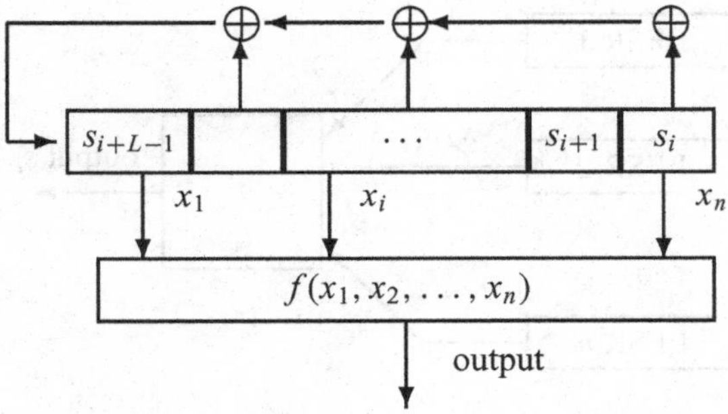

Figure 8.4. Filter Model.

An alternative model is the *filter model*, which uses a single LFSR (of a longer length). A filtered LFSR outputs $f(x_1, \ldots, x_n)$ where f is some n-variable Boolean function, called a filtering function, and where $x_1, \ldots, x_n$ are the bits contained in some flip-flops of the LFSR; see Figure 8.4.

Such a system is equivalent to the combiner model using n copies of the LFSR. However, the attacks, even when they apply to both systems, do not work similarly (a first obvious difference is that the lengths of the LFSRs are different in the two models). Consequently, the criteria that the involved Boolean functions must satisfy because of these attacks may be different for the two models and we shall have to distinguish between the two models when describing the attacks and the related criteria.

Other pseudorandom generators exist. A *feedback shift register* has the same structure as an LFSR, but the leftmost flip-flop is fed with $f(x_{i_1}, \ldots, x_{i_n})$ where $n \leq L$ and $x_{i_1}, \ldots, x_{i_n}$ are bits contained in the flip-flops of the FSR, and where f is some n-variable Boolean function. The linear complexity of the produced sequence can then be near 2^L, see [199] for general FSRs and [97] for FSRs with quadratic feedback function f. The linear complexity is difficult to study in general. Nice results similar to those on the m-sequences exist in the case of FCSRs (feedback with carry shift registers); see [8, 167, 168, 218].

Boolean functions also play an important role in block ciphers. Every block cipher admits as input a binary vector $(x_1, \ldots, x_n)$ (a block of plaintext) and outputs a binary vector $(y_1, \ldots, y_m)$; the coordinates $y_1, \ldots, y_m$ are the outputs to Boolean functions (depending on the key) whose common input is $(x_1, \ldots, x_n)$; see Figure 8.5.

But the number n of variables of these Boolean functions being large (most often, more than 100), these functions cannot be analyzed. Boolean functions on fewer variables are in fact involved in the ciphers. All known block ciphers are the iterations of a number of rounds.

We give in Figures 8.6 and 8.7 a description of the rounds of the DES and of the AES. The input to a DES round is a binary string of length 64, divided into two strings of 32-bits-each (in the figure, they enter the round, from above, on the left and on the right); confusion (see the later discussion of what this term means)

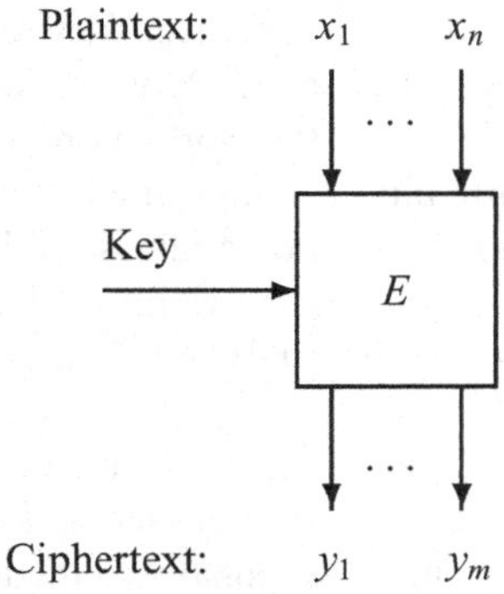

Figure 8.5. Block Cipher.

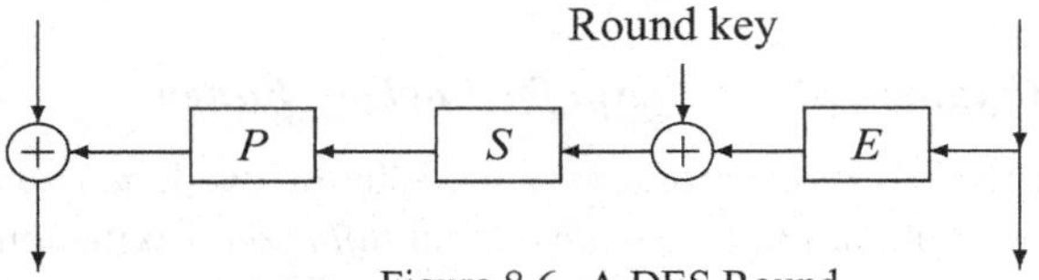

Figure 8.6. A DES Round.

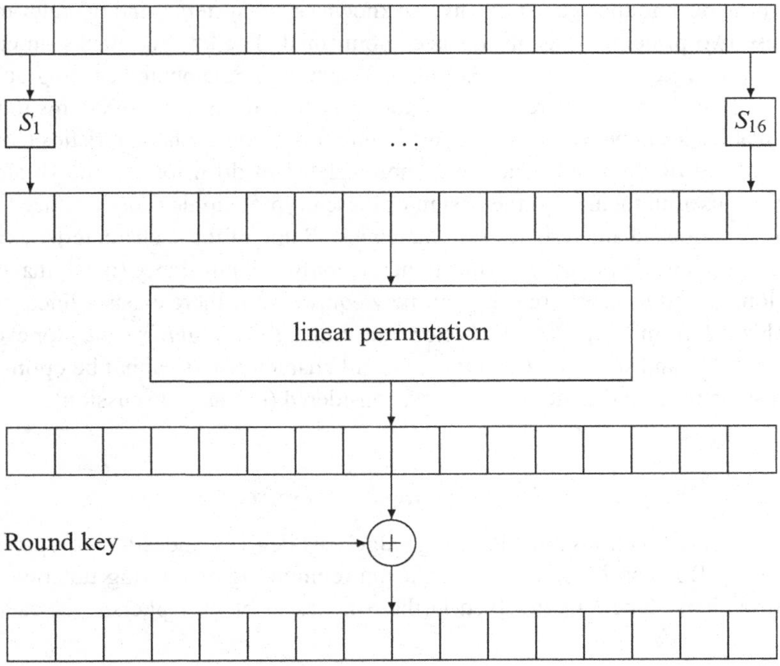

Figure 8.7. An AES Round.

is achieved by the S-box, which is a nonlinear transformation of a binary string of 48 bits[20] into a 32-bit-long one. So, 32 Boolean functions on 48 variables are involved. But, in fact, this nonlinear transformation is the concatenation of eight sub-S-boxes, which transform binary strings of 6 bits into 4-bit-long ones. So, 32 (that is, 8×4) Boolean functions on 6 variables are involved. In the (standard) AES round, the input is a 128-bit-long string, divided into 16 strings of 8 bits each; the S-box is the concatenation of 16 sub-S-boxes corresponding to 16×8 Boolean functions on 8 variables.

A block cipher being considered, the individual properties of all the involved Boolean functions can be studied (see Subsection 8.4.1), but this is not sufficient. The whole sub-S-boxes must be globally studied (see Chapter 9, "Vectorial Boolean Functions for Cryptography").

8.4.1 Cryptographic Criteria for Boolean Functions

The design of conventional cryptographic systems relies on two fundamental principles introduced by Shannon [329]: *confusion* and *diffusion*. Confusion aims at concealing any algebraic structure in the system. It is closely related to the complexity[21] of the involved Boolean functions. Diffusion consists in spreading out the influence of any minor modification of the input data or of the key over all outputs. These two principles were stated more than half a century ago. Since then, many attacks have been found against the diverse known cryptosystems, and the relevance of these two principles has always been confirmed. The known attacks on each cryptosystem lead to criteria [275, 299, 333] that the implemented cryptographic functions must satisfy. More precisely, the resistance of the cryptosystems to the known attacks can be quantified through some fundamental characteristics (some more related to confusion, and some more related to diffusion) of the Boolean functions used in them; also the design of these cryptographic functions needs to consider various characteristics simultaneously. Some of these characteristics are affine invariants: that is, they are invariant under affine equivalence (recall that two functions f and g on $\mathbb{F}_2^n$ are called affinely equivalent if there exists a linear isomorphism L from $\mathbb{F}_2^n$ to $\mathbb{F}_2^n$ and a vector a such that $f(x) = g(L(x) + a)$ for every input $x \in \mathbb{F}_2^n$) and some are not. Of course, all characteristics cannot be optimum in the same time, and trade-offs must be considered (see later discussion).

8.4.1.1 The Algebraic Degree

Cryptographic functions must have high algebraic degrees. Indeed, all cryptosystems using Boolean functions for confusion (combining or filtering functions in stream ciphers, functions involved in the S-boxes of block ciphers, ...) can be

[20] The E-box has expanded the 32-bit-long string into a 48-bit-long one.
[21] That is, the cryptographic complexity, which is different from circuit complexity, for instance.

attacked if the functions have low degrees. For instance, in the case of combining functions, if n LFSRs having lengths $L_1, \ldots, L_n$ are combined by the function

$$f(x) = \bigoplus_{I \in \mathcal{P}(N)} a_I \left(\prod_{i \in I} x_i \right),$$

where $\mathcal{P}(N)$ denotes the power set of $N = \{1, \ldots, n\}$, then (see [315]) the sequence produced by f has linear complexity

$$\mathcal{L} \leq \sum_{I \in \mathcal{P}(N)} a_I \left(\prod_{i \in I} L_i \right)$$

(and $\mathcal{L}$ equals this number under the sufficient condition that the sequences output by the LFSRs are m-sequences and the lengths of these LFSRs are pairwise coprime); see [364]. In the case of the filter model, we have a less precise result [314]: if L is the length of the LFSR and if the feedback polynomial is primitive, then the linear complexity of the sequence satisfies

$$\mathcal{L} \leq \sum_{i=0}^{d^\circ f} \binom{L}{i}.$$

Moreover, if L is a prime, then

$$\mathcal{L} \geq \binom{L}{d^\circ f},$$

and the fraction of functions f of given algebraic degree that output a sequence of linear complexity equal to $\sum_{i=0}^{d^\circ f} \binom{L}{i}$ is at least $e^{-1/L}$. In both models, the algebraic degree of f (recall that this is the largest size of I such that $a_I = 1$) has to be high so that $\mathcal{L}$ can have high value (the number of those nonzero coefficients a_I, in the ANF of f, such that I has large size, can also play a role, but clearly a less important one). In the case of block ciphers, using Boolean functions of low degrees makes the higher differential attack [215, 227] effective.

When n tends to infinity, random Boolean functions have almost surely algebraic degrees at least $n - 1$ because the number of Boolean functions of algebraic degrees at most $n - 2$ equals $2^{\sum_{i=0}^{n-2} \binom{n}{i}} = 2^{2^n - n - 1}$ and is negligible with respect to the number 2^{2^n} of all Boolean functions. But we shall see that the functions of algebraic degrees $n - 1$ or n do not allow achieving some other characteristics (balancedness, resiliency, ...).

We have seen in Subsection 8.2.1 that the algebraic degree is an affine invariant.

8.4.1.2 The Nonlinearity

In order to provide confusion, cryptographic functions must lie at large Hamming distance to all affine functions. Let us explain why. We shall say that there is a correlation between a Boolean function f and a linear function ℓ if $d_H(f, \ell)$ is different from 2^{n-1}. Because of Parseval's relation (8.23) applied to the sign function f_χ and of Relation (8.14), any Boolean function has correlation with some linear functions of its input. But this correlation should be small: the existence of affine approximations of the Boolean functions involved in a cryptosystem allows in various situations (block ciphers, stream ciphers) attacks to be built on this system (see [173, 270]).

In the case of stream ciphers, these attacks are the so-called *fast correlation attacks* [46, 108, 156, 200, 274]: let g be a linear approximation of f (or $f \oplus 1$; then we change f into $f \oplus 1$) whose distance to f is smaller than 2^{n-1}. Then, denoting by $Pr[E]$ the probability of an event E;

$$p = Pr[f(x_1, \ldots, x_n) \neq g(x_1, \ldots, x_n)] = \frac{d_H(f, g)}{2^n} = \frac{1}{2} - \varepsilon,$$

where $\varepsilon > 0$. The pseudorandom sequence s then corresponds to the transmission with errors of the sequence σ that would be produced by the same model, but with g instead of f. Attacking the cipher can be done by correcting the errors as in the transmission of the sequence σ over a noisy channel. Assume that we have N bits $s_u, \ldots, s_{u+N-1}$ of the pseudorandom sequence s; then $Pr[s_i \neq \sigma_i] \approx p$. The set of possible sequences $\sigma_u, \ldots, \sigma_{u+N-1}$ is a vectorspace, that is, a linear code of length N and dimension at most L. We then use a decoding algorithm to recover $\sigma_u, \ldots, \sigma_{u+N-1}$ from $s_u, \ldots, s_{u+N-1}$, and since g is linear, the linear complexity of the sequence σ is small and we obtain by the Berlekamp-Massey algorithm the initialization of the LFSR. We can then compute the whole sequence s.

There are several ways to perform the decoding. The method exposed in [274] and improved by [108] is as follows. We call a *parity check polynomial* any polynomial $a(x) = 1 + \sum_{j=1}^{r} a_j x^j$ $(a_r \neq 0)$ that is a multiple of the feedback polynomial of an LFSR generating the sequence σ_i. Denoting by $\sigma(x)$ the generating function $\sum_{i \geq 0} \sigma_i x^i$, the product $a(x)\sigma(x)$ is a polynomial of degree less than r. We use for the decoding a set of parity check polynomials satisfying three conditions: their degrees are bounded by some integer m, the number of nonzero coefficients a_j in each of them is at most some number t (*i.e.*, each polynomial has Hamming weight at most $t + 1$), and for every $j = 1, \ldots, m$, at most one polynomial has nonzero coefficient a_j. Each parity check polynomial $a(x) = 1 + \sum_{j=1}^{r} a_j x^j$ gives a linear relation $\sigma_i = \sum_{j=1}^{r} a_j \sigma_{i-j} = \sum_{j=1,\ldots,r \,/\, a_j \neq 0} \sigma_{i-j}$ for every $i \geq m$, and the relations corresponding to different polynomials involve different indices $i - j$. If we replace the (unknown) σ_is by the s_is, then some of these relations become false, but it is possible by using the method of Gallager [160] to compute a sequence z_i such that $Pr(z_i = \sigma_i) > 1 - p$. Then it can be proved that iterating this process converges to the sequence σ (with a speed that depends on m, t, and p).

In the case of block ciphers, we shall see in Chapter 9 that the Boolean functions involved in their S-boxes must also lie at large Hamming distances to affine functions, to allow resistance to the linear attacks.

The *nonlinearity* of f is the minimum Hamming distance between f and affine functions. The larger is the nonlinearity, the larger is p in the fast algebraic attack and the less efficient is the attack. Hence, the nonlinearity must be high (in a sense that will be clarified later), and we shall see that this condition happens to be necessary against other attacks as well. A high nonlinearity is surely one of the most important cryptographic criteria.

The nonlinearity is an affine invariant, by definition, because $d_H(f \circ L, \ell \circ L) = d_H(f, \ell)$, for every function f and ℓ and for every affine automorphism L, and because $\ell \circ L$ ranges over the whole set of affine functions when ℓ does.

It can be computed through the Walsh transform: let $\ell_a(x) = a_1 x_1 \oplus \cdots \oplus a_n x_n = a \cdot x$ be any linear function; according to Relation (8.14), we have $d_H(f, \ell_a) = 2^{n-1} - \frac{1}{2}\widehat{\mathrm{f}_\chi}(a)$, and we deduce $d_H(f, \ell_a \oplus 1) = 2^{n-1} + \frac{1}{2}\widehat{\mathrm{f}_\chi}(a)$; the nonlinearity of f is therefore equal to

$$nl(f) = 2^{n-1} - \frac{1}{2} \max_{a \in \mathbb{F}_2^n} |\widehat{\mathrm{f}_\chi}(a)|. \tag{8.35}$$

Hence, a function has high nonlinearity if all of its Walsh values have low magnitudes.

Parseval's Relation (8.23) applied to f_χ gives $\sum_{a \in \mathbb{F}_2^n} \widehat{\mathrm{f}_\chi}^{\,2}(a) = 2^{2n}$ and implies that the mean of $\widehat{\mathrm{f}_\chi}^{\,2}(a)$ equals 2^n. The maximum of $\widehat{\mathrm{f}_\chi}^{\,2}(a)$ being greater than or equal to its mean (equality occurs if and only if $\widehat{\mathrm{f}_\chi}^{\,2}(a)$ is constant), we deduce that $\max_{a \in \mathbb{F}_2^n} |\widehat{\mathrm{f}_\chi}(a)| \geq 2^{n/2}$. This implies

$$nl(f) \leq 2^{n-1} - 2^{n/2-1}. \tag{8.36}$$

This bound, valid for every Boolean function and tight for every even n, as we shall see, will be called the *covering radius bound* (since this is the value of the covering radius of the Reed-Muller code of order 1 if n is even; indeed, in the case of the Reed-Muller code of order 1, the covering radius coincides with the maximum nonlinearity of Boolean functions). The covering radius bound can be improved when we restrict ourselves, to sub-classes of functions (e.g., resilient and correlation-immune functions; see Section 8.7). A Boolean function is considered as highly nonlinear if its nonlinearity lies near the upper bound corresponding to the class of functions to which it belongs. The meaning of "near" depends on the framework; see [203]. Olejár and Stanek [288] have shown that, when n tends to infinity, random Boolean functions on $\mathbb{F}_2^n$ almost surely have nonlinearity greater than $2^{n-1} - \sqrt{n}\, 2^{\frac{n-1}{2}}$ (this is easy to prove by counting – or more precisely by upper bounding – the number of functions whose nonlinearities are bounded above by a given number; see [65]). Rodier [308] later showed more precisely that, asymptotically, almost all Boolean functions have nonlinearity between $2^{n-1} - 2^{n/2-1}\sqrt{n}\left(\sqrt{2\ln 2} + \frac{4\ln n}{n}\right)$ and $2^{n-1} - 2^{n/2-1}\sqrt{n}\left(\sqrt{2\ln 2} - \frac{5\ln n}{n}\right)$ and therefore located in the neighborhood of $2^{n-1} - 2^{n/2-1}\sqrt{2n\ln 2}$.

Equality occurs in (8.36) if and only if $|\widehat{\mathrm{f}_\chi}(a)|$ equals $2^{n/2}$ for every vector a. The corresponding functions are called *bent functions*. They exist only for even values of n, because $2^{n-1} - 2^{n/2-1}$ must be an integer (in fact, they exist for every n even; see Section 8.6). The whole of Section 8.6 is devoted to bent functions.

For n odd, Inequality (8.36) cannot be tight. The maximum nonlinearity of n-variable Boolean functions then lies between $2^{n-1} - 2^{\frac{n-1}{2}}$ (which can always be achieved *e.g.*, by quadratic functions; see Subsection 8.5.2) and $2^{n-1} - 2^{n/2-1}$. It has been shown in [177, 283] that it equals $2^{n-1} - 2^{\frac{n-1}{2}}$ when $n = 1, 3, 5, 7$, and in [294, 295], by Patterson and Wiedemann,[22] that it is strictly greater than $2^{n-1} - 2^{\frac{n-1}{2}}$ if $n \geq 15$ (a review on what was known in 1999 on the best nonlinearities of functions on odd numbers of variables was given in [154]; see also [28, 237]). This value $2^{n-1} - 2^{\frac{n-1}{2}}$ is called the *quadratic bound* because, as we already mentioned, such nonlinearity can be achieved by quadratic functions. It is also called the *bent-concatenation bound* because it can also be achieved by the concatenation of two bent functions in $n-1$ variables. Very recently it has been proved in [210] (see also [261]) that the best nonlinearity of Boolean functions in odd numbers of variables is strictly greater than the quadratic bound for any $n > 7$.

The nonlinearity of a Boolean function f equals the minimum distance of the linear code $R(1, n) \cup (f \oplus R(1, n))$. More generally, the minimum distance of a code defined as the union of cosets $f \oplus R(1, n)$ of the Reed-Muller code of order 1, where f ranges over a set $\mathcal{F}$, equals the minimum nonlinearity of the functions $f \oplus g$, where f and g are distinct and range over $\mathcal{F}$. This observation allows constructing good nonlinear codes such as Kerdock codes (see Subsection 8.6.10).

Bent functions not being balanced (*i.e.*, their values not being uniformly distributed; see later discussion), they are improper for use in cryptosystems[23] (see later discussion). For this reason, even when they exist (for n even), it is also necessary to study those functions that have large but not optimal nonlinearities, say between $2^{n-1} - 2^{\frac{n-1}{2}}$ and $2^{n-1} - 2^{n/2-1}$, among which some balanced functions exist. The maximum nonlinearity of balanced functions is unknown for any $n \geq 8$.

Two relations were first observed in [357, 360] between the nonlinearity and the derivatives of Boolean functions (here we give simpler proofs): applying Relation (8.27), relating the values of the Walsh transform of a function on a flat $a + E$ to the autocorrelation coefficients of the function on a flat $b + E^\perp$, to all linear hyperplanes $E = \{0, e\}^\perp$, $e \neq 0$, to all vectors a, and to $b = 0$, and using that $\max_{u \in E} \widehat{\mathrm{f}_{\chi}}^2(u) \geq \frac{1}{|E|} \sum_{u \in E} \widehat{\mathrm{f}_{\chi}}^2(u)$, we deduce

$$nl(f) \leq 2^{n-1} - \frac{1}{2}\sqrt{2^n + \max_{e \neq 0} |\mathcal{F}(D_e f)|}.$$

[22] It has been later proved (see [141, 325] and [216, 266]) that balanced functions with nonlinearity strictly greater than $2^{n-1} - 2^{\frac{n-1}{2}}$, and with algebraic degree $n-1$, or satisfying $PC(1)$, exist for every odd $n \geq 15$.

[23] As soon as n is large enough (say $n \geq 20$), the difference $2^{n/2-1}$ between their weights and the weight 2^{n-1} of balanced functions is very small with respect to this weight. However, according to [13, Theorem 6], 2^n bits of the pseudorandom sequence output by f are enough to distinguish it from a random sequence. Nevertheless, we shall see in Section 8.6 that highly nonlinear functions can be built from bent functions.

And the obvious relation $w_H(f) \geq \frac{1}{2}\, w_H(D_e f)$, valid for every $e \in \mathbb{F}_2^n$, leads when applied to the functions $f \oplus \ell$, where ℓ is affine, to the lower bound:

$$nl(f) \geq 2^{n-2} - \frac{1}{4} \min_{e \neq 0} |\mathcal{F}(D_e f)|. \tag{8.37}$$

Another lower bound on the nonlinearity is a consequence of Remark 2 after Theorem 8.16: if f admits a maximal odd weighting subspace E of dimension $d \geq 2$, then for every affine function ℓ, the function $f \oplus \ell$ also admits E as a maximal odd weighting subspace (because the restriction of ℓ to E and to any of its superspaces has an even weight) and thus has nonlinearity at least 2^{n-d}.

The rth-Order Nonlinearity. Changing one or a few bits in the output to a low degree Boolean function (that is, in its truth table) gives a function with high degree and does not fundamentally modify the robustness of the system using this function (explicit attacks using approximations by low-degree functions exist for block ciphers but not for all stream ciphers, however; see, e.g., [219]). A relevant criterion is the *nonlinearity profile*, that is, the sequence of the Hamming distances to the Reed-Muller code of order r, for all values of $r < n$. This distance is called the rth-order nonlinearity (and if r is not specified, the *higher order nonlinearity*) of f and denoted by $nl_r(f)$. This criterion is related to the maximum correlation of the Boolean function with respect to a subset of variables, or equivalently, to the minimal distance of the function to functions depending on a subset of variables (which plays a role with respect to the correlation attack; see Subsection 8.4.1.7), because a function depending on k variables has algebraic degree at most k. Hence the rth-order nonlinearity is a lower bound to the distance to functions depending of at most k variables. The former is much more difficult to study than the latter.

The best known asymptotic upper bound on $nl_r(f)$ is

$$2^{n-1} - \frac{\sqrt{15}}{2} \cdot (1+\sqrt{2})^{r-2} \cdot 2^{n/2} + O(n^{r-2})$$

(see [91], where a nonasymptotic – and more complex – bound is also given). Counting the number of functions whose rth-order nonlinearities are bounded above by a given number allows proving that, when n tends to infinity, there exist functions with rth order nonlinearity greater than $2^{n-1} - \sqrt{\sum_{i=0}^{r} \binom{n}{i}}\, 2^{\frac{n-1}{2}}$. But this does not help obtaining explicit functions with nonweak rth-order nonlinearity.

Computing the rth-order nonlinearity of a given function with algebraic degree strictly greater than r is a hard task for $r > 1$ (in the case of the first order, we have seen that much is known in theory and also algorithmically because the nonlinearity is related to the Walsh transform, which can be computed by the algorithm of the fast Fourier transform; but for $r > 1$, very little is known). Even the second-order nonlinearity is known only for a few peculiar functions and for functions in small numbers of variables. A nice algorithm due to G. Kabatiansky and C. Tavernier and improved and implemented by Fourquet and Tavernier [157] works well for $r = 2$ and $n \leq 11$ (in some cases, $n \leq 13$), only. It can be applied for higher orders, but it is then efficient only for very small numbers of variables. No better algorithm is known. Proving lower bounds on the rth-order nonlinearity of functions (and therefore proving their good behavior with respect to this criterion) is also a quite difficult task, even for the second order. Until recently, there had been only one attempt, by Iwata and Kurosawa [198], to construct functions with lower bounded rth-order nonlinearity. But the obtained value, $2^{n-r-3}(r+5)$, of the lower bound was small. Also, lower bounds on the rth-order nonlinearity by means of the algebraic immunity of Boolean functions have been derived (see Section 8.9), but they are small, too. In [72] is introduced a method for efficiently lower bounding the nonlinearity profile of a given function in the case lower bounds exist for the $(r-1)$th-order nonlinearities of the derivatives of f:

Proposition 8.19. *Let f be any n-variable function and r a positive integer smaller than n. We have*

$$nl_r(f) \geq \frac{1}{2} \max_{a \in F_2^n} nl_{r-1}(D_a f)$$

and

$$nl_r(f) \geq 2^{n-1} - \frac{1}{2}\sqrt{2^{2n} - 2\sum_{a \in F_2^n} nl_{r-1}(D_a f)}.$$

The first bound is easy to prove, and the second one comes from the equalities

$$nl_r(f) = 2^{n-1} - \frac{1}{2} \max_{h \in \mathcal{BF}_n \,/\, d^\circ f \leq r} \left| \sum_{x \in \mathbb{F}_2^n} (-1)^{f(x) \oplus h(x)} \right|$$

and

$$\left(\sum_{x \in \mathbb{F}_2^n} (-1)^{f(x) \oplus h(x)} \right)^2 = \sum_{a \in F_2^n} \sum_{x \in F_2^n} (-1)^{D_a f(x) \oplus D_a h(x)}.$$

Lower bounds for the second-order nonlinearities of some functions (known for being highly nonlinear) are deduced in [72], as well as bounds for the whole nonlinearity profile of the multiplicative inverse function $tr_n(x^{2^n-2})$ (used in the S-box of the AES with $n = 8$; see Chapter 9): the rth-order nonlinearity of this function is approximately bounded below by $2^{n-1} - 2^{(1-2^{-r})n}$ and therefore asymptotically equivalent to 2^{n-1}, for every fixed r. Note that the extension of the Weil bound recalled in Subsection 8.5.6 is efficient for lower bounding the rth-order nonlinearity of the inverse function only for $r = 1$. Indeed, already for $r = 2$, the univariate degree of a quadratic function in trace representation form can be bounded above by $2^{\lfloor n/2 \rfloor} + 1$ only, and this gives a bound in 2^n on the maximum magnitude of the Walsh transform and therefore no information on the nonlinearity.

8.4.1.3 Balancedness and Resiliency

Cryptographic functions must be *balanced functions* (their output must be uniformly – that is, equally – distributed over $\{0, 1\}$) for avoiding statistical dependence between the input and the output (which can be used in attacks). Notice that f is balanced if and only if $\widehat{f_\chi}(0) = \mathcal{F}(f) = 0$.

A stronger condition is necessary in the filtering model of pseudorandom generators, in order to avoid so-called *distinguishing attacks*. These attacks are able to distinguish the pseudorandom sequence $(s_i)_{i\in\mathbb{N}}$ from a random sequence. A way of doing so is to observe that the distribution of the sequences $(s_{i+\gamma_1}, \ldots, s_{i+\gamma_n})$ is not uniform, where $\gamma_1, \ldots, \gamma_n$ are the positions where the input bits to the filtering function are chosen. J. Golić [163] has observed that if the feedback polynomial of the LFSR is primitive and if the filtering function has the form $x_1 \oplus g(x_2, \ldots, x_n)$ or $g(x_1, \ldots, x_{n-1}) \oplus x_n$, then this property is satisfied. A. Canteaut [39] has proved that this condition on the function is also necessary. For choosing a filtering function, we shall have to choose a function g satisfying the cryptographic criteria listed in the present section, and use f defined by means of g in one of the two ways just described.

There is an additional condition to balancedness in the case of the combiner model: any combining function $f(x)$ must stay balanced if we keep constant some number of coordinates x_i of x.

Definition 8.3. *Let n be a positive integer and $m < n$ a nonnegative integer. An n-variable function f is called an m-*resilient function[24] *if any of its restrictions obtained by fixing at most*[25] *m of its input coordinates x_i is balanced.*

This definition resiliency was introduced by Siegenthaler[26] in [333]; it is related to an attack on the combiner model,[27] called *correlation attack*: if f is not m-resilient, then there exists a correlation between the output to the function and (at most) m coordinates of its input; if m is small, a divide-and-conquer attack due to Siegenthaler [334] uses this weakness for attacking a system using f as combining function; in the original attack by Siegenthaler, all the possible initializations of the m LFSRs corresponding to these coordinates are tested (in other words, an exhaustive search of the initializations of these specific LFSRs is done); when we

[24] More generally, a (not necessarily balanced) combining function whose output distribution probability is unaltered when any m (or, equivalently, at most m) of the inputs are kept constant is called an mth-order *correlation-immune function*. Similarly to resiliency, correlation immunity is characterized by the set of zero values in the Walsh spectrum of the function: f is mth-order correlation-immune if and only if $\widehat{f_\chi}(u) = 0$, *i.e.*, $\widehat{f}(u) = 0$, for all $u \in \mathbb{F}_2^n$ such that $1 \le w_H(u) \le m$. The notion of correlation-immune function is related to the notion of orthogonal array (see [34]). Only resilient functions are of interest as cryptographic functions (but Boolean correlation-immune functions play a role with respect to vectorial resilient functions; see Chapter 9).

[25] Or exactly m, this is equivalent.

[26] The term of *resiliency* was, in fact, introduced in [109], in relationship to another cryptographic problem.

[27] This attack has no equivalent for the filter model, where first-order resiliency seems sufficient; see more precisely in [170] the status of resiliency in the filter generator.

arrive at the correct initialization of these LFSRs, we observe a correlation (before that, the correlation is negligible, as for random pairs of sequences); now that the initializations of the m LFSRs is known, those of the remaining LFSRs can be found with an independent exhaustive search. The fast correlation attacks that we saw earlier can be more efficient if the Boolean combining function is not highly nonlinear. More precisely, Canteaut and Trabbia in [46] and Canteaut in [37] show that, to make the correlation attack on the combiner model with an m-resilient combining function as inefficient as possible, the coefficient $\widehat{\mathrm{f}_\chi}(u)$ of the function has to be small for every vector u of Hamming weight higher than but close to m. This condition is satisfied if the function is highly nonlinear. Hence we see that nonlinearity plays a role with respect to all the main attacks.

Note that when we say that a function f is m-resilient, we do not mean that m is the maximum value of k such that f is k-resilient. We will call this maximum value the *resiliency order* of f.

Resiliency has been characterized by Xiao and Massey through the Fourier and the Walsh transforms.

Theorem 8.20. ([174].) *Any n-variable Boolean function f is m-resilient if and only if $\widehat{\mathrm{f}_\chi}(u) = 0$ for all $u \in \mathbb{F}_2^n$ such that $w_H(u) \leq m$. Equivalently, f is m-resilient if and only if it is balanced and $\widehat{f}(u) = 0$ for all $u \in \mathbb{F}_2^n$ such that $0 < w_H(u) \leq m$.*

We give here a first direct proof of this fact: we apply Relation (8.28) to $E = \{x \in \mathbb{F}_2^n / \ x_i = 0, \ \forall i \in I\}$ where I is any set of indices of size m; the sum of E and $E^\perp = \{x \in \mathbb{F}_2^n / \ x_i = 0, \ \forall i \notin I\}$ is direct and equals $\mathbb{F}_2^n$; hence we can take $E' = E^\perp$ and we get $\sum_{u \in E^\perp} \widehat{\mathrm{f}_\chi}^2(u) = |E^\perp| \sum_{a \in E^\perp} \mathcal{F}^2(h_a)$, where h_a is the restriction of f to $a + E$, that is, the restriction obtained by fixing the coordinates of x whose indices belong to I to the corresponding coordinates of a. The number $\mathcal{F}(h_a)$ is null if and only if h_a is balanced, and clearly, all the numbers $\mathcal{F}(h_a)$, $a \in E^\perp$ are null if and only if all the numbers $\widehat{\mathrm{f}_\chi}(u)$, $u \in E^\perp$ are null. Because this is valid for every mutiindex I of size m, this completes the proof.

An alternate proof of this same result is obtained by applying the Poisson summation formula (8.17) to $\varphi = \mathrm{f}_\chi$, $a = 0$ and $E = \{x \in \mathbb{F}_2^n / \ x_i = 0, \ \forall i \notin I\}$, b ranging over $\mathbb{F}_2^n$. We obtain that f is m-resilient if and only if, for every b and every I of size m, we have $\sum_{u \in \mathbb{F}_2^n / \ u_i = 0, \ \forall i \notin I} (-1)^{b \cdot u} \, \widehat{\mathrm{f}_\chi}(u) = 0$, and it can easily be shown that this is equivalent to $\widehat{\mathrm{f}_\chi}(u) = 0$ for every u of weight at most m.

Theorem 8.20 shows that f is m-resilient if and only if its support has size 2^{n-1} and dual distance at least $m + 1$. Indeed, if C denotes the support of f, the dual distance of C equals the number $\min\{w_H(b);\ b \neq 0, \sum_{x \in C} (-1)^{b \cdot x} \neq 0\}$, according to Relation (8.34) and to the observation that follows it; we have, for every vector b: $\sum_{x \in C} (-1)^{b \cdot x} = \widehat{f}(b)$ and therefore, for every $b \neq 0$: $\sum_{x \in C} (-1)^{b \cdot x} = -\frac{1}{2} \widehat{\mathrm{f}_\chi}(b)$. More generally, f is mth-order correlation immune if and only if its support has dual distance at least $m + 1$. This had been observed by Delsarte in [129, 130] (see also in a paper by J. Massey [269] a generalization of this result to arrays over finite fields and other related nice results).

An easily provable related property is that, if G is the generator matrix of an $[n, k, d]$ linear code, then for every k-variable balanced function g, the n-variable function $f(x) = g(x \times G^t)$ is $(d-1)$-resilient [128] (but such a function has nonzero linear structures; see later discussion).

Contrary to the algebraic degree, to the nonlinearity, and to the balancedness, the resiliency order is not an affine invariant, except for the null order (and for the order n, but the set of n-resilient functions is empty, because of Parseval's relation). It is invariant under any translation $x \mapsto x + b$, according to Proposition 8.7 and Theorem 8.20. The symmetry group of the set of m-resilient functions and the orbits under its action have been studied in [194]).

The whole of Section 8.7 is devoted to resilient functions.

8.4.1.4 Strict Avalanche Criterion and Propagation Criterion

The *strict avalanche criterion* (SAC) was introduced by Webster and Tavares [349] and this concept was generalized into the *propagation criterion* (PC) by Preneel et al. [299] (see also [300]). The SAC, and its generalizations, are based on the properties of the derivatives of Boolean functions. These properties describe the behavior of a function whenever some coordinates of the input are complemented. Thus, they are related to the property of diffusion of the cryptosystems using the function. They must be satisfied at high levels, in particular by the Boolean functions involved in block ciphers. Let f be a Boolean function on $\mathbb{F}_2^n$ and $E \subset \mathbb{F}_2^n$. The function f satisfies the *propagation criterion PC with respect to E* if, for all $a \in E$, the derivative $D_a f(x) = f(x) \oplus f(a + x)$ (see Definition 8.2) is balanced. It satisfies $PC(l)$ if it satisfies PC with respect to the set of all nonzero vectors of weights at most l. In other words, f satisfies $PC(l)$ if the autocorrelation coefficient $\mathcal{F}(D_a f)$ is null for every $a \in \mathbb{F}_2^n$ such that $1 \leq w_H(a) \leq l$. Criterion SAC corresponds to $PC(1)$.

It is needed, for some cryptographic applications, to have Boolean functions that still satisfy $PC(l)$ when a certain number k of coordinates of the input x are kept constant (whatever these coordinates are and whatever are the constant values chosen for them). We say that such functions satisfy the *propagation criterion $PC(l)$ of order k*. This notion, introduced in [299], is a generalization of the strict avalanche criterion of order k, $SAC(k)$ (which is equivalent to $PC(1)$ of order k), introduced in [155]. Obviously, if a function f satisfies $PC(l)$ of order $k \leq n - l$, then it satisfies $PC(l)$ of order k' for any $k' \leq k$.

There exists another notion, which is similar to $PC(l)$ of order k, but stronger [299, 301] (see also [60]): a Boolean function satisfies the *extended propagation criterion $EPC(l)$ of order k* if every derivative $D_a f$, with $a \neq 0$ of weight at most l, is k-resilient.

All of these criteria are not affine invariants, in general.

A weakened version of the PC criterion has been studied in [222].

8.4.1.5 Nonexistence of Nonzero Linear Structure

We shall call the *linear kernel* of f the set of those vectors e such that $D_e f$ is a constant function. The linear kernel of any Boolean function is an $\mathbb{F}_2$-subspace of $\mathbb{F}_2^n$. Any element e of the linear kernel of f is said to be a *linear structure* of f. Nonlinear cryptographic functions used in block ciphers should have no nonzero linear structure (see [148]). The existence of nonzero (involuntary) linear structures, for the functions implemented in stream ciphers, is a potential risk that should also be avoided, despite the fact that such existence has not been used in attacks so far.

Proposition 8.21. *An n-variable Boolean function admits a nonzero linear structure if and only if it is linearly equivalent to a function of the form* $f(x_1,\ldots,x_n)=g(x_1,\ldots,x_{n-1})\oplus\varepsilon\, x_n$ *where* $\varepsilon\in\mathbb{F}_2$. *More generally, its linear kernel has dimension at least k if and only if it is linearly equivalent to a function of the form* $f(x_1,\ldots,x_n)=g(x_1,\ldots,x_{n-k})\oplus\varepsilon_{n-k+1}\,x_{n-k+1}\oplus\cdots\oplus\varepsilon_n\,x_n$ *where* $\varepsilon_{n-k+1},\ldots,\varepsilon_n\in\mathbb{F}_2$.

Indeed, if we compose f on the right with a linear automorphism L such that $L(0,\ldots,0,1)=e$ is a nonzero linear structure, we have then $D_{(0,\ldots,0,1)}(f\circ L)(x)=f\circ L(x)\oplus f\circ L(x+(0,\ldots,0,1))=f\circ L(x)\oplus f(L(x)+e)=D_e f(L(x))$. The case of dimension k is similar.

Note that, according to Proposition 8.21, if f admits a nonzero linear structure, then the nonlinearity of f is bounded above by $2^{n-1}-2^{\frac{n-1}{2}}$ (this implies that the functions obtained by Patterson and Wiedemann cannot have nonzero linear structure), because it equals twice that of g and since, g being an $(n-1)$-variable function, it has nonlinearity bounded above by $2^{n-2}-2^{\frac{n-1}{2}-1}$. Similarly, if k is the dimension of the linear kernel of f, we have straighforwardly $nl(f)\le 2^{n-1}-2^{\frac{n+k-2}{2}}$ [40].

Another characterization of linear structures [146, 228] (see also [42]) is a direct consequence of Relation (8.27), relating the values of the Walsh transform of a function on a flat $a+E$ to the autocorrelation coefficients of the function on a flat $b+E^\perp$, with $b=0$ and $E=\{0,e\}^\perp$, that is, $\sum_{u\in a+E}\widehat{\mathrm{f}_\chi}^2(u)=2^{n-1}(2^n+(-1)^{a\cdot e}\mathcal{F}(D_e f))$.

Proposition 8.22. *Let f be any n-variable Boolean function. The derivative* $D_e f$ *equals the null function (resp. the function 1) if and only if the support* $S_{\widehat{\mathrm{f}_\chi}}=\{u\in\mathbb{F}_2^n\,/\,\widehat{\mathrm{f}_\chi}(u)\neq 0\}$ *of* $\widehat{\mathrm{f}_\chi}$ *is included in* $\{0,e\}^\perp$ *(resp. its complement).*

This is a direct consequence of the relation deduced from (8.27), with $a\cdot e=1$ if $D_e f$ is null and $a\cdot e=0$ if $D_e f=1$. Notice that, if $D_e f$ is the constant function 1 for some $e\in\mathbb{F}_2^n$, then f is balanced (indeed, the relation $f(x+e)=f(x)\oplus 1$ implies that f takes the values 0 and 1 equally often). Thus, a nonbalanced function f has no nonzero linear structure if and only if there is no nonzero vector e such that $D_e f$ is null. According to Proposition 8.22, this is equivalent to saying that the support of its Walsh transform has rank n. A similar characterization exists for balanced functions by replacing the function $f(x)$ by a nonbalanced function $f(x) \oplus b \cdot x$. It is deduced in [104] (see more in [344]) that resilient functions of high orders must have linear structures.

The existence/nonexistence of nonzero linear structures is clearly an affine invariant. But, contrary to the other criteria, it is an all-or-nothing criterion. Meier and Staffelbach introduced in [275] a related criterion, leading to a characteristic (that is, a criterion that can be satisfied at levels quantified by numbers): a Boolean function on $\mathbb{F}_2^n$ being given, its *distance to linear structures* is its distance to the set of all Boolean functions admitting nonzero linear structures (among which we have all affine functions – hence, this distance is bounded above by the nonlinearity – but also other functions, such as all nonbent quadratic functions). This distance is always bounded above by 2^{n-2}. More precisely, it equals[28] $2^{n-2} - \frac{1}{4}\max_{e \in \mathbb{F}_2^{n*}} |\mathcal{F}(D_e f)|$, because a function g, which admits some nonzero vector e as a linear structure, and which lies at minimum distance from f among all such functions, can be obtained by choosing an affine hyperplane H such that $\mathbb{F}_2^n = H \cup (e + H)$ and defining $g(x) = f(x)$ for every $x \in H$ and $g(x) = f(x + e) \oplus \varepsilon$ for every $x \in (e + H)$, where ε is chosen in $\mathbb{F}_2$; the Hamming distance between f and this function g equals $|\{x \in e + H / \, D_e f(x) = \varepsilon \oplus 1\}| = \frac{1}{2}|\{x \in \mathbb{F}_2^n / \, D_e f(x) = \varepsilon \oplus 1\}| = \frac{1}{2}\left(2^{n-1} - \frac{(-1)^\varepsilon}{2}\mathcal{F}(D_e f)\right)$. Recall that $\Delta_f(e) = \mathcal{F}(D_e f)$ is the autocorrelation function of f. We see (according to Theorem 8.28) that the distance of f to linear structures equals 2^{n-2} if and only if f is bent.

8.4.1.6 Algebraic Immunity

A new kind of attack, called *algebraic attack*, has been introduced recently (see [112, 116, 150]). Algebraic attacks recover the secret key, or at least the initialization of the system, by solving a system of multivariate algebraic equations. The idea that the key bits can be characterized as the solutions of a system of multivariate equations comes from C. Shannon [329]. In practice, for cryptosystems that are robust against the usual attacks such as the Berlekamp-Massey attack, this system is too complex to be solved (its equations being highly nonlinear). However, in the case of stream ciphers, we can get a very overdefined system (i.e., a system with a number of linearly independent equations much greater than the number of unknowns). Let us consider the combiner or the filter model, with a linear part (the n LFSRs in the case of the combiner model, the single LFSR in the case of the filter model) of size N and with an n-variable Boolean function f as combining or filtering function; then there exists a linear permutation $L : \mathbb{F}_2^N \mapsto \mathbb{F}_2^N$ and a linear mapping $L' : \mathbb{F}_2^N \mapsto \mathbb{F}_2^n$ such that, denoting by $u_1, \ldots, u_N$ the initialisation of the LFSR and by $(s_i)_{i \geq 0}$ the pseudorandom

[28] Note that this again proves Relation (8.37).

sequence output by the generator, we have, for every $i \geq 0$:

$$s_i = f(L' \circ L^i(u_1, \dots, u_N)).$$

The number of equations can then be much larger than the number of unknowns. This makes less complex the resolution of the system by using the Groebner basis (see [150]), and even allows linearizing the system (i.e., obtaining a system of linear equations by replacing every monomial of degree greater than 1 by a new unknown); the resulting linear system has, however, too many unknowns and cannot be solved. Nevertheless, Courtois and Meier have had a simple but very efficient idea. Assume that there exist functions $g \neq 0$ and h of low degrees (say, of degrees at most d) such that $f \times g = h$ (where $f \times g$ denotes the Hadamard product of f and g, whose support is the intersection of the supports of f and g, we shall omit writing $\times$ in the sequel). We have then, for every $i \geq 0$:

$$s_i\, g(L' \circ L^i(u_1, \dots, u_N)) = h(L' \circ L^i(u_1, \dots, u_N)).$$

This equation in $u_1, \dots, u_N$ has degree at most d, because L and L' are linear, and the system of equations obtained after linearization can then be solved by Gaussian elimination.

Low-degree relations have been shown to exist for several well-known constructions of stream ciphers, which were immune to all previously known attacks.

Note that if we only know the existence of a nonzero low-degree multiple h of f, then the support of h being included in that of f, we have $(f \oplus 1)h = 0$, and taking $g = h$, we have the desired relation $fg = h$ (the paper [116] mentioned the existence of low-degree multiples of f for making the attack feasible). It is a simple matter to see also that the existence of functions $g \neq 0$ and h, of degrees at most d, such that $fg = h$ is equivalent to the existence of a function $g \neq 0$ of degree at most d such that $fg = 0$ or $(f \oplus 1)g = 0$. Indeed, $fg = h$ implies $f^2 g = fh$, that is (since $f^2 = f$), $f(g \oplus h) = 0$, which gives the desired equality if $g \neq h$ by replacing $g \oplus h$ by g, and if $g = h$ then $fg = h$ is equivalent to $(f \oplus 1)g = 0$. A function g such that $fg = 0$ is called an *annihilator* of f. The set of all annihilators is equal to the ideal of all the multiples of $f \oplus 1$.

Let g be a function of degree d. Let the ANF of g equal $a_0 + \sum_{i=1}^{n} a_i x_i + \sum_{1 \leq i < j \leq n} a_{i,j} x_i x_j + \dots + \sum_{1 \leq i_1 \leq \dots \leq i_d \leq n} a_{i_1,\dots i_d} x_{i_1} \dots x_{i_d}$. Note that g is an annihilator of f if and only if $f(x) = 1$ implies $g(x) = 0$. Hence, g is an annihilator of f if and only if the coefficients in its ANF satisfy the system of homogeneous linear equations that translates this fact. In this system, we have $\sum_{i=0}^{d} \binom{n}{i}$ number of variables (the coefficients of the monomials of degrees at most d) and $w_H(f)$ many equations.

The minimum degree of $g \neq 0$ such that $fg = 0$ (i.e., such that g is an annihilator of f) or $(f \oplus 1)g = 0$ (i.e., such that g is a multiple of f) is called the (standard) *algebraic immunity* of f and denoted by $AI(f)$. This important characteristic is an affine invariant. As shown in [116], the algebraic immunity of any n-variable function is bounded above[29] by $\lceil n/2 \rceil$. Indeed, the sum of the number of monomials of degrees at most $\lceil n/2 \rceil$ and of the (equal) number of the products between f and these monomials being greater than 2^n, these functions are necessarily linearly dependent elements of the 2^n-dimensional vectorspace of all Boolean functions. This linear dependence gives two functions g and h of degrees at most $\lceil n/2 \rceil$ such that $fg = h$ and $(g, h) \neq (0, 0)$, that is, $g \neq 0$.

Let us now see what the consequences are of the existence of this attack on the design of stream ciphers: let an n-variable function f, with algebraic immunity $\lceil n/2 \rceil$ be used for instance as a filtering function on an LFSR) of length $N \geq 2k$, where k is the length of the key (otherwise, it is known that the system is not robust against an attack called time-memory-data trade-off attack). Then the complexity of an algebraic attack using one annihilator of degree $\lceil n/2 \rceil$ is roughly $7\left(\binom{N}{0} + \cdots + \binom{N}{\lceil n/2 \rceil}\right)^{\log_2 7} \approx 7\left(\binom{N}{0} + \cdots + \binom{N}{\lceil n/2 \rceil}\right)^{2.8}$ (see [116]). Let us choose $k = 128$ (which is usual) and $N = 256$; then the complexity of the algebraic attack is at least 2^{80} (which is nowadays considered just sufficient complexity) for $n \geq 13$; and it is greater than the complexity of an exhaustive search, that is, 2^{128}, for $n \geq 15$. If the attacker knows several linearly independent annihilators of degree $\lceil n/2 \rceil$, then the number of variables must be enhanced! In practice, the number of variables will have to be near 20 (but this then poses a problem of efficiency of the stream cipher).

A high value of $AI(f)$ is not a sufficient property for a resistance to algebraic attacks, because of fast algebraic attacks, which work if one can find g of low degree and $h \neq 0$ of reasonable degree such that $fg = h$; see [4, 116, 176] (note, however, that fast algebraic attacks need more data than standard ones). This has been exploited in [114] to present an attack on a stream cipher called SFINKS. As before, when the number of monomials of degrees at most e, plus the number of monomials of degrees at most d, is strictly greater than 2^n – that is, when $d^\circ g + d^\circ h \geq n$, there exist g of degree at most e and h of degree at most d such that $fg = h$. An n-variable function f is then optimal with respect to fast algebraic attacks if there do not exist two functions $g \neq 0$ and h such that $fg = h$ and $d^\circ g + d^\circ h < n$. Because $fg = h$ implies $fh = ffg = fg = h$, we see that h is then an annihilator of $f \oplus 1$, and if $h \neq 0$, its degree is then at least equal to the algebraic immunity of f. This means that having a high algebraic immunity is a necessary condition not only for a resistance to standard algebraic attacks but also for a resistance to fast algebraic attacks.

The pseudorandom generator must also resist algebraic attacks on the augmented function [153], that is (considering now f as a function in N variables, to simplify description), the vectorial function $F(x)$ whose output equals the vector $(f(x), f(L(x)), \ldots, f(L^{m-1}(x)))$, where L is the (linear) update function of the linear part of the generator. Algebraic attacks can be more efficient when applied

[29] Consequently, it is bounded above by $\lceil k/2 \rceil$ if, up to affine equivalence, it depends only on k variables, and by $\lceil k/2 + 1 \rceil$ if it has a linear kernel of dimension $n - k$, because it is then equivalent, according to Proposition 8.21, to a function in k variables plus an affine function.

to the augmented function rather than to the function f itself. The efficiency of the attack depends not only on the function f, but also on the update function (and naturally also on the choice of m), because for two different update functions L and L', the vectorial functions $F(x)$ and $F'(x) = (f(x), f(L'(x)), \dots, f(L'^{m-1}(x))$ are not linearly equivalent (neither equivalent in the more general sense called CCZ-equivalence, that is, affine equivalence of the graphs of the functions; see Chapter 9). Testing the behavior of a function with respect to this attack is therefore a long-term project (all possible update functions have to be investigated).

Finally, a powerful attack on the filter generator has been introduced by S. Rønjom and T. Helleseth in [310], which also adapts the idea of algebraic attacks due to Shannon, but in a different way. The complexity of the attack is in about $\sum_{i=0}^{d} \binom{N}{i}$ operations, where d is the algebraic degree of the filter function and N is the length of the LFSR. It needs about $\sum_{i=0}^{d} \binom{N}{i}$ consecutive bits of the keystream output by the pseudorandom generator. Because d is supposed to be close to the number n of variables of the filter function, the number $\sum_{i=0}^{d} \binom{N}{i}$ is comparable to $\binom{N}{n}$, whereas in the case of a standard algebraic attack with the method due to Courtois and Meier, the complexity of the attack is in $O\left(\left(\sum_{i=0}^{AI(f)} \binom{N}{i}\right)^{\omega}\right)$ operations, where $\omega \approx 3$ is the exponent of the Gaussian reduction[30] and $AI(f)$ is the algebraic immunity of the filter function, and it needs about $\sum_{i=0}^{AI(f)} \binom{N}{i}$ consecutive bits of the keystream. Because $AI(f)$ is supposed to be close to $\lceil n/2 \rceil$, we can see that denoting by C the complexity of the Courtois-Meier attack and by C' the amount of data it needs, the complexity of the Rønjom-Helleseth attack roughly equals $C^{2/3}$ and the amount of data it needs is roughly C'^2. From the viewpoint of complexity, it is more efficient and from the viewpoint of data it is less efficient.

The whole of Section 8.9 is devoted to the algebraic immunity of Boolean functions.

8.4.1.7 Other Criteria

- The second moment of the autocorrelation coefficients

$$\mathcal{V}(f) = \sum_{e \in \mathbb{F}_2^n} \mathcal{F}^2(D_e f) \tag{8.38}$$

has been introduced by Zhang and Zheng [356] for measuring the *global avalanche criterion* (GAC) and also called the *sum-of-squares indicator*. The *absolute indicator* is by definition $\max_{e \in \mathbb{F}_2^n,\, e \neq 0} \mid \mathcal{F}(D_e f) \mid$. Both indicators are clearly affine invariants. In order to achieve good diffusion, cryptographic functions should have low sum-of-squares indicators and absolute indicators. Obviously, we have $\mathcal{V}(f) \geq 2^{2n}$, since $\mathcal{F}^2(D_0 f) = 2^{2n}$. Note that every lower

[30] As already seen, it can be taken equal to $\log_2 7 \approx 2.8$ and the coefficient in the O can be taken equal to 7, according to Strassen [337]; a still better exponent is due to Coppersmith and Winograd but the multiplicative constant is then inefficiently high for our framework.

bound of the form $\mathcal{V}(f) \geq V$ straightforwardly implies that the absolute indicator is bounded below by $\sqrt{\frac{V-2^{2n}}{2^n-1}}$. The functions achieving $\mathcal{V}(f) = 2^{2n}$ are those functions whose derivatives $D_e f(x)$, $e \neq 0$, are all balanced. We shall see in Section 8.6 that these are the bent functions. If f has a k-dimensional linear kernel, then $\mathcal{V}(f) \geq 2^{2n+k}$ (with equality if and only if f is partially bent; see later discussion).

Note that, according to Relation (8.26) applied to $D_e f$ for every e, we have

$$\mathcal{V}(f) = \sum_{a,e \in \mathbb{F}_2^n} \mathcal{F}(D_a D_e f),$$

where $D_a D_e f(x) = f(x) \oplus f(x+a) \oplus f(x+e) \oplus f(x+a+e)$ is the *second order derivative* of f.

Note also that, according to Relation (8.21) applied to $\varphi(e) = \psi(e) = \mathcal{F}(D_e f)$, we have, for any n-variable Boolean function f:

$$\forall a \in \mathbb{F}_2^n, \sum_{e \in \mathbb{F}_2^n} \widehat{f_\chi}^2(e)\widehat{f_\chi}^2(a+e) = 2^n \sum_{e \in \mathbb{F}_2^n} \mathcal{F}^2(D_e f)(-1)^{e \cdot a},$$

as shown in [41] (indeed, the Fourier transform of φ equals $\widehat{f_\chi}^2$, according to Relation (8.25)), and thus, for $a = 0$,

$$\sum_{e \in \mathbb{F}_2^n} \widehat{f_\chi}^4(e) = 2^n\, \mathcal{V}(f). \tag{8.39}$$

We have

$$\sum_{e \in \mathbb{F}_2^n} \widehat{f_\chi}^4(e) \leq \left(\sum_{e \in \mathbb{F}_2^n} \widehat{f_\chi}^2(e)\right)\left(\max_{e \in \mathbb{F}_2^n} \widehat{f_\chi}^2(e)\right) \leq 2^n \max_{e \in \mathbb{F}_2^n} \widehat{f_\chi}^4(e).$$

According to Parseval's relation $\sum_{e \in \mathbb{F}_2^n} \widehat{f_\chi}^2(e) = 2^{2n}$, we deduce, using Relation (8.39): $\max_{e \in \mathbb{F}_2^n} \widehat{f_\chi}^2(e) \geq \frac{\mathcal{V}(f)}{2^n} \geq \sqrt{\mathcal{V}(f)}$; thus, according to Relation (8.35) relating the nonlinearity to the Walsh transform, we have (as first shown in [357, 360]):

$$nl(f) \leq 2^{n-1} - 2^{-n/2-1}\sqrt{\mathcal{V}(f)} \leq 2^{n-1} - \frac{1}{2}\sqrt[4]{\mathcal{V}(f)}.$$

Denoting again by $N_{\widehat{f_\chi}}$ the cardinality of the support $\{a \in \mathbb{F}_2^n / \widehat{f_\chi}(a) \neq 0\}$ of the Walsh transform of f, Relation (8.39) also implies the following relation, first observed in [360]: $\mathcal{V}(f) \times N_{\widehat{f_\chi}} \geq 2^{3n}$. Indeed, using for instance the Cauchy-Schwartz inequality, we see that $\left(\sum_{a \in \mathbb{F}_2^n} \widehat{f_\chi}^2(a)\right)^2 \leq \left(\sum_{a \in \mathbb{F}_2^n} \widehat{f_\chi}^4(a)\right) \times N_{\widehat{f_\chi}}$ and we have $\sum_{a \in \mathbb{F}_2^n} \widehat{f_\chi}^2(a) = 2^{2n}$, according to Parseval's relation (8.23). Clearly, the functions satisfying $nl(f) = 2^{n-1} - 2^{-n/2-1}\sqrt{\mathcal{V}(f)}$ (resp. $\mathcal{V}(f) \times N_{\widehat{f_\chi}} = 2^{3n}$) are the functions whose Walsh transforms take at most one nonzero magnitude. These functions are called *plateaued functions* (see Subsection 8.6.8 for further properties of plateaued functions). The functions satisfying $nl(f) = 2^{n-1} - \frac{1}{2}\sqrt[4]{\mathcal{V}(f)}$ are (also clearly) the bent functions.

Constructions of balanced Boolean functions with low absolute indicators and high nonlinearities have been studied in [259].

- The *maximum correlation* of an n-variable Boolean function f with respect to a subset I of $N = \{1, \ldots, n\}$ equals by definition (see [355])

$$C_f(I) = \max_{g \in \mathcal{BF}_{I,n}} \frac{\mathcal{F}(f \oplus g)}{2^n},$$

where $\mathcal{BF}_{I,n}$ is the set of n-variable Boolean functions depending on $\{x_i, i \in I\}$ only. According to Relation (8.13), the distance from f to $\mathcal{BF}_{I,n}$ equals $2^{n-1}(1 - C_f(I))$. As we saw earlier, denoting the size of I by r, this distance is bounded below by the rth-order nonlinearity.

The maximum correlation of any combining function with respect to any subset I of small size should be small (*i.e.*, its distance to $\mathcal{BF}_{I,n}$ should be high). It is straightforward to prove, by decomposing the sum $\mathcal{F}(f \oplus g)$, that $C_f(I)$ equals $\sum_{j=1}^{2^{|I|}} \frac{|\mathcal{F}(h_j)|}{2^n}$, where $h_1, \ldots, h_{2^{|I|}}$ are the restrictions of f obtained by keeping constant the x_i's for $i \in I$, and to see that the distance from f to $\mathcal{BF}_{I,n}$ is achieved by the functions g taking value 0 (resp. 1) when the corresponding value of $\mathcal{F}(h_j)$ is positive (resp. negative), and that we have $C_f(I) = 0$ if and only if all h_j's are balanced (thus, f is m-resilient if and only if $C_f(I) = 0$ for every set I of size at most m). Also, according to the Cauchy-Schwartz inequality, we have $\left(\sum_{j=1}^{2^{|I|}} |\mathcal{F}(h_j)|\right)^2 \leq 2^{|I|} \sum_{j=1}^{2^{|I|}} \mathcal{F}^2(h_j)$, and Relation (8.28) directly implies the following inequality observed in [37]:

$$C_f(I) \leq 2^{-n} \left(\sum_{u \in \mathbb{F}_2^n /\, u_i = 0,\ \forall i \notin I} \widehat{\mathrm{f}_\chi}^2(u) \right)^{\frac{1}{2}} \leq 2^{-n+\frac{|I|}{2}} \left(2^n - 2nl(f)\right) \quad (8.40)$$

or equivalently:

$$d_H(f, \mathcal{BF}_{I,n}) \geq 2^{n-1} - \frac{1}{2} \left(\sum_{\substack{u \in \mathbb{F}_2^n / \\ supp(u) \subseteq I}} \widehat{\mathrm{f}_\chi}^2(u) \right)^{\frac{1}{2}} \geq 2^{n-1} - 2^{\frac{|I|}{2}-1} \max_{u \in \mathbb{F}_2^n} \widehat{\mathrm{f}_\chi}(u)|.$$

This inequality shows that, contrary to the case of approximations by functions of algebraic degrees at most r (higher order nonlinearity), it is sufficient that the first-order nonlinearity of a combining function be high for avoiding close approximations of f by functions of $\mathcal{BF}_{I,n}$ (when I has small size).

An affine invariant criterion related to the maximum correlation and also related to the "distance to linear structures" is the following: the distance to the Boolean functions g such that the space $\{e \in \mathbb{F}_2^n / D_e g = 0\}$ has dimension at least k (the functions of $\mathcal{BF}_{I,n}$ can be viewed as n-variable functions g such that the set $\{e \in \mathbb{F}_2^n / D_e g = 0\}$ contains $\mathbb{F}_2^{N \setminus I}$). The results on the maximum correlation just given generalize to this criterion [37].

- The main cryptographic *complexity criteria* for a Boolean function are the algebraic degree and the nonlinearity, but other criteria have also been studied: the minimum number of terms in the algebraic normal forms of all affinely equivalent functions, called the *algebraic thickness* (studied in [65] and first evoked in [275]); the maximum dimension k of those flats E such that the restriction of f to E is constant (f is then called a *k-normal function*) or is affine (f is called a *k-weakly-normal function*) [65] (see Subsection 8.5.4); and the number of nonzero coefficients of the Walsh transform [301, 313]. It has been shown in [65, 288, 313] that (asymptotically) almost all Boolean functions have high complexities with respect to all these criteria.

  For every even integer k such that $4 \leq k \leq 2^n$, the kth-order *nonhomomorphicity* [359] of a Boolean function equals the number of k-tuples $(u_1, \ldots, u_k)$ of vectors of $\mathbb{F}_2^n$ such that $u_1 + \cdots + u_k = 0$ and $f(u_1) \oplus \cdots \oplus f(u_k) = 0$. It is a simple matter to show (more directly than in [359]) that it equals $2^{(k-1)n-1} + 2^{-n-1} \sum_{u \in \mathbb{F}_2^n} \widehat{\mathrm{f}_\chi}^k(u)$. This parameter should be small (but no related attack exists on stream ciphers). It is maximum and equals $2^{(k-1)n}$ if and only if the function is affine. It is minimum and equals $2^{(k-1)n-1} + 2^{\frac{nk}{2}-1}$ if and only if the function is bent, and some relationship obviously exists between nonhomomorphicity and nonlinearity.

Conclusion of This Subsection. As we can see, there are numerous cryptographic criteria for Boolean functions. The ones that must be necessarily satisfied are balancedness, a high algebraic degree, a high nonlinearity, a high algebraic immunity, and a good resistance to fast algebraic attacks. It is difficult but not impossible to find functions satisfyingly good trade-offs between all these criteria (see Section 8.9). It is not clear whether it is possible to achieve additionally resiliency of a sufficient order, which is necessary for the combiner model. Hence, the filter model may be more appropriate (future research will determine this). Once we know the foregoing criteria are satisfied by some function f (except resiliency), it is a simple matter to render f 1-resilient by composing it with a linear automorphism (we just need for this that there exist n linearly independent vectors at which the Walsh transform vanishes). First-order resiliency is useful for resisting some distinguishing (less dreadful) attacks.

8.5 Classes of Functions for Which Restrictions on the Possible Values of the Weights, Walsh Spectra and Nonlinearities Can Be Proved

8.5.1 *Affine Functions*

The weights and the Walsh spectra of affine functions are peculiar: the Walsh transform of the function $\ell(x) = a \cdot x \oplus \varepsilon$ takes null value at every vector $u \neq a$ and takes value $2^n (-1)^\varepsilon$ at a.

Concatenating affine functions gives the so-called Maiorana-McFarland functions: for every n-variable function f, if we order all the binary words of length n in lexicographic order, with the bit of higher weight on the right (for instance), then the truth table of f is the concatenation of the restrictions of f obtained by setting the values of the (say) s last bits of the input and letting the others freely range over $\mathbb{F}_2$. If all these restrictions are affine, then f is called a Maiorana-McFarland function. These Maiorana-McFarland functions are studied in Section 8.6 (Subsection 8.6.4, for bent functions) and Section 8.7 (Subsection 8.7.5, for resilient functions). The computations of their weights, Walsh spectra, and nonlinearities are easier than for general Boolean functions and in some cases can be completely determined.

8.5.2 Quadratic Functions

The behavior of the functions of $R(2, n)$, called *quadratic functions*, is also peculiar. Recall that Relation (8.26) states that, for every Boolean function f:

$$\mathcal{F}^2(f) = \sum_{b \in \mathbb{F}_2^n} \mathcal{F}(D_b f).$$

If f is quadratic, then $D_b f$ is affine for every $b \in \mathbb{F}_2^n$ and is therefore either balanced or constant. Because $\mathcal{F}(g) = 0$ for every balanced function g, we deduce

$$\mathcal{F}^2(f) = 2^n \sum_{b \in \mathcal{E}_f} (-1)^{D_b f(0)}, \tag{8.41}$$

where $\mathcal{E}_f$ is the set of all $b \in \mathbb{F}_2^n$ such that $D_b f$ is constant. The set $\mathcal{E}_f$ is the linear kernel of f (see Subsection 8.4.1). In the case of quadratic functions, it also equals the kernel $\{x \in \mathbb{F}_2^n / \forall y \in \mathbb{F}_2^n,\ \varphi_f(x, y) = 0\}$ of the symplectic (*i.e.*, bilinear, symmetric, and null over the diagonal) form associated to f: $\varphi_f(x, y) = f(0) \oplus f(x) \oplus f(y) \oplus f(x + y)$. The restriction of the function $b \mapsto D_b f(0) = f(b) \oplus f(0)$ to this vectorspace is linear, as can be easily checked; we deduce that $\mathcal{F}^2(f)$ equals $2^n\, |\mathcal{E}_f|$ if this linear form on $\mathcal{E}_f$ is null, that is, if f is constant on $\mathcal{E}_f$, and is null otherwise. According to Relation (8.13), this proves the following theorem.

Theorem 8.23. *Any quadratic function f is balanced if and only if its restriction to its linear kernel $\mathcal{E}_f$ (i.e., the kernel of its associated symplectic form) is not constant. If it is not balanced, then its weight equals $2^{n-1} \pm 2^{\frac{n+k}{2}-1}$ where k is the dimension of $\mathcal{E}_f$.*

Note that Theorem 8.23 implies that f is balanced if and only if there exists $b \in \mathbb{F}_2^n$ such that the derivative $D_b f(x) = f(x) \oplus f(x + b)$ equals the constant function 1 (take b in $\mathcal{E}_f$ such that $f(b) \neq f(0)$). For general Boolean functions, this condition is sufficient for f being balanced, but it is not necessary.

Theorem 8.23 applied to $f \oplus \ell$, where ℓ is a linear function such that $f \oplus \ell$ is not balanced (such a function ℓ always exists, according to Parseval's relation) shows that the codimension of $\mathcal{E}_f$ must be even (this codimension is the *rank of* φ_f).

The weight of a quadratic function can be any element of the set $\{2^{n-1}\} \cup \{2^{n-1} \pm 2^i;\ i = \lceil \frac{n}{2} \rceil - 1, \ldots, n-1\}$. Its nonlinearity can be any element of the set $\{2^{n-1} - 2^i;\ i = \frac{n}{2} - 1, \ldots, n-1\}$, and if f has weight $2^{n-1} \pm 2^i$, then for every affine function l, the weight of the function $f \oplus l$ belongs to the set $\{2^{n-1} - 2^i, 2^{n-1}, 2^{n-1} + 2^i\}$.

Determining whether the weight is $2^{n-1} - 2^i$ or $2^{n-1} + 2^i$ (when the function is not balanced) and more generally studying the sign of the Walsh transform is in general much more difficult than determining the value of i, or equivalently the magnitude of the Walsh transform. In [226] the sign of the values of the Walsh transform of Gold and Kasami functions is studied. The former are quadratic (the latter are not but they are related to quadratic functions; see Chapter 9). In [164], the result of [226] is generalized: for every AB power function x^d over $\mathbb{F}_{2^n}$ (see definition in Chapter 9) whose restriction to any subfield of $\mathbb{F}_{2^n}$ is also AB, the value $\sum_{x \in \mathbb{F}_{2^n}} (-1)^{tr_n(x^d + x)}$ equals $2^{\frac{n+1}{2}}$ if $n \equiv \pm 1$ [mod 8] and $-2^{\frac{n+1}{2}}$ if $n \equiv \pm 3$ [mod 8].

Every quadratic nonaffine function f has a monomial of degree 2 in its ANF, and we can assume without loss of generality that, up to a nonsingular linear transformation, this monomial is $x_1 x_2$. The function then has the form $x_1 x_2 \oplus x_1 f_1(x_3, \ldots, x_n) \oplus x_2 f_2(x_3, \ldots, x_n) \oplus f_3(x_3, \ldots, x_n)$ where f_1, f_2 are affine functions and f_3 is quadratic. Then, $f(x)$ equals $(x_1 \oplus f_2(x_3, \ldots, x_n))(x_2 \oplus f_1(x_3, \ldots, x_n)) \oplus f_1(x_3, \ldots, x_n) f_2(x_3, \ldots, x_n) \oplus f_3(x_3, \ldots, x_n)$ and is therefore affinely equivalent to the function $x_1 x_2 \oplus f_1(x_3, \ldots, x_n) f_2(x_3, \ldots, x_n) \oplus f_3(x_3, \ldots, x_n)$. Applying this method recursively shows the following.

Theorem 8.24. *Every quadratic nonaffine function is affinely equivalent to* $x_1 x_2 \oplus \cdots \oplus x_{2l-1} x_{2l} \oplus x_{2l+1}$ *(where* $l \le \frac{n-1}{2}$*) if it is balanced, to* $x_1 x_2 \oplus \cdots \oplus x_{2l-1} x_{2l}$ *(where* $l \le n/2$*) if it has weight smaller than* 2^{n-1} *and to* $x_1 x_2 \oplus \cdots \oplus x_{2l-1} x_{2l} \oplus 1$ *(where* $l \le n/2$*) if it has weight greater than* 2^{n-1}.

This allows precisely describing the weight distribution of $R(2, n)$ [257].

Remark. *Let* f_1, f_2 *and* f_3 *be any Boolean functions on* $\mathbb{F}_2^n$. *Define the function on* $\mathbb{F}_2^{n+2}$: $f(x, y_1, y_2) = y_1 y_2 \oplus y_1 f_1(x) \oplus y_2 f_2(x) \oplus f_3(x)$. *Then we have*

$$\mathcal{F}(f) = \sum_{x \in \mathbb{F}_2^n / \, y_1, y_2 \in \mathbb{F}_2} (-1)^{(y_1 \oplus f_2(x))(y_2 \oplus f_1(x)) \oplus f_1(x) f_2(x) \oplus f_3(x)}$$

$$= \sum_{x \in \mathbb{F}_2^n / \, y_1, y_2 \in \mathbb{F}_2} (-1)^{y_1 y_2 \oplus f_1(x) f_2(x) \oplus f_3(x)} = 2 \sum_{x \in \mathbb{F}_2^n} (-1)^{f_1(x) f_2(x) \oplus f_3(x)}.$$

So, starting with a function $g = f_1 f_2 \oplus f_3$, *we can relate* $\mathcal{F}(g)$ *to* $\mathcal{F}(f)$, *on two more variables, in which the term* $f_1 f_2$ *has been replaced by* $y_1 y_2 \oplus y_1 f_1(x) \oplus y_2 f_2(x)$. *Applying this repeatedly ("breaking" this way all the monomials of degrees at least 4), this allows showing easily (see [51]) that, for every Boolean* *function g on $\mathbb{F}_2^n$, there exists an integer m and a Boolean function f of algebraic degree at most 3 on $\mathbb{F}_2^{n+2m}$ whose Walsh transform takes value $\widehat{f_\chi}(0) = 2^m\, \widehat{g_\chi}(0)$ at 0. As we have mentioned, this proves that the functions of algebraic degree 3 can have weights much more diverse than functions of degrees at most 2.*

The trace representation of quadratic functions is $tr_n\left(\beta_\emptyset + \sum_{i=0}^{\frac{n-1}{2}} \beta_i\, x^{2^i+1}\right)$ for n odd and $tr_n\left(\beta_\emptyset + \sum_{i=0}^{\frac{n}{2}-1} \beta_i\, x^{2^i+1}\right) + tr_{\frac{n}{2}}(\gamma x^{2^{n/2}+1})$ for n even, where the β_is belong to $\mathbb{F}_{2^n}$ and γ belongs to $\mathbb{F}_{2^{n/2}}$. For n odd, the quadratic functions of nonlinearity $2^{n-1} - 2^{\frac{n-1}{2}}$ (called *semibent functions*) of the form $tr_n(\sum_{i=1}^{(n-1)/2} c_i x^{2^i+1})$ have been studied in [105]. Further functions of this kind have been given and studied in [197, 217].

Concatenating quadratic functions gives a superclass of the class of Maiorana-McFarland functions, studied in [63] and presented in Section 8.7 (Subsection 8.7.5.2).

8.5.3 Indicators of Flats

As we have already seen, a Boolean function f is the indicator of a flat A of codimension r if and only if it has the form $f(x) = \prod_{i=1}^{r}(a_i \cdot x \oplus \varepsilon_i)$ where $a_1, \ldots, a_r \in \mathbb{F}_2^n$ are linearly independent and $\varepsilon_1, \ldots, \varepsilon_r \in \mathbb{F}_2$. Then f has weight 2^{n-r}. Moreover, set $a \in \mathbb{F}_2^n$. If a is linearly independent of $a_1, \ldots, a_r$, then the function $f(x) \oplus a \cdot x$ is balanced (and hence $\widehat{f_\chi}(a) = 0$), because it is linearly equivalent to a function of the form $g(x_1, \ldots, x_r) \oplus x_{r+1}$. If a is linearly dependent on $a_1, \ldots, a_r$, say $a = \sum_{i=1}^{r} \eta_i\, a_i$, then $a \cdot x$ takes constant value $\bigoplus_{i=1}^{r} \eta_i\,(a_i \cdot x) = \bigoplus_{i=1}^{r} \eta_i\,(\varepsilon_i \oplus 1)$ on the flat; hence, $\widehat{f}(a) = \sum_{x\in A}(-1)^{a\cdot x}$ equals then $2^{n-r}(-1)^{\bigoplus_{i=1}^{r} \eta_i\,(\varepsilon_i\oplus 1)}$. Thus, if $a = \sum_{i=1}^{r} \eta_i\, a_i \neq 0$, then we have $\widehat{f_\chi}(a) = -2^{n-r+1}(-1)^{\bigoplus_{i=1}^{r} \eta_i\,(\varepsilon_i\oplus 1)}$, and we have $\widehat{f_\chi}(0) = 2^n - 2|A| = 2^n - 2^{n-r+1}$.

Note that the nonlinearity of f equals 2^{n-r} and is bad as soon as $r \geq 2$. But indicators of flats can be used to design Boolean functions with good nonlinearities: concatenating sums of indicators of flats and of affine functions gives another superclass of the Maiorana-McFarland functions, studied in [67] and presented in Section 8.7 (Subsection 8.7.5.2).

Note. As recalled in Section 8.3.1, the functions of $R(r, n)$ whose weights occur in the range $[2^{n-r}; 2^{n-r+1}[$ have been characterized by Kasami and Tokura [207]; any such function is the product of the indicator of a flat and of a quadratic function or is the sum (modulo 2) of two indicators of flats. The Walsh spectra of such functions can also be precisely computed.

8.5.4 Normal Functions

Let E and E' be subspaces of $\mathbb{F}_2^n$ such that $E \cap E' = \{0\}$ and whose direct sum equals $\mathbb{F}_2^n$. Denote by k the dimension of E. For every $a \in E'$, let h_a be the restriction of f to the coset $a + E$. Then, Relation (8.28) in Proposition 8.13

implies

$$\max_{u\in\mathbb{F}_2^n} \widehat{\mathrm{f}_\chi}^2(u) \geq \sum_{a\in E'} \mathcal{F}^2(h_a)$$

(indeed, the maximum of $\widehat{\mathrm{f}_\chi}^2(u)$ is greater than or equal to its mean). Hence we have $\max_{u\in\mathbb{F}_2^n} \widehat{\mathrm{f}_\chi}^2(u) \geq \mathcal{F}^2(h_a)$ for every a. Applying this property to $f \oplus \ell$, where ℓ is any linear function, and using Relation (8.35) relating the nonlinearity of a function to the maximum magnitude of its Walsh transform, we deduce:

$$\forall a \in E',\ nl(f) \leq 2^{n-1} - 2^{k-1} + nl(h_a). \tag{8.42}$$

This bound was first proved (in a different way) by Zheng et al. in [361]. The present proof is from [41]. Relation (8.42) can also be deduced from the Poisson summation formula (8.17) applied to the sign function of f, and in which the roles of E and $E^\perp$ are exchanged: let us choose $b \in \mathbb{F}_2^n$ such that $\left|\sum_{x\in a\oplus E}(-1)^{f(x)\oplus b\cdot x}\right|$ is maximum, that is, equals $\left(2^k - 2nl(h_a)\right)$. Then

$$\left|\sum_{u\in b\oplus E^\perp} (-1)^{a\cdot u}\, \widehat{\mathrm{f}_\chi}(u)\right| = |E^\perp| \left(2^k - 2nl(h_a)\right).$$

Then the mean of $(-1)^{a\cdot u}\widehat{\mathrm{f}_\chi}(u)$, when u ranges over $b \oplus E^\perp$, is equal to $\pm\left(2^k - 2nl(h_a)\right)$. Thus, the maximum magnitude of $\widehat{\mathrm{f}_\chi}(u)$ is greater than or equal to $2^k - 2nl(h_a)$. This implies Relation (8.42). These two methods, for proving (8.42) lead to two different necessary conditions for the case of equality (see [65]).

Relation (8.42) implies in particular that, if the restriction of f to a k-dimensional flat of $\mathbb{F}_2^n$ is affine (say equals ℓ), then $nl(f) \leq 2^{n-1} - 2^{k-1}$, and that, if equality occurs, then $f \oplus \ell$ is balanced on every other coset of this flat.

Definition 8.4. *A function is called k-weakly normal (resp. k-normal) if its restriction to some k-dimensional flat is affine (resp. constant).*

H. Dobbertin introduced this terminology by calling normal the functions that we call $n/2$-normal here (we shall also call normal the $n/2$-normal functions, in the sequel). He used this notion for constructing balanced functions with high nonlinearities (see Subsection 8.7.5.1). It is proved in [65] that, for every $\alpha > 1$, when n tends to infinity, random Boolean functions are almost surely $[\alpha \log_2 n]$-nonnormal. This means that almost all Boolean functions have high complexity with respect to this criterion. As usual, the proof of existence of nonnormal functions does not give examples of such functions. Alon, Goldreich, Hastad, and Peralta give in [2] several constructions of functions that are nonconstant on flats of dimension $n/2$. This is not explicitly mentioned in the paper. What is shown is that the functions are not constant on flats defined by equations $x_{i_1} = a_1, \ldots, x_{i_{n/2}} = a_{n/2}$. As the proof still works when composing the function by an affine automorphism, it implies the result.

There are also explicit constructions that work for dimensions $(1/2-\varepsilon)n$, for some small $\varepsilon>0$ very recently found by Jean Bourgain [23].

Functions that are nonconstant on flats of dimensions n^δ for every $\delta>0$ are also given in [14]. These constructions are very good asymptotically (but may not be usable to obtain functions in explicit numbers of variables).

As far as we know, no construction is known below n^δ.

8.5.5 Functions Admitting Partial Covering Sequences

The notion of covering sequence of a Boolean function has been introduced in [94].

Definition 8.5. *Let f be an n-variable Boolean function. An integer-valued*[31] *sequence $(\lambda_a)_{a\in\mathbb{F}_2^n}$ s called a* covering sequence *of f if the integer-valued function $\sum_{a\in\mathbb{F}_2^n}\lambda_a D_a f(x)$ takes a constant value. This constant value is called the* level of a covering sequence. *If the level is nonzero, we say that the covering sequence is a* nontrivial covering sequence.

Note that the sum $\sum_{a\in\mathbb{F}_2^n}\lambda_a D_a f(x)$ involves both kinds of additions: the addition $\sum$ in $\mathbb{Z}$ and the addition $\oplus$ in $\mathbb{F}_2$ (which is concealed inside $D_a f$). It was shown in [94] that any function admitting a nontrivial covering sequence is balanced (see Theorem 8.25 for a proof) and that any balanced function admits the constant sequence 1 as covering sequence (the level of this sequence is 2^{n-1}).

A characterization of covering sequences by means of the Walsh transform was also given in [94]: denote again by $S_{\widehat{f_\chi}}$ the support $\{u\in\mathbb{F}_2^n \mid \widehat{f_\chi}(u)\neq 0\}$ of $\widehat{f_\chi}$; then f admits an integer-valued sequence $\lambda=(\lambda_a)_{a\in\mathbb{F}_2^n}$ as covering sequence if and only if the Fourier transform $\widehat{\lambda}$ of the function $a\mapsto\lambda_a$ takes a constant value on $S_{\widehat{f_\chi}}$. Indeed, replacing $D_a f(x)$ by $\frac{1}{2}-\frac{1}{2}(-1)^{D_a f(x)}=\frac{1}{2}-\frac{1}{2}(-1)^{f(x)}(-1)^{f(x+a)}$ in the equality $\sum_{a\in\mathbb{F}_2^n}\lambda_a D_a f(x)=\rho$, we see that f admits the covering sequence λ with level ρ if and only if, for every $x\in\mathbb{F}_2^n$, we have $\sum_{a\in\mathbb{F}_2^n}\lambda_a(-1)^{f(x+a)}=\left(\sum_{a\in\mathbb{F}_2^n}\lambda_a-2\rho\right)(-1)^{f(x)}$; since two integer-valued functions are equal if and only if their Fourier transforms are equal, the characterization follows, thanks to the straightforward relation $\sum_{a,x\in\mathbb{F}_2^n}\lambda_a(-1)^{f(x+a)+x\cdot b}=\left(\sum_{a\in\mathbb{F}_2^n}\lambda_a(-1)^{a\cdot b}\right)\widehat{f_\chi}(b)=\widehat{\lambda}(b)\widehat{f_\chi}(b)$.

Knowing a covering sequence (trivial or not) of a function f allows knowing that all the vectors a such that $f(x)\oplus a\cdot x$ is nonbalanced belong to the set $\widehat{\lambda}^{-1}(\mu)$, where $\mu=\widehat{\lambda}(0)-2\rho$ is the constant value of $\widehat{\lambda}$ on $S_{\widehat{f_\chi}}$; hence, if f admits a covering sequence $\lambda=(\lambda_a)_{a\in\mathbb{F}_2^n}$ with level ρ (resp. with level $\rho\neq 0$), then f is k-th order correlation-immune (resp. k-resilient) where $k+1$ is the minimum Hamming weight of nonzero $b\in\mathbb{F}_2^n$ such that $\widehat{\lambda}(b)=\mu$. Conversely, if f is k-th order correlation-immune (resp. k-resilient) and if it is not $(k+1)$-th order correlation-immune (resp. $(k+1)$-resilient), then there exists at least one

[31] or real-valued, or complex-valued; but taking real or complex sequences instead of integer-valued ones has no practical sense.

(non-trivial) covering sequence $\lambda = (\lambda_a)_{a\in\mathbb{F}_2^n}$ with level ρ such that $k+1$ is the minimum Hamming weight of $b \in \mathbb{F}_2^n$ satisfying $\widehat{\lambda}(b) = \widehat{\lambda}(0) - 2\rho$.

A particularly simple covering sequence is the indicator of the set of vectors of weight one. The functions which admit this covering sequence are called regular; they are $(\rho - 1)$-resilient (where ρ is the level); more generally, any function, admitting as covering sequence the indicator of a set of vectors whose supports are disjoint, has this same property. See further properties in [94].

But knowing a covering sequence for f gives no information on the nonlinearity of f, since it gives only information on the support of the Walsh transform, not on the nonzero values it takes. In [68] is weakened the definition of covering sequence, so that it can help computing the (nonzero) values of the Walsh transform.

Definition 8.6. *Let f be a Boolean function on $\mathbb{F}_2^n$. A partial covering sequence for f is a sequence $(\lambda_a)_{a\in\mathbb{F}_2^n}$ such that $\sum_{a\in F_2^n} \lambda_a D_a f(x)$ takes two values ρ and ρ' (distinct or not) called the* levels *of the sequence. The partial covering sequence is called* nontrivial *if one of the constants is nonzero.*

A simple example of nontrivial partial covering sequence is as follows: let $\mathcal{E}$ be any set of derivatives of f. Assume that $\mathcal{E}$ contains a nonzero function and is stable under addition (*i.e.*, is a nontrivial $\mathbb{F}_2$-vectorspace). Then $\sum_{g\in\mathcal{E}} g$ takes on values 0 and $\frac{|\mathcal{E}|}{2}$. Thus, if $\mathcal{E} = \{D_a f / \ a \in E\}$ (where we choose E minimum, so that any two different vectors of the set E give different functions of $\mathcal{E}$), then 1_E is a nontrivial partial covering sequence.

The interest of nontrivial partial covering sequences is that they allow simplifying the computation of the weight and of the Walsh transform of f.

Theorem 8.25. *Let $(\lambda_a)_{a\in F_2^n}$ be a partial covering sequence of a Boolean function f, of levels ρ and ρ'.*

Let $A = \{x \in F_2^n \ / \ \sum_{a\in F_2^n} \lambda_a D_a f(x) = \rho'\}$ (assuming that $\rho' \neq \rho$; otherwise, when λ is in fact a covering sequence of level ρ, we set $A = \emptyset$).

Then, for every vector $b \in F_2^n$, we have

$$\left(\widehat{\lambda}(b) - \widehat{\lambda}(0) + 2\,\rho\right)\,\widehat{\mathrm{f}_\chi}(b) = 2\,(\rho - \rho')\sum_{x\in A}(-1)^{f(x)\oplus b\cdot x}.$$

Proof. By definition, we have, for every $x \in F_2^n$,

$$\sum_{a\in F_2^n} \lambda_a D_a f(x) = \rho'\,1_A(x) + \ \rho\,1_{A^c}(x)$$

and therefore,

$$\begin{aligned}\sum_{a\in F_2^n} \lambda_a(-1)^{D_a f(x)} &= \sum_{a\in F_2^n} \lambda_a(1 - 2\,D_a f(x)) \\ &= \sum_{a\in F_2^n} \lambda_a - 2\,\rho'\,1_A(x) - 2\,\rho\,1_{A^c}(x).\end{aligned}$$

We deduce

$$\sum_{a\in F_2^n} \lambda_a(-1)^{f(x+a)} = (-1)^{f(x)}\left(\sum_{a\in F_2^n} \lambda_a - 2\,\rho'\,1_A(x) - 2\,\rho\,1_{A^c}(x)\right). \qquad (8.43)$$

The Fourier transform of the function $(-1)^{f(x+a)}$ maps every vector $b \in F_2^n$ to the value $\sum_{x\in F_2^n}(-1)^{f(x+a)\oplus x\cdot b} = \sum_{x\in F_2^n}(-1)^{f(x)\oplus(x+a)\cdot b} = (-1)^{a\cdot b}\, \widehat{f}_\chi(b)$. Hence, taking the Fourier transform of both terms of equality (8.43), we get

$$\left(\sum_{a\in F_2^n} \lambda_a(-1)^{a\cdot b}\right)\widehat{f}_\chi(b) =$$

$$\left(\sum_{a\in F_2^n} \lambda_a\right)\widehat{f}_\chi(b) - 2\,\rho' \sum_{x\in A}(-1)^{f(x)\oplus b\cdot x} - 2\,\rho \sum_{x\in A^c}(-1)^{f(x)\oplus b\cdot x},$$

that is,

$$\widehat{\lambda}(b)\,\widehat{f}_\chi(b) = \widehat{\lambda}(0)\,\widehat{f}_\chi(b) - 2\,\rho\,\widehat{f}_\chi(b) + 2\,(\rho-\rho') \sum_{x\in A}(-1)^{f(x)\oplus b\cdot x}.$$

Hence,

$$\left(\widehat{\lambda}(b) - \widehat{\lambda}(0) + 2\,\rho\right)\widehat{f}_\chi(b) = 2\,(\rho-\rho') \sum_{x\in A}(-1)^{f(x)\oplus b\cdot x}.$$

■

Hence, if $\rho \neq 0$, we have in particular an information on the weight of f:

$$2^n - 2w_H(f) = \widehat{f}_\chi(0) = \left(1 - \frac{\rho'}{\rho}\right)\sum_{x\in A}(-1)^{f(x)}.$$

Examples are given in [68] of computations of the weights or Walsh spectra of some Boolean functions (quadratic functions, Maiorana-McFarland's functions and their extensions, and other examples of functions), using Theorem 8.25.

8.5.6 Functions with Low Univariate Degree

The following Weil's theorem is very well known in finite field theory (*cf.* [247, Theorem 5.38]).

Theorem 8.26. *Let q be a prime power and $f \in \mathbb{F}_q[x]$ a univariate polynomial of degree $d \geq 1$ with $gcd(d,q) = 1$. Let χ be a nontrivial character of $\mathbb{F}_q$. Then*

$$\left|\sum_{x\in\mathbb{F}_q} \chi(f(x))\right| \leq (d-1)\,q^{1/2}.$$

For $q = 2^n$, this *Weil's bound* means that, for every nonzero $a \in \mathbb{F}_{2^n}$: $\left|\sum_{x\in\mathbb{F}_{2^n}}(-1)^{tr_n(af(x))}\right| \leq (d-1)\,2^{n/2}$. And since adding a linear function $tr_n(bx)$

to the function $tr_n(af(x))$ corresponds to adding $(b/a)x$ to $f(x)$ and does not change its univariate degree, we deduce that, if $d > 1$ is odd and $a \neq 0$, then

$$nl(tr_n(af)) \geq 2^{n-1} - (d-1)\,2^{n/2-1}.$$

An extension of the Weil bound to the character sums of functions of the form $f(x) + g(1/x)$ (where $1/x = x^{2^n-2}$ takes value 0 at 0), among which are the so-called *Kloosterman sums* $\sum_{x\in\mathbb{F}_{2^n}}(-1)^{tr_n(1/x+ax)}$, has been first obtained by Carlitz and Uchiyama [96] and extended by Shanbhag, Kumar, and Helleseth [327]: if f and g have odd univariate degrees, then

$$\sum_{x\in\mathbb{F}_{2^n}}(-1)^{tr_n(f(1/x)+g(x))} \leq (d^\circ f + d^\circ g)2^{n/2}.$$

8.6 Bent Functions

We recall the definition of bent functions.

Definition 8.7. *A Boolean function f on $\mathbb{F}_2^n$ (n even) is called bent if its Hamming distance to the set $R(1, n)$ of all n-variable affine functions (the nonlinearity of f) equals $2^{n-1} - 2^{n/2-1}$ (the covering radius of the Reed-Muller code of order 1).*

Equivalently, as seen in Subsection 8.4.1, f is bent if and only if $\widehat{f_\chi}$ takes on values $\pm 2^{n/2}$ only (this characterization is independent of the choice of the inner product on $\mathbb{F}_2^n$, since any other inner product has the form $\langle x, s\rangle = x \cdot L(s)$, where L is an auto-adjoint linear automorphism, that is, an automorphism whose associated matrix is symmetric). Hence, f is bent if and only if its distance to any affine function equals $2^{n-1} \pm 2^{n/2-1}$. Note that, for any bent function f, half of the elements of the Reed-Muller code of order 1 lie at distance $2^{n-1} + 2^{n/2-1}$ from f and half lie at distance $2^{n-1} - 2^{n/2-1}$ (indeed, if ℓ lies at distance $2^{n-1} + 2^{n/2-1}$ from f, then $\ell \oplus 1$ lies at distance $2^{n-1} - 2^{n/2-1}$ and VICE VERSA). In fact, the condition on $\widehat{f_\chi}$ can be weakened, without losing the property of being necessary and sufficient.

Lemma 8.27. *Any n-variable (n even ≥ 2) Boolean function f is bent if and only if, for every $a \in \mathbb{F}_2^n$, $\widehat{f_\chi}(a) \equiv 2^{n/2}\left[\,mod\ 2^{n/2+1}\right]$, or equivalently $\widehat{f}(a) \equiv 2^{n/2-1}\left[\,mod\ 2^{n/2}\right]$.*

Proof. This necessary condition is also sufficient, since, if it is satisfied, then writing $\widehat{f_\chi}(a) = 2^{n/2}\lambda_a$, where λ_a is odd for every a, Parseval's relation (8.23) implies $\sum_{a\in\mathbb{F}_2^n}\lambda_a^2 = 2^n$, which implies that $\lambda_a^2 = 1$ for every a. ∎

A slightly different viewpoint is that of bent sequences[32] but we shall not adopt it here because it most often gives no extra insight into the problems. The nonlinearity being an affine invariant, so is the notion of bent function. Clearly, if f is bent and ℓ is affine, then $f \oplus \ell$ is bent. A class of bent functions is called a *complete class of functions* if it is globally invariant under the action of the general affine group and the addition of affine functions.

Thanks to Relation (8.25) and to the fact that the Fourier transform of a function is constant if and only if the function equals δ_0 times some constant, we see that any function f is bent if and only if, for any nonzero vector a, the Boolean function $D_a f(x) = f(x) \oplus f(x+a)$ is balanced. In other words, we have the following theorem.

Theorem 8.28. *Any n-variable Boolean function (n even[33]) is bent if and only if it satisfies $PC(n)$.*

For this reason, bent functions are also called *perfect nonlinear functions*[34]. Equivalently, f is bent if and only if the $2^n \times 2^n$ matrix $H = [(-1)^{f(x+y)}]_{x,y \in \mathbb{F}_2^n}$ is a Hadamard matrix (i.e., satisfies $H \times H^t = 2^n\, I$, where I is the identity matrix), and if and only if the support of f is a *difference set*[35] of the elementary Abelian 2-group $\mathbb{F}_2^n$ [136, 204] (other types of difference sets exist; see, e.g., [139]). This implies that the Cayley graph G_f (see Subsection 8.2.2.2) is strongly regular (see [18] for more precision).

The functions whose derivatives $D_a f$, $a \in H$, $a \neq 0$ are all balanced, where H is a linear hyperplane of $\mathbb{F}_2^n$, are characterized in [40, 41] for every n; they are all bent if n is even. The functions whose derivatives $D_a f$, $a \in E$, $a \neq 0$ are all balanced, where E is a vector subspace of $\mathbb{F}_2^n$ of dimension $n-2$, are also characterized in these two papers.

Bent functions have the property that, for every even positive integer w, the sum $\sum_{a \in \mathbb{F}_2^n} \widehat{\mathrm{f}_\chi}^{\,w}(a)$ is minimum. Such sums (for even or odd w) play a role with respect

[32] For each vector X in $\{-1,1\}^{2^n}$, define: $\hat{X} = \frac{1}{\sqrt{2^n}} H_n X$, where H_n is the Walsh-Hadamard matrix, recursively defined by

$$H_n = \begin{bmatrix} H_{n-1} & H_{n-1} \\ H_{n-1} & -H_{n-1} \end{bmatrix}, \ H_0 = [1].$$

The vectors X such that $\hat{X}$ belongs to $\{-1,1\}^{2^n}$ are called bent sequences. They are the images by the character $\chi = (-1)^{\cdot}$ of the bent functions on $\mathbb{F}_2^n$.

[33] In fact, according to the observations above, "n even" is implied by "f satisfies $PC(n)$"; functions satisfying $PC(n)$ do not exist for odd n.

[34] The characterization of Theorem 8.28 leads to a generalization of the notion of bent function to non-binary functions. In fact, several generalizations exist [3, 220, 256] (see [77] for a survey); the equivalence between being bent and being perfect nonlinear is no more valid if we consider functions defined over residue class rings (see [79]).

[35] Thus, bent functions are also related to designs, since any difference set can be used to construct a symmetric design, see [11], pages 274–8. The notion of difference set is anterior to that of a bent function, but it had not been much studied in the case of elementary 2-groups before the introduction of bent functions.

to fast correlation attacks [39, 46] (when these sums have small magnitude for low values of w, this contributes to a good resistance to fast correlation attacks).

A last way of looking at bent functions deals with *linear codes*: let f be any n-variable Boolean function (n even). Denote its support $\{x \in \mathbb{F}_2^n \mid f(x) = 1\}$ by S_f and write $S_f = \{u_1, \dots, u_{w_H(f)}\}$. Consider a matrix G whose columns are all the vectors of S_f, without repetition, and let C be the linear code generated by the rows of this matrix. Thus, C is the set of all the vectors $U_v = (v \cdot u_1, \dots, v \cdot u_{w_H(f)})$, where v ranges over $\mathbb{F}_2^n$. Then Proposition 8.29 follows.

Proposition 8.29. *Let n be any even positive integer. Any n-variable Boolean function f is bent if and only if the linear code C defined earlier has dimension n (i.e., G is a generator matrix of C) and has exactly two nonzero Hamming weights: 2^{n-2} and $w_H(f) - 2^{n-2}$.*

Indeed, $w_H(U_v)$ equals $\sum_{x \in \mathbb{F}_2^n} f(x) \times v \cdot x = \sum_{x \in \mathbb{F}_2^n} f(x) \frac{1-(-1)^{v \cdot x}}{2} = \frac{\widehat{f}(0) - \widehat{f}(v)}{2}$. Hence, according to Relation (8.12), $w_H(U_v)$ equals $2^{n-2} + \frac{\widehat{f_\chi}(v) - \widehat{f_\chi}(0)}{4}$, for every nonzero vector v. Thus, C has dimension n and has the two nonzero Hamming weights 2^{n-2} and $w_H(f) - 2^{n-2}$ if and only if, for every $v \neq 0$, U_v is nonzero and $\widehat{f_\chi}(v) = \widehat{f_\chi}(0)$ or $\widehat{f_\chi}(v) = \widehat{f_\chi}(0) + 4w_H(f) - 2^{n+1} = \widehat{f_\chi}(0) - 2\widehat{f_\chi}(0) = -\widehat{f_\chi}(0)$. If f is bent, then this condition is clearly satisfied. Conversely, according to Parseval's Relation (8.23), if this condition is satisfied, then $\widehat{f_\chi}(v)$ equals $\pm 2^{n/2}$ for every v, *i.e.*, f is bent.

There exist two other characterizations [350] dealing with C:

(1) C has dimension n and C has exactly two weights, whose sum equals $w_H(f)$;
(2) The length $w_H(f)$ of C is even, C has exactly two weights, and one of these weights is 2^{n-2}.

8.6.1 The Dual

If f is bent, then the *dual function* $\widetilde{f}$ of f, defined on $\mathbb{F}_2^n$ by

$$\widehat{f_\chi}(u) = 2^{n/2}(-1)^{\widetilde{f}(u)},$$

is also bent and its own dual is f itself. Indeed, the inverse Fourier transform property (8.19) applied to $\varphi = f_\chi$ (the sign function of f) gives, for every vector a, $\sum_{u \in \mathbb{F}_2^n} (-1)^{\widetilde{f}(u) \oplus a \cdot u} = 2^{n/2} f_\chi(a) = 2^{n/2}(-1)^{f(a)}$.

Let f and g be two bent functions, then Relation (8.22) applied with $\varphi = f_\chi$ and $\psi = g_\chi$ shows that

$$\mathcal{F}(\widetilde{f} \oplus \widetilde{g}) = \mathcal{F}(f \oplus g). \tag{8.44}$$

Thus, $f \oplus g$ and $\widetilde{f} \oplus \widetilde{g}$ have the same weight and the mapping $f \mapsto \widetilde{f}$ is an isometry.

According to Proposition 8.7, for every $a, b \in \mathbb{F}_2^n$ and for every bent function f, the dual of the function $f(x + b) \oplus a \cdot x$ equals $\widetilde{f}(x + a) \oplus b \cdot (x + a) = \widetilde{f}(x +$

$a) \oplus b \cdot x \oplus a \cdot b$. Denoting $b \cdot x$ by $\ell_b(x)$, Relation (8.44), applied with $g(x) = f(x+b) \oplus a \cdot x$, gives $\mathcal{F}(D_a\widetilde{f} \oplus \ell_b) = (-1)^{a \cdot b}\mathcal{F}(D_b f \oplus \ell_a)$, and applied with $g(x) = f(x) \oplus \ell_a(x)$ and with $f(x+b)$ in the place of $f(x)$, it gives the following property, first observed in [60] (and rediscovered in [42]):

$$\mathcal{F}(D_a\widetilde{f} \oplus \ell_b) = \mathcal{F}(D_b f \oplus \ell_a) \tag{8.45}$$

(from these two relations, we deduce that, if $a \cdot b = 1$, then $\mathcal{F}(D_a\widetilde{f} \oplus \ell_b) = \mathcal{F}(D_b f \oplus \ell_a) = 0$). Notice that, for every a and b, we have $D_b f = \ell_a \oplus \varepsilon$ if and only if $D_a\widetilde{f} = \ell_b \oplus \varepsilon$).

Moreover, if a pair of Boolean functions f and f' satisfies the relation $\mathcal{F}(D_a f' \oplus \ell_b) = \mathcal{F}(D_b f \oplus \ell_a)$, then these functions are bent (indeed, taking $a = 0$ shows that $D_b f$ is balanced for every $b \neq 0$ and taking $b = 0$ shows that $D_a f'$ is balanced for every $a \neq 0$) and are then the duals of each other up to the addition of a constant. Indeed, summing up the relation $\mathcal{F}(D_a f' \oplus \ell_b) = \mathcal{F}(D_b f \oplus \ell_a)$ for b ranging over $\mathbb{F}_2^n$ shows that $f'(0) \oplus f'(a) = \widetilde{f}(0) \oplus \widetilde{f}(a)$ for every a, since we have $\sum_{x,b \in \mathbb{F}_2^n} (-1)^{f'(x) \oplus f'(x+a) \oplus b \cdot x} = 2^n(-1)^{f'(0) \oplus f'(a)}$, and $\sum_{x,b \in \mathbb{F}_2^n} (-1)^{f(x) \oplus f(x+b) \oplus a \cdot x} = \widehat{\mathrm{f}_\chi}(0) \times \widehat{\mathrm{f}_\chi}(a)$.

The NNF of $\widetilde{f}$ can be deduced from the NNF of f. Indeed, using equality $\widetilde{f} = \frac{1-(-1)^{\widetilde{f}}}{2}$, we have $\widetilde{f} = \frac{1}{2} - 2^{-n/2-1}\widehat{\mathrm{f}_\chi} = \frac{1}{2} - 2^{n/2-1}\delta_0 + 2^{-n/2}\widehat{f}$ (according to Relation (8.12)). Now applying Relation (8.30) (expressing the value of the Fourier transform by means of the coefficients of the NNF) to $\varphi = f$, we deduce that if $\sum_{I \in \mathcal{P}(N)} \lambda_I x^I$ is the NNF of f, then

$$\widetilde{f}(x) = \frac{1}{2} - 2^{n/2-1}\delta_0(x) + (-1)^{w_H(x)} \sum_{I \in \mathcal{P}(N) \,|\, supp(x) \subseteq I} 2^{n/2-|I|}\lambda_I.$$

Changing I into $N \setminus I$ in this relation and observing that $supp(x)$ is included in $N \setminus I$ if and only if $x_i = 0, \forall i \in I$, we obtain the NNF of $\widetilde{f}$ by expanding the following relation:

$$\widetilde{f}(x) = \frac{1}{2} - 2^{n/2-1}\prod_{i=1}^{n}(1-x_i) + (-1)^{w_H(x)} \sum_{I \in \mathcal{P}(N)} 2^{|I|-n/2}\lambda_{N \setminus I}\prod_{i \in I}(1-x_i).$$

We deduce Proposition 8.30 (as shown in [86]).

Proposition 8.30. *Let f be any n-variable bent function (n even). For every $I \neq N$ such that $|I| > n/2$, the coefficient of x^I in the NNF of $\widetilde{f}$ (resp. of f) is divisible by $2^{|I|-n/2}$.*

Reducing this equality modulo 2 proves Rothaus' bound (see later discussion) and that, for $n \geq 4$ and $|I| = n/2$, the coefficient of x^I in the ANF of $\widetilde{f}$ equals the coefficient of $x^{N \setminus I}$ in the ANF of f. Using Relation (8.9), the preceding equality can be related to the main result of [191] (but this result by Hou was stated in a complex way).

The Poisson summation formula (8.17) applied to $\varphi = f_\chi$ gives (see [53]) that for every vector subspace E of $\mathbb{F}_2^n$, and for every elements a and b of $\mathbb{F}_2^n$, we have:

$$\sum_{x\in a+E} (-1)^{\widetilde{f}(x)\oplus b\cdot x} = 2^{-n/2}|E|(-1)^{a\cdot b} \sum_{x\in b+E^\perp} (-1)^{f(x)\oplus a\cdot x}. \tag{8.46}$$

Self-dual bent functions are studied in [76].

8.6.2 Bent Functions of Low Algebraic Degrees

Obviously, no affine function can be bent. All the quadratic bent functions are known: according to the properties recalled in Subsection 8.5.2, any such function

$$f(x) = \bigoplus_{1\le i<j\le n} a_{i,j}\, x_i\, x_j \oplus h(x) \ (h \text{ affine}, a_{i,j} \in \mathbb{F}_2)$$

is bent if and only if one of the following equivalent properties is satisfied:

(i) Its Hamming weight is equal to $2^{n-1} \pm 2^{n/2-1}$;
(ii) Its associated symplectic form: $\varphi_f : (x, y) \mapsto f(0) \oplus f(x) \oplus f(y) \oplus f(x + y)$ is nondegenerate (*i.e.*, has kernel $\{0\}$);
(iii) The skew-symmetric matrix $M = (m_{i,j})_{i,j\in\{1,\ldots,n\}}$ over $\mathbb{F}_2$, defined by: $m_{i,j} = a_{i,j}$ if $i < j$, $m_{i,j} = 0$ if $i = j$, and $m_{i,j} = a_{j,i}$ if $i > j$, is regular (*i.e.*, has determinant 1); indeed, M is the matrix of the bilinear form φ_f;
(iv) $f(x)$ is equivalent, up to an affine nonsingular transformation, to the function $x_1x_2 \oplus x_3x_4 \oplus \cdots \oplus x_{n-1}x_n \oplus \varepsilon$ $(\varepsilon \in \mathbb{F}_2)$.

It is interesting to charaterize quadratic bent functions in the trace representation. This leads, for instance, to the Kerdock code; see Subsection 8.6.10 where the bent functions leading to this code are given.

Let us study for example the case of the *Gold function* $tr_n(vx^{2^i+1})$, where $gcd(i, n) = 1$. It is bent if and only if there is no nonzero $x \in \mathbb{F}_{2^n}$ such that $tr_n(vx^{2^i}y + vxy^{2^i}) = 0$ for every $y \in \mathbb{F}_{2^n}$, *that is*, the equation $vx^{2^i} + (vx)^{2^{n-i}} = 0$ has no nonzero solution. Raising this equation to the 2^ith power gives $v^{2^i}x^{2^{2i}} + vx = 0$ and, $2^i - 1$ being coprime with $2^n - 1$, it is equivalent, after dividing by vx (when $x \neq 0$) and taking the $(2^i - 1)$th root, to $vx^{2^i+1} \in \mathbb{F}_2$. Hence, the function $tr_n(vx^{2^i+1})$ is bent if and only if v is not the $(2^i + 1)$th power of an element of $\mathbb{F}_{2^n}$: that is, (since $gcd(2^i + 1, 2^n - 1) = 3$), v is not the third power of an element of $\mathbb{F}_{2^n}$. The same result can be proven similarly with the *Kasami function* $tr_n(x^{2^{2i}-2^i+1})$.

Another example of a quadratic bent function in the trace representation that uses two trace functions, the trace function tr_n on the whole field $\mathbb{F}_{2^n}$ and the trace function $tr_{\frac{n}{2}}$ on the subfield $\mathbb{F}_{2^{n/2}}$, is $f(x) = tr_n(\sum_{i=1}^{\frac{n}{2}-1} x^{2^i+1}) \oplus tr_{\frac{n}{2}}(x^{2^{n/2}+1})$.

A third example has not yet appeared in the literature (as far as we know): let n be coprime with 3 and i be coprime with n, then the function $f(x, y) = tr_{\frac{n}{2}}(x^{2^i+1} + y^{2^i+1} + xy)$, $x, y \in \mathbb{F}_{2^{n/2}}$ is bent. Indeed, its associated symplectic form equals the function $((x, y), (x', y')) \to f(0, 0) \oplus f(x, y) \oplus f(x', y') \oplus$

$f(x+x', y+y') = tr_{\frac{n}{2}}(x^{2^i}x' + xx'^{2^i} + y^{2^i}y' + yy'^{2^i} + xy' + x'y)$. The kernel of this symplectic form equals

$$\left\{(x, y) \in \mathbb{F}_{2^{n/2}}^2 \;/\; \left\{\begin{array}{l} x^{2^i} + x^{2^{n-i}} + y = 0 \\ y^{2^i} + y^{2^{n-i}} + x = 0 \end{array}\right.\right\};$$

this set is reduced to $\{(0, 0)\}$, since denoting $z = x + y$ we have $z^{2^i} + z^{2^{n-i}} + z = 0$, which implies $z^{2^{2i}} = z^{2^i} + z$ and, therefore $z^{2^{3i}} = z$, that is, $z \in \mathbb{F}_{2^{3i}}$ and therefore $z \in \mathbb{F}_2$, and since 1 is not a solution, $z = 0$. Then x and y must be null.

Open problem: Characterize the bent functions of algebraic degrees at least 3 (that is, classify them under the action of the general affine group). This has been done for $n \leq 6$ in [312] (see also [301] where the number of bent functions is computed for these values of n). For $n = 8$, it has been done in [190], for functions of algebraic degrees at most 3 only; all of these functions have at least one affine derivative $D_a f$, $a \neq 0$ (it has been proved in [42] that this happens for $n \leq 8$ only). The determination of all bent eight-variable functions has been completed very recently; see [233]. Hans Dobbertin (with G. Leander) has presented in the posthumous paper [143] a nice approach for generating new bent functions by recursively gluing so-called $\mathbb{Z}$-bent functions.

8.6.3 Bound on Algebraic Degree

The algebraic degree of any Boolean function f being equal to the maximum size of the multiindex I such that x^I has an odd coefficient in the NNF of f, Proposition 8.30 gives the following result.

Proposition 8.31. *Let n be any even integer greater than or equal to 4. The algebraic degree of any bent function on* $\mathbb{F}_2^n$ *is at most* $n/2$.

In the case that $n = 2$, the bent functions have degree 2, because they have odd weight (in fact, they are the functions of odd weights).

The bound of Proposition 8.31 (which is obviously also true for $\widetilde{f}$) was first proved in [312] and will be called *Rothaus' bound* in the sequel. It can also be proved (see later discussion) by using a similar method as in the proof of Proposition 8.15. This same method also allows obtaining a bound, shown in [192], relating the gaps between $n/2$ and the algebraic degrees of f and $\widetilde{f}$:

Proposition 8.32. *The algebraic degrees of any n-variable bent function and of its dual satisfy:*

$$n/2 - d^\circ f \geq \frac{n/2 - d^\circ \widetilde{f}}{d^\circ \widetilde{f} - 1}. \tag{8.47}$$

Proof of Proposition 8.32 and Second Proof of Proposition 8.31. Let us denote by d (resp. by $\widetilde{d}$) the algebraic degree of f (resp. of $\widetilde{f}$) and consider a term x^I of degree d in the ANF of f. The Poisson summation formula (8.18) applied to $\varphi = \mathrm{f}_{\chi}$ (or Relation (8.46) with $a = b = 0$) and to the vectorspace $E = \{u \in \mathbb{F}_2^n / \forall i \in I,\ u_i = 0\}$ gives $\sum_{u \in E}(-1)^{\tilde{f}(u)} = 2^{n/2-d} \sum_{x \in E^\perp} \mathrm{f}_{\chi}(x)$. The orthogonal $E^\perp$ of E equals $\{u \in \mathbb{F}_2^n / \forall i \notin I,\ u_i = 0\}$. According to Relation (8.3), the restriction of f to $E^\perp$ has odd weight w, thus $\sum_{x \in E^\perp} \mathrm{f}_{\chi}(x) = 2^d - 2w$ is not divisible by 4. Hence, $\sum_{u \in E}(-1)^{\tilde{f}(u)}$ is not divisible by $2^{n/2-d+2}$. We deduce the proof of Proposition 8.31: suppose that $d > n/2$, then $\sum_{u \in E}(-1)^{\tilde{f}(u)}$ is not even, a contradiction with the fact that E has an even size. We prove now Proposition 8.32: according to McEliece's theorem (or Ax's theorem), $\sum_{u \in E}(-1)^{\tilde{f}(u)}$ is divisible by $2^{\lceil \frac{n-d}{d} \rceil}$. We deduce the inequality $n/2 - d + 2 > \lceil \frac{n-d}{d} \rceil$, that is, $n/2 - d + 1 \geq \frac{n-d}{d}$, which is equivalent to (8.47). ■

Using Relation (8.7) instead of Relation (8.3) gives a more precise result than Proposition 8.31, first shown in [86], which will be given in Subsection 8.6.6.

Proposition 8.32 can also be deduced from Proposition 8.30 and from some divisibility properties, shown in [86], of the coefficients of the NNFs of Boolean functions of algebraic degree d.

More on the algebraic degree of bent functions can be said for homogeneous functions (whose ANF contain monomials of fixed degree); see [278].

8.6.4 Constructions

There does not exist for $n \geq 10$ a classification of bent functions under the action of the general affine group. In order to understand better the structure of bent functions, we can try to design constructions of bent functions. It is useful also to deduce constructions of highly nonlinear balanced functions. Some of the known constructions of bent functions are direct, that is, do not use as building blocks previously constructed bent functions. We will call these direct constructions *primary constructions*. The others, sometimes leading to recursive constructions, will be called *secondary constructions*.

8.6.4.1 Primary Constructions

(1) The *Maiorana-McFarland original class* $\mathcal{M}$ (see [136, 272]) is the set of all the Boolean functions on $\mathbb{F}_2^n = \{(x, y); x, y \in \mathbb{F}_2^{n/2}\}$, of the form

$$f(x, y) = x \cdot \pi(y) \oplus g(y) \tag{8.48}$$

where π is any permutation on $\mathbb{F}_2^{n/2}$ and g any Boolean function on $\mathbb{F}_2^{n/2}$ ("$\cdot$" here denotes an inner product in $\mathbb{F}_2^{n/2}$). Any such function is bent. More precisely, the bijectivity of π is a necessary and sufficient condition[36]

[36] It is, because the input has been cut in two pieces x and y of the same length; it is also possible to cut them in pieces of different lengths (see Proposition 8.33), and bentness is then obviously not characterized by the bijectivity of π.

for f being bent, according to Relation (8.49) later, applied with $r = n/2$. The dual function $\widetilde{f}(x, y)$ equals $y \cdot \pi^{-1}(x) \oplus g(\pi^{-1}(x))$, where π^{-1} is the inverse permutation of π. The completed class of $\mathcal{M}$ (that is, the smallest possible complete class including $\mathcal{M}$) contains all the quadratic bent functions (according to Line 4 of the characterization of quadratic bent functions given in Subsection 8.6.2; take $\pi = id$ and g constant in (8.48)).

As we saw already in Subsection 8.5.1, the fundamental idea of Maiorana-McFarland's construction consists in *concatenating affine functions*. If we order all the binary words of length n in lexicographic order, with the bit of higher weight on the right, then the truth table of f is the concatenation of the restrictions of f obtained by setting the value of y and letting x freely range over $\mathbb{F}_2^{n/2}$. These restrictions are affine. In fact, Maiorana-McFarland's construction is a particular case of a more general construction of bent functions [64] (see the next proposition), which is, properly speaking, a secondary construction for $r < n/2$ and which is the original Maiorana-McFarland construction for $r = n/2$ (this is why we give it in this subsection).

Proposition 8.33. *Let $n = r + s$ $(r \leq s)$ be even. Let ϕ be any mapping from $\mathbb{F}_2^s$ to $\mathbb{F}_2^r$ such that, for every $a \in \mathbb{F}_2^r$, the set $\phi^{-1}(a)$ is an $(n - 2r)$-dimensional affine subspace of $\mathbb{F}_2^s$. Let g be any Boolean function on $\mathbb{F}_2^s$ whose restriction to $\phi^{-1}(a)$ (viewed as a Boolean function on $\mathbb{F}_2^{n-2r}$ via an affine isomorphism between $\phi^{-1}(a)$ and this vectorspace) is bent for every $a \in \mathbb{F}_2^r$, if $n > 2r$ (no condition on g being imposed if $n = 2r$). Then the function $f_{\phi,g} = x \cdot \phi(y) \oplus g(y)$ is bent on $\mathbb{F}_2^n$.*

Proof. This is a direct consequence of the equality (valid for every ϕ and every g):

$$\widehat{\mathrm{f}_{\phi,\mathrm{g}}}_{\chi}(a, b) = 2^r \sum_{y \in \phi^{-1}(a)} (-1)^{g(y) \oplus b \cdot y}, \tag{8.49}$$

which comes from the fact that every function $x \mapsto f_{\phi,g}(x, y) \oplus a \cdot x \oplus b \cdot y$ being affine, and thus constant or balanced, it contributes for a nonzero value in the sum $\sum_{x \in \mathbb{F}_2^r, y \in \mathbb{F}_2^s} (-1)^{f_{\phi,g}(x,y) \oplus x \cdot a \oplus y \cdot b}$ only if $\phi(y) = a$. According to Relation (8.49), the function $f_{\phi,g}$ is bent if and only if $r \leq n/2$ and $\sum_{y \in \phi^{-1}(a)} (-1)^{g(y) \oplus b \cdot y} = \pm 2^{n/2-r}$ for every $a \in \mathbb{F}_2^r$ and every $b \in \mathbb{F}_2^s$. The hypothesis in Proposition 8.33 is a sufficient condition for that (but it is not a necessary one). ∎

This construction is pretty general: the choice of any partition of $\mathbb{F}_2^s$ in 2^r flats of dimension $(n - 2r)$ and of an $(n - 2r)$-variable bent function on each of these flats leads to an n-variable bent function.

Obviously, every Boolean function can be represented (in several ways) in the form $f_{\phi,g}$ for some values of $r \geq 1$ and s and for some mapping ϕ from $\mathbb{F}_2^s$ to $\mathbb{F}_2^r$ and Boolean function g on $\mathbb{F}_2^s$. It has been observed in [256] that, if a bent function has this form, then ϕ is balanced (*i.e.,* is uniformly distributed over $\mathbb{F}_2^r$). This is a direct consequence of the fact that, for every nonzero $a \in \mathbb{F}_2^r$, the Boolean function $a \cdot \phi$ is balanced, because it equals the derivative $D_{(a,0)} f_{\phi,g}$, and of the characterization of balanced vectorial functions given in Chapter 9.

It is shown in [24] that every bent function in 6 variables is affinely equivalent to a function of the Maiorana-McFarland class.

Remark. *There exist $n/2$-dimensional vector spaces of n-variable Boolean functions whose nonzero elements are all bent. The Maiorana-McFarland construction easily allows constructing such vector spaces. A result by K. Nyberg (see Chapter 9) shows that k-dimensional vector spaces of n-variable Boolean functions whose nonzero elements are all bent cannot exist for $k > n/2$.*

(2) The *Partial Spreads class $\mathcal{PS}$*, introduced in [136] by J. Dillon, is the set of all the sums (modulo 2) of the indicators of $2^{n/2-1}$ or $2^{n/2-1}+1$ "disjoint" $n/2$-dimensional subspaces of $\mathbb{F}_2^n$ ("disjoint" meaning that any two of these spaces intersect in 0 only, and therefore that their sum is direct and equals $\mathbb{F}_2^n$). The bentness of such function is a direct consequence of Theorem 8.41. This is why we omit the proof of this fact here. According to this same theorem, the dual of such a function has the same form, all the $n/2$-dimensional spaces E being replaced by their orthogonals. Note that the Boolean functions equal to the sums of the indicators of "disjoint" $n/2$-dimensional subspaces of $\mathbb{F}_2^n$ share with quadratic functions the property of being bent if and only if they have the weight of a bent function (which is $2^{n-1} \pm 2^{n/2-1}$). J. Dillon denotes by $\mathcal{PS}^-$ (resp. $\mathcal{PS}^+$) the class of those bent functions for which the number of $n/2$-dimensional subspaces is $2^{n/2-1}$ (resp. $2^{n/2-1}+1$). All the elements of $\mathcal{PS}^-$ have algebraic degree $n/2$ exactly, but not all those of $\mathcal{PS}^+$ (which contains, for instance, all the quadratic functions, if $n/2$ is even). It is an open problem to characterize the algebraic normal forms of the elements of class $\mathcal{PS}$, and it is not a simple matter to construct, practically, elements of this class. J. Dillon exhibits in [136] a subclass of $\mathcal{PS}^-$, denoted by $\mathcal{PS}_{ap}$, whose elements (which we shall call *Dillon's functions*) are defined in an explicit form: $\mathbb{F}_2^{n/2}$ is identified to the Galois field $\mathbb{F}_{2^{n/2}}$ (an inner product in this field being defined as $x \cdot y = tr_{\frac{n}{2}}(xy)$, where $tr_{\frac{n}{2}}$ is the trace function from $\mathbb{F}_{2^{n/2}}$ to $\mathbb{F}_2$; we know that the notion of bent function is independent of the choice of the inner product); the space $\mathbb{F}_2^n \approx \mathbb{F}_{2^{n/2}} \times \mathbb{F}_{2^{n/2}}$, viewed[37] as a two-dimensional $\mathbb{F}_{2^{n/2}}$-vectorspace, is equal to the "disjoint" union of its $2^{n/2}+1$ lines through the origin; these lines are $n/2$-dimensional $\mathbb{F}_2$-subspaces of $\mathbb{F}_2^n$. Choosing any $2^{n/2-1}$ of the lines and taking them different from those of equations

[37] Let ω be an element of $\mathbb{F}_{2^n} \setminus \mathbb{F}_{2^{n/2}}$; the pair $(1, \omega)$ is a basis of the $\mathbb{F}_{2^{n/2}}$-vectorspace $\mathbb{F}_{2^n}$; hence, we have $\mathbb{F}_{2^n} = \mathbb{F}_{2^{n/2}} + \omega\mathbb{F}_{2^{n/2}}$.

$x = 0$ and $y = 0$ leads, by definition, to an element of $\mathcal{PS}_{ap}$, that is, to a function of the form $f(x, y) = g\left(x\, y^{2^{n/2}-2}\right)$, *i.e.*, $g\left(\frac{x}{y}\right)$ with $\frac{x}{y} = 0$ if $y = 0$, where g is a balanced Boolean function on $\mathbb{F}_2^{n/2}$ that vanishes at 0. The complements $g\left(\frac{x}{y}\right) \oplus 1$ of these functions are the functions $g(\frac{x}{y})$ where g is balanced and does not vanish at 0; they belong to the class $\mathcal{PS}^+$. In both cases, the dual of $g(\frac{x}{y})$ is $g(\frac{y}{x})$ (this is a direct consequence of Theorem 8.41). The elements of the class $\mathcal{PS}^{\#}_{ap}$, of those Boolean functions over $\mathbb{F}_{2^n}$ that can be obtained from those of $\mathcal{PS}_{ap}$ by composition by the transformations $x \in \mathbb{F}_{2^n} \mapsto \delta x$, $\delta \neq 0$, and by addition of a constant,[38] are those Boolean functions f of weight $2^{n-1} \pm 2^{n/2-1}$ on $\mathbb{F}_{2^n}$ such that, denoting by α a primitive element of this field, $f(\alpha^{2^{n/2}+1}x) = f(x)$ for every $x \in \mathbb{F}_{2^n}$. It is proved in [81, 136] that these functions are the functions of weight $2^{n-1} \pm 2^{n/2-1}$ which can be written as $\sum_{i=1}^{r} tr_n(a_i x^{j_i})$ for $a_i \in \mathbb{F}_{2^n}$ and j_i a multiple of $2^{n/2} - 1$ with $j_i \leq 2^n - 1$.

(3) Dobbertin gives in [141] the construction of a class of bent functions that contains both $\mathcal{PS}_{ap}$ and $\mathcal{M}$. The elements of this class are the functions f defined by $f(x, \phi(y)) = g\left(\frac{x+\psi(y)}{y}\right)$, where g is a balanced Boolean function on $\mathbb{F}_{2^{n/2}}$ and ϕ, ψ are two mappings from $\mathbb{F}_{2^{n/2}}$ to itself such that, if T denotes the affine subspace of $\mathbb{F}_{2^{n/2}}$ spanned by the support of the function $\widehat{g_\chi}$ (where $g_\chi = (-1)^g$), then, for any a in $\mathbb{F}_{2^{n/2}}$, the functions ϕ and ψ are affine on $aT = \{ax, x \in T\}$. The mapping ϕ must additionally be one to one. The elements of this class do not have an explicit form, but Dobbertin gives two explicit examples of bent functions constructed this way. In both, ϕ is a power function (see later discussion).

(4) If $n/2$ is odd, then it is possible to deduce a bent Boolean function on $\mathbb{F}_2^n$ from any almost bent function from $\mathbb{F}_2^{n/2}$ to $\mathbb{F}_2^{n/2}$. A vectorial Boolean function $F : \mathbb{F}_2^m \to \mathbb{F}_2^m$ is called almost bent if all of the component functions $v \cdot F$, $v \neq 0$ in $\mathbb{F}_2^m$, are plateaued with amplitude $2^{\frac{m+1}{2}}$ (see in Subsection 8.6.8 the definition of these terms). The function $\gamma_F(a, b)$, $a, b \in \mathbb{F}_2^m$, equal to 1 if the equation $F(x) + F(x + a) = b$ admits solutions, with $a \neq 0$ in $\mathbb{F}_2^m$, and equal to 0 otherwise, is then bent (see the proof of this result in Chapter 9). This gives new bent functions related to the almost bent functions listed in this same chapter. However, determining the ANF or the univariate representation of γ_F is an open problem when F is a Kasami, Welch, or Niho almost-bent function.

(5) Some infinite classes of bent functions have also been obtained, thanks to the identification between the vectorspace $\mathbb{F}_2^n$ and the field $\mathbb{F}_{2^n}$, as *power functions* (which can also be called *monomial functions*), that is, functions of the form $tr_n(ax^i)$, where tr_n is the trace function on $\mathbb{F}_{2^n}$ and where a and x belong to this same field. The known values of i for which there exists at

[38] The functions of $\mathcal{PS}_{ap}$ are among them those satisfying $f(0) = f(1) = 0$.

least one a such that $tr_n(ax^i)$ is bent are (up to conjugacy $i \to 2i$ [mod $2^n - 1$]) the Gold exponents $i = 2^j + 1$, where $\frac{n}{gcd(j,n)}$ is even (the corresponding function $tr_n(ax^i)$ belongs to the Maiorana-McFarland class; it is bent if and only if $a \notin \{x^i,\ x \in \mathbb{F}_{2^n}\}$; the condition "$\frac{n}{gcd(j,n)}$ even" is imposed to allow the existence of such a); the Dillon exponents [135] of the form $j \cdot (2^{n/2} - 1)$, where $gcd(j, 2^{n/2} + 1) = 1$ (the function $tr_n(ax^i)$, where $a \in \mathbb{F}_{2^{n/2}}$ and $i = j\,(2^{n/2} - 1)$ is then bent if and only if the Kloosterman sum $\sum_{x \in \mathbb{F}_{2^{n/2}}} (-1)^{tr_{\frac{n}{2}}(1/x+ax)}$ is null,[39] where $1/0 = 0$ and where $tr_{\frac{n}{2}}$ is the trace function on the field $F_{2^{n/2}}$; this equivalence was first proved by Dillon [136]; more recently, Leander [240] has found another proof that gives more insight; a small error in his proof has been corrected in [101]; the function $tr_n(ax^i)$ belongs then to the PS_{ap} class); the Kasami exponents $i = 2^{2j} - 2^j + 1$, where $gcd(j, n) = 1$ (the corresponding function $tr_n(ax^i)$ is bent if and only if $a \notin \{x^3,\ x \in \mathbb{F}_{2^n}\}$, see [139]) and two exponents more recently found: $i = (2^{n/4} + 1)^2$ where n is divisible by 4 and not by 8 (see [240] where the Gold and Dillon exponents are also revisited, see also [103]; the set of all as such that the corresponding function $tr_n(ax^i)$ is bent is not known; an example of such a is the fifth power of a primitive element of $\mathbb{F}_{16}$; the function belongs to the Maiorana-McFarland class) and $i = 2^{n/3} + 2^{n/6} + 1$, where n is divisible by 6 [43] (the corresponding function $tr_n(ax^i)$ is bent if and only if the value $a + a^{2^{n/6}} + a^{2^{2n/6}} + a^{2^{3n/6}} + \cdots + a^{2^{n/2-n/6}}$ at a of the trace function from $\mathbb{F}_{2^{n/2}}$ to $\mathbb{F}_{2^{n/6}}$ is null; it belongs to the Maiorana-McFarland class). Note that a still simpler bent function (but which is not expressed by means of the function tr_n itself) is $f(x) = tr_{\frac{n}{2}}(x^{2^{n/2}+1})$, that is, $f(x) = x^{2^{n/2}+1} + \left(x^{2^{n/2}+1}\right)^2 + \left(x^{2^{n/2}+1}\right)^{2^2} + \cdots + \left(x^{2^{n/2}+1}\right)^{2^{n/2-1}}$. The symplectic form $\varphi_f(x, y)$ associated to f equals $tr_n(y^{2^{n/2}}x)$; its kernel is therefore trivial and f is bent.

Some other functions are defined as the sums of a few power functions, see [101, 136, 139, 144, 145, 197, 239, 354].

Note that power functions and sums of power functions represent for the designer of the cryptosystem using them the interest of being more easily computable than general functions (which allows using them with more variables while keeping a good efficiency). Power functions have the peculiarity that, denoting the set $\{x^i;\ x \in \mathbb{F}_{2^n}^*\}$ by U, two functions $tr_n(ax^i)$ and $tr_n(bx^i)$ such that $a/b \in U$ are linearly equivalent. It is not clear whether this is more an advantage for the designer or for the attacker of a system using a nonlinear balanced function derived from such a bent function.

Obviously, a power function $tr_n(ax^i)$ can be bent only if the mapping $x \to x^i$ is not one to one (otherwise, the function would be balanced, a contradiction),

[39] The existence of a such that the Kloosterman sum is null had been conjectured by Dillon. It has been proved by Lachaud and Wolfmann [225] who proved that the values of such Kloosterman sums are all the numbers divisible by 4 in the range $[-2^{n/4+1} + 1; 2^{n/4+1} + 1]$.

that is, if i is not coprime with $2^n - 1$. It has been proved in [39] that i must be coprime either with $2^{n/2} - 1$ or with $2^{n/2} + 1$.

Finally, bent functions have been also obtained by Dillon and McGuire [140] as the restrictions of functions on $\mathbb{F}_{2^{n+1}}$, with $n+1$ odd, to a hyperplane of this field: these functions are the Kasami functions $tr_n\left(x^{2^{2k}-2^k+1}\right)$, and the hyperplane has equation $tr_n(x) = 0$. The restriction is bent under the condition that $n + 1 = 3k \pm 1$.

Remark. *The bent sequences given in [352] are particular cases of the constructions given above (using also some of the secondary constructions given below).*

In [98] homogeneous bent functions (i.e. bent functions whose ANFs are the sums of monomials of the same degree) are constructed on 12 (and fewer) variables by using the invariant theory (which makes the computer searches feasible).

8.6.4.2 Secondary Constructions

We have already seen in Proposition 8.33 a secondary construction based on the Maiorana-McFarland construction. We now describe the others that have been found so far.

(1) The first secondary construction given by J. Dillon and O. Rothaus in [136, 312] is very simple: let f be a bent function on $\mathbb{F}_2^n$ (n even); and g a bent function on $\mathbb{F}_2^m$ (m even); then the function h defined on $\mathbb{F}_2^{n+m}$ by $h(x, y) = f(x) \oplus g(y)$ is bent. Indeed, we clearly have $\widehat{\mathrm{h}_\chi}(a, b) = \widehat{\mathrm{f}_\chi}(a) \times \widehat{\mathrm{g}_\chi}(b)$. This construction, called the *direct sum*, unfortunately has no great interest from a cryptographic point of view, because it produces *decomposable functions* (a Boolean function is called decomposable if it is equivalent to the sum of two functions that depend on two disjoint subsets of coordinates; such a peculiarity is easy to detect and can be used for designing divide-and-conquer attacks, as pointed out by J. Dillon in [137]).

(2) A more interesting result, by the same authors, is the following: if g, h, k and $g \oplus h \oplus k$ are bent on $\mathbb{F}_2^n$ (n even), then the function defined at every element (x_1, x_2, x) of $\mathbb{F}_2^{n+2}$ ($x_1, x_2 \in \mathbb{F}_2$, $x \in \mathbb{F}_2^n$) by

$$f(x_1, x_2, x) =$$

$$g(x)h(x) \oplus g(x)k(x) \oplus h(x)k(x) \oplus [g(x) \oplus h(x)]x_1 \oplus [g(x) \oplus k(x)]x_2 \oplus x_1x_2$$

is bent (this is a particular case of Theorem 8.35). No general class of bent functions has been deduced from this *Rothaus construction*.

(3) Two classes of bent functions have been derived in [53] from Maiorana-McFarland's class, by adding to some functions of this class the indicators of some vector subspaces:
   - The class $\mathcal{D}_0$ whose elements are the functions of the form $f(x, y) = x \cdot \pi(y) \oplus \delta_0(x)$ (recall that δ_0 is the Dirac symbol; the ANF of $\delta_0(x)$ is

$\prod_{i=1}^{n/2}(x_i \oplus 1))$. The dual of such a function f is the function $y \cdot \pi^{-1}(x) \oplus \delta_0(y)$. It is proved in [53] that this class is not included[40] in the completed versions $\mathcal{M}^\#$ and $\mathcal{PS}^\#$ of classes $\mathcal{M}$ and $\mathcal{PS}$ (*i.e.*, the smallest possible classes including them) and that every bent function in six variables is affinely equivalent to a function of this class, up to the addition of an affine function. Class $\mathcal{D}_0$ is a subclass of the class denoted by $\mathcal{D}$, whose elements are the functions of the form $f(x, y) = x \cdot \pi(y) \oplus 1_{E_1}(x)1_{E_2}(y)$, where π is any permutation on $\mathbb{F}_2^{n/2}$ and where E_1, E_2 are two linear subspaces of $\mathbb{F}_2^{n/2}$ such that $\pi(E_2) = E_1{}^\perp$ (1_{E_1} and 1_{E_2} denote their indicators). The dual of f belongs to the completed version of this same class.

- The class $\mathcal{C}$ of all the functions of the form $x \cdot \pi(y) \oplus 1_L(x)$, where L is any linear subspace of $\mathbb{F}_2^{n/2}$ and π any permutation on $\mathbb{F}_2^{n/2}$ such that, for any element a of $\mathbb{F}_2^{n/2}$, the set $\pi^{-1}(a + L^\perp)$ is a flat. It is a simple matter to see, as shown in [44], that, under the same hypothesis on π, if g is a Boolean function whose restriction to every flat $\pi^{-1}(a + L^\perp)$ is affine, then the function $x \cdot \pi(y) \oplus 1_L(x) \oplus g(y)$ is also bent.

The fact that any function in class $\mathcal{D}$ or class $\mathcal{C}$ is bent comes from the following theorem proved in [53], which has its own interest.

Theorem 8.34. *Let $b + E$ be any flat in $\mathbb{F}_2^n$ (E being a linear subspace of $\mathbb{F}_2^n$). Let f be any bent function on $\mathbb{F}_2^n$. The function $f^\star = f \oplus 1_{b+E}$ is bent if and only if one of the following equivalent conditions is satisfied:*

(i) *For any a in $\mathbb{F}_2^n \setminus E$, the function $D_a f$ is balanced on $b + E$.*

(ii) *The restriction of the function $\widetilde{f}(x) \oplus b \cdot x$ to any coset of $E^\perp$ is either constant or balanced.*

If f and $f^\star$ are bent, then E has dimension greater than or equal to $n/2$ and the algebraic degree of the restriction of f to $b + E$ is at most $\dim(E) - n/2 + 1$.

If f is bent, if E has dimension $n/2$, and if the restriction of f to $b + E$ has algebraic degree at most $\dim(E) - n/2 + 1 = 1$, that is, *is affine, then conversely $f^\star$ is bent, too.*

Proof. Recall that a function is bent if and only if it satisfies $PC(n)$. The equivalence between condition *1* and the bentness of $f^\star$ then comes from the fact that $\mathcal{F}(D_a f^\star)$ equals $\mathcal{F}(D_a f)$ if $a \in E$, and equals $\mathcal{F}(D_a f) - 4\sum_{x\in b+E}(-1)^{D_a f(x)}$ otherwise.

We have $\widehat{f_\chi}(a) - \widehat{f^\star_\chi}(a) = 2\sum_{x\in b+E}(-1)^{f(x)\oplus a\cdot x}$. Using Relation (8.46), applied with $E^\perp$ in the place of E, we deduce that for every $a \in \mathbb{F}_2^n$:

$$\sum_{u\in a+E^\perp}(-1)^{\widetilde{f}(u)\oplus b\cdot u} = 2^{\dim(E^\perp)-n/2-1}(-1)^{a\cdot b}\left(\widehat{f_\chi}(a) - \widehat{f^\star_\chi}(a)\right),$$

[40] It is easy to show that a function f does not belong to $\mathcal{M}^\#$ by showing that there does not exist an $n/2$-dimensional vector-subspace E of $\mathbb{F}_2^n$ such that $D_a D_b f$ is null for every $a, b \in E$; it is much more difficult to show that it does not belong to $\mathcal{PS}^\#$.

and $\widehat{f_\chi}(a) - \widehat{f^\star_\chi}(a)$ takes value 0 or $\pm 2^{n/2+1}$ for every a if and only if condition *2* is satisfied. So condition *2* is necessary and sufficient, according to Lemma 8.27 (at the beginning of Section 8.6).

Let us now assume that f and $f^\star$ are bent. Then $1_{b+E} = f^\star \oplus f$ has algebraic degree at most $n/2$, according to Rothaus' bound, and thus $\dim(E) \geq n/2$. The values of the Walsh transform of the restriction of f to $b+E$ being equal to those of $\frac{1}{2}\left(\widehat{f_\chi} - \widehat{f^\star_\chi}\right)$, they are divisible by $2^{n/2}$ and thus the restriction of f to $b+E$ has algebraic degree at most $\dim(E) - n/2 + 1$, according to Proposition 8.15.

If f is bent, if E has dimension $n/2$, and if the restriction of f to $b+E$ is affine, then the relation $\widehat{f_\chi}(a) - \widehat{f^\star_\chi}(a) = 2\sum_{x\in b+E}(-1)^{f(x)\oplus a\cdot x}$ shows that $f^\star$ is bent too, according to Lemma 8.27. ∎

Remark.

- *Relation (8.46) applied to $E^\perp$ in the place of E, where E is some $n/2$-dimensional subspace, shows straightforwardly that, if f is a bent function on $\mathbb{F}_2^n$, then $f(x) \oplus a \cdot x$ is constant on $b+E$ if and only if $\widetilde{f}(x) \oplus b \cdot x$ is constant on $a + E^\perp$. The same relation shows that $f(x) \oplus a \cdot x$ is then balanced on every other coset of E and $\widetilde{f}(x) \oplus b \cdot x$ is balanced on every other coset of $E^\perp$. Notice that Relation (8.46) also shows that $f(x) \oplus a \cdot x$ cannot be constant on a flat of dimension strictly greater than $n/2$ (i.e., that f cannot be k-weakly normal with $k > n/2$).*
- *Let f be bent on $\mathbb{F}_2^n$. Let a and a' be two linearly independent elements of $\mathbb{F}_2^n$. Let us denote by E the orthogonal of the subspace spanned by a and a'. According to condition (2) of Theorem 8.34, the function $f \oplus 1_E$ is bent if and only if $D_a D_{a'} \widetilde{f}$ is null (indeed, a 2-variable function is constant or balanced if and only if it has even weight, and $\widetilde{f}$ has even weight on any coset of the vector subspace spanned by a and a' if and only if, for every vector x, we have $f(x) \oplus f(x+a) \oplus f(x+a') \oplus f(x+a+a') = 0$). This result has been restated in [42] and used in [44] to design (potentially) new bent functions.*

(4) Other classes of bent functions have been deduced from a construction given in [56], which generalizes the secondary constructions given earlier in (1) and (2).

Theorem 8.35. *Let n and m be two even positive integers. Let f be a Boolean function on $\mathbb{F}_2^{n+m} = \mathbb{F}_2^n \times \mathbb{F}_2^m$ such that, for any element y of $\mathbb{F}_2^m$, the function on $\mathbb{F}_2^n$*

$$f_y : x \mapsto f(x, y)$$

is bent.

Then f is bent if and only if, for any element s of $\mathbb{F}_2^n$, the function

$$\varphi_s : y \mapsto \widetilde{f_y}(s)$$

is bent on $\mathbb{F}_2^m$. If this condition is satisfied, then the dual of f is the function $\widetilde{f}(s,t) = \widetilde{\varphi_s}(t)$ (taking as inner product in $\mathbb{F}_2^n \times \mathbb{F}_2^m$: $(x,y)\cdot(s,t) = x\cdot s \oplus y\cdot t$).

This very general result is easy to prove, using that, for every $s \in \mathbb{F}_2^n$,

$$\sum_{x\in\mathbb{F}_2^n} (-1)^{f(x,y)\oplus x\cdot s} = 2^{n/2}(-1)^{\widetilde{f_y}(s)} = 2^{n/2}(-1)^{\varphi_s(y)},$$

and thus that

$$\widehat{\mathrm{f}}_{\chi}(s,t) = 2^{n/2}\sum_{y\in\mathbb{F}_2^m}(-1)^{\varphi_s(y)\oplus y\cdot t}.$$

This construction has also been considered in a particular case by Adams and Tavares [1] under the name of bent-based functions, and later studied by J. Seberry and X.-M. Zhang in [323] in special cases, too.

A case of application of this construction is nicely simple (see [66]): let f_1 and f_2 be two n-variable bent functions (n even) and let g_1 and g_2 be two m-variable bent functions (m even). Define[41]

$$h(x,y) = f_1(x) \oplus g_1(y) \oplus (f_1 \oplus f_2)(x)\,(g_1 \oplus g_2)(y);\ x \in \mathbb{F}_2^n,\ y \in \mathbb{F}_2^m$$

(this construction $(f_1, f_2, g_1, g_2) \mapsto h$ will appear again later to construct resilient functions; see Theorem 8.55). For every y, the function h_y of Theorem 8.35 equals f_1 plus a constant or f_2 plus a constant (depending on the values of y) and thus is bent; φ_s equals g_1 plus a constant or g_2 plus a constant (depending on the values of s), and thus is bent, too. According to Theorem 8.35, h is then bent. Its dual $\widetilde{h}$ is obtained from $\widetilde{f_1}$, $\widetilde{f_2}$, $\widetilde{g_1}$, and $\widetilde{g_2}$ by the same formula by which h is obtained from f_1, f_2, g_1, and g_2:

$$\widetilde{h}(s,t) = \widetilde{f_1}(s) \oplus \widetilde{g_1}(t) \oplus (\widetilde{f_1} \oplus \widetilde{f_2})(s)(\widetilde{g_1} \oplus \widetilde{g_2})(t).$$

What is interesting in this particular case (sometimes called the *indirect sum*) is that we only assume the bentness of f_1, f_2, g_1, and g_2 for deducing the bentness of h; no extra condition is needed, contrary to the general construction.

Another simple application of Theorem 8.35, called the *extension of Maiorana-McFarland type*, is given in [78]: let π be a permutation on $\mathbb{F}_2^{n/2}$ and $f_{\pi,g}(x,y) = x\cdot\pi(y) \oplus g(y)$ a related Maiorana-McFarland bent function. Let $(h_y)_{y\in\mathbb{F}_2^{n/2}}$ be a collection of bent functions on $\mathbb{F}_2^m$ for some

[41] h is the concatenation of the four functions f_1, $f_1 \oplus 1$, f_2 and $f_2 \oplus 1$, in an order controlled by $g_1(y)$ and $g_2(y)$.

even integer m. Then the function $(x, y, z) \in \mathbb{F}_2^{n/2} \times \mathbb{F}_2^{n/2} \times \mathbb{F}_2^m \to h_y(z) \oplus f_{\pi,g}(x, y)$ is bent.

Several classes have been deduced from Theorem 8.35 in [56], and later in [192].

- Let n and m be two even positive integers. The elements of $\mathbb{F}_2^{n+m}$ are written (x, y, z, τ), where x, y are elements of $\mathbb{F}_2^{n/2}$ and z, τ are elements of $\mathbb{F}_2^{m/2}$. Let π and π' be permutations on $\mathbb{F}_2^{n/2}$ and $\mathbb{F}_2^{m/2}$ (respectively) and h a Boolean function on $\mathbb{F}_2^{m/2}$. Then, the following Boolean function on $\mathbb{F}_2^{n+m}$ is bent:

$$f(x, y, z, \tau) = x \cdot \pi(y) \oplus z \cdot \pi'(\tau) \oplus \delta_0(x)h(\tau)$$

(recall that $\delta_0(x)$ equals 1 if $x = 0$ and is null otherwise). It is possible to prove, see [56], that such a function does not belong, in general, to the completed version of class $\mathcal{M}$. It is also easy to prove that f does not belong, in general, to the completed version of class $\mathcal{D}_0$, because any element of $\mathcal{D}_0$ has algebraic degree $\frac{n+m}{2}$, and it is a simple matter to produce examples of functions f whose algebraic degree is smaller than $\frac{n+m}{2}$.

- Let n and m be two even positive integers. We identify $\mathbb{F}_2^{n/2}$ (resp. $\mathbb{F}_2^{m/2}$) with the Galois field $\mathbb{F}_{2^{n/2}}$ (resp. with $\mathbb{F}_{2^{m/2}}$). Let k be a Boolean function on $\mathbb{F}_{2^{n/2}} \times \mathbb{F}_{2^{m/2}}$ such that, for any element x of $\mathbb{F}_{2^{n/2}}$, the function $z \mapsto k(x, z)$ is balanced on $\mathbb{F}_{2^{m/2}}$, and for any element z of $\mathbb{F}_{2^{m/2}}$, the function $x \mapsto k(x, z)$ is balanced on $\mathbb{F}_{2^{n/2}}$. Then the function

$$f(x, y, z, \tau) = k\left(\frac{x}{y}, \frac{z}{\tau}\right)$$

is bent on $\mathbb{F}_2^{n+m}$.

- Let r be a positive integer. We identify $\mathbb{F}_2^r$ with $\mathbb{F}_{2^r}$. Let π and π' be two permutations on $\mathbb{F}_{2^r}$ and g a balanced Boolean function on $\mathbb{F}_{2^r}$. The following Boolean function on $\mathbb{F}_2^{4r} = (\mathbb{F}_2^r)^4$:

$$f(x, y, z, \tau) = z \cdot \pi'\left[\tau + \pi\left(\frac{x}{y}\right)\right] \oplus \delta_0(z)g\left(\frac{x}{y}\right)$$

is a bent function.

(5) X.-D. Hou and P. Langevin, in [196], made a very simple observation that leads to potentially new bent functions.

Proposition 8.36. *Let f be a Boolean function on $\mathbb{F}_2^n$, n even. Let σ be a permutation on $\mathbb{F}_2^n$. We denote its coordinate functions by $\sigma_1, \ldots, \sigma_n$ and we assume that, for every $a \in \mathbb{F}_2^n$,*

$$d_H(f, \bigoplus_{i=1}^{n} a_i \sigma_i) = 2^{n-1} \pm 2^{n/2-1}.$$

Then $f \circ \sigma^{-1}$ is bent.

Indeed, the Hamming distance between $f \circ \sigma^{-1}$ and the linear function $\ell_a(x) = a \cdot x$ equals $d_H(f, \bigoplus_{i=1}^n a_i\, \sigma_i)$.

Hou and Langevin deduced that, if h is an affine function on $\mathbb{F}_2^n$, if f_1, f_2 and g are Boolean functions on $\mathbb{F}_2^n$, and if the following function is bent:

$$f(x_1, x_2, x) = x_1\, x_2\, h(x) \oplus x_1\, f_1(x) \oplus x_2\, f_2(x) \oplus g(x) /\ x \in \mathbb{F}_2^n,\ x_1,\ x_2 \in \mathbb{F}_2,$$

then the function

$$f(x_1, x_2, x) \oplus (h(x) \oplus 1)\, f_1(x) f_2(x) \oplus f_1(x) \oplus (x_1 \oplus h(x) \oplus 1) f_2(x) \oplus x_2\, h(x)$$

is bent.

They also deduced that, if f is a bent function on $\mathbb{F}_2^n$ whose algebraic degree is at most 3, and if σ is a permutation on $\mathbb{F}_2^n$ such that for every $i = 1, \ldots, n$, there exists a subset U_i of $\mathbb{F}_2^n$ and an affine function h_i such that

$$\sigma_i(x) = \bigoplus_{u \in U_i} (f(x) \oplus f(x+u)) \oplus h_i(x),$$

then $f \circ \sigma^{-1}$ is bent.

Finally, X.-D. Hou [192] deduced that if $f(x, y)$ $(x, y \in \mathbb{F}_2^{n/2})$ is a Maiorana-McFarland function of the particular form $x \cdot y \oplus g(y)$ and if $\sigma_1, \ldots, \sigma_n$ are all of the form $\bigoplus_{1 \le i < j \le n/2} a_{i,j} x_i\, y_j \oplus b \cdot x \oplus c \cdot y \oplus h(y)$, then $f \circ \sigma^{-1}$ is bent. He gave several examples of application of this result.

(6) Note that the construction of (5) does not increase the number of variables, contrary to most other secondary constructions. Another secondary construction without extension of the number of variables was introduced in [69]. It is based on the following result.

Proposition 8.37. *Let f_1, f_2 and f_3 be three Boolean functions on $\mathbb{F}_2^n$. Denote by s_1 the Boolean function equal to $f_1 \oplus f_2 \oplus f_3$ and by s_2 the Boolean function equal to $f_1 f_2 \oplus f_1 f_3 \oplus f_2 f_3$. Then we have $f_1 + f_2 + f_3 = s_1 + 2s_2$. This implies the following equality between the Fourier transforms: $\widehat{f_1} + \widehat{f_2} + \widehat{f_3} = \widehat{s_1} + 2\widehat{s_2}$ and the similar equality between the Walsh transforms:*

$$\widehat{\mathrm{f_1}_\chi} + \widehat{\mathrm{f_2}_\chi} + \widehat{\mathrm{f_3}_\chi} = \widehat{\mathrm{s_1}_\chi} + 2\,\widehat{\mathrm{s_2}_\chi}. \tag{8.50}$$

Proof. The fact that $f_1 + f_2 + f_3 = s_1 + 2s_2$ (the sums being computed in $\mathbb{Z}$ and not modulo 2) can be checked easily. The linearity of the Fourier transform with respect to the addition in $\mathbb{Z}$ implies then $\widehat{f_1} + \widehat{f_2} + \widehat{f_3} = \widehat{s_1} + 2\widehat{s_2}$. The equality $f_1 + f_2 + f_3 = s_1 + 2s_2$ also directly implies $\mathrm{f_1}_\chi + \mathrm{f_2}_\chi + \mathrm{f_3}_\chi = \mathrm{s_1}_\chi + 2\mathrm{s_2}_\chi$, thanks to the equality $\mathrm{f}_\chi = 1 - 2f$ valid for every Boolean function, which implies Relation (8.50). ■

Proposition 8.37 leads to the following double construction of bent functions.

Corollary 8.38. *Let f_1, f_2 and f_3 be three n-variable bent functions, n even. Denote by s_1 the function $f_1 \oplus f_2 \oplus f_3$ and by s_2 the function $f_1 f_2 \oplus f_1 f_3 \oplus f_2 f_3$. Then:*

- *If s_1 is bent and if $\tilde{s_1} = \tilde{f_1} \oplus \tilde{f_2} \oplus \tilde{f_3}$, then s_2 is bent, and $\tilde{s_2} = \tilde{f_1}\tilde{f_2} \oplus \tilde{f_1}\tilde{f_3} \oplus \tilde{f_2}\tilde{f_3}$.*
- *If $\widehat{s_{2_\chi}}(a)$ is divisible by $2^{n/2}$ for every a (e.g., if s_2 is bent, or if it is quadratic, or more generally if it is plateaued; see the definition in Subsection 8.6.8), then s_1 is bent.*

Proof.

- If s_1 is bent and if $\tilde{s_1} = \tilde{f_1} \oplus \tilde{f_2} \oplus \tilde{f_3}$, then, for every a, Relation (8.50) implies

$$\widehat{s_{2_\chi}}(a) = \left[(-1)^{\tilde{f_1}(a)} + (-1)^{\tilde{f_2}(a)} + (-1)^{\tilde{f_3}(a)} - (-1)^{\tilde{f_1}(a)\oplus\tilde{f_2}(a)\oplus\tilde{f_3}(a)}\right] 2^{\frac{n-2}{2}}$$
$$= (-1)^{\tilde{f_1}(a)\tilde{f_2}(a)\oplus\tilde{f_1}(a)\tilde{f_3}(a)\oplus\tilde{f_2}(a)\tilde{f_3}(a)}\, 2^{n/2}.$$

  Indeed, as we already saw with the relation $f_{1_\chi} + f_{2_\chi} + f_{3_\chi} = s_{1_\chi} + 2s_{2_\chi}$, for every bit ε, η, and τ, we have $(-1)^\varepsilon + (-1)^\eta + (-1)^\tau - (-1)^{\varepsilon\oplus\eta\oplus\tau} = 2\,(-1)^{\varepsilon\eta\oplus\varepsilon\tau\oplus\eta\tau}$.
- If $\widehat{s_{2_\chi}}(a)$ is divisible by $2^{n/2}$ for every a, then the number $\widehat{s_{1_\chi}}(a)$, which is equal to $\left[(-1)^{\tilde{f_1}(a)} + (-1)^{\tilde{f_2}(a)} + (-1)^{\tilde{f_3}(a)}\right] 2^{n/2} - 2\,\widehat{s_{2_\chi}}(a)$, according to Relation (8.50), is congruent with $2^{n/2}$ modulo $2^{n/2+1}$ for every a. This is sufficient to imply that s_1 is bent, according to Lemma 8.27 (at the beginning of Section 8.6). ■

(7) A construction related to the notion of normal extension of a bent function can be found in Proposition 8.51.

8.6.4.3 *Decompositions of Bent Functions*

The following theorem, proved in [41], is a direct consequence of Relation (8.28), applied to $f \oplus \ell$ where ℓ is linear, and to a linear hyperplane E of $\mathbb{F}_2^n$, and of the well-known (easy to prove) fact that, for every even integer $n \geq 4$, the sum of the squares of two integers equals 2^n (resp. 2^{n+1}) if and only if one of these squares is null and the other one equals 2^n (resp. both squares equal 2^n).

Theorem 8.39. *Let n be an even integer, $n \geq 4$, and let f be an n-variable Boolean function. Then the following properties are equivalent.*

(i) *f is bent.*
(ii) *For every (resp. for some) linear hyperplane E of $\mathbb{F}_2^n$, the Walsh transforms of the restrictions h_1, h_2 of f to E and to its complement (viewed as*

Boolean functions on $\mathbb{F}_2^{n-1}$) take values $\pm 2^{n/2}$ and 0 only, and the disjoint union of their supports equals the whole space $\mathbb{F}_2^{n-1}$.

Hence, a simple way of obtaining a plateaued function in an odd number of variables and with optimal nonlinearity is to take the restriction of a bent function to an affine hyperplane. Note that we have also (see [41]) that, if a function in an odd number of variables is such that, for some nonzero $a \in \mathbb{F}_2^n$, every derivative $D_u f$, $u \neq 0$, $u \in a^\perp$, is balanced, then its restriction to the linear hyperplane $a^\perp$ or to its complement is bent.

It is also proved in [41] that the Walsh transforms of the four restrictions of a bent function to an $(n-2)$-dimensional vector subspace E of $\mathbb{F}_2^n$ and to its cosets have the same sets of magnitudes. It is a simple matter to see that, denoting by a and b two vectors such that $E^\perp$ is the linear space spanned by a and b, these four restrictions are bent if and only if $D_a D_b \widetilde{f}$ takes on constant value 1.

More on decomposing bent functions can be found in [41, 42, 100].

8.6.5 On the Number of Bent Functions

The class of bent functions produced by the original Maiorana-McFarland construction is far the widest class, compared to the classes obtained from the other primary constructions.

The number of bent functions of the form (8.48) equals $(2^{n/2})! \times 2^{2^{n/2}}$, which is asymptotically equivalent to $\left(\frac{2^{n/2+1}}{e}\right)^{2^{n/2}} \sqrt{2^{n/2+1}\pi}$ (according to Stirling's formula), whereas the only other important construction of bent functions, $\mathcal{PS}_{ap}$, leads to only $\binom{2^{n/2}}{2^{n/2-1}} \approx \frac{2^{2^{n/2}+\frac{1}{2}}}{\sqrt{\pi 2^{n/2}}}$ functions. However, the number of provably bent Maiorana-McFarland functions seems negligible with respect to the total number of bent functions. The number of (bent) functions that are affinely equivalent to Maiorana-McFarland functions is unknown; it is at most equal to the number of Maiorana-McFarland functions times the number of affine automorphisms, which equals $2^n(2^n-1)(2^n-2)\cdots(2^n-2^{n-1})$. It seems also negligible with respect to the total number of bent functions. The problem of determining an efficient lower bound on the number of n-variable bent functions is open.

Rothaus' inequality recalled in Subsection 8.6.3 (Proposition 8.31) states that any bent function has algebraic degree at most $n/2$. Thus, the number of bent functions is at most

$$2^{1+n+\ldots+\binom{n}{n/2}} = 2^{2^{n-1}+\frac{1}{2}\binom{n}{n/2}}.$$

We shall call this upper bound the *naive bound.* For $n = 6$, the number of bent functions is known and is approximately equal to $2^{32.3}$ (see [301]), which is much less than what gives the naive bound: 2^{42}. For $n = 8$, the number is also known: it was first shown in [234] that it is inferior to $2^{129.2}$. It has been very recently calculated by Langevin, Leander et al. [233] and equals approximately $2^{106.3}$ (the naive bound gives 2^{163}). Hence, picking at random an eight-variable

Boolean function of algebraic degree bounded above by 4 does not allow obtaining bent functions (but more clever methods exist; see [81, 127]). An upper bound improving upon the naive bound has been found in [89]. It is exponentially better than the naive bound because it divides it by approximately $2^{2^{n/2}-n/2-1}$. But it seems to be still far from the exact number of bent functions: for $n = 6$ it gives roughly 2^{38} (to be compared with $2^{32.3}$) and for $n = 8$ it gives roughly 2^{152} (to be compared with $2^{106.3}$).

8.6.6 Characterizations of Bent Functions

8.6.6.1 Characterization through the NNF

Proposition 8.40. *Let $f(x) = \sum_{I\in\mathcal{P}(N)} \lambda_I x^I$ be the NNF of a Boolean function f on $\mathbb{F}_2^n$. Then f is bent if and only if:*

(1) *For every I such that $n/2 < |I| < n$, the coefficient λ_I is divisible by $2^{|I|-n/2}$;*
(2) *λ_N (with $N = \{1, \ldots, n\}$) is congruent with $2^{n/2-1}$ modulo $2^{n/2}$.*

Proof. According to Lemma 8.27, f is bent if and only if, for every $a \in \mathbb{F}_2^n$, $\widehat{f}(a) \equiv 2^{n/2-1} \left[\bmod 2^{n/2}\right]$. We deduce that, according to Relation (8.30) applied with $\varphi = f$, Conditions (*1*) and (*2*) imply that f is bent.

Conversely, Condition (*1*) *is necessary, according to Proposition 8.30. Condition* (*2*) is also necessary because $\widehat{f}(1, \ldots, 1) = (-1)^n \lambda_N$ (from Relation (8.30)). ■

Proposition 8.40 and Relation (8.9) imply some restrictions on the coefficients of the ANFs of bent functions, observed and used in [89] (and also partially observed by Hou and Langevin in [196]).

Proposition 8.40 can be seen as a (much) stronger version of Rothaus' bound, since the algebraic degree of a Boolean function whose NNF is $f(x) = \sum_{I\in\mathcal{P}(N)} \lambda_I x^I$ equals the maximum size of I, such that λ_I is odd.

8.6.6.2 Geometric Characterization

Proposition 8.40 also allows proving the following characterization.

Theorem 8.41. ([84].) *Let f be a Boolean function on $\mathbb{F}_2^n$. Then f is bent if and only if there exist $n/2$-dimensional subspaces $E_1, \ldots, E_k$ of $\mathbb{F}_2^n$ (there is no constraint on the number k) and integers $m_1, \ldots, m_k$ (positive or negative) such that, for any element x of $\mathbb{F}_2^n$:*

$$f(x) \equiv \sum_{i=1}^{k} m_i 1_{E_i}(x) - 2^{n/2-1}\delta_0(x) \quad \left[\, mod\ 2^{n/2} \right]. \tag{8.51}$$

If we have $f(x) = \sum_{i=1}^{k} m_i 1_{E_i}(x) - 2^{n/2-1}\delta_0(x)$ *then the dual of* f *equals* $\widetilde{f}(x) = \sum_{i=1}^{k} m_i 1_{E_i^{\perp}}(x) - 2^{n/2-1}\delta_0(x)$.

Proof. [Sketch] Relation (8.51) is a sufficient condition for f being bent, according to Lemma 8.27 (at the beginning of Section 8.6) and to Relation (8.16). This same Relation (8.16) also implies the last sentence of Theorem 8.41. Conversely, if f is bent, then Proposition 8.40 allows us to deduce Relation (8.51), by expressing all the monomials x^I by means of the indicators of subspaces of dimension at least $n - |I|$ (indeed, the NNF of the indicator of the subspace $\{x \in \mathbb{F}_2^n / x_i = 0, \forall i \in I\}$ being equal to $\prod_{i \in I}(1 - x_i) = \sum_{J \subseteq I}(-1)^{|J|}x^J$, the monomial x^I can be expressed by means of this indicator and of the monomials x^J, where J is strictly included in I) and by using Lemma 8.42 (note that $d \geq n - |I|$ implies $|I| - n/2 \geq n/2 - d$ and that $\prod_{i \in N}(1 - x_i) = \delta_0(x)$). ■

Lemma 8.42. *Let* F *be any* d*-dimensional subspace of* $\mathbb{F}_2^n$. *There exist* $n/2$*-dimensional subspaces* $E_1, \ldots, E_k$ *of* $\mathbb{F}_2^n$ *and integers* $m, m_1, \ldots, m_k$ *such that, for any element* x *of* $\mathbb{F}_2^n$:

$$2^{n/2-d}\, 1_F(x) \equiv m + \sum_{i=1}^{k} m_i 1_{E_i}(x) \left[\ mod\ 2^{n/2}\right]\ \textit{if}\ d < n/2,\ \textit{and}$$

$$1_F(x) \equiv \sum_{i=1}^{k} m_i 1_{E_i}(x) \left[\ mod\ 2^{n/2}\right]\ \textit{if}\ d > n/2.$$

The class of those functions f which satisfy the relation obtained from (8.51) by withdrawing "[mod $2^{n/2}$]" is called the *generalized partial spread* class and denoted by $\mathcal{GPS}$ (it includes $\mathcal{PS}$). The dual $\widetilde{f}$ of such function f of $\mathcal{GPS}$ equaling $\widetilde{f}(x) = \sum_{i=1}^{k} m_i 1_{E_i^{\perp}}(x) - 2^{n/2-1}\delta_0(x)$, it belongs to $\mathcal{GPS}$ too.

There is no uniqueness of the representation of a given bent function in the form (8.51). There exists another characterization, shown in [85], in the form $f(x) = \sum_{i=1}^{k} m_i 1_{E_i}(x) \pm 2^{n/2-1}\delta_0(x)$, where $E_1, \ldots, E_k$ are vector subspaces of $\mathbb{F}_2^n$ of dimensions $n/2$ or $n/2 + 1$ and where $m_1, \ldots, m_k$ are integers (positive or negative). There is not a unique way, either, to choose these spaces E_i. But it is possible to define some subclass of $n/2$-dimensional and $(n/2 + 1)$-dimensional spaces such that there is uniqueness, if the spaces E_i are chosen in this subclass.

P. Guillot proved subsequently in [171] that, up to composition by a translation $x \mapsto x + a$, every bent function belongs to $\mathcal{GPS}$.

8.6.6.3 Characterization by Second-Order Covering Sequences

Proposition 8.43. ([92].) *A Boolean function* f *defined on* $\mathbb{F}_2^n$ *is bent if and only if*

$$\forall x \in \mathbb{F}_2^n,\ \sum_{a,b \in \mathbb{F}_2^n} (-1)^{D_a D_b f(x)} = 2^n. \tag{8.52}$$

Proof. If we multiply both terms of Relation (8.52) by $f_\chi(x) = (-1)^{f(x)}$, we obtain the (equivalent) relation

$$\forall x \in \mathbb{F}_2^n, f_\chi \otimes f_\chi \otimes f_\chi(x) = 2^n f_\chi(x);$$

indeed, we have $f_\chi \otimes f_\chi \otimes f_\chi(x) = \sum_{b\in\mathbb{F}_2^n}\left(\sum_{a\in\mathbb{F}_2^n}(-1)^{f(a)\oplus f(a+b)}\right)(-1)^{f(b+x)} = \sum_{a,b\in\mathbb{F}_2^n}(-1)^{f(a+x)\oplus f(a+b+x)\oplus f(b+x)}$. According to the bijectivity of the Fourier transform and to Relation (8.20), this is equivalent to

$$\forall u \in \mathbb{F}_2^n, \widehat{f_\chi}^3(u) = 2^n \widehat{f_\chi}(u).$$

Thus, we have $\sum_{a,b\in\mathbb{F}_2^n}(-1)^{D_aD_bf(x)} = 2^n$ if and only if, for every $u \in \mathbb{F}_2^n$, $\widehat{f_\chi}(u)$ equals $\pm\sqrt{2^n}$ or 0. According to Parseval's relation, the value 0 cannot be achieved by $\widehat{f_\chi}$ and this is therefore equivalent to the bentness of f. ■

Relation (8.52) is equivalent to the relation $\sum_{a,b\in\mathbb{F}_2^n}(1 - 2D_aD_bf(x)) = 2^n$, that is, $\sum_{a,b\in\mathbb{F}_2^n} D_aD_bf(x) = 2^{2n-1} - 2^{n-1}$, and hence to the fact that f admits the *second-order covering sequence* with all-1 coefficients and with level $2^{2n-1} - 2^{n-1}$.

It is shown similarly in [92] that the relation similar to (8.52) but with any integer in the place of 2^n characterizes the class of plateaued functions (see Subsection 8.6.8).

A characterization of bent functions through Cayley graphs also exists; see [18].

8.6.7 Subclasses: Hyperbent Functions

In [353], A. Youssef and G. Gong studied the Boolean functions f on the field $\mathbb{F}_{2^n}$ (n even) whose Hamming distances to all functions $tr_n(a\,x^i)\oplus\varepsilon$ ($a \in \mathbb{F}_{2^n}$, $\varepsilon \in \mathbb{F}_2$), where tr_n denotes the trace function from $\mathbb{F}_{2^n}$ to $\mathbb{F}_2$ and where i is coprime with $2^n - 1$, equal $2^{n-1} \pm 2^{n/2-1}$. These functions are bent, because every affine function has the form $tr_n(a\,x)\oplus\varepsilon$. They are called *hyperbent functions*. The (equivalent) condition that $\sum_{x\in\mathbb{F}_{2^n}}(-1)^{f(x)\oplus tr_n(a\,x^i)}$ equals $\pm 2^{n/2}$ for every $a \in \mathbb{F}_{2^n}$ and every i coprime with $2^n - 1$, seems difficult to satisfy, because it is equivalent to the fact that the function $f(x^i)$ is bent for every such i. However, Youssef and Gong showed in [353] that hyperbent functions exist. Their result is equivalent to the following proposition (see [81]).

Proposition 8.44. *All the functions of class $\mathcal{PS}_{ap}$ are hyperbent.*

Let us give a direct proof of this fact here.

Proof. Let ω be any element in $\mathbb{F}_{2^n}\setminus\mathbb{F}_{2^{n/2}}$. The pair $(1,\omega)$ is a basis of the $\mathbb{F}_{2^{n/2}}$-vectorspace $\mathbb{F}_{2^n}$. Hence, we have $\mathbb{F}_{2^n} = \mathbb{F}_{2^{n/2}} + \omega\mathbb{F}_{2^{n/2}}$. Moreover, every element y of $\mathbb{F}_{2^{n/2}}$ satisfies $y^{2^{n/2}} = y$, and therefore $tr_n(y) = y + y^2 + \cdots + y^{2^{n/2-1}} + y + y^2 + \cdots + y^{2^{n/2-1}} = 0$. Consider the inner product in $\mathbb{F}_{2^n}$ defined by $y \cdot y' = tr_n(y\,y')$; the subspace $\mathbb{F}_{2^{n/2}}$ is then its own orthogonal. Hence, according to Relation (8.16), any sum of the form $\sum_{y \in \mathbb{F}_{2^{n/2}}} (-1)^{tr_n(\lambda y)}$ is null if $\lambda \notin \mathbb{F}_{2^{n/2}}$ and equals $2^{n/2}$ if $\lambda \in \mathbb{F}_{2^{n/2}}$.

Let us consider any element of the class $\mathcal{PS}_{ap}$, choosing a balanced Boolean function g on $\mathbb{F}_2^{n/2}$, vanishing at 0, and defining $f(y' + \omega\, y) = g\left(\frac{y'}{y}\right)$, with $\frac{y'}{y} = 0$ if $y = 0$. For every $a \in \mathbb{F}_{2^n}$, we have

$$\sum_{x \in \mathbb{F}_{2^n}} (-1)^{f(x) \oplus tr_n(a\,x^i)} = \sum_{y, y' \in \mathbb{F}_{2^{n/2}}} (-1)^{g\left(\frac{y'}{y}\right) \oplus tr_n(a\,(y' + \omega y)^i)}.$$

Denoting $\frac{y'}{y}$ by z, we see that

$$\sum_{y \in \mathbb{F}^*_{2^{n/2}},\, y' \in \mathbb{F}_{2^{n/2}}} (-1)^{g\left(\frac{y'}{y}\right) \oplus tr_n(a\,(y' + \omega y)^i)} = \sum_{z \in \mathbb{F}_{2^{n/2}},\, y \in \mathbb{F}^*_{2^{n/2}}} (-1)^{g(z) \oplus tr_n(a\,y^i\,(z + \omega)^i)}.$$

The remaining sum

$$\sum_{y' \in \mathbb{F}_{2^{n/2}}} (-1)^{g(0) \oplus tr_n(a\,y'^i)} = \sum_{y' \in \mathbb{F}_{2^{n/2}}} (-1)^{tr_n(a\,y')}$$

(this equality being due to the fact that the mapping $x \to x^i$ is one-to-one) equals $2^{n/2}$ if $a \in \mathbb{F}_{2^{n/2}}$ and is null otherwise.

Thus, $\sum_{x \in \mathbb{F}_{2^n}} (-1)^{f(x) \oplus tr_n(a\,x^i)}$ equals

$$\sum_{z \in \mathbb{F}_{2^{n/2}}} (-1)^{g(z)} \sum_{y \in \mathbb{F}_{2^{n/2}}} (-1)^{tr_n(a(z+\omega)^i\,y)} - \sum_{z \in \mathbb{F}_{2^{n/2}}} (-1)^{g(z)} + 2^{n/2} 1_{\mathbb{F}_{2^{n/2}}}(a).$$

The sum $\sum_{z \in \mathbb{F}_{2^{n/2}}} (-1)^{g(z)}$ is null because g is balanced.

The sum $\sum_{z \in \mathbb{F}_{2^{n/2}}} (-1)^{g(z)} \sum_{y \in \mathbb{F}_{2^{n/2}}} (-1)^{tr_n(a(z+\omega)^i\,y)}$ equals $\pm 2^{n/2}$ if $a \notin \mathbb{F}_{2^{n/2}}$, because we prove in the next lemma that there exists then exactly one $z \in \mathbb{F}_{2^{n/2}}$ such that $a(z+\omega)^i \in \mathbb{F}_{2^{n/2}}$, and this sum is null if $a \in \mathbb{F}_{2^{n/2}}$ (this can be checked, if $a = 0$ thanks to the balancedness of g, and if $a \neq 0$ because y ranges over $\mathbb{F}_{2^{n/2}}$ and $a(z+\omega)^i \notin \mathbb{F}_{2^{n/2}}$). This completes the proof. ■

Lemma 8.45. *Let n be any positive integer. Let a and ω be two elements of the set $\mathbb{F}_{2^n} \setminus \mathbb{F}_{2^{n/2}}$, and let i be coprime with $2^n - 1$. There exists a unique element $z \in \mathbb{F}_2^{n/2}$ such that $a(z+\omega)^i \in \mathbb{F}_2^{n/2}$.*

Proof. Let j be the inverse of i modulo $2^n - 1$. We have $a(z+\omega)^i \in \mathbb{F}_2^{n/2}$ if and only if $z \in \omega + a^{-j} \times \mathbb{F}_2^{n/2}$. The sets $\omega + a^{-j} \times \mathbb{F}_2^{n/2}$ and $\mathbb{F}_2^{n/2}$ are two flats whose directions $a^{-j} \times \mathbb{F}_2^{n/2}$ and $\mathbb{F}_2^{n/2}$ are subspaces whose sum is direct and equals $\mathbb{F}_{2^n}$. Hence, they have a unique vector in their intersection. ■

Relationships between the notion of hyperbent function and cyclic codes are studied in [81]. It is proved that every hyperbent function $f : \mathbb{F}_{2^n} \to \mathbb{F}_2$ can be represented as $f(x) = \sum_{i=1}^{r} tr_n(a_i x^{t_i}) \oplus \varepsilon$, where $a_i \in \mathbb{F}_{2^n}$, $\varepsilon \in \mathbb{F}_2$ and $w_2(t_i) = n/2$. Consequently, all hyperbent functions have algebraic degree $n/2$.

In [101] it is proved that, for every even n, every $\lambda \in \mathbb{F}_{2^{n/2}}^*$ and every $r \in]0; \frac{n}{2}[$ such that the cyclotomic cosets of 2 modulo $2^{n/2}+1$ containing respectively $2^r - 1$ and $2^r + 1$ have size n and such that the function $tr_{\frac{n}{2}}\left(\lambda x^{2^r+1}\right)$ is balanced on $\mathbb{F}_{2^{n/2}}$; the function $tr_n\left(\lambda\left(x^{(2^r-1)(2^{n/2}-1)} + x^{(2^r+1)(2^{n/2}-1)}\right)\right)$ is bent (*i.e.*, hyperbent) if and only if the function $tr_{\frac{n}{2}}\left(x^{-1} + \lambda x^{2^r+1}\right)$ is also balanced on $\mathbb{F}_{2^{n/2}}$.

Remark. *In [55] were determined those Boolean functions on $\mathbb{F}_2^n$ such that, for a given even integer k $(2 \le k \le n-2)$, any of the Boolean functions on $\mathbb{F}_2^{n-k}$, obtained by keeping constant k coordinates among $x_1, \ldots, x_n$, is bent (*i.e., *those functions that satisfy the propagation criterion of degree $n-k$ and order k; see Section 8.8). These are the four symmetric bent functions (see Section 8.10). They were called hyperbent in [55], but we keep this term for the notion introduced by Youssef and Gong.*

8.6.8 Superclasses: Partially Bent Functions, Partial Bent Functions, and Plateaued Functions

We have seen that bent functions can never be balanced, which makes them improper for a direct cryptographic use. This has led to research on superclasses of the class of bent functions, whose elements can have high nonlinearities but can also be balanced (and possibly be m-resilient with large m or satisfy $PC(l)$ with large l). A first superclass having these properties has been obtained as the set of those functions that achieve a bound expressing some trade-off between the number of nonbalanced derivatives (*i.e.*, of nonzero autocorrelation coefficients) of a Boolean function and the number of nonzero values of its Walsh transform. This bound, given in the next proposition, was conjectured in [300] and proved later in [52].

Proposition 8.46. *Let n be any positive integer. Let f be any Boolean function on $\mathbb{F}_2^n$. Let us denote the cardinalities of the sets $\{b \in \mathbb{F}_2^n \mid \mathcal{F}(D_b f) \neq 0\}$ and $\{b \in \mathbb{F}_2^n \mid \widehat{f_\chi}(b) \neq 0\}$ by N_{Δ_f} and $N_{\widehat{f_\chi}}$, respectively. Then*

$$N_{\Delta_f} \times N_{\widehat{f_\chi}} \ge 2^n. \qquad (8.53)$$

Moreover, $N_{\Delta_f} \times N_{\widehat{f_\chi}} = 2^n$ if and only if, for every $b \in \mathbb{F}_2^n$, the derivative $D_b f$ is either balanced or constant. This property is also equivalent to the fact that there exist two linear subspaces E (of even dimension) and E' of $\mathbb{F}_2^n$, whose direct sum equals $\mathbb{F}_2^n$, and Boolean functions g, bent on E, and h, affine on E', such that

$$\forall x \in E,\ \forall y \in E',\ f(x+y) = g(x) \oplus h(y). \qquad (8.54)$$

Proof. Inequality (8.53) comes directly from Relation (8.25): since the value of the autocorrelation coefficient $\mathcal{F}(D_b f)$ lies between -2^n and 2^n for every b, we have $N_{\Delta_f} \ge 2^{-n} \sum_{b \in \mathbb{F}_2^n} (-1)^{u \cdot b} \mathcal{F}(D_b f) = 2^{-n} \widehat{f_\chi}^2(u)$, for every $u \in \mathbb{F}_2^n$, and thus

$N_{\Delta_f} \geq 2^{-n} \max_{u \in \mathbb{F}_2^n} \widehat{\mathrm{f}_{\chi}}^2(u)$. And we have

$$N_{\widehat{\mathrm{f}_{\chi}}} \geq \frac{\sum_{u \in \mathbb{F}_2^n} \widehat{\mathrm{f}_{\chi}}^2(u)}{\max_{u \in \mathbb{F}_2^n} \widehat{\mathrm{f}_{\chi}}^2(u)} = \frac{2^{2n}}{\max_{u \in \mathbb{F}_2^n} \widehat{\mathrm{f}_{\chi}}^2(u)}$$

This proves Inequality (8.53). This inequality is an equality if and only if both inequalities above are equalities, that is, if and only if, for every b, the autocorrelation coefficient $\mathcal{F}(D_b f)$ equals 0 or $2^n(-1)^{u_0 \cdot b}$, where $\max_{u \in \mathbb{F}_2^n} \widehat{\mathrm{f}_{\chi}}^2(u) = \widehat{\mathrm{f}_{\chi}}^2(u_0)$ and if f is plateaued. The condition that $D_b f$ is either balanced or constant, for every b, is sufficient to imply that f has the form (8.54): E' is the linear kernel of f, and the restriction of f to E has balanced derivatives. Conversely, any function of the form (8.54) is such that Relation (8.53) is an equality. ■

These functions such that $N_{\Delta_f} \times N_{\widehat{\mathrm{f}_{\chi}}} = 2^n$ are called *partially bent functions*. Every quadratic function is partially bent. Partially bent functions share with quadratic functions almost all of their nice properties (a Walsh spectrum that is easier to calculate, potential good nonlinearity, and good resiliency order); see [52]. In particular, the values of the Walsh transform equal 0 or $\pm 2^{dim(E')+dim(E)/2}$.

The following generalization of Relation (8.53) was obtained in [304].

Proposition 8.47. *Let φ be any nonzero n-variable pseudo-Boolean function. Let $N_\varphi = |\{x \in \mathbb{F}_2^n / \varphi(x) \neq 0\}|$ and $N_{\widehat{\varphi}} = |\{u \in \mathbb{F}_2^n / \widehat{\varphi}(u) \neq 0\}|$. Then $N_\varphi \times N_{\widehat{\varphi}} \geq 2^n$.*

Equality occurs if and only if there exists a number λ and a flat F of $\mathbb{F}_2^n$ such that $\varphi(x) = \lambda(-1)^{u \cdot x}$ if $x \in F$ and $\varphi(x) = 0$ otherwise.

Proof. Denoting by 1_φ the indicator of the support $\{x \in \mathbb{F}_2^n / \varphi(x) \neq 0\}$ of φ, and replacing $\varphi(x)$ by $1_\varphi(x)\,\varphi(x)$ in the definition of $\widehat{\varphi}$, gives, for every $u \in \mathbb{F}_2^n$; $\widehat{\varphi}(u) = \sum_{x \in \mathbb{F}_2^n} 1_\varphi(x)\varphi(x)(-1)^{u \cdot x}$. Then, applying the Cauchy-Schwartz inequality gives $\widehat{\varphi}^2(u) \leq N_\varphi \sum_{x \in \mathbb{F}_2^n} \varphi^2(x) = 2^{-n} N_\varphi \sum_{v \in \mathbb{F}_2^n} \widehat{\varphi}^2(v)$ (according to Parseval's relation (8.12)). Hence, $\widehat{\varphi}^2(u) \leq 2^{-n} N_\varphi \times N_{\widehat{\varphi}} \max_{v \in \mathbb{F}_2^n} \widehat{\varphi}^2(v)$. Choosing u such that $\widehat{\varphi}^2(u)$ is maximum gives the desired inequality, since, according to Parseval's inequality, and φ being nonzero, this maximum cannot be null.

Equality occurs if and only if all of the inequalities above are equalities, that is, $\widehat{\varphi}^2(v)$ takes only one nonzero value (say μ) and there exists a number λ such that, for every u such that $\widehat{\varphi}^2(u) = \mu$, we have $\varphi(x) \neq 0 \Rightarrow \varphi(x) = \lambda(-1)^{u \cdot x}$. This is equivalent to the condition stated at the end of Proposition 8.47. ■

Partially bent functions must not be mistaken for *partial bent functions*, studied by P. Guillot in [172]. By definition, the Fourier transforms of partial bent functions take exactly two values[42] λ and $\lambda + 2^{n/2}$ on $\mathbb{F}_2^{n*}$ (n even). Rothaus' bound on the degree generalizes to partial bent functions. The dual $\widetilde{f}$ of f, defined by $\widetilde{f}(u) = 0$ if $\widehat{f}(u) = \lambda$ and $\widetilde{f}(u) = 1$ if $\widehat{f}(u) = \lambda + 2^{n/2}$, is also partial bent, and its dual

[42] Partial bent functions are the indicators of partial difference sets.

is f. Two kinds of partial bent functions f exist: those such that $\widehat{f}(0) - f(0) = -\lambda(2^{n/2} - 1)$ and those such that $\widehat{f}(0) - f(0) = (2^{n/2} - \lambda)(2^{n/2} + 1)$. This can be proved by applying Parseval's relation (8.23). The sum of two partial bent functions of the same kind, whose supports have at most the zero vector in common, is partial bent. A potential reason in interest in partial bent functions is the possibility of using them as building blocks for constructing bent functions.

In spite of their good properties, partially bent functions, when they are not bent, have by definition nonzero linear structures and so do not give full satisfaction. The class of *plateaued* functions, already encountered in Subsection 8.4.1 (and sometimes called *three-valued functions*), is a natural extension of that of partially bent functions. They have been first studied by Zheng and Zhang in [360]. A function is called plateaued if its squared Walsh transform takes at most one nonzero value, which is, if its Walsh transform takes at most three values 0 and $\pm\lambda$ (where λ is some positive integer, which we call the *amplitude* of the plateaued function). Bent functions are plateaued and, according to Parseval's relation (8.23), a plateaued function is bent if and only if its Walsh transform never takes the value 0.

Note that, according to Parseval's relation again, denoting as before by $N_{\widehat{f}_\chi}$ the cardinality of the support $\{a \in \mathbb{F}_2^n / \widehat{f}_\chi(a) \neq 0\}$ of the Walsh transform of a given n-variable Boolean function f, we have $N_{\widehat{f}_\chi} \times \max_{a \in \mathbb{F}_2^n} \widehat{f}_\chi^2(a) \geq 2^{2n}$ and therefore, according to Relation (8.35) relating the nonlinearity to the Walsh transform, $nl(f) \leq 2^{n-1}\left(1 - 1/\sqrt{N_{\widehat{f}_\chi}}\right)$. Equality is achieved if and only if f is plateaued.

Still because of Parseval's relation, the amplitude λ of any plateaued function must be of the form 2^r where $r \geq n/2$ (since $N_{\widehat{f}_\chi} \leq 2^n$). Hence, the values of the Walsh transform of a plateaued function are divisible by $2^{n/2}$ if n is even and by $2^{(n+1)/2}$ if n is odd. The class of plateaued functions contains those functions that achieve the best possible trade-offs between resiliency, nonlinearity, and algebraic degree: the order of resiliency and the nonlinearity of any Boolean function are bounded by Sarkar et al.'s bound (see Section 8.7), and the best compromise between those two criteria is achieved by plateaued functions only; the third criterion – the algebraic degree – is then also optimum. Other properties of plateaued functions can be found in [41]. Plateaued functions can be characterized by second-order covering sequences (see [92]).

Proposition 8.48. *A Boolean function f on $\mathbb{F}_2^n$ is plateaued if and only if there exists λ such that, for every $x \in \mathbb{F}_2^n$:*

$$\sum_{a,b \in \mathbb{F}_2^n} (-1)^{D_a D_b f(x)} = \lambda^2. \tag{8.55}$$

The proof is very similar to that of Proposition 8.52, and λ is necessarily the amplitude of the plateaued function. Indeed, a function f is plateaued with amplitude λ if and only if, for every $u \in \mathbb{F}_2^n$, we have, $\widehat{f}_\chi(u)\left(\widehat{f}_\chi^2(u) - \lambda^2\right) = 0$, that is, $\widehat{f_\chi}^3(u) - \lambda^2 \widehat{f_\chi}(u) = 0$. Applying the Fourier transform to both terms of this equality and using Relation (8.20), we see that this is equivalent to the fact that, for every $a \in \mathbb{F}_2^n$, we have

$$\sum_{x,y\in\mathbb{F}_2^n} (-1)^{f(x)\oplus f(y)\oplus f(x+y+a)} = \lambda^2(-1)^{f(a)}.$$

The fact that quadratic functions are plateaued is a direct consequence of Proposition 8.48, because their second-order derivatives are constant; also, Proposition 8.48 gives more insight into the relationship between the nonlinearity of a quadratic function and the number of its nonzero second-order derivatives.

P. Langevin proved in [230] that, if f is a plateaued function, then the coset $f \oplus R(1,n)$ of the Reed-Muller code of order 1, is an *orphan of* $R(1,n)$. The notion of orphan has been introduced in [178] (with the term "urcoset" instead of orphan) and studied in [30]. A coset of $R(1,n)$ is an orphan if it is maximum with respect to the following partial order relation: $g \oplus R(1,n)$ is smaller than $f \oplus R(1,n)$ if there exists in $g \oplus R(1,n)$ an element g_1 of weight $nl(g)$ (that is, of minimum weight in $g \oplus R(1,n)$), and in $f \oplus R(1,n)$ an element f_1 of weight $nl(f)$, such that $supp(g_1) \subseteq supp(f_1)$. Clearly, if f is a function of maximum nonlinearity, then $f \oplus R(1,n)$ is an orphan of $R(1,n)$ (the converse is false, because plateaued functions with nonoptimum nonlinearity exist). The notion of orphans can be used in algorithms searching for functions with high nonlinearities.

8.6.9 Normal and Nonnormal Bent Functions

As observed in [53] (see Theorem 8.34), if a bent function f is normal (resp. weakly normal), that is, constant (resp. affine) on an $n/2$-dimensional flat $b + E$ (where E is a subspace of $\mathbb{F}_2^n$), then its dual $\widetilde{f}$ is such that $\widetilde{f}(u) \oplus b \cdot u$ is constant on $E^\perp$ (resp. on $a + E^\perp$, where a is a vector such that $f(x) \oplus a \cdot x$ is constant on E). Thus, $\widetilde{f}$ is weakly normal. Moreover, we have already seen that f (resp. $f(x) \oplus a \cdot x$) is balanced on each of the other cosets of the flat. H. Dobbertin used this idea to construct balanced functions with high nonlinearities from normal bent functions (see Subsection 8.7.5.1).

The existence of nonnormal (and even non–weakly normal) bent functions, that is, bent functions that are nonconstant (resp. nonaffine) on every $n/2$-dimensional flat, has been shown, contradicting a conjecture made by several authors that such bent functions did not exist. It is proved in [139] that the so-called Kasami function defined over $\mathbb{F}_{2^n}$ by $f(x) = tr_n\left(ax^{2^{2k}-2^k+1}\right)$, with $gcd(k,n) = 1$, is bent if n is not divisible by 3 and if $a \in \mathbb{F}_{2^n}$ is not a cube. As shown in [44], if $a \in \mathbb{F}_4 \setminus \mathbb{F}_2$ and $k = 3$, then for $n = 10$, the function $f(x) \oplus tr_n(b)$ is nonnormal for some b, and for $n = 14$, the function f is not weakly normal. Cubic bent functions on eight variables are all normal, as shown in [100].

The direct sum (see the definition in Subsection 8.6.4) of two normal functions is obviously a normal function, whereas the direct sum of two nonnormal functions can be normal. What about the sum of a normal bent function and of a nonnormal bent function? This question has been studied in [78]. To this aim, a notion more general than normality has been introduced as follows.

Definition 8.8. *Let $U \subseteq V$ be two vector spaces over $\mathbb{F}_2$. Let $\beta : U \rightarrow \mathbb{F}_2$ and $f : V \rightarrow \mathbb{F}_2$ be bent functions. Then we say that f is a* normal extension *of β, in symbols $\beta \preceq f$, if there is a direct decomposition $V = U \oplus W_1 \oplus W_2$ such that*

(i) $\beta(u) = f(u + w_1)$ *for all* $u \in U$, $w_1 \in W_1$,
(ii) $\dim W_1 = \dim W_2$.

The relation $\preceq$ is transitive, and if $\beta \preceq f$, then the same relation exists between the duals: $\widetilde{\beta} \preceq \widetilde{f}$.

A bent function is normal if and only if $\varepsilon \preceq f$, where $\varepsilon \in \mathbb{F}_2$ is viewed as a Boolean function over the vector space $\mathbb{F}_2^0 = \{0\}$.

Examples of normal extensions are given in [78] (including the construction of Theorem 8.35 and its particular cases, the indirect sum, and the extension of the Maiorana-McFarland type).

The clarification about the sum of a normal bent function and of a nonnormal bent function comes from the two following propositions:

Proposition 8.49. *Let $f_i : V_i \rightarrow \mathbb{F}_2$, $i = 1, 2$, be bent functions. The direct sum $f_1 \oplus f_2$ is normal if and only if bent functions β_1 and β_2 exist such that f_i is a normal extension of β_i $(i = 1, 2)$ and either β_1 and β_2 or β_1 and $\beta_2 \oplus 1$ are linearly equivalent.*

Proposition 8.50. *Suppose that $\beta \preceq f$ for bent functions β and f. If f is normal, then β also is normal.*

Hence, since the direct sum of a bent function β and of a normal bent function g is a normal extension of β, the direct sum of a normal and a nonnormal bent function is always nonnormal.

Normal extension leads to a secondary construction of bent functions.

Proposition 8.51. *Let β be a bent function on U and f a bent function on $V = U \times W \times W$. Assume that $\beta \preceq f$. Let*

$$\beta' : U \rightarrow \mathbb{F}_2$$

be any bent function. Modify f by setting for all $x \in U$, $y \in W$

$$f'(x, y, 0) = \beta'(x),$$

while $f'(x, y, z) = f(x, y, z)$ for all $x \in U$, $y, z \in W$, $z \neq 0$. Then f' is bent and we have $\beta' \preceq f'$.

8.6.10 Kerdock Codes

For every even n, the *Kerdock code* $\mathcal{K}_n$ [211] is a supercode of $R(1,n)$ (*i.e.*, contains $R(1,n)$ as a subset) and is a subcode of $R(2,n)$. More precisely, $\mathcal{K}_n$ is a union of cosets $f_u \oplus R(1,n)$ of $R(1,n)$, where the functions f_u are quadratic (one of them is null and all the others have algebraic degree 2). The difference $f_u \oplus f_v$ between two distinct functions f_u and f_v being bent, $\mathcal{K}_n$ has minimum distance $2^{n-1} - 2^{n/2-1}$ (n even), which is the best possible minimum distance for a code equal to a union of cosets of $R(1,n)$, according to the covering radius bound. The size of $\mathcal{K}_n$ equals 2^{2n}. This is the best possible size for such minimum distance (see [129]). We now describe how the construction of Kerdock codes can be simply stated.

8.6.10.1 Construction of the Kerdock Code

The function

$$f(x) = \bigoplus_{1\le i<j\le n} x_i x_j \tag{8.56}$$

(which can also be defined as $f(x) = \binom{w_H(x)}{2}$ [mod 2]) is bent[43] because the kernel of its associated symplectic form $\varphi(x,y) = \bigoplus_{1\le i\neq j\le n} x_i y_j$ equals $\{0\}$. Thus, the linear code $R(1,n) \cup (f \oplus R(1,n))$ has minimum distance $2^{n-1} - 2^{n/2-1}$.

We want to construct a code of size 2^{2n} with this same minimum distance.

We use the structure of field to this aim. We recalled in Subsection 8.2.1 some properties of the field $\mathbb{F}_{2^m}$ (where m is any positive integer). In particular, we saw that there exists $\alpha \in \mathbb{F}_{2^m}$ (called a primitive element) such that $\mathbb{F}_{2^m} = \{0, \alpha, \alpha^2, \cdots, \alpha^{2^m-1}\}$. Moreover, there exists α, a primitive element, such that $(\alpha, \alpha^2, \alpha^{2^2}, \cdots, \alpha^{2^{m-1}})$ is a basis of the vectorspace $\mathbb{F}_{2^m}$. Such a basis is called a *normal basis*. If m is odd, then there exists a self-dual normal basis, that is, a normal basis such that $tr_m(\alpha^{2^i+2^j}) = 1$ if $i = j$ and $tr_m(\alpha^{2^i+2^j}) = 0$ otherwise, where tr_m is the trace function over $\mathbb{F}_{2^m}$.

The *consequence*: for all $x = x_1\alpha + \cdots + x_m\alpha^{2^{m-1}}$ in $\mathbb{F}_{2^m}$, we have

$$tr_m(x) = \bigoplus_{i=1}^{m} x_i \qquad tr_m(x^{2^j+1}) = \bigoplus_{i=1}^{m} x_i x_{i+j}$$

(where x_{i+j} is replaced by x_{i+j-m} if $i + j > m$).

[43] We see in Section 8.10 that its is, up to the addition of affine functions, the sole symmetric bent function.

The function f of Relation (8.56), viewed as a function $f(x, x_n)$ on $\mathbb{F}_{2^m} \times \mathbb{F}_2$, where $m = n - 1$ is odd – say $m = 2t + 1$ – can now be written as[44]

$$f(x, x_n) = tr_m\left(\sum_{j=1}^{t} x^{2^j+1}\right) \oplus x_n tr_m(x).$$

Notice that the associated symplectic form associated to f equals $tr_m(x)tr_m(y) \oplus tr_m(xy) \oplus x_n tr_m(y) \oplus y_n tr_m(x)$.

Let us denote $f(ux, x_n)$ by $f_u(x, x_n)$ ($u \in \mathbb{F}_{2^m}$). Then $\mathcal{K}_n$ is defined as the union, when u ranges over $\mathbb{F}_{2^m}$, of the cosets $f_u \oplus R(1, n)$.

$\mathcal{K}_n$ contains 2^{n+1} affine functions and $2^{2n} - 2^{n+1}$ quadratic bent functions. Its minimum distance equals $2^{n-1} - 2^{n/2-1}$ because the sum of two distinct functions f_u and f_v is bent. Indeed, the kernel of the associated symplectic form equals the set of all ordered pairs (x, x_n) verifying $tr_m(ux)tr_m(uy) \oplus tr_m(u^2xy) \oplus x_n tr_m(uy) \oplus y_n tr_m(ux) = tr_m(vx)tr_m(vy) \oplus tr_m(v^2xy) \oplus x_n tr_m(vy) \oplus y_n tr_m(vx)$ for every $(y, y_n) \in \mathbb{F}_{2^m} \times \mathbb{F}_2$, that is, $utr_m(ux) + u^2x + x_n u = vtr_m(vx) + v^2x + x_n v$ and $tr_m(ux) = tr_m(vx)$; it is a simple matter to show that it equals $\{(0, 0)\}$.

Open problem. Other examples of codes having the same parameters exist [205]. All are equal to subcodes of the Reed-Muller code of order 2, up to affine equivalence. We do not know how to obtain the same parameters with non-quadratic functions. This would be useful for cryptographic purposes as well as for the design of sequences for code division multiple access (CDMA) in telecommunications.

Remark. *The Kerdock codes are not linear. However, they share some nice properties with linear codes: the distance distribution between any codeword and all the other codewords does not depend on the choice of the codeword (we say that the Kerdock codes are distance-invariant; this results in the fact that their distance enumerators are equal to their weight enumerators); and, as proved by Semakov and Zinoviev [326], the weight enumerators of the Kerdock codes satisfy a relation similar to Relation (8.33), in which* C *is replaced by* $\mathcal{K}_n$ *and* $C^\perp$ *is replaced by the so-called Preparata code of the same length (we say that the Kerdock codes and the Preparata codes are formally dual). An explanation of this astonishing property was recently obtained [175]: the Kerdock code is stable under an addition inherited of the addition in* $\mathbb{Z}_4 = \mathbb{Z}/4\mathbb{Z}$ *(we say it is* $\mathbb{Z}_4$*-linear) and the MacWilliams identity still holds in this different framework. Such an explanation had been an open problem for two decades.*

[44] Obviously, this expression can be taken as the definition of f.

8.7 Resilient Functions

We have seen in Subsection 8.4.1 that the combining functions in stream ciphers must be m-resilient with large m. As any cryptographic functions, they must also have high algebraic degrees and high nonlinearities.

Notation: By an $(n, m, d, \mathcal{N})$- function, we mean an n-variable, m-resilient function having algebraic degree at least d and nonlinearity at least $\mathcal{N}$.

There are necessary trade-offs between n, m, d, and $\mathcal{N}$.

8.7.1 Bound on Algebraic Degree

Siegenthaler's bound states that any m-resilient function $(0 \leq m < n-1)$ has algebraic degree less than or equal to $n - m - 1$ and that any $(n-1)$-resilient function is affine.[45] This can be proved directly by using Relation (8.3) and the original definition of resiliency given by Siegenthaler (Definition 8.3), since the bit $\bigoplus_{x\in\mathbb{F}_2^n/\, supp(x)\subseteq I} f(x)$ equals the parity of the weight of the restriction of f obtained by setting to 0 the coordinates of x that lie outside I. Note that instead of using the original Siegenthaler definition in the proof of Siegenthaler's bound, we can also use the characterization by Xiao and Massey, recalled in Theorem 8.20, together with the Poisson summation formula (8.18) applied to $\varphi = f$ and with $E^\perp = \{x \in \mathbb{F}_2^n \mid supp(x) \subseteq I\}$, where I has size strictly greater than $n - m - 1$. But this gives a less simple proof. Siegenthaler's bound is also a direct consequence of a characterization of resilient functions[46] through their NNFs and of the fact that the algebraic degrees of Boolean functions are smaller than or equal to their numerical degrees:

Proposition 8.52. ([87].) *Let n be any positive integer and $m < n$ a non-negative integer. A Boolean function f on $\mathbb{F}_2^n$ is m-resilient if and only if the NNF of the function $f(x) \oplus x_1 \oplus \cdots \oplus x_n$ has degree at most $n - m - 1$.*

Proof. Let us denote by $g(x)$ the function $f(x) \oplus x_1 \oplus \cdots \oplus x_n$. For each vector $a \in \mathbb{F}_2^n$, we denote by $\overline{a}$ the componentwise complement of a equal to $a + (1, \ldots, 1)$. We have $\widehat{f_\chi}(a) = \widehat{g_\chi}(\overline{a})$. Thus, f is m-resilient if and only if, for each vector u of weight greater than or equal to $n - m$, the number $\widehat{g_\chi}(u)$ is null. Consider the NNF of g:

$$g(x) = \sum_{I\in\mathcal{P}(N)} \lambda_I x^I.$$

[45] Siegenthaler also proved that any n-variable mth-order correlation-immune function has degree at most $n - m$. This can be shown by using similar methods as for resilient functions. Moreover, if such function has weight divisible by 2^{m+1}, then it satisfies the same bound as m-resilient functions.

[46] A similar characterization of correlation-immune functions can be found in [62].

According to Relations (8.30), (8.31), and (8.12) applied to g, we have for nonzero u

$$\widehat{g_\chi}(u) = (-1)^{w_H(u)+1} \sum_{I \in \mathcal{P}(N)\,|\,supp(u) \subseteq I} 2^{n-|I|+1} \lambda_I$$

and for nonempty I

$$\lambda_I = 2^{-n}(-2)^{|I|-1} \sum_{u \in \mathbb{F}_2^n\,|\,I \subseteq supp(u)} \widehat{g_\chi}(u).$$

We deduce that $\widehat{g_\chi}(u)$ is null for every vector u of weight greater than or equal to $n-m$ if and only if the NNF of g has degree at most $n-m-1$. ■

Proposition 8.52 can also be proved by using the Xiao-Massey characterization (again) and Relation (8.8) relating the values of the coefficients of the NNF to the values of the function, applied to $g(x) = f(x) \oplus x_1 \oplus \cdots \oplus x_n$.

Proposition 8.52 was used by X.-D. Hou in [193] for constructing resilient functions. Siegenthaler's bound gives an example of the trade-offs that must be accepted in the design of combiner generators.[47] Sarkar and Maitra showed in [318] that the values of the Walsh transform of an n-variable, m-resilient (resp. mth-order correlation-immune) function are divisible by 2^{m+2} (resp. 2^{m+1}) if $m \leq n-2$ (a proof of a slightly more precise result is given in the next subsection, at Theorem 8.53).[48] This *Sarkar-Maitra divisibility* bound (which implies in particular that the weight of any mth-order correlation-immune function is divisible by 2^m) also allows us to deduce Siegenthaler's bound, thanks to Proposition 8.15 applied with $k = m+2$ (resp. $k = m+1$).

8.7.2 Bounds on the Nonlinearity

Sarkar-Maitra's divisibility bound, recalled at the end of the previous subsection, provided a nontrivial upper bound on the nonlinearity of resilient functions, independently obtained by Tarannikov [342] and by Zheng and Zhang [363]: the nonlinearity of any m-resilient function ($m \leq n-2$) is bounded above by $2^{n-1} - 2^{m+1}$. This bound is tight, at least when $m \geq 0.6\, n$; see [342, 343].[49] We call it *Sarkar et al.'s bound*. Notice that, if an m-resilient function f achieves nonlinearity $2^{n-1} - 2^{m+1}$, then f is plateaued. Indeed, the distances between f and affine functions lie then between $2^{n-1} - 2^{m+1}$ and $2^{n-1} + 2^{m+1}$ and must be therefore equal to $2^{n-1} - 2^{m+1}$, 2^{n-1} and $2^{n-1} + 2^{m+1}$ because of the divisibility result of

[47] One approach to avoid such a trade-off is to allow memory in the nonlinear combination generator, that is, to replace the combining function by a finite state machine; see [276].

[48] More is proved in [62, 93]; in particular: if the weight of an mth-order correlation-immune is divisible by 2^{m+1}, then the values of its Walsh Transform are divisible by 2^{m+2}.

[49] Also Zheng and Zhang [363], showed that the upper bound on the nonlinearity of correlation-immune functions of high orders is the same as the upper bound on the nonlinearity of resilient functions of the same orders. The distances between resilient functions and Reed-Muller codes of orders greater than 1 have also been studied by Kurosawa et al. and by Borissov et al. [22, 221].

Sarkar and Maitra. Thus, the Walsh transform of f takes three value, 0 and $\pm 2^{m+2}$. Moreover, it is proved in [342] that such a function f also achieves Siegenthaler's bound (and as proved in [260], achieves a minimum sum-of-squares indicator). These last properties can also be deduced from a more precise divisibility bound shown later in [62].

Theorem 8.53. *Let f be any n-variable m-resilient function ($m \le n-2$) and let d be its algebraic degree. The values of the Walsh transform of f are divisible by $2^{m+2+\lfloor \frac{n-m-2}{d} \rfloor}$. Hence the nonlinearity of f is divisible by $2^{m+1+\lfloor \frac{n-m-2}{d} \rfloor}$.*

The approach for proving this result was first to use the numerical normal form (see [62]). Later, a second proof using only the properties of the Fourier transform was given in [93].

Proof. The Poisson summation formula (8.18) applied to $\varphi = f_\chi$ and to the vectorspace $E = \{u \in \mathbb{F}_2^n / \forall i \in N,\ u_i \le v_i\}$, where v is some vector of $\mathbb{F}_2^n$, whose orthogonal equals $E^\perp = \{u \in \mathbb{F}_2^n / \forall i \in N,\ u_i \le v_i \oplus 1\}$, gives $\sum_{u \in E} \widehat{f_\chi}(u) = 2^{w_H(v)} \sum_{x \in E^\perp} f_\chi(x)$. It is then a simple matter to prove the result by induction on the weight of v, starting with the vectors of weight $m+1$ (because it is obvious for the vectors of weights at most m), and using McEliece's divisibility property (see Subsection 8.3.1). ■

A similar proof shows that the values of the Walsh transform of any mth-order correlation-immune function are divisible by $2^{m+1+\lfloor \frac{n-m-1}{d} \rfloor}$ (and by $2^{m+2+\lfloor \frac{n-m-2}{d} \rfloor}$ if its weight is divisible $2^{m+1+\lfloor \frac{n-m-2}{d} \rfloor}$; see [93]).

Theorem 8.53 directly gives a more precise upper bound on the nonlinearity of any m-resilient function of degree d: this nonlinearity is bounded above by $2^{n-1} - 2^{m+1+\lfloor \frac{n-m-2}{d} \rfloor}$. This gives a simpler proof that it can be equal to $2^{n-1} - 2^{m+1}$ only if $d = n - m - 1$, that is, if Siegenthaler's bound is achieved. Moreover, the proof just given also shows that the nonlinearity of any m-resilient n-variable Boolean function is bounded above by $2^{n-1} - 2^{m+1+\lfloor \frac{n-m-2}{d} \rfloor}$ where d is the minimum algebraic degree of the restrictions of f to the subspaces $\{u \in \mathbb{F}_2^n / \forall i \in N,\ u_i \le v_i \oplus 1\}$ such that v has weight $m+1$ and $\widehat{f_\chi}(v) \ne 0$.

If $2^{n-1} - 2^{m+1}$ is greater than the best possible nonlinearity of all balanced functions (and in particular if it is greater than the covering radius bound), then, obviously, a better bound exists. In the case of n even, the best possible nonlinearity of all balanced functions being strictly smaller than $2^{n-1} - 2^{n/2-1}$, Sarkar and Maitra deduce that $nl(f) \le 2^{n-1} - 2^{n/2-1} - 2^{m+1}$ for every m-resilient function f with $m \le n/2 - 2$. In the case of n odd, they state that $nl(f)$ is less than or equal to the highest multiple of 2^{m+1}, which is less than or equal to the best possible nonlinearity of all Boolean functions. But a potentially better upper bound can be given, whatever the parity of n. Indeed, Sarkar-Maitra's divisibility bound shows that $\widehat{f_\chi}(a) = \varphi(a) \times 2^{m+2}$ where $\varphi(a)$ is integer-valued. But Parseval's relation (8.23) and the fact that $\widehat{f_\chi}(a)$ is null for every vector a of weight $\le m$

imply

$$\sum_{a\in\mathbb{F}_2^n/\ w_H(a)>m} \varphi^2(a) = 2^{2n-2m-4}$$

and, thus,

$$\max_{a\in\mathbb{F}_2^n} |\varphi(a)| \geq \sqrt{\frac{2^{2n-2m-4}}{2^n - \sum_{i=0}^m \binom{n}{i}}} = \frac{2^{n-m-2}}{\sqrt{2^n - \sum_{i=0}^m \binom{n}{i}}}.$$

Hence, we have $\max_{a\in\mathbb{F}_2^n} |\varphi(a)| \geq \left\lceil \frac{2^{n-m-2}}{\sqrt{2^n - \sum_{i=0}^m \binom{n}{i}}} \right\rceil$, and this implies

$$nl(f) \leq 2^{n-1} - 2^{m+1} \left\lceil \frac{2^{n-m-2}}{\sqrt{2^n - \sum_{i=0}^m \binom{n}{i}}} \right\rceil. \tag{8.57}$$

When n is even and $m \leq n/2 - 2$, this number is always less than or equal to the number $2^{n-1} - 2^{n/2-1} - 2^{m+1}$ (given by Sarkar and Maitra), because $(2^{n-m-2})/\sqrt{2^n - \sum_{i=0}^m \binom{n}{i}}$ is strictly greater than $2^{n/2-m-2}$ and $2^{n/2-m-2}$ is an integer, and, thus, $\left\lceil (2^{n-m-2})/\sqrt{2^n - \sum_{i=0}^m \binom{n}{i}} \right\rceil$ is at least $2^{n/2-m-2} + 1$. And when n increases, the right hand-side of Relation (8.57) is strictly smaller than $2^{n-1} - 2^{n/2-1} - 2^{m+1}$ for an increasing number of values of $m \leq n/2 - 2$ (but this improvement does not appear when we compare the values we obtain with this bound to the values indicated in the table given by Sarkar and Maitra in [318], because the values of n they consider in this table are small).

When n is odd, it is difficult to say whether inequality (8.57) is better than the bound given by Sarkar and Maitra, because their bound involves a value that is unknown for $n \geq 9$ (the best possible nonlinearity of all balanced Boolean functions). In any case, this makes (8.57) more usable.

We know (see [257], page 310) that $\sum_{i=0}^m \binom{n}{i} \geq (2^{nH_2(m/n)})/\sqrt{8m(1-m/n)}$, where $H_2(x) = -x\log_2(x) - (1-x)\log_2(1-x)$ is the so-called *binary entropy function* and satisfies $H_2(\frac{1}{2} - x) = 1 - 2x^2 \log_2 e + o(x^2)$. Thus, we have

$$nl(f) \leq 2^{n-1} - 2^{m+1} \left\lceil \frac{2^{n-m-2}}{\sqrt{2^n - \frac{2^{nH_2(m/n)}}{\sqrt{8m(1-m/n)}}}} \right\rceil. \tag{8.58}$$

8.7.3 Bound on the Maximum Correlation with Subsets of N

An upper bound on the maximum correlation of m-resilient functions with respect to subsets I of N can be directly deduced from Relation (8.40) and from Sarkar et al.'s bound. Note that we get an improvement by using that the support of $\widehat{f_\chi}$, restricted to the set of vectors $u \in \mathbb{F}_2^n$ such that $u_i = 0,\ \forall i \notin I$, contains at most

$\sum_{i=m+1}^{|I|} \binom{|I|}{i}$ vectors. In particular, if $|I| = m + 1$, the maximum correlation of f with respect to I equals $2^{-n}\,|\widehat{\mathrm{f}_\chi}(u)|$, where u is the vector of support I; see [37, 46, 355]. The optimal number of LFSRs that should be considered together in a correlation attack on a cryptosystem using an m-resilient combining function is $m + 1$; see [37].

8.7.4 Relationship with Other Criteria

The relationships between resiliency and other criteria have been studied in [104, 260, 345, 362]. For instance, m-resilient $PC(l)$ functions can exist only if $m + l \leq n - 1$. This is a direct consequence of Relation (8.27), relating the values of the Walsh transform of a function on a flat $a + E$ to the autocorrelation coefficients of the function on a flat $b + E^\perp$, applied with $a = b = 0$, $E = \{x \in \mathbb{F}_2^n;\ x_i = 0,\ \forall i \in I\}$ and $E^\perp = \{x \in \mathbb{F}_2^n;\ x_i = 0,\ \forall i \notin I\}$, where I has size $n - m$: if $l \geq n - m$, then the right-hand term of (8.27) is nonzero while the left-hand term is null. Equality $m + l = n - 1$ is possible only if $l = n - 1$, n is odd and $m = 0$ [104, 362]. The known upper bounds on the nonlinearity (see Section 8.7) can then be improved with the same argument.

The definition of resiliency was weakened in [26, 88, 222] in order to relax some of the trade-offs just recalled without weakening the cryptosystem against the correlation attack.

Resiliency is related to the notion of corrector (useful for the generation of random sequences having good statistical properties) introduced by P. Lacharme in [224].

8.7.5 Constructions

High-order resilient functions with high algebraic degrees, high nonlinearities, and good immunity to algebraic attacks are needed for applications in stream ciphers using the combiner model. But designing constructions of Boolean functions meeting all these cryptographic criteria is still a challenge nowadays (although we would need numerous such functions in order to be able to choose from among them functions satisfying additional design criteria). The primary constructions (which allow designing resilient functions without using known ones) lead potentially to wider classes of functions than secondary constructions (recall that the number of Boolean functions on $n - 1$ variables is only equal to the square root of the number of n-variable Boolean functions). But the known primary constructions of such Boolean functions do not lead to very large classes of functions. In fact, only one reasonably large class of Boolean functions is known, whose elements can be analyzed with respect to the cryptographic criteria recalled in Subsection 8.4.1. So we observe some imbalance in the knowledge on cryptographic functions for stream ciphers: much is known on the properties of resilient functions, but little is known on how constructing them. Examples of m-resilient functions achieving the best possible nonlinearity $2^{n-1} - 2^{m+1}$ (and thus the best algebraic degree) have been obtained for $n \leq 10$ in [291, 317, 318] and for every $m \geq 0.6\, n$ [342, 343] (n then being not limited). But $n \leq 10$ is too small for applications, and $m \geq 0.6\, n$ is too large (because of Siegenthaler's bound), and almost nothing is known about the immunity of these functions to algebraic attacks. Moreover, these examples give very limited numbers of functions (they are often defined recursively or obtained after a computer search), and many of these functions have cryptographic weaknesses such as linear structures (see [104, 260]). Balanced Boolean functions with high nonlinearities were obtained by C. Fontaine in [154] and by E. Filiol and C. Fontaine in [152], who made a computer investigation – but for $n = 7, 9$, which is too small – of the corpus of *idempotent functions*. These functions, whose ANFs are invariant under the cyclic shifts of the coordinates x_i, were later called *rotation symmetric* (see Subsection 8.10.6).

8.7.5.1 Primary Constructions

Maiorana-McFarland's Construction. An extension of the class of bent functions that we called the Maiorana-McFarland original class was given in [34], based on the same principle of concatenating affine functions[50] (we saw this generalization, which we shall call the *Maiorana-McFarland general construction*, in Section 8.6): let r be a positive integer smaller than n; we denote $n - r$ by s; let g be any Boolean function on $\mathbb{F}_2^s$, and let ϕ be a mapping from $\mathbb{F}_2^s$ to $\mathbb{F}_2^r$. Then, we define the function

$$f_{\phi,g}(x, y) = x \cdot \phi(y) \oplus g(y) = \bigoplus_{i=1}^{r} x_i \phi_i(y) \oplus g(y), \quad x \in \mathbb{F}_2^r, \; y \in \mathbb{F}_2^s, \tag{8.59}$$

where $\phi_i(y)$ is the ith coordinate function of $\phi(y)$.

For every $a \in \mathbb{F}_2^r$ and every $b \in \mathbb{F}_2^s$, we have seen in Subsection 8.6.4 that

$$\widehat{\mathrm{f}_{\phi,\mathrm{g}}}_{\chi}(a, b) = 2^r \sum_{y \in \phi^{-1}(a)} (-1)^{g(y) \oplus b \cdot y}. \tag{8.60}$$

This can be used to design resilient functions: if every element in $\phi(\mathbb{F}_2^s)$ has Hamming weight strictly greater than k, then $f_{\phi,g}$ is m-resilient with $m \geq k$ (in particular, if $\phi(\mathbb{F}_2^s)$ does not contain the null vector, then $f_{\phi,g}$ is balanced). Indeed, if $w_H(a) \leq k$ then $\phi^{-1}(a)$ is empty in Relation (8.60); hence, if $w_H(a) + w_H(b) \leq k$, then $\widehat{\mathrm{f}_{\phi,\mathrm{g}}}_{\chi}(a, b)$ is null. The k-resiliency of $f_{\phi,g}$ under this hypothesis can also be deduced from the facts that any affine function $x \in \mathbb{F}_2^r \mapsto a \cdot x \oplus \varepsilon$ ($a \in \mathbb{F}_2^r$ nonzero, $\varepsilon \in \mathbb{F}_2$) is $(w_H(a) - 1)$-resilient, and that any Boolean function equal to the concatenation of k-resilient functions is a k-resilient function (see secondary construction III discussed later).

Degree. The algebraic degree of $f_{\phi,g}$ is at most $s + 1 = n - r + 1$. It equals $s + 1$ if and only if ϕ has algebraic degree s (*i.e.* if at least one of its coordinate functions has algebraic degree s). If we assume that every element in $\phi(\mathbb{F}_2^s)$

[50] These functions were also studied under the name linear-based functions in [1, 352].

has Hamming weight strictly greater than k, then ϕ can have algebraic degree s only if $k \leq r-2$, because if $k = r-1$, then ϕ is constant. Thus, if $m = k$, then the algebraic degree of $f_{\phi,g}$ reaches Siegenthaler's bound $n-k-1$ if and only if either $k = r-2$ and ϕ has algebraic degree $s = n-k-2$, or $k = r-1$ and g has algebraic degree $s = n-k-1$. There are cases where $m > k$ (see [63, 64, 118]). An obvious one is when each set $\phi^{-1}(a)$ has even size and the restriction of g to this set is balanced: then $m \geq k+1$.

Nonlinearity. Relations (8.35), relating the nonlinearity to the Walsh transform, and (8.60) lead straightforwardly to a general lower bound on the nonlinearity of Maiorana-McFarland's functions (first observed in [324]):

$$nl(f_{\phi,g}) \geq 2^{n-1} - 2^{r-1} \max_{a\in\mathbb{F}_2^r} |\phi^{-1}(a)| \tag{8.61}$$

(where $|\phi^{-1}(a)|$ denotes the size of $\phi^{-1}(a)$). A more recent upper bound

$$nl(f_{\phi,g}) \leq 2^{n-1} - 2^{r-1} \left\lceil \sqrt{\max_{a\in\mathbb{F}_2^r} |\phi^{-1}(a)|} \right\rceil \tag{8.62}$$

obtained in [63] strengthens a bound previously obtained in [106, 107] that stated $nl(f_{\phi,g}) \leq 2^{n-1} - 2^{r-1}$.

Proof of (8.62). The sum

$$\sum_{b\in\mathbb{F}_2^s} \left(\sum_{y\in\phi^{-1}(a)} (-1)^{g(y)+b\cdot y} \right)^2 = \sum_{b\in\mathbb{F}_2^s} \left(\sum_{y,z\in\phi^{-1}(a)} (-1)^{g(y)+g(z)+b\cdot(y+z)} \right)$$

equals $2^s|\phi^{-1}(a)|$ (because the sum $\sum_{b\in\mathbb{F}_2^s}(-1)^{b\cdot(y+z)}$ is null if $y \neq z$). The maximum of a set of values always being greater than or equal to its mean, we deduce

$$\max_{b\in\mathbb{F}_2^s} \left| \sum_{y\in\phi^{-1}(a)} (-1)^{g(y)+b\cdot y} \right| \geq \sqrt{|\phi^{-1}(a)|}$$

and thus, according to Relation (8.60),

$$\max_{a\in\mathbb{F}_2^r; b\in\mathbb{F}_2^s} |\widehat{\mathrm{f}}_{\chi\,\phi,g}(a,b)| \geq 2^r \left\lceil \sqrt{\max_{a\in\mathbb{F}_2^r} |\phi^{-1}(a)|} \right\rceil.$$

Relation (8.35) completes the proof. ■

This new bound allowed characterizing the Maiorana-McFarland's functions $f_{\phi,g}$ such that $w_H(\phi(y)) > k$ for every y and achieving nonlinearity $2^{n-1} - 2^{k+1}$: the inequality

$$nl(f_{\phi,g}) \leq 2^{n-1} - \frac{2^{r+\frac{s}{2}-1}}{\sqrt{\sum_{i=k+1}^{r} \binom{r}{i}}}$$

implies either that $r = k+1$ or that $r = k+2$.

If $r = k + 1$, then ϕ is the constant $(1, \ldots, 1)$ and $n \leq k + 3$. Either $s = 1$ and $g(y)$ is then any function in one variable, or $s = 2$ and g is then any function of the form $y_1 y_2 \oplus \ell(y)$ where ℓ is affine (thus, f is quadratic).

If $r = k + 2$, then ϕ is injective, $n \leq k + 2 + \log_2(k + 3)$, g is any function on $n - k - 2$ variables, and $d^\circ f_{\phi,g} \leq 1 + \log_2(k + 3)$.

A simple example of k-resilient Maiorana-McFarland functions such that $nl(f_{\phi,g}) = 2^{n-1} - 2^{k+1}$ (and thus achieving Sarkar et al.'s bound) can be given for any $r \geq 2^s - 1$ and for $k = r - 2$ (see [63]). And, for every even $n \leq 10$, Sarkar et al.'s bound with $m = n/2 - 2$ can be achieved by Maiorana-McFarland functions. Also, functions with high nonlinearities but not achieving Sarkar et al.'s bound exist in Maiorana-McFarland's class (for instance, for every $n \equiv 1 [\bmod 4]$, there exist such $\frac{n-1}{4}$-resilient functions on $\mathbb{F}_2^n$ with nonlinearity $2^{n-1} - 2^{\frac{n-1}{2}}$).

Generalizations of Maiorana-McFarland's construction were introduced in [63] and [92]; the latter generalization was further generalized into a class introduced in [67]. A motivation for introducing such generalizations is that Maiorana-McFarland functions have the weakness that $x \mapsto f_{\phi,g}(x, y)$ is affine for every $y \in \mathbb{F}_2^s$ and have high divisibilities of their Fourier spectra (indeed, if we want to ensure that f is m-resilient with large value of m, then we need to choose r large; then the Walsh spectrum of f is divisible by 2^r according to Relation (8.60); there is also a risk that this property can be used in attacks, as it was used in [47] to attack block ciphers). The functions constructed in [63, 92] are concatenations of quadratic functions instead of affine functions. This makes them harder to study than Maiorana-McFarland functions. But they are more numerous and more general. Two classes of such functions have been studied.

- The functions of the first class are defined as:

$$f_{\psi,\phi,g}(x, y) = \bigoplus_{i=1}^{t} x_{2i-1}x_{2i}\ \psi_i(y) \oplus x \cdot \phi(y) \oplus g(y),$$

  with $x \in \mathbb{F}_2^r$, $y \in \mathbb{F}_2^s$, where $n = r + s$, $t = \lfloor \frac{r}{2} \rfloor$, and where $\psi : \mathbb{F}_2^s \to \mathbb{F}_2^t$, $\phi : \mathbb{F}_2^s \to \mathbb{F}_2^r$ and $g : \mathbb{F}_2^s \to \mathbb{F}_2$ can be chosen arbitrarily;
- The functions of the second class are defined as

$$f_{\phi_1,\phi_2,\phi_3,g}(x, y) = (x \cdot \phi_1(y))\,(x \cdot \phi_2(y)) \oplus x \cdot \phi_3(y) \oplus g(y),$$

  with $x \in \mathbb{F}_2^r$, $y \in \mathbb{F}_2^s$, where ϕ_1, ϕ_2, and ϕ_3 are three functions from $\mathbb{F}_2^s$ into $\mathbb{F}_2^r$ and g is any Boolean function on $\mathbb{F}_2^s$. The size of this class roughly equals $\left[(2^r)^{2^s}\right]^3 \times 2^{2^s} = 2^{(3r+1)2^s}$ (the exact number, which is unknown, is smaller because the same function can be represented in this form in several ways) and is larger than the size of the first class, roughly equal to $(2^t)^{2^s} \times (2^r)^{2^s} \times 2^{2^s} = 2^{(t+r+1)2^s}$.

The second construction was generalized in [67]. The functions of this generalized class are the concatenations of functions equal to the sums of r-variable affine functions and of flat indicators:

$$\forall(x,y)\in\mathbb{F}_2^r\times\mathbb{F}_2^s,\ f(x,y)=\prod_{i=1}^{\varphi(y)}(x\cdot\phi_i(y)\oplus g_i(y)\oplus 1)\oplus x\cdot\phi(y)\oplus g(y),$$

where φ is a function from $\mathbb{F}_2^s$ into $\{0,1,\ldots,r\}$, $\phi_1,\ldots,\phi_r$ and ϕ are functions from $\mathbb{F}_2^s$ into $\mathbb{F}_2^r$ such that, for every $y\in\mathbb{F}_2^s$, the vectors $\phi_1(y),\ldots,\phi_{\varphi(y)}(y)$ are linearly independent, and $g_1,\ldots,g_r$ and g are Boolean functions on $\mathbb{F}_2^s$. There exist formulas for the Walsh transforms of the functions of these classes, which result in sufficient conditions for their resiliency and in bounds on their nonlinearities (see [63, 67]).

Other Constructions. We first make a *preliminary observation*. Let $k<n$. For any k-variable function g, any surjective linear mapping $L:\mathbb{F}_2^n\to\mathbb{F}_2^k$ and any element s of $\mathbb{F}_2^n$, the function $f(x)=g\circ L(x)\oplus s\cdot x$ is $(d-1)$-resilient, where d is the Hamming distance between s and the linear code C whose generator matrix equals the matrix of L. Indeed, for any vector $a\in\mathbb{F}_2^n$ of Hamming weight at most $d-1$, the vector $s+a$ does not belong to C. This implies that the Boolean function $f(x)\oplus a\cdot x$ is linearly equivalent to the function $g(x_1,\ldots,x_k)\oplus x_{k+1}$, becaue we may assume without loss of generality that L is systematic (*i.e.*, has the form $[Id_k|N]$); it is therefore balanced. But such a function f, having nonzero linear structures, does not give full satisfaction.

A construction derived from $\mathcal{PS}_{ap}$ construction is introduced in [57] to obtain resilient functions: let k and r be positive integers and $n\ge r$; we denote $n-r$ by s; the vectorspace $\mathbb{F}_2^r$ is identified to the Galois field $\mathbb{F}_{2^r}$. Let g be any Boolean function on $\mathbb{F}_{2^r}$ and ϕ an $\mathbb{F}_2$-linear mapping from $\mathbb{F}_2^s$ to $\mathbb{F}_{2^r}$; set $a\in\mathbb{F}_{2^r}$ and $b\in\mathbb{F}_2^s$ such that, for every y in $\mathbb{F}_2^s$ and every z in $\mathbb{F}_{2^r}$, $a+\phi(y)$ is nonzero and $\phi^*(z)+b$ has weight greater than k, where ϕ^* is the adjoint of ϕ (satisfying $u\cdot\phi(x)=\phi^*(u)\cdot x$ for every x and u, that is, having for matrix the transpose of that of ϕ). Then, the function

$$f(x,y)=g\left(\frac{x}{a+\phi(y)}\right)\oplus b\cdot y,\ \text{where } x\in\mathbb{F}_{2^r},\ y\in\mathbb{F}_2^s, \tag{8.63}$$

is m-resilient with $m\ge k$. There exist bounds on the nonlinearities of these functions (see [64]), similar to those existing for Maiorana-McFarland's functions. But this class has many fewer elements than Maiorana-McFarland's class, because ϕ is linear.

Dobbertin's Construction. In [141] is given a nice generalization of a method, introduced by Seberry et al. in [325], for modifying bent functions into balanced functions with high nonlinearities. He observes that most known bent functions on $\mathbb{F}_2^n$ (n even) are normal (that is, constant on at least one $n/2$-dimensional flat). Up to affine equivalence, we can then assume that $f(x,y)$, $x\in\mathbb{F}_2^{n/2}$, $y\in\mathbb{F}_2^{n/2}$

is such that $f(x,0)=\varepsilon$ ($\varepsilon \in \mathbb{F}_2$) for every $x \in \mathbb{F}_2^{n/2}$ and that $\varepsilon = 0$ (otherwise, consider $f \oplus 1$).

Proposition 8.54. *Let $f(x,y)$, $x \in \mathbb{F}_2^{n/2}$, $y \in \mathbb{F}_2^{n/2}$, be any bent function such that $f(x,0)=0$ for every $x \in \mathbb{F}_2^{n/2}$, and let g be any balanced function on $\mathbb{F}_2^{n/2}$. Then the Walsh transform of the function $h(x,y) = f(x,y) \oplus \delta_0(y)g(x)$, where δ_0 is the Dirac symbol, satisfies*

$$\widehat{\mathrm{h}_\chi}(u,v) = 0 \textit{ if } u = 0 \textit{ and } \widehat{\mathrm{h}_\chi}(u,v) = \widehat{\mathrm{f}_\chi}(u,v) + \widehat{\mathrm{g}_\chi}(u) \textit{ otherwise.} \quad (8.64)$$

Proof. We have $\widehat{\mathrm{h}_\chi}(u,v) = \widehat{\mathrm{f}_\chi}(u,v) - \sum_{x\in\mathbb{F}_2^{n/2}}(-1)^{u\cdot x} + \sum_{x\in\mathbb{F}_2^{n/2}}(-1)^{g(x)\oplus u\cdot x} = \widehat{\mathrm{f}_\chi}(u,v) - 2^{n/2}\delta_0(u) + \widehat{\mathrm{g}_\chi}(u)$. The function g being balanced, we have $\widehat{\mathrm{g}_\chi}(0) = 0$. And $\widehat{\mathrm{f}_\chi}(0,v)$ equals $2^{n/2}$ for every v, because f is null on $\mathbb{F}_2^{n/2} \times \{0\}$ and according to Relation (8.46) applied to $E = \{0\} \times \mathbb{F}_2^{n/2}$ and $a = b = 0$ (or see the remark after Theorem 8.34). ■

We deduce that

$$\max_{u,v\in\mathbb{F}_2^{n/2}} |\widehat{\mathrm{h}_\chi}(u,v)| \le \max_{u,v\in\mathbb{F}_2^{n/2}} |\widehat{\mathrm{f}_\chi}(u,v)| + \max_{u\in\mathbb{F}_2^{n/2}} |\widehat{\mathrm{g}_\chi}(u)|,$$

that is, that $2^n - 2nl(h) \le 2^n - 2nl(f) + 2^{n/2} - 2nl(g)$, that is,

$$nl(h) \ge nl(f) + nl(g) - 2^{n/2-1} = 2^{n-1} - 2^{n/2} + nl(g).$$

Applying this principle recursively (if $n/2$ is even, g can be constructed in the same way), we see that if $n = 2^k n'$ (n' odd), Dobbertin's method allows reaching the nonlinearity $2^{n-1} - 2^{n/2-1} - 2^{\frac{n}{4}-1} - \cdots - 2^{n'-1} - 2^{\frac{n'-1}{2}}$ because we know that, for every odd n', the nonlinearity of functions on $\mathbb{F}_2^{n'}$ can be as high as $2^{n'-1} - 2^{\frac{n'-1}{2}}$, and that balanced (quadratic) functions can achieve this value. If $n' \le 7$, then this value is the best possible and $2^{n-1} - 2^{n/2-1} - 2^{\frac{n}{4}-1} - \cdots - 2^{n'-1} - 2^{\frac{n'-1}{2}}$ is therefore the best known nonlinearity of balanced functions in general. For $n' > 7$, the best nonlinearity of balanced n'-variable functions is larger than $2^{n'-1} - 2^{\frac{n'-1}{2}}$ (see the paragraph devoted to nonlinearity in Section 8.4.1) and $2^{n-1} - 2^{n/2-1} - 2^{\frac{n}{4}-1} - \cdots - 2^{2n'-1} - 2^{n'} + nl(g)$, where g is an n'-variable balanced function, can therefore reach higher values.

Unfortunately, according to Relation (8.64), Dobbertin's construction cannot produce m-resilient functions with $m > 0$ because, g being a function defined on $\mathbb{F}_2^{n/2}$, there cannot exist more than one vector a such that $\widehat{\mathrm{g}_\chi}(a)$ equals $\pm 2^{n/2}$.

8.7.5.2 Secondary Constructions

There exist several simple secondary constructions that can be combined to obtain resilient functions achieving the bounds of Sarkar et al. and Siegenthaler. We list them next in chronological order. As we shall see in the end, they all are particular cases of a single general one.

I. **Direct Sum of Functions**

A. *Adding a Variable.* Let f be an r-variable t-resilient function. The Boolean function on $\mathbb{F}_2^{r+1}$

$$h(x_1, \ldots, x_r, x_{r+1}) = f(x_1, \ldots, x_r) \oplus x_{r+1}$$

is $(t+1)$-resilient [333]. If f is an $(r, t, r-t-1, 2^{r-1} - 2^{t+1})$ function,[51] then h is an $(r+1, t+1, r-t-1, 2^r - 2^{t+2})$ function, and thus achieves Siegenthaler's and Sarkar et al.'s bounds. But h has the linear structure $(0, \ldots, 0, 1)$.

B. Generalization. If f is an r-variable t-resilient function ($t \geq 0$) and if g is an s-variable m-resilient function ($m \geq 0$), then the function:

$$h(x_1, \ldots, x_r, x_{r+1}, \ldots, x_{r+s}) = f(x_1, \ldots, x_r) \oplus g(x_{r+1}, \ldots, x_{r+s})$$

is $(t+m+1)$-resilient. This comes from the easily provable relation $\widehat{\mathrm{h}_\chi}(a, b) = \widehat{\mathrm{f}_\chi}(a) \times \widehat{\mathrm{g}_\chi}(b)$, $a \in \mathbb{F}_2^r$, $b \in \mathbb{F}_2^s$. We also have $d^\circ h = \max(d^\circ f, d^\circ g)$ and, thanks to Relation (8.35) relating the nonlinearity to the Walsh transform, $\mathcal{N}_h = 2^{r+s-1} - \frac{1}{2}(2^r - 2\mathcal{N}_f)(2^s - 2\mathcal{N}_g) = 2^r\mathcal{N}_g + 2^s\mathcal{N}_f - 2\mathcal{N}_f\mathcal{N}_g$. Such a decomposable function does not give full satisfaction. Moreover, h has low algebraic degree, in general. And if $\mathcal{N}_f = 2^{r-1} - 2^{t+1}$ $(t \leq r-2)$ and $\mathcal{N}_g = 2^{s-1} - 2^{m+1}$ $(m \leq s-2)$, *that is*, if $\mathcal{N}_f$ and $\mathcal{N}_g$ have maximum possible values, then $\mathcal{N}_h = 2^{r+s-1} - 2^{t+m+3}$ and h does not achieve Sarkar's and Maitra's bound (note that this is not in contradiction with the properties of the construction recalled in **I.A**, because the function $g(x_{r+1}) = x_{r+1}$ is 1-variable, 0-resilient, and has null nonlinearity).

Function h has no nonzero linear structure if and only if f and g both have no nonzero linear structure.

II. **Siegenthaler's Construction.** Let f and g be two Boolean functions on $\mathbb{F}_2^r$. Let us consider the function

$$h(x_1, \ldots, x_r, x_{r+1}) = (x_{r+1} \oplus 1) f(x_1, \ldots, x_r) \oplus x_{r+1} g(x_1, \ldots, x_r)$$

on $\mathbb{F}_2^{r+1}$. Note that the truth table of h can be obtained by concatenating the truth tables of f and g. Then

$$\widehat{\mathrm{h}_\chi}(a_1, \ldots, a_r, a_{r+1}) = \widehat{\mathrm{f}_\chi}(a_1, \ldots, a_r) + (-1)^{a_{r+1}} \widehat{\mathrm{g}_\chi}(a_1, \ldots, a_r). \qquad (8.65)$$

Thus:

(1) If f and g are m-resilient, then h is m-resilient [333]; moreover, if for every $a \in \mathbb{F}_2^r$ of Hamming weight $m+1$, we have $\widehat{\mathrm{f}_\chi}(a) + \widehat{\mathrm{g}_\chi}(a) = 0$, then h is $(m+1)$-resilient. Note that the construction recalled in **I.A** corresponds to $g = f \oplus 1$ and satisfies this condition. Another possible choice of a function g satisfying this condition (first pointed out

[51] Recall that, by an $(n, m, d, \mathcal{N})$- function, we mean an n-variable, m-resilient function having algebraic degree at least d and nonlinearity at least $\mathcal{N}$.

in [34]) is $g(x) = f(x_1 \oplus 1, \ldots, x_r \oplus 1) \oplus \varepsilon$, where $\varepsilon = m$ [mod 2], because $\widehat{g_\chi}(a) = \sum_{x \in \mathbb{F}_2^r} (-1)^{f(x) \oplus \varepsilon \oplus (x \oplus (1,\ldots,1)) \cdot a} = (-1)^{\varepsilon + w_H(a)} \widehat{f_\chi}(a)$. It leads to a function h that also has a nonzero linear structure (namely, the vector $(1, \ldots, 1)$).

(2) The value $\max_{a_1,\ldots,a_{r+1} \in \mathbb{F}_2} |\widehat{h_\chi}(a_1, \ldots, a_r, a_{r+1})|$ is bounded above by the number $\max_{a_1,\ldots,a_r \in \mathbb{F}_2} |\widehat{f_\chi}(a_1, \ldots, a_r)| + \max_{a_1,\ldots,a_r \in \mathbb{F}_2} |\widehat{g_\chi}(a_1, \ldots, a_r)|$; this implies $2^{r+1} - 2\mathcal{N}_h \le 2^{r+1} - 2\mathcal{N}_f - 2\mathcal{N}_g$, that is $\mathcal{N}_h \ge \mathcal{N}_f + \mathcal{N}_g$.
   (a) If f and g achieve maximum possible nonlinearity $2^{r-1} - 2^{m+1}$ and if h is $(m+1)$-resilient, then the nonlinearity $2^r - 2^{m+2}$ of h is the best possible.
   (b) If f and g are such that, for every vector a, at least one of the numbers $\widehat{f_\chi}(a)$, $\widehat{g_\chi}(a)$ is null (in other words, if the supports of the Walsh transforms of f and g are disjoint), then we have $\max_{a_1,\ldots,a_{r+1} \in \mathbb{F}_2} |\widehat{h_\chi}(a_1, \ldots, a_r, a_{r+1})| = \max\left(\max_{a_1,\ldots,a_r \in \mathbb{F}_2} |\widehat{f_\chi}(a_1, \ldots, a_r)|; \max_{a_1,\ldots,a_r \in \mathbb{F}_2} |\widehat{g_\chi}(a_1, \ldots, a_r)|\right)$. Hence, we have $2^{r+1} - 2\mathcal{N}_h = 2^r - 2\min(\mathcal{N}_f, \mathcal{N}_g)$, and $\mathcal{N}_h$ therefore equals $2^{r-1} + \min(\mathcal{N}_f, \mathcal{N}_g)$; thus, if f and g achieve the best possible nonlinearity $2^{r-1} - 2^{m+1}$, then h achieves the best possible nonlinearity $2^r - 2^{m+1}$;

(3) If the monomials of highest degree in the algebraic normal forms of f and g are not all the same, then $d^\circ h = 1 + \max(d^\circ f, d^\circ g)$. Note that this condition is not satisfied in the two cases indicated in **1**, for which h is $(m+1)$-resilient.

(4) For every $a = (a_1, \ldots, a_r) \in \mathbb{F}_2^r$ and every $a_{r+1} \in \mathbb{F}_2$, we have, denoting $(x_1, \ldots, x_r)$ by x, $D_{(a,a_{r+1})} h(x, x_{r+1}) = D_a f(x) \oplus a_{r+1}(f \oplus g)(x) \oplus x_{r+1} D_a(f \oplus g)(x) \oplus a_{r+1} D_a(f \oplus g)(x)$. If $d^\circ(f \oplus g) \ge d^\circ f$, then $D_{(a,1)}h$ is nonconstant, for every a. And if, additionally, there does not exist $a \ne 0$ such that $D_a f$ and $D_a g$ are constant and equal to each other, then h admits no nonzero linear structure.

This construction allows obtaining the following:

- From any two m-resilient functions f and g having disjoint Walsh spectra, achieving nonlinearity $2^{r-1} - 2^{m+1}$ and such that $d^\circ(f \oplus g) = r - m - 1$, an m-resilient function h having algebraic degree $r - m$ and having nonlinearity $2^r - 2^{m+1}$, that is, achieving Siegenthaler's and Sarkar et al.'s bounds; note that this construction increases (by 1) the algebraic degrees of f and g.
- From any m-resilient function f achieving algebraic degree $r - m - 1$ and nonlinearity $2^{r-1} - 2^{m+1}$, a function h having resiliency order $m + 1$ and nonlinearity $2^r - 2^{m+2}$, that is, achieving Siegenthaler's and Sarkar et al.'s bounds and having same algebraic degree as f (but having nonzero linear structures).

So it allows us, when combining these two methods, to keep the best trade-offs between resiliency order, algebraic degree and nonlinearity, and to increase by 1 the degree and the resiliency order.

Generalization: Let $(f_y)_{y\in\mathbb{F}_2^s}$ be a family of r-variable m-resilient functions; then the function on $\mathbb{F}_2^{r+s}$ defined by $f(x, y) = f_y(x)$ $(x \in \mathbb{F}_2^r, y \in \mathbb{F}_2^s)$ is m-resilient. Indeed, we have $\widehat{f_\chi}(a, b) = \sum_{y\in\mathbb{F}_2^s}(-1)^{b\cdot y}\,\widehat{f_{y\,\chi}}(a)$. The function f corresponds to the concatenation of the functions f_y; hence, this secondary construction can be viewed as a generalization of Maiorana-McFarland's construction (in which the functions f_y are m-resilient affine functions).

More on the resilient functions, achieving high nonlinearities, and constructed by using, among others, the secondary constructions above (as well as algorithmic methods) can be found in [216, 290].

III. **Tarannikov's Elementary Construction.** Let g be any Boolean function on $\mathbb{F}_2^r$. We define the Boolean function h on $\mathbb{F}_2^{r+1}$ by $h(x_1, \ldots, x_r, x_{r+1}) = x_{r+1} \oplus g(x_1, \ldots, x_{r-1}, x_r \oplus x_{r+1})$. By the change of variable $x_r \leftarrow x_r \oplus x_{r+1}$, we see that the Walsh transform $\widehat{h_\chi}(a_1, \ldots, a_{r+1})$ is equal to

$$\sum_{x_1,\ldots,x_{r+1}\in\mathbb{F}_2} (-1)^{a\cdot x\oplus g(x_1,\ldots,x_r)\oplus a_r x_r\oplus(a_r\oplus a_{r+1}\oplus 1)x_{r+1}},$$

where $a = (a_1, \ldots, a_{r-1})$ and $x = (x_1, \ldots, x_{r-1})$; if $a_{r+1} = a_r$ then this value is null, and if $a_r = a_{r+1} \oplus 1$ then it equals $2\,\widehat{g_\chi}(a_1, \ldots, a_{r-1}, a_r)$. Thus:

(1) $\mathcal{N}_h = 2\,\mathcal{N}_g$.
(2) If g is m-resilient, then h is m-resilient. If, additionally, $\widehat{g_\chi}(a_1, \ldots, a_{r-1}, 1)$ is null for every vector $(a_1, \ldots, a_{r-1})$ of weight at most m, then for every such vector $\widehat{g_\chi}(a_1, \ldots, a_{r-1}, a_r)$ is null for every a_r and h is then $(m + 1)$-resilient, because if $a_r = a_{r+1} \oplus 1$ then (a_r, a_{r+1}) has weight 1. Note that, in such a case, if g has nonlinearity $2^{r-1} - 2^{m+1}$ then the nonlinearity of h, which equals $2^r - 2^{m+2}$, also then achieves Sarkar et al.'s bound. The condition that $\widehat{g_\chi}(a_1, \ldots, a_{r-1}, 1)$ is null for every vector $(a_1, \ldots, a_{r-1})$ of weight at most m is achieved if g does not actually depend on its last input bit; but the construction is then a particular case of the construction recalled in **I.A**. The condition is also achieved if g is obtained from two m-resilient functions, by using Siegenthaler's construction (recalled in **II**), according to Relation (8.65).
(3) $d^\circ f = d^\circ g$ if $d^\circ g \geq 1$.
(4) h has the nonzero linear structure $(0, \ldots, 0, 1, 1)$.

In [342], Tarannikov combined this construction with the constructions recalled in **I** and **II** to build a more complex secondary construction, which allows us to increase at the same time the resiliency order and the algebraic degree of the functions, and which leads to an infinite sequence of functions achieving Siegenthaler's and Sarkar et al.'s bounds. Increasing then, by using the construction recalled in **I.A**, the set of ordered pairs (r, m) for which such functions can be constructed, he deduced the existence of r-variable m-resilient functions achieving Siegenthaler's and Sarkar et al.'s bounds for any number of variables r and any resiliency order m such that $m \geq \frac{2r-7}{3}$ and $m > \frac{r}{2} - 2$ (but these functions have nonzero linear structures). In [291], Pasalic et al. slightly modified this more complex Tarannikov construction into a construction that we shall call *Tarannikov et al.'s construction*, which allowed, when iterating it together with the construction recalled in **I.A**, a slight relaxation of the condition on m into $m \geq \frac{2r-10}{3}$ and $m > \frac{r}{2} - 2$.

IV. **Indirect Sum of Functions.** Tarannikov et al.'s construction was, in its turn, generalized into the following construction. All the secondary constructions previously listed are particular cases of it.

Theorem 8.55. ([66].) *Let r and s be positive integers, and let t and m be nonnegative integers such that $t < r$ and $m < s$. Let f_1 and f_2 be two r-variable t-resilient functions. Let g_1 and g_2 be two s-variable m-resilient functions. Then the function*

$$h(x, y) = f_1(x) \oplus g_1(y) \oplus (f_1 \oplus f_2)(x)\,(g_1 \oplus g_2)(y), \quad x \in \mathbb{F}_2^r, y \in \mathbb{F}_2^s,$$

is an $(r+s)$-variable $(t+m+1)$-resilient function. If f_1 and f_2 are distinct and if g_1 and g_2 are distinct, then the algebraic degree of h equals $\max(d^\circ f_1, d^\circ g_1, d^\circ(f_1 \oplus f_2) + d^\circ(g_1 \oplus g_2))$; otherwise, it equals $\max(d^\circ f_1, d^\circ g_1)$. The Walsh transform of h takes the value

$$\widehat{\mathrm{h}_\chi}(a, b) = \frac{1}{2}\widehat{\mathrm{f_1}_\chi}(a)\left[\widehat{\mathrm{g_1}_\chi}(b) + \widehat{\mathrm{g_2}_\chi}(b)\right] + \frac{1}{2}\widehat{\mathrm{f_2}_\chi}(a)\left[\widehat{\mathrm{g_1}_\chi}(b) - \widehat{\mathrm{g_2}_\chi}(b)\right]. \quad (8.66)$$

If the Walsh transforms of f_1 and f_2 have disjoint supports and if the Walsh transforms of g_1 and g_2 have disjoint supports, then

$$\mathcal{N}_h = \min_{i,j \in \{1,2\}} \left(2^{r+s-2} + 2^{r-1}\mathcal{N}_{g_j} + 2^{s-1}\mathcal{N}_{f_i} - \mathcal{N}_{f_i}\mathcal{N}_{g_j}\right). \quad (8.67)$$

In particular, if f_1 and f_2 are two $(r, t, -, 2^{r-1} - 2^{t+1})$ functions with disjoint Walsh supports, if g_1 and g_2 are two $(s, m, -, 2^{s-1} - 2^{m+1})$ functions with disjoint Walsh supports, and if $f_1 \oplus f_2$ has degree $r - t - 1$ and $g_1 \oplus g_2$ has algebraic degree $s - m - 1$, then h is a $(r+s, t+m+1, r+s-t-m-2, 2^{r+s-1} - 2^{t+m+2})$ function and thus achieves Siegenthaler's and Sarkar et al.'s bounds.

Proof. We have

$$\widehat{\mathrm{h}_\chi}(a,b) = \sum_{y\in\mathbb{F}_2^s/\, g_1\oplus g_2(y)=0} \left(\sum_{x\in\mathbb{F}_2^r}(-1)^{f_1(x)\oplus a\cdot x}\right)(-1)^{g_1(y)\oplus b\cdot y}$$

$$+ \sum_{y\in\mathbb{F}_2^s/\, g_1\oplus g_2(y)=1} \left(\sum_{x\in\mathbb{F}_2^r}(-1)^{f_2(x)\oplus a\cdot x}\right)(-1)^{g_1(y)\oplus b\cdot y}$$

$$= \widehat{\mathrm{f_1}_\chi}(a) \sum_{\substack{y\in\mathbb{F}_2^s/\\ g_1\oplus g_2(y)=0}} (-1)^{g_1(y)\oplus b\cdot y} + \widehat{\mathrm{f_2}_\chi}(a) \sum_{\substack{y\in\mathbb{F}_2^s/\\ g_1\oplus g_2(y)=1}} (-1)^{g_1(y)\oplus b\cdot y}$$

$$= \widehat{\mathrm{f_1}_\chi}(a) \sum_{y\in\mathbb{F}_2^s}(-1)^{g_1(y)\oplus b\cdot y}\left(\frac{1+(-1)^{(g_1\oplus g_2)(y)}}{2}\right)$$

$$+ \widehat{\mathrm{f_2}_\chi}(a) \sum_{y\in\mathbb{F}_2^s}(-1)^{g_1(y)\oplus b\cdot y}\left(\frac{1-(-1)^{(g_1\oplus g_2)(y)}}{2}\right).$$

We deduce Relation (8.66). If (a, b) has weight at most $t + m + 1$, then a has weight at most t or b has weight at most m; hence, we have $\widehat{\mathrm{h}_\chi}(a, b) = 0$. Thus, h is $t + m + 1$-resilient.

If $f_1 \oplus f_2$ and $g_1 \oplus g_2$ are nonconstant, then the algebraic degree of h equals $\max(d^\circ f_1, d^\circ g_1, d^\circ(f_1 \oplus f_2) + d^\circ(g_1 \oplus g_2))$ because the terms of highest degree in $(g_1 \oplus g_2)(y)(f_1 \oplus f_2)(x)$, in $f_1(x)$ and in $g_1(y)$ cannot cancel each other. We deduce from Relation (8.66) that if the supports of the Walsh transforms of f_1 and f_2 are disjoint, as well as those of g_1 and g_2, then

$$\max_{(a,b)\in\mathbb{F}_2^r\times\mathbb{F}_2^s} |\widehat{\mathrm{h}_\chi}(a,b)| = \frac{1}{2}\max_{i,j\in\{1,2\}}\left(\max_{a\in\mathbb{F}_2^r}|\widehat{\mathrm{f_i}}(a)|\max_{b\in\mathbb{F}_2^s}|\widehat{\mathrm{g_j}}(b)|\right),$$

and according to Relation (8.35), relating the nonlinearity to the Walsh transform, this implies

$$2^{r+s} - 2nl(h) = \frac{1}{2}\max_{i,j\in\{1,2\}}\left((2^r - 2nl(f_i))(2^s - 2nl(g_j))\right),$$

which is equivalent to Relation (8.67). ∎

Note that function h, defined this way, is the concatenation of the four functions f_1, $f_1 \oplus 1$, f_2 and $f_2 \oplus 1$, in an order controled by $g_1(y)$ and $g_2(y)$.

Examples of pairs (f_1, f_2) (or (g_1, g_2)) satisfying the hypotheses of Theorem 8.55 can be found in [66].

V. **Constructions without Extension of the Number of Variables.** Proposition 8.37 leads to the following construction.

Proposition 8.56. ([69].) *Let n be any positive integer and k any nonnegative integer such that $k \le n$.Let f_1, f_2, and f_3 be three kth-order correlation immune (resp.*

k-resilient) functions. Then the function $s_1 = f_1 \oplus f_2 \oplus f_3$ *is kth-order correlation immune (resp. k-resilient) if and only if the function* $s_2 = f_1 f_2 \oplus f_1 f_3 \oplus f_2 f_3$ *is kth-order correlation immune (resp. k-resilient). Moreover,*

$$nl(s_2) \geq \frac{1}{2}\left(nl(s_1) + \sum_{i=1}^{3} nl(f_i)\right) - 2^{n-1}, \tag{8.68}$$

and if the Walsh supports of f_1, f_2, *and* f_3 *are pairwise disjoint (that is, if at most one value* $\widehat{\chi_{f_i}}(s)$, $i = 1, 2, 3$ *is nonzero, for every vector s), then*

$$nl(s_2) \geq \frac{1}{2}\left(nl(s_1) + \min_{1\leq i\leq 3} nl(f_i)\right). \tag{8.69}$$

Proof. Relation (8.50) and the fact that, for every (nonzero) vector a of weight at most k, we have $\widehat{\mathrm{f_i}_{\chi}}(a) = 0$ for $i = 1, 2, 3$ imply that $\widehat{\mathrm{s_1}_{\chi}}(a) = 0$ if and only if $\widehat{\mathrm{s_2}_{\chi}}(a) = 0$. Relations (8.68) and (8.69) are also direct consequences of Relation (8.50) and of Relation (8.35) relating the nonlinearity to the Walsh transform. ■

Note that this secondary construction is proper to allow achieving high algebraic immunity with s_2, given functions with lower algebraic immunities f_1, f_2, f_3, and s_1, because the support of s_2 can be made more complex than those of these functions. This is done without changing the number of variables and keeping similar resiliency order and nonlinearity.

Remark . *Let g and h be two Boolean functions on* $\mathbb{F}_2^n$ *with disjoint supports and let f be equal to* $g \oplus h = g + h$. *Then, f is balanced if and only if* $w_H(g) + w_H(h) = 2^{n-1}$. *By linearity of the Fourier transform, we have* $\widehat{f} = \widehat{g} + \widehat{h}$. *Thus, if g and h are mth-order correlation-immune, then f is m-resilient. For every nonzero* $a \in \mathbb{F}_2^n$, *we have* $|\widehat{f_{\chi}}(a)| = 2\,|\widehat{f}(a)| \leq 2\,|\widehat{g}(a)| + 2\,|\widehat{h}(a)| = |\widehat{g_{\chi}}(a)| + |\widehat{h_{\chi}}(a)|$. *Thus, assuming that f is balanced, we have* $nl(f) \geq nl(g) + nl(h) - 2^{n-1}$. *The algebraic degree of f is bounded above by (and can be equal to) the maximum of the algebraic degrees of g and h.*

Most of the secondary constructions of bent functions described in Section 8.6.4 can be altered into constructions of correlation-immune and resilient functions; see [57].

8.7.6 On the Number of Resilient Functions

It is important to ensure that the selected criteria for the Boolean functions, supposed to be used in some cryptosystems, do not restrict the choice of the functions too severely. Hence, the set of functions should be enumerated. But this enumeration is unknown for most criteria, and the case of resilient functions is not an exception in this matter. We recall later what is known. As for bent functions, the class of balanced or resilient functions produced by Maiorana-McFarland's construction is by far the widest class, compared to the classes obtained from the other usual constructions, and the number of provably balanced or resilient

Maiorana-McFarland's functions seems negligible with respect to the total number of functions with the same properties. For balanced functions, this can be checked: for every positive r, the number of balanced Maiorana-McFarland's functions (8.59) obtained by choosing ϕ such that $\phi(y) \neq 0$, for every y, equals $(2^{r+1}-2)^{2^s}$ and is less than or equal to $2^{2^{n-1}}$ (because $r \geq 1$). It is quite negligible with respect to the number $\binom{2^n}{2^{n-1}} \approx (2^{2^n+\frac{1}{2}})/\sqrt{\pi 2^n}$ of all balanced functions on $\mathbb{F}_2^n$. The number of m-resilient Maiorana-McFarland's functions obtained by choosing ϕ such that $w_H(\phi(y)) > m$ for every y equals $\left[2\sum_{i=m+1}^{r}\binom{r}{i}\right]^{2^{n-r}}$ and is probably also very small compared to the number of all m-resilient functions. But this number is unknown.

The exact numbers of m-resilient functions is known for $m \geq n-3$ (see [34], where $(n-3)$-resilient functions are characterized) and $(n-4)$-resilient functions have been characterized [25, 74].

As for bent functions, an upper bound comes directly from the Siegenthaler bound on the algebraic degree: the number of m-resilient functions is bounded above by $2^{\sum_{i=0}^{n-m-1}\binom{n}{i}}$. This bound is the so-called naive bound. In 1990, Yang and Guo published an upper bound on the number of first-order correlation-immune (and thus resilient) functions. At the same time, Denisov obtained a much stronger result (see later discussion), but his result being published in Russian, it was not known internationally. His paper was translated into English two years later but is not widely known. This explains why several papers appeared with weaker results. Park, Lee, Sung, and Kim [293] improved on Yang-Guo's bound. Schneider [322] proved that the number of m-resilient n-variable Boolean functions is less than

$$\prod_{i=1}^{n-m}\left(\frac{2^i}{2^{i-1}}\right)^{\binom{n-i-1}{m-1}},$$

but this result was known; see [158]. A general upper bound on the number of Boolean functions whose distances to affine functions are all divisible by 2^m was obtained in [89]. It implies an upper bound on the number of m-resilient functions that improves upon previous bounds for about half the values of (n, m) (it is better for m large). This bound divides the naive bound by approximately $2^{\sum_{i=0}^{n-m-1}\binom{m-1}{i}-1}$ if $m \geq n/2$ and by approximately $2^{2^{2m+1}-1}$ if $m < n/2$.

An upper bound on m-resilient functions ($m \geq n/2-1$) partially improving upon this latter bound was obtained for $n/2-1 \leq m < n-2$ in [83]: the number of n-variable m-resilient functions is lower than

$$2^{\sum_{i=0}^{n-m-2}\binom{n}{i}} + \frac{\binom{n}{n-m-1}}{2^{\binom{m+1}{n-m-1}+1}}\prod_{i=1}^{n-m}\left(\frac{2^i}{2^{i-1}}\right)^{\binom{n-i-1}{m-1}}.$$

The expressions of these bounds seem difficult to compare mathematically. Tables were computed in [83].

The problem of counting resilient functions is related to counting integer solutions of a system of linear equations; see [280].

An asymptotic formula for the number of m-resilient (and also for mth-order correlation-immune functions), where m is very small compared to n – namely $m = o(\sqrt{n})$ – was given by O. Denisov in [131]. This formula was not correct for $m \geq 2$, and a correction was given by the same author in [132] (as well as a simpler proof): the number of m-resilient functions is equivalent to

$$\exp_2\left(2^n - \frac{n-m}{2}\binom{n}{m} - \sum_{i=0}^{m}\binom{n}{i}\log_2\sqrt{\pi/2}\right).$$

For large resiliency orders, Y. Tarannikov and D. Kirienko showed in [344] that, for every positive integer m, there exists a number $p(m)$ such that for $n > p(m)$, any $(n-m)$-resilient function $f(x_1, \ldots, x_n)$ is equivalent, up to permutation of its input coordinates, to a function of the form $g(x_1, \ldots, x_{p(m)}) \oplus x_{p(m)+1} \oplus \cdots \oplus x_n$. It is then a simple matter to deduce that the number of $(n-m)$-resilient functions equals $\sum_{i=0}^{p(m)} A(m,i)\binom{n}{i}$, where $A(m,i)$ is the number of i-variable $(i-m)$-resilient functions that depend on all inputs $x_1, x_2, \ldots, x_i$ nonlinearly. Hence, it is equivalent to $\frac{A(m,p(m))}{p(m)!} n^{p(m)}$ for m constant when n tends to infinity, and it is at most $A_m\, n^{p(m)}$, where A_m depends on m only. It is proved in [345] that $3 \cdot 2^{m-2} \leq p(m) \leq (m-1)2^{m-2}$ and in [344] that $p(4) = 10$; hence the number of $(n-4)$-resilient functions equals $(1/2)n^{10} + O(n^9)$.

8.8 Functions Satisfying the Strict Avalanche and Propagation Criteria

In this section, we are interested in the functions (and more particularly, in the balanced functions) that achieve $PC(l)$ for some $l < n$ (the functions achieving $PC(n)$ are the bent functions and they cannot be balanced).

8.8.1 PC(l) Criterion

It is shown in [59, 60, 180] that, if n is even, then $PC(n-2)$ implies $PC(n)$; so we can find balanced n-variable $PC(l)$ functions for n even only if $l \leq n-3$. For odd $n \geq 3$, it is also known that the functions that satisfy $PC(n-1)$ are those functions of the form $g(x_1 \oplus x_n, \ldots, x_{n-1} \oplus x_n) \oplus \ell(x)$, where g is bent and ℓ is affine, and that the $PC(n-2)$ functions are those functions of a similar form, but where, for at most one index i, the term $x_i \oplus x_n$ may be replaced by x_i or by x_n (other equivalent characterizations exist [60]).

The only known upper bound on the algebraic degrees of $PC(l)$ functions is $n-1$. A lower bound on the nonlinearity of functions satisfying the propagation criterion exists [357] and can be very easily proved: if there exists an l-dimensional subspace F such that, for every nonzero $a \in F$, the derivative $D_a f$ is balanced, then $nl(f) \geq 2^{n-1} - 2^{n-\frac{1}{2}l-1}$; Relation (8.27), relating the values of the Walsh transform of a function on a flat $a + E$ to the autocorrelation coefficients of the function on a flat $b + E^\perp$, applied to any $a \in \mathbb{F}_2^n$, with $b = 0$ and $E = F^\perp$, shows indeed that every value $\widehat{\mathrm{f}_\chi}^2(u)$ is bounded above by 2^{2n-l}; it implies that $PC(l)$ functions have nonlinearities bounded below by $2^{n-1} - 2^{n-\frac{1}{2}l-1}$. Equality can occur only if $l = n - 1$ (n odd) and $l = n$ (n even).

The maximum correlation of Boolean functions satisfying $PC(l)$ (and in particular, of bent functions) can be directly deduced from Relations (8.40) and (8.27); see [37].

8.8.1.1 Characterizations

There exist characterizations of the propagation criterion. A first obvious one is that, according to Relation (8.24), the Wiener-Khintchine theorem, f satisfies $PC(l)$ if and only if $\sum_{u\in\mathbb{F}_2^n}(-1)^{a\cdot u}\,\widehat{\mathrm{f}_\chi}^2(u) = 0$ for every nonzero vector a of weight at most l. A second one follows.

Proposition 8.57. ([60].) *Any n-variable Boolean function f satisfies $PC(l)$ if and only if, for every vector u of weight at least $n - l$, and every vector v,*

$$\sum_{w\preceq u} \widehat{\mathrm{f}_\chi}^2(w + v) = 2^{n+w_H(u)}.$$

This is a direct consequence of Relation (8.27). A third characterization is given in Subsection 8.8.2 below (apply it to $k = 0$).

8.8.1.2 Constructions

Maiorana-McFarland's construction can be used to produce functions satisfying the propagation criterion: the derivative $D_{(a,b)}(x, y)$ of a function of the form (8.59) being equal to $x \cdot D_b\,\phi(y) \oplus a \cdot \phi(y + b) \oplus D_b g(y)$, the function satisfies $PC(l)$ under the sufficient condition that:

(1) For every nonzero $b \in \mathbb{F}_2^s$ of weight less than or equal to l, and every vector $y \in \mathbb{F}_2^s$, the vector $D_b\,\phi(y)$ is nonzero (or equivalently, every set $\phi^{-1}(u)$, $u \in \mathbb{F}_2^r$, either is empty or is a singleton or has minimum distance strictly greater than l).
(2) Every linear combination of at least one and at most l coordinate functions of ϕ is balanced (this condition corresponds to the case $b = 0$).

Constructions of such functions have been given in [59, 60, 223].

According to Proposition 8.57, Dobbertin's construction cannot produce functions satisfying $PC(l)$ with $l \geq n/2$. Indeed, if u is for instance the vector with $n/2$ first coordinates equal to 0, and with $n/2$ last coordinates equal to 1, we have, according to Relation (8.64): $\widehat{\mathrm{h}_\chi}^2(w) = 0$ for every $w \preceq u$.

8.8.2 PC(l) of Order k and EPC(l) of Order k Criteria

According to the characterization of resilient functions and to the definitions of PC and EPC criteria, we have the following proposition.

Proposition 8.58. ([301].) *A function f satisfies $EPC(l)$ (resp. $PC(l)$) of order k if and only if, for any vector a of Hamming weight smaller than or equal to l and any vector b of Hamming weight smaller than or equal to k, if $(a, b) \neq (0, 0)$ (resp. if $(a, b) \neq (0, 0)$ and if a and b have disjoint supports) then:*

$$\sum_{x \in \mathbb{F}_2^n} (-1)^{f(x) \oplus f(x+a) \oplus b \cdot x} = 0.$$

A recent paper [305] gives the following characterization.

Proposition 8.59. *Any n-variable Boolean function f satisfies $EPC(l)$ (resp. $PC(l)$) of order k if and only if, for every vector u of weight at least $n - l$, and every vector v of weight at least $n - k$ (resp. of weight at least $n - k$ and such that $\overline{v}$ and $\overline{u}$ have disjoint supports),*

$$\sum_{w \preceq u} \widehat{\mathrm{f}_\chi}(w) \widehat{\mathrm{g}_\chi}(w) = 2^{w_H(u) + w_H(v)},$$

where g is the restriction of f to the vectorspace $\{x \in \mathbb{F}_2^n / x \preceq v\}$.

This can be proved by applying the Poisson summation formula (8.17) to the function $(a, b) \mapsto \widehat{D_a \mathrm{f}_\chi}(b)$.

Preneel showed in [299] that $SAC(k)$ functions have algebraic degrees at most $n - k - 1$ (indeed, all of their restrictions obtained by fixing k input coordinates have algebraic degrees at most $n - k - 1$). In [252], the criterion $SAC(n - 3)$ was characterized through the ANF of the function, and its properties were further studied. A construction of $PC(l)$ of order k functions based on Maiorana-McFarland's method is given in [223] (the mapping ϕ being linear and constructed from linear codes) and generalized in [59, 60] (the mapping ϕ being not linear and constructed from nonlinear codes). A construction of n-variable balanced functions satisfying $SAC(k)$ and having algebraic degree $n - k - 1$ is given, for $n - k - 1$ odd, in [223] and, for $n - k - 1$ even, in [317] (where balancedness and nonlinearity are also considered).

It is shown in [60] that, for every positive even $l \leq n - 4$ (with $n \geq 6$) and every odd l such that $5 \leq l \leq n - 5$ (with $n \geq 10$), the functions that satisfy $PC(l)$ of order $n - l - 2$ are the functions of the form

$$\bigoplus_{1 \leq i < j \leq n} x_i x_j \oplus h(x_1, \ldots, x_n)$$

where h is affine.

8.9 Algebraic Immune Functions

We recalled in Section 8.4.1 the different algebraic attacks on stream ciphers and the related criteria of resistance for the Boolean functions used in their pseudo-random generators. We now study these criteria in more detail and describe the known functions satisfying them.

8.9.1 General Properties of the Algebraic Immunity and Its Relationship with Some Other Criteria

We have seen that the algebraic immunity of any n-variable Boolean function is bounded above by $\lceil n/2 \rceil$ and that the functions used in stream ciphers must have an algebraic immunity close to this maximum.

8.9.1.1 Algebraic Immunity of Random Functions

Random functions behave well with respect to algebraic immunity:[52] it has been proved in [133] that, for all $a < 1$, when n tends to infinity, $AI(f)$ is almost surely greater than $\frac{n}{2} - \sqrt{\frac{n}{2} \ln\left(\frac{n}{2a \ln 2}\right)}$.

8.9.1.2 Algebraic Immunity of Monomial Functions

It has been shown in [284] that, if the number of runs $r(d)$ of 1s in the binary expansion of the exponent d of a power function $tr_n(ax^d)$ (that is, the number of full subsequences of consecutive 1s) is less than $\sqrt{n}/2$, then the algebraic immunity is bounded above by $r(d)\lfloor\sqrt{n}\rfloor + \left\lceil \frac{n}{\lfloor\sqrt{n}\rfloor} \right\rceil - 1$. Note that this bound is better than the general bound $\lceil n/2 \rceil$ for only a negligible part of power mappings, but, however, it concerns all of those whose exponents have a constant 2-weight or a constant number of runs – the power functions studied as potential S-boxes in block ciphers enter in this framework (see Chapter 9). Moreover, the bound is further improved when n is odd and the function is almost bent (see the present chapter for a definition): the algebraic immunity of such functions is bounded above by $2\,\lfloor\sqrt{n}\rfloor$.

8.9.1.3 Functions in Odd Numbers of Variables with Optimal Algebraic Immunity

In [38], A. Canteaut observed the following property.

Proposition 8.60. *If an n-variable balanced function f, with n odd, admits no nonzero annihilator of algebraic degree at most* $(n-1)/2$*, then it has optimum algebraic immunity* $(n+1)/2$.

[52] No result is known on the behavior of random functions against fast algebraic attacks.

This means that we do not need to check also that $f \oplus 1$ has no nonzero annihilator of algebraic degree at most $(n-1)/2$ for showing that f has optimum algebraic immunity. Indeed, consider the Reed-Muller code of length 2^n and of order $(n-1)/2$. This code is self-dual (i.e., is its own dual), according to Theorem 8.18. Let G be a generator matrix of this code. Each column of G is labeled by the vector of F_2^n obtained by keeping its coordinates of indices $2, \ldots, n+1$. Saying that f has no nonzero annihilator of algebraic degree at most $(n-1)/2$ is equivalent to saying that the matrix obtained by selecting those columns of G corresponding to the elements of the support of f has full rank $\sum_{i=0}^{\frac{n-1}{2}} \binom{n}{i} = 2^{n-1}$. By hypothesis, f has weight 2^{n-1}. Because the order of the columns in G can be freely chosen, we assume for simplicity that the columns corresponding to the support of f are the 2^{n-1} first ones. Then we have $G = (A \mid B)$ where A is an invertible $2^{n-1} \times 2^{n-1}$ matrix (and the matrix $G' = A^{-1} \times G = (Id \mid A^{-1} \times B)$ is also a generator matrix). In terms of coding theory, the support of the function is an *information set*. Then the complement of the support of f is also an information set (*i.e.*, B is also invertible): otherwise, there would exist a vector $(z \mid 0)$, $z \neq 0$, in the code, and this is clearly impossible because G and G' are also parity-check matrices of the code.

8.9.1.4 Relationship between Normality and Algebraic Immunity

If an n-variable function f is k-normal, then its algebraic immunity is at most $n-k$, because the fact that $f(x) = \varepsilon \in F_2$ for every $x \in A$, where A is a k-dimensional flat, implies that the indicator of A is an annihilator of $f + \varepsilon$. This bound is tight because, being symmetric, the majority function is $\lfloor n/2 \rfloor$-normal for every n (see later discussion) and has algebraic immunity $\lceil n/2 \rceil$. Obviously, $AI(f) \leq \ell$ does not imply conversely that f is $(n-\ell)$-normal, because when n tends to infinity, for every $a > 1$, n-variable Boolean functions are almost surely non-$(a \log_2 n)$-normal [65] (note that $k \sim a \log_2 n$ implies that $n - k \sim n$), and the algebraic immunity is always bounded above by $n/2$.

8.9.1.5 Relationship between Algebraic Immunity, Weight, and Nonlinearity

It can be easily shown that $\sum_{i=0}^{AI(f)-1} \binom{n}{i} \leq w_H(f) \leq \sum_{i=0}^{n-AI(f)} \binom{n}{i}$: the left-hand-side inequality must be satisfied because, otherwise, the number $w_H(f)$ of equations in the linear system expressing that a function of algebraic degree at most $AI(f) - 1$ is an annihilator of f would have a number of equations smaller than its number of unknowns (*i.e.*, the number of coefficients in its algebraic normal form) and it would therefore have nontrivial solutions, a contradiction. The right-hand-side inequality is obtained from the other one by replacing f by $f \oplus 1$. This implies that a function f such that $AI(f) = (n+1)/2$ (n odd) must be balanced.

It has been shown in [121] and [75] that low nonlinearity implies low algebraic immunity (but high algebraic immunity does not imply high nonlinearity): it can be easily proved that, for every function h of algebraic degree r, we have $AI(f) - r \leq AI(f \oplus h) \leq AI(f) + r$, and this implies

$$nl(f) \geq \sum_{i=0}^{AI(f)-2} \binom{n}{i}$$

and, more generally,

$$nl_r(f) \geq \sum_{i=0}^{AI(f)-r-1} \binom{n}{i}.$$

These bounds were improved in all cases for the first-order nonlinearity into

$$nl(f) \geq 2 \sum_{i=0}^{AI(f)-2} \binom{n-1}{i}$$

by Lobanov [255], and in most cases for the rth-order nonlinearity into

$$nl_r(f) \geq 2 \sum_{i=0}^{AI(f)-r-1} \binom{n-r}{i}$$

(in fact, the improvement was slightly stronger than this, but more complex) in [70]. Another improvement:

$$nl_r(f) \geq \sum_{i=0}^{AI(f)-r-1} \binom{n}{i} + \sum_{i=AI(f)-2r}^{AI(f)-r-1} \binom{n-r}{i}$$

(which always improves upon the bound of [75] and improves upon the bound of [70] for low values of r) was subsequently obtained by Mesnager in [279].

8.9.2 The Problem of Finding Functions Achieving High Algebraic Immunity and High Nonlinearity

We know that functions achieving optimal or suboptimal algebraic immunity and in the same time high algebraic degree and high nonlinearity must exist thanks to the results of [133, 308]. But knowing that almost all functions have high algebraic immunity does not mean that constructing such functions is easy.

The bounds of [70] and [279] seen earlier are weak[53] and Lobanov's bound, which is tight, does not ensure that the nonlinearity is high enough:

- For n even and $AI(f) = \frac{n}{2}$, it gives $nl(f) \geq 2^{n-1} - 2\binom{n-1}{n/2-1} = 2^{n-1} - \binom{n}{n/2}$, which is much smaller than the best possible nonlinearity $2^{n-1} - 2^{n/2-1}$ and, more problematically, much smaller than the asymptotic almost sure nonlinearity of Boolean functions, which is, when n tends to ∞, located in the neighbourhood of $2^{n-1} - 2^{n/2-1}\sqrt{2n \ln 2}$ as we saw. Until recently, the

[53] Their interest is to be valid for every function with given algebraic immunity.

best nonlinearity reached by the known functions with optimum AI was that of the majority function and of the iterative construction (see more details below on these functions): $2^{n-1} - \binom{n-1}{n/2} = 2^{n-1} - \frac{1}{2}\binom{n}{n/2}$. This was a little better than what gives Lobanov's bound, but insufficient.

- For n odd and $AI(f) = (n+1)/2$, Lobanov's bound gives $nl(f) \geq 2^{n-1} - \binom{n-1}{(n-1)/2} \simeq 2^{n-1} - \frac{1}{2}\binom{n}{(n-1)/2}$, which is a little better than in the n even case, but still far from the average nonlinearity of Boolean functions. Until recently, the best known nonlinearity was that of the majority function and matched this bound.

Efficient algorithms have been given in [5, 134] for computing the algebraic immunity, and tables are given in [5].

8.9.3 The Functions with High Algebraic Immunity Found So Far and Their Parameters

Sporadic Functions. Balanced highly nonlinear functions in up to 20 variables (derived from power functions) with high algebraic immunities have been exhibited in [82] and [5].

Infinite Classes of Functions. The majority function (first proposed by J. D. Key, T. P. McDonough, and V. C. Mavron in the context of the erasure channel [213] – rediscovered by Dalai et al. in the context of algebraic immunity [124]), defined as $f(x) = 1$ if $w_H(x) \geq n/2$ and $f(x) = 0$ otherwise, has optimum algebraic immunity.[54] It is a symmetric function (which can represent a weakness), and its nonlinearity is insufficient. Some variants also have optimum algebraic immunity.

A nice iterative construction of an infinite class of functions with optimum algebraic immunity was given in [122] and further studied in [75]; however, the functions it produces are neither balanced nor highly nonlinear.

All of these functions are weak against fast algebraic attacks, as shown in [5].

More numerous functions with optimum algebraic immunity were given in [71]. Among them are functions with better nonlinearities. However, the method of [71] did not allow high nonlinearities to be reached (see [95]) and some functions constructed in [245, 246] seem still worse from this viewpoint. Hence, the question of designing infinite classes of functions achieving all the necessary criteria remained open after these papers.

A function with optimum algebraic immunity, apparently (according to computer investigations) good immunity to fast algebraic attacks, provably much better nonlinearity than the functions mentioned earlier, and, in fact, according to computer investigations, quite sufficient nonlinearity has been exhibited very recently in [80, 151].

[54] Changing $w_H(x) \geq n/2$ into $w_H(x) > n/2$ or $w_H(x) \leq n/2$ or $w_H(x) < n/2$ changes the function into an affinely equivalent one, up to addition of the constant 1, and therefore does not change the AI.

Theorem 8.61. *Let n be any positive integer and α a primitive element of the field $\mathbb{F}_{2^n}$. Let f be the balanced Boolean function on $\mathbb{F}_{2^n}$ whose support equals $\{0, 1, \alpha, \dots, \alpha^{2^{n-1}-2}\}$. Then f has optimum algebraic immunity $\lceil n/2 \rceil$.*

Proof. Let g be any Boolean function of algebraic degree at most $\lceil n/2 \rceil - 1$. Let $g(x) = \sum_{i=0}^{2^n-1} g_i x^i$ be its univariate representation in the field $\mathbb{F}_{2^n}$, where $g_i \in \mathbb{F}_{2^n}$ is null if the 2-weight $w_2(i)$ of i is at least $\lceil n/2 \rceil$ (which implies in particular that $g_{2^n-1} = 0$).

If g is an annihilator of f, then we have $g(\alpha^i) = 0$ for every $i = 0, \dots, 2^{n-1} - 2$, that is, the vector $(g_0, \dots, g_{2^n-2})$ belongs to the Reed-Solomon code over $\mathbb{F}_{2^n}$ of zeroes $1, \alpha, \dots, \alpha^{2^{n-1}-2}$ (see [257]). According to the BCH bound, if g is nonzero, then this vector has Hamming weight at least 2^{n-1}. We briefly recall how this lower bound can be simply proved in our framework. By definition, we have

$$\begin{pmatrix} g(1) \\ g(\alpha) \\ g(\alpha^2) \\ \vdots \\ g(\alpha^{2^n-2}) \end{pmatrix} = \begin{pmatrix} 1 & 1 & 1 & \cdots & 1 \\ 1 & \alpha & \alpha^2 & \cdots & \alpha^{2^n-2} \\ 1 & \alpha^2 & \alpha^4 & \cdots & \alpha^{2(2^n-2)} \\ \vdots & \vdots & \vdots & \cdots & \vdots \\ 1 & \alpha^{2^n-2} & \alpha^{2(2^n-2)} & \cdots & \alpha^{(2^n-2)(2^n-2)} \end{pmatrix} \times \begin{pmatrix} g_0 \\ g_1 \\ g_2 \\ \vdots \\ g_{2^n-2} \end{pmatrix}$$

which implies (because $\sum_{k=0}^{2^n-2} \alpha^{(i-j)k}$ equals 1 if $i = j$ and 0 otherwise),

$$\begin{pmatrix} g_0 \\ g_1 \\ g_2 \\ \vdots \\ g_{2^n-2} \end{pmatrix} = \begin{pmatrix} 1 & 1 & 1 & \cdots & 1 \\ 1 & \alpha^{-1} & \alpha^{-2} & \cdots & \alpha^{-(2^n-2)} \\ 1 & \alpha^{-2} & \alpha^{-4} & \cdots & \alpha^{-2(2^n-2)} \\ \vdots & \vdots & \vdots & \cdots & \vdots \\ 1 & \alpha^{-(2^n-2)} & \alpha^{-2(2^n-2)} & \cdots & \alpha^{-(2^n-2)(2^n-2)} \end{pmatrix} \times \begin{pmatrix} g(1) \\ g(\alpha) \\ g(\alpha^2) \\ \vdots \\ g(\alpha^{2^n-2}) \end{pmatrix}$$

$$= \begin{pmatrix} 1 & 1 & \cdots & 1 \\ \alpha^{-(2^{n-1}-1)} & \alpha^{-2^{n-1}} & \cdots & \alpha^{-(2^n-2)} \\ \vdots & \vdots & \cdots & \vdots \\ \alpha^{-(2^{n-1}-1)(2^n-2)} & \alpha^{-2^{n-1}(2^n-2)} & \cdots & \alpha^{-(2^n-2)(2^n-2)} \end{pmatrix} \times \begin{pmatrix} g(\alpha^{2^{n-1}-1}) \\ g(\alpha^{2^{n-1}}) \\ \vdots \\ g(\alpha^{2^n-2}) \end{pmatrix}.$$

Suppose that at least 2^{n-1} of the g_is are null. Then, $g(\alpha^{2^{n-1}-1}), \dots, g(\alpha^{2^n-2})$ satisfy a homogeneous system whose matrix is obtained from the latter matrix just shown by erasing $2^{n-1} - 1$ rows. This is a $2^{n-1} \times 2^{n-1}$ Vandermonde matrix and its determinant is therefore nonnull. This implies that $g(\alpha^{2^{n-1}-1}), \dots, g(\alpha^{2^n-2})$ and therefore g must then be null. Hence, the vector $(g_0, \dots, g_{2^n-2})$ has weight at least 2^{n-1}.

Moreover, suppose that the vector $(g_0, \dots, g_{2^n-2})$ has Hamming weight 2^{n-1} exactly. Then n is odd and

$$g(x) = \sum_{\substack{0 \le i \le 2^n-2 \\ w_2(i) \le (n-1)/2}} x^i;$$

but this contradicts the fact that $g(0) = 0$. We deduce that the vector $(g_0, \ldots, g_{2^n-2})$ has Hamming weight strictly greater than 2^{n-1}, leading to a contradiction with the fact that g has algebraic degree at most $\lceil n/2 \rceil - 1$, since the number of integers of 2-weight at most $\lceil n/2 \rceil - 1$ is not strictly greater than 2^{n-1}.

Let g now be a nonzero annihilator of $f \oplus 1$. The vector $(g_0, \ldots, g_{2^n-2})$ then belongs to the Reed-Solomon code over $\mathbb{F}_{2^n}$ of zeros $\alpha^{2^{n-1}-1}, \ldots, \alpha^{2^n-2}$. According to the BCH bound (which can be proven as before), this vector has then Hamming weight strictly greater than 2^{n-1}. We arrive at the same contradiction. Hence, there does not exist a nonzero annihilator of f or $f \oplus 1$ of algebraic degree at most $\lceil n/2 \rceil - 1$, and f has then (optimum) algebraic immunity $\lceil n/2 \rceil$. ■

It is shown in [80] that the univariate representation of f equals

$$1 + \sum_{i=1}^{2^n-2} \frac{\alpha^i}{(1 + \alpha^i)^{1/2}} x^i \tag{8.70}$$

where $u^{1/2} = u^{2^{n-1}}$, which shows that f has algebraic degree $n - 1$ (which is optimum for a balanced function) and that we have

$$nl(f) \geq 2^{n-1} - n \cdot \ln 2 \cdot 2^{\frac{n}{2}} - 1.$$

It could be checked, for small values of n, that the exact value of $nl(f)$ is much better than what gives this lower bound and seems quite sufficient for resisting fast correlation attacks (for these small values of n, it behaves as $2^{n-1} - 2^{n/2}$). Finally, the function seems to show good immunity against fast algebraic attacks: the computer investigations made using Algorithm 2 of [5] suggest the following properties:

- No nonzero function g of algebraic degree at most e and no function h of algebraic degree at most d exist such that $fg = h$, when $(e, d) = (1, n - 2)$ for n odd and $(e, d) = (1, n - 3)$ for n even. This has been checked for $n \leq 12$, and we conjecture it for every n.
- For $e > 1$, pairs (g, h) of algebraic degrees (e, d) such that $e + d < n - 1$ were never observed. Precisely, the nonexistence of such pairs could be checked exhaustively for $n \leq 9$ and $e < n/2$, for $n = 10$ and $e \leq 3$, and for $n = 11$ and $e \leq 2$. This suggests that this class of functions, even if not always optimal against fast algebraic attacks, has very good behavior.

Hence, the functions of this class gather all the properties needed for allowing the stream ciphers using them as filtering functions to resist all the main attacks (the Berlekamp-Massey and Rønjom-Helleseth attacks, fast correlation attacks, standard and fast algebraic attacks). They are the only functions found so far for which such properties could be shown. There remains at least one attack against which the resistance of the functions should be evaluated: the algebraic attack on the augmented function (this obliges us to consider all possible update functions of the linear part of the pseudo-random generator).

The construction of Proposition 8.37 allows increasing the complexity of Boolean functions while keeping their high nonlinearities and may allow increasing their algebraic immunity as well.

8.10 Symmetric Functions

A Boolean function is called a *symmetric function* if it is invariant under the action of the symmetric group (*i.e.*, if its output is invariant under permutation of its input bits). Its output then depends only on the Hamming weight of the input. So, in other words, f is symmetric if and only if there exists a function $f^\#$ from $\{0, 1, \ldots, n\}$ to $\mathbb{F}_2$ such that $f(x) = f^\#(w_H(x))$.
Such functions are of some interest to cryptography, as they allow nonlinear functions on large numbers of variables to be implemented in an efficient way. Let us consider, for example, an LFSR filtered by a 63-variable symmetric function f, whose input is the content of an interval of 63 consecutive flip-flops of the LFSR. This device may be implemented with a cost similar to that of a 6-variable Boolean function, thanks to a 6-bit counter calculating the weight of the input to f (this counter is incremented if a 1 is shifted in the interval and decremented if a 1 is shifted out). However, the pseudorandom sequence obtained this way has correlation with transitions (sums of consecutive bits), and a symmetric function should not take all its inputs in a full interval. In fact, it is not yet completely clarified whether the advantage of allowing many more variables and the cryptographic weaknesses these symmetric functions may introduce result in an advantage for the designer or for the attacker, in more sophisticated devices.

8.10.1 Representation

Let $r = 0, \ldots, n$ and let φ_r be the Boolean function whose support is the set of all vectors of weight r in $\mathbb{F}_2^n$. Then, according to Relation (8.8) relating the values of the coefficients of the NNF to the values of the function, the coefficient of x^I, $I \in \mathcal{P}(N)$, in the NNF of φ_r, is $\lambda_I = (-1)^{|I|-r}\binom{|I|}{r}$.

Any symmetric function f being equal to $\bigoplus_{r=0}^n f^\#(r)\,\varphi_r$, it is therefore equal to $\sum_{r=0}^n f^\#(r)\,\varphi_r$, because the functions φ_r have disjoint supports. The coefficient of x^I in its NNF equals then $\sum_{r=0}^{n} f^\#(r)(-1)^{|I|-r}\binom{|I|}{r}$ and depends only on the size of I. The NNF of f is then

$$f(x) = \sum_{i=0}^{n} c_i\, S_i(x), \text{ where } c_i = \sum_{r=0}^{n} f^\#(r)(-1)^{i-r}\binom{i}{r} \tag{8.71}$$

and where $S_i(x)$ is the ith elementary symmetric pseudo-Boolean function whose NNF is $\sum_{I\in\mathcal{P}(N)/\,|I|=i} x^I$. The degree of the NNF of f equals $\max\{i\,/\,c_i \neq 0\}$.

We clearly have $S_i(x) = \binom{w_H(x)}{i} = (w_H(x)(w_H(x)-1)\cdots(w_H(x)-i+1))/i!$. According to Relation (8.71), we see that the univariate function $f^\#(z)$ admits the polynomial representation $\sum_{i=0}^n c_i\binom{z}{i} = \sum_{i=0}^n c_i(z(z-1)\cdots(z-i+1))/i!$ in one variable z, whose degree equals the degree of the NNF of f. Because this degree is at most n, and the values taken by this polynomial at $n+1$ points are set, this polynomial representation is unique.

Denoting by $\sigma_i(x)$ the reduction of $S_i(x)$ modulo 2, $\sigma_i(x)$ equals 1 if and only if $\binom{w_H(x)}{i}$ is odd, that is, according to Lucas' theorem [257], if and only if the binary expansion of i is covered by that of $w_H(x)$. Reducing Relation (8.71) modulo 2 and writing that $j \preceq i$ when the binary expansion of i covers that of j (*i.e.*, $j = \sum_{l=1}^{\log_2 n} j_l\, 2^{l-1}$, $i = \sum_{l=1}^{\log_2 n} i_l\, 2^{l-1}$, $j_l \le i_l$, $\forall l = 1, \ldots, \log_2 n$), we deduce from Lucas' theorem again that the ANF of f is

$$f(x) = \bigoplus_{i=0}^{n} \lambda_i\, \sigma_i(x), \text{ where } \lambda_i = \bigoplus_{j \preceq i} f^\#(j). \qquad (8.72)$$

Conversely (because the Möbius transform is involutive as we saw), $f^\#(i) = \bigoplus_{j\preceq i} \lambda_j$.

Note that a symmetric Boolean function f has algebraic degree 1 if and only if it equals $\bigoplus_{i=1}^n x_i$ or $\bigoplus_{i=1}^n x_i \oplus 1$, that is, if the binary function $f^\#(r)$ equals r [mod 2] or $r+1$ [mod 2], and that it is quadratic if and only if it equals $\bigoplus_{1\le i<j\le n} x_i x_j$ (introduced to generate the Kerdock code) plus a symmetric function of algebraic degree at most 1, that is, if the function $f^\#(r)$ equals $\binom{r}{2}$ [mod 2] or $\binom{r}{2}+r$ [mod 2] or $\binom{r}{2}+1$ [mod 2] or $\binom{r}{2}+r+1$ [mod 2]. Hence, f has algebraic degree 1 if and only if $f^\#$ satisfies $f^\#(r+1) = f^\#(r)\oplus 1$, and it has degree 2 if and only if $f^\#$ satisfies $f^\#(r+2) = f^\#(r)\oplus 1$. As observed in [48], the algebraic degree of a symmetric function f is at most $2^t - 1$, for some positive integer t, if and only if the sequence $(f^\#(r))_{r\ge 0}$ is periodic with period 2^t. This is a direct consequence of (8.72). Here again, it is not clear whether this is a greater advantage for the designer of a cryptosystem using such a symmetric function f (since, to compute the image of a vector x by f, it is enough to compute the number of nonzero coordinates $x_1, \ldots, x_t$ only) or for the attacker.

8.10.2 Fourier and Walsh Transforms

By linearity, the Fourier transform of any symmetric function $\sum_{r=0}^n f^\#(r)\,\varphi_r$ equals $\sum_{r=0}^n f^\#(r)\,\widehat{\varphi_r}$.

For every vector $a \in \mathbb{F}_2^n$, denoting by ℓ the Hamming weight of a, we have

$$\widehat{\varphi_r}(a) = \sum_{x\in\mathbb{F}_2^n \mid w_H(x)=r} (-1)^{a\cdot x} = \sum_{j=0}^{n} (-1)^j \binom{\ell}{j}\binom{n-\ell}{r-j},$$

denoting by j the size of $supp(a)\cap supp(x)$. The polynomials $K_{n,r}(X) = \sum_{j=0}^n (-1)^j \binom{X}{j}\binom{n-X}{r-j}$ are called *Krawtchouk polynomials*. They are characterized by their generating series,

$$\sum_{r=0}^{n} K_{n,r}(\ell)z^r = (1-z)^\ell(1+z)^{n-\ell},$$

and have nice resulting properties (see, *e.g.*, [95, 257]).

From the Fourier transform, we can deduce the Walsh transform thanks to Relation (8.12).

8.10.3 Nonlinearity

If n is even, then the restriction of every symmetric function f on $\mathbb{F}_2^n$ to the $n/2$-dimensional flat

$$A = \{(x_1, \ldots, x_n) \in \mathbb{F}_2^n\ ;\ x_{i+n/2} = x_i \oplus 1, \forall i \le n/2\}$$

is constant, because all the elements of A have the same weight $n/2$. Thus, f is $n/2$-normal[55] (see Definition 8.4). But Relation (8.42) gives nothing more than the covering radius bound (8.36). The symmetric functions that achieve this bound, *that is*, which are bent, were first characterized by P. Savicky in [321]: the bent symmetric functions are the four symmetric functions of algebraic degree 2 already described: $f_1(x) = \bigoplus_{1\le i<j\le n} x_i x_j$, $f_2(x) = f_1(x) \oplus 1$, $f_3(x) = f_1(x) \oplus x_1 \oplus \cdots \oplus x_n$, and $f_4(x) = f_3(x) \oplus 1$. A stronger result can be proved in a very simple way [169].

Theorem 8.62. *For every positive even n, the $PC(2)$ n-variable symmetric functions are the functions f_1, f_2, f_3, and f_4 just given.*

Proof. Let f be any $PC(2)$ n-variable symmetric function, and let $i < j$ be two indices in the range $[1; n]$. Let us denote by x' the following vector: $x' = (x_1, \ldots, x_{i-1}, x_{i+1}, \ldots, x_{j-1}, x_{j+1}, \ldots, x_n)$. Because $f(x)$ is symmetric, it has the form $x_i x_j g(x') \oplus (x_i \oplus x_j) h(x') \oplus k(x')$. Let us denote by $e_{i,j}$ the vector of weight 2 whose nonzero coordinates stand at positions i and j. The derivative $D_{e_{i,j}} f$ of f with respect to $e_{i,j}$ equals $(x_i \oplus x_j \oplus 1)g(x')$. Because this derivative is balanced, by hypothesis, then g must be equal to the constant function 1 (indeed, if $g(x') = 1$, then $(x_i \oplus x_j \oplus 1)g(x')$ equals 1 for half of the inputs and otherwise it equals 1 for none). Hence, the degree-2 part of the ANF of f equals $\bigoplus_{1\le i<j\le n} x_i x_j$. ∎

Some more results on the propagation criterion for symmetric functions can be found in [48].

If n is odd, then the restriction of any symmetric function f to the $(n+1)/2$-dimensional flat

$$A = \{(x_1, \ldots, x_n) \in \mathbb{F}_2^n\ ;\ x_{i+\frac{n-1}{2}} = x_i \oplus 1, \forall i \le n/2\}$$

[55] Obviously, this is more generally valid for every function that is constant on the set $\{x \in \mathbb{F}_2^n;\ w_H(x) = n/2\}$.

is affine, because the weight function w_H is constant on the hyperplane of A of equation $x_n = 0$ and on its complement. Thus, f is $(n+1)/2$, weakly-normal. According to Relation (8.42), this implies that its nonlinearity is upper bounded by $2^{n-1} - 2^{\frac{n-1}{2}}$. It also allows showing that the only symmetric functions achieving this bound are the same as the four functions f_1, f_2, f_3, and f_4 shown earlier, but with n odd (this was first proved by Maitra and Sarkar [264], in a more complex way). Indeed, Relation (8.42) implies the following result.

Theorem 8.63. ([65].) *Let n be any positive integer and let f be any symmetric function on $\mathbb{F}_2^n$. Let l be any integer satisfying $0 < l \le n/2$. Denote by h_l the symmetric Boolean function on $n-2l$ variables defined by $h_l(y_1, \ldots, y_{n-2l}) = f(x_1, \ldots, x_l, x_1 \oplus 1, \ldots, x_l \oplus 1, y_1, \ldots, y_{n-2l})$, where the values of $x_1, \ldots, x_l$ are arbitrary (equivalently, h_l can be defined by $h_l^\#(r) = f^\#(r+l)$, for every $0 \le r \le n-2l$). Then $nl(f) \le 2^{n-1} - 2^{n-l-1} + 2^l nl(h_l)$.*

Proof. Let $A = \{(x_1, \ldots, x_n) \in \mathbb{F}_2^n \mid x_{i+l} = x_i \oplus 1, \forall i \le l\}$. For every element x of A, we have $f(x) = h_l(x_{2l+1}, \ldots, x_n)$. Let us consider the restriction g of f to A as a Boolean function on $\mathbb{F}_2^{n-l}$, say $g(x_1, \ldots, x_l, x_{2l+1}, \ldots, x_n)$. Then, since $g(x_1, \ldots, x_l, x_{2l+1}, \ldots, x_n) = h_l(x_{2l+1}, \ldots, x_n)$, g has nonlinearity $2^l\, nl(h_l)$. According to Relation (8.42) applied with $h_a = g$, we have $nl(f) \le 2^{n-1} - 2^{n-l-1} + 2^l nl(h_l)$. ■

Then, the characterizations recalled earlier of the symmetric functions achieving best possible nonlinearity can be straightforwardly deduced. Moreover:

- If, for some integer l such that $0 \le l < \lfloor \frac{n-1}{2} \rfloor$, the nonlinearity of an n-variable symmetric function f is strictly greater than $2^{n-1} - 2^{n-l-1} + 2^l \left(2^{n-2l-1} - 2^{\lfloor \frac{n-2l-1}{2} \rfloor} - 1\right) = 2^{n-1} - 2^{\lfloor \frac{n-1}{2} \rfloor} - 2^l$, then, thanks to these characterizations and to Theorem 8.63, the function h_l must be quadratic, and $f^\#$ satisfies $f^\#(r+2) = f^\#(r) \oplus 1$, for all $l \le r \le n-2-l$ (this property has been observed in [48, Theorem 6], but proved slightly differently).
- If the nonlinearity of f is strictly greater than $2^{n-1} - 2^{\lfloor \frac{n-1}{2} \rfloor} - 2^{l+1}$, then h_l either is quadratic or has odd weight, that is, either $f^\#$ satisfies $f^\#(r+2) = f^\#(r) \oplus 1$ for all $l \le r \le n-2-l$, or h_l has odd weight.

Further properties of the nonlinearities of symmetric functions can be found in [48, 65].

8.10.4 Resiliency

There exists a conjecture on symmetric Boolean functions and, equivalently, on functions defined over $\{0, 1, \ldots, n\}$ and valued in $\mathbb{F}_2$: if f is a nonconstant symmetric Boolean function, then the numerical degree of f (that is, the degree of the polynomial representation in one variable of $f^\#$) is greater than or equal to $n-3$. It is a simple matter to show that this numerical degree is greater than or equal to $n/2$ (otherwise, the polynomial ${f^\#}^2 - f^\#$ would have degree at most n, and being null at $n+1$ points, it would equal the null polynomial, a contradiction with the fact that f is assumed not to be constant), but the gap between $n/2+1$ and $n-3$ is open. According to Proposition 8.52, the conjecture is equivalent to saying that there does not exist any symmetric 3-resilient function. Proving this conjecture is also a problem on binomial coefficients, because the numerical degree of f is bounded above by d if and only if, for every k such that $d < k \leq n$,

$$\sum_{r=0}^{k}(-1)^r\binom{k}{r}f^{\#}(r) = 0. \tag{8.73}$$

Hence, the conjecture is equivalent to saying that Relation (8.73), with $d = n-4$, has no binary solution $f^{\#}(0), \ldots, f^{\#}(n)$.

J. von zur Gathen and J. R. Roche [161] observed that all symmetric n-variable Boolean functions have numerical degrees greater than or equal to $n-3$, for any $n \leq 128$ (they exhibited Boolean functions with numerical degree $n-3$; see also [166]).

The same authors observed also that, if the number $m = n+1$ is a prime, then all nonconstant n-variable symmetric Boolean functions have numerical degree n (and therefore, considering the function $g(x) = f(x) \oplus x_1 \oplus \cdots \oplus x_n$ and applying Proposition 8.52, all nonaffine n-variable symmetric Boolean functions are not 0-resilient, that is, are unbalanced): indeed, the binomial coefficient $\binom{n}{r}$ being congruent with $\frac{(-1)(-2)\cdots(-r)}{1\cdot 2\cdots r} = (-1)^r$, modulo m, the sum $\sum_{r=0}^{n}(-1)^r\binom{n}{r}f^{\#}(r)$ is congruent with $\sum_{r=0}^{n} f^{\#}(r)$, modulo m; and Relation (8.73) with $k = n$ implies then that $f^{\#}$ must be constant.

Notice that, applying Relation (8.73) with $k = p-1$, where p is the largest prime less than or equal to $n+1$, shows that the numerical degree of any symmetric nonconstant Boolean function is greater than or equal to $p-1$ (or equivalently that no symmetric nonaffine Boolean function is $(n-p+1)$-resilient): otherwise, reducing (8.73) modulo p, we would have that the string $f^{\#}(0), \ldots, f^{\#}(k)$ is constant, and $f^{\#}$ having univariate degree less than or equal to k, the function $f^{\#}$, and thus f itself, would be constant.

More results on the balancedness and resiliency/correlation immunity of symmetric functions can be found in [20, 282, 351] and more recently in [48, 320].

8.10.5 Algebraic Immunity

We have seen that the majority function, which is symmetric, has optimum algebraic immunity. In the case n is odd, it is the only symmetric function having such a property, up to the addition of a constant (see [302], which completed a partial result of [244]). In the case n is even, other symmetric functions exist (up to the addition of a constant and to the transformation $x \to \overline{x} = (x_1 \oplus 1, \ldots, x_n \oplus 1)$) with this property; more precision and more results on the algebraic immunity of symmetric functions can be found in [27, 251] and the references therein.

8.10.6 The Superclasses of Rotation Symmetric and Matriochka Symmetric Functions

A superclass of symmetric functions, called idempotent or rotation symmetric functions (see Subsection 8.7.5), has been investigated from the viewpoints of bentness and correlation immunity (see, e.g., [152, 335]). Recently, it was proved in [210], thanks to a further investigation of these functions, that the best nonlinearity of Boolean functions in odd numbers of variables is strictly greater than the quadratic bound if and only if $n > 7$. Indeed, a function of nonlinearity 241 could be found (while the quadratic bound gives 240, and the covering radius bound 244), and using direct sum with quadratic functions, it then gave 11-variable functions of nonlinearity 994 (while the quadratic bound gives 992 and the covering radius bound 1000), and 13-variable functions of nonlinearity 4036 (while the quadratic bound gives 4032 and the covering radius bound 4050). Still more recently, it was checked that 241 is the best nonlinearity of 9-variable rotation symmetric functions, but that 9-variable functions whose truth tables (or equivalently ANFs) are invariant under cyclic shifts by 3 steps and under inversion of the order of the input bits can reach nonlinearity 242, which led to 11-variable functions of nonlinearity 996 and 13-variable functions of nonlinearity 4040. Balanced functions in 13 variables beating the quadratic bound could also be found. However, this construction gives worse nonlinearity than the Patterson-Widemann functions for 15 variables (whose nonlinearity is 16276).

In [238] is introduced the class of Matriochka symmetric functions, which are the sums of symmetric functions whose sets of variables are different and nested. Contrary to symmetric functions, they do not depend on the single weight of the input but on the sequence of the weights of the corresponding subinputs, and contrary to rotation symmetric functions, they are not invariant under cyclic shifts of the input coordinates. They can be almost as quickly computable as symmetric functions. Their cryptographic parameters require further study.

Acknowledgment

We thank Caroline Fontaine for her careful reading of a previous draft of this chapter.

References

[1] C. M. Adams and S. E. Tavares. Generating and counting binary bent sequences, *IEEE Transactions on Information Theory* 36(5), pp. 1170–1173, 1990.

[2] N. Alon, O. Goldreich, J. Hastad, and R. Peralta. Simple constructions of almost *k*-wise independent random variables. *Random Stuctures and Algorithms* 3(3), pp. 289–304, 1992.

[3] A. S. Ambrosimov. Properties of bent functions of *q*-valued logic over finite fields. *Discrete Mathematics Applications* 4(4), pp. 341–50, 1994.

[4] F. Armknecht. Improving fast algebraic attacks. *Proceedings of Fast Software Encryption 2004, Lecture Notes in Computer Science* 3017, pp. 65–82, 2004.

[5] F. Armknecht, C. Carlet, P. Gaborit, S. Künzli, W. Meier, and O. Ruatta. Efficient computation of algebraic immunity for algebraic and fast algebraic attacks. *Proceedings of EUROCRYPT 2006, Lecture Notes in Computer Science* 4004, pp. 147–64, 2006.

[6] F. Armknecht and M. Krause. Algebraic attacks on combiners with memory. *Proceedings of CRYPTO 2003, Lecture Notes in Computer Science* 2729, pp. 162–75, 2003.

[7] F. Armknecht and M. Krause. Constructing single- and multi-output Boolean functions with maximal immunity. *Proceedings in ICALP 2006, Lecture Notes in Computer Science* 4052, pp. 180–91, 2006.

[8] F. Arnault and T. P. Berger. Design and properties of a new pseudorandom generator based on a filtered FCSR automaton. *IEEE Transactions on Computers* 54(11), pp. 1374–83, 2005.

[9] G. Ars and J.-C. Faugère. Algebraic immunities of functions over finite fields. *Proceedings of the Conference BFCA 2005*, Publications des universités de Rouen et du Havre, pp. 21–38, 2005.

[10] E. F. Assmus. On the Reed-Muller codes. *Discrete Mathematics* 106/107, pp. 25–33, 1992.

[11] E. F. Assmus and J. D. Key. *Designs and Their Codes*, Cambridge University Press, Cambridge, UK, 1992.

[12] J. Ax. Zeroes of polynomials over finite fields. *American Journal of Mathematics* 86, pp. 255–61, 1964.

[13] T. Baignères, P. Junod, and S. Vaudenay. How far can we go beyond linear cryptanalysis? *Proceedings of ASIACRYPT 2004, Lecture Notes in Computer Science* 3329, pp. 432–50, 2004.

[14] B. Barak, G. Kindler, R. Shaltiel, B. Sudakov, and A. Wigderson. Simulating Independence: New constructions of condensers, Ramsey graphs, dispersers, and extractors. *Proceedings of the 37th ACM STOC*, 2005. Preprint available at http://www.math.ias.edu/ boaz/Papers/BKSSW.html

[15] E. Berlekamp, *Algebraic Coding Theory*. McGraw-Hill, New York, 1968.

[16] E. R. Berlekamp and N. J. A. Sloane. Restrictions on the weight distributions of the Reed-Muller codes. *Information and Control* 14, pp. 442–6, 1969.

[17] E. R. Berlekamp and L. R. Welch. Weight distributions of the cosets of the (32,6) Reed-Muller code. *IEEE Transactions on Information Theory* 18(1), pp. 203–7, 1972.

[18] A. Bernasconi and B. Codenotti. Spectral analysis of Boolean functions as a graph eigenvalue problem. *IEEE Transactions on Computers* 48(3), pp. 345–51, 1999.

[19] A. Bernasconi and I. Shparlinski. Circuit complexity of testing square-free numbers. *Proceedings of STACS 99, 16th Annual Symposium on Theoretical Aspects of Computer Science, Lecture Notes in Computer Science* 1563, pp. 47–56, 1999.

[20] J. Bierbrauer, K. Gopalakrishnan, and D. R. Stinson. Bounds for resilient functions and orthogonal arrays. *Proceedings of CRYPTO '94, Lecture Notes in Computer Science* 839, pp. 247–56, 1994.

[21] Y. Borissov, N. Manev, and S. Nikova. On the non-minimal codewords of weight $2d_{min}$ in the binary Reed-Muller code. *Proceedings of the Workshop on Coding and Cryptography* 2001, published by *Electronic Notes in Discrete Mathematics*, Elsevier, vo. 6, pp. 103–110, 2001. A revised version has been published in *Discrete Applied Mathematics* 128 (Special Issue "International Workshop on Coding and Cryptography (2001)"), pp. 65–74, 2003.

[22] Y. Borissov, A. Braeken, S. Nikova, and B. Preneel. On the covering radii of binary Reed-Muller codes in the set of resilient Boolean functions. *IEEE Transactions on Information Theory* 51(3), pp. 1182–89, 2005.

[23] J. Bourgain. On the construction of affine extractors. *Geometric & Functional Analysis GAFA* 17(1), pp. 33–57, 2007.

[24] A. Braeken, Y. Borisov, S. Nikova, and B. Preneel. Classification of Boolean functions of 6 variables or less with respect to cryptographic properties. *Proceedings of ICALP 2005, Lecture Notes in Computer Science* 3580, pp. 324–34, 2005.

[25] A. Braeken, Y. Borisov, S. Nikova, and B. Preneel. Classification of cubic $(n-4)$-resilient Boolean functions. *IEEE Transactions on Information Theory* 52(4), pp. 1670–6, 2006.

[26] A. Braeken, V. Nikov, S. Nikova, and B. Preneel. On Boolean functions with generalized cryptographic properties. *Proceedings of INDOCRYPT'2004, Lecture Notes in Computer Science* 3348, pp. 120–35, 2004.

[27] A. Braeken and B. Preneel. On the algebraic immunity of symmetric Boolean functions. *Proceedings of Indocrypt 2005, Lecture Notes in Computer Science* 3797, pp. 35–48, 2005. Some false results of this reference have been corrected in Braeken's PhD thesis entitled "Cryptographic properties of Boolean functions and S-boxes," available at URL http://homes.esat.kuleuven.be/ abraeken/thesisAn.pdf.

[28] J. Bringer, V. Gillot, and P. Langevin. Exponential sums and Boolean functions. *Proceedings of the Conference BFCA 2005*, Publications des universités de Rouen et du Havre, pp. 177–85, 2005.

[29] E. Brier and P. Langevin. Classification of cubic Boolean functions of 9 variables. *Proceedings of 2003 IEEE Information Theory Workshop*, Paris, 2003.

[30] R. A. Brualdi, N. Cai, and V. S. Pless. Orphans of the first order Reed-Muller codes. *IEEE Transactions on Information Theory* 36, pp. 399–401, 1990.

[31] P. Camion and A. Canteaut. Construction of t-resilient functions over a finite alphabet. *Proceedings of EUROCRYPT'96, Lecture Notes in Computer Science* 1070, pp. 283–93, 1996.

[32] P. Camion and A. Canteaut. Generalization of Siegenthaler inequality and Schnorr-Vaudenay multipermutations. *Proceedings of CRYPTO'96, Lecture Notes in Computer Science* 1109, pp. 372–86, 1996.

[33] P. Camion and A. Canteaut. Correlation-immune and resilient functions over finite alphabets and their applications in cryptography. *Designs, Codes and Cryptography* 16, 1999.

[34] P. Camion, C. Carlet, P. Charpin, and N. Sendrier. On correlation-immune functions, *Proceedings of CRYPTO'91, Lecture Notes in Computer Science* 576, pp. 86–100, 1991.

[35] A. Canteaut. On the weight distributions of optimal cosets of the first-order Reed-Muller code. *IEEE Transactions on Information Theory*, 47(1), pp. 407–13, 2001.

[36] A. Canteaut. Cryptographic functions and design criteria for block ciphers. *Proceedings of INDOCRYPT 2001, Lecture Notes in Computer Science* 2247, pp. 1–16, 2001.

[37] A. Canteaut. On the correlations between a combining function and functions of fewer variables. *Proceedings of the Information Theory Workshop '02*, Bangalore, 2002.

[38] A. Canteaut. Open problems related to algebraic attacks on stream ciphers. *Proceedings of Workshop on Coding and Cryptography* WCC 2005, pp. 1–10, 2005. See also a revised version in *Lecture Notes in Computer Science* 3969, pp. 120–34, 2006. Paper available at http://www-rocq.inria.fr/codes/Anne.Canteaut/Publications/canteaut06a.pdf

[39] A. Canteaut. Analysis and design of symmetric ciphers. Habilitation for Directing Theses, University of Paris 6, 2006.

[40] A. Canteaut, C. Carlet, P. Charpin, and C. Fontaine. Propagation characteristics and correlation-immunity of highly nonlinear Boolean functions. *Proceedings of EUROCRYPT'2000, Lecture Notes in Computer Science* 187, pp. 507–22 (2000).

[41] A. Canteaut, C. Carlet, P. Charpin, and C. Fontaine. On cryptographic properties of the cosets of $R(1, m)$. *IEEE Transactions on Information Theory* 47(4), pp. 1494–513, 2001.

[42] A. Canteaut and P. Charpin. Decomposing bent functions. *IEEE Transactions on Information Theory* 49, pp. 2004–19, 2003.

[43] A. Canteaut, P. Charpin, and G. Kyureghyan. A new class of monomial bent functions. *Finite Fields and Application* 14(1), pp. 221–41, 2008.

[44] A. Canteaut, M. Daum, H. Dobbertin, and G. Leander. normal and non-normal bent functions. *Proceedings of the Workshop on Coding and Cryptography 2003*, pp. 91–100, 2003.

[45] A. Canteaut and E. Filiol. Ciphertext only reconstruction of stream ciphers based on combination generators. *Proceedings of Fast Software Encryption 2000, Lecture Notes in Computer Science* 1978, pp. 165–80, 2001.

[46] A. Canteaut and M. Trabbia. Improved fast correlation attacks using parity-check equations of weight 4 and 5, *Advanced in Cryptology-EUROCRYPT 2000. Lecture Notes in Computer Science* 1807, pp. 573–88, 2000.

[47] A. Canteaut and M. Videau. Degree of composition of highly nonlinear functions and applications to higher order differential cryptanalysis. *Advances in Cryptology, EUROCRYPT 2002, Lecture Notes in Computer Science* 2332, pp. 518–33, 2002.

[48] A. Canteaut and M. Videau. Symmetric Boolean functions. *IEEE Transactions on Information Theory* 51(8), pp. 2791–811, 2005.

[49] Jean Robert du Carlet. La Cryptographie, contenant une très subtile manière d'escrire secrètement, composée par Maistre Jean Robert Du Carlet, 1644. A manuscript exists at the Bibliothèque Nationale (Très Grande Bibliothèque), Paris.

[50] C. Carlet. *Codes de Reed-Muller, codes de Kerdock et de Preparata*. PhD thesis. Publication of LITP, Institut Blaise Pascal, Université Paris 6, 90.59, 1990.

[51] C. Carlet. A transformation on Boolean functions, its consequences on some problems related to Reed-Muller codes. *Proceedings of EUROCODE '90, Lecture Notes in Computer Science* 514, pp. 42–50, 1991.

[52] C. Carlet. Partially-bent functions, *Designs Codes and Cryptography*, 3, pp. 135–145, 1993, and *Proceedings of CRYPTO' 92, Lecture Notes in Computer Science* 740, pp. 280–91, 1993.

[53] C. Carlet. Two new classes of bent functions. In *Proceedings of EUROCRYPT '93, Lecture Notes in Computer Science* 765, pp. 77–101, 1994.

[54] C. Carlet. Generalized Partial Spreads, *IEEE Transactions on Information Theory*, 41(5), pp. 1482–1487, 1995.

[55] C. Carlet. Hyper-bent functions. *PRAGOCRYPT '96, Czech Technical University Publishing House*, pp. 145–55, 1996.

[56] C. Carlet. A construction of bent functions. *In Finite Fields and Applications, London Mathematical Society*, Lecture Series 233. Cambridge University Press, Cambridge, UK, pp. 47–58, 1996.

[57] C. Carlet. More correlation-immune and resilient functions over Galois fields and Galois rings. *Advances in Cryptology, EUROCRYPT' 97, Lecture Notes in Computer Science* 1233, pp. 422–33, 1997.

[58] C. Carlet. On Kerdock codes, American Mathematical Society. *Proceedings of the conference Finite Fields and Applications Fq4, Contemporary Mathematics* 225, pp. 155–63, 1999.

[59] C. Carlet. On the propagation criterion of degree ℓ and order k. *Proceedings of EUROCRYPT'98, Lecture Notes in Computer Science* 1403, pp. 462–74, 1998.

[60] C. Carlet. On cryptographic propagation criteria for Boolean functions. *Information and Computation*, vol. 151, Academic Press, pp. 32–56, 1999.

[61] C. Carlet. Recent results on binary bent functions. *Proceedings of the International Conference on Combinatorics, Information Theory and Statistics; Journal of Combinatorics, Information and System Sciences*, 25(1–4), pp. 133–49, 2000.

[62] C. Carlet. On the coset weight divisibility and nonlinearity of resilient and correlation-immune functions, *Proceedings of SETA'01 (Sequences and Their Applications 2001), Discrete Mathematics and Theoretical Computer Science*, pp. 131–44, 2001.

[63] C. Carlet. A larger class of cryptographic Boolean functions via a study of the Maiorana-McFarland construction. *Proceedings of CRYPTO 2002, Lecture Notes in Computer Science* 2442, pp. 549–64, 2002.

[64] C. Carlet. On the confusion and diffusion properties of Maiorana-McFarland's and extended Maiorana-McFarland's functions. *Special Issue "Complexity Issues in Coding and Cryptography," dedicated to Prof. Harald Niederreiter on the occasion of his 60th birthday, Journal of Complexity* 20, pp. 182–204, 2004.

[65] C. Carlet. On the degree, nonlinearity, algebraic thickness and non-normality of Boolean functions, with developments on symmetric functions. *IEEE Transactions on Information Theory* 50, pp. 2178–85, 2004.

[66] C. Carlet. On the secondary constructions of resilient and bent functions. In *Proceedings of the Workshop on Coding, Cryptography and Combinatorics 2003. Birkhäuser Verlag* pp. 3–28, 2004.

[67] C. Carlet. Concatenating indicators of flats for designing cryptographic functions. *Design, Codes and Cryptography* 36(2), pp. 189–202, 2005.

[68] C. Carlet. Partial covering sequences: a method for designing classes of cryptographic functions. *Proceedings of the conference First Symposium on Algebraic Geometry and Its Applications (SAGA'07), Tahiti, 2007, World Scientific, Singapore; Series on Number Theory and Its Applications*, Vol. 5, pp. 366–387, 2008.

[69] C. Carlet. On bent and highly nonlinear balanced/resilient functions and their algebraic immunities. *Proceedings of AAECC 16, Lecture Notes in Computer Science* 3857, pp. 1–28, 2006. This paper is an extended version of the paper entitled "Improving the algebraic immunity of resilient and nonlinear functions and constructing bent function," IACR ePrint Archive http://eprint.iacr.org/ 2004/276.

[70] C. Carlet. On the higher order nonlinearities of algebraic immune functions. *Proceedings of CRYPTO 2006, Lecture Notes in Computer Science* 4117, pp. 584–601, 2006.

[71] C. Carlet. A method of construction of balanced functions with optimum algebraic immunity. To appear in the Proceedings of the International Workshop on Coding and Cryptography, The Wuyi Mountain, Fujiang, China, June 11–15, 2007, World Scientific, in its series Coding and Cryptology, 2008. A preliminary version is available on IACR ePrint Archive, http://eprint.iacr.org/, 2006/149.

[72] C. Carlet. Recursive lower bounds on the nonlinearity profile of Boolean functions and their applications. *IEEE Transactions on Information Theory*, 54(3), pp. 1262–72, 2008.

[73] C. Carlet. On the higher order nonlinearities of Boolean functions and S-boxes, and their generalizations. *Proceedings of SETA 2008, Lecture Notes in Computer Science* 5203, pp. 345–67, 2008.

[74] C. Carlet and P. Charpin. Cubic Boolean functions with highest resiliency. *IEEE Transactions on Information Theory* 51(2), pp. 562–71, 2005.

[75] C. Carlet, D. Dalai, K. Gupta, and S. Maitra. Algebraic immunity for cryptographically significant Boolean functions: Analysis and construction. *IEEE Transactions on Information Theory* 52(7), pp. 3105–21, 2006.

[76] C. Carlet, L. E. Danielsen, M. G. Parker, and P. Solé self dual bent functions. Proceedings of the conference BFCA 2008, Copenhagen, to appear in Lecture Notes in Computer Science.

[77] C. Carlet and C. Ding. Highly nonlinear mappings. *Special Issue "Complexity Issues in Coding and Cryptography," dedicated to Prof. Harald Niederreiter on the occasion of his 60th birthday, Journal of Complexity* 20, pp. 205–44, 2004.

[78] C. Carlet, H. Dobbertin, and G. Leander. Normal extensions of bent functions. *IEEE Transactions on Information Theory* 50(11), pp. 2880–5, 2004.

[79] C. Carlet and S. Dubuc. On generalized bent and q-ary perfect nonlinear functions. *Proceedings of Finite Fields and Applications Fq5*, Augsburg, Germany, Springer, pp. 81–94, 2000.

[80] C. Carlet and K. Feng. An infinite class of balanced functions with optimum algebraic immunity, good immunity to fast algebraic attacks and good nonlinearity. To appear in the Proceedings of ASIACRYPT 2008, Lecture Notes in Computer Science.

[81] C. Carlet and P. Gaborit. Hyper-bent functions and cyclic codes. *Journal of Combinatorial Theory, Series A*, 113(3), 466–82, 2006.

[82] C. Carlet and P. Gaborit. On the construction of balanced Boolean functions with a good algebraic immunity. *Proceedings of International Symposium on Information Theory, ISIT*, Adelaide, Australia, 2005. A longer version of this paper has been published in

the *Proceedings of BFCA 2005*, Publications des universités de Rouen et du Havre, pp. 1–20, 2005.

[83] C. Carlet and A. Gouget. An upper bound on the number of m-resilient Boolean functions. *Proceedings of ASIACRYPT 2002, Lecture Notes in Computer Science* 2501, pp. 484–96, 2002.

[84] C. Carlet and P. Guillot. A characterization of binary bent functions, *Journal of Combinatorial Theory, Series A*, 76(2), pp. 328–35, 1996.

[85] C. Carlet and P. Guillot. An alternate characterization of the bentness of binary functions, with uniqueness. *Designs, Codes and Cryptography* 14, pp. 133–40, 1998.

[86] C. Carlet and P. Guillot. A new representation of Boolean functions, *Proceedings of AAECC'13, Lecture Notes in Computer Science* 1719, pp. 94–103, 1999.

[87] C. Carlet and P. Guillot. Bent, resilient functions and the Numerical Normal Form. *DIMACS Series in Discrete Mathematics and Theoretical Computer Science,* 56, pp. 87–96, 2001.

[88] C. Carlet, P. Guillot, and S. Mesnager. On immunity profile of Boolean functions. *Proceedings of SETA 2006 (International Conference on Sequences and their Applications), Lecture Notes in Computer Science* 4086, pp. 364–75, 2006.

[89] C. Carlet and A. Klapper. Upper bounds on the numbers of resilient functions and of bent functions. This paper was meant to appear in an issue of Lecture Notes in Computer Sciences dedicated to Philippe Delsarte, Editor Jean-Jacques Quisquater. But this issue finally never appeared. A shorter version has appeared in the *Proceedings of the 23rd Symposium on Information Theory in the Benelux*, Louvain-La-Neuve, Belgium, 2002.

[90] C. Carlet and S. Mesnager. On the supports of the Walsh transforms of Boolean functions. *Proceedings of the conference BFCA 2005*, Publications des universités de Rouen et du Havre, pp. 65–82, 2005.

[91] C. Carlet and S. Mesnager. Improving the upper bounds on the covering radii of binary Reed-Muller codes. *IEEE Transactions on Information Theory* 53, pp. 162–73, 2007.

[92] C. Carlet and E. Prouff. On plateaued functions and their constructions. *Proceedings of Fast Software Encryption 2003, Lecture Notes in Computer Science* 2887, pp. 54–73, 2003.

[93] C. Carlet and P. Sarkar. Spectral domain analysis of correlation immune and resilient Boolean functions. *Finite fields and Applications* 8, pp. 120–30, 2002.

[94] C. Carlet and Y. V. Tarannikov. Covering sequences of Boolean functions and their cryptographic significance. *Designs, Codes and Cryptography* 25, pp. 263–79, 2002.

[95] C. Carlet, X. Zeng, C. Lei, and L. Hu. Further properties of several classes of Boolean functions with optimum algebraic immunity. *Proceedings of the First International Conference on Symbolic Computation and Cryptography* SCC 2008, LMIB, pp. 42–54, 2008.

[96] L. Carlitz and S. Uchiyama. Bounds for exponential sums. *Duke Mathematical Journal* 1, pp. 37–41, 1957.

[97] A. H. Chan and R. A. Games. On the quadratic spans of De Bruijn sequences. *IEEE Transactions on Information Theory* 36(4), pp. 822–9, 1990.

[98] C. Charnes, M. Rötteler, and T. Beth. Homogeneous bent functions, invariants, and designs. *Designs, Codes and Cryptography* 26, pp. 139–54, 2002.

[99] P. Charpin. *Open problems on cyclic codes.* In Handbook of Coding Theory, Part 1, chapter 11, Amsterdam, 1998.

[100] P. Charpin. Normal Boolean functions. *Special Issue "Complexity Issues in Coding and Cryptography," dedicated to Prof. Harald Niederreiter on the occasion of his 60th birthday, Journal of Complexity* 20, pp. 245–65, 2004.

[101] P. Charpin and G. Gong. Hyperbent functions, Kloosterman sums and Dickson polynomials. *IEEE Transactions on Information Theory* 54(9), pp. 4230–8, 2008.

[102] P. Charpin, T. Helleseth, and V. Zinoviev. Propagation characteristics of $x \to 1/x$ and Kloosterman sums. *Finite Fields and Applications* 13(2), 366–81, 2007.
[103] P. Charpin and G. Kyureghyan. Cubic monomial bent functions: A subclass of $\mathcal{M}$. *SIAM Journal on Discrete Mathematics* 22(2), pp. 650–65, 2008.
[104] P. Charpin and E. Pasalic. On propagation characteristics of resilient functions. *Proceedings of SAC 2002, Lecture Notes in Computer Science* 2595, pp. 356–65, 2002.
[105] P. Charpin, E. Pasalic, and C. Tavernier. On bent and semi-bent quadratic Boolean functions. *IEEE Transactions on Information Theory* 51(12), pp. 4286–98, 2005.
[106] S. Chee, S. Lee, K. Kim, and D. Kim. Correlation immune functions with controlable nonlinearity. *ETRI Journal* 19(4), pp. 389–401, 1997.
[107] S. Chee, S. Lee, D. Lee, and S. H. Sung. On the correlation immune functions and their nonlinearity. *Proceedings of Asiacrypt'96, Lecture Notes in Computer Science* 1163, pp. 232–43.
[108] V. Chepyzhov and B. Smeets. On a fast correlation attack on certain stream ciphers. *Proceedings of EUROCRYPT'91, Lecture Notes in Computer Science* 547, pp. 176–85, 1992.
[109] B. Chor, O. Goldreich, J. Hastad, J. Freidmann, S. Rudich, and R. Smolensky. The bit extraction problem or t-resilient functions. *Proceedings of the 26th IEEE Symposium on Foundations of Computer Science*, pp. 396–407, 1985.
[110] G. Cohen, I. Honkala, S. Litsyn, and A. Lobstein. *Covering Codes*. North-Holland, Amsterdam, 1997.
[111] N. Courtois. Higher Order Correlation Attacks, XL algorithm and Cryptanalysis of Toyocrypt. *Proceedings of ICISC 2002, Lecture Notes in Computer Science* 2587, pp. 182–99, 2003.
[112] N. Courtois. Fast algebraic attacks on stream ciphers with linear feedback. *Proceedings of CRYPTO 2003, Lecture Notes in Computer Science* 2729, pp. 177–94, 2003.
[113] N. Courtois. Algebraic attacks on combiners with memory and several outputs. *Proceedings of ICISC 2004, Lecture Notes in Computer Science* 3506, pp. 3–20, 2005.
[114] N. Courtois. Cryptanalysis of SFINKS. *Proceedings of ICISC 2005*. Also available at IACR ePrint Archive http://eprint.iacr.org/, Report 2005/243, 2005.
[115] N. Courtois. General principles of algebraic attacks and new design criteria for components of symmetric ciphers. *AES 4 Conference, Lecture Notes in Computer Science* 3373, 2004.
[116] N. Courtois and W. Meier. Algebraic attacks on stream ciphers with linear feedback. *Proceedings of EUROCRYPT 2003, Lecture Notes in Computer Science* 2656, pp. 346–59, 2002.
[117] Y. Crama and P. L. Hammer. *Boolean Functions: Theory, Algorithms, and Applications.* Cambridge University Press, Cambridge, UK, 2010.
[118] T. W. Cusick. On constructing balanced correlation immune functions. *Proceedings of SETA '98 (Sequences and their Applications 1998). Discrete Mathematics and Theoretical Computer Science*, pp. 184–90, 1999.
[119] T. W. Cusick, C. Ding, and A. Renvall, *Stream Ciphers and Number Theory,* North-Holland Mathematical Library 55. North-Holland/Elsevier, Amsterdam, 1998.
[120] D. M. Cvetkovic, M. Doob, and H. Sachs. *Spectra of Graphs*. Academic Press, 1979.
[121] D. K. Dalai, K. C. Gupta, and S. Maitra. Results on algebraic immunity for cryptographically significant Boolean functions. *Proceedings of Indocrypt 2004, Lecture Notes in Computer Science* 3348, pp. 92–106, 2004.
[122] D. K. Dalai, K. C. Gupta, and S. Maitra. Cryptographically significant Boolean functions: Construction and analysis in terms of algebraic immunity. *Fast Software Encryption 2005, Lecture Notes in Computer Science* 3557, pp. 98–111, 2005.
[123] D. K. Dalai, K. C. Gupta, and S. Maitra. Notion of algebraic immunity and its evaluation related to fast algebraic attacks. *Proceedings of the conference BFCA 2006*, Publications des universités de Rouen et du Havre, pp. 107–24, 2006.

[124] D. K. Dalai, S. Maitra, and S. Sarkar. Basic theory in construction of Boolean functions with maximum possible annihilator immunity. *Designs, Codes and Cryptography* 40(1), pp. 41–58, July 2006. IACR ePrint Archive http://eprint.iacr.org/, No. 2005/229, 15 July, 2005.

[125] D. Dalai and S. Maitra. Balanced Boolean functions with (more than) maximum algebraic immunity. *Proceedings of the Workshop on Coding and Cryptography* (in the memory of Hans Dobbertin) WCC 2007, pp. 99–108, 2007.

[126] M. Daum, H. Dobbertin, and G. Leander. An algorithm for checking normality of Boolean functions. *Proceedings of the Workshop on Coding and Cryptography 2003*, pp. 133–42, 2003.

[127] M. Daum, H. Dobbertin, and G. Leander. Short description of an algorithm to create bent functions. Private communication.

[128] E. Dawson and C.-K. Wu. Construction of correlation immune Boolean functions. *Proceedings of ICICS 1997*, pp. 170–80, 1997.

[129] P. Delsarte. *An algebraic approach to the association schemes of coding theory*. PhD thesis. Université Catholique de Louvain, 1973.

[130] P. Delsarte. Four fundamental parameters of a code and their combinatorial significance. *Information and Control* 23(5), pp. 407–38, 1973.

[131] O. Denisov. An asymptotic formula for the number of correlation-immune of order k Boolean functions. *Discrete Mathematics Applications* 2(4), pp. 407–26, 1992. Translation of a Russian article in Diskretnaya Matematika 3, pp. 25–46, 1990.

[132] O. Denisov. A local limit theorem for the distribution of a part of the spectrum of a random binary function. *Discrete Mathematics and Applications* 10(1), pp. 87–102, 2000.

[133] F. Didier. A new upper bound on the block error probability after decoding over the erasure channel. *IEEE Transactions on Information Theory* 52, pp. 4496–503, 2006.

[134] F. Didier. Using Wiedemann's algorithm to compute the immunity against algebraic and fast algebraic attacks. *Proceedings of Indocrypt 2006, Lecture Notes in Computer Science* 4329, pp. 236–50.

[135] J. Dillon. A survey of bent functions. *NSA Technical Journal Special Issue*, pp. 191–215, 1972.

[136] J. F. Dillon. *Elementary Hadamard difference sets*. PhD Thesis, University of Maryland, 1974.

[137] J. F. Dillon. Elementary Hadamard difference sets. *Proceedings of the Sixth Southeastern Conference on Combinatorics, Graph Theory and Computing*, Winnipeg Utilitas Math, pp. 237–49, 1975.

[138] J. Dillon. More DD difference sets. To appear in *Designs, Codes and Cryptography* (online), 2008.

[139] J. F. Dillon and H. Dobbertin. New cyclic difference sets with Singer parameters. *Finite Fields and Their Applications* 10, pp. 342–89, 2004.

[140] J. F. Dillon and G. McGuire. Near bent functions on a hyperplane. *Finite Fields and Their Applications* 14(3), pp. 715–20, 2008.

[141] H. Dobbertin. Construction of bent functions and balanced Boolean functions with high nonlinearity. *Proceedings of Fast Software Encryption, Second International Workshop*, Lecture Notes in Computer Science 1008, pp. 61–74, 1995.

[142] H. Dobbertin, P. Felke, T. Helleseth, and P. Rosenthal. Niho type cross-correlation functions via Dickson polynomials and Kloosterman sums. *IEEE Transactions on Information Theory* 52(2), pp. 613–27, 2006.

[143] H. Dobbertin and G. Leander. Bent functions embedded into the recursive framework of $\mathbb{Z}$-bent functions. To appear in a Special Issue of Designs, Codes and Cryptography, 2008. Available online.

[144] H. Dobbertin, and G. Leander. A survey of some recent results on bent functions. *Proceeding of SETA 2004, Lecture Notes in Computer Science* 3486, pp. 1–29, 2005.

[145] H. Dobbertin, G. Leander, A. Canteaut, C. Carlet, P. Felke, and P. Gaborit. Construction of Bent Functions via Niho Power Functions. *Journal of Combinatorial Theory, Series A*, 113(5), pp. 779–98, 2006.

[146] S. Dubuc. Characterization of linear structures. *Designs, Codes and Cryptography* 22, pp. 33–45, 2001.

[147] I. Dumer, G. Kabatiansky, and C. Tavernier. List decoding of Reed-Muller codes up to the Johnson bound with almost linear complexity. *Proceedings of ISIT 2006*. Seattle, WA.

[148] J. H. Evertse. Linear structures in block ciphers. *Proceedings of EUROCRYPT '87, Lecture Notes in Computer Science* 304, pp. 249–66, 1988.

[149] J. C. Faugère. Fast Gröbner. Algebraic cryptanalysis of HFE and filter generators. *Proceedings of the Workshop on Coding and Cryptography 2003*, pp. 175–6, 2003.

[150] J.-C. Faugère and G. Ars. An algebraic cryptanalysis of nonlinear filter generators using Gröbner bases. *Rapport de Recherche INRIA* 4739, 2003.

[151] K. Feng, Q. Liao, and J. Yang: Maximal values of generalized algebraic immunity. *Designs, Codes and Cryptography* 50, pp. 243–252, 2009.

[152] E. Filiol and C. Fontaine. Highly nonlinear balanced Boolean functions with a good correlation-immunity. *Proceedings of EUROCRYPT '98, Lecture Notes in Computer Science* 1403, pp. 475–88, 1998.

[153] S. Fischer and W. Meier. Algebraic immunity of S-boxes and augmented functions. *Proceedings of Fast Software Encryption 2007, Lecture Notes in Computer Science* 4593, pp. 366–381, 2007.

[154] C. Fontaine. On some cosets of the first-order Reed-Muller code with high minimum weight. *IEEE Transactions on Information Theory* 45(4), pp. 1237–43, 1999.

[155] R. Forré. The strict avalanche criterion: Spectral properties of Boolean functions and an extended definition. *Proceedings of CRYPTO '88, Lecture Notes in Computer Science* 403, pp. 450–68, 1989.

[156] R. Forré. A fast correlation attack on nonlinearly feedforward filtered shift register sequences. *Proceedings of EUROCRYPT '89, Lecture Notes in Computer Science* 434, pp. 586–95, 1990.

[157] R. Fourquet and C. Tavernier. List decoding of second order Reed-Muller and its covering radius implications. *Proceedings of the Workshop on Coding and Cryptography 2007* WCC, pp. 147–156, 2007.

[158] J. Friedman. On the bit extraction problem. *Proceedings of the 33rd IEEE Symposium on Foundations of Computer Science*, pp. 314–19, 1992.

[159] THE BOOLEAN PLANET. Webpage on the equivalence classes of Boolean functions in at most 6 variables, maintained by J. Fuller at URL http://www.booleanfunction.com/.

[160] R. G. Gallager. *Low Density Parity Check Codes*. MIT Press, Cambridge, MA, 1963.

[161] J. von zur Gathen and J. R. Roche. Polynomials with two values. *Combinatorica* 17(3), pp. 345–62, 1997.

[162] J. Golić. Fast low order approximation of cryptographic functions. *Proceedings of EUROCRYPT '96, Lecture Notes in Computer Science* 1070, pp. 268–82, 1996.

[163] J. Golić. On the security of nonlinear filter generators. *Proceedings of Fast Software Encryption'96, Lecture Notes in Computer Science* 1039, pp. 173–88, 1996.

[164] F. Göloglu and A. Pott. Results on the crosscorrelation and autocorrelation of sequences. *Proceedings of Sequences and Their Applications – SETA 2008, Lecture Notes in Computer Science* 5203, pp. 95–105, 2008.

[165] S. W. Golomb. *Shift Register Sequences*. Aegean Park Press, Laguna Hills, CA 1982.

[166] K. Gopalakrishnan, D. G. Hoffman, and D. R. Stinson. A note on a conjecture concerning symmetric resilient functions. *Information Processing Letters* 47(3), pp. 139–43, 1993.

[167] M. Goresky and A. Klapper. Fibonacci and Galois representation of feedback with carry shift registers. *IEEE Transactions on Information Theory* 48, pp. 2826–36, 2002.

[168] M. Goresky and A. Klapper. Periodicity and distribution properties of combined FCSR sequences. *Proceedings of Sequences and Their Applications – SETA 2006, Lecture Notes in Computer Science* 4086, pp. 334–41, 2006.

[169] A. Gouget. On the propagation criterion of Boolean functions. *Proceedings of the Workshop on Coding, Cryptography and Combinatorics 2003, pp. 153–68. Birkhäuser Verlag*, Basel, Switzerland, 2004.

[170] A. Gouget and H. Sibert. Revisiting correlation-immunity in filter generators. *Proceedings of SAC 2007, Lecture Notes in Computer Science* 4876, pp. 378–95, 2007.

[171] P. Guillot. Completed GPS Covers All Bent Functions. *Journal of Combinatorial Theory*, Series A 93, pp. 242–60, 2001.

[172] P. Guillot. Partial bent functions. *Proceedings of the World Multiconference on Systemics, Cybernetics and Informatics, SCI 2000*, 2000.

[173] Xiao Guo-Zhen, C. Ding, and W. Shan. *The stability theory of stream ciphers*, Lecture Notes in Computer Science 561, 1991.

[174] Xiao Guo-Zhen and J. L. Massey. A spectral characterization of correlation-immune combining functions. *IEEE Transactions on Information Theory* 34(3), pp. 569–71, 1988.

[175] A. R. Hammons Jr., P. V. Kumar, A. R. Calderbank, N. J. A. Sloane, and P. Solé. The Z_4-linearity of Kerdock, Preparata, Goethals and related codes. *IEEE Transactions on Information Theory*, 40, pp. 301–19, 1994.

[176] P. Hawkes and G. Rose. Rewriting variables: The complexity of fast algebraic attacks on stream ciphers. *Proceedings of CRYPTO 2004, Lecture Notes in Computer Science* 3152, pp. 390–406, 2004.

[177] T. Helleseth, T. Kløve, and J. Mykkelveit. On the covering radius of binary codes. *IEEE Transactions on Information Theory* 24(5), pp. 627–8, 1978.

[178] T. Helleseth and H. F. Mattson Jr. On the cosets of the simplex code. *Discrete Mathematics* 56, pp. 169–89, 1985.

[179] S. Hirose and K. Ikeda. Nonlinearity criteria of Boolean functions. *KUIS Technical Report*, KUIS-94–0002, 1994.

[180] S. Hirose and K. Ikeda. Complexity of Boolean functions satisfying the propagation criterion. *Proceedings of the 1995 Symposium on Cryptography and Information Security*, SCIS95-B3.3, 1995.

[181] I. Honkala and A. Klapper. Bounds for the multicovering radii of Reed-Muller codes with applications to stream ciphers. *Designs, Codes and Cryptography* 23, pp. 131–45, 2001.

[182] X.-D. Hou. Some results on the covering radii of Reed-Muller codes. *IEEE Transactions on Information Theory* 39(2), pp. 366–78, 1993.

[183] X.-D. Hou. Classification of cosets of the Reed-Muller code $R(m-3,m)$. *Discrete Mathematics*, 128, pp. 203–24, 1994.

[184] X.-D. Hou. The covering radius of $R(1,9)$ in $R(4,9)$. *Designs, Codes and Cryptography* 8(3), pp. 285–92, 1995.

[185] X.-D. Hou. $AGL(m,2)$ acting on $R(r,m)/R(s,m)$. *Journal of Algebra* 171, pp. 921–38, 1995.

[186] X.-D. Hou. Covering radius of the Reed-Muller code $R(1,7)$ – a simpler proof. *Journal of Combinatorial Theory*, Series A 74, pp. 337–41, 1996.

[187] X.-D. Hou. $GL(m,2)$ acting on $R(r,m)/R(r-1,m)$. *Discrete Mathematics* 149, pp. 99–122, 1996.

[188] X.-D. Hou. On the covering radius of $R(1,m)$ in $R(3,m)$. *IEEE Transactions on Information Theory* 42(3), pp. 1035–7, 1996.

[189] X.-D. Hou. The Reed-Muller code $R(1,7)$ is normal. *Designs, Codes and Cryptography* 12, pp. 75–82, 1997.

[190] X.-D. Hou. Cubic bent functions. *Discrete Mathematics* vol. 189, pp. 149–61, 1998.

[191] X. D. Hou. On the coefficients of binary bent functions. *Proceedings of the American Mathematical Society* 128(4), pp. 987–996, 2000.

[192] X.-D. Hou. New constructions of bent functions. *Proceedings of the International Conference on Combinatorics, Information Theory and Statistics; Journal of Combinatorics, Information and System Sciences* 25(1–4), pp. 173–89, 2000.

[193] X.-D. Hou. On binary resilient functions. *Designs, Codes, and Cryptography* 28(1), pp. 93–112, 2003.

[194] X.-D. Hou. Group Actions on Binary Resilient Functions. *Applied Algebra Engineerings, Communication and Computing* 14(2), pp. 97–115, 2003.

[195] X.-D. Hou. A note on the proof of a theorem of Katz. *Finite Fields and Their Applications* 11, pp. 316–19, 2005.

[196] X.-D. Hou and P. Langevin. Results on bent functions, *Journal of Combinatorial Theory Series A* 80, pp. 232–46, 1997.

[197] H. Hu and D. Feng. On quadratic bent functions in polynomial forms. *IEEE Transactions on Information Theory* 53, pp. 2610–15, 2007.

[198] T. Iwata and K. Kurosawa. Probabilistic higher order differential attack and higher order bent functions. *Proceedings of ASIACRYPT'99, Lecture Notes in Computer Science* 1716, pp. 62–74, 1999.

[199] C. J. A. Jansen and D. E. Boekee. The shortest feedback shift register that can generate a given sequence. *Proceedings of CRYPTO'89, Lecture Notes in Computer Science* 435, pp. 90–9, 1990 (this paper refers to the classified PhD thesis of C. J. A. Jansen entitled "Investigations on nonlinear streamcipher systems: construction and evaluation methods," Philips).

[200] T. Johansson and F. Jönsson. Improved fast correlation attack on stream ciphers via convolutional codes. *Proceedings of EUROCRYPT '99, Lecture Notes in Computer Science* 1592, pp. 347–62, 1999.

[201] T. Johansson and F. Jönsson. Fast correlation attacks based on turbo code techniques. *Advances in Cryptology – CRYPTO '99, Lecture Notes in Computer Science* 1666, pp. 181–97, 1999.

[202] T. Johansson and F. Jönsson. Fast correlation attacks through reconstruction of linear polynomials. *Advances in Cryptology – CRYPTO 2000, Lecture Notes in Computer Science* 1880. pp. 300–15, 2000.

[203] F. Jönsson. *Some results on fast correlation attacks*. PhD thesis, Lund University. 2002.

[204] D. Jungnickel. *Difference sets.* In Contemporary Design Theory: A Collection of Surveys, J. Dinitz and D. R. Stinson eds. John Wiley & Sons, New York, 1992.

[205] W. Kantor. An exponential number of generalized Kerdock codes. *Information and Control* 53, pp. 74–80, 1982.

[206] T. Kasami. The weight enumerators for several classes of subcodes of the second order binary Reed-Muller codes. *Information and Control* 18, pp. 369–94, 1971.

[207] T. Kasami and N. Tokura. On the weight structure of the Reed Muller codes, *IEEE Transactions on Information Theory* 16, pp. 752–9, 1970.

[208] T. Kasami, N. Tokura, and S. Azumi. On the weight enumeration of weights less than $2.5d$ of Reed-Muller codes. *Information and Control* 30, 380–95, 1976.

[209] N. Katz. On a theorem of Ax. *American Journal of Mathematics* 93, pp. 485–99, 1971.

[210] S. Kavut, S. Maitra, and M. D. Yücel. Search for Boolean functions with excellent profiles in the rotation symmetric class. *IEEE Transactions on Information Theory* 53(5), pp. 1743–51, 2007.

[211] A. M. Kerdock. A class of low-rate non linear codes. *Information and Control* 20, pp. 182–7, 1972.

[212] A. Kerckhoffs. La cryptographie militaire. *Journal des Sciences Militaires*, 1883.

[213] J. D. Key, T. P. McDonough, and V. C. Mavron. Information sets and partial permutation decoding for codes from finite geometries. *Finite Fields and Their Applications* 12(2), pp. 232–47, 2006.

[214] J. Kahn, G. Kalai, and N. Linial. The influence of variables on Boolean functions. *IEEE 29th Symp. on Foundations of Computer Science*, pp. 68–80, 1988.

[215] L. R. Knudsen. Truncated and higher order differentials. *Proceedings of Fast Software Encryption, Second International Workshop, Lecture Notes in Computer Science* 1008, pp. 196–211, 1995.

[216] K. Khoo and G. Gong. New constructions for resilient and highly nonlinear Boolean functions. *Proceedings of 8th Australasian Conference, ACISP 2003, Wollongong, Austrialia, Lecture Notes in Computer Science* 2727, pp. 498–509, 2003.

[217] K. Khoo, G. Gong, and D. Stinson. A new characterization of semi-bent and bent functions on finite fields. *Designs, Codes and Cryptography* 38(2), pp. 279–95, 2006.

[218] A. Klapper and M. Goresky. Feedback shift registers, 2-Adic span, and combiners with memory. *Journal of Cryptology* 10, pp. 111–47. 1997.

[219] L. R. Knudsen and M. P. J. Robshaw. Non-linear approximations in linear cryptanalysis. *Proceedings of EUROCRYPT '96, Lecture Notes in Computer Science* 1070, pp. 224–36, 1996.

[220] P. V. Kumar, R. A. Scholtz, and L. R. Welch. Generalized bent functions and their properties, *Journal of Combinatorial Theory Series A* 40, pp. 90–107, 1985.

[221] K. Kurosawa, T. Iwata, and T. Yoshiwara. New covering radius of Reed-Muller codes for t-resilient functions. *Proceedings of Selected Areas in Cryptography, 8th Annual International Workshop, Lecture Notes in Computer Science* 2259, pp. 75ff, 2001.

[222] K. Kurosawa and R. Matsumoto. Almost security of cryptographic Boolean functions. *IEEE Transactions on Information Theory* 50(11), pp. 2752–61, 2004.

[223] K. Kurosawa and T. Satoh. Design of $SAC/PC(\ell)$ of order k Boolean functions and three other cryptographic criteria. *Advances in Cryptology, EUROCRYPT' 97, Lecture Notes in Computer Science* 1233, pp. 434–49, 1997.

[224] P. Lacharme. Post processing functions for a physical random number generator. *Proceedings of Fast Software Encryption 2008, Lecture Notes in Computer Science* 5086, pp. 334–42, 2008.

[225] G. Lachaud and J. Wolfmann. The weights of the orthogonals of the extended quadratic binary Goppa codes. *IEEE Transactions Information Theory* 36, pp. 686–92, 1990.

[226] J. Lahtonen, G. McGuire, and H. Ward. Gold and Kasami-Welch functions, quadratic forms and bent functions. *Advances in Mathematics of Communications* 1, pp. 243–50, 2007.

[227] X. Lai. Higher order derivatives and differential cryptanalysis. *Proceedings of the Symposium on Communication, Coding and Cryptography, in honor of J. L. Massey on the occasion of his 60th birthday*. 1994.

[228] X. Lai. Additive and linear structures of cryptographic functions. *Proceedings of Fast Software Encryption, Second International Workshop, Lecture Notes in Computer Science* 1008, pp. 75–85, 1995.

[229] P. Langevin. Covering radius of $RM(1, 9)$ in $RM(3, 9)$. *Eurocode'90, Lecture Notes in Computer Science* 514, pp. 51–59, 1991.

[230] P. Langevin. On the orphans and covering radius of the Reed-Muller codes. *Proceedings of AAECC 9, Lecture Notes in Computer Science* 539, pp. 234–40, 1991.

[231] P. Langevin. On generalized bent functions. *CISM Courses and Lectures 339 (Eurocode)*, pp. 147–57, 1992.

[232] P. Langevin and G. Leander. Monomial Bent Functions and Stickelberger's Theorem. *Finite Fields and Their Applications*, 14, pp. 727–42, 2008.

[233] P. Langevin. Private communication, 2008. See also the web page http://langevin.univ-tln.fr/project/quartics/ maintained by P. Langevin, G. Leander, P. Rabizzoni, P. Veron, and J.-P. Zanotti.

[234] P. Langevin, P. Rabizzoni, P. Veron, and J.-P. Zanotti. On the number of bent functions with 8 variables. *Proceedings of the conference BFCA 2006*, pp. 125–136. Publications des universités de Rouen et du Havre, 2006.

[235] P. Langevin and P. Solé. Kernels and defaults. American Mathematical Society (*Proceedings of the conference Finite Fields and Applications* Fq4), *Contemporary Mathematics* 225, pp. 77–85, 1999.

[236] P. Langevin and P. Véron. On the nonlinearity of power functions. *Designs, Codes and Cryptography* 37, pp. 31–43, 2005.

[237] P. Langevin and J.-P. Zanotti. Nonlinearity of some invariant Boolean functions. *Designs, Codes and Cryptography* 36, pp. 131–46, 2005.

[238] C. Lauradoux and M. Videau. Matriochka symmetric Boolean functions. *Proceedings of International Symposium on Information Theory, ISIT 2008*.

[239] G. Leander. Bent functions with 2^r Niho exponents. *Proceedings of the Workshop on Coding and Cryptography 2005*, pp. 454–61, 2005.

[240] G. Leander. Monomial bent functions. *Proceedings of the Workshop on Coding and Cryptography* 2005, Bergen, pp. 462–70, 2005, and *IEEE Transactions on Information Theory* 52(2), pp. 738–43, 2006.

[241] G. Leander. Another class of non-normal bent functions. *Proceedings of the conference BFCA 2006*, pp. 87–98. Publications des universités de Rouen et du Havre, 2006.

[242] R. J. Lechner. *Harmonic analysis of switching functions*. In Recent Developments in Switching Theory. A. Mukhopadhyay, ed. Academic Press, New York, 1971.

[243] S. Leveiller, G. Zemor, P. Guillot, and J. Boutros. A new cryptanalytic attack for PN-generators filtered by a Boolean function. *Proceedings of Selected Areas of Cryptography 2002, Lecture Notes in Computer Science* 2595, pp. 232–49 (2003).

[244] N. Li and W. Qi. Symmetric Boolean functions depending on an odd number of variables with maximum algebraic immunity. *IEEE Transactions on Information Theory* 52(5), pp. 2271–3, 2006.

[245] N. Li and W.-Q. Qi. Construction and analysis of Boolean functions of $2t + 1$ variables with maximum algebraic immunity. *Proceedings of Asiacrypt 2006, Lecture Notes in Computer Science* 4284, pp. 84–98, 2006.

[246] N. Li, L. Qu, W.-F. Qi, G. Feng, C. Li, and D. Xie. On the construction of Boolean functions with optimal algebraic immunity. *IEEE Transactions on Information Theory* 54(3), pp. 1330–4, 2008.

[247] R. Lidl and H. Niederreiter. *Finite fields*. In Encyclopedia of Mathematics and Its Applications, Vol. 20. Addison-Wesley, Reading, MA, 1983.

[248] S. Ling and C. Xing. *Coding Theory*. Cambridge University Press, Cambridge, UK, 2004.

[249] N. Linial, Y. Mansour, and N. Nisan. Constant depth circuits, Fourier transform, and learnability. *Journal of the Association for Computing Machinery* 40(3), pp. 607–20, 1993.

[250] J. H. van Lint. *Introduction to Coding Theory*. Springer, New York, 1982.

[251] F. Liu and K. Feng. On the 2^m-variable symmetric Boolean functions with maximum algebraic immunity 2^{m-1}. *Proceedings of the Workshop on Coding and Cryptography 2007* WCC, pp. 225–32, 2007.

[252] S. Lloyd. Properties of binary functions. *Proceedings of EUROCRYPT '90, Lecture Notes in Computer Science* 473, pp. 124–139, 1991.

[253] S. Lloyd. Counting binary functions with certain cryptographic properties. *Journal of Cryptology* 5, pp. 107–131, 1992.

[254] S. Lloyd. Balance, uncorrelatedness and the strict avalanche criterion. *Discrete Applied Mathematics* 41, pp. 223–33, 1993.

[255] M. Lobanov. Tight bound between nonlinearity and algebraic immunity. IACR ePrint Archive http://eprint.iacr.org/ 2005/441.

[256] O. A. Logachev, A. A. Salnikov, and V. V. Yashchenko. Bent functions on a finite Abelian group. *Discrete Mathematics Applications* 7(6), pp. 547–64, 1997.

[257] F. J. MacWilliams and N. J. Sloane. *The Theory of Error-Correcting Codes*. North-Holland, Amsterdam, 1977.

[258] J. A. Maiorana. A classification of the cosets of the Reed-Muller code $R(1, 6)$. *Mathematics of Computation* 57(195), pp. 403–14, 1991.
[259] S. Maitra. Highly nonlinear balanced Boolean functions with very good autocorrelation property. *Proceedings of the Workshop on Coding and Cryptography 2001, Electronic Notes in Discrete Mathematics*, 6, pp. 355–64, 2001.
[260] S. Maitra. Autocorrelation properties of correlation immune Boolean functions. *Proceedings of INDOCRYPT 2001, Lecture Notes in Computer Science* 2247, pp. 242–53, 2001.
[261] S. Maitra, S. Kavut, and M. Yucel. Balanced Boolean function on 13-variables having nonlinearity greater than the bent concatenation bound. Proceedings of the conference BFCA 2008, Copenhagen, to appear.
[262] S. Maitra and E. Pasalic. Further constructions of resilient Boolean functions with very high nonlinearity. *IEEE Transactions on Information Theory* 48(7), pp. 1825–34, 2002.
[263] S. Maitra and P. Sarkar. Enumeration of correlation-immune Boolean functions. *Proceedings of ACISP 1999*, pp. 12–25, 1999.
[264] S. Maitra and P. Sarkar. Maximum nonlinearity of symmetric Boolean functions on odd number of variables. *IEEE Transactions on Information Theory* 48, pp. 2626–30, 2002.
[265] S. Maitra and P. Sarkar. Highly nonlinear resilient functions optimizing Siegenthaler's inequality. *Proceedings of CRYPTO '99, Lecture Notes in Computer Science* 1666, pp. 198–215, 1999.
[266] S. Maitra and P. Sarkar. Modifications of Patterson-Wiedemann functions for cryptographic applications. *IEEE Transactions on Information Theory* 48, pp. 278–84, 2002.
[267] S. Maity and S. Maitra. Minimum distance between bent and 1-resilient Boolean functions. *Proceedings of Fast Software Encryption 2004, Lecture Notes in Computer Science* 3017, pp. 143–60, 2004.
[268] J. L. Massey. Shift-register analysis and BCH decoding. *IEEE Transactions on Information Theory*, 15, pp. 122–7, 1969.
[269] J. L. Massey. Randomness, arrays, differences and duality. *IEEE Transactions on Information Theory* 48, pp. 1698–1703, 2002.
[270] M. Matsui. Linear cryptanalysis method for DES cipher. *Proceedings of EUROCRYPT'93, Lecture Notes in Computer Science* 765, pp. 386–97, 1994.
[271] R. J. McEliece. Weight congruence for p-ary cyclic codes. *Discrete Mathematics* 3, pp. 177–92, 1972.
[272] R. L. McFarland. A family of noncyclic difference sets, *Journal of Combinatonal Theory Series A*, 15, pp. 1–10, 1973.
[273] W. Meier, E. Pasalic, and C. Carlet. Algebraic attacks and decomposition of Boolean functions. *Advances in Cryptology, EUROCRYPT 2004, Lecture Notes in Computer Science* 3027, pp. 474–91, 2004.
[274] W. Meier and O. Staffelbach. Fast correlation attacks on stream ciphers. *Advances in Cryptology, EUROCRYPT '88, Lecture Notes in Computer Science* 330, pp. 301–14, 1988.
[275] W. Meier and O. Staffelbach. Nonlinearity Criteria for Cryptographic Functions. *Advances in Cryptology, EUROCRYPT' 89, Lecture Notes in Computer Science* 434, pp. 549–62, 1990.
[276] W. Meier and O. Staffelbach. Correlation properties of combiners with memory in stream ciphers. *Advances in Cryptology, EUROCRYPT '90, Lecture Notes in Computer Science* 473, pp. 204–13, 1990.
[277] A. Menezes, P. van Oorschot, and S. Vanstone. In *Handbook of Applied Cryptography*. CRC Press Series on Discrete Mathematics and Its Applications, 1996.
[278] Q. Meng, H. Zhang, M. Yang, and J. Cui. On the degree of homogeneous bent functions. *Discrete Applied Mathematics* 155(5), pp. 665–9, 2007.
[279] S. Mesnager. Improving the lower bound on the higher order nonlinearity of Boolean functions with prescribed algebraic immunity. *IEEE Transactions on Information*

Theory 54(8), pp. 3656–62, 2008. Preliminary version available at IACR ePrint Archive http://eprint.iacr.org/, 2007/117.

[280] S. Mesnager. On the number of resilient Boolean functions. *Proceedings of the conference. First Symposium on Algebraic Geometry and Its Applications (SAGA'07), Tahiti, 2007. Series on Number Theory and Its Applications* 5, pp. 419–33, 2008.

[281] W. Millan, A. Clark, and E. Dawson. Heuristic design of cryptographically strong balanced Boolean functions. *EUROCRYPT '98, Advances in Cryptology, Lecture Notes in Computer Science* 1403, 1998.

[282] C. J. Mitchell. Enumerating Boolean functions of cryptographic signifiance. *Journal of Cryptology* 2(3), pp. 155–70, 1990.

[283] J. Mykkelveit. The covering radius of the [128,8] Reed-Muller code is 56. *IEEE Transactions on Information Theory*, 26(3), pp. 359–62, 1980.

[284] Y. Nawaz, G. Gong, and K. Gupta. Upper bounds on algebraic immunity of power functions. *Proceeding of Fast Software Encryption 2006, Lecture Notes in Computer Science* 4047, pp. 375–89, 2006.

[285] N. Nisan and M. Szegedy. On the degree of Boolean functions as real polynomials. *Computational Complexity* 4, pp. 301–13, 1994.

[286] K. Nyberg. Constructions of bent functions and difference sets, *EUROCRYPT '90, Advances in Cryptology, Lecture Notes in Computer Science* 473, pp. 151–60, 1991.

[287] L. O'Connor and A. Klapper. Algebraic nonlinearity and its applications to cryptography. *Journal of Cryptology* 7, pp. 213–27, 1994.

[288] D. Olejár and M. Stanek. On cryptographic properties of random Boolean functions. *Journal of Universal Computer Science* 4(8), pp. 705–17, 1998.

[289] J. D. Olsen, R. A. Scholtz, and L. R. Welch. Bent function sequences. *IEEE Transactions on Information Theory* 28(6), 1982.

[290] E. Pasalic. *On Boolean functions in symmetric-key ciphers*. PhD thesis, 2003.

[291] E. Pasalic, T. Johansson, S. Maitra, and P. Sarkar. New constructions of resilient and correlation immune Boolean functions achieving upper bounds on nonlinearity. *Proceedings of the Workshop on Coding and Cryptography* 2001, *Electronic Notes in Discrete Mathematics* 6, pp. 425–34, 2001.

[292] E. Pasalic and S. Maitra. A Maiorana-McFarland type construction for resilient Boolean functions on n variables (n even) with nonlinearity $> 2^{n-1} - 2^{n/2} + 2^{n/2-2}$. *Proceedings of the Workshop on Coding and Cryptography 2003*, pp. 365–74, 2003.

[293] S. M. Park, S. Lee, S. H. Sung, and K. Kim. Improving bounds for the number of correlation-immune Boolean functions. *Information Processing Letters* 61, pp. 209–12, 1997.

[294] N. J. Patterson and D. H. Wiedemann. The covering radius of the $[2^{15}, 16]$ Reed-Muller code is at least 16276. *IEEE Transactions on Information Theory* 29, pp. 354–6, 1983.

[295] N. J. Patterson and D. H. Wiedemann. Correction to [294]. *IEEE Transactions on Information Theory* 36(2), p. 443, 1990.

[296] J. Pieprzyk and X.-M. Zhang. Computing Möbius transforms of Boolean functions and characterizing coincident Boolean functions. *Proceedings of the conference BFCA 2007*. Publications des universités de Rouen et du Havre, 2007.

[297] V. S. Pless, W. C. Huffman, eds., R. A. Brualdi, assistant editor. *Handbook of Coding Theory*, Elsevier, Amsterdam, 1998.

[298] A. Pott. *Finite Geometry and Character Theory*. Lecture Notes in Mathematics, 1601. Springer Verlag, Berlin, 1995.

[299] B. Preneel, W. Van Leekwijck, L. Van Linden, R. Govaerts, and J. Vandevalle. Propagation characteristics of Boolean functions, *Proceedings of EUROCRYPT '90, Lecture Notes in Computer Sciences* 473, pp. 161–73, 1991.

[300] B. Preneel, R. Govaerts, and J. Vandevalle. Boolean functions satisfying higher order propagation criteria, *Proceedings of EUROCRYPT'91, Lecture Notes in Computer Science* 547, pp. 141–52, 1991.

[301] B. Preneel. *Analysis and Design of Cryptographic Hash Functions*, PhD thesis, Katholieke Universiteit Leuven, K. Mercierlaan 94, 3001 Leuven, Belgium, U. D. C. 621.391.7, 1993.

[302] L. Qu, C. Li, and K. Feng. A note on symmetric Boolean functions with maximum algebraic immunity in odd number of variables. *IEEE Transactions on Information Theory* 53, pp. 2908–10, 2007.

[303] M. Quisquater. *Applications of character theory and the Möbius inversion principle to the study of cryptographic properties of Boolean functions*. PhD thesis, Katholieke Universiteit Leuven, 2004.

[304] M. Quisquater, B. Preneel and J. Vandewalle. A new inequality in discrete Fourier theory. *IEEE Transactions on Information Theory* 49, pp. 2038–40, 2003.

[305] M. Quisquater, B. Preneel, and J. Vandewalle. Spectral characterization of cryptographic Boolean functions satisfying the (extended) propagation criterion of degree l and order *k*. *Information Processing Letters* 93(1), pp. 25–8, 2005.

[306] C. Riera and M. G. Parker. Generalised bent criteria for Boolean functions (I). *IEEE Transactions on Information Theory* 52(9), pp. 4142–59, 2006.

[307] C. R. Rao. Factorial experiments derived from combinatorial arrangements of arrays. *Journal of the Royal Statistical Society* 9, pp. 128–39, 1947.

[308] F. Rodier. Asymptotic nonlinearity of Boolean functions. *Designs, Codes and Cryptography* 40(1), pp. 59–70, 2006.

[309] S. Ronjom, M. Abdelraheem, and L. E. Danielsen. Online database of Boolean Functions. http://www.ii.uib.no/ mohamedaa/odbf/index.html

[310] S. Rønjom and T. Helleseth. A new attack on the filter generator. *IEEE Transactions on Information theory* 53(5), pp. 1752–8, 2007.

[311] S. Rønjom and T. Helleseth. Attacking the filter generator over $GF(2^m)$. *Proceedings of the International Workshop on the Arithmetic of Finite Fields, WAIFI 2007, Lecture Notes in Computer Science* 4547, pp. 264–75, 2007.

[312] O. S. Rothaus. On "bent" functions. *Journal of Combinatorial Theory* A20, pp. 300–5, 1976.

[313] B. V. Ryazanov. On the distribution of the spectral complexity of Boolean functions. *Discrete Mathematics Applications* 4(3), pp. 279–88, 1994.

[314] R. A. Rueppel. *Analysis and Design of Stream Ciphers*. Com. and Contr. Eng. Series, Tokyo 1986

[315] R. A. Rueppel and O. J. Staffelbach. Products of linear recurring sequences with maximum complexity. *IEEE Transactions on Information Theory* 33(1), 1987.

[316] P. Sarkar. The filter-combiner model for memoryless synchronous stream ciphers. *Proceedings of CRYPTO 2002, Lecture Notes in Computer Science* 2442, pp. 533–48, 2002.

[317] P. Sarkar and S. Maitra. Construction of nonlinear Boolean functions with important cryptographic properties. *Proceedings of EUROCRYPT 2000, Lecture Notes in Computer Science* 1807, pp. 485–506, 2000.

[318] P. Sarkar and S. Maitra. Nonlinearity bounds and constructions of resilient Boolean functions. *Proceedings of CRYPTO 2000, Lecture Notes in Computer Science* 1880, pp. 515–32, 2000.

[319] P. Sarkar and S. Maitra. Construction of nonlinear resilient Boolean functions using "small" affine functions. *IEEE Transactions on Information Theory* 50(9), pp. 2185–93, 2004.

[320] P. Sarkar and S. Maitra. Balancedness and correlation immunity of symmetric Boolean functions. *Discrete Mathematics* 307, pp. 2351–8, 2007.

[321] P. Savicky. On the bent Boolean functions that are symmetric. *European Journal of Combinatorics* 15, pp. 407–10, 1994.

[322] M. Schneider. A note on the construction and upper bounds of correlation-immune functions. *Proceedings of the 6th IMA Conference, Lecture Notes In Computer Science* 1355, pp. 295–306, 1997. An extended version appeared under the title "On the

construction and upper bounds of balanced and correlation-immune functions", *Selected Areas in Cryptography*, pp. 73–87, 1997.

[323] J. Seberry and X.-M. Zhang. Constructions of bent functions from two known bent functions. *Australasian Journal of Combinatorics* 9, pp. 21–35, 1994.

[324] J. Seberry, X.-M. Zhang, and Y. Zheng. On constructions and nonlinearity of correlation immune Boolean functions. *Advances in Cryptology – EUROCRYPT '93, Lecture Notes in Computer Science* 765, pp. 181–99, 1994.

[325] J. Seberry, X.-M. Zhang, and Y. Zheng. Nonlinearly balanced Boolean functions and their propagation characteristics. *Advances in Cryptology – CRYPTO '93*, pp. 49–60, 1994.

[326] N. V. Semakov and V. A. Zinoviev. Balanced codes and tactical configurations. *Problems of Information Transfer* 5(3), pp. 22–8, 1969.

[327] A. Shanbhag, V. Kumar, and T. Helleseth. An upper bound for the extended Kloosterman sums over Galois rings. *Finite Fields and Their Applications* 4, pp. 218–38, 1998.

[328] C. E. Shannon. A mathematical theory of communication. *Bell System Technical Journal* 27, pp. 379–423, 1948.

[329] C. E. Shannon. Communication theory of secrecy systems. *Bell System Technical Journal* 28, pp. 656–715, 1949.

[330] C. E. Shannon. The synthesis of two-terminal switching circuits. *Bell System Technical Journal* 28, pp. 59–98, 1949.

[331] I. Shparlinski. Bounds on the Fourier coefficients of the weighted sum function. *Information Processing Letters* 103(3), pp. 83–7, 2007.

[332] I. Shparlinski and A. Winterhof. On the nonlinearity of linear recurrence sequences. *Applied Mathematics Letters* 19, pp. 340–4, 2006.

[333] T. Siegenthaler. Correlation-immunity of nonlinear combining functions for cryptographic applications. *IEEE Transactions on Information Theory* 30(5), pp. 776–80, 1984.

[334] T. Siegenthaler. Decrypting a class of stream ciphers using ciphertext only. *IEEE Transactions on Computers*, C-34(1), pp. 81–5, 1985.

[335] P. Stanica, S. Maitra, and J. Clark. Results on rotation symmetric bent and correlation immune Boolean functions. *Proceedings of Fast Software Encryption 2004, Lecture Notes in Computer Science* 3017, pp. 161–77, 2004.

[336] P. Stanica and S. H. Sung. Boolean functions with five controllable cryptographic properties. *Designs, Codes and Cryptography* 31, pp. 147–57, 2004.

[337] V. Strassen. Gaussian elimination is not optimal. *Numerische Mathematik* 13, pp. 354–6, 1969.

[338] I. Strazdins. Universal affine classification of Boolean functions. *Acta Applicandae Mathematicae* 46, pp. 147–67, 1997.

[339] T. Sugita, T. Kasami, and T. Fujiwara. Weight distributions of the third and fifth order Reed-Muller codes of length 512. Nara Institute of Science Technical Report, 1996.

[340] S. H. Sung, S. Chee, and C. Park. Global avalanche characteristics and propagation criterion of balanced Boolean functions. *Information Processing Letters* 69, pp. 21–24, 1999.

[341] H. Tapia-Recillas and G. Vega. An upper bound on the number of iterations for transforming a Boolean function of degree greater than or equal than 4 to as function of degree 3. *Designs, Codes and Cryptography* 24, pp. 305–12, 2001.

[342] Y. V. Tarannikov. On resilient Boolean functions with maximum possible nonlinearity. *Proceedings of INDOCRYPT 2000, Lecture Notes in Computer Science* 1977, pp. 19–30, 2000.

[343] Y. V. Tarannikov. New constructions of resilient Boolean functions with maximum nonlinearity. *Proceedings of FSE 2001, 8th International Workshop, FSE 2001, Lecture Notes in Computer Science* 2355, pp. 66–77, 2001.

[344] Y. V. Tarannikov and D. Kirienko. Spectral analysis of high order correlation immune functions. *Proceedings of 2001 IEEE International Symposium on Information Theory*, p. 69, 2001 (full preliminary version at IACR ePrint Archive http://eprint.iacr.org/).

[345] Y. V. Tarannikov, P. Korolev, and A. Botev. Autocorrelation coefficients and correlation immunity of Boolean functions. *Proceedings of Asiacrypt 2001, Lecture Notes in Computer Science* 2248, pp. 460–79, 2001

[346] F. Didier and J. Tillich. Computing the Algebraic Immunity Efficiently. *Proceedings of Fast Software Encryption 2006, Lecture Notes in Computer Science* 4047, pp. 359–74, 2006.

[347] S. Tsai. Lower bounds on representing Boolean functions as polynomials in $\mathbb{Z}_m^{\star}$. *SIAM Journal on Discrete Mathematics* 9(1), pp. 55–62, 1996.

[348] S. F. Vinokurov and N. A. Peryazev. An expansion of Boolean function into a sum of products of subfunctions. *Discrete Mathematics Applications* 3(5), pp. 531–3, 1993.

[349] A. F. Webster and S. E. Tavares. On the design of S-boxes. In *Proceedings of CRYPTO'85, Lecture Notes in Computer Science* 219, pp. 523–34, 1985.

[350] J. Wolfmann. Bent functions and coding theory. In *Difference Sets, Sequences and their Correlation Properties,* A. Pott, P. V. Kumar, T. Helleseth and D. Jungnickel, eds., pp. 393–417. Kluwer, Amsterdam, 1999.

[351] Y. X. Yang and B. Guo. Further enumerating Boolean functions of cryptographic signifiance. *Journal of Cryptology* 8(3), pp. 115–22, 1995.

[352] R. Yarlagadda and J. E. Hershey. Analysis and synthesis of bent sequences. *IEE proceedings, Part E, Computers and Digital Techniques*, Vol. 136, pp. 112–23, 1989.

[353] A. M. Youssef and G. Gong. Hyper-bent functions. *Proceedings of EUROCRYPT 2001, Lecture Notes in Computer Science* 2045, pp. 406–19, 2001.

[354] N. Y. Yu and G. Gong. Constructions of quadratic bent functions in polynomial forms. *IEEE Transactions on Information Theory* 52(7), pp. 3291–9, 2006.

[355] M. Zhang. Maximum correlation analysis of nonlinear combining functions in stream ciphers. *Journal of Cryptology* 13(3), pp. 301–13, 2000.

[356] X.-M. Zhang and Y. Zheng. GAC – the criterion for global avalanche characteristics of cryptographic functions. *Journal of Universal Computer Science* 1(5), pp. 320–37, 1995.

[357] X.-M. Zhang and Y. Zheng. Auto-correlations and new bounds on the nonlinearity of Boolean functions. *Proceedings of EUROCRYPT'96, Lecture Notes in Computer Science* 1070, pp. 294–306, 1996.

[358] X.-M. Zhang and Y. Zheng. Characterizing the structures of cryptographic functions satisfying the propagation criterion for almost all vectors. *Designs, Codes and Cryptography* 7(1), pp. 11–134, 1996.

[359] X.-M. Zhang and Y. Zheng. The nonhomomorphicity of Boolean functions. *Proceedings of SAC 1998, Lecture Notes in Computer Science* 1556, pp. 280–95, 1999.

[360] Y. Zheng and X. M. Zhang. Plateaued functions. *Proceedings of ICICS'99, Lecture Notes in Computer Science* 1726, pp. 284–300, 1999.

[361] Y. Zheng, X.-M. Zhang, and H. Imai. Restriction, terms and nonlinearity of Boolean functions. *Theoretical Computer Science* 226(1–2), pp. 207–23, 1999.

[362] Y. Zheng and X.-M. Zhang. On relationships among avalanche, nonlinearity and correlation immunity. *Proceedings of Asiacrypt 2000, Lecture Notes in Computer Science* 1976, pp. 470–83, 2000.

[363] Y. Zheng and X.-M. Zhang. Improving upper bound on the nonlinearity of high order correlation immune functions. *Proceedings of Selected Areas in Cryptography 2000, Lecture Notes in Computer Science* 2012, pp. 262–74, 2001.

[364] N. Zierler and W. H. Mills. Products of linear recurring sequences. *Journal of Algebra* 27, pp. 147–57, 1973.

9

Vectorial Boolean Functions for Cryptography

Claude Carlet

This chapter is dedicated to the memory of Hans Dobbertin.

9.1 Introduction

This chapter deals with multi-output Boolean functions viewed from a cryptographic viewpoint, that is, functions from the vector space $\mathbb{F}_2^n$, of all binary vectors of length n, to the vector space $\mathbb{F}_2^m$, for some positive integers n and m, where $\mathbb{F}_2$ is the finite field with two elements.[1] Obviously, these functions include the (single-output) Boolean functions that correspond to the case $m = 1$. The present chapter follows the chapter "Cryptography and Error-Correcting Codes" (dedicated to Boolean functions), to which we refer for all the definitions and properties that will be needed in the present chapter. As in this previous chapter, additions of bits performed in characteristic 0 (that is, in $\mathbb{Z}$, *i.e.*, not modulo 2) will be denoted by $+$, and additions modulo 2 (in $\mathbb{F}_2$) will be denoted by $\oplus$. The multiple sums will be denoted by $\sum_i$ when they are calculated in characteristic 0 and by $\bigoplus_i$ when they are calculated modulo 2. These two different notations are necessary because some representations of (vectorial) Boolean functions live in characteristic 2 and some representations of the same functions live in characteristic 0. However, the additions of elements of the finite field $\mathbb{F}_{2^n}$ will be denoted by $+$, as is usual in mathematics, despite the fact they are performed in characteristic 2. So, for simplicity (because $\mathbb{F}_2^n$ will often be identified with $\mathbb{F}_{2^n}$) and because there will be no ambiguity, we shall also denote by $+$ the addition of vectors of $\mathbb{F}_2^n$ when $n > 1$.

Let n and m be two positive integers. The functions from $\mathbb{F}_2^n$ to $\mathbb{F}_2^m$ are called *(n, m)-functions*. Such function F being given, the Boolean functions $f_1, \ldots, f_m$ defined, at every $x \in \mathbb{F}_2^n$, by $F(x) = (f_1(x), \ldots, f_m(x))$, are called the *coordinate functions* of F. When the numbers m and n are not specified, (n, m)-functions

[1] Denoted by $\mathcal{B}$ in some chapters of the present collection.

are called *multioutput Boolean functions*, *vectorial Boolean functions*, or *S-boxes*[2] (this last term is the most often used in cryptography, but is dedicated to the vectorial functions whose role is to provide confusion into the system; see the subsection on the cryptographic criteria for Boolean functions in Chapter 8 for the meaning of this term).

S-boxes are parts of iterative *block ciphers*, and they play a central role in their robustness. Iterative block ciphers are the iterations of a transformation depending on a key over each block of plaintext. The iterations are called rounds, and the key used in an iteration is called a round key. The round keys are computed from the secret key (called the master key) by a key scheduling algorithm. The rounds consist of vectorial Boolean functions combined in different ways involving the round key. Figures displaying the location of the S-boxes in the two main block ciphers, DES and AES, can be found in Chapter 8.

The main attacks on block ciphers, which will result in design criteria, are the following.

The *differential attack*, introduced by Biham and Shamir [11], assumes the existence of ordered pairs (α, β) of binary strings of the same length as the blocks (which are binary strings, too), such that, a block m of plaintext being randomly chosen, the bitwise difference $c + c'$ (recall that we use $+$ to denote the bitwise addition/difference in $\mathbb{F}_2^n$) between the ciphertexts c and c' corresponding to the plaintexts m and $m + \alpha$ has a larger probability to be equal to β than if c and c' were binary strings randomly chosen. Such an ordered pair (α, β) is called a differential; the larger the probability of the differential is, the more efficient is the attack. The related criterion on an (n, m)-function F used as an S-box in the round functions of the cipher is that the output to its derivatives $D_a(x) = F(x) + F(x + a); x, a \in \mathbb{F}_2^n$, must be as uniformly distributed as possible (except for the case $a = 0$, obviously). There are several ways to mount the differential cryptanalysis. The most common (and most efficient) one is to use differentials for the *reduced cipher*, that is, the input to the last round (*i.e.*, the cipher obtained from the original one by removing its last round); this allows (*see Figure 9.1*) to distinguish, in a *last round attack*, the reduced cipher from a random permutation. The existence of such *distinguisher* allows recovering the key used in the last round (either by an exhaustive search, which is efficient if this key is shorter than the master key, or by using specificities of the cipher, allowing replacement of the exhaustive search by, for instance, solving algebraic equations).

The *linear attack*, introduced by Matsui [122], is based on an idea from [144]. Its most common version is also an attack on the reduced cipher. It uses as distinguishers triples (α, β, γ) of binary strings such that, a block m of plaintext and a key k being randomly chosen, the bit $\alpha \cdot m \oplus \beta \cdot c \oplus \gamma \cdot k$, where "$\cdot$" denotes the usual inner product, has a probability different from 1/2 of being null. The more distant from 1/2 the probability is, the more efficient is the attack. The

2 "S" for "Substitution".

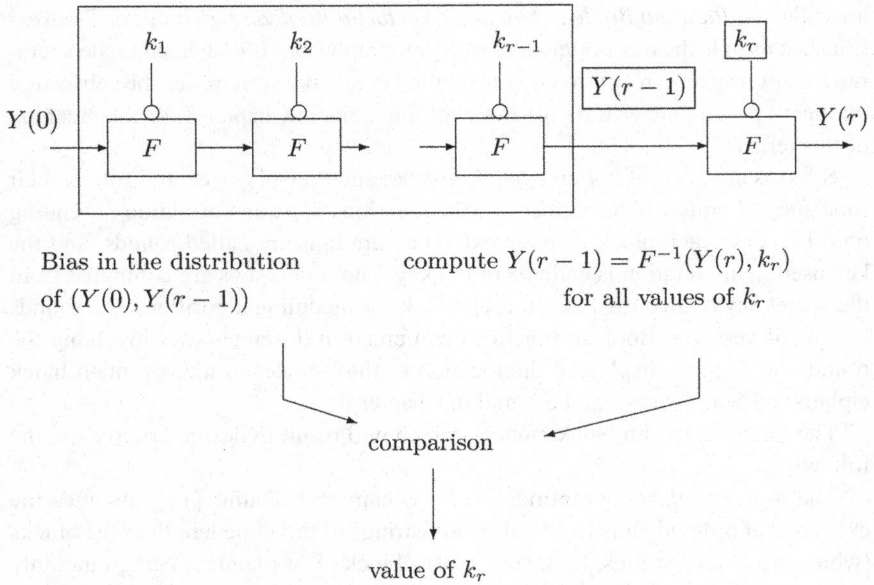

Figure 9.1. Last round attacks.

related criterion on the S-boxes used in the round functions of the cipher deals with the so-called *component functions*, which are the linear combinations, with non all-zero coefficients, of the coordinate functions of the S-box (their set is the vector space spanned by the coordinate functions, deprived of the null function if the coordinate functions are $\mathbb{F}_2$-linearly independent). The nonlinearities (see definition in Chapter 8 or see the later discussion) of all these component functions must be as high as possible. The design of the AES has been partly founded on studies (by K. Nyberg and others) of the notions of nonlinearity (for the resistance to linear attacks) and differential uniformity (for the resistance to differential attacks). This has allowed the AES to use S-boxes working on bytes (it would not have been possible to find a good 8-bit-to-8-bit S-box by a computer search as had been done for the 6-bit-to-4-bit S-boxes of the DES).

The *higher order differential attack* [115, 109] exploits the fact that the algebraic degree of the S-box F is low, or more generally that there exists a low-dimensional vector subspace V of $\mathbb{F}_2^n$ such that the function $D_V F(x) = \sum_{v \in V} F(x + v)$ is constant.

The *interpolation attack* [101] is efficient when the degree of the univariate polynomial representation of the S-box over $\mathbb{F}_{2^n}$ – see the next section – is low or when the distance of the S-box to the set of low univariate degree functions is small.

Algebraic attacks also exist on block ciphers (see, *e.g.*, [68]), exploiting the existence of multivariate equations involving the input to the S-box and its output (an example of such equation is $x^2y = x$ in the case of the AES), but their efficiency has to be more precisely studied: the number of variables in the resulting system of equations, which equals the global number of data bits and of key bits in all rounds of the cipher, is much larger than for stream ciphers, and the resulting systems of equations are not as overdefined as for stream ciphers. However, the AES allowing bilinear relations between the input and the output bits to the S-boxes,[3] this may represent a thread.

In the *pseudorandom generators* of *stream ciphers*, (n, m)-functions can be used to combine the outputs to n linear feedback shift registers (LFSRs), or to filter the content of a single one, then generating m bits at each clock cycle instead of only one, which increases the speed of the cipher (but risks decreasing its robustness). The attacks, described in Chapter 8, are obviously also efficient on these kinds of ciphers. They are in fact often more efficient (see Subsection 9.3.3).

9.2 Generalities on Vectorial Boolean Functions

9.2.1 The Walsh Transform

We shall call *Walsh transform* of an (n, m)-function F the function which maps any ordered pair $(u, v) \in \mathbb{F}_2^n \times \mathbb{F}_2^m$ to the value at u of the Walsh transform of the component[4] function $v \cdot F$, that is, $\sum_{x \in \mathbb{F}_2^n} (-1)^{v \cdot F(x) \oplus u \cdot x}$.

If we denote by G_F the graph $\{(x, y) \in \mathbb{F}_2^n \times \mathbb{F}_2^m / y = F(x)\}$ of F, and by 1_{G_F} its indicator (taking value 1 on G_F and 0 outside), then we have $\sum_{x \in \mathbb{F}_2^n} (-1)^{v \cdot F(x) \oplus u \cdot x} = \widehat{1_{G_F}}(u, v)$, where $\widehat{1_{G_F}}$ is the Fourier transform of the Boolean function 1_{G_F} (see the definition of the Fourier transform in the previous chapter). This observation gives more insight into the nature of the Walsh transform, and moreover, it gives a convenient notation for denoting it.

Observation. *The Walsh transform of any vectorial function is the Fourier transform of the indicator of its graph.*

There is a simple way of expressing the value of the Walsh transform of the composition of two vectorial functions by means of those of the functions.

Proposition 9.1. *If we write the values of the function $\widehat{1_{G_F}}$ in a $2^m \times 2^n$ matrix (in which the term located at the row indexed by $v \in \mathbb{F}_2^m$ and at the column indexed by $u \in \mathbb{F}_2^n$ equals $\widehat{1_{G_F}}(u, v)$), then the matrix corresponding to the composition $F \circ H$ of F, where H is an (r, n)-function, equals the product (in the same order) of the matrices associated to F and H, divided by 2^n.*

[3] It is possible to avoid such relations when the number of input/output bits is 8, if nonpower S-boxes are used (but this may have a cost in terms of speed).

[4] Properly speaking, we can use the term *component function* only for $v \neq 0$, so we are abusing it here.

Proof. For every $w \in \mathbb{F}_2^r$ and every $v \in \mathbb{F}_2^m$, we have

$$\sum_{u\in\mathbb{F}_2^n} \widehat{1_{G_F}}(u,v)\widehat{1_{G_H}}(w,u) = \sum_{u\in\mathbb{F}_2^n; x\in\mathbb{F}_2^r; y\in\mathbb{F}_2^n} (-1)^{v\cdot F(y)\oplus u\cdot y\oplus u\cdot H(x)\oplus w\cdot x}$$
$$= 2^n \sum_{x\in\mathbb{F}_2^r; y\in\mathbb{F}_2^n \,/\, y=H(x)} (-1)^{v\cdot F(y)\oplus w\cdot x}$$
$$= 2^n \widehat{1_{G_{F\circ H}}}(w,v),$$

because $\sum_{u\in\mathbb{F}_2^n}(-1)^{u\cdot y\oplus u\cdot H(x)}$ equals 2^n if $y = H(x)$ and is null otherwise. ■

Remark. *Because of Proposition 9.1, it could seem more convenient to exchange the positions of u and v in $\widehat{1_{G_F}}(u,v)$, in order to have the row index first. However, it seems to us more natural to respect the order (input, output).*

We shall term the *Walsh spectrum* of F the multiset of all the values of the Walsh transform of F, *that is*, $\sum_{x\in\mathbb{F}_2^n}(-1)^{v\cdot F(x)\oplus u\cdot x}$, where $u \in \mathbb{F}_2^n$, $v \in \mathbb{F}_2^{m*}$ (where $\mathbb{F}_2^{m*} = \mathbb{F}_2^m \setminus \{0\}$). We shall term the *extended Walsh spectrum* of F the multiset of their absolute values, and the *Walsh support* of F the set of those (u, v) such that $\sum_{x\in\mathbb{F}_2^n}(-1)^{v\cdot F(x)\oplus u\cdot x} \neq 0$.

Remark. *We have*

$$\sum_{x\in\mathbb{F}_2^n}(-1)^{v\cdot F(x)\oplus u\cdot x} = \sum_{b\in\mathbb{F}_2^m} \widehat{\varphi_b}(u)(-1)^{v\cdot b} \tag{9.1}$$

where $\widehat{\varphi_b}$ is the discrete Fourier transform of the indicator function φ_b of the pre-image $F^{-1}(b) = \{x \in \mathbb{F}_2^n / F(x) = b\}$, defined by $\varphi_b(x) = 1$ if $F(x) = b$ and $\varphi_b(x) = 0$ otherwise.

9.2.2 The Different Ways of Representing Vectorial Functions

9.2.2.1 The Algebraic Normal Form

The notion of *algebraic normal form* of Boolean functions can easily be extended to (n, m)-functions. Because each coordinate function of such a function F is uniquely represented as a polynomial on n variables, with coefficients in $\mathbb{F}_2$ and in which every variable appears in each monomial with degree 0 or 1, the function F itself is uniquely represented as a polynomial of the same form with coefficients in $\mathbb{F}_2^m$, or more precisely as an element of $\mathbb{F}_2^m[x_1, \ldots, x_n]/(x_1^2 \oplus x, \ldots, x_n^2 \oplus x)$:

$$F(x) = \sum_{I\in\mathcal{P}(N)} a_I \left(\prod_{i\in I} x_i\right) = \sum_{I\in\mathcal{P}(N)} a_I\, x^I, \tag{9.2}$$

where $\mathcal{P}(N)$ denotes the power set of $N = \{1, \ldots, n\}$, and a_I belongs to $\mathbb{F}_2^m$ (according to our convention on the notation for additions, we used $\sum$ to denote the sum in $\mathbb{F}_2^m$, but recall that, coordinate by coordinate, this sum is a $\bigoplus$). This polynomial is again called the algebraic normal form (ANF) of F. Keeping the ith coordinate of each coefficient in this expression gives back the ANF of the ith coordinate function of F. Moreover, according to the relations recalled in Chapter 8, a_I equals $\sum_{x\in\mathbb{F}_2^n/\ supp(x)\subseteq I} F(x)$ (this sum being calculated in $\mathbb{F}_2^m$) and conversely, we have $F(x) = \sum_{I\subseteq supp(x)} a_I$.

The *algebraic degree* of the function is by definition the global degree of its ANF: $d^\circ F = \max\{|I|/\ a_I \neq (0,\ldots,0);\ I \in \mathcal{P}(N)\}$. It therefore equals the maximal algebraic degree of the coordinate functions of F. It also equals the maximal algebraic degree of the component functions of F. It is a right and left *affine invariant* (that is, its value does not change when we compose F, on the right or on the left, by an affine automorphism). Another notion of degree is also relevant to cryptography (and is also affine invariant): the minimum algebraic degree of all the component functions[5] of F, often called the *minimum degree*.

9.2.2.2 The Representation as a Univariate Polynomial over $\mathbb{F}_{2^n}$

A second representation of (n, m)-functions exists when $m = n$: we endow $\mathbb{F}_2^n$ with the structure of the field $\mathbb{F}_{2^n}$, as explained in Chapter 8 (see "The Trace Representation" in Subsection 8.2.1); any (n, n)-function F then admits a unique *univariate polynomial representation* over $\mathbb{F}_{2^n}$, of degree at most $2^n - 1$:

$$F(x) = \sum_{j=0}^{2^n-1} \delta_j x^j\ ,\quad \delta_j \in \mathbb{F}_{2^n}\ . \tag{9.3}$$

Indeed, the mapping that maps any such polynomial to the corresponding (n, n)-function is $\mathbb{F}_{2^n}$-linear and has kernel $\{0\}$, because a nonzero univariate equation of degree at most $2^n - 1$ over a field cannot have more than $2^n - 1$ solutions. The dimensions of the vector spaces over $\mathbb{F}_{2^n}$ of, respectively, all such polynomials and all (n, n)-functions both being equal to 2^n, this mapping is bijective. Note that the univariate representation (9.3) of F can be obtained by expanding and simplifying the expression

$$\sum_{a\in\mathbb{F}_{2^n}} F(a)(1 + (x + a)^{2^n-1}).$$

The way to obtain the ANF from this univariate polynomial is similar to the case of Boolean functions seen in the previous chapter. We recall it for self-completeness: for every binary vector $x \in \mathbb{F}_2^n$, we can also denote by x the element $\sum_{i=1}^n x_i\alpha_i$ of $\mathbb{F}_{2^n}$, where $(\alpha_1,\ldots,\alpha_n)$ is a basis of the $\mathbb{F}_2$-vector space $\mathbb{F}_{2^n}$. Let us write the *binary expansion* of every integer $j \in [0; 2^n - 1]$: $\sum_{s=0}^{n-1} j_s 2^s$, $j_s \in \{0, 1\}$.

[5] Not just the coordinate functions; the notion would then not be affine invariant.

We have

$$F(x) = \sum_{j=0}^{2^n-1} \delta_j \left(\sum_{i=1}^{n} x_i \alpha_i\right)^j$$
$$= \sum_{j=0}^{2^n-1} \delta_j \left(\sum_{i=1}^{n} x_i \alpha_i\right)^{\sum_{s=0}^{n-1} j_s 2^s}$$
$$= \sum_{j=0}^{2^n-1} \delta_j \prod_{s=0}^{n-1} \left(\sum_{i=1}^{n} x_i \alpha_i^{2^s}\right)^{j_s}$$

because the mapping $x \to x^2$ is $\mathbb{F}_2$-linear over $\mathbb{F}_{2^n}$ and $x_i \in \mathbb{F}_2$. Expanding these last products, simplifying, and decomposing again over the basis $(\alpha_1, \ldots, \alpha_n)$ gives the ANF of F.

It is then possible to read the algebraic degree of F directly on the univariate polynomial representation: let us denote by $w_2(j)$ the number of nonzero coefficients j_s in the binary expansion $\sum_{s=o}^{n-1} j_s 2^s$ of j, that is, $w_2(j) = \sum_{s=0}^{n-1} j_s$. The number $w_2(j)$ is called the *2-weight* of j. Then, the function F has algebraic degree $\max_{j=0,\ldots,2^n-1/\,\delta_j \neq 0} w_2(j)$. Indeed, according to the foregoing equalities, the algebraic degree of F is clearly bounded above by this number, and it cannot be strictly smaller, because the number $2^n \sum_{i=0}^{d} \binom{n}{i}$ of those (n, n)-functions of algebraic degrees at most d equals the number of those univariate polynomials $\sum_{j=0}^{2^n-1} \delta_j x^j$, $\delta_j \in \mathbb{F}_{2^n}$, such that $\max_{j=0,\ldots,2^n-1/\,\delta_j \neq 0} w_2(j) \leq d$.

In particular, F is $\mathbb{F}_2$-linear (resp. affine) if and only if $F(x)$ is a *linearized polynomial* over $\mathbb{F}_{2^n}$: $\sum_{j=0}^{n-1} \delta_j x^{2^j}$, $\delta_j \in \mathbb{F}_{2^n}$ (resp. a linearized polynomial plus a constant).

If m is a divisor of n, then any (n, m)-function F can be viewed as a function from $\mathbb{F}_{2^n}$ to itself, because $\mathbb{F}_{2^m}$ is a subfield of $\mathbb{F}_{2^n}$. Hence, the function admits a univariate polynomial representation. Note that this unique polynomial can be represented in the form $tr_{n/m}(\sum_{j=0}^{2^n-1} \delta_j x^j)$, where $tr_{n/m}(x) = x + x^{2^m} + x^{2^{2m}} + x^{2^{3m}} + \cdots + x^{2^{n-m}}$ is the trace function from $\mathbb{F}_{2^n}$ to $\mathbb{F}_{2^m}$. Indeed, there exists a function G from $\mathbb{F}_{2^n}$ to $\mathbb{F}_{2^n}$ such that F equals $tr_{n/m} \circ G$ (for instance, $G(x) = \lambda F(x)$, where $tr_{n/m}(\lambda) = 1$). But there is no uniqueness of G in this representation.

9.2.2.3 *The Multidimensional Walsh Transform*

K. Nyberg defined in [132] a polynomial representation, called the *multidimensional Walsh transform*. Let us define

$$\mathcal{W}(F)(z_1, \ldots, z_m) = \sum_{x \in \mathbb{F}_2^n} \prod_{j=1}^{m} z_j^{f_j(x)} \in \mathbb{Z}[z_1, \ldots, z_m]/(z_1^2 - 1, \ldots, z_m^2 - 1),$$

where $f_1, \ldots, f_m$ are the coordinate functions of F. The multidimensional Walsh transform maps every linear (n, m)-function L to the polynomial $\mathcal{W}(F + L)$

$(z_1, \ldots, z_m)$. This is a representation with uniqueness of F, because for every L, the knowledge of $\mathcal{W}(F + L)$ is equivalent to that of the evaluation of $\mathcal{W}(F + L)$ at $(\chi_1, \ldots, \chi_m)$ for every choice of χ_j, $j = 1, \ldots, m$, in the set $\{-1, 1\}$ of roots of the polynomial $z_j^2 - 1$. For such a choice, let us define the vector $v \in \mathbb{F}_2^m$ by $v_j = 1$ if $\chi_j = -1$ and $v_j = 0$ otherwise. For every $j = 1, \ldots, m$, let us denote by a_j the vector of $\mathbb{F}_2^n$ such that the jth coordinate of $L(x)$ equals $a_j \cdot x$. We then denote by u the vector $\sum_{j=1}^m v_j a_j \in \mathbb{F}_2^n$. Then this evaluation equals $\sum_{x \in \mathbb{F}_2^n} (-1)^{v \cdot F(x) \oplus u \cdot x}$. We see that the correspondence between the multidimensional Walsh transform and the Walsh transform is the correspondence between a multivariate polynomial of $\mathbb{Z}[z_1, \ldots, z_m]/(z_1^2 - 1, \ldots, z_m^2 - 1)$ and its evaluation over $\{(z_1, \ldots, z_m) \in \mathbb{Z}^m \,/\, z_1^2 - 1 = \cdots = z_m^2 - 1 = 0\} = \{-1, 1\}^m$. Consequently, the multidimensional Walsh transform satisfies a relation equivalent to Parseval's relation (see [132]).

9.2.3 Balanced Functions

As for Boolean functions, balancedness plays an important role for vectorial Boolean functions in cryptography. An (n, m)-function F is called *balanced* if it takes every value of $\mathbb{F}_2^m$ the same number 2^{n-m} of times. By definition, F is balanced if every function φ_b has Hamming weight 2^{n-m}.

Obviously, the balanced (n, n)-functions are the permutations on $\mathbb{F}_2^n$.

9.2.3.1 Characterization through the Component Functions

The balanced S-boxes (and among them, the permutations) can be nicely characterized by the balancedness of their component functions.

Proposition 9.2. [120] *An (n, m)-function is balanced if and only if its component functions are balanced, that is, if and only if, for every nonzero $v \in \mathbb{F}_2^m$, the Boolean function $v \cdot F$ is balanced.*

Proof. The relation

$$\sum_{v \in \mathbb{F}_2^m} (-1)^{v \cdot (F(x)+b)} = \begin{cases} 2^m \text{ if } F(x) = b \\ 0 \text{ otherwise} \end{cases} = 2^m \, \varphi_b(x), \tag{9.4}$$

is valid for every (n, m)-function F, every $x \in \mathbb{F}_2^n$ and every $b \in \mathbb{F}_2^m$, since the function $v \mapsto v \cdot (F(x) + b)$ being linear, it is either balanced or null. Thus:

$$\sum_{x \in \mathbb{F}_2^n; v \in \mathbb{F}_2^m} (-1)^{v \cdot (F(x)+b)} = 2^m \, |F^{-1}(b)| = 2^m \, w_H(\varphi_b), \tag{9.5}$$

where w_H denotes the Hamming weight as in the previous chapter. Hence, the discrete Fourier transform of the function $v \mapsto \sum_{x \in \mathbb{F}_2^n} (-1)^{v \cdot F(x)}$ equals the function $b \mapsto 2^m \, |F^{-1}(b)|$. We know (see the previous chapter) that a pseudo-Boolean function has constant Fourier transform if and only if it is null at every nonzero vector.

We deduce that F is balanced if and only if the function $v \mapsto \sum_{x\in\mathbb{F}_2^n}(-1)^{v\cdot F(x)}$ is null on $\mathbb{F}_2^{m*}$. ∎

9.2.4 Generalizations to Vectorial Functions of Notions on Boolean Functions

The most important notion on Boolean functions is the nonlinearity. We devote Section 9.3 to its generalization to S-boxes. We also devote a section (9.4) to the notion of resiliency of vectorial functions.

9.2.4.1 Covering Sequences

The notion of covering sequence of a balanced Boolean function has been generalized to vectorial functions, and the properties of this generalization have been studied in [58].

9.2.4.2 Algebraic Immunity

The notion of *algebraic immunity* of S-boxes has been studied in [1, 2]. As recalled in the introduction, the existence of multivariate relations of low degrees between the input bits and the output bits may be exploited in algebraic attacks [68] (but contrary to the case of stream ciphers, the system of equations is generally not overdefined). Several notions of algebraic immunity of an S-box F have been related to these attacks. We first recall the definition of *annihilator*, and we give the definition of the algebraic immunity of a set.

Definition 9.1. *We call* annihilator *of a subset E of $\mathbb{F}_2^n$ any n-variable Boolean function vanishing on E. We call algebraic immunity of E, and we denote by $AI(E)$, the minimum algebraic degree of all the nonzero annihilators of E.*

The algebraic immunity of a Boolean function f (see the previous chapter) equals by definition $\min(AI(f^{-1}(0)), AI(f^{-1}(1)))$.

The first generalization of algebraic immunity to S-boxes is its direct extension.

Definition 9.2. *The* basic algebraic immunity $AI(F)$ *of any (n, m)-function F is the minimum algebraic immunity of all the pre-images $F^{-1}(z)$ of elements z of $\mathbb{F}_2^m$ by F.*

Note that $AI(F)$ also equals the minimum algebraic immunity of all the indicators φ_z of the preimages $F^{-1}(z)$ because, the algebraic immunity being a nondecreasing function over sets, we have for every $z \in \mathbb{F}_2^m$

$$AI(\mathbb{F}_2^n \setminus F^{-1}(z)) \geq AI(F^{-1}(z')), \ \forall z \neq z'.$$

This notion is of interest only for sufficiently small values of m (for instance, for S-boxes used in stream ciphers); see the later discussion. A second notion of algebraic immunity of S-boxes, more relevant when m is comparable to n (which is the case of S-boxes used in block ciphers) has been called the *graph algebraic immunity* and is defined as follows.

Definition 9.3. *The graph algebraic immunity of any* (n,m)*-function* F *is the algebraic immunity of the graph* $\{(x, F(x));\ x \in \mathbb{F}_2^n\}$ *of the S-box.*

This second notion will be denoted by $AI_{gr}(F)$.

Two other notions have been studied in [2], but it is proved in [119] that they are in fact different expressions for the same $AI(F)$ and $AI_{gr}(F)$.

A third notion, which we shall call the *component algebraic immunity*, also seems natural.

Definition 9.4. *The component algebraic immunity of any* (n,m)*-function* F *is the minimal algebraic immunity of the component functions* $v \cdot F$ ($v \neq 0$ *in* $\mathbb{F}_2^m$) *of the S-box. We shall denote it by* $AI_{comp}(F)$.

Properties. It has been observed in [1] that, for any (n,m)-function F, we have $AI(F) \leq AI_{gr}(F) \leq AI(F) + m$. The left-hand-side inequality is straightforward (by restricting an annihilator of the graph to a value of y such that the annihilator does not vanish for every x), and the right-hand-side inequality comes from the fact that, because there exists z and a nonzero annihilator $g(x)$ of $F^{-1}(z)$ of algebraic degree $AI(F)$, the function $g(x)\prod_{i=1}^{m}(y_j \oplus z_j \oplus 1)$ is an annihilator of algebraic degree $AI(F) + m$ of the graph of F.

It has been also observed in [1] that, denoting by d the smallest integer such that $\sum_{i=0}^{d}\binom{n}{i} > 2^{n-m}$, we have $AI(F) \leq d$ (indeed, there is at least one z such that $|F^{-1}(z)| \leq 2^{n-m}$, the annihilators of $F^{-1}(z)$ are the solutions of $|F^{-1}(z)|$ linear equations in $\sum_{i=0}^{d}\binom{n}{i}$ unknowns – which are the coefficients of the ANF of an unknown annihilator of degree at most d – and the number of equations being strictly smaller than the number of unknowns, the system must have nontrivial solutions). It has been proved in [88] (among other results) that this bound is tight. Note that it shows that the basic algebraic immunity has no relevance when m is not small enough: we need $m \leq n - \log_2(n+1)$ for $AI(F)$ being possibly greater than 1; more generally, we know (see [121], page 310) that $\sum_{i=0}^{d}\binom{n}{i} \geq \frac{2^{nH_2(d/n)}}{\sqrt{8d(1-d/n)}}$, where $H_2(x) = -x\log_2(x) - (1-x)\log_2(1-x)$; hence, for $AI(F)$ being possibly greater than a number k, we must have $m \leq n\,(1 - H_2(k/n)) + \frac{1}{2}\,(3 + \log_2(k(1-k/n)))$.

Finally, it has also been proved in [1] that, denoting by D the smallest integer such that $\sum_{i=0}^{D}\binom{n+m}{i} > 2^n$, we have $AI_{gr}(F) \leq D$ (the proof is similar, considering annihilators in $n+m$ variables – the input coordinates and the output coordinates – of the graph), but it is not known whether this bound is tight (it is shown in [1] that it is tight for $n \leq 14$ and partially for $n = 15$).

Because the algebraic immunity of any Boolean function is bounded above by its algebraic degree, the component algebraic immunity of any vectorial function is bounded above by its minimum degree and therefore by its algebraic degree:

$$AI_{comp}(F) \leq d^\circ F.$$

We also have:

$$AI_{comp}(F) \geq AI(F),$$

because $AI_{comp}(F)$ equaling the algebraic immunity of the Boolean function $v \cdot F$ for some $v \neq 0$, it equals $AI(F^{-1}(H))$ for some affine hyperplane H of $\mathbb{F}_2^m$, and AI is a nondecreasing function over sets. We have

$$AI_{comp}(F) \geq AI_{gr}(F) - 1$$

because of the following:

- If g is a nonzero annihilator of $v \cdot F$, $v \neq 0$, then the product $h(x, y) = g(x)\,(v \cdot y)$ is a nonzero annihilator of the graph of F.
- If g is a nonzero annihilator of $v \cdot F \oplus 1$, then $h(x, y) = g(x)\,(v \cdot y) \oplus g(x)$ is a nonzero annihilator of the graph of F.

9.3 Highly Nonlinear Vectorial Boolean Functions

9.3.1 Nonlinearity of S-Boxes in Block Ciphers

A generalization to (n, m)-functions of the notion of nonlinearity of Boolean functions has been introduced and studied by Nyberg [127] and further studied by Chabaud and Vaudenay [60].

Definition 9.5. *The* nonlinearity $nl(F)$ *of an* (n, m)*-function* F *is the minimum nonlinearity of all the component functions* $x \in \mathbb{F}_2^n \mapsto v \cdot F(x)$, $v \in \mathbb{F}_2^m$, $v \neq 0$.

In other words, $nl(F)$ equals the minimum Hamming distance between all the component functions of F and all affine functions on n variables. As we saw in the introduction, this generalization quantifies the level of resistance of the S-box to the linear attack.

The nonlinearity of S-boxes is clearly a right and left affine invariant (that is, it does not change when we compose F by affine automorphisms), and the nonlinearity of an S-box F does not change if we add to F an affine function. Moreover, if A is a surjective linear (or affine) function from $\mathbb{F}_2^p$ (where p is some positive integer) into $\mathbb{F}_2^n$, then it is easily shown that $nl(F \circ A) = 2^{p-n} nl(F)$, since by affine invariance, we can assume without loss of generality that A is a projection.

According to the equality relating the nonlinearity of a Boolean function to the maximal magnitude of its Walsh transform, we have

$$nl(F) = 2^{n-1} - \frac{1}{2} \max_{v\in\mathbb{F}_2^{m*};\ u\in\mathbb{F}_2^n} \left| \sum_{x\in\mathbb{F}_2^n} (-1)^{v\cdot F(x)\oplus u\cdot x} \right| \tag{9.6}$$

$$= 2^{n-1} - \frac{1}{2} \max_{v\in\mathbb{F}_2^{m*};\ u\in\mathbb{F}_2^n} \left| \widehat{1_{G_F}}(u,v) \right| .$$

Note that "$\max_{v\in\mathbb{F}_2^{m*};\ u\in\mathbb{F}_2^n}$" can be replaced by "$\max_{(u,v)\in\mathbb{F}_2^n\times\mathbb{F}_2^m;(u,v)\neq(0,0)}$," because we have $\sum_{x\in\mathbb{F}_2^n}(-1)^{u\cdot x} = 0$ for every nonzero u. Hence, if $n = m$ and if F is a permutation, then F and its inverse F^{-1} have the same nonlinearity (change the variable x into $F^{-1}(x)$).

Relation with Linear Codes. As observed in [52, 147], there is a relationship between the maximal possible nonlinearity of (n, m)-functions and the possible parameters of the linear supercodes of the Reed-Muller code of order 1. Let C be a linear $[2^n, K, D]$ binary code including the Reed-Muller code $RM(1, n)$ as a subcode. Let $(b_1, \ldots, b_K)$ be a basis of C completing a basis $(b_1, \ldots, b_{n+1})$ of $RM(1, n)$. The n-variable Boolean functions corresponding to the vectors $b_{n+2}, \ldots, b_K$ are the coordinate functions of an $(n, K - n - 1)$-function whose nonlinearity is D. Conversely, if $D > 0$ is the nonlinearity of some (n, m)-function, then the linear code equal to the union of the cosets $v \cdot F + RM(1, n)$, where v ranges over $\mathbb{F}_2^m$, has parameters $[2^n, n + m + 1, D]$. Existence and nonexistence results[6] on highly nonlinear vectorial functions are deduced in [147], and upper bounds on the nonlinearity of (n, m)-functions are derived in [54].

9.3.1.1 The Covering Radius Bound; Bent/Perfect Nonlinear Functions

The covering radius bound being valid for every n-variable Boolean function (see the previous chapter), it is *a fortiori* valid for every (n, m)-function:

$$nl(F) \leq 2^{n-1} - 2^{n/2-1}. \tag{9.7}$$

Definition 9.6. *An* (n, m) *function is called* bent *if it achieves the covering radius bound* (9.7) *with equality.*

The notion of bent vectorial function is invariant under composition on the left and on the right by affine automorphisms and by addition of affine functions. Clearly, an (n, m)-function is bent if and only if all of the component functions $v \cdot F$, $v \neq 0$ of F are bent (*i.e.*, achieve the same bound[7]). Hence, the algebraic degree of any bent (n, m)-function is at most $n/2$. Note also that, because any n-variable Boolean function f is bent if and only if all of its derivatives $D_a f(x) = f(x) \oplus f(x + a)$, $a \neq 0$, are balanced, an (n, m)-function F is bent

[6] Using the linear programming bound due to Delsarte.

[7] In other words, the existence of a bent (n, m)-function is equivalent to the existence of an m-dimensional vector space of n-variable Boolean bent functions.

if and only if, for every $v \in \mathbb{F}_2^m$, $v \neq 0$, and every $a \in \mathbb{F}_2^n$, $a \neq 0$, the function $v \cdot (F(x) + F(x+a))$ is balanced. According to Proposition 9.2, this implies the following.

Proposition 9.3. *An (n, m)-function is bent if and only if all of its* derivatives *$D_aF(x) = F(x) + F(x+a)$, $a \in \mathbb{F}_2^{n*}$, are balanced.*

For this reason, bent functions are also called *perfect nonlinear*[8]; they then also contribute to an optimum resistance to the differential attack (see introduction) of those cryptosystems in which they are involved (but they are not balanced). They can be used to design *authentication schemes* (or codes); see [61].

Thanks to the observations made in Subsection 9.2.2 (where we saw that the evaluation of the multidimensional Walsh transform corresponds in fact to the evaluation of the Walsh transform), it is a simple matter to characterize the bent functions as those functions whose squared expression of the multidimensional Walsh transform at L is the same for every L.

Note that, according to the results recalled in Chapter 8, if a bent (n, m)-function F is normal in the sense that it is null on (say) an $n/2$-dimensional vector space E, then F is balanced on any translate of E. Indeed, for every $v \neq 0$ in $\mathbb{F}_2^m$ and every $u \in \mathbb{F}_2^n \setminus E$, the function $v \cdot F$ is balanced on $u + E$.

Existence of Bent (n, m)-Functions. Since bent n-variable Boolean functions exist only if n is even, bent (n, m)-functions exist only under this same hypothesis. But, as shown by Nyberg in [126], this condition is not sufficient for the existence of bent (n, m)-functions. Indeed, we have seen in Relation (9.5) that, for every (n, m)-function F and any element $b \in \mathbb{F}_2^m$, the size of $F^{-1}(b)$ is equal to $2^{-m} \sum_{x \in \mathbb{F}_2^n; v \in \mathbb{F}_2^m} (-1)^{v \cdot (F(x)+b)}$. Assuming that F is bent and denoting, for every $v \in \mathbb{F}_2^{n*}$, by $\widetilde{v \cdot F}$ the dual of the bent Boolean function $x \mapsto v \cdot F(x)$, we have, by definition: $\sum_{x \in \mathbb{F}_2^n} (-1)^{v \cdot F(x)} = 2^{n/2} (-1)^{\widetilde{v \cdot F}(0)}$. The size of $F^{-1}(b)$ equals then $2^{n-m} + 2^{n/2-m} \sum_{v \in \mathbb{F}_2^{n*}} (-1)^{\widetilde{v \cdot F}(0) \oplus v \cdot b}$. Because the sum $\sum_{v \in \mathbb{F}_2^{n*}} (-1)^{\widetilde{v \cdot F}(0) \oplus v \cdot b}$ has an odd value ($\mathbb{F}_2^{n*}$ having an odd size), we deduce that, if $m \leq n$ then $2^{n/2-m}$ must be an integer. And it is also easily shown that $m > n$ is impossible. Hence the following proposition.

Proposition 9.4. *Bent (n, m)-functions exist only if n is even and $m \leq n/2$.*

We shall see later that, for every ordered pair (n, m) satisfying this condition, bent functions do exist.

Open problem: Find a better bound than the covering radius bound for:

- n odd and $m < n$ (we shall see that for $m \geq n$, the Sidelnikov-Chabaud-Vaudenay bound, and other bounds if m is large enough, are better);
- n even and $n/2 < m < n$ (idem).

[8] We shall see that perfect nonlinear (n, n)-functions do not exist; but they do exist in other characteristics than 2 (see, e.g., [53]); they are then often called *planar*.

Primary Constructions of Bent Functions. The two main classes of bent Boolean functions described in Chapter 8 lead to two classes of bent (n, m)-functions (this was first observed by Nyberg in [126]). We endow $\mathbb{F}_2^{n/2}$ with the structure of the field $\mathbb{F}_{2^{n/2}}$. We identify $\mathbb{F}_2^n$ with $\mathbb{F}_{2^{n/2}} \times \mathbb{F}_{2^{n/2}}$.

- Let us define $F(x, y) = L(x\,\pi(y)) + H(y)$, where the product $x\,\pi(y)$ is calculated in $\mathbb{F}_{2^{n/2}}$, where L is any linear or affine mapping from $\mathbb{F}_{2^{n/2}}$ onto $\mathbb{F}_2^m$, π is any permutation of $\mathbb{F}_{2^{n/2}}$ and H is any $(n/2, m)$-function. This gives a bent function that we shall call *strict Maiorana-McFarland's bent* (n, m)-function. More generally, we obtain bent functions (that we can call *general Maiorana-McFarland's bent* (n, m)-functions) by taking for $F = (f_1, \ldots, f_m)$ any (n, m)-function such that, for every $v \in \mathbb{F}_2^{m*}$, the Boolean function $v \cdot F = v_1 f_1 \oplus \cdots \oplus v_m f_m$ belongs, up to linear equivalence, to the original Maiorana-McFarland class of bent functions. The function $L(x\,\pi(y)) + H(y)$ has this property, since the function $v \cdot L(z)$ being a nonzero linear function, it equals $tr_{\frac{n}{2}}(\lambda\, z)$ for some $\lambda \neq 0$, where $tr_{\frac{n}{2}}(x) = x + x^2 + x^{2^2} + \cdots + x^{2^{\frac{n}{2}-1}}$ is the (absolute) trace function from $\mathbb{F}_{2^{n/2}}$ to $\mathbb{F}_2$.
  An example of general Maiorana-McFarland's bent function is given in [138]: the ith coordinate of this function is defined as $f_i(x, y) = tr_{\frac{n}{2}}(x\,\phi_i(y)) \oplus g_i(y)$, $x, y \in \mathbb{F}_{2^{n/2}}$, where g_i is any Boolean function on $\mathbb{F}_{2^{n/2}}$ and where
  $$\phi_i(y) = \begin{cases} 0 \text{ if } y = 0 \\ \alpha^{dec(y)+i-1} \text{ otherwise} \end{cases},$$
  where α is a primitive element of $\mathbb{F}_{2^{n/2}}$ and $dec(y) = 2^{n/2-1}y_1 + 2^{n/2-2}y_2 + \cdots + y_{n/2}$. This function belongs to the general Maiorana-McFarland class of bent functions because the mapping
  $$y \to \begin{cases} 0 \text{ if } y = 0 \\ \alpha^{dec(y)} \text{ otherwise} \end{cases}$$
  is a permutation from $\mathbb{F}_2^{n/2}$ to $\mathbb{F}_{2^{n/2}}$, and for every $y \neq 0$, we have $\sum_{i=1}^{n/2} v_i \alpha^{dec(y)+i-1} = \alpha^{dec(y)} \sum_{i=1}^{n/2} v_i \alpha^{i-1}$ and $\sum_{i=1}^{n/2} v_i \alpha^{i-1} \neq 0$ when $v \neq 0$.
  Modifications of the Maiorana-McFarland bent functions have been proposed in [129], using the classes $\mathcal{C}$ and $\mathcal{D}$ of bent Boolean functions recalled in the previous chapter.
- Defining $F(x, y) = G(xy^{2^n-2}) = G(\frac{x}{y})$ (with $\frac{x}{y} = 0$ if $y = 0$), where G is a balanced $(n/2, m)$-function, also gives a bent (n, m)-function: for every $v \neq 0$, the function $v \cdot F$ belongs to the class $\mathcal{PS}_{ap}$ of Dillon's functions (seen in the previous chapter), according to Proposition 9.2.

Remark. *The functions above are given as defined over $\mathbb{F}_{2^{n/2}} \times \mathbb{F}_{2^{n/2}}$. In the case they are valued in $\mathbb{F}_{2^{n/2}}$, we may want to see them as functions from $\mathbb{F}_{2^n}$ to itself and wish to express them in the univariate representation. If $n/2$*

is odd, this is quite easy: we have then $\mathbb{F}_{2^{n/2}} \cap \mathbb{F}_4 = \mathbb{F}_2$ and we can choose the basis $(1, w)$ of the 2-dimensional vector space $\mathbb{F}_{2^n}$ over $\mathbb{F}_{2^{n/2}}$, where w is a primitive element of $\mathbb{F}_4$. Then $w^2 = w + 1$ and $w^{2^{n/2}} = w^2$ since $n/2$ is odd. A general element of $\mathbb{F}_{2^n}$ has the form $X = x + wy$ where $x, y \in \mathbb{F}_{2^{n/2}}$ and we have $X^{2^{n/2}} = x + w^2 y = X + y$ and therefore $y = X + X^{2^{n/2}}$, and $x = w^2 X + wX^{2^{n/2}}$. For instance, the univariate representation of the simplest Maiorana-McFarland function, that is the function $(x, y) \to xy$, is $(X + X^{2^{n/2}})(w^2 X + wX^{2^{n/2}})$, that is, up to linear terms: $X^{1+2^{n/2}}$.

- We have already observed that constructing a bent (n, m)-function corresponds to finding an m-dimensional vector space of functions whose nonzero elements are all bent. An example (found by the author in common with G. Leander) of such a construction is the following: let n be divisible by 2 but not by 4. Then $\mathbb{F}_{2^{n/2}}$ consists only of cubes (because $gcd(3, 2^{n/2} - 1) = 1$). If we choose some $w \in \mathbb{F}_{2^n}$ that is not a cube, then all the nonzero elements of the vector space $U = w\,\mathbb{F}_{2^{n/2}}$ are noncubes. Then if $F(X) = X^d$ where $d = 2^i + 1$ (d is called a Gold exponent, see later discussion) or $2^{2i} - 2^i + 1$ (d is then called a Kasami exponent) and $gcd(n, i) = 1$, all the functions $tr_n(vF(X))$, where $v \in U^*$, are bent (see the section on bent functions in Chapter 8). This leads to the bent $(n, n/2)$-functions $X \in \mathbb{F}_{2^n} \to (tr_n(\beta_1 wX^d), \ldots, tr_n(\beta_{n/2} wX^d) \in \mathbb{F}_2^{n/2}$, where $(\beta_1, \ldots, \beta_{n/2})$ is a basis of $\mathbb{F}_{2^{n/2}}$ over $\mathbb{F}_2$. Let us see how these functions can be represented as functions from $\mathbb{F}_{2^n}$ to $\mathbb{F}_{2^{n/2}}$. Let us choose a basis $(\alpha_1, \ldots, \alpha_{n/2})$ of $\mathbb{F}_{2^{n/2}}$ orthogonal to $(\beta_1, \ldots, \beta_{n/2})$, that is, such that $tr_{\frac{n}{2}}(\alpha_i \beta_j) = \delta_{i,j}$. Because the two bases are orthogonal, for every $y \in \mathbb{F}_{2^{n/2}}$, we have $y = \sum_{j=1}^{n/2} \alpha_j tr_{\frac{n}{2}}(\beta_j y)$. For every $X \in \mathbb{F}_{2^n}$, the image of X by the function equals $\sum_{j=1}^{n/2} \alpha_j tr_n(\beta_j wX^d) = \sum_{j=1}^{n/2} \alpha_j tr_{\frac{n}{2}}(\beta_j(wX^d + (wX^d)^{2^{n/2}})) = wX^d + (wX^d)^{2^{n/2}}$. Let us see now how, in the case of the Gold exponent, it can be represented as a function from $\mathbb{F}_{2^{n/2}} \times \mathbb{F}_{2^{n/2}}$ to $\mathbb{F}_{2^{n/2}}$: we express X in the form $x + wy$ where $x, y \in \mathbb{F}_{2^{n/2}}$ and if n is not a multiple of 3, we can take for w a primitive element of $\mathbb{F}_4$ (otherwise, all elements of F_4 are cubes and we must then take w outside F_4), for which we have then $w^2 = w + 1$, $w^{2^i} = w^2$ (since i is necessarily odd) and $w^{2^i+1} = w^3 = 1$. We have then $X^d = x^{2^i+1} + wx^{2^i} y + w^2 xy^{2^i} + y^{2^i+1}$ and $wX^d + (wX^d)^{2^{n/2}} = (w + w^2)x^{2^i+1} + (w^2 + w)x^{2^i} y + (w^3 + w^3)xy^{2^i} + (w + w^2)y^{2^i+1} = x^{2^i+1} + x^{2^i} y + y^{2^i+1}$.

  It would be nice to be able to do the same for the Kasami function.

Note that, in the case of the Gold functions x^d with $d = 2^i + 1$, we can extend the construction to the case where i is not coprime with n. The exact condition for $tr_n(vX^d)$ to be bent is then (as we saw in the previous chapter) that $\frac{n}{gcd(i,n)}$ is even and $v \notin \{x^d,\ x \in \mathbb{F}_{2^n}\}$.

Secondary Constructions. Given any bent (n, m)-function F, any chopped function obtained by deleting some coordinates of F (or more generally by composing it on the left with any surjective affine mapping) is obviously still bent.

But there exist other more useful secondary constructions (that is, constructions of new bent functions from known ones). In [47] the following secondary construction of bent Boolean functions is given (recalled in Chapter 8): let r and s be two positive integers with the same parity and such that $r \leq s$, and let $n = r + s$; let ϕ be a mapping from $\mathbb{F}_2^s$ to $\mathbb{F}_2^r$ and g a Boolean function on $\mathbb{F}_2^s$. Let us assume that ϕ is balanced and, for every $a \in \mathbb{F}_2^r$, the set $\phi^{-1}(a)$ is an $(s-r)$-dimensional affine subspace of $\mathbb{F}_2^s$. Let us assume additionally if $r < s$ that the restriction of g to $\phi^{-1}(a)$ (viewed as a Boolean function on $\mathbb{F}_2^{n-2r}$ via an affine isomorphism between $\phi^{-1}(a)$ and this vector space) is bent. Then the function $f_{\phi,g}(x, y) = x \cdot \phi(y) \oplus g(y)$, $x \in \mathbb{F}_2^r$, $y \in \mathbb{F}_2^s$, where "$\cdot$" is an inner product in $\mathbb{F}_2^r$, is bent on $\mathbb{F}_2^n$. This generalizes directly to vectorial functions in the following proposition.

Proposition 9.5. *Let r and s be two positive integers with the same parity and such that $r \leq \frac{s}{3}$. Let ψ be any (balanced) mapping from $\mathbb{F}_2^s$ to $\mathbb{F}_{2^r}$ such that, for every $a \in \mathbb{F}_{2^r}$, the set $\psi^{-1}(a)$ is an $(s-r)$-dimensional affine subspace of $\mathbb{F}_2^s$. Let H be any (s,r)-function whose restriction to $\psi^{-1}(a)$ (viewed as an $(s-r,r)$-function via an affine isomorphism between $\psi^{-1}(a)$ and $\mathbb{F}_2^{s-r}$) is bent for every $a \in \mathbb{F}_{2^r}$. Then the function $F_{\psi,H}(x, y) = x\,\psi(y) + H(y)$, $x \in \mathbb{F}_{2^r}$, $y \in \mathbb{F}_2^s$, is a bent function from $\mathbb{F}_2^{r+s}$ to $\mathbb{F}_{2^r}$.*

Indeed, taking $x \cdot y = tr_r(xy)$ for inner product in $\mathbb{F}_{2^r}$, for every $v \in \mathbb{F}_{2^r}^*$, the function $tr_r(v\,F_{\psi,H}(x, y))$ is bent, according to the result of [47] just recalled, with $\phi(y) = v\,\psi(y)$ and $g(y) = tr_r(v\,H(y))$. The condition $r \leq \frac{s}{3}$, more restrictive than $r \leq s$, is meant so that $r \leq \frac{s-r}{2}$, which is necessary, according to Proposition 9.4, for allowing the restrictions of H to be bent. The condition on ψ being easily satisfied,[9] it is then a simple matter to choose H. Hence, this construction is quite effective (but only for designing bent (n, m)-functions such that $m \leq n/4$, because $r \leq \frac{s}{3}$ is equivalent to $r \leq \frac{r+s}{4}$).

In [46] another secondary construction of bent Boolean functions is given, which is very general and can be adapted to vectorial functions as follows.

Proposition 9.6. *Let r and s be two positive even integers and m a positive integer such that $m \leq r/2$. Let H be a function from $\mathbb{F}_2^n = \mathbb{F}_2^r \times \mathbb{F}_2^s$ to $\mathbb{F}_2^m$. Assume that, for every $y \in \mathbb{F}_2^s$, the function $H_y : x \in \mathbb{F}_2^r \to H(x, y)$ is a bent (r, m)-function. For every nonzero $v \in \mathbb{F}_2^m$ and every $a \in \mathbb{F}_2^r$ and $y \in \mathbb{F}_2^s$, let us denote by $f_{a,v}(y)$ the value at a of the dual of the Boolean function $v \cdot H_y$, that is, the binary value such that $\sum_{x\in\mathbb{F}_2^r}(-1)^{v\cdot H(x,y)\oplus a\cdot x} = 2^{r/2}(-1)^{f_{a,v}(y)}$. Then H is bent if and only if, for every nonzero $v \in \mathbb{F}_2^m$ and every $a \in \mathbb{F}_2^r$, the Boolean function $f_{a,v}$ is bent.*

Indeed, we have, for every nonzero $v \in \mathbb{F}_2^m$ and every $a \in \mathbb{F}_2^r$ and $b \in \mathbb{F}_2^s$,

$$\sum_{\substack{x\in\mathbb{F}_2^r\\ y\in\mathbb{F}_2^s}}(-1)^{v\cdot H(x,y)\oplus a\cdot x\oplus b\cdot y} = 2^{r/2}\sum_{y\in\mathbb{F}_2^s}(-1)^{f_{a,v}(y)\oplus b\cdot y}.$$

[9] Note that it does not make ψ necessarily affine.

An example of application of Proposition 9.6 is when we choose every H_y in the Maiorana-McFarland class: $H_y(x, x') = x\, \pi_y(x') + G_y(x')$, $x, x' \in \mathbb{F}_{2^{r/2}}$, where π_y is bijective for every $y \in \mathbb{F}_2^s$. According to the results recalled in the previous chapter on the duals of Maiorana-McFarland's functions, for every $v \in \mathbb{F}_{2^{r/2}}^*$ and every $a, a' \in \mathbb{F}_{2^{r/2}}$, we then have $f_{(a,a'),v}(y) = tr_{\frac{r}{2}}\left(a'\, \pi_y^{-1}\left(\frac{a}{v}\right) + v\, G_y\left(\pi_y^{-1}\left(\frac{a}{v}\right)\right)\right)$, where $tr_{\frac{r}{2}}$ is the trace function from $\mathbb{F}_{2^{r/2}}$ to $\mathbb{F}_2$. Then H is bent if and only if, for every $v \in \mathbb{F}_{2^{r/2}}^*$ and every $a, a' \in \mathbb{F}_{2^{r/2}}$, the function $y \to tr_{\frac{r}{2}}\left(a'\, \pi_y^{-1}(a) + v\, G_y(\pi_y^{-1}(a))\right)$ is bent on $\mathbb{F}_2^s$. A simple possibility for achieving this is for $s = r/2$ to choose π_y^{-1} such that, for every a, the mapping $y \to \pi_y^{-1}(a)$ is an affine automorphism of $\mathbb{F}_{2^{r/2}}$ (e.g., $\pi_y^{-1}(a) = \pi_y(a) = a + y$) and to choose G_y such that, for every a, the function $y \to G_y(a)$ is bent.

An obvious corollary of Proposition 9.6 is that the so-called *direct sum of bent functions* gives bent functions: we define $H(x, y) = F(x) + G(y)$, where F is any bent (r, m)-function and G any bent (s, m)-function, and we then have $f_{a,v}(y) = v \cdot F(a) \oplus v \cdot G(y)$, which is a bent Boolean function for every a and every $v \neq 0$. Hence, H is bent.

Remark. *The direct sum of bent Boolean functions has been generalized into the indirect sum (see the previous chapter). The direct sum of bent vectorial functions cannot be similarly generalized into a secondary construction of bent vectorial functions, as is. As mentioned in [48], we can identify $\mathbb{F}_2^m$ with $\mathbb{F}_{2^m}$ and define $H(x, y) = F_1(x) + G_1(y) + (F_1(x) + F_2(x))(G_1(y) + G_2(y))$, where F_1 and F_2 are (r, m)-functions and G_1 and G_2 are (s, m)-functions. However, in general, Proposition 9.6 cannot be applied as is. Indeed, taking (as usual) for inner product in $\mathbb{F}_{2^m}$: $u \cdot v = tr_m(uv)$, then $v \cdot H_y(x)$ equals*

$$tr_m(v\, F_1(x)) \oplus tr_m(v\, G_1(y)) + tr_m\left(v\, (F_1(x) + F_2(x))\, (G_1(y) + G_2(y))\right),$$

which does not enter, in general, in the framework of the construction of Boolean functions called "indirect sum." Note that the function $f_{a,v}$ exists under the sufficient condition that, for every nonzero ordered pair $(v, w) \in \mathbb{F}_{2^m} \times \mathbb{F}_{2^m}$, the function $tr_m(v\, F_1(x)) + tr_m(w\, F_2(x))$ is bent (which is equivalent to saying that the $(r, 2m)$-function (F_1, F_2) is bent). There are particular cases where the construction works.

Open problem: Find new constructions of bent (perfect nonlinear) functions.

9.3.1.2 The Sidelnikov-Chabaud-Vaudenay Bound

Because bent (n, m)-functions do not exist if $m > n/2$, this leads to asking the question whether better upper bounds than the covering radius bound can be proved in this case. Such a bound was (in a way) rediscovered[10] by Chabaud and Vaudenay in [60], as shown in the following theorem.

Theorem 9.7. *Let n and m be any positive integers such that $m \geq n-1$. Let F be any (n,m)-function. Then*

$$nl(F) \leq 2^{n-1} - \frac{1}{2}\sqrt{3 \times 2^n - 2 - 2\frac{(2^n-1)(2^{n-1}-1)}{2^m-1}}.$$

Proof. Recall that

$$nl(F) = 2^{n-1} - \frac{1}{2} \max_{v \in \mathbb{F}_2^{m*};\, u \in \mathbb{F}_2^n} \left| \sum_{x \in \mathbb{F}_2^n} (-1)^{v \cdot F(x) \oplus u \cdot x} \right|.$$

We have

$$\max_{\substack{v \in \mathbb{F}_2^{m*} \\ u \in \mathbb{F}_2^n}} \left(\sum_{x \in \mathbb{F}_2^n} (-1)^{v \cdot F(x) \oplus u \cdot x} \right)^2 \geq \frac{\sum_{\substack{v \in \mathbb{F}_2^{m*} \\ u \in \mathbb{F}_2^n}} \left(\sum_{x \in \mathbb{F}_2^n} (-1)^{v \cdot F(x) \oplus u \cdot x} \right)^4}{\sum_{\substack{v \in \mathbb{F}_2^{m*} \\ u \in \mathbb{F}_2^n}} \left(\sum_{x \in \mathbb{F}_2^n} (-1)^{v \cdot F(x) \oplus u \cdot x} \right)^2}. \tag{9.8}$$

Parseval's relation (see the previous chapter) states that, for every $v \in \mathbb{F}_2^m$,

$$\sum_{u \in \mathbb{F}_2^n} \left(\sum_{x \in \mathbb{F}_2^n} (-1)^{v \cdot F(x) \oplus u \cdot x} \right)^2 = 2^{2n}. \tag{9.9}$$

Using the fact (already used in the proof of Proposition 9.2) that any character sum $\sum_{x \in E} (-1)^{\ell(x)}$ associated to a linear function ℓ over any $\mathbb{F}_2$-vector space E is nonzero if and only if ℓ is null on E, we have:

$$\begin{aligned}
&\sum_{v \in \mathbb{F}_2^m,\, u \in \mathbb{F}_2^n} \left(\sum_{x \in \mathbb{F}_2^n} (-1)^{v \cdot F(x) \oplus u \cdot x} \right)^4 \\
&= \sum_{x,y,z,t \in \mathbb{F}_2^n} \left[\sum_{v \in \mathbb{F}_2^m} (-1)^{v \cdot (F(x)+F(y)+F(z)+F(t))} \right] \left[\sum_{u \in \mathbb{F}_2^n} (-1)^{u \cdot (x+y+z+t)} \right] \\
&= 2^{n+m} \left| \left\{ (x,y,z,t) \in \mathbb{F}_2^{4n} / \left\{ \begin{array}{ll} x+y+z+t & = 0 \\ F(x)+F(y)+F(z)+F(t) = 0 \end{array} \right. \right\} \right| \\
&= 2^{n+m} |\{(x,y,z) \in \mathbb{F}_2^{3n} / F(x)+F(y)+F(z)+F(x+y+z) = 0\}| \qquad (9.10) \\
&\geq 2^{n+m} |\{(x,y,z) \in \mathbb{F}_2^{3n} / x = y \text{ or } x = z \text{ or } y = z\}|. \qquad (9.11)
\end{aligned}$$

[10] We write "re-discovered" because a bound on sequences due to Sidelnikov [141] is equivalent to the bound obtained by Chabaud and Vaudenay for power functions, and its proof is in fact valid for all functions.

Clearly, $|\{(x,y,z)/\, x=y \text{ or } x=z \text{ or } y=z\}| = 3\cdot|\{(x,x,y)/\, x,y\in\mathbb{F}_2^n\}| - 2\cdot|\{(x,x,x)/\, x\in\mathbb{F}_2^n\}| = 3\cdot 2^{2n} - 2\cdot 2^n$. Hence, according to Relation (9.8):

$$\max_{v\in\mathbb{F}_2^{m*};\ u\in\mathbb{F}_2^n}\left(\sum_{x\in\mathbb{F}_2^n}(-1)^{v\cdot F(x)\oplus u\cdot x}\right)^2 \geq$$

$$\frac{2^{n+m}(3\cdot 2^{2n} - 2\cdot 2^n) - 2^{4n}}{(2^m-1)\,2^{2n}} = 3\times 2^n - 2 - 2\frac{(2^n-1)(2^{n-1}-1)}{2^m-1},$$

and this gives the desired bound, according to Relation (9.6). ∎

The condition $m \geq n-1$ is assumed in Theorem 9.7 to make nonnegative the expression located under the square root. Note that for $m = n-1$, this *Sidelnikov-Chabaud-Vaudenay bound* coincides with the covering radius bound. For $m \geq n$, it strictly improves upon it. For $m > n$, the square root in it cannot be an integer (see [60]). Hence, the Sidelnikov-Chabaud-Vaudenay bound can be tight only if $n = m$ with n odd. We see in the next subsection that, under this condition, it is actually tight.

Other bounds were obtained in [54] and improve, when m is sufficiently greater than n (which makes them less interesting, cryptographically), upon the covering radius bound and the Sidelnikov-Chabaud-Vaudenay bound (examples are given).

9.3.1.3 Almost Bent and Almost Perfect Nonlinear Functions

Almost Bent Functions

Definition 9.7. ([60].) *The (n,n)-functions F which achieve the bound of Theorem 9.7 with equality – that is, such that $nl(F) = 2^{n-1} - 2^{\frac{n-1}{2}}$ (n odd) – are called* almost bent *(AB).*

Remark. *The term* almost bent *is a little misleading. It gives the feeling that these functions are not quite optimal. But they are. Recall that, according to Nyberg's result (Proposition 9.4), (n,n)-bent functions do not exist.*

According to Inequality (9.8), the AB functions are those (n,n)-functions such that, for every $u, v \in \mathbb{F}_2^n$, $v \neq 0$, the sum $\sum_{x\in\mathbb{F}_2^n}(-1)^{v\cdot F(x)\oplus u\cdot x} = \widehat{1_{G_F}}(u,v)$ equals 0 or $\pm 2^{\frac{n+1}{2}}$ (indeed, the maximum of a sequence of nonnegative integers equals the ratio of the sum of their squares over the sum of their values if and only if these integers take at most one nonzero value). Note that this condition does not depend on the choice of the inner product.

There exists a bound on the algebraic degree of AB functions, similar to the bound for bent functions.

Proposition 9.8. ([52].) *Let F be any (n,n)-function ($n \geq 3$). If F is AB, then the algebraic degree of F is less than or equal to $(n+1)/2$.*

This is a direct consequence of the fact that the Walsh transform of any function $v \cdot F$ is divisible by $2^{\frac{n+1}{2}}$ and the fact, recalled in Chapter 8, that if the Walsh transform values of an n-variable Boolean function are divisible by 2^k, then the algebraic degree of the function is at most $n - k + 1$. Note that the divisibility also plays a role with respect to the algebraic degree of the composition of two vectorial functions: in [45] it was proved that, if the Walsh transform values of a vectorial function $F : \mathbb{F}_2^n \to \mathbb{F}_2^n$ are divisible by 2^ℓ then, for every vectorial function $F' : \mathbb{F}_2^n \to \mathbb{F}_2^n$, the algebraic degree of $F' \circ F$ is at most equal to the algebraic degree of F' plus $n - \ell$. This means that using AB power functions as S-boxes in block ciphers may not be a good idea. Suboptimal functions (such as the multiplicative inverse function discussed later) may be better (as usual in cryptography).

Almost Perfect Nonlinear Functions Inequality (9.11) is an equality if and only if the relation $F(x) + F(y) + F(z) + F(x + y + z) = 0$ can be achieved only when $x = y$ or $x = z$ or $y = z$. There are several equivalent ways of characterizing this property:

- The restriction of F to any two-dimensional flat (*i.e.*, affine subspace) of $\mathbb{F}_2^n$ is nonaffine (indeed, the set $\{x, y, z, x + y + z\}$ is a flat and it is two-dimensional if and only if $x \neq y$ and $x \neq z$ and $y \neq z$. Saying that $F(x) + F(y) + F(z) + F(x + y + z) = 0$ is equivalent to saying that the restriction of F to this flat is affine, because we know that a function F is affine on a flat A if and only if, for every x, y, z in A, we have $F(x + y + z) = F(x) + F(y) + F(z)$).
- For every distinct nonzero (that is, $\mathbb{F}_2$-linearly independent) vectors a and a', the second-order derivative $D_a D_{a'} F(x) = F(x) + F(x + a) + F(x + a') + F(x + a + a')$ takes only nonzero values.
- The equation $F(x) + F(x + a) = F(y) + F(y + a)$ (obtained from $F(x) + F(y) + F(z) + F(x + y + z) = 0$ by denoting $x + z$ by a) can be achieved only for $a = 0$ or $x = y$ or $x = y + a$.
- For every $a \in \mathbb{F}_2^{n*}$ and every $b \in \mathbb{F}_2^n$, the equation $F(x) + F(x + a) = b$ has at most two solutions (that is, zero or two solutions, because if it has one solution x, then it has $x + a$ for a second solution).

Definition 9.8. ([133, 8, 128].) *An (n, n)-function F is called* almost perfect nonlinear *(APN) if, for every $a \in \mathbb{F}_2^{n*}$ and every $b \in \mathbb{F}_2^n$, the equation $F(x) + F(x + a) = b$ has zero or two solutions; that is, equivalently, if the restriction of F to any two-dimensional flat (*i.e., *affine subspace) of $\mathbb{F}_2^n$ is nonaffine.*

Remark. *Here again, the term* almost perfect nonlinear *is a little misleading, giving the feeling that these functions are almost optimal, whereas they are optimal.*

According to the proof of Sidelnikov-Chabaud-Vaudenay's bound given earlier, every AB function is APN (this was first observed by Chabaud and Vaudenay). In fact, this implication can be more precisely changed into a characterization of AB functions (see Proposition 9.9) involving the notion of a plateaued function.

Definition 9.9. *An (n,m)-function is called* plateaued *if, for every nonzero $v \in \mathbb{F}_2^m$, the component function $v \cdot F$ is plateaued, that is, there exists a positive integer λ_v (called the amplitude of the plateaued Boolean function $v \cdot F$) such that the values of its Walsh transform $\sum_{x\in\mathbb{F}_2^n}(-1)^{v\cdot F(x)\oplus u\cdot x}$, $u \in \mathbb{F}_2^n$, all belong to the set $\{0, \pm\lambda_v\}$.*

Then, because of Parseval's relation (9.9), 2^{2n} equals λ_v^2 times the size of the set $\{u \in \mathbb{F}_2^n / \sum_{x\in\mathbb{F}_2^n}(-1)^{v\cdot F(x)\oplus u\cdot x} \neq 0\}$, and λ_v then equals a power of 2 whose exponent is greater than or equal to $n/2$ (because this size is at most 2^n). The extreme case $\lambda_v = 2^{n/2}$ corresponds to the case where $v \cdot F$ is bent. Every *quadratic* function (that is, every function of algebraic degree 2) is plateaued; see Chapter 8.

It was proved in [33] that no power plateaued bijective (n,n)-function exists[11] when n is a power of 2 and in [123] that no such function exists with Walsh spectrum $\{0, \pm 2^{n/2+1}\}$ when n is divisible by 4.

Proposition 9.9. *Every AB function is APN. More precisely, any vectorial function $F : \mathbb{F}_2^n \to \mathbb{F}_2^n$ is AB if and only if F is APN and the functions $v \cdot F$, $v \neq 0$, are plateaued with the same amplitude.*

This comes directly from Relations (9.8) and (9.11). *We shall see later, in Proposition 9.17, that if n is odd, the condition "with the same amplitude" is in fact not necessary.*

Note that, according to Relations (9.10) and (9.11) and to the two lines following them, APN (n,n)-functions F are characterized by the fact that the power sums of degree 4 of the values of their Walsh transform take the minimal value $3 \cdot 2^{4n} - 2 \cdot 2^{3n}$, that is, F is APN if and only if

$$\sum_{v\in\mathbb{F}_2^n, u\in\mathbb{F}_2^n} \left(\sum_{x\in\mathbb{F}_2^n} (-1)^{v\cdot F(x)\oplus u\cdot x} \right)^4 = 3 \cdot 2^{4n} - 2 \cdot 2^{3n} \tag{9.12}$$

or equivalently, replacing $\sum_{u\in\mathbb{F}_2^n}\left(\sum_{x\in\mathbb{F}_2^n}(-1)^{u\cdot x}\right)^4$ by its value 2^{4n} and using Parseval's relation (9.9), we arrive at the following proposition.

Proposition 9.10. *Any (n,n)-function F is APN if and only if*

$$\sum_{\substack{v\in\mathbb{F}_2^{n*} \\ u\in\mathbb{F}_2^n}} \left(\sum_{x\in\mathbb{F}_2^n} (-1)^{v\cdot F(x)\oplus u\cdot x} \right)^2 \left(\left(\sum_{x\in\mathbb{F}_2^n} (-1)^{v\cdot F(x)\oplus u\cdot x} \right)^2 - 2^{n+1} \right) = 0. \tag{9.13}$$

This characterization will have nice consequences in the sequel.

[11] A conjecture by T. Helleseth states that there is no power permutation having three Walsh transform values when n is a power of 2.

Note that, similarly as for the power sum of degree 4, the power sum $\sum_{v\in\mathbb{F}_2^n,u\in\mathbb{F}_2^n}\left(\sum_{x\in\mathbb{F}_2^n}(-1)^{v\cdot F(x)\oplus u\cdot x}\right)^3$ of degree 3 equals

$$2^{2n}\left|\{(x,y)\in\mathbb{F}_2^{2n}/\,F(x)+F(y)+F(x+y)=0\}\right|.$$

Applying (with $z=0$) the property that, for every APN function F, the relation $F(x)+F(y)+F(z)+F(x+y+z)=0$ can be achieved only when $x=y$ or $x=z$ or $y=z$, we then have, for every APN function such that $F(0)=0$,

$$\sum_{v\in\mathbb{F}_2^n,u\in\mathbb{F}_2^n}\left(\sum_{x\in\mathbb{F}_2^n}(-1)^{v\cdot F(x)\oplus u\cdot x}\right)^3=3\cdot 2^{3n}-2\cdot 2^{2n}. \tag{9.14}$$

But this property is not characteristic (except for quadratic functions; see later discussion) of APN functions among those (n,n)-functions such that $F(0)=0$, because it is only characteristic of the fact that $\sum_{x\in E}F(x)\neq 0$ for every two-dimensional vector subspace E of $\mathbb{F}_2^n$.

The APN property is a particular case of a notion introduced by Nyberg [126, 127]: an (n,m)-function F is called *differentially δ-uniform* if, for every nonzero $a\in\mathbb{F}_2^n$ and every $b\in\mathbb{F}_2^m$, the equation $F(x)+F(x+a)=b$ has at most δ solutions. The number δ is then bounded below by 2^{n-m} and equals 2^{n-m} if and only if F is perfect nonlinear. The behavior of δ for general S-boxes has been studied in [146].

The smaller δ is, the better is the contribution of F to a resistance to differential cryptanalysis. When $m=n$, the smallest possible value of δ is 2, because we already saw that if x is a solution of equation $F(x)+F(x+a)=b$, then $x+a$ is also a solution. Hence, APN functions contribute to a maximal resistance to differential cryptanalysis when $m=n$ and AB functions contribute to a maximal resistance to both linear and differential cryptanalyses.

Note that if F is a quadratic (n,n)-function, the equation $F(x)+F(x+a)=b$ is a linear equation. It admits then at most two solutions for every nonzero a and every b if and only if the related homogeneous equation $F(x)+F(x+a)+F(0)+F(a)=0$ admits at most two solutions for every nonzero a. Hence, F is APN if and only if the associated bilinear symmetric $(2n,n)$-function $\varphi_F(x,y)=F(0)+F(x)+F(y)+F(x+y)$ never vanishes when x and y are $\mathbb{F}_2$-linearly independent vectors of $\mathbb{F}_2^n$. For functions of higher degrees, the fact that $\varphi_F(x,y)$ (which is no longer bilinear) never vanishes when x and y are linearly independent is only a necessary condition for APNness.

A subclass of APN functions (and a potential superclass of APN quadratic permutations), called crooked functions, has been considered in [6] and further studied in [32, 71, 111]. All known crooked functions are quadratic. It can be proved [112] that every power crooked function is a Gold function (see the later definition).

Other Characterizations of AB and APN Functions

- A necessary condition dealing with *quadratic terms in the ANF of any APN function* has been observed in [8]. Given any APN function F (quadratic or not), every quadratic term $x_i x_j$ $(1 \leq i < j \leq n)$ must appear with a nonnull coefficient in the algebraic normal form of F. Indeed, we know that the coefficient of any monomial $\prod_{i \in I} x^i$ in the ANF of F equals $a_I = \sum_{x \in \mathbb{F}_2^n / \, supp(x) \subseteq I} F(x)$ (this sum being calculated in $\mathbb{F}_2^n$). Applied for instance to $I = \{n-1, n\}$, this gives $a_I = F(0, \ldots, 0, 0, 0) + F(0, \ldots, 0, 0, 1) + F(0, \ldots, 0, 1, 0) + F(0, \ldots, 0, 1, 1)$, and F being APN, this vector cannot be null. Note that, because the notion of almost perfect nonlinearity is affinely invariant (see later discussion), this condition must be satisfied by all of the functions $L' \circ F \circ L$, where L' and L are affine automorphisms of $\mathbb{F}_2^n$. Extended this way, the condition becomes necessary and sufficient (indeed, for every distinct x, y, z in $\mathbb{F}_2^n$, there exists an affine automorphism L of $\mathbb{F}_2^n$ such that $L(0, \ldots, 0, 0, 0) = x$, $L(0, \ldots, 0, 1, 0) = y$ and $L(0, \ldots, 0, 0, 1) = z$).
- The properties of APNness and ABness can be translated in terms of Boolean functions, as observed in [52].

Proposition 9.11. *Let F be any (n, n)-function. For every $a, b \in \mathbb{F}_2^n$, let $\gamma_F(a, b)$ equal 1 if the equation $F(x) + F(x + a) = b$ admits solutions, with $a \neq 0$. Otherwise, let $\gamma_F(a, b)$ be null. Then, F is APN if and only if γ_F has weight $2^{2n-1} - 2^{n-1}$, and F is AB if and only if γ_F is bent. The dual of γ_F is then the indicator of the Walsh support of F, deprived of* $(0, 0)$.

Proof.

(1) If F is APN, then for every $a \neq 0$, the mapping $x \mapsto F(x) + F(x + a)$ is two-to-one (that is, the size of the preimage of any vector equals 0 or 2). Hence, γ_F has weight $2^{2n-1} - 2^{n-1}$. The converse is also straightforward.

(2) We now assume that F is APN. For every $u, v \in \mathbb{F}_2^n$, replacing $(-1)^{\gamma_F(a,b)}$ by $1 - 2\gamma_F(a, b)$ in the character sum $\sum_{a,b \in \mathbb{F}_2^n} (-1)^{\gamma_F(a,b) \oplus u \cdot a \oplus v \cdot b}$ leads to $\sum_{a,b \in \mathbb{F}_2^n} (-1)^{u \cdot a \oplus v \cdot b} - 2 \sum_{a,b \in \mathbb{F}_2^n} \gamma_F(a, b)(-1)^{u \cdot a \oplus v \cdot b}$. Denoting by δ_0 the Dirac symbol ($\delta_0(u, v) = 1$ if $u = v = 0$ and 0 otherwise), we deduce that the Walsh transform of γ_F equals $2^{2n}\ \delta_0(u, v) - \sum_{x \in \mathbb{F}_2^n, a \in \mathbb{F}_2^{n*}} (-1)^{u \cdot a \oplus v \cdot (F(x) + F(x+a))} = 2^{2n}\ \delta_0(u, v)\ - (\sum_{x,y \in \mathbb{F}_2^n} (-1)^{v \cdot F(x) \oplus v \cdot F(y) \oplus u \cdot x \oplus u \cdot y})$ $+ 2^n = 2^{2n}\ \delta_0(u, v) - (\sum_{x \in \mathbb{F}_2^n} (-1)^{v \cdot F(x) \oplus u \cdot x})^2 + 2^n$. Hence, F is AB if and only if the value of this Walsh transform equals $\pm 2^n$ at every $(u, v) \in \mathbb{F}_2^n \times \mathbb{F}_2^n$, *that is*, if γ_F is bent. Moreover, if γ_F is bent, then for every $(u, v) \neq 0$, we have $\widetilde{\gamma_F}(u, v) = 0$, that is, $\sum_{a,b \in \mathbb{F}_2^n} (-1)^{\gamma_F(a,b) \oplus u \cdot a \oplus v \cdot b} = 2^n$ if and only if $\sum_{x \in \mathbb{F}_2^n} (-1)^{v \cdot F(x) \oplus u \cdot x} = 0$. Hence, the dual of γ_F is the indicator of the Walsh support of F, deprived of $(0, 0)$. ■

- Obviously, an (n, n)-function F is APN if and only if, for every $(a, b) \neq (0, 0)$, the system

$$\begin{cases} x + y & = a \\ F(x) + F(y) = b \end{cases}$$

admits zero or two solutions. As shown by van Dam and Fon-Der-Flaass in [71], it is AB if and only if the system

$$\begin{cases} x + y + z & = a \\ F(x) + F(y) + F(z) = b \end{cases}$$

admits $3 \cdot 2^n - 2$ solutions if $b = F(a)$ and $2^n - 2$ solutions otherwise. This can easily be proved by using the facts that F is AB if and only if, for every $v \in \mathbb{F}_2^{n*}$ and every $u \in \mathbb{F}_2^n$, we have $\left(\sum_{x\in\mathbb{F}_2^n}(-1)^{v\cdot F(x)\oplus u\cdot x}\right)^3 = 2^{n+1}\sum_{x\in\mathbb{F}_2^n}(-1)^{v\cdot F(x)\oplus u\cdot x}$, and that two pseudo-Boolean functions (that is, two functions from $\mathbb{F}_2^n$ to $\mathbb{Z}$) are equal to each other if and only if their discrete Fourier transforms are equal to each other: the value at (a, b) of the Fourier transform of the function of (u, v) equal to $\left(\sum_{x\in\mathbb{F}_2^n}(-1)^{v\cdot F(x)\oplus u\cdot x}\right)^3$ if $v \neq 0$, and to 0 otherwise equals

$$\sum_{\substack{u\in\mathbb{F}_2^n \\ v\in\mathbb{F}_2^n}}\left(\sum_{x\in\mathbb{F}_2^n}(-1)^{v\cdot F(x)\oplus u\cdot x}\right)^3(-1)^{a\cdot u\oplus b\cdot v} - 2^{3n} =$$

$$2^{2n}\left|\left\{(x, y, z) \in \mathbb{F}_2^{3n} \,/\, \begin{cases} x + y + z = a \\ F(x) + F(y) + F(z) = b \end{cases}\right\}\right| - 2^{3n},$$

and the value of the Fourier transform of the function that is equal to $2^{n+1}\sum_{x\in\mathbb{F}_2^n}(-1)^{v\cdot F(x)\oplus u\cdot x}$ if $v \neq 0$, and to 0 otherwise, equals

$$2^{3n+1}\left|\left\{x \in \mathbb{F}_2^n \,/\, \begin{cases} x = a \\ F(x) = b \end{cases}\right\}\right| - 2^{2n+1}.$$

This proves the result. Note that $3 \cdot 2^n - 2$ is the number of triples (x, x, a), (x, a, x) and (a, x, x) where x ranges over $\mathbb{F}_2^n$. Hence the condition when $F(a) = b$ means that these particular triples are the only solutions of the system

$$\begin{cases} x + y + z & = a \\ F(x) + F(y) + F(z) = F(a). \end{cases}$$

This is equivalent to saying that F is APN and we can therefore replace the first condition of van Dam and Fon-Der-Flaass by "F is APN." Denoting $c = F(a) + b$, we have the following.

Proposition 9.12. *Let n be any positive integer and F any APN (n, n)-function. Then F is AB if and only if, for every $c \neq 0$ and every a in $\mathbb{F}_2^n$, the equation $F(x) + F(y) + F(a) + F(x + y + a) = c$ has $2^n - 2$ solutions.*

Let us denote by $\mathcal{A}_2$ the set of two-dimensional flats of $\mathbb{F}_2^n$ and by Φ_F the mapping $A \in \mathcal{A}_2 \to \sum_{x \in A} F(x) \in \mathbb{F}_2^n$. Proposition 9.12 is equivalent to saying that an APN function is AB if and only if, for every $a \in \mathbb{F}_2^n$, the restriction of Φ_F to those flats that contain a is a $(2^{n-1} - 1)/3$-to-1 function. Hence we have the following corollary.

Corollary 9.13. *Any (n, n)-function F is APN if and only if Φ_F is valued in $\mathbb{F}_2^{n*} = \mathbb{F}_2^n \setminus \{0\}$, and F is AB if and only if, additionally, the restriction of $\Phi_F : \mathcal{A}_2 \to \mathbb{F}_2^{n*}$ to those flats that contain a vector a is a balanced function, for every $a \in \mathbb{F}_2^n$.*

Note that, for every APN function F and any two distinct vectors a and a', the restriction of Φ_F to those flats that contain a and a' is injective, because for two such distinct flats $A = \{a, a', x, x + a + a'\}$ and $A' = \{a, a', x', x' + a + a'\}$, we have $\Phi_F(A) + \Phi_F(A') = F(x) + F(x + a + a') + F(x') + F(x' + a + a') = \Phi_F(\{x, x + a + a', x', x' + a + a'\}) \neq 0$. But this restriction of Φ_F cannot be surjective because the number of flats containing a and a' equals $2^{n-1} - 1$, which is less than $2^n - 1$.

Remark. *Other characterizations can be derived with the same method as in the proof of the result of van Dam and Fon-Der-Flaass. For instance, F is AB if and only if, for every $v \in \mathbb{F}_2^{n*}$ and every $u \in \mathbb{F}_2^n$, we have $\left(\sum_{x \in \mathbb{F}_2^n} (-1)^{v \cdot F(x) \oplus u \cdot x}\right)^4 = 2^{n+1} \left(\sum_{x \in \mathbb{F}_2^n} (-1)^{v \cdot F(x) \oplus u \cdot x}\right)^2$. By again applying the Fourier transform and dividing by 2^{2n}, we deduce that F is AB if and only if, for every (a, b), we have*

$$\left|\left\{(x, y, z, t) \in \mathbb{F}_2^{4n} / \begin{cases} x + y + z + t = a \\ F(x) + F(y) + F(z) + F(t) = b \end{cases}\right\}\right| - 2^{2n} =$$

$$2^{n+1} \left|\left\{(x, y) \in \mathbb{F}_2^{2n} / \begin{cases} x + y = a \\ F(x) + F(y) = b \end{cases}\right\}\right| - 2^{n+1}.$$

Hence, F is AB if and only if the system

$$\begin{cases} x + y + z + t & = a \\ F(x) + F(y) + F(z) + F(t) & = b \end{cases}$$

admits $3 \cdot 2^{2n} - 2^{n+1}$ solutions if $a = b = 0$ (this is equivalent to saying that F is APN), $2^{2n} - 2^{n+1}$ solutions if $a = 0$ and $b \neq 0$ (note that this condition corresponds to adding all the conditions of Proposition 9.12 with c fixed to b and with a ranging over $\mathbb{F}_2^n$), and $2^{2n} + 2^{n+2}\gamma_F(a, b) - 2^{n+1}$ solutions if $a \neq 0$ (indeed, F is APN; note that this gives a new property of AB functions).

- A relationship has been observed in [52]) (see also [147, 54]) between the properties, for an (n, n)-function, of being APN or AB and properties of related codes:

Proposition 9.14. *Let F be any (n, n)-function such that $F(0) = 0$. Let H be the matrix*

$$\begin{bmatrix} 1 & \alpha & \alpha^2 & \dots & \alpha^{2^n-2} \\ F(1) & F(\alpha) & F(\alpha^2) & \dots & F(\alpha^{2^n-2}) \end{bmatrix},$$

where α is a primitive element of the field $\mathbb{F}_{2^n}$, and where each symbol stands for the column of its coordinates with respect to a basis of the $\mathbb{F}_2$-vector space $\mathbb{F}_{2^n}$. Let C_F be the linear code admitting H for parity-check matrix. Then, F is APN if and only if C_F has minimum distance 5, and F is AB if and only if $C_F^{\perp}$ (i.e., the code admitting H for generator matrix) has weights $0, 2^{n-1} - 2^{\frac{n-1}{2}}, 2^{n-1}$, and $2^{n-1} + 2^{\frac{n-1}{2}}$.

Proof. Because H contains no zero column, C_F has no codeword of Hamming weight 1, and because all columns of H are distinct vectors, C_F has no codeword of Hamming weight 2. Hence,[12] C_F has minimum distance at least 3. This minimum distance is also at most 5 (this is known; see [52]). The fact that C_F has no codeword of weight 3 or 4 is by definition equivalent to the APNness of F, because a vector $(c_0, c_1, \dots, c_{2^n-2}) \in \mathbb{F}_2^{2^n-1}$ is a codeword if and only if

$$\begin{cases} \sum_{i=0}^{2^n-2} c_i\alpha^i = 0 \\ \sum_{i=0}^{2^n-2} c_i F(\alpha^i) = 0 \end{cases}$$

The inexistence of codewords of weight 3 is then equivalent to the fact that $\sum_{x\in E} F(x) \neq 0$ for every two-dimensional vector subspace E of $\mathbb{F}_{2^n}$, and the inexistence of codewords of weight 4 is equivalent to the fact that $\sum_{x\in A} F(x) \neq 0$ for every two-dimensional flat A not containing 0. The characterization of ABness through the weights of $C_F^{\perp}$ comes directly from the characterization of AB functions by their Walsh transform values, and from the fact that the weight of the Boolean function $v \cdot F(x) \oplus u \cdot x$ equals $2^{n-1} - \frac{1}{2}\widehat{1_{G_F}}(u, v)$. ■

Remark.

(1) Any subcode of dimension $2^n - 1 - 2n$ of the $[2^n - 1, n, 3]$ Hamming code is a code C_F for some function F.

(2) Proposition 9.14 assumes that $F(0) = 0$. If we want to express the APNness of any (n, n)-function, another matrix can be considered as in [21]: the

[12] We can also say that, C_F being a subcode of the Hamming code (see the definition of the Hamming code in Chapter 8), it has minimum distance at least 3.

$(2n+1) \times (2^n - 1)$ *matrix*

$$\begin{bmatrix} 1 & 1 & 1 & 1 & \dots & 1 \\ 0 & 1 & \alpha & \alpha^2 & \dots & \alpha^{2^n-2} \\ F(0) & F(1) & F(\alpha) & F(\alpha^2) & \dots & F(\alpha^{2^n-2}) \end{bmatrix}.$$

Then F is APN if and only if the code $\widetilde{C_F}$ admitting this parity-check matrix has parameters $[2^n, 2^n - 1 - 2n, 6]$. To prove this, note first that this code does not change if we add a constant to F (contrary to C_F). Hence, by adding the constant $F(0)$, we can assume that $F(0) = 0$. Then, the code $\widetilde{C_F}$ is the extended code of C_F (obtained by adding to each codeword of C_F a first coordinate equal to the sum modulo 2 of its coordinates). Because $F(0) = 0$, we can apply Proposition 9.14, and it is clear that C_F is a $[2^n - 1, 2^n - 1 - 2n, 5]$ code if and only if $\widetilde{C_F}$ is a $[2^n, 2^n - 1 - 2n, 6]$ code (we know that C_F cannot have minimum distance greater than 5, as recalled in [52]).

As shown in [52], using Parseval's relation and Relations (9.12) and (9.14), it can be proved that the weight distribution of $C_F^\perp$ is unique[13] for every AB (n, n)-function F such that $F(0) = 0$: there is one codeword of null weight, $(2^n - 1)(2^{n-2} + 2^{\frac{n-3}{2}})$ codewords of weight $2^{n-1} - 2^{\frac{n-1}{2}}$, $(2^n - 1)(2^{n-2} - 2^{\frac{n-3}{2}})$ codewords of weight $2^{n-1} + 2^{\frac{n-1}{2}}$, and $(2^n - 1)(2^{n-1} + 1)$ codewords of weight 2^{n-1}. We shall see that the function $x \to x^3$ over the field $\mathbb{F}_{2^n}$ is an AB function. The code $C_F^\perp$ corresponding to this function is known in coding theory as the dual of the 2-error-correcting BCH code of length $2^n - 1$.

If F is APN on $\mathbb{F}_{2^n}$ and null at 0, and $n > 2$, it can also be proved that the code $C_F^\perp$ has dimension $2n$. Equivalently, let us prove that the code whose generator matrix equals $\left[F(1)\ F(\alpha)\ F(\alpha^2) \dots\ F(\alpha^{2^n-2}) \right]$, and which can therefore be seen as the code $\{tr_n(vF(x));\ v \in \mathbb{F}_{2^n}\}$, has dimension n and intersects the simplex code $\{tr_n(ux);\ u \in \mathbb{F}_{2^n}\}$ (whose generator matrix is equal to $\left[1\ \alpha\ \alpha^2 \dots\ \alpha^{2^n-2} \right]$) only in the null vector. Slightly more generally, we have the following proposition.

Proposition 9.15. *Let F be an APN function in $n > 2$ variables. Then the nonlinearity of F cannot be null and, assuming that $F(0) = 0$, the code $C_F^\perp$ has dimension $2n$.*

Proof. Suppose there exists $v \neq 0$ such that $v \cdot F$ is affine. Without loss of generality (by composing F with an appropriate linear automorphism and adding an affine function to F), we can assume that $v = (0, \dots, 0, 1)$ and that $v \cdot F$ is null. Then, every derivative of F is 2-to-1 and has null last coordinate. Hence, for every $a \neq 0$ and every b, the equation $D_aF(x) = b$ has no solution if $b_n = 1$

[13] Being able to determine such weight distribution is rare (when the code does not contain the all-1 vector): it is equivalent to determining the Walsh value distribution of the function, and we have seen in the previous chapter that this is much more difficult in general than just determining the distribution of the absolute values, which for an AB function is easily deduced from the single Parseval relation.

and it has two solutions if $b_n = 0$. The $(n, n-1)$ function obtained by erasing the last coordinate of $F(x)$ has therefore balanced derivatives; hence it is a bent $(n, n-1)$-function, a contradiction with Nyberg's result, because $n - 1 > n/2$. ■

Note that for $n = 2$, the nonlinearity can be null. An example is the function $(x_1, x_2) \to (x_1x_2, 0)$.
J. Dillon (private communication) observed that the property of Proposition 9.15 implies that, for every nonzero $c \in \mathbb{F}_{2^n}$, the equation $F(x) + F(y) + F(z) + F(x + y + z) = c$ must have a solution (that is, the function Φ_F introduced after Proposition 9.12 is onto $\mathbb{F}_2^{n*}$). Indeed, otherwise, for every Boolean function $g(x)$, the function $F(x) + g(x)\,c$ would be APN. But this is contradictory with Proposition 9.15 if we take $g(x) = v_0 \cdot F(x)$ (that is, $g(x) = tr_n(v_0F(x))$ if we have identified $\mathbb{F}_2^n$ with the field $\mathbb{F}_{2^n}$) with $v_0 \notin c^\perp$, since we have then $v_0 \cdot [F(x) + g(x)\,c] = v_0 \cdot F(x) \oplus g(x)\,(v_0 \cdot c) = 0$.

There is a connection between AB functions and the so-called *uniformly packed codes* [3].

Definition 9.10. *Let C be any binary code of length N, with minimum distance $d = 2e + 1$ and covering radius ρ. For any $x \in \mathbb{F}_2^N$, let us denote by $\zeta_j(x)$ the number of codewords of C at distance j from x. The code C is called uniformly packed if there exist real numbers $h_0,\ h_1, \ldots,\ h_\rho$ such that, for any $x \in \mathbb{F}_2^N$, the following equality holds:*

$$\sum_{j=0}^{\rho} h_j\, \zeta_j(x) = 1.$$

As shown in [4], this is equivalent to saying that the covering radius of the code equals its external distance (*i.e.*, the number of different nonzero distances between the codewords of its dual). Then, as shown in [52], we have the following proposition.

Proposition 9.16. *Let F be any polynomial of the form (9.3), where n is odd. Then F is AB if and only if C_F is a uniformly packed code of length $N = 2^n - 1$ with minimum distance $d = 2e + 1 = 5$ and covering radius $\rho = e + 1 = 3$.*

- We have seen that all AB functions are APN. The converse is false, in general. But if n is odd and if F is APN, then, as shown in [42, 39], there exists a nice necessary and sufficient condition, for F being AB: the weights of $C_F^\perp$ are all divisible by $2^{\frac{n-1}{2}}$ (see also [43], where the divisibilities for several types of such codes are calculated, where tables of exact divisibilities are computed, and where proofs are given that a great many power functions are not AB). In other words:

Proposition 9.17. *Let F be an APN (n, n)-function, n odd. Then F is AB if and only if all the values $\sum_{x\in\mathbb{F}_2^n}(-1)^{v\cdot F(x)\oplus u\cdot x}$ of the Walsh spectrum of F are divisible by $2^{\frac{n+1}{2}}$.*

Proof. The condition is clearly necessary. Conversely, assume that F is APN and that all the values $\sum_{x\in\mathbb{F}_2^n}(-1)^{v\cdot F(x)\oplus u\cdot x}$ are divisible by $2^{\frac{n+1}{2}}$. Writing $\left(\sum_{x\in\mathbb{F}_2^n}(-1)^{v\cdot F(x)\oplus u\cdot x}\right)^2 = 2^{n+1}\lambda_{u,v}$, where all $\lambda_{u,v}$s are integers, Relation (9.13) then implies

$$\sum_{v\in\mathbb{F}_2^{n*},u\in\mathbb{F}_2^n} (\lambda_{u,v}^2 - \lambda_{u,v}) = 0, \tag{9.15}$$

and because all the integers $\lambda_{u,v}^2 - \lambda_{u,v}$ are nonnegative ($\lambda_{u,v}$ being an integer), we deduce that $\lambda_{u,v}^2 = \lambda_{u,v}$ for every $v \in \mathbb{F}_2^{n*}$, $u \in \mathbb{F}_2^n$, *that is*, $\lambda_{u,v} \in \{0, 1\}$. ■

Hence, if an APN function F is plateaued, or more generally if $F = F_1 \circ F_2^{-1}$ where F_2 is a permutation and where the linear combinations of the component functions of F_1 and F_2 are plateaued, then F is AB. Indeed, the sum $\sum_{x\in\mathbb{F}_2^n}(-1)^{v\cdot F(x)\oplus u\cdot x} = \sum_{x\in\mathbb{F}_2^n}(-1)^{v\cdot F_1(x)\oplus u\cdot F_2(x)}$ is then divisible by $2^{\frac{n+1}{2}}$.

This allows us to easily deduce the AB property of Gold and Kasami functions (see their definitions later) from their APN property, because the Gold functions are quadratic and the Kasami functions are equal, when n is odd, to $F_1 \circ F_2^{-1}$, where $F_1(x) = x^{2^{3i}+1}$ and $F_2(x) = x^{2^i+1}$ are quadratic.[14]

In the Case n Even. If F is APN, then there must exist $v \in \mathbb{F}_2^{n*}$, $u \in \mathbb{F}_2^n$ such that $\sum_{x\in\mathbb{F}_2^n}(-1)^{v\cdot F(x)\oplus u\cdot x}$ is not divisible by $2^{(n+2)/2}$. Indeed, suppose that all the Walsh values of F have such divisibility. Then, again denoting $\left(\sum_{x\in\mathbb{F}_2^n}(-1)^{v\cdot F(x)\oplus u\cdot x}\right)^2 = 2^{n+1}\lambda_{u,v}$, we have Relation (9.15). All the values $\lambda_{u,v}^2 - \lambda_{u,v}$ are nonnegative integers and (for each $v \neq 0$) at least one value is strictly positive, a contradiction. If all the Walsh values of F are divisible by $2^{n/2}$ (e.g., if F is plateaued), then we deduce that there must exist $v \in \mathbb{F}_2^{n*}$, $u \in \mathbb{F}_2^n$ such that $\sum_{x\in\mathbb{F}_2^n}(-1)^{v\cdot F(x)\oplus u\cdot x}$ is congruent with $2^{n/2}$ modulo $2^{n/2+1}$. Hence, if F is plateaued, there must exist $v \in \mathbb{F}_2^{n*}$ such that the Boolean function $v \cdot F$ is bent. *Note that this implies that F cannot be a permutation*, according to Proposition 9.2 and since a bent Boolean function is never balanced. More precisely, when F is plateaued and APN, the numbers $\lambda_{u,v}$ involved in Equation (9.15) can be divided into two categories: those such that the function $v \cdot F$ is bent (for each such v, we have $\lambda_{u,v} = 1/2$ for every u and therefore $\sum_{u\in\mathbb{F}_2^n}(\lambda_{u,v}^2 - \lambda_{u,v}) = -2^{n-2}$); and those such that $v \cdot F$ is not bent (then $\lambda_{u,v} \in \{0, 2^i\}$ for some $i \geq 1$ depending on v, and therefore $\lambda_{u,v}^2 = 2^i\lambda_{u,v}$ and we have, thanks to Parseval's relation applied to the Boolean function $v \cdot F$, $\sum_{u\in\mathbb{F}_2^n}(\lambda_{u,v}^2 - \lambda_{u,v}) = (2^i - 1)\sum_{u\in\mathbb{F}_2^n}\lambda_{u,v} = (2^i - 1)\frac{2^{2n}}{2^{n+1}} = (2^i - 1)2^{n-1} \geq 2^{n-1}$). Equation (9.15) implies then that the number B of those v such that $v \cdot F$ is bent satisfies $-B\,2^{n-2} + (2^n - 1 - B)\,2^{n-1} \leq 0$, which implies that *the number of bent functions among the functions* $v \cdot F$ *is at least* $\frac{2}{3}(2^n - 1)$.

[14] It is conjectured that the component functions of the Kasami functions are plateaued for every n even, too. This is already proved in Theorem 11 of [76], when n is not divisible by 6.

In the case of the Gold functions $F(x) = x^{2^i+1}$, $gcd(i, n) = 1$ (see Subsection 9.3.1.7), the number of bent functions among the functions $tr_n(vF(x))$ equals $\frac{2}{3}(2^n - 1)$. Indeed, according to the results recalled in the section on bent functions of the previous chapter, the function $tr_n(vF(x))$ is bent if and only if v is not the third power of an element of $\mathbb{F}_{2^n}$.

Note that, given an APN plateaued function F, saying that the number of bent functions among the functions $tr_n(vF(x))$ equals $\frac{2}{3}(2^n - 1)$ is equivalent to saying, according to the preceding observations, that there is no v such that $\lambda_{u,v} = \pm 2^i$ with $i > \frac{n}{2} + 1$, that is, F has nonlinearity $2^{n-1} - 2^{n/2}$, and it is also equivalent to saying that F has the same extended Walsh spectrum as the Gold functions.

The fact that an APN function F has same extended Walsh spectrum as the Gold functions can be characterized by using a similar method as for proving Proposition 9.12: this situation happens if and only if, for every $v \in \mathbb{F}_2^{n*}$ and every $u \in \mathbb{F}_2^n$, we have $\widehat{1_{G_F}}(u, v) \in \{0, \pm 2^{\frac{n}{2}}, \pm 2^{\frac{n+2}{2}}\}$ (where $\widehat{1_{G_F}}(u, v) = \sum_{x \in \mathbb{F}_2^n} (-1)^{v \cdot F(x) \oplus u \cdot x}$), that is,

$$\widehat{1_{G_F}}(u, v) \left(\widehat{1_{G_F}}^2(u, v) - 2^{n+2} \right) \left(\widehat{1_{G_F}}^2(u, v) - 2^n \right) = 0,$$

or, equivalently, $\widehat{1_{G_F}}^5(u, v) - 5 \cdot 2^n\, \widehat{1_{G_F}}^3(u, v) + 2^{2n+2} \widehat{1_{G_F}}(u, v) = 0$. Applying the Fourier transform and dividing by 2^{2n}, this is equivalent to the fact that

$$\left| \left\{ (x_1, \ldots, x_5) \in \mathbb{F}_2^{5n} / \left\{ \begin{array}{l} \sum_{i=0}^5 x_i = a \\ \sum_{i=0}^5 F(x_i) = b \end{array} \right\} \right| - 2^{3n} -$$

$$5 \cdot 2^n \left(\left| \left\{ (x_1, \ldots, x_3) \in \mathbb{F}_2^{3n} / \left\{ \begin{array}{l} \sum_{i=0}^3 x_i = a \\ \sum_{i=0}^3 F(x_i) = b \end{array} \right\} \right| - 2^n \right) +$$

$$2^{2n+2} \left(\left| \left\{ x \in \mathbb{F}_2^n / \left\{ \begin{array}{l} x = a \\ F(x) = b \end{array} \right\} \right| - 2^{-n} \right) = 0$$

for every $a, b \in \mathbb{F}_2^n$. A necessary condition is (taking $b = F(a)$ and using that F is APN) that, for every $a, b \in \mathbb{F}_2^n$, we have

$$\left| \left\{ (x_1, \ldots, x_5) \in \mathbb{F}_2^{5n} / \left\{ \begin{array}{l} \sum_{i=0}^5 x_i = a \\ \sum_{i=0}^5 F(x_i) = b \end{array} \right\} \right| =$$

$$2^{3n} + 5 \cdot 2^n (3 \cdot 2^n - 2 - 2^n) - 2^{2n+2}(1 - 2^{-n}) =$$

$$2^{3n} + 3 \cdot 2^{2n+1} - 3 \cdot 2^{n+1}.$$

There exist APN quadratic functions whose Walsh spectra are different from the Gold functions. K. Browning *et al.* [21] have exhibited such a function in six variables: $F(x) = x^3 + \alpha^{11} x^5 + \alpha^{13} x^9 + x^{17} + \alpha^{11} x^{33} + x^{48}$, where α is a primitive element in the field. For this function, we get the following spectrum:

46 functions $tr_6(vF(x))$ are bent, 16 are plateaued with amplitude 16, and 1 is plateaued with amplitude 32.

9.3.1.4 The Particular Case of Power Functions

We have seen that the notion of AB function being independent of the choice of the inner product, we can identify $\mathbb{F}_2^n$ with the field $\mathbb{F}_{2^n}$ and take $x \cdot y = tr_n(xy)$ for inner product (where tr_n is the trace function from this field to $\mathbb{F}_2$). This allows us to consider those particular (n, n)-functions that have the form $F(x) = x^d$, called *power functions* (and, sometimes, *monomial functions*).

When F is a power function, it is enough to check the APN property for $a = 1$ only, because changing, for every $a \neq 0$, the variable x into ax in the equation $F(x) + F(x + a) = b$ gives $F(x) + F(x + 1) = \frac{b}{F(a)}$. Moreover, checking the AB property $\sum_{x\in\mathbb{F}_{2^n}}(-1)^{tr_n(vF(x)+ux)} \in \{0, \pm 2^{\frac{n+1}{2}}\}$, for every $u, v \in \mathbb{F}_{2^n}$, $v \neq 0$, is enough for $u = 0$ and $u = 1$ (and every $v \neq 0$), because changing x into x/u (if $u \neq 0$) in this sum gives $\sum_{x\in\mathbb{F}_{2^n}}(-1)^{tr_n(v'F(x)+x)}$, for some $v' \neq 0$. If F is a permutation, then checking the AB property is also enough for $v = 1$ and every u, since changing x into $\frac{x}{F^{-1}(v)}$ in this sum gives $\sum_{x\in\mathbb{F}_{2^n}}(-1)^{tr_n\left(F(x)+\frac{u}{F^{-1}(v)}x\right)}$.

Also, when F is an APN power function, we have additional information on its bijectivity. It was proved in [52] that, when n is even, no APN function exists in a class of permutations including power permutations, that we describe now. Let $k = (2^n - 1)/3$ (which is an integer, because n is even) and let α be a primitive element of the field $\mathbb{F}_{2^n}$. Then $\beta = \alpha^k$ is a primitive element of $\mathbb{F}_4$. Hence, $\beta^2 + \beta + 1 = 0$. For every j, the element $(\beta + 1)^j + \beta^j = \beta^{2j} + \beta^j$ equals 1 if j is coprime with 3 (because β^j is then also a primitive element of $\mathbb{F}_4$) and is null otherwise. Let $F(x) = \sum_{j=0}^{2^n-1}\delta_j x^j$, ($\delta_j \in \mathbb{F}_{2^n}$) be an (n, n)-function. According to the preceding observations, β and $\beta + 1$ are the solutions of the equation $F(x) + F(x + 1) = \sum_{gcd(j,3)=1}\delta_j$. Also, the equation $F(x) + F(x + 1) = \sum_{j=1}^{2^n-1}\delta_j$ admits 0 and 1 for solutions. Thus, we obtain the following.

Proposition 9.18. *Let n be even and let $F(x) = \sum_{j=0}^{2^n-1}\delta_j x^j$ be any APN (n, n)-function, then $\sum_{j=1}^{k}\delta_{3j} \neq 0$, $k = \frac{2^n-1}{3}$. If F is a power function, then it cannot be a permutation.*

H. Dobbertin gives in [85] a result valid only for power functions but slightly more precise, and he completes it in the case that n is odd.

Proposition 9.19. *If a power function $F(x) = x^d$ over $\mathbb{F}_{2^n}$ is APN, then for every $x \in \mathbb{F}_{2^n}$, we have $x^d = 1$ if and only if $x^3 = 1$, that is, $F^{-1}(1) = \mathbb{F}_4 \cap \mathbb{F}_{2^n}^*$. If n is odd, then $gcd(d, 2^n - 1)$ equals 1 and, if n is even, then $gcd(d, 2^n - 1)$ equals 3. Consequently, APN power functions are permutations if n is odd, and are three-to-one if n is even.*

Proof. Let $x \neq 1$ be such that $x^d = 1$. There is a (unique) y in $\mathbb{F}_{2^n}$, $y \neq 0, 1$, such that $x = (y + 1)/y$. The equality $x^d = 1$ then implies $(y + 1)^d + y^d = 0 =$

$(y^2+1)^d+(y^2)^d$. By the APN property and because $y^2 \neq y$, we conclude $y^2+y+1=0$. Thus, y, and therefore x, are in $\mathbb{F}_4$ and $x^3=1$. Conversely, if $x \neq 1$ is an element of $\mathbb{F}_{2^n}^*$ such that $x^3=1$, then 3 divides 2^n-1 and n must be even. Moreover, d must also be divisible by 3 (indeed, otherwise, the restriction of x^d to $\mathbb{F}_4$ would coincide with the function $x^{gcd(d,3)}=x$ and therefore would be linear, a contradiction). Hence, we have $x^d=1$. The rest is straightforward. ■

A. Canteaut proves in [40] that for n even, if a power function $F(x)=x^d$ on $\mathbb{F}_{2^n}$ is not a permutation (*i.e.*, if $gcd(d,2^n-1)>1$), then the nonlinearity of F is bounded above by $2^{n-1}-2^{n/2}$ (she also studies the case of equality). Indeed, denoting $gcd(d,2^n-1)$ by d_0, then for every $v \in \mathbb{F}_{2^n}$, the sum $\sum_{x\in\mathbb{F}_{2^n}}(-1)^{tr_n(vx^d)}$ equals $\sum_{x\in\mathbb{F}_{2^n}}(-1)^{tr_n(vx^{d_0})}$, which implies that $\sum_{v\in\mathbb{F}_{2^n}}\left(\sum_{x\in\mathbb{F}_{2^n}}(-1)^{tr_n(vx^d)}\right)^2$ equals $2^n\,|\{(x,y),\ x,y\in\mathbb{F}_{2^n},\ x^{d_0}=y^{d_0}\}|$. The number of elements in the image of $\mathbb{F}_{2^n}^*$ by the mapping $x\to x^{d_0}$ is $(2^n-1)/d_0$, and every element of this image has d_0 preimages. Hence, $\sum_{v\in\mathbb{F}_{2^n}^*}\left(\sum_{x\in\mathbb{F}_{2^n}}(-1)^{tr_n(vx^d)}\right)^2$ equals $2^n[(2^n-1)d_0+1]-2^{2n}=2^n(2^n-1)(d_0-1)$ and $\max_{v\in\mathbb{F}_{2^n}^*}\left(\sum_{x\in\mathbb{F}_{2^n}}(-1)^{tr_n(vx^d)}\right)^2\geq 2^n(d_0-1)\geq 2^{n+1}$.

The possible values of the sum $\sum_{x\in\mathbb{F}_{2^n}}(-1)^{tr_n(vx^d)}$ are determined in [7] for APN power functions in an even number of variables.

If F is a power function, then the linear codes C_F and $C_F^\perp$ (viewed in Proposition 9.14) are *cyclic codes*, that is, are invariant under cyclic shifts of their coordinates (see [121] and Chapter 8). Indeed, $(c_0,\ldots,c_{2^n-2})$ belongs to C_F if and only if $c_0+c_1\alpha+\cdots+c_{2^n-2}\alpha^{2^n-2}=0$ and $c_0+c_1\alpha^d+\cdots+c_{2^n-2}\alpha^{(2^n-2)d}=0$; this implies (by multiplying these equations by α and α^d, respectively) $c_{2^n-2}+c_0\alpha+\cdots+c_{2^n-3}\alpha^{2^n-2}=0$ and $c_{2^n-2}+c_0\alpha^d+\cdots+c_{2^n-3}\alpha^{(2^n-2)d}=0$. Recall that, representing each codeword $(c_0,c_1,\ldots,c_{2^n-2})$ by the element $\sum_{i=0}^{2^n-2}c_iX^i$ of the algebra $\mathbb{F}_2[X]/(X^{2^n-1}+1)$, the code is then an ideal of this algebra, and it equals the set of all those polynomials of degrees at most 2^n-2 that are multiples (as elements of the algebra and more strongly as polynomials) of a polynomial, called a generator polynomial, dividing $X^{2^n-1}+1$, which is the unique element of minimal degree in the code. In other words, $\sum_{i=0}^{2^n-2}c_iX^i$ is a codeword if and only if the roots in $\mathbb{F}_{2^n}$ of the generator polynomial are also roots of $\sum_{i=0}^{2^n-2}c_iX^i$. The roots of the generator polynomial are of the form $\{\alpha^i,\ i\in I\}$ where $I\subseteq\mathbb{Z}/(2^n-1)\mathbb{Z}$ is a union of cyclotomic classes of 2 modulo 2^n-1. The set I is called the *defining set* of the code. In the case of C_F, the defining set I is precisely the union of the two cyclotomic classes containing 1 and d.

A very efficient bound on the minimum distance of cyclic codes, also recalled in the previous chapter, is the *BCH bound* [121]: if I contains a string $\{l+1,\ldots,l+k\}$ of length k in $\mathbb{Z}/(2^n-1)\mathbb{Z}$, then the cyclic code has minimum distance greater than or equal to $k+1$. This bound shows, for instance, in an original way that the function $x^{2^{\frac{n-1}{2}}+1}$, n odd, is AB: by definition, the defining set I of C_F equals the union of the cyclotomic classes of 1 and $2^{\frac{n-1}{2}}+1$, that is,

$$\{1, 2, \ldots, 2^{n-1}\} \cup$$

$$\{2^{\frac{n-1}{2}}+1, 2^{\frac{n+1}{2}}+2, \ldots, 2^{n-1}+2^{\frac{n-1}{2}}, 2^{\frac{n+1}{2}}+1, 2^{\frac{n+3}{2}}+2, \ldots, 2^{n-1}+2^{\frac{n-3}{2}}\}.$$

The defining set of $C_F^\perp$ then equals $\mathbb{Z}/(2^n-1)\mathbb{Z} \setminus \{-i;\, i \notin I\}$ (this property is valid for every cyclic code; see [121]). Because there is no element equal to $2^{n-1}+2^{\frac{n-1}{2}}+1, \ldots, 2^n-1$ in I, the defining set of $C_F^\perp$ then contains a string of length $2^{n-1}-2^{\frac{n-1}{2}}-1$. Hence, the nonzero codewords of this code have weights greater than or equal to $2^{n-1}-2^{\frac{n-1}{2}}$. This is not sufficient for concluding that the function is AB (because we need also to know that the complements of the extended codewords have weight at least $2^{n-1}-2^{\frac{n-1}{2}}$), but we can apply the previous reasoning to the cyclic code $C_F^\perp \cup ((1, \ldots, 1)+C_F^\perp)$: the defining set of the dual of this code being equal to that of C_F, plus 0, the defining set of the code itself equals that of $C_F^\perp$ less 0, which gives a string of length $2^{n-1}-2^{\frac{n-1}{2}}-2$ instead of $2^{n-1}-2^{\frac{n-1}{2}}-1$. Hence, the complements of the codewords of $C_F^\perp$ have weights at least $2^{n-1}-2^{\frac{n-1}{2}}-1$, and because for these codewords, the corresponding Boolean function takes value 1 at the zero vector (which is not taken into account in the corresponding codeword), this allows us to deduce that all functions $tr_n(vx^{2^{\frac{n-1}{2}}+1}+ux) \oplus \varepsilon$, $v \neq 0$, $\varepsilon \in \mathbb{F}_2$, have weights between $2^{n-1}-2^{\frac{n-1}{2}}$ and $2^{n-1}+2^{\frac{n-1}{2}}$, that is, F is AB. The powerful *McEliece theorem* (see, *e.g.*, [121]) gives the exact divisibility of the codewords of cyclic codes. Translated in terms of vectorial functions, it says that if d is relatively prime to 2^n-1, the exponent e_d of the greatest power of 2 dividing all the Walsh coefficients of the power function x^d is given by $e_d = \min\{w_2(t_0)+w_2(t_1),\ 1 \le t_0,\ t_1 < 2^n-1;\ t_0+t_1 d \equiv 0 \ [\text{mod } 2^n-1]\}$. It can be used in relationship with Proposition 9.17. This led to the proof, by Canteaut, Charpin, and Dobbertin, of a several-decades-old conjecture due to Welch (see later discussion).

Note finally that, if F is a power function, then the Boolean function γ_F seen in Proposition 9.11 is within the framework of Dobbertin's triple construction [77].

9.3.1.5 Notions of Equivalence Respecting the APN and AB Properties

The right and left compositions of an APN (resp. AB) function by an affine permutation are APN (resp. AB). Two functions are called *affine equivalent* if one is equal to the other, composed by such affine permutations.

Adding an affine function to an APN (resp. AB) function respects its APN (resp. AB) property. Two functions are called *extended affine equivalent* (EA-equivalent) if one is affine equivalent to the other, added with an affine function.

The inverse of an APN (resp. AB) permutation is APN (resp. AB) but is in general not EA-equivalent to it. There exists a notion of equivalence between (n, n)-functions that respects APNness and ABness and for which any permutation is equivalent to its inverse. As we shall see, this equivalence relation is still more general than EA-equivalence between functions, up to replacing the functions by their inverses when they are permutations.

Definition 9.11. *Two (n,n)-functions F and G are called* CCZ-equivalent[15] *if their graphs $C_F = \{(x,y) \in \mathbb{F}_2^n \times \mathbb{F}_2^n \mid y = F(x)\}$ and $C_G = \{(x,y) \in \mathbb{F}_2^n \times \mathbb{F}_2^n \mid y = G(x)\}$ are affine equivalent, that is, if there exists an affine automorphism $L = (L_1, L_2)$ of $\mathbb{F}_2^n \times \mathbb{F}_2^n$ such that $y = F(x) \Leftrightarrow L_2(x,y) = G(L_1(x,y))$.*

As observed in [21], given two (n,n)-functions F and G such that $F(0) = G(0) = 0$, there exists a linear automorphism[16] that maps G_F to G_G if and only if the codes C_F and C_G (see the definition of these codes in Proposition 9.14) are equivalent (that is, are equal up to some permutation of the coordinates of their codewords). Indeed, the graph G_F of F equals the (unordered) set of columns in the parity-check matrix of the code C_F, plus an additional point equal to the all-zero vector. Hence, the existence of a linear automorphism that maps G_F onto G_G is equivalent to the fact that the parity-check matrices[17] of the codes C_F and C_G are equal up to multiplication (on the left) by an invertible matrix and to permutation of the columns. Because two codes with given parity-check matrices are equal if and only if these matrices are equal up to multiplication on the left by an invertible matrix, this completes the proof. It is nicely deduced in [21] that two functions F and G taking any values at 0 are CCZ-equivalent if and only if the codes $\widetilde{C_F}$ and $\widetilde{C_G}$ (see the definition of these codes in the remark – line (2) – following Proposition 9.14) are equivalent.

The notion of CCZ-equivalence can be similarly defined for functions from $\mathbb{F}_2^n$ to $\mathbb{F}_2^m$.

Given a function $F : \mathbb{F}_2^n \to \mathbb{F}_2^m$ and an affine automorphism $L = (L_1, L_2)$ of $\mathbb{F}_2^n \times \mathbb{F}_2^m$, the image of the graph of F by L is the graph of a function if and only if the function $F_1(x) = L_1(x, F(x))$ is a permutation of $\mathbb{F}_2^n$. Indeed, if F_1 is a permutation, then $L(G_F)$ equals the graph of the function $G = F_2 \circ F_1^{-1}$; and conversely, denoting $F_2(x) = L_2(x, F(x))$, the image of the graph of F by L equals $\{(F_1(x), F_2(x)); x \in \mathbb{F}_2^n\}$, and since L is a permutation, if $F_1(x) = F_1(x')$ for some $x \neq x'$, then $F_2(x) \neq F_2(x')$, and $L(G_F)$ is not the graph of a function.

Proposition 9.20. *If two (n,n)-functions F and G are CCZ-equivalent, then F is APN (resp. AB) if and only if G is APN (resp. AB). Moreover, denoting by $L = (L_1, L_2)$ an affine automorphism between the graphs of F and G, the function γ_F (see Proposition 9.11) equals $\gamma_G \circ \mathcal{L}$, where $\mathcal{L}$ is the linear automorphism such that $L = \mathcal{L} + cst$.*

15 This notion has been introduced in [52] and later named CCZ-equivalence in [23, 24]; it could be also called graph-equivalence.

16 Note that this is a subcase of CCZ-equivalence – in fact, a strict subcase as shown in [21].

17 This is true also for the generator matrices of the codes.

Proof. We have seen that $G = F_2 \circ F_1^{-1}$, where $F_1(x) = L_1(x, F(x))$ and $F_2(x) = L_2(x, F(x))$. The value $\gamma_G(a, b)$ equals 1 if and only if $a \neq 0$ and there exists (x, y) in $\mathbb{F}_2^n \times \mathbb{F}_2^n$ such that $F_1(x) + F_1(y) = a$ and $F_2(x) + F_2(y) = b$, that is, $\mathcal{L}(x, F(x)) + \mathcal{L}(y, F(y)) = \mathcal{L}(x + y, F(x) + F(y)) = (a, b)$. Thus, γ_G is equal to $\gamma_F \circ \mathcal{L}^{-1}$. The function γ_G is therefore bent (resp. has weight $2^{2n-1} - 2^{n-1}$) if and only if γ_F is bent (resp. has weight $2^{2n-1} - 2^{n-1}$). Proposition 9.11 completes the proof. ■

All the transformations respecting the APN (resp. AB) property that we have seen previously to Proposition 9.20 are particular cases of this general one:

- If $L_1(x, y)$ depends only on x, then writing $L_1(x, y) = L_1(x)$ and $L_2(x, y) = L'(x) + L''(y)$, the function $F_1(x) = L_1(x)$ is a permutation (because, L being onto $\mathbb{F}_2^n \times \mathbb{F}_2^m$, L_1 must be onto $\mathbb{F}_2^n$) and we have $F_2 \circ F_1^{-1}(x) = L' \circ L_1^{-1}(x) + L'' \circ F \circ L_1^{-1}(x)$. This corresponds to EA-equivalence.
- If $(L_1, L_2)(x, y) = (y, x)$, then $F_2(x) = x$ and $F_1(x) = F(x)$. If F is a permutation, then F_1 is a permutation, and $F_2 \circ F_1^{-1}$ is equal to F^{-1}.

It was proved in [23, 24] that CCZ-equivalence is strictly more general than EA-equivalence between the functions or their inverses (when they exist), by exhibiting (see later discussion) APN functions that are CCZ-equivalent to the APN function $F(x) = x^3$ on $\mathbb{F}_{2^n}$, but that are provably EA-inequivalent to it and (for n odd) to its inverse.

Note, however, that if we reduce ourselves to bent functions, then CCZ-equivalence and EA-equivalence coincide: let F be a bent (n, m)-function (n even, $m \leq n/2$) and let (without loss of generality) L_1 and L_2 be two linear functions from $\mathbb{F}_2^n \times \mathbb{F}_2^m$ to (respectively) $\mathbb{F}_2^n$ and $\mathbb{F}_2^m$, such that (L_1, L_2) and $L_1(x, F(x))$ are permutations. For every vector v in $\mathbb{F}_2^m$, the function $v \cdot L_1(x, F(x))$ is necessarily unbent because, if $v = 0$, then it is null, and if $v \neq 0$, then it is balanced, according to Proposition 9.2. Let us denote $L_1(x, y) = L'(x) + L''(y)$. We then have $F_1(x) = L'(x) + L'' \circ F(x)$. The adjoint operator L''' of L'' (satisfying by definition $v \cdot L''(y) = L'''(v) \cdot y$, that is, the linear function having for matrix the transpose of the matrix of L'') is then the null function, because if $L'''(v) \neq 0$, then $v \cdot F_1(x) = v \cdot L'(x) \oplus L'''(v) \cdot F(x)$ is bent. This means that L'' is null and L_1 then depends only on x, which corresponds to EA-equivalence.

Proving the CCZ-inequivalence between two functions is mathematically (and also computationally) difficult, unless some CCZ-invariant parameters can be proved different for the two functions. Examples of direct proofs of CCZ-inequivalence using only the definition can be found in [26, 27].

Examples of CCZ-invariant parameters are the following (see [21] and [87] where they are introduced and used):

- The extended Walsh spectrum.
- The equivalence class of the code $\widetilde{C_F}$ (under the relation of equivalence of codes), according to the result of [21] recalled after Definition 9.11, and all

the invariants related to this code (the weight enumerator of $\widetilde{C_F}$, the weight enumerator of its dual – but it corresponds to the extended Walsh spectrum of the function – the automorphism group, etc., which coincide with some of the following invariants).

- The Γ-rank: let $\mathcal{G} = \mathbb{F}_2[\mathbb{F}_2^n \times \mathbb{F}_2^n]$ be the so-called group algebra of $\mathbb{F}_2^n \times \mathbb{F}_2^n$ over $\mathbb{F}_2$, consisting of the formal sums $\sum_{g \in \mathbb{F}_2^n \times \mathbb{F}_2^n} a_g\, g$, where $a_g \in \mathbb{F}_2$. If S is a subset of $\mathbb{F}_2^n \times \mathbb{F}_2^n$, then it can be identified with the element $\sum_{s \in S} s$ of $\mathcal{G}$. The dimension of the ideal of $\mathcal{G}$ generated by the graph $G_F = \{(x, F(x));\ x \in \mathbb{F}_2^n\}$ of F is called the *Γ-rank* of F. The Γ-rank equals (see [87]) the rank of the matrix M_{G_F} whose term indexed by $(x, y) \in \mathbb{F}_2^n \times \mathbb{F}_2^n$ and by $(a, b) \in \mathbb{F}_2^n \times \mathbb{F}_2^n$ equals 1 if $(x, y) \in (a, b) + G_F$ and equals 0 otherwise.
- The Δ-rank, that is, the dimension of the ideal of $\mathcal{G}$ generated by the set $D_F = \{(a, F(x) + F(x + a));\ a, x \in \mathbb{F}_2^n;\ a \neq 0\}$ (recall that, according to Proposition 9.11, this set has size $2^{2n-1} - 2^{n-1}$ and is a difference set when F is AB). The Δ-rank equals the rank of the matrix M_{D_F} whose term indexed by (x, y) and by (a, b) equals 1 if $(x, y) \in (a, b) + D_F$ and equals 0 otherwise.
- The order of the automorphism group of the design $dev(G_F)$, whose points are the elements of $\mathbb{F}_2^n \times \mathbb{F}_2^n$ and whose blocks are the sets $(a, b) + G_F$ (and whose incidence matrix is M_{G_F}), that is, of all those permutations on $\mathbb{F}_2^n \times \mathbb{F}_2^n$ that map every such block to a block.
- The order of the automorphism group of the design $dev(D_F)$, whose points are the elements of $\mathbb{F}_2^n \times \mathbb{F}_2^n$ and whose blocks are the sets $(a, b) + D_F$ (and whose incidence matrix is M_{D_F}).
- The order of the automorphism group $\mathcal{M}(G_F)$ of the so-called multipliers of G_F, that is, the permutations π of $\mathbb{F}_2^n \times \mathbb{F}_2^n$ such that $\pi(G_F)$ is a translate $(a, b) + G_F$ of G_F. This order is easier to compute and it allows us in some cases to easily prove CCZ-inequivalence. As observed in [21], $\mathcal{M}(G_F)$ is the automorphism group of the code $\widetilde{C_F}$.
- The order of the automorphism group $\mathcal{M}(D_F)$.

CCZ-equivalence does not preserve crookedness or the algebraic degree.

9.3.1.6 The Known AB Functions

Power Functions. Until recently, the only known examples of AB functions were (up to EA-equivalence) the power functions $x \mapsto x^d$ on the field $\mathbb{F}_{2^n}$ (n odd) corresponding to the following values of d, and the inverses of these power functions:

- $d = 2^i + 1$ with $\gcd(i, n) = 1$ and $1 \le i \le \frac{n-1}{2}$ (proved by Gold; see [90, 128]). The condition $1 \le i \le \frac{n-1}{2}$ (here and below) is not necessary, but we mention it because the other values of i give EA-equivalent functions. These power functions are called *Gold functions*.
- $d = 2^{2i} - 2^i + 1$ with $\gcd(i, n) = 1$ and $2 \le i \le \frac{n-1}{2}$ (the AB property of this function is equivalent to a result by Kasami [106], historically due to

Welch, but never published by him; see another proof in [80]). These power functions are called *Kasami functions* (some authors call them *Kasami-Welch functions*).

- $d = 2^{(n-1)/2} + 3$ (conjectured by Welch and proved by Canteaut, Charpin and Dobbertin; see [81, 42, 43]). These power functions are called *Welch functions*.
- $d = 2^{(n-1)/2} + 2^{(n-1)/4} - 1$, where $n \equiv 1 \pmod 4$ (conjectured by Niho, proved by Hollman and Xiang, after the work by Dobbertin, see [82, 98]).
- $d = 2^{(n-1)/2} + 2^{(3n-1)/4} - 1$, where $n \equiv 3 \pmod 4$ (idem). The power functions in these two last cases are called *Niho functions*.

The almost bentness of these functions can be deduced from their almost perfect nonlinearity (see below) by using Proposition 9.17 (and McEliece's theorem in the cases of the Welch and Niho functions; the proofs are then not easy). The direct proof that the Gold function is AB is easy by using the properties of quadratic functions recalled in the previous chapter, in the subsection devoted to quadratic functions. The value at a of the Walsh transform of the Gold Boolean function $tr_n(x^{2^i+1})$ equals $\pm 2^{\frac{n+1}{2}}$ if $tr_n(a) = 1$ and is null otherwise, because $tr_n(x^{2^i} y + xy^{2^i}) = tr_n((x^{2^i} + x^{2^{n-i}})y)$ is null for every y if and only if $x^{2^{2i}} + x = 0$, that is, if and only if $x \in \mathbb{F}_2$ (since $gcd(2^{2i} - 1, 2^n - 1) = 1$), and since $tr_n(x^{2^i+1} + ax)$ is constant on $\mathbb{F}_2$ if and only if $tr_n(a) = 1$. This easily gives the magnitude (but not the sign, which is studied in [114]) of the Walsh transform of the vectorial Gold function, this function being a permutation (see Subsection 9.3.1.4).

The inverse of x^{2^i+1} is x^d, where

$$d = \sum_{k=0}^{\frac{n-1}{2}} 2^{2ik},$$

and x^d therefore has the algebraic degree $\frac{n+1}{2}$ [128].

It was proved in [75] (Theorem 7) and [76] (Theorem 15) that, if $3i$ is congruent with 1 mod n, then the Walsh support of the Kasami Boolean function $tr_n(x^{2^{2i}-2^i+1})$ equals the support of the Gold Boolean function $tr_n(x^{2^i+1})$ (*i.e.* the set $\{x \in \mathbb{F}_{2^n} \mid tr_n(x^{2^i+1}) = 1\}$) if n is odd and equals the set $\{x \in \mathbb{F}_{2^n} \mid tr_{n/2}(x^{2^i+1}) = 0\}$ if n is even, where $tr_{n/2}$ is the trace function from $\mathbb{F}_{2^n}$ to the field $\mathbb{F}_{2^2}$: $tr_{n/2}(x) = x + x^4 + x^{4^2} + \cdots + x^{4^{n/2-1}}$. When n is odd, this gives the magnitude (but not the sign) of the Walsh transform of the vectorial Kasami function, this function being a permutation. Note that this also gives information on the autocorrelation of the Kasami Boolean function: according to the Wiener-Khintchine theorem (see the previous chapter), the Fourier transform of the function $a \to \mathcal{F}(D_a f) = \sum_{x \in \mathbb{F}_2^n} (-1)^{D_a f(x)}$, where f is the Kasami Boolean function, equals the square of the Walsh transform of f. According to Dillon's and Dobbertin's result recalled earlier and because we know that the Kasami function is almost bent when n is odd, the value at b of the square of the Walsh transform of f equals then 2^{n+1} if $tr_n(x^{2^i+1}) = 1$ and equals zero otherwise. Hence, by applying the inverse

Fourier transform (that is, by applying the Fourier transform again and dividing by 2^n), $\mathcal{F}(D_a f)$ equals twice the Fourier transform of the function $tr_n(x^{2^i+1})$. We deduce that, except at the zero vector, $\mathcal{F}(D_a f)$ equals the opposite of the Walsh transform of the function $tr_n(x^{2^i+1})$.

It is proved in [27] that Gold functions are pairwise CCZ-inequivalent and that they are in general CCZ-inequivalent to Kasami and Welch functions.

We have seen that the Walsh value distribution of AB functions is known. A related result of [114] is generalized in [91]: for every AB power function x^d over $\mathbb{F}_{2^n}$ whose restriction to any subfield of $\mathbb{F}_{2^n}$ is also AB, the value $\sum_{x \in \mathbb{F}_{2^n}} (-1)^{tr_n(x^d+x)}$ equals $2^{\frac{n+1}{2}}$ if $n \equiv \pm 1$ [mod 8] and $-2^{\frac{n+1}{2}}$ if $n \equiv \pm 3$ [mod 8].

Remark. *There is a close relationship between AB power functions and* sequences *used for radars and for spread-spectrum communications. A binary sequence that can be generated by an LFSR, or equivalently that satisfies a linear recurrence relation $s_i = a_1 s_{i-1} \oplus \cdots \oplus a_n s_{i-n}$, is called an m-*sequence *or a* maximum-length sequence *if its period equals $2^n - 1$, which is the maximal possible value. Such a sequence has the form $tr_n(\lambda \alpha^i)$, where $\lambda \in \mathbb{F}_{2^n}$ and α is some primitive element of $\mathbb{F}_{2^n}$, and where tr_n is the trace function on $\mathbb{F}_{2^n}$. Consequently, its auto-correlation values $\sum_{i=0}^{2^n-2} (-1)^{s_i \oplus s_{i+t}}$ $(1 \leq t \leq 2^n - 2)$ are equal to -1, that is, are optimum. Such a sequence can be used for radars and for code division multiple access (CDMA) in telecommunications, because it allows sending a signal that can be easily distinguished from any time-shifted version of itself. Finding an AB power function x^d on the field $\mathbb{F}_{2^n}$ allows to have a d-decimation[18] $s'_i = tr_n(\lambda \alpha^{di})$ of the sequence, whose cross-correlation values $\sum_{i=0}^{2^n-2} (-1)^{s_i \oplus s'_{i+t}}$ $(0 \leq t \leq 2^n - 2)$ with the sequence s_i have minimum overall magnitude[19] [94]. The cross-correlation is then called a* preferred cross-correlation *function; see [33]. The conjectures that the power functions above were AB have been stated (before being proved later) in the framework of sequences for this reason.*

It has been conjectured by Hans Dobbertin that the preceding list of power AB functions is complete. See [117] about this conjecture.

Nonpower Functions. It was first conjectured that all AB functions are equivalent to power functions and to permutations. These two conjectures were later disproved, in a first step by exhibiting AB functions that are EA-inequivalent to power functions and to permutations, but that are by construction CCZ-equivalent to the Gold function $x \to x^3$, and in a second step by finding AB functions that are CCZ-inequivalent to power functions[20] (at least for some values of n):

[18] Another m-sequence if d is co-prime with $2^n - 1$.

[19] This allows, in code division multiple access, to give different signals to different users.

[20] The question of knowing whether all AB functions are CCZ-equivalent to permutations remains open, as far as we know.

Functions CCZ-Equivalent to Power Functions. Two examples of linear permutations over $\mathbb{F}_{2^n} \times \mathbb{F}_{2^n}$ transforming the graph of the Gold function $x \to x^3$ into the graph of a function have been found in [23], giving new classes of AB functions:

- The function $F(x) = x^{2^i+1} + (x^{2^i} + x)\, tr_n(x^{2^i+1} + x)$, where $n > 3$ is odd and $gcd(n, i) = 1$, is AB. It is provably EA-inequivalent to any power function and it is EA-inequivalent to any permutation (at least for $n = 5$).
- For n odd and divisible by m, $n \neq m$, and $\gcd(n, i) = 1$, the following function from $\mathbb{F}_{2^n}$ to $\mathbb{F}_{2^n}$:

$$x^{2^i+1} + tr_{n/m}(x^{2^i+1}) + x^{2^i}\, tr_{n/m}(x) + x\; tr_{n/m}(x)^{2^i} +$$
$$[tr_{n/m}(x)^{2^i+1} + tr_{n/m}(x^{2^i+1}) + tr_{n/m}(x)]^{\frac{1}{2^i+1}}(x^{2^i} + tr_{n/m}(x)^{2^i} + 1) +$$
$$[tr_{n/m}(x)^{2^i+1} + tr_{n/m}(x^{2^i+1}) + tr_{n/m}(x)]^{\frac{2^i}{2^i+1}}(x + tr_{n/m}(x)),$$

  where $tr_{n/m}$ denotes the trace function $tr_{n/m}(x) = \sum_{i=0}^{n/m-1} x^{2^{mi}}$ from $\mathbb{F}_{2^n}$ to $\mathbb{F}_{2^m}$, is an AB function of algebraic degree $m + 2$ that is provably EA-inequivalent to any power function; the question of knowing whether it is EA-inequivalent to any permutation is open.

Open problem: Find classes of AB functions by using CCZ-equivalence with Kasami (resp. Welch, Niho) functions.

- Though the AB functions constructed in [23] cannot be obtained from power functions by applying only EA-equivalence and inverse transformation, L. Budaghyan shows in [22] that AB functions EA-inequivalent to power functions can be constructed by only applying EA-equivalence and inverse transformation to power AB functions.

Functions CCZ-Inequivalent to Power Functions. The problem of knowing whether there exist AB functions that are CCZ-inequivalent to power functions remained open after the introduction of the two functions just shown. Also, it was conjectured that any quadratic APN function is EA-equivalent to Gold functions, and this problem remained open. A paper by Edel, Kyureghyan, and Pott [86] introduced two quadratic APN functions from $\mathbb{F}_{2^{10}}$ (resp. $\mathbb{F}_{2^{12}}$) to itself. The first one was proved to be CCZ-inequivalent to any power function.

These two (quadratic) functions were isolated, and this left open the question of knowing whether a whole infinite class of APN/AB functions being not CCZ-equivalent to power functions could be exhibited.

- The following new class of AB functions was found in [25, 26].

Proposition 9.21. *Let s and k be positive integers with* $\gcd(s, 3k) = 1$ *and* $t \in \{1, 2\}$, $i = 3 - t$. *Furthermore, let* $d = 2^{ik} + 2^{tk+s} - (2^s + 1)$,

$$g_1 = \gcd(2^{3k} - 1, d/(2^k - 1)),$$
$$g_2 = \gcd(2^k - 1, d/(2^k - 1)).$$

If $g_1 \neq g_2$, *then the function*

$$F : \mathbb{F}_{2^{3k}} \to \mathbb{F}_{2^{3k}}$$
$$x \mapsto \alpha^{2^k-1} x^{2^{ik}+2^{tk+s}} + x^{2^s+1}$$

where α *is primitive in* $\mathbb{F}_{2^{3k}}$ *is AB when k is odd and APN when k is even.*

It could be proved in [25, 26] that some of these functions are EA-inequivalent to power functions and CCZ-inequivalent to some AB power functions, and this was sufficient to deduce that they are CCZ-inequivalent to all power functions for some values of n.

Proposition 9.22. *Let* s *and* $k \geq 4$ *be positive integers such that* $s \leq 3k - 1$, $\gcd(k, 3) = \gcd(s, 3k) = 1$, *and* $i = sk$ *[mod 3]*, $t = 2i$ *[mod 3]*, $n = 3k$. *If* $a \in \mathbb{F}_{2^n}$ *has the order* $2^{2k} + 2^k + 1$, *then the function* $F(x) = x^{2^s+1} + ax^{2^{ik}+2^{tk+s}}$ *is an AB permutation on* $\mathbb{F}_{2^n}$ *when n is odd and is APN when n is even. It is EA-inequivalent to power functions and CCZ-inequivalent to Gold and Kasami mappings.*

- The following proposition has been shown in [29].

Proposition 9.23. *For every odd positive integer, the function* $x^3 + tr_n(x^9)$ *is AB on* $\mathbb{F}_{2^n}$ *(and it is APN for n even).*

This function is the only example, with the function x^3, of a function that is AB for any odd n (if we consider it as the same function for every n, which is not quite true because the trace function depends on n). It is CCZ-inequivalent to any Gold function on $\mathbb{F}_{2^n}$ if $n \geq 7$.

Open problem: Find infinite classes of AB functions CCZ-inequivalent to power functions and to quadratic functions.

9.3.1.7 The Known APN Functions

We now list the known APN functions (in addition to the AB functions listed previously).

Power Functions. The so-called *multiplicative inverse permutation* (or simply *inverse function*) $x \mapsto F(x) = x^{2^n-2}$ (which equals $\frac{1}{x}$ if $x \neq 0$, and 0 otherwise) is APN if n is odd [8, 128]. Indeed, the equation $x^{2^n-2} + (x+1)^{2^n-2} = b$ ($b \neq 0$, since the inverse function is a permutation) admits 0 and 1 for solutions if and only if $b = 1$; and it (also) admits (two) solutions different from 0 and 1 if and only if there exists $x \neq 0, 1$ such that $\frac{1}{x} + \frac{1}{x+1} = b$, that is, $x^2 + x = \frac{1}{b}$. It is well known that such existence is equivalent to the fact that $tr_n\left(\frac{1}{b}\right) = 0$. Hence, F is APN if and only if $tr_n(1) = 1$, that is, if n is odd.

Consequently, the functions $x \mapsto x^{2^n-2^i-1}$, which are linearly equivalent to F (through the linear isomorphism $x \mapsto x^{2^i}$), are also APN, if n is odd.

If n is even, then the equation $x^{2^n-2} + (x+1)^{2^n-2} = b$ admits at most two solutions if $b \neq 1$ and admits four solutions (the elements of $\mathbb{F}_4$) if $b = 1$, which means that F opposes a good (but not optimal) resistance against differential cryptanalysis. Its nonlinearity equals $2^{n-1} - 2^{n/2}$ when n is even, and it equals the highest even number bounded above by this number, when n is odd (see [59]; Lachaud and Wolfmann proved in [113] that the set of values of its Walsh spectrum equals the set of all integers $s \equiv 0$ [mod 4] in the range $[-2^{n/2+1} + 1; 2^{n/2+1} + 1]$; see more in [97]). Knowing whether there exist (n, n)-functions with nonlinearity strictly greater than this value when n is even is an open question (even for power functions). These are some of the reasons why the function $x \mapsto x^{2^n-2}$ has been chosen for the S-boxes of the AES.

Until recently, the only known examples of APN functions were (up to affine equivalence and to the addition of an affine function) power functions $x \mapsto x^d$. We next list the known values of d for which we obtain APN functions, without repeating the cases where the functions are AB:

- $d = 2^n - 2$, n odd (inverse function);
- $d = 2^i + 1$ with $\gcd(i, n) = 1$, n even and $1 \leq i \leq \frac{n-2}{2}$ (*Gold functions*, see [90, 128]);
- $d = 2^{2i} - 2^i + 1$ with $\gcd(i, n) = 1$, n even and $2 \leq i \leq \frac{n-2}{2}$ (*Kasami functions*; see [102], see also [80]);
- $d = 2^{\frac{4n}{5}} + 2^{\frac{3n}{5}} + 2^{\frac{2n}{5}} + 2^{\frac{n}{5}} - 1$, with n divisible by 5 (*Dobbertin functions*, see [83]). It has been shown by Canteaut, Charpin, and Dobbertin [43] that this function cannot be AB: they showed that $C_F^{\perp}$ contains words whose weights are not divisible by $2^{\frac{n-1}{2}}$.

The proof that the Gold functions are APN (whatever the parity of n) is easy: the equality $F(x) + F(x+1) = F(y) + F(y+1)$ is equivalent to $(x+y)^{2^i} = (x+y)$, and thus implies that $x + y = 0$ or $x + y = 1$, because i and n are coprime. Hence, any equation $F(x) + F(x+1) = b$ admits at most two solutions.

The proofs that the Kasami and Dobbertin functions are APN are difficult. They come down to showing that some mappings are permutations. H. Dobbertin, in [84], gives a nice general method for this.

The Gold and Kasami functions, for n even, have the best known nonlinearity when n is even, too [90, 106], but not the Dobbertin functions. See [43] for a list of all known *permutations* with best known nonlinearity. See also [78].

Inverse and Dobbertin functions are CCZ-inequivalent to all other known APN functions because of their peculiar Walsh spectra.

It is proven in [19] that there exists no APN function CCZ-inequivalent to power mappings on $\mathbb{F}_{2^n}$ for $n \leq 5$.

The exponents d such that the function x^d is APN on infinitely many fields $\mathbb{F}_{2^n}$ have been called *exceptional* by J. Dillon (see, *e.g.*, [21]). We have seen that a power function x^d is APN if and only if the function $x^d + (x+1)^d + 1$ (we write "+1" so that 0 is a root, which simplifies presentation) is 2-to-1. For every

(n, n)-function F over $\mathbb{F}_{2^n}$, there clearly always exists a polynomial P such that $F(x) + F(x+1) + F(1) = P(x + x^2)$. J. Dillon observed that, in the cases of the Gold and Kasami functions, the polynomial P is an exceptional polynomial (*i.e.*, is a permutation over infinitely many fields $\mathbb{F}_{2^n}$); from there comes the term. In the case of the Gold function x^{2^i+1}, we have $P(x) = x + x^2 + x^{2^2} + \cdots + x^{2^{i-1}}$, which is a linear function over the algebraic closure of $\mathbb{F}_2$ having kernel $\{x \in \mathbb{F}_{2^i} \,/\, tr_i(x) = 0\}$ and is therefore a permutation over $\mathbb{F}_{2^n}$ for every n coprime with i. In the case of the Kasami function, $P(x) = ((tr_i(x))^{2^i+1})/x^{2^i}$ is the Müller-Cohen-Matthews polynomial [66]. It is conjectured that the Gold and Kasami exponents are the only exceptional exponents.

Nonpower Functions. As for AB functions, it had been conjectured that all APN functions were EA-equivalent to power functions, and this conjecture was proven false.

Functions CCZ-Equivalent to Power Functions. Also using the stability properties recalled in Subsection 9.3.1.5, two more infinite classes of APN functions have been introduced in [23] and disprove the preceding conjecture:

- The function $F(x) = x^{2^i+1} + (x^{2^i} + x + 1)\, tr_n(x^{2^i+1})$, where $n \geq 4$ is even and $gcd(n, i) = 1$ is APN and is EA-inequivalent to any power function.
- For n even and divisible by 3, the function $F(x)$ equal to

$$[x + tr_{n/3}(x^{2(2^i+1)} + x^{4(2^i+1)}) + tr_n(x)\, tr_{n/3}(x^{2^i+1} + x^{2^{2i}(2^i+1)})]^{2^i+1},$$

  where $gcd(n, i) = 1$, is APN, and is EA-inequivalent to any known APN function.

Open problem: Find classes of APN functions by using CCZ-equivalence with Kasami (resp. Welch, Niho, Dobbertin, inverse) functions.

Functions CCZ-Inequivalent to Power Functions

- As recalled earlier, the paper [86] introduced two quadratic APN functions from $\mathbb{F}_{2^{10}}$ (resp. $\mathbb{F}_{2^{12}}$) to itself. The first one, $F(x) = x^3 + ux^{36}$, where $u \in \mathbb{F}_4 \setminus \mathbb{F}_2$, was proved to be CCZ-inequivalent to any power function by computing its Δ-rank.

  The functions viewed in Proposition 9.21 are APN when n is even and generalize the second function: $F(x) = x^3 + \alpha^{15} x^{528}$, where α is a primitive element of $\mathbb{F}_{2^{12}}$; some of them can be proven CCZ-inequivalent to Gold and Kasami mappings, as seen in Proposition 9.22. A similar class but with n divisible by 4 was later given in [28]. As observed by J. Bierbrauer, a common framework exists for the class of Proposition 9.22 and this new class.

Theorem 9.24. *Let:*

- $n = tk$ *be a positive integer, with* $t \in \{3, 4\}$, *and* s *be such that* t, s, k *are pairwise coprime and such that* t *is a divisor of* $k + s$; *and*

- *α be a primitive element of $\mathbb{F}_{2^n}$ and $w = \alpha^e$, where e is a multiple of $2^k - 1$, coprime with $2^t - 1$.*

Then the function

$$F(x) = x^{2^s+1} + wx^{2^{k+s}+2^{k(t-1)}}$$

is APN.

For $n \geq 12$, these functions are EA-inequivalent to power functions and CCZ-inequivalent to Gold and Kasami mappings [26].

In particular, for $n = 12, 20, 24, 28$ they are CCZ-inequivalent to all power functions.

- Proposition 9.22 has been generalized[21] in [15, 16] by C. Bracken, E. Byrne, N. Markin, and G. McGuire:

$$F(x) = u^{2^k} x^{2^{2k}+2^{k+s}} + ux^{2^s+1} + vx^{2^{2k}+1} + wu^{2^k+1}x^{2^{k+s}+2^s}$$

is APN on $\mathbb{F}_{2^{3k}}$, when $3 \mid k+s$, $(s, 3k) = (3, k) = 1$ and u is primitive in $\mathbb{F}_{2^{3k}}$, $v \neq w^{-1} \in \mathbb{F}_{2^k}$.

The same authors in the same paper obtained another class of APN functions:

$$F(x) = bx^{2^s+1} + b^{2^k}x^{2^{k+s}+2^k} + cx^{2^k+1} + \sum_{i=1}^{k-1} r_i x^{2^{i+k}+2^i},$$

where k, s are odd and coprime, $b \in \mathbb{F}_{2^{2k}}$ is not a cube, $c \in \mathbb{F}_{2^{2k}} \setminus \mathbb{F}_{2^k}$, $r_i \in \mathbb{F}_{2^k}$ is APN on $\mathbb{F}_{2^{2k}}$.

The extended Walsh spectrum of these functions is the same as for the Gold function; see [13]. But it is proved in [17] that at least some of these functions are inequivalent to Gold functions.

- As already mentioned, the construction of AB functions of Proposition 9.23 gives APN functions for n even: *for any positive integer n, the function $x^3 + tr_n(x^9)$ is APN on $\mathbb{F}_{2^n}$.*

  This function is CCZ-inequivalent to any Gold function on $\mathbb{F}_{2^n}$ if $n \geq 7$.

  The extended Walsh spectrum of this function is the same as for the Gold functions as shown in [12].

- An idea of J. Dillon [74] was that (n, n)-functions (over $\mathbb{F}_{2^n}$) of the form

$$F(x) = x(Ax^2 + Bx^q + Cx^{2q}) + x^2(Dx^q + Ex^{2q}) + Gx^{3q},$$

where $q = 2^{n/2}$, n even, have good chances to be differentially 4-uniform. This idea was exploited and pushed further in [30], which gave new APN functions.

Proposition 9.25. *Let n be even and i be coprime with $n/2$. Set $q = 2^{n/2}$ and let $c, b \in \mathbb{F}_{2^n}$ be such that $c^{q+1} = 1$, $c \notin \{\lambda^{(2^i+1)(q-1)}, \lambda \in \mathbb{F}_{2^n}\}$, $cb^q + b \neq 0$.*

[21] Note that Proposition 9.21 covers a larger class of APN functions than Proposition 9.22.

Then the function

$$F(x) = x^{2^{2i}+2^i} + bx^{q+1} + cx^{q(2^{2i}+2^i)}$$

is APN on $\mathbb{F}_{2^n}$.
Such vectors b, c *do exist if and only if* $gcd(2^i + 1, q + 1) \neq 1$. *For* $n/2$ *odd, this is equivalent to saying that* i *is odd.*

The extended Walsh spectrum of these functions is the same as that of the Gold functions [154].

- Another class was obtained in this same paper [30] with the same idea.

Proposition 9.26. *Let* n *be even and* i *be co-prime with* $n/2$. *Set* $q = 2^{n/2}$ *and let* $c \in \mathbb{F}_{2^n}$ *and* $s \in \mathbb{F}_{2^n} \setminus \mathbb{F}_q$. *If the polynomial*

$$X^{2^i+1} + cX^{2^i} + c^q X + 1$$

is irreducible over $\mathbb{F}_{2^n}$, *then the function*

$$F(x) = x(x^{2^i} + x^q + cx^{2^i q}) + x^{2^i}(c^q x^q + sx^{2^i q}) + x^{(2^i+1)q}$$

is APN on $\mathbb{F}_{2^n}$.

It was checked with a computer that some of the functions of the present class and of the previous one are CCZ-inequivalent to power functions for $n = 6$. It remains open to prove the same property for every even $n \geq 6$.

Open problem: The APN power functions listed above are not permutations when n is even. The question of knowing whether there exist APN permutations when n is even is open. This question was first raised (at least in printed form) in [130]. We have seen that the answer is "no" for all plateaued functions (this was first observed in this same paper [130] when all the component functions of F are partially bent; Nyberg generalized there a result given without a complete proof in [139], which was valid only for quadratic functions). We also saw in Subsection 9.3.1.4 that the answer is "no" for a class of functions including power functions. And X.-d. Hou proved in [100] that it is also "no" for those functions whose univariate representation coefficients lie in $\mathbb{F}_{2^{n/2}}$; he showed that this problem is related to a conjecture on the symmetric group of $\mathbb{F}_{2^n}$.

- We now introduce a method for constructing APN functions from bent functions, which leads to two classes (we do not yet know if these classes are new) and should lead to others. Let B be a bent $(n, n/2)$-function and let G be a function from $\mathbb{F}_2^n$ to $\mathbb{F}_2^{n/2}$. Let

$$F : x \in \mathbb{F}_2^n \to (B(x), G(x)) \in \mathbb{F}_2^{n/2} \times \mathbb{F}_2^{n/2}.$$

F is APN if and only if, for every nonzero $a \in \mathbb{F}_2^n$, and for every $c \in \mathbb{F}_2^{n/2}$ and $d \in \mathbb{F}_2^{n/2}$, the system of equations

$$\begin{cases} B(x) + B(x+a) = c \\ G(x) + G(x+a) = d \end{cases}$$

has 0 or 2 solutions.
Because B is bent, the number of solutions of the first equation equals $2^{n/2}$ for every $a \neq 0$. We need to find functions G such that, among these $2^{n/2}$ solutions, only 0 or 2 additionally satisfy the second equation.

Obviously, the condition on G depends on the choice of B. We take the Maiorana-McFarland function defined on $\mathbb{F}_{2^{n/2}} \times \mathbb{F}_{2^{n/2}}$ by $B(x, y) = xy$, where xy is the product of x and y in the field $\mathbb{F}_{2^{n/2}}$. We then write (a, b) with $a, b \in \mathbb{F}_{2^{n/2}}$ instead of $a \in \mathbb{F}_2^n$. Changing c into $c + ab$, the preceding system of equations becomes

$$\begin{cases} bx + ay & = c \\ G(x, y) + G(x+a, y+b) & = d \end{cases}.$$

It is straightforward to check that F is APN if and only if, for every nonzero ordered pair (a, b) in $\mathbb{F}_{2^{n/2}} \times \mathbb{F}_{2^{n/2}}$ and every c, d in $\mathbb{F}_{2^{n/2}}$, denoting $G_{a,b,c}(x) = G(ax, bx + c)$:

(i) For every $y \in \mathbb{F}_{2^{n/2}}$, the function $x \in \mathbb{F}_{2^{n/2}} \to G(x, y)$ is APN (this condition corresponds to the case $b = 0$);
(ii) For every $x \in \mathbb{F}_{2^{n/2}}$, the function $y \in \mathbb{F}_{2^{n/2}} \to G(x, y)$ is APN (this condition corresponds to the case $a = 0$);
(iii) For every $(a, b) \in \mathbb{F}_{2^{n/2}} \times \mathbb{F}_{2^{n/2}}$ such that $a \neq 0$ and $b \neq 0$, and for every $c, d \in \mathbb{F}_{2^{n/2}}$, the equation $G_{a,b,c}(x) + G_{a,b,c}(x+1) = d$ has 0 or 2 solutions.

We give two classes of APN functions obtained with this method:

- Let us choose $G(x, y) = sx^{2^i+1} + ty^{2^i+1} + ux^{2^i}y + vxy^{2^i}$, where $(n/2, i) = 1$ and $s, t, u, v \in \mathbb{F}_{2^{n/2}}, s \neq 0, t \neq 0$. Then, because the Gold function x^{2^i+1} is APN, the function $x \in \mathbb{F}_{2^{n/2}} \to G(x, y)$ is APN for every $y \in \mathbb{F}_{2^{n/2}}$ (the other terms being affine in x), and the function $y \in \mathbb{F}_{2^{n/2}} \to G(x, y)$ is APN for every $x \in \mathbb{F}_{2^{n/2}}$. The function $G_{a,b,c}(x)$ equals $(sa^{2^i+1} + tb^{2^i+1} + ua^{2^i}b + vab^{2^i})x^{2^i+1}$, plus an affine function. Then, if the polynomial $sX^{2^i+1} + t + uX^{2^i} + vX$ has no zero in $\mathbb{F}_{2^{n/2}}$ (for instance, if it is irreducible over $\mathbb{F}_{2^{n/2}}$), we have $sa^{2^i+1} + tb^{2^i+1} + ua^{2^i}b + vab^{2^i} \neq 0$ for every $a \neq 0$ and every $b \neq 0$ (dividing this expression by b^{2^i+1} and taking $X = a/b$), and the equation $G_{a,b,c}(x) + G_{a,b,c}(x+1) = d$ has at most two solutions, because the function $x \to x^{2^i+1}$ is APN. Thus, F is then APN. If we take, for instance, $G(x, y) = x^3 + xy^2 + y^3$ with $(n/2, 3) = 1$, then the preceding polynomial equals $X^3 + X + 1$ and has no zero (which implies that it is irreducible because it has degree 3), because $X^3 = X + 1$ implies $X^8 = X^2(X^2 + 1) = X(X^3 + X) = X$, which in its turn implies $X \in \mathbb{F}_2$, since $(n/2, 3) = 1$, and $X^3 + X + 1$ has no zero in $\mathbb{F}_2$.

- Now take $G(x, y) = x^{2^i+1} + \lambda\, y^{2^i+1}$ where $(i, n/2) = 1$ and λ is not a cube (λ can exist only if $n/2$ is even, i.e., n is divisible by 4). Then conditions (i) and (ii) are clearly satisfied and, since $G_{a,b,c}(x)$ equals $(a^{2^i+1} + \lambda\, b^{2^i+1})\, x^{2^i+1}$ plus an affine function, condition (iii) is also satisfied because $a^{2^i+1} + \lambda\, b^{2^i+1}$ never vanishes; so, the function is APN.
- It is also easy to construct differentially 4-uniform functions this way. For instance the functions $(x, y) \to (xy, x^3 + y^5)$, $(x, y) \to (xy, x^3 + y^6)$ and $(x, y) \to (xy, x^5 + y^6)$. More interestingly there are non-quadratic differentially 4-uniform functions: the function $(x, y) \to (xy, (x^3 + w)(y^3 + w'))$, where w and w' belong to $\mathbb{F}_{2^{n/2}} \setminus \{x^3,\ x \in \mathbb{F}_{2^{n/2}}\}$, with $n/2$ even (for allowing the existence of such elements), and $(x, y) \to (xy, x^3(y^2 + y + 1) + y^3)$, with $n/2$ odd (so that $y^2 + y + 1$ is never null).

Open problem: Find infinite classes of APN functions CCZ-inequivalent to power functions and to quadratic functions.

Observation. *A classification under CCZ-inequivalence of all APN functions up to dimension five and a (nonexhaustive) list of CCZ-inequivalent functions in dimension 6 have been given in [19]. One of the functions in dimension 6 is CCZ-inequivalent to power functions and to quadratic functions, as proved by Edel and Pott in [87] (this had not been seen by the authors of [19]). This function is*

$$x^3 + \alpha^{17}(x^{17} + x^{18} + x^{20} + x^{24}) + tr_2(x^{21}) + tr_3(\alpha^{18}x^9)$$
$$+\alpha^{14}\, tr_6\,(\alpha^{52}x^3 + \alpha^6x^5 + \alpha^{19}x^7 + \alpha^{28}x^{11} + \alpha^2x^{13}).$$

Some constructions of differentially 4-uniform functions have been given in [124], in connection with commutative semifields. A semifield is a finite algebraic structure $(E, +, \circ)$ such that (1) $(E, +)$ is an abelian group, (2) the operation $\circ$ is distributive on the left and on the right with respect to $+$, (3) there is no nonzero divisor of 0 in E, and (4) E contains an identity element with respect to $\circ$. This structure has been very useful for constructing planar functions in odd characteristic. In characteristic 2, it may lead to new APN functions by considering for instance the function $(x \circ x) \circ x$ in a classical semifield (there are examples of these whose underlying abelian group is the additive group of $\mathbb{F}_{2^n}$: the Albert semifields, in which the multiplication is $x \circ y = xy + \beta(xy)^\sigma$, where $x \to x^\sigma$ is an automorphism of the field $\mathbb{F}_{2^n}$ that is not a generator and $\beta \notin \{x^{\sigma+1};\ x \in \mathbb{F}_{2^n}\}$; and the Knuth semifield, where the multiplication is $x \circ y = xy + (xtr(y) + ytr(x))^2$, where tr is a trace function from $\mathbb{F}_{2^n}$ to a suitable subfield).

More open problems:

(1) Find secondary constructions of APN and AB functions.

(2) Derive more constructions of APN/AB functions from perfect nonlinear functions, and *vice versa*.
(3) Classify APN functions, or at least their extended Walsh spectra, or at least their nonlinearities.

Observations. *For n odd, the known APN functions have three possible spectra:*

- *The spectrum of the AB functions which gives a nonlinearity of* $2^{n-1} - 2^{\frac{n-1}{2}}$.
- *The spectrum of the inverse function, which takes any value divisible by 4 in* $[-2^{n/2+1} + 1; 2^{n/2+1} + 1]$ *and gives a nonlinearity close to* $2^{n-1} - 2^{n/2}$.
- *The spectrum of the Dobbertin function which is more complex (it is divisible by* $2^{n/5}$ *and not divisible by* $2^{2n/5+1}$*); its nonlinearity seems to be bounded below by approximately* $2^{n-1} - 2^{3n/5-1} - 2^{2n/5-1}$ *– maybe equal – but this has to be proven (or disproven).*

For n even, the spectra may be more diverse:

- *The Gold functions, whose component functions are bent for a third of them and have nonlinearity* $2^{n-1} - 2^{n/2}$ *for the rest of them; the Kasami functions, which have the same extended spectra.*
- *The Dobbertin function (same observation as before).*
- *As soon as* $n \geq 6$*, we find (quadratic) functions with different spectra.*

The nonlinearities also seem bounded below by approximately $2^{n-1} - 2^{3n/5-1} - 2^{2n/5-1}$ (but this has to be proven . . . or disproven). Note that the question of classifying APN functions is open even when we restrict ourselves to quadratic APN functions (even classifying their Walsh spectra is open for even numbers of variables). Already for $n = 6$ there are at least nine mutually CCZ-inequivalent quadratic APN polynomials that are CCZ-inequivalent to power functions [21].

Open question. The nonlinearities of the known APN functions do not seem to be very weak; is this situation general to all APN functions or specific to the APN functions found so far?

Observation. *We have seen in Proposition 9.15 that an APN function cannot have null nonlinearity. We can improve upon this lower bound under some hypothesis:*

Proposition 9.27. ([51].) *Let F be an APN function in* $n > 2$ *variables. For all real numbers a and b such that* $a \leq b$*, let* $N_{a,b}$ *be the number of ordered pairs* $(u, v) \in \mathbb{F}_2^n \times \mathbb{F}_2^{n*}$ *such that* $\widehat{1_{G_F}}^2(u, v) \in]2^n + a; 2^n + b[$*, where* $\widehat{1_{G_F}}(u, v) = \sum_{x \in \mathbb{F}_2^n} (-1)^{v \cdot F(x) \oplus u \cdot x}$*. Then*

$$nl(F) \geq 2^{n-1} - \frac{1}{2}\sqrt{2^n + \frac{1}{2}(b + a + \sqrt{\Delta_{a,b}})},$$

where $\Delta_{a,b} = (N_{a,b} + 1)(b - a)^2 + a\, b\, 2^{n+2}(2^n - 1) + 2^{4n+2} - 2^{3n+2}$.

Proof. Relation (9.13) shows that for all real numbers a, b we have

$$\sum_{\substack{u\in\mathbb{F}_2^n,\\ v\in\mathbb{F}_2^{n*}}} (\widehat{1_{G_F}}^2(u,v) - 2^n - a)(\widehat{1_{G_F}}^2(u,v) - 2^n - b) = (2^{3n} + a\,b\,2^n)(2^n - 1), \tag{9.16}$$

because $\sum_{u\in\mathbb{F}_2^n, v\in\mathbb{F}_2^{n*}} (\widehat{1_{G_F}}^2(u,v) - 2^n) = 0$ and $\sum_{u\in\mathbb{F}_2^n, v\in\mathbb{F}_2^{n*}} (\widehat{1_{G_F}}^2(u,v) - 2^n)^2 = \sum_{u\in\mathbb{F}_2^n, v\in\mathbb{F}_2^{n*}} \widehat{1_{G_F}}^2(u,v)(\widehat{1_{G_F}}^2(u,v) - 2^{n+1}) + 2^{3n}(2^n - 1)$.

The expression $(x-a)(x-b)$ is nonnegative outside $]a, b[$; it takes its minimum at $x = (b+a)/2$, and this minimum equals $-(b-a)^2/4$. We deduce that we have $(\widehat{1_{G_F}}^2(u,v) - 2^n - a)(\widehat{1_{G_F}}^2(u,v) - 2^n - b) \geq -(b-a)^2/4$ for these $N_{a,b}$ ordered pairs and $(\widehat{1_{G_F}}^2(u,v) - 2^n - a)(\widehat{1_{G_F}}^2(u,v) - 2^n - b) \geq 0$ for all the others. Hence $-(b-a)^2/4 \leq (\widehat{1_{G_F}}^2(u,v) - 2^n - a)(\widehat{1_{G_F}}^2(u,v) - 2^n - b) \leq 2^{4n} - 2^{3n} + a\,b\,2^n(2^n - 1) + N_{a,b}(b-a)^2/4$ for any $(u,v) \in \mathbb{F}_2^n \times \mathbb{F}_2^{n*}$, that is, $(\widehat{1_{G_F}}^2(u,v) - 2^n)^2 - (b+a)(\widehat{1_{G_F}}^2(u,v) - 2^n) + ab - (2^{4n} - 2^{3n} + a\,b\,2^n(2^n - 1) + N_{a,b}(b-a)^2/4) \leq 0$, which implies

$$\frac{1}{2}\left(b + a - \sqrt{\Delta_{a,b}}\right) \leq \widehat{1_{G_F}}^2(u,v) - 2^n \leq \frac{1}{2}\left(b + a + \sqrt{\Delta_{a,b}}\right),$$

where $\Delta_{a,b} = (b+a)^2 - 4(a\,b - 2^{4n} + 2^{3n} - a\,b\,2^n(2^n - 1) - N_{a,b}(b-a)^2/4) = (N_{a,b}+1)(b-a)^2 + a\,b\,2^{n+2}(2^n - 1) + 2^{4n+2} - 2^{3n+2}$. This implies that the nonlinearity of F is bounded below by

$$2^{n-1} - \frac{1}{2}\sqrt{2^n + \frac{1}{2}\left(b + a + \sqrt{\Delta_{a,b}}\right)}.$$

■

Consequences:

- Taking $b = -a = 2^n$, we see that if $\widehat{1_{G_F}}^2(u,v)$ does not take values in the range $]0; 2^{n+1}[$, then F is AB (this was known according to Relation (9.13)).
- More generally, taking $a = -2^{2n}/b$ (b necessarily greater than or equal to 2^n, because we see later that otherwise this would contradict the Sidelnikov-Chabaud-Vaudenay bound), we see that if $\widehat{1_{G_F}}^2(u,v)$ does not take values in the range $]2^n - \frac{2^{2n}}{b}; 2^n + b[$, the nonlinearity of F is bounded below by $2^{n-1} - \frac{1}{2}\sqrt{2^n + b}$. For instance (for $b = 2^{n+1}$), if $\widehat{1_{G_F}}^2(u,v)$ does not take values in the range $]2^{n-1}; 3\cdot 2^n[$, the nonlinearity of F is bounded below by $2^{n-1} - \frac{1}{2}\sqrt{3\cdot 2^n}$.

As observed by G. Leander (private communication), if n is odd and F is an APN power (n,n)-function, then because we know that F is a bijection and thus all functions $v \cdot F$ have the same Walsh spectrum, we have, according to Relation

(9.12),

$$\max_{v\neq 0,u} \widehat{1_{G_F}}^4(u,v) \leq \frac{\sum_{v\neq 0,u} \widehat{1_{G_F}}^4(u,v)}{2^n-1} = 2^{3n+1}.$$

Thus we have $\max_{v\neq 0,u} |\widehat{1_{G_F}}(u,v)| \leq 2^{\frac{3n+1}{4}}$ and

$$nl(F) \geq 2^{n-1} - 2^{\frac{3n-3}{4}}. \tag{9.17}$$

In fact, the lower bound of Proposition 9.27 can then be improved: denoting by $N'_{a,b}$ the number $N_{a,b}/2^n - 1$ of elements u of $\mathbb{F}_2^n$ such that $\widehat{1_{G_F}}^2(u,v) \in]2^n + a; 2^n + b[$ (which is the same for every $v \in \mathbb{F}_2^{n*}$), we have

$$nl(F) \geq 2^{n-1} - \frac{1}{2}\sqrt{2^n + \frac{1}{2}\left(b + a + \sqrt{\Delta'_{a,b}}\right)},$$

where $\Delta'_{a,b} = (N'_{a,b}+1)(b-a)^2 + a\,b\,2^{n+2} + 2^{3n+2}$. This does not improve the lower bound in the cases considered earlier, but it does in general. Note that, for $a = b = -2^n$, it gives (9.17).

Concluding Remark. As we can see, very few functions usable as S-boxes have emerged so far. The Gold functions, all the other recently found quadratic functions, and the Welch functions have algebraic degrees too low to make them widely chosen for the design of new S-boxes. The Kasami functions themselves seem too closely related to quadratic functions. The inverse function has many very nice properties: large Walsh spectrum and good nonlinearity, differential uniformity of order at least 4, fast implementation. But it has a potential weakness, which has not yet led to efficient attacks, but may in the future: denoting its input by x and its output by y, the bilinear expression xy equals 1 for every nonzero x. The candidates for future block ciphers not using the inverse function as an S-box are the Niho and Dobbertin functions. The Niho functions exist only in odd numbers of variables (which is not convenient for implementation in software, but is not a real problem in hardware), and the Dobbertin function needs n to be divisible by 5 (idem). The nonlinearity is also a concern. So further studies seem indispensable for the future designs of SP networks. This is the main open problem.

9.3.1.8 Lower Bounds on the Nonlinearity of S-Boxes by Means of Their Algebraic Immunity

As proved in [50], Lobanov's bound, recalled in the previous chapter for Boolean functions, generalizes to (n, m)-functions as follows:

$$nl(F) \geq 2^m \sum_{i=0}^{AI(F)-2} \binom{n-1}{i},$$

where $AI(f)$ is the basic algebraic immunity of F.

Note that, applying Lobanov's bound to the component functions of F, we obtain

$$nl(F) \geq 2 \sum_{i=0}^{AI_{comp}(F)-2} \binom{n-1}{i},$$

where $AI_{comp}(F)$ is the component algebraic immunity of F. The inequality $AI_{comp}(F) \geq AI_{gr}(F) - 1$ then implies

$$nl(F) \geq 2 \sum_{i=0}^{AI_{gr}(F)-3} \binom{n-1}{i},$$

where $AI_{gr}(F)$ is the graph algebraic immunity of F.

9.3.2 Higher Order Nonlinearities

For every positive integer r, the rth-order nonlinearity of a vectorial function F is the minimum rth-order nonlinearity of its component functions (recall that, as defined in the previous chapter, the rth-order nonlinearity of a Boolean function equals its minimum Hamming distance to functions of algebraic degrees at most r). As proved in [50], the bounds recalled in Chapter 8 for Boolean functions generalize to (n, m)-functions as follows:

$$nl_r(F) \geq 2^m \sum_{i=0}^{AI(F)-r-1} \binom{n-r}{i}$$

and

$$nl_r(F) \geq 2^{m-1} \sum_{i=0}^{AI(F)-r-1} \binom{n}{i} + 2^{m-1} \sum_{i=AI(F)-2r}^{AI(F)-r-1} \binom{n-r}{i}$$

(the first of these two bounds can be slightly improved as for Boolean functions).

Applying the bounds valid for Boolean functions to the component functions of F, we also have

$$nl_r(F) \geq 2 \sum_{i=0}^{AI_{comp}(F)-r-1} \binom{n-r}{i}$$

and

$$nl_r(F) \geq \sum_{i=0}^{AI_{comp}(F)-r-1} \binom{n}{i} + \sum_{i=AI_{comp}(F)-2r}^{AI_{comp}(F)-r-1} \binom{n-r}{i}.$$

The inequality $AI_{comp}(F) \geq AI_{gr}(F) - 1$ then implies

$$nl_r(F) \geq 2 \sum_{i=0}^{AI_{gr}(F)-r-2} \binom{n-r}{i}$$

and

$$nl_r(F) \geq \sum_{i=0}^{AI_{gr}(F)-r-2} \binom{n}{i} + \sum_{i=AI_{gr}(F)-2r-1}^{AI_{gr}(F)-r-2} \binom{n-r}{i}.$$

In the definition of $nl_r(F)$, we consider approximations by Boolean functions of algebraic degrees at most r of the component functions of F, that is, of the functions equal to F composed on the left by nonzero linear Boolean functions on $\mathbb{F}_2^m$ (and taking nonconstant affine functions instead does not change the value). We can also consider F composed by functions of higher degrees.

Definition 9.12. *For every S-box $F : F_2^n \to \mathbb{F}_2^m$, for every positive integer $s \leq m$ and $t \leq n+m$, and every nonnegative integer $r \leq n$, we define*

$$\begin{aligned} nl_{s,r}(F) &= \min\{nl_r(f \circ F);\ f \in \mathcal{B}_m,\ d^\circ f \leq s,\ f \neq cst\} \\ &= \min\{d_H(g, f \circ F);\ f \in \mathcal{B}_m,\ d^\circ f \leq s,\ f \neq cst,\ g \in \mathcal{B}_n,\ d^\circ g \leq r\} \end{aligned}$$

and

$$NL_t(F) = \min\{w_H(h(x, F(x)));\ h \in \mathcal{B}_{n+m},\ d^\circ h \leq t,\ h \neq cst\},$$

where d_H denotes the Hamming distance and $\mathcal{B}_m$ the set of m-variable Boolean functions, as in the previous chapter.

Definition 9.12 obviously excludes $f = cst$ and $h = cst$ because the knowledge of the distance $d_H(g, f \circ F)$ or of the weight $w_H(h(x, F(x)))$ when f or h is constant gives no information specific to F and usable in an attack against a stream or block cryptosystem using F as an S-box.

Clearly, for every S-box F and every integer $t \leq t'$, $s \leq s'$ and $r \leq r'$, we have $NL_t(F) \geq NL_{t'}(F)$ and $nl_{s,r}(F) \geq nl_{s',r'}(F)$. Note also that, for every vectorial function F, we have $NL_1(F) = nl(f)$.

T. Shimoyama and T. Kaneko exhibited, in [140], several quadratic functions h and pairs (f, g) of quadratic functions showing that the nonlinearities NL_2 and $nl_{2,2}$ of some sub-S-boxes of the DES are null (and therefore that the global S-box of each round of the DES has the same property). They deduced a "higher-order nonlinear" attack (an attack using the principle of the linear attack by Matsui but with nonlinear approximations) that needs 26% less data than Matsui's attack. This improvement is not very significant, practically, but some recent studies, not yet published, seem to show that the notions of NL_t and $nl_{s,r}$ can be related to potentially more powerful attacks. Note that we have $NL_{\max(s,r)}(F) \leq nl_{s,r}(F)$ by taking $h(x, y) = g(x) + f(y)$ (because $f \neq cst$ implies then $h \neq cst$), and the inequality can be strict if $s > 1$ or $r > 1$ because a function $h(x, y)$ of low degree and such that $w_H(h(x, F(x)))$ is small can exist while no such function exists with separated variables x and y, that is, of the form $g(x) + f(y)$. This is the case, for instance, of the S-box of the AES for $s = 1$ and $r = 2$ (see later discussion).

We now study bounds on these parameters. We begin with an easy one coming from the existence of n-variable Boolean functions of algebraic degree s and Hamming weight 2^{n-s}.

Proposition 9.28. ([50].) *For every positive integers m, n, $s \leq m$, and $r \leq n$ and every (n,m)-function F, we have $NL_s(F) \leq 2^{n-s}$ and $nl_{s,r}(F) \leq 2^{n-s}$. These inequalities are strict if F is not balanced (that is, if its output is not uniformly distributed over $\mathbb{F}_2^m$).*

The bound $nl_{s,r}(F) \leq 2^{n-s}$ is asymptotically almost tight (in a sense that will be made precise in Proposition 9.30) for permutations when $r \leq s \leq 0.227\,n$.

9.3.2.1 Existence of Permutations with Lower Bounded Higher Order Nonlinearities

Proposition 9.29. ([50].) *Let n and s be positive integers and let r be a non-negative integer. Let D be the greatest integer such that*

$$\sum_{t=0}^{D} \binom{2^n}{t} \leq \frac{\binom{2^n}{2^{n-s}}}{2^{\sum_{i=0}^{s}\binom{n}{i}+\sum_{i=0}^{r}\binom{n}{i}}}.$$

There exist (n,n)-permutations F whose higher order nonlinearity $nl_{s,r}(F)$ is strictly greater than D.

Proof. For every integer $i \in [0, 2^n]$ and r, let us denote by $A_{r,i}$ the number of codewords of Hamming weight i in the Reed-Muller code of order r. Given a number D, a permutation F, and two Boolean functions f and g, if we have $d_H(f \circ F, g) \leq D$, then F^{-1} maps the support $supp(f)$ of f onto the symmetric difference $supp(g)\Delta E$ between $supp(g)$ and a set E of size at most D (equal to the symmetric difference between $F^{-1}(supp(f))$ and $supp(g)$). And F^{-1} maps $\mathbb{F}_2^n \setminus supp(f)$ onto the symmetric difference between $\mathbb{F}_2^n \setminus supp(g)$ and E. Given f, g, and E and denoting by i the size of $supp(f)$ (with $0 < i < 2^n$, since $f \neq cst$), the number of permutations whose restriction to $supp(g)\Delta E$ is a one-to-one function onto $supp(f)$ and whose restriction to $(\mathbb{F}_2^n \setminus supp(g))\Delta E$ is a one-to-one function onto $\mathbb{F}_2^n \setminus supp(f)$ equals $i!\,(2^n - i)!$. We deduce that the number of permutations F such that $nl_{s,r}(F) \leq D$ is bounded above by

$$\sum_{t=0}^{D} \binom{2^n}{t} \sum_{0<i<2^n} \sum_{j=1}^{2^n} A_{s,i} A_{r,j}\, i!\,(2^n - i)!$$

Because the nonconstant codewords of the Reed-Muller code of order s have weights between 2^{n-s} and $2^n - 2^{n-s}$, we deduce that the probability $P_{s,r,D}$ that a permutation F chosen at random (with uniform probability) satisfies $nl_{s,r}(F) \leq D$

is bounded above by

$$\sum_{t=0}^{D}\binom{2^n}{t}\sum_{j=0}^{2^n}A_{r,j}\sum_{2^{n-s}\le i\le 2^n-2^{n-s}}A_{s,i}\frac{i!\,(2^n-i)!}{2^n!}=$$

$$\sum_{t=0}^{D}\binom{2^n}{t}\sum_{j=0}^{2^n}A_{r,j}\sum_{2^{n-s}\le i\le 2^n-2^{n-s}}\frac{A_{s,i}}{\binom{2^n}{i}}$$

$$<\frac{\left(\sum_{t=0}^{D}\binom{2^n}{t}\right)2^{\sum_{i=0}^{s}\binom{n}{i}+\sum_{i=0}^{r}\binom{n}{i}}}{\binom{2^n}{2^{n-s}}}. \tag{9.18}$$

If this upper bound is at most 1, then we deduce that $P_{s,r,D}<1$, and this proves that there exist permutations F from $\mathbb{F}_2^n$ to itself whose higher order nonlinearity $nl_{s,r}(F)$ is strictly greater than D. This completes the proof. ■

Let us now see what happens when n tends to ∞. Let $H_2(x)=-x\log_2(x)-(1-x)\log_2(1-x)$ be the binary entropy function.

Proposition 9.30. ([50].) *Let $\frac{s_n}{n}$ tend to a limit $\rho\le 0.227$ when n tends to ∞. If $r_n\le\mu\, n$ for every n, where $1-H_2(\mu)>\rho$ (e.g., if r_n/s_n tends to a limit strictly smaller than 1), then for every $\rho'>\rho$, almost all permutations F of $\mathbb{F}_2^n$ satisfy $nl_{s_n,r_n}(F)\ge 2^{(1-\rho')n}$.*

Proof. We know (see, e.g., [121], p. 310) that, for every integer n and every $\lambda\in[0,1/2]$, we have $\sum_{i\le\lambda n}\binom{n}{i}\le 2^{nH_2(\lambda)}$. According to the Stirling formula, we have also, when i and j tend to ∞: $i!\sim i^ie^{-i}\sqrt{2\pi i}$ and $\binom{i+j}{i}\sim\frac{(\frac{i+j}{i})^i(\frac{i+j}{j})^j}{\sqrt{2\pi}}\sqrt{\frac{i+j}{ij}}$. For $i+j=2^n$ and $i=2^{n-s_n}$, this gives

$$\binom{2^n}{2^{n-s_n}}\sim\frac{(2^{s_n})^{2^{n-s_n}}}{\sqrt{2\pi}(1-2^{-s_n})^{2^n-2^{n-s_n}}}\sqrt{\frac{2^{s_n}}{2^n-2^{n-s_n}}}$$

$$=\frac{2^{s_n2^{n-s_n}}}{\sqrt{2\pi}\,2^{(2^n-2^{n-s_n})\ln(1-2^{-s_n})\log_2 e}}\sqrt{\frac{2^{s_n}}{2^n-2^{n-s_n}}}.$$

We then deduce from Inequality (9.18)

$$\log_2 P_{s_n,r_n,D_n}=O\Bigg(2^n\Bigg[H_2\left(\frac{D_n}{2^n}\right)+2^{-n(1-H_2(s_n/n))}+2^{-n(1-H_2(r_n/n))}$$

$$-2^{-s_n+\log_2(s_n)}-2^{-s_n}(1-2^{-s_n})\log_2 e\Bigg]\Bigg)$$

(we omit $-\frac{s_n}{2^{n+1}}+\frac{n}{2^{n+1}}\log_2(1-2^{-s_n})$ inside the brackets because it is negligible).

For $\rho\le 0.227$, we have $1-H_2(\rho)>\rho$. If $\lim\frac{s_n}{n}=\rho\le 0.227$, then there exists $\rho'>\rho$ such that $1-H_2(\rho')>\rho'$ and such that asymptotically we have $s_n\le\rho' n$; hence $2^{-n(1-H_2(s_n/n))}$ is negligible with respect to 2^{-s_n}. And if $r_n\le\mu\, n$

where $1 - H_2(\mu) > \rho$, then we have $2^{-n(1-H_2(r_n/n))} = o(2^{-s_n})$, and for $D_n = 2^{(1-\rho')n}$, where ρ' is any number strictly greater than ρ, we have $H_2\left(\frac{D_n}{2^n}\right) = H_2\left(2^{-\rho' n}\right) = \rho' n\, 2^{-\rho' n} - (1 - 2^{-\rho' n})\log_2(1 - 2^{-\rho' n}) = o(2^{-\rho n}) = o(2^{-s_n})$. We obtain that, asymptotically, $nl_{s_n,r_n}(F) > 2^{(1-\rho')n}$ for every $\rho' > \rho$. ■

9.3.2.2 The Inverse S-Box

For $F_{inv}(x) = x^{2^n-2}$ and $f_{inv}(x) = tr_n(F_{inv}(x))$, we have $nl_r(F_{inv}) = nl_r(f_{inv})$ as for any power permutation. Recall that, for $r = 1$, this parameter equals $2^{n-1} - 2^{n/2}$ when n is even and is close to this number when n is odd, and that for $r > 1$, it is approximately bounded below by $2^{n-1} - 2^{(1-2^{-r})n}$ (see more in [49]). We have $NL_2(F_{inv}) = 0$, because we have $w_H(h(x, F_{inv}(x))) = 0$ for the bilinear function $h(x, y) = tr_n(axy)$ where a is any nonzero element of null trace and xy denotes the product of x and y in $\mathbb{F}_{2^n}$. Indeed, we have $x\, F_{inv}(x) = 1$ for every nonzero x. As observed in [68], we have also $w_H(h(x, F_{inv}(x))) = 0$ for the bilinear functions $h(x, y) = tr_n(a(x + x^2 y))$ and $h(x, y) = tr_n(a(y + y^2 x))$ where a is now any nonzero element, and for the quadratic functions $h(x, y) = tr_n(a(x^3 + x^4 y))$ and $h(x, y) = tr_n(a(y^3 + y^4 x))$. These properties are the core properties used in the tentative algebraic attack on the AES by Courtois and Pieprzyk [68].

It is proved in [50] that, for every ordered pair (s, r) of strictly positive integers, we have

- $nl_{s,r}(F_{inv}) = 0$ if $r + s \geq n$;
- $nl_{s,r}(F_{inv}) > 0$ if $r + s < n$;

and that, in particular, for every ordered pair (s, r) of positive integers such that $r + s = n - 1$, we have $nl_{s,r}(F_{inv}) = 2$. The other values are unknown when $r + s < n$, except for small values of n.

9.3.3 Nonlinearity of S-Boxes in Stream Ciphers

The classical notion of nonlinearity (see Definition 9.5) and its generalizations given in Subsection 9.3.2 have been introduced in the framework of block ciphers. Because of the iterative structure of these ciphers, the knowledge of a nonlinear combination by a function f of the output bits of an S-box F, such that $f \circ F$ has a low (higher order) nonlinearity, does not necessarily lead to an attack, unless the degree of f is low. This is why, in Definition 9.12, the degree of f is also specified. We recall in Figures 9.2 and 9.3 how vectorial functions can be used in the pseudorandom generators of stream ciphers to speed up the ciphers.

Because the structure of these pseudorandom generators is not iterative, all of the m binary sequences produced by an (n, m)-function can be combined by a linear or nonlinear (but nonconstant) m-variable Boolean function f to perform (fast) correlation attacks. Consequently, a second generalization to (n, m)-functions of the notion of nonlinearity has been introduced (in [57], but the definition was based on the observations of Zhang and Chan in [153]).

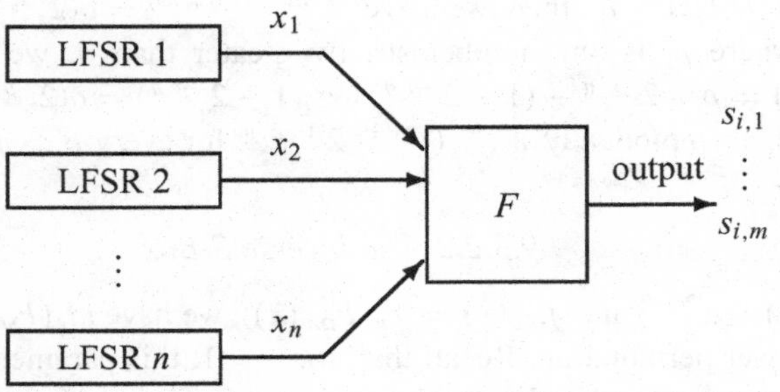

Figure 9.2. Combiner model.

Definition 9.13. *Let F be an (n, m)-function. The* unrestricted nonlinearity $unl(F)$ *of F is the minimum Hamming distance between all nonconstant affine functions and all Boolean functions $g \circ F$, where g is any nonconstant Boolean function on m variables.*

If $unl(F)$ is small, then one of the linear or nonlinear (nonconstant) combinations of the output bits to F has high correlation to a nonconstant affine function of the input, and a (fast) correlation attack is feasible.

Remark.

(1) *In Definition 9.13, the considered affine functions are nonconstant, because the minimum distance between all Boolean functions $g \circ F$ (g nonconstant) and all constant functions equals $\min_{b \in \mathbb{F}_2^m} |F^{-1}(b)|$ (each number $|F^{-1}(b)|$ is indeed equal to the distance between the null function and $g \circ F$, where g equals the indicator of the singleton $\{b\}$); it is therefore an indicator of*

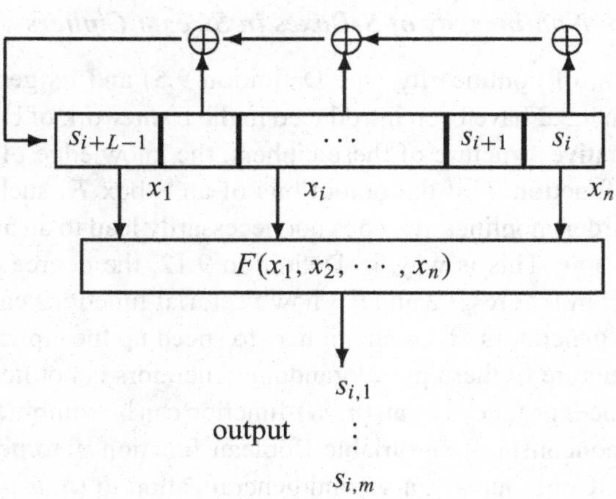

Figure 9.3. Filter model.

the balancedness of F. It is bounded above by 2^{n-m} (and it equals 2^{n-m} if and only if F is balanced).

(2) *We can replace "nonconstant affine functions" by "nonzero linear functions" in the statement of Definition 9.13 (replacing g by $g \oplus 1$, if necessary).*

(3) *Thanks to the fact that the affine functions considered in Definition 9.13 are nonconstant, we can relax the condition that g is nonconstant: the distance between a constant function and a nonconstant affine function equals 2^{n-1}, and $unl(F)$ is clearly always smaller than 2^{n-1}.*

The unrestricted nonlinearity of any (n, m)-function F is obviously unchanged when F is right-composed with an affine invertible mapping. Moreover, if A is a surjective linear (or affine) function from $\mathbb{F}_2^p$ (where p is some positive integer) into $\mathbb{F}_2^n$, then it is easily shown that $unl(F \circ A) = 2^{p-n} unl(F)$. Also, for every (m, p)-function ϕ, we have $unl(\phi \circ F) \geq unl(F)$ (indeed, the set $\{g \circ \phi,\ g \in \mathcal{BF}_p\}$, where $\mathcal{BF}_p$ is the set of p-variable Boolean functions, is included in $\mathcal{BF}_m$), and if ϕ is a permutation on $\mathbb{F}_2^m$, then we have $unl(\phi \circ F) = unl(F)$ (by applying the preceding inequality to $\phi^{-1} \circ F$).

A further generalization of the Zhang-Chan attack, called the *generalized correlation attack*, was introduced in [55]: considering implicit equations that are linear in the input variable x and of any degree in the output variable $z = F(x)$, the following probability is considered, for any nonconstant function g and every functions $w_i : \mathbb{F}_2^m \to \mathbb{F}_2$:

$$Pr\,[g(z) + w_1(z)x_1 + w_2(z)x_2 + \cdots + w_n(z)x_n = 0], \tag{9.19}$$

where $z = F(x)$ and where x uniformly ranges over $\mathbb{F}_2^n$.

The knowledge of such approximation g with a probability significantly higher than 1/2 leads to an attack, because $z = F(x)$ corresponding to the output keystream is known, and therefore $g(z)$ and $w_i(z)$ are known for all $i = 1, \ldots, n$.

This led to a new notion of generalized nonlinearity.

Definition 9.14. *Let $F : \mathbb{F}_2^n \to \mathbb{F}_2^m$. The* generalized Hadamard transform *$\hat{F} : (\mathbb{F}_2^{2^m})^{n+1} \to \mathbb{R}$ is defined as*

$$\hat{F}(g(\cdot), w_1(\cdot), \ldots, w_n(\cdot)) = \sum_{x \in \mathbb{F}_2^n} (-1)^{g(F(x)) + w_1(F(x))x_1 + \cdots + w_n(F(x))x_n},$$

where the input is an $(n+1)$-tuple of Boolean functions $g, w_i : \mathbb{F}_2^m \to \mathbb{F}_2$, $i = 1, \ldots, n$.

Let $\mathcal{W}$ be the set of all n-tuple functions $w(\cdot) = (w_1(\cdot), \ldots, w_n(\cdot))$, where w_i is an m-variable Boolean function and such that $w(z) = (w_1(z), \ldots, w_n(z)) \neq (0, \ldots, 0)$ for all $z \in \mathbb{F}_2^m$.

The generalized nonlinearity *is defined as*

$$gnl(F) = \min\{\min_{0 \neq u \in \mathbb{F}_2^m} (w_H(u \cdot F), 2^n - w_H(u \cdot F)), nonlin_{gen} F\},$$

where

$$nonlin_{gen}F = 2^{n-1} - \frac{1}{2}\max_{g\in\mathcal{G},w\in\mathcal{W}} \hat{F}(g(\cdot), w_1(\cdot), \ldots, w_n(\cdot)). \qquad (9.20)$$

The generalized nonlinearity is clearly not greater than the other nonlinearity measures and provides linear approximations with better bias for (fast) correlation attacks.

9.3.3.1 Relations to the Walsh Transforms and Lower Bounds

The unrestricted nonlinearity of F can be related to the values of the discrete Fourier transforms of the functions φ_b, and a lower bound (observed in [153]) depending on $nl(F)$ can be directly deduced.

Proposition 9.31. *For every (n, m)-function, we have*

$$unl(F) = 2^{n-1} - \frac{1}{2}\max_{u\in\mathbb{F}_2^{n*}} \sum_{b\in\mathbb{F}_2^m} |\widehat{\varphi_b}(u)| \qquad (9.21)$$

and

$$unl(F) \geq 2^{n-1} - 2^{m/2}\left(2^{n-1} - nl(F)\right). \qquad (9.22)$$

The lower bound (9.22) is far from giving a good idea of the best possible unrestricted nonlinearities: even if $nl(F)$ is close to the nonlinearity of bent functions, that is, $2^{n-1} - 2^{n/2-1}$, it implies that $unl(F)$ is approximately greater than $2^{n-1} - 2^{\frac{n+m}{2}-1}$, whereas there exist balanced $(n, n/2)$-functions F such that $unl(F) = 2^{n-1} - 2^{n/2}$ (see later discussion).

Proposition 9.32. ([55].) *Let $F : \mathbb{F}_2^n \to \mathbb{F}_2^m$ and let $w(\cdot)$ denote the n-tuple of m-bit Boolean functions $(w_1(\cdot), \ldots, w_n(\cdot))$. Then*

$$nonlin_{gen}F = 2^{n-1} - 1/2 \sum_{z\in\mathbb{F}_2^m} \max_{w(z)\in\mathbb{F}_2^n - \{0\}} |\widehat{\varphi_b}(w(z))|$$

$$= 2^{n-1} - \frac{1}{2^{m+1}} \sum_{z\in\mathbb{F}_2^m} \max_{\substack{0\neq w(z)\in\\ \mathbb{F}_2^n}} \left| \sum_{v\in\mathbb{F}_2^m} (-1)^{v\cdot z} \widehat{1_{G_F}}(w(z), v)\right|,$$

where $\widehat{1_{G_F}}$ denotes the Walsh transform. Hence,

$$gnl(F) \geq 2^{n-1} - (2^m - 1)\left(2^{n-1} - nl(F)\right).$$

9.3.3.2 Upper Bounds

If F is balanced, the minimum distance between the component functions $v \cdot F$ and the affine functions cannot be achieved by constant affine functions, because $v \cdot F$, which is a Boolean balanced function, has distance 2^{n-1} to constant functions. Hence the following proposition.

Proposition 9.33. (Covering Radius Bound.) *For every balanced S-box F, we have*

$$unl(F) \leq nl(F). \tag{9.23}$$

This implies $unl(F) \leq 2^{n-1} - 2^{n/2-1}$.

Another upper bound,

$$unl(F) \leq 2^{n-1} - \frac{1}{2}\left(\frac{2^{2m} - 2^m}{2^n - 1} + \sqrt{\frac{2^{2n} - 2^{2n-m}}{2^n - 1} + \left(\frac{2^{2m} - 2^m}{2^n - 1} - 1\right)^2 - 1}\right),$$

has been obtained in [57]. It improves upon the covering radius bound only for $m \geq n/2 + 1$, and the question of knowing whether it is possible to improve upon the covering radius bound for $m \leq n/2$ is open. In any case, this improvement will not be dramatic, at least for $m = n/2$, because it is shown (by using Relation (9.21)) in this same paper that the balanced function

$$F(x, y) = \begin{cases} \frac{x}{y} & \text{if } y \neq 0 \\ x & \text{if } y = 0 \end{cases}$$

satisfies $unl(F) = 2^{n-1} - 2^{n/2}$ (see other examples of S-boxes in [107], whose unrestricted nonlinearities seem low, however). It is pretty astonishing that an S-box with such high unrestricted nonlinearity exists; but it can be shown that this balanced function does not contribute to a good resistance to algebraic attacks and has null generalized nonlinearity (see later discussion).

Proposition 9.34. *Let $F : \mathbb{F}_2^n \to \mathbb{F}_2^m$. Then the following inequality holds:*

$$nonlin_{gen} F \leq 2^{n-1} - -\frac{1}{4}\sum_{z \in \mathbb{F}_2^m} \sqrt{\frac{2^{n+2}|F^{-1}(z)| - 4|F^{-1}(z)|^2}{2^n - 1}}.$$

Furthermore, if $F(x)$ is balanced, then we have

$$gnl(F) \leq 2^{n-1} - -2^{n-1}\sqrt{\frac{2^m - 1}{2^n - 1}}.$$

This upper bound is much lower than the covering radius bound $2^{n-1} - 2^{n/2-1}$ and than the upper bound given earlier for UN_F.

It is proved in [56] that the balanced function

$$F(x, y) = \begin{cases} \frac{x}{y} & \text{if } y \neq 0 \\ x & \text{if } y = 0 \end{cases}$$

has null generalized nonlinearity. Hence, a vectorial function may have very high unrestricted nonlinearity and have zero generalized nonlinearity. Some functions with good generalized nonlinearity are given in [56]:

(i) $F(x) = tr_{n/m}(x^k)$ where $k = 2^i + 1$, $\gcd(i, n) = 1$, is a Gold exponent;
(ii) $F(x) = tr_{n/m}(x^k)$ where $k = 2^{2i} - 2^i + 1$ is a Kasami exponent such that $3i \equiv 1$ [mod] n,

where m divides n and n is odd, and where $tr_{n/m}$ is the trace function from $\mathbb{F}_{2^n}$ to $\mathbb{F}_{2^m}$, have generalized nonlinearity satisfying $gnl(F) \geq 2^{n-1} - 2^{(n-1)/2+m-1}$.

9.4 Resilient Vectorial Boolean Functions

Resilient Boolean functions were studied in Chapter 8. The notion, when extended to vectorial functions, is relevant, in cryptology, to quantum cryptographic key distribution [5] and to pseudorandom sequence generation for stream ciphers.

Definition 9.15. *Let n and m be two positive integers. Let t be an integer such that $0 \leq t \leq n$. An (n, m)-function $F(x)$ is called* tth-order correlation-immune *if its output distribution does not change when at most t coordinates x_i of x are kept constant. It is called* t-resilient *if it is balanced and tth-order correlation-immune: that is, if it stays balanced when at most t coordinates x_i of x are kept constant.*

This notion has a relationship with another notion that also plays a role in cryptography: an (n, m)-function F is called a *multipermutation* (see [145]) if any two ordered pairs $(x, F(x))$ and $(x', F(x'))$, such that $x, x' \in \mathbb{F}_2^n$ are distinct, differ in at least $m + 1$ distinct positions (that is, collide in at most $n - 1$ positions). Such (n, m)-function then ensures a perfect diffusion. An (n, m)-function is a multipermutation if and only if the indicator of its graph $\{(x, F(x));\ x \in \mathbb{F}_2^n\}$ is an nth-order correlation-immune Boolean function (see [35]).

Because S-boxes must be balanced, we focus on resilient functions, but most of the following results can also be stated for correlation-immune functions.

We call an (n, m) function that is t-resilient an (n, m, t)-function. Clearly, if such a function exists, then $m \leq n - t$, because balanced (n, m)-functions can exist only if $m \leq n$. This bound is weak (it is tight if and only if $m = 1$ or $t = 1$). It is shown in [65] (see also [9]) that, if an (n, m, t)-function exists, then $m \leq n - \log_2 \left[\sum_{i=0}^{t/2} \binom{n}{i}\right]$ if t is even and $m \leq n - \log_2 \left[\binom{n-1}{(t-1)/2} + \sum_{i=0}^{(t-1)/2} \binom{n}{i}\right]$ if t is odd. This can be deduced from a classical bound on orthogonal arrays, due to Rao [136]. But, as shown in [9] (see also [118]), potentially better bounds can be deduced from the linear programming bound due to Delsarte [72]: $t \leq \left\lfloor \frac{2^{m-1} n}{2^m - 1} \right\rfloor - 1$ and $t \leq 2 \left\lfloor \frac{2^{m-2}(n+1)}{2^m - 1} \right\rfloor - 1$.

Note that composing a t-resilient (n, m)-function by a permutation on $\mathbb{F}_2^m$ does not change its resiliency order (this obvious result was first observed in [151]).

Also, the t-resiliency of S-boxes can be expressed by means of the t-resiliency and tth-order correlation immunity of Boolean functions.

Proposition 9.35. *Let F be an (n,m) function. Then F is t-resilient if and only if one of the following conditions is satisfied:*

(1) *for every nonzero vector $v \in \mathbb{F}_2^m$, the Boolean function $v \cdot F(x)$ is t-resilient,*
(2) *for every balanced m-variable Boolean function g, the n-variable Boolean function $g \circ F$ is t-resilient.*

Equivalently, F is t-resilient if and only if, for every vector $u \in \mathbb{F}_2^n$ such that $w_H(u) \le t$, one of the following conditions is satisfied:

(i) $\sum_{x\in\mathbb{F}_2^n}(-1)^{v\cdot F(x)+u\cdot x} = 0$, *for every $v \in \mathbb{F}_2^{m*}$,*
(ii) $\sum_{x\in\mathbb{F}_2^n}(-1)^{g(F(x))+u\cdot x} = 0$, *for every balanced m-variable Boolean function g.*

Finally, F is t-resilient if and only if, for every vector $b \in \mathbb{F}_2^m$, the Boolean function φ_b is tth-order correlation-immune and has weight 2^{n-m}.

Proof. According to the characterization recalled in the previous chapter, condition 1 (resp. condition 2) is equivalent to the fact that condition (i) (resp. condition (ii)) is satisfied for every vector $u \in \mathbb{F}_2^n$ such that $w_H(u) \le t$. Let us now prove that the t-resiliency of F implies condition 2, which implies condition 1, which implies that, for every vector $b \in \mathbb{F}_2^m$, the Boolean function φ_b is tth-order correlation-immune and has weight 2^{n-m}, which implies that F is t-resilient. If F is t-resilient, then for every balanced m-variable Boolean function g, the function $g \circ F$ is t-resilient, by definition; hence condition 2 is satisfied. This clearly implies condition 1, because the function $g(x) = v \cdot x$ is balanced for every nonzero vector v. Relation (9.4) then implies that, for every nonzero vector $u \in \mathbb{F}_2^n$ such that $w_H(u) \le t$ and for every $b \in \mathbb{F}_2^m$, we have $\widehat{\varphi_b}(u) = 2^{-m}\sum_{x\in\mathbb{F}_2^n, v\in\mathbb{F}_2^m}(-1)^{v\cdot(F(x)+b)+u\cdot x} = 2^{-m}\sum_{x\in\mathbb{F}_2^n}(-1)^{u\cdot x} = 0$. Hence, condition 1 implies that φ_b is tth-order correlation-immune for every b. Also, according to Proposition 9.2, condition 1 implies that F is balanced: *that is*, φ_b has weight 2^{n-m}, for every b. These two conditions obviously imply, by definition, that F is t-resilient. ■

Consequently, the t-resiliency of vectorial functions is invariant under the same transformations as for Boolean functions.

9.4.1 Constructions

9.4.1.1 Linear or Affine Resilient Runctions

The construction of t-resilient linear functions is easy: Bennett et al. [5] and Chor et al. [65] established the connection between linear resilient functions and linear codes (correlation-immune functions being related to orthogonal arrays; see

[37, 36], we could in fact refer to Delsarte [73] for this relationship). There exists a linear (n, m, t)-function if and only if there exists a binary linear $[n, m, t+1]$ code.

Proposition 9.36. *Let G be a generating matrix for an $[n, m, d]$ binary linear code. We define $L : \mathbb{F}_2^n \mapsto \mathbb{F}_2^m$ by the rule $L(x) = x \times G^T$, where G^T is the transpose of G. Then L is an $(n, m, d-1)$-function.*

Indeed, for every nonzero $v \in \mathbb{F}_2^m$, the vector $v \cdot L(x) = v \cdot (x \times G^t)$ has the form $x \cdot u$ where $u = v \times G$ is a nonzero codeword. Hence, u has weight at least d and the linear function $v \cdot L$ is $(d-1)$-resilient, because it has at least d independent terms of degree 1 in its ANF.

The converse of Proposition 9.36 is clearly also true.

Proposition 9.36 is still trivially true if L is affine instead of linear, that is, $L(x) = x \times G^t + a$, where a is a vector of $\mathbb{F}_2^k$.

Stinson [142] considered the equivalence between resilient functions and what he called large sets of orthogonal arrays. According to Proposition 9.35, an (n, m)-function is t-resilient if and only if there exists a set of 2^m disjoint binary arrays of dimensions $2^{n-m} \times n$, such that, in any t columns of each array, every one of the 2^t elements of $\mathbb{F}_2^t$ occurs in exaclty 2^{n-m-t} rows and no two rows are identical.

The construction of (n, m, t)-functions by Proposition 9.36 can be generalized by considering nonlinear codes of length n (that is subsets of $\mathbb{F}_2^n$) and of size 2^{n-m} whose dual distance $d^\perp$ equals $t+1$ (see [143]). In the case of Proposition 9.36, C is the dual of the code of generating matrix G. As recalled in the previous chapter, the *dual distance of a code C* of length n is the smallest nonzero integer i such that the coefficient of the monomial $X^{n-i}Y^i$ in the polynomial $\sum_{x,y \in C}(X+Y)^{n-w_H(x+y)}(X-Y)^{w_H(x+y)}$ is nonzero (when the code is linear, the dual distance is equal to the minimum Hamming distance of the dual code, according to MacWilliams' identity). Equivalently, according to the calculations made in Chapter 8 for proving the MacWilliams identity and to Proposition 9.35, the dual distance is the number $d^\perp$ such that the indicator of C is $d^\perp$-th order correlation immune. The nonlinear code needs also to be *systematic* (that is, there must exist a subset I of $\{1, \cdots, n\}$ called an *information set* of C, necessarily of size $n-m$ because the code has size 2^{n-m}, such that every possible tuple occurs in exactly one codeword within the specified coordinates x_i; $i \in I$) to allow the construction of an $(n, m, d^\perp - 1)$-function: the image of a vector $x \in \mathbb{F}_2^n$ is the unique vector y of $\mathbb{F}_2^n$ such that $y_i = 0$ for every $i \in I$ and such that $x \in y + C$ (in other words, to calculate y, we first determine the unique codeword c of C that matches with x on the information set and we have $y = x + c$). It is deduced in [143] that, for every $r \geq 3$, a $(2^{r+1}, 2^{r+1} - 2r - 2, 5)$-resilient function exists (the construction is based on the Kerdock code), and that no affine resilient function with such good parameters exists.

9.4.1.2 Maiorana-MacFarland Resilient Functions

The idea of designing resilient vectorial functions by generalizing the Maiorana-MacFarland construction is natural. One can find a first reference to such construction in a paper by Nyberg [126], but for generating perfect nonlinear functions. This technique has been used by Kurosawa et al. [110], Johansson and Pasalic [104], Pasalic and Maitra [134], and Gupta and Sarkar [92] to produce functions having high resiliency and high nonlinearity.[22]

Definition 9.16. *The class of* Maiorana-McFarland (n, m)*-functions is the set of those functions* F *that can be written in the form*

$$F(x, y) = x \times \begin{pmatrix} \varphi_{11}(y) & \cdots & \varphi_{1m}(y) \\ \vdots & \ddots & \vdots \\ \varphi_{r1}(y) & \cdots & \varphi_{rm}(y) \end{pmatrix} + H(y),\ (x, y) \in \mathbb{F}_2^r \times \mathbb{F}_2^s \tag{9.24}$$

where r *and* s *are two integers satisfying* $r + s = n$*,* H *is any* (s, m)*-function and, for every index* $i \leq r$ *and every index* $j \leq m$*,* φ_{ij} *is a Boolean function on* $\mathbb{F}_2^s$*.*

The concatenation of t-resilient functions being still t-resilient, if the transpose matrix of the matrix involved in Equation (9.24) is the generator matrix of a linear $[r, m, d]$-code for every vector y ranging over $\mathbb{F}_2^s$, then the (n, m)-function F is $(d - 1)$-resilient.
Any Maiorana-McFarland (n, m)-function F can be written in the form

$$F(x, y) = \left(\bigoplus_{i=1}^{r} x_i \varphi_{i1}(y) \oplus h_1(y), \ldots, \bigoplus_{i=1}^{r} x_i \varphi_{im}(y) \oplus h_m(y) \right) \tag{9.25}$$

where $H = (h_1, \ldots, h_m)$.

After denoting, for every $i \leq m$, by ϕ_i the (s, r)-function that admits the Boolean functions $\varphi_{1i}, \ldots, \varphi_{ri}$ for coordinate functions, we can rewrite Relation (9.25) as

$$F(x, y) = (x \cdot \phi_1(y) \oplus h_1(y), \ldots, x \cdot \phi_m(y) \oplus h_m(y)) \,. \tag{9.26}$$

Resiliency. As a direct consequence of Proposition 9.36, we have (equivalently to what was written earlier in terms of codes) the following proposition.

Proposition 9.37. *Let* n*,* m*,* r *and* s *be three integers such that* $n = r + s$*. Let* F *be a Maiorana-McFarland* (n, m)*-function defined as in Relation (9.26) and such that, for every* $y \in \mathbb{F}_2^s$*, the family* $(\phi_i(y))_{i \leq m}$ *is a basis of an* m*-dimensional subspace of* $\mathbb{F}_2^r$ *having* $t + 1$ *for minimum Hamming weight, then* F *is at least* t*-resilient.*

[22] But, as recalled in Section 9.3.3, this notion of nonlinearity is not relevant to S-boxes for stream ciphers. The unrestricted nonlinearity of resilient functions and of Maiorana-MacFarland functions must be further studied.

Nonlinearity. According to the known facts about the Walsh transform of the Boolean Maiorana-MacFarland functions, the nonlinearity $nl(F)$ of any Maiorana-McFarland (n, m)-function defined as in Relation (9.26) satisfies

$$nl(F) = 2^{n-1} - 2^{r-1} \max_{(u,u') \in \mathbb{F}_2^r \times \mathbb{F}_2^s, v \in \mathbb{F}_2^{m*}} \left| \sum_{y \in E_{u,v}} (-1)^{v \cdot H(y) + u' \cdot y} \right| \tag{9.27}$$

where $E_{u,v}$ denotes the set $\{y \in \mathbb{F}_2^s;\ \sum_{i=1}^m v_i \phi_i(y) = u\}$.

The bounds proved in Chapter 8, for the nonlinearities of Maiorana-McFarland Boolean functions, imply that the nonlinearity $nl(F)$ of a Maiorana-McFarland (n, m)-function defined as in Relation (9.26) satisfies

$$2^{n-1} - 2^{r-1} \max_{u \in \mathbb{F}_2^r, v \in \mathbb{F}_2^{m*}} |E_{u,v}| \leq nl(F) \leq 2^{n-1} - 2^{r-1} \left\lceil \sqrt{\max_{u \in \mathbb{F}_2^r, v \in \mathbb{F}_2^{m*}} |E_{u,v}|} \right\rceil .$$

If, for every element y, the vector space spanned by the vectors $\phi_1(y), \ldots, \phi_m(y)$ admits m for dimension and has a minimum Hamming weight strictly greater than k (so that F is t-resilient with $t \geq k$), then we have

$$nl(F) \leq 2^{n-1} - 2^{r-1} \left\lceil \frac{2^{s/2}}{\sqrt{\sum_{i=k+1}^{r} \binom{r}{i}}} \right\rceil . \tag{9.28}$$

The nonlinearity can be exactly calculated in two situations (at least): if, for every vector $v \in \mathbb{F}_2^{m*}$, the (s, r)-function $y \mapsto \sum_{i \leq m} v_i \phi_i(y)$ is injective, then F admits $2^{n-1} - 2^{r-1}$ for nonlinearity; and if, for every vector $v \in \mathbb{F}_2^{m*}$, this same function takes exactly two times each value of its image set, then F admits $2^{n-1} - 2^r$ for nonlinearity.

Johansson and Pasalic described in [104] a way to specify the vectorial functions $\phi_1, \ldots, \phi_m$ so that this kind of condition is satisfied. Their result can be generalized in the following form.

Lemma 9.38. *Let C be a binary linear $[r, m, t+1]$ code. Let $\beta_1, \ldots, \beta_m$ be a basis of the $\mathbb{F}_2$-vector space $\mathbb{F}_{2^m}$, and let L_0 be a linear isomorphism between $\mathbb{F}_{2^m}$ and C. Then the functions $L_i(z) = L_0(\beta_i z)$, $i = 1, \ldots, m$, have the property that, for every vector $v \in \mathbb{F}_2^{m*}$, the function $z \in \mathbb{F}_{2^m} \mapsto \sum_{i=1}^m v_i L_i(z)$ is a bijection from $\mathbb{F}_{2^m}$ into C.*

Proof. For every vector v in $\mathbb{F}_2^m$ and every element z of $\mathbb{F}_{2^m}$, we have $\sum_{i=1}^m v_i L_i(z) = L_0\left(\left(\sum_{i=1}^m v_i \beta_i\right) z\right)$. If the vector v is nonzero, then the element $\sum_{i=1}^m v_i \beta_i$ is nonzero. Hence, the function $z \in \mathbb{F}_{2^m} \mapsto \sum_{i=1}^m v_i L_i(z)$ is a bijection. ■

Because the functions $L_1, L_2, \ldots, L_m$ vanish at $(0, \ldots, 0)$, they do not satisfy the hypothesis of Proposition 9.37 (i.e., the vectors $L_1(z), \ldots, L_m(z)$ are not linearly independent for every $z \in \mathbb{F}_{2^m}$). A solution to derive a family of vectorial functions also satisfying the hypothesis of Proposition 9.37 is then to right-compose the functions L_i with a same injective (or two-to-one) function π from $\mathbb{F}_2^s$ into $\mathbb{F}_{2^m}^*$. Then, for every nonzero vector $v \in \mathbb{F}_2^{m*}$, the function $y \in \mathbb{F}_2^s \mapsto \sum_{i=1}^m v_i L_i[\pi(y)]$ is injective from $\mathbb{F}_2^s$ into C^*.

This gives the following construction:[23]

Given two integers m and r ($m < r$), construct an $[r, m, t+1]$-code C such that t is as large as possible (Brouwer gives in [20] a precise overview of the best known parameters of codes). Then, define m linear functions $L_1, \ldots, L_m$ from $\mathbb{F}_{2^m}$ into C as in Lemma 9.38. Choose an integer s strictly lower than m (resp. lower than or equal to m) and define an injective (resp. two-to-one) function π from $\mathbb{F}_2^s$ into $\mathbb{F}_{2^m}^*$. Choose any (s, m)-function $H = (h_1, \ldots, h_m)$ and denote $r + s$ by n. Then the (n, m)-function F whose coordinate functions are defined by $f_i(x, y) = x \cdot [L_i \circ \pi](y) \oplus h_i(y)$ is t-resilient and admits $2^{n-1} - 2^{r-1}$ (resp. $2^{n-1} - 2^r$) for nonlinearity.

All the primary constructions presented in [104, 110, 134, 127] are based on this principle. Also, the recent construction of (n, m, t)-functions defined by Gupta and Sarkar in [92] is also a particular application of this construction, as shown in [58].

9.4.1.3 Other Constructions

Constructions of highly nonlinear resilient vectorial functions, based on elliptic curves theory and on the trace of some power functions $x \mapsto x^d$ on finite fields, have been designed respectively by Cheon [63] and by Khoo and Gong [108]. However, it is still an open problem to design highly nonlinear functions with high algebraic degrees and high resiliency orders with Cheon's method. Besides, the number of functions that can be designed by these methods is very small.

Zhang and Zheng proposed in [151, 152] a secondary construction consisting of the composition $F = G \circ L$ of a linear resilient (n, m, t)-function L with a highly nonlinear (m, k)-function. F is obviously t-resilient and admits $2^{n-m} nl(G)$ for nonlinearity, where $nl(G)$ denotes the nonlinearity of G and its degree is the same as that of G. Taking for function G the inverse function $x \mapsto x^{-1}$ on the finite field $\mathbb{F}_{2^m}$ studied by Nyberg in [128] (and later used for designing the S-boxes of the AES), Zhang and Zheng obtained t-resilient functions having a nonlinearity greater than or equal to $2^{n-1} - 2^{n-m/2}$ and having $m - 1$ for algebraic degree. But the linear (n, m)-functions involved in the construction of Zhang and Zheng introduce a weakness: their *unrestricted nonlinearity* being null, this kind of function cannot be used as a multioutput combination function in stream ciphers. Nevertheless, this drawback can be avoided by concatenating such functions (recall that the concatenation of t-resilient functions gives t-resilient functions, and a good nonlinearity can be obtained by concatenating functions with disjoint Walsh

[23] Another construction based on Lemma 9.38 is given by Johansson and Pasalic in the same paper [104]. It involves a family of *nonintersecting codes*, that is, a family of codes having the same parameters (same length, same dimension, and same minimum distance) and whose pairwise intersections are reduced to the null vector. However, this construction is often worse for large resiliency orders, as shown in [58].

supports). In this way we obtain a modified Maiorana-McFarland construction, which should be investigated.
Other secondary constructions of resilient vectorial functions can be derived from the secondary constructions of resilient Boolean functions (see, *e.g.*, [36, 48]).

Acknowledgment

We thank Lilya Budaghyan, Anne Canteaut, and Gregor Leander for useful comments and Caroline Fontaine for her careful reading of a previous draft of this chapter.

References

[1] F. Armknecht and M. Krause. Constructing single- and multi-output boolean functions with maximal immunity. *Proceedings of ICALP 2006, Lecture Notes in Computer Science* 4052, pp. 180–91, 2006.

[2] G. Ars and J.-C. Faugère. Algebraic immunities of functions over finite fields. *Proceedings of the conference BFCA 2005*, pp. 21–38, 2005.

[3] L. A. Bassalygo, G. V. Zaitsev, and V. A. Zinoviev, Uniformly packed codes. *Problems of Information Transmission* 10(1), pp. 9–14, 1974.

[4] L.A. Bassalygo and V.A. Zinoviev. Remarks on uniformly packed codes. *Problems of Information Transmission* 13(3), pp. 22–5, 1977.

[5] C. H. Bennett, G. Brassard, and J. M. Robert. Privacy amplification by public discassion. *SIAM Journal on Computing* 17, pp. 210–29, 1988.

[6] T. Bending and D. Fon-Der-Flaass. Crooked functions, bent functions and distance regular graphs. *Electronic Journal of Combinatorics* 5, Research paper 34 (electronic), 14 pages, 1998.

[7] T. Berger, A. Canteaut, P. Charpin, and Y. Laigle-Chapuy. On almost perfect nonlinear functions. *IEEE Transactions on Information Theory* 52(9), pp. 4160–70, 2006.

[8] T. Beth and C. Ding, On almost perfect nonlinear permutations. *Proceedings of Eurocrypt' 93, Lecture Notes in Computer Science* 765, pp. 65–76, 1994.

[9] J. Bierbrauer, K. Gopalakrishnan, and D.R. Stinson. Orthogonal arrays, resilient functions, error-correcting codes, and linear programming bounds. *SIAM Journal on Discrete Mathematics* 9(3), pp. 424–52, 1996.

[10] J. Bierbrauer and G. Kyureghyan. Crooked binomials. *Designs, Codes and Cryptography* 46(3), pp. 269–301, 2008.

[11] E. Biham and A. Shamir. Differential cryptanalysis of DES-like cryptosystems. *Journal of Cryptology* 4(1), pp. 3–72, 1991.

[12] C. Bracken, E. Byrne, N. Markin, and G. McGuire. On the Walsh Spectrum of a New APN Function. *Proceedings of IMA conference on Cryptography and Coding, Lecture Notes in Computer Science* 4887, pp. 92–8, 2007.

[13] C. Bracken, E. Byrne, N. Markin, and G. McGuire. Determining the nonlinearity of a new family of APN functions. *Proceedings of AAECC-17 Conference, Lecture Notes in Computer Science* 4851, pp. 72–9, 2007.

[14] C. Bracken, E. Byrne, N. Markin, and G. McGuire. An infinite family of quadratic quadrinomial APN functions. arXiv:0707.1223v1, 2007.

[15] C. Bracken, E. Byrne, N. Markin, and G. McGuire. New families of quadratic almost perfect nonlinear trinomials and multinomials. *Finite Fields and Their Applications* 14, pp. 703–14, 2008.

[16] C. Bracken, E. Byrne, N. Markin, and G. McGuire. A few more quadratic APN functions. arXiv:0804.4799v1, 2007.

[17] C. Bracken, E. Byrne, G. McGuire, and G. Nebe. Automorphisms of some APN functions and an inequivalence result. Preprint, 2008.

[18] C. Bracken and G. Leander. New families of functions with differential uniformity of 4. Proceedings of the conference BFCA 2008, Copenhagen, to appear.

[19] M. Brinkmann and G. Leander. On the classification of APN functions up to dimension five. To appear in *Designs, Codes and Cryptography* 49(1–3), 2008. Revised and extended version of a paper with the same title in the Proceedings of the Workshop on Coding and Cryptography (in the memory of Hans Dobbertin), WCC 2007, pp. 39–48, 2007.

[20] A. E. Brouwer. Bounds on the minimum distance of linear codes (Table of the best known codes). Available at http://www.win.tue.nl/ aeb/voorlincod.html.

[21] K. Browning, J. F. Dillon, R. E. Kibler, and M. McQuistan. APN polynomials and related codes. To appear in a special volume of Journal of Combinatorics, Information and System Sciences, honoring the 75th birthday of Prof. D. K. Ray-Chaudhuri, 2008.

[22] L. Budaghyan. The simplest method for constructing APN polynomials EA-inequivalent to power functions. *Proceedings of the International Workshop on the Arithmetic of Finite Fields, WAIFI 2007, Lecture Notes in Computer Science* 4547, pp. 177–88, 2007.

[23] L. Budaghyan, C. Carlet, and A. Pott. New classes of almost bent and almost perfect nonlinear polynomials. *Proceedings of the Workshop on Coding and Cryptography 2005*, Bergen, pp. 306–15, 2005.

[24] L. Budaghyan, C. Carlet, and A. Pott. New classes of almost bent and almost perfect nonlinear functions. *IEEE Transactions on Information Theory* 52(3), pp. 1141–52, March 2006. This is a completed version of [23].

[25] L. Budaghyan, C. Carlet, P. Felke, and G. Leander. An infinite class of quadratic APN functions which are not equivalent to power functions. *Proceedings of IEEE International Symposium on Information Theory (ISIT) 2006*.

[26] L. Budaghyan, C. Carlet, and G. Leander. Two classes of quadratic APN binomials inequivalent to power functions. *IEEE Transactions on Information Theory* 54(9), pp. 4218–29, 2008. This paper is a completed and merged version of [25] and [28].

[27] L. Budaghyan, C. Carlet, and G. Leander. On inequivalence between known power APN functions. Proceedings of the conference BFCA 2008, Copenhagen, to appear.

[28] L. Budaghyan, C. Carlet, and G. Leander. Another class of quadratic APN binomials over $\mathbb{F}_{2^n}$: The case n divisible by 4. *Proceedings of the Workshop on Coding and Cryptography, WCC 2007*, pp. 49–58, 2007.

[29] L. Budaghyan, C. Carlet, and G. Leander. Constructing new APN functions from known ones. Finite Fields and Applications 15(2), pp. 150–159, 2009.

[30] L. Budaghyan and C. Carlet. Classes of Quadratic APN Trinomials and Hexanomials and Related Structures. *IEEE Transactions on Information Theory* 54(5), pp. 2354–57, 2008.

[31] L. Budaghyan and A. Pott. On differential uniformity and nonlinearity of functions. Discrete Mathematics 309(2), pp. 371–384, 2009.

[32] E. Byrne and G. McGuire. On the non-existence of crooked functions on finite fields. *Proceedings of the Workshop on Coding and Cryptography, WCC 2005*, pp. 316–24, 2005.

[33] A.R. Calderbank, G. McGuire, B. Poonen, and M. Rubinstein, *On a conjecture of Helleseth regarding pairs of binary m-sequences*. IEEE Transactions on Information Theory 42, pp. 988–90 (1996).

[34] P. Camion and A. Canteaut. Construction of t-resilient functions over a finite alphabet, *Proceedings of EUROCRYPT'96, Lecture Notes in Computer Sciences* 1070, pp. 283–93, 1996.

[35] P. Camion and A. Canteaut. Generalization of Siegenthaler inequality and Schnorr-Vaudenay multipermutations. *Proceedings of CRYPTO'96, Lecture Notes in Computer Science* 1109, pp. 372–86, 1996.

[36] P. Camion and A. Canteaut. Correlation-immune and resilient functions over finite alphabets and their applications in cryptography. *Designs, Codes and Cryptography* 16, pp. 121–49, 1999.

[37] P. Camion, C. Carlet, P. Charpin, and N. Sendrier. On correlation-immune functions, *Proceedings of CRYPTO'91, Lecture Notes in Computer Science* 576, pp. 86–100, 1991.
[38] A. Canteaut. Differential cryptanalysis of Feistel ciphers and differentially uniform mappings. *Proceedings of Selected Areas on Cryptography, SAC'97*, pp. 172–84, Ottawa, Canada, 1997.
[39] A. Canteaut. Cryptographic functions and design criteria for block ciphers. *Proceedings of INDOCRYPT 2001, Lecture Notes in Computer Science* 2247, pp. 1–16, 2001.
[40] A. Canteaut. Analysis and design of symmetric ciphers. Habilitation for Directing Theses, University of Paris 6, 2006.
[41] A. Canteaut, P. Charpin, and H. Dobbertin. A new characterization of almost bent functions. *Proceedings of Fast Software Encryption 99, Lecture Notes in Computer Science* 1636, pp. 186–200, 1999.
[42] A. Canteaut, P. Charpin, and H. Dobbertin. Binary m-sequences with three-valued crosscorrelation: A proof of Welch's conjecture. *IEEE Transactions on Information Theory* 46(1), pp. 4–8, 2000.
[43] A. Canteaut, P. Charpin, and H. Dobbertin. Weight divisibility of cyclic codes, highly nonlinear functions on $GF(2^m)$ and crosscorrelation of maximum-length sequences. *SIAM Journal on Discrete Mathematics* 13(1), pp. 105–38, 2000.
[44] A. Canteaut, P. Charpin, and M. Videau. Cryptanalysis of block ciphers and weight divisibility of some binary codes. *Information, Coding and Mathematics (Workshop in Honor of Bob McEliece's 60th Birthday)*. Kluwer, pp. 75–97, 2002.
[45] A. Canteaut and M. Videau. Degree of composition of highly nonlinear functions and applications to higher order differential cryptanalysis. *Proceedings of EUROCRYPT 2002, Lecture Notes in Computer Science* 2332, pp. 518–33, 2002.
[46] C. Carlet. A construction of bent functions. In *Finite Fields and Applications, pp. 47–58. London Mathematical Society*, Lecture Series 233, Cambridge University Press, Cambridge, UK, 1996.
[47] C. Carlet. On the confusion and diffusion properties of Maiorana-McFarland's and extended Maiorana-McFarland's functions. *Special issue "Complexity Issues in Coding and Cryptography," dedicated to Prof. Harald Niederreiter on the occasion of his 60th birthday, Journal of Complexity* 20, pp. 182–204, 2004.
[48] C. Carlet. On the secondary constructions of resilient and bent functions. *Proceedings of the Workshop on Coding, Cryptography and Combinatorics 2003, pp. 3–28, Birkhäuser Verlag*, 2004.
[49] C. Carlet. Recursive lower bounds on the nonlinearity profile of Boolean functions and their applications. *IEEE Transactions on Information Theory* 54(3), pp. 1262–72, 2008.
[50] C. Carlet. On the higher order nonlinearities of Boolean functions and S-boxes, and their generalizations. *Proceedings of SETA 2008, Lecture Notes in Computer Science* 5203, pp. 345–67, 2008.
[51] C. Carlet. On almost perfect nonlinear functions. To appear in the *Special Section on Signal Design and its Application in Communications, IEICE Transactions on Fundamentals* E91-A(12), 2008.
[52] C. Carlet, P. Charpin, and V. Zinoviev. Codes, bent functions and permutations suitable for DES-like cryptosystems. *Designs, Codes and Cryptography* 15(2), pp. 125–56, 1998.
[53] C. Carlet and C. Ding. Highly nonlinear mappings. *Special issue "Complexity Issues in Coding and Cryptography," dedicated to Prof. Harald Niederreiter on the occasion of his 60th birthday, Journal of Complexity* 20, pp. 205–44, 2004.
[54] C. Carlet and C. Ding. Nonlinearities of S-boxes. *Finite Fields and Its Applications* 13(1), pp. 121–35, 2007.
[55] C. Carlet, K. Khoo, C.-W. Lim, and C.-W. Loe. Generalized correlation analysis of vectorial Boolean functions. *Proceedings of FSE 2007. Lecture Notes in Computer Science* 4593, pp. 382–98, 2007.
[56] C. Carlet, K. Khoo, C.-W. Lim, and C.-W. Loe. On an improved correlation analysis of stream ciphers using multi-output Boolean functions and the related generalized

notion of nonlinearity. *Advances in Mathematics of Communications* 2(2), pp. 201–21, 2008.

[57] C. Carlet and E. Prouff. On a new notion of nonlinearity relevant to multi-output pseudo-random generators. *Proceedings of Selected Areas in Cryptography 2003, Lecture Notes in Computer Science* 3006, pp. 291–305, 2004.

[58] C. Carlet and E. Prouff. Vectorial functions and covering sequences. *Proceedings of Finite Fields and Applications, Fq7, Lecture Notes in Computer Science* 2948, pp. 215–48, 2004.

[59] L. Carlitz and S. Uchiyama. Bounds for exponential sums. *Duke Mathematical Journal* 1, pp. 37–41, 1957.

[60] F. Chabaud and S. Vaudenay. Links between differential and linear cryptanalysis. *Proceedings of EUROCRYPT'94, Lecture Notes in Computer Science* 950, pp. 356–65, 1995.

[61] S. Chanson, C. Ding, and A. Salomaa. Cartesian authentication codes from functions with optimal nonlinearity. *Theoretical Computer Science* 290, pp. 1737–52, 2003.

[62] P. Charpin and E. Pasalic. Highly nonlinear resilient functions through disjoint codes in projecting spaces. *Designs, Codes and Cryptography* 37, pp. 319–46, 2005.

[63] J. H. Cheon. Nonlinear vector resilient functions. *Proceedings of CRYPTO 2001, Lecture Notes in Computer Science* 2139, pp. 458–69, 2001.

[64] J. H. Cheon and D. H. Lee. Resistance of S-boxes against algebraic attacks. *Proceedings of FSE 2004, Lecture Notes in Computer Science* 3017, pp. 83–94, 2004.

[65] B. Chor, O. Goldreich, J. Hastad, J. Friedman, S. Rudich, and R. Smolensky. The bit extraction problem or t-resilient functions. *Proceedings of the 26th IEEE Symposium on Foundations of Computer Science*, pp. 396–407, 1985.

[66] S. D. Cohen and R. W. Matthews. A class of exceptional polynomials. *Transactions of the AMS* 345, pp. 897–909, 1994.

[67] N. Courtois, B. Debraize, and E. Garrido. On exact algebraic [non-]immunity of S-boxes based on power functions. IACR e-print archive 2005/203.

[68] N. Courtois and J. Pieprzyk. Cryptanalysis of block ciphers with overdefined systems of equations. *Proceedings of ASIACRYPT 2002, Lecture Notes in Computer Science* 2501, pp. 267–87, 2003.

[69] T. W. Cusick and H. Dobbertin, Some new 3-valued crosscorrelation functions of binary sequences, *IEEE Transactions on Information Theory* 42, pp. 1238–40, 1996.

[70] J. Daemen and V. Rijmen. AES proposal: Rijndael. Available at http://csrc.nist.gov/encryption/aes/rijndael/Rijndael.pdf, 1999.

[71] E. R. van Dam and D. Fon-Der-Flaass. Codes, graphs, and schemes from nonlinear functions. *European Journal of Combinatorics* 24(1), pp. 85–98, 2003.

[72] P. Delsarte. Bounds for unrestricted codes, by linear programming. *Philips Research Reports* 27, pp. 272–89, 1972.

[73] P. Delsarte. An algebraic approach to the association schemes of coding theory. PhD thesis, Université Catholique de Louvain, 1973.

[74] J. F. Dillon. APN polynomials and related codes. Banff Conference, November 2006.

[75] J. F. Dillon. Multiplicative difference sets via additive characters. *Designs, Codes and Cryptography* 17, pp. 225–35, 1999.

[76] J. F. Dillon and H. Dobbertin. New cyclic difference sets with Singer parameters. *Finite Fields and Applications* 10, pp. 342–89, 2004.

[77] H. Dobbertin. Construction of bent functions and balanced Boolean functions with high nonlinearity. *Proceedings of Fast Software Encryption, Second International Workshop, Lecture Notes in Computer Science* 1008, pp. 61–74, 1995.

[78] H. Dobbertin. One-to-one highly nonlinear power functions on $GF(2^n)$. *Applicable Algebra in Engineering, Computing and Communication* 9(2), pp. 139–52, 1998.

[79] H. Dobbertin. Kasami power functions, permutation polynomials and cyclic difference sets. *Proceedings of the NATO-A.S.I. Workshop "Difference Sets, Sequences and Their Correlation Properties,"* Bad Windsheim, pp. 133–58, Kluwer Verlag, 1998.

[80] H. Dobbertin. Another proof of Kasami's theorem. *Designs, Codes and Cryptography* 17, pp. 177–80, 1999.

[81] H. Dobbertin, Almost perfect nonlinear power functions on $GF(2^n)$: The Welch case, *IEEE Transactions on Information Theory* 45(4), pp. 1271–5, 1999.

[82] H. Dobbertin, Almost perfect nonlinear power functions on $GF(2^n)$: The Niho case. *Information and Computation* 151, pp. 57–72, 1999.

[83] H. Dobbertin. Almost perfect nonlinear power functions on GF(2^n): A new case for n divisible by 5. *Proceedings of Finite Fields and Applications* Fq5, Augsburg, Germany, Springer, pp. 113–21, 2000.

[84] H. Dobbertin. Uniformly representable permutation polynomials. *Proceedings of Sequences and Their Applications, SETA 01, Discrete Mathematics and Theoretical Computer Science*, pp. 1–22, Springer, 2002.

[85] H. Dobbertin. Private communication, 1998.

[86] Y. Edel, G. Kyureghyan, and A. Pott. A new APN function which is not equivalent to a power mapping. *IEEE Transactions on Information Theory* 52(2), pp. 744–7, 2006.

[87] Y. Edel and A. Pott. A new almost perfect nonlinear function that is not quadratic. Preprint, 2008.

[88] K. Feng, Q. Liao, and J. Yang. Maximal values of generalized algebraic immunity. To appear in Designs, Codes and Cryptography.

[89] J. Friedman. The bit extraction problem. *Proceedings of the 33th IEEE Symposium on Foundations of Computer Science*, pp. 314–19, 1992.

[90] R. Gold, Maximal recursive sequences with 3-valued recursive crosscorrelation functions. *IEEE Transactions on Information Theory*. 14, pp. 154–6, 1968.

[91] F. Göloglu and A. Pott. Results on the crosscorrelation and autocorrelation of sequences. *Proceedings of Sequences and their Applications – SETA 2008 – Lecture Notes in Computer Science* 5203, pp. 95–105, 2008.

[92] K. Gupta and P. Sarkar. Improved construction of nonlinear resilient S-boxes. *Proceedings of ASIACRYPT 2002, Lecture Notes in Computer Science* 2501, pp. 466–83, 2002.

[93] K. Gupta and P. Sarkar. Construction of perfect nonlinear and maximally nonlinear multiple-output Boolean functions satisfying higher order strict avalanche criteria. *IEEE Transactions on Information Theory* 50, pp. 2886–94, 2004.

[94] T. Helleseth and P. V. Kumar. Sequences with low correlation. In *Handbook of Coding Theory*, V. Pless and W. C. Huffman, eds. Vol. II, pp. 1765–854, Elsevier, Amsterdam, 1998.

[95] T. Helleseth and D. Sandberg. Some power mappings with low differential uniformity. *Applicable Algebra in Engineering, Computing and Communication* 8, pp. 363–70, 1997.

[96] T. Helleseth, C. Rong, and D. Sandberg. New families of almost perfect nonlinear power mappings. *IEEE Transactions on Information Theory* 45, pp. 475–85, 1999.

[97] T. Helleseth and V. Zinoviev. On $\mathbb{Z}_4$-linear Goethals codes and Kloosterman sums. *Designs, Codes and Cryptography* 17, pp. 269–88, 1999.

[98] H. Hollman and Q. Xiang. A proof of the Welch and Niho conjectures on crosscorrelations of binary m-sequences. *Finite Fields and Their Applications* 7, pp. 253–86, 2001.

[99] K. Horadam. *Hadamard Matrices and their Applications*. Princeton University Press, Princeton, NJ, 2006.

[100] X.-d. Hou. Affinity of permutations of $\mathbb{F}_2^n$. *Proceedings of the Workshop on Coding and Cryptography* WCC 2003, pp. 273–80, 2003. Completed version in *Discrete Applied Mathematics* 154(2), pp. 313–25, 2006.

[101] T. Jakobsen and L.R. Knudsen. The interpolation attack on block ciphers. *Proceedings of Fast Software Encryption'97, Lecture Notes in Computer Science* 1267, pp. 28–40, 1997.

[102] H. Janwa and R. Wilson, Hyperplane sections of Fermat varieties in P^3 in char. 2 and some applications to cyclic codes. *Proceedings of AAECC-10, Lecture Notes in Computer Science* 673, pp. 180–194, 1993.

[103] D. Jedlicka. APN monomials over $GF(2^n)$ for infinitely many n. Preprint.

[104] T. Johansson and E. Pasalic. A construction of resilient functions with high nonlinearity. *Proceedings of the IEEE International Symposium on Information Theory*, Sorrente, Italy, 2000.

[105] D. Jungnickel and A. Pott. Difference sets: An introduction. In *Difference Sets, Sequences and Their Autocorrelation Properties*, A. Pott, P. V. Kumar, T. Helleseth, and D. Jungnickel, eds., pp. 259–95, Kluwer, Amsterdam, 1999.

[106] T. Kasami. The weight enumerators for several classes of subcodes of the second order binary Reed-Muller codes, *Information and Control* 18, pp. 369–94, 1971.

[107] K. Khoo, G. Gong, and D. Stinson. Highly nonlinear S-boxes with reduced bound on maximum correlation. *Proceedings of 2003 IEEE International Symposium on Information Theory*, 2003. Available at http://www.cacr.math.uwaterloo.ca/techreports/2003/corr2003–12.ps

[108] K. Khoo and G. Gong. New constructions for resilient and highly nonlinear Boolean functions. *Proceedings of 8th Australasian Conference, ACISP 2003, Lecture Notes in Computer Science* 2727, pp. 498–509, 2003.

[109] L. Knudsen. Truncated and higher order differentials. *Proceedings of Fast Software Encryption, Second International Workshop, Lecture Notes in Computer Science* 1008, pp. 196–211, 1995.

[110] K. Kurosawa, T. Satoh, and K. Yamamoto. Highly nonlinear t-resilient functions. *Journal of Universal Computer Science* 3(6), pp. 721–29, 1997.

[111] G. Kyureghyan. Differentially affine maps. *Proceedings of the Workshop on Coding and Cryptography, WCC 2005*, pp. 296–305, 2005.

[112] G. Kyureghyan. The only crooked power functions are $x^{2^k+2^l}$. *European Journal of Combinatorics* 28(4), pp. 1345–50, 2007.

[113] G. Lachaud and J. Wolfmann. The weights of the orthogonals of the extended quadratic binary Goppa codes. *IEEE Transactions on Information Theory* 36, pp. 686–92, 1990.

[114] J. Lahtonen, G. McGuire, and H. Ward. Gold and Kasami-Welch functions, quadratic forms and bent functions. *Advances in Mathematics of Communications* 1, pp. 243–50, 2007.

[115] X. Lai. Higher order derivatives and differential cryptanalysis. *Proceedings of the Symposium on Communication, Coding and Cryptography, in honor of J. L. Massey on the occasion of his 60th birthday*, 1994.

[116] P. Langevin and P. Véron. On the nonlinearity of power functions. *Designs, Codes and Cryptography* 37, pp. 31–43, 2005.

[117] G. Leander and P. Langevin. On exponents with highly divisible Fourier coefficients and conjectures of Niho and Dobbertin. *Proceedings of the First Symposium on Algebraic Geometry and Its Applications (SAGA'07), Tahiti, 2007, Vol. 5, pp. 410–418, Series on Number Theory and Its Applications.* World Scientific, Singapore, 2008.

[118] V. I. Levenshtein. Split orthogonal arrays and maximum independent resilient systems of functions. *Designs, Codes and Cryptography* 12(2), pp. 131–60, 1997.

[119] Q. Liao and K. Feng. A note on algebraic immunity of vectorial Boolean functions. Preprint.

[120] R. Lidl and H. Niederreiter. *Finite fields*, In Encyclopedia of Mathematics and Its Applications, Vol. 20, Addison-Wesley, Reading, MA, 1983.

[121] F. J. MacWilliams and N. J. Sloane. *The theory of error-correcting codes*, North-Holland, Amsterdam, 1977.

[122] M. Matsui. Linear cryptanalysis method for DES cipher. *Proceedings of EUROCRYPT'93, Lecture Notes in Computer Science* 765, pp. 386–97, 1994.

[123] G. McGuire and A.R. Calderbank. Proof of a conjecture of Sarwate and Pursley regarding pairs of binary m-sequences. *IEEE Transactions on Information Theory* 41(4), pp. 1153–5, 1995.

[124] N. Nakagawa and S. Yoshiara. A construction of differentially 4-uniform functions from commutative semifields of characteristic 2. *Proceedings of the International Workshop on the Arithmetic of Finite Fields, WAIFI 2007, Lecture Notes in Computer Science* 4547, pp. 134–46, 2007.

[125] Y. Nawaz, K. Gupta, and G. Gong. Efficient Techniques to find algebraic immunity of S-boxes based on power mappings. *Proceedings of the Workshop on Coding and Cryptography 2007* WCC, pp. 237–46, 2007.

[126] K. Nyberg. Perfect non-linear S-boxes. *Proceedings of EUROCRYPT '91, Lecture Notes in Computer Science* 547, pp. 378–86, 1992.

[127] K. Nyberg. On the construction of highly nonlinear permutations. *Proceedings of EUROCRYPT '92, Lecture Notes in Computer Science* 658, pp. 92–8, 1993.

[128] K. Nyberg. Differentially uniform mappings for cryptography. *Proceedings of EUROCRYPT '93, Lecture Notes in Computer Science* 765, pp. 55–64, 1994.

[129] K. Nyberg. New bent mappings suitable for fast implementation. *Proceedings of Fast Software Encryption 1993, Lecture Notes in Computer Science* 809, pp. 179–84, 1994.

[130] K. Nyberg. S-boxes and round functions with controllable linearity and differential uniformity. *Proceedings of Fast Software Encryption 1994, Lecture Notes in Computer Science* 1008, pp. 111–30, 1995.

[131] K. Nyberg. Correlation theorems in crytptanalysis. *Discrete Applied Mathematics* 111 (Special Issue on Coding and Cryptography), pp. 177–88, 2000.

[132] K. Nyberg. Multidimensional Walsh transform and a characterization of bent functions. *Proceedings of Information Theory Workshop ITW 2007*, Bergen, Norway, July 2007.

[133] K. Nyberg and L. R. Knudsen. Provable security against differential cryptanalysis. *Proceedings of CRYPTO' 92, Lecture Notes in Computer Science* 740, pp. 566–74, 1993.

[134] E. Pasalic and S. Maitra. Linear codes in generalized construction of resilient functions with very high nonlinearity. *IEEE Transactions on Information Theory* 48, pp. 2182-91, 2002, completed version of a paper published in the *Proceedings of Selected Areas in Cryptography, SAC 2001, Lecture Notes in Computer Science* 2259, pp. 60–74, 2002.

[135] E. Prouff. DPA attacks and S-boxes. *Proceedings of Fast Software Encryption 2005, Lecture Notes in Computer Science* 3557, pp. 424–42, 2005.

[136] C. R. Rao. Factorial experiments derivable from combinatorial arrangements of arrays. *Journal of the Royal Statistical Society* 9, pp. 128–39, 1947.

[137] F. Rodier Bounds on the degrees of APN polynomials. Proceedings of the conference BFCA 2008, Copenhagen, to appear.

[138] T. Satoh, T. Iwata, and K. Kurosawa. On cryptographically secure vectorial Boolean functions. *Proceedings of Asiacrypt 1999, Lecture Notes in Computer Science* 1716, pp. 20–8, 1999.

[139] J. Seberry, X.-M. Zhang, and Y. Zheng. Nonlinearity characteristics of quadratic substitution boxes. *Proceedings of Selected Areas in Cryptography (SAC '94)*. This paper appeared under the title "Relationship among nonlinearity criteria" in the *Proceedings of EUROCRYPT '94, Lecture Notes in Computer Science* 950, pp. 376–88, 1995.

[140] T. Shimoyama and T. Kaneko. Quadratic relation of S-box and its application to the linear attack of full round DES. *Proceedings of CRYPTO 98, Lecture Notes in Computer Science* 1462, pp. 200–11, 1998.

[141] V. M. Sidelnikov. *On the mutual correlation of sequences*. Soviet Mathematics Doklad 12, pp. 197–201, 1971.

[142] D. R. Stinson. Resilient functions and large sets of orthogonal arrays. *Congressus Numerantium* 92, pp. 105–10, 1993.

[143] D. R. Stinson and J. L. Massey. An infinite class of counterexamples to a conjecture concerning nonlinear resilient functions. *Journal of Cryptology* 8(3), pp. 167–73, 1995.

[144] A. Tardy-Corfdir and H. Gilbert. A known plaintext attack on feal-4 and feal-6. In *Proceedings of CRYPTO'91, Lecture Notes in Computer Science* 576, pp. 172–81, 1991.

[145] S. Vaudenay. On the need for multipermutations: cryptanalysis of MD4 and SAFER. *Proceedings of Fast Software Encryption, Lecture Notes in Computer Science* 1008, pp. 286–97, 1995.

[146] J. F. Voloch. Symmetric cryptography and algebraic curves. *Proceedings of the first Symposium on Algebraic Geometry and its Applications" (SAGA'07), Tahiti, 2007, published by Series on Number Theory and Its Applications*, Vol. 5, pp. 135–141, World Scientific, 2008.

[147] T. Wadayama, T. Hada, K. Wagasugi, and M. Kasahara. Upper and lower bounds on the maximum nonlinearity of n-input m-output Boolean functions. *Designs, Codes and Cryptography* 23, pp. 23–33, 2001.

[148] A. F. Webster and S. E. Tavares. On the design of S-boxes. In *Proceedings of CRYPTO'85, Lecture Notes in Computer Science* 219, pp. 523–34, 1985.

[149] I. Wegener, *The Complexity of Boolean Functions*. John Wiley & Sons, Chichester, 1987.

[150] L. Welch, Lower bounds on the maximum cross correlation of signals. *IEEE Transactions on Information Theory* 20(3), pp. 397–9, 1974.

[151] X.-M. Zhang and Y. Zheng. On nonlinear resilient functions. *Proceedings of EURO-CRYPT '95, Lecture Notes in Computer Science* 921, pp. 274–88, 1995.

[152] X.-M. Zhang and Y. Zheng. Cryptographically resilient functions. *IEEE Transactions on Information Theory* 43, pp. 1740–7, 1997.

[153] M. Zhang and A. Chan. Maximum correlation analysis of nonlinear S-boxes in stream ciphers. *Proceedings of CRYPTO 2000, Lecture Notes in Computer Science* 1880, pp. 501–14, 2000.

[154] Y. Zhou and C. Li. The Walsh spectrum of a new family of APN functions. *Proceedings of WSPC*, 2008.

Part IV

Graph Representations and Efficient Computation Models

10

Binary Decision Diagrams

Beate Bollig, Martin Sauerhoff, Detlef Sieling, and Ingo Wegener

10.1 Introduction

Let $f\colon D^n \to R$ be a finite function: that is, D and R are finite sets. Such a function can be represented by the table of all $(a, f(a))$, $a \in D^n$, which always has an exponential size of $|D|^n$. Therefore, we are interested in representations that for many important functions are much more compact. The best-known representations are circuits and decision diagrams. Circuits are a hardware model reflecting the sequential and parallel time to compute $f(a)$ from a (see Chapter 11). *Decision diagrams* (DDs), also called *branching programs* (BPs), are nonuniform programs for computing $f(a)$ from a based on only two types of instructions represented by nodes in a graph (see also Figure 10.1):

- *Decision nodes*: depending on the value of some input variable x_i the next node is chosen.
- *Output nodes* (also called *sinks*): a value from R is presented as output.

A decision diagram is a directed acyclic graph consisting of decision nodes and output nodes. Each node v represents a function f_v defined in the following way. Let $a = (a_1, \ldots, a_n) \in D^n$. At decision nodes, choose the next node as described before. The value of $f_v(a)$ is defined as the value of the output node that is finally reached when starting at v. Hence, for each node each input $a \in D^n$ activates a unique *computation path* that we follow during the computation of $f_v(a)$. An edge $e = (v, w)$ of the diagram is called *activated* by a if the computation path starting at v runs via e. A graph-theoretical path is called *inconsistent* if it is not activated by any input.

The obvious resources consumed by the DD are the *size* of the DD, defined as the number of its nodes, and the *length* of the DD, defined as the length of the longest computation path of the DD. Note that the length may be smaller than the depth, which is the length of the longest graph-theoretical path. The size and the length of a function f are defined as the smallest size and length, resp., of a DD representing f at some node. As an example, we consider the function *hidden*

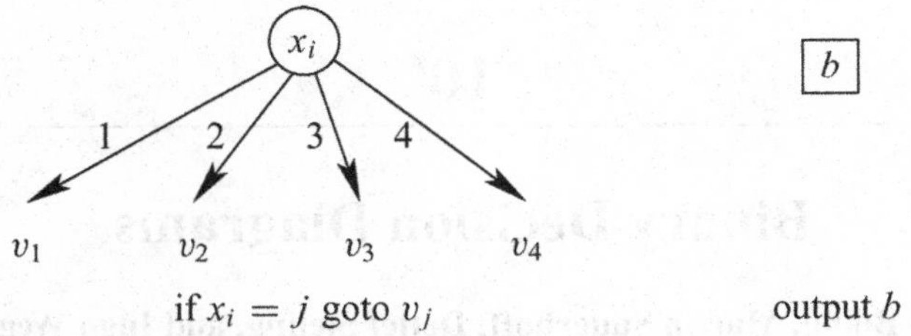

Figure 10.1. A decision node for $D = \{1, 2, 3, 4\}$ and an output node.

weighted bit, $\text{HWB}_n : \{0, 1\}^n \rightarrow \{0, 1\}$, defined by $\text{HWB}_n(x) = x_s$ where s is the number of ones in the input and $x_0 = 0$. Figure 10.2 shows two DDs for HWB_4.

It is obvious that each function $f : D^n \rightarrow R$ can be represented by a DD of length n, implying that DDs are not adequate to prove time lower bounds. In Section 10.6, we investigate the relationship between DD-based complexity measures and other complexity measures. Then it turns out that the logarithm of the DD size is closely related to the space complexity. Hence, DDs are adequate for proving space lower bounds and time-space trade-offs. Such bounds are easier to obtain for large domains D, implying that the Boolean case $D = \{0, 1\}$ and $R = \{0, 1\}$ is of special interest. In the following, we consider the Boolean case (if nothing else is mentioned), and DDs for the Boolean case are called *binary decision diagrams* (BDDs).

Lower bounds on the BDD complexity of explicitly defined Boolean functions are difficult to prove. In Section 10.7, we describe the largest known lower bound,

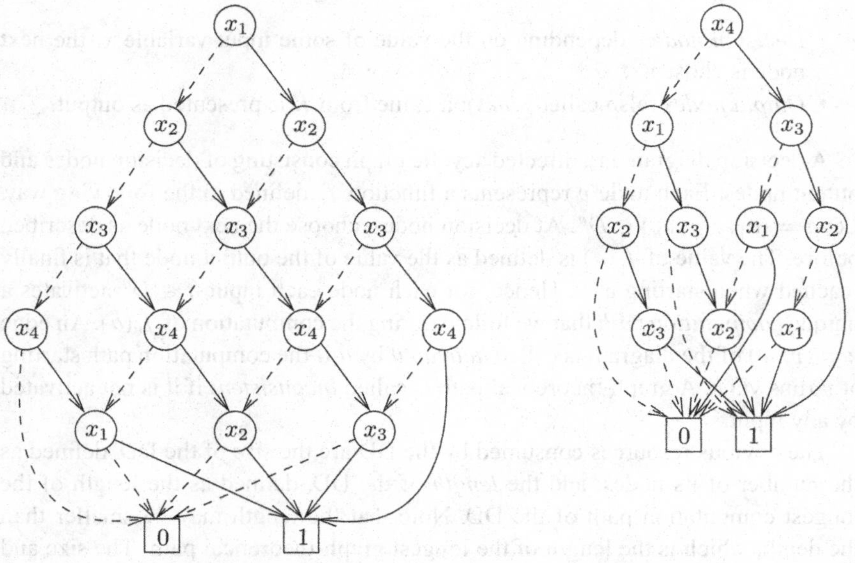

Figure 10.2. Two DDs for HWB on four Boolean variables. Dotted lines represent edges labeled by 0 and solid ones, edges labeled by 1.

whose size is only $\Omega(n^2/\log^2 n)$. This result has not been improved in more than 40 years, which motivates the investigation of restricted BDDs.

A BDD is called *s-oblivious* for a sequence $s = (s_1, \ldots, s_\ell)$ of (not necessarily different) variables if the node set can be partitioned into $\ell + 1$ levels such that all nodes on level $i \leq \ell$ are decision nodes for the variable s_i, all nodes on level $\ell + 1$ are output nodes, and each edge leaving level i reaches some level $j > i$. Lower bounds for oblivious BDDs are presented in Section 10.8. The special case that s is a permutation $(x_{\pi(1)}, \ldots, x_{\pi(n)})$ of the input variables is called an *ordered binary decision diagram* (OBDD) and, if the permutation π is fixed, π-OBDD. Then $(x_{\pi(1)}, \ldots, x_{\pi(n)})$ is the *variable order* of the π-OBDD. If s repeats the variable order π k times, the resulting BDDs are called k-π-OBDDs and, for arbitrary π, k-OBDDs. If s is the concatenation of k (perhaps different) variable orders corresponding to $(\pi_1, \ldots, \pi_k)$, the resulting BDDs are called $(\pi_1, \ldots, \pi_k)$-IBDDs or simply k-IBDDs. Oblivious BDDs and, in particular, π-OBDDs allow the efficient realization of many important operations on Boolean functions. Hence, we may use them not only as programs for Boolean functions but also for the manipulation of Boolean functions, that is, as a dynamic data structure for Boolean functions. Indeed, π-OBDDs are the most often applied data structure for Boolean functions, with many applications in various areas, among them verification, model checking, CAD tools, optimization, and graph algorithms. In Section 10.2, the requirements for data structures for Boolean functions are discussed, and in Section 10.3, the corresponding efficient algorithms for π-OBDDs are presented.

The other type of restrictions is length restrictions. In the *read-k model*, the access to the variables is restricted. It is essential to distinguish *syntactic* from *semantic* restrictions. In a syntactic read-k BDD graph, theoretical paths contain at most k decision nodes labeled with x_i for each variable x_i. In a semantic read-k BDD, only computation paths have to fulfill the read-k restriction. Syntactic restrictions simplify lower bound proofs. In the case of $k = 1$, called *free binary decision diagrams* (FBDDs), the syntactic and the semantic restriction coincide, since inconsistent paths are impossible. Defining graph orders as canonical generalization of variable orders, one obtains G-FBDDs (generalizing π-OBDDs) whose algorithmic properties are discussed in Section 10.4. Only length-restricted BDDs where the length of all computation paths is bounded allow the proof of general time-space trade-offs. In Section 10.9, such results are presented.

Deterministic models of computation such as BDDs and their restrictions can be generalized to *randomized* and *nondeterministic models*. This is done by allowing *randomized nodes* with two outgoing edges where one of these edges is activated independently from all other random decisions with probability $1/2$. In the Boolean case, this leads to an acceptance and a rejection probability for each input a. We may distinguish the model of zero error, the model of one-sided error, and the model of two-sided error. The permitted error probability is a critical value for certain BDD models, because not all models allow an efficient *probability amplification*. The case of unbounded but nontrivial probability for one-sided errors is the case of *existential nondeterminism* (OR-nondeterminism), which also can be described in the usual way. An input a is accepted if it activates at least one accepting computation path. We also investigate *universal* (AND-) nondeterminism and *parity* (EXOR-) nondeterminism. Lower bounds for nondeterministic BDDs are discussed together with their deterministic counterparts in Section 10.8. Those restricted OR-π-OBDDs that have good algorithmic properties are called *partitioned* BDDs (PBDDs) and are discussed in Section 10.5. Results on randomized BDD variants are presented in Section 10.10.

This chapter is a brief introduction to the rich world of BDDs and BPs, including aspects of complexity theory, algorithms, data structures, and applications. A comprehensive monograph is available (Wegener [52]).

10.2 Data Structures for Boolean Functions

In many applications, such as symbolic verification, test pattern generation, symbolic simulation, analysis of circuits and automata, and logical synthesis, representations of Boolean functions are needed that are small for many and, in particular, for important functions and simultaneously allow the efficient execution of important operations. In order to look for suitable data structures, one first has to clarify which operations have to be supported. Bryant [15] has presented a list of problems for which efficient algorithms should exist.

- EVALUATION PROBLEM. Given a representation G for some Boolean function f and an input a, compute $f(a)$.
- SATISFIABILITY TEST. Given a representation G for some Boolean function f, decide whether $f(a) = 1$ for some input a.
- SATISFIABILITY COUNT. Given a representation G for some Boolean function f, compute $|f^{-1}(1)|$.
- SATISFIABILITY ALL. Given a representation G for some Boolean function f, compute a list of all $a \in f^{-1}(1)$.
- MINIMIZATION (OR REDUCTION). Given a representation G for some Boolean function f, compute a minimal-size representation G' for f within the class of representations described by the chosen data structure. If G' is unique (up to isomorphism), the computation of G' is called *reduction* of G.
- SYNTHESIS. Given representations G_1 and G_2 for some Boolean functions g and h and some binary Boolean operation $\otimes$, compute a representation G for $f := g \otimes h$.
- EQUALITY TEST. Given representations G_1 and G_2 for some Boolean functions g and h, decide whether g and h are equal.
- REPLACEMENT BY CONSTANTS. Given a representation G for some Boolean function g, some variable x_i, and some constant $c \in \{0, 1\}$, compute a representation G' for $f := g_{|x_i=c}$ defined by $f(a) = g(a_1, \ldots, a_{i-1}, c, a_{i+1}, \ldots, a_n)$.
- REPLACEMENT BY FUNCTIONS. Given representations G_1 and G_2 for some Boolean functions g and h and some variable x_i, compute a representation G' for $f := g_{|x_i=h}$ defined by $f = (\overline{h} \wedge g_{|x_i=0}) \vee (h \wedge g_{|x_i=1})$.

- EXISTENTIAL QUANTIFICATION. Given a representation G for some function g and some variable x_i, compute a representation G' for $f := g_{|x_i=0} \vee g_{|x_i=1}$.
- UNIVERSAL QUANTIFICATION. Given a representation G for some function g and some variable x_i, compute a representation G' for $f := g_{|x_i=0} \wedge g_{|x_i=1}$.

In order to motivate some of the listed operations, we discuss a typical computer-aided design application (see also the monograph [19] as well as Chapters 15 and 16 in the present volume for different viewpoints on this topic). The physical construction of a new VLSI circuit for a Boolean function f is quite expensive. Hence, it is necessary to verify the correctness of a circuit design. *Verification of combinational circuits* is the problem to prove the equality of a specification, such as an already verified circuit, and a new realization. Let us denote the function realized by the new design f'. Verification can be done by transforming the specification and the realization into the chosen representation type and by an equality test for the results. If $f \neq f'$, we like to analyze the set of inputs a where $f(a) \neq f'(a)$. For the equality test, we may design a representation of $g = f \oplus f'$. The equality test for f and f' is equivalent to the satisfiablity test for g. If $g^{-1}(1)$ is not empty, the size of $g^{-1}(1)$ measures the number of inputs for which f' computes the wrong value. If the number of 1-inputs for g is small, it may be useful to list $g^{-1}(1)$ in order to correct the new design.

Several representations for Boolean functions are known. Many data structures such as circuits, formulas, and branching programs allow succinct representations of many Boolean functions, but not all operations can be performed in polynomial time. For example, the satisfiability test is NP-complete for these models, and the circuit minimization problem seems to be even harder. For other data structures such as disjunctive or conjunctive normal forms or decision trees, simple functions require representations of exponential size.

Ordered binary decision diagrams (OBDDs), introduced by Bryant [15], are a compromise. If the order of the variables is fixed, efficient algorithms for operations on OBDDs have to work only under the assumption that all functions are represented by OBDDs respecting the same order π of the variables. Efficient polynomial algorithms on π-OBDDs exist for all operations mentioned earlier (see Section 10.3). Although π-OBDDs are a nice data structure with many advantages, there are several functions that only have OBDDs of exponential size but that appear to be simple enough to be represented in small poynomial size. This observation has led to several extensions of the OBDD model. In Section 10.4 and Section 10.5, we discuss graph-driven BDDs and partitioned BDDs, which are less restricted than OBDDs but nevertheless have good algorithmic properties.

10.3 OBDDs and π-OBDDs

The first variant of binary decision diagrams that was suggested as a data structure for Boolean functions was π-OBDDs (Bryant [15]), and they are still the most popular. In this section we discuss the properties of π-OBDDs that make them suitable for this purpose. We furthermore present some theoretical results on π-OBDDs. For details on the efficient implementation of algorithms on π-OBDDs, we refer to Somenzi [50].

Examples of functions with small π-OBDDs are all symmetric functions, all outputs of the addition of two numbers, or the bitwise equality test of two n-bit strings. In Figure 10.3, π-OBDDs for the functions $x_1 \oplus x_2 \oplus x_3$ and for the bitwise equality test of $(x_0, \ldots, x_{n-1})$ and $(y_0, \ldots, y_{n-1})$ are shown. On the other hand, there are several functions that only have OBDDs of exponential size, such as the hidden weighted bit function HWB from Section 10.1 or the middle bit of the multiplication of two n-bit numbers (Bryant [16]; see Section 10.8); As a tool for obtaining the π-OBDD size of some function we may look at the functions f_v represented at the internal nodes v of the π-OBDD. For the examples in Figure 10.3 these functions are indicated at the nodes, where $[A]$ takes the value 1 if the expression A is true and otherwise the value 0. The π-OBDD size can be obtained by the following lemma due to Sieling and Wegener [46]. To simplify the presentation the lemma is only shown for the variable order $x_1, \ldots, x_n$, that is, for $\pi = \text{id}$. Remember that a function f essentially depends on x_i if the two subfunctions $f_{|x_i=0}$ and $f_{|x_i=1}$ are different.

Lemma 10.1. *Let S_i be the set of subfunctions $f_{|x_1=c_1,\ldots,x_{i-1}=c_{i-1}}$ that essentially depend on x_i and where $c_1, \ldots, c_{i-1} \in \{0, 1\}$. A minimal-size id-OBDD for f contains exactly $|S_i|$ nodes labeled by x_i.*

The proof of the lemma also shows that for each function in S_i there is an internal node computing this function and that the successors of this node are uniquely determined. This implies the following result, which was first proved by Bryant [15].

Theorem 10.2. *For each function f, the minimal-size π-OBDD is unique up to isomorphism.*

The question arises how to obtain the minimal-size π-OBDD for some function f from an arbitrary π-OBDD for f. We assume that there are no nodes that are not reachable from the node representing f, which we therefore may call the source of the π-OBDD for f. Bryant [15] observed that the following two simple reduction rules suffice to minimize π-OBDDs: by the deletion rule, nodes where the 0- and 1-successor coincide are deleted and the incoming edges are redirected to the successor. The merging rule allows two nodes v and w with the same label, the same 0-successors, and the same 1-successors to be merged: that is, v is deleted and the incoming edges are redirected to w. Similarly, sinks with the same label can be merged. The reduction rules are illustrated in Figure 10.4. Bryant proved that independent of the order of the application of these rules, we eventually obtain the minimal-size π-OBDD, which is also called the reduced π-OBDD. In order to obtain an efficient reduction algorithm, it is useful to apply the reduction rules bottom-up, because each application of a reduction rule may allow new applications of reduction rules only for preceding nodes. The reduction algorithm

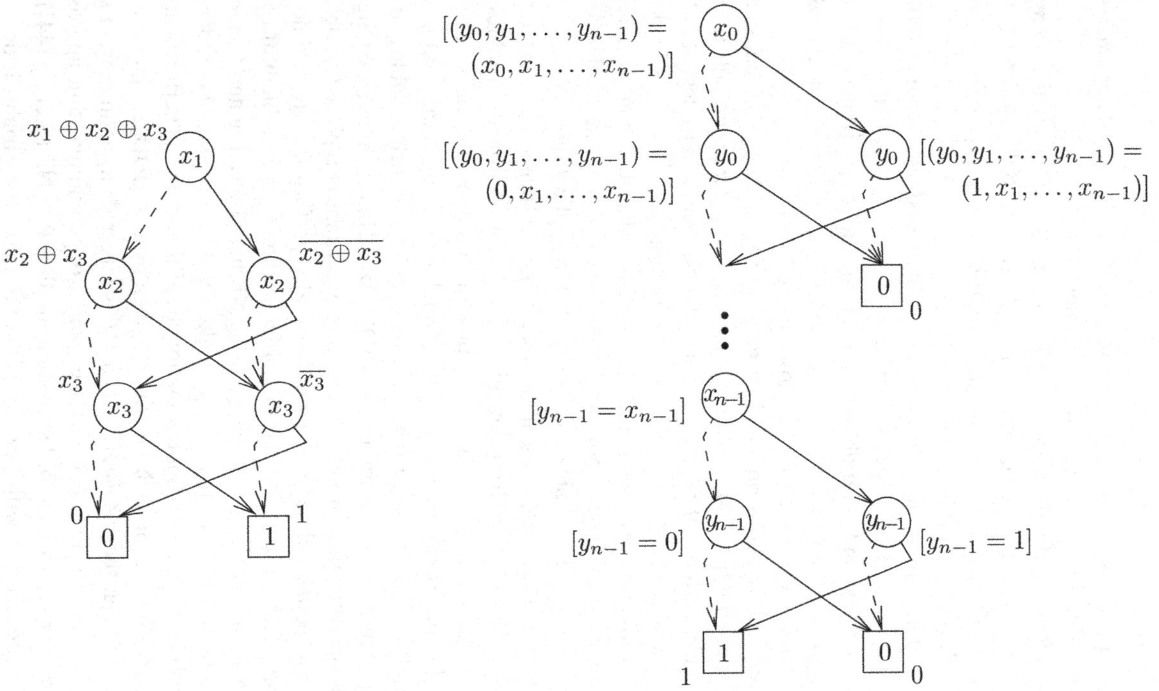

Figure 10.3. Examples of OBDDs for the functions $x_1 \oplus x_2 \oplus x_3$ and the bitwise equality test of $(x_0, \ldots, x_{n-1})$ and $(y_0, \ldots, y_{n-1})$.

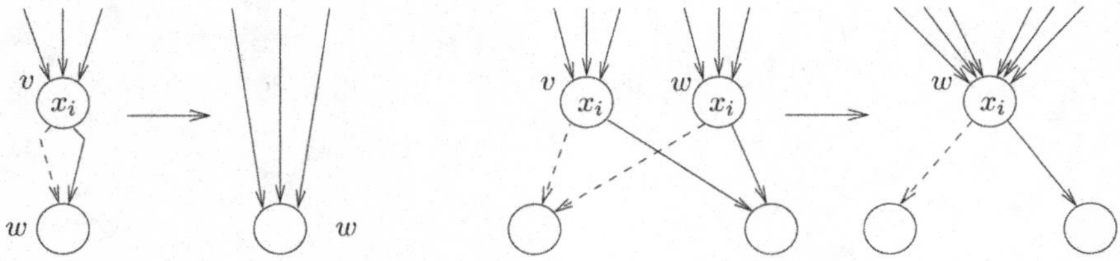

Figure 10.4. The reduction rules for OBDDs.

of Bryant [15] has a run time of $O(|G|\log|G|)$ and was later improved to a linear-time algorithm by Sieling and Wegener [47].

The uniqueness of reduced π-OBDDs and the efficient reduction of π-OBDDs have consequences for the EQUALITY TEST of two π-OBDDs: it suffices to reduce the given π-OBDDs and test them for isomorphy, which is possible in linear time because of the labels of the edges.

A careful investigation of the bitwise equality test shows that its π-OBDDs are only small for a variable order π such as $x_0, y_0, x_1, y_1, \ldots$, that is, for a variable order where the x- and y-variables with the same index are arranged near together. For example, for the variable order $x_0, \ldots, x_{n-1}, y_0, \ldots, y_{n-1}$, the π-OBDD size is exponential. This can easily be shown using Lemma 10.1 because the number of different subfunctions obtained by replacing $x_0, \ldots, x_{n-1}$ by constants is 2^n. Intuitively the π-OBDD has to store $x_0, \ldots, x_{n-1}$ in order to compute the function, which requires exponential size. A more general approach to describe this effect is communication complexity theory; see Section 10.8. Wegener [52] has even shown that for almost all choices of the variable order, the functions addition and bitwise equality test have exponential π-OBDD size. This shows that algorithms for choosing a good variable order are needed in applications.

An algorithm for computing an optimal variable order was presented by Friedman and Supowit [21]. However, this algorithm works on truth tables and has an exponential run time. Several heuristics for improving the variable order were presented in the literature. An example of such a heuristic is the sifting algorithm due to Rudell [38]. This algorithm successively chooses each variable and searches for the position leading to minimal OBDD size. This is done by trying all possible positions for the variable. The sifting algorithm is reported to obtain good results in reasonable time. However, it may also perform poorly, and no efficient algorithm for optimizing the variable order is known. The problem of computing an optimal variable order was shown to be NP-hard by Bollig and Wegener [11]. In applications, efficient approximation algorithms for the variable order problem would also be helpful. An approximation algorithm with the performance ratio c is an algorithm that, for each function f given by an OBDD, computes a variable order π such that the π-OBDD size for f is larger than the minimum size by a factor of at most c. However, even the existence of polynomial-time approximation algorithms for the variable order problem implies P = NP (Sieling [44]). Hence, we have to be satisfied with heuristics for optimizing the variable order.

A satisfying assignment of a function f given by a π-OBDD can be found by searching for a path from the source to a 1-sink. SATISFIABILITY ALL can be solved by enumerating all such paths where we have to take into account that variables not tested on such a path can take both values 0 and 1. Furthermore, an algorithm for SATISFIABILITY COUNT may use a labeling procedure that for each node v stores the number of assignments to the variables such that the corresponding computation path leads through v. The run time for SATISFIABILITY and SATISFIABILITY COUNT is linear with respect to the input size, and for SATISFIABILITY ALL the run time is linear with respect to the input and output size. We remark that these algorithms depend on the property that in π-OBDDs each variable may be tested at most once on each computation path such that computation paths and graph-theoretical paths coincide. For many variants of BDDs without this property, the satisfiability operations are NP-hard.

In applications such as hardware verification, π-OBDDs have to be computed for functions represented by circuits. We first remark that this problem is NP-hard because SATISFIABILITY is NP-hard for circuits but can be done in linear time for π-OBDDs. The general approach for the transformation of circuits into π-OBDDs works in the following way: We run through the circuit in some topological order. For the functions represented at the inputs, that is, projections on variables, it is simple to construct π-OBDDs. For a function represented at the output of a gate, it is possible to compute a π-OBDD by combining the π-OBDDs representing the functions at the input of the gate with SYNTHESIS. In the following we briefly discuss SYNTHESIS.

SYNTHESIS of π-OBDDs $G_1 = (V_1, E_1)$ and $G_2 = (V_2, E_2)$ is mainly based on computing a product graph of the given π-OBDDs. The node set of the resulting π-OBDD is some subset of $V_1 \times V_2$, and reaching an x_i-node $(v_1, v_2) \in V_1 \times V_2$ for some assignment a to $x_1, \ldots, x_{i-1}$ means that in G_1 the computation path for the partial assignment a ends at the node v_1 and in G_2 at the node v_2. It follows that the number of nodes in the resulting π-OBDD is bounded above by $|V_1||V_2|$, and it is even possible to compute this π-OBDD in time $O(|V_1||V_2|)$. The resulting π-OBDD is not necessarily reduced. We remark that in implementations the reduction is integrated into the synthesis algorithm in order to avoid the construction of large nonreduced OBDDs.

The computation of the product graph is possible only if the given OBDDs G_1 and G_2 for f_1 and f_2 have the same variable order. Otherwise the construction of an OBDD for, as an example, $f_1 \wedge f_2$ is even NP-hard (Fortune et al. [20]). Nevertheless, it is a common technique to change the variable order during the computation of an OBDD from a circuit in order to avoid large intermediate OBDDs (Bryant [17]).

REPLACEMENT BY CONSTANTS can be performed by redirecting the edges leading to each x_i-node v to the c-successor of v. If the source is labeled by x_i, the c-successor is defined as the new source of the OBDD. The quantification operations are combinations of REPLACEMENT BY CONSTANTS and SYNTHESIS. Finally, to perform REPLACEMENTS BY FUNCTIONS, we define the ternary operation $\text{ite}(a, b, c) =$

$ab \vee \bar{a}c$ (if a then b else c) and compute a π-OBDD for $g_{|x_i=h} = \text{ite}(h, g_{|x_i=1}, g_{|x_i=0})$ by a generalization of SYNTHESIS to ternary operators.

Altogether, for many of the important operations on Boolean functions there are efficient algorithms working on π-OBDDs, and π-OBDDs are of reasonable size for many important functions. However, in applications the operations on π-OBDDs, in particular SYNTHESIS, have to be applied successively such that we cannot bound the size of intermediate π-OBDDs. Hence, π-OBDDs are a heuristic approach for solving practically important problems.

10.4 Graph-Driven BDDs

Several important and also quite simple functions have exponential OBDD size. Therefore, more general representations with good algorithmic behavior are necessary. With a restricted variant of FBDDs such as the restriction of OBDDs to π-OBDDs, we obtain a new data structure for Boolean functions. Gergov and Meinel [22] and Sieling and Wegener [48] have independently generalized the concept of variable orders to graph orders.

A *graph order* is a BDD with a single sink, where on each path from the source to the sink all variables appear exactly once. A *graph-driven* FBDD G' according to a graph order G, or G-FBDD for short, is an FBDD with the following property: if for an input a, a variable x_i appears on the computation path of a in G' before the variable x_j, then x_i also appears on the computation path of a in G before x_j.

In graph-driven BDDs (according to a fixed order), for each input the variables are tested in the same order, whereas (different from OBDDs) for different inputs different orders may be used. It is not difficult to see that any FBDD G' is a graph-driven FBDD for a suitably chosen graph order: we merge the sinks of G' to a single sink. Then we run through G' top-down. Let $Var(v)$ be the set of all variables tested on some path between the source and v. Replace each edge (v, w), where v is labeled by x_i, with a list of tests for the variables in $Var(w) \setminus (Var(v) \cup \{x_i\})$ in order to obtain G. It follows that graph, driven BDDs have the same expressive power as FBDDs, that is, all Boolean functions with polynomial-size FBDDs can be represented by graph-driven BDDs of polynomial size.

For the operations EVALUATION, SATISFIABILITY, SATISFIABILITY ALL, and SATISFIABILITY COUNT, the OBDD algorithms (see Section 10.3) also work for FBDDs, but efficient algorithms are not known for all operations. Blum, Chandra, and Wegman [8] have proved that the EQUALITY TEST for FBDDs is contained in co-RP, that is, the inequality can be tested probabilistically with one-sided error in polynomial time but no deterministic polynomial-time algorithm is known. Furthermore, SYNTHESIS may lead to an exponential blow-up, and the problem REPLACEMENT BY FUNCTIONS is as hard as the synthesis problem.

Because efficient algorithms exist for all operations on π-OBDDs and a fixed variable order π, we hope for efficient algorithms on G-FBDDs and a fixed graph order G. First, we look at the functions that have to be represented at the internal nodes of a G-FBDD. Let v be a node in the graph order G and let w.l.o.g.

$Var(v) = \{x_1, \ldots, x_{i-1}\}$, and $\mathcal{A}(v) \subseteq \{0, 1\}^{i-1}$ be the set of partial assignments $(a_1, \ldots, a_{i-1})$ to the variables $x_1, \ldots, x_{i-1}$ such that v is reached for all inputs a starting with $(a_1, \ldots, a_{i-1})$. We define $\mathcal{F}_v = \{f_{|x_1=a_1,\ldots,x_{i-1}=a_{i-1}} | (a_1, \ldots, a_{i-1}) \in \mathcal{A}(v)\}$ and denote the graph order that is the subgraph of G with source v by $G(v)$. A G-FBDD representing a function f has to contain a $G(v)$-driven FBDD for each subfunction $f_v \in \mathcal{F}_v$.

Sieling and Wegener [48] have proved that there is (up to isomorphism) a unique G-FBDD of minimal size for each function f, that is, G-FBDDs are a canonical representation of Boolean functions, and the operation Reduction is well defined. Like π-OBDDs, a G-FBDD is reduced iff neither the deletion rule nor the merging rule is applicable. Sieling and Wegener [48] have designed a linear-time reduction algorithm for G-FBDDs. The difficulty is to decide how to proceed bottom-up. In the case of OBDDs, the variable order helps to investigate the x_i-node before the x_j-node if x_j precedes x_i in the variable order. A graph order can combine many variable orders. Therefore, the bottom-up application of the reduction rules has to be guided quite carefully in order to guarantee the linear run time.

The representation by reduced G-FBDDs implies a linear-time Equality Test because the equality check is a simple isomorphism check.

For the Synthesis of two G-FBDDs $G_1 = (V_1, E_1)$ and $G_2 = (V_2, E_2)$, we simultaneously run through $G = (V, E)$, G_1, and G_2. The node set of the resulting G-FBDD is some subset of $V \times V_1 \times V_2$. Therefore, the size of the resulting G-FBDD can be bounded above by $|V||V_1||V_2|$, and the result can be computed in time $O(|V||V_1||V_2|)$. This algorithm does not create reduced G-FBDDs. Because the application of BDDs in practice is limited more by restrictions of the available storage space than by restrictions of the available time, the reduction is integrated into the synthesis process in implementations.

Replacement by Constants and Quantification may cause an exponential blow-up of the size of G-FBDDs. Since $f_{|x_i=c} = (\overline{c} \wedge f_{|x_i=0}) \vee (c \wedge f_{|x_i=1})$, the same holds for Replacement by Functions. As a consequence, G-FBDDs cannot be used efficiently if our application needs one of these operations. But if we know in advance which variables are used in these operations, we may work with a graph order with some additional properties. A graph order G is called x_i-oblivious if for each x_i-node its 0-successor coincides with its 1-successor. The Replacement and Quantification problems can be solved efficiently for variables for which the considered graph order is oblivious.

Similarly to the variable ordering problem for OBDDs, we are faced with the graph ordering problem for FBDDs, that is, the problem of finding a suitable graph order. Sieling [45] has shown that the existence of polynomial-time approximation schemes for optimizing the graph order implies P=NP. The known heuristics do not lead to satisfactory results. The only graph ordering algorithm tested in experiments is due to Bern, Meinel, and Slobodová [7]. Their approach creates graph orders of the following kind. For a parameter d, the graph order starts with a complete binary tree of depth d. For each leaf of this tree, a variable order of the remaining $n - d$ variables follows.

10.5 Partitioned Binary Decision Diagrams

Nondeterminism is a powerful compexity theoretical concept, and nondeterministic representations of Boolean functions can be much smaller than their deterministic counterparts. For example, the function HWB (see Section 10.1) has exponential OBDD size but can be represented by OR-OBDDs of size $O(n^3)$. The output bit x_i, $1 \le i \le n$, is guessed; afterward it is verified whether $x_1 + \cdots + x_n = i$ and $x_i = 1$. A disadvantage is that the simple operation NOT causes an exponential size blow-up for some functions. In order to obtain representation types with good algorithmic behavior, we have to consider restrictions where, in particular, NEGATION is not difficult. Partitioned binary decision diagrams (PBDDs), introduced by Jain, Bitner, Fussell, and Abraham [26] and more intensively studied by Narayan, Jain, Fujita, and Sangiovanni-Vincentelli [33], have the desired properties. PBDDs are a generalized OBDD model allowing a restricted use of nondeterminism and different variable orders. They are restricted enough such that most of the essential operations can be performed efficiently, and they allow polynomial-size representations for more functions than OBDDs.

We define (k, w, π)-PBDDs where $k = k(n)$ is the number of parts, $w = (w_1, \ldots, w_k)$ is the vector of so-called window functions, and $\pi = (\pi_1, \ldots, \pi_k)$ is the vector of variable orders. A necessary condition for the window functions is that their disjunction is the constant 1. Figure 10.5 describes a (k, w, π)-PBDD. One of the k parts is chosen nondeterministically. The ith part represents $f \wedge w_i$ by a π_i-OBDD. The function represented by the (k, w, π)-PBDD equals

$$(f \wedge w_1) + \cdots + (f \wedge w_k) = f \wedge (w_1 + \cdots + w_k) = f,$$

explaining the necessary condition for the window functions. The window functions are called disjoint if $w_i \wedge w_j = 0$ for $i \neq j$. Then $w_1^{-1}(1), \ldots, w_k^{-1}(1)$ are a partition of the input space justifying the notion of partitioned BDDs. Then

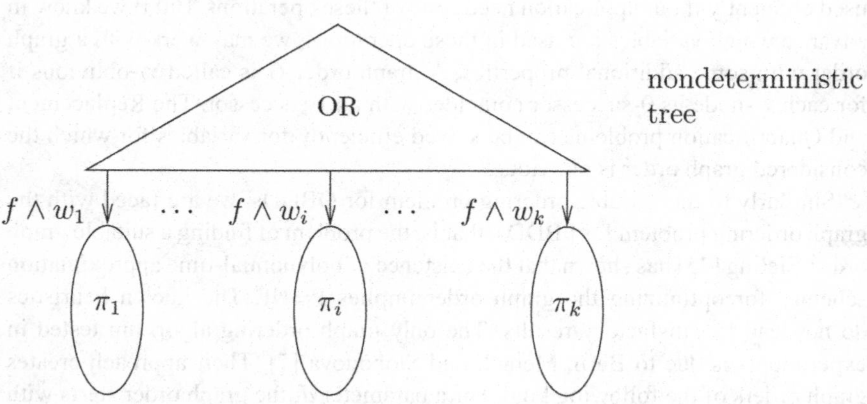

Figure 10.5. A partitioned (nondeterministic) BDD with k parts representing $f \wedge w_1, \ldots, f \wedge w_k$.

at most one part computes 1, which is known in complexity theory as unique nondeterminism.

The size of a (k, w, π)-PBDD consisting of the π_i-OBDDs G_i, $1 \le i \le k$, is the sum of the sizes of its parts G_i. If only the number k of parts is fixed, we use the notion of k-PBDDs. Because the algorithms on PBDDs need π_i-OBDDs for the corresponding window functions w_i, π and w should be chosen such that the π_i-OBDD size of w_i is small.

As in the case of OBDDs, we look for efficient algorithms only if the number of parts k, the vector of window functions w, and the vector of variable orders are fixed. The representation by (k, w, π)-PBDDs is canonical and the REDUCTION of a (k, w, π)-PBDD G is possible in time $O(|G|)$ because it can be done individually for all parts. Moreover, there exists an efficient equality check for two functions f and g because we can use the EQUALITY TEST for π-OBDDs (see Section 10.3) to check pairwise the equality of the corresponding parts. SYNTHESIS of (k, w, π)-PBDDs G_f and G_g, representing the functions f and g respectively, can be computed in time $O(|G_f||G_g|)$ for the operations AND, OR, and EXOR. It is sufficient to apply the π_i-OBDD synthesis algorithm to the ith parts of G_f and G_g, $1 \le i \le k$. Generally, $\overline{f \wedge w_i} \neq \overline{f} \wedge w_i$, but

$$(\overline{f \wedge w_i}) \wedge w_i = (\overline{f} \vee \overline{w_i}) \wedge w_i = \overline{f} \wedge w_i.$$

Hence, NEGATION can be performed as negation of each part followed by an AND-SYNTHESIS with the corresponding window function. A function f is satisfiable iff one of the functions $f \wedge w_i$ is satisfiable. SATISFIABILITY COUNT is difficult if the window functions are nondisjoint, because many parts may compute 1 for an input. For disjoint window functions, the number of satisfying inputs is the sum of the number of satisfying inputs for the parts.

The quantification operations are based on REPLACEMENT BY CONSTANTS, but this operation causes problems for (k, w, π)-PBDDs. Let g_n test whether a Boolean matrix consists of rows with exactly one 1-entry, and let h_n be the analogous function for the columns. The function $f_n = sg_n \vee \overline{s}h_n$ can be represented by 2-PBDDs and the window functions s and $\overline{s}$. For the first part we use a rowwise variable order, and for the second part a columnwise variable order. For the replacement of s by 1, we obtain the function g_n. Then we have to represent $\overline{s}g_n$ in the second part by a columnwise variable order that needs exponential size.

The heuristics for the generation of window functions often construct window functions that are the minterms with respect to a small set V of variables. Then the replacement of a variable $x_i \in X_n \setminus V$ by a constant is easy. This holds more generally if the window functions do not essentially depend on x_i. Then $f_{|x_i=c} \wedge w_j = (f \wedge w_j)_{|x_i=c}$, and the replacement can be done for each part in the usual way (see Section 10.3). Afterwards, QUANTIFICATION is a binary synthesis operation.

Why are the window functions necessary? The model of k-PBDDs in its general form has several practical drawbacks. NEGATION may lead to an exponential blow-up. Even if the k variable orders are fixed, it is difficult to check whether two k-PBDDs G' and G'' represent the same function. An input a may lead to 1 in the

ith part of G', but only in the jth part of G'', where $i \neq j$. Furthermore, it is easy to prove that the EQUALITY TEST is co-NP complete already for 2-PBDDs.

Since the number of parts and the window functions have to be fixed if we work with (k, w, π)-PBDDs, one may ask whether the choice of the number of parts and the corresponding window functions is crucial for small size PBDDs. Bollig and Wegener [12] have proved that PBDDs of polynomial size form a tight hierarchy with respect to the number k of their parts. They have shown that k-PBDDs may be exponentially larger than $(k+1)$-PBDDs for the same function if $k = o((\log n / \log\log n)^{1/2})$. Sauerhoff [40] has improved their result up to $k = O((n/\log^{1+\varepsilon} n)^{1/4})$, where $\varepsilon > 0$ is an arbitrarily small constant.

Sometimes we find appropriate window functions because of our knowledge of the structural properties of the functions: for example, for HWB we use the window functions w_i, $0 \leq i \leq n$, which compute 1 iff $x_1 + \cdots + x_n = i$. Good heuristic algorithms for the automatic creation of appropriate window functions have been developed using methods known as functional partitioning (Jain et al. [26] and Lai et al. [32]). Heuristic algorithms for which experimental results have proved the practical usefulness have also been presented by Jain, Mohanram, Moundanos, Wegener, and Lu [27].

10.6 BDD Size versus Other Complexity Measures

Because BDDs are a nonuniform model of computation, we discuss relations between the size of BDDs and complexity measures for nonuniform models of computation, namely nonuniform space complexity, circuit size, and formula size. We recall that for considering nonuniform models of computation we decompose the function $f\colon \{0,1\}^* \to \{0,1\}$ into a sequence $f = (f_n)$, where $f_n\colon \{0,1\}^n \to \{0,1\}$, and for each function of the sequence we may design a different algorithm or circuit. A *nonuniform Turing machine* is a Turing machine with an advice tape, whose contents are initialized only depending on the input length such that it may perform different algorithms for different input lengths. For more details we refer to Johnson [28].

The first result concerns the relation between BDD size and nonuniform sequential space complexity. Let $\mathrm{BDD}(f_n)$ denote the minimum size of a BDD computing f_n. Let $\mathrm{S}(f_n)$ be the minimum space of a nonuniform Turing machine computing f_n. The following relationship between $\mathrm{BDD}(f_n)$ and $\mathrm{S}(f_n)$ was obtained by Cobham [18] and Pudlák and Žák [37].

Theorem 10.3. *Let $f = (f_n)$ be a sequence of Boolean functions. If $\mathrm{S}(f_n) \geq \log n$, then $\mathrm{BDD}(f_n) = 2^{O(\mathrm{S}(f_n))}$. If $\mathrm{BDD}(f_n) \geq n$, then $\mathrm{S}(f_n) = O(\log \mathrm{BDD}(f_n))$.*

Proof. The theorem follows from simulations between BDDs and nonuniform space-bounded Turing machines. For the first statement, we observe that the number of configurations of a nonuniform $\mathrm{S}(f_n)$ space-bounded Turing machine on input length n is at most $2^{c\,\mathrm{S}(f_n)}$ for some constant c. For each such configuration, the BDD for input length n contains a node. Accepting configurations are 1-sinks and rejecting configurations are 0-sinks. In each nonhalting configuration C, the Turing machine reads one input bit x_i and, according to the value $d \in \{0, 1\}$ of x_i, the successor configuration C_d is reached. Then C is simulated by a BDD node labeled by x_i and with the successors corresponding to C_0 and C_1.

For the second statement we construct a nonuniform Turing machine from a given sequence of BDDs of size $\mathrm{BDD}(f_n)$. For input length n, the advice tape contains a coding of the BDD for input length n. The simulation of the computation of the BDD for the given input merely requires storing a pointer to the node reached. Such a pointer can be stored in space $O(\log \mathrm{BDD}(f_n))$. ■

Now we consider the relation between BDD size and circuit and formula size. Let $\mathrm{C}(f_n)$ and $\mathrm{L}(f_n)$ denote the minimum size of circuits and formulas over the basis {AND, OR, NOT} for f_n, resp., where it is common to count only the number of AND- and OR-gates and to neglect the NOT-gates. Let $\mathrm{L}^*(f_n)$ denote the minimum size of formulas for f_n over the basis consisting of all Boolean functions with two inputs.

Theorem 10.4. *Let $f = (f_n)$ be a sequence of Boolean functions. Then*

$$\frac{\mathrm{C}(f_n)}{3} \le \mathrm{BDD}(f_n) \le \mathrm{L}(f_n) + 3$$

and for each $\varepsilon > 0$,

$$\mathrm{BDD}(f_n) = O(\mathrm{L}^*(f_n)^{1+\varepsilon}).$$

Again the first inequalities can be proved by simulations. In order to simulate a BDD by a circuit, we replace each c-sink of the BDD by the constant input c, and we replace each node labeled by x_i and with the successors a and b by a multiplexer circuit. This circuit computes the function $a\bar{x}_i \vee bx_i$. Furthermore, we assume that the edges are now directed from the constant inputs to the former source of the BDD. Then we obtain a circuit computing the function represented by the BDD, where the output is the former source of the BDD. Each internal node is replaced by one OR-gate and two AND-gates. Hence, the size of the circuit is bounded above by three times the size of the BDD.

The simulation of a formula by a BDD can be performed using the SYNTHESIS operation, which, however, is performed in a way different from that for π-OBDDs. Let two BDDs G_1 and G_2 for functions f_1 and f_2 be given, where we may assume that each of the BDDs only contains one 0-sink and one 1-sink. Then a BDD for $f_1 \wedge f_2$ can be obtained by replacing the 1-sink of G_1 by a copy of G_2 and merging the 0-sinks of G_1 and G_2. Similarly, a BDD for $f_1 \vee f_2$ can be obtained by replacing the 0-sink of G_1 by a copy of G_2 and merging the 1-sinks of G_1 and G_2. Obviously the size of the resulting BDD is bounded by $|G_1| + |G_2| - 2$. A BDD for $\bar{f}_1$ can be obtained from G_1 by exchanging the sinks. For $f_1 \oplus f_2$ the situation is different: We replace the 0-sink of G_1 by a copy of G_2 and the 1-sink of G_1 by a BDD for $\bar{f}_2$, which we obtain from G_2 by exchanging the sinks. Alternatively, we may exchange the roles of G_1 and G_2.

Now we may apply the algorithm given in Section 10.3 for the transformation of circuits into BDDs to the given formula. For formulas over the basis {AND, OR, NOT}, the size bound follows by a simple induction. For formulas with arbitrary binary gates, we may first replace each gate in a straightforward way by an AND- or EXOR-gate combined with some NOT-gates. In the foregoing simulation of an EXOR-gate combining f_1 and f_2, we need two copies of G_1 or G_2. Sauerhoff, Wegener, and Werchner [42] showed how to perform this simulation in order to obtain a BDD for f_n of size bounded by $\alpha(\mathrm{L}^*(f_n)+1)^\beta$, where $\alpha \le 1.360$ and $\beta = \log_4(3+\sqrt{5}) < 1.195$. The stronger result shown in Theorem 10.4 was obtained by Giel [23] using much more involved arguments.

10.7 Lower Bounds for BDDs

In the last section, we have seen that lower bounds on the BDD size imply lower bounds of the same size on the {AND, OR, NOT}-formula size and lower bounds of almost the same size on the formula size for the full binary basis. A method due to Nechiporuk [34] still yields the largest lower bounds for BDDs as well as for general binary formulas (see Chapter 11).The idea is that a BDD for f of limited size can realize only a limited number of subfunctions of f that are obtained by replacing the variables outside a chosen set $S \subseteq X := \{x_1, \ldots, x_n\}$ by constants. Hence, functions with many different subfunctions need BDDs of not too small a size. Before we prove Nechiporuk's bound, we show that BDDs of a fixed, given size cannot represent too many functions. The bound of the following lemma is sufficient for our purposes.

Lemma 10.5. *The number of functions $f\colon \{0,1\}^n \to \{0,1\}$ of BDD size s is at most $n^{s-2}((s-1)!)^2$.*

Proof. It suffices to count the syntactically different BDDs of size s. Each BDD for nonconstant functions has two sinks and can be described by an ordered list of $s-2$ instructions. For each instruction, we have the choice between n variables. For the ith instruction and each successor, we have the choice between $s-i$ nodes, namely the instructions $i+1, \ldots, s-2$ and two sinks. Hence, the number of syntactically different BDDs of size s is bounded by $n^{s-2}((s-1)!)^2$. ■

In the following, an *S-subfunction* of f is a subfunction obtained by replacing the variables outside S by constants.

Theorem 10.6. *Let $f\colon \{0,1\}^n \to \{0,1\}$ essentially depend on all n variables and let $S_1, \ldots, S_k$ be disjoint subsets of the variable set X. Then*

$$\mathrm{BDD}(f) = \Omega\left(\sum_{1\le i\le k,\, s_i\ge 3} (\log s_i)/\log\log s_i\right),$$

where s_i is the number of S_i-subfunctions of f.

Proof. For technical reasons, we assume that $s_i > 16$ for all i. It is obvious that the sum of the terms in the claimed lower bound for which this is not fulfilled is at most $O(n)$, while on the other hand trivially $\mathrm{BDD}(f) \geq n$, because f essentially depends on all variables.

We fix an optimal BDD G representing f. Let t_i be the number of nodes labeled by S_i-variables. This implies that $\mathrm{BDD}(f) \geq t_1 + \cdots + t_k + 2$. We prove that there is a constant $c > 0$ such that for all i with $t_i \geq 3$, we have $t_i \geq c \cdot (\log s_i)/\log\log s_i$. We justify at the end why this is sufficient to prove the claim. Obviously, $t_i \geq |S_i|$, because f essentially depends on all variables. Moreover, for each S_i-subfunction, we obtain a BDD from G whose size is $t_i + 2$. For this, we replace the variables outside S_i by the appropriate constants and add dummy nodes afterward if needed. By Lemma 10.5 and using that $|S_i| \leq t_i$ and $t_i \geq 3$,

$$s_i \leq |S_i|^{t_i}((t_i + 1)!)^2 \leq t_i^{t_i}((t_i + 1)!)^2 \leq t_i^{3t_i}.$$

The assumption that $t_i \leq (1/6) \cdot (\log s_i)/\log\log s_i$ now leads to a contradiction, implying a lower bound of the claimed type on t_i for $t_i \geq 3$. This proves the theorem, because $t_i \leq 2$ implies that $s_i \leq 2^{2^{|S_i|}} \leq 2^{2^{t_i}} \leq 16$, and thus i corresponds to a term we have excluded at the beginning. ■

What is the largest possible size of Nechiporuk's lower bound? There are $2^{n-|S_i|}$ different assignments to the variables outside S_i and $2^{2^{|S_i|}}$ functions on S_i. Hence,

$$(\log s_i)/\log\log s_i \leq \min\{(n - |S_i|)/\log(n - |S_i|), 2^{|S_i|}/|S_i|\}.$$

This implies that each set S_i contributes not more than $n/\log n$ to the lower bound. However, a large contribution is only possible for a large set S_i. By elementary calculus, it follows that Nechiporuk's lower bound is bounded above by $O(n^2/\log^2 n)$.

There is a simple function called ISA modeling indirect storage access where Nechiporuk's bound is of size $\Omega(n^2/\log^2 n)$. ISA is defined on $n + k$ variables $x_0, \ldots, x_{n-1}, y_0, \ldots, y_{k-1}$, where $n = 2^m$ and $k = m - \lfloor \log m \rfloor$. The vector $y = (y_{k-1}, \ldots, y_0)$ is interpreted as a binary number $|y|$. If $|y| \geq \lfloor n/m \rfloor$, $\mathrm{ISA}(x, y) := 0$. Otherwise, the x-block $x(y) := (x_{|y|\cdot m}, \ldots, x_{|y|\cdot m+m-1})$ is considered as an address. Then $\mathrm{ISA}(x, y) := x_{|x(y)|}$ is the contents of the storage cell addressed by $x(y)$. It is easy to see that even the FBDD size of ISA is bounded above by $O(n^2/\log n)$. We read y, then $x(y)$ and, if not already done, $x_{|x(y)|}$. These are at most $2m - \lfloor \log m \rfloor + 1$ variables leading to an upper bound of $2^{2m-\lfloor \log m \rfloor+1} = O(n^2/\log n)$. We apply Nechiporuk's bound to ISA and define $S_i := \{x_{im}, \ldots, x_{im+m-1}\}, 0 \leq i \leq \lfloor n/m \rfloor - 1$. The number of S_i-subfunctions of ISA is bounded below by 2^{n-m}, because for $|y| = i$ all 2^{n-m} assignments to the x-variables outside S_i lead to different S_i-subfunctions. Nechiporuk's bound consists of $\lfloor n/m \rfloor$ terms of size $(n - m)/\log(n - m)$, implying the lower bound $\Omega(n^2/\log^2 n)$.

We conclude by mentioning an example for a more sophisticated application of Nechiporuk's method to a natural function. Using this method, Wegener and

Woelfel [53] have obtained that the middle bit of n-bit integer multiplication, that is, the bit at position $n-1$ of the product of two n-bit integers, requires branching program size $\Omega(n^{3/2}/\log n)$. They have also shown that Nechiporuk's method cannot give larger bounds than $O(n^{5/3}/\log n)$.

10.8 Lower Bounds for Oblivious BDDs

Each BDD with s decision nodes is an oblivious BDD with s levels, because we can use the topological order of the decision nodes and can define the ith level as the set containing only the ith decision node. Hence, this BDD is even an (sn)-π-OBDD for each variable order π. However, oblivious BDDs with a small number of levels are strongly restricted BDDs. Nevertheless, there are many important functions that have short representations by small-size and small-length oblivious BDDs:

- All bits of n-bit addition can be represented by an OBDD of size $9n-5$.
- The multiplexer or direct storage access function has OBDD size $2n+1$.
- Symmetric functions have quadratic OBDD size.
- The *equality test* ($\text{EQ}(x, y) = 1$ iff $x = y$) and the *inner product* ($\text{IP}(x, y) = x_1y_1 \oplus \cdots \oplus x_ny_n$) have linear OBDD size.
- The test PERM whether an $n \times n$ Boolean matrix is a permutation matrix can be represented by a 2-IBDD of linear size (choose a rowwise and a columnwise variable order).

In fact, Bollig, Range, and Wegener [9] have proved that the bounds just given for addition and for the multiplexer function are the *exact* OBDD sizes, that is, they have proved matching lower bounds. (To be precise, the case $n = 1$ is an exception for n-bit addition, where the exact OBDD size is 6.)

Furthermore, nondeterminism is a powerful concept for oblivious BDDs.

- HWB (see Section 10.1) and ISA (see Section 10.7) can be represented by nondeterministic OBDDs of polynomial size. Guess the output bit and verify that the guess was correct and the output bit equals 1. Because we may also check whether the output bit equals 0, and because we obtain at most one accepting path, these bounds hold for all three types of nondeterminism.
- The test whether a matrix is a permutation matrix is easy for AND-OBDDs, because it is sufficient to check whether each row and each column contains exactly one 1-entry.

These upper bounds motivate the investigation of lower-bound techniques. The most common approach is based on the theory of communication complexity (see Chapter 11, Hromkovič [24], Kushilevitz and Nisan [31]). The communication game is defined for a "distributed" function $f\colon \{0,1\}^n \times \{0,1\}^m \rightarrow \{0,1\}$, where Alice holds the first n variables and Bob the last m variables. Before knowing the specific values of their variables they agree on a communication protocol. The communication can be stopped if one of the players knows $f(a, b)$. The length of a protocol is the largest number of bits (with respect to all inputs) exchanged between the two players. The *communication complexity* $C(f)$ of f with respect to a given partition of the variables is the minimal length of all communication protocols for f. In the following we prove a simple upper bound on the communication complexity of f depending on parameters of oblivious BDDs representing f. This implies that we can apply lower bounds on the communication complexity of Boolean functions to obtain lower bounds on the size of oblivious BDDs.

We identify the variables given to Alice with A and the variables given to Bob with B. A level of an oblivious BDD G is "owned" by the player holding the variable that is the label of the decision nodes of this level. A *layer* of G is a maximal block of consecutive levels owned by the same player. The *layer depth* $\mathrm{ld}(G)$ equals the number of layers of G.

Lemma 10.7. *Let G be an oblivious BDD representing $f\colon \{0,1\}^n \times \{0,1\}^m \to \{0,1\}$. Then*

$$C(f) \le (\mathrm{ld}(G) - 1) \cdot \lceil \log |G| \rceil$$

and the communication protocol works in $\mathrm{ld}(G) - 1$ *rounds.*

Proof. The player holding the variables of the first layer, w.l.o.g. Alice, starts the communication. She knows the variables tested in the first layer. Hence, she can compute the first node on the computation path which lies outside her layers. Her first message equals the number of this node. Then the communication is continued in the same way by Bob. After at most $\mathrm{ld}(G) - 1$ messages of length $\lceil \log |G| \rceil$ each, the player owning the last layer can compute $f(a, b)$. ■

The result of this lemma can be restated as a lower bound on the size of oblivious BDDs G, namely,

$$|G| \ge 2^{C(f)/(\mathrm{ld}(G)-1)-1}.$$

We have stated and proved the lemma for deterministic BDDs and deterministic communication protocols. We can state and prove it in the same way for nondeterministic protocols and all considered modes of nondeterminism. Nondeterministic protocols are defined in the obvious way. For example, an EXOR-nondeterministic protocol accepts (a, b) if the number of accepted bit sequences describing possible communications is odd, and the corresponding complexity measure is called $C_{\mathrm{EXOR}}(f)$. The theory of communication complexity provides methods for proving large lower bounds on the deterministic and even nondeterministic communication complexity of important functions. However, we only obtain large lower bounds on $|G|$ if $\mathrm{ld}(G)$ is not too large.

First, we discuss the basic lower-bound techniques. The *communication matrix* M consists of 2^n rows corresponding to the possible inputs of Alice and of 2^m columns for the inputs of Bob. The entry at position (a, b) equals $f(a, b)$. A *rectangle* R is defined as the submatrix consisting of all entries of M that belong to a certain subset of the rows and a certain subset of the columns. It is called *monochromatic* if all entries have the same color (or value). Let t, t_0, and t_1 denote the minimal number of monochromatic, 0-colored, and 1-colored rectangles, resp., which are a disjoint cover of M, the zeros of M, and the ones of M, resp. Let t^*, t_0^*, and t_1^* be the corresponding numbers of not necessarily disjoint covers. Then

$$C(f) \geq \log t,\ C_{\mathrm{OR}}(f) \geq \log t_1^*,\ \text{and } C_{\mathrm{AND}}(f) \geq \log t_0^*.$$

The fooling set method is a method to get lower bounds on t^*, t_0^*, and t_1^*. A *c-fooling set* is a set S_c of entries (a, b) such that $f(a, b) = c$ for all $(a, b) \in S_c$ and $f(a, b') \neq c$ or $f(a', b) \neq c$ for all distinct $(a, b), (a', b') \in S_c$. Then $t_c^* \geq |S_c|$, because no c-rectangle can cover two elements of S_c. Moreover, let $R_1, \ldots, R_k$ be disjoint rectangles covering all 1-entries of M. Let M_i be the $2^n \times 2^m$-matrix where all entries of R_i equal 1 and all other entries equal 0. Then M is the sum of all M_i and, by the subadditivity of the rank operator and the fact that all M_i have rank 1, we get the lower bound rank(M) on $C(f)$. This does not hold for nondisjoint covers with the exception of the case of EXOR-nondeterminism and the field $\mathbb{Z}_2$. This leads to the lower bound $\mathrm{rank}_{\mathbb{Z}_2}(M)$ on $C_{\mathrm{EXOR}}(f)$.

The next step is to investigate the layer depth. For a k-π-OBDD we can give the first r, $1 \leq r \leq n$, variables according to the variable order to Alice and the remaining variables to Bob. Then the layer depth equals $2k$, and it is sufficient to prove lower bounds on the length of protocols with $2k - 1$ rounds. For lower bounds for k-OBDDs, we have to consider all variable orders and the corresponding partitions of the variables between Alice and Bob.

For k-IBDDs and general oblivious BDDs, the situation is more difficult. The idea is to find not too small disjoint subsets A and B of the variable set such that the layer depth with respect to A and B is small. Then we have the freedom to choose "good" assignments to the variables outside $A \cup B$ such that the communication complexity of the resulting subfunction is large. Alon and Maass [4] have used a Ramsey-type argument to prove that, for oblivious BDDs with kn levels, A and B can be chosen such that they contain at least $n/2^{4k+1}$ variables each and such that the layer depth is bounded by $4k + 1$. This approach works only if $k = o(\log n)$. We have no complete control over which variables survive. Hence, we obtain lower bounds only for functions where subfunctions on many subsets of variables and arbitrary partitions of the variables between Alice and Bob are difficult.

An example is the middle bit of integer multiplication where we are interested in the bit at position $n - 1$ of the product of two n-bit numbers x and y. We decide that no y-variable survives and we assign constants to the y-bits such that only two y-bits equal 1. This implies that $x \cdot y = x2^i + x2^j$ for some $i, j \in \{0, \ldots, n-1\}$, $i \neq j$. By the pigeonhole principle, there exist i and j such that $x2^i$ and $x2^j$ have many bit positions $k \leq n - 1$ such that Alice owns bit k of one of the terms and Bob owns bit k of the other term. All other positions $\ell \leq n - 1$ are fixed such that one term contains 1 and the other 0. Then we are basically in the situation of the addition of two numbers a and b such that Alice owns a and Bob owns b, and we are interested in the carry bit. The communication matrix is of size $2^r \times 2^r$ for some r and contains 1s at positions (i, j) such that $i + j \geq 2^r, 0 \leq i, j \leq 2^r - 1$. The matrix has rank $2^r - 1$ over $\mathbb{R}$ as well as over $\mathbb{Z}_2$. It contains a 0-fooling set of size 2^r (all (i, j) where $i + j = 2^r$) and a 1-fooling set of size $2^r - 1$ (all (i, j) where $i + j = 2^r + 1$). Choosing the right parameters (Bryant [16], Gergov and Meinel [22]) this leads to the bound $2^{\Omega(n/k^3 2^{4k})}$ for oblivious BDDs of length $2kn$ and all types of nondeterminism.

In order to separate the different modes of nondeterminism, we again apply results from communication complexity:

- The equality test $\mathrm{EQ}(x, y)$ (Alice owns x and Bob owns y) has linear communication complexity for OR- and EXOR-nondeterminism, but only logarithmic communication complexity for AND-nondeterminism; for the negation $\overline{\mathrm{EQ}}$ the roles of OR and AND are interchanged.
- The inner product function $\mathrm{IP}(x, y)$ has linear communication complexity for OR- and AND-nondeterminism but only logarithmic communication complexity for EXOR-nondeterminism.

However, these results hold only for the bad partition of the input where Alice gets x and Bob gets y. All these functions have linear OBDD size for the interleaved variable order $x_1, y_1, \ldots, x_n, y_n$. A general technique working with so-called bit masks a and b generalizes the lower bounds on communication complexity to all balanced partitions of the input. For $f_n\colon \{0, 1\}^n \times \{0, 1\}^n \to \{0, 1\}$ we define the generalized function $f_n^* : \{0, 1\}^{4n} \to \{0, 1\}$. The input is (a, b, x, y). We obtain x' as vector of all x_i where $a_i = 1$, similarly y' based on y and b. If x' and y' have different length, $f_n^*(a, b, x, y) = 0$. If they have the same length m, then $f_n^*(a, b, x, y) = f_m(x', y')$. Then we obtain for each type of nondeterminism a function ($\overline{\mathrm{EQ}}^*$ for OR, EQ^* for AND, IP^* for EXOR) such that the function has polynomial-size OBDDs for all variable orders and the chosen type of nondeterminism but exponential size for oblivious BDDs of linear length and the other two types of nondeterminism.

Finally, we compare k-OBDDs and k-IBDDs. The permutation matrix test PERM can be represented by linear-size 2-IBDDs but needs size $2^{\Omega(n/k)}$ for k-OBDDs. The lower bound method for k-OBDDs works, because we can argue about all variables, and it breaks down for 2-IBDDs, because, as shown earlier, we have to fix too many variables. Moreover, the number k is of importance. The *pointer jumping function* is defined as follows. The function describes a graph on $2n + 1$ vertices $u, v_0, \ldots, v_{n-1}, w_0, \ldots, w_{n-1}, n = 2^k$, and for each vertex v^* there are k Boolean variables describing a number $j \in \{0, \ldots, n - 1\}$. If $v^* \in \{u, w_0, \ldots, w_{n-1}\}$, the pointer from v^* leads to v_j, for $v^* \in \{v_0, \ldots, v_{n-1}\}$ the pointer from v^* leads to w_j. Moreover, there are n Boolean variables describing the colors of $v_0, \ldots, v_{n-1}$. The function $\mathrm{PJ}_{n,k}$ outputs the color of the vertex reached by the unique path of length $2k + 1$ starting at u. Obviously, the k-OBDD size of $\mathrm{PJ}_{n,k}$ is bounded by $O(kn^2)$. Based on lower bounds on the length of communication protocols with a fixed number of rounds due to Nisan and Wigderson [35], Bollig, Sauerhoff, Sieling, and Wegener [10] have proved that $\mathrm{PJ}_{n,k}$ has exponential $(k-1)$-IBDD size if $k \leq c \log \log n$ for a suitable constant $c > 0$. Using bit masks, an analogous result has been obtained even for all $k \leq c' \log n$ for a suitable constant $c' > 0$ by Hromkovič and Sauerhoff [25].

Altogether, oblivious BDDs allow a much more compact representation than OBDDs, and powerful lower-bound techniques based on communication complexity are available.

10.9 Lower Bounds for Length-Restricted DDs

In this section, we consider general DDs, not just BDDs. Presently available methods allow us to prove exponential size lower bounds for DDs whose length is bounded above by a function of order $O(n \log^2 n)$. These results have the nice feature that they can be interpreted as length-size trade-offs for unrestricted DDs or also as time-space trade-offs for general sequential models of computation such as register machines. The proof methods for length-restricted DDs, of which we give only a brief outline here because of the highly technical nature of the subject, have been developed in a succession of papers by Beame, Jayram, and Saks [5], Ajtai [2, 3], and Beame, Saks, Sun, and Vee [6]. These are in turn based on a long history of methods for much more restricted BDDs, e. g., oblivious BDDs (see Section 10.8), read-once BDDs (Wegener [51], Žák [54]) and syntactic read-k BDDs for $k > 1$ (Borodin, Razborov, and Smolensky [14], Okol'nishnikova [36]).

A good starting point for the proof methods is again communication complexity theory. In Section 10.8, we saw that the minimum number t_1^* of 1-colored rectangles required to cover the 1s in the communication matrix of f provides a lower bound on the nondeterministic communication complexity of f and in turn on the size of nondeterministic oblivious BDDs. Here it is more appropriate to regard rectangles as sets of the form $R = R_A \times R_B$, where R_A and R_B are sets of partial assignments to the variables in the sets A and B given to Alice and Bob, resp. Then t_1^* can be characterized as the minimum number of rectangles $R \subseteq f^{-1}(1)$ required to cover $f^{-1}(1)$. Lower bounds on the number t_1^* can be proved by the *rectangle size method* from communication complexity:

(1) Choose a probability distribution μ on the inputs.
(2) Prove a lower bound on $\mu(f^{-1}(1))$ and an upper bound on the measure (density) $\mu(R)$ of each rectangle $R \subseteq f^{-1}(1)$, say $\mu(R) \leq \beta$ for each such R.

Then, obviously, $t_1^* \geq \mu(f^{-1}(1))/\beta$. The fooling set method is the special case where μ is the uniform distribution over a 1-fooling set $S_1 \subseteq f^{-1}(1)$ and $\beta = 1/|S_1|$.

Essentially, an extension of the rectangle size method is also behind the proof method for length-restricted DDs. We develop a simple version of that method (based on the paper of Beame et al. [5]) and comment on further developments later on. The main work is to show how a BDD can be translated into a rectangle cover of $f^{-1}(1)$. We first generate a cover of $f^{-1}(1)$ by functions that can be represented by DDs with a simple structure, so-called *decision forests*. Then we further partition the sets of inputs accepted by each of these decision forests into rectangles. In the following, we consider functions $f\colon D^n \to \{0,1\}$, D any finite set, that are defined on the variable set $X = \{x_1, \ldots, x_n\}$.

Definition 10.1. *A* decision tree *is a DD where each node has indegree at most* 1. *The function of a decision tree T is also denoted by T. An* (r,k)-decision forest *is a set of decision trees $F = \{T_1, \ldots, T_r\}$ where for each input a and for each T_i, the number of variable accesses on the path activated by a in T_i is at most $\lceil kn/r \rceil$. The function computed by F is $F = \bigwedge_{1 \le i \le r} T_i$.*

Lemma 10.8. *Let r, k be positive integers with $r \le kn$. Let G be a DD of length at most kn for any function $f\colon D^n \to \{0,1\}$. Then there is a collection of at most $|G|^{r-1}$ (r,k)-decision forests whose sets of accepted inputs form a partition of $f^{-1}(1)$.*

Proof. Modify the given DD G of length $\ell \le kn$ as follows: first, use $\ell + 1$ copies $G_0, \ldots, G_\ell$ of G and redirect each edge originating in copy G_i to the copy of its successor node in G_{i+1}. Then replace the sinks in G_i with nodes testing an arbitrary variable and having the respective sinks in G_{i+1} as successors. The new graph G' still computes f, edges lead only from the nodes of G_i to the nodes of G_{i+1}, and each path in G' has length exactly ℓ. For different nodes v and w in G', let $f_{v,w}(a) = 1$ if there is a path in G' activated by a leading from v to w. For $i = 1, \ldots, r-1$, let C_i be the set of nodes of $G_{i\lceil \ell/r \rceil}$. Let v_0 be the source of G' and v_r the 1-sink. Then f is the disjunction of the functions $F_{v_1,\ldots,v_{r-1}} = \bigwedge_{0 \le i \le r-1} f_{v_i,v_{i+1}}$ over all $(v_1, \ldots, v_{r-1}) \in C_1 \times \cdots \times C_{r-1}$, the sets of inputs accepted by these functions are pairwise disjoint, and there are at most $|C_1| \cdots |C_{r-1}| \le |G|^{r-1}$ terms in this disjunction. Furthermore, because each of the functions $f_{v_i,v_{i+1}}$ can obviously be computed by a decision tree where at most $\lceil kn/r \rceil$ variables are read during each computation, each $F_{v_1,\ldots,v_{r-1}}$ is computed by an (r,k)-decision forest. This proves the claim. ∎

The idea behind the decomposition into decision trees is that we can now observe the behavior of the DD in a sufficiently coarse way by just looking at the points in time where the computations have been cut (after reading roughly $i \cdot kn/r$ variables, $i = 0, \ldots, r$). Next, we want to find rectangles in a decision forest. For that, we assign each of the decision trees to Alice and Bob randomly with probability $1/2$ for both of them. Typically, many variables read during a computation will occur in both Alice's and Bob's trees. But we can nevertheless hope that there will be some variables that occur exclusively in Alice's trees and some that will be only in Bob's trees, resp. This is because each of the trees reads only kn/r variables during a computation, and thus there are not too many pairs of variables that occur together in one of the trees. We make this precise now.

Definition 10.2. *For a decision forest F and a subforest $F' \subseteq F$, let $\mathtt{core}_F(a, F')$ be the set of variables read* exclusively *in the trees in F' during the computation for a, called F'*-core of *a (in F).*

We usually omit the index F because we consider only one decision forest at a time. The random assignment of decision trees in a decision forest F to Alice and Bob yields a random partition (F_A, F_B) of F. We show that $\mathtt{core}(a, F_A)$ and $\mathtt{core}(a, F_B)$ will be not too small with high probability.

Lemma 10.9. *Let F be an (r, k)-decision forest with $r \leq n$ and let $a \in \{0, 1\}^n$ be an input. Let (F_A, F_B) be a random partition of F as described previously. Then $\mathtt{core}(a, F_A)$ and $\mathtt{core}(a, F_B)$ have the same expected size $\mu(a) \geq n/2^{k+1}$ and $\big||\mathtt{core}(a, F_A)| - \mu(a)\big| \geq \mu(a)/2$ with probability at most $4(k+1)^2 2^{2(k+1)}/r$, analogously for F_B.*

Proof. By symmetry, it suffices to consider F_A. Let $t(i)$ be the number of trees in F that access variable x_i during the computation for a. Let Z_i be the indicator variable for the event $x_i \in \mathtt{core}(a, F_A)$ and let $Z = Z_1 + \cdots + Z_n$. We have $\Pr\{Z_i = 1\} = 2^{-t(i)}$ for all i and $\mu(a) = E(Z) = \sum_{1 \leq i \leq n} 2^{-t(i)}$. Because F is an (r, k)-decision tree, $t(1) + \cdots + t(n) \leq r\lceil kn/r \rceil \leq (k+1)n$. It is easy to see that the minimum of the convex function $\sum_{1 \leq i \leq n} 2^{-t(i)}$ of the $t(i)$ as variables under the constraint $t(1) + \cdots + t(n) \leq (k+1)n$ is $n/2^{k+1}$. This proves the first part of the claim. For the second part, we first upper bound the variance of Z,

$$V(Z) = \sum_{1 \leq i,j \leq n} (\Pr\{Z_i = 1 \wedge Z_j = 1\} - \Pr\{Z_i = 1\} \cdot \Pr\{Z_j = 1\}).$$

If there is no tree that reads both the variables x_i and x_j, then Z_i and Z_j are independent and the respective term in the preceding sum is zero. We crudely estimate all other terms with 1 and count their number. For a fixed i, there are at most $t(i) \cdot \lceil kn/r \rceil$ variables x_j that are read together with x_i in one of the trees. Thus, altogether, there are at most $\sum_{1 \leq i \leq n} t(i)\lceil kn/r \rceil \leq (k+1)^2 n^2/r$ nonzero terms in the sum that is also an upper bound for $V(Z)$. By Chebyshev's inequality, we obtain

$$\Pr\big\{|Z - E(Z)| \geq E(Z)/2\big\} \leq 4V(Z)/E(Z)^2 \leq 4(k+1)^2 2^{2(k+1)}/r. \quad \blacksquare$$

Let F be a fixed decision forest. Set $r = 16(k+1)^2 2^{2(k+1)}$. Then for each input a, the foregoing lemma yields a fixed partition (F_A, F_B) of F such that for $A := \mathtt{core}(a, F_A)$ and $B := \mathtt{core}(a, F_B)$, we have $|A|, |B| \geq m$ with $m = n/2^{k+2}$. We are now essentially in the same situation as for oblivious BDDs in Section 10.8, where we had obtained suitably large subsets A and B of the variables for the two players.

As for oblivious BDDs, we want to fix the variables outside A and B. We first group together inputs for which the same sets A and B are suitable (there is no longer one fixed choice for all inputs here). For two disjoint sets of variables A, B, let $Q = Q(A, B, F_A, F_B)$ be the set of inputs $a \in F^{-1}(1)$ with $A \subseteq \mathtt{core}(a, F_A)$ and $B \subseteq \mathtt{core}(a, F_B)$. Each input $a \in F^{-1}(1)$ is contained in such a set with $|A| = |B| = m$ by Lemma 10.9. Notice that usually there will be more than one such set covering a given input. Let c be any partial assignment to all variables in

$X - (A \cup B)$, and let Q_c be the set of all partial assignments to $A \cup B$ that together with c form an assignment in Q. We show that Q_c is a (possibly empty) rectangle. For assignments u and v to disjoint sets of variables, let uv denote the joint assignment to the union of these sets. Let $Q_{c,A}$ be the set of all assignments a to A such that, for all assignments b_0 to B, $F_A(ab_0c) = 1$ and $\texttt{core}(ab_0c, F_A) \supseteq A$, and let $Q_{c,B}$ be the set of all assignments b to B such that, for all assignments a_0 to A, $F_B(a_0bc) = 1$ and $\texttt{core}(a_0bc, F_B) \supseteq B$. Then $Q_c = Q_{c,A} \times Q_{c,B}$. Thus, we can first cover all inputs in $F^{-1}(1)$ by the sets $Q(A, B, F_A, F_B)$ and then partition each of these sets into rectangles. It remains to quantify this approach.

We first extend our notion of rectangles to include fixed input parts that do not belong to any of the two players and introduce a name for sets like $Q(A, B, F_A, F_B)$.

Definition 10.3. *Let $A, B \subseteq X$ be disjoint sets with $|A| = |B| = m$. A set of inputs R is called* m-rectangle with respect to (A, B) *if there is an assignment c to $X - (A \cup B)$ and there are sets of assignments R_A and R_B to A and B, resp., such that $R = R_A \times R_B \times \{c\}$. A set of inputs Q is called* m-pseudorectangle with respect to (A, B) *if for each assignment c to $X - (A \cup B)$, $Q_c \times \{c\}$ is an m-rectangle with respect to (A, B).*

Lemma 10.10. *Let $k \le n$, $r = 16(k+1)^2 2^{2(k+1)} \le n$, and $m = n/2^{k+2}$. Let F be an (r, k)-decision forest. Then there is a family of at most $2^{4(k+2)m+r}$ m-pseudorectangles that cover $F^{-1}(1)$.*

Proof. It only remains to prove the upper bound on the number of pseudorectangles. Each input $a \in F^{-1}(1)$ is contained in a set $Q(A, B, F_1, F_2)$ with $|A| = |B| = m$ due to Lemma 10.9, and there are at most $2^r \binom{n}{m}^2$ such sets. Using that $\binom{n}{m} \le 2^{H(m/n)n}$ with $H(x) = -(x \log x + (1-x)\log(1-x))$ and the Taylor series approximation $H(x) \le -2x \log x$ for $x \le 1/2$, we get $2^r \binom{n}{m}^2 \le 2^{4(k+2)m+r}$. ■

Analogously to the rectangle size method for the uniform distribution, we obtain the following connection between the density of 1-colored rectangles and the size of DDs.

Theorem 10.11. *Let $2 \le k \le n$, $r = 16(k+1)^2 2^{2(k+1)} \le n$, and $m = n/2^{k+2}$. Let G be a DD for $f\colon D^n \to \{0, 1\}$ of length at most kn. Then there is an m-rectangle $R \subseteq f^{-1}(1)$ such that $|R|/|D|^{2m} \ge 2^{-4(k+2)m-r} \cdot (1/|G|)^r \cdot |f^{-1}(1)|/|D|^n$.*

Proof. Applying Lemma 10.8 and Lemma 10.10, we obtain a cover of $f^{-1}(1)$ by at most $2^{4(k+2)m+r} \cdot |G|^r$ m-pseudorectangles. Each of these pseudo rectangles can be partitioned into m-rectangles as described earlier. By averaging, it follows that there is at least one such rectangle with the required density. ■

Beame, Jayram, and Saks [5] have applied this theorem to *quadratic form functions*. We consider one concrete example. Let $n = 2^d$ and let S be the $n \times n$-Sylvester matrix defined by $S_{a,b} = (-1)^{a^{\top}b}$ for $a, b \in \{0, 1\}^d$. Let S^* be the matrix obtained from S by replacing the diagonal with zeros. For any odd prime power q and $x \in \mathbb{Z}_q^n$, let $\mathrm{SQF}_{q,n}(x) = 1$ if $x^\top S^* x \equiv 0 \bmod q$.

The matrix S has the remarkable property that each of its submatrices has large rank compared to its size. In Beame, Jayram, and Saks [5], a bound for the submatrix rank of S and elementary algebra have been used to prove that, for any m-rectangle $R \subseteq \mathrm{SQF}_{q,n}^{-1}(1)$, $|R|/|D|^{2m} \leq |D|^{-m^2/n}$. Furthermore, it is easy to see that even after fixing $n-2$ variables of x, the function $x \mapsto (x^\top S^* x) \bmod q$ can still attain any value in $\mathbb{Z}_q$. Thus, $|\mathrm{SQF}_{q,n}^{-1}(1)| \geq q^{n-2}$. Plugging these facts into Theorem 10.11, we get the following theorem.

Theorem 10.12. *For any constant $\varepsilon > 0$, there is a constant $c > 0$ such that for k, n, q with $\log\log q \geq ck$ and $n \geq 16(k+1)^2 2^{2(k+1)}$, any DD for $\mathrm{SQF}_{q,n}$ of length kn requires size at least $2^{n \log^{1-\varepsilon} q}$.*

To illustrate this result further, we choose q as the smallest prime greater than or equal to n and set k to roughly $\log\log n$. Furthermore, we use the well-known correspondence between length and size of DDs and time and space for register machines (see, e. g., Borodin and Cook [13]).

Corollary 10.13. *For any constant $\varepsilon > 0$, there is a constant $c' > 0$ such that each deterministic algorithm for $\mathrm{SQF}_{q,n}$ on a register machine with space $n \log^{1-\varepsilon} n$ requires time at least $c' n \log\log n$.*

Theorem 10.11 makes sense only for large domains D that grow with the input length. In the Boolean case $D = \{0, 1\}$, the number of rectangles eats up even the best possible upper bound of 2^{-2m} on the rectangle density. By a further, even more sophisticated variant of the rectangle method, Ajtai [2, 3] and Beame, Saks, Sun, and Vee [6] have managed to overcome this problem. They have been able to show the existence of a rectangle $R = R_A \times R_B \times \{c\} \subseteq f^{-1}(1)$, where R_A and R_B are sets of assignments to disjoint sets A and B, with the following properties:

(i) $|A|, |B| \geq m$ for some large m.
(ii) There is a constant ε with $0 < \varepsilon < 1$ such that both densities $|R_A|/|D|^{|A|}$ and $|R_B|/|D|^{|B|}$ satisfy the lower bound $2^{-\varepsilon m} \cdot (1/|G|^r) \cdot |f^{-1}(1)|/|D|^n$.

Being able to bound the densities of *both* parts of the rectangle separately makes the approach applicable to a larger class of functions. Actually, Beame, Saks, Sun, and Vee [6] have even managed to show that all but a small fraction of the accepted inputs of f can be covered by rectangles with the properties (i) and (ii), and, furthermore, that these rectangles do not overlap too much. Although one only needs a single rectangle for the deterministic case, the stronger result also allows the method to be used for proving exponential lower bounds on the size of randomized length-restricted BDDs (see next section).

The first result for deterministic BDDs (and, therefore, Boolean inputs) of linear length was achieved by Ajtai [2] for a quadratic form function similar to SQF but based on modified Hankel matrices, for which Ajtai obtained an especially strong bound on the rank of submatrices. Furthermore, he proved [3] deterministic lower bounds for the practically interesting functions *element distinctness* (check whether there are two identical numbers in a list) and *Hamming closeness* (check whether there are two vectors in a list that have small Hamming distance). These last two results are again for the "large domain case." Finally, also by a variant of the method presented here for variables with values from a "large domain," Sauerhoff and Woelfel [43] have obtained a time-space trade-off lower bound for the middle bit of nk-bit integer multiplication for branching programs where the input variables encode blocks of k consecutive bits of the input for not too small k.

10.10 Randomized BDDs

Many practically relevant complexity theoretical questions regarding randomized algorithms are still open today: for example, we do not even know whether randomization helps at all to solve more problems in polynomial time compared to deterministic algorithms. BDDs allow study of such questions in the scenario where space is the primary resource. We look at examples of upper and lower bounds for randomized oblivious BDDs and discuss a proof method and results for randomized length-restricted BDDs, thus far the most general restricted type of BDDs that can be handled by lower bound methods.

For OBDDs, it is easy to prove that allowing randomization can indeed lead to exponential savings in size (Ablayev and Karpinski [1]). As a simple example, we consider the bit mask version of the equality function, EQ_n^*, from Section 10.8. Recall that nondeterministic oblivious OBDDs of linear length for EQ_n^* require exponential size.

Theorem 10.14. *The complement of* EQ_n^* *can be computed by a randomized OBDD of polynomial size with one-sided error probability* $1/n$.

Proof. The construction is based on the so-called *fingerprinting technique.* To compare objects from a large universe we map these objects to "fingerprints" or "hash codes" from a small universe that are efficiently comparable. Using randomization, we can ensure that the probability of different objects having the same fingerprint is small.

The function EQ_n^* is defined on vectors $a, b, x, y \in \{0, 1\}^n$. Recall that $\mathrm{EQ}_n^*(a, b, x, y) = 1$ if the sub-vectors of x' of x and y' of y chosen by a and b, resp., are equal. The randomized OBDD G for $\overline{\mathrm{EQ}_n^*}$ uses the variable order $a_1, x_1, b_1, y_1, \ldots, a_n, x_n, b_n, y_n$ and works as follows. By a tree of randomized nodes at the top of the OBDD, a prime from the set of the first n^2 primes is chosen. For a prime p, a subprogram computes the fingerprints $h_{x',p} = \left(\sum_{1\le i\le n} x_i' 2^{i-1}\right) \bmod p$ and $h_{y',p} = \left(\sum_{1\le i\le n} y_i' 2^{i-1}\right) \bmod p$. This can be done by storing the intermediate results with p^2 nodes per level in the OBDD. The OBDD accepts an input if $h_{x',p} \neq h_{y',p}$.

If $x' = y'$, $h_{x',p} = h_{y',p}$ for each p and the OBDD G correctly rejects the input. Let $x' \neq y'$. Since $\left|\sum_{1\le i\le n} x'_i 2^{i-1} - \sum_{1\le i\le n} y'_i 2^{i-1}\right| \le 2^n - 1$, there are fewer than n primes p with $h_{x',p} = h_{y',p}$. Hence, $\Pr\{G(a,b,x,y)=1\} < n/n^2 = 1/n$. Using that each of the first n^2 primes is at most of size $O(n^2 \log n)$ by the prime number theorem, it is easy to prove that the OBDD is of polynomial size. ■

By the same technique, the complement of the permutation matrix test PERM can be shown to have randomized OBDDs of polynomial size with small one-sided error (Sauerhoff [39]), while we know that PERM requires exponential size even for nondeterministic FBDDs (Jukna [29] and Krause, Meinel, and Waack [30]).

The indirect storage access function ISA is a typical example for a function that is hard for deterministic OBDDs (recall the definition from Section 10.7). Because variables may be read only once in a fixed order, and because each x-variable may occur as a bit in $x(y)$ as well as the output bit, a deterministic OBDD has to store a large number of bits and thus requires exponential size. The function can be easily computed by a nondeterministic OBDD of polynomial size (see Section 10.8). Nevertheless, randomization does not help here (Sauerhoff [41]).

Theorem 10.15. *Each randomized OBDD with arbitrary two-sided error ε, $0 < \varepsilon < 1/2$, for* ISA_n *has exponential size.*

Proof. Consider a randomized OBDD for ISA_n with an arbitrary variable order given as a list. Cut the list in two parts such that the first part contains $\ell = b - 1$ of the x-variables, where $b = \Theta(n/\log n)$ is the number of blocks $x(y)$. Then there is at least one block $x(y_0)$ that lies completely in the second part. Give the first and last part of the list to Alice and Bob, resp., and set the y-variables to y_0 in the randomized OBDD. Then the two players can solve the following *direct storage access problem*: Alice has an ℓ-bit "memory" vector (her x-variables), Bob has an "address" in this memory (encoded in $x(y)$), and Bob has to output the addressed bit after receiving only a single message from Alice (encoded as the number of an OBDD node). In communication complexity theory, it is proved that Alice essentially has to tell Bob her complete memory contents for solving this task: that is, $\Omega(\ell) = \Omega(n/\log n)$ bits of communication are required even in the randomized case. This amount of information has to be encoded by nodes in the OBDD, leading to size $2^{\Omega(n/\log n)}$. ■

Applying suitable lower bounds for one-round communication protocols as in the proof of Theorem 10.15, several other functions have been shown to require exponentially large randomized OBDDs. For general oblivious BDDs, lower bounds for randomized communication protocols with several rounds are required. As one practical example, we consider the multiplication function from Section 10.8. By the same ideas as in that section, each randomized oblivious BDD of length $2kn$ for the multiplication of n-bit numbers yields a randomized k-round communication protocol for the carry bit problem of input length $r = \Theta(n/((k+1)^2 2^{4k}))$ described in Section 10.8. Smirnov [49] has shown that each randomized k-round communication protocol for the latter problem with error bounded by a constant smaller than $1/2$ requires $\Omega(r^{1/k} \log r)$ bits of communication. This implies an exponential lower bound on the size of randomized oblivious BDDs of linear length for the multiplication function.

Upper and lower bound results have also been proved for randomized variants of FBDDs and syntactic read-k BDDs for $k > 1$. We skip these results and devote the rest of the section to randomized length-restricted DDs.

We sketch a variant of the proof method from Section 10.9 that also works for the randomized case. We again consider a function $f\colon D^n \to \{0, 1\}$. We say that a deterministic DD G *approximates* f *with error* ε if the output of G agrees with f on at least a $(1 - \varepsilon)$-fraction of all inputs with respect to the uniform distribution on the inputs. By a generally applicable counting argument, we may consider approximating DDs rather than randomized DDs.

Lemma 10.16. *Let G be a randomized DD that computes f with two-sided error ε. Then there is a deterministic DD G' with $|G'| \le |G|$ that approximates f with error at most ε.*

Proof. Let $G_1, \ldots, G_N$ be the different deterministic DDs obtained by selecting one outgoing edge of each randomized node in G and removing the other one. Let G_i also denote the function represented by G_i. For each input $a \in D^n$, let $d(i, a) = 1$ if $G_i(a) \ne f(a)$ and 0 otherwise. Then $\frac{1}{N}\sum_{1\le i\le N} d(i, a) \le \varepsilon$ for each a due to the error bound of G, and thus

$$|D|^{-n} \sum_{a\in D^n} \frac{1}{N} \sum_{1\le i\le N} d(i, a) = \frac{1}{N} \sum_{1\le i\le N} |D|^{-n} \sum_{a\in D^n} d(i, a) \le \varepsilon.$$

This implies that there is an i_0 with $|D|^{-n} \sum_{a\in D^n} d(i_0, a) \le \varepsilon$. ■

In Section 10.9, we have considered a deterministic DD for f and have obtained a cover of the accepted inputs of f by pseudorectangles. We have shown that the number of pseudorectangles is not too large and have argued that this implies the existence of one dense pseudorectangle that contains only accepted inputs. Here we consider a deterministic DD G approximating the given function f with error ε, and our aim is to produce a rectangle that is dense and contains only few inputs that are not accepted by f. Let G also denote the function represented by G.

Because G is deterministic, Lemma 10.8 yields a decomposition of $G^{-1}(1)$ into disjoint sets of inputs accepted by decision forests. Instead of covering the inputs accepted by each decision forest by overlapping pseudorectangles as in Section 10.9, we directly partition these sets into rectangles (we omit the details how this can be done). This yields a partition of the whole set $G^{-1}(1)$ into rectangles. By averaging arguments, it can then be shown that at least half of all accepted inputs of G can be partitioned into dense rectangles (again we omit the details). Because of the error bound of G, f is 0 for at most $\varepsilon|D|^n$ inputs in the obtained rectangle partition. On the other hand, these rectangles contain at least $(|f^{-1}(1)| - \varepsilon|D|^n)/2$ accepted inputs altogether. Hence, again by averaging, there is at least one dense rectangle R such that f is 0 for at most a $2\varepsilon/(|f^{-1}(1)| \cdot |D|^{-n} - \varepsilon)$-fraction of the inputs in R. Altogether, we arrive at the following analog of the method from Section 10.9 for approximating DDs.

Theorem 10.17. *Let k, n, r be suitable integer parameters. Let G be a deterministic DD of length at most kn that approximates $f\colon D^n \to \{0, 1\}$ with error ε. Then there is an m-rectangle R with respect to variable sets A and B such that*

(i) $|R|/|D|^{|A|}, |R|/|D|^{|B|} \geq 2^{-12(k+1)m} \cdot (1/|G|^r) \cdot (\eta - \varepsilon)$, *where* $\eta = |f^{-1}(1)|/|D|^n$ *and*
(ii) *f equals 0 for at most a $2\varepsilon/(\eta - \varepsilon)$-fraction of the inputs in R.*

We apply this theorem to the following variant of the Sylvester matrix function from Section 10.9. Let p be a prime, let $M \subseteq \mathbb{Z}_p$ with $|M| = \lfloor p/2 \rfloor$, and let S be the $n \times n$-Sylvester matrix as in Section 10.9. Define the function BSQF (*balanced* SQF) for $x \in \mathbb{Z}_p^n$ by $\mathrm{BSQF}_{p,M,n}(x) = a$ if $x^\top Sx \in M$ and $1 - a$ otherwise, where $a \in \{0, 1\}$ is chosen such that $|\,\mathrm{BSQF}_{p,M,n}^{-1}(1)| \geq 1/2$. Beame, Saks, Sun, and Vee [6] have shown that for each m-rectangle $R \subseteq \mathbb{Z}_p^n$ with $|R|/|D|^{2m} \geq |D|^{-m^2/(2n)+3}$ and any $c \in \mathbb{Z}_p$ the fraction of inputs $x \in R$ with $x^\top Sx \equiv c \bmod p$ is at least $1/(4p)$. It follows that each dense rectangle contains a large fraction of inputs from $\mathrm{BSQF}_{p,M,n}^{-1}(0)$. For $p \geq n$, Theorem 10.17 and simple calculations yield the following result.

Theorem 10.18. *For any constant $\varepsilon > 0$, each deterministic DD of size $2^{n^{1-\varepsilon}}$ that approximates $\mathrm{BSQF}_{p,M,n}$ with error at most $1/50$ has length $\Omega(n \log\log n)$.*

Similar results have been obtained by Beame, Saks, Sun, and Vee [6] for quadratic form functions with other matrices. The best lower bound on the length is of order $\Omega(n \log n)$ for DDs of size $2^{n^{1-\varepsilon}}$ with error $1/100$. As a more practically relevant example, we finally mention their particularly nice result for the computation of the element distinctness function on register machines, which is a direct consequence of an analogous DD result. Let $\mathrm{ED}_n(x_1, \ldots, x_n) = 1$ if the numbers $x_1, \ldots, x_n \in \{1, \ldots, n^2\}$ are pairwise distinct and 0 otherwise.

Theorem 10.19. *For any $\varepsilon > 0$ there is a constant $c_\varepsilon > 0$ such that any randomized algorithm that computes ED_n with error $n^{-\varepsilon}$ and runs on a register machine with space $n^{1-\varepsilon}$ requires time at least $c_\varepsilon n\sqrt{\log n/\log\log n}$.*

All of these results are for functions on a large domain D growing with the input length. In the Boolean case $D = \{0, 1\}$, Beame, Saks, Sun, and Vee [6] have used a more sophisticated version of the method described earlier (see also the end of Section 10.9) to obtain time-space trade-offs for randomized and approximating BDDs representing quadratic form functions based on Hankel matrices.

References

[1] F. Ablayev and M. Karpinski. On the power of randomized branching programs. In *Proceedings of 23rd ICALP, Lecture Notes in Computer Science 1099*, pp. 348–56, 1996.

[2] M. Ajtai. A non-linear time lower bound for boolean branching programs. In *Proceedings of 40th Symposium on Foundations of Computer Science*, pp. 60–70, 1999.

[3] M. Ajtai. Determinism versus non-determinism for linear time RAMs with memory restrictions. *Journal of Computer and System Sciences 65*, pp. 2–37, 2002.

[4] N. Alon and W. Maass. Meanders and their applications in lower bound arguments. *Journal of Computer and System Sciences 37*, pp. 118–29, 1988.

[5] P. Beame, T. S. Jayram, and M. Saks. Time-space tradeoffs for branching programs. *Journal of Computer and System Sciences 63*, pp. 542–72, 2001.

[6] P. Beame, M. Saks, X. Sun, and E. Vee. Time-space trade-off lower bounds for randomized computation of decision problems. *Journal of the ACM 50*, pp. 154–95, 2003.

[7] J. Bern, C. Meinel, and A. Slobodová. Some heuristics for generating treelike FBDD types. *IEEE Transactions on Computer-Aided Design of Integrated Circuits and Systems 15*, pp. 127–30, 1996.

[8] M. Blum, A. K. Chandra, and M. N. Wegman. Equivalence of free boolean graphs can be decided probabilistically in polynomial time. *Information Processing Letters 10*, pp. 80–2, 1980.

[9] B. Bollig, N. Range, and I. Wegener. Exact OBDD bounds for some fundamental functions. In *Proceedings of 34th SOFSEM, Lecture Notes in Computer Science 4910*, pp. 174–85, 2008.

[10] B. Bollig, M. Sauerhoff, D. Sieling, and I. Wegener. Hierarchy theorems for kOBDDs and kIBDDs. *Theoretical Computer Science 205*, pp. 45–60, 1998.

[11] B. Bollig and I. Wegener. Improving the variable ordering of OBDDs is NP-complete. *IEEE Transactions on Computers 45*, pp. 993–1002, 1996.

[12] B. Bollig and I. Wegener. Complexity theoretical results on partitioned (nondeterministic) binary decision diagrams. *Theory of Computing Systems 32*, pp. 487–503, 1999.

[13] A. Borodin and S. Cook. A time-space tradeoff for sorting on a general sequential model of computation. *SIAM Journal on Computing 11*, pp. 287–97, 1982.

[14] A. Borodin, A. A. Razborov, and R. Smolensky. On lower bounds for read-k-times branching programs. *Computational Complexity 3*, pp. 1–18, 1993.

[15] R. E. Bryant. Graph-based algorithms for boolean function manipulation. *IEEE Transactions on Computers 35*, pp. 677–91, 1986.

[16] R. E. Bryant. On the complexity of VLSI implementations and graph representations of boolean functions with application to integer multiplication. *IEEE Transactions on Computers 40*, pp. 205–13, 1991.

[17] R. E. Bryant. Binary decision diagrams and beyond: enabling technologies for formal verification. In *Proceedings of International Conference on Computer-Aided Design*, pp. 236–43, 1995.

[18] A. Cobham. The recognition problem for the set of perfect squares. In *Proceedings of 7th Symposium on Switching and Automata Theory*, pp. 78–87, 1966.

[19] Y. Crama and P.L. Hammer. *Boolean Functions: Theory, Algorithms, and Applications.* Cambridge University Press, Cambridge, UK, 2010.

[20] S. Fortune, J. Hopcroft, and E. M. Schmidt. The complexity of equivalence and containment for free single variable program schemes. In *Proceedings of 5th ICALP, Lecture Notes in Computer Science 62*, pp. 227–40, 1978.

[21] S. J. Friedman and K. J. Supowit. Finding the optimal variable ordering for binary decision diagrams. *IEEE Transactions on Computers* 39, pp. 710–13, 1990.

[22] J. Gergov and C. Meinel. Efficient boolean manipulation with OBDDs can be extended to FBDDs. *IEEE Transactions on Computers 43*, pp. 1197–209, 1994.

[23] O. Giel. Branching program size is almost linear in formula size. *Journal of Computer and System Sciences 63*, pp. 222–35, 2001.

[24] J. Hromkovič. Communication Complexity and Parallel Computing. Springer-Verlag, New York, 1997.

[25] J. Hromkovič and M. Sauerhoff. On the power of nondeterminism and randomness for oblivious branching programs. *Theory of Computing Systems 36, pp. 159–82*, 2003.

[26] J. Jain, J. Bitner, D. S. Fussell, and J. A. Abraham. Functional partitioning for verification and related problems. *Brown MIT VLSI Conf.*, 45, pp. 210–26. MIT Press, Cambridge, MA, 1992.

[27] J. Jain, K. Mohanram, D. Moundanos, I. Wegener, and Y. Lu. Analysis of composition, and how to obtain smaller canonical graphs. In *Proceedings of 37th Design Automation Conference*, pp. 681–6, 2000.

[28] D. S. Johnson. A catalog of complexity classes. In *Handbook of Theoretical Computer Science, Vol. A*, J. van Leeuwen, ed., Chapter 2, pp. 67–161. Elsevier, Amsterdam, 1990.

[29] S. P. Jukna. Entropy of contact circuits and lower bounds on their complexity. *Theoretical Computer Science 57*, pp. 113–29, 1988.

[30] M. Krause, C. Meinel, and S. Waack. Separating the eraser Turing machine classes L_e, NL_e, co-NL_e and P_e. *Theoretical Computer Science 86*, pp. 267–75, 1991.

[31] E. Kushilevitz and N. Nisan. Communication Complexity. Cambridge University Press, Cambridge, UK, 1997.

[32] Y.-P. Lai, M. Pedram, and S. B. K. Vrudhula. BDD based decomposition of logic functions with application to FPGA synthesis. In *Proceedings of 30th Design Automation Conference*, pp. 642–7, 1993.

[33] A. Narayan, J. Jain, M. Fujita, and A. Sangiovanni-Vincentelli. Partitioned ROBDDs – a compact, canonical, and efficiently manipulable representation for Boolean functions. In *Proceedings of International Conference on Computer-Aided Design*, pp. 547–54, 1996.

[34] È. I. Nechiporuk. A boolean function. *Soviet Mathematics Doklady 7*, pp. 999–1000, 1966.

[35] N. Nisan and A. Wigderson. Rounds in communication complexity revisited. *SIAM Journal on Computing 22*, pp. 211–9, 1993.

[36] E. A. Okol'nishnikova. On lower bounds for branching programs. *Siberian Advances in Mathematics 3*, pp. 152–66, 1993.

[37] P. Pudlák and S. Žák. Space complexity of computations. Preprint, University of Prague, 1983.

[38] R. Rudell. (1993). Dynamic variable ordering for ordered binary decision diagrams. In *Proceedings of International Conference on Computer-Aided Design*, 42–47.

[39] M. Sauerhoff. Lower bounds for randomized read-*k*-times branching programs. In *Proceedings of 15th STACS, Lecture Notes in Computer Science 1373*, pp. 105–15, 1998.

[40] M. Sauerhoff. An improved hierarchy result for partitioned BDDs. *Theory of Computing Systems 33*, pp. 313–29, 2000.

[41] M. Sauerhoff. On the size of randomized OBDDs and read-once branching programs for *k*-stable functions. *Computational Complexity 10*, pp. 155–78, 2001.

[42] M. Sauerhoff, I. Wegener, and R. Werchner. Relating branching program size and formula size over the full binary basis. In *Proceedings of 16th STACS, Lecture Notes in Computer Science 1563*, pp. 57–67, 1999.

[43] M. Sauerhoff and P. Woelfel. Time-space tradeoff lower bounds for integer multiplication and graphs of arithmetic functions. In *Proceedings of 35th STOC*, pp. 186–95, 2003.

[44] D. Sieling. The nonapproximability of OBDD minimization. *Information and Computation 172*, 103–38, 2002.

[45] D. Sieling. The complexity of minimizing and learning OBDDs and FBDDs. *Discrete Applied Mathematics 122*, pp. 263–82, 2002.

[46] D. Sieling and I. Wegener. *NC*-algorithms for operations on binary decision diagrams. *Parallel Processing Letters 3*, pp. 3–12, 1993.

[47] D. Sieling and I. Wegener. Reduction of OBDDs in linear time. *Information Processing Letters 48*, pp. 139–44, 1993.

[48] D. Sieling and I. Wegener. Graph driven BDDs – a new data structure for boolean functions. *Theoretical Computer Science 141*, pp. 283–310, 1995.

[49] D. V. Smirnov. *Shannon's information methods for lower bounds for probabilistic communication complexity*. Master's thesis, Moscow University, 1988.

[50] F. Somenzi. Efficient manipulation of decision diagrams. *Software Tools for Technology Transfer 3*, pp. 171–81, 2001.

[51] I. Wegener. On the complexity of branching programs and decision trees for clique functions. *Journal of the ACM 35*, pp. 461–71, 2000.

[52] I. Wegener. Branching Programs and Binary Decision Diagrams – Theory and Applications. SIAM Monographs on Discrete Mathematics and Applications, 2000.

[53] I. Wegener and P. Woelfel. New results on the complexity of the middle bit of multiplication. *Computational Complexity 16*(3), 298–323, 2007.

[54] S. Žák. An exponential lower bound for one-time-only branching programs. In *Proceedings of 11th MFCS, Lecture Notes in Computer Science 176*, pp. 562–6, 1984.

11

Circuit Complexity

Matthias Krause and Ingo Wegener

11.1 Introduction

The theory on efficient algorithms and complexity theory is software oriented. Their hardware-oriented counterpart is the theory on *combinational circuits* or, simply, circuits. The main difference is that circuits are a *nonuniform* model. A circuit is designed for one Boolean function $f \in B_{n,m}$, that is, $f: \{0, 1\}^n \rightarrow \{0, 1\}^m$. However, most circuit designs lead to sequences of circuits realizing a sequence of functions. Typical adders are sequences of adders, one for each input length. If there is an efficient algorithm computing for each n the circuit for input length n, the circuit family is called *uniform.* However, for basic functions like arithmetic functions or storage access the circuit model is more adequate than software models. Moreover, circuits are a very simple and natural computation model reflecting all aspects of efficiency.

A circuit model needs a *basis* of elementary functions that can be realized by simple gates. In the basic circuit model, a basis is a finite set. Then a circuit for input size n is a finite sequence of instructions or gates and, therefore, a straight-line program: the ith instruction consists of a function g from the chosen basis and, if $g \in B_j := B_{j,1}$, a sorted list $I_{i,1}, \ldots, I_{i,j}$ of inputs. The constants 0 and 1, the variables $x_1, \ldots, x_n$, and the results $r_1, \ldots, r_{i-1}$ of the first $i - 1$ instructions are possible inputs. If we consider $0, 1, x_1, \ldots, x_n$ as Boolean functions, the semantics of the circuit is defined by

$$r_i := g(I_{i,1}, \ldots, I_{i,j}).$$

The circuit realizes $f = (f_1, \ldots, f_m) \in B_{n,m}$ if $f_k = r_{i(k)}$ for each k and some $i(k)$. The *size* of the circuit equals the number of its gates. A circuit has an obvious representation by a directed acyclic graph with the sources $0, 1, x_1, \ldots, x_n$. The ith instruction is represented as a vertex that is reached by edges from $I_{i,1}, \ldots, I_{i,j}$. The *depth* of the circuit equals the length of the longest path in its graph representation. The *circuit size* $C_\Omega(f)$ of f with respect to the basis Ω is the smallest size of an Ω-circuit representing f; the depth $D_\Omega(f)$ of f is defined in a similar way.

It is obvious that circuit size is a measure for the cost of realizing hardware for f (e.g., wires, wire length, area, and power consumption are ignored). It is also obvious that *circuit depth* is a measure for the parallel time to evaluate f. Based on the so-called *parallel computation thesis*, parallel time is related to storage space.

Historically, mathematicians and, in particular, logicians were the first to start a discussion of computability and decidability (Hilbert, Gödel, Church, Turing, and others). However, complexity theory started only in the 1960s, whereas much earlier Shannon [63, 64] discussed a circuit-based complexity theory. Lower-bound arguments were discussed as early as the 1950s, for example, by Lupanov [47].

One reason for the importance of the circuit model is its robustness. A basis Ω is called *complete* if each Boolean function can be represented by an Ω-circuit; typical complete bases are the binary basis B_2 and the basis U_2, which equals B_2 without EXOR and its negation. However, it is sufficient to have (binary) AND, OR, and (unary) NOT, AND and NOT, OR and NOT, AND and EXOR, or even only NAND (see Chapter 1). The ternary function ite defined by

$$\text{ite}(x, y, z) = xy \vee \overline{x}z = (\text{if } x \text{ then } y \text{ else } z)$$

also forms a complete basis. It is a simple fact that circuit size and circuit depth with respect to different complete bases are related by a multiplicative constant. The only important incomplete basis is the monotone basis consisting of AND and OR. Exactly the monotone Boolean functions can be realized by monotone circuits.

The reader may wonder that the fan-out of the gates of the circuit is not limited. Hoover, Klawe, and Pippenger [31] have shown that general circuits can be simulated by circuits with fan-out bounded by 2 such that size and depth grow by a small constant factor only. The special case of fan-out 1 (tree circuits with multiple leaves for the inputs) is the case of *formulas*.

In order to show the power of circuits, efficient circuits for the fundamental arithmetic functions are presented in Section 11.2. We are especially interested in circuits realizing small size and small depth simultaneously. In Section 11.3, we compare circuit size and depth with other complexity measures and, afterward, we focus on lower bounds. The situation in the general case seems hopeless. Only bounds of linear size are known. The corresponding methods are discussed in Section 11.4. The situation for formulas is only a little better: the best-known lower bounds are of polynomial size. They are presented in Section 11.5. Exponential lower bounds are available for monotone circuits; see Section 11.6. We also discuss lower-bound techniques based on communication complexity.

For the case of general circuits, we have no good lower bounds on circuit size and no good lower bounds on circuit depth. However, for upper bounds we are interested in circuits with small size *and* depth. Perhaps, it is easier to prove that small size is not possible if the depth is limited. The complexity class P/poly of all sequences of Boolean functions with polynomial circuit size considers circuits of unlimited depth. If the depth is restricted by $O(\log^k n)$, we obtain the complexity class NC^k (called *Nick's class* because of the fundamental paper of Nick Pippenger

[53]). However, we are not able to prove for explicitly defined Boolean functions that they are not contained in NC^1. In order to restrict the depth even further, we have to allow gates with unbounded fan-in. The class AC^k (alternating class) contains all Boolean functions computable in depth $O(\log^k n)$ (constant depth for $k = 0$) and polynomial size by circuits allowing negation and unbounded fan-in AND and OR. It was a breakthrough in the 1980s to prove that simple functions are not in AC^0 (see Section 11.7). The choice of the basis is essential in the case of unbounded fan-in circuits. Other bases such as unbounded AND and EXOR, AND and MOD_p (testing whether its input contains a number of ones which is a multiple of p), and threshold functions (testing whether its input contains at least or at most a certain number of ones) are also investigated in Sections 11.7 and 11.8. Section 11.9 briefly discusses barriers to proving lower bounds, and Section 11.10 is devoted to further lower-bound techniques based on voting polynomials and spectral techniques. Finally, in Section 11.11, the most challenging open problems are listed. The classical results are contained in the monograph by Wegener [73]. Clote and Kranakis [18] present several more recent results.

11.2 Efficient Circuits for Arithmetic Functions

Most people believe that the so-called school methods for addition, multiplication, and division they have learned in primary school are "optimal." Perhaps they are optimal in the sense that they are simple enough that people can learn them (and perhaps remember them), but they are not the best solutions with respect to hardware design.

It is obvious that the school method for addition has linear size and also linear depth. Indeed, it is a good sequential solution but a poor parallel solution. Sklansky [66] described an adder with logarithmic depth and polynomial size. His idea was to perform three additions of numbers of half the length in parallel:

- The less significant halves of the numbers,
- The most significant halves of the numbers without a carry, and
- The most significant halves of the numbers with a carry bit.

Afterward, the less significant half of the sum and its carry are known. Using this carry, it is possible to choose the correct most significant bits in constant depth and linear size. The depth bound is obvious. This recursive approach leads to a size of $\Theta(n^{\log 3})$. By an iterative approach avoiding multiple solutions of the same subproblem, the size can be decreased to $\Theta(n \log n)$.

Ladner and Fischer [45] have shown that prefix computation is the core of addition. Each position of the given two numbers x and y is of one of three types, namely;

- Eliminate (E), if $x_i + y_i = 0$,
- Propagate (P), if $x_i + y_i = 1$,
- Generate (G), if $x_i + y_i = 2$,

describing whether a carry is eliminated ($E(c) = 0$), a carry is propagated ($P(c) = c$), or a carry is generated ($G(c) = 1$). The idea is to encode the type T_i of position i. The prefix problem is the simultaneous computation of all $c_i = T_i \circ \cdots \circ T_0(0)$. Finally, the sum bits s_j can be computed easily, because $s_n = c_{n-1}$ and $s_j = x_j \oplus y_j \oplus c_{j-1}$ for $0 \le j \le n-1$. The prefix problem can be solved with less than $4n$ "$\circ$"-operations in depth $\lceil \log n \rceil$. Using a good encoding of T_i the operation "$\circ$" can be realized with three gates in depth 2 leading to an adder of size less than $15n$ and depth $2\lceil \log n \rceil + 2$ (over the basis B_2). This trick also works for subtraction if we use the representation of two's-complement numbers.

The school method of multiplication computes all $x_i y_j$ and then has the problem of adding n numbers. In its pure form size and depth are $\Theta(n^2)$. We may use a balanced binary tree to parallelize the addition of n numbers to $\lceil \log n \rceil$ parallel addition steps. Using Ladner-Fischer adders, this results in depth $\Theta(\log^2 n)$. It is easy to see that addition needs depth $\Omega(\log n)$ and that addition of n numbers needs $\Omega(\log n)$ parallel steps. However, Wallace [70] has observed that addition in redundant number representation can be done in constant depth. The *Wallace tree* consists of CSA-gates (carry-save-adders). Each CSA-gate works on three binary numbers and produces two numbers with the same sum using linear size and constant depth. For each position a full adder is applied. The first number consists of all sum bits and the second number of all carry bits with an additional bit of value 0 at position 0. At each level the number of terms is reduced approximately by a factor of 2/3. Only for the last two terms is an efficient adder used. This leads to size $\Theta(n^2)$ and optimal depth $\Theta(\log n)$.

Karatsuba and Ofman [34] showed how to reduce the size for multiplication by divide-and-conquer. Let $x = (x', x'')$ and $y = (y', y'')$ be partitioned into two blocks of size $n/2$. If we identify x with the number represented by x,

$$x \cdot y = x' \cdot y' \cdot 2^n + (x' \cdot y'' + x'' \cdot y') \cdot 2^{n/2} + x'' \cdot y''.$$

After having computed $x' \cdot y'$ and $x'' \cdot y''$ the middle term can be computed by one further multiplication:

$$x' \cdot y'' + x'' + y' = (x' + x'') \cdot (y' + y'') - x' \cdot y' - x'' \cdot y''.$$

This multiplier has size $\Theta(n^{\log 3})$ and is still far from being optimal. Schönhage and Strassen [62] still hold the record with their multiplier, which simultaneously realizes logarithmic depth and quasilinear size, more exactly size $\Theta(n \log n \log\log n)$. The method is too complicated to be presented in detail. It is based on another redundant number representation, namely radix-4 representation. The computations are performed in the Fermat ring $\mathbb{Z}_m$ where $m = 4^n + 1$ and $x = (x_n, \ldots, x_0)$ where $x_i \in \{-3, -2, -1, 0, 1, 2, 3\}$ represents the sum of all $x_i \cdot 4^i$. This allows addition in constant depth and linear size. The same holds for multiplication by powers of 2. The transformation into numbers in binary representation is possible in linear size and logarithmic depth. One main idea is that multiplication of numbers resembles multiplication of polynomials and includes convolution, more precisely,

$$\left(\sum x_i \cdot 2^i\right) \cdot \left(\sum y_j \cdot 2^j\right) = \sum p_k 2^k,$$

where

$$p_k = \sum_{i+j=k} x_i y_j$$

and $p = (p_0, \ldots, p_{2n-2})$ is the convolution of a and b. Because the algorithm works recursively, it is essential to shorten the intermediate results and to compute only the negative envelope $q = (q_0, \ldots, q_{n-1})$ of p where $q_k := p_k - p_{n+k}$ and $p_{2n-1} = 0$. Also, the negative envelope of convolution can be computed with the FFT (fast Fourier transform) algorithm.

Let $n = 2^k$. In order to use radix-4 representations, we consider the problem of computing $x \cdot y \bmod (2^n + 1)$. Let $b = 2^{\lfloor k/2 \rfloor}$. The input numbers x and y are partitioned into b blocks of length $l := n/b$ each. These blocks are interpreted as coefficients of a polynomial. Because x and y are the values of the polynomials at 2^l, it is sufficient to multiply the polynomials and to evaluate the result at 2^l. For this purpose the negative envelope of convolution is computed. This is done $\bmod(2^l + 1)$ and $\bmod b$. Because $\gcd(2^{2l} + 1, b) = 1$, the final result can be obtained by Chinese remaindering. (For details, see Wegener [73].)

The school method for division leads to circuits of size $\Theta(n^2)$ and depth $\Theta(n \log n)$ if the Ladner-Fischer circuit for subtraction is used. Division is equivalent to the computation of the inverse of the divisor and multiplying it with the dividend. Moreover, we are satisfied with an approximative result where the n most significant bits are correct. It is sufficient to investigate the computation of the inverse of a number z where $1/2 \le z < 1$. The function $f(x) = x^{-1} - z$ is convex and has z^{-1} as unique solution of $f(x) = 0$. Hence, we can apply Newton's method to approximate z^{-1}. This leads to the sequence $z_{i+1} = 2z_i - z \cdot z_i^2$. For $\varepsilon_i := z^{-1} - z_i > 0$ and $z_0 := 1$, it is easy to prove that $\varepsilon_i > 0$ and $\varepsilon_{i+1} < \varepsilon_i^2$. Hence, z_k with $k = \lceil \log n \rceil + 1$ is a good approximation of z^{-1}. However, a correct calculation leads to very long numbers and bad circuit size. A careful design starts with short numbers. In each phase the result is rounded down and the number of considered bits increases. Using Schönhage-Strassen multipliers, we obtain a divider of size $\Theta(n \log n \log\log n)$ and depth $\Theta(\log^2 n)$.

Beame, Cook, and Hoover [13] were the first to present a divider of depth $\Theta(\log n)$ that was uniform only in a weak sense. More recently, Hesse [28] constructed strongly uniform dividers of this kind. Again, it is enough to describe how z^{-1} can be computed approximatively. Let $x = 1 - z$. Then $0 < x \le 1/2$ and z^{-1} can be approximated by the product of all $(1 + x^{2^k})$, $0 \le k \le \lceil \log n \rceil$, even if we work with numbers whose length is bounded by $2n$. This implies that the main problem is to multiply n numbers of bit length n. It is sufficient to compute the product $\bmod M$ where $M = (2^n - 1)^n$. This can be done by Chinese remaindering, that is, computing the product $\bmod p_j$ for distinct prime numbers whose product is larger than M. The prime number theorem ensures that it is sufficient to work with $r = O(n^2/\log n)$ primes each of size $O(n^2)$. Hence, p_j is small enough that we can use table-lookup methods.

Let $x_1, \ldots, x_n$ be the numbers we want to multiply, and let p be their product. Then it is easy to compute $y_{ij} \equiv x_i \bmod p_j$. The problem is to compute z_i, the product of all y_{ij}, $1 \leq j \leq n$, mod, p_j that equals $p \bmod p_j$. Finally, it is easy to compute p from all $p \bmod p_j$ by the technique of Chinese remaindering. For the computation of z_i we use the fact that $\mathbb{Z}_{p_j}$ has a generator g_j that can be computed in advance, because it does not depend on the input. (This is a problem that makes the circuit only weakly uniform.) It is a fact that $y_{ij} \equiv g_j^{\text{ind}(y_{ij})} \bmod p_j$ for the so-called index $\text{ind}(y_{ij})$ of y_{ij}. The table of all $(y, \text{ind}(y))$ is computed in advance. Then $\text{ind}(y_{ij})$ can be computed by table lookup. This reduces the multiplication of all y_{ij}, $1 \leq j \leq n$, to the addition of all $\text{ind}(y_{ij})$, $1 \leq j \leq n$, $\text{mod}(p_j - 1)$. Finally, by table lookup, the result $I(j)$ is retransformed into $p \bmod p_j$.

However, in applications, n is not large enough such that the asymptotically best circuits are the best ones. The Schönhage-Strassen multiplier is applied for the multiplication of very large numbers, such as in cryptography. Dividers like the famous Pentium divider (famous because of its wrong implementation) follow an approach that is based on the school method with many improvements. The main ideas were described by Atkins [10]. The result is again computed in radix-4 representation. In order to simplify the computation of the new remainder, the quotient contains only entries from $\{-2, -1, 0, +1, +2\}$. Then the multiplication with the divisor is easier than for the values -3 or $+3$. The main idea is that it is sufficient to use seven bits of the remainder (including the sign bit) and five bits of the divisor (including the leading bit, which always equals 1) in order to find a correct value for the next position of the quotient. This correct value can be found by table lookup. This divider has good properties, although the size equals $\Theta(n^2)$ and the depth $\Theta(n)$.

This short section on the design of efficient circuits proves that the design of circuits with small size *and* depth is a fascinating subject – even for functions we believe we understand, such as the arithmetic functions.

11.3 Circuit Size, Circuit Depth, and Other Complexity Measures

In this section, we compare circuit-based complexity measures with other complexity measures.

Branching programs (see Chapter 10) can be interpreted as special circuits. We have seen that the function ite, defined by $ite(x, y, z) = xy \vee \overline{x}z$, is a complete basis. Branching programs are circuits over the basis *ite* with the restriction that the first input has to be an input variable. Because polynomial branching program size corresponds to logarithmic space and polynomial circuit size corresponds to polynomial time, one expects the existence of functions representable by polynomial-size circuits, but not by polynomial-size branching programs.

Barrington [11] has proved an interesting characterization of the complexity class NC^1. This class equals the class of all (sequences of) Boolean functions representable by polynomial-size branching programs of width 5. The nodes of the branching program are partitioned into levels whose size is bounded by 5, and all edges leaving nodes from the ith level reach nodes on the $(i+1)$-th level. It is straightforward to transform such a branching program into an NC^1-circuit, but the other direction is based on a clever algebraic characterization.

We have seen that circuits of fan-out 1, also called formulas, play a special role. Fan-out 1 is not necessary in applications. However, the tree structure of formulas is of theoretical interest. Moreover, formula size is closely related to circuit depth. More precisely, circuit depth and the logarithm of formula size have the same asymptotic behavior (see Spira [68] for complete bases and Wegener [72] for the monotone case).

It is easy to replace a circuit of depth d by a formula of the same depth. If the fan-in of the basis equals k, the size of the constructed formula is bounded above by k^d, implying that the logarithm (of base k) of the formula size is a lower bound for the circuit depth. For the other direction, we reconstruct a given formula such that the depth gets small. We consider the case of a binary basis and, therefore, binary trees representing f by a formula of size s. Such a formula contains a subformula whose size is between $s/3$ and $2s/3$. Let g be the function represented by this subformula, and let f_c be the function that we obtain if we replace the subtree for g in the formula for f by the constant c. The formula size of f_c is bounded by $2s/3$. Because $f = ite(g, f_1, f_0)$, we can use this as the main idea of a recursive approach where we construct recursively small-depth formulas for g, f_1, and f_0. For each complete basis, *ite* can be realized in constant depth. In the monotone case, by construction, $f_0 \leq f_1$, and we can replace *ite* by the monotone function $monite(x, y, z) = xy \vee z$.

Now we discuss the relations between sequences of circuits representing Boolean functions $f_n \in B_n$ and Turing machines for the corresponding language, which is the union of all $f_n^{-1}(1)$.

This first idea is to relate circuit size and computation time. The main step is to simulate a Turing machine by an oblivious one where the head position at time t can depend on the input length n but not on the input x itself. Efficient simulations are due to Schnorr [61] and Pippenger and Fischer [54]. The time increases by a factor of only $O(\log s(n))$, where $s(n)$ is the space used by the Turing machine. Each step of an oblivious Turing machine can be simulated by a circuit of constant size leading to a simulation of a Turing machine with resource bounds $t(n)$ for time and $s(n)$ for space by a circuit of size $O(t(n) \log s(n))$.

The second idea is to relate circuit depth and space (Borodin [15]). A Turing machine using space $s(n) \geq \log n$ has $c(n) = 2^{O(s(n))}$ possible configurations. Therefore, the one-step behavior can be represented by a $c(n) \times c(n)$ transition matrix. The t-step behavior can be obtained as the tth power of the configuration matrix, and this can be computed by iterative squaring. Altogether, we obtain a circuit of depth $O(s(n) \cdot \log t(n)) = O(s(n)^2)$. This result confirms the parallel computation hypothesis.

Sequences of circuits can realize nonrecursive languages. Circuits are a nonuniform computation model and can be simulated by nonuniform Turing machines. These Turing machines obtain extra information for free that may depend on the input length but not on the input itself. The description of a circuit or formula is suitable extra information. Circuits of size $s(n)$ can be simulated by $O(s(n)^2)$ time-bounded nonuniform Turing machines, and circuits of depth $d(n)$ by $O(d(n))$ space-bounded nonuniform Turing machines.

Finally, the power of nonuniform computation is confirmed by the following derandomization result (Adleman [1]). If the language L corresponding to the sequence f_n of Boolean functions is contained in BPP (polynomial probabilistic time, two-sided error bounded by $1/3$), then f_n can be realized by polynomial-size circuits that are by definition deterministic. The first step is to apply the well-known probability amplification technique by using a majority decision rule on $O(n)$ independent trials of the algorithm. By Chernoff bounds, it follows that the error probability can be reduced to 2^{-2n}. The next step is to apply the pigeonhole principle to prove the existence of an assignment to the random bits leading to an error-free computation. However, nobody knows how to compute this assignment efficiently.

The results of this section underline the central role of circuit size and circuit depth.

11.4 Lower Bounds on Circuit Size

It is easy to prove the existence of Boolean functions whose circuit size is (for large n) at least $2^n/n$ (Shannon [64]) and that this result is the best possible (Lupanov [47]). However, we are interested in lower bounds for "concrete" functions or "explicitly defined" functions excluding diagonalization techniques. Here we consider sequences of functions as explicitly defined if the corresponding language belongs to NP. For a discussion of other notions of explicitness, see Wegener [73].

There is no proof of a nonlinear lower bound on the circuit size of explicitly defined Boolean functions $f_n : \{0, 1\}^n \to \{0, 1\}^n$. We cite some interesting results. Red'kin [60] has proved that the circuit size of binary addition ($2n$ inputs and $n + 1$ outputs) equals $5n - 3$. Hence, a sequence of a half-adder and $n - 1$ full-adders is size-optimal. The record for one-output functions and the full binary basis is $3n - o(n)$ (Blum [14]) and for the basis U_2 after all $5n - o(n)$ (Iwama and Morizumi [33]). All these bounds are based on the gate elimination method. The idea is to to assign constants to some input variables such that a certain number of gates is eliminated (because they are useless, e.g., since they realize a function realized somewhere else in the circuit) and such that the resulting function is of the same type. This idea has been refined such that certain parameters of the circuit are considered. It is sufficient to eliminate a smaller number of gates if one can guarantee that the parameters are changed in a "good direction." This good direction indicates that we can later eliminate some more gates.

As a very simple application of this technique, we prove a lower bound of size $2n - 3$ on the circuit size of threshold-2. This function checks whether the input contains at least two ones. For $n = 2$, this is the function $x_1 \wedge x_2$ and the bound is trivial. For $n > 2$, we consider the first gate of an optimal circuit for threshold-2. This implies that it works on two different variables x_i and x_j. By case inspection, we see that the first gate realizes

$$\big[(x_i \oplus a) \wedge (x_j \oplus b)\big] \oplus c \text{ or } x_i \oplus x_j \oplus c$$

for some constants a, b, and c. We consider the graph representation of the circuit. If x_i or x_j has a fan-out of at least 2, we replace this variable by 0 and can eliminate at least two gates. The resulting subfunction is a threshold-2 function, and the bound follows by induction. Hence, we are in the case that x_i and x_j have fan-out 1. If the first gate is of the first type described earlier, $x_i = a$ makes the circuit independent of x_j while threshold-2 depends for $x_i = a$ essentially on x_j. If the first gate is of the second type described earlier, we obtain for $x_i = x_j = 0$ and $x_i = x_j = 1$ the same output of the circuit, which again is a contradiction of the assumption that the circuit realizes threshold-2. Already this simple example shows that the gate-elimination technique leads to tedious case inspections.

11.5 Lower Bounds on Formula Size and Circuit Depth

Formulas are fan-out restricted circuits, and we may hope that it is easier to obtain lower bounds on the formula size of explicitly defined Boolean functions. There are superlinear lower bounds but, for complete bases, no superpolynomial lower bounds and, therefore, no superlogarithmic lower bounds on the circuit depth. Here we discuss the largest known lower bounds for the bases B_2 and U_2, and we present communication games whose complexity equals circuit depth and monotone circuit depth, respectively. By counting arguments it is easy to prove that the formula size of almost all Boolean functions is bounded below by $2^n(1 - o(1))/\log n$), and Lupanov [48] has proved an upper bound of $2^n(2 + o(1))/\log n$ on the formula size of each Boolean function.

For the complete basis B_2, the best lower bound is still due to Nechiporuk [52]. The lower bound method follows the same idea as the lower bound method for the branching program size (see Chapter 10). We consider the tree representation of formulas. Here it is easier to measure the formula size by the number of leaves of the tree, which is only by one larger than the "real" formula size. For $f \in B_n$, let T be the tree of an optimal formula. Let $S_1, \ldots, S_k$ be disjoint subsets of the variable set, and let l_i be the number of leaves labeled by variables from S_i. An S_i-subfunction of f is a function obtained by an assignment of constants to the variables outside S_i. Let s_i be the number of S_i-subfunctions of f. Nechiporuk's lower bound is stated as

$$L(f) = \Omega(\sum_{1 \leq i \leq k} \log s_i)$$

where $L(f)$ denotes the formula size of f. It is sufficient to prove that $l_i \geq (1/4) \cdot \log s_i$, since $L(f) + 1 \geq l_1 + \cdots + l_k$. This claim is proved by measuring the influence of the S_i-leaves on the formula. Let V_i be the set of inner nodes of T where both subtrees contain S_i-leaves. Then $|V_i| = l_i - 1$. We consider paths starting at S_i-leaves or V_i-nodes, ending at V_i-nodes or the root, and containing no V_i-node as inner node. Because the number of these paths ending in a V_i-node is bounded above by 2 and there may be one more path ending at the root, the number p_i of these paths is bounded above by $2|V_i| + 1$. Let us fix an assignment of the variables outside S_i. If g is computed at the first node of one of the considered paths, then g, $\overline{g}$, 0, or 1 is computed at its last edge. Because we obtain all S_i-subfunctions in this way and because the S_i-leaves are not influenced by these assignments,

$$s_i \leq 4^{p_i} \leq 4^{2|V_i|+1} \leq 4^{2l_i},$$

implying the claim that $l_i \geq (1/4) \cdot \log s_i$. It can be shown that Nechiporuk's lower bound is at most of size $n^2/\log n$ and the bound $\Omega(n^2/\log n)$ can be proved for the indirect storage access function in the same way as the $\Omega(n^2/\log^2 n)$ bound for the branching program size of this function (see Chapter 10).

The complete basis U_2 does not contain EXOR and its negation. By De Morgan rules, we may consider formula trees where the leaves are labeled by literals, namely positive or negated variables, and all gates are of type AND or OR. The main difference between AND/OR and EXOR is the following. Fixing the input of an AND/OR gate by the appropriate constant, the output of the gate is a constant. This is not possible for EXOR-gates. Hence, one can hope to work with the following approach. If we choose a good assignment to certain variables, the formula size shrinks by a large factor, but the resulting subfunction still is not too easy. Because this does not work for EXOR-gates, it is a good idea to investigate the parity function, that is, the EXOR of n variables. Subbotovskaya [69] used this approach to prove an $\Omega(n^{3/2})$ bound on the U_2-formula size of parity. Khrapchenko [37] improved this to the optimal bound $n^2 - 1$. His bound can be considered as a further development of Subbotovskaya's method. However, people preferred to consider the result as a purely combinatorial one. It can be stated as follows. For some $A \subseteq f^{-1}(1)$, and $B \subseteq f^{-1}(0)$, let $H(A, B)$ be the set of pairs $(a, b) \in A \times B$ whose Hamming distance equals 1. Then the U_2-formula size of f is bounded below by $|H(A, B)|^2/(|A| \cdot |B|) - 1$, which equals $n^2 - 1$ for $A = f^{-1}(1)$, $B = f^{-1}(0)$, and the parity function f.

It was a long time before Andreev [8] came up with a larger bound on the U_2-formula size. He reevaluated Subbotovskaya's method. Moreover, he came up with a function that is the right choice for the method of shrinking formulas by applying random restrictions. These random restrictions keep x_i as a variable with probability $p \in (0, 1)$ and assign it the value 0 or 1 with probability $(1 - p)/2$ each. The function is defined on $2n$ variables where $n = 2^k$. The first n variables are partitioned into k blocks of size n/k each. Let a_i be the parity of the variables of the ith block. Then we have k values that are the inputs of a function g whose value table is stored in the other n variables of the input. Andreev's function computes $g(a)$. This function has two properties that support lower bound proofs. We may replace the second group of variables by the value table of a difficult function, namely one of formula size $\Omega(n/\log\log n)$. Then we apply an appropriate random restriction to the other n variables. If at least one variable of each block survives, the remaining formula has a size of at least $\Omega(n/\log\log n)$. If the given formula has size s and is shrunk to αs, then $\alpha s = \Omega(n/\log\log n)$ and we get the lower bound $\Omega(\alpha^{-1}n/\log\log n)$ on s. Andreev [8] obtained a bound of $\Omega(n^{5/2-o(1)})$ on the U_2-formula size of his function. Several authors improved this result before Håstad [27] obtained the best possible shrinking lemma and the bound $\Omega(n^{3-o(1)})$ on the U_2-formula size of Andreev's function.

Karchmer and Wigderson [35] established a tight connection between circuit depth and the communication complexity (see Kushilevitz and Nisan [44] and Chapter 10) of relations. For a Boolean function f, Alice gets some input $a \in f^{-1}(1)$ and Bob gets some input $b \in f^{-1}(0)$. They have to agree on a communication protocol, which for all such pairs (a, b) implies that Alice and Bob agree on an index i such that $a_i \neq b_i$. If we do not care about constant factors, we can work with circuits whose inputs are literals and whose gates are of type AND and OR. The idea of the following communication protocol is that the players start at the output gate and follow a reversed path to some input such that the following invariant holds. For each gate on this path and the function g realized at this gate, $g(a) \neq g(b)$. At the input x_i or $\overline{x}_i$ is realized and i is a correct result of the protocol. For monotone circuits, the same idea leads to an input x_i, and the resulting index even has the property that $a_i = 1$ and $b_i = 0$ (and not vice versa). Hence, lower bounds on the communication complexity of this relation (its monotone counterpart) yield lower bounds of the same size for the (monotone) circuit depth. It is also not too difficult to design from a communication protocol for the considered (monotone) relation a (monotone) circuit whose depth is bounded by the length of the given protocol.

Nontrivial bounds have been obtained only in the monotone case, for example:

- An $\Omega(n)$-bound on the monotone circuit depth of the function on $\binom{n}{2}$ variables deciding for a graph on n vertices whether it has a matching with at least $n/3$ edges (Raz and Wigderson [56])
- An $\Omega(\log^2 n)$-bound on the monotone circuit depth for the s-t-connectivity problem in directed graphs (Karchmer and Wigderson [35])
- A $\log^{2-o(1)} n$-bound on the monotone circuit depth for the decision problem whether an undirected graph is connected (Goldmann and Håstad [21])

A quite general proof method has been presented by Raz and McKenzie [55]. They obtain a new proof of the $\Omega(\log^2 n)$-bound on the monotone circuit depth for s-t-connectivity and an $\Omega(k \log n)$ bound for the problem to decide whether a graph on n vertices contains a k-clique. Moreover, they have obtained a hierarchy result, namely that monotone circuits of polynomial size and depth $O(\log^{i+1} n)$ compute more functions than monotone circuits of polynomial size and depth $O(\log^i n)$.

11.6 Lower Bounds on the Size of Monotone Circuits

In the last section we considered lower bounds on the depth of monotone circuits. Here we investigate the size of monotone circuits. However, size s implies a depth of at least $\log s$.

Monotone circuits allow local replacement rules. Hence, one has some knowledge of the structure of optimal monotone circuits. In the beginning, lower bounds have been proved by proving that different prime implicants need different gates to be created. This was improved by Wegener [71] who estimated the contribution of each gate to the computation of the prime implicants of the considered function. This led to bounds of size $n^2/\log n$ for functions with n outputs. These methods tried to be too precise. For functions of exponential size, it is possible to underestimate the complexity by a smaller exponential factor in order to obtain an exponential lower bound. Applying these ideas, Razborov [58] proved the first nonpolynomial lower bound on the monotone circuit size of an explicitly defined monotone function, namely a clique function. Alon and Boppana [4] improved the bound to an exponential one, and Andreev [7] obtained a slightly larger bound for another function.

The lower-bound method works as follows. Monotone functions f are identified with $f^{-1}(1)$. Because $(f \wedge g)^{-1}(1) = f^{-1}(1) \cap g^{-1}(1)$ and $(f \vee g)^{-1}(1) = f^{-1}(1) \cup g^{-1}(1)$, monotone computations are computations in the lattice of all $f^{-1}(1)$, f monotone, with the operations $\cap$ and $\cup$. The idea of Razborov was to define an "approximating lattice" such that lower bounds on the lattice complexity are lower bounds on the monotone circuit size and such that the lattice has enough structure to allow the proof of large lower bounds.

The approximating lattice L has to be a subset of all $f^{-1}(1)$, f monotone, and it has to contain the counterparts of the circuit inputs, namely, $\emptyset$, $\{0,1\}^n$, $x_i^{-1}(1)$, $1 \le i \le n$. Moreover, we need two operations called meet "$\sqcap$" and join "$\sqcup$" that have to fulfill the conditions $M \sqcap N \subseteq M \cap N$ and $M \sqcup N \supseteq M \cup N$. The "approximation mistakes" are denoted by $\delta_\sqcap(M, N) := (M \cap N) \setminus (M \sqcap N)$ and $\delta_\sqcup(M, N) := (M \sqcup N) \setminus (M \cup N)$. Finally, the lattice complexity of the monotone function f is the minimal t such that there exist lattice sets $M, M_1, N_1, \ldots, M_t, N_t$ such that $f^{-1}(1)$ and M are close with respect to these sets, more precisely,

$$M \subseteq f^{-1}(1) \cup \bigcup_{1 \le i \le t} \delta_\sqcup(M_i, N_i)$$

and

$$f^{-1}(1) \subseteq M \cup \bigcup_{1 \le i \le t} \delta_{\sqcap}(M_i, N_i).$$

It is not very difficult to prove that the lattice complexity of f is a lower bound on the monotone circuit size of f. The difficult part is to define an "appropriate" lattice and to prove a large lower bound on the lattice complexity of the considered function (see Wegener [73]). Several authors have pointed out that these bounds are not adequate for circuits over complete bases.

11.7 Constant-Depth Unbounded-Fan-In Circuits

We have seen that our lower-bound methods for circuits are poor. Because good circuits have small size *and* small depth, we are also interested in lower bounds on the size of depth-restricted circuits. Again, our methods are too poor to prove that functions cannot be realized by NC^1-circuits. Because functions depending essentially on n variables need depth $\lceil \log n \rceil$, smaller depth bounds make no sense for bases of constant fan-in. This changes for unbounded-fan-in circuits.

The simplest model allows negation and unbounded-fan-in AND- and OR-gates and circuits of bounded depth d. These circuits can be easily transformed into the following normal form. All literals are inputs of the circuit, the circuit is leveled, and all gates on even levels are of one type and all gates on odd levels are of the other type. Σ^k-circuits have k levels and the output gate is an OR-gate, whereas Π^k-circuits also have k levels ending with an AND-gate. Σ^2-circuits are sums-of-monomials and their minimization is (or at least was) an important problem in hardware design (see Chapter 16).

The parity function has the largest possible Σ^2-size and is a good candidate to prove that Σ^k-circuits, k not too large, need exponential size. We obtain a polynomial upper bound for depth $\lceil \log n / \log \log n \rceil + 1$. We start with a balanced tree of depth $d = \lceil \log n / \log \log n \rceil$ where each gate is an EXOR-gate of fan-in $\lceil \log n \rceil$. Such a gate can be replaced by a Σ^2- or Π^2-circuit of size bounded by $2n$. Using a Σ^2-circuit for the last gate and alternating levelwise between Σ^2-circuits and Π^2-circuits, we obtain a polynomial-size circuit of depth $2d$. However, the levels $2i$ and $2i + 1$ are of the same type and can be merged, leading to a depth of $d + 1$. This bound is very close to optimal, as was proved by Håstad [25], who improved lower bounds of Furst, Saxe, and Sipser [20], Ajtai [2], and Yao [75]. His result can be stated in the following way. The depth of polynomial-size unbounded-fan-in parity circuits is not smaller than $(\log n)/(\log \log n + c)$ for some constant c.

The proof is by induction on the depth where the case of depth 2 is obvious. For technical reasons it is better to work with circuits where the fan-in of the first level (called 1-fan-in) is bounded by some parameter. This can be achieved in the beginning by adding a dummy level with 1-fan-in 1. The first two levels are Σ^2- or Π^2-circuits of bounded 1-fan-in. If we replace them by Π^2- resp.

Σ^2-circuits for the same functions, levels 2 and 3 are of the same type and can be merged. However, the size can increase exponentially and the 1-fan-in is no longer bounded. The idea is to replace certain variables by appropriate constants in order to guarantee small circuits with small 1-fan-in. Nobody knows how to compute such a restriction. Hence, the well-known probabilistic method is applied: namely, a random restriction is applied and it is proved that a good restriction exists. We state Håstad's switching lemma in its original form, although some improved variants exist. Let p be the probability that x_i is kept as a variable, and let α be the unique solution of

$$(1 + 4p/((1 + p)\alpha))^t = 1 + (1 + 2p/((1 + p)\alpha))^t.$$

Let S be a Π^2-circuit (Σ^2-circuit) of 1-fan-in t computing g, and let γ be a random restriction with parameter p. Then with probability at least $1 - \alpha^s$, the random subfunction g_γ of s can be described by a Σ^2-circuit (Π^2-circuit) of 1-fan-in s. This lemma is applied with well-chosen parameters until we obtain a Σ^2- or Π^2-circuit realizing a parity function on a not-too-small number m of variables. The assumption of too small size of the given circuit leads to a 1-fan-in of less than m for the resulting Σ^2- or Π^2-circuit, which is impossible for such parity circuits. The argument works for all functions where all prime implicants and prime clauses are long. More precisely, we obtain a bound $2^{\Omega(n^{1/(k-1)})}$ for the size of Σ^k- and Π^k-circuits and all functions where the length of each prime implicant and each prime clause is at least $n - (9/100)n^{1/(k-1)}$ (Wegener [74]).

This success motivated the investigation of other types of unbounded fan-in circuits. Here it is essential which gates are allowed. The parity function can be computed with one gate if unbounded-fan-in EXOR gates are available. Hence, it is interesting to allow such gates. Then OR-gates can be replaced easily and negation can be realized by $x \oplus 1$. This leads to unbounded-fan-in circuits with AND- and EXOR-gates. Again we can assume that the circuits are leveled and the gate types alternate between the levels.

The situation with AND-, OR-, and NOT-gates is a combinatorial one. We can argue with implicants, clauses, subcubes, and their unions and intersections. It is much more difficult to investigate EXOR in this setting. However, ($\{0, 1\}$, EXOR, AND) is the field $\mathbb{Z}_2$, and it seems to be appropriate to apply algebraic methods: that is, we consider Boolean functions as $\mathbb{Z}_2$-polynomials. Only AND-gates can increase the degree of functions. It is easy to obtain a polynomial of large degree; namely $x_1 x_2 \cdots x_n$ is of maximal degree n and can be computed at one AND-gate. However, this function can be approximated by a degree-0 polynomial, namely the constant 0. The quality of approximations is based on the Hamming distance of functions, namely $H(f, g) := |\{a \mid f(a) \neq g(a)\}|$. Razborov [57] has proved that all functions computable with small-depth unbounded-fan-in EXOR/AND-circuits have a good approximation by a polynomial of low degree. More precisely: Let $r \geq 1$, and let k be the number of AND-levels of an unbounded-fan-in EXOR/AND-circuit realizing f. Let d be the minimal distance between f

and a polynomial of degree m^k. Then the size of the circuit is at least $d/2^{n-m}$. This is proved by induction on k.

In order to apply this result, we look for a "simple" function that has large distance from all polynomials of small degree. Majority is such a function. It computes 1 iff the input contains at least as many ones as zeros. Razborov [57] investigated the distance between majority and polynomials of bounded degree. This led to a lower bound of $2^{\Omega(n^{1/(2k-1)})}$ on the size of unbounded-fan-in EXOR/AND-circuits that realize majority with k AND-levels.

Smolensky [67] generalized this result to the case where EXOR (or MOD-2) is replaced by MOD-p for some prime p. He also proved large lower bounds if MOD-q for some prime q has to be realized with gates of type AND and MOD-p for some prime $p \neq q$. The situation for lower bounds gets much more difficult if we allow gates of type AND, MOD-p, and MOD-q for different primes p and q (this is equivalent to gates of type AND and MOD-pq).

11.8 Threshold Circuits of Constant Depth

In the previous section we investigated the class AC^0 (alternating class of depth $O(\log^0 n) = O(1)$) of all sequences $f = (f_n)$ of Boolean functions realizable by constant-depth polynomial-size circuits with NOT-, AND-, and OR-gates of unbounded fan-in. Because the parity function PAR is not contained in AC^0, we have then allowed EXOR-gates or, more generally MOD-m-gates leading to the class $ACC^0[m]$ (alternating counting class of depth $O(1)$) of all $f = (f_n)$ realizable by constant-depth polynomial-size circuits with AND- and MOD-m-gates of unbounded fan-in. Because the majority function MAJ is not contained in $ACC^0[p]$ for primes p, the next step is to allow MAJ-gates. It is more convenient to allow threshold gates where the positive threshold function $T^n_{\geq k}$ computes 1 if the number of incoming ones is at least k and the negative threshold function $T^n_{\leq k}$ can be defined as $1 - T^n_{\geq k+1}$. Threshold gates can simulate NOT, AND, and OR. Therefore, we investigate threshold circuits containing only threshold gates. (This topic is also handled from different angles by Bruck in Chapter 12 and by Anthony in Chapter 13.)

In the case of AC^0 or $ACC^0[m]$, the circuit size is defined as the number of gates, although the number of edges can be significantly larger. However, it makes no sense to connect two gates by more than 2 resp. m edges, and therefore, the number of edges is polynomially bounded iff the number of gates is. This is different for threshold circuits. One threshold gate is enough to compute the carry bit CAR_n of the sum of two n-bit numbers. This holds because the carry bit equals 1 iff the weighted sum of all $(x_i + y_i) \cdot 2^i$ is at least 2^n. However, this gate has $2 \cdot (2^n - 1)$ incoming edges. Hence, we are interested in the complexity class TC^0_k (threshold class of depth k, k a constant) containing all $f = (f_n)$ with depth-k threshold circuits with a polynomial number of *edges* and in the class LT^0_k containing all $f = (f_n)$ with depth-k threshold circuits with a polynomial number of *gates*. It is convenient to replace multiple edges by edges weighted by integers.

The gate computes 1 iff the weighted sum of the incoming values is not smaller resp. not larger than the given threshold t, which also is allowed to be an integer. The edge size is replaced by the weight, where an edge with weight w contributes $|w|$ to the weight of the gate and the weight of the circuit.

First, we discuss the relations between the TC- and the LT-classes. Afterward, we investigate the computational power of small-depth threshold circuits of polynomial size. The results imply that it is difficult to obtain good lower bounds. Then the known lower bound techniques are presented. Finally, we discuss some relations to the $\mathrm{ACC}^0[m]$-classes and to cryptography.

The example of the carry bit CAR shows that exponential weight seems to be useful. What about even larger weights? Muroga [49] showed that each threshold gate with n inputs can be replaced by a threshold gate whose weight is bounded by $2^{O(n \log n)}$ (see also Chapter 13). Much later, Håstad [26] could prove that this bound is the best possible. This implies that TC^0, the union of all TC_k^0, is contained in NC^1.

By definition, $\mathrm{TC}_k^0 \subseteq \mathrm{LT}_k^0$. In order to simulate LT_k^0-circuits by TC^0-circuits, we may try a gate-by-gate simulation. Using the result of Muroga [49], it is enough to compute the sum of polynomially many $O(n \log n)$-bit numbers in order to simulate an arbitrary threshold gate. For each bit position, the sum of the bits has length $O(\log n)$ and can be computed bit by bit by threshold gates of polynomial weight in the obvious way. Hence, we can decrease the number of terms logarithmically in depth 1. Finally, addition of two numbers can be done in depth 3 even in AC^0-circuits. This implies a simulation in depth $O(\log^* n)$ for each layer of the LT_k^0-circuit. This easy result is by no means optimal. Goldmann, Håstad, and Razborov [22] could prove that $\mathrm{LT}_k^0 \subseteq \mathrm{TC}_{k+1}^0$. Their proof of the existence of a TC_{k+1}^0-circuit was improved by Goldmann and Karpinski [23] by an explicit construction of such a circuit. However, the size of their TC_2^0-circuit simulating a threshold gate with n inputs was $\Theta(n^{20} \log^{20} n)$. Nowadays, the best bound of $O(n^8)$ is due to Hofmeister [30].

In order to understand the power of threshold circuits, we investigate the parity function. It is not contained in LT_1^0, because a threshold gate can realize only functions that are monotone (positively or negatively) with respect to each input. However, the following construction shows that $\mathrm{PAR} \in \mathrm{TC}_2^0$. The first level contains $2\lceil n/2 \rceil$ gates, each with one edge from each input. For each odd $m \leq n$ we realize $T_{\geq m}^n(x)$ and $T_{\leq m}^n(x)$. If x contains an even number of ones, each such pair of gates realizes one 1. If x contains an odd number of r ones, the same holds for all $m \neq r$ and the pair $T_{\geq r}^n(x)$ and $T_{\leq r}^n(x)$ realizes two ones. Hence, PAR_n is realized by checking whether the gates on the first level realize at least $\lceil n/2 \rceil + 1$ ones. This idea can be extended to prove that all symmetric functions are contained in TC_2^0. The corresponding circuit needs only $O(n)$ gates and weight $O(n^2)$. A natural question in this context is whether there are more efficient ways for representing PAR by constant-depth threshold circuits, for instance with linear weight. A negative answer was given by Impagliazzo, Paturi, and Saks [32]. They showed that any depth d threshold circuit computing PAR_n

has weight at least $n^{1+c\cdot\alpha^{-d}}$, where $c > 0$ and $\alpha \leq 3$ are constants independent of n and d.

We know that CAR $\in \mathrm{LT}_1^0 \subseteq \mathrm{TC}_2^0$ and the same result can be proven for the comparison COM of two numbers. This can be used to compute the rank of a number among n numbers of bit length n implying that sorting SOR is contained in TC_3^0.

Arithmetic functions are representable by TC^0-circuits of small depth:

- Addition is in TC_2^0 (Alon and Bruck [5]). This holds even for multiple addition, that is, the addition of n numbers of bit length n.
- Multiplication, division, modular powering (x^y mod z), multiple multiplication in $\mathbb{Z}_p$ for primes p of bit length $O(\log n)$ or in finite abelian groups is in TC_3^0 (Siu, Bruck, Kailath, and Hofmeister [65], Hofmeister [29]).
- Multiple multiplication in $\mathbb{Z}$ and in abelian groups is in TC_4^0.
- Multiple multiplication in finite solvable groups is in TC^0 (Barrington and Thérien [12]).

However, multiple multiplication in finite nonsolvable groups is in TC^0 only if $\mathrm{TC}^0 = \mathrm{NC}^1$ (Barrington [11]). Most people believe that TC^0 is a proper subclass of NC^1.

There is a common idea behind many of these upper bounds. A function f on n inputs is called *polynomially bounded* if there are weights w_i, $1 \leq i \leq n$, and polynomially many integer intervals I_j such that $f(x) = 1$ is equivalent to $w_1x_1 + \cdots + w_nx_n \in I_j$ for some j. The simulation results of Goldmann, Håstad, and Razborov [22] can be generalized to circuits whose gates realize polynomially bounded functions. Such functions support the design of small-depth circuits. Each bit of the sum of n numbers of bit length n is a polynomially bounded function allowing the application of Chinese remaindering. The preceding arguments have shown that small-depth polynomial-size threshold circuits are surprisingly powerful. This implies that it should be difficult to come up with powerful lower bound techniques. Indeed, no proof is known that a function is contained in NP $-\ \mathrm{TC}_3^0$. For many functions it is easy to prove that they are not contained in TC_1^0, because all functions in TC_1^0 are monotone with respect to each variable. Hence, the only interesting case which has been solved is the case of depth 2.

Hajnal, Maass, Pudlák, Szegedy, and Turán [24] have shown that the inner product in $\mathbb{Z}_2$ called IP is in $\mathrm{TC}_3^0 - \mathrm{TC}_2^0$. $\mathrm{IP}_n(x, y)$ is the EXOR-sum of all x_iy_i, $1 \leq i \leq n$. Because PAR $\in \mathrm{TC}_2^0$, it is obvious that IP $\in \mathrm{TC}_3^0$. The idea for a lower bound for depth 2 is the following one. We assume that a depth-2 weight-$w(n)$ threshold circuit realizes IP_n. By some manipulations we can assume that IP_n is realized at a MAJ-gate with $w'(n) = O(w(n))$ inputs where $w'(n)$ is even and the number of incoming ones is always different from $w'(n)/2$. If $\mathrm{IP}_n(a, b) = 1$, there are more incoming ones than incoming zeros while, for $\mathrm{IP}_n(a, b) = 0$, there are more incoming zeros than incoming ones. On average, the input functions to the output gate are more likely to equal IP_n than to differ from IP_n. There has to be at least one input function of the output gate that is $1/(2w'(n))$-correlated to IP_n. Two functions f and g are called ε-correlated if, for a randomly chosen input a, the event $f(a) = g(a)$ has a probability of at least $1/2 + \varepsilon$. Now a lower bound follows by proving that IP_n and threshold functions cannot be $1/(2w'(n))$-correlated. For this purpose it is useful to represent the function table of IP_n as matrix whose rows correspond to the partial input a and whose columns correspond to the partial input b. This matrix is closely related to the well-known Hadamard matrix: more precisely, we obtain the Hadamard matrix if we replace 1 by -1 and 0 by 1. All rows or columns of this matrix are orthogonal with respect to $\mathbb{Z}_2$. Therefore, it is possible to apply arguments from linear algebra to prove that IP_n and threshold functions can be $\varepsilon(n)$-correlated only for exponentially small $\varepsilon(n)$. Krause [38] has used these arguments to show that multiple multiplication in $\mathbb{Z}$ has no TC_3^0-circuits based on Chinese remaindering.

The proof method just described can be reformulated in the language of communication complexity. The matrix representing the function is then called a communication matrix, and we assume that Alice holds a, Bob holds b, and they have to compute $\mathrm{IP}_n(a, b)$. A depth-2 small-weight threshold circuit leads to a short randomized two-sided error communication protocol. The lower bound follows by proving that such protocols do not exist (see Kushilevitz and Nisan [44]).

Krause [39] has presented another approach to prove that functions are not contained in TC_2^0. His approach is also based on the communication matrix of the given function and some distribution of the input variables between Alice and Bob. Here it is convenient to replace 1 by -1 and 0 by 1. This matrix is called $M(f)$. The lower bound technique is based on the following observation. If f can be realized by a depth-2 threshold circuit of weight $w(n)$, then there exists a matrix M of the same number of rows and columns as $M(f)$ with the following properties: all entries are nonzero integers, they are positive exactly for the 1-entries of $M(f)$ (M and $M(f)$ are called *signum-consistent*), and the rank is bounded by $O(w(n))$. Hence, we are interested in lower bounds on the rank of matrices with these properties. Such bounds can be obtained via the operator norm of $M(f)$. The operator norm of an $N \times N$-matrix A is the largest length (L_2-norm) of some vector Ax where x has length 1 and is denoted by $||A||_2$. Improving a result of Krause [39], Forster [19] proved that the rank of all matrices M that are signum-consistent to f is bounded below by $2^n/||M(f)||_2$. This approach gives at present the largest lower bounds for depth-2 threshold circuits.

We mention a result that underlines the power of small-depth threshold circuits in comparison to $\mathrm{ACC}^0[m]$-circuits of arbitrary constant depth. If $f = (f_n) \in \mathrm{ACC}^0[m]$, then f_n can be realized by depth-3 threshold circuits where the number of incoming edges for the gates on the first level is polylogarithmically bounded and the weight is bounded by some function $2^{w(n)}$ for some polylogarithmically bounded $w(n)$. This has been shown by Allender [3] for primes m and by Yao [76] for the general case.

Another argument for the computational power of the complexity class TC_3^0 is based on results from learning theory. The scenario is the following. One does not know f, but one knows that f belongs to some complexity class C. Based on a random set of pairs $(a, f(a))$, one has to construct a function g that is "close" to f. It is known that subexponential learning algorithms exist for AC^0 (Linial, Mansour, and Nisan [46]). Kearns and Valiant [36] have shown that the existence of such algorithms for TC^0_3 would disprove widely believed cryptographic hardness assumptions. This result is based on the observation that the encryption operators of famous cryptosystems (RSA, Rabin, El-Gamal) can be realized by TC^0_3-circuits. Thus, subexponential learning algorithm for TC^0_3 would break these systems.

11.9 Barriers to Proving Lower Bounds

Apparently, there is a connection between the resistance of certain complexity classes against effective lower bound proofs and the hardness of certain cryptographical primitives. Pioneering work in this direction was carried out by Razborov and Rudich [59]. The typical fashion of a lower bound argument is to determine a combinatorial property

$$P = \left(P_n : \{0,1\}^{2^n} \longrightarrow \{0,1\}\right)$$

and to give a proof that Boolean functions $f = (f_n)$ having this property (i.e., there is some n_0 with $P_n(f_n) = 1$ for $n \geq n_0$) do not belong to C. Razborov and Rudich [59] observed that all known lower bound arguments are *natural proofs*, which, roughly speaking, means that P_n has small circuits (in the input length 2^n), and that $P_n(f) = 1$ with probability close to 1 on a random Boolean function f. The main result of Razborov and Rudich [59] is that the existence of natural proofs against P/poly disproves the existence of exponentially hard pseudorandom bit generators and, thus, contradicts widely believed cryptographic hardness assumptions. This makes the existence of natural lower bound arguments against P/poly unlikely.

For getting similar arguments for complexity classes below P/poly one can use the following observation of Krause and Lucks [40]. All known effective lower bound arguments can be reformulated as efficient adversary algorithms for distinguishing an unknown function in the given complexity class C from a random Boolean function. This implies that for those complexity classes C that contain cryptographically hard pseudorandom function generators (i.e., collections F of functions in C for which there is no efficient adversary algorithm that distinguishes a randomly chosen function from F from a truly random function), effective lower bound arguments should not exist. Naor and Reingold [51] have constructed a pseudorandom function generator in TC^0 that is cryptographically hard under the condition that the Decisional Diffie Hellman Conjecture, a widely believed assumption on the hardness of the discrete logarithm, is true. Krause and Lucks [40] could show that Naor and Reingold's generator even belongs to TC^0_4. Altogether, we have good cryptographical reasons for being not able to show $P \neq NP$ via circuit lower bounds or to separate TC^0_4 from NP.

11.10 Voting Polynomials

Our lower bound techniques are powerful enough to obtain bounds of exponential size for depth-2 threshold circuits but not powerful enough for depth-3 threshold circuits. This motivates the investigation of complexity classes contained in TC_3^0 (or LT_3^0) but not in TC_2^0. One such model are depth-2 circuits with EXOR-gates on the first level and a threshold gate on the second level. This circuit model can be easily transformed into the model of voting polynomials. We consider functions $f\colon \{1,-1\}^n \to \{1,-1\}$ instead of $g\colon \{0,1\}^n \to \{0,1\}$ where 1 is replaced by -1 and 0 is replaced by 1. The EXOR-sum of some Boolean variables is the same as the product of the corresponding $\{1,-1\}$-valued variables. Hence, on the first level the circuit computes $\{1,-1\}$-monomials that are square-free (because $1^2 = (-1)^2 = 1$). The monomials get weights on the edges to the output gate; hence, we get a polynomial, namely the sum of the weighted monomials (all weights are integer valued). Because we can easily replace the output gate by an output gate with the threshold value 0 such that the polynomial never takes the value 0, the monomials vote for the output, that is, $f(a) = 1$ iff the majority vote is positive and $f(a) = -1$ iff the majority vote is negative. More formally, $f\colon \{1,-1\}^n \to \{1,-1\}$ can be represented as the sign of a square-free polynomial $p(x)$. This computation model leads to the following three complexity measures:

- The length, that is, the number of gates on the first level or the number of monomials of the voting polynomial,
- The weight of all edges leading to the output gate, and
- The degree of the voting polynomial, that is, the largest fan-in of a gate on Level 1.

This implies that threshold functions have voting polynomials of degree 1 and EXOR-functions have voting polynomials of length 1. The complexity class $\widehat{\mathrm{PT}}_1$ of all functions with polynomial-weight voting polynomials is contained in TC_2^0, and the complexity class PT_1 of all functions with polynomial-length voting polynomials is contained in $\mathrm{LT}_2^0 \subseteq \mathrm{TC}_3^0$. This follows by simulating the EXOR-gates by TC_2^0-circuits and by a trick to save one level of the resulting circuit.

The problem is to prove lower bounds on the complexity measures related to voting polynomials. This can be done by a spectral analysis on the function tables of functions $f\colon \{1,-1\}^n \to \{1,-1\}$. Each function is considered as a 2^n-dimensional vector (representing all $f(a)$, $a \in \{1,-1\}^n$). In this vector space an inner product (f, f') is defined as the sum of all $f(a) \cdot f'(a)$, $a \in \{1,-1\}^n$. It is well known that the set of the 2^n monomials m_a, $a \in \{0,1\}^n$, is an orthonormal basis of this vector space. The monomial m_a is in the $\{1,-1\}$-representation the product of all x_i where $a_i = 1$. The spectral coefficients of f are the numbers (f, m_a), and it is well known that f equals the sum of all $(f, m_a) \cdot m_a$. The sum of all $2^n \cdot (f, m_a) \cdot m_a$, $a \in \{0,1\}^n$, is a voting polynomial for f. Based on the size of the spectral coefficients, one obtains properties of *all* voting polynomials for f. We refer to Chapter 12 by Bruck for more details on spectral analysis.

Bruck [16] showed that functions f whose voting polynomial length is bounded by t have a spectral coefficient whose absolute value is at least $1/t$. It is easy to prove that the absolute value of all spectral coefficients of IP_n equals 2^{-n}, implying that $\mathrm{IP} \notin \mathrm{PT}_1$. The EXOR-sum of all $x_i x_j$, $i \neq j$, is a symmetric function and, therefore, in TC_2^0. Computing the spectral coefficients leads to the result that this function is in $\mathrm{TC}_2^0 - \mathrm{PT}_1$.

The L_1-norm of f is defined as the sum of all $|(f, m_a)|$, $a \in \{0, 1\}^n$. It is easy to see that $1 \leq L_1(f) \leq 2^{n/2}$ for all f. Again Bruck [16] presented a randomized construction proving the existence of voting polynomials for f whose weight is bounded by $O(n \cdot L_1(f)^2)$. The idea is to choose randomly the sum of $O(n \cdot L_1(f)^2)$ monomials $\pm m_a$ where m_a gets a positive sign iff $(f, m_a) > 0$. Moreover, $\pm m_a$ is chosen with a probability of $|(f, m_a)|/L_1(f)$. Then it is possible to prove that the probability of obtaining a voting polynomial for f is positive. These results imply the existence of polynomial-weight voting polynomials for COM and CAR. Explicit constructions were obtained by Alon and Bruck [5].

Voting polynomials are not allowed to refuse to vote, that is, a voting polynomial p has the property that $p(a) \neq 0$ for all inputs a. A weaker notion is to consider all polynomials p that are different from the constant 0. They weakly represent f if they vote correctly for all a where $p(a) \neq 0$. The weak degree of f is the minimal degree of a polynomial that weakly represents f. Aspnes, Beigel, Furst, and Rudich [9] have proved that functions representable by polynomial-size constant-depth circuits where all gates besides the output gate are of type AND, OR, and NOT and the output gate is allowed to be a threshold gate of unbounded weight have a not very large weak degree. Because the parity function has the maximal possible weak degree, namely n, it cannot be represented by the class of circuits described above.

We have seen that $\widehat{\mathrm{PT}}_1$ contains functions not contained in AC^0. However, $\mathrm{AC}_2^0 \subseteq \widehat{\mathrm{PT}}_1$ (Alon and Bruck [5]). For the proof of this claim we investigate the function ROW_n deciding whether a Boolean $n \times n$-matrix contains at least one 1-entry in each row, more formally,

$$\mathrm{ROW}_n(X) = \bigwedge_{1 \leq i \leq n} \bigvee_{1 \leq j \leq n} x_{ij}.$$

Without giving a formal definition, $\mathrm{ROW} = (\mathrm{ROW}_n)$ can be considered as AC_2^0-complete function. The voting polynomial degree of ROW_n is $\Omega(n)$ as already observed by Minsky and Papert [50]. However, Alon and Bruck [5] have proved the existence of voting polynomials of weight $O(n^4)$ for ROW_n. Their probabilistic arguments are based on the following observation. Let us choose randomly a row number $i \in \{1, \ldots, n\}$ and a subset $A \subseteq \{1, \ldots, n\}$. The random function $g_{i,A}$ is the EXOR-sum of all x_{ij}, $j \in A$. For each Boolean matrix the random value of $g_{i,A}$ equals 1 with probability $1/2$. Restricting the Boolean matrices to those matrices X where $\mathrm{ROW}_n(X) = 0$, this probability is bounded above by $(1 - 1/n)/2$, because with probability $1/n$ the 0-row is chosen. This difference between random matrices and random matrices with a 0-row is exploited to prove the result. Alon, Goldreich, Håstad, and Peralta [6] present an explicit construction of a polynomial-weight voting polynomial for ROW.

Krause and Pudlák [41] have shown that AC_3^0 is not even contained in PT_1. They have investigated the following variant ROW_n^* of ROW_n. There are three input matrices U, V, and W. Let $x_{ij} := \overline{u}_{ij}v_{ij} + u_{ij}w_{ij}$. Then $ROW_n^*(U, V, W) := ROW_n(X)$. By definition, $ROW^* \in AC_3^0$. The main result is that the voting polynomial length of ROW_n^* is at least 2^d where d is the voting polynomial degree of ROW_n and therefore $2^{\Omega(n)}$. This result holds for all f^* obtained from f as ROW^* from ROW.

We have compared AC_2^0 and AC_3^0 with $\widehat{PT}_1$ and PT_1. The same problem can be considered for LT_1^0 and TC_2^0. The function $T_{n,n}$ that outputs 1 if the sum of all $2^{j-1}x_{ij}$, $1 \le i, j \le n$, is at least $n2^{n-1}$ is a decision variant of multiple addition. It is contained in LT_1^0 and TC_2^0 but not in $\widehat{PT}_1$ as shown by Goldmann, Håstad, and Razborov [22]. Let $T_{n,n}^*$ be the function $T_{n,n}$ applied to $x_{ij} := y_{ij}z_{ij}$ where y_{ij} and z_{ij} are the input variables of $T_{n,n}^*$. Then the methods discussed in Section 11.8 lead to the result that $T^* = (T_{n,n}^*)$ is not contained in TC_2^0. However, this function is contained in PT_1. Another variant of $T_{n,n}$ has been investigated by Krause and Pudlák [42]. Let $T_{n,n}^*$ be the function $T_{n,n}$ applied to disjunctions x_{ij} of n variables x_{ijk}. Krause and Pudlák [42] have proved a general result implying that $T_{n,n}^*$ has exponential voting polynomial length.

11.11 Open Problems

Finally, we give a list of the most challenging open problems where all considered functions have to be explicitly defined:

- Prove for some functions that they are not realizable by circuits of fan-in 2, linear size, and logarithmic depth.
- Prove for some functions that they are not realizable by formulas of size $O(n^2/\log n)$.
- Prove for some functions that they are not realizable by polynomial-size unbounded fan-in circuits of AND- and MOD-6-gates.
- Prove for some functions that they are not contained in LT_2^0.

References

[1] L. Adleman. Two theorems on random polynomial time. Proceedings of 19th Symposium on Foundations of Computer Science (FOCS), pp. 75–83, 1978.

[2] M. Ajtai. Σ_1^1-formulae on finite structures. Annals of Pure and Applied Logic 24, pp. 1–48, 1983.

[3] E. Allender. A note on the power of threshold circuits. Proceedings of the 30th Symposium on Foundations of Computer Science (FOCS), pp. 580–4, 1989.

[4] N. Alon and R. B. Boppana. The monotone circuit complexity of boolean functions. Combinatorica 7, pp. 1–22, 1987.

[5] N. Alon and J. Bruck. Explicit constructions of depth two majority circuits for comparison and addition. Technical Report RJ 8300 (75661). IBM San Jose, 1991.

[6] N. Alon, O. Goldreich, J. Håstad, and R. Peralta. Simple constructions of almost k-wise independent random variables. Random Structures and Algorithms 3, pp. 289–304, 1992.

[7] A. E. Andreev. On a method for obtaining lower bounds for the complexity of individual monotone functions. Soviet Mathematics Doklady 31, pp. 530–4, 1985.

[8] A. E. Andreev. On a method for obtaining more than quadratic effective lower bounds for the complexity of π-schemes. Moscow University Mathematical Bulletin 42, pp. 63–6, 1987.

[9] J. Aspnes, R. Beigel, M. Furst, and S. Rudich. The expressive power of voting polynomials. Proceedings of the 23rd Symposium of Theory of Computing (STOC), pp. 402–9, 1991.

[10] D. E. Atkins. Higher-radix division using estimates of the divisor and partial remainder. IEEE Transaction on Computers 17, pp. 925–34, 1968.

[11] D. A. Barrington. Bounded-width polynomial-size branching programs recognize exactly those languages in NC^1. Journal of Computer and System Sciences 38, pp. 150–64, 1989.

[12] D. A. Barrington and D. Thérien. Finite monoids and the fine structure of NC^1. Journal of the ACM 35, pp. 599–616, 1988.

[13] P. Beame, S. A. Cook, and J. Hoover. Log depth circuits for division and related problems. SIAM Journal on Computing 15, pp. 994–1003, 1986.

[14] N. Blum. A boolean function requiring $3n$ network size. Theoretical Computer Science 28, pp. 337–45, 1984.

[15] A. Borodin. On relating time and space to size and depth. SIAM Journal on Computing 6, pp. 733–44, 1977.

[16] J. Bruck. Harmonic analysis of polynomial threshold functions. SIAM Journal on Discrete Mathematics 3, pp. 168–77, 1990.

[17] J. Bruck and R. Smolensky. Polynomial threshold functions, AC^0-functions, and spectral norms. Proceedings of the 31st Symposium on Foundations of Computer Science (FOCS), pp. 632–41, 1990.

[18] P. Clote and E. Kranakis. *Boolean Functions and Computational Models*. Springer, New York, 2002.

[19] J. Forster. A linear lower bound on the unbounded error probabilistic communication complexity. Journal of Computer and System Sciences 65, pp. 612–25, 2002.

[20] M. L. Furst, J. B. Saxe, and M. Sipser. Parity circuits and the polynomial time hierarchy. Math. Systems Theory 17, pp. 13–27, 1984.

[21] M. Goldmann and J. Håstad. Monotone circuits for connectivity have depth $(\log n)^{2-o(1)}$. SIAM Journal on Computing 27, pp. 1283–94, 1998.

[22] M. Goldmann, J. Håstad, and A. A. Razborov. Majority gates versus general weighted threshold gates. Journal of Computational Complexity 2, pp. 277–300, 1992.

[23] M. Goldmann and M. Karpinski. Simulating threshold circuits by majority circuits. Proceedings of the 25th Symposium on Theory of Computing (STOC), pp. 551–60, 1993.

[24] A. Hajnal, W. Maass, P. Pudlák, M. Szegedy, and G. Turán. Threshold circuits of bounded depth. Proceedings of the 28th Symposium on Foundations of Computer Science (FOCS), pp. 99–110, 1987.

[25] J. Håstad. Almost optimal lower bounds for small depth circuits. In Randomness and Computation. Advances in Computing Research 5, S. Micali, ed., pp. 143–70. JAI Press, Stamford, CT, 1989.

[26] J. Håstad. On the size of weights for threshold gates. SIAM Journal on Discrete Mathematics 7, pp. 484–92, 1994.

[27] J. Håstad. The shrinkage exponent is 2. SIAM Journal on Computing 27, pp. 48–64, 1998.

[28] W. Hesse. Division is in uniform TC^0. Proceedings of the 28th International Colloquium on Automata, Languages and Programming (ICALP), Lecture Notes on Computer Science 2076, pp. 104–14, 2001.

[29] T. Hofmeister. Depth-efficient threshold circuits for arithmetic functions. In *Theoretical Advances in Neural Computation and Learning*, V. Roychowdhury, K. Y. Siu, and A. Orlitski, eds. Kluwer Acadamic, Dordrecht, 1994.

[30] T. Hofmeister. A note on the simulation of exponential threshold weights. Proceedings of the 2nd International Computing and Combinatorics Conference (COCOON), Lecture Notes on Computer Science 1090, pp. 136–41, 1996.

[31] H. J. Hoover, M. M. Klawe, and N. J. Pippenger. Bounding fan-out in logical networks. Journal of the ACM 31, pp. 13–18, 1984.

[32] R. Impagliazzo, R. Paturi, and M. E. Saks. Size-depth tradeoffs for threshold circuits. SIAM Journal on Computing 26, pp. 693–707, 1997.

[33] K. Iwama and H. Morizumi. An explicit lower bound of $5n - o(n)$ for boolean circuits. Proceedings of the 27th Symposium on Mathematical Foundations of Computer Science (MFCS), Lecture Notes on Computer Science 2420, pp. 353–64, 2002.

[34] A. Karatsuba and Y. Ofman. Multiplication of multidigit numbers on automata. Soviet Physics Doklady 7, pp. 595–6, 1969.

[35] M. Karchmer and A. Wigderson. Monotone circuits for connectivity require super-logarithmic depth. SIAM Journal on Discrete Mathematics 3, pp. 255–65, 1990.

[36] M. Kearns and L. Valiant. Cryptographic limitations on learning boolean formulae and finite automata. Journal of the ACM 41, pp. 67–95, 1994.

[37] V. M. Khrapchenko. A method of obtaining lower bounds for the complexity of π-schemes. Mathematical Notes of the Academy of Science of the USSR 11, pp. 474–9, 1972.

[38] M. Krause. On realizing iterated multiplication by small depth threshold circuits. Proceedings of 12th Symposium on Theoretical Aspects of Computer Science (STACS), Lecture Notes on Computer Science 900, pp. 83–94, 1995.

[39] M. Krause. Geometric arguments yield better bounds for threshold circuits and distributed computing. Theoretical Computer Science 156, pp. 99–117, 1996.

[40] M. Krause and S. Lucks. On the minimal hardware complexity of pseudorandom function generators. Proceedings of 18th Symposium on Theoretical Aspects of Computer Science (STACS), Lecture Notes on Computer Science 2010, pp. 419–30, 2001.

[41] M. Krause and P. Pudlák. On the computational power of depth-2 circuits with threshold and modulo gates. Theoretical Computer Science 174, pp. 137–56, 1997.

[42] M. Krause and P. Pudlák. Computing boolean functions by polynomials and threshold circuits. Computational Complexity 7, pp. 346–70, 1998.

[43] M. Krause and S. Waack. Variation ranks of communication matrices and lower bounds for depth two circuits having symmetric gates with unbounded fan-in. Mathematical System Theory 28, pp. 553–564, 1995.

[44] E. Kushilevitz and N. Nisan. *Communication Complexity*. Cambridge University Press, Cambridge; UK, 1997.

[45] R. E. Ladner and M. J. Fischer. Parallel prefix computation. Journal of the ACM 27, pp. 831–38, 1980.

[46] N. Linial, Y. Mansour, and N. Nisan. Constant depth circuits, Fourier transforms, and learnability. Journal of the ACM 40, pp. 607–20, 1993.

[47] O. B. Lupanov. A method of circuit synthesis. Izveska VUZ Radiofiz 1, pp. 120–40, 1958.

[48] O. B. Lupanov. Complexity of formula realization of functions of logical algebra. Problems of Cybernetics 3, pp. 782–811, 1962.

[49] S. Muroga. *Threshold Logic and Its Applications*. (1971.) Wiley-Interscience, New York, 1971.

[50] M. Minsky and S. Papert. *Perceptrons*. MIT Press, Cambridge, MA: 1988 (expanded edition; first edition 1968).

[51] M. Naor and O. Reingold. Number-theoretic constructions of efficient pseudo-random functions. Proceedings of the 38th Symposium on Foundations of Computer Science (FOCS), pp. 458–67, 1997.

[52] É. I. Nechiporuk. A boolean function. Soviet Mathematics Doklady 7, pp. 999–1000, 1966.

[53] N. Pippenger. On simultaneous resource bounds. Proceedings of the 20th Symposium on Foundations of Computer Science (FOCS), pp. 307–11, 1979.

[54] N. Pippenger and M. J. Fischer. Relations among complexity measures. Journal of the ACM 26, pp. 361–81, 1979.
[55] R. Raz and P. McKenzie. Separation of the monotone NC hierarchy. Combinatorica 19, pp. 403–35, 1999.
[56] R. Raz and A. Wigderson. Monotone circuits for matching require linear depth. Journal of the ACM 39, pp. 736–44, 1992.
[57] A. A. Razborov. Lower bounds on the size of bounded depth networks over a complete basis with logical addition. Mathematical of Notes of the Academy of Sciences of the USSR 41, pp. 333–38, 1987.
[58] A. A. Razborov. Lower bounds for monotone complexity of boolean functions. American Mathematical Society Translations 147, pp. 75–84, 1990.
[59] A. A. Razborov and S. Rudich. Natural proofs. Journal of Computer and System Sciences 55, pp. 24–35, 1997.
[60] N. P. Red'kin. Minimal realization of a binary adder. Problems of Cybernetics. 38, pp. 181–216, 1981.
[61] C. P. Schnorr. The network complexity and the Turing machine complexity of finite functions. Acta Informatica 7, pp. 95–107, 1976.
[62] A. Schönhage and V. Strassen. Schnelle Multiplikation grosser Zahlen. Computing 7, pp. 281–92, 1971.
[63] C. E. Shannon. A symbolic analysis of relay and switching circuits. Transactors of the AIEE 57, pp. 713–23, 1938.
[64] C. E. Shannon. The synthesis of two-terminal switching circuits. Bell Systems Technical Journal 28, pp. 59–98, 1949.
[65] K. Siu, J. Bruck, T. Kailath, and T. Hofmeister. Depth efficient neural networks for division and related problems. IEEE Transactions on Information Theory 39, pp. 946–56, 1993.
[66] J. Sklansky. Conditional-sum addition logic. IRE Trans. Elect. Comp. 9. pp. 226–31, 1960.
[67] R. Smolensky. Algebraic methods in the theory of lower bounds for boolean circuit complexity. Proceedings of the 19th Symposium on Theory of Computing (STOC), pp. 77–82, 1987.
[68] P. M. Spira. On time-hardware complexity tradeoffs for boolean functions. 4. Hawaii Symposium on System Sciences, pp. 525–27, 1971.
[69] B. A. Subbotovskaya. Realizations of linear functions by formulas using $+, *, -$. Soviet Mathematics Doklady 2, pp. 110–12, 1961.
[70] C. S. Wallace. A suggestion for a fast multiplier. IEEE Transactions on Computers 13, pp. 14–17, 1964.
[71] I. Wegener. Boolean functions whose monotone complexity is of size $n^2 / \log n$. Theoretical Computer Science 21, pp. 213–24, 1982.
[72] I. Wegener. Relating monotone formula size and monotone depth of boolean functions. Information Processing Letters 16, pp. 41–42, 1983.
[73] I. Wegener. *The Complexity of Boolean Functions*. Wiley, New York, 1987.
[74] I. Wegener. The range of new lower bound techniques for WRAMs and bounded depth circuits. Journal of Information Processing and Cybernetics (EIK) 23, pp. 537–43, 1987b.
[75] A. C. Yao. Separating the polynomial-time hierarchy with oracles. Proceedings of the 26th Symposium on Foundations of Computer Science (FOCS), pp. 1–10, 1985.
[76] A. C. Yao. On ACC and threshold circuits. Proceedings of 31st Symposium on Foundations of Computer Science (FOCS), pp. 619–628, 1990.

12

Fourier Transforms and Threshold Circuit Complexity

Jehoshua Bruck

12.1 Introduction

There exists a large gap between the empirical evidence of the computational capabilities of neural networks and our ability to systematically analyze and design those networks. Although it is well known that classical Fourier analysis is a very effective mathematical tool for the design and analysis of linear systems, such a tool was not available for artificial neural networks, which are inherently nonlinear. In the late 1980s, the spectral analysis tool was introduced in the domain of discrete neural networks. The application of the spectral technique led to a number of new insights and results, including lower and upper bounds on the complexity of computing with neural networks as well as methods for constructing optimal (in terms of performance) feedforward networks for computing various arithmetic functions.

The focus of the presentation in this chapter is on an elementary description of the basic techniques of Fourier analysis and its applications in threshold circuit complexity. Our hope is that this chapter will serve as background material for those who are interested in learning more about the progress and results in this area. We also provide extensive bibliographic notes that can serve as pointers to a number of research results related to spectral techniques and threshold circuit complexity.

The chapter is organized as follows. In the next section we prove the representation theorem and describe an important set of basic properties of the spectral representation. In Section 12.3 we present a number of examples of computing the spectrum of Boolean functions. In Section 12.4 we describe a key connection between the complexity of computing with threshold circuits and the spectrum. In particular, we state the spectral characterization theorem for polynomial threshold functions and describe some of the key ideas in the proof. In Section 12.5 we show, as an example, results related to the trade-off between the size of the weights and the depth of a neural network. Finally, in Section 12.6 we provide an overview of the relevant publications.

12.2 The Spectral Representation of Boolean Functions

There are a number of methods for representing Boolean functions. Examples are truth tables, DNFs (disjunctive normal forms), and CNFs (conjunctive normal forms). In this section we describe the spectral representation of Boolean functions. (This representation is also mentioned in several other chapters, e.g., in Chapters 1, 8, 11, and it is closely related to the pseudo-Boolean representation discussed in [14].)

12.2.1 The Spectrum

The main idea in the spectral representation is to specify functions as polynomials that compute the functions over the field of rational numbers. The coefficients of the polynomials are called the spectrum of the corresponding functions.

Throughout this chapter a Boolean function f of n variables is defined as a mapping

$$f : \{1, -1\}^n \to \{1, -1\}.$$

Note that we use the multiplicative representation of $\{0, 1\}$ via the transformation $a \mapsto (-1)^a$.

For example, let $f = x_1 \oplus x_2$; that is, f is the exclusive-OR (XOR) of two variables. It is easy to check that the polynomial representation of f is $f(x_1, x_2) = x_1 x_2$.

Notice that for every Boolean function f, the polynomial representation is linear in each of its variables because $x^2 = 1$ for $x \in \{-1, 1\}$.

Let a_α, $\alpha \in \{0, 1\}^n$, be the coefficient of X^α in the polynomial representation where $X^\alpha = x_1^{\alpha_1} x_2^{\alpha_2} \ldots x_n^{\alpha_n}$. For example, for $\alpha = 11$ we have $X^\alpha = x_1 x_2$, and for $\alpha = 00$ we have $X^\alpha = 1$.

We denote the set of 2^n coefficients of the polynomial representation of a Boolean function f with n variables by

$$A_n(f) = \{a_\alpha \mid \alpha \in \{0, 1\}^n\}.$$

This set of coefficients is called the spectrum of the function.

Hence, every Boolean function can be written as

$$f(X) = \sum_{\alpha \in \{0,1\}^n} a_\alpha X^\alpha.$$

For example, for $f(x_1, x_2) = x_1 x_2$,

$$A_2(f) = \{a_{00} = 0, a_{10} = 0, a_{01} = 0, a_{11} = 1\}.$$

In the following discussion $A_n(f)$ denotes both the set of spectral coefficients and a vector consisting of the coefficients in a lexicographic order according to α. This abuse of notation will be made clear given the context.

12.2.2 The Representation Theorem

Next we prove that for any Boolean function f the spectrum $A_n(f)$ is indeed unique, and we present a method for computing it. The main concept used in the proof of the theorem is that of a Sylvester-type Hadamard matrix that is defined by the following recursion:

$$H_1 = [1]$$
$$H_2 = \begin{bmatrix} 1 & 1 \\ 1 & -1 \end{bmatrix}$$
$$H_{2^{n+1}} = \begin{bmatrix} H_{2^n} & H_{2^n} \\ H_{2^n} & -H_{2^n} \end{bmatrix}. \tag{12.1}$$

For example,

$$H_4 = \begin{bmatrix} 1 & 1 & 1 & 1 \\ 1 & -1 & 1 & -1 \\ 1 & 1 & -1 & -1 \\ 1 & -1 & -1 & 1 \end{bmatrix}.$$

Theorem 12.1. (Representation.) *Let $f(X)$ be a Boolean function of n variables, $X = (x_1, x_2, \cdots, x_n)$. Let $\underline{f_n}$ be a vector that consists of the values of f written in lexicographic order (in the variables X). Then:*

(i) *f can be written uniquely as*

$$f(X) = \sum_{\alpha \in \{0,1\}^n} a_\alpha X^\alpha.$$

(ii) *The set of coefficients of f (the spectrum), namely $A_n(f)$, can be computed as follows:*

$$A_n(f) = \frac{1}{2^n} H_{2^n} \underline{f_n}.$$

Proof. The proof is constructive and uses induction. The idea is to compute $A_n(f)$ by solving a system of linear equations. Let us start by computing the spectrum of f with $n = 1$, namely, f is a function of a single variable:

$$f(x_1) = a_0 + a_1 x_1$$

and

$$f(1) = a_0 + a_1$$
$$f(-1) = a_0 - a_1.$$

Clearly,

$$\underline{f_1} = A_2(f) H_2$$

and by the orthogonality of the Hadamard matrix,

$$A_2 = \frac{1}{2} H_2 \underline{f_2}.$$

The foregoing can be generalized to n variables by induction on n. Assume it is true for n, namely,

$$\underline{f_n} = H_{2^n} A_n(f).$$

For $(n+1)$, consider the different values of x_{n+1} and get that

$$\begin{aligned} \underline{f_{n+1}} &= \begin{bmatrix} H_{2^n} & H_{2^n} \\ H_{2^n} & -H_{2^n} \end{bmatrix} A_{n+1}(f) \\ &= H_{2^{n+1}} A_{n+1}(f). \end{aligned}$$

Hadamard matrices are nonsingular; thus, for any given f a unique polynomial representation defined by the set of coefficients (the spectrum) always exists. ■

Example 12.1.

- *Consider the AND function of two variables. Namely,* $f(x_1, x_2) = x_1 \wedge x_2$. *Then*
  $f(1,1) = 1$, $f(1,-1) = 1$, $f(-1,1) = 1$ *and* $f(-1,-1) = -1$.
  By Theorem 12.1,

$$\frac{1}{4}\begin{bmatrix} 1 & 1 & 1 & 1 \\ 1 & -1 & 1 & -1 \\ 1 & 1 & -1 & -1 \\ 1 & -1 & -1 & 1 \end{bmatrix}\begin{bmatrix} 1 \\ 1 \\ 1 \\ -1 \end{bmatrix} = \begin{bmatrix} 0.5 \\ 0.5 \\ 0.5 \\ -0.5 \end{bmatrix}. \tag{12.2}$$

  Hence,

$$f(x_1, x_2) = \frac{1}{2}(1 + x_1 + x_2 - x_1 x_2).$$

- *Consider the OR function of two variables. Namely,* $f(x_1, x_2) = x_1 \vee x_2$. *Then*
  $f(1,1) = 1$, $f(1,-1) = -1$, $f(-1,1) = -1$ *and* $f(-1,-1) = -1$.
  By Theorem 12.1,

$$\frac{1}{4}\begin{bmatrix} 1 & 1 & 1 & 1 \\ 1 & -1 & 1 & -1 \\ 1 & 1 & -1 & -1 \\ 1 & -1 & -1 & 1 \end{bmatrix}\begin{bmatrix} 1 \\ -1 \\ -1 \\ -1 \end{bmatrix} = \begin{bmatrix} -0.5 \\ 0.5 \\ 0.5 \\ 0.5 \end{bmatrix}. \tag{12.3}$$

  Hence,

$$f(x_1, x_2) = \frac{1}{2}(-1 + x_1 + x_2 + x_1 x_2).$$

We note here that the foregoing method is applicable not only to Boolean functions but also to any pseudo-Boolean function $f : \{1, -1\}^n \rightarrow \mathbb{R}$.

12.2.3 Basic Properties of the Spectral Representation

Here we present a number of basic properties of the spectral representation. In all the lemmas that follow, f denotes a Boolean function and $A_n(f) = \{a_\alpha \mid \alpha \in \{0, 1\}^n\}$ is its spectral representation.

Lemma 12.2. (Orthogonality.)

$$\sum_{X \in \{1,-1\}^n} X^\alpha = \begin{cases} 2^n & \textit{if } \alpha = \textit{all-0 vector} \\ 0 & \textit{else.} \end{cases}$$

Proof. The result follows from the fact that every X^α corresponds to a row in a Sylvester-type Hadamard matrix. Every row but the first has the same number of 1s and -1s (proved by induction). ■

Lemma 12.3. (Zero-Value.)

$$\sum_{\alpha \in \{0,1\}^n} a_\alpha = f(1, 1, \ldots, 1) \in \{1, -1\}.$$

Proof. Let $f(X) = \sum_{\alpha \in \{0,1\}^n} a_\alpha X^\alpha$. The result is obtained by plugging in $X = (1, 1, \ldots, 1)$. Notice that as a result,

$$|\sum_{\alpha \in \{0,1\}^n} a_\alpha| = 1.$$

■

Lemma 12.4. (Moments.)

$$\sum_{X \in \{1,-1\}^n} f(X) X^\alpha = 2^n a_\alpha$$

for all $\alpha \in \{0, 1\}^n$.

Proof. The result follows from the orthogonality lemma, by considering the terms of $f(X)X^\alpha$. ■

Lemma 12.5. (Correlation.) *If* $f(X) = \sum_{\alpha \in \{0,1\}^n} a_\alpha X^\alpha$ *and* $\hat{f}(X) = \sum_{\alpha \in \{0,1\}^n} \hat{a}_\alpha X^\alpha$ *are two Boolean functions of n variables, then*

$$\sum_{X \in \{1,-1\}^n} f(X)\hat{f}(X) = 2^n \sum_{\alpha \in \{0,1\}^n} a_\alpha \hat{a}_\alpha.$$

Proof. The result follows by application of the moments lemma. ■

Lemma 12.6. (Parseval's Property.)

$$\sum_{X \in \{1,-1\}^n} f^2(X) = 2^n \sum_{\alpha \in \{0,1\}^n} a_\alpha^2 = 2^n.$$

Proof. Because $f(X) \in \{1, -1\}$, one identity is trivial. The second is a special case of the correlation lemma.
Hence, $\sum_{\alpha \in \{0,1\}^n} a_\alpha^2 = 1$. The power spectrum of a Boolean function sums to 1. ■

Example 12.2. *Consider the AND and OR functions of two variables:*

$$AND(x_1, x_2) = \frac{1}{2}(1 + x_1 + x_2 - x_1x_2),$$

$$OR(x_1, x_2) = \frac{1}{2}(-1 + x_1 + x_2 + x_1x_2).$$

The zero value property is

$$AND(1, 1) = \frac{1}{2}(1 + 1 + 1 - 1) = 1,$$

$$OR(1, 1) = \frac{1}{2}(-1 + 1 + 1 + 1) = 1.$$

The Parseval property (the power spectrum sums to 1) is

$$1/4 + 1/4 + 1/4 + 1/4 = 1.$$

12.3 Computing the Spectrum

The spectrum of a Boolean function can be computed using the Hadamard matrix (see Theorem 12.1). However, this technique requires computation using the explicit representation (truth table) of a Boolean function, which is exponential in the number of variables. In this section we present more practical techniques for computing the spectrum of Boolean functions. These techniques will be useful in the following sections where we show how the spectrum of Boolean functions is related to their circuit complexity.

12.3.1 Simple Functions

We first show what is the spectrum of three basic functions, namely, XOR, AND, and OR. We then show how we can use this knowledge to compute the spectrum of Boolean functions that are represented in CNF or DNF.

Proposition 12.7. *Let $XOR(X) = XOR(x_1, x_2, \ldots, x_n)$ be the XOR function of n variables. Then $XOR(X) = x_1x_2 \cdots x_n$.*

Proof. $x_1x_2 \cdots x_n$ is 1 if and only if the number of -1s in X is even. ■

Proposition 12.8. *Let $AND(X) = AND(x_1, x_2, \ldots, x_n)$ be the AND function of n variables and let its spectrum be $A_n(AND) = \{a_\alpha \mid \alpha \in \{0,1\}^n\}$. Then*

$$a_\alpha = \begin{cases} 1 - 2^{1-n} & \text{if } |\alpha| = 0 \\ 2^{1-n} & \text{if } |\alpha| \text{ is odd} \\ -2^{1-n} & \text{if } |\alpha| \neq 0 \text{ is even.} \end{cases}$$

Proof. Consider the following polynomial:

$$g(X) = \frac{1}{2^n} \prod_{i=1}^{n} (1 - x_i).$$

Clearly $g(X) = 1$ if and only if X is the all-(-1) vector. In all other cases, $g(X) = 0$. Hence,

$$AND(X) = 1 - 2g(X).$$

The theorem is proved by considering the coefficients of the polynomial $1 - 2g(X)$. ■

Proposition 12.9. *Let $OR(X) = OR(x_1, x_2, \ldots, x_n)$ be the OR function of n variables and let its spectrum be $A_n(OR) = \{a_\alpha \mid \alpha \in \{0,1\}^n\}$. Then*

$$a_\alpha = \begin{cases} -1 + 2^{1-n} & \text{if } |\alpha| = 0 \\ 2^{1-n} & \text{if } |\alpha| \text{ is odd} \\ 2^{1-n} & \text{if } |\alpha| \text{ is even.} \end{cases}$$

Proof. The proof follows from Proposition 12.8 and the fact that

$$OR(X) = -AND(-X).$$

■

Example 12.3.

$$AND(x_1, x_2, x_3) = \frac{1}{4}(3 + x_1 + x_2 + x_3 - x_1x_2 - x_1x_3 - x_2x_3 + x_1x_2x_3)$$

$$OR(x_1, x_2, x_3) = \frac{1}{4}(-3 + x_1 + x_2 + x_3 + x_1x_2 + x_1x_3 + x_2x_3 + x_1x_2x_3)$$

Given the fact that we know the representation of AND and OR, we can compute the representation of other functions by using substitutions. For example, let

$$f(x_1, x_2, x_3) = x_1 \wedge (x_2 \vee x_3).$$

Since

$$AND(x_1, x_2) = \frac{1}{2}(1 + x_1 + x_2 - x_1x_2)$$

and

$$OR(x_1, x_2) = \frac{1}{2}(-1 + x_1 + x_2 + x_1x_2),$$

we get that

$$\begin{aligned} f(x_1, x_2, x_3) &= x_1 \wedge (x_2 \vee x_3) \\ &= \frac{1}{2}(1 + x_1 + \frac{1}{2}(-1 + x_2 + x_3 + x_2x_3) - \frac{1}{2}x_1(-1 + x_2 + x_3 + x_2x_3)) \\ &= \frac{1}{4}(1 + 3x_1 + x_2 + x_3 + x_2x_3 - x_1x_2 - x_1x_3 - x_1x_2x_3). \end{aligned}$$

12.3.2 Symmetric Functions

A Boolean function $f(X)$ is symmetric if and only if any permutation π of the input variables does not change the function. Namely, for an arbitrary permutation π,

$$f(x_1, x_2, \ldots, x_n) = f(x_{\pi(1)}, x_{\pi(2)}, \ldots, x_{\pi(n)}).$$

For example, the functions AND, OR, and XOR are symmetric functions. Notice that a symmetric function is a function of the number of logical 1s in the input.

In the previous subsection we computed the spectrum of the functions AND, OR, and XOR. Notice that for those functions the spectral coefficient that corresponds to the term $x_1x_2 \cdots x_n$ is nonzero. Namely, the degree of the polynomial representation of these three functions is n.

For any function f, we call degree of f, and we denote by $d(f)$, the degree of the polynomial representation of f. In this subsection we explore the general question, namely, what is the minimal degree of a symmetric function with n variables?

We first prove a lower bound.

Proposition 12.10. *Let $f(X)$ be a nonconstant symmetric function of n variables. Then $d(f) > \lfloor n/2 \rfloor$.*

Proof. Notice that $f^2(X) = 1$ (this is true for any f). Namely, the degree of f^2 is 0. Because $f(X)$ is symmetric, the polynomial $f(X)$ is invariant under permutations on the inputs. Namely, all the terms of degree $d(f)$ (in general of same degree) have identical coefficients. Thus, if $d(f) \leq \lfloor n/2 \rfloor$ then the degree of f^2 is $2d(f)$, which implies that $d(f) = 0$ and that f is constant. ■

Next we prove an upper bound.

Proposition 12.11. *There exists symmetric functions f with n variables for which $d(f) = n - 1$.*

Proof. A useful trick is to assume that f gets the values 0 and 1 instead of 1 and -1, respectively. The reason for that is that the transformation $(1 - f)/2$ preserves the degree. From now on we assume that $f(X) \in \{0, 1\}$ is a symmetric function with $X \in \{1, -1\}^n$.

The coefficient in $f(X)$ that corresponds to the term $x_1 x_2 x_3 \cdots x_n$ is

$$\sum_{X \in \{1,-1\}^n} x_1 x_2 x_3 \cdots x_n f(X).$$

Assume f is 1 for Xs that have exactly i (-1)s with $i \in I$, where $I \subset \{0, 1, \ldots n\}$. Then

$$\sum_{X \in \{1,-1\}^n} x_1 x_2 x_3 \cdots x_n f(X) = \sum_{i \in I} (-1)^i \binom{n}{i}.$$

Hence, a symmetric Boolean function defined by the set I has degree at most $n - 1$ if and only if

$$\sum_{i \in I} (-1)^i \binom{n}{i} = 0. \qquad (12.4)$$

For n odd, all functions that are defined by I, where I contains both i and $(n - i)$, have degree at most $n - 1$. That follows from the fact that for n odd

$$(-1)^i \binom{n}{i} + (-1)^{n-i} \binom{n}{n-i} = 0.$$

An example is the function that is 1 for X the all-1 vector and X the all-(-1) vector and 0 otherwise.

The case where n is even is a bit more difficult. Here we need to involve three terms to get a solution to Equation 12.4. It is easy to prove that $I = \{i, i+1, n-i\}$ is a solution to Equation 12.4 for $n = 3i + 2$. An example is the function with $n = 8$ variables, which is 1 if exactly 2, 3, or 6 of its variables are (-1) and 0 otherwise. In this case Equation 12.4 is

$$\binom{8}{2} - \binom{8}{3} + \binom{8}{6} = 0.$$

■

It is still an open problem to compute the exact lower bound on $d(f)$.

12.3.3 The Complete Quadratic Function

The complete quadratic (CQ) function is the function that consists of the XOR of all the $\binom{n}{2}$ possible ANDs between pairs of variables. Namely,

$$CQ(X) = (x_1 \wedge x_2) \oplus (x_1 \wedge x_3) \oplus \cdots \oplus (x_{n-1} \wedge x_n). \qquad (12.5)$$

For example,

$$CQ(x_1, x_2, x_3) = (x_1 \wedge x_2) \oplus (x_1 \wedge x_3) \oplus (x_2 \wedge x_3).$$

The interesting property of the CQ function is that it is symmetric and has a "flat" spectrum. This property is useful in the later sections for proving lower bounds.

Next we prove that the CQ function is symmetric by considering an equivalent definition.

Proposition 12.12.

$$CQ(X) = \begin{cases} 1 & \textit{if the number of } -1s \textit{ in } X \textit{ equals } 0 \textit{ or } 1 \textit{ mod } 4 \\ -1 & \textit{otherwise.} \end{cases}$$

Proof. Suppose there are m -1's in X. Because a pair in Equation (12.5) is -1 iff both variables are -1, we have exactly $\binom{m}{2}$ pairs which are -1. Hence, the value of CQ(X) is determined by the evenness of $\binom{m}{2}$, and the result follows. ∎

First we calculate the spectrum for the case when n is even.

Proposition 12.13. *Let* $\{a_\alpha \mid \alpha \in \{0, 1\}^n\}$ *be the spectral representation of* $CQ(X)$. *Assume that* n *is even. Then*

$$| a_\alpha | = 2^{-\frac{n}{2}}, \qquad \forall \alpha \in \{0, 1\}^n.$$

Proof. The proof is by induction on n. For $n = 2$, we have

$$CQ(x_2, x_2) = \frac{1}{2}(1 + x_1 + x_2 - x_1 x_2).$$

Assume true for n and show that the statement is true for $(n + 2)$. We use the same notation as in Section 12.2, namely, $\underline{f_n}$ represents the vector with the values of CQ and A_n represents the vector of the spectral coefficients of CQ. Using Proposition 12.12, it can be shown that $\underline{f_{n+2}}$ can be expressed as a function of $\underline{f_n}$:

$$\underline{f_{n+2}} = \begin{bmatrix} \underline{f_n} \\ \underline{\hat{f}_n} \\ \underline{\hat{f}_n} \\ -\underline{f_n} \end{bmatrix}$$

where

$$\underline{\hat{f}_n} = \hat{X}_n \circ \underline{f_n},$$

$\hat{X}_n$ is the vector representation of XOR(X) $= x_1 x_2 \cdots x_n$, and "$\circ$" is bitwise multiplication.

Hence, by Theorem 12.1,

$$A_{n+2} = \frac{1}{2^{n+2}} H_{2^{n+2}} \underline{f_{n+2}}$$

$$= \frac{1}{2^{n+2}} \begin{bmatrix} H_{2^n} & H_{2^n} & H_{2^n} & H_{2^n} \\ H_{2^n} & -H_{2^n} & H_{2^n} & -H_{2^n} \\ H_{2^n} & H_{2^n} & -H_{2^n} & -H_{2^n} \\ H_{2^n} & -H_{2^n} & -H_{2^n} & H_{2^n} \end{bmatrix} \begin{bmatrix} \underline{f_n} \\ \underline{\hat{f}_n} \\ \underline{\hat{f}_n} \\ -\underline{f_n} \end{bmatrix}$$

$$= \frac{1}{2} \begin{bmatrix} \hat{A}_n \\ A_n \\ A_n \\ -\hat{A}_n \end{bmatrix},$$

where $\hat{A}_n$ is the reflection of A_n. Hence, if the result is true for n, it is also true for $n+2$. ■

Example 12.4. *Using the foregoing recursive description of the spectrum of $CQ(X)$, we can calculate A_4 from A_2:*

$$A_2^T = \frac{1}{2}(1, 1, 1, -1)$$

and

$$A_4^T = \frac{1}{4}(-1, 1, 1, 1, \quad 1, 1, 1, -1, \quad 1, 1, 1, -1, \quad 1, -1, -1, -1).$$

The foregoing is true for n even. For n odd we have the following proposition.

Proposition 12.14. *Let $\{a_\alpha \mid \alpha \in \{0,1\}^n\}$ be the spectral representation of $CQ(X)$. Assume that n is odd. Then*

$$\mid a_\alpha \mid = 0 \text{ or } 2^{-\frac{n-1}{2}} \qquad \forall \alpha \in \{0,1\}^n.$$

The proof is similar to the even case. We use induction on n and can write the recursive description of the spectrum.

Example 12.5. *Let $n = 3$. Then*

$$CQ(x_1, x_2, x_3) = \frac{1}{2}(x_1 + x_2 + x_3 - x_1 x_2 x_3).$$

12.3.4 The Comparison Function

In this section, we focus on the *COMPARISON* function of two n-bit numbers.

Definition. *Let $X = (x_1, \ldots, x_n)$, $Y = (y_1, \ldots, y_n) \in \{1, -1\}^n$. We consider X and Y as two n-bit numbers representing $\sum_{i=1}^n x_i \cdot 2^i$ and $\sum_{i=1}^n y_i \cdot 2^i$, respectively.*

The *COMPARISON* function is defined as

$$C(X, Y) = 1 \text{ iff } X \geq Y.$$

In other words,

$$C(X, Y) = sgn\left\{\sum_{i=1}^{n} 2^i (x_i - y_i) + 1\right\}.$$

We write a recursion for the spectral representation of the *COMPARISON* function $C(X, Y)$. If C_n is the polynomial corresponding to the function of $x_n, ..., x_1$ and $y_n, ..., y_1$, it is easy to see that

$$C_n = \frac{x_n - y_n}{2} + \frac{1 + x_n y_n}{2} C_{n-1}.$$

For example,

$$C_1 = \frac{x_1 - y_1}{2} + \frac{1 + x_1 y_1}{2},$$

and

$$C_2 = \frac{x_2 - y_2}{2} + \frac{1 + x_2 y_2}{2} C_1.$$

12.4 Threshold Functions and Spectral Norms

A Boolean function $f(X)$ is a *threshold function* if

$$f(X) = sgn\,(F(X)) = \begin{cases} 1 & \text{if } F(X) > 0 \\ -1 & \text{if } F(X) < 0, \end{cases}$$

where

$$F(X) = \sum_{\alpha \in \{0,1\}^n} w_\alpha X^\alpha,$$

$$X^\alpha \stackrel{\text{def}}{=} \prod_{i=1}^{n} x_i^{\alpha_i},$$

and n is the number of variables. Recall that a *Boolean function* is defined as $f : \{1, -1\}^n \longrightarrow \{1, -1\}$; namely, 0 and 1 are represented by 1 and -1, respectively. It is also assumed, without loss of generality, that $F(X) \neq 0$ for all X.

A *threshold gate* is a gate that computes a threshold function. We define LT_1 to be the class of functions that are computable by a single linear threshold gate, where a linear threshold gate is a threshold gate that includes only the constant and linear terms. We can also define the class of functions that can be computed by circuits that consist of gates. The size of a circuit is the number of gates in the circuit. The depth of a circuit is the longest path (in terms of the number of gates) from an input to an output of a circuit. Let LT_k be the set of functions that can be computed by a depth-k linear threshold circuit with size bounded by a polynomial in the number of inputs.

We can also define elements and circuits based on threshold functions with other restriction (besides being linear).

It follows from Theorem 12.1 that any Boolean function can be computed by a single threshold gate if we allow the number of monomials in $F(X)$ to be as large as 2^n. This stimulates the following natural question: what happens when the number of monomials (terms) in $F(X)$ is bounded by a polynomial in n?

The question can be formulated by defining a new complexity class of Boolean functions. Let PT_1, for *polynomial threshold* functions, be the set of all Boolean functions that can be computed by a single threshold gate where the number of monomials is bounded by a polynomial in n.

Another interesting practical restriction is the size of the weights (coefficients) of the threshold function. In particular, we define $\widehat{LT}_1$ (similarly $\widehat{PT}_1$) to be the set of functions that can be computed by a linear threshold gate in which the weights are bounded by a polynomial in the size of the inputs. The corresponding circuit classes are $\widehat{LT}_k$ and $\widehat{PT}_k$.

12.4.1 Characterization of Polynomial Threshold Functions

The main goal in this subsection is to characterize PT_1 using the spectral representation of Boolean functions. This characterization provides a link between threshold circuit complexity and spectral techniques.

We are interested in the L_1 and L_∞ norms associated with the spectrum of f. Namely,

$$L_1(f) = \sum_{\alpha \in \{0,1\}^n} |a_\alpha|$$

and

$$L_\infty(f) = \max_{\alpha \in \{0,1\}^n} |a_\alpha|.$$

Note that for any Boolean function f, $L_1(f) \geq 1$, $L_2(f) = 1$, and $L_\infty(f) \leq 1$.

Example 12.6. *Consider the function $f(x_1, x_2) = x_1 \wedge x_2$. Then*

$$f(x_1, x_2) = \frac{1}{2}(1 + x_1 + x_2 - x_1 x_2).$$

Notice that $L_1(f) = 2$, $L_2(f) = 1$, and $L_\infty(f) = 1/2$.

The main result here is a characterization of the functions in PT_1 using spectral norms. Formally, let PL_1 be the class of Boolean functions for which the spectral norm L_1 is bounded by a polynomial in n and PL_∞ the class of Boolean functions for which L_∞^{-1} is bounded by a polynomial in n.

Theorem 12.15. (Characterization.)

$$PL_1 \subset PT_1 \subset PL_\infty.$$

The characterization theorem is also interesting because of the following connection with circuits.

Proposition 12.16.

$$PT_1 \subset LT_2.$$

Hence, $PL_1 \subset LT_2$.

We can use this theorem and the results on the spectrum of Boolean functions to prove the following proposition.

Proposition 12.17. *The complete quadratic function is not in PT_1.*

Proof. It follows from Proposition 12.13 that $L_\infty(CQ(x_1, x_2, \ldots, x_n)) = 2^{-\frac{n}{2}}$. Namely, $CQ \notin PL_\infty$ and by Theorem 12.15, $CQ \notin PT_1$. ■

Proposition 12.18. *The $COMPARISON$ function is in PT_1.*

Proof. It follows from the results in Subsection 12.3.4 that the polynomial representation of $COMPARISON$ can be written in a recursive form:

$$C_n = \frac{x_n - y_n}{2} + \frac{1 + x_n y_n}{2} C_{n-1}.$$

Notice that $L_1(C_1) = 2$. Also, notice that L_1 is increasing by 1 as n grows by one. Hence, $L_1(C_n) = n + 1$. Namely, $C_n \in PL_1$ and by Theorem 12.15, $C_n \in PT_1$. ■

The proof of Theorem 12.15 has two parts. The proof that

$$PL_1 \subset PT_1$$

is based on probabilistic arguments and is not constructive. The proof that

$$PT_1 \subset PL_\infty$$

is based on algebraic techniques. In the next subsection we present some of the ideas related to the algebraic techniques.

12.4.2 Necessary and Sufficient Conditions for Threshold Functions

The following concept is very useful in the proof using the algebraic technique.

Definition. *A Boolean function $f(X)$ is S-threshold, for a given set $S \subset \{0, 1\}^n$, iff there exist weights such that $f(X) = sgn(\sum_{\alpha \in S} w_\alpha X^\alpha)$.*

One of the main ideas in the proof of Theorem 12.15 is a necessary and sufficient condition for a function to be an S-threshold function, for arbitrary S.

The following simple lemma enables us to express the relation between $f(X)$ and $F(X)$ in a global way. That is, instead of having 2^n conditions we have only one.

Lemma 12.19. *Let $f(X)$ be a Boolean function and let $F(X) \neq 0$ for all $X \in \{1, -1\}^n$; then*

$$f(X) = sgn(F(X)) \qquad \forall X \in \{1, -1\}^n \tag{12.6}$$

iff

$$\sum_{X \in \{1,-1\}^n} | F(X) | = \sum_{X \in \{1,-1\}^n} f(X)F(X). \tag{12.7}$$

Proof. Suppose there exists an X_1 that violates (12.6); that implies ($F(X) \neq 0$ for all X) that

$$| F(X_1) | > f(X_1)F(X_1).$$

Hence, (12.7) is also not true because a violation in Equation (12.6) can only decrease the value on the right-hand side of (12.7). Clearly, if (12.6) is true, so is (12.7). ∎

The necessary and sufficient condition follows from (12.7) by using the polynomial representation of a Boolean function.

Theorem 12.20. *Fix $S \subseteq \{0, 1\}^n$. Let $F(X) = \sum_{\alpha \in S} w_\alpha X^\alpha$. Let $f(X)$, $X \in \{-1, 1\}^n$, be a Boolean function with spectral representation $\{a_\alpha \mid \alpha \in \{0, 1\}^n\}$. Then*

$$f(X) = sgn(F(X)) \qquad \forall X \in \{1, -1\}^n$$

iff

$$\sum_{X \in \{1,-1\}^n} | F(X) | = 2^n \sum_{\alpha \in S} w_\alpha a_\alpha. \tag{12.8}$$

Proof. By Lemma 12.19 it is enough to show that

$$\sum_{X \in \{1,-1\}^n} f(X)F(X) = 2^n \sum_{\alpha \in S} w_\alpha a_\alpha.$$

We write $f(X)$ as a polynomial,

$$f(X) = \sum_{\alpha \in \{0,1\}^n} a_\alpha X^\alpha,$$

and find that the constant term of $f(X)F(X)$ is $\sum_{\alpha \in S} w_\alpha a_\alpha$. Hence, the result follows from the orthogonality lemma. ∎

12.4.3 Applications of the Conditions

Theorem 12.20 is interesting because it suggests that an S-threshold function is fully characterized by the set of spectral coefficients that correspond to S. This theorem is the key for proving that

$$PT_1 \subset PL_\infty.$$

We omit the details of this proof and instead focus on presenting three other applications. The first application (Theorem 12.21) is a generalization of the known result that XOR is not in LT_1; the second application provides a "sampling theorem" (Theorem 12.22); and the third application (Theorem 12.24) is an upper bound on the number of S-threshold functions.

Theorem 12.21. *Let*

$$XOR(X) = \begin{cases} -1 & \textit{if the number of } -1\textit{s in } X \textit{ is odd} \\ 1 & \textit{otherwise.} \end{cases}$$

Then XOR is not an S-threshold function for any S such that the all-1 vector is not in S, that is, $(1, 1, \ldots, 1) \notin S$.

Proof. Note that $\text{XOR}(X) = x_1 x_2 \cdots x_n$; hence, for a set S that does not include the all-1 vector, we have $\sum_{\alpha \in S} w_\alpha a_\alpha = 0$, and the condition

$$\sum_{X \in \{1,-1\}^n} |F(X)| = 2^n \sum_{\alpha \in S} w_\alpha a_\alpha = 0$$

cannot be satisfied because $F(X) \neq 0$. ∎

A nice interpretation of this result is as follows: the X^α form a basis for the space of Boolean functions. Hence, by the definition of a basis, an element of the basis cannot be represented as a linear combination of the other elements in the basis. What Theorem 12.21 is saying is that it is also impossible to represent an element of the basis as the *sign* of a linear combination of the other elements in the basis.

The next theorem has the qualities of a "sampling theorem." It shows that certain restricted sets of threshold functions can be fully characterized using only part of the spectrum.

Theorem 12.22. *Let f_1 and f_2 be Boolean functions with $\{a_\alpha^i \mid \alpha \in \{0, 1\}^n\}$, $i = 1, 2$, as their spectral representation, respectively. Assume that at least one of the functions is an S-threshold function. Then $f_1(X) = f_2(X)$ for all $X \in \{1, -1\}^n$ iff $a_\alpha^1 = a_\alpha^2$ for all $\alpha \in S$.*

Proof. Suppose $f_1 = f_2$. By the uniqueness of the spectral representation (Theorem 12.1), we get the "only if." Assume with loss of generality that f_1 is an S-threshold function. Now suppose $a_\alpha^1 = a_\alpha^2$ for all $\alpha \in S$. By Theorem 12.20 and the assumption that f_1 is S-threshold, we get that there exists a set of weights that satisfies (12.8) for both f_1 and f_2. Hence, f_2 is also S-threshold and $F_1(X) = F_2(X)$ for all $X \in \{1, -1\}^n$. Thus, $f_1 = f_2$. ∎

Corollary 12.23. *Consider the set $S \subseteq \{0, 1\}^n$. Let f_1 and f_2 be Boolean functions of n variables. If $a_\alpha^1 = a_\alpha^2$ for all $\alpha \in S$, then either both f_1 and f_2 are S-threshold or both are not S-threshold.*

One application of the foregoing is to counting the number of different S-threshold functions.

Theorem 12.24. *Fix $S \subseteq \{0, 1\}^n$. There are at most $2^{(n+1)|S|}$ different S-threshold functions.*

Proof. It can be shown that for all $\alpha \in \{0, 1\}$, a_α can assume at most $(2^n + 1)$ different values. Hence, there are at most $2^{(n+1)|S|}$ different sets of $|S|$ spectral coefficients. Thus, the result follows from Corollary 12.23 . ∎

The preceding is a generalization of a known upper bound on the number of linear threshold functions for which $|S| = n + 1$.

12.5 Small Weights and Circuit Complexity

In this section we show how the result regarding the *COMPARISON* function in the previous section can be used to prove an upper bound on computing with threshold circuits with small weights.

12.5.1 Large versus Small Weights

Although most linear threshold functions require exponentially large weights, we can always simulate them by three layers of $\widehat{LT}_1$ elements.

Theorem 12.25.

$$LT_1 \subsetneq \widehat{LT}_3.$$

Proof. The proof of Theorem 12.25 is divided into three parts. First, we show how any LT_1 function namely, the computation of $f(X) = \operatorname{sgn}(F(X))$, can be reduced to the *COMPARISON* function in two layers. Second, by Proposition 12.18, the final result can be obtained using another two layers. Finally, we see how the second and the third layers can be combined into one layer so that all together only three layers are needed. The strict inclusion follows from the well-known fact that the XOR function is not computable in LT_1, whereas XOR $\in \widehat{LT}_2$.

(i) Reduction to COMPARISON
We show how we can reduce the computation of $F(X)$ to a sum of two numbers using two layers. First observe that by considering the binary representation of the weights w_i, we can introduce more variables and assign some constant values

to the renamed variables in such a way that any linear threshold function can be assumed to be of the following generic form:

$$f(X) = \operatorname{sgn}(F(X)), \quad \text{where}$$
$$F(X) = \sum_{i=1}^{n \log n} 2^i (x_{1_i} + x_{2_i} + \cdots + x_{n_i}).$$

We can further assume that n is *odd* by noting that $f(X)$ is not changed if we add one to $F(X)$ and the fact that $2^{n \log n} - 2^{n \log n - 1} - \cdots - 2 - 1 = 1$. For the sake of convenience of presentation, we assume $n \log n$ to be integer, where log denotes logarithm to the base 2. Let

$$s_i = x_{1_i} + x_{2_i} + \cdots + x_{n_i} \quad \text{for } i = 1, \ldots, n. \tag{12.9}$$

Note that $|s_i| \leq n$. Now partition the sum $F(X) = \sum_{i=1}^{n \log n} s_i 2^i$ into n consecutive blocks of $l = \lceil \log n \rceil$ summands each, so that

$$F(X) = \sum_{i=1}^{l} s_i 2^i + \sum_{i=l+1}^{2l} s_i 2^i + \ldots + \sum_{i=n \log n - l + 1}^{n \log n} s_i 2^i$$
$$= \sum_{j=0}^{n-1} \left(\sum_{i=jl+1}^{(j+1)l} s_i 2^i \right) = \sum_{j=0}^{n-1} \left(\sum_{k=1}^{l} s_{jl+k} 2^{k-1} \right) 2^{jl+1}$$
$$= \sum_{j=0}^{n-1} \tilde{s}_j 2^{jl+1} \quad \text{where } \tilde{s}_j = \sum_{k=1}^{l} s_{jl+k} 2^{k-1}.$$

Note that

$$|\tilde{s}_j| \leq \sum_{k=1}^{l} |s_{jl+k}| 2^{k-1} \leq n(2^l - 1) < 2^{2\lceil \log n \rceil}. \tag{12.10}$$

Further, notice that every *odd* number z such that $|z| < 2^{2\lceil \log n \rceil}$ can be expressed in ± 1 *binary representation* with $2\lceil \log n \rceil$ bits:

$$z = \sum_{i=0}^{2\lceil \log n \rceil - 1} z_i 2^i \quad \text{where } z_i = \pm 1. \tag{12.11}$$

Because n is assumed to be odd without loss of generality, it follows that $\tilde{s}_j$ is also odd and therefore can be represented as $2\lceil \log n \rceil$ bits of ± 1 as in expression (12.11). Now observe that because of (12.10), there is no overlapping in the ± 1 binary representation between $\tilde{s}_j 2^{jl+1}$ and $\tilde{s}_{j+2} 2^{(j+2)l+1} = 2^{2\lceil \log n \rceil} (\tilde{s}_{j+2} 2^{jl+1})$. Thus, we can compute the ± 1 binary representation of each $\tilde{s}_j$ for j odd in parallel and concatenate the resulting bits together to obtain the ± 1 binary representation of

$$s_{odd} = \sum_{j\ odd} \tilde{s}_j 2^{jl+1}.$$

We can obtain the ± 1 binary representation of

$$s_{even} = \sum_{j\ even} \tilde{s}_j 2^{jl+1}$$

in a similar fashion. Obviously, $F(X)$ is the sum of s_{odd} and s_{even}. It remains to show how to compute the ± 1 binary representation of each $\tilde{s}_j$. Observe that each $\tilde{s}_j$ is a polynomially bounded linear combination of $\log n \times n$ input variables. Let b_k be the kth bit in the ± 1 binary representation of $\tilde{s}_j$. Then there exists a set of numbers $\{k_1, \ldots, k_l\}$ such that $b_k = 1$ iff $\tilde{s}_j \in \{k_1, \ldots, k_l\}$. Let

$$y_{k_m} = \text{sgn}\{2(\tilde{s}_j - k_m) + 1\}\ ;\ \tilde{y}_{k_m} = \text{sgn}\{2(k_m - \tilde{s}_j) + 1\} \quad \text{for } m = 1, \ldots, l. \tag{12.12}$$

Then

$$b_k = \left\{ \sum_{m=1}^{l} (y_{k_m} + \tilde{y}_{k_m}) - 1 \right\}. \tag{12.13}$$

Because $\tilde{s}_j$ is polynomially bounded, there are only polynomially many different k_ms. The first layer of our circuit consists of $\widehat{LT}_1$ elements that compute the values of y_{k_m}s and $\tilde{y}_{k_m}$s. The second layer takes as inputs y_{k_m}s and $\tilde{y}_{k_m}$s and outputs $sgn\{\sum_{m=1}^{l}(y_{k_m} + \tilde{y}_{k_m}) - 1\}$. Hence the ± 1 binary representation of each $\tilde{s}_j$ can be computed in two layers.

(ii) Another two layers to compute *COMPARISON*
Now since the computation of $F(X)$ can be reduced to a sum of two $O(n \log n)$-bit numbers in ± 1 binary representation, it follows from Propositions 12.18 and 12.16 that we only need two more layers to compute the function.

(iii) Combining the second and third layers
So far we have used four layers of $\widehat{LT}_1$ elements to simulate an LT_1 element: two layers to reduce the computing of a general LT_1 function to that of a *COMPARISON*, then another two layers to compute the *COMPARISON*. Now we see how to combine the second and the third layer together so that $LT_1 \subset \widehat{LT}_3$.

Notice from (12.13) that the inputs b_k to the third layer are linear combinations of the outputs from the first layer. Thus, it is redundant to compute the $sgn(\ldots)$ after computing the linear combination. Therefore, we can directly feed the outputs from the first layer to the third layer without the use of the second layer. ∎

The result stated in Theorem 12.25 implies that a depth-d threshold circuit with exponentially large weights can be simulated by a depth-$3d$ threshold circuit with polynomially large weights. In fact, this result was improved: it was shown that it is possible to obtain an optimal depth simulation ($LT_d \subseteq \widehat{LT}_{d+1}$).

12.6 Bibliographic Notes

- The ideas of representations using various orthogonal functions over fields and groups is very old. In fact, the Sylvester-type Hadamard matrix used in the representation theorem of Section 12.2 was constructed by Sylvester in his 1867 paper [44]. The spectral representation of Boolean functions was first used by Muller in 1954 to construct the well-known Reed-Muller codes [33] that were later used in the *Mariner 9* spacecraft in 1972 to protect the transmission of images [28]. Muller as well as Niniomya [35] (in his PhD thesis) utilized the spectral representation to classify Boolean functions under certain symmetries. The MacWilliams identities that form a key result in coding theory were derived using spectral techniques applied to the Boolean domain [28]. Lechner's chapter in [26] is an excellent source for more details on this early work related to harmonic analysis of Boolean functions, and MacWilliams and Sloane's book [28] is an excellent source for studying the connections between the spectral techniques and the theory of error-correcting codes.
- The earliest connection made between threshold functions and the spectrum is the work by Chow on the characterization of linear threshold functions using the spectral coefficients [13] (Chow was not explicitly using the spectral language to describe his results). An early work on Boolean complexity and the spectrum was reported in [21], mainly focusing on an elementary interpretation of the relations between the properties of the function and its spectral coefficients.
- The area of threshold functions was very active as branch of pattern recognition and logic circuits in the 1960s and early 1970s [20, 34]. The connection between brain activity and threshold logic was suggested in [30]. There was a decay in interest in part because of the perception that small-depth neural networks have limited capabilities for certain pattern recognition tasks [32] and because of the strong success of AND/OR logic technology.
- New interest in neural networks emerged as a result of new insights regarding their dynamical properties [19] and their ability to learn by adjusting the weights [29]. See [1, 3, 24, 31] for a description of early studies regarding the capabilities of neural networks.
- The area of circuit complexity (as part of computer science) is focusing on studying complexity classes that are defined by circuits with various sets of gates. The book by Savage [40], the chapter by Boppana and Sipser [6], and the book by Wegener [45] are excellent general sources on this topic.
- The seminal work by Razborov [37] that showed that a majority gate can not be computed by a constant-depth polynomial size circuit that consists of AND, OR, NOT, and XOR gates resulted in a new focus in the area of circuit complexity, namely, the study of threshold circuits [17, 22, 36, 39, 46].
- An early study on the connection between neural networks and error-correcting codes suggested the use of the Sylvester-type Hadamard matrix

in this domain [9]. The first study on harmonic analysis of neural networks was reported in [8]. Studies reporting the application the spectral techniques to circuit complexity [7] and learning [27] were reported at about the same time.

- The characterization theorem and the related results in Section 12.4 are from [8] and [11]. Further work on spectral classification of threshold circuits is reported in [38].
- Related research activity include studies on applications of spectral techniques to learning [4, 23, 25], circuit complexity [5, 41], and neural networks [10].
- Studies on threshold circuit complexity, in particular issues related to computing with small weights and computing of arithmetic function, appeared in [2, 12, 15, 16, 18, 42, 43]. In particular, the results in Section 12.5 are from [42] and a constructive design for a depth-2 circuit for the *COMPARISON* function appears in [2]. A novel construction technique that improves the results in Section 12.5 and shows that a single extra layer is sufficient to transform a general threshold circuit to one with only "small" weights is reported in [16].
- Some of the topics in this chapter are treated in accompanying chapters in this volume. Specifically, for additional material and insights, the reader is referred to Chapter 8 by Claude Carlet, to Chapter 11 by Matthias Krause and Ingo Wegener, and to Chapter 13 by Martin Anthony.

References

[1] Y. S. Abu-Mostafa and J. M. St. Jacques. *Information capacity of the Hopfield model.* IEEE Transactions on Information Theory 31(4), pp. 461–4, 1985.

[2] N. Alon and J. Bruck. *Explicit constructions of depth-2 majority circuits for comparison and addition.* SIAM Journal on Discrete Mathematics 7(1), pp. 1–8, 1994.

[3] E. B. Baum. *On the capabilities of multilayer perceptrons.* Journal of Complexity 4, pp. 193–215, 1988.

[4] M. Bellare. *A technique for upper bounding the spectral norm with applications to learning.* ACM Workshop on Computational Learning Theory, 5, pp. 62–70, 1992.

[5] A. Bernasconi and B. Codenotti. *Spectral analysis of Boolean functions as a graph eigenvalue problem*, IEEE Transactions on Computers 48(3), pp. 345–51, 1999.

[6] R. Boppana and M. Sipser. The complexity of finite functions. In Handbook of Theoretical Computer Science, Vol. A, Algorithms and Complexity, J. Van Leeuwen, ed. MIT Press, Cambridge, MA, 1990.

[7] Y. Brandman, A. Orlitsky, and J. Hennessy. *A spectral lower bound technique for the size of decision trees and two-level AND/OR circuits.* IEEE Transactions on Computers 39(2), pp. 282–7, 1990.

[8] J. Bruck. *Harmonic analysis of polynomial threshold functions.* SIAM Journal on Discrete Mathematics 3(2), pp. 168–177, 1990.

[9] J. Bruck and M. Blaum. *Neural networks, error-correcting codes and polynomials over the binary n-cube.* IEEE Transactions on Information Theory 35, pp. 976–87, 1989.

[10] J. Bruck and V. P. Roychowdhury. *On the number of spurious memories in the Hopfield model.* IEEE Transactions on Information Theory 36(2), pp. 393–7, 1990.

[11] J. Bruck and R. Smolensky. *Polynomial threshold functions, AC^0 functions and spectral norms*. SIAM Journal on Computing 21(1), pp. 33–42, 1992.

[12] A. K. Chandra, L. Stockmeyer, and U. Vishkin. *Constant depth reducibility*, SIAM Journal on Computing 13, pp. 423–39, 1984.

[13] C. K. Chow. *On the characterization of threshold functions.* Symposium on Switching Circuit Theory and Logical Design, pp. 34–8, 1961.

[14] Y. Crama and P. L. Hammer. *Boolean Functions: Theory, Algorithms, and Applications*. Cambridge University Press, Cambridge, UK, 2009.

[15] M. Goldmann, J. Håstad, and A. Razborov. *Majority gates vs. general weighted threshold gates,* Structure in Complexity Theory Conference 7, pp. 2–13, 1992.

[16] M. Goldmann and M. Karpinski. *Simulating threshold circuits by majority circuits.* ACM Symposium on Theory of Computing 25, pp. 551–60, 1993.

[17] A. Hajnal, W. Maass, P. Pudlak, M. Szegedy, and G. Turan. *Threshold circuits of bounded depth*. IEEE Symposium on the Foundations of Computer Science 28, pp. 99–110, 1987.

[18] T. Hofmeister, W. Hohberg, and S. Köhling. *Some notes on threshold circuits and multiplication in depth 4*. Information Processing Letters 39, pp. 219–25, 1991.

[19] J. J. Hopfield. *Neural networks and physical systems with emergent collective computational abilities*. Proceedings of the National Academy of Sciences of the United States of America 79, pp. 2554–8, 1982.

[20] S. Hu. *Threshold Logic*. University of California Press, 1965.

[21] S. L. Hurst, D. M. Miller, and J. C. Muzio. *Spectral methods of Boolean function complexity*. Electronics Letters 18(33), pp. 572–4, 1982.

[22] R. Impagliazzo, R. Paturi, and M. Saks. *Size-depth trade-offs for threshold circuits*, ACM Symposium on Theory of Computing, 25, pp. 541–50, 1993.

[23] J. Jackson. *An efficient membership-query algorithm for learning DNF with respect to the uniform distribution*. Journal of Computer and System Sciences 55(3), pp. 414–40, 1997.

[24] T. Kohonen. *Self-Organization and Associative Memory*. Springer-Verlag, Berlin, 1989.

[25] E. Kushilevitz and Y. Mansour. *Learning decision trees using the Fourier spectrum.* ACM Symposium on Theory of Computing 23, pp. 455–64, 1991.

[26] R. J. Lechner. *Harmonic analysis of switching functions*. In Recent Development in Switching Theory, A. Mukhopadhyay, ed. Academic Press, New York, 1971.

[27] N. Linial, Y. Mansour, and N. Nisan. *Constant depth circuits, Fourier transforms, and learnability*. Journal of the Association for Computing Machinery 40(3), pp. 607–20, 1993.

[28] F. J. MacWilliams and N. J. A. Sloane. *The Theory of Error-Correcting Codes*. North-Holland, New York, 1977.

[29] J. L. McClelland, D. E. Rumelhart, and the PDP Research Group. *Parallel Distributed Processing: Explorations in the Microstructure of Cognition, Vol. 1*. MIT Press, Cambridge, MA, 1986.

[30] W. S. McCulloch and W. Pitts. *A logical calculus of ideas imminent in nervous activity*. Bulletin of Mathematical Biophysics 5, pp. 115–33, 1943.

[31] R. J. McEliece, E. C. Posner, E. R. Rodemich, and S. S. Venkatesh. *The capacity of the Hopfield associative memory.* IEEE Transactions on Information Theory 33(4), pp. 461–82, 1987.

[32] M. Minsky and S. Papert. *Perceptrons: An Introduction to Computational Geometry*. MIT Press, Cambridge, MA, 1969.

[33] D. E. Muller. *Application of Boolean algebra to switching circuit design and to error detection*. IRE Transactions on Electronic Computers 3, pp. 6–12, 1954.

[34] S. Muroga. *Threshold Logic and Its Applications*. John Wiley & Sons, New York, 1971.

[35] I. Ninomiya. *A theory of coordinate representation of switching functions*. Memoirs of the Faculty of Engineering, Nagoya University 10, pp. 175–90, 1958.

[36] I. Parberry and G. Schnitger. *Parallel computation with threshold functions*. Journal of Computer and System Sciences 36(3), pp. 278–302, 1988.

[37] A. Razborov. *Lower bounds on the size of bounded-depth networks over a complete basis with logical addition.* Mathematical Notes of the Academy of Sciences of the USSR 41, pp. 333–8, 1987.

[38] A. Razborov. *On small depth threshold circuits.* Scandinavian Workshop on Algorithm Theory 3, Lecture Notes in Computer Science 621, pp. 42–52, 1992.

[39] J. Reif, *On threshold circuits and polynomial computation.* Structure in Complexity Theory Symposium, pp. 118–23, 1987.

[40] J. E. Savage. *The Complexity of computing.* John Wiley & Sons, New York, 1976.

[41] M. Sitharam. *Evaluating spectral norms for constant depth circuits with symmetric gates.* Computational Complexity 5, pp. 167–89, 1995.

[42] K. Y. Siu and J. Bruck. *On the power of threshold circuits with small weights.* SIAM Journal of Discrete Mathematics 4(3), pp. 423–35, 1991.

[43] K. Y. Siu, J. Bruck, T. Kailath, and T. Hofmeister. *Depth efficient neural networks for division and related problems.* IEEE Transactions on Information Theory 39(3), pp. 946–56, 1993.

[44] J. J. Sylvester. *Thoughts on Inverse Orthogonal Matrices, Simultaneous Sign Successions, and Tessellated Pavements in Two or More Colors*, Philosophical Magazine 34, pp. 461–75, 1867.

[45] I. Wegener. *The Complexity of Boolean Functions.* John Wiley & Sons, New York, 1987.

[46] A. C. Yao. *On ACC and threshold circuits.* IEEE Symposium on the Foundations of Computer Science 31, pp. 619–27, 1990.

13

Neural Networks and Boolean Functions

Martin Anthony

13.1 Introduction

There has recently been much interest in "artificial neural networks," machines (or models of computation) based loosely on the ways in which the brain is believed to work. Neurobiologists are interested in using these machines as a means of modeling biological brains, but much of the impetus comes from their applications. For example, engineers wish to create machines that can perform "cognitive" tasks, such as speech recognition, and economists are interested in financial time series prediction using such machines.

In this chapter we focus on individual "artificial neurons" and feed-forward artificial neural networks. We are particularly interested in cases where the neurons are linear threshold neurons, sigmoid neurons, polynomial threshold neurons, and spiking neurons. We investigate the relationships between types of artificial neural network and classes of Boolean function. In particular, we ask questions about the type of Boolean functions a given type of network can compute, and about how extensive or expressive the set of functions so computable is.

13.2 Artificial Neural Networks

13.2.1 Introduction

It appears that one reason why the human brain is so powerful is the sheer complexity of connections between neurons. In computer science parlance, the brain exhibits huge parallelism, with each neuron connected to many other neurons. This has been reflected in the design of artificial neural networks. One advantage of such parallelism is that the resulting network is robust: in a serial computer, a single fault can make computation impossible, whereas in a system with a high degree of parallelism and many computation paths, a small number of faults may be tolerated with little or no upset to the computation. There are many good general texts on neural networks, such as [7, 17]. Here we briefly describe the aspects of neural networks that we are interested in from a Boolean functions point of view.

Generally speaking, we can say that an artificial neural network consists of a directed graph with *computation units* (or *neurons*) situated at the vertices. One or more of these computation units are specified as *output units*. These are the units with zero out-degree in the directed graph. We consider networks in which there is only one output unit. Additionally, the network has *input units*, which receive signals from the outside world. Each unit produces an output, which is transmitted to other units along the arcs of the directed graph. The outputs of the input units are simply the input signals that have been applied to them. The computation units have *activation functions* determining their outputs. The degree to which the output of one computation unit influences those of its neighbors is determined by the weights assigned to the network. This description is quite abstract at this stage, but we shall concretize it shortly by focusing on particular types of network.

13.2.2 Neurons

The building blocks of feed-forward networks are *computation units* (or *neurons*). In isolation, a computation unit has some number, k, of *inputs* and is capable of taking on a number of *states*, each described by a vector $w = (w_0, w_1, \ldots, w_p) \in \mathbb{R}^p$ of p real numbers, known as *weights* or *parameters*. Here, p, the number of parameters of the unit, will depend on k. If the unit is a *linear threshold* unit or *sigmoid unit*, then $p = k + 1$ and, in these cases, it is useful to think of the weights $w_1, w_2, \ldots, w_k$ as being assigned to each of the k inputs. For *spiking neurons* and *polynomial threshold units*, the number of parameters will be greater than $k + 1$. The different types of neurons we consider are best described by defining how they process their inputs.

Generally, when in the state described by $w \in \mathbb{R}^p$, and on receiving input $x = (x_1, x_2, \ldots, x_k)$, the computation unit produces as output an *activation* $g(w, x)$, where $g : \mathbb{R}^p \times \mathbb{R}^k \to \mathbb{R}$ is a fixed function. We may regard the unit as a parameterized function class. That is, we may write $g(w, x) = g_w(x)$, where, for each state w, $g_w : \mathbb{R}^k \to \mathbb{R}$ is the function computed by the unit on the inputs x.

Linear Threshold Units

For a linear threshold unit, the function g takes a particularly simple form:

$$g(w, x) = \text{sgn}\,(w_0 + w_1 x_1 + \cdots + w_k x_k)\,,$$

where sgn is the sign function, given by

$$\text{sgn}(z) = \begin{cases} 1 & \text{if } z \geq 0 \\ 0 & \text{if } z < 0. \end{cases}$$

Thus, when the state of the unit is given by $w = (w_0, w_1, \ldots, w_k)$, the output is either 1 or 0, and it is 1 precisely when

$$w_0 + w_1x_1 + \cdots + w_kx_k \geq 0,$$

which may be written as

$$w_1x_1 + \cdots + w_kx_k \geq \theta,$$

where $\theta = -w_0$ is known as the *threshold*. In other words, the computation unit gives output 1 (in biological parlance, it *fires*) if and only if the weighted sum of its inputs is at least the threshold θ. If the inputs to the threshold unit are restricted to $\{0, 1\}^n$, then the set of Boolean functions it computes is precisely the *threshold functions*.

Sigmoid Units

For a (standard) sigmoid unit, we have

$$g(w, x) = \sigma(w_0 + w_1x_1 + \cdots + w_kx_k),$$

where the "activation function" $\sigma(z) = 1/(1 + e^{-z})$ is the *standard sigmoid function*. Writing $\theta = -w_0$, as we did previously for the linear threshold unit, we see that the output of the sigmoid unit is $\sigma(\sum_{i=1}^{k} w_ix_i - \theta)$. If the weighted sum $\sum_{i=1}^{k} w_ix_k$ is much larger than the threshold, then the output is close to 1; if it is much less than the threshold, the output is close to 0; and if it is very close to the threshold, then the output is close to $1/2$. In fact, the sigmoid function can be thought of as a "smoothed" version of the sign function, sgn, because σ maps from $\mathbb{R}$ into the interval (0, 1), is differentiable and satisfies

$$\lim_{z\to-\infty} \sigma(z) = 0, \quad \lim_{z\to\infty} \sigma(z) = 1.$$

Note that, whereas the linear threshold unit has output in $\{0, 1\}$, the output of a sigmoid unit lies in the interval (0, 1) of real numbers.

Polynomial Threshold Units

The linear threshold and sigmoid units both work with $w_1x_1 + \cdots + w_kx_k$, a linear combination of the inputs to the unit, but we can generalize from this and consider instead units that use a nonlinear combination of the x_i. For example, when $k = 3$, imagine a unit that computes the quadratic expression

$$w_1x_1 + w_2x_2 + w_3x_3 + w_4x_1^2 + w_5x_2^2 + w_6x_3^2 + w_7x_1x_2 + w_8x_1x_3 + w_9x_2x_3,$$

for some contants w_i, $(1 \leq i \leq 9)$, and then compares this with a threshold value θ. Such a unit is a *polynomial threshold unit* of degree 2. We now set up a description of this generalization of linear threshold units. We denote by $[n]^m$ the set of all selections, in which repetition is allowed, of at most m objects from the set $[n] = \{1, 2, \ldots, n\}$. Thus, $[n]^m$ is a collection of "multisets." For example, $[3]^2$ consists of the multisets

$$\emptyset, \{1\}, \{1, 1\}, \{2\}, \{2, 2\}, \{3\}, \{3, 3\}, \{1, 2\}, \{1, 3\}, \{2, 3\}.$$

A *polynomial threshold unit of degree m* (also termed a *sigma-pi unit* [38, 45, 49]) has $p = \binom{n+m}{m}$ parameters w_S, one for each multiset $S \in [n]^m$. For $S \in [n]^m$ and $x = (x_1, x_2, \ldots, x_n) \in \mathbb{R}^n$, let x_S denote the product of the x_i for $i \in S$ (with repetitions as required). For example, $x_{\{1,2,3\}} = x_1x_2x_3$ and $x_{\{1,1,2\}} = x_1^2x_2$. When $S = \emptyset$, the empty set, we interpret x_S as the constant 1. The output of the unit is given by

$$g_w(x) = g(w, x) = \mathrm{sgn}\left(\sum_{S \in [n]^m} w_S x_S\right).$$

Of course, when $m = 1$, we obtain a linear threshold unit. But for $m > 1$, a polynomial threshold unit can compute functions that a linear threshold unit is incapable of computing. Furthermore (and this will prove useful later), note that if we restrict the inputs x_i to belong to $\{0, 1\}$ then we do not need terms of the form $w_S x_S$ where the multiset S contains repeated members: this is simply because if $x_i \in \{0, 1\}$ then $x_i^r = x_i$ for all $r > 1$.

Consider, for example, the case $n = m = 2$, and suppose we take

$$w_\emptyset = -\frac{1}{2}, \ w_{\{1\}} = w_{\{2\}} = 1, \ w_{\{1,2\}} = -2,$$

with the remaining weights $w_{\{1,1\}}$ and $w_{\{2,2\}}$ equal to 0. Then

$$g_w(x) = \mathrm{sgn}\left(-\frac{1}{2} + x_1 + x_2 - 2x_1x_2\right).$$

It is easy to verify that, as a Boolean function on $\{0, 1\}^2$, g is the exclusive-or function, which is not computable by a linear threshold unit.

Spiking Neurons

A very interesting class of artificial neurons are the *spiking neurons*. A number of results on the capabilities of these neurons and networks of them have been obtained by Maass and Schmitt [26, 27, 29, 42]. In this chapter we present some results from [29, 42] concerning spiking neurons of a simplified type. The type of neuron considered is a "Type A" spiking neuron with "binary encoding" [29]. For biological motivation for this model, see [29] and the references cited there. The key difference between this type of neuron and the ones considered so far is the introduction of a time variable. In the three types of neuron discussed so far, a weighted sum is immediately computed and the output of the neuron depends directly on that weighted sum. Here, however, *delays* in the inputs to the neuron are modeled by assuming not only that to each input there is associated a weight w_i, but also a *delay* d_i. It is assumed that the weighted input corresponding to input unit i is only "active" during the time interval $[d_i, d_i + 1)$. If, at any time, the sum of the currently active weighted inputs is at least the threshold value, then the neuron fires; otherwise it does not. Formally, with k inputs and in state

$$w = (w_0, w_1, w_2, \ldots, w_k, d_1, d_2, \ldots, d_k),$$

the output of the spiking neuron is given by

$$g(w, x) = \text{sgn}\left(w_0 + \max_{t \geq 0} \sum_{i=1}^{k} w_i x_i \chi_{[d_i, d_i+1)}(t)\right),$$

where $\chi_{[d_i, d_i+1)}$, the characteristic function of the time interval $[d_i, d_i + 1)$, is given by

$$\chi_{[d_i, d_i+1)}(t) = \begin{cases} 1 & \text{if } d_i \leq t < d_i + 1 \\ 0 & \text{otherwise.} \end{cases}$$

Observe that if all delays d_i are fixed at 0, then the spiking neuron behaves just like the linear threshold neuron with weights $(w_0, w_1, \ldots, w_k)$.

13.2.3 Networks

As mentioned in the general description just given, a neural network is formed when we place units at the vertices of a directed graph, with the arcs of the digraph representing the flows of signals between units. Some of the units are termed *input units*: these do not receive signals from other units, but instead they take their signals from the outside environment. Units that do not transmit signals to other units are termed *output units*. The network is said to be a *feed-forward network* if the underlying directed graph is acyclic (that is, it has no directed cycles). This feed-forward condition means that the units can be labeled with integers in such a way that if there is a connection from the computation unit labeled i to the computation unit labeled j, then $i < j$. We will often be interested in *multilayer* networks. In such networks, the units may be grouped into *layers*, labeled $0, 1, 2, \ldots, \ell$, in such a way that the input units form layer 0, these feed into the computation units, and if there is a connection from a computation unit in layer r to a computation unit in layer s, then we must have $s > r$. Note, in particular, that there are no connections between any two units in a given layer. We call such a network an ℓ-layer network. (Strictly speaking, it has $\ell + 1$ layers, but one of these consists entirely of input units, and it is the number of layers of computation units that is usually important.) Any feed-forward network is a multilayer network (because we could just take the layers to consist of single computation units), but we shall often be interested in feed-forward networks with a small number of layers. It is easy to see that the smallest ℓ for which such a layering is possible is the *depth* of the network, defined as the length of the largest directed path in the underlying directed graph.

We shall primarily be interested in single polynomial threshold units and spiking neurons, and in one-output feed-forward networks in which the computation units are linear threshold units or sigmoid units. A threshold or sigmoid network with n input units is capable of computing a number of functions from $\mathbb{R}^n$ to $\mathbb{R}$, or (simply restricting the input signals to be $\{0, 1\}$-valued) from $\{0, 1\}^n \to \mathbb{R}$. The precise function computed depends on the state of each computation unit. Recall that for the threshold and sigmoid neurons, if a unit has k inputs, then the state is a vector of $k + 1$ real numbers: one of these numbers (w_0 or its negative, the threshold θ in the description given earlier) can be thought of as being attached to the unit itself, and the other k can be thought of as describing the weight attached to each of the k arcs feeding into the unit. Suppose that the network has N computation units, labeled $1, 2, \ldots, N$, and that computation unit i has k_i inputs. Then the total number of weights in the network is

$$\sum_{i=1}^{N}(k_i + 1) = N + \sum_{i=1}^{N} k_i = N + E,$$

where E denotes the total number of arcs in the digraph. We may therefore say that the *state of the network* as a whole is described by a vector w of $W = N + E$ real numbers. When there are n input units and one output unit, the network computes, for each state w, a function $h_w : \mathbb{R}^n \to \mathbb{R}$. The set of functions *computable* by the network when the weight vector can be chosen from a subset Ω of $\mathbb{R}^W$ is $\{h_w : w \in \Omega\}$. (Often, Ω will simply be $\mathbb{R}^W$, but one may want, for example, to restrict the sizes of the allowable weights, in which case Ω will be a strict subset of $\mathbb{R}^W$.)

Linear threshold networks have long been studied and were the subject of much work in "threshold logic" in the 1960s; see the books by Hu [18], Muroga [33], and Crama and Hammer [10], and the papers cited there. A single linear threshold unit may be regarded as a linear threshold network, and this simplest of all neural networks is often called the *perceptron*, though that term is also used more generally [31]. Questions concerning the type of function computable by a polynomial threshold unit have been worked on by a number of researchers and were considered in [9, 31, 35]. For more recent results, see the survey article by Saks [39]: this provides an excellent overview of much of the theoretical work on functions computable by threshold and polynomial threshold units and related areas (some of which are touched on later in this chapter). See also [48].

In the rest of this chapter, we concentrate on two main issues. First, how many and what type of Boolean functions can be computed by neural networks of particular types? Second, what is the expressive power (as measured by the VC-dimension of Chapter 6) of neural network classes? (The latter is a question important for learning theory; see Chapter 6.)

13.3 Computing Boolean Functions by Neural Networks

13.3.1 Linear Threshold Units

We have noted that the Boolean functions computed by the single linear threshold unit are precisely the Boolean threshold functions. Recall that f is a (Boolean) threshold defined on $\{0, 1\}^n$ if there are $w \in \mathbb{R}^n$ and $\theta \in \mathbb{R}$ such that

$$f(x) = \begin{cases} 1 & \text{if } \langle w, x\rangle \geq \theta \\ 0 & \text{if } \langle w, x\rangle < \theta, \end{cases}$$

where $\langle w, x\rangle = w^T x$ is the standard inner product of w and x. Given such w and θ, we say that f is represented by $[w, \theta]$, and we write $f \leftarrow [w, \theta]$. The vector w is known as the weight-vector, and θ is known as the threshold. We denote the class of threshold functions on $\{0, 1\}^n$ by T_n. Note that any $f \in T_n$ will satisfy $f \leftarrow [w, \theta]$ for ranges of w and θ.

Asummability and Linear Separability

Geometrically, a Boolean function f is a threshold function if the true and false points are separable by a hyperplane; that is, f is *linearly separable*. Such functions can also be characterized by the asummability property, as follows (see, e.g., [10]).

Theorem 13.1. *The Boolean function f is a threshold function if and only if it is* asummable, *meaning that for any $k \in \mathbb{N}$, for any sequence $x_1, x_2, \ldots, x_k$ of (not necessarily distinct) true points of f and any sequence $y_1, y_2, \ldots, y_k$ of (not necessarily distinct) false points of f,*

$$\sum_{i=1}^{k} x_i \neq \sum_{i=1}^{k} y_i.$$

Asummability can be seen to be equivalent to the nonintersection of the convex hulls of the sets true points and false points of f. (It can be seen quite directly to be equivalent to the assertion that there is no point that is simultaneously a rational convex combination of true points and a rational convex combination of false points. This, in turn, is equivalent to the nonintersection of the convex hulls.) By the separating hyperplanes theorem, asummability is therefore equivalent to linear separability.

Number of Functions Computed

A classical result, which dates back to work by Schläfli in the last century [41] and which also appears in [9], is that the maximum number of connected regions into which $\mathbb{R}^d$ can be partitioned by N hyperplanes passing through the origin is bounded above by

$$C(N, d) = 2\sum_{k=0}^{d-1} \binom{N-1}{k}.$$

(Here, we apply the usual convention that $\binom{a}{b} = 0$ if $b > a$, and $\binom{0}{b} = 1$.) From this, it is possible to obtain the following result [9].

Theorem 13.2. *Suppose that $S \subseteq \mathbb{R}^n$ is finite. Then the number of different functions $f : S \to \{0, 1\}$ computable by a linear threshold unit on domain S is at most*

$$2\sum_{k=0}^{n}\binom{|S|-1}{k}.$$

Taking $S = \{0, 1\}^n$, the set of computable functions is just T_n, the set of Boolean threshold functions, so we obtain

$$|T_n| \le 2\sum_{k=0}^{n}\binom{2^n-1}{k} \le 2^{n^2}.$$

It is clear that T_n is a vanishingly small fraction of all Boolean functions on $\{0, 1\}^n$, as might be expected. Since the 1960s (see Muroga's book [33]), a lower bound on $|T_n|$ of the form $2^{(n^2/2)(1+o(1))}$ has been known. More recently, Zuev [50] showed that, for sufficiently large n, $\log_2 |T_n| > n^2 (1 - 10/\ln n)$. So the upper bound is asymptotically of the right order.

Sizes of Weights

A weight-vector and threshold are said to be *integral* if the threshold and each entry of the weight-vector are integers. Any Boolean threshold function can be represented by an integral weight-vector and threshold. To see this, note first that, by the discreteness of $\{0, 1\}^n$, any Boolean threshold function can be represented by a rational threshold and weight-vector. Scaling these by a suitably large integer yields integral threshold and weight-vector representing the same function. A natural question is how large the integer weights (including the threshold) have to be. An upper bound is as follows [33].

Theorem 13.3. *For any Boolean threshold function f on $\{0, 1\}^n$, there is an integral weight-vector w and an integral threshold θ such that $t \leftarrow [w, \theta]$ and such that*

$$\max\{|\theta|, |w_1|, \ldots, |w_n|\} \le (n+1)n^{n/2}.$$

It is easy to show that exponential-sized integer weights are sometimes necessary just by a simple counting argument. A result of Muroga [33] alluded to earlier says that there are at least $2^{n(n-1)/2}$ threshold functions on $\{0, 1\}^n$. For $B \in \mathbb{N}$, the number of pairs (w, θ) of integer weight-vector and threshold that satisfy $|w_i| \le B$ for $i = 1, 2, \ldots, n$, and $|\theta| \le B$ is at most B^{n+1}. So, for example, the number of threshold functions representable with integer weights and threshold bounded in magnitude by $2^{n/6}$ is no more than $2^{n(n+1)/6}$. But this is less than $2^{n(n-1)/2}$ for $n \ge 2$, so there must be some threshold functions in which, using integer weights, we would need weights greater than $2^{n/6}$ in magnitude. This simple argument given earlier establishes the need for large weights, but it does not provide a concrete example of a threshold function requiring such large weights. Specific examples of such functions have long been known (see [32, 33]). We now present an example function that, although it is not the simplest possible, will be useful later and has been of much interest in analyzing the performance of the perceptron learning algorithm [5, 15].

Consider, for n even, the Boolean function f_n on variables with formula

$$f_n = u_n \wedge (u_{n-1} \vee (u_{n-2} \wedge (u_{n-3} \vee (\ldots (u_2 \wedge u_1)) \ldots).$$

Thus, for example,

$$f_6 = u_6 \wedge (u_5 \vee (u_4 \wedge (u_3 \vee (u_2 \wedge u_1)))).$$

It can be shown (see [1, 37], for example) that if w is any integral weight-vector in a threshold representation of f_n, then $w_i \geq F_i$ for $1 \leq i \leq n$, where F_i is the ith Fibonacci number. Because

$$F_n \geq \frac{\sqrt{5}}{6}\left(\frac{1+\sqrt{5}}{2}\right)^n,$$

for all n, this function requires integer weights exponential in n.

The general upper bound on integral weights given in Theorem 13.3 is $(n+1)n^{n/2}$, whereas the specific lower bound exhibited by the function f_n is (merely) exponential in n. The question arises as to whether the general upper bound is loose and could potentially be considerably improved. In fact, however, the upper bound is quite tight. Specifically, Håstad [16] has proved that there are constants $k > 0$ and $c > 1$ such that, for n a power of 2, there is a threshold function f on $\{0, 1\}^n$ such that any integral weight-vector representing f has a weight at least $kc^{-n}n^{n/2}$.

Test Sets for Linear Threshold Functions

For $f \in T_n$, we say that a set $S \subseteq \{0, 1\}^n$ is a test set for f if, when $h \in T_n$ and h classifies the inputs in S in the same way as f does, then h is necessarily equal to f, among all threshold functions. In other words, S is a test set for f if the inputs in S serve to specify uniquely the function f. Denote by $\sigma(f)$ the cardinality of the smallest test set for t. This parameter is useful in considering the complexity of "teaching" linear threshold functions; see [4, 12]. The following result was obtained in [4].

Theorem 13.4. *Suppose $f \in T_n$ and suppose that $k \geq 1$ is such that any weight-vector realizing f has at least k nonzero weights and that there is a weight-vector realizing f that has exactly k nonzero weights. Then*

$$2^{n-k}(k+1) \leq \sigma(f) \leq 2^{n-k}\binom{k+1}{\lfloor \frac{k+1}{2} \rfloor},$$

and equality is possible in both of these inequalities.

Despite the fact that the testing number can be exponential, it can be shown [4] that the average, or expected, testing number of a function in T_n is at most n^2.

Fixing attention for the moment on the case $k = n$ just given, it has been shown [4] that there is a large family of threshold functions – the nested functions – each having minimum possible testing number. Let us recursively define a Boolean function to be canonically nested as follows: both functions of 1 variable are canonically nested, and t_n, a function of n variables, is canonically nested if $t_n = u_n \star t_{n-1}$ or $t_n = \bar{u}_n \star t_{n-1}$, where $\star$ is $\vee$ (the OR connective) or $\wedge$ (the AND connective) and t_{n-1} is a canonically nested function of $n - 1$ variables. (Here, we mean that t_{n-1} acts on the first $n - 1$ entries of its argument.) We say that a function f is *nested* if, by permuting (or relabeling) the variables, we obtain a canonically nested function. One may relate nested functions to particular types of *decision lists* (discussed in Chapter 14). It is straightforward to see that any nested function can be realized as a 1-decision list of length n in which, for each i between 1 and n, precisely one term of the form (u_i, b) or $(\bar{u}_i, b)$ occurs (for some $b \in \{0, 1\}$) (and *vice versa*). It is easily seen that any nested function is a threshold function. Examples of nested functions include the functions f_n described earlier. It turns out [4] that all nested functions (regarded as threshold functions) have the smallest possible testing numbers, since each has testing number $n + 1$.

13.3.2 Polynomial Threshold Units

We now consider the Boolean functions computable by a polynomial threshold unit. For $n \in \mathbb{N}$ and $d \leq n$, let $[n]^{(d)}$ denote all subsets of $[n] = \{1, 2, \ldots, n\}$ of cardinality at most d. A Boolean function f defined on $\{0, 1\}^n$ is a *polynomial threshold function* of *degree d* if there are real numbers w_S, one for each $S \in [n]^{(d)}$, such that

$$f(x) = \text{sgn}\left(\sum_{S \in [n]^{(d)}} w_S x_S \right),$$

where the notation is as defined earlier. The set of polynomial threshold functions on $\{0, 1\}^n$ of degree d will be denoted by $\mathcal{P}(n, d)$. The class $\mathcal{P}(n, 1)$ is, of course, simply as the set of threshold functions T_n on $\{0, 1\}^n$. (Note that we have used the earlier observation that, for Boolean inputs to the polynomial threshold unit, no powers of x_i other than 0 or 1 are needed; so S ranges over subsets rather than multisubsets of $[n]$.)

Asummability and Polynomial Separability

We have already observed that a function is a linear threshold function if and only if the true points can be separated from the false points by a hyperplane. For a polynomial threshold function of degree m, we have the corresponding geometrical characterization that the true points can be separated from the false points by a surface whose equation is a polynomial of degree m.

It is possible to relate such polynomial separation to linear separation in a higher-dimensional space [9, 48]. For $x \in \{0,1\}^n$, we define the *m-augment*, $x^{(m)}$, of x to be the $\{0,1\}$-vector of length $\sum_{i=1}^{m} \binom{n}{i}$ whose entries are x_S for $\emptyset \neq S \in [n]^{(m)}$ in some prescribed order. To be precise, we shall suppose the entries are in order of increasing degree and that terms of the same order are listed in lexicographic (dictionary) order. Thus, for example, when $n = 5$ and $m = 2$,

$$x^{(5)} = (x_1, x_2, x_3, x_4, x_5, x_1x_2, x_1x_3, x_1x_4, x_1x_5,\\ x_2x_3, x_2x_4, x_2x_5, x_3x_4, x_3x_5, x_4x_5).$$

We observe that a Boolean function f is a polynomial threshold function of degree m if and only if there is some linear threshold function h_f, defined on $\{0,1\}$ vectors of length $r = \sum_{i=1}^{m} \binom{n}{i}$, such that

$$f(x) = 1 \Longleftrightarrow h_f(x^{(m)}) = 1;$$

that is, if and only if the m-augments of the true points of f and the m-augments of the false points of f can be separated by a hyperplane in the higher-dimensional space $\mathbb{R}^r$, where $r = \sum_{i=1}^{m} \binom{n}{i}$.

The m-augments can be used to provide an asummability criterion similar to Theorem 13.1.

We say that f is m-asummable if for any $k \in \mathbb{N}$, for any sequence $x_1, x_2, \ldots, x_k$ of (not necessarily distinct) true points of f, and any sequence $y_1, y_2, \ldots, y_k$ of (not necessarily distinct) false points of f,

$$\sum_{i=1}^{k} x_i^{(m)} \neq \sum_{i=1}^{k} y_i^{(m)}.$$

Note that if f is m-asummable, then f is m'-asummable for any $m' > m$. The following result holds [48].

Theorem 13.5. *The Boolean function f is a threshold function of degree m if and only if f is m-asummable.*

Number of Polynomial Threshold Functions

We can obtain an upper bound on the number of polynomial threshold functions of a given degree by using Theorem 13.2, together with the fact that a Boolean function is a polynomial threshold function of degree m if and only if the m-augments of true points and the m-augments of the false points are linearly separable.

Theorem 13.6. *The number, $|\mathcal{P}(n, m)|$ of polynomial threshold functions of degree m on $\{0, 1\}^n$ satisfies*

$$|\mathcal{P}(n, m)| \leq 2 \sum_{k=0}^{\sum_{i=1}^{m} \binom{n}{i}} \binom{2^n - 1}{k}$$

for all m, n with $1 \leq m \leq n$.

It is fairly easy to deduce from this that $\log_2 |T(n, m)|$ is at most $n\binom{n}{m} + O(n^m)$ as $n \to \infty$, with $m = o(n)$.

Saks [39] observed that $|\mathcal{P}(n, m)| \geq |\mathcal{P}(n - 1, m)||\mathcal{P}(n - 1, m - 1)|$, for $2 \leq m \leq n - 1$. The next theorem follows [1, 39] from this.

Theorem 13.7. *The number, $|\mathcal{P}(n, m)|$, of polynomial threshold functions of degree m on $\{0, 1\}^n$ satisfies $|\mathcal{P}(n, m)| \geq 2^{\binom{n}{m+1}}$. for all m, n with $1 \leq m \leq n - 1$.*

Note that this lower bound is not at all tight for $m > n/2$. However, for constant m it provides a good match for the upper bound of Theorem 13.6. Taken together, the results imply that, for fixed m, for some positive constants c, k, $\log_2 |T(n, m)|$ is, between cn^{m+1} and kn^{m+1}.

Threshold Order

A Boolean function is said to be a k-DNF function if it has a DNF formula in which each term is of degree at most k. It is easy to see that any k-DNF f on $\{0, 1\}^n$ is in $\mathcal{P}(n, k)$, as follows. Given a term $T_j = u_{i_1} u_{i_2} \dots u_{i_r} \bar{u}_{j_1} \bar{u}_{j_2} \dots \bar{u}_{j_s}$ of the DNF, we form the expression

$$A_j = x_{i_1} x_{i_2} \dots x_{i_r} (1 - x_{j_1})(1 - x_{j_2}) \dots (1 - x_{j_s}).$$

We do this for each term $T_1, T_2, \dots, T_l$ and expand the algebraic expression $A_1 + A_2 + \dots + A_l$ according to the normal rules of algebra, until we obtain a linear combination of the form $\sum_{S \in [n]^{(k)}} w_S x_S$. Then, since $f(x) = 1$ if and only if $A_1 + A_2 + \dots + A_l > 0$, it follows that

$$f(x) = \text{sgn}\left(\sum_{S \in [n]^{(k)}} w_S x_S \right),$$

so $f \in \mathcal{P}(n, k)$. Thus, any k-DNF function is also a polynomial threshold function of degree at most k.

Generally, given a Boolean function f, the *threshold order* [31, 48] of f is the least k such that $f \in \mathcal{P}(n, k)$. We mention that there are always (exactly) two functions with threshold order n, namely the parity function PARITY_n (defined by $\text{PARITY}_n(x) = 1$ if and only if x has an odd number of entries equal to 1) and its complement; see [48].

A very precise behavior of the "expected" threshold order has been conjectured by Wang and Williams [48]. Roughly speaking, the conjecture says that, for large even numbers n, almost all the Boolean functions on $\{0, 1\}^n$ have threshold order equal to $n/2$; and that for large odd n, almost every function has threshold order $(n-1)/2$ or $(n+1)/2$, with an equal split between these. To make this precise, we introduce some notation. Let $\sigma(n,k)$ denote the proportion of Boolean functions of n variables with threshold order k; thus,

$$\sigma(n,k) = \frac{|\mathcal{P}(n,k)| - |\mathcal{P}(n,k-1)|}{2^{2^n}}.$$

Wang and Williams conjectured that for even values of n, $\sigma(n, n/2) \to 1$ as $n \to \infty$ and that for odd values of n, $\sigma(n, (n-1)/2) \to 1/2$ and $\sigma(n, (n+1)/2) \to 1/2$ as $n \to \infty$. The following observation [3] provides a partial proof of this.

Theorem 13.8. *For $k = k(n) \leq \lfloor n/2 \rfloor - 1$, $\sigma(n, k(n)) \to 0$ as $n \to \infty$. Furthermore, for all odd n, $\sigma((n, \lfloor \frac{n}{2} \rfloor) \leq 1/2$.*

This result shows, among other things, that the representational power of $\mathcal{P}(n,k)$ is limited unless k is of the same order as n. In particular, it might be said that the "typical" Boolean function has threshold order at least $\lfloor n/2 \rfloor$.

Some progress has been made on the remaining parts of the conjecture. As reported in [39], Alon, using a result of Gotsman [13] on the harmonic analysis of Boolean functions, showed that there is a fixed constant $\varepsilon > 0$ such that almost all Boolean functions of n variables have threshold order less than $(1-\varepsilon)n$; that is, $\sigma(n, (1-\varepsilon)n) \to 0$ as $n \to \infty$. This was improved upon by Samorodnitsky [40] and, independently, by O'Donnell and Servedio [36], who show (again, using harmonic analysis) that almost all Boolean functions have threshold order at most $\frac{n}{2} + O(\sqrt{n \log n})$.

13.3.3 Linear Threshold Networks

We now move on to consider the representation of Boolean functions by feed-forward linear threshold networks (which we refer to as threshold networks for the sake of brevity). Single linear threshold units have very limited computational abilities, but we can easily see that any Boolean function can be represented by a threshold network with one hidden layer. It is natural to ask how small a threshold network can be used for particular functions or types of functions. Questions like this bring us into the realm of circuit complexity (in which threshold networks are usually referred to as *threshold circuits*), a large area that we only very briefly touch on here but that is more extensively handled by Krause and Wegener in Chapter 11 and by Bruck in Chapter 12.

The existence of a DNF formula for every Boolean function can be used to show that any Boolean function can be computed by a two-layer feed-forward threshold network.

Theorem 13.9. *There is a two-layer threshold network capable of computing any Boolean function.*

Proof. Suppose that $f : \{0, 1\}^n$, and let ϕ be the DNF formula obtained as the disjunction of the prime implicants of f. Suppose $\phi = T_1 \vee T_2 \vee \cdots \vee T_k$, where each T_j is a term of the form $T_j = (\bigwedge_{i \in P_j} u_i) \bigwedge (\bigwedge_{j \in N_j} \bar{u}_j)$, for some disjoint subsets P_j, N_j of $\{1, 2, \ldots, n\}$. Suppose that the network has 2^n hidden units, and let us set the weights to and from all but the first k of these to equal 0, and the corresponding thresholds equal to 1 (so the effect is as if these units were absent). Then for each of the first k units, let the weight-vector $\alpha^{(j)}$ from the inputs to unit j correspond directly to T_j, in that $\alpha_i^{(j)} = 1$ if $i \in P_j$, $\alpha_i^{(j)} = -1$ if $i \in N_j$, and $\alpha_i^{(j)} = 0$ otherwise. We take the threshold on unit j to be $|P|$, the weight on the connection between the unit and the output unit to be 1, and the threshold on the output unit to be $1/2$. It is clear that unit j outputs 1 on input x precisely when x satisfies T_j, and that the output unit computes the "or" of all the outputs of the hidden units. Thus, the output of the network is the disjunction of the terms T_j and hence equals f. ■

A *universal network* for Boolean functions on $\{0, 1\}^n$ is a threshold network that is capable of computing every Boolean function of n variables. Theorem 13.9 shows that the two-layer threshold network with n inputs, 2^n units in the hidden layer, and one output unit, is universal. The question arises as to whether there is a universal network with fewer threshold units. By an easy counting argument, one can obtain a lower bound on the size of *any* universal network, regardless of its structure. In particular (see [34, 44]), any universal network (regardless of how many layers it has) must have at least $\Omega\left(2^{n/2}/\sqrt{n}\right)$ threshold units. Moreover, any two-layer universal network for Boolean functions must have at least $\Omega(2^n/n^2)$ threshold units.

Much work in circuit complexity has gone into consideration of the sizes of threshold network needed to compute particular Boolean functions. Of particular interest has been the parity function parity$_n$. Many sophisticated techniques have been used to produce lower bounds on the sizes of networks (with particular numbers of layers, for example) capable of computing parity; see [44]. One such result is that any two-layer threshold network capable of computing parity$_n$ must have $\Omega(\sqrt{n})$ units.

13.3.4 Spiking Neurons

We have observed that if all delays on a spiking neuron are set to zero, then the neuron behaves exactly like a linear threshold unit. So the spiking neuron is at least as powerful as the linear threshold unit, and the set S_n of Boolean functions it computes is at least as large as T_n. However, S_n is not significantly larger than T_n, for as shown by Maass and Schmitt [29], $\log_2 |S_n| \leq n^2 + O(\log n)$, whereas, as noted earlier, $\log_2 |T_n|$ is $n^2(1 + o(1))$.

By the way the neuron acts, the weighted signal from input i is "active" (if at all) on the time interval $[d_i, d_i + 1)$ and the output of the neuron is 1 if and only if the sum of active weighted inputs exceeds the threshold, at some time. By partitioning the time axis into intervals on which the same weighted inputs are active, it can be seen [29] that there are at most $2n - 1$ intervals on which the sum of active weighted inputs is constant. (For, there are at most $2n$ times at which the set of active weighted inputs can change.) Hence, the neuron fires if one of these $2n - 1$ sums exceeds the threshold. Thus, we obtain the result from [29] that any function in S_n can be expressed as a disjunction of at most $2n - 1$ threshold functions. Schmitt [42] improved this to $n - 1$. Hammer *et al.* [14] defined the *threshold number* of a Boolean function to be the smallest number of threshold functions of which it is a conjunction (a number that is well defined and at most 2^{n-1} by a result of Jeroslow [19]). Thus, this result may be rephrased as saying that any function in S_n has threshold number at most $n - 1$. That there are functions in S_n quite different from threshold functions has been indicated by Schmitt [42], who showed that there is a function in S_n with threshold number at least $\lfloor n/2 \rfloor$ (whereas, of course, any function in T_n has threshold number 1). (He also showed, however, that there is some function of threshold number 2 that is not in S_n.)

Further differences between the spiking neuron and the threshold (and polynomial threshold) unit emerge when the threshold order of computable functions is considered [42]. Whereas the threshold order of any function in T_n is 1, there are functions in S_n with threshold order $n^{1/3}/4^{1/3}$. This shows, additionally, that the functions in S_n cannot be computed by a polynomial threshold unit of any fixed degree. (Schmitt also shows that some Boolean function of threshold order 2 is not in S_n.)

A Boolean function is a μ-DNF function if it has a DNF formula in which each variable appears, either negated or not, at most once. Maass and Schmitt [29] showed that any μ-DNF function can be computed by a spiking neuron and that, by contrast, there are μ-DNF functions that cannot be computed by a linear threshold unit.

13.4 Expressive Power of Neural Networks

13.4.1 Growth Function and VC-Dimension

Definitions

We saw in Chapter 6 that the sample complexity of learning can be quantified fairly precisely by the Vapnik-Chervonenkis dimension, or VC-dimension, of the class of functions being used as hypotheses. In this sense, the VC-dimension is a useful way of measuring the expressive power of a set of functions. In this section, we examine the growth functions and VC-dimensions of the sets of functions computable by certain types of neural networks.

We start by recalling what is meant by the growth function and VC-dimension. Suppose that H is a set of functions from a set X to $\{0, 1\}$. (So, when H is the set of functions computable by an n-input neural network, X will be $\mathbb{R}^n$ or – the case of most interest to us – $\{0, 1\}^n$.) For a finite subset S of X, $\Pi_H(S)$

denotes the cardinality of the set of functions $H|_S$ obtained by restricting H to domain S. For $m \in \mathbb{N}$, $\Pi_H(m)$ is defined to be the maximum of $\Pi_H(S)$ over all subsets of cardinality m. For all m, $\Pi_H(m) \leq 2^m$. The *Vapnik-Chervonenkis dimension* [8, 47] of H is defined as the maximum m (possibly infinite, in the case where the domain is $\mathbb{R}^n$) such that $\Pi_H(m) = 2^m$. We say that $S \subseteq X$ is *shattered* by H, or that H *shatters* S, if $\Pi_H(S) = 2^{|S|}$: that is, if H gives all possible classifications of the points of S. Thus, S is shattered by H if for each subset R of S, there is some function f_R in H such that for $1 \leq i \leq m$, $f_R(x_i) = 1 \iff x_i \in R$.

The neural networks considered in this chapter compute a class of $\{0, 1\}$-valued functions. So we can define the VC-dimension of a neural network to be the VC-dimension of the set of functions computable by the network. For a network $\mathcal{N}$ with n inputs, we denote by VCdim$(\mathcal{N}, \mathbb{R}^n)$ the VC-dimension of the class of functions from $\mathbb{R}^n \to \{0, 1\}^n$ computed by $\mathcal{N}$, and VCdim$(\mathcal{N}, \{0, 1\}^n)$ denotes the VC-dimension of the corresponding class of Boolean functions. In this chapter, we are primarily interested in the VC-dimension of the set of Boolean functions computable by the network.

VC-Dimension and Linear Dimension

There is a useful connection between linear (vector-space) dimension and the VC-dimension [11]. Suppose $\mathcal{V}$ is a set of real functions defined on some set X. For $f, g \in \mathcal{V}$ and $\lambda \in \mathbb{R}$, we can form the function $f + g : X \to \mathbb{R}$ by *pointwise addition* and the function $\lambda f : X \to \mathbb{R}$ by *pointwise scalar multiplication*, as follows:

$$(f + g)(x) = f(x) + g(x), \quad (\lambda f)(x) = \lambda f(x), \quad (x \in X).$$

If $\mathcal{V}$ is closed under these operations, then it is a vector space of functions. Then, in $\mathcal{V}$, we say that the set $\{f_1, f_2, \ldots, f_k\}$ of functions is *linearly dependent* if there are constants λ_i $(1 \leq i \leq k)$, not all zero, such that, *for all* $x \in X$,

$$\lambda_1 f_1(x) + \lambda_2 f_2(x) + \cdots + \lambda_k f_k(x) = 0;$$

that is, some nontrivial linear combination of the functions is the zero function on X. The vector space $\mathcal{V}$ is finite-dimensional, of linear dimension d, if the maximum cardinality of a linearly independent set of functions in $\mathcal{V}$ is d. We have the following result, due to Dudley [11].

Theorem 13.10. *Let $\mathcal{V}$ be a real vector space of real-valued functions defined on a set X. Suppose that $\mathcal{V}$ has linear dimension d. For any $f \in \mathcal{V}$, define the $\{0, 1\}$-valued function f_+ on X by*

$$f_+(x) = \begin{cases} 1 & \text{if } f(x) \geq 0 \\ 0 & \text{if } f(x) < 0, \end{cases}$$

and let $\mathrm{sgn}(\mathcal{V}) = \{f_+ : f \in \mathcal{V}\}$. *Then the VC-dimension of* $\mathrm{sgn}(\mathcal{V})$ *is d.*

13.4.2 Linear Threshold Units

The VC-dimension of the single linear threshold unit can be bounded fairly directly using Theorem 13.10, for the class of functions in question is precisely $\text{sgn}(\mathcal{V})$, where $\mathcal{V}$ is the set of affine functions, of the form $x \mapsto w_0 + w_1x_1 + w_2x_2 + \cdots + w_nx_n$, for some constants $w_0, w_1, \ldots, w_n$. The set $\mathcal{V}$ is easily seen to be a vector space of linear dimension $n+1$ and hence has VC-dimension $n+1$. In fact, this is so even if we restrict the inputs to $\{0,1\}^n$.

Theorem 13.11. *The VC-dimension of T_n, the set of (Boolean) threshold functions, is $n+1$.*

Proof. We have already indicated why the VC-dimension of the set of functions computable by the threshold unit on $\mathbb{R}^n$ is $n+1$. Certainly, we must therefore have $\text{VCdim}(T_n)$ no more than $n+1$, because T_n is a restriction to the Boolean domain, $\{0,1\}^n$, of this class. So the result will follow if we show that $\text{VCdim}(T_n) \geq n+1$. We do this by proving that a particular subset of $\{0,1\}^n$ of cardinality $n+1$ is shattered by the T_n. Let $\mathbf{0}$ denote the all-0 vector and, for $1 \leq i \leq n$, let e_i be the point with a 1 in the ith coordinate and all other coordinates 0. We shall show that T_n shatters the set $S = \{\mathbf{0}, e_1, e_2, \ldots, e_n\}$. Suppose that R is any subset of S and, for $i = 1, 2, \ldots, n$, let

$$w_i = \begin{cases} 1, & \text{if } e_i \in R \\ -1, & \text{if } e_i \notin R, \end{cases}$$

and let

$$\theta = \begin{cases} -1/2, & \text{if } \mathbf{0} \in R \\ 1/2, & \text{if } \mathbf{0} \notin R. \end{cases}$$

Then it is straightforward to verify that if h_R is the function computed by the threshold unit when the weight-vector is $w = (w_1, w_2, \ldots, w_n)$ and the threshold is θ, then the set of positive examples of h_R in S is precisely R. Therefore S is shattered by T_n and, consequently, $\text{VCdim}(T_n) \geq n+1$. ■

Theorem 13.2 shows that

$$\Pi_H(m) \leq 2\sum_{k=0}^{n} \binom{m-1}{k}.$$

This upper bound is easily seen to equal 2^m for $m \leq n+1$ and to be less than 2^m for $m > n+1$, from which it follows also that $\text{VCdim}(T_n) \leq n+1$.

13.4.3 Polynomial Threshold Units

We now bound the VC-dimension of the class $\mathcal{P}(n, m)$ of (Boolean) polynomial threshold functions of degree m. Recall that such a function takes the form

$$f(x) = \text{sgn}\left(\sum_{S \in [n]^m} w_S x_S\right),$$

for some $w_S \in \mathbb{R}$, where $[n]^{(m)}$ is the set of subsets of at most m elements from $\{1, 2, \ldots, n\}$ and x_S denotes the product of the x_i for $i \in S$. For $m \leq n$, let $C(n, m) = \{x_S : S \in [n]^{(m)}\}$, regarded as a set of real functions on domain $\{0, 1\}^n$.

Theorem 13.12. *For all n, m with $m \leq n$, $C(n, m)$ is a linearly independent set of real functions defined on $\{0, 1\}^n$.*

Proof. Let $n \geq 1$ and suppose that for some constants c_S and for all $x \in \{0, 1\}^n$,

$$A(x) = \sum_{S \in [n]^{(m)}} c_S x_S = 0.$$

Set x to be the all-0 vector to deduce that $c_\emptyset = 0$. Let $1 \leq k \leq m$ and assume, inductively, that $c_S = 0$ for all $S \subseteq [n]$ with $|S| < k$. Let $S \subseteq [n]$ with $|S| = k$. Setting $x_i = 1$ if $i \in S$ and $x_j = 0$ if $j \notin S$, we deduce that $A(x) = c_S = 0$. Thus for all S of cardinality k, $c_S = 0$. Hence $c_S = 0$ for all S, and the functions are linearly independent. ■

It is therefore immediate, from Theorem 13.10, that for all n, m with $m \leq n$,

$$\text{VCdim}(\mathcal{P}(n, m)) = \sum_{i=0}^{m} \binom{n}{i}.$$

A similar analysis will determine the VC-dimension of the set of functions from $\mathbb{R}^n$ to $\{0, 1\}$ computable by the polynomial threshold unit. In this case, the set of functions of degree m is $\text{sgn}(\mathcal{V})$, where $\mathcal{V}$ is the vector space with basis x_S for all $\binom{n+m}{m}$ multisets of at most m elements from $[n]$. So the VC-dimension in this case is $\binom{n+m}{m}$. To sum up, we have the following results.

Theorem 13.13. *Let $\mathcal{N}$ be a single n-input polynomial threshold unit of degree m. Then, for all $m, n \in \mathbb{N}$,*

$$\text{VCdim}(\mathcal{N}, \mathbb{R}^n) = \binom{n+m}{m},$$

and for all n, m with $m \leq n$,

$$\text{VCdim}(\mathcal{N}, \{0, 1\}^n) = \sum_{i=0}^{m} \binom{n}{i}.$$

We have only considered single polynomial threshold units here, but clearly networks could be formed from such units. The VC-dimensions of the resulting networks (and of further generalizations of these types of network) have been bounded by Schmitt [43].

13.4.4 Linear Threshold Networks

We now provide a bound on the VC-dimension of feed-forward linear threshold networks. This is a slightly weaker version (with an easier proof, from [25]) of a bound due to Baum and Haussler [6].

Theorem 13.14. *Suppose that $\mathcal{N}$ is a feed-forward linear threshold network having a total of W variable weights and thresholds, and n inputs. Then*

$$\text{VCdim}(\mathcal{N}, \{0, 1\}^n) \leq \text{VCdim}(\mathcal{N}, \mathbb{R}^n) < 6W \log_2 W.$$

Proof. Let $X = \mathbb{R}^n$ and suppose that $S \subseteq X$ is of cardinality m. Let H be the set of functions computable by $\mathcal{N}$. We bound the growth function of H by bounding $\Pi_H(S)$ independently of S. Denote by N the number of computation units (that is, the number of linear threshold neurons) in the network. Because the network is a feed-forward network, the computation units may be labeled with the integers $1, 2, \ldots, N$ so that if the output of threshold unit i feeds into unit j, then $i < j$. Consider any particular threshold unit, i. Denote the in-degree of i by d_i. By Theorem 13.2, the number of different ways in which a set of m points can be classified by unit i is at most $2 \sum_{k=0}^{d_i} \binom{m-1}{k}$, which is certainly at most m^{d_i+2} for $m \geq d_i + 1$. It follows that, (if $m > \max_i d_i + 1$) the number of classifications $\Pi_H(S)$ of S by the network is bounded by

$$m^{d_1+2} m^{d_2+2} \ldots m^{d_N+2},$$

which, because $W = d_1 + d_2 + \cdots + d_N + N$, the total number of weights and thresholds, is at most m^{W+N}. Because $W \geq N$ (there being a threshold for each threshold unit), this is at most m^{2W}. Now, $m^{2W} < 2^m$ if $m = 6W \log_2 W$, from which it follows that the VC-dimension of the network is less than $6W \log_2 W$. ■

With more careful bounding [6], the VC-dimension can be bounded above by $2W \log_2(eN)$. This upper bound is of order $W \ln N$, where W is the total number of weights and thresholds: that is, the total number of variable parameters determining the state of the network. We have already seen that the linear threshold unit on n inputs has VC-dimension $n + 1$, which is exactly the number of variable parameters in this case. We have also seen that for polynomial threshold functions, the VC-dimension is precisely the number of variable parameters. The question therefore arises as to whether the $O(W \ln N)$ bound is of the best possible order or whether in this case, too, the VC-dimension is of order W. In fact, the $\ln N$ factor cannot, in general, be removed, as the following result of Maass [24] shows.

Theorem 13.15. *Let W be any positive integer greater than* 32. *Then there is a three-layer feed-forward linear threshold network $\mathcal{N}_W$ with at most W weights and thresholds, for which* $\text{VCdim}(\mathcal{N}_W, \{0, 1\}^n) > (1/132) W \log_2 (N/16)$, *where N is the number of computation units.*

13.4.5 Sigmoid Networks

Bounding the VC-dimension of sigmoid networks is rather more complicated than for threshold networks. Finiteness of the VC-dimension of sigmoid networks was established by Macintyre and Sontag [30], using deep results from logic. This in itself was a significant result, because it had previously been shown by Sontag [46] that for small networks with activation function other than the standard sigmoid, σ, the VC-dimension could be infinite. (Indeed, there are activation functions very similar to the standard sigmoid, for which the VC-dimension of a very small corresponding network is infinite; see [2].) The following result of Karpinski and Macintyre [20, 21] provides concrete, polynomial, upper bounds on the VC-dimension of sigmoid networks. The proof, which is quite involved, brings together techniques from logic and algebraic geometry. (See also [2].)

Theorem 13.16. *Let $\mathcal{N}$ be a feed-forward sigmoid network. Suppose that the total number of adjustable weights and thresholds is W, that the number of inputs is n, and that there are N computation units. Then*

$$\text{VCdim}(\mathcal{N}, \mathbb{R}^n) \leq (WN)^2 + 11WN\log_2(18WN^2).$$

Note that this bound, which is $O(W^4)$, is polynomial in the number of weights and thresholds. It has been shown by Koiran and Sontag [22, 23] that the VC-dimension of (unbounded depth) sigmoid nets can be as large as W^2. There is thus, generally, a strict separation between the VC-dimension of threshold networks (with VC-dimension $O(W \ln W)$) and sigmoid networks.

13.4.6 Spiking Neurons

Recall that S_n denotes the set of Boolean functions computable by the n-input spiking neuron. Maass and Schmitt [29] obtained the following result on the VC-dimension of a single spiking neuron.

Theorem 13.17. *The VC-dimension of S_n, the set of functions computable by a spiking neuron on $\{0, 1\}^n$, is $O(n \log n)$ and $\Omega(n \log n)$. Moreover, this lower bound is also true for a subclass of S_n in which the weights and threshold are kept fixed and only the delay parameters are varied.*

Thus, although, as noted earlier, there are not significantly many more Boolean functions computable by the spiking neuron than by the threshold unit, the spiking neuron is considerably more expressive. For, the VC-dimension of the linear threshold unit is $n + 1$, whereas the VC-dimension of the spiking neuron is $\Theta(n \log n)$.

The VC-dimension of feed-forward networks of spiking neurons has also been investigated. Maass and Schmitt [29] proved that for each n, there is a network of this type with $O(n)$ edges, for which, varying only the delays (and leaving weights and threshold fixed), the resulting class of functions defined on $\{0, 1\}^n$ has VC-dimension $\Omega(n^2)$. Note that, here, only the delays are variable and there are $O(n)$ of these. Thus the VC-dimension is at least quadratic in the number of variable delays. Recall that any linear threshold network has VC-dimension $O(W \log W)$ where W is the number of weights and thresholds. Thus, the VC-dimension of a network of spiking neurons with a given number of adjustable delays (and weights and threshold fixed) can be larger than the VC-dimension of a threshold network with the same number of adjustable weights and thresholds. Maass and Schmitt also showed that any such network has a VC-dimension (over inputs from $\mathbb{R}^n$) that is at most $O(E^2)$, where E is the number of edges in the underlying digraph (that is, the number of network connections). So the VC-dimension is at most quadratic in the number of variable weights, thresholds, and delays, and their lower bound is asymptotically tight.

References

[1] M. Anthony. *Discrete mathematics of neural networks: Selected topics*. SIAM Monographs on Discrete Mathematics and Applications DT08. SIAM, Philadelphia, 2001.

[2] M. Anthony and P. L. Bartlett. *Neural Network Learning: Theoretical Foundations*. Cambridge University Press, Cambridge, UK, 1999.

[3] M. Anthony. Classification by polynomial surfaces. *Discrete Applied Mathematics* 61, pp. 91–103, 1996.

[4] M. Anthony, G. Brightwell, and J. Shawe-Taylor. On specifying Boolean functions by labelled examples. *Discrete Applied Mathematics* 61, pp. 1–25, 1995.

[5] M. Anthony and J. Shawe-Taylor. Using the perceptron algorithm to find consistent hypotheses. *Combinatorics, Probability and Computing* 4(2), pp. 385–7, 1993.

[6] E. Baum and D. Haussler. What size net gives valid generalization? *Neural Computation* 1(1), pp. 151–60, 1989.

[7] C. M. Bishop. *Neural Networks for Pattern Recognition*. Oxford University Press, Oxford, 1995.

[8] A. Blumer, A. Ehrenfeucht, D. Haussler, and M. K. Warmuth. Learnability and the Vapnik-Chervonenkis dimension. *Journal of the ACM* 36(4), pp. 929–65, 1989.

[9] T. M. Cover. Geometrical and statistical properties of systems of linear inequalities with applications in pattern recognition. *IEEE Transactions on Electronic Computers* EC-14, pp. 326–34, 1965.

[10] Y. Crama and P. L. Hammer. *Boolean Functions: Theory, Algorithms, and Applications*, Cambridge University Press, Cambridge, UK, 2010.

[11] R. M. Dudley. Central limit theorems for empirical measures. Annals of Probability 6, pp. 899–929, 1978.

[12] S. A. Goldman and M. J. Kearns. On the complexity of teaching. *Journal of Computer and System Sciences* 50(1), pp. 20–31, 1995.

[13] C. Gotsman. On Boolean functions, polynomials and algebraic threshold functions. Technical report TR-89-18, Department of Computer Science, Hebrew University, 1989.

[14] P. L. Hammer, T. Ibaraki, and U. N. Peled. Threshold numbers and threshold completions. *Annals of Discrete Mathematics* 11, pp. 125–45, 1981.

[15] S. E. Hampson and D. J. Volper. Linear function neurons: structure and training. Biological Cybernetics 53, pp. 203–17, 1986.

[16] J. Håstad. On the size of weights for threshold gates. *SIAM Journal on Discrete Mathematics* 7(3), pp. 484–92, 1994.

[17] J. Hertz, A. Krogh, and R. G. Palmer. *Introduction to the Theory of Neural Computation*. Addison-Wesley, Redwood City, CA, 1991.

[18] S.-T. Hu. Threshold Logic, University of California Press, Berkeley, 1965.

[19] R. G. Jeroslow. On defining sets of vertices of the hypercube by linear inequalities. *Discrete Mathematics* 11, pp. 119–24, 1975.

[20] M. Karpinski and A. J. Macintyre. Polynomial bounds for VC dimension of sigmoidal neural networks. In *Proceedings of the 27th Annual ACM Symposium on Theory of Computing*, pp. 200–8. ACM Press, New York, 1995.

[21] M. Karpinski and A. J. Macintyre. Polynomial Bounds for VC dimension of sigmoidal and general Pfaffian neural networks. *Journal of Computer and System Sciences*, 54, pp. 169–76, 1997.

[22] P. Koiran and E. D. Sontag. Neural networks with quadratic VC dimension. *Journal of Computer and System Sciences* 54(1), pp. 190–8, 1997.
[23] P. Koiran and E. D. Sontag. Neural networks with quadratic VC dimension. In *Advances in Neural Information Processing Systems 8*, D. S. Touretzky, M. C. Mozer, and M. E. Hasselmo, eds., pp. 197–203. MIT Press, Cambridge, MA, 1996.
[24] W. Maass. Bounds for the computational power and learning complexity of analog neural nets. In *Proceedings of the 25th Annual ACM Symposium on the Theory of Computing*, pp. 335–44. ACM Press, New York, 1993.
[25] W. Maass. On the complexity of learning in feedforward neural nets. Manuscript, Institute for Theoretical Computer Science, Technische Universitaet Graz, 1993.
[26] W. Maass. On the relevance of time in neural computation and learning. In *Proceedings of the 8th International Workshop on Algorithmic Learning Theory, ALT '97*, M. Li and A. Maruoka, eds. Springer, Berlin, 1997.
[27] W. Maass. Networks of spiking neurons: The third generation of neural network models. *Neural Networks* 10, pp. 1659–71, 1997.
[28] W. Maass. Lower bounds for the computational power of networks of spiking neurons. *Neural Computation* 8, pp. 1–40, 1996.
[29] W. Maass and M. Schmitt. On the complexity of learning for spiking neurons with temporal coding. Information and Computation 153, pp. 26–46, 1999.
[30] A. Macintyre and E. D. Sontag. Finiteness results for sigmoidal "neural" networks. In *Proceedings of the 25th Annual ACM Symposium on the Theory of Computing*, pp. 325–34, ACM Press, New York, 1993.
[31] M. Minsky and S. Papert. Perceptrons. MIT Press, Cambridge, MA, 1969. (Expanded edition 1988.)
[32] S. Muroga. Lower bounds of the number of threshold functions and a maximum weight. *IEEE Transactions on Electronic Computers* 14, pp. 136–48, 1965.
[33] S. Muroga, Threshold Logic and Its Applications. Wiley, New York, 1971.
[34] E. I. Nechiporuk. The synthesis of networks from threshold elements. *Problemy Kibernetiki* 11, pp. 49–62, 1964.
[35] N. J. Nilsson. Learning Machines. McGraw-Hill, New York, 1965.
[36] R. O'Donnell and R. A. Servedio. Extremal properties of polynomial threshold functions. *Journal of Computer and System Sciences* 74, pp. 298–312, 2008. [Earlier version in Eighteenth Annual Conference on Computational Complexity (CCC)], pp. 3–12, 2003.
[37] I. Parberry. *Circuit Complexity and Neural Networks*. Foundations of Computing Series. MIT Press, Cambridge, MA, 1994.
[38] D. E. Rumelhart, G. E. Hinton, and J. L. McClelland. A general framework for parallel distributed processing. In D. E. Rumelhart and J. L. McClelland, eds. *Parallel Distributed Processing: Explorations in the Microstructure of Cognition*, Vol. 1. MIT Press, Cambridge, MA, 1987.
[39] M. Saks. Slicing the hypercube. In Surveys in Combinatorics, K. Walker, ed. Cambridge University Press, Cambridge, UK, 1993.
[40] A. Samorodnitsky. Unpublished manuscript (personal communication), 2000.
[41] L. Schläfli. *Gesammelte Mathematische Abhandlungen I*. Birkhäuser, Basel, 1950.
[42] M. Schmitt. On computing Boolean functions by a spiking neuron. *Annals of Mathematics and Artificial Intelligence* 24, pp. 181–91, 1998.
[43] M. Schmitt. On the complexity of computing and learning with multiplicative neural networks. *Neural Computation* 14(2), pp. 241–301, 2002.
[44] K.-Y. Siu, V. Roychowdhury, and T. Kailath. *Discrete Neural Computation: A Theoretical Foundation*. Prentice Hall Information and System Sciences Series. Prentice Hall, Englewood Cliffs, NJ, 1995.
[45] W. Softky and C. Koch. Single-cell models. In The *Handbook of Brain Theory and Neural Networks*, M. A. Arbib, ed., pp. 879–84. MIT Press, Cambridge, MA, 1995.
[46] E. D. Sontag. Feedforward nets for interpolation and classification. *Journal of Computer and System Sciences* 45, pp. 20–48, 1992.

[47] V. N. Vapnik and A. Y. Chervonenkis. On the uniform convergence of relative frequencies of events to their probabilities. *Theory of Probability and Its Applications* 16(2), pp. 264–80, 1971.

[48] C. Wang and A. C. Williams. The threshold order of a Boolean function. Discrete Applied Mathematics 31, pp. 51–69, 1991.

[49] R. J. Williams. The logic of activation functions. In D. E. Rumelhart and J. L. McClelland, eds. *Parallel Distributed Processing: Explorations in the Microstructure of Cognition*, Vol. 1. MIT Press, Cambridge, MA, 1987.

[50] Y. A. Zuev. Asymptotics of the logarithm of the number of threshold functions of the algebra of logic. *Soviet Mathematics Doklady* 39, pp. 512–13, 1989.

14

Decision Lists and Related Classes of Boolean Functions

Martin Anthony

14.1 Introduction

Decision lists provide a useful way of representing Boolean functions. Just as every Boolean function can be represented by a DNF formula, we shall see that every Boolean function can also be represented by a decision list. This representation is sometimes more compact. By placing restrictions on the type of decision list considered, we obtain some interesting subclasses of Boolean functions. As we shall see, these subclasses have some interesting properties, and certain algorithmic questions can be settled for them.

14.2 Decision Lists

14.2.1 Definition

Suppose that K is any set of Boolean functions on $\{0, 1\}^n$, n fixed. We shall usually suppose (for the sake of simplicity) that K contains the identically-1 function **1**. A Boolean function f with the same domain as K is said to be a *decision list based on* K if it can be evaluated as follows. Given an example y, we first evaluate $f_1(y)$ for some fixed $f_1 \in K$. If $f_1(y) = 1$, we assign a fixed value c_1 (either 0 or 1) to $f(y)$; if not, we evaluate $f_2(y)$ for a fixed $f_2 \in K$, and if $f_2(y) = 1$ we set $f(y) = c_2$, otherwise we evaluate $f_3(y)$, and so on. If y fails to satisfy any f_i then $f(y)$ is given the default value 0.

The evaluation of a decision list f can therefore be thought of as a sequence of `'if then else'` commands:

`if` $f_1(y) = 1$ `then` set $f(y) = c_1$
    `else if` $f_2(y) = 1$ `then` set $f(y) = c_2$
        ...
        ...
        `else if` $f_r(y) = 1$ `then` set $f(y) = c_r$
            `else` set $f(y) = 0$.

We define $DL(K)$, the class of *decision lists based on* K, to be the set of finite sequences

$$f = (f_1, c_1),\ (f_2, c_2), \ldots,\ (f_r, c_r),$$

such that $f_i \in K$ and $c_i \in \{0, 1\}$ for $1 \leq i \leq r$. The values of f are defined by $f(y) = c_j$ where $j = \min\{i \mid f_i(y) = 1\}$, or 0 if there are no j such that $f_j(y) = 1$. We call each f_j a *test* (or, following Krause [15], a *query*) and the pair (f_j, c_j) a *term* of the decision list.

Decision lists were introduced by Rivest [19], where a key concern was to develop a learning algorithm for them. (This is discussed later in this chapter.)

Note that we do not always draw a strong distinction between a decision list as a Boolean function, and a decision list as a *representation* of a Boolean function. Strictly speaking, of course, a decision list is a representation of a Boolean function, just as a DNF formula is.

There is no loss of generality in requiring that all tests f_i occurring in a decision list be distinct. This observation enables us to obtain the following bound:

$$|DL(K)| \leq \sum_{i=0}^{|K|} \binom{|K|}{i} i!2^i \leq 2^{|K|}|K|! \sum_{i=0}^{|K|} \frac{1}{(|K| - i)!} = 2^{|K|}|K|! \sum_{i=0}^{|K|} \frac{1}{i!}$$
$$\leq e2^{|K|}|K|!.$$

(Each decision list of length i is formed by choosing i functions of K, in a particular order, and assigning a $c_i \in \{0, 1\}$ to each.)

Example 14.1. *Suppose that* $K = M_{3,2}$, *the set of monomials (that is, simple conjunctions or terms) of length at most two in three Boolean variables. Consider the decision list*

$$(x_2, 1),\ (x_1\bar{x}_3, 0),\ (\bar{x}_1, 1).$$

Those examples for which x_2 *is satisfied are assigned the value 1: these are 010, 011, 110, 111. Next the remaining examples for which* $x_1\bar{x}_3$ *is satisfied are assigned the value 0: the only such example is 100. Finally, the remaining examples for which* $\bar{x}_1$ *is satisfied are assigned the value 1: this accounts for 000 and 001, leaving only the example 101 which is assigned the value 0.*

Suppose that $K = M_{n,k}$ is the set of monomials (or terms) consisting of at most k literals, so each test is a simple conjunction of degree at most k. Then, following Rivest [19], $DL(K)$ is usually denoted k-DL, and we call such decision lists *k-decision lists*. (Krause [15] defines a k-decision list to be one in which each test involves at most k variables, but such a decision list can be transformed into one in which the tests are in $M_{n,k}$.)

Note that when $K = M_{n,k}$, we have $|K| \leq (2n)^k$, and hence

$$|k\text{-DL}| = |DL(M_{n,k})| \leq 2^{(2n)^k} e\left((2n)^k\right)! = 2^{O(n^k \log n)},$$

for fixed k.

Later, we will want to consider K being the set of threshold functions, but unless it is explicitly said so, K will either be $M_{n,k}$ for some fixed k, or simply M_n, the set of all monomials on n variables.

14.2.2 Special Types of Decision List

By restricting the types of decision list considered, subclasses of decision list arise. We have already witnessed this, when we considered the k-decision lists; these arise from restricting the degree of each test in the decision list to be no more than k. Later in this chapter we look more closely at the very special case in which $k = 1$.

Rather than restrict the degree of each individual test, a restriction could be placed on the total number of terms in the decision list: an r-term decision list is (a function representable by) a decision list in which the number of terms is no more than r. We can also combine structural restrictions on decision lists. For example, the r-term k-DLs are those k-DLs in which the number of terms is at most r. (So, here, there is a restriction both on the number of terms and on the degree of each test.)

As observed by Guijarro, Lavín, and Raghavan [10], any function representable by a decision list with few terms (but possibly high degree) is also representable by one with terms of low degree (but possibly many terms).

Theorem 14.1. ([10].) *Suppose that $f : \{0, 1\}^n \to \{0, 1\}$ is (representable by) a decision list with r terms. Then f is also (representable by) an r-decision list.*

Proof. Suppose the decision list

$$f = (f_1, c_1), (f_2, c_2), \ldots, (f_r, c_r)$$

is given, where, as we may assume, $c_r = 1$. We construct an r-decision list g representing the same function. First, for each choice of a literal from each of $f_1, \ldots, f_r$, we have a term of g of the form $(T, 0)$, where T is the conjunction of the negations of these literals. We take all such terms, in any order, as the first set of terms of g. Note that each such T is of degree no more than r. For example, if f is the decision list

$$(x_2, 1),\ (\bar{x}_1 x_4, 0),\ (x_1 x_3, 1),$$

then take the first four terms of g, in any order, to be

$$(\bar{x}_2 x_1 \bar{x}_1, 0), (\bar{x}_2 \bar{x}_4 \bar{x}_1, 0), (\bar{x}_2 x_1 \bar{x}_3, 0), (\bar{x}_2 \bar{x}_4 \bar{x}_3, 0).$$

(In fact, in this case the first term is vacuous and can be deleted.) Next, we consider in turn each of the terms (f_i, c_i) as i is decreased from r to 2. Corresponding to f_i, we form terms (T, c_i) of g by choosing a literal from each preceding term $f_1, \ldots, f_{i-1}$ in f, and forming the conjunction of the negations of these. (Note that these tests have degree no more than $i - 1$, which is less than r.) If $c_1 = 0$ we are then done; otherwise, we add $(\mathbf{1}, 1)$ as the last term of g. For example, for the example decision list, a final suitable g is as follows:

$$(\bar{x}_2\bar{x}_4\bar{x}_1, 0), (\bar{x}_2x_1\bar{x}_3, 0), (\bar{x}_2\bar{x}_4\bar{x}_3, 0), (\bar{x}_2x_1, 1), (\bar{x}_2\bar{x}_4, 1), (\bar{x}_2, 0), (\mathbf{1}, 1).$$

(We have deleted the redundant first term created earlier). ■

Additionally, Bshouty [4] has obtained the following result, which gives a better dependence on r (at the expense of some dependence on n).

Theorem 14.2. ([4].) *Suppose that $f : \{0, 1\}^n \to \{0, 1\}$ is (representable by) a decision list with r terms. Then f is also (representable by) a k-decision list, where $k = 4\sqrt{n \ln n \ln(r + 1)}$.*

The proof of Bshouty's theorem (which is omitted here) relates decision lists to certain type of decision tree (in which the leaves are k-decision lists).

Another special type of decision list arises when the tests are required to be positive monomials. Guijarro et al. [10] refer to such decision lists as *monotone term decision lists*. Here, we instead call them *positive-term decision lists*. Note that, even though the tests are positive, the overall function computed by the decision list need not be positive. Guijarro *et al.* studied a number of aspects of this class, and, as we shall see later, discovered that there are efficient algorithms for many problems associated with the class.

14.3 Representation of Boolean Functions as Decision Lists

14.3.1 DNF and Decision List Representations

We first state a relationship between k-decision lists and special classes of Boolean functions. For any $k \leq n$, k-DNF denotes the Boolean functions that have a DNF formula in which each term is of degree at most k; dually, k-CNF denotes the set of functions having a CNF representation in which each clause involves at most k literals.

The following result, noted by Rivest [19], is easily obtained.

Theorem 14.3. *Let K be any set of Boolean functions. The disjunction of any set of functions in K is a decision list based on K. Explicitly, $f_1 \vee f_2 \vee \cdots \vee f_r$ is represented by the decision list*

$$(f_1, 1), (f_2, 1), \ldots, (f_r, 1).$$

It follows immediately from this that any k-DNF function, as the disjunction of terms of degree at most k, is also a k-decision list. It is easy to see, however, that, for $0 < k < n$, there are k-decision lists that are not k-DNF functions. For example, the function f with formula $x_1x_2 \ldots x_n$ is certainly not a k-DNF with $k < n$. (This is quite apparent: it has just one true point, whereas any k-DNF, $k < n$, has at least $2^{n-k} \geq 2$ true points, because any one of its terms does.) However, f can be expressed as the following 1-decision list:

$$(\bar{x}_1, 0), (\bar{x}_2, 0), \ldots, (\bar{x}_n, 0), (\mathbf{1}, 1).$$

If K contains the identically-1 function $\mathbf{1}$, then $DL(K)$ is closed under complementation, because

$$(f_1, 1 - c_1), (f_2, 1 - c_2), \ldots (f_r, 1 - c_r), (\mathbf{1}, 1)$$

is the complement of

$$(f_1, c_1), (f_2, c_2), \ldots, (f_r, c_r).$$

Thus, in contrast to the DNF and CNF representations, the decision list representations of a function and its negation are of the same size (but, possibly, for a difference of one additional term).

In particular, because k-CNF functions are the complements of k-DNF functions, and because k-DL contains k-DNF, we have that k-DL also contains k-CNF. (Here, by identifying it as a monomial with no literals, we also use the fact that the identically-1 function belongs to $K = M_{n,k}$.) In fact, we have the following result, due to Rivest [19]. The fact that the containment is strict demonstrates that the k-decision list representation is, in fact, more powerful than k-DNF and k-CNF representations.

Theorem 14.4. ([19].) *For $n \geq 2$ and $k \geq 1$,*

$$k\text{-DNF} \cup k\text{-CNF} \subseteq k\text{-DL},$$

and the containment is strict for $n > 2$ and $0 < k < n$.

Proof. The containment has been established in the arguments just given. It remains to show that the containment is strict. We use the fact that if a Boolean function has a prime implicant of degree s, then it does not belong to k-DNF for any $k < s$. We deal first with the case $k = 1$. Consider the function f represented by the 1-decision list

$$(x_1, 0), (x_2, 1), (x_3, 1).$$

Because f has $\bar{x}_1 x_2$ as a prime implicant, it is not in 1-DNF, and because the complement $\bar{f}$ has $\bar{x}_2 \bar{x}_3$ as an implicant, $\bar{f}$ is not in 1-DNF, and hence f is not in 1-CNF. Now suppose $n > 2$, that $1 < k < n$, and that k is odd. (The case of even k can be treated similarly.) Let g_k denote the parity function on the first k variables $x_1, x_2, \ldots, x_k$ (that is, the exclusive-or of them), regarded as a function on $\{0, 1\}^n$. Through its DNF representation, g_k can be represented by a k-decision list, ℓ. Consider the function $f : \{0, 1\}^n \to \{0, 1\}$ represented by the k-decision list

$$(\bar{x}_1 x_{k+1}, 0), (x_1 x_{k+1}, 1), \ell.$$

Then, f has degree-$(k + 1)$ prime implicant $x_1 \bar{x}_2 \bar{x}_3 \ldots \bar{x}_k \bar{x}_{k+1}$ and so is not in k-DNF. Furthermore, the complement of f is not in k-DNF (and hence f

is not in k-CNF) because the complement has degree-$(k+1)$ prime implicant $\bar{x}_1\bar{x}_2\ldots\bar{x}_k\bar{x}_{k+1}$. ■

As an interesting example of decision list representation, consider the function f_n, for even n, with formula

$$f_n = x_1 \wedge (x_2 \vee (x_3 \wedge (\cdots (x_{n-1} \wedge x_n) \cdots).$$

For example,

$$f_6 = x_1 \wedge (x_2 \vee (x_3 \wedge (x_4 \vee (x_5 \wedge x_6)))).$$

The function f_n is difficult to represent in DNF or CNF form: it is easily seen that both f_n and its complement have prime implicants of degree at least $n/2$, so f_n cannot be represented by a k-DNF or a k-CNF formula when $k < n/2$. However, f_n is easily represented by a 1-decision list, for

$$f_n = (\bar{x}_1, 0), (x_2, 1), (\bar{x}_3, 0), \ldots, (x_{n-2}, 1), (\bar{x}_{n-1}, 0), (\bar{x}_n, 0), (\mathbf{1}, 1),$$

where we regard **1** as being represented by the empty monomial (with no literals). Note that this example almost demonstrates the strictness part of Theorem 14.4 and is, moreover, not just in k-DL, but in 1-DL.

The inclusion of k-DNF in k-DL shows that any Boolean function can be represented by a decision list in which the tests are of sufficiently high degree; that is, n-DL is the set of all Boolean functions on $\{0, 1\}^n$. So, in this sense, the decision list representation is universal. Simple counting will show that most Boolean functions need decision lists of high degree. (As we see later, using a result on polynomial threshold functions, almost every Boolean function needs tests of degree at least $k \geq \lfloor n/2 \rfloor$ in any decision list representation.)

Explicit examples can also be given of functions that have reasonably compact representations as decision lists, but that have very long DNF or CNF representations.

Let COMP_n denote the function from $\{0, 1\}^{2n} \to \{0, 1\}$ given by

$$\text{COMP}_n(x, y) = 1 \text{ if and only if } \langle x \rangle > \langle y \rangle,$$

where, for $x \in \{0, 1\}^n$, $\langle x \rangle$ is the integer whose binary representation is x. (Thus, $\langle x \rangle = \sum_{i=1}^{n} 2^{n-i} x_i$.) Then, as noted in [15], for example, COMP_n can be represented by a short decision list but has no polynomial-sized DNF or CNF formula. (It is not in the circuit complexity class AC_2^0.) A 2-DL representation of COMP_n is

$$(\bar{x}_1 y_1, 0), (x_1 \bar{y}_1, 1), (\bar{x}_2 y_2, 0), (x_2 \bar{y}_2, 1), \ldots, (\bar{x}_n y_n, 0), (x_n \bar{y}_n, 1).$$

14.3.2 Universality of Positive-Term Decision Lists

As we have seen, every Boolean function has a decision list representation, obtained from any DNF representation of the function. If, moreover, the function is positive and we use a positive DNF representation, then we can obtain a positive-term decision list representation. But it is fairly easy to see that any Boolean function (positive or not), can be represented by a positive-term decision list, as the following result of Guijarro et al. [10] establishes.

Theorem 14.5. ([10].) *Every Boolean function can be represented as a positive-term decision list.*

Proof. Suppose f is a given Boolean function, and construct a positive-term decision list as follows. For $y \in \{0, 1\}^n$, let T_y be the (positive) conjunction of all literals x_i which are true on y (meaning that $y_i = 1$). Then, the first term is $(T_{11\ldots1}, f(11\ldots1))$; that is, $(x_1x_2\ldots x_n, f(11\ldots1))$. The next $n-1$ terms consist (in any order) of all terms $(T_y, f(y))$ for those y having $n-1$ ones. We continue in this way, dealing with the y of weight $n-2$, and so on, until we reach $y = 00\ldots0$, so the final term of the decision list (if it is needed) is $(\mathbf{1}, f(00\ldots0))$. Clearly, f is a positive-term decision list, and it computes f. ■

Note that the construction in this proof will result in a very long decision list (2^n terms) of high degree (n). Some subsequent reduction in size of the list may be possible, but the question naturally arises as to how long a positive-term decision list representation of a Boolean function is compared to, say, the standard DNF and CNF representations. Guijarro et al. observed that there is a sequences of functions (f_n) such that the shortest length of a positive-term decision list representing f_n is exponential in the number of terms in the shortest DNF representation of f_n, and that a corresponding result also holds for CNF representations. On the other hand, Guijarro et al. (invoking results of Ehrenfeucht and Haussler [6] and Fredman and Khachiyan [9]) also prove the following.

Theorem 14.6. ([10].) *For a Boolean function f, let $|f|_{dnf}$ and $|f|_{cnf}$ denote, respectively, the number of terms (clauses) in the shortest DNF (CNF) formulas representing f, and let $|f|$ be the larger of these two measures. Then there is a positive-term decision list representing f and having no more than $|f|^{\log^2 |f|}$ terms.*

14.4 Algorithmic Aspects of Decision Lists

14.4.1 Membership Problems

Recall that, for a class $\mathcal{C}$ of functions, the (*functional*) membership problem for $\mathcal{C}$ is as follows.

MEMBERSHIP ($\mathcal{C}$)
Instance: A DNF formula ϕ.
Question: Does the function f represented by ϕ belong to $\mathcal{C}$?

A useful general result due to Hegedűs and Megiddo [13] shows that MEMBERSHIP ($\mathcal{C}$) is NP-complete for all classes $\mathcal{C}$ satisfying certain properties. The following definition describes these properties.

Definition 14.1. *Suppose that $\mathcal{C} = \{\mathcal{C}_n\}$ is a class of Boolean functions. (Here, $\mathcal{C}_n$ maps from $\{0, 1\}^n$.) We say that a class $\mathcal{C}$* has the projection property *if*

(i) *$\mathcal{C}$ is closed under restrictions (so, all restrictions of a function in $\mathcal{C}$ also belong to $\mathcal{C}$);*
(ii) *for every $n \in \mathbb{N}$, the identically-1 function $\mathbf{1}$ belongs to $\mathcal{C}_n$;*
(iii) *there exists $k \in \mathbb{N}$ such that some Boolean function on $\{0, 1\}^k$ does* not *belong to $\mathcal{C}_k$.*

Then, we have the following result.

Theorem 14.7. ([13].) *Suppose that $\mathcal{C}$ is a class of Boolean functions having the projection property. Then* MEMBERSHIP*($\mathcal{C}$) is NP-hard.*

For any $k < n$, the class k-DL is easily seen to have the projection property, and hence the membership problem MEMBERSHIP(k-DL) is NP-hard. (In fact, it is co-NP-complete; see [8].) (The same is true of positive-term decision lists.)

However, in the case of 1-decision lists, Eiter, Ibaraki, and Makino [8] have established that the membership problem can be solved if the DNF is positive.

Theorem 14.8. *Deciding whether a* positive *DNF formula ϕ represents a function in 1-DL can be solved in polynomial time.*

14.4.2 Extending and Learning Decision Lists

It is often important to determine, given a set T of points labeled "true" and a set F of points labeled "false," whether there is a Boolean function in a certain class $\mathcal{C}$ that is an *extension* of the *partially defined Boolean function pdBf*(T, F). In other words, the problem is to determine whether there is $f \in \mathcal{C}$ such that $f(y) = 0$ for $y \in F$ and $f(y) = 1$ for $y \in T$. In many applications, it is also important to produce such an extension efficiently. Rivest [19] developed the following *learning algorithm*. This takes as input a sequence (or sample) $\mathbf{s} = ((y_1, b_1), \ldots, (y_m, b_m))$ of labeled points of $\{0, 1\}^n$ (where $y_i \in \{0, 1\}^n$ and $b_i \in \{0, 1\}$) and finds, if one exists, a decision list in $DL(K)$ that is an extension of the sample (or, if we like, of the pdBf corresponding to the sample). (We use an ordered sample of labeled points rather than simply two subsets T, F because this is more natural in many learning contexts. However, it should be noted that the decision list output by the following algorithm does not depend on the ordering of the labeled points in the sample.)

The extension (or learning) algorithm may be described as follows. At each step in the construction of the required decision list, some of the examples have been deleted, while others remain. The procedure is to run through K seeking a function $g \in K$ and a bit c such that, for all remaining points y_i, whenever $g(y_i) = 1$, then b_i is the constant Boolean value c. The pair (g, c) is then selected as the next term of the sequence defining the decision list, and all the examples satisfying g are deleted. The procedure is repeated until all the examples in $\mathbf{s}$ have been deleted.

Let $\{g_1, g_2, \ldots, g_p\}$ be an enumeration of K. The algorithm is as follows.

```
set I = {1, 2, ..., m};  j:= 1;
repeat
if  for all i ∈ I,  g_j(y_i) = 1 implies b_i = c
          then begin select (g_j, c);
                     delete from I all i for which g_j(y_i) = 1;
                     j:= 1 end
          else j:= j+1;
until  I = ∅
```

Note, of course, that the way in which K is enumerated has an effect on the decision list output by the algorithm: different orderings of the functions in K potentially lead to different decision lists, as the following example demonstrates.

Example 14.2. *Suppose we want to find a 2-decision list on five variables that is an extension of the pdBf described by the following labelled sample:*

$$\begin{aligned}\mathbf{s} &= ((y_1, b_1), (y_2, b_2), (y_3, b_3), (y_4, b_4), (y_5, b_5), (y_6, b_6)) \\ &= ((10000, 0), (01110, 0), (11000, 0), (10101, 1), (01100, 1), (10111, 1)).\end{aligned}$$

Suppose we list the functions of $K = M_{5,2}$ in lexicographic (or dictionary) order, based on the ordering $x_1, x_2, x_3, x_4, x_5, \bar{x}_1, \bar{x}_2, \bar{x}_3, \bar{x}_4, \bar{x}_5$ of the literals. The first few entries in the list are the identically-1 monomial $\mathbf{1}$, x_1, x_1x_2, x_1x_3. Then the algorithm operates as follows. To begin, we select the first item from the list that satisfies the required conditions. Clearly $\mathbf{1}$ will not do, because all the examples satisfy it but some have label 0 and some have label 1. x_1 also will not do, because (for example) y_1 and y_4 both satisfy it but $b_1 \neq b_4$. However, x_1x_2 is satisfied only by y_3, and $b_3 = 0$, so we select $(x_1x_2, 0)$ as the first term in the decision list and delete y_3. The subsequent steps are as follows:

- *Select $(x_1x_3, 1)$, delete y_4 and y_6;*
- *Select $(x_1, 0)$, delete y_1;*
- *Select $(x_2x_4, 0)$, delete y_2;*
- *Select $(\mathbf{1}, 1)$, delete y_5.*

In this case the output decision list is therefore

$$(x_1x_2, 0),\ (x_1x_3, 1),\ (x_1, 0),\ (\bar{x}_1x_4, 0),\ (\mathbf{1}, 1).$$

Suppose instead that the functions in $K = M_{5,2}$ were enumerated instead in such a way that the smaller monomials came first: that is, we started with $\mathbf{1}$, then listed (in lexicographic order) all monomials of length 1, then all of length 2:

$$\mathbf{1},\ x_1,\ x_2,\ x_3,\ x_4,\ x_5,\ \bar{x}_1,\ \bar{x}_2,\ \bar{x}_3,\ \bar{x}_4,\ \bar{x}_5,\ x_1x_2,\ x_1x_3,\ x_1x_4, \ldots.$$

In this case, the decision list output by the algorithm is the 1-decision list

$$(x_5, 1), (x_1, 0), (x_4, 0), (x_2, 1).$$

This is simpler, in the sense that it is a 1-decision list rather than a 2-decision list.

It is easily verified that both decision lists are indeed extensions of the pdBf given by the sample.

Correctness of the algorithm in general is easily established [1, 19].

Theorem 14.9. *Suppose that K is a set of Boolean functions containing the identically-1 function,* **1**. *Suppose that* **s** *is a sample of labeled elements of* $\{0, 1\}^n$. *If there is an extension in $DL(K)$ of the partially defined Boolean function described by* **s**, *then the foregoing algorithm will produce such an extension.*

The extension algorithm is also efficient: when $K = M_{n,k}$, so that the class $DL(K)$ is k-DL, then the algorithm is easily seen to have running time $O(mn^{k+1})$ for fixed k. There is no guarantee, however, that the algorithm will necessarily produce a decision list that is nearly as short as it could be, as Hancock *et al.* [12] have shown.

Eiter, Ibaraki, and Makino [8] considered 1-decision lists in some detail and were able to find an improved (that is, faster) extension algorithm. In fact, rather than a running time of $O(mn^2)$, their algorithm has linear running time $O(mn)$. They also develop a *polynomial delay algorithm* for generating *all* 1-decision list extensions of a pdBf (when such extensions exist). Such an algorithm outputs, one by one and without repetition, all extensions of the pdBf in such a way that the running time between outputs is polynomial in nm. (This is a reasonable requirement: to ask for the *total* time to generate all extensions to be polynomial in mn would be inappropriate because the number of extensions may well be exponential in mn.)

14.4.3 Algorithmic Issues for Positive-Term Decision Lists

For positive-term decision lists, Guijarro et al. [10] have shown that a number of problems that are intractable for general decision lists become efficiently solvable.

The following result is useful. It shows that the question of whether two positive-term decision lists are equivalent (that is, represent the same function) can be resolved quite simply.

Theorem 14.10. ([10].) *There is an algorithm with running time $O(n(p+q)pq)$ that, given two positive-term decision lists on n variables, involving p and q tests, decides whether or not the decision lists represent the same function.*

Proof. Suppose the decision lists are L_1, L_2. For any test T from L_1 and any test S from L_2, let $y(T, S) \in \{0, 1\}^n$ have 1s in precisely those positions i for which x_i appears in T or S. Let f_1, f_2 be the functions computed by L_1 and L_2. Suppose $f_1 \neq f_2$ and let z be such that $f_1(z) \neq f_2(z)$. Let (T, c) be the first term of L_1 "activated" by z (meaning that T is the first test in L_1 passed by z). The first term of L_2 activated by z is then necessarily of the form $(S, 1-c)$. Then, as can easily be seen, $f_1(y(T, S)) = c$ and $f_2(y(T, S)) = 1 - c$. Thus there exist tests T and S of

L_1, L_2, respectively, such that $f_1(y(T, S)) \neq f_2(y(T, S))$. Conversely, of course, if such T, S exist, then $f_1 \neq f_2$. So it suffices to check, for each of the pq pairs (T, S), whether $f_1(y(T, S)) = f_2(y(T, S))$, and, for each pair, this can be done in $O(n(p + q))$ time. ■

As Guijarro et al. observe, the existence of efficient algorithms for other problems follows from this result. For example, to check whether a term is redundant (unnecessary), simply remove it and check the equivalence of the new decision list with the original. Furthermore, to check whether a positive-term decision list represents a positive function, one can remove from the list all redundant terms (using the redundancy-checking method just described) and check whether the remaining terms all have label 1; they do so if and only if the function is positive.

14.5 Properties of 1-Decision Lists

14.5.1 Threshold Functions and **1***-Decision Lists*

Recall that a Boolean function t defined on $\{0, 1\}^n$ is a *threshold function* if there are $w \in \mathbb{R}^n$ and $\theta \in \mathbb{R}$ such that

$$t(x) = \begin{cases} 1 & \text{if } \langle w, x \rangle \geq \theta \\ 0 & \text{if } \langle w, x \rangle < \theta, \end{cases}$$

where $\langle w, x \rangle = w^T x$ is the standard inner product of w and x (see, e.g., [5]). Given such w and θ, we say that t is represented by $[w, \theta]$, and we write $t \leftarrow [w, \theta]$. The vector w is known as the weight-vector, and θ is known as the threshold. We denote the class of threshold functions on $\{0, 1\}^n$ by T_n. Note that any $t \in T_n$ will satisfy $t \leftarrow [w, \theta]$ for ranges of w and θ.

We have the following connection between 1-decision lists and threshold functions [7] (see also [2]).

Theorem 14.11. *Any 1-decision list is a threshold function.*

Proof. We prove this by induction on the number of terms in the decision list. Note that the identically-1 function $\mathbf{1}$ is regarded as a monomial of length 0. Suppose, for the base case of the induction, that a decision list has just one term and is of the form $(x_i, 1)$, or $(\bar{x}_i, 1)$, or $(\mathbf{1}, 1)$, where $\mathbf{1}$ is the identically-1 function. (Note that if it were of the form $(x_i, 0)$, $(\bar{x}_i, 0)$, or $(\mathbf{1}, 0)$, then, because a decision list outputs 0 by default, the term is redundant, and the decision list computes the identically-0 function, which is certainly a threshold function.) In the first case, the function may be represented as a threshold function by taking the weight-vector to be $(0, \ldots, 0, 2, 0, \ldots, 0)$, where the nonzero entry is in position i, and by taking the threshold to be 1. In the second case, we may take weight-vector $(0, \ldots, 0, -2, 0, \ldots, 0)$ and threshold -1. In the third case, the function is the identically-1 function, and we may take as weight-vector the all-0 vector and threshold 0. Assume, as the inductive hypothesis, that any decision list of length r is a threshold function, and suppose we have a decision list of length $r+1$,

$$f = (\ell_1, c_1), (\ell_2, c_2), \ldots, (\ell_{r+1}, c_{r+1}),$$

where each ℓ_i is a literal, possibly negated. We assume, without any loss of generality (for one can simply rename the variables or, equivalently, permute the entries of the weight vector), that $\ell_1 = x_1$ or $\bar{x}_1$. By the induction hypothesis, the decision list

$$(\ell_2, c_2), \ldots, (\ell_{r+1}, c_{r+1})$$

is a threshold function. Suppose it is represented by weight-vector $w = (w_1, \ldots, w_n)$ and threshold θ, and let $\|w\|_1 = \sum_{i=1}^{n} |w_n|$ be the 1-norm of w. There are four possibilities for (ℓ_1, c_1), as follows:

$$(x_1, 1), \quad (x_1, 0), \quad (\bar{x}_1, 1), \quad (\bar{x}_1, 0).$$

Denoting by e_1 the vector $(1, 0, \ldots, 0)$, and letting $M = \|w\|_1 + |\theta| + 1$, we claim that the decision list f is a threshold function represented by the weight-vector w' and threshold θ', where, respectively,

$$\begin{aligned} w' &= w + Me_1, & \theta' &= \theta, \\ w' &= w - Me_1, & \theta' &= \theta, \\ w' &= w - Me_1, & \theta' &= \theta - M, \\ w' &= w + Me_1, & \theta' &= \theta + M. \end{aligned}$$

This claim is easy to verify in each case. Consider, for example, the third case. For $x \in \{0, 1\}^n$,

$$\langle w', x \rangle = \langle w - Me_1, x \rangle = \langle w, x \rangle - Mx_1,$$

and therefore $\langle w', x \rangle \geq \theta' = \theta - M$ if and only if

$$\langle w, x \rangle - Mx_1 \geq \theta - M.$$

If $x_1 = 0$ (in which case the decision list outputs 1), this inequality becomes $\langle w, x \rangle \geq \theta - M$. Now, for any $x \in \{0, 1\}^n$, $-\|w\|_1 \leq \langle w, x \rangle \leq \|w\|_1$, and

$$\theta - M = \theta - (\|w\|_1 + |\theta| + 1) = -\|w\|_1 - 1 + (\theta - |\theta|) \leq -\|w\|_1 - 1 < -\|w\|_1,$$

so in this case the inequality is certainly satisfied, and the output of the threshold function is 1, equal to the output of the decision list. Now suppose that $x_1 = 1$. Then the inequality

$$\langle w, x \rangle - Mx_1 \geq \theta - M$$

becomes $\langle w, x \rangle - M \geq \theta - M$, which is $\langle w, x \rangle \geq \theta$. But, by the inductive assumption, the decision list

$$f' = (\ell_2, c_2), \ldots, (\ell_{r+1}, c_{r+1})$$

is a threshold function represented by the weight-vector w and threshold θ. So in this case, the output of the threshold function is 1 if and only if the output of decision list f' is 1, which is exactly how f calculates its output in this case. So we see that this representation is indeed correct. The other cases can be verified similarly. ■

14.5.2 Characterizations of 1-Decision Lists

Eiter et al. [8] obtain some results relating the class of 1-decision lists closely to other classes of Boolean function. To describe their results, we need to recall a few more definitions concerning Boolean functions (see [5] for additional information about these classes of functions).

For i between 1 and n, let $e_i \in \{0, 1\}^n$ have ith entry 1 and all other entries 0. Then, recall that a Boolean function is said to be 2-*monotonic* if for each pair i and j between 1 and n, *either*

$$\text{for all } x \text{ with } x_i = 0 \text{ and } x_j = 1,\ f(x) \leq f(x + e_i - e_j),$$

or

$$\text{for all } x \text{ with } x_i = 0 \text{ and } x_j = 1,\ f(x) \geq f(x + e_i - e_j).$$

It is easily seen that threshold functions are 2-monotonic. (However, there are 2-monotonic functions that are not threshold functions.)

A Boolean function is *read-once* if there is a Boolean formula representing f in which each variable appears at most once. (So, for each j, the formula contains either x_j at most once, or $\bar{x}_j$ at most once, but not both.)

A DNF formula is *Horn* if each term contains at most one negated literal, and a function is said to be Horn if it has a representation as a Horn DNF. Given a class H of Boolean functions, the *renaming closure* of H is the set of all Boolean functions obtained from H by replacing every occurrence of some literals by their complements. (For example, we might replace every x_1 by $\bar{x}_1$ and every $\bar{x}_1$ by x_1.)

Eiter et al. [8] obtained the following characterizations (among others).

Theorem 14.12. *The class 1-DL coincides with the following classes:*

- *The class of all 2-monotonic read-once functions;*
- *The class of all read-once threshold functions;*
- *The renaming closure of the class of functions f such that f and $\bar{f}$ are Horn.*

In particular, whereas Theorem 14.11 establishes that 1-decision lists are threshold functions, the theorem just given states that the 1-decision lists are precisely those threshold functions that are also read-once.

Eiter et al. show that these characterizations do not extend to k-DL for $k > 1$. They do, however, have one characterization of k-DL. Given a class H of Boolean functions, the class of *nested differences* of H is defined as the set of all Boolean functions of the form

$$h_1 \setminus (h_2 \setminus (\ldots (h_{l-1} \setminus h_l)))$$

for $h_1, h_2, \ldots, h_l \in H$. Eiter et al. prove that k-DL coincides with the nested differences of clauses with at most k-literals. (They show, too, that k-DL is the set of nested differences of k-CNF.)

For $k > 1$, k-decision lists are not necessarily threshold functions, but they can be described in terms of *polynomial threshold functions*, a convenient generalization of threshold functions that we met in Chapter 12 and Chapter 13. Theorem 14.11 shows that 1-decision lists are threshold functions. An analogous argument establishes the following.

Theorem 14.13. *Any k-decision list is a polynomial threshold function of degree k.*

It was noted in Chapter 13 that almost every Boolean function has threshold order at least $\lfloor n/2 \rfloor$. This, together with Theorem 14.13, means that almost every Boolean function will need $k \geq \lfloor n/2 \rfloor$ in order to be in k-DL.

14.6 Threshold Decision Lists

14.6.1 Definition and Geometrical Interpretation

We now consider the class of decision lists in which the individual tests are themselves threshold functions. We call such decision lists *threshold decision lists*, but they have also been called *neural* decision lists [16] and *linear* decision lists [21].

Suppose we are given the points in $\{0, 1\}^n$, each one labeled by the value of a Boolean function f. Of course, because there are very few threshold functions, it is unlikely that the true and false points of f can be separated by a hyperplane. One possible alternative is to try a separator of higher degree, representing f as a polynomial threshold function. Here we consider a different approach, in which we successively "chop off" like points (that is, points all possessing the same label). We first use a hyperplane to separate off a set of points all having the same classification (either all are true points or all are false points). These points are then removed from consideration and the procedure is iterated until no points remain. For simplicity, we assume that at each stage, no data point lies on the hyperplane. This procedure is similar in nature to one of Jeroslow [14], but at each stage in his procedure, only positive examples may be "chopped off" (not positive *or* negative). We give one example for illustration.

Example 14.3. *Suppose the function f is the parity function, so that the true points are precisely those with an odd number of 1s. We first find a hyperplane such that all points on one side of the plane are either positive or negative. It is clear that all we can do at this first stage is chop off one of the points, because the nearest neighbors* *of any given point have the opposite classification. Let us suppose that we decide to chop off the origin. We may take as the first hyperplane the plane with equation* $y_1 + y_2 + \cdots + y_n = 1/2$. *(Of course, there are infinitely many other choices of hyperplane that would achieve the same effect with respect to the data points.) We then ignore the origin and consider the remaining points. We can next chop off all neighbors of the origin, all the points that have precisely one entry equal to* 1. *All of these are positive points, and the hyperplane* $y_1 + y_2 + \cdots + y_n = 3/2$ *will separate them from the other points. These points are then deleted from consideration. We may continue in this manner. The procedure iterates n times, and at stage i in the procedure we chop off all data points having precisely* $(i-1)$ *1s, by using the hyperplane* $y_1 + y_2 + \cdots + y_n = i - 1/2$, *for example. (These hyperplanes are in fact all parallel, but this is not necessary.)*

Note that, by contrast, Jeroslow's method [14] (described earlier) requires 2^{n-1} iterations in this example, because at each stage it can only chop off one positive point.

We may regard the chopping procedure as deriving a representation of the function by a threshold decision list. If, at stage i of the procedure, the hyperplane with equation $\sum_{i=1}^{n} w_i y_i = \theta$ chops off true (false) points, and these lie on the side of the hyperplane with equation $\sum_{i=1}^{n} w_i y_i > \theta$, then we take as the ith term of the threshold decision list the pair $(f_i, 1)$ (respectively, $(f_i, 0)$), where $f_i \leftarrow [w, \theta]$; otherwise we take the ith term to be $(g_i, 1)$ (respectively, $(g_i, 0)$), where $g_i \leftarrow [-w, -\theta]$.

If one applies this construction to the series of hyperplanes resulting from the Jeroslow method, a restricted form of decision list results – one in which all terms are of the form $(f_i, 1)$. But, as we saw earlier, such a decision list is quite simply the *disjunction* $f_1 \vee f_2 \vee \ldots$. The problem of decomposing a function into the disjunction of threshold functions has also been considered by Hammer, Ibaraki, and Peled [11] and by Zuev and Lipkin [22]. Hammer et al. defined the *threshold number* of a Boolean function to be the minimum s such that f is a disjunction of s threshold functions, and they showed that there is a positive function with threshold number $\binom{n}{n/2}/n$. Zuev and Lipkin [22] showed that almost all positive functions have a threshold number of this order, and that almost all Boolean functions have a threshold number that is $\Omega(2^n/n)$ and $O(2^n \ln n/n)$.

The decision lists arising from the chopping procedure are more general than disjunctions of threshold functions, just as k-decision lists are more general than k-DNF. Such threshold decision lists may provide a more compact representation of the function. (That is, because fewer hyperplanes might be used, the decision list could be smaller.)

14.6.2 Algorithmics and Heuristics of the Chopping Procedure

The chopping procedure just described was in some ways merely a device to help us see that threshold decision lists have a fairly natural geometric interpretation.

Furthermore, because all points of $\{0, 1\}^n$ are labeled, it is clear that the method, if implemented, would generally be inefficient. However, if only some of the points are labeled, so that we have a partially defined Boolean function, then the chopping procedure might constitute a heuristic for building a threshold decision list extension of the pdBf. This was considered by Marchand and Golea [16]. (See also [17].) Marchand and Golea derive a greedy heuristic for constructing a sequence of "chops." This relies on an incremental heuristic for the NP-hard problem of finding at each stage a hyperplane that chops off as many remaining points as possible. Reports on the experimental performance of their method can be found in the papers cited.

14.7 Threshold Network Representations

We now show how we can make use of the chopping procedure to find a threshold network (the simplest type of artificial neural network) representing a given Boolean function. We use linear threshold networks having just one hidden layer. Such a network will consist of k "hidden nodes," each of which computes a threshold function of the n inputs. The (binary-valued) outputs of these hidden nodes are then used as the inputs to the output node, which calculates a threshold function of these. Thus, the neural network computes a threshold function of the outputs of the k threshold functions computed by the hidden nodes. If the threshold function computed by the output node is described by weight-vector $\beta \in \mathbb{R}^k$ and threshold ϕ, and the threshold function computed by hidden node i is $f_i \leftarrow [w^{(i)}, \theta^{(i)}]$, then the threshold network as a whole computes the function $f : \mathbb{R}^n \rightarrow \{0, 1\}$ given by

$$f(y) = 1 \Longleftrightarrow \sum_{i=1}^{k} \beta_i f_i(y) > \phi;$$

that is,

$$f(y_1 y_2 \dots y_n) = \operatorname{sgn}\left(\sum_{i=1}^{k} \beta_k \operatorname{sgn}\left(\sum_{j=1}^{n} w_j^{(i)} y_j - \theta^{(i)}\right) - \phi\right),$$

where $\operatorname{sgn}(x) = 1$ if $x > 0$ and $\operatorname{sgn}(x) = 0$ if $x < 0$.

It is well known that any Boolean function can be represented by a linear threshold network with one hidden layer (albeit with potentially a large number of nodes in the hidden layer). The standard way of doing so, based on the function's disjunctive normal form, was demonstrated in Chapter 13.

The threshold decision list representation of a function gives rise to a different method of representing Boolean functions by threshold networks.

We have seen that a 1-decision list is a threshold function and that a k-decision list is a polynomial threshold function of degree k. In an easy analog of this, we see that any threshold decision list is a threshold function of threshold functions. But a threshold function of threshold functions is nothing more than a two-layer threshold network of the type considered here. So, by representing a function by a threshold decision list and then representing this as a threshold function over the threshold functions in the decision list, we obtain another method for finding a threshold network representation of a Boolean function. It is clear that the resulting representation is in general different from the standard DNF-based one. For example, the standard representation of the parity function on $\{0, 1\}^n$ will require a neural network with 2^{n-1} hidden units, whereas the representation derived from the procedure described here will require only n hidden units.

Marchand and Golea [16] drew attention to (essentially) this link between threshold decision lists and threshold networks. (Their networks were, however, slightly different, in that they had connections between nodes in the hidden layer.)

14.8 Representational Power of Threshold Decision Lists

14.8.1 A Function with Long Threshold Decision List Representation

Turán and Vatan [21] gave a specific example of a function with a necessarily long threshold decision list representation. The inner-product modulo 2 function $\mathrm{IP}_2 : \{0, 1\}^{2n} \to \{0, 1\}$ is given by $\mathrm{IP}_2(x, y) = \bigoplus_{i=1}^{n} x_i y_i$, for $x, y \in \{0, 1\}^n$, where $\bigoplus$ denotes addition modulo 2. Turán and Vatan proved the following.

Theorem 14.14. *In any threshold decision list representation of* IP_2, *the number* s *of terms satisfies* $s \geq 2^{n/2} - 1$.

14.8.2 Multilevel Threshold Functions

We saw in the earlier example that the parity function can be represented by a threshold decision list with n terms. We also noted that the hyperplanes in that example were parallel. By demanding that the hyperplanes be parallel, we obtain a special subclass of threshold decision lists, known as the *multilevel threshold functions*. These have been considered in a number of papers, such as [3, 18, 20].

We define the class of s-level threshold functions to be the set of Boolean functions representable by a threshold decision list of length at most s and having the test hyperplanes parallel to each other.

Geometrically, a Boolean function is an s-level threshold function if there are s parallel hyperplanes with the property that the $s + 1$ regions defined by these hyperplanes each contain only true points or only false points. Equivalently (following Bohossian and Bruck [3]), f is an s-level threshold function if there is a weight-vector $w = (w_1, w_2, \ldots, w_n)$ such that

$$f(x) = F\left(\sum_{i=1}^{n} w_i x_i\right),$$

where the function $F : \mathbb{R} \to \{0, 1\}$ is piecewise constant with at most $s + 1$ pieces.

Bohossian and Bruck observed that any Boolean function is a 2^n-level threshold function, an appropriate weight-vector being $w = (2^{n-1}, 2^{n-2}, \ldots, 2, 1)$. For that reason, they paid particular attention to the question of whether a function can be computed by a multilevel threshold function where the number of levels is polynomial. A related question considered by Bohossian and Bruck is whether a function can be computed by such a function, with polynomial weights (in addition to the restriction that the number of levels be polynomial).

It was explained earlier that, through the chopping procedure, a threshold decision list and, subsequently, a threshold network could be produced representing a given Boolean function. The translation from threshold decision list to threshold network is established by an analog of Theorem 14.11. From the proof of that theorem, it emerges that the weights in the resulting threshold network are, necessarily, exponential in size. It is often useful to focus on networks of the same structure, which is to say, having one "hidden" layer, but that are restricted to have integer weights polynomial in n. (Any such network can, insofar as it is regarded as computing a Boolean function, be assumed to have integer weights: we can simply scale up rational weights appropriately; and there is never a need for irrational weights because the domain is discrete.) The class of functions that can be computed by threshold networks with one hidden layer (that is, of depth 2) is denoted LT_2, and the subset of those in which the (integer) weights can be polynomial in the number of inputs (or variables), n, is denoted $\hat{\mathrm{LT}}_2$. Let LTM denote the set of Boolean functions f (or, more precisely, the set of sequences of Boolean functions (f_n) where f maps from $\{0, 1\}^n$) that can be computed by a multilevel threshold function with a polynomial number of levels. Then the following inclusion is valid.

Theorem 14.15. ([3].) *Let* LTM *denote the set of Boolean functions realisable by multilevel threshold functions with a polynomial number of levels. Then* $\mathrm{LTM} \subseteq \hat{\mathrm{LT}}_2$.

Bohossian and Bruck also obtain "separation" results, which show that there are functions in $\hat{\mathrm{LT}}_2$ but not in LTM; and that there are functions that are in LTM, but that are not representable with polynomial-sized weights.

14.9 Conclusions

In this chapter, we have looked at decision lists, a powerful and versatile way of representing Boolean functions. Decision lists have a number of interesting properties. There are, moreover, efficient algorithms for certain problems associated with classes of decision list. Allowing decision lists to be based on threshold functions allows greater generality and draws connections with threshold networks. There are still many open problems concerning decision lists, particularly threshold decision lists.

References

[1] M. Anthony and N. L. Biggs. *Computational Learning Theory: An Introduction*, Cambridge University Press, Cambridge, UK, 1992.

[2] M. Anthony, G. Brightwell, and J. Shawe-Taylor. On specifying Boolean functions by labelled examples. *Discrete Applied Mathematics* 61, pp. 1–25, 1995.

[3] V. Bohossian and J. Bruck. Multiple threshold neural logic. In *Advances in Neural Information Processing, Vol. 10: NIPS 1997*, Michael Jordan, Michael Kearns, and Sara Solla, eds. MIT Press, Cambridge, MA, 1998.

[4] N. Bshouty. A subexponential exact learning algorithm for DNF using equivalence queries. *Information Processing Letters* 59(1), pp. 37–9, 1996.

[5] Y. Crama and P. L. Hammer, *Boolean Functions: Theory, Algorithms, and Applications*. Cambridge University Press, Cambridge, UK, 2010.

[6] A. Ehrenfeucht and D. Haussler. Learning decision trees from random examples. *Information and Computation* 82, pp. 231–46, 1989.

[7] A. Ehrenfeucht, D. Haussler, M. Kearns, and L. Valiant. A general lower bound on the number of examples needed for learning. *Information and Computation* 82, pp. 247–61, 1989.

[8] T. Eiter, T. Ibaraki, and K. Makino. Decision lists and related Boolean functions. *Theoretical Computer Science* 270(1–2), pp. 493–524, 2002.

[9] M. Fredman and L. Khachiyan. On the complexity of dualization of monotone disjunctive normal form. *Journal of Algorithms* 21, pp. 618–28, 1996.

[10] D. Guijarro, V. Lavín, and V. Raghavan. Monotone term decision lists. *Theoretical Computer Science* 256, pp. 549–75, 2001.

[11] P. L. Hammer, T. Ibaraki, and U. N. Peled. Threshold numbers and threshold completions. *Annals of Discrete Mathematics* 11, pp. 125–45, 1981.

[12] T. Hancock, T. Jiang, M. Li, and J. Tromp. Lower bounds on learning decision lists and trees. *Information and Computation* 126(2), pp. 114–22, 1996.

[13] T. Hegedűs and N. Megiddo. On the geometric separability of Boolean functions. *Discrete Applied Mathematics* 66, pp. 205–18, 1996.

[14] R. G. Jeroslow. On defining sets of vertices of the hypercube by linear inequalities. *Discrete Mathematics* 11, pp. 119–24, 1975.

[15] M. Krause. On the computational power of Boolean decision lists. In Proceedings of the 19th Annual Symposium of Theoretical Aspects of Computer Science (STACS), 2002. Springer Lecture Notes in Computer Science 2285, pp. 372–83. Springer, New York, 2002.

[16] M. Marchand and M. Golea. On learning simple neural concepts: From halfspace intersections to neural decision lists. *Network: Computation in Neural Systems* 4, pp. 67–85, 1993.

[17] M. Marchand, M. Golea and P. Ruján. A convergence theorem for sequential learning in two-layer perceptrons. *Europhysics Letters* 11, p. 487, 1990.

[18] S. Olafsson and Y. S. Abu-Mostafa. The capacity of multilevel threshold functions. *IEEE Transactions on Pattern Analysis and Machine Intelligence* 10(2), pp. 277–81, 1988.

[19] R. R. Rivest. Learning decision lists. Machine Learning 2(3), pp. 229–46, 1987.

[20] R. Takiyama. The separating capacity of a multi-threshold element. *IEEE Transactions on Pattern Analysis and Machine Intelligence* 7, pp. 112–16, 1985.

[21] G. Turán and F. Vatan. Linear decision lists and partitioning algorithms for the construction of neural networks. Foundations of Computational Mathematics: selected papers of a conference held at Rio de Janeiro, pp. 414–23. Springer, Berlin, 1997.

[22] A. Zuev and L. I. Lipkin. Estimating the efficiency of threshold representations of Boolean functions. *Cybernetics* 24, pp. 713–23, 1988. (Translated from Kibernetika (Kiev) 6, pp. 29–37, 1988.)

Part V

Applications in Engineering

15

Hardware Equivalence and Property Verification

J.-H. Roland Jiang and Tiziano Villa

15.1 Introduction to Formal Verification

15.1.1 The Problem of Verification

Synthesis and verification are two basic steps in designing a digital electronic system, which may involve both hardware and software components. Synthesis aims to produce an implementation that satisfies the specification while minimizing some cost objectives, such as circuit area, code size, timing, and power consumption. Verification deals with the certification that the synthesized component is correct.

In system design, hardware synthesis and verification are more developed than the software counterparts and will be our focus. The reason for this asymmetric development is threefold. First, hardware design automation is better driven by industrial needs; after all, hardware costs are more tangible. Second, the correctness and time-to-market criteria of hardware design are in general more stringent. As a result, hardware design requires rigorous design methodology and high automation. Third, hardware synthesis and verification admit simpler formulation and are better studied.

There are various types of hardware verification, according to design stages, methodologies, and objectives. By design stages, verification can be deployed in high-level design from specification, called *design verification*; during synthesis transformation, called *implementation verification*; or after circuit manufacturing, called *manufacture verification*.

Manufacture verification is also known as *testing*. There is a whole research and engineering community devoted to it. In hardware testing, we would like to know if some defects appear in a manufactured circuit by testing the conformance between it and its intended design. In this case, the circuit – usually a sequential circuit, modeled abstractly as a finite-state machine (FSM) – is treated as a black box. The temporal behavior of the FSM can only be observed through output responses to input stimuli. In contrast, in design and implementation verification, a state transition system is seen as a white box. Its transition relation is known beforehand, and thus not only the output sequences but also the state traces are known for given input sequences.

By methodologies, verification can be *informal* or *formal*. Informal verification shows the presence, but not the absence, of design errors; formal verification shows both the presence and absence. Verifying the correctness of the design itself can be performed in different contexts. A common design practice at different stages is *simulation*: that is, stimulating the circuit with input sequences to check if the actual output sequences correspond to the expected ones [28, 50, 59, 65, 90]. The confidence of the outcome depends on the view of the circuit (e.g., digital blocks, logic gates, or transistors) and the size of the simulated input space. For instance, it is reasonable to simulate an electronic circuit using a numerical simulator such as SPICE, to characterize the behavior of a circuit at the transistor level corresponding to a few logic gates. For register-transfer level (RTL) and gate-level circuits, there are logic simulators, but it is usually impossible to exercise the complete input space. Therefore simulation is commonly conceived as an informal approach to verification.

Since the 1980s, theoretical and practical breakthroughs have made possible *formal verification*, the automatic process of checking formally whether a design satisfies some requirements. "Formally" means that we establish that the circuit satisfies the requirements by a mathematical analysis of the system, exploring the full set of behaviors, which is equivalent to an exhaustive simulation. A variety of formalisms have been introduced and studied to support a rigorous analysis, and they are based on automata and logical calculus. Automata may be regular or ω-regular, according to whether we care about computed sequences of finite or infinite lengths.

A very important class of formal methods goes under the name of *model checking* [12], where a finite-state system is represented by a labeled state transition graph, the so-called Kripke structure, where labels of a state are the values of atomic propositions in that state (for example, the values of the registers). Properties of the system are expressed as formulas in temporal logic, of which the state transition system is to be a "model." So, model checking consists of traversing the graph of the transition system and of verifying that it satisfies the formula representing the property: that is, that the system is a model of the property. We do not discuss here infinite-state systems (software programs); there are, however, vibrant research efforts devoted to verification of such systems, too.

By objectives, from the point-of-view of the design flow, formal verification may be divided into *property checking* (is what I designed what I wanted?) and *equivalence checking* (is what I produced, at a certain design stage, the same as what I had at the previous design stage?).

The objective of equivalence checking is to ensure that, in a design developed through phases that refine previous stages, the refinement process preserves the original behavior. So equivalence checking is a sanity check to guarantee that we are maintaining the design integrity through the synthesis and optimization phases. In theory these transformations should be correct by construction; however, the same tools that perform them should be formally verified, and the process would be endless. So it is a good practice to check conformance of input and output of each synthesis phase.

In property verification, given a transition system, we test whether the informal description of the specification has been captured correctly by a formal description with a hardware description language (an HDL), by checking whether the system satisfies some property. Properties capture functionality, timing and temporal relations, safety (is it possible for a bad event to happen?), liveness (does an expected good event eventually take place?), and fairness (does every request eventually receive service?).

Safety properties are the most common to express. For instance, do the traffic light controllers of a crossroad make sure that intersecting road lines have mutually exclusive rights of the way? As an example of liveness, in communication systems with lossy channels, we make certain types of fairness assumptions about the loss of messages: that is, if a message is transmitted infinitely many times, then it will be received infinitely often. In such systems, in general, we can only prove liveness properties. Fairness assumptions are often imposed upon liveness properties, but are not needed for safety properties. In fact, the extent of the role played by liveness and fairness properties is debated. It was suggested that liveness property checking can be converted into safety property checking [79].

In this chapter we focus on safety properties because they are the most commonly used in hardware verification problems. Two important hardware verification problems, sequential equivalence checking and resettability verification, fall into this category. They assert that something bad never happens during the evolution of the system, and their violation can be detected by analyzing finite executions of the system. They can be posed as checking that some property p holds in the reachable states R of system S starting from the initial state(s). This checking justifies the crucial role played by *reachability analysis* in the verification of safety properties.

Notice that hardware designs may be specified hierarchically; this is also consistent with how a human designer operates. In order to formally verify a design, it must first be converted into a simpler "verifiable" format. The design is specified as a set of interacting systems; each has a finite number of configurations, called states. States and transition between states constitute FSMs. The entire system is an FSM, which can be obtained by composing the FSMs associated with each component. Hence, the first step in verification consists of obtaining a complete FSM description of the system. Given a present state (or current configuration), the next state (or successive configuration) of an FSM can be written as a function of its present state and inputs (transition function or transition relation).

15.1.2 Model Checking and Temporal Logics

In this subsection we briefly survey model checking of properties expressed in temporal logics. A temporal logic is used to express the ordering of events in time by means of operators that specify properties such as "property p will eventually hold." There are various versions of temporal logics, such as linear temporal logic

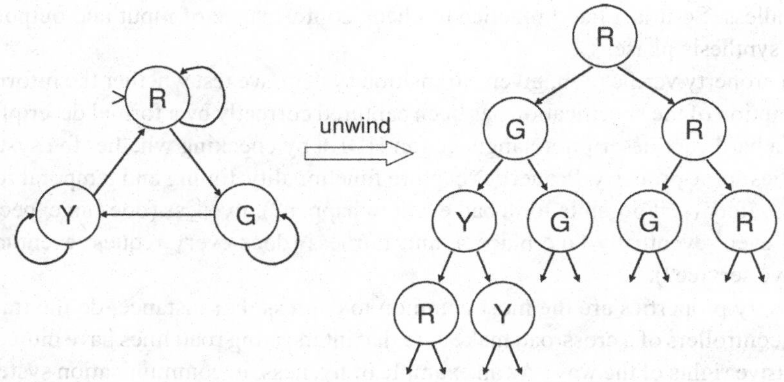

Figure 15.1. Unwinding of state transition graph.

(LTL), computation tree logic (CTL), and other variants. Different temporal logics can have different expressive powers. For instance, LTL and CTL are incomparable. That is, there are temporal formulas in LTL not expressible in CTL, and vice versa. Here we focus on CTL to exemplify temporal logics.

Computation trees are derived from state transition graphs. The graph structure is unwound into an infinite tree rooted at the initial state. Figure 15.1 shows an example of unwinding a graph into a tree. Paths in this tree represent all possible computations of the system being modeled. Formulas in CTL refer to the computation tree derived from the model. CTL is classified as a branching time logic (in contrast to a linear time logic, such as LTL) because it has operators that describe the branching structure of this tree.

Formulas in CTL are built from atomic propositions (where each proposition corresponds to a variable in the model), standard Boolean connectives of propositional logic (e.g., AND, OR, XOR, NOT), and temporal operators. Each temporal operator consists of two parts[1]: a path quantifier **A** or **E** followed by a temporal modality **F**, **G**, **X**, or **U**. All temporal operators are interpreted relative to an implicit "current state." There are in general many execution paths (sequences of state transitions) of the system starting at the current state. The path quantifier indicates whether the modality defines a property that should be true of all those possible paths (denoted by universal path quantifier **A**) or whether the property need only hold on some path (denoted by existential path quantifier **E**). The temporal modalities describe the ordering of events in time along an execution path and have the following intuitive meaning:

(i) **F** ϕ (reads "ϕ holds sometime in the future") is true on a path if there exists a state in the path where formula ϕ is true.

[1] A formula that contains any temporal modality of **F**, **G**, **X**, and **U** without an associated path quantifier **A** or **E** is *not* a legal CTL formula.

(ii) **G** ϕ (reads "ϕ holds globally") is true on a path if ϕ is true at every state in the path.
(iii) **X** ϕ (reads "ϕ holds in the next state") is true on a path if ϕ is true in the state reached immediately after the current state in the path.
(iv) ϕ **U** ψ (reads "ϕ holds until ψ holds," called "strong until"[2]) is true on a path if ψ is true in some state in the path, and ϕ holds in all preceding states.

The state of a system consists of the values stored in all registers. Each formula of the logic is either true or false in a given state; its truth is evaluated from the truth of its subformulas in a recursive fashion, until one reaches atomic propositions that are either true or false in a given state. A formula is satisfied by a system if it is true for all the initial states of the system. If the property does not hold, the model checker will produce a counterexample, that is, an execution path that witnesses the failure. An efficient algorithm for automatic model checking used also in VIS has been described by Clarke et al. [11]. The following table shows examples of evaluations of formulas on the computation tree of Figure 15.1.

Formula	T/F
EG (RED)	true
E (RED **U** GREEN)	true
AF (GREEN)	false

15.1.3 Tools and Representations

Tools for automatic formal verification have been based on various theoretical approaches, of which model checking and language containment have been the most useful in hardware verification. In temporal logic model checking, the properties to be checked are expressed as formulas in a temporal logic, and the system is expressed as a finite-state system. In particular, CTL model checking is a formalism pioneered by Clarke et al. [10] to verify whether a finite-state system satisfies properties expressed as formulas in CTL, a branching-time temporal logic. SMV [55], a system developed at Carnegie Mellon University, belongs to this class of tools. Symbolic trajectory evaluation [31] uses a restricted form of linear temporal logic to express properties as bounded-length sequences of circuit states. Verification is done by using an extension of symbolic simulation.

Certain properties are not expressible in CTL, but they can be expressed as ω-automata. The second approach, language containment, requires the description of the system and properties as ω-automata and verifies correctness by checking that the language of the system is contained in the language of the property. Note that certain types of CTL properties involving existential quantification are not

[2] "Weak until" is when ϕ holds forever, i.e., ψ is not required to hold at some state in the future.

expressible by ω-automata. COSPAN [49], a system developed at Bell Labs, offers language containment checking.

VIS [5, 6], a system developed jointly at the University of California at Berkeley and the University of Colorado at Boulder, emphasizes model checking, but it also offers to the user a limited form of language containment (language emptiness) checking.

The computation engines inside these tools represent and operate with discrete functions, which represent sets of elements rather than individual elements. Such techniques are often referred to as *implicit/symbolic computation*, in contrast to *explicit/enumerative computation*.

Implicit representations allow operating on sets of states, instead of individual states. A typical example is the computation of the set of reachable states from the initial states. Using binary decision diagrams (BDDs) as an underlying data structure, there are cases when Boolean formulas and operations on them can be performed efficiently even on instances of huge size, where explicit enumeration would fail. The reason is that the size of BDD representations is not linear in the size of the represented sets. A paper summarized this state of affair with the catchy title "Symbolic model checking: 10^{20} states and beyond" [8]. The other side of the coin is that this is not guaranteed to happen, because there may be no such good representation with BDDs, or if it exists, we may be unable to find it (the size of BDDs depends on the order of the support variables). Issues relating to BDD ordering, partition of the system representation, and new types of decision diagrams have been explored to represent compactly important sets of Boolean and discrete functions; see, for instance, Chapter 10 by Bollig, Sauerhoff, Sieling, and Wegener in this volume.

In recent years, the role played by BDD computations has been restricted because of their lack of robustness; at the same time, improvements of SAT solvers (which check the satisfiability of formulas with conjunctions of clauses) made them the preferred computational engines and allowed a more judicious balance in the usage of different function representations. We refer, for instance, to Crama and Hammer [17] or to Chapter 4 by Franco and to Chapter 5 by Hooker in this volume for additional information on SAT algorithms. In the rest of the chapter, we clarify the rationale behind these technical developments.

We can summarize the prevailing data representation and typical design size in successive generation of tools as follows:

(i) Explicit representation, 10^4 states, 1980s;
(ii) Implicit BDD-based representation, 10^{30} states, 1990s;
(iii) Implicit SAT-based representation, 10^{100} states, 2000s.

In this survey we cannot provide a comprehensive coverage of hardware verification. We focus on equivalence checking, which is a special case of safety property checking. Even in the restricted domain, we have to leave out some interesting topics and recent advances. An in-depth discussion on aspects of sequential equivalence checking will be given, because it is the most widely encountered case, still a subject of active theoretical and experimental investigation.

In Section 15.2 we introduce basic terminology and technical background. In Section 15.3 we discuss the computational complexity of hardware equivalence checking. In Section 15.4 we introduce combinational equivalence checking. In Section 15.5 we describe the basics of sequential equivalence checking, which is further expanded in Section 15.6, introducing bounded and unbounded model checking. Section 15.7 discusses how to bridge the gap between combinational and sequential equivalence checking. Section 15.8 covers resettability verification and other variants of equivalence checking. Section 15.9 concludes by mentioning advanced topics and active research areas.

15.2 Preliminaries

In the sequel, as a notational convention, a vector (or an ordered set) $\boldsymbol{v} = (v_1, \ldots, v_n)$ is specified by a boldface letter. The cardinality of $\boldsymbol{v}$ is denoted as $|\boldsymbol{v}|$, and the (unordered) set $\{v_1, \ldots, v_n\}$ is denoted as $\{\boldsymbol{v}\}$.

15.2.1 Characteristic Functions

Progress since the 1980s in expanding the capability of formal verification can be attributed to the introduction of the data structure called a reduced ordered binary decision diagram (ROBDD; [7] and Chapter 10 in this volume), which turned out to be effective in representing and manipulating Boolean functions. ROBDDs allowed computations on sets to be reduced to Boolean computations on their *characteristic functions*. A characteristic function, in our discussion, is a (total) function $\chi_A : U \to \mathbb{B}$, where U is a finite set and $\mathbb{B} = \{\textbf{false}, \textbf{true}\}$ or $\{0, 1\}$, such that $\chi_A(s) = 1$ if and only if $s \in A$. (In some applications, it can be extended to multiple-valued output, e.g., {**true**, **false**, **undefined**}.) When represented in computers, a characteristic function is often a Boolean formula where multiple-valued symbols are encoded and expressed in binary form. It serves as a predicate indicating a membership relation. That is, the function χ_A answers the query whether an element $e \in U$ is in $A \subseteq U$. Essentially, any finite set A can be represented by a characteristic function χ_A such that an element $e \in A$ if and only if $\chi_A(e) = \textbf{true}$. Thus set operations (e.g., intersection $\cap$, union $\cup$, and complement) are in effect Boolean operations (e.g., conjunction $\wedge$, disjunction $\vee$, and negation F, respectively) over characteristic functions (**false** and **true** can be seen as characteristic functions for the empty set $\emptyset$ and universal set U, respectively). In the sequel, we do not distinguish a set (respectively a set operation) and its corresponding characteristic function (respectively its corresponding Boolean operation).

By dealing with characteristic functions, we are able to manipulate sets of elements simultaneously rather than manipulate elements individually. For instance, the intersection of two sets A and B can be done by performing $\chi_A \wedge \chi_B$ rather than examining, for each element $e \in A$, whether e is in B as well. Such approaches, capable of manipulating sets of elements simultaneously, are known as (*implicit*) *symbolic algorithms* in contrast to the traditional (*explicit*) *enumerative algorithms*. Although BDDs and symbolic algorithms were once almost synonyms, more recently other data structures were developed as alternatives to BDDs. Notably, Boolean reasoning engines using SAT (satisfiability solving over conjunctive normal forms) and AIGs (And-Inverter Graphs), for instance, are gaining popularity in hardware verification.

15.2.2 State Transition Systems

We are concerned with a type of state transition system called a *finite-state machine* (FSM).

Definition 15.1. (Finite-State Machine.) *An FSM* $\mathcal{M}$ *is a tuple* $(Q, Q^0, \Sigma, \Omega, \boldsymbol{\delta}, \boldsymbol{\lambda})$*, where* Q *is a finite set of states,* $Q^0 \subseteq Q$ *is the set of initial states,* Σ *and* Ω *are the input and output alphabets, respectively, and* $\boldsymbol{\delta} : \Sigma \times Q \to Q$ *(respectively* $\boldsymbol{\lambda} : \Sigma \times Q \to \Omega$*) is the transition function (respectively output function).*

(In our discussion, we focus on Mealy-type FSMs, whose output valuations depend on both input and state variables. The results can be applied to Moore-type FSMs as well, whose output valuations depend only on state variables.) In the sequel, we assume that $\boldsymbol{\delta}$ and $\boldsymbol{\lambda}$ are total functions. That is, the FSMs to be discussed are deterministic and completely specified.

In modern hardware designs, state, input and output symbols are encoded by binary representations; transition and output functions are encoded by Boolean functional vectors. Let $\boldsymbol{s} = (s_1, \ldots, s_n)$ and $\boldsymbol{s}' = (s'_1, \ldots, s'_n)$ be the vectors of (binary-valued) current- and next-state variables, respectively; let $\boldsymbol{x} = (x_1, \ldots, x_k)$ and $\boldsymbol{y} = (y_1, \ldots, y_l)$ be the vectors of (binary-valued) input and output variables, respectively. That is, vector $\boldsymbol{s}$ encodes the states Q, $\boldsymbol{x}$ encodes Σ, and $\boldsymbol{y}$ encodes Ω. With notation $[\![\boldsymbol{v}]\!]$ representing the set of all possible valuations (or interpretations) on variables $\boldsymbol{v}$, we have $Q = [\![\boldsymbol{s}]\!]$, $\Sigma = [\![\boldsymbol{x}]\!]$ and $\Omega = [\![\boldsymbol{y}]\!]$. Also, let $\boldsymbol{\delta} = (\delta_1, \ldots, \delta_n)$ and $\boldsymbol{\lambda} = (\lambda_1, \ldots, \lambda_l)$. Then $\delta_i : [\![\boldsymbol{x}]\!] \times [\![\boldsymbol{s}]\!] \to [\![s'_i]\!]$ for $i = 1, \ldots, n$ and $\lambda_j : [\![\boldsymbol{x}]\!] \times [\![\boldsymbol{s}]\!] \to [\![y_j]\!]$ for $j = 1, \ldots, l$. Given two states q_0 and q_1 of an FSM with $q_1 = \boldsymbol{\delta}(\sigma, q_0)$ for some $\sigma \in \Sigma$, q_1 is called the *successor* of q_0 under σ, denoted as $Succ_\sigma(q_0)$, and q_0 is called the *predecessor* of q_1 under σ, denoted as $Pred_\sigma(q_1)$.

An FSM (in the six-tuple form) can be alternatively represented with a graph.

Definition 15.2. (State Transition Graph.) *Given an FSM* $\mathcal{M} = (Q, Q^0, \Sigma, \Omega, \boldsymbol{\delta}, \boldsymbol{\lambda})$*, it can be represented with a* state transition graph *(STG)* $G = (V, E)$*, where any state* $q \in Q$ *is modeled as a vertex* $q \in V$ *and any transition* $q_t = \boldsymbol{\delta}(\sigma, q_s)$ *is modeled as a directed edge* $(q_s, q_t) \in E$ *labeled '*σ/ω*' for* $\omega = \boldsymbol{\lambda}(\sigma, q)$*. Also, for initial states, their corresponding vertices are identified.*

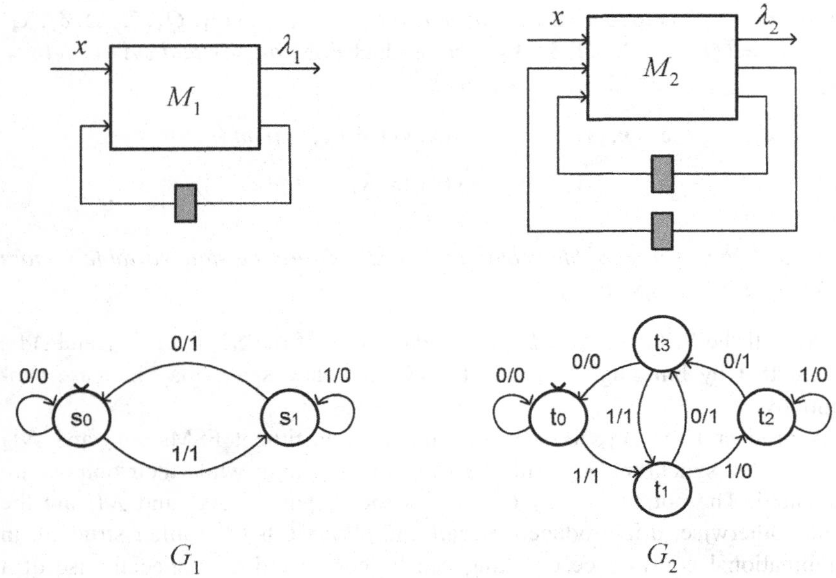

Figure 15.2. Two FSMs $\mathcal{M}_1$ and $\mathcal{M}_2$ (in schematic diagrams) with their corresponding STGs G_1 and G_2 are shown at left and right, respectively. In a schematic circuit diagram, detailed functions are omitted and only connections are shown, and stateholding elements are denoted with shaded boxes. An initial state in an STG is identified with an arrowhead: for example, state s_0 in G_1.

Example 15.1. *Figure 15.2 shows the schematic diagrams of two FSMs $\mathcal{M}_1$ and $\mathcal{M}_2$ and their respective corresponding STGs G_1 and G_2.*

Because the behavior of an FSM is described with transition functions, its state transitions are deterministic. That is, given any input assignment and any current state of the FSM, there is a unique next-state. A more general description is to use *transition relations* instead of *transition functions*. In this case, nondeterministic transitions can be handled. Essentially, a transition function can be converted to a transition relation. We may consider the Cartesian product $\Sigma \times Q$ as the new state space. By treating input variables as part of the state variables, we can write $T((\boldsymbol{x}, \boldsymbol{s}), (\boldsymbol{x}', \boldsymbol{s}')) = \bigwedge_i (s_i' \equiv \delta_i(\boldsymbol{x}, \boldsymbol{s}))$, where the equality sign "$\equiv$" means exclusive-nor, often denoted as $\overline{\oplus}$, in Boolean operation. Because input variables $\boldsymbol{x}'$ of the next time frame are unconstrained, we write $T(\boldsymbol{x}, \boldsymbol{s}, \boldsymbol{s}')$ in place of $T((\boldsymbol{x}, \boldsymbol{s}), (\boldsymbol{x}', \boldsymbol{s}'))$ for short. Consequently, FSMs can be described with transition relations in addition to transition functions. In the sequel, we may alternatively use the more general transition relation notation and describe an FSM with the tuple $(Q, Q^0, \Sigma, \Omega, T, \boldsymbol{\lambda})$.

To study the behavior between two FSMs subject to the same input stimuli, we define a product machine as follows.

Definition 15.3. (Product FSM.) *Given two FSMs* $\mathcal{M}_1 = (Q_1, Q_1^0, \Sigma, \Omega, \delta_1, \boldsymbol{\lambda}_1)$ *and* $\mathcal{M}_2 = (Q_2, Q_2^0, \Sigma, \Omega, \delta_2, \boldsymbol{\lambda}_2)$*, the* product FSM *of* $\mathcal{M}_1$ *and* $\mathcal{M}_2$ *is* $\mathcal{M}_\times = (Q_1 \times Q_2, Q_1^0 \times Q_2^0, \Sigma, \mathbb{B}, \boldsymbol{\delta}_\times, \boldsymbol{\lambda}_\times)$ *with*

$$\boldsymbol{\delta}_\times(\boldsymbol{x}, (\boldsymbol{s}_1, \boldsymbol{s}_2)) = (\boldsymbol{\delta}_1(\boldsymbol{x}, \boldsymbol{s}_1), \boldsymbol{\delta}_2(\boldsymbol{x}, \boldsymbol{s}_2)) \text{ and}$$
$$\boldsymbol{\lambda}_\times(\boldsymbol{x}, (\boldsymbol{s}_1, \boldsymbol{s}_2)) = \bigwedge_i (\lambda_{1i}(\boldsymbol{x}, \boldsymbol{s}_1) \equiv \lambda_{2i}(\boldsymbol{x}, \boldsymbol{s}_2)),$$

where $\boldsymbol{x}$ *is the input variable vector, and* $\boldsymbol{s}_1$ *and* $\boldsymbol{s}_2$ *are the state variable vectors of* $\mathcal{M}_1$ *and* $\mathcal{M}_2$*, respectively.*

We call the STG $G_\times$ of $\mathcal{M}_\times$ the *product STG* of the STGs of $\mathcal{M}_1$ and $\mathcal{M}_2$. Notice that by building the product FSM, the state space may increase substantially.

A product FSM $\mathcal{M}_\times$ is such that its two constituent FSMs $\mathcal{M}_1$ and $\mathcal{M}_2$ are executed synchronously with the same input stimuli while their outputs are compared. The output of $\mathcal{M}_\times$ equals 1 if the outputs of $\mathcal{M}_1$ and $\mathcal{M}_2$ are the same. Otherwise, 0 is produced. Recall and observe that the miter structure in combinational equivalence checking can be considered as a special case of a product FSM (with a single state).

Example 15.2. *Figure 15.3 shows the product FSM* $\mathcal{M}_\times$ *of the FSMs* $\mathcal{M}_1$ *and* $\mathcal{M}_2$ *in Figure 15.2, and the corresponding STG of* $\mathcal{M}_\times$*.*

To model the disjoint union of two FSMs, we state the following definition.

Definition 15.4. (Multiplexed FSM.) *Given two FSMs* $\mathcal{M}_1 = (Q_1, Q_1^0, \Sigma, \Omega, \delta_1, \boldsymbol{\lambda}_1)$ *and* $\mathcal{M}_2 = (Q_2, Q_2^0, \Sigma, \Omega, \delta_2, \boldsymbol{\lambda}_2)$*, the* multiplexed FSM *of* $\mathcal{M}_1$

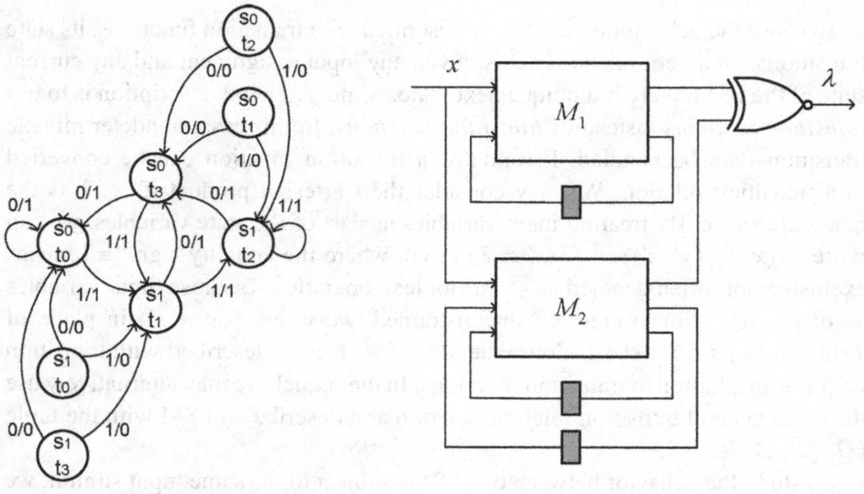

Figure 15.3. The product FSM of $\mathcal{M}_1$ and $\mathcal{M}_2$ of Figure 15.2, and its corresponding STG.

and $\mathcal{M}_2$ *is* $\mathcal{M}_{\uplus} = (Q_1 \uplus Q_2, Q_1^0 \uplus Q_2^0, \Sigma, \Omega, \boldsymbol{\delta}_{\uplus}, \boldsymbol{\lambda}_{\uplus})$ *with*

$$\boldsymbol{\delta}_{\uplus}(\boldsymbol{x}, (\boldsymbol{s}_\alpha, \alpha)) = \begin{cases} \boldsymbol{\delta}_1(\boldsymbol{x}, \boldsymbol{s}_1) & \text{if } \alpha = 1 \\ \boldsymbol{\delta}_2(\boldsymbol{x}, \boldsymbol{s}_2) & \text{if } \alpha = 2 \end{cases}, \text{ and}$$

$$\boldsymbol{\lambda}_{\uplus}(\boldsymbol{x}, (\boldsymbol{s}_\alpha, \alpha)) = \begin{cases} \boldsymbol{\lambda}_1(\boldsymbol{x}, \boldsymbol{s}_1) & \text{if } \alpha = 1 \\ \boldsymbol{\lambda}_2(\boldsymbol{x}, \boldsymbol{s}_2) & \text{if } \alpha = 2 \end{cases}$$

where symbol $\uplus$ *denotes the* disjoint union *operator,* $\alpha = \{1, 2\}$ *acts as a machine indicator,* $\boldsymbol{x}$ *is the input variable vector, and* $\boldsymbol{s}_1$ *and* $\boldsymbol{s}_2$ *are the state variable vectors of* $\mathcal{M}_1$ *and* $\mathcal{M}_2$*, respectively.*

We call the STG $G_{\uplus}$ of $\mathcal{M}_{\uplus}$ the *disjoint union STG* of the STGs of $\mathcal{M}_1$ and $\mathcal{M}_2$. Notice that the state space of a multiplexed FSM is the disjoint union of those of its two constituent FSMs, and it is usually much smaller than the state space of the product FSM.

Suppose that $\mathcal{M}_1$ and $\mathcal{M}_2$ have m_1 and m_2 registers respectively and that $m_2 \geq m_1$. To minimize the state variables of the multiplexed machine, we arbitrarily pair every next state variable of $\mathcal{M}_1$ with one of $\mathcal{M}_2$. The pair is then multiplexed before being fed to a register, whose output is then demultiplexed to recover the current state variables for $\mathcal{M}_1$ and $\mathcal{M}_2$. In addition, one self-looped auxiliary state variable is added, α, which controls all multiplexers, as indicated by the dotted lines in Figure 15.4; the value α stays constant as its initial value.

A multiplexed FSM includes its two constituent FSMs and acts as $\mathcal{M}_1$ or $\mathcal{M}_2$ depending on the machine indicator α. Although a multiplexed FSM is not as intuitive as a product FSM, its usefulness in equivalence verification is demonstrated later in Section 15.5.

Example 15.3. *Figure 15.4 shows the multiplexed FSM* $\mathcal{M}_{\uplus}$ *of* $\mathcal{M}_1$ *and* $\mathcal{M}_2$ *of Figure 15.2, and the corresponding STG of* $\mathcal{M}_{\uplus}$*. Note that the pairing of transition*

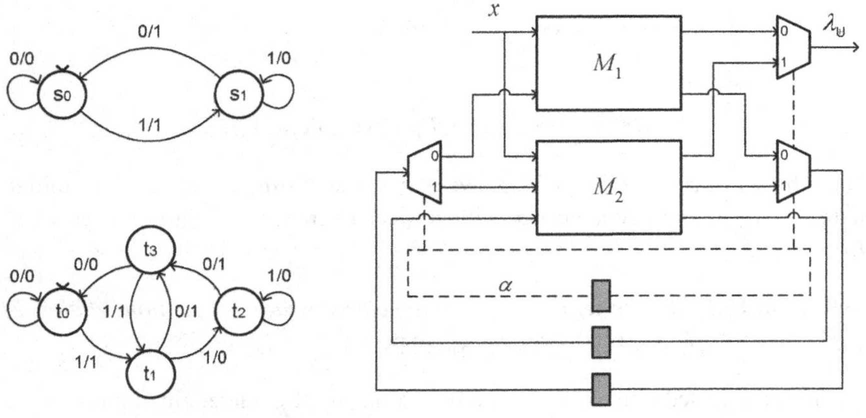

Figure 15.4. The multiplexed FSM of $\mathcal{M}_1$ and $\mathcal{M}_2$ of Figure 15.2, and its corresponding STG.

functions between $\mathcal{M}_1$ and $\mathcal{M}_2$ for the inputs of a multiplexor can be arbitrary, whereas that of output functions needs to be the corresponding pairs. When $\alpha = 0$ (respectively $\alpha = 1$), $\mathcal{M}_{\uplus}$ behaves like $\mathcal{M}_1$ (respectively $\mathcal{M}_2$) of Figure 15.2. In the STG of $\mathcal{M}_{\uplus}$, states with $\alpha = 0$ are $\{s_0, s_1\}$, and those with $\alpha = 1$ are $\{t_0, \ldots, t_3\}$.

15.2.3 Image and Preimage Computation

There are two important operations that are key ingredients to formal hardware verification. The *image* of state set $C \subseteq \Sigma \times Q$, which depends on both input variables $\boldsymbol{x}$ and current-state variables $\boldsymbol{s}$, with respect to transition relation T is computed as

$$Img(C, T) = [\exists \boldsymbol{x}, \boldsymbol{s}.(T(\boldsymbol{x}, \boldsymbol{s}, \boldsymbol{s}') \wedge C(\boldsymbol{x}, \boldsymbol{s}))]_{\boldsymbol{s}' \leftarrow \boldsymbol{s}},$$

where the subscript denotes the substitution of $\boldsymbol{s}$ in place of $\boldsymbol{s}'$ in order that the resulting image will depend on current-state variables rather than next-state variables. (Rigorously speaking, $C(\boldsymbol{x}, \boldsymbol{s})$ means $\chi_C(\boldsymbol{x}, \boldsymbol{s})$.) It characterizes the set of states that can be reached under the inputs and current states constrained by C. On the other hand, the *preimage* of $C^\dagger$ over next-state variables with respect to transition relation T can be computed as

$$PreImg(C^\dagger, T) = \exists \boldsymbol{x}, \boldsymbol{s}'.(T(\boldsymbol{x}, \boldsymbol{s}, \boldsymbol{s}') \wedge C^\dagger(\boldsymbol{s}')).$$

It characterizes the set of states that can reach $C^\dagger$ in one step under some input assignments to variables $\boldsymbol{x}$.

Given initial states I and an input sequence $\boldsymbol{\sigma} = \sigma_1, \ldots, \sigma_n$ for $\sigma_i \in \Sigma$, the image computation can be applied to obtain the destination states D with respect to a transition relation T as follows.

$$\begin{aligned} D_0 &= I(\boldsymbol{s}) \\ &\vdots \\ D_n &= Img(\sigma_n(\boldsymbol{x}) \wedge D_{n-1}(\boldsymbol{s}), T(\boldsymbol{x}, \boldsymbol{s}, \boldsymbol{s}')) \end{aligned}$$

Then the set of states D_n equals D. We denote the destination states D of initial states I under input sequence $\boldsymbol{\sigma}$ with respect to transition relation T as $D = Img_{\boldsymbol{\sigma}}(I, T)$.

Definition 15.5. *A state set $C \subseteq Q$ is called closed under a transition relation T if $\{C \cup Img(C, T)\} \subseteq C$. Otherwise, C is* open.

That is, a closed state set C cannot transition out of C under any inputs.

Let $\boldsymbol{\sigma}_1$ and $\boldsymbol{\sigma}_2$ be two input sequences. We denote their concatenation as $\boldsymbol{\sigma}_1 \circ \boldsymbol{\sigma}_2$, meaning that $\boldsymbol{\sigma}_1$ is applied first and then followed by $\boldsymbol{\sigma}_2$.

15.3 Computational Complexity of Equivalence Checking

15.3.1 Complexity of Combinational Equivalence Checking

Combinational equivalence checking (CEC) is basically tautology checking of Boolean expressions.

CEC of Boolean Circuits
Instance: Two combinational Boolean circuits C_1 and C_2.
Question: Are the output values of C_1 and C_2 the same for each input vector? That is, are C_1 and C_2 equivalent?

Theorem 15.1. *CEC is coNP-complete.*

The proof is by reduction from validity, that is, given an instance of validity checking of formula ϕ, we build polynomially an instance of CEC as follows: C_1 is the natural circuit representation of the formula ϕ and C_2 is any circuit representing the function constant 1. Then C_1 is equivalent to C_2 if and only if ϕ is valid. If a formula ϕ is not valid, then it can be disqualified succinctly by providing a truth assignment that does not satisfy ϕ. On the other hand, a valid ϕ has no such disqualification. VALIDITY is also called TAUTOLOGY. Checking the VALIDITY of a Boolean expression in disjunctive normal form (DNF) is coNP-complete. Any language $\mathcal{L}$ in coNP is reducible to VALIDITY. A string x is in $\mathcal{L}$ if and only if the formula φ_x in DNF by the reduction from $\mathcal{L}$ to VALIDITY is valid. Indeed, if $\mathcal{L}$ is in coNP, then its complement $\overline{\mathcal{L}}$ is in NP. Hence there is a reduction from $\overline{\mathcal{L}}$ to SAT. A string x is in $\overline{\mathcal{L}}$ if and only if the formula φ_x in conjunctive normal form (CNF) by the reduction is satisfiable.

15.3.2 Complexity of Sequential Equivalence Checking

The complexity measure of sequential equivalence checking (SEC) depends on how state-transition systems are represented. When they are represented explicitly by state-transition graphs, the corresponding SEC is tractable in terms of the size of the graphs. When they are represented by sequential circuits (with logic gates and memory elements), the corresponding SEC is intractable in the size of the circuits.

SEC of Deterministic Automata (explicit graph representation)
Instance: Two deterministic finite automata A_1 and A_2.
Question: Are the output languages produced by A_1 and A_2 the same?

The problem is in P (solvable in polynomial time in the size of the input graph), because we can perform state minimization of A_1 and A_2. It is well known that, given a finite automaton or finite transducer, state minimization yields a unique minimum automaton up to state renaming. So the SEC problem is reduced to verifying the isomorphism of the two state-minimized automata. Notice that graph isomorphism is not an easy problem; however, comparing the graphs of the two minimized automata is easy because they are special graphs with a unique starting state and unique edge labeling for every state.

The SEC problem becomes harder when the state-transition system is not given explicitly via a graph, but implicitly via a circuit (or a program) that encodes the graph. So a sequential circuit can be seen as a succinct representation of a graph $G = (V, E)$ with states V and (labeled) transitions $E \subseteq V \times V$. An edge $(s, t) \in E$ with label a denotes that t is the next state of s under input a. In this way we are encoding an exponentially large graph into a small circuit, but the reachability problem on the succinct graph representation jumps in complexity from P to PSPACE.

SEC of Sequential Circuits (implicit graph representation)
Instance: Two sequential Boolean circuits C_1 and C_2 with the same input and output variables, and with respective initial states s_1 and s_2.
Question: Are the output sequences of C_1 and C_2 the same for every sequence of input vectors?

The negation of the foregoing problem can be stated as follows.

SNEC of Sequential Circuits (implicit graph representation)
Instance: Two sequential Boolean circuits C_1 and C_2 with the same input and output variables, and with respective initial states s_1 and s_2.
Question: Is there a sequence of input vectors such that C_1 and C_2 differ in at least one output value?

Theorem 15.2. *SNEC is PSPACE-complete.*

Proof. We show first that SNEC is in PSPACE, because the following polynomial-space algorithm decides the language. For convenience, consider the product machine of C_1 and C_2 with initial state $s = (s_1, s_2)$. The algorithm first guesses the length l of a path from s providing a counter-example to equivalence, and then guesses its $l - 2$ inner vertices and the final vertex t (vertices are states of the product machine).

(i) Guess a number $l \in \{2, 3, \ldots, 2^n\}$, where n is the the number of storage elements in the product machine (i.e., the sum of the number of storage elements in C_1 and C_2), and so 2^n is an upper bound on the length of a counterexample to equivalence.
(ii) Set $k = 1$.
(iii) Repeat until k is equal to $l - 1$:
   (a) Guess $v_k \in \mathbb{B}^n$ (a state of the product machine) and an input vector i.

(b) Simulate the product machine of C_1 and C_2 from state v_{k-1} under i:
If v_k is not the next state of v_{k-1} (with $v_0 = s$) under i "reject",
else if C_1 and C_2 differ at v_k in at least one output "accept" ($v_k = t$)
else if $k = l - 1$ "reject".

(c) Set $k = k + 1$.

The foregoing simulation of the product machine takes polynomial space. ■

SNEC is PSPACE-hard. Let $\mathcal{L}$ be an arbitrary language in PSPACE and $M_{\mathcal{L}}$ be a Turing machine[3] deciding $\mathcal{L}$ using space no more than $poly(n)$, polynomial in the size n of the input given to $M_{\mathcal{L}}$. We show that $\mathcal{L} \leq_p$ SNEC, namely, $\mathcal{L}$ is polynomial-time reducible to SNEC.

The idea of the reduction is to build in polynomial time a circuit that is an implicit representation of the configuration graph of $M_{\mathcal{L}}$. The vertices of the configuration graph are represented by the encoded vectors stored in the memory elements of the sequential circuit. The number of memory elements is a polynomial in n because each configuration can grow at most as a polynomial in n.

Given an input string w to $M_{\mathcal{L}}$ with $|w| = n$ (size of the input), the possible configurations of $M_{\mathcal{L}}$ when running on w are of length at most $poly(n)$. Each configuration can be represented as a sequence of $poly(n)$ cells where each cell contains either a symbol from the work alphabet of $M_{\mathcal{L}}$ or a symbol from the state set. Therefore, each cell can be represented by a constant number c of bits (that does not depend on the input length), and the entire configuration can be represented by a sequence of $m = c \cdot poly(n)$ bits. In addition, we define a vector of m bits that does not represent a configuration, but instead means "accepted," and another vector of m bits meaning "rejected."

Given w, let us see in detail how to build an instance of SNEC.

First we construct a sequential circuit C_1^w that emulates a run of $M_{\mathcal{L}}$ on w. This sequential circuit has m storage elements (to store the configurations of $M_{\mathcal{L}}$ running on w) and logic circuitry computing the next configuration yielded by the current configuration. The circuitry will implement with logic gates the transition relation of the Turing machine $M_{\mathcal{L}}$; in practice, a lookup table will do. Two states correspond to the vectors for "accepted" and "rejected," respectively. The initial state s_1 of C_1^w corresponds to the initial configuration of $M_{\mathcal{L}}$ on w. Moreover, C_1^w outputs 1 in every state. We can build in polynomial time the sequential circuit C_1^w that works in polynomial space. Notice that this sequential circuit is autonomous, that is, it has no inputs except the clock.

Finally, to complete the construction of the instance of SNEC, we build C_2^w as a copy of C_1^w, which differs from C_1^w only in the fact that it outputs 0 in the

[3] We assume that $M_{\mathcal{L}}$ is a deterministic Turing machine, due to the theorem by Savitch [76], stating that a nondeterministic Turing machine using $f(n)$ space can be simulated by a deterministic Turing machine using $f^2(n)$ space for any polynomial function f. So the simulation preserves polynomial space.

accepted state of C_2^w. As before, the initial state s_2 of C_2^w corresponds to the initial configuration of $M_{\mathcal{L}}$ on w.

In summary, we established PSPACE-hardness, because

(i) The overall reduction can be done in polynomial time.
(ii) $M_{\mathcal{L}}$ accepts w if and only if, starting respectively from s_1 and s_2, C_1 and C_2 reach a state where they differ in at least one output value.

The fact that SNEC is PSPACE-complete implies by definition that its complement problem SEC is coPSPACE-complete. By a theorem of Immerman [35] and Szelepscenyi [84], it holds that NSPACE = coNSPACE; by a theorem of Savitch [76] it holds that PSPACE = NSPACE. So it follows that PSPACE = coNSPACE = coPSPACE.

Corollary 15.3. *SEC is PSPACE-complete.*

It was proved recently in [40] that even a restricted case of sequential equivalence checking where the circuits differ only by retiming and resynthesis transformations is still PSPACE-complete (by reduction from finite function generation [26]).

15.4 Combinational Equivalence Checking

Given two combinational Boolean circuits C_1 and C_2, the problem is to check whether the corresponding outputs of the two circuits are the same for every input vector $\boldsymbol{i} = (i_1, \ldots, i_n)$. For single-output circuits, their equivalence can be expressed as

$$\forall \boldsymbol{i}.(C_1(\boldsymbol{i}) \equiv C_2(\boldsymbol{i})). \tag{15.1}$$

For multioutput circuits with outputs $C_{1,j}$ and $C_{2,j}$, where $j = 1, \ldots, m$, their equivalence can be expressed as

$$\forall \boldsymbol{i}. \bigwedge_{j=1}^{m} (C_{1,j}(\boldsymbol{i}) \equiv C_{2,j}(\boldsymbol{i})). \tag{15.2}$$

The tautology checking of formulas (15.1) and (15.2) can be achieved through the "miter" circuit construction [4] as shown in Figures 15.5 and 15.6, respectively. A counterexample to equivalence exists if and only if there is an input vector that makes the output of a miter circuit false. Observe that the miter circuit of CEC is similar to the product machine of SEC.

Tautology checking can be alternatively negated for satisfiability checking. Technically speaking, it makes no difference whether we consider tautology checking and satisfiability checking. One can dually define the miter circuit construction by replacing the XNOR-gates with XOR-gates and by replacing the final AND-gate with an OR-gate. Then the previous statements are dualized

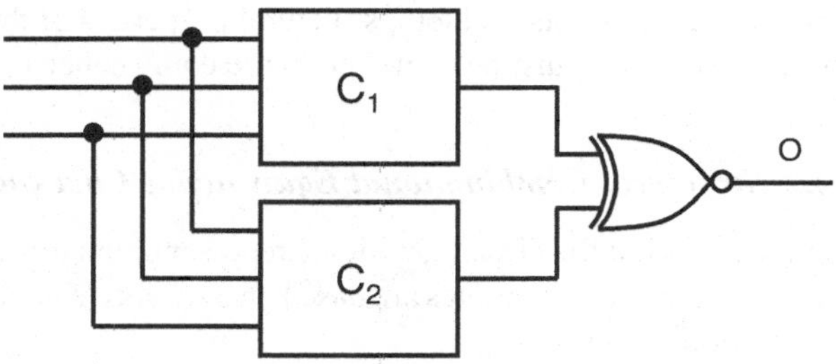

Figure 15.5. Single-output miter circuit from networks C_1 and C_2.

in that circuit equivalence is the same as the miter circuit being satisfiable at the output.

We new review the foundations of the two basic approaches to combinational equivalence checking: functional techniques and structural techniques. Surveys of CEC techniques can be found in [37] for early work, and in [25, 47] for more recent work.

15.4.1 Functional Combinational Equivalence Checking

In functional combinational equivalence checking, the functions realized by the two circuits are compared directly by representing them with canonical representations (they guarantee a unique form for a given function). Since Bryant's influential paper [7], ROBDDs have been the most widely used canonical data structure for Boolean functions. For instance, given the miter construction in Figure 15.6, one builds the ROBDD for $O = \bigwedge_j (C_{1,j} \equiv C_{2,j})$. If the BDD represents the constant 1, then C_1 and C_2 are equivalent. Otherwise, C_1 and C_2 are not. Notice that the ROBDD representing the constant-1 function has a unique representation by the constant-1 terminal node.

A major problem is that ROBDDs may be unable to represent circuits with a large number of primary inputs, and in general their growth is not bounded *a priori*

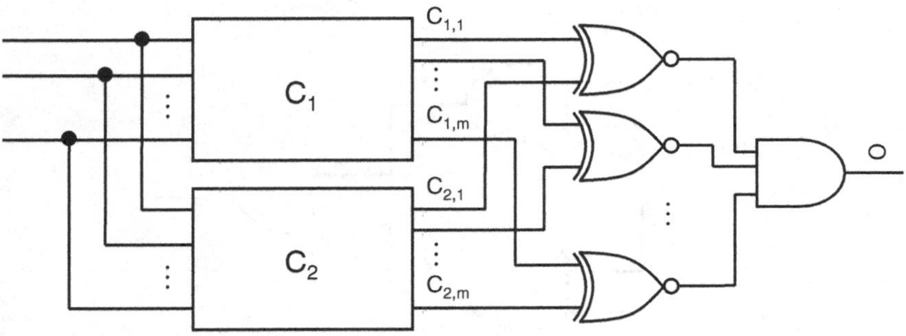

Figure 15.6. Multioutput miter circuit from networks C_1 and C_2.

by tight bounds. To overcome this problem, structural properties of the multilevel representations of the circuits must be exploited in the equivalence check.

15.4.2 Structural Combinational Equivalence Checking

Structural techniques exploit the circuit gate-level representation to verify equivalence. Two common structural techniques are based on SAT solving and Automatic Test Pattern Generation (ATPG) [48].

15.4.2.1 Combinational Equivalence by SAT

Given the miter of Figure 15.5, its output produces 0 under some input truth assignment if and only if the two circuits C_1 and C_2 represent different Boolean functions. A SAT solver can be used to search for a counterexample to equivalence.

To apply SAT we must translate a circuit structure into a CNF, by introducing new variables for all circuit nodes and adding relational constraints in CNF to model the functionality of each gate (as originally proposed by Tseitin [88]; see also Crama and Hammer [17] for a description of this technique). The translation is linear in the number of gates if the fan-in sizes of XOR-gates and XNOR-gates are upper bounded by a constant.

Example 15.4. *Consider the circuit of Figure 15.7. By Tseitin's circuit-to-CNF conversion, in addition to the primary input variables a, b, and c, variables x, y, and f are added to represent the logic values of the gate outputs. The CNFs of the upper AND-gate, lower AND-gate, and OR-gate are*

$$(\overline{a} \vee \overline{b} \vee x)(a \vee \overline{x})(b \vee \overline{x}), \tag{15.3}$$

$$(\overline{b} \vee \overline{c} \vee y)(b \vee \overline{y})(c \vee \overline{y}), \text{ and} \tag{15.4}$$

$$(x \vee y \vee \overline{f})(\overline{x} \vee f)(\overline{y} \vee f), \text{ respectively.} \tag{15.5}$$

The CNF of the whole circuit is simply the conjunction of formulas (15.3), (15.4), and (15.5), namely,

$$(\overline{a} \vee \overline{b} \vee x)(a \vee \overline{x})(b \vee \overline{x})(\overline{b} \vee \overline{c} \vee y)(b \vee \overline{y})(c \vee \overline{y})(x \vee y \vee \overline{f})(\overline{x} \vee f)(\overline{y} \vee f).$$

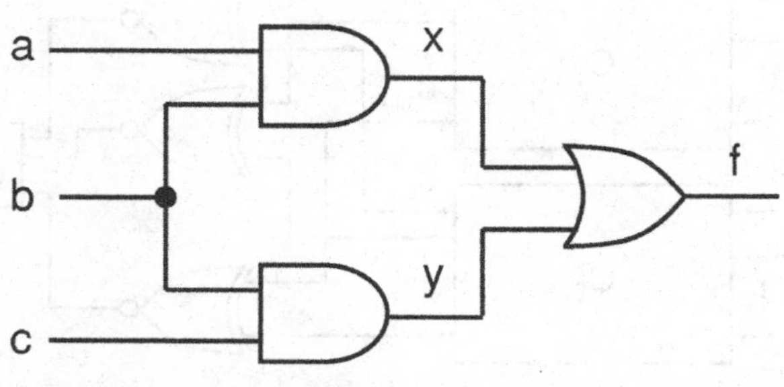

Figure 15.7. A combinational circuit for circuit-to-CNF conversion.

Even though SAT is NP-complete, much progress has been made in engineering methods that solve instances of practical interest. A modern SAT solver may deploy a mixture of *backtrack search*, *implication* (*Boolean constraint propagation*), *conflict analysis*, *resolution*, *nonchronological back-tracking* (*conflict-based learning*), and *two-literal watching*. We refer to the specialized literature for more details on this topic; see, for example the monograph [17], and the chapters by Franco and Hooker in the present volume.

15.4.2.2 Combinational Equivalence by ATPG

Automatic Test Pattern Generation (ATPG) has the goal to generate a test for every fault in a circuit. Under the stuck-at-fault model, for a test to exist there must be a set of input assignments such that the fault under test can be:

(i) Activated, that is, the faulty stuck-at-1 signal is set to 0 and stuck-at-0 signal to 1, and
(ii) Observed at the primary outputs, that is, the faulty effect at the fault location propagates to at least one primary output.

The connection between ATPG and equivalence checking is that, given the miter circuit of Figure 15.6, if the stuck-at-1 fault at O is not testable, then C_1 is equivalent to C_2. Otherwise the test patterns are counterexamples to the assertion of equivalence.

Plain SAT- and ATPG-based CEC tend to fail if the size of the miter circuit is large. A key idea to alleviate this problem is to exploit similarities between C_1 and C_2 for miter circuit reduction. The similarities are likely to exist when many internal nodes in C_1 and C_2 are equivalent. It is expected that the circuits to be compared share structural and functional similarities, because usually CEC is applied to verify the correctness of incremental synthesis steps. So, for each node n_1 in a given list of good candidate nodes in C_1, and for each node n_2 in a given list of good candidate nodes in C_2, one builds the miter circuit $z = (n_1 \equiv n_2)$. If z stuck-at-1 is not testable or SAT does not find a falsifying assignment, then n_1 and n_2 are equivalent. Otherwise, a counterexample is found and they are inequivalent. Identification of the internal equivalences is intrinsically a difficult problem. It is usually done by scanning the circuit from inputs to outputs, with a variety of methods from local analysis to comparison of local BDDs.

In identifying equivalent nodes, if n_1 and n_2 turn out to be equivalent, then there can be different strategies to simplify the miter circuit. One (safe) method is to merge n_1 and n_2 by replacing n_2 with n_1. Another more aggressive (but unsafe) method is to substitute a free variable for both n_1 and n_2 in the miter circuit. The latter may result in false negatives, that is, C_1 and C_2 may be declared inequivalent even if they are equivalent, because the newly added variables are unconstrained and allow truth assignments that are infeasible because of the constraints imposed by the subcircuits of n_1 and n_2. To fix this problem, one has to carefully partition the circuit and, in the worst case, enlarge the partition to include the primary inputs. Or, one can apply range-preserving parametric representation for miter circuit rewriting.

There is a close relationship between ATPG and SAT [48]. On the one hand, ATPG can be seen as SAT solving directly over circuits rather than CNF formulas. On the other hand, ATPG can be reformulated as standard SAT solving by expressing the constraints of fault activation and observation as a CNF formula. ATPG and SAT have their own strengths and weaknesses. ATPG has better circuit information for Boolean reasoning; SAT has a simpler and more generic data structure for Boolean reasoning. In practice, efficient hybrid methods deploying features of both ATPG and SAT have been developed.

15.4.3 Trading Off Canonicity versus Structure

Between the two extremes of functional equivalence checking with canonical representations, like ROBDDs, which do not scale well, and structural equivalence checking directly on the network, a middle ground has been found with "intermediate" representations that efficiently store the multilevel network by using structural hashing to detect common subnetworks.

One such data structure is AND-Inverter graphs (AIGs), which are Boolean networks composed of two-input AND-gates and inverters represented as flipped bits on the edges [46]. AIGs are a multilevel logic representation whose construction time and size are linearly proportional to the size of the circuit. AIGs are not canonical. That is, a Boolean function has many AIG representations. For example, function $F = abc$ can be represented as $((ab)c)$, $(a(bc))$, $((ac)(bc))$, and so on. However, when comparing circuits C_1 and C_2, structural graph hashing can be applied to their AIGs to identify structurally identical subgraphs in the miter structure. The AIG representation of the miter is constructed from the inputs to the outputs in topological order, as is done with BDDs; before adding a new AND vertex, a hash-lookup checks for the existence of a structurally equivalent vertex, and, if it exists, uses it to realize the current AND [44].

Because AIGs are not canonical, internal nodes of an AIG may have equivalent functionality. This increases the number of AIG nodes and makes reasoning on the graph structure inefficient and time consuming. An algorithm to detect and merge equivalent nodes (called functional reduction) during the process of AIG construction was presented in [60]; this construction yields a functional representation called Functionally Reduced AIGs (FRAIGs). FRAIGs are "semicanonical" because no two nodes in a FRAIG have the same function in terms of the primary inputs, but the same function may have different FRAIG structures. The construction uses simulation and SAT.

It is the small set of base functions (two-input AND-gates and inverters) that makes AIGs structure hashing efficient and makes them preferable in scalable Boolean function manipulation.

There are other noncanonical representations, such as Boolean Expression Diagrams (BEDs) and extended BDDs (XBDDs), that use richer sets of base functions. The BED data structure [34] is obtained by extending the ROBDD representation with operator vertices, besides variable vertices; a variable vertex (inherited from standard BDDs) has as attributes a variable and two child vertices, whereas an operator vertex has as attributes a binary Boolean operator and two child vertices. Any Boolean circuit can be transformed into a BED by a translation that is linear in size. For instance, because there are combinational circuits that implement multiplication by using only a quadratic number of gates, there are BEDs of this size representing them. Moreover, BEDs can recognize and share identical subexpressions. The other side of the coin is that BEDs are not canonical. In contrast, XBDDs [68] are obtained by adding structural variables that can be both universally and existentially quantified. Quantifications are described as annotations on pointers leading to nodes with structural variables.

15.5 Sequential Equivalence Checking

In the design process of a hardware system, a design may be transformed (manually by circuit designers or automatically by software tools) back and forth to explore an optimal implementation with respect to some design criteria. These transformations may introduce errors in the design. It is crucial to ensure that the final optimized design is indeed equivalent to the original one. Therefore, equivalence verification is one of the most important problems in ensuring the correctness of hardware designs.

When two circuits under comparison contain no state-holding elements (registers), the corresponding verification problem is called *combinational equivalence checking*; in contrast, when the circuits under comparison contain registers, the corresponding verification problem is called *sequential equivalence checking*. In fact, combinational equivalence checking can be seen as a special case of sequential equivalence checking, because a combinational circuit can be modeled as a single-state finite state machine. We argued in Section 15.3 that the computational complexity of combinational equivalence checking is coNP-complete, whereas that of sequential checking is PSPACE-complete. Even though combinational equivalence is likely to be *intractable* [26] (not solvable in deterministic polynomial time) in theory, it is considered to be efficiently solvable in practice in most design instances (except for arithmetic circuits). As a matter of fact, the equivalence of combinational circuits of multimillion logic gates has been demonstrated: for example, in [45]. This apparent contradiction arises from the fact that, in real-world designs, two circuits under comparison mostly possess structural similarities to a large extent. This structural information provides hints assisting equivalence checking. Nevertheless, structural similarities in sequential equivalence checking may not be as helpful as in the combinational case. As a result, there is still no efficient solution to sequential equivalence checking even in practical applications. In this section we study approaches to general sequential equivalence checking; in the next two sections we discuss some reduction techniques to make sequential equivalence checking more akin to combinational equivalence checking.

To begin with, we state the following definition.

Definition 15.6. (k-Step State Equivalence.) *Two states q_1 and q_2 of an FSM $(Q, Q^0, \Sigma, \Omega, T, \boldsymbol{\lambda})$ are* k-step equivalent, *denoted as $q_1 \sim^k q_2$, if they satisfy the relation $\sim^k$ that is defined inductively as follows:*

$$q_1 \sim^0 q_2 : \quad \forall \sigma \in \Sigma.(\boldsymbol{\lambda}(\sigma, q_1) \equiv \boldsymbol{\lambda}(\sigma, q_2))$$

$$\vdots$$

$$q_1 \sim^k q_2 : \quad (q_1 \sim^{k-1} q_2) \wedge \forall \sigma \in \Sigma.(Img_\sigma(q_1, T) \sim^{k-1} Img_\sigma(q_2, T)).$$

It can be easily checked that $\sim^k$ is reflexive, symmetric, and transitive and thus forms an equivalence relation. Moreover, when viewed as a set, $\sim^k$ is monotonically decreasing, that is, $\sim^k \supseteq \sim^{k+1}$, for $k = 0, 1, \ldots$, because $(q_1, q_2) \in \sim^{k+1}$ implies $(q_1, q_2) \in \sim^k$. In particular, if $\sim^k = \sim^{k+1}$ for some k, then $\sim^{k+i} = \sim^{k+i+1}$ for all $i \geq 0$. Such $\sim^k$, denoted as $\sim^*$, is known as the greatest fixed point of the k-step state equivalence of an FSM.

The partition induced by $\sim^*$ is one instance of the so-called *stable partition.*

Definition 15.7. (Stable Partition.) *A partition π is* stable *with respect to a transition relation T if, for any q_1, q_2 in the same block of π and $(q_1, q_1') \in T$, there exists q_2', with $(q_2, q_2') \in T$, in the same block as q_1'.*

The foregoing fixed-point computation is a process of *stabilization* for a given arbitrary initial partition (not necessarily the partition $\sim^0$ induced by the observational equivalence due to the outputs $\boldsymbol{\lambda}$). In fact, the fixed point derived in this way is the *coarsest stable partition* that refines the initial partition. (A partition π_1 *refines* another partition π_2 if any two elements in a block of π_1 are also in a block of π_2.)

Relation $\sim^*$ is equivalent to the state equivalence defined next.

Definition 15.8. (State Equivalence.) *Two states q_1 and q_2 of an FSM $\mathcal{M} = (Q, Q^0, \Sigma, \Omega, T, \boldsymbol{\lambda})$ are* equivalent, *denoted as $q_1 \sim q_2$, if $\boldsymbol{\lambda}(\sigma', Img_{\boldsymbol{\sigma}}(q_1, T)) = \boldsymbol{\lambda}(\sigma', Img_{\boldsymbol{\sigma}}(q_2, T))$ for every $\sigma' \in \Sigma$ and every finite (including empty) input sequence $\boldsymbol{\sigma} \in \Sigma^*$.*

That is, starting from either q_1 or q_2 with $q_1 \sim q_2$, an FSM behaves indistinguishably from its input-output behavior. It can be shown that $\sim$ and $\sim^*$ are identical. As a matter of fact, $\sim^*$ can be seen as an operational definition of $\sim$. It allows a finite computation of state equivalence. (As a side note, it is possible to define other notions of state equivalence, whose distinguishing capabilities may induce a preorder $\preceq$ on different equivalence definitions. For instance, equivalence relation $\sim_S$, with $q_1 \sim_S q_2$ if and only if $q_1 = q_2$, is the most distinguishing state equivalence relation, and thus is on the top of the preorder. On the other hand, equivalence relation $\sim_U$, with $q_1 \sim_U q_2$ for any states q_1, q_2, is the least distinguishing state equivalence relation and thus is at the bottom of the preorder. In comparison, we see that $\sim_U \preceq \sim \preceq \sim_S$ among other possible definitions of state equivalences.)

The state equivalence of an FSM can be straightforwardly generalized to state equivalence among multiple FSMs by treating the disjoint state spaces as a single large state space. On the other hand, we may define FSM equivalence as follows.

Definition 15.9. (FSM Equivalence.) *Two FSMs $\mathcal{M}_1$ and $\mathcal{M}_2$ are* equivalent *if starting from their respective initial states, their input-output behaviors are indistinguishable.*

Given two FSMs, the problem of sequential equivalence checking asks if these two machines are indistinguishable from their output sequences under any input sequence. Notice that even if any state of one FSM has a corresponding equivalent state in the other, the two FSMs may be inequivalent unless they start from equivalent initial states.

Based on their underlying data structures, verification approaches can be divided into two classes: that of explicit enumerative algorithms and that of implicit symbolic algorithms. The former perform manipulations directly over state transition graphs (e.g., reachability analysis based on traversal on state transition graphs); the latter, on the other hand, perform manipulations over characteristic functions representing state sets, where characteristic functions are realized by some abstract data types, such as ROBDD, CNF, or AIG (e.g., reachability analysis based on Boolean manipulations).

Before effective data structures and algorithms were available for Boolean manipulations in the late 1980s, early verification algorithms performed explicit enumeration over state transition graphs. Since the late 1980s, verification algorithms have relied on implicit manipulations over Boolean formulas and characteristic functions. Because STGs (in a graph representation) and FSMs (in a six-tuple representation) point to the same state transition system, verification can be done both explicitly over STGs and implicitly over FSMs.

Before delving into the algorithms, we introduce some FSM constructs (and thus their STG counterparts) that will be useful later on.

Definition 15.10. (Quotient FSM.) *Given an FSM $\mathcal{M} = (Q, Q^0, \Sigma, \Omega, \delta, \lambda)$, its* quotient FSM $\mathcal{M}_{/\sim} = (Q_{/\sim}, Q^0_{/\sim}, \Sigma, \Omega, \delta_{/\sim}, \lambda_{/\sim})$ *(with respect to relation $\sim$) is obtained by collapsing equivalent states $C = \{q_i \in Q \mid q_i \sim q \text{ for some } q \in Q\}$ into its arbitrary representative state, denoted as $q_C \in Q_{/\sim}$, of the equivalence class. In addition, for any $\sigma \in \Sigma$,*

$$\begin{aligned} q_C \in Q^0_{/\sim} &\Leftrightarrow C \cap Q^0 \neq \emptyset, \\ q_{C_2} = \delta_{/\sim}(\sigma, q_{C_1}) &\Leftrightarrow q_2 = \delta(\sigma, q_1) \text{ for some } q_1 \in C_1, q_2 \in C_2, \text{ and} \\ \omega = \lambda_{/\sim}(\sigma, q_C) &\Leftrightarrow \omega = \lambda(\sigma, q) \text{ for some } q \in C. \end{aligned}$$

(Recall the assumption that all initial states in an FSM have to be equivalent. So under this assumption there is a single initial state in any quotient FSM.) Also, we refer to the STG $G_{/\sim}$ of $\mathcal{M}_{/\sim}$ as the *quotient* STG of $\mathcal{M}$.

We can think of $\mathcal{M}_{/\sim}$ as a *reachability-preserving abstraction* of $\mathcal{M}$ with respect to the partition $\sim$ imposed by the observational equivalence of output functions. It is an *abstraction* in the sense that some detailed state information is hidden and the FSM is simplified. The abstraction is reachability-preserving in that, for any $\sigma \in \Sigma$, if $\exists\, q_1 \in C_1 .\ \delta(\sigma, q_1) \in C_2$, then $\forall\, q_1 \in C_1 .\ \delta(\sigma, q_1) \in C_2$. The significance of this kind of abstraction is that, for a given property to be verified that can be formulated as a reachability problem, there exists a reachable state violating the property in the original transition system if and only if there exists a reachable bad state in the abstracted counterpart. Hence, the abstraction is both sound and complete for safety property checking. For instance, in $\mathcal{M}_{/\sim}$, $\boldsymbol{\lambda}$ induces an initial partition $\sim^0$ (which distinguishes states with different output observations, that is, our concerned property) over the state space Q. We may study whether some state in one block of $\sim^0$ is reachable from a state in another block. The answer is the same, no matter whether the analysis is conducted over $\mathcal{M}$ or $\mathcal{M}_{/\sim}$. This kind of abstraction is helpful in simplifying sequential equivalence checking and other general safety property checking.

15.5.1 Explicit Graph Algorithms

In discussing explicit graph algorithms, we are concerned with STGs (instead of FSMs in the six-tuple representation). We introduce three STG-based approaches for sequential equivalence checking. One relies on analyzing reachability on a product STG. Another relies on checking the isomorphism between two quotient STGs. The third one, though similar to the previous one, relies on building the quotient of a disjoint union STG.

15.5.1.1 Reachability Analysis in the Product of Two STGs

In reachability analysis, one wants to assert that bad states of an STG are unreachable from its initial states. In the context of equivalence checking of two STGs G_1 and G_2, we are concerned with their product STG $G_\times$. A state of $G_\times$ is bad if any of its outgoing edges contains a 0, instead of a 1, in the output label.

Example 15.5. *In Figure 15.3, states* $\{(s_0, t_1), (s_0, t_2), (s_1, t_0), (s_1, t_3)\}$ *are bad.*

Proposition 15.4. *Two STGs are equivalent if and only if any output label on the edges in the reachable subspace of their product STG is 1.*

Reachability analysis on an STG with n vertices and m edges can be done, for example by a breadth-first traversal, in linear time complexity $O(m + n)$. Assume the input alphabet is of size k, i.e. $|\Sigma| = k$. Then $m = kn$ and the foregoing complexity can be rewritten as $O(kn)$. Therefore, to traverse the product space of two STGs $G_1 = (V_1, E_1)$ with $|V_1| = n_1$ and $G_2 = (V_2, E_2)$ with $|V_2| = n_2$, the time complexity is of $O(kn_1n_2)$.

15.5.1.2 Isomorphism of the Quotients of Two STGs

In addition to the previous approach based on reachability, the equivalence of two STGs can be alternatively formulated based on the partition induced by state equivalence. Essentially, we have the following proposition.

Proposition 15.5. *The quotients of equivalent STGs are canonical (i.e. unique up to an isomorphism).*

Proof. For two equivalent STGs G_1 and G_2, there exists a bijection between the equivalence classes of states and their initial states must be in corresponding equivalence classes. Otherwise, there exists an input sequence that can drive G_1 and G_2 into inequivalent states. Because states in an equivalent class are merged in computing a quotient, there exists a bijection between equivalent states of $G_{1/\sim}$ and $G_{2/\sim}$. That is, $G_{1/\sim}$ and $G_{2/\sim}$ are isomorphic. The proposition follows. ∎

Therefore, to check the equivalence of two FSMs, one can first canonicalize the corresponding STGs into their quotient (state-minimized) forms and then check their graph isomorphism. Two FSMs are equivalent if and only if their quotient STGs are isomorphic (including the matching of initial states) in the reachable state subspace.

The computation of equivalent states is, in fact, already implicit in Definition 15.6. Figure 15.8 sketches a procedure computing equivalent states. The

ComputeEquivalentStates
input: an STG G
output: the coarsest partition π of equivalent states of G
begin
$i := 0$
let $\pi^{(i)}$ be the partition of states induced by $\sim^0$
repeat
  $i := i + 1$
  $\pi^{(i)} := \emptyset$
  for each block $B_j \in \pi^{(i-1)}$
    refine B_j into a partition $\pi_{B_j} = \{C_k\}$ with
      $q_1, q_2 \in C_k \Leftrightarrow \forall \sigma \in \Sigma, \exists B_l \in \pi^{(i-1)}.Succ_\sigma(q_1), Succ_\sigma(q_2) \in B_l$
    $\pi^{(i)} := \pi^{(i)} \cup \pi_{B_j}$
until $|\pi^{(i)}| = |\pi^{(i-1)}|$
return $\pi^{(i)}$
end

Figure 15.8. An algorithm that computes the coarsest partition of the state equivalence of a given STG.

initial partition is induced by $\sim^0$. The partition is then iteratively refined. A block B_j in the current partition $\pi^{(i)}$ is split into several subblocks such that two states $q_1, q_2 \in B_j$ are in different subblocks if and only if $\exists \sigma \in \Sigma$ such that the successor of q_1 and that of q_2 under σ go to different blocks in $\pi^{(i)}$. The procedure terminates when no more refinements can be made. Any block in the final partition represents an equivalence class of states. Given a state equivalence partition, the corresponding quotient STG can then be constructed by merging equivalent states.

For an STG with n vertices and m edges, an analysis shows that the foregoing algorithm for computing equivalent states may take up to n iterations. Because in each iteration we may need to traverse m edges, the total time complexity is of $O(mn)$. An improved algorithm that constructs the coarsest stable partition in $O(m \log n)$ time can be found in [32]. (The result was extended in [64] to relational coarsest partition and may deal with nondeterministic transitions beyond STGs.) Moreover, constructing the quotient STG for a given partition can be done in $O(m + n)$ time. Consequently, the total time complexity in constructing a quotient STG can be achieved in $O(m \log n)$ time. On the other hand, the isomorphism checking of two quotient STGs is of linear time complexity in the graph size. (The isomorphism checking here is much easier than general graph isomorphism checking because the quotient STGs are deterministic in their state transitions and the correspondence of their two initial states is known.) Consequently, equivalence checking of two STGs, $G_1 = (V_1, E_1)$ with $|V_1| = n_1, |E_1| = m_1$ and $G_2 = (V_2, E_2)$ with $|V_2| = n_2, |E_2| = m_2$, is of time complexity $O(m_1 \log n_1 + m_2 \log n_2)$. In terms of the alphabet size $|\Sigma| = k$, the complexity can be reexpressed as $O(k(n_1 \log n_1 + n_2 \log n_2))$.

15.5.1.3 State Equivalence in the Disjoint Union of Two STGs

In the previous approach, the quotients of two STGs are derived individually and then compared. Here we show that the quotient computation can be computed once, and further that isomorphism checking is unnecessary.

Again, the same procedure of Figure 15.8 is used. However, this time, rather than minimizing STGs G_1 and G_2 individually twice, we apply the procedure once on the disjoint union $G_\uplus$ of the two constituent STGs. In the end, the algorithm yields a partition that contains the state equivalence information not only within the individual STGs but also the equivalence information between them.

Proposition 15.6. *Two STGs G_1 and G_2 are equivalent if and only if their initial states are in the same equivalence class of their disjoint union STG.*

Even though the computation is simplified, the time complexity is essentially the same, that is, $O(k(n_1 \log n_1 + n_2 \log n_2))$, as the previous algorithm based on isomorphism checking of two quotient STGs.

In the three preceding enumerative algorithms, STGs are the underlying data structure representing state transition systems. It should be noted that the size of an STG (in terms of the number of vertices and edges) can be exponentially larger than an FSM in the six-tuple representation (in terms of the number of bits or Boolean circuit representing the six-tuple). Therefore the early enumerative algorithms may not be effective because of the state explosion problem, namely that the number of states is exponentially larger than the number of state variables. In what follows, we introduce symbolic algorithms, which manipulate Boolean formulas instead of enumerating through transition graphs.

15.5.2 Implicit Symbolic Algorithms

The previous graph enumerative algorithms are not scalable to large instances. The reasons are twofold: firstly, representing STGs is memory-space consuming because every state and transition needs to be represented explicitly; second, enumeration over STGs is time consuming because no parallelism can be exploited, that is, transitions need to be enumerated separately. These shortcomings are overcome in the symbolic algorithms to be introduced.

In symbolic algorithms, state transition systems are represented with FSMs (in the six-tuple representation) or sequential circuits. Enumeration is done through Boolean manipulations. Therefore, sets of states and transitions can be manipulated simultaneously.

To design a symbolic algorithm, its underlying data structure must be effective in representing Boolean formulas and in supporting Boolean manipulations. ROBDD (or BDD for short) is one such data structure. Hereafter, unless otherwise stated, BDD is meant to be ROBDD. There are several unique properties of BDDs that make them particularly suitable for verification:

(i) Most Boolean functions can be compactly represented by BDDs.
(ii) Boolean manipulations over BDDs are efficient (polynomial in BDD sizes).
(iii) BDD representation is canonical. Thus, checking equivalence of two BDDs (with the same variable ordering) takes constant time.

Despite these nice properties, BDD sizes are in general unpredictable and sensitive to different variable orderings. Because of this intrinsic feature, BDD-based algorithms suffer from nonrobustness even though good heuristics exist for finding good variable orderings. All together, the BDD is still one of the most important data structures for model-checking algorithms, and, in the following discussion, we assume it to be the data structure of choice.

15.5.2.1 Reachability Analysis of Product Machine

As in the case of explicit graph algorithms, we may formulate the equivalence checking problem of two FSMs $\mathcal{M}_1$ and $\mathcal{M}_2$ as a reachability problem over the product FSM $\mathcal{M}_\times$. Similarly to Proposition 15.4, we have the following proposition.

Algorithm: ForwardStateTraversal

input: initial states I and a transition relation T
output: reachable states R
begin
$i := 0$
$R^{(0)} := I$
repeat
$i := i + 1$
$R^{(i)} := R^{(i-1)} \cup Img(R^{(i-1)}, T)$
until $R^{(i)} = R^{(i-1)}$
return $R^{(i)}$
end

Figure 15.9. An algorithm that performs forward reachability analysis for given initial states and a transition relation.

Proposition 15.7. *Two FSMs* $\mathcal{M}_1 = (Q_1, Q_1^0, \Sigma, \Omega, T_1, \lambda_1)$ *and* $\mathcal{M}_2 = (Q_2, Q_2^0, \Sigma, \Omega, T_2, \lambda_2)$ *are equivalent if and only if no bad state* $(q_1, q_2) \in Q_1 \times Q_2$ *with* $\lambda_1(\sigma, q_1) \neq \lambda_2(\sigma, q_2)$ *for some* $\sigma \in \Sigma$ *is reachable from the initial states* $Q_1^0 \times Q_2^0$ *of the product FSM.*

Because we are concerned with emptiness of the set intersection of the reachable states and the bad states of the product FSM, reachability analysis over the product machine forms the computation core of sequential equivalence checking. Essentially, reachability analysis can be conducted in two directions, forward [14] and backward [23]. Figures 15.9 and 15.10 sketch the procedures for forward and backward reachability analysis, respectively.

Example 15.6. *Figure 15.11 shows the effective reached state sets of the product FSM of Figure 15.3 with forward and backward state traversals. The two*

Algorithm: BackwardStateTraversal

input: final states F and a transition relation T
output: reachable states R
begin
$i := 0$
$R^{(0)} := F$
repeat
$i := i + 1$
$R^{(i)} := R^{(i-1)} \cup PreImg(R^{(i-1)}, T)$
until $R^{(i)} = R^{(i-1)}$
return $R^{(i)}$
end

Figure 15.10. An algorithm that performs backward reachability analysis for given target states and a transition relation.

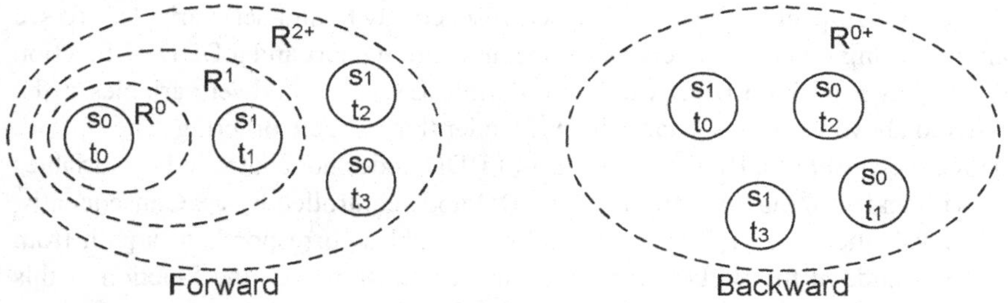

Figure 15.11. The reached state sets of the product FSM of Figure 15.3 with forward and backward state traversals.

constituent FSMs $\mathcal{M}_1$ and $\mathcal{M}_2$ are equivalent because in forward traversal, the reachable state sets are disjoint with the bad states $\{(s_0, t_1), (s_0, t_2), (s_1, t_0), (s_1, t_3)\}$ (whose transitions may produce "0" at the output), and similarly, in backward traversal, the states reachable from bad states are disjoint with the initial states. In this example, the forward traversal requires more iterations than the backward traversal. In the symbolic algorithms, the reached state sets are represented with characteristic functions.

The most sophisticated computation involved in a reachability analysis is the existential quantification in the image computation (in Step 5 of the algorithms in Figures 15.9 and 15.10). Extensive research efforts have been made in the 1990s to extend the capability of image computation. One of the most important and successful techniques is the so-called *quantification scheduling* [87], which determines the order of quantification over a set of variables. The objective is to make quantifications as local as possible such that the intermediate Boolean formulas are kept small.

Unlike that of an explicit algorithm over STGs, the complexity measure of an implicit algorithm is harder to quantify because it depends heavily on BDD sizes, which cannot be predicted well. On the other hand, because the procedures of Figures 15.9 and 15.10 in effect perform breadth-first search, the number of iterations is upper bounded by the forward and backward diameters, respectively. (The forward diameter is the length of the longest shortest path starting from an initial state to any state; the backward diameter is the length of the longest shortest path starting from any state to a target state.) Notice that the forward and backward diameter may differ substantially. Hence, it is sometimes beneficial to combine both forward and backward analysis in the same procedure.

15.5.2.2 State Partitioning of Multiplexed Machine

An alternative to symbolic equivalence checking is through the state equivalence formulation [38].

A *decomposition chart* of a Boolean function $\phi : [\![\boldsymbol{v}_r]\!] \times [\![\boldsymbol{v}_c]\!] \to \mathbb{B}$ is a two-dimensional truth table with rows indexed by $[\![\boldsymbol{v}_r]\!]$ and columns indexed by $[\![\boldsymbol{v}_c]\!]$.

We call $\boldsymbol{v}_r$ (respectively $\boldsymbol{v}_c$) the free-set (respectively bound-set) variables. To see an interesting connection between a decomposition chart and a BDD of function ϕ, we play a trick on BDD variable ordering. Let the bound-set variables $\boldsymbol{v}_c$ be ordered above the free-set variables $\boldsymbol{v}_r$. Under this variable ordering criterion, we define the *cutset* of a BDD to be the set of BDD nodes not controlled by variables $\boldsymbol{v}_c$ with an incoming edge from some BDD node controlled by $\boldsymbol{v}_c$. Consequently, every valuation $c \in [\![\boldsymbol{v}_c]\!]$ of the bound-set variables corresponds to a path from the root node of the BDD to a node in the cutset. Moreover, the function of this node in the cutset is $\phi(\boldsymbol{v}_r, c)$, that is, the function of column c (or called the *column pattern* of c) in the decomposition chart. Because of the BDD property that no two nodes in a BDD possess the same function (since the ordered BDD is reduced), we know that, for each column pattern in a decomposition chart, there is a unique corresponding BDD node in the cutset. Thus, the number of distinct column patterns equals the size of the cutset.

Example 15.7. *Consider a Boolean function $f : \mathbb{B}^4 \to \mathbb{B}$ over variables $\boldsymbol{v} = (v_1, v_2, v_3, v_4)$ with*

$$f(\boldsymbol{v}) = v_2 v_4.$$

Let $\boldsymbol{v}_c = (v_1, v_2)$ and $\boldsymbol{v}_r = (v_3, v_4)$ be the bound-set and free-set variables, respectively. Figure 15.12 (a) and (b) show the decomposition chart and the BDD of f. The function of the BDD node controlled by v_4 corresponds to the column pattern of the first and third columns. The function of the zero-terminal node of the BDD corresponds to the column pattern of the second and fourth columns. Valuations $\{(0,0),(1,0)\}$ of (v_1, v_2) (the indices for the first and third columns, respectively) are in one equivalence class; valuations $\{(0,1),(1,1)\}$ of (v_1, v_2) (the indices for

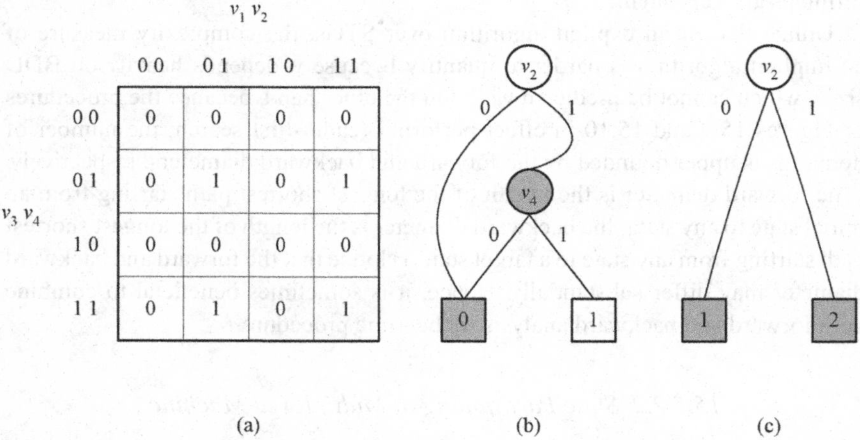

Figure 15.12. (a) The decomposition chart of function $f(\boldsymbol{v}) = v_2 v_4$ with bound-set and free-set variables $\{v_1, v_2\}$ and $\{v_3, v_4\}$, respectively. (b) The BDD of $f(\boldsymbol{v})$ with the bound-set variables ordered above the free-set variables. The nodes in the cutset are shaded. (c) An MTBDD abstraction.

the second and fourth columns, respectively) are in the other equivalence class. Observe that valuations $\{(0,0),(1,0)\}$ *(respectively* $\{(0,1),(1,1)\}$*) correspond to the else-branch (respectively then-branch) of the root of the BDD of (b). The decomposition chart and the BDD are equivalent representations.*

To see how this is useful, suppose we are asked to compute the partition over $[\![\boldsymbol{v}_c]\!]$ induced by the observational output of ϕ for all $[\![\boldsymbol{v}_r]\!]$. That is, we need to characterize the equivalence classes among elements in $[\![\boldsymbol{v}_c]\!]$ such that two elements $c_1, c_2 \in [\![\boldsymbol{v}_c]\!]$ are in the same equivalence class if and only if $\phi(r, c_1) = \phi(r, c_2)$ for all $r \in [\![\boldsymbol{v}_r]\!]$. Equivalently, in the decomposition chart of ϕ, two columns c_1 and c_2 having the same pattern (i.e., $\phi(r, c_1) = \phi(r, c_2)$ for all row index $r \in [\![\boldsymbol{v}_r]\!]$) are in the same equivalence class. Characterizing such equivalence classes can be done implicitly by constructing a BDD with $\boldsymbol{v}_c$ ordered above $\boldsymbol{v}_r$. It is because the paths leading to a node in the cutset of the BDD correspond to an equivalence class. Hence, the cutset and the paths leading to it encode all the information we need. Essentially, the size of the cutset equals the number of equivalence classes in the partition. Note that the BDD structure below the cutset is immaterial, and only the structure above (and including) the cutset is important. Therefore, we may possibly abstract the BDD with a multiterminal binary decision diagram (MTBDD). In the abstraction, each node in the cutset of the BDD is replaced with a distinct terminal node in the MTBDD while the structure of the BDD above the cutset is preserved in the MTBDD. Thus, the MTBDD can be seen as a function that maps an element to the index of its corresponding equivalence class. Note that even though the decomposition chart and the BDD are equivalent representations of equivalence classes, the latter can be much more succinct than the former.

Example 15.8. *The BDD of Figure 15.12 (b) can be abstracted with the MTBDD of (c) when the primary concern is the equivalence classes rather than the characteristics of the equivalence classes. In the abstraction, the zero-terminal node of (b) is replaced with the one-terminal node in (c), and the cutset node controlled by* v_4 *of (b) is replaced with the two-terminal node in (c). Notice that the BDD structure above the cutset of (b) is preserved in the MTBDD (above the terminal nodes) of (c).*

We may need also to compute the partition induced by a set of functions $\{\phi_1(\boldsymbol{v}_r, \boldsymbol{v}_c), \dots, \phi_n(\boldsymbol{v}_r, \boldsymbol{v}_c)\}$ instead of just a single function. We may think that the partition is induced by the column patterns in the decomposition chart formed by stacking the decomposition charts of $\phi_1, \dots, \phi_n$ altogether. To apply the implicit computation using BDDs, we may construct a single *hyperfunction* $\Phi : [\![\eta]\!] \times [\![\boldsymbol{v}_r]\!] \times [\![\boldsymbol{v}_c]\!] \to \mathbb{B}$ with

$$\Phi(\eta, \boldsymbol{v}_r, \boldsymbol{v}_c) = \bigwedge_{i=1}^{n} ((\eta \equiv i) \wedge \phi_i(\boldsymbol{v}_r, \boldsymbol{v}_c)),$$

where η is an n-valued variable (which can be encoded with a vector of Boolean variables in constructing the BDD of Φ). In the BDD of Φ, let variables η and $\boldsymbol{v}_r$ be the free set, and variables $\boldsymbol{v}_c$ be the bound set. Accordingly, the cutset of the BDD characterizes all the equivalence classes induces by $\phi_1, \ldots, \phi_n$.

Example 15.9. *To compute the partition induced by* $f(\boldsymbol{v}) = v_2 v_4$ *and* $g(\boldsymbol{v}) = v_3(v_1 \vee v_2)$ *with respect to bound set variables* $\{v_1, v_2\}$*, we construct the hyperfunction* h *of* f *and* g *as* $h(\eta, \boldsymbol{v}) = \overline{\eta} f(\boldsymbol{v}) \vee \eta g(\boldsymbol{v}) = \overline{\eta} v_2 v_4 \vee \eta v_3(v_1 \vee v_2)$*. Let* $\{v_1, v_2\}$ *and* $\{\eta, v_3, v_4\}$ *be the bound-set and free-set variables, respectively. Figure 15.13 shows the decomposition chart, BDD, and MTBDD of* $h(\eta, \boldsymbol{v})$ *in (a), (b), and (c), respectively.*

To see how the BDD analog of the decomposition chart is useful in equivalence verification, we first show that it can be exploited to characterize the state equivalence relation of an FSM. We take advantage of the BDD-based characterization of equivalence classes to compute equivalent states of an FSM $\mathcal{M} = ([\![\boldsymbol{s}]\!], Q^0, [\![\boldsymbol{x}]\!], \Omega, \boldsymbol{\delta}, \boldsymbol{\lambda})$. Essentially we need to compute the partition π^* from π^0, where π^i denotes the partition induced by $\sim^i$. To compute π^0 we build

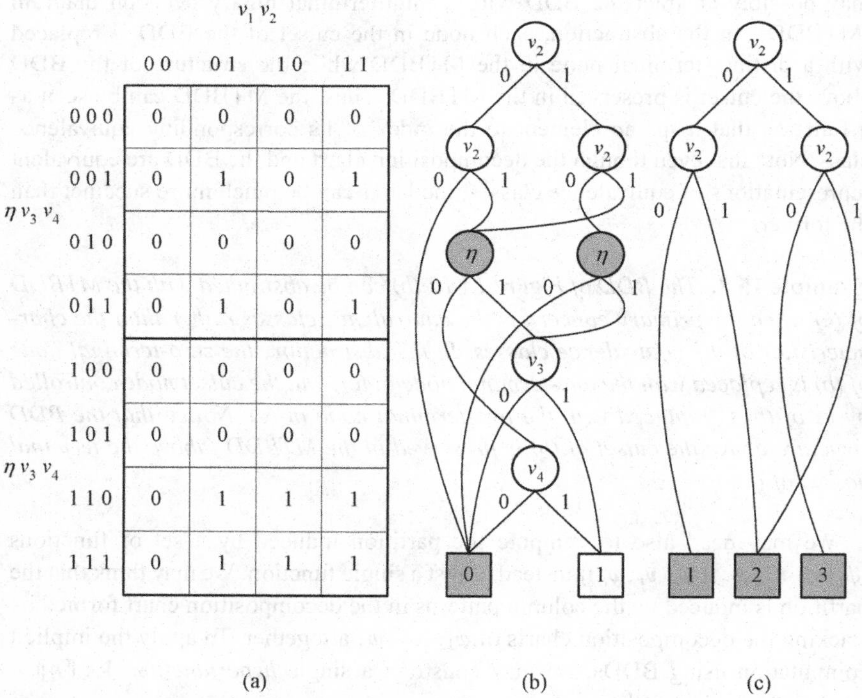

Figure 15.13. (a) The decomposition chart of function $h(\eta, \boldsymbol{v}) = \overline{\eta} v_2 v_4 \vee \eta v_3(v_1 \vee v_2)$ with bound-set and free-set variables $\{v_1, v_2\}$ and $\{\eta, v_3, v_4\}$, respectively. (b) The BDD of $h(\eta, \boldsymbol{v})$ with the bound-set variables ordered above the free-set variables. The nodes in the cutset are shaded. (c) An MTBDD abstraction.

the BDD of a hyperfunction for the output functions $\boldsymbol{\lambda}(\boldsymbol{x}, \boldsymbol{s})$ by setting $\boldsymbol{s}$ as the bound-set variables and all others as the free-set variables. Accordingly, the cutset C^0 of the BDD characterizes π^0. The BDD can then be simplified and abstracted into an MTBDD, which defines a function $\chi^0 : [\![\boldsymbol{s}]\!] \to \{1, \ldots, |C^0|\}$. For $q \in [\![\boldsymbol{s}]\!]$, $\chi^0(q)$ gives the equivalence class to which state q belongs.

By Definition 15.6, to compute π^{k+1} from π^k for $k \geq 0$, observe that two states $q_1, q_2 \in [\![\boldsymbol{s}]\!]$ are in the same equivalence class of π^{k+1} if and only if

(i) $\chi^k(q_1) = \chi^k(q_2)$, and
(ii) $\chi^k(Succ_\sigma(q_1)) = \chi^k(Succ_\sigma(q_2))$ for all $\sigma \in [\![\boldsymbol{x}]\!]$.

Assuming χ^k of π^k has been derived, we need to compute the equivalence classes induced by the functions $\chi^k(\boldsymbol{s})$ and $\chi^k(\boldsymbol{\delta}(\boldsymbol{x}, \boldsymbol{s}))$. Hence, we need to build the BDD for the hyperfunction of $\chi^k(\boldsymbol{s})$ and $\chi^k(\boldsymbol{\delta}(\boldsymbol{x}, \boldsymbol{s}))$ by setting state variables $\boldsymbol{s}$ as the only bound-set variables. (Note that the multiple-valued function χ^k can be rewritten in terms of a vector of Boolean functions, and thus can be represented with a BDD through a hyperfunction construction. Also, $\chi^k(\boldsymbol{\delta}(\boldsymbol{x}, \boldsymbol{s}))$ can be obtained through the BDD composition operation by substituting the variable s'_i of $\chi^k(\boldsymbol{s}')$ with the corresponding transition function $\delta_i(\boldsymbol{x}, \boldsymbol{s})$.) Again, the cutset C^{k+1} of the BDD gives us the equivalence classes of π^{k+1}, from which we may obtain the MTBDD of χ^{k+1}. The iteration terminates when the number of equivalence classes does not increase, that is, $|C^{k+1}| = |C^k|$. Upon termination, the MTBDD of χ^* characterizes the partition π^*.

For checking the equivalence of two FSMs $\mathcal{M}_1$ and $\mathcal{M}_2$ (as in the development of the corresponding explicit algorithm), the previous procedure can be extended to characterize the state equivalence of the multiplexed FSMs $\mathcal{M}_\uplus$ of $\mathcal{M}_1$ and $\mathcal{M}_2$. (Note that the auxiliary variable α is treated as a state variable and thus is considered as a bound-set variable.) Upon termination, we check if the initial states of $\mathcal{M}_1$ and $\mathcal{M}_2$ are in the same equivalence class. By Proposition 15.6, $\mathcal{M}_1$ and $\mathcal{M}_2$ are equivalent if and only if their initial states are in the same equivalence class. A detailed exposition can be found in [38].

Example 15.10. *Figure 15.14 shows the state space partitioning of the multiplexed FSM of Figure 15.4. The two constituent FSMs $\mathcal{M}_1$ and $\mathcal{M}_2$ are equivalent*

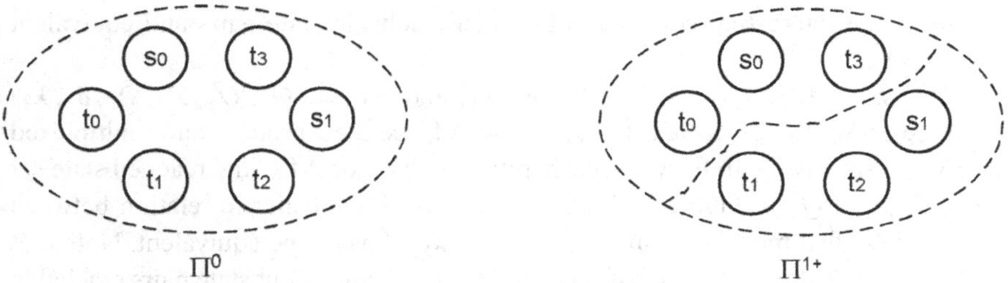

Figure 15.14. The state space partitioning of the multiplexed FSM of Figure 15.4.

because their initial states s_0 and t_0 are in the same equivalence class in the final partition π^1. In the symbolic algorithm, the equivalence classes of a partition are represented with a BDD.

In the computation of state equivalence, since every equivalence class is represented with a BDD node, the verification capability of the approach may be limited. In practice, the approach may handle designs with up to millions of equivalence classes. Hence, reducing the peak amount of equivalence classes may extend the verification capability and improve the efficiency of the algorithm. An effective way to do so is by divide-and-conquer and handling separately the partition induced by every output function. In fact, two FSMs are equivalent if and only if their initial states are in one equivalence class for the partition induced by every output function. Actually, the number of equivalence classes induced by all output functions may be exponentially larger than the one induced by any single function. (This reduction can be considered as one instance of the so-called *cone-of-influence reduction* [12].) Another way to reduce the number of equivalence classes is to restrict state partitioning to the subspace of reachable states. When the (exact or overapproximated) reachable state sets of $\mathcal{M}_1$ and $\mathcal{M}_2$ are known, the state equivalence can be characterized within these subspaces by the BDD *constrain* operator [15, 38], which removes from consideration useless equivalence classes. (Note that, unlike verification in the disjoint union space, verification in the product space may not benefit much from reachability information of the individual FSMs. Moreover, reachability analysis on individual FSMs may be much easier than analysis on their product FSM.)

Whereas reachability analysis of a product FSM is usually unpredictable, verification in the disjoint union space may be more robust. After all, two equivalent FSMs must have the same number of equivalence classes in their respective reachable state spaces. On the other hand, one limitation of the above BDD-based characterization of equivalence classes is that it is applicable only to partitions induced by functions rather than by relations; in comparison, verification based on reachability analysis is more flexible and can handle nondeterministic transitions straightforwardly.

15.5.2.3 Connections between Reachability and Equivalence

There is an interesting connection between reachable state pairs and equivalent state pairs.

Assume $\mathcal{M}_1 = (Q_1, Q_1^0, \Sigma_1, \Omega_1, \boldsymbol{\delta}_1, \boldsymbol{\lambda}_1)$ and $\mathcal{M}_2 = (Q_2, Q_2^0, \Sigma_2, \Omega_2, \boldsymbol{\delta}_2, \boldsymbol{\lambda}_2)$ are the FSMs to be verified. Let $\mathcal{M}_\times$ and $\mathcal{M}_\uplus$ be their product and multiplexed FSMs, respectively. In forward reachability analysis of $\mathcal{M}_\times$, any reached state set $R^{(i)} \subseteq Q_1 \times Q_2$ in Figure 15.9 can be seen as an equivalence relation between Q_1 and Q_2 that must be maintained for the two FSMs to be equivalent. Note that forward reachability analysis of $\mathcal{M}_\times$ characterizes equivalent state pairs reachable only from the initial states of $\mathcal{M}_\times$ (but not all equivalent state pairs in $Q_1 \times Q_2$).

In contrast, in backward reachability analysis of $\mathcal{M}_\times$, the complement of $R^{(i)} \subseteq Q_1 \times Q_2$ in Figure 15.10 is equal to $\sim^i \subseteq (Q_1 \uplus Q_2) \times (Q_1 \uplus Q_2)$ of $\mathcal{M}_\uplus$ except for ignoring the equivalence among individual FSMs, that is,

$$(Q_1 \times Q_2) \backslash R^{(i)} = \sim^i \backslash \{Q_1 \times Q_1 \cup Q_2 \times Q_2\}.$$

Example 15.11. *Consider the reached state set with backward state traversal in Figure 15.11,*

$$R^* = (s0, t1), (s0, t2), (s1, t0), (s1, t3),$$

and the state space partition of the related multiplexed FSM in Figure 15.14,

$$\sim^* = (s0, t0), (s0, t3), (s1, t1), (s1, t2), (t0, t3)(t1, t2).$$

It holds that $(Q_1 \times Q_2) \backslash R^\star = \sim^\star \backslash \{Q_1 \times Q_1 \cup Q_2 \times Q_2\}$.

Note that backward reachability analysis of $\mathcal{M}_\times$ characterizes all equivalent state pairs in $Q_1 \times Q_2$ whereas state partitioning of $\mathcal{M}_\uplus$ considers all equivalent state pairs in $Q_1 \times Q_2 \cup Q_1 \times Q_1 \cup Q_2 \times Q_2$. Consequently, the number of iterations needed in state partitioning of $\mathcal{M}_\uplus$ may be greater than that in backward reachability analysis of $\mathcal{M}_\times$. Nevertheless, when restricted to the reachable subspace of $\mathcal{M}_1$ and $\mathcal{M}_2$, state partitioning of $\mathcal{M}_\uplus$ terminates at the same iteration as backward reachability analysis of $\mathcal{M}_\times$ because the unreachable state space is not partitioned. In fact, backward reachability analysis can be seen as state space partitioning over the product space, in contrast to state space partitioning over the disjoint union space. (Notice that $R^{(i)}$ in forward reachability analysis does not follow k-step state equivalence, so the number of iterations may not be comparable with backward reachability analysis.)

15.6 Safety Property Checking through Time-Frame Expansion

In Section 15.5, we see that sequential equivalence checking can be formulated as reachability analysis, where we want to assert that the underlying product machine always outputs **true** under any state reachable from the initial state set. It is in fact an instance of safety property checking, where temporal properties of a state transition system can be checked through reachability analysis. In this section, we broaden our discussion a bit and speak of safety property checking. The discussion is in turn applicable to sequential equivalence checking.

The capacity of ROBDD-based verification algorithms is typically limited to designs with at most hundreds of registers. This limitation is due to the memory explosion problem because BDDs may not represent some characteristic functions efficiently or may grow too large due to quantifier elimination (e.g., in image computation), even though dynamic variable reordering [74], quantification scheduling, and some other techniques are helpful.

In addition to BDDs, satisfiability (SAT) solving over Boolean formulas in the conjunctive normal form (CNF) is another core engine for Boolean reasoning.

SAT solvers based on the the DPLL algorithm (Davis-Putnam-Logemann-Loveland) [18, 19] were recently engineered into a very effective technology, due to breakthroughs including conflict-based nonchronological back-tracking [58], two watched literals [91], fast Boolean constraint propagation [62], and so forth. As a consequence, more and more verification problems are reformulated as SAT solving to exploit the capability of SAT solvers.

In SAT solving, we are asked if a CNF Boolean formula $\phi(\boldsymbol{x})$ is *satisfiable* (true under some valuation, or interpretation, of the Boolean variables $\boldsymbol{x}$, i.e., $\exists \boldsymbol{x}.\phi(\boldsymbol{x})$). Therefore, a SAT solver semantically performs existential quantification. On the other hand, a SAT solver can be exploited to test if a Boolean formula $\phi(\boldsymbol{x})$ is *valid* (true under every valuation of $\boldsymbol{x}$, i.e., $\forall \boldsymbol{x}.\phi(\boldsymbol{x})$) by testing the satisfiability of the complement $F\phi(\boldsymbol{x})$. That is, $\phi(\boldsymbol{x})$ is valid if and only if $F\phi(\boldsymbol{x})$ is unsatisfiable. However, the main problem of validity checking is whether the complement can be represented efficiently, because a SAT solver expects $F\phi(\boldsymbol{x})$ to be in CNF. In general, complementing a CNF into another CNF may suffer from an exponential blow-up in the Boolean formula size. Fortunately, in hardware verification, the problem with the complement can sometimes be overcome. For instance, to test if the output of a circuit always produces 1 regardless of the valuations of the input variables, we may add an inverter at the output. By translating the new circuit into a CNF, we may apply SAT solving. The original circuit always produces 1 if and only if the CNF of the new circuit is unsatisfiable. Translating a circuit into a CNF can be done in polynomial time, and the size of the resultant CNF is polynomial in the size of the circuit [67]. (The polynomial conversion is possible because of the introduction of new local variables. Existentially quantifying out these variables results in an equivalent Boolean formula depending only on the original variables.) Nonetheless, image computation using SAT is less straightforward than using BDDs. In fact, a genuinely hard task for SAT solving is to solve quantified Boolean formulas alternating existential and universal quantifications: solving quantified Boolean formulas (QBFs) is PSPACE-complete [83], whereas SAT [13] is NP-complete. After all, SAT solvers are good at spotting *one* satisfying valuation instead of *all* satisfying valuations.

In contrast to BDD-based algorithms, SAT-based verification algorithms are less memory hungry. However, long run time may be the main bottleneck in SAT solving. In hardware verification, SAT may be preferable to BDDs in that a verification task is less likely to abort because of memory limitations, and it catches more design bugs as time flows, so it is more suitable in hardware debugging than BDDs. However, because SAT is not effective in dealing with QBFs, SAT is not as applicable to general model checking as BDDs. To take full advantage of the strengths of SAT solving, the approach to model checking needs to be modified as discussed later.

15.6.1 Bounded Model Checking

To verify whether a state transition system respects a temporal safety property, it can be translated into a set of QBFs. (In fact, any PSPACE-complete problem can be converted into QBF solving.) For instance, the termination condition of a reachability analysis of Figure 15.9 or 15.10 can be expressed with a QBF. Essentially, $R^{(i+1)} \Rightarrow R^{(i)}$ (equivalent to the termination check $R^{(i+1)} = R^{(i)}$) is valid if and only if $R^{(i+1)} \wedge FR^{(i)}$, or equivalently $Img(R^{(i)}, T) \wedge FR^{(i)}$, is unsatisfiable. Recursively reexpressing $R^{(i)}$ with $R^{(i-1)} \vee Img(R^{(i-1)}, T)$ yields a QBF, where existential and universal quantifications alternate due to the complement of $R^{(i)}$ and the existential quantification in image computation. However, notice that SAT solvers are not good at solving QBFs.

Example 15.12. *Given the initial state set I and the transition relation T, the termination check $R^{(2)} = R^{(1)}$ can be translated to the QBF*

$$\forall \boldsymbol{s}^0, \boldsymbol{s}^1, \boldsymbol{s}^2, \exists \boldsymbol{t}^0.[(I(\boldsymbol{s}^0) \wedge T(\boldsymbol{s}^0, \boldsymbol{s}^1) \wedge T(\boldsymbol{s}^1, \boldsymbol{s}^2)) \Rightarrow (I(\boldsymbol{s}^2) \vee (I(\boldsymbol{t}^0) \wedge T(\boldsymbol{t}^0, \boldsymbol{s}^2)))],$$

where $T(\boldsymbol{s}, \boldsymbol{s}')$ is an abbreviation for $\exists \boldsymbol{x}.T(\boldsymbol{x}, \boldsymbol{s}, \boldsymbol{s}')$ (or, alternatively, it results from considering input variables $\boldsymbol{x}$ as part of the states variables). The QBF asserts that for all states $\boldsymbol{s}^0, \boldsymbol{s}^1, \boldsymbol{s}^2$, if $\boldsymbol{s}^0$ is an initial state, and there are transitions respectively from $\boldsymbol{s}^0$ to $\boldsymbol{s}^1$ and from $\boldsymbol{s}^1$ to $\boldsymbol{s}^2$, then either $\boldsymbol{s}^2$ is an initial state or $\boldsymbol{s}^2$ can be reached from an initial state $\boldsymbol{t}^0$. In other words, if a state is reached at the second iteration, it must be an initial state or have been reached at the first iteration. That is, no new state is reached at the second iteration.

To take full advantage of SAT solving, we had better restrict the formulation to quantifier-free Boolean formulas. One possible solution is to ignore the termination condition in reachability analysis, and to ask if the considered transition system satisfies the property under any execution of input sequences whose length is upper-bounded by k. By sacrificing "completeness," instead of QBF solving one will be reduced to solving quantifier-free Boolean formulae, as we will see shortly. This bounded-length model checking is known as *bounded model checking* (BMC) [2]. Note that, for finite-state systems, BMC is indeed complete, however, under a rather useless upper bound on k, such as, the number of states of the transition system. Although BMC is in theory complete for finite-state systems, in almost all practical cases it runs out of computational resources long before the completeness bound is reached. Therefore, it is considered incomplete in practice.

BMC is very similar to sequential ATPG, developed earlier in the testing community, with some minor difference in their data structures. Let I, T, and P be the initial state set, transition relation, and temporal property, respectively. (In the context of sequential equivalence checking, the temporal property $P(\boldsymbol{s})$ to be verified is $\lambda_\times(\boldsymbol{s}) \equiv 1$.) BMC checks the satisfiability of the following Boolean formulas in order:

$$\begin{aligned}
\mathrm{Bmc}^0 &: I(\boldsymbol{s}^0) \wedge FP(\boldsymbol{s}^0),\\
\mathrm{Bmc}^1 &: I(\boldsymbol{s}^0) \wedge T(\boldsymbol{s}^0, \boldsymbol{s}^1) \wedge FP(\boldsymbol{s}^1),\\
&\vdots\\
\mathrm{Bmc}^k &: I(\boldsymbol{s}^0) \wedge T(\boldsymbol{s}^0, \boldsymbol{s}^1) \wedge \cdots \wedge T(\boldsymbol{s}^{k-1}, \boldsymbol{s}^k) \wedge FP(\boldsymbol{s}^k),
\end{aligned}$$

where the state variables are annotated with time indices in superscript. (Here we treat primary input variables as part of the state variables.) The procedure either stops at a satisfiable Boolean formula Bmc^i or proceeds otherwise with a new formula Bmc^{i+1}. Boolean formula Bmc^i is satisfiable if the temporal property P is violated and a counterexample of length i is found. Consequently, BMC can be used to locate bugs under some length bound constrained mainly by computational resources.

To "visualize" BMC with circuits, the feedback loop of a sequential circuit is eliminated and replaced with a series of concatenated replicas of the combinational core of the circuit. A sequential circuit is expanded into a combinational one with several time frames. Thereby, BMC converts a sequential verification problem into a combinational one (which can be solved using SAT straightforwardly) through the so-called *time-frame expansion*. In effect, the corresponding STG of the original circuit is unwound into a tree through the expansion. Thus, states at level i of the tree correspond to states at the ith time frame reached from the initial states; BMC tries to locate a state at level i that violates the property P under verification.

Example 15.13. *Figure 15.15 (a) shows the time-frame expansion of the FSM $\mathcal{M}_1$ of Figure 15.2. When $\mathcal{M}_1$ is converted to its time-frame expansion, the corresponding STG is unwound as shown in Figure 15.15 (b). Because circuits can be efficiently converted to CNFs as mentioned earlier, BMC using SAT applies.*

Figure 15.16 sketches the procedure of BMC. Let D be the target states that violate property P, that is, the FP-states. The procedure checks if there is some target state reachable from the initial states I under a transition path of length i. Because the Boolean formulas of two consecutive iterations Bmc^i and Bmc^{i+1} are almost the same, intuitively many of the learned clauses induced by the conflict analysis in Bmc^i are applicable for solving Bmc^{i+1}. Therefore, one important issue for BMC to be effective is how to efficiently reuse some of the learned clauses from the previous iterations and remove those that become invalid in the current iteration. It brings up the notion of *incremental SAT solving*. More details can be found in, for example, [20, 89].

15.6.2 Unbounded Model Checking

The main weakness of BMC is the termination criterion: it is unknown how many time frames are needed to conclude for the absence of bugs. The length of time-frame expansion is mostly set by the limit of computational resources. Shortly after the introduction of BMC, several techniques, for example, [56, 57, 80], were developed to cope with the termination problem in order to guarantee automatic termination of SAT-based model checking whenever a proof is established, just like in BDD-based model checking. Removing the boundedness limitation of BMC yields the so-called *unbounded model checking* (UMC).

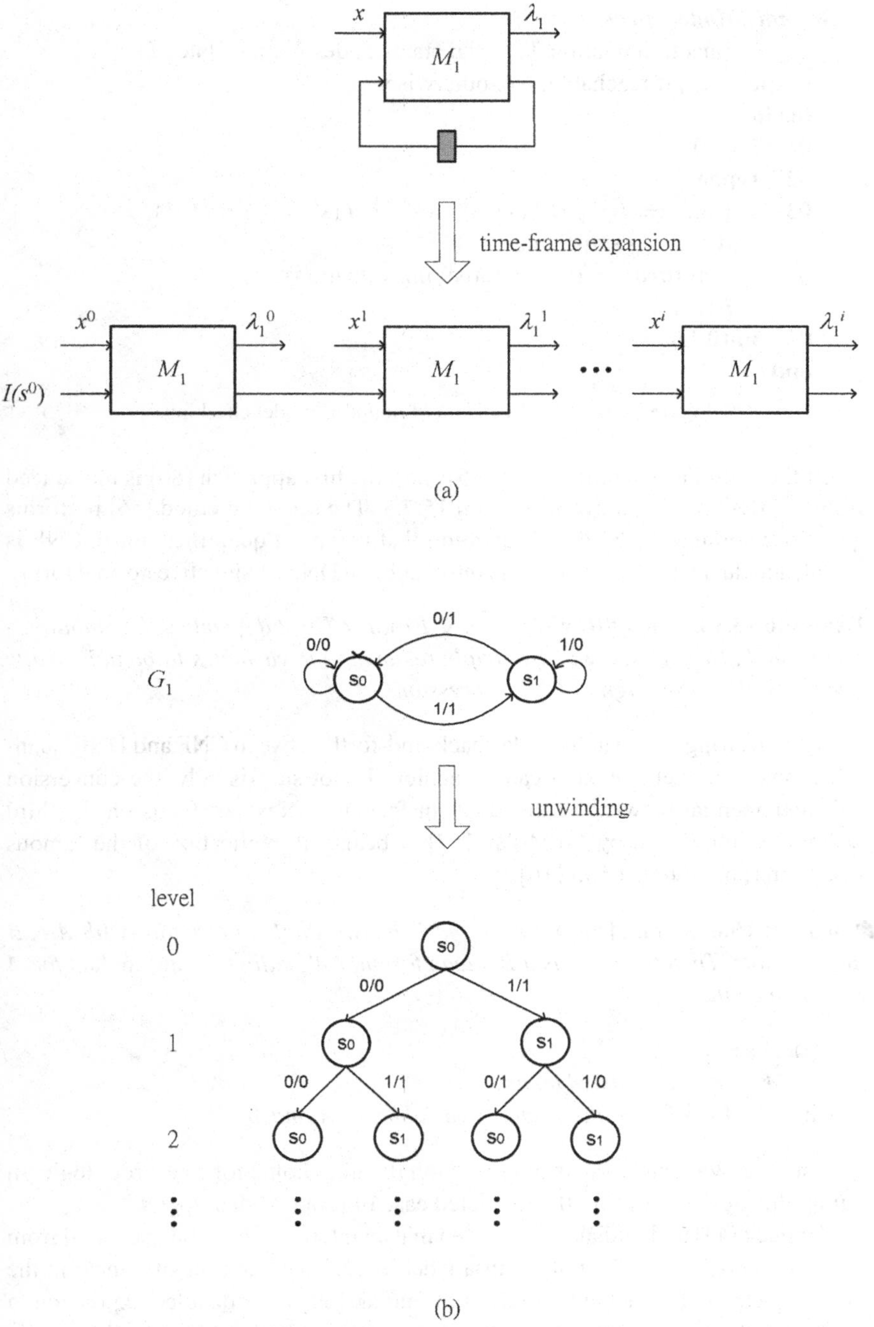

Figure 15.15. (a) Time-frame expansion of the FSM $\mathcal{M}_1$ of Figure 15.2. (b) The effective unwinding of the STG G_1.

BoundedModelChecking

input: transition relation T, initial states I, destination states D
output: Yes, if reachable; No, otherwise
begin
  $i := 0$
  repeat
    $\mathrm{Bmc}^i := I(s^0) \wedge T(s^0, s^1) \wedge \cdots \wedge T(s^{i-1}, s^i) \wedge D(s^i)$
    if Bmc^i is satisfiable
      return Yes (with a satisfying valuation)
    $i := i + 1$
  until false
end

Figure 15.16. The procedure of bounded model checking.

Of the techniques reported in [56, 57, 80], the first approach [80] is introduced from a different perspective in Section 15.7.3. The second method [56] performs quantifier elimination by the observation that universal quantification in CNF is trivial, and dually so is existential quantification in DNF (disjunctive normal form).

Example 15.14. *The QBF* $\forall b.[(a \vee b \vee Fc)(a \vee Fb \vee d)]$ *equals the quantifier-free formula* $(a \vee Fc)(a \vee d)$ *by simply removing the variables to be universally quantified from the original CNF expression.*

By converting Boolean formulas back-and-forth between CNF and DNF, quantifications in model checking can be achieved. Not surprisingly, the conversion risks exponential blow-up in the Boolean formulas. Next we focus on the third technique, introduced by McMillan [57], a beautiful application of the famous Craig interpolation theorem [16].

Theorem 15.8. (Craig [57].) *Let A and B be two Boolean formulas with $A \wedge B$ unsatisfiable. Then there exists a Boolean formula A', called the* interpolant *for A and B, such that*

(i) $A \Rightarrow A'$,
(ii) *$A' \wedge B$ is unsatisfiable, and*
(iii) *A' only refers to the common variables of A and B.*

Note that, whereas Craig's interpolation theorem holds for first-order logic, in our application we use it in the restricted case of propositional logic.

Krajíček [43] and Pudlák [69] showed that an interpolant can be generated from a *resolution refutation* [73] of an unsatisfiable CNF with complexity linear in the proof. (Interested readers are referred to Pudlák [69] for a detailed exposition; a different but equivalent construct can be found in [57].) In fact, modern DPLL-style SAT solvers (such as GRASP [58], Chaff [62], BerkMin [27]) can be exploited to generate resolution refutations for unsatisfiable CNFs, and thus interpolants

UnboundedModelChecking

input: transition relation T, initial states I, destination states D
output: Yes, if reachable; No, otherwise
begin
  if $I \wedge D$ is satisfiable
    return Yes (with a satisfying valuation)
  $i := 0$
  repeat
    $R := I(s^{-1})$
    $A := R(s^{-1}) \wedge T(s^{-1}, s^0)$
    $B := T(s^0, s^1) \wedge \cdots \wedge T(s^{i-1}, s^i) \wedge (D(s^0) \vee \cdots \vee D(s^i))$
    if $A \wedge B$ is satisfiable
      return Yes (with a satisfying valuation)
    repeat
      generate interpolant $A'(s^0)$ for A and B
      if $A'(s^0) \Rightarrow R(s^0)$
        return No
      $R := R(s^{-1}) \vee A'(s^{-1})$
      $A := R(s^{-1}) \wedge T(s^{-1}, s^0)$
    until $A \wedge B$ is satisfiable
    $i := i + 1$
  until false
end

Figure 15.17. An algorithm that performs unbounded model checking based on interpolation.

(with respect to prespecified clause sets A and B). In what follows, we suppose that SAT solvers capable of generating interpolants are available.

Figure 15.17 shows the algorithm of interpolation-based unbounded model checking, where all Boolean formulas are assumed to be in CNF. Given the transition relation T, it checks whether the destination states D are reachable from the initial states I. Variable vectors in parentheses are instantiated with time indices in superscript as in BMC. Observe that the outer loop of the algorithm is the same essentially as the BMC steps of Figure 15.16, whereas the inner loop performs overapproximated reachability analysis.

To understand the overapproximated image computation in the inner loop, we sketch the main idea. Recall that reachability analysis looks for the fixed point $R^* = \vee_i R^{(i)}$, where $R^{(i)} = R^{(i-1)} \vee Img(R^{(i-1)}, T)$ and $R^{(0)} = I$, the initial state set. A destination state set D is reachable from I if and only if $R^* \wedge D \neq$ **false**. Replacing the exact image computation Img with an overapproximation image operator Img' (i.e., $Img(C, T) \Rightarrow Img'(C, T)$ for any state set C and transition relation T) results in an overapproximated reachability computation. To preserve the accuracy of such approximated reachability analysis, we define the *adequacy* of an image overapproximation as follows.

Definition 15.11. (Adequacy.) *Given a state transition system with transition relation T, an overapproximation R' of the image of a state set C, that is, $R' \supseteq Img(C, T)$, is* adequate *with respect to the destination states D when, if C cannot reach D, R' cannot reach D.*

So if an overapproximation image operator Img' is adequate with respect to D, then D is reachable from I if and only if $R'^{*} \wedge D \neq$ **false**, where $R'^{*} = \vee_i R'^{(i)}$ with $R'^{(i)} = R'^{(i-1)} \vee Img'(R'^{(i-1)}, T)$ and $R'^{(0)} = I$. The question is how to find an adequate Img' operator. As a step to our goal, we first define an image operator that is adequate only for k steps.

Definition 15.12. (k-Adequacy.) *Given a state transition system with transition relation T, an overapproximation R' of the image of a state set C, that is, $R' \supseteq Img(C, T)$, is* k-adequate *with respect to the destination states D when, if C cannot reach D, R' cannot reach D in k or fewer steps.*

We will argue that in the k-th iteration of the outer loop of the UMC algorithm in Figure 15.17, step 11 is a k-adequate image operation. (Notice that, unlike exact image computation, no existential quantification is needed in the computation.) In addition, if k is such that all states reaching D are within distance k, k-adequacy becomes general adequacy, because the overapproximated reachable states computed from I under such a k-adequate image operator are disjoint from the states reaching D. The conclusion is that even under such an approximation, reachability analysis with respect to I and D returns a correct answer, within a finite number of steps.

Figure 15.18 illustrates the computation of a few steps of the inner loop of the interpolation-based UMC at the kth iteration of the outer loop, where $R'(s^{-1}) = I(s^{-1}) \vee A'(s^{-1})$, $R''(s^{-1}) = R'(s^{-1}) \vee A''(s^{-1})$, and $T^k(s^0, s^1, \ldots, s^k)$ is a shorthand for $T(s^0, s^1) \wedge T(s^1, s^2) \wedge \cdots \wedge T(s^{k-1}, s^k)$. By Theorem 15.8, for unsatisfiable $A \wedge B$ with $A = R(s^{-1}) \wedge T(s^{-1}, s^0)$ and $B = T(s^0, s^1) \wedge T(s^1, s^2) \wedge \cdots \wedge T(s^{k-1}, s^k) \wedge (D(s^0) \vee \cdots \vee D(s^k))$, we know that A' depends only on variables s^0 (which are common to A and B), that $A \Rightarrow A'$ implies A' is an over-approximation of $Img(R, T)$, and that the unsatisfiability of $A' \wedge B$ implies that states A' cannot reach D within k steps. Consequently, A' is a k-adequate overapproximation of the image of R. (Note that Boolean formula B changes only outside the inner loop so k-adequacy is maintained for the image computations at the kth iteration of the outer loop.) As an example, R' (respectively R'') of Figure 15.18 is essentially a k-adequate overapproximation of $R^{(1)}$ (respectively $R^{(2)}$) in the exact forward reachability analysis of Figure 15.9. The inner loop of Figure 15.17 iterates until the overapproximated reached states R can reach the destination states D in $k+1$ transitions, that is, $A \wedge B$ is satisfiable at step 16. Because R is an overapproximated reached state set, the satisfying valuation of step 16 may be a spurious counterexample. Thus, another iteration of the outer loop is needed to strengthen the k-adequacy of image computation by increasing i. It is worth mentioning that, if in the ith iteration of the outer loop of Figure 15.17 the inner loop is executed j times, then certainly step 9 will not be executed until the $(i+j)$th iteration of the

$$\underbrace{I(s^{-1}) \wedge T(s^{-1}, s^0)}_{A} \quad \wedge \quad \underbrace{T^k(s^0, s^1, \ldots, s^k) \wedge (D(s^0) \vee \cdots \vee D(s^k))}_{B}$$

$$\downarrow$$

$$A'(s^0)$$

$$\swarrow$$

$$\underbrace{R'(s^{-1}) \wedge T(s^{-1}, s^0)}_{A} \quad \wedge \quad \underbrace{T^k(s^0, s^1, \ldots, s^k) \wedge (D(s^0) \vee \cdots \vee D(s^k))}_{B}$$

$$\downarrow$$

$$A''(s^0)$$

$$\swarrow$$

$$\underbrace{R''(s^{-1}) \wedge T(s^{-1}, s^0)}_{A} \quad \wedge \quad \underbrace{T^k(s^0, s^1, \ldots, s^k) \wedge (D(s^0) \vee \cdots \vee D(s^k))}_{B}$$

$$\vdots$$

Figure 15.18. The image overapproximations $A', A'', \ldots$ computed by the inner loop (Steps 10–16) at the kth iteration of the outer loop of the UMC procedure of Figure 15.17.

outer loop because it implies I cannot reach D in $(i + j)$ transitions. Therefore, step 17 can be placed alternatively in the inner loop between steps 15 and 16 of the procedure.

We comment on steps 7, 12, and 14 of Figure 15.17. For step 7, translating the Boolean formula into CNF may be difficult because of the disjunctions. Nevertheless, the aforementioned polynomial-size translation from circuit to CNF [67] can be applied, provided that the circuit representing the characteristic function of D is available. Note that the disjunctions of $D(s^i)$ with $D(s^0), \ldots, D(s^{i-1})$ are necessary for k-adequacy to hold for the overapproximated image computation. (In BMC, these disjunctions are redundant in Boolean formula Bmc^i because in checking Bmc^i we assume Bmc^j is unsatisfiable for any $j = 0, \ldots, i$.) For step 12, checking the validity of $A' \Rightarrow R$ is identical to checking the unsatisfiability of $A' \wedge FR$. Because the interpolant A' is in circuit representation when generated from a refutation [57, 69] and the circuit for the characteristic function of R can be built, the circuit representing $A' \wedge FR$ can be built as well and translated into CNF in polynomial time for a SAT solver to solve. For step 14, similar to step 7, the disjunction of A' and R can be translated into CNF by the circuit-to-CNF conversion from the circuits representing the characteristic functions of A' and R.

The UMC procedure of Figure 15.17 terminates at steps 2, 9, or 13. At step 2, it returns Yes because the initial states I and destination states D intersect. At step 9, it returns Yes because at least an initial state reaches the destination set in $i + 1$ transitions. At step 13, it returns No because there is no newly reached state in the overapproximated image A' and thus the reached state set R is closed, and moreover it is disjoint from D.

The number i of iterations of the outer loop of Figure 15.17 is bounded from above by the backward diameter from destination states D (i.e., the length of the longest shortest path that leads to a state in D). The reason is that, when i is greater than or equal to this diameter, Boolean formula B is satisfiable under any state reaching D, that is, B contains all the states reaching D. As long as I cannot reach D, the computed interpolant in every inner-loop iteration must be disjoint from B and thus is adequate. As the overapproximated reached state set R grows monotonically, the computation must reach a fixed point, and the algorithm terminates eventually at step 13. The upper bound defined by the backward diameter is tight and is the same as the number of iterations needed in BDD-based backward reachability analysis. However, unlike BDD-based exact reachability analysis, the interpolation-based UMC may terminate in fewer iterations than this bound because of the overapproximated image computation.

In BMC, the proof of a satisfiable Bmc^i is useful in providing a trace that activates a design error. On the contrary, the proof of an unsatisfiable Bmc^i is rather useless and is a waste in some sense. In comparison, the interpolation-based UMC extracts useful information out of the refutation through the Craig Interpolation Theorem. On the other hand, when BDD-based and SAT-based model checking approaches are compared, it is observed that SAT-based methods generate only proofs relevant to the properties to be checked and thus are often more effective than BDD-based methods, which may generate irrelevant proofs as well as relevant ones.

15.7 Bridging the Gap between Combinational and Sequential Equivalence Checking

The aforementioned verification methods of Sections 15.5 and 15.6 make no assumptions about structural similarities between circuits to be compared. They are applicable to equivalence verification of designs with completely different implementations. However, it is this generality that limits their capability and scalability to verify large designs. The generality is often unnecessary because in most cases the designs being compared possess structural similarities to some extent. Although the similarities may not be sufficient to demonstrate the equivalence between two designs, their identification may substantially simplify verification tasks. As an extreme example, when the relation between state encodings of the two designs to be compared is known, the equivalence verification becomes pure combinational checking, and the complexity reduces from a PSPACE-complete problem (sequential equivalence checking) to a coNP-complete problem (combinational equivalence checking). This dramatic complexity reduction motivates the exploitation of circuit similarities in equivalence verification.

To be precise, we define equivalent signals of sequential circuits as a metric for structural similarities, where the term *signal* means a wire in a circuit.

Definition 15.13. (Signal Equivalence.) *Given a circuit implementation of an FSM, two signals in the circuit are* equivalent *(i.e., they are* corresponding signals*)* *if their values are all identical or all complementary to each other in any execution of the FSM.*

In fact, this notion of correspondence can be extended straightforwardly to signals between two different circuits. For instance, one may construct a product machine of the two circuits and seek equivalent signals inside it.

Given two sequential circuits, especially when one is optimized from the other through logic synthesis tools, their circuit structures, though they may look different, may possess a large number of equivalent signals. The identification of these equivalent signals may not be trivial. The labels (names) of signals may have been changed or lost during logic optimization, for instance, because of the creation of new signals, elimination of existing signals, and so on. There is little hope of obtaining corresponding signals simply by name matching. Moreover, even signals with the same name are not necessarily equivalent, and they need to be proved as well. Identifying corresponding signals without relying on pure name matching has its practical importance. In this section, we introduce some effective techniques that identify circuit similarities for sequential circuits without relying on name matching.

15.7.1 Inductive Register Correspondence

Among signals whose correspondences are to be identified, signals at register outputs (state variables) are the most important to demonstrate the equivalence between two sequential circuits. We call the equivalence of signals at register outputs the *register correspondence*. It can be envisaged that, if two sequential circuits differ only combinationally, there exists a register correspondence between these two circuits, that is, every state transition function of one circuit has an equivalent transition function of the other. If the register correspondence is established, then the underlying sequential equivalence checking problem is reduced to a combinational one.

By Definition 15.13, to compute equivalent signals requires knowledge of the reachable state set of an FSM. As reachability analysis is hard, we turn to seek approximation approaches to the identification of equivalent signals. In fact, induction can be used as an effective way of detecting register correspondences [22, 24]. Although the characterization is incomplete, it captures an interesting class of register correspondences.

Definition 15.14. (Inductive Register Correspondence.) *An* inductive register correspondence *of an FSM* $\mathcal{M} = (\llbracket \boldsymbol{s} \rrbracket, I, \llbracket \boldsymbol{x} \rrbracket, \Omega, \boldsymbol{\delta}, \boldsymbol{\lambda})$ *is an equivalence relation* $\stackrel{\text{rc}}{\equiv} \subseteq \{\boldsymbol{s}\} \times \{\boldsymbol{s}\}$ *over the state variables that satisfies, for all valuations of* $\boldsymbol{x}$, $\boldsymbol{s}$,

$$\textit{Base case:} \quad I(\boldsymbol{s}) \Rightarrow R_{\stackrel{\text{rc}}{\equiv}}(\boldsymbol{s}), \ \textit{and} \tag{15.6}$$

$$\textit{Inductive case:} \quad R_{\stackrel{\text{rc}}{\equiv}}(\boldsymbol{s}) \Rightarrow R_{\stackrel{\text{rc}}{\equiv}}(\boldsymbol{\delta}(\boldsymbol{x}, \boldsymbol{s})), \tag{15.7}$$

where

$$R_{\overset{\mathrm{rc}}{\equiv}}(\boldsymbol{s}) = \bigwedge_{(s_i,s_j)\in \overset{\mathrm{rc}}{\equiv}} s_i \equiv s_j.$$

It can be checked that, for state variables s_i and s_j satisfying the foregoing conditions, their valuations are always identical throughout the execution of the underlying FSM.

Notice that the equivalence relations $\overset{\mathrm{rc}}{\equiv}$ satisfying 15.6 and 15.7 may not be unique. However, there exists a unique *maximum* inductive register correspondence. In fact, the equivalence relations satisfying Equations (15.6) and (15.7) form a bounded partially ordered set (or poset) $P = (Z, \subseteq)$, where Z is the set of equivalence relations satisfying Equations (15.6) and (15.7). Because relation $\subseteq$ is a partial order (satisfying reflexivity, antisymmetry, and transitivity) on Z, $P = (Z, \subseteq)$ is a poset. In particular, the empty set $\emptyset$ is a lower bound for Z; it corresponds to an equivalence relation with no equivalent state variables. On the other hand, there is an upper bound for Z. This is because, if $\overset{\mathrm{rc}}{\equiv}_1$ and $\overset{\mathrm{rc}}{\equiv}_2$ are two elements in Z, then $\overset{\mathrm{rc}}{\equiv}_1 \cup \overset{\mathrm{rc}}{\equiv}_2$ forms another valid inductive register correspondence and is in Z as well. It certifies that a (locally) *maximal* inductive register correspondence is also a (globally) *maximum* one.

The signal correspondence of Definition 15.14 can be slightly relaxed by further considering complementary state variables s_i and s_j satisfying Equations (15.6) and (15.7) with

$$R_{\overset{\mathrm{rc}}{\equiv}}(\boldsymbol{s}) = \bigwedge_{(s_i,s_j)\in \overset{\mathrm{rc}}{\equiv}} s_i \equiv \overline{s_j}.$$

Because the valuations of s_i and s_j are always complementary to each other in the execution of the underlying FSM, they can be seen as corresponding signals as well.

The procedure sketched in Figure 15.19 computes the maximum inductive register correspondence, where the function call *InitialValue*(s) obtains the initial value of state variable s. Here we assume that the initial value of any register is a fixed deterministic value. That is, there is a single initial state in the underlying FSM. (Note that, for a state variable with nondeterministic initial values, it may not have equivalent state variables by Definition 15.14.) For the sake of simplicity, the computed register correspondence does not include state variables with complementary values, although this extension can easily be added. On the other hand, as mentioned earlier, the procedure can be used to compute the register correspondence between two sequential circuits because it can take the product machine of the two circuits as its input.

ComputeInductiveRegisterCorrespondence

input: a sequential circuit $\mathcal{M}$ with transition functions $\boldsymbol{\delta}$ and input and state variables $\boldsymbol{x}$ and $\boldsymbol{s}$, respectively
output: the inductive register correspondence of $\mathcal{M}$
begin

$i := 0$

$\overset{\mathrm{rc}\,(i)}{=} := \{(s_p, s_q) \mid \mathit{InitialValue}(s_p) = \mathit{InitialValue}(s_q)\}$

repeat

$\quad i := i + 1$

$\quad R := \bigwedge_{(s_p, s_q) \in \overset{\mathrm{rc}\,(i-1)}{=}} (s_p \equiv s_q)$

$\quad \overset{\mathrm{rc}\,(i)}{=} := \{(s_p, s_q) \in \overset{\mathrm{rc}\,(i-1)}{=} \mid \forall \boldsymbol{x}, \boldsymbol{s}.[R(\boldsymbol{s}) \Rightarrow (\delta_p(\boldsymbol{x}, \boldsymbol{s}) \equiv \delta_q(\boldsymbol{x}, \boldsymbol{s}))]\}$

until $\overset{\mathrm{rc}\,(i)}{=} \; = \; \overset{\mathrm{rc}\,(i-1)}{=}$

return $\overset{\mathrm{rc}\,(i)}{=}$

end

Figure 15.19. An algorithm that characterizes the inductive register correspondence for a given sequential circuit.

Example 15.15. *Consider the FSM* $\mathcal{M} = (\mathbb{B}^4, (0,0,0,0) \in \mathbb{B}^4, \mathbb{B}, \Omega, \boldsymbol{\delta}, \boldsymbol{\lambda})$, *where* $\boldsymbol{\delta} = (\delta_1, \ldots, \delta_4)$ *with*

$$\delta_1(x, \boldsymbol{s}) = \overline{x}\, s_1 \vee x\, s_3,$$
$$\delta_2(x, \boldsymbol{s}) = \overline{x}\, s_2 \vee x\, s_3 \vee (s_1 \oplus s_2),$$
$$\delta_3(x, \boldsymbol{s}) = \overline{x}\, s_3 \vee x\, \overline{s_1}\, \overline{s_3}, \text{ and}$$
$$\delta_4(x, \boldsymbol{s}) = (\overline{x}\, s_4 \vee x\, \overline{s_2}\, \overline{s_4})(\overline{s_2} \vee \overline{s_3}).$$

The inductive register correspondence of $\mathcal{M}$ *can be identified through the procedure of Figure 15.19. It can be checked that the register correspondences at different iterations are*

$$\overset{\mathrm{rc}\,(0)}{=} = \{(s_1, s_2), (s_1, s_3), (s_1, s_4), (s_2, s_3), (s_2, s_4), (s_3, s_4)\},$$
$$\overset{\mathrm{rc}\,(1)}{=} = \{(s_1, s_2)\}, \text{ and}$$
$$\overset{\mathrm{rc}\,(2)}{=} = \{(s_1, s_2)\}.$$

Thus, the procedure terminates in two iterations and identifies s_1 *and* s_2 *as corresponding state variables. Because* s_1 *and* s_2 *have the same valuations throughout the execution of* $\mathcal{M}$, *they are equivalent signals.*

Theorem 15.9. *The procedure of Figure 15.19 terminates and returns the maximum inductive register correspondence.*

Proof. The termination is easily seen because $\overset{\text{rc}}{\equiv}^{(i)}$ is initially finite and decreases monotonically until $\overset{\text{rc}}{\equiv}^{(i)}$ equals $\overset{\text{rc}}{\equiv}^{(i-1)}$ with cardinality no less than zero.

The computed register correspondence upon termination follows Definition 15.14 because steps 1 and 6 of the procedure in Figure 15.19 are satisfied. On the other hand, we show that the returned register correspondence is maximum. Let S be the set of state variables. Then $\mathcal{L} = (\mathcal{P}(S \times S), \subseteq, S \times S, \emptyset)$ forms a *complete lattice* because relation $\subseteq$ is a partial order on $\mathcal{P}(S \times S)$ (the power set of $S \times S$), and $S \times S$ and $\emptyset$ are the greatest and least elements, respectively, of $\mathcal{P}(S \times S)$. Also, observe that step 6 in the procedure of Figure 15.19 defines a mapping $f : (S \times S) \rightarrow (S \times S)$, which is monotone, that is, for all $a, b \in \mathcal{P}(S \times S)$, $a \subseteq b$ implies $f(a) \subseteq f(b)$. Because the procedure iteratively removes inequivalent state-variable pairs through f, it computes a greatest fixed point by the Knaster-Tarski theorem [86]. ■

The procedure *ComputeInductiveRegisterCorrespondence* is effective and scalable to large designs, for example, with tens of thousands of registers. The main reason of this practicality is that an equivalence relation among state variables, instead of states, is computed. As a result, the state explosion problem does not occur. However, it should be noted that not all register correspondences can be characterized inductively. In fact, the relation $\overset{\text{rc}}{\equiv}$ can be seen as an overapproximation of the reachable state set R of the underlying sequential circuit $\mathcal{M}$, i.e., $\overset{\text{rc}}{\equiv} \supseteq R$, because $\overset{\text{rc}}{\equiv}$ must be an invariant over R. The procedure in Figure 15.19 can be seen as a kind of reachability analysis as the inductive register correspondence $\overset{\text{rc}}{\equiv}^{(i)}$ at the ith iteration is gradually refined, that is, $\overset{\text{rc}}{\equiv}^{(i)} \supseteq \overset{\text{rc}}{\equiv}^{(i+1)}$, and approximates R more and more accurately.

Example 15.16. *Continue Example 15.15. State variables s_3 and s_4 are, in fact, equivalent signals by Definition 15.13. However, they cannot be detected by the inductive procedure of Figure 15.19. The reason is that the valuations of s_3 and s_4 are the same in the reachable state space, with characteristic function $R(\boldsymbol{s}) = \overline{s_1}\,\overline{s_2}\,\overline{s_3}\,\overline{s_4} \vee \overline{s_1}\,\overline{s_2}\,s_3\,s_4 \vee s_1\,s_2\,\overline{s_3}\,\overline{s_4}$, of $\mathcal{M}$. However, it is not true in the overapproximated state space, with characteristic function $R_{\overset{\text{rc}}{\equiv}}(\boldsymbol{s}) = s_1 s_2 \vee \overline{s_1}\,\overline{s_2}$, because δ_3 and δ_4 are not equivalent under $R_{\overset{\text{rc}}{\equiv}}(\boldsymbol{s})$.*

The algorithm of inductive register correspondence can be applied for sequential equivalence checking of two FSMs $\mathcal{M}_1$ and $\mathcal{M}_2$ by constructing their product machine $\mathcal{M}_\times$ and analyzing register correspondence on $\mathcal{M}_\times$. It checks if under the computed overapproximated reachable state set $R_{\overset{\text{rc}}{\equiv}}$ the output of $\mathcal{M}_\times$ always produces constant one: that is, for all valuations of $\boldsymbol{x}, \boldsymbol{s}$,

$$R_{\overset{\text{rc}}{\equiv}}(\boldsymbol{s}) \Rightarrow \lambda_\times(\boldsymbol{x}, \boldsymbol{s}) \equiv 1. \tag{15.8}$$

If the preceding formula is valid, then $\mathcal{M}_1$ and $\mathcal{M}_2$ are indeed equivalent. Otherwise, $\mathcal{M}_1$ and $\mathcal{M}_2$ may or may not be equivalent. This yields a so-called *false negative*, which means that a negative answer may be problematic. Therefore, inductive register correspondence is incomplete for sequential equivalence checking. As a matter of fact, its proving power is very limited and is only complete for special occasions, such as when, the FSMs under comparison are transformed combinationally.

15.7.1.1 Quasiinductive Functional Dependency

In inductive register correspondence $\overset{\text{rc}}{\equiv}$, every two state variables s_i and s_j are related with the identity function or the complement of the identity function. That is, $s_i \equiv s_j$ or $s_i \equiv \overline{s_j}$. In fact, more general *functional dependencies* [39] can be considered, where a state variable s_i can be expressed as a function θ over other state variables $s_1, \ldots, s_{i-1}, s_{s+1}, \ldots, s_n$, that is, $s_i = \theta(s_1, \ldots, s_{i-1}, s_{s+1}, \ldots, s_n)$. Similar to inductive register correspondence, functional dependencies may be characterized in an inductive manner. However, the computation may risk nontermination unless some artificial conditions are further imposed. A detailed exposition on combinational and sequential functional dependencies can be found in [39]. Based on Craig interpolation and incremental SAT solving, recent work [51] further extends the scalability of the exploration of combinational functional dependencies, where designs with hundreds of thousands of gates are handled effectively.

15.7.2 Inductive Signal Correspondence

In register correspondence, we search an equivalence relation over signals at register outputs. In fact, there is no need to restrict ourselves to the correspondence of registers. Given a circuit implementation of an FSM, we may compute the correspondences for all signals (including intermediate signals) of the circuit [21]. At first glance it may seem that there is no advantage in computing correspondences among intermediate signals. However, it turns out that identifying such correspondences helps to tighten the induction hypothesis and thus to tighten the approximation of the reachable state set.

Similar to Definition 15.14, we have the following definition.

Definition 15.15. (Inductive Signal Correspondence.) *An* inductive signal correspondence *of a circuit implementing FSM* $\mathcal{M} = ([\![\boldsymbol{s}]\!], I, [\![\boldsymbol{x}]\!], \Omega, \boldsymbol{\delta}, \boldsymbol{\lambda})$ *is an equivalence relation* $\overset{\text{sc}}{\equiv} \subseteq F \times F$ *over the set* F *of circuit signals that satisfies, for all valuations of* $\boldsymbol{x}, \boldsymbol{s}$,

$$\text{Base case: } I(\boldsymbol{s}) \Rightarrow R_{\overset{\text{sc}}{\equiv}}(\boldsymbol{x}, \boldsymbol{s}), \text{ and} \tag{15.9}$$

$$\text{Inductive case: } R_{\overset{\text{sc}}{\equiv}}(\boldsymbol{x}, \boldsymbol{s}) \Rightarrow R'_{\overset{\text{sc}}{\equiv}}(\boldsymbol{x}, \boldsymbol{s}), \tag{15.10}$$

where

$$R_{\overset{\text{sc}}{\equiv}}(\boldsymbol{x}, \boldsymbol{s}) = \bigwedge_{(f_i, f_j) \in \overset{\text{sc}}{\equiv}} f_i(\boldsymbol{x}, \boldsymbol{s}) \equiv f_j(\boldsymbol{x}, \boldsymbol{s}), \text{ and}$$

$$R'_{\overset{\text{sc}}{\equiv}}(\boldsymbol{x}, \boldsymbol{s}) = \bigwedge_{(f_i, f_j) \in \overset{\text{sc}}{\equiv}} \forall \boldsymbol{x}'.[f_i(\boldsymbol{x}', \boldsymbol{\delta}(\boldsymbol{x}, \boldsymbol{s})) \equiv f_j(\boldsymbol{x}', \boldsymbol{\delta}(\boldsymbol{x}, \boldsymbol{s}))].$$

ComputeInductiveSignalCorrespondence

input: a sequential circuit (with signals F) implementing $\mathcal{M} = ([\![\boldsymbol{s}]\!], I, [\![\boldsymbol{x}]\!], \Omega, \boldsymbol{\delta}, \boldsymbol{\lambda})$

output: the inductive register correspondence of $\mathcal{M}$

begin

$i := 0$

$\overset{\text{sc}\,(i)}{\equiv} := \{(f_p, f_q) \mid \forall \boldsymbol{x}, \boldsymbol{s}.[(f_p(\boldsymbol{x}, \boldsymbol{s}) \equiv f_q(\boldsymbol{x}, \boldsymbol{s})) \wedge I(\boldsymbol{s})]\}$

repeat

$i := i + 1$

$R := \bigwedge_{(f_p, f_q) \in \overset{\text{sc}\,(i-1)}{\equiv}} (f_p(\boldsymbol{x}, \boldsymbol{s}) \equiv f_q(\boldsymbol{x}, \boldsymbol{s}))$

$R' := \bigwedge_{(f_p, f_q) \in \overset{\text{sc}\,(i-1)}{\equiv}} (f_p(\boldsymbol{x}', \boldsymbol{\delta}(\boldsymbol{x}, \boldsymbol{s})) \equiv f_q(\boldsymbol{x}', \boldsymbol{\delta}(\boldsymbol{x}, \boldsymbol{s})))$

$\overset{\text{sc}\,(i)}{\equiv} := \{(f_p, f_q) \in \overset{\text{sc}\,(i-1)}{\equiv} \mid \forall \boldsymbol{x}, \boldsymbol{x}', \boldsymbol{s}.[R(\boldsymbol{x}, \boldsymbol{s}) \Rightarrow R'(\boldsymbol{x}, \boldsymbol{x}', \boldsymbol{s})]\}$

until $\overset{\text{sc}\,(i)}{\equiv} = \overset{\text{sc}\,(i-1)}{\equiv}$

return $\overset{\text{sc}\,(i)}{\equiv}$

end

Figure 15.20. An algorithm that characterizes the inductive signal correspondence for a given sequential circuit.

In the foregoing definition, the function of a signal $f_i \in F$ is denoted as $f_i(\boldsymbol{x}, \boldsymbol{s})$. Note that the functions of all signals are expressed in terms of primary-input and state variables. With almost the same procedure, we may compute corresponding signals as shown in Figure 15.20. Comparing $R_{\overset{\text{rc}}{\equiv}}$ and $R_{\overset{\text{sc}}{\equiv}}$, we see that $R_{\overset{\text{sc}}{\equiv}}(\boldsymbol{x}, \boldsymbol{s})$ imposes a strictly stronger precondition for the implication in Equation (15.10). With the stronger inductive hypothesis, it is possible to identify more register correspondence not detectable in $\overset{\text{rc}}{\equiv}$.

As a modern integrated-circuit design may contain millions of logic gates, considering the correspondences of all signals is formidable. In implementing the computation of inductive signal correspondence, it is desirable to first screen out obvious inequivalent signal pairs. For that purpose, simulation is often adopted and is shown to be fast and effective.

The foregoing algorithms in computing register and signal correspondences assume that the initial values of registers are known and unique. However, when the initial values are not known or not unique *a priori*, the computation needs to be modified to take nondeterministic initialization into account. Different notions of conformance, such as [9], may be defined depending on the imposed initial conditions. (Initialization issues are discussed in Section 15.8.) Nevertheless, the essential computation based on induction is the same.

Similar to $\overset{\text{rc}}{\equiv}$, signal correspondence $\overset{\text{sc}}{\equiv}$ can be used for checking the equivalence of two sequential circuits $\mathcal{M}_1$ and $\mathcal{M}_2$. In effect, these two FSMs are equivalent if, for all valuations of $\boldsymbol{x}$, $\boldsymbol{s}$,

$$R_{\overset{\text{sc}}{\equiv}}(\boldsymbol{x}, \boldsymbol{s}) \Rightarrow \lambda_{\times}(\boldsymbol{x}, \boldsymbol{s}) \equiv 1. \qquad (15.11)$$

Similar to the case of register correspondence, the foregoing condition is not sufficient to show the inequivalence of two FSMs. That is, false negatives may occur as well. Nevertheless, the effectiveness in proving of signal correspondence is strictly stronger than that of register correspondence. In particular, it is complete for verifying circuits optimized by retiming [52, 53]. By exploiting signal correspondence among different timeframes, inductive signal correspondence can further be made complete for equivalence checking of circuits optimized with retiming plus resynthesis plus retiming [41]. Although verification via signal correspondence is incomplete in general, signal correspondence can be used as a preprocessing step to simplify the circuits to be compared before reachability analysis is applied.

15.7.3 Inductive Safety Property Checking

Signal correspondence captures the equivalence relation between two signals. However, the equivalence relation is not the only criterion to be exploited. For instance, among many other possibilities, the relation of implication, $\Rightarrow$, between two signals can be computed, as was suggested in [3]. It yields an even stronger induction hypothesis than equivalence relation, $\Leftrightarrow$, since equivalent signals $f_1 \Leftrightarrow f_2$ can be captured by $(f_1 \Rightarrow f_2) \wedge (f_2 \Rightarrow f_1)$ but $f_1 \Rightarrow f_2$ cannot be written in terms of $\Leftrightarrow$. A further generalization is to consider general temporal properties as the candidates to be proved by induction. In fact, $R_{\underline{\underline{\mathrm{rc}}}}$ and $R_{\underline{\underline{\mathrm{sc}}}}$ are just two special instances of temporal properties of interest. The inductive proof of a general safety property P can be formalized as follows:

$$\text{Base case:}\quad I(\boldsymbol{s}) \Rightarrow P(\boldsymbol{s}),\ \text{and} \qquad (15.12)$$

$$\text{Inductive case:}\quad P(\boldsymbol{s}) \wedge T(\boldsymbol{s}, \boldsymbol{s}') \Rightarrow P(\boldsymbol{s}'), \qquad (15.13)$$

where initial state set I may be of cardinality greater than 1, and transition relation T may contain nondeterministic transitions. (Recall that we may treat primary input variables as part of the state variables. Thereby, T can be expressed in terms of state variables only.) By Definition 15.5, note that the induction aims to certify that P characterizes a state set closed under the transition relation T. (It is easily seen that the reachable state set R of reachability analysis satisfies Equations (15.12) and (15.13).)

Observe that proving safety properties by the foregoing induction is incomplete. When Equations (15.12) and (15.13) are satisfied, P holds at any state reachable from I under T. In contrast, when Equation (15.12) is violated, P is not true for some initial state, and a true counterexample is generated. When Equation (15.13) is violated instead, P may or may not hold for all reachable states, and a spurious counterexample, that is, a false negative, may be generated. Examining the causes, we see that the induction fails (one of Equations (15.12) and (15.13) is violated) if and only if, in the entire state space, there exists a P-state (a state that satisfies P)

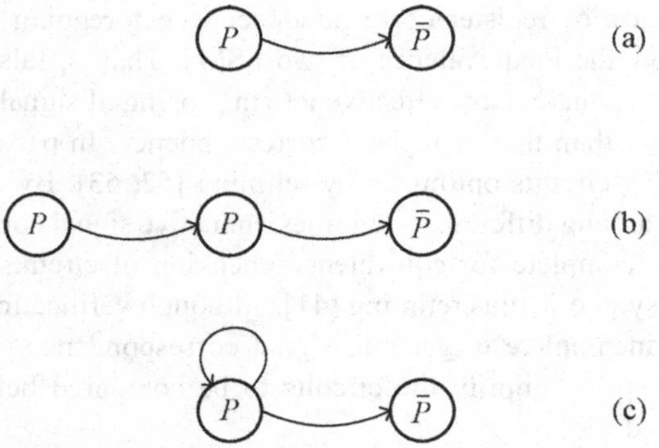

Figure 15.21. (a) A transition condition that makes an 1-step induction fail. (b) A transition condition that makes a 2-step induction fail. (c) A transition condition that makes a general k-step induction fail.

that can transition to a $\overline{P}$-state (a state that satisfies $\overline{P}$) as shown in Figure 15.21 (a). In particular, a false negative can be generated if and only if, in the unreachable state space, there exists a transition of Figure 15.21 (a). Because the entire state space is an overapproximation of the reachable state space, inductive proofs are in general incomplete.

To make induction complete for property checking, k-step induction [80] was introduced to strengthen the induction hypothesis by increasing the transition length k when a false negative occurs. Formally speaking, a k-step induction consists of

$$\text{Base case:}\quad I(\boldsymbol{s}^0) \wedge T^k(\boldsymbol{s}^0, \ldots, \boldsymbol{s}^k) \Rightarrow P^k(\boldsymbol{s}^0, \ldots, \boldsymbol{s}^k), \text{ and} \qquad (15.14)$$

$$\text{Inductive case:}\quad P^k(\boldsymbol{s}^0, \ldots, \boldsymbol{s}^k) \wedge T^{k+1}(\boldsymbol{s}^0, \ldots, \boldsymbol{s}^{k+1}) \Rightarrow P(\boldsymbol{s}^{k+1}), \qquad (15.15)$$

where a state-variable vector $\boldsymbol{s}$ is annotated with a superscript i indicating that $\boldsymbol{s}$ is instantiated for the ith timeframe, $T^i(\boldsymbol{s}^0, \ldots, \boldsymbol{s}^i)$ is the shorthand for $T(\boldsymbol{s}^0, \boldsymbol{s}^1) \wedge T(\boldsymbol{s}^1, \boldsymbol{s}^2) \wedge \cdots \wedge T(\boldsymbol{s}^{i-1}, \boldsymbol{s}^i)$, and $P^i(\boldsymbol{s}^0, \ldots, \boldsymbol{s}^i)$ is the shorthand for $P(\boldsymbol{s}^0) \wedge \cdots \wedge P(\boldsymbol{s}^i)$. Because for $k = 0$ Equations (15.14) and (15.15) reduce to Equations (15.12) and (15.13), respectively, ordinary induction is a special case of k-step induction.

It is not hard to see that a false negative caused by Figure 15.21 (a) is ruled out in a 2-step induction. Similarly, a false negative caused by Figure 15.21 (b) (which may appear in a 2-step induction) is ruled out in a 3-step induction. Increasing the length of transition paths along which P is satisfied makes sure that the condition that fails the k-step induction does not appear in the $(k + i)$-step induction for $i = 1, 2, \ldots$ However, a false negative with an infinite path length is possible, for example, due to the transition of Figure 15.21 (c); it cannot be ruled out by increasing k. Thus, k-step induction risks nontermination.

Fortunately, this problem can be overcome by adding the following uniqueness criterion, in addition to Equations (15.14) and (15.15):

$$\bigwedge_{i \le j \le k} s^i \not\equiv s^j \left(= \bigwedge_{i \le j \le k} \bigvee_{l} s_l^i \not\equiv s_l^j \right) \tag{15.16}$$

for the state variables. Effectively, in the k-step induction, only simple paths (on which a state never appears twice) are considered. With the uniqueness constraint, the termination of the k-step induction is guaranteed.

To summarize, the basic procedure for proving a safety property by the k-step induction (with the uniqueness criterion) proceeds as follows. We start from $k = 1$, that is, the normal induction. In a k-step induction, three cases may happen. If Equations (15.14) and (15.15) are both satisfied, the property is proven to be true for the underlying transition system. If Equation (15.14) is violated, the property does not hold for the underlying state transition system. In these two cases, the induction terminates. On the other hand, if Equation (15.15) is violated, we increase k by 1 and repeat the induction. In fact, k is upper-bounded by the length of the longest simple path from an initial state in the underlying STG, which is not smaller and, in fact, can be exponentially larger than the forward diameter (the upper bound for forward reachability analysis). An example of such a worst case is a complete graph.

The induction criteria (including Equations (15.14), (15.15), and the uniqueness criterion) can be formulated as SAT solving, especially, incremental SAT solving [20]. Therefore, in addition to the interpolation-based approach in Section 15.6.2, k-step induction is an alternative solution to unbounded model checking of safety properties.

15.7.4 Some Reduction Techniques

15.7.4.1 Reencoding States for Similarity

The aforementioned computation of register correspondence and that of functional dependency identify the functional connections among registers.

Consider the special case of two state-minimized equivalent FSMs. When their corresponding equivalent states are encoded in the same way, they are in fact combinationally equivalent. Once their register correspondence is established, a combinational check suffices to show sequential equivalence. On the other hand, when the corresponding equivalent states are encoded differently, no register correspondence can be concluded in general. In this case, more powerful techniques, such as the identification of functional dependency, are needed.

When two FSMs $\mathcal{M}_1$ and $\mathcal{M}_2$ have a one-to-one equivalent state mapping, there exists some reencoding able to produce identical next-state and output functions, thus again reducing sequential to combinational equivalence. (The mapping need not hold for the entire state sets of $\mathcal{M}_1$ and $\mathcal{M}_2$ and can hold merely for their exact or overapproximated reachable state sets.) How to find such reencoding is a problem that was addressed in the work by Quer et al. [70]. Even when complete reencoding cannot be found, because of computational expensiveness or nonexistence, partial reencoding may increase the similarity between the two FSMs and simplify their product machine. Notice that the number of states of a product machine may be larger by an exponential factor with respect to the number of states of the constituent machines, even though by experimental evidence it is usually larger by only a small constant factor. The technique was also suggested to work for sequential circuits with different number of registers, to ease sequential equivalence checking.

State minimization of a sequential circuit without explicit state space exploration was studied in [38, 54, 85], where equivalence classes are represented with BDDs.

15.7.4.2 Exposing Registers for Observability

One can view the process to check the equivalence of FSMs $\mathcal{M}_1$ and $\mathcal{M}_2$ through the product machine $\mathcal{M}_\times$ as constructing a relation on pairs of states $R \subseteq Q_1 \times Q_2$ such that $(q_1, q_2) \in R$ if they are simultaneously reachable, and they produce the same outputs under the same inputs, namely,

$$(q_1, q_2) \in R \Leftrightarrow ((q_1, q_2) \text{ reachable in } \mathcal{M}_\times \wedge \forall \sigma \in \Sigma.(\lambda_1(\sigma, q_1) \equiv \lambda_2(\sigma, q_2))).$$

It can be proved that the two FSMs are equivalent if and only if the relation R forms a *bisimulation relation*, which for every $(q_1, q_2) \in R$ and every $\sigma \in \Sigma$ satisfies

(i) $(q_1^0, q_2^0) \in R$ for $q_1^0 \in Q_1^0$ and $q_2^0 \in Q_2^0$,
(ii) $\delta_1(\sigma, q_1) = q_1'$ implies $\exists q_2' \in Q_2.(\delta_2(\sigma, q_2) = q_2' \wedge (q_1', q_2') \in R)$, and
(iii) $\delta_2(\sigma, q_2) = q_2'$ implies $\exists q_1' \in Q_1.(\delta_1(\sigma, q_1) = q_1' \wedge (q_1', q_2') \in R)$.

In other words, the two FSMs are equivalent if and only if first, their initial states are equivalent; second, for every state q_1' (q_2') to which q_1 (q_2) can go, there is a state q_2' (q_1') to which q_2 (q_1) can go under the same input; and third, q_1' and q_2' are equivalent. In order to obtain R, we need to find the set of simultaneously reachable state pairs by state traversal of $\mathcal{M}_\times$.

By exposing register outputs as observable pseudoprimary outputs, state equivalence is simplified to 1-step state equivalence (recall Definition 15.6). The idea was exploited in [1] to check the bisimulation property of a state relation between $\mathcal{M}_1$ and $\mathcal{M}_2$ under 1-step state equivalence.

Consider constructing, instead of the relation R, another relation $R^\dagger \subseteq Q_1 \times Q_2$, called a *1-equivalence relation* in [1], with

$$(q_1, q_2) \in R^\dagger \Leftrightarrow (q_1 \text{ reachable in } \mathcal{M}_1 \;\wedge\; q_2 \text{ reachable in } \mathcal{M}_2 \\ \wedge\, \forall \sigma \in \Sigma.(\lambda_1(\sigma, q_1) \equiv \lambda_2(\sigma, q_2))).$$

That is, two states $q_1 \in Q_1$ and $q_2 \in Q_2$ with $(q_1, q_2) \in R^\dagger$ if they are *individually* reachable in their constituent FSMs and produce the same outputs under the same inputs. Therefore, $R^{\dagger}$ differs from R in that it holds between pairs of states *individually* reachable in the two FSMs. So to compute $R^{\dagger}$, we need not find out which states are *simultaneously* reachable by traversing the product machine; we only need to know the set of reachable states of each individual FSM.

A question to ask is whether or not still the proposition holds that $\mathcal{M}_1$ and $\mathcal{M}_2$ are equivalent if and only if the relation $R^{\dagger}$ is a bisimulation relation. Because $R \subseteq R^{\dagger}$, it is immediate that, if $R^{\dagger}$ satisfies the bisimulation property, then so does R. Hence, the checking is sufficient to verify the equivalence of $\mathcal{M}_1$ and $\mathcal{M}_2$. However, because $R^{\dagger}$ is an overapproximation of R, there can be a false negative, that is, a case when the check fails on a pair of states that are not simultaneously reachable.

To rule out such false negatives, in [1] the *complete-1-distinguishability* (C-1-D) property, defined as any two states can be distinguished at the outputs with an input sequence of length 1, is imposed upon the states of $\mathcal{M}_1$ and $\mathcal{M}_2$. By restricting $\mathcal{M}_1$ to obey the C-1-D property, it can be proved that, if $\mathcal{M}_1$ and $\mathcal{M}_2$ are equivalent, then $R^{\dagger}$ is a bisimulation relation. Conversely, if $R^{\dagger}$ is not a bisimulation relation, then $\mathcal{M}_1$ and $\mathcal{M}_2$ are certainly inequivalent. The C-1-D property avoids the false negatives because, when the property holds for $\mathcal{M}_1$, for each state q_2 in $\mathcal{M}_2$ there is exactly one state q_1 in $\mathcal{M}_1$ such that $(q_1, q_2) \in R^{\dagger}$, and also, if q_2 is reachable in $\mathcal{M}_2$, then (q_1, q_2) is reachable in $\mathcal{M}_{\times}$. So, when $\mathcal{M}_1$ has the C-1-D property, the set of pairs of equivalent separately reachable states is the same as the set of pairs of equivalent simultaneously reachable states.

The C-1-D property can be viewed as a condition under which backward reachability analysis attains convergence in one iteration, because backward traversal iteratively refines the partition obtained initially by 1-step state equivalence. (Connections between reachability analysis and state equivalence are detailed in Section 15.5.2.3.)

The preceding discussion can be straightforwardly generalized to k-equivalence relation and complete-k-distinguishability. To make sure that $\mathcal{M}_1$ is 1- or k-distinguishable, some registers have to be exposed as observable primary outputs. It imposes restrictions on circuit synthesis. When all registers are exposed, the circuit can only be synthesized combinationally, keeping intact register boundaries. Therefore the method can be seen as intermediate between pure combinational synthesis and verification versus unrestricted sequential synthesis and verification.

15.8 A Hierarchy of Classes of Sequential Equivalence

In the previous exposition, we assumed that FSMs can start from their pre-specified initial states, so that equivalence checking reduces to verifying the equivalence of the initial states. This initialization assumption, however, should be questioned, because in reality a circuit may not begin with the desired initial state, and this fact has to be enforced somehow. We now discuss what happens when there are no known initial states. In fact, the way an FSM is initialized affects the definition of sequential equivalence. We first study the initialization problem of FSMs and then discuss some variants of sequential equivalence.

15.8.1 Resettability and Alignability

In hardware implementation of an FSM, it is not immediate that the implemented FSM can be prepared in its designated initial states. This is because, when a circuit is powered up, the latched values in the state-holding elements (registers) are uncontrollable and can be of value either 0 or 1 indefinitely. Therefore, the circuit can be in an arbitrary state. An initialization (or reset) mechanism is needed to drive the circuit into some designated initial states before the FSM can operate correctly. A simple approach to rectifying the problem of uncertain power-up states is to add reset circuitry for every register. (A register is of *explicit reset* if some reset circuitry is associated with it. Otherwise, a register is of *implicit reset.*) Once a circuit is powered up, a reset signal is activated to trigger the reset mechanism for a register to latch its right initial value. However, this approach may incur heavy area penalties, especially for a design with an excessive number of registers, because the reset signal needs to be connected to all the registers. Another solution (without incurring any hardware overhead) to the reset problem is to apply some input sequence enforcing registers to have desired values. On the other hand, a hybrid solution with *partial reset* is possible, where only a subset of the registers is selected for explicit reset.

It is noteworthy that, if we treat the additional reset circuitry as an integral part of a design and thus the reset signal as part of the primary inputs, explicit reset can be seen as a special case of implicit reset. Consequently, we may only need to focus on implicit reset as we do hereafter. Because the input sequences for implicit reset may not always exist, we study the resettability of an FSM. Moreover, for a resettable FSM, we would like to obtain an input sequence that resets the FSM.

To reflect the situation that a circuit when powered up may not be in designated initial states, we modify the definition of an FSM and speak of a *hardware finite-state machine* instead.

Definition 15.16. (Hardware Finite-State Machine.) *A* hardware finite-state machine *(HFSM) is a tuple* $(Q, \Sigma, \Omega, \delta, \lambda)$ *or* $(Q, \Sigma, \Omega, T, \lambda)$, *similar to the definition of an FSM except for the absence of an initial state set.*

We define a *hardware state transition graph* (HSTG) for an HFSM in the same way as an STG for an FSM except that initial states are not identified. An HFSM (HSTG) is more primitive than an FSM (STG) in the sense that there is no notion of initial states.

Given an HFSM and some target initial state, we may study whether the HFSM can be brought to this state through some input sequence from any power-up states.

Definition 15.17. (Strict Resettability.) *An HFSM* $\mathcal{H} = (Q, \Sigma, \Omega, T, \lambda)$ *is called* strictly resettable *if there exists an input sequence* $\sigma \in \Sigma^*$ *(known as the* reset sequence *or* synchronization sequence*) and a state* $q_0 \in Q$ *(known as the* reset state*) such that* $Img_{\boldsymbol{\sigma}}(q, T) = q_0$*, for any* $q \in Q$.

From the foregoing definition, an FSM $\mathcal{M}$ with a single initial state can be derived from an HFSM $\mathcal{H}$ if $\mathcal{H}$ is resetable with respect to the initial state of $\mathcal{M}$. Notice that a reset sequence is *universal* in the sense that it applies to arbitrary power-up states. By initializing an HFSM to a single reset state, Definition 15.17 may seem somewhat restricted. When in particular two states are equivalent, it does not matter if the HFSM is reset to anyone of them. It brings up a relaxed definition of resettability.

Definition 15.18. (Essential Resettability.) *An HFSM is called* essentially resettable *if, under any power-up states, it can be initialized to a set of equivalent reset states* I *through a common reset sequence.*

In hardware design, FSMs are supposed to behave deterministically after reset. Therefore, unless the reset states I are all equivalent, the HFSM $\mathcal{H}$ may behave differently depending on which state in I it is reset to. It is the reason that we assume all initial states must be equivalent.

Strict resettability and essential resettability can be connected as follows.

Proposition 15.10. *An HFSM is essentially resettable if and only if its quotient (i.e., state-minimized) HFSM is strictly resettable.*

(A quotient HFSM is defined the same as a quotient FSM except for the ignorance of initial states.) The proof can be shown straightforwardly by the deferred Lemma 15.15 and is omitted. Essential resettability is more adequate than strict resettability in the context of hardware design. In the sequel, when resettability is mentioned, we mean essential resettability unless otherwise stated. Equivalently, we may assume an HFSM is in its quotient form and speak of strict resettability.

Algorithms for resettability analysis, like those for reachability analysis, are heavily influenced by the data structures used to realize the computations. Following the historical development, we begin with an enumerative approach to resettability analysis as well as reset sequence generation, and then we proceed with an algorithm that lends itself to a symbolic implementation.

15.8.1.1 Explicit Graph Algorithm for Resettability Analysis

The reset problem can be resolved based on explicit graph enumeration [42]. Essentially, the resettability of an HFSM can be easily understood through a tree construction.

Definition 15.19. (Synchronization Tree.) *The* synchronization tree *of an HFSM* $\mathcal{H} = (Q, \Sigma, \Omega, T, \lambda)$ *is an infinite tree* $G = (V, E)$*, where a vertex* $v \in V$ *corresponds to a state set* $Q_v \subseteq Q$ *and an edge labeled with* $\sigma \in \Sigma$ *from vertex* u *to* *vertex v signifies* $Q_v = Img_\sigma(Q_u, T)$. *The unique root vertex* $r \in V$ *corresponds to the universal state set* Q, *that is,* $Q_r = Q$.

A synchronization tree enumerates all possible input sequences and lists the corresponding possible states of the considered HFSM.

Example 15.17. *Figure 15.22 shows an example, where T is the synchronization tree of the given HSTG G. Under an empty input sequence (at power-up), the HFSM may be in any state as indicated in the root node; under the input sequence '0', it is possible to be in any state of* $\{q_0, q_1, q_3\}$ *as indicated in the left successor node of the root.*

Because we are concerned with finite-state sets, labels on vertices will repeat eventually. A finite construction suffices for an exhaustive enumeration.

The relation between a synchronization tree and the resettability of an HFSM can be stated as follows.

Proposition 15.11. *An HFSM is resettable with respect to a state set* $I \subseteq Q$ *if and only if its corresponding synchronization tree has a path from the root vertex r to some vertex v with its corresponding state set* $Q_v \subseteq \{q \in Q \mid \exists q_i \in I.q \sim q_i\}$. *Also, the ordered labels on the edges along this path give the corresponding reset sequence.*

(Assume that all the states in I are equivalent.) In hardware design, short reset sequences are preferable to long ones.

Example 15.18. *The HSTG of Figure 15.22 is resettable to any reset state because input sequences '1,1,0', '1,1,1', '1,1,0,0', and '1,1,0,1' are reset sequences for reset states* q_3, q_2, q_1, *and* q_0, *respectively.*

Whereas the synchronization tree gives an exact characterization of resettability, there are some interesting necessary conditions for an HFSM to be resettable that may be used for fast screening of resettability. Whether an HFSM is resettable depends on the structure of its HSTG, in particular, the *strongly connected components* (SCCs).

Definition 15.20. (Strongly Connected Component.) *A subgraph G of an STG or HSTG is a* strongly connected component *if every (ordered) vertex pair* (u, v) *of G is connected by some path from u to v. An SCC is called* closed *if its vertex set corresponds to a closed state set. Otherwise it is* open.

Note that there may exist an open SCC contained by a closed SCC, but not the converse, that is, a closed SCC cannot be contained by an open or closed SCC except for itself. If fact, any two distinct closed SCCs must be disjoint.

Example 15.19. *In Figure 15.23, the open SCC induced by vertex* q_8 *is contained in the closed SCC induced by vertices* q_6, q_7, q_8.

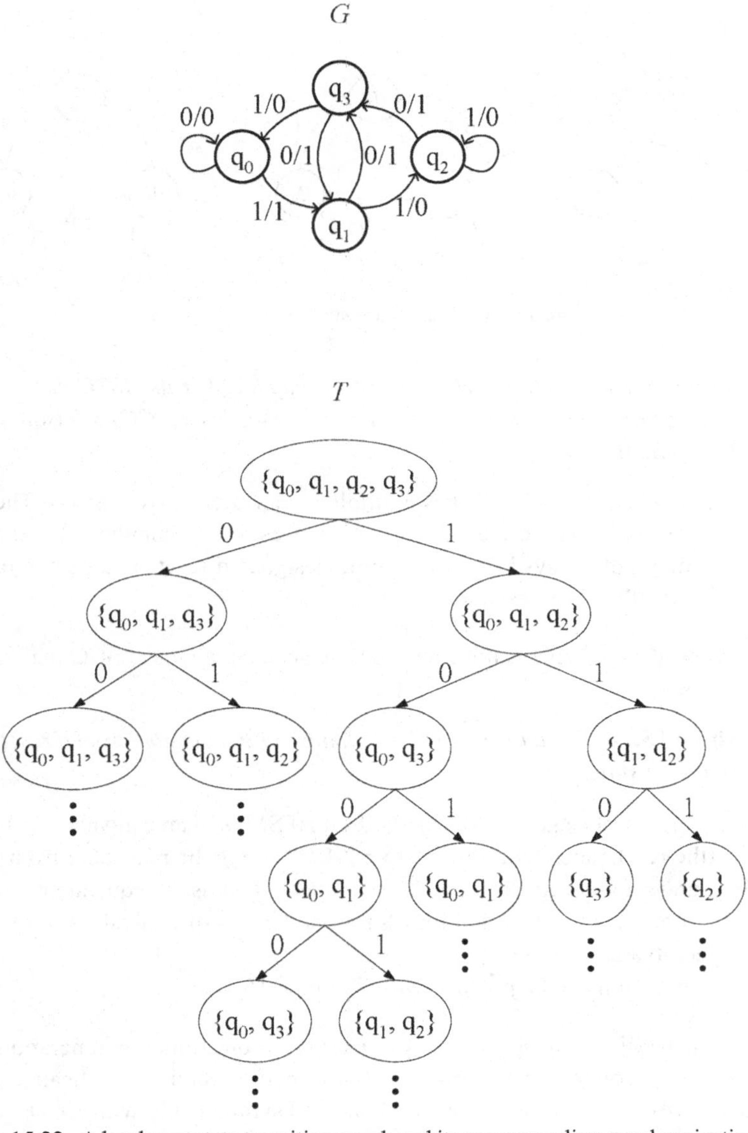

Figure 15.22. A hardware state transition graph and its corresponding synchronization tree.

Proposition 15.12. *The reset states of an HFSM $\mathcal{H}$ must be inside a closed SCC, or equivalent to some state in a closed SCC of the HSTG of $\mathcal{H}$.*

Proof. For the sake of contradiction, assume that the reset states of $\mathcal{H}$ are not in a closed SCC and inequivalent to any state in a closed SCC. Then, when powered up in a state in a closed SCC, $\mathcal{H}$ cannot be reset to the designated reset states. It contradicts with the assumption that $\mathcal{H}$ is resettable. ∎

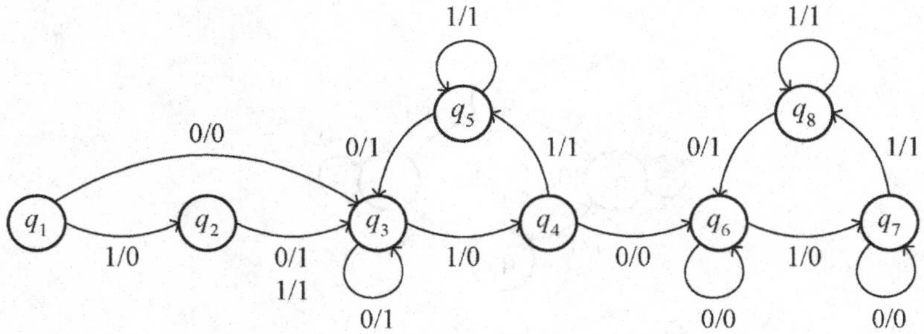

Figure 15.23. A hardware state transition graph.

Proposition 15.13. *An HFSM $\mathcal{H}$ is resettable only if either its HSTG has a single closed SCC or all the state subgraphs induced by the closed SCCs are isomorphic after state minimization.*

Proof. Assume the HSTG of $\mathcal{H}$ has multiple inequivalent closed SCCs. Then $\mathcal{H}$, once powered up in some closed SCC, cannot escape to another closed SCC. Hence, $\mathcal{H}$ may not always be reset to some designated reset state. It contradicts the assumption that $\mathcal{H}$ is resettable. ■

However, the converse is not true, because states in a closed SCC may not be resettable at all.

Proposition 15.14. *For a resettable HFSM, any state reachable from a reset state must be a reset state.*

Proof. After a reset sequence $\boldsymbol{\sigma}$ is applied, an HFSM is in an equivalence class of states I (the reset states with respect to $\boldsymbol{\sigma}$). Let state $q^\dagger$ be reachable from some state in I under input sequence $\boldsymbol{\sigma}^\dagger$. Then, all such $q^\dagger$ must be equivalent because states in I are all equivalent and thus form another set of equivalent reset states with respect to a new reset sequence $\boldsymbol{\sigma} \circ \boldsymbol{\sigma}^\dagger$.

This proposition can be justified with Example 15.18. ■

The resettability analysis discussed earlier relies on explicit enumeration over the synchronization tree. An alternative approach to resetability verification relies on reachability analysis and can be implemented symbolically with BDDs [66].

15.8.1.2 Implicit Symbolic Algorithm for Resettability Analysis

Resettability analysis can be reformulated in terms of the alignability of a pair of states.

Definition 15.21. (State Alignment.) *Two states q_1 and q_2 of an HFSM $\mathcal{H} = (Q, \Sigma, \Omega, T, \boldsymbol{\lambda})$ are* alignable *if there exists an* aligning sequence $\boldsymbol{\sigma} \in \Sigma^*$ *of q_1 and q_2 such that $Img_{\boldsymbol{\sigma}}(q_1, T) \sim Img_{\boldsymbol{\sigma}}(q_2, T)$.*

Lemma 15.15. *For an HFSM $\mathcal{H} = (Q, \Sigma, \Omega, T, \lambda)$, if two states $q_1, q_2 \in Q$ are alignable under some input sequence $\boldsymbol{\sigma} \in \Sigma^*$, then states $q_3, q_4 \in Q$, with $q_3 \sim q_1$ and $q_4 \sim q_2$, are alignable as well under $\boldsymbol{\sigma}$.*

Proof. Because q_1 and q_2 are alignable under $\boldsymbol{\sigma}$, $Img_{\boldsymbol{\sigma}}(q_1, T) \sim Img_{\boldsymbol{\sigma}}(q_2, T)$. On the other hand, $q_3 \sim q_1$ implies $Img_{\boldsymbol{\sigma}}(q_3, T) \sim Img_{\boldsymbol{\sigma}}(q_1, T)$. Similarly, $Img_{\boldsymbol{\sigma}}(q_4, T) \sim Img_{\boldsymbol{\sigma}}(q_2, T)$. It follows that $Img_{\boldsymbol{\sigma}}(q_3, T) \sim Img_{\boldsymbol{\sigma}}(q_4, T)$. That is, q_3 and q_4 are alignable under $\boldsymbol{\sigma}$. ■

Consequently, we may speak of alignments over state equivalence classes rather than individual states. To compute reset sequences, it is more convenient to first reduce an HSTG to its quotient (state-minimized) HSTG such that no state equivalence needs to be checked later on.

The following theorem shows that a universal reset sequence can be built from concatenating local aligning sequences that align pairwise states.

Theorem 15.16. *An HFSM is resettable if and only if every of its state pairs is alignable under some finite input sequence.*

Proof. ($\Rightarrow$) By the definition of HFSM resettability, all state pairs are alignable (under a unique reset sequence).

($\Leftarrow$) For an HFSM with any pair of its states alignable, a universal sequence that aligns all state pairs can be constructed from local aligning sequences. Figures 15.24 and 15.25 show a procedure that determines whether a given HFSM is resettable and returns a reset sequence if it exists. The correctness of the procedure is proved in Theorem 15.17. ■

The procedure can be implemented with BDD-based computations [66].

Algorithm: ComputeResetSequence
input: a state-minimized HFSM $\mathcal{H} = (Q, \Sigma, \Omega, T, \lambda)$
output: a reset sequence
begin
$i := 0$
$S^{(i)} := Q$
repeat
  $i := i + 1$
  $\sigma_i := ComputeAligningSequence(s_a, s_b, \mathcal{H})$ for some $s_a, s_b \in S^{(i-1)}$ and $s_a \neq s_b$
  if $\sigma_i = \emptyset$
    return HFSM unresettable
  $S^{(i)} := Img_{\sigma_i}(S^{(i-1)}, T)$
until $|S^{(i)}| < 2$
return $\sigma_1 \circ \cdots \circ \sigma_i$
end

Figure 15.24. An algorithm that computes a universal reset sequence for a given HFSM.

Algorithm: ComputeAligningSequence

```
input: two distinct states q_a, q_b and their underlying HFSM H
output: an input sequence that aligns the given state pair
begin
    let T_× be the transition relation of the product machine H × H
    i := 0
    R^(0) := (q_a, q_b)
    repeat
        i := i + 1
        R^(i) := R^(i−1) ∪ Img(R^(i−1), T_×)
        D := R^(i) ∩ {(q, q) | q ∈ Q}
    until D ≠ ∅ or R^(i) = R^(i−1)
    if D = ∅
        return ∅
    n := i
    let (q_d^(i), q_d^(i)) ∈ D
    repeat
        let (q_d^(i−1), q_d^(i−1)) ∈ {R^(i−1) ∩ PreImg((q_d^(i), q_d^(i)), T_×)}
        σ_i := arg_x[T_×(x, (q_d^(i−1), q_d^(i−1)), (q_d^(i), q_d^(i)))]
        i := i − 1
    until i = 0
    return σ_1 ∘ ··· ∘ σ_n
end
```

Figure 15.25. An algorithm that computes an aligning sequence for a given state pair under a given transition relation.

Theorem 15.17. *The procedure in Figures 15.24 and 15.25 terminates and returns a correct answer upon termination.*

Proof. The algorithm *ComputeResetSequence* in Figure 15.24 implicitly assumes $|Q| \geq 2$. Otherwise, the given HFSM $\mathcal{H}$ is readily in its reset state when powered up, and thus there is no need to compute the reset sequence. ■

In the iterative computation, two states not aligned yet are selected for alignment. If the two states are unalignable, they provide a certificate of nonresettability of the given HFSM, and the program terminates. Otherwise, the set $S^{(i)}$ in Figure 15.24 decreases monotonically because at each iteration a state pair is aligned. Because $|S^{(i-1)}| - |S^{(i)}| \geq 1$ and $S^{(0)} = Q$, there exists some j such that $|S^{(j)}| < 2$, and thus the procedure terminates.

To see that the computed input sequence is indeed a reset sequence, consider first the algorithm *ComputeAligningSequence*. The procedure consists of two disjoint loops. The first loop performs forward reachability to check if state pair (q_a, q_b) in the product machine of $\mathcal{H} \times \mathcal{H}$ can reach some state $(q, q) \in Q \times Q$. (The product of two HFSMs is defined similarly to that of two FSMs, except for ignoring the initial state set.) If the answer is negative, states q_a and q_b are not alignable, and an empty input sequence is returned. Otherwise, the input sequence for (q_a, q_b) to reach (q, q) is computed by backward reachability analysis in the second loop. The computed aligning sequences are concatenated and returned by algorithm *ComputeResetSequence*. The overall input sequence drives the given HFSM $\mathcal{H}$ to a unique state (i.e. the corresponding reset state) upon termination. That unique state is the only state in the final image $S^{(i)} := Img_{\sigma_i}(S^{(i-1)}, T)$ produced by the algorithm at the last cycle, when the condition $|S^{(i)}| < 2$ holds.

If the resettability of an HFSM is the only primary concern but not the reset sequences, then we may test the resettability more effectively based on the following corollary.

Corollary 15.18. *An HFSM* $\mathcal{H} = (Q, \Sigma, \Omega, T, \lambda)$ *is strictly resettable if and only if any state of the product HFSM* $\mathcal{H} \times \mathcal{H} = (Q \times Q, \Sigma, \Omega, T_\times, \lambda_\times)$ *can reach some state in* $S = \{(q, q) \mid q \in Q\}$*, that is, the backward reachable state set of* S *equals the universal set.*

Proof. The corollary follows from Theorem 15.16. ■

A similar argument holds for essential resettability by modifying S to be the set $\{(q_1, q_2) \mid q_1, q_2 \in Q, q_1 \sim q_2\}$.

15.8.2 A Landscape of Sequential Equivalences

Depending on the equivalence criteria of HFSMs in the initialization phase, we may have different notions of equivalence.

15.8.2.1 Alignment Equivalence

Definition 15.21 and Theorem 15.16 can be generalized to the alignment of two different HFSMs.

Definition 15.22. (HFSM Alignment and Alignment Equivalence.) *Two HFSMs* $\mathcal{H}_1 = (Q_1, \Sigma, \Omega, T_1, \lambda_1)$ *and* $\mathcal{H}_2 = (Q_2, \Sigma, \Omega, T_2, \lambda_2)$ *are* alignable *if* $\forall q_1 \in Q_1, q_2 \in Q_2, \exists \sigma \in \Sigma^*. Img_\sigma(q_1, T_1) \sim Img_\sigma(q_2, T_2)$*. In this case, we say that* $\mathcal{H}_1$ *and* $\mathcal{H}_2$ *are* alignment equivalent*, denoted as* $\mathcal{H}_1 \cong \mathcal{H}_2$*.*

Similar to Theorem 15.16, we have the following.

Theorem 15.19. *Two HFSMs* $\mathcal{H}_1 = (Q_1, \Sigma, \Omega, T_1, \lambda_1)$ *and* $\mathcal{H}_2 = (Q_2, \Sigma, \Omega, T_2, \lambda_2)$ *are alignable (or alignment equivalent) if and only if* $\exists \sigma \in \Sigma^*, \forall q_1 \in Q_1, q_2 \in Q_2, Img_\sigma(q_1, T_1) \sim Img_\sigma(q_2, T_2)$*.*

Proof. ($\Leftarrow$) This direction follows from Definition 15.22.

($\Rightarrow$) A universal aligning sequence can be constructed in a way similar to the procedure *ComputeResetSequence* in Figure 15.24. ■

In other words, $\mathcal{H}_1 \cong \mathcal{H}_2$ if they share a common reset sequence and behave equivalently after reset. Alignment equivalence is more stringent than FSM equivalence because it requires that two FSMs share a common reset sequence in addition to indistinguishable input-output behavior after reset. However, note that, by the definition of alignment equivalence of two HFSMs, their input-output behaviors during the reset phase need not be identical.

Theorem 15.20. *Alignment equivalence $\cong$ is symmetric and transitive, but not necessarily reflexive in general.*

Proof. For $\mathcal{H}_1 \cong \mathcal{H}_2$, any state pair (q_1, q_2), with q_1 of $\mathcal{H}_1$ and q_2 of $\mathcal{H}_2$, is alignable and independent of the order of $\mathcal{H}_1$ and $\mathcal{H}_2$. Therefore, alignment equivalence is symmetric.

To prove the transitivity, assume $\mathcal{H}_1 \cong \mathcal{H}_2$ and $\mathcal{H}_1 \cong \mathcal{H}_3$. We show $\mathcal{H}_2 \cong \mathcal{H}_3$. There exists a reset sequence $\boldsymbol{\sigma}_{1,2}$ aligning $\mathcal{H}_1$ and $\mathcal{H}_2$ and a reset sequence $\boldsymbol{\sigma}_{1,3}$ aligning $\mathcal{H}_1$ and $\mathcal{H}_3$. Let $\boldsymbol{\sigma} = \boldsymbol{\sigma}_{1,2} \circ \boldsymbol{\sigma}_{1,3}$. We claim that $\boldsymbol{\sigma}$ is a reset sequence aligning any pair of $\mathcal{H}_1$, $\mathcal{H}_2$, and $\mathcal{H}_3$. To see it, consider first the alignment of $\mathcal{H}_1$ and $\mathcal{H}_2$ under $\boldsymbol{\sigma}$. After $\boldsymbol{\sigma}_{1,2}$ is applied, $\mathcal{H}_1$ and $\mathcal{H}_2$ are aligned and the current states of $\mathcal{H}_1$ and $\mathcal{H}_2$ are in a single equivalence class. Now applying another input sequence $\boldsymbol{\sigma}_{1,3}$ on $\mathcal{H}_1$ and $\mathcal{H}_2$ does not drive equivalent states to nonequivalent states. Therefore, the new current states of $\mathcal{H}_1$ and $\mathcal{H}_2$ after applying $\boldsymbol{\sigma}$ remain in a single equivalence class. That is, for any $q_1 \in Q_1$ and $q_2 \in Q_2$, let $q_1' = Img_{\boldsymbol{\sigma}}(q_1, T_1)$ and $q_2' = Img_{\boldsymbol{\sigma}}(q_2, T_2)$. Then $q_1' \sim q_2'$. On the other hand, $\boldsymbol{\sigma}$ must be an alignment sequence for $\mathcal{H}_1$ and $\mathcal{H}_3$ because, no matter what state pair $(q_1^\dagger, q_3^\dagger)$ is reached after applying $\boldsymbol{\sigma}_{1,2}$ to $\mathcal{H}_1$ and $\mathcal{H}_3$, it can always be aligned by $\boldsymbol{\sigma}_{1,3}$. Thus, for any $q_3 \in Q_3$, let $q_3' = Img_{\boldsymbol{\sigma}}(q_3, T_3)$. Then $q_1' \sim q_3'$, from which $q_2' \sim q_3'$, and thus $\mathcal{H}_2 \cong \mathcal{H}_3$. ■

Alignment equivalence is not necessarily reflexive, because if an HFSM is not essentially resettable at all, it cannot be alignment equivalent to itself.

Theorem 15.21. *An HFSM is alignment equivalent to itself if and only if it is essentially resettable.*

Proof. Let $\mathcal{H}^\dagger$ be another copy of an HFSM $\mathcal{H}$.

($\Rightarrow$) Suppose $\mathcal{H}$ is alignment equivalent to itself. Then any state pair $(q_1, q_2^\dagger)$ for q_1 of $\mathcal{H}$ and $q_2^\dagger$ of $\mathcal{H}^\dagger$ can be aligned by a universal reset sequence. Because every state of one HFSM, say $q_2^\dagger$ of $\mathcal{H}^\dagger$, has a corresponding equivalent state of the other, say q_2 of $\mathcal{H}$, then state pair (q_1, q_2) of $\mathcal{H}$ can be aligned in $\mathcal{H}$ with the same reset sequence by Lemma 15.15. Thus $\mathcal{H}$ must be essentially resettable.

($\Leftarrow$) If $\mathcal{H}$ is essentially resettable, then any state pair (q_1, q_2) of $\mathcal{H}$ can be aligned by some universal reset sequence $\boldsymbol{\sigma}$. Consider the disjoint union state space of $\mathcal{H}$ and $\mathcal{H}^\dagger$. Every state of one HFSM has a corresponding equivalent state of the other, say $q_2 \sim q_2^\dagger$ for state $q_2^\dagger$ of $\mathcal{H}^\dagger$. Again by Lemma 15.15, $(q_1, q_2^\dagger)$ has the same reset sequence $\boldsymbol{\sigma}$. Hence $\mathcal{H}$ and $\mathcal{H}^\dagger$ are alignment equivalent. ■

In other words, alignment equivalence $\cong$ is reflexive only for essentially resettable HFSMs. Therefore, alignment equivalence $\cong$ is an equivalence relation among essentially resetable HFSMs.

Corollary 15.22. *If* $\mathcal{H}_1 \cong \mathcal{H}_2$*, then* $\mathcal{H}_1 \cong \mathcal{H}_1$.

Proof. From Theorem 15.20, $\mathcal{H}_1 \cong \mathcal{H}_2$ implies $\mathcal{H}_2 \cong \mathcal{H}_1$ by symmetry. Further, $\mathcal{H}_1 \cong \mathcal{H}_1$ holds by transitivity. ∎

For two HFSMs, suppose that any state in one HFSM has a corresponding equivalent state in the other. Even then, it is not necessarily the case that there exists an input sequence that aligns these two HFSMs. This is similar to the fact that the converse condition of Proposition 15.13 does not hold. However, the next theorem shows that then unalignability implies that both of the two HFSMs are unresettable.

Theorem 15.23. *Given two HFSMs, assume that any state in one machine has some corresponding equivalent state in the other. Then these two HFSMs are alignable if and only if one of them (equivalently, each of them) is essentially resettable.*

Proof. Let Q and $Q^\dagger$ be the state sets of HFSMs $\mathcal{H}$ and $\mathcal{H}^\dagger$, respectively. Assume that any state of $\mathcal{H}$ has a corresponding equivalent state of $\mathcal{H}^\dagger$, and vice versa.

($\Rightarrow$) Assume $\mathcal{H} \cong \mathcal{H}^\dagger$. Then, by Lemma 15.15, any state pair $(q_1, q_2) \in Q \times Q$ of $\mathcal{H}$ must be alignable because there always exists $(q_1, q_2^\dagger) \in Q \times Q^\dagger$, with $q_2^\dagger \sim q_2$, alignable between $\mathcal{H}$ and $\mathcal{H}^\dagger$. Therefore, $\mathcal{H}$ is essentially resettable. Similarly, we know $\mathcal{H}^\dagger$ is essentially resettable as well.

($\Leftarrow$) Without loss of generality, assume $\mathcal{H}$ is resettable. Then, by Lemma 15.15, any state pair $(q_1, q_2^\dagger) \in Q \times Q^\dagger$ must be alignable between $\mathcal{H}$ and $\mathcal{H}^\dagger$ because there always exists $(q_1, q_2) \in Q \times Q$, with $q_2^\dagger \sim q_2$, alignable in $\mathcal{H}$. Therefore, $\mathcal{H}$ and $\mathcal{H}^\dagger$ are alignable. ∎

Theorem 15.24. *Consider the closed SCCs of the HSTGs of two alignment equivalent HFSMs. Suppose that power-up states are in the SCCs. Then any reset sequence for one HFSM is also a reset sequence for the other.*

Proof. In the SCCs, any state in one HFSM must have an equivalent state in the other. Otherwise, these two HFSMs are not alignment equivalent. On the other hand, any of the HFSMs is alignment equivalent to itself by Corollary 15.22, and thus resettable by Theorem 15.21. From Lemma 15.15, these two HFSMs must share the same set of reset sequences in the SCCs. ∎

To study the structure of a reset sequence, we employ the following definition.

Definition 15.23. *A state q of an HFSM is* dangling *if it has no predecessor states (i.e., no states can transition to q) or all of its predecessor states are dangling. Otherwise, it is* nondangling.

Example 15.20. *Consider Figure 15.23. States q_1 and q_2 are dangling, and all others are nondangling.*

Any reset sequence $\boldsymbol{\sigma}$ can be decomposed into three subsequences such that $\boldsymbol{\sigma} = \boldsymbol{\sigma}_1 \circ \boldsymbol{\sigma}_2 \circ \boldsymbol{\sigma}_3$ with:

(i) Subsequence $\boldsymbol{\sigma}_1$ drives an HFSM out of dangling paths. In this phase, as long as $|\boldsymbol{\sigma}_1|$ is no less than the length of the longest dangling path, any input sequence is valid. (A shorter $\boldsymbol{\sigma}_1$ may require detailed knowledge about the transitions of the dangling paths of an HSTG.)
(ii) Subsequence $\boldsymbol{\sigma}_2$ drives an HFSM out of open SCCs and into a closed SCC.
(iii) Subsequence $\boldsymbol{\sigma}_3$ enforces an HFSM entering a reset state (in a closed SCC).

Notice that subsequences $\boldsymbol{\sigma}_1$ and $\boldsymbol{\sigma}_2$ may be empty.

Example 15.21. *Continuing Figure 15.23, observe that the HSTG has a reset sequence $\boldsymbol{\sigma}$ = '1, 1, 0, 1, 0, 0, 1, 1' with reset state q_8. As analyzed previously, $\boldsymbol{\sigma}$ can be decomposed into $\boldsymbol{\sigma}_1$ = '1, 1', $\boldsymbol{\sigma}_2$ = '0, 1, 0', and $\boldsymbol{\sigma}_3$ = '0, 1, 1'.*

For two alignment equivalent HFSMs, any state in a closed SCC of one HFSM must have an equivalent state in a closed SCC of the other. Therefore, by Theorem 15.24, subsequence $\boldsymbol{\sigma}_3$ is a reset subsequence common to all alignment equivalent HFSMs. Two alignment equivalent HFSMs can always have a common suffix.

In certain classes of circuit transformations (e.g., see [40]), only dangling paths of an HSTG can be changed. In that case, (non-)resettability is preserved.

15.8.2.2 Reset-Independent Equivalences

Recall that FSM equivalence is an equivalence relation over FSMs, where initial states are prespecified; alignment equivalence is an equivalence relation only over essentially resettable HFSMs. Here we study some other notions of equivalence over HFSMs that are independent of initial states or resettability.

Definition 15.24. (Strict Equivalence.) *Two HFSMs are* strictly equivalent *if, for every state in one machine, there is an equivalent state in the other.*

This definition corresponds to the definition of "FSM equivalence" in [30, 42], where the "FSM" is meant to be our "HFSM." Here we name it differently to avoid confusion.

If two HFSMs are strictly equivalent, they are equivalent according to Definition 15.9 as long as the initial states are properly specified. On the other hand, FSM equivalence (of Definition 15.9) implies HFSM strict equivalence under restriction to the reachable state subspace. It can be checked that HFSM strict equivalence forms an equivalence relation over resettable and nonresettable HFSMs.

Another equivalence definition [81], more relaxed than HFSM strict equivalence, is based on the notion of safe replacement.

Definition 15.25. (Safe Replacement.) *An HFSM* $\mathcal{H}_1 = (Q_1, \Sigma, \Omega, T_1, \lambda_1)$ *is a* safe replacement *for* $\mathcal{H}_2 = (Q_2, \Sigma, \Omega, T_2, \lambda_2)$*, denoted as* $\mathcal{H}_1 \sqsubseteq \mathcal{H}_2$*, if* $\forall q_1 \in Q_1, \boldsymbol{\sigma} \in \Sigma^*, \exists q_2 \in Q_2$ *such that* $\lambda_1(\sigma_1, q_1) = \lambda_2(\sigma_1, q_2)$ *and* $\lambda_1(\sigma_i, Img_{\boldsymbol{\sigma}_{i-1}}(q_1, T_1)) = \lambda_2(\sigma_i, Img_{\boldsymbol{\sigma}_{i-1}}(q_2, T_2))$ *for* $i = 2, \ldots, |\boldsymbol{\sigma}|$*, where* $\boldsymbol{\sigma}_j = \sigma_1\sigma_2 \ldots \sigma_j$ *is a substring of* $\boldsymbol{\sigma}$*.*

By comparing the foregoing definition with Definition 15.8 of equivalent states, one notices that the difference is in the quantification order $\forall\boldsymbol{\sigma}\exists q_2$ here versus $\exists q_2\forall\boldsymbol{\sigma}$ in Definition 15.8. In Definition 15.25, a state $q_1 \in Q_1$ need not have an equivalent state $q_2 \in Q_2$, but for every input sequence there may be a different state of Q_2 that behaves like q_1 under that input sequence. However, if $\mathcal{H}_2$ is replaced with $\mathcal{H}_1$ for $\mathcal{H}_1 \sqsubseteq \mathcal{H}_2$, the underlying environment will not experience any new response because the input-output behavior of $\mathcal{H}_1$ when powered up can always be simulated by $\mathcal{H}_2$.

Example 15.22. *Let* G_1 *and* G_2 *of Figure 15.26 be the HSTGs of HFSMs* $\mathcal{H}_1$ *and* $\mathcal{H}_2$*, respectively. Then* $\mathcal{H}_1 \sqsubseteq \mathcal{H}_2$*. This is because states* q_b*,* q_c*, and* q_d *of* G_1 *are equivalent to* q_6*,* q_7*, and* q_8 *of* G_2*, respectively. In addition,* q_a *of* G_1 *and* q_4 *of* G_2

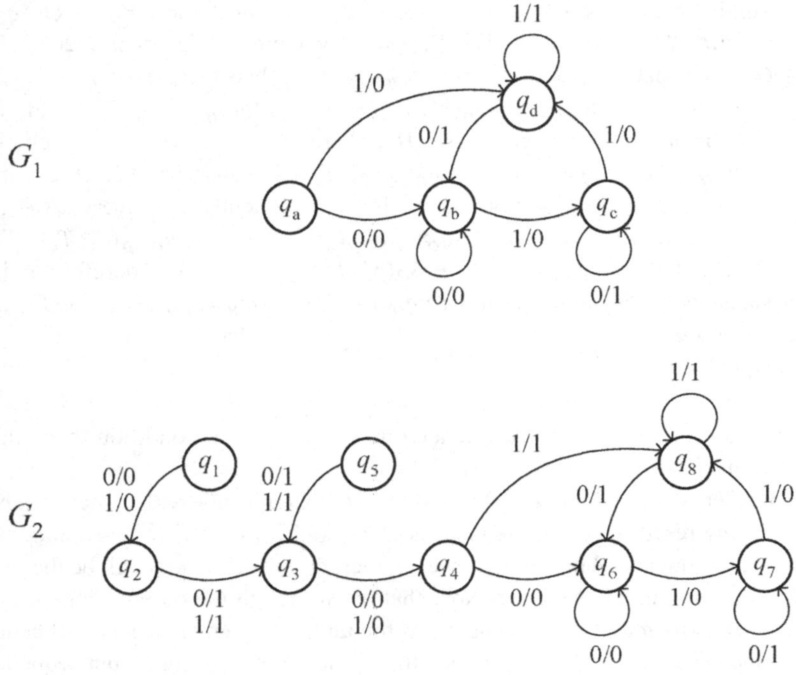

Figure 15.26. Two HSTGs G_1 and G_2, where given any state in G_1 and any input sequence $\boldsymbol{\sigma}$, there exists a state in G_2 that generates the same output sequence under $\boldsymbol{\sigma}$.

are equivalent under input 0, whereas q_a of G_1 and q_7 of G_2 are equivalent under input 1.

Definition 15.26. (Safe-Replacement Equivalence.) *Two HFSMs $\mathcal{H}_1$ and $\mathcal{H}_2$ are* safe-replacement equivalent *if $\mathcal{H}_1 \sqsubseteq \mathcal{H}_2$ and $\mathcal{H}_2 \sqsubseteq \mathcal{H}_1$.*

Safe-replacement equivalence is reflexive, symmetric, and transitive and thus forms an equivalence relation. From the logical validity $\exists\forall \Rightarrow \forall\exists$, it follows that strict equivalence implies safe-replacement equivalence. Moreover, regardless of resettability, safe-replacement equivalence is the coarsest condition for the environment not being able to distinguish the replacement of one HFSM with another.

When restricted to *resettable* HFSMs, we can connect safe replacement and alignment equivalence as follows.

Theorem 15.25. *If HFSM $\mathcal{H}_1 = (Q_1, \Sigma, \Omega, T_1, \lambda_1)$ is a safe replacement of a resettable HFSM $\mathcal{H}_2 = (Q_2, \Sigma, \Omega, T_2, \lambda_2)$, then $\mathcal{H}_1$ and $\mathcal{H}_2$ are alignment equivalent.*

Proof. We show that safe replacement preserves reset sequences, that is, any reset sequence of $\mathcal{H}_2$ is also a reset sequence of $\mathcal{H}_1$. In addition, $\mathcal{H}_1$ and $\mathcal{H}_2$ are aligned to equivalent states under the same reset sequence.

Assuming that $\boldsymbol{\sigma}$ is a reset sequence of $\mathcal{H}_2$, we must show $\forall q_1 \in Q_1, q_2 \in Q_2 . Img_{\boldsymbol{\sigma}}(q_1, T_1) \sim Img_{\boldsymbol{\sigma}}(q_2, T_2)$. Suppose by contradiction that there exists $q_1' \in Q_1$ such that $Img_{\boldsymbol{\sigma}}(q_1', T_1) \not\sim Img_{\boldsymbol{\sigma}}(q_2, T_2)$. Then there exist $\boldsymbol{\rho} \in \Sigma^*$ and $\sigma' \in \Sigma$ such that $\lambda_1(\sigma', Img_{\boldsymbol{\sigma}\circ\boldsymbol{\rho}}(q_1', T_1)) \neq \lambda_2(\sigma', Img_{\boldsymbol{\sigma}\circ\boldsymbol{\rho}}(q_2, T_2))$. But, because $\mathcal{H}_1$ is a safe replacement of $\mathcal{H}_2$, there is a state $q_2' \in Q_2$ such that $\lambda_1(\sigma', Img_{\boldsymbol{\sigma}\circ\boldsymbol{\rho}}(q_1', T_1)) = \lambda_2(\sigma', Img_{\boldsymbol{\sigma}\circ\boldsymbol{\rho}}(q_2', T_2))$. Because $\boldsymbol{\sigma}$ is a reset sequence of $\mathcal{H}_2$, so is $\boldsymbol{\sigma} \circ \boldsymbol{\rho}$ by Proposition 15.14. Consequently, for every $q_2 \in Q_2$, $\lambda_2(\sigma', Img_{\boldsymbol{\sigma}\circ\boldsymbol{\rho}}(q_2, T_2)) = \lambda_2(\sigma', Img_{\boldsymbol{\sigma}\circ\boldsymbol{\rho}}(q_2', T_2)) = \lambda_1(\sigma', Img_{\boldsymbol{\sigma}\circ\boldsymbol{\rho}}(q_1', T_1))$. This contradicts $\lambda_1(\sigma', Img_{\boldsymbol{\sigma}\circ\boldsymbol{\rho}}(q_1', T_1)) \neq \lambda_2(\sigma', Img_{\boldsymbol{\sigma}\circ\boldsymbol{\rho}}(q_2, T_2))$. Therefore, $\boldsymbol{\sigma}$ is a reset sequence of $\mathcal{H}_1$ as well, and $Img_{\boldsymbol{\sigma}}(q_1, T_1) \sim Img_{\boldsymbol{\sigma}}(q_2, T_2)$ holds for any $q_1 \in Q_1$ and $q_2 \in Q_2$. By Theorem 15.19, it follows that $\mathcal{H}_1$ and $\mathcal{H}_2$ are alignment equivalent. ∎

Thus, safe-replacement equivalence imposes a stronger condition than alignment equivalence over resettable HFSMs.

Consider two resettable HFSMs $\mathcal{H}_1$ and $\mathcal{H}_2$ that are safe-replacement equivalent. During reset, the output sequences of $\mathcal{H}_1$ and $\mathcal{H}_2$ may differ even under the same reset sequence. After reset, however, their output sequences will be the same under the same input sequence. Note that, even though the output sequences of the two HFSMs may differ during reset, the underlying environment still cannot tell if one HFSM is replaced with the other. (These inconsistent output sequences are purely due to the nondeterminism at power-up. Even for the same HFSM, the output sequence during reset at this time may differ from that at next time.)

Example 15.23. *Consider the corresponding HFSM $\mathcal{H}$ of the HSTG G of Figure 15.22. Suppose two copies of $\mathcal{H}$ are powered up in different states, say q_0 and q_1. (These two copies are safe-replacement equivalent.) Under the reset sequence '1,1,1', they produce output sequences '1,0,0' and '0,0,0', respectively, and both are reset to state q_2.*

Safe replacement is sometimes more stringent than necessary. In some cases, delayed replacements [82] are allowed, where the new HFSM replacing the old one can be clocked for several cycles before the original reset sequence is applied. Here we do not care about input-output behavior in the first n clock cycles.

Definition 15.27. (Delayed HFSM.) *The* n-cycle delayed HFSM *of an HFSM $\mathcal{H} = (Q, \Sigma, \Omega, T, \lambda)$, denoted as $\mathcal{H}^n$, is the same as $\mathcal{H}$ except for restricting its state set to $\{q' \mid \exists\, q \in Q, \sigma \in \Sigma^n.\, q' = Img_\sigma(q, T)\}$.*

The purpose of delaying an HFSM for n clock cycles is to let the HFSM get rid of some dangling states before the original reset sequence is applied.

Definition 15.28. (Delay Replacement.) *An HFSM $\mathcal{H}_1 = (Q_1, \Sigma, \Omega, T_1, \lambda_1)$ is an* n-cycle delayed safe replacement *(or simply* n-cycle delay replacement*) for $\mathcal{H}_2 = (Q_2, \Sigma, \Omega, T_2, \lambda_2)$ if ${\mathcal{H}_1}^n \sqsubseteq \mathcal{H}_2$.*

It is easily seen that ${\mathcal{H}_1}^n \sqsubseteq \mathcal{H}_2$ implies ${\mathcal{H}_1}^{n+1} \sqsubseteq \mathcal{H}_2$. Moreover, delay replacement is a relaxed definition of safe replacement because a safe replacement is also a 0-cycle delayed replacement. On the other hand, we may connect delay replacement and alignment equivalence over resettable HFSMs as follows.

Proposition 15.26. *If HFSM $\mathcal{H}_1$ is an n-cycle delay replacement of a resetable HFSM $\mathcal{H}_2$, then $\mathcal{H}_1$ and $\mathcal{H}_2$ are alignment equivalent.*

Proof. Suppose σ_2 is a reset sequence of $\mathcal{H}_2$. Because $\mathcal{H}_1$ is an n-cycle delay replacement of $\mathcal{H}_2$, then, for any input sequence σ_1 with $|\sigma_1| = n$ and for any $q_1 \in Q_1$, $Img_{\sigma_1}(q_1, T_1)$ is in the state set Q_1^n of $\mathcal{H}_1^n$; moreover, $\mathcal{H}_1^n$ is a safe replacement of $\mathcal{H}_2$, but safe replacement preserves reset sequences by Theorem 15.25, and so $\sigma = \sigma_1 \circ \sigma_2$ is a reset sequence for $\mathcal{H}_1$ and $\forall q_1 \in Q_1, q_2 \in Q_2.\ Img_\sigma(q_1, T_1) \sim Img_{\sigma_2}(q_2, T_2)$. However, observe that $\sigma_1 \circ \sigma_2$ must also be a reset sequence for $\mathcal{H}_2$ because the possible states of $\mathcal{H}_2$ after applying σ_1 are a subset of Q_2 that can still be reset by σ_2. Thus, $\forall q_1 \in Q_1, q_2 \in Q_2.\ Img_\sigma(q_1, T_1) \sim Img_\sigma(q_2, T_2)$, and $\mathcal{H}_1 \cong \mathcal{H}_2$. ■

Similar to the definition of safe-replacement equivalence, we may define delay-replacement equivalence as follows.

Definition 15.29. (Delay-Replacement Equivalence.) *Two HFSMs $\mathcal{H}_1$ and $\mathcal{H}_2$ are* delay-replacement equivalent *if ${\mathcal{H}_1}^m \sqsubseteq \mathcal{H}_2$ and ${\mathcal{H}_2}^n \sqsubseteq \mathcal{H}_1$ for some positive integers m and n.*

Figure 15.27 summarizes the hierarchical structure among the aforementioned equivalence relations over *resettable* HFSMs, where FSM equivalence (of

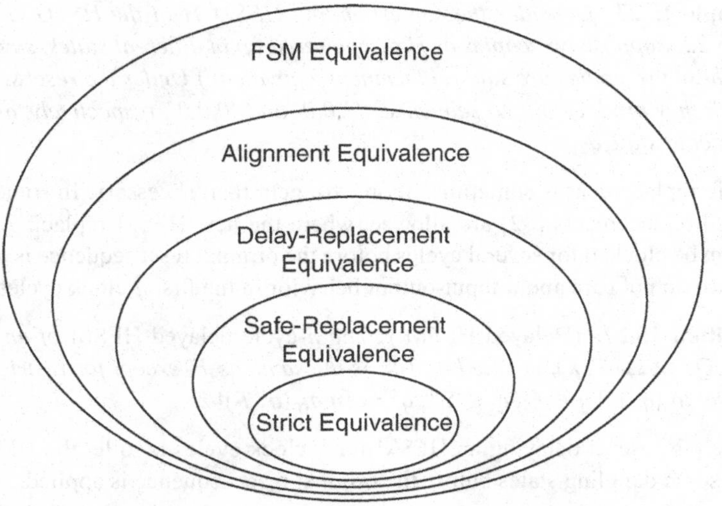

Figure 15.27. The hierarchical structure formed by the inclusion relation among various notions of equivalence over resettable HFSMs.

Definition 15.9) disregards any inconsistent behavior before initialization (in fact, the two circuits under comparison need not be initialized in the same way) and cares only about equivalence after initialization. The Euler diagram shows that, if two resetable HFSMs are equivalent under a contained relation, then they are equivalent under the containing relation. There are other notions of equivalence, mainly developed in the testing community. A review can be found in [61].

15.9 Conclusions

This chapter provided the foundations of combinational and sequential hardware equivalence. For space reasons, interesting topics were not covered: arithmetic circuit verification, property checking for RTL codes, verification under different levels of abstraction, don't cares, and so on.

External sequential don't cares (and combinational don't cares as a special case) arise as incomplete specification that can be used for better optimization of the final circuit. They cannot be disregarded during verification to avoid false negatives, that is, situations where a reported difference is actually due to a don't-care sequence, and thus it cannot happen. Sequential don't cares were discussed in [33], whose authors describe a software package AQUILA, a sequential equivalence checker that is able to handle them. AQUILA is based on an array of techniques centered around ATPG analysis. The way in which sequential don't cares are incorporated can be understood from the revised miter construction shown in Figure 15.28. It can be seen that the external don't-care set is represented by an additional sequential network with only one primary output *dc* (part of the specification). If an input

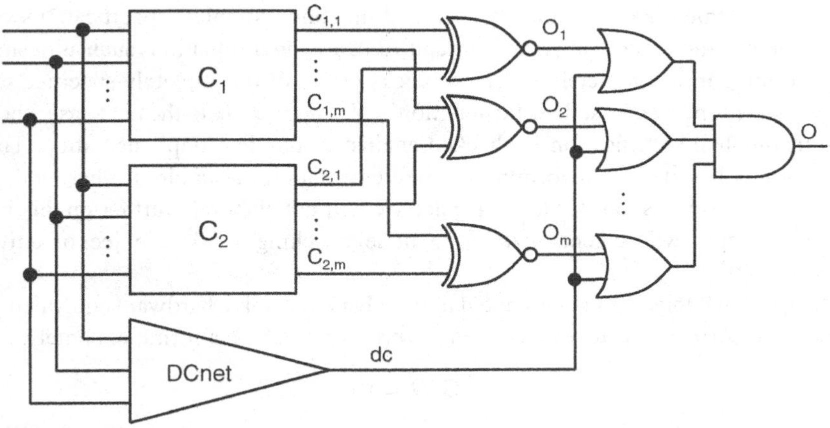

Figure 15.28. Multioutput miter circuit from networks C_1 and C_2, augmented by a don't-care sequential network *DCnet*.

sequence is a don't-care sequence, then the value of the output signal dc is 1; otherwise it is 0. For example, they mention the case of a sequential multiplier that produces one care output for every certain number of clock cycles; that is, the don't-care network might produce the output sequence 11111110, which means that only in the eighth cycle is the output a care output.

The basic fact is that the stuck-at-1 fault at the output O in Figure 15.28 is untestable if and only if signal $C_{1,k}$ is equal to signal $C_{2,k}$, $k = 1, \dots, m$, for all care input sequences. Indeed, to enforce the output O to 0, at least one of the OR-gates should produce a 0, which requires $dc = 0$ (care sequence) and the XNOR gate to 0 (distinguishing sequence). So a sequential ATPG program can either prove the equivalence or find a distinguishing sequence for a pair of output signals.

In addition to external don't cares, there are internal don't cares that represent impossible value combinations at some state variables, corresponding to unreachable states. Provided one has a cheap way of finding them, they can be exploited to ease the task of verification by avoiding traversing the unreachable state space.

Other discussions of don't cares in combinational and sequential equivalence checking can be found in [75] and in [71, 72], where it is remarked that don't cares can be seen as extending equivalence checking into a problem of inclusion checking, meaning that the implementation is included in the specification if there is an assignment of the don't-care conditions that makes them equal.

An even more radical extension, under the name of *black box equivalence checking*, supposes that the specification is known, but only parts of the implementation are completed or known (black boxes being the unfinished or unknown parts), so that inequivalence is declared when the implementation differs from the specification for all possible replacements of the black boxes (see [29, 36] and [77]). Then Scholl and Becker [77, 78] studied the related problem when boxes implement an incompletely specified function (complete specifications vs. incomplete implementations) and variants of it, arising in situations such as design partitioning in incompletely specified blocks or use of incompletely specified intellectual property cores; their formulation is also applicable to the dual case when an incomplete specification is checked against a complete implementation. The solution is based on transforming the implementation into a circuit that models the incompleteness, but for lack of space we will not elaborate further on this interesting topic, whose generalization to model checking is still an object of active research [63].

Notice that the effort documented in this chapter to solve hardware equivalence checking corresponds to verifying on the product machine a formula, namely,

$$\mathbf{AG}(O = 0)$$

in CTL. It is read as: for every path and at every node in the path, it is true that the output O is 0. In general the **AG** operator asserts that a formula is true for all possible evolutions in the future. As mentioned in Section 15.1.2, however, this is just one simple formula, out of the many that can be built in CTL or other temporal logics, using the full power of path quantifiers and temporal modalities. A discussion of general model checking requires telling another story.

References

[1] P. Ashar, A. Gupta, and S. Malik. Using complete-1-distinguishability for FSM equivalence checking. *ACM Transactions on Design Automation of Electronic Systems* 6(4), pp. 569–90, 2001.

[2] A. Biere, A. Cimatti, E. Clarke, and Y. Zhu. Symbolic model checking without BDDs. In *Proceedings of Tools and Algorithms for the Construction and Analysis of Systems*, pp. 193–207, 1999.

[3] P. Bjesse and K. Claessen. SAT-based verification without state space traversal. In *Proceedings of Formal Methods in Computer-Aided Design*, 2000.

[4] D. Brand. Verification of large synthesized designs. In *Proceedings of the International Conference on Computer-Aided Design*, pp. 534–7, November 1993.

[5] R. Brayton, G. Hachtel, A. Sangiovanni-Vincentelli, F. Somenzi, A. Aziz, S.-T. Cheng, S. Edwards, S. Khatri, Y. Kukimoto, A. Pardo, S. Qadeer, R. Ranjan, S. Sarwary, T. Shiple, G. Swamy, and T. Villa. VIS: A system for verification and synthesis. In *Proceedings of the Conference on Computer-Aided Verification*, R. Alur and T. Henzinger, eds. *Lecture Notes in Computer Science*, Vol. 1102, pp. 332–4, 1996.

[6] R. Brayton, G. Hachtel, A. Sangiovanni-Vincentelli, F. Somenzi, A. Aziz, S.-T. Cheng, S. Edwards, S. Khatri, Y. Kukimoto, A. Pardo, S. Qadeer, R. Ranjan, S. Sarwary, T. Shiple, G. Swamy, and T. Villa. VIS. In *Proceedings of the Conference on Formal Methods in Computer-Aided Design*, M. Srivas and A. Camilleri, eds. *Lecture Notes in Computer Science*, Vol. 1166, pp. 248–56, 1996.

[7] R. E. Bryant. Graph-based algorithms for boolean function manipulation. *IEEE Transactions on Computers*, pp. 677–91, 1986.

[8] J. R. Burch, E. M. Clarke, K. L. McMillan, D. L. Dill, and L. J. Hwang. Symbolic model checking: 10^{20} states and beyond. *Information and Computation* 98(2), pp. 142–70, 1992.

[9] J. Burch and V. Singhal. Robust latch mapping for combinational equivalence checking. In *Proceedings of the International Conference on Computer-Aided Design*, 1998.

[10] E. M. Clarke, E. A. Emerson, and A. P. Sistla. Automatic verification of finite-state concurrent systems using temporal logic specifications. *ACM Transactions on Programming Languages and Systems* 8(2), pp. 244–63, 1986.
[11] E. M. Clarke, O. Grumberg, K. L. McMillan, and X. Zhao. Efficient generation of counterexamples and witnesses in symbolic model checking. In *32nd Design Automation Conference (DAC 95)*, pp. 427–32, San Francisco, 1995.
[12] E. M. Clarke, O. Grumberg, and D. A. Peled. *Model Checking*. MIT Press, Cambridge, MA, 1999.
[13] S. Cook. The complexity of theorem-proving procedures. In *Proceedings of the IEEE Symposium on the Foundations of Computer Science*, pp. 151–8, 1971.
[14] O. Coudert, C. Berthet, and J. C. Madre. Verification of synchronous sequential machines based on symbolic execution. In *Proceedings of the International Workshop on Automatic Verification Methods for Finite State Systems*, 1989.
[15] O. Coudert and J. C. Madre. A unified framework for the formal verification of sequential circuits. In *Proceedings International Conference on Computer-Aided Design*, pp. 126–9, 1990.
[16] W. Craig. Linear reasoning: A new form of the Herbrand-Gentzen theorem. *Journal of Symbolic Logic* 22(3), pp. 250–68, 1957.
[17] Y. Crama and P. L. Hammer. *Boolean Functions: Theory, Algorithms, and Applications*. Cambridge University Press, Cambridge, UK, 2010.
[18] M. Davis, G. Logemann, and D. Loveland. A machine program for theorem proving. *Communications of the ACM* 5, pp. 394–7, 1962.
[19] M. Davis and H. Putnam. A computing procedure for quantification theory. *Journal of the ACM* 7, pp. 201–15, 1960.
[20] N. Eén and N. Sörensson. Temporal induction by incremental SAT solving. In *Proceedings of Bounded Model Checking*, 2003.
[21] C. A. J. van Eijk. Sequential equivalence checking based on structural similarities. *IEEE Transactions of Computer-Aided Design* 19(7), pp. 814–19, 2000.
[22] C. A. J. van Eijk and J. A. G. Jess. Detection of equivalent state variables in finite state machine verification. In *Proceedings of the International Workshop on Logic Synthesis*, 1995.
[23] T. Filkorn. A method for symbolic verification of synchronous circuits. In *Proceedings of the International Symposium on Computer Hardware Description Languages and Their Applications*, pp. 249–59, 1991.
[24] T. Filkorn. *Symbolische Methoden für die Verifikation endlicher Zustandssysteme*. PhD dissertation, Institut für Informatik der Technischen Universität München, 1992.
[25] M. Fujita, S. Komatsu, and H. Saito. Formal verification techniques for digital systems. In *Dependable Computing Systems*, H. B. Diab, S. Hassoun, and A. Y. Zomaya, eds. J. Wiley, New York, 2005.
[26] M. R. Garey and D. S. Johnson. *Computers and Intractability: A Guide to the Theory of NP-Completeness*. W. H. Freeman, New York, 1979.
[27] E. Goldberg and Y. Novikov. BerkMin: A fast and robust SAT-solver. In *Proceedings of Design Automation and Test in Europe*, pp. 142–9, 2002.
[28] J. B. Gosling. *Simulation in the Design of Digital Electronic Systems*. Cambridge University Press, Cambridge, UK, 1993.
[29] W. Guenther, N. Drechsler, R. Drechsler, and B. Becker. Verification of designs containing black boxes. In *EUROMICRO*, pp. 100–5, 2000.
[30] J. Hartmanis and R. E. Stearns. *Algebraic Structure Theory of Sequential Machines*. Prentice-Hall, Upper Saddle River, NJ, 1966.
[31] S. Hazelhurst and C.-J. Seger. Symbolic trajectory evaluation. In *Formal Hardware Verification*, T. Kropf, ed., *Lecture Notes in Computer Science*, 1287, pp. 3–78, 1997.
[32] J. Hopcroft. An $n \log n$ algorithm for minimizing states in a finite automaton. In *Theory of Machines and Computations*, Z. Kohavi and A. Paz, eds., pp. 189–96. Academic Press, New York, 1971.

[33] S.-Y. Huang, K.-T. Cheng, K.-C. Chen, C.-Y. Huang, and F. Brewer. Aquila: An equivalence checking system for large sequential designs. *IEEE Transactions on Computers* 49(5), pp. 443–64, 2000.

[34] H. Hulgaard, P. F. Williams, and H. R. Andersen. Equivalence checking of combinational circuits using boolean expression diagrams. *IEEE Transactions on Computer-Aided Design* 18(7), pp. 903–17, 1999.

[35] N. Immerman. Nondeterministic space is closed under complementation. *SIAM Journal on Computing*, 17(5), pp. 935–8, 1988.

[36] A. Jain, V. Boppana, R. Mukherjee, J. Jain, M. Fujita, and M. Hsiao. Testing, verification, and diagnosis in the presence of unknowns. In *VLSI Test Symposium*, pp. 263–9, 2000.

[37] J. Jain, A. Narayan, M. Fujita, and A. Sangiovanni-Vincentelli. A survey of techniques for formal verification of combinational circuits. In *Proceedings of the International Conference on Computer Design*, pp. 445–54, 1997.

[38] J.-H. Jiang and R. Brayton. On the verification of sequential equivalence. *IEEE Transactions on Computer-Aided Design* 6, pp. 686–97, 2003.

[39] J.-H. Jiang and R. Brayton. Functional dependency for verification reduction. In *Proceedings of the International Conference on Computer Aided Verification*, pp. 268–80, 2004.

[40] J.-H. Jiang and R. Brayton. Retiming and resynthesis: A complexity perspective. *IEEE Transactions on Computer-Aided Design* 25(12), pp. 2674–86, 2006.

[41] J.-H. R. Jiang and W.-L. Hung. Inductive equivalence checking under retiming and resynthesis. In *Proceedings of the International Conference on Computer-Aided Design*, pp. 326–33, 2007.

[42] Z. Kohavi. *Switching and Finite Automata Theory*. McGraw-Hill, New York, 1978.

[43] J. Krajíček. Interpolation theorems, lower bounds for proof systems, and independence results for bounded arithmetic. *Journal of Symbolic Logic* 62(2), pp. 457–86, 1997.

[44] A. Kuehlmann and C. A. J. van Eijk. Combinational and sequential equivalence checking. In *Logic Synthesis and Verification*, R. Brayton, S. Hassoun, and T. Sasao, eds., pp. 343–72. Kluwer, Dordrecht, 2001.

[45] A. Kuehlmann and F. Krohm. Equivalence checking using cuts and heaps. In *Proceedings of the Design Automation Conference*, pp. 263–8, 1997.

[46] A. Kuehlmann, V. Paruthi, F. Krohm, and M. K. Ganai. Robust boolean reasoning for equivalence checking and functional property verification. *IEEE Transactions on Computer-Aided Design* 21(12), pp. 1377–94, 2002.

[47] W. Kunz and A. Biere. SAT and ATPG: Boolean engines for formal hardware verification. In *Proceedings of the International Conference on Computer-Aided Design*, pp. 782–5, 2002.

[48] W. Kunz, J. Marques-Silva, and S. Malik. SAT and ATPG: Algorithms for boolean decision problems. In *Logic Synthesis and Verification*, R. Brayton, S. Hassoun, and T. Sasao, eds., pp. 309–41. Kluwer, Dordrecht, 2001.

[49] R. P. Kurshan. *Computer-Aided Verification of Coordinating Processes*. Princeton University Press, Princeton, NJ, 1994.

[50] W. K. C. Lam. *Hardware Design Verification: Simulation and Formal Method-Based Approaches*. Prentice Hall Professional Technical Reference, Upper Saddle River, NJ, 2005.

[51] C.-C. Lee, J.-H. R. Jiang, C.-Y. Huang, and A. Mishchenko. Scalable exploration of functional dependency by interpolation and incremental SAT solving. In *Proceedings of the International Conference on Computer-Aided Design*, pp. 227–33, 2007.

[52] C. Leiserson and J. Saxe. Optimizing synchronous systems. *Journal of VLSI and Computer Systems* 1(1), pp. 41–67, 1983.

[53] C. Leiserson and J. Saxe. Retiming synchronous circuitry. *Algorithmica* 6, pp. 5–35, 1991.

[54] B. Lin and A. R. Newton. Implicit manipulation of equivalence classes using binary decision diagrams. In *International Conference on Computer Design*, pp. 81–5. IEEE Computer Society, 1991.

[55] K. L. McMillan. *Symbolic Model Checking*. Kluwer Academic Publishers, Dordrecht, 1993.

[56] K. L. McMillan. Applying SAT methods in unbounded symbolic model checking. In *Proceedings of Computer Aided Verification*, pp. 250–64, 2002.

[57] K. L. McMillan. Interpolation and SAT-based model checking. In *Proceedings of Computer Aided Verification*, pp. 1–13, 2003.

[58] J. Marques-Silva and K. Sakallah. GRASP: A search algorithm for propositional satisfiability. *IEEE Transactions on Computers* 48(5), pp. 506–21, 1999.

[59] A. Miczo. *Digital Logic Testing and Simulation*. J. Wiley, New York, 2003.

[60] A. Mishchenko, S. Chatterjee, J.-H. R. Jiang, and R. Brayton. FRAIGs: A unifying representation for logic synthesis and verification. Technical report, Electronics Research Laboratory, University of California, Berkeley, March 2005.

[61] M. Mneimneh and K. Sakallah. Principles of sequential-equivalence verification. *IEEE Design & Test of Computers*, pp. 248–57, 2005.

[62] M. Moskewicz, C. Madigan, L. Zhang, and S. Malik. Chaff: Engineering an efficient SAT solver. In *Proceedings of the Design Automation Conference*, pp. 530–5, 2001.

[63] T. Nopper and C. Scholl. Approximate symbolic model checking for incomplete designs. In *Formal Methods in Computer-Aided Design*, pp. 290–305, 2004.

[64] R. Paige and R. Tarjan. Three partition refinement algorithms. *SIAM Journal on Computing* 16(6), pp. 973–89, 1987.

[65] L. Pillage. *Electronic Circuit and System Simulation Methods (SRE)*. McGraw-Hill, New York, 1998.

[66] C. Pixley, S. Jeong, and G. Hatchel. Exact calculation of synchronization sequences based on binary decision diagrams. *IEEE Transactions on Computer-Aided Design* 13, pp. 1024–34, 1994.

[67] D. Plaisted and S. Greenbaum. A structure preserving clause form translation. *Journal of Symbolic Computation* 2, pp. 293–304, 1986.

[68] B. Plessier, G. Hachtel, and F. Somenzi. Extended BDDs: trading off canonicity for structure in verification algorithms. *Formal Methods in System Design* 4(2), pp. 167–85, 1994.

[69] P. Pudlák. Lower bounds for resolution and cutting plane proofs and monotone computations. *Journal of Symbolic Logic* 62(2), pp. 981–98, 1997.

[70] S. Quer, G. Cabodi, P. Camurati, L. Lavagno, E. Sentovich, and R. K. Brayton. Verification of similar FSMs by mixing incremental re-encoding, reachability analysis, and combinational checks. *Formal Methods in System Design* 17(2), pp. 107–34, 2000.

[71] S. Rahim, B. Rouzeyre, and L. Torres. A flip-flop matching engine to verify sequential optimizations. *Computing and Informatics* 23(5–6), pp. 437–60, 2004.

[72] S. Rahim, B. Rouzeyre, L. Torres, and J. Rampon. Matching in the presence of don't cares and redundant sequential elements for sequential equivalence checking. In *HLDVT '03: Proceedings of the Eighth IEEE International Workshop on High-Level Design Validation and Test*, p. 129, IEEE Computer Society, 2003.

[73] J. Robinson. A machine-oriented logic based on the resolution principle. *Journal of the ACM* 12(1), pp. 23–41, 1965.

[74] R. Rudell. Dynamic variable ordering for binary decision diagrams. In *Proceedings of the International Conference on Computer-Aided Design*, pp. 42–7, 1993.

[75] S. Safarpour, G. Fey, A. Veneris, and R. Drechsler. Utilizing don't care states in SAT-based bounded sequential problems. In *GLSVSLI '05: Proceedings of the 15th ACM Great Lakes Symposium on VLSI*, pp. 264–9, ACM Press, New York, 2005.

[76] W. J. Savitch. Relationships between nondeterministic and deterministic space complexities. *Journal of Computer and System Sciences* 4(2), pp. 177–92, 1970.

[77] C. Scholl and B. Becker. Checking equivalence for partial implementations. In *Design Automation Conference*, pp. 238–43, 2001.
[78] C. Scholl and B. Becker. Checking equivalence for circuits containing incompletely specified boxes. In *International Conference on Computer Design*, pp. 56–63, 2002.
[79] V. Schuppan and A. Biere. Efficient reduction of finite state model checking to reachability analysis. *International Journal on Software Tools for Technology Transfer (STTT)* 5(2–3), 2004.
[80] M. Sheeran, S. Singh, and G. Stålmarck. Checking safety properties using induction and a SAT-solver. In *Proceedings of Formal Methods in Computer-Aided Design*, 2000.
[81] V. Singhal and C. Pixley. The verification problem for safe replaceability. In *Proceedings of Computer Aided Verification*, pp. 311–23, 1994.
[82] V. Singhal, C. Pixley, A. Aziz, and R. Brayton. Exploiting power-up delay for sequential optimization. In *Proceedings of the European Design Automation Conference*, pp. 54–9, 1995.
[83] L. Stockmeyer and A. Meyer. Word problems requiring exponential time. In *Proceedings of the ACM Symposium on the Theory of Computing*, pp. 1–9, 1973.
[84] R. Szelepcsenyi. The method of forced enumeration for nondeterministic automata. *Acta Informatica* 26(3), pp. 279–84, 1988.
[85] T. Tamisier. Computing the observable equivalence relation of a finite state machine. In *ICCAD '93: Proceedings of the 1993 IEEE/ACM International Conference on Computer-Aided Design*, pp. 184–7. IEEE Computer Society Press, 1993.
[86] A. Tarski. A lattice-theoretic fixpoint theorem and its applications. *Pacific Journal of Mathematics* 5, pp. 285–309, 1955.
[87] H. Touati, H. Savoj, B. Lin, R. Brayton, and A. Sangiovanni-Vincentelli. Implicit enumeration of finite state machines using BDDs. In *Proceedings of the International Conference on Computer-Aided Design*, pp. 130–3, 1990.
[88] G. Tseitin. On the complexity of derivation in propositional calculus. *Studies in Constructive Mathematics and Mathematical Logic*, pp. 466–83, 1970.
[89] J. Whittemore, J. Kim, and K. Sakallah. SATIRE: A new incremental satisfiability engine. In *Proceedings of the Design Automation Conference*, pp. 542–5, 2001.
[90] B. P. Zeigler, T. G. Kim, and H. Praehofer. *Theory of Modeling and Simulation*. Academic Press, New York, 2000.
[91] H. Zhang. SATO: An efficient propositional prover. In *Proceedings of the International Conference on Automated Deduction*, pp. 272–5, 1997.

16

Synthesis of Multilevel Boolean Networks

Tiziano Villa, Robert K. Brayton, and Alberto L. Sangiovanni-Vincentelli

16.1 Boolean Networks

16.1.1 Introduction

Two-level logic minimization has been a success story both in terms of theoretical understanding (see, e.g., [15]) and availability of practical tools (such as ESPRESSO) [2, 14, 31, 38]. However, two-level logic is not suitable to implement large Boolean functions, whereas multilevel implementations allow a better trade-off between area and delay. Multilevel logic synthesis has the objective to explore multilevel implementations guided by some function of the following metrics:

(i) The **area** occupied by the logic gates and interconnect (e.g., approximated by **literals**, which correspond to **transistors** in technology-independent optimization);
(ii) The **delay** of the longest path through the logic;
(iii) The **testability** of the circuit, measured in terms of the percentage of faults covered by a specified set of test vectors, for an appropriate fault model (e.g., single stuck faults, multiple stuck faults);
(iv) The **power** consumed by the logic gates and wires.

Often good implementations must satisfy simultaneously upper or lower constraints placed on these parameters and look for good compromises among the cost functions.

It is common to classify optimization as technology-independent versus technology-dependent, where the former represents a circuit by a network of abstract nodes, whereas the latter represents a circuit by a network of the actual gates available in a given library or programmable architecture. A common paradigm is to first try technology-independent optimization and then map the optimized circuit into the final library (technology mapping). In some cases it has been found advantageous to iterate this process; then it is called technology semiindependent optimization. Splitting multilevel optimization into two steps has been suggested by the complexity of the problem but pays the penalty of selecting suboptimal solutions. It must be said that, contrary to the two-level and three-level cases, there is no established theory of optimum multilevel implementations, so here we are in the domain of heuristics.

Example 16.1. *Given the specification*

$$\begin{aligned} w &= ab \vee \overline{a}\overline{b} \\ \textit{if } w \textit{ then } z &= cd \vee \overline{a}\overline{d}; u = cd \vee \overline{a}\overline{d} \vee e(f \vee b) \\ \textit{else } z &= e(f \vee b); u = (cd \vee \overline{a}\overline{d})e(f \vee b) \end{aligned}$$

a straighforward multilevel implementation is

$$\begin{aligned} w &= ab \vee \overline{a}\overline{b} \\ z &= w(cd \vee \overline{a}\overline{d}) \vee \overline{w}e(f \vee b) \\ u &= w(cd \vee \overline{a}\overline{d} \vee e(f \vee b)) \vee \overline{w}(cd \vee \overline{a}\overline{d})e(f \vee b) \end{aligned}$$

A more succinct multilevel implementation is

$$\begin{aligned} w &= ab \vee \overline{a}\overline{b} \\ t &= cd \vee \overline{a}\overline{d} \\ s &= e(f \vee b) \\ z &= wt \vee \overline{w}s \\ u &= w(t \vee s) \vee \overline{w}ts \end{aligned}$$

Interesting work on the efficient decomposition of Boolean functions into networks of digital gates had been carried on since the 1960s [19, 20, 23, 30, 45]. But it was only in the 1980s that a combination of more powerful theory and computers gave wings to the field, culminating in the development of modern logic synthesis packages, often originated in academia and then engineered into industrial tools by a number of electronic design automation (EDA) companies. The first industrially employed logic synthesis tool was LSS [16–18], developed at IBM and based on local transformations. A pioneering package from academia was MIS [4, 5], developed at Berkeley in the mid-1980s. Two students who played a pivotal role in the theory and implementation of MIS were R. Rudell [6, 39] and A. Wang [6, 22, 44], who then joined a successful start-up in EDA. In parallel, a research group at the University of Colorado, Boulder, led by G. Hachtel, developed the package BOLD [1]. A few years later a Berkeley team developed SIS [40, 41], a more advanced and complete package (including also sequential synthesis), which became the de-facto standard for comparing algorithms in multilevel logic synthesis. The current champion is the package ABC [42], developed at UCB by a team whose main architect is A. Mishchenko; ABC achieves high scalability by using FRAIGs, that is, a representation based on functionally reduced two-input AND gates and inverters [36].

In this chapter we provide a rigorous survey of the basics of modern multilevel logic synthesis, developed since the beginning of the 1980s under the influence of seminal papers on factoring and division by R. Brayton and C. McMullen, at the IBM Yorktown Research Labs [3, 7, 8]. We first describe the setting of the problem based on the abstraction of a Boolean network and on factored forms of Boolean functions. Then we describe Boolean division, for completely specified and incompletely specified Boolean functions, and heuristic algorithms to perform it. Next we describe algebraic division, introducing weak algebraic division, and we discuss how to find common algebraic divisors by restricting the search to kernels and other subsets of divisors. Finally we show how division can be the engine of the main operations to restructure a Boolean network, namely factoring, decomposition, substitution, and extraction.

Textbook expositions of multilevel logic synthesis can be found in [21, 24, 25, 34].

16.1.2 Network Representation

Let us consider the network representation first. A model used in multilevel logic synthesis is a network of nodes that are single-output functions, where each node is abstracted as a sum-of-products or a factored form, see Figure 16.1.

Definition 16.1. *A* **Boolean network** *is a three-tuple $\mathcal{N} = (V, E, \mathbf{f})$, consisting of a directed acyclic graph (DAG) $G = (V, E)$ and a collection of Boolean (or logic) functions* **f**.

The set of nodes is $V = V^I \cup V^{int} \cup V^O$, where V^I is the set of source nodes, V^{int} is the set of internal nodes, V^O is the set of sink nodes. There may be an arc from any node in V^I to any node in V^{int}, and from any node in V^{int} to any node in V^{int}, as long as the graph is acyclic. For every node in V^O there is a unique node in V^{int} from which there is an arc to it. The network is characterized as follows:

(i) *A primary input $x_i, i = 1, \ldots, m$, is associated with each node $i \in V^I, i = 1, \ldots, m$.*
(ii) *An internal variable $y_i, i = 1, \ldots, n$, and a representation of a completely specified Boolean function $f_i, i = 1, \ldots, n$, are associated with each internal node $i \in V^{int}, i = 1, \ldots, n$. It holds that $y_i = f_i(y_{i_1}, \ldots, y_{i_n}, x_{i_1}, \ldots, x_{i_m})$, $i = 1, \ldots, n$, where $y_{i_1}, \ldots, y_{i_n}$ are the internal variables associated with internal nodes for which there is an arc to node i, and where $x_{i_1}, \ldots, x_{i_m}$ are the primary inputs associated with source nodes from which there is an arc to node i.*
(iii) *A primary output $z_i, i = 1, \ldots, p$, and a completely specified output function $f_i, i = n+1, \ldots, n+p$, are associated with each node $i \in V^O, i = 1, \ldots, p$. It holds that $f_i = f_k$, where f_k is the function associated with the unique internal node k from which there is an arc to node i.*

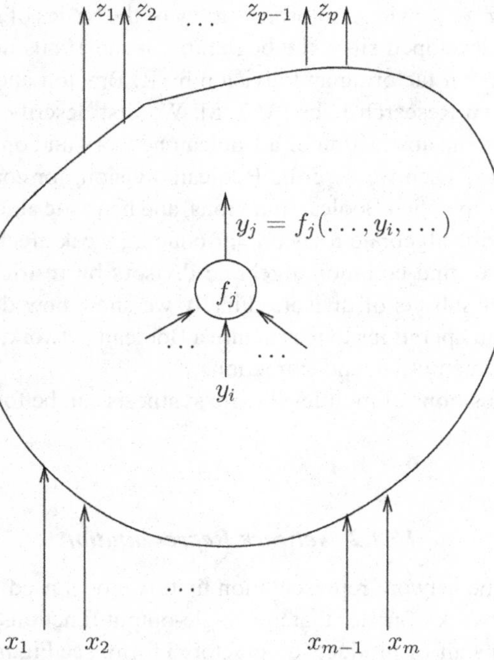

Figure 16.1. Structure of a Boolean network. External inputs are $x_1, \ldots, x_m$; external outputs are $y_1, \ldots, y_p$. A node f_j with output signal y_j computes the function $y_j = f_j(y_{j_1}, \ldots, y_{j_n}, x_{j_1}, \ldots, x_{j_m})$, where some local inputs may be external inputs.

The set of primary inputs is represented by the vector of variables $\mathbf{x} = (x_1, \ldots, x_m)$. *The set of internal variables is represented by the vector of variables* $\mathbf{y} = (y_1, \ldots, y_n)$. *The set of primary outputs is represented by the vector of variables* $\mathbf{z} = (z_1, \ldots, z_p)$. *The set of functions is represented by the vector of functions* $\mathbf{f} = (f_1, \ldots, f_n, f_{n+1}, \ldots, f_{n+p})$

There is an external don't-care set *associated with the primary outputs and defined by the vector of completely specified functions* $\mathbf{d} = (d_1(\mathbf{x}), \ldots, d_p(\mathbf{x}))$, *where* $d_i(\mathbf{x}) = 1$ *if and only the primary output* z_i *under input* $\mathbf{x}$ *is unspecified, that is, it may be either 0 or 1.*

Notice that if $\mathbf{x}$ is a don't-care input for the primary output z_i, it means that we are free to synthesize a network that computes any value for z_i under the input $\mathbf{x}$, and we can choose the value that makes the implementation more cost-effective.

Definition 16.2. *A* **literal** *is a variable or its complement, such as,* x *or* $\overline{x}$. *A* **cube** *is a set* C *of literals, such as, in set notation* $\{x, y, \overline{z}\}$; *a cube can be viewed as representing the conjunction of its literals, such as,* $xy\overline{z}$. *An* **expression** *or* **cover** F *is a set of cubes, such as, in set notation* $\{\{\overline{x}\}, \{x, y, \overline{z}\}\}$; *an expression represents the disjunction of its cubes, also called sum-of-product (SOP) or disjunctive normal form (DNF), for example,* $\overline{x} \vee xy\overline{z}$.

Definition 16.3. *The* **support** *of an expression F, $sup(F)$, is the set of variables that appear in F, that is,*

$$sup(F) = \{x \mid \exists \text{ cube } C \text{ in } F \text{ such that } x \in C \text{ or } \overline{x} \in C\}.$$

For example, $sup(xy \vee \overline{xz}) = \{x, y, z\}$.

Definition 16.4. *A node j is a* **fan-in** *node of a node i if function f_i depends on variable y_j explicitly, that is, there is an arc from j to i. The set of all fan-ins of a node i is denoted by*

$$FI(i) = \begin{cases} \emptyset & i \text{ is a primary input} \\ \{j \mid j \in sup(i)\} & \text{otherwise.} \end{cases}$$

Definition 16.5. *The* **transitive fan-in** *of a node i, denoted by $TFI(i)$, is defined recursively as*

$$TFI(i) = \begin{cases} \emptyset & i \text{ is a primary input} \\ FI(i) \cup \bigcup_{j \in FI(i)} TFI(j) & \text{otherwise.} \end{cases}$$

Definition 16.6. *A node j is a* **fan-out** *node of a node i if function f_j depends on variable y_i explicitly, that is, there is an arc from i to j. The set of all fan-outs of a node i is denoted by*

$$FO(i) = \begin{cases} \emptyset & i \text{ is a primary output} \\ \{j \mid i \in sup(j)\} & \text{otherwise} \end{cases}$$

Definition 16.7. *The* **transitive fan-out** *of a node i, denoted by $TFO(i)$, is defined recursively as*

$$TFO(i) = \begin{cases} \emptyset & i \text{ is a primary output} \\ FO(i) \cup \bigcup_{j \in FO(i)} TFO(j) & \text{otherwise} \end{cases}$$

An example of a Boolean network is given in Figure 16.2.

The given notion of Boolean network is abstract enough that it can be used for both technology-independent and technology-dependent representations. What makes the difference is the type of node representation. Nodes may be abstract functions of many sorts in a technology-independent representation, whereas they are valid gates from a library in a technology-dependent representation. In the former case nodes may be classified as follows:

- General node: each node is the representation of an **arbitrary** Boolean function; then a theory is easier to develop because there are no arbitrary restrictions dependent on the technology. This is the choice of SIS, with the restriction that, in SIS, nodes must be single-output with the following possible choices of function representation:
  - Sum-of-products form
  - Factored form
  - Binary decision diagram (BDD)

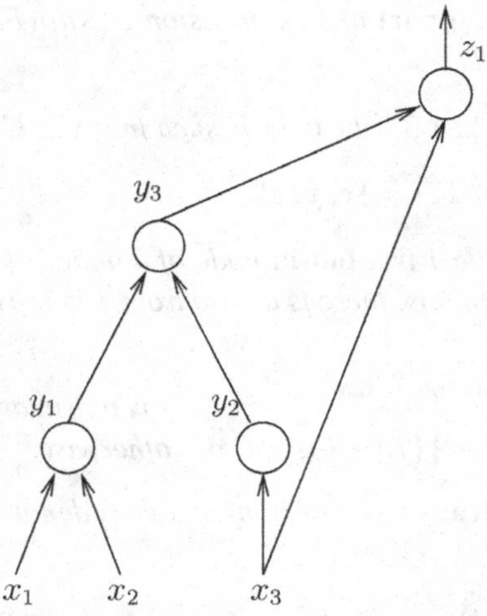

Figure 16.2. An example of Boolean network. The nodes compute the following functions: $y_1 = x_1x_2$, $y_2 = \overline{x}_3$, $y_3 = y_1y_2$, $z_1 = y_3 \vee x_3$.

- Generic node: every node in the network is a simple generic node, like a 2-input NAND gate; some manipulations are much faster using this structure, but the network is finely decomposed in a particular way, and some natural structures may be lost. Recently, And-Inverter graphs (AIGs) have seen increasing use [29, 35].
- Discrete node. A node can be one of a small set of logic functions, such as AND, OR, NOT, DECODE, ADD or other complex blocks of logic manipulated as a single unit, also allowing multiple output nodes; this is used mostly in rule-based systems and a theory for such networks seems more difficult.

In the sequel we develop technology-independent multilevel optimization based on general nodes represented as in SIS by sum-of-products (SOPs) and factored forms. SOPs are easy to manipulate, and there are many algorithms for their minimization, spawned by a rigorous theory of optimum two-level forms. However, their main disadvantage is that they do not reliably represent the logic complexity of a function. Consider

$$f = ad \vee ae \vee bd \vee be \vee cd \vee ce$$
$$\overline{f} = \overline{a}\overline{b}\overline{c} \vee \overline{d}\overline{e}$$

These differ in their implementation by only one inverter, but their SOPs differ by many products and literals. So SOPs are not a good estimator of progress during logic minimization.

For ROBDDs, we refer to Chapter 10 in this collection. Here it suffices to say that they represent a function and its complement with the same complexity, such as factored forms; however, they are not a good estimator for the complexity of implementation, because they are effectively networks of muxes restricted to be controlled by primary input variables. BDDs are a good replacement for truth tables, because they are canonical for a given ordering, and operations on them are well defined and efficient, with true support (dependency) exposed explicitly. Finding the best ordering is a difficult problem; moreover, there are functions such as multipliers for which there is no ordering that yields a small BDD.

As will be seen in the formal definition, factored forms are recursively defined as sums or products of factored forms down to the terminal cases of single literals. An example of a factored form is $(ab \vee \overline{b}c)(c \vee \overline{d}(e \vee a\overline{c})) \vee (d \vee e)(fg)$. Factored forms represent a function and its complement with the same complexity:

$$f = ad \vee ae \vee bd \vee be \vee cd \vee ce$$
$$\overline{f} = \overline{a}\overline{b}\overline{c} \vee \overline{d}\overline{e}$$
$$f = (a \vee b \vee c)(d \vee e).$$

They are good estimators of complexity of logic implementation and do not blow up easily; in many design styles the implementation of a function corresponds directly to its factored form (e.g., complex-gate CMOS, where factored form literal count correlates to transistor count that correlates to area, but area depends on wiring, too). A disadvantage is that there are not as many algorithms available for manipulation; hence, usually they must be converted into SOPs before operating on them.

16.1.3 Factored Forms

Definition 16.8. *An* **algebraic expression** *F is a sum-of-products representation of a Boolean function that is minimal with respect to single cube containment, that is, such that every proper subset of F represents another function than F.*

Example 16.2. *$ab \vee cd$ is an algebraic expression, but $a \vee ab$ and $ab \vee abc \vee cd$ are not algebraic expressions (e.g., $a \vee ab$ is ruled out because factoring would yield $a(1 \vee b)$, where the 1 indicates single-cube containment).*

Definition 16.9. *The* **product** *of two expressions F and G is a set defined by*

$$FG = \{cd \mid c \in F \text{ and } d \in G \text{ and } cd \neq \emptyset\}.$$

Example 16.3. $(a \vee b)(ce \vee d \vee \overline{a}) = ace \vee ad \vee bce \vee bd \vee \overline{a}b.$
The result is not necessarily an algebraic expression, for example, $(a \vee b)(a \vee c) = aa \vee ac \vee ab \vee bc$.

Definition 16.10. *FG is an* **algebraic product** *if F and G are algebraic expressions and have disjoint support (that is, they have no input variables in common); otherwise, it is a* **Boolean product**.

Example 16.4. *$(a \vee b)(c \vee d) = ac \vee ad \vee bc \vee bd$ is an algebraic product, but $(a \vee b)(a \vee c) = aa \vee ac \vee ab \vee bc$ is a Boolean product.*

Lemma 16.1. *The algebraic product of two expressions F and G is an algebraic expression.*

Proof. By contradiction. Suppose there exists $c_i d_j \subseteq c_k d_l$ for some i, j, k, l, with either $i \neq k$ or $j \neq l$. Because $sup(F) \cap sup(G) = \emptyset$, then $c_i \subseteq c_k$ and $d_j \subseteq d_l$, against the hypothesis that F and G are algebraic expressions. ■

A factored form is a parenthesized expression.

Definition 16.11. *A* **factored form** *can be defined recursively by the following rules:*

A factored form is either a product or a sum where:

- *A* product *is either a single literal or a product of factored forms.*
- *A* sum *is either a single literal or a sum of factored forms.*

In effect, a factored form is a product of sums of products of ... or a sum of products of sums of ... factored forms.

Any Boolean function can be represented by a factored form, and any factored form is a representation of some Boolean function.

Example 16.5.

- *Examples of factored forms are*

$$\begin{array}{l} x \\ \overline{y} \\ ab\overline{c} \\ a \vee \overline{b}c \\ ((\overline{a} \vee b)cd \vee e)(a \vee \overline{b}) \vee \overline{e}. \end{array}$$

- *The following is not a factored form, because complementation is allowed only on literals in factored forms:*

$$\overline{(a \vee b)c}.$$

- *The following are three equivalent factored forms (factored forms are not unique):*

$$\begin{array}{l} ab \vee c(a \vee b) \\ bc \vee a(b \vee c) \\ ac \vee b(a \vee c). \end{array}$$

Definition 16.12. *The* **factorization value** *of an algebraic factorization $F = G_1G_2 \vee R$, where G_1, G_2, and R are algebraic expressions, is defined to be*

$$\begin{aligned} fact\_val(F, G_2) &= lits(F) - (lits(G_1) + lits(G_2) + lits(R)) \\ &= (|G_1| - 1)lits(G_2) + (|G_2| - 1)lits(G_1). \end{aligned}$$

Here, for any algebraic expression P, $lits(P)$ = number of literals in P, and $|P|$ is the number of cubes in P.

The factorization value is the number of literals saved by doing one level of factoring. It is important that the expressions are algebraic.

Example 16.6. *The algebraic expression*

$$F = ae \vee af \vee ag \vee bce \vee bcf \vee bcg \vee bde \vee bdf \vee bdg$$

can be expressed in the factored form

$$F = (a \vee b(c \vee d))(e \vee f \vee g).$$

Note that only seven, rather than 24, literals are required.

If $G_1 = (a \vee bc \vee bd)$, and $G_2 = (e \vee f \vee g)$, then $R = \emptyset$, and $fact\_val(F, G_2) = (2)(3) + (2)(5) = 16$. Note that the given factored form saved 17 rather than 16 literals. The extra literal that was saved comes from recursively applying the formula to the factored form of G_1.

The following facts are true:

- Factored forms are more compact representations of Boolean functions than the traditional sum-of-products forms. For example, the following factored form has 10 literals:

  $$(a \vee b)(c \vee d(e \vee f(j \vee i \vee h \vee g))),$$

  but when represented as a sum-of-products form, it requires 42 literals:

  $$ac \vee ade \vee adfg \vee adfh \vee adfi \vee adfj \vee bc \vee bde \vee bdfg \vee$$
  $$bdfh \vee bdfi \vee bdfj.$$

- Every sum-of-products form can be viewed as a factored form.

When measured in terms of number of inputs, there are functions whose size is exponential in sum-of-products forms, but polynomial in factored forms.

Example 16.7. *Consider the Achilles' heel function represented as product-of-sums:*

$$\prod_{i=1}^{i=n/2} (x_{2i-1} \vee x_{2i}).$$

There are n literals in this factored form and $(n/2) \times 2^{n/2}$ literals in the smallest sum-of-products form. (See, e.g., [15] for additional examples of this nature.)

Factored forms are useful in estimating area and delay in a multilevel logic synthesis and optimization system. In most design styles (for example, complex-gate CMOS design) the implementation of a function corresponds directly to its factored form.

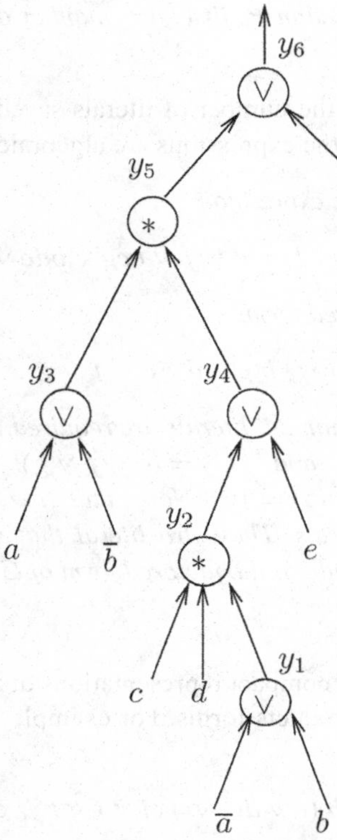

Figure 16.3. Factoring tree for Example 16.8. The tree represents the function $y_6 = ((\overline{a} \vee b)cd \vee e)(a \vee \overline{b}) \vee \overline{e}$.

Factored forms can also be graphically represented as labeled trees, called factoring trees, in which each internal node including the root has a label either "+" (or "∨") for disjunction, or "∗" for conjunction, and each leaf has a label that is either a variable or its complement.

Example 16.8. *Figure 16.3 shows the computation tree of the factored form* $((\overline{a} \vee b)cd \vee e)(a \vee \overline{b}) \vee \overline{e}$.

Definition 16.13. *The* **size** *of a factored form F (denoted $\rho(F)$) is the number of literals in the factored form. A factored form is* **optimum** *if no other equivalent factored form has fewer literals.*

Example 16.9. $\rho((a \vee b)c\overline{a}) = 4$ *and* $\rho((a \vee b \vee cd)(\overline{a} \vee \overline{b})) = 6$.

Definition 16.14. *A factored form F is* **positive unate** *in x if x appears in F but $\overline{x}$ does not.*
A factored form F is **negative unate** *in x if $\overline{x}$ appears in F and x does not.*
A factored form F is **unate** *in x if it is either positive unate or negative unate in x.*

Example 16.10. *$(a \vee \overline{b})c \vee \overline{a}$ is positive unate in c, negative unate in b, and not unate in a.*

Definition 16.15. *The* **cofactor** *of a factored form F with respect to a literal x_1 (or $\overline{x}_1$) is the factored form $F_{x_1} = F|_{x_1=1}(x)$ (or $F_{\overline{x}_1} = F|_{\overline{x}_1=1}(x)$) obtained by*

(i) *Replacing all occurrences of x_1 with 1 and all occurrences of $\overline{x}_1$ with 0, and*
(ii) *Simplifying the factored form using the following identities of Boolean algebra:*

$$\begin{aligned} 1x &= x \\ 1 \vee x &= 1 \\ 0x &= 0 \\ 0 \vee x &= x. \end{aligned}$$

The cofactor of a factored form F with respect to a cube c is a factored form, denoted by F_c, obtained by successively cofactoring F with respect to each literal in c.

After constant propagation (all constants are removed), part of the factored form may appear as $G \vee G$. In general, G is another factored form. In fact, the two Gs may have different factored forms. Identifying these equivalent factored forms to apply the simplification $G \vee G = G$ is a nontrivial task.

Example 16.11. *Let $F = (x \vee \overline{y} \vee z)(\overline{x}u \vee \overline{z}\,\overline{y}(v \vee \overline{u}))$ and $c = v\overline{z}$. Then*

$$\begin{aligned} F_{\overline{z}} &= (x \vee \overline{y})(\overline{x}u \vee \overline{y}(v \vee \overline{u})) \\ F_c &= (x \vee \overline{y})(\overline{x}u \vee \overline{y}) \end{aligned}$$

Note that cofactoring does not preserve algebraic expressions: $F = abc \vee bcd$, $F_a = bc \vee bcd$.

Sum-of-products forms are used as the internal representation of Boolean functions in most multilevel logic optimization systems. The advantage is that good algorithms for manipulating SOPs are available. Disadvantages are twofold:

(i) The quality of solutions of SOP algorithms is unpredictable: they may accidentally generate a function whose sum-of-products form is too large.
(ii) Factoring algorithms have to be used constantly to provide an estimate for the size of the Boolean network, so that time spent in factoring may become significant; therefore there is a need for quick but still good factoring methods.

A possible solution to overcome these disadvantages is to avoid sum-of-products forms by using factored forms as internal representation. However, this is not practical, unless we know how to perform logic operations on the factored forms directly without converting them to sum-of-products forms. Extensions to factored forms of the most common logic operations have been partially provided, but more research is needed.

16.1.4 Incompletely Specified Boolean Functions

Definition 16.16. *An incompletely specified function* $\mathcal{F} = (f, d, r) : B^n \to \{0, 1, \star\}$ *(*$\star$ *stands for* don't-care *value) is a triple of completely specified Boolean functions* f, d, *and* r, *respectively the onset, don't care, and offset functions; that is,* $f(x) = 1$ *if and only if* $\mathcal{F}(x) = 1$, $d(x) = 1$ *if and only if* $\mathcal{F}(x) = *$, $r(x) = 1$ *if and only if* $\mathcal{F}(x) = 0$, *where* f, d, r *are a partition of* B^n, *that is, they are disjoint and together they cover all of* B^n. *When the don't-care function* d *is empty, we have a completely specified Boolean function.*

An equivalent view is to say that an incompletely specified function $\mathcal{F} = (f, d, r)$ is a collection of completely specified functions g such that $f \subseteq g \subseteq f \vee d$. A cover for such a g (i.e., a SOP expression representing g) is said to be a **cover** of $\mathcal{F} = (f, d, r)$, and sometimes – with slight abuse of notation – g itself is said to be a cover of $\mathcal{F} = (f, d, r)$.

For minimization of incompletely specified Boolean functions there is a huge literature [38] and there are practical software packages [2]. Often in the text we refer to applying minimization to functions associated with nodes of a Boolean network, such as the operation *minimize* (a.k.a. *simplification* or *simplify*), or *minimize with don't cares*, the latter when we underline that there is a nonempty don't-care function d. Because each node has – among others – a representation of the associated function as a SOP, by minimization we mean applying any technique that will reduce the number of product terms and literals of the expression representing the function. According to the context and the practical requirements, we may be interested only in "light" minimization, for instance at least ensuring that the final cover has no single-cube containment, or in producing an irredundant cover (in which no cube can be removed without uncovering at least a point of the onset), or in obtaining an absolute minimum. In practice we are often interested in an irredundant cover (minimal cover, which is a local optimum), as it can be obtained by the program ESPRESSO when run in the heuristic mode. In pseudocode descriptions we may denote a call to ESPRESSO with a pseudoinstruction such as ESPRESSO (*F,D,R*), where F, D, and R are respectively covers of the onset, don't-care set, and offset of the function to be minimized.

16.1.5 Manipulation of Boolean Networks

The basic techniques to restructure Boolean networks can be summarized as:

- Structural operations (they change the topological structure of the network), divided as:
  - Algebraic;
  - Boolean.
- Node simplification (they change the functions of nodes):
  - Computation of don't cares due (1) to the dependencies that the network topology imposes on the fan-in signals (satisfiability or controllability don't

cares), and (2) to the limited sensitivity of the outputs to vectors of inputs and internal nodes (observability don't cares);
  - Node minimization by exploiting the computed don't cares.
- Phase assignment of the internal and output signals.

Given an initial network, the restructuring problem is to find the **best** equivalent network. In this chapter, for lack of space, we almost exclusively discuss structural operations.

Example 16.12. *Given the functions f_1 and f_2, we apply to them the operations* **minimize** *(node minimization performed by an algorithm like in* ESPRESSO *[2]),* **factor**, *and* **decompose**, *whose precise meaning we discuss later.*

$$
\begin{aligned}
f_1 &= abcd \vee abce \vee a\overline{b}c\overline{d} \vee a\overline{b}\overline{c}\overline{d} \vee a\overline{b}\overline{c}\overline{d} \vee \overline{a}c \vee cdf \\
&\quad \vee ab\overline{c}\overline{d}\overline{e} \vee a\overline{b}\overline{c}d\overline{f} \\
f_2 &= bdg \vee \overline{b}dfg \vee \overline{b}\,\overline{d}g \vee b\overline{d}eg
\end{aligned}
$$

minimize:

$$
\begin{aligned}
f_1 &= bcd \vee bce \vee \overline{b}\,\overline{d} \vee \overline{b}f \vee \overline{a}c \vee ab\overline{c}\overline{d}\overline{e} \vee a\overline{b}\overline{c}d\overline{f} \\
f_2 &= bdg \vee dfg \vee \overline{b}\,\overline{d}g \vee \overline{d}eg
\end{aligned}
$$

factor:

$$
\begin{aligned}
f_1 &= c(b(d \vee e) \vee \overline{b}(\overline{d} \vee f) \vee \overline{a}) \vee a\overline{c}(b\overline{d}\overline{e} \vee \overline{b}d\overline{f}) \\
f_2 &= g(d(b \vee f) \vee \overline{d}(\overline{b} \vee e))
\end{aligned}
$$

decompose:

$$
\begin{aligned}
f_1 &= c(\overline{a} \vee x) \vee a\overline{c}\,\overline{x} \\
f_2 &= gx \\
x &= d(b \vee f) \vee \overline{d}(\overline{b} \vee e)
\end{aligned}
$$

The basic structural operations that can be used to change the internal topology of a Boolean network are now briefly introduced on the basis of examples only. Operative definitions can be found in the manual of the multilevel synthesis package SIS [40]. Detailed definitions and descriptions are given in the follow-up of the chapter, especially in Section 16.5.

Example 16.13. *The following steps show the application of basic operations to restructure Boolean networks:*

(i) *Decomposition of a single function:*

$$
\begin{aligned}
f &= abc \vee abd \vee \overline{a}\,\overline{c}\overline{d} \vee \overline{b}\overline{c}\overline{d} \\
&\Downarrow \\
f &= xy \vee \overline{x}\,\overline{y} \\
x &= ab \\
y &= c \vee d.
\end{aligned}
$$

(ii) *Extraction, that is, decomposition of multiple functions:*

$$
\begin{aligned}
f &= (az \vee b\overline{z})cd \vee e\\
g &= (az \vee b\overline{z})\overline{e}\\
h &= cde\\
&\Downarrow\\
f &= xy \vee e\\
g &= x\overline{e}\\
h &= ye\\
x &= az \vee b\overline{z}\\
y &= cd.
\end{aligned}
$$

(iii) *Factoring, that is, series-parallel decomposition:*

$$
\begin{aligned}
f &= ac \vee ad \vee bc \vee bd \vee e\\
&\Downarrow\\
f &= (a \vee b)(c \vee d) \vee e
\end{aligned}
$$

(iv) *Substitution, that is, making the value of a function available into another function:*

$$
\begin{aligned}
g &= a \vee b\\
f &= a \vee bc\\
&\Downarrow\\
f &= g(a \vee c).
\end{aligned}
$$

(v) *Collapsing (also called elimination), that is, replacing a function by its expression:*

$$
\begin{aligned}
f &= ga \vee \overline{g}b\\
g &= c \vee d\\
&\Downarrow\\
f &= ac \vee ad \vee b\overline{c}\overline{d}\\
g &= c \vee d.
\end{aligned}
$$

In general, "elimination" is a term that we use for partial collapsing.

In comparing factoring versus decomposition, we remark that the former is restricted to a series-parallel graph restructuring of the given function, whereas the latter may produce a nontree structure, so it is similar to BDD collapsing of common nodes and using negative pointers. But decomposition is not canonical, so there is not a perfect identification of common nodes.

We next introduce the value of a node to define a cost function that measures the effect of an operation to restructure a Boolean network. The value of a node is the difference in literal cost of the network without the node (node eliminated or no factoring) and with the node (node factored out). For a node j, let n_j be the number of times that literal y_j and/or $\overline{y}_j$ appears in the factored network, and let l_j be the number of literals in the factored form of node j (we can treat y_j and $\overline{y}_j$ as the

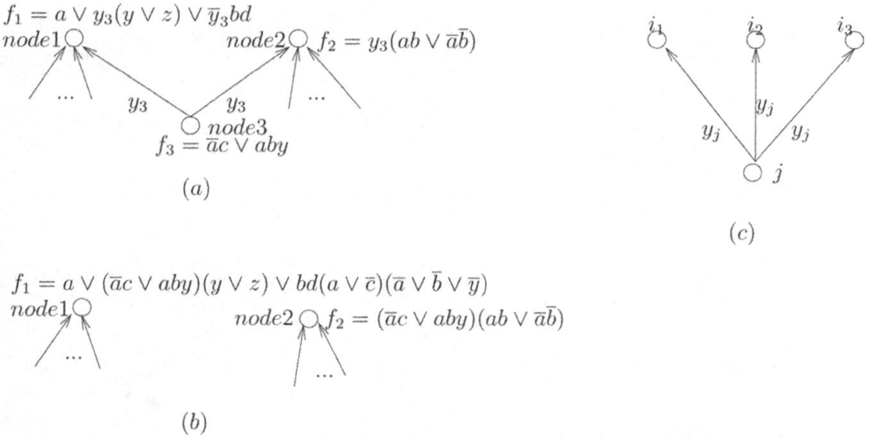

Figure 16.4. Illustration of Example 16.14. (a) Fragment of a Boolean network before elimination. (b) The same fragment after elimination. (c) A node j with l_j literals whose output fans out as wire y_j to nodes i_1, i_2, i_3.

same because $\rho(f_j) = \rho(\overline{f_j})$); moreover, let c be the cost of the rest of the nework. The literal cost of the network without node j (elimination, no factoring) is

$$l_j \sum_{i \in FANOUT(j)} n_i + c,$$

and the literal cost of the network with node j (no elimination, factoring) is

$$l_j + \sum_{i \in FANOUT(j)} n_i + c.$$

Their difference (cost change due to elimination) is

$$\begin{aligned} value(j) &= (l_j \sum_{i \in FANOUT(j)} n_i + c) - (l_j + \sum_{i \in FANOUT(j)} n_i + c) \\ &= ((\sum_{i \in FANOUT(j)} n_i) - 1)(l_j - 1) - 1. \end{aligned}$$

Example 16.14. *Figure 16.4 (a) shows a network fragment before elimination; its literal cost is* $5 + 7 + 5 + c = 17 + c$. *Figure 16.4 (b) shows the same network fragment after elimination; its literal cost is* $9 + 15 + c = 24 + c$, *for a difference of* $24 + c - (17 + c) = 7$. *This is exactly the value computed by the formula* $value(node3) = (1 + 2 - 1)(5 - 1) - 1$, *given* $n_1 = 1, n_2 = 2, l_3 = 5$.

The value might be different if we were to eliminate, simplify, and then refactor. Figure 16.5 (a) shows a network fragment with values annotated for nodes l, m, n; *Figure 16.5 (b) shows the value of node* n *after* eliminating -1 *of nodes* l *and* m. *For instance, in (a)* $value(l) = -1$, *because the network without node* l *has value less by 1 than the network with node* l; *in (b) node* n *has value* 3 *because the value of the network with node* n *is* $18 + c$ *and with node* n *it is* $15 + c$, *a difference*

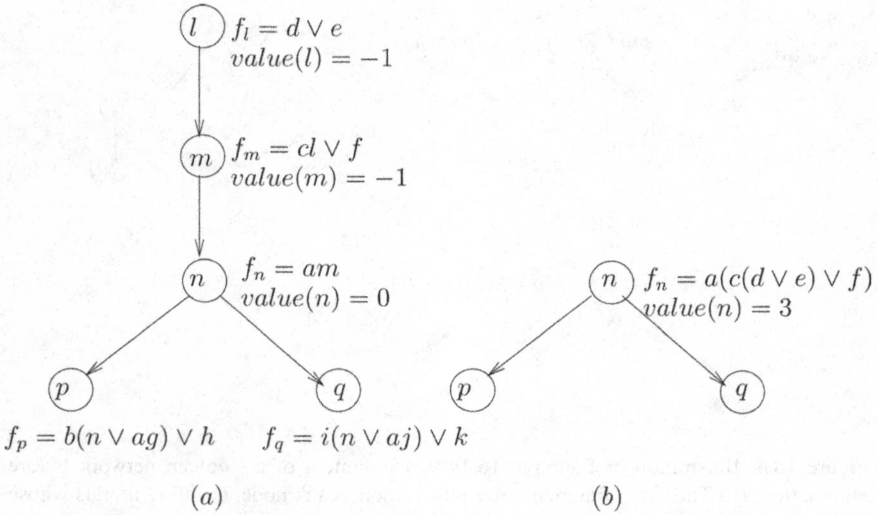

Figure 16.5. Illustration of Example 16.14. (a) Fragment of a Boolean network with some node values. (b) The same fragment after elimination of nodes with negative value.

of $18 + c - (15 + c) = 3$. *In general the operation* eliminate x *eliminates all the nodes whose removal does not increment the value of the network by more than* x. *The special case* eliminate -1 *eliminates the nodes whose removal decreases the value of the network by at least 1, and in particular it removes nodes that are used only once (as it can be verified by the formula accounting for the value). The order of elimination may be important.*

Note that "**Division**" plays a key role in all of these operations. Their essence can be abstracted as addressing the following two problems:

(i) Find good common subfunctions.
(ii) Carry out the division.

In the next section we survey techniques to handle these problems.

16.2 Boolean Division

Division is central in the operations to restructure a Boolean network. To be able to use it, we must answer the following questions:

(i) What exactly does "division" mean in the context of Boolean functions?
(ii) What is a "divisor," that is, what do we divide by?
(iii) How do we perform division?
(iv) How do we apply division to restructure a network, such as in factoring, resubstitution, extraction?

Definition 16.17. *A Boolean function g is a* **divisor** *of a Boolean function f if there exist Boolean functions h and r such that $f = gh \vee r$, with $gh \neq \emptyset$. In this case, h is called a* **quotient** *and r is called a* **remainder** *of the* **division** *of f by g. Note that h and r may not be unique.*

The function g is said to be a **factor** *of f if in addition $r = \emptyset$, that is, if $f = gh$.*

We provide algorithms for division when the functions are represented by covers.

Definition 16.18. *Let F, G, H, R be covers respectively of the functions f, g, h, r in Definition 16.17.*

If GH is restricted to be an algebraic product, G is an **algebraic divisor** *of F. Otherwise, G is a* **Boolean divisor** *of G.*

If GH is restricted to be an algebraic product and $R = \emptyset$, G is an **algebraic factor** *of F. Otherwise, G is a* **Boolean factor** *of F.*

If GH is restricted to be an algebraic product, H is an **algebraic quotient**, *denoted $F//G$. Otherwise, H is a (nonunique)* **Boolean quotient** *denoted $F \div G$.*

If H is an algebraic quotient, the operation $(F, G) \rightarrow (H, R)$ is called **algebraic division**, *denoted by $//$; otherwise it is called* **Boolean division**, *denoted by $\div$.*

We reserve the notation F/G for the more useful "weak division" that is a type of algebraic division defined later.

Example 16.15. *Consider:*

$$F = ad \vee ae \vee bcd \vee j$$
$$G_1 = a \vee bc$$
$$G_2 = a \vee b.$$

Examples of algebraic division are:

(i) $F//(bc) = d$;
(ii) $F//a = d \vee e$; *also* $F//a = d$ *or* $F//a = e$, *that is, algebraic division is not unique;*
(iii) $H_1 \equiv F//G_1 = d,\ R_1 = ae \vee j$.

Examples of Boolean division are:

(i) $H_2 \equiv F \div G_2 = (a \vee c)d,\ R_2 = ae \vee j$, *that is,* $F = (a \vee b)(a \vee c)d \vee ae \vee j$;
(ii) $G_1 \div G_2 = a \vee c$, *i.e.,* $G_1 = (a \vee b)(a \vee c)$.

16.2.1 Boolean Division for Completely Specified Functions

Lemma 16.2. *A Boolean function g is a Boolean factor of a Boolean function f if and only if $f \subseteq g$ (i.e., $f\overline{g} = \emptyset$, i.e. $\overline{g} \subseteq \overline{f}$).*

Proof. $\Rightarrow$: If g is a Boolean factor of f, then there exists h such that $f = gh$. Hence, $f \subseteq g$ (as well as $f \subseteq h$).

$\Leftarrow$: Assume $f \subseteq g$; since $f = fg \vee f\overline{g}$, then $f = fg = g(f \vee r)$ where r is any function $r \subseteq \overline{g}$. Thus $f = gh$, where $h = f \vee r$. ■

Note that:

(i) $h = f$ works fine for the proof.
(ii) Given f and g, h is not unique. Getting a small quotient h is the same as getting a small $f \vee r$. Because $rg = \emptyset$, this is the same as minimizing (simplifying) f with don't care (DC) = $\overline{g}$.

Example 16.16. *Let $f = (a \vee b)(a \vee c) \vee \overline{d}b$. Because $a \vee b \subseteq f$, then by Lemma 16.2 $g = a \vee b$ is a factor of f, that is, $f = (a \vee b)h$. We have $f = fg \vee f\overline{g} = fg = ((a \vee b)(a \vee c) \vee \overline{d}b)(a \vee b)$. To get a small h we can minimize f and obtain $f = (a \vee bc \vee \overline{d}b)(a \vee b)$. The minimization of f is performed by representing $f = (a \vee b)(a \vee c) \vee \overline{d}b = a \vee ab \vee ac \vee bc \vee \overline{d}b$ by the minimal SOP representation $f = a \vee bc \vee \overline{d}b$.*

Actually, according to the proof, h can be any function of the form $f \vee r$, where r is orthogonal to g. Hence we can get a smaller h by minimizing f with $DC = \overline{g} = \overline{a}\overline{b}$, which gives $h = a \vee c \vee \overline{d}$ and so $f = (a \vee c \vee \overline{d})(a \vee b)$, where $g = a \vee b$ is the divisor and $h = f \vee r = a \vee c \vee \overline{d}$ is the quotient. The minimization of f with $DC = \overline{g} = \overline{a}\overline{b}$ is performed by applying two-level logic minimization to the incompletely specified function $\mathcal{F}$ whose onset is $a \vee bc \vee \overline{d}b$ and whose don't-care set is $\overline{a}\overline{b}$, yielding the optimal SOP representation $h = a \vee c \vee \overline{d}$, where the minimization procedure used the don't-care set $\overline{a}\overline{b}$ by adding to the onset the terms in the don't-care set given by $r = \overline{a}\overline{b}\,\overline{d} \vee \overline{a}\overline{b}c \subseteq \overline{a}\overline{b}$.

Lemma 16.3. *g is a Boolean divisor of f if and only if $fg \neq \emptyset$.*

Proof. $\Rightarrow$: $f = gh \vee r$, $gh \neq \emptyset$ implies $fg = gh \vee gr$. Because $gh \neq \emptyset$, $fg \neq \emptyset$.

$\Leftarrow$: Letting $r = f\overline{g}$, we can write $f = fg \vee f\overline{g} = fg \vee r$. If $fg \neq \emptyset$, then this means that g is a Boolean divisor of f, in view of Definition 16.17. ■

Hence f has **many** Boolean divisors. We are looking for a g such that $f = gh \vee r$, where g, h, r are **simple** functions. From $f = fg \vee f\overline{g} = gh \vee r$, given g and h, we could minimize $f\overline{g}$ with $DC = (fg) = gh$ to get a small r, and instead given g and r we could minimize $hg = fg$ with $DC = (f\overline{g}) = r$ to get a small h (or also with $DC = r \vee \overline{g}$ since $(r \vee \overline{g})g = rg$). The problem is how to minimize simultaneously all the unknown terms to obtain a small f, that is, a representation of f with fewer literals.

16.2.2 Boolean Division for Incompletely Specified Functions

The following results extend Boolean division to incompletely specified functions.

Definition 16.19. *A completely specified Boolean function g is a* **Boolean divisor** *of $\mathcal{F} = (f, d, r)$ if there exist completely specified Boolean functions h, e such* *that*

$$f \subseteq gh \vee e \subseteq f \vee d, \text{ that is, } f = gh \vee e \bmod d,$$

and

$$gh \not\subseteq d, \text{ that is, } gh \neq 0 \bmod d.$$

g is a **Boolean factor** *of* $\mathcal{F} = (f, d, r)$ *if there exists h such that*

$$f \subseteq gh \subseteq f \vee d, \text{ that is, } f = gh \bmod d.$$

Lemma 16.4. $f \subseteq g$ *if and only if g is a Boolean factor of* $\mathcal{F} = (f, d, r)$.

Proof. $\Rightarrow$: Let k be any function such that $kg \subseteq d$ (e.g., $k = \emptyset$) and let $h = f \vee k$. Since $f \subseteq g$,

$$f = fg \subseteq gh = fg \vee kg \subseteq f \vee d,$$

and thus g is a Boolean factor of $\mathcal{F}$.

$\Leftarrow$: If g is a Boolean factor of $\mathcal{F}$, then $f \subseteq gh \subseteq g$. ■

Note that we would like to find the simplest h; because h has the form $f \vee k$, where $kg \subseteq d$ or $k \subseteq d \vee \overline{g}$ ($kg \subseteq d$, $kg \vee \overline{g} \subseteq d \vee \overline{g}$, $kg \vee k\overline{g} \subseteq d \vee \overline{g}$, $k \subseteq d \vee \overline{g}$), we can simplify $hf \vee k$ by minimizing f with $DC = d \vee \overline{g}$. So the steps would be:

(1) Choose $g \supseteq f$ and let G be a cover of g.
(2) Simplify $(f, d \vee \overline{g}, rg)$ to obtain H.
(3) GH is a cover of $\mathcal{F}$.

Lemma 16.5. $fg \neq \emptyset$ *if and only if g is a Boolean divisor of* $\mathcal{F} = (f, d, r)$.

Proof. $\Rightarrow$: Assume $fg \neq \emptyset$.

Define $\mathcal{F}_h = (fg, d \vee f\overline{g}, r)$. Now $g \supseteq fg$, so by Lemma 16.4 there exists h such that $fg \subseteq gh \subseteq fg \vee d \vee f\overline{g} = f \vee d$.

Let e be a cover of $\mathcal{F}_e = (f\overline{g}, d \vee fg, r)$, that is, $f\overline{g} \subseteq e \subseteq f\overline{g} \vee d \vee fg = f \vee d$.

Thus $f = fg \vee f\overline{g} \subseteq gh \vee e \subseteq f \vee d$.

From $fg \subseteq gh$ it follows that $fg \subseteq fgh$, and $\emptyset \neq fg \subseteq ghf$ implies $ghf \neq \emptyset$. Because $fd = \emptyset$ by definition and $ghf \neq \emptyset$, then $gh \not\subseteq d$, verifying the conditions for Boolean division.

$\Leftarrow$: Suppose there exists h such that $gh \not\subseteq d$ and $f \subseteq gh \vee e \subseteq f \vee d$.

From $gh \vee e \subseteq f \vee d$ it follows $gh \subseteq f \vee d$ and, because $gh \not\subseteq d$ implies $gh \neq \emptyset$, then $ghf \neq \emptyset$. Hence $fg \neq \emptyset$. ■

In summary, the steps to find H and E would be:

(1) Choose g such that $fg \neq \emptyset$, and let G be a cover of g.
(2) Simplify $(fg, d \vee f\overline{g}, r)$ to obtain H.
(3) Simplify $(f\overline{g}, d \vee fg, r)$ to obtain E.
(4) $GH \vee E$ is a cover of $\mathcal{F}$.

A correct variant of step 2 is:

(2) Simplify $(fg, d \vee \overline{g}, r)$ to obtain H, because $fg \subseteq h \subseteq fg \vee d \vee \overline{g}$ implies $fg \subseteq gh \subseteq fg \vee dg \subseteq f \vee d$.

A correct variant of step 3 is:

(3) Simplify $(f\overline{g} \vee f\overline{h}, d \vee fgh, r)$ to obtain H, because $f\overline{g} \vee f\overline{h} \subseteq e \subseteq f\overline{g} \vee f\overline{h} \vee d \vee fgh = f \vee d$.

Because there are many divisors g (only $fg \neq \emptyset$ is needed), the following questions are the **unsolved problems of common Boolean divisors**:

(1) Given (f, d, r), find a good g $(fg \neq \emptyset)$ such that both $(fg, d \vee f\overline{g}, r)$ and $(f\overline{g}, d \vee fg, r)$ simplify "nicely."
(2) Given two functions (or more) (f_1, d_1, r_1) and (f_2, d_2, r_2), find a common g "good and simple" such that step 1 applies for both.

Lemma 16.6. *Suppose g is an algebraic divisor of F, a cover of $\mathcal{F} = (f, d, r)$. If $f \not\subseteq e$,* where e is the remainder in the algebraic division, that is, $F = gh \vee e$, *then g is a Boolean divisor of $\mathcal{F}$.*

Proof. Assume $F = gh \vee e, gh \neq 0, f \not\subseteq e$. Because $f \subseteq gh \vee e$ and $f \not\subseteq e$, then $fgh \neq \emptyset$, implying that $fg \neq \emptyset$. Therefore, by Lemma 16.5, g is a Boolean divisor of $\mathcal{F}$. ■

Lemma 16.7. *If g is an algebraic factor of F, a cover of $\mathcal{F} = (f, d, r)$, then g is a Boolean factor of $\mathcal{F}$.*

Proof. Assume $F = gh$. Because $f \subseteq F$, then

$$f \subseteq gh \Rightarrow f \subseteq g.$$

By Lemma 16.4, g is a Boolean factor of $\mathcal{F}$. ■

Lemma 16.8. *Suppose g is an algebraic divisor of F, a prime cover of $\mathcal{F} = (f, d, r)$. Then $\overline{g}$ is a Boolean divisor of $\tilde{\mathcal{F}} = (r, d, f)$.*

Proof. We need to show that $\overline{g}r \neq \emptyset$ by Lemma 16.5. Now $F = gh \vee e$, so $\overline{F} = \overline{e}\,\overline{g} \vee \overline{e}\overline{h}$. But $d \vee r \supseteq \overline{F} \supseteq r$, thus $\overline{F}r = r = \overline{e}\,\overline{g}r \vee \overline{e}\overline{h}r$.

Suppose by contradiction that $\overline{g}r = \emptyset$, then $r = \overline{e}\,\overline{h}r$, that is, $\overline{e}\,\overline{h} \supset r$, and so $e \vee h \subseteq \overline{r} = f \vee d$.

But, from $F = gh \vee e \subseteq h \vee e \subseteq f \vee d$, then $e \vee h \subseteq f \vee d$ implies that the cubes of gh were not prime, reaching a contradiction. ■

16.2.3 Performing Boolean Division

Given $\mathcal{F} = (f, d, r)$, and a divisor g, the problem is to find a cover for $\mathcal{F}$ in the form $GH \vee E$ where H, E are minimal in some sense, such as the minimum factored form. A variant is to find a cover in the form $gH_1 \vee \overline{g}H_0 \vee E$, where again H_1, H_0, E are minimal.

There is a method for performing the Boolean division operation (i.e., finding H and E) based on ESPRESSO, even though a minimum SOP may not be the best objective. Informally the steps of the method are:

(i) Create a **new** variable x to "represent" g.
(ii) Form the don't-care set $\tilde{d} = x\overline{g} \vee \overline{x}g$. (Because $x = g$, we don't care if $x \neq g$.)
(iii) Minimize $(f\,\overline{(\tilde{d})}, d \vee \tilde{d}, r\,\overline{(\tilde{d})})$ to get $\tilde{f}$. Note that $(f\,\overline{(\tilde{d})}, d \vee \tilde{d}, r\,\overline{(\tilde{d})})$ is a partition.
(iv) Return (H, E) where H are the terms of $\tilde{f}$ containing x but with x removed, and E is the remainder of $\tilde{f}$ (i.e., the terms not containing x).

We can say that the method returns $(H = \tilde{f}/x, E)$, where f/x denotes "weak division," a maximal form of algebraic division introduced formally in Definition 16.24.

Example 16.17. *Consider f with cover $F = a \vee bc$ and g with cover $G = a \vee b$. The Boolean division $F \div G$ can be performed as follows:*

- $\tilde{d} = x\overline{a}\overline{b} \vee \overline{x}(a \vee b)$ *where* $x = g = (a \vee b)$.
- *Minimize* $(a \vee bc)\overline{(\tilde{d})} = (a \vee bc)(\overline{x}\;\overline{a}\overline{b} \vee x(a \vee b)) = xa \vee xbc$ *with* $DC = x\overline{a}\overline{b} \vee \overline{x}(a \vee b)$.
- *A minimum cover is* $a \vee bc$, *but it does not use* x *or* $\overline{x}$*!*
- *Force* x *in the cover. It yields* $F = xa \vee xc$ *and, making it a prime cover; finally we get* $F = a \vee xc = a \vee (a \vee b)c$.

Heuristic: Try to find an answer **involving** *x and that uses the* **fewest variables** *(or* **literals***).*

We provide two algorithms based on the previous method. Assume that F is a cover for $\mathcal{F} = (f, d, r)$, D is a cover for d, and g is a divisor. The first algorithm, shown in Figure 16.6, finds H, E such that $xH \vee E$ is a cover for (f, d, r), where x is a literal denoting the divisor g.

The second algorithm, shown in Figure 16.7, finds H_1, H_0, E such that $xH_1 \vee \overline{x}H_0 \vee E$ is a cover for (f, d, r), where x is a literal denoting the divisor g and $\overline{x}$ is a literal denoting the divisor $\overline{g}$. The second algorithm is a slight variation of the first one: it uses $\overline{x}$ also while dividing, by skipping the step F_2 = remove $\overline{x}$ from F_1.

The given algorithms use an operation MinLiteral_Support (or MinVariable_Support), which finds a prime cover with the smallest literal (or variable) support, that is, a prime cover with the smallest number of literals (variables) that appear in at least a prime.

We remark that in any cover of primes of a completely specified function, the minimum support is always the same and is the set of variables on which the function effectively depends, that is, the essential variables of the function.

$$
\begin{aligned}
(H, E) &\leftarrow \text{Boolean\_Division}(F, D, g)\\
D_1 &= D \vee x\overline{g} \vee \overline{x}g \quad \text{(don't care)}\\
F_1 &= F\overline{D_1} \quad \text{(on-set)}\\
R_1 &= \overline{(F_1 \vee D_1)} = \overline{F}_1\overline{D}_1 = \overline{F}\,\overline{D}_1 \quad \text{(off-set)}\\
F_2 &= \text{remove } \overline{x} \text{ from } F_1.\\
F_3 &= \text{MinLiteral\_Support}(F_2, R_1, x)\\
&\quad \text{(minimum literal support including } x)\\
F_4 &= \text{ESPRESSO}(F_3, D_1, R_1)\\
H &= F_4/x \quad \text{(quotient)}\\
E &= F_4 - \{xH\} \quad \text{(remainder)}
\end{aligned}
$$

Figure 16.6. An algorithm for Boolean division. Given covers F of (f, d, r) and D of d and a divisor g, it finds H, E such that $xH \vee E$ is a cover of (f, d, r), where x is a literal denoting the divisor g.

However, this is not true when there is a don't-care set d, and so the result does not apply to our context, where $d \neq \emptyset$ by construction. Given a cover F of $\mathcal{F} = (f, d, r)$, and the notations

$$
\begin{aligned}
v\_sup(F) &= \{v | v \in c \text{ or } \overline{v} \in c \text{ for some } c \in F\},\\
\ell\_sup(F) &= \{\ell | \ell \in c \text{ for some } c \in F\},
\end{aligned}
$$

the following lemma states that the minimum support sets v_sup and ℓ_sup are unique for all prime covers.

$$
\begin{aligned}
(H_1, H_0, e) &\leftarrow \text{Boolean\_Division}(F, D, g)\\
D_1 &= D \vee x\overline{g} \vee \overline{x}g \quad \text{(don't care)}\\
F_1 &= F\overline{D_1} \quad \text{(on-set)}\\
R_1 &= \overline{(F_1 \vee D_1)} = \overline{F}_1\overline{D}_1 = \overline{F}\,\overline{D}_1 \quad \text{(off-set)}\\
F_3 &= \text{MinLiteral\_Support}(F_1, R_1, x, \overline{x})\\
&\quad \text{(minimum literal support including } x, \overline{x})\\
F_4 &= \text{ESPRESSO}(F_3, D_1, r_1)\\
H_1 &= F_4/x\\
H_0 &= F_4/\overline{x}\\
E &= F_4 - \{xH_1\} - \{\overline{x}H_0\}
\end{aligned}
$$

Figure 16.7. An algorithm for Boolean division. Given covers F of (f, d, r) and D of d and a divisor g, it finds H, E such that $xH_1 \vee \overline{x}H_0 \vee E$ is a cover of (f, d, r), where x is a literal denoting the divisor g and $\overline{x}$ is a literal denoting the divisor $\overline{g}$.

Lemma 16.9. *For a completely specified Boolean function, that is, for $\mathcal{F} = (f, d, r)$ with $d = \emptyset$, then $v\_sup(F_1) = v\_sup(F_2)$ and $\ell\_sup(F_1) = \ell\_sup(F_2)$ for all prime covers F_1, F_2 of $\mathcal{F}$.*

Proof. Let F_1 and F_2 be two prime covers of $\mathcal{F}$, with F_1 independent of x, but let x be in F_2. Consider a prime cube $xc \in F_2$, where c is a cube covering the minterms $\{m_1, m_2, \ldots, m_k\}$ (defined over all variables except x). Because xm_i is an implicant of $\mathcal{F}$, it is covered by F_1, because $d = \emptyset$ and so all minterms in every implicant must be covered by any cover. However, F_1 is independent of x, and so xm_i covered by F_1 implies $\overline{x}m_i$ covered by F_1. Hence $\overline{x}m_i$ is also an implicant of $\mathcal{F}$. Hence m_i is an implicant of $\mathcal{F}$, and this holds for all $i = 1, \ldots, k$. Therefore c (which is the collection of all $m_i, i = 1, \ldots, k$) is an implicant of $\mathcal{F}$, and so xc can be raised to c, contradicting the fact that xc is a prime. ■

The procedures to minimize the variable or literal support are based on the notion of blocking matrices, used in two-level logic minimization to expand cubes to primes [2, 38]. Recall [2] that the operation of expansion of an implicant is performed by removing one or more literals from its expression as a cube, which corresponds to increasing its size by a factor of 2 per deleted literal, and therefore to covering more minterms. It is legitimate to expand an implicant with respect to a literal, if the expanded cube does not cover minterms of the offset. Expanding a cover means expanding the single terms of the collection.

Given $\mathcal{F} = (f, d, r)$, let $F = \{c^1, c^2, \ldots, c^k\}$ be a cover of $\mathcal{F}$ and $R = \{r^1, r^2, \ldots, r^m\}$ be a cover of r. A blocking matrix B^i of a cube c^i of F keeps track of all variables that make cube c^i disjoint from each cube of R.

Definition 16.20. *The* **blocking matrix** B^i *of cube* c^i *of* F *is defined as*

$$(B^i)_{qj} = \begin{cases} 1 & \text{if} (c^i)_j \cap (r^q)_j = \emptyset \\ 0 & \text{otherwise} \end{cases} \tag{16.1}$$

for $q = 1, \ldots, m$ *and* $j = 1, \ldots, n$.

A row cover of a blocking matrix B^i of a cube c^i of F is a set of variables that make cube c^i disjoint from each cube of R.

Definition 16.21. *A* **row cover** *of a 0-1 matrix B is a set of column indices $S = \{j_1, \ldots, j_v\}$ such that $\forall q$, $\exists j \in S$, such that $B_{qj} = 1$.*

Theorem 16.10. *Let S be a row cover of B^i and suppose $|S|$ is minimum. Construct the cube*

$$(\widetilde{c}^i)_j = \begin{cases} (c^i)_j & \text{if } j \in S \\ \{0, 1\} = 2 & \text{otherwise.} \end{cases} \tag{16.2}$$

Then $\widetilde{c}^i \supseteq c^i$ is the largest prime implicant of $\mathcal{F}$ containing c^i.

(i) Construct blocking matrix B^i for each c^i.
(ii) Form superblocking matrix B.
(iii) Find a minimum cover S of B, where $S = \{j_1, j_2, \ldots, j_v\}$.
(iv) Build $\widetilde{F} \leftarrow \{\widetilde{c}^1, \widetilde{c}^2, \ldots, \widetilde{c}^k\}$, where $(\widetilde{c}^i)_j = \begin{cases} (c^i)_j & \text{if } j \in S \\ \{0,1\} = 2 & \text{otherwise.} \end{cases}$

Figure 16.8. Steps of the algorithm MinVariable_Support to compute the minimum variable support.

Definition 16.22. *The* **super-blocking matrix** *B of cover $F = \{c^1, c^2, \ldots, c^k\}$ is defined as*

$$B = \begin{bmatrix} B^1 \\ B^2 \\ \vdots \\ B^k \end{bmatrix}$$

where B^i is the blocking matrix of cube c^i of F.

Theorem 16.11. *Let S be a minimum row cover of the superblocking matrix B. The set $\{x_j \mid j \in S\}$ is a minimum set of variables that appear in any cover of $\mathcal{F}$ obtained by expanding F.*

Proof. Expand treats each $c^i \in F$ and builds B^i. Let $\hat{c}^i$ be any prime containing c^i. Then the variables in $\hat{c}^i$ "cover" B^i. Thus the union of the set of variables in $\hat{c}^i$ taken over all i covers B. Hence, this set cannot be smaller than a minimum cover of B. ■

Note that in general, there could exist another cover of $\mathcal{F}$ that has fewer variables, but one not obtained by expanding F.

In summary, given $\mathcal{F} = (f, d, r)$, $F = \{c^1, c^2, \ldots, c^k\}$ cover of $\mathcal{F}$, and $R = \{r^1, r^2, \ldots, r^m\}$ cover of r, the algorithm to find an expanded cover with the fewest variables is outlined in Figure 16.8.

Given $\mathcal{F} = (f, d, r)$, $F = \{c^1, c^2, \ldots, c^k\}$ cover of $\mathcal{F}$, $R = \{r^1, r^2, \ldots, r^m\}$ cover of r, the minimum literal support is computed in a similar way, defining the literal blocking matrix as an extension of the standard (variable) blocking matrix.

Definition 16.23. *The* **literal blocking matrix** *$\hat{B}^i$ of cube c^i of F over B^n is defined as*

$$(\hat{B}^i)_{qj} = \begin{cases} 1 & \textit{if } v_j \in c^i \textit{ and } \overline{v}_j \in r^q \\ 0 & \textit{otherwise,} \end{cases}$$

$$(\hat{B}^i)_{q,j+n} = \begin{cases} 1 & \textit{if } \overline{v}_j \in c^i \textit{ and } v_j \in r^q \\ 0 & \textit{otherwise.} \end{cases}$$

Example 16.18. *Given $c^i = a\bar{d}\bar{e}$, $r^q = \bar{a}ce$, the literal blocking matrix $\hat{B}^i_q$ of c^i is*

$$\hat{B}^i_q = \begin{array}{cccccccccc} a & b & c & d & e & \bar{a} & \bar{b} & \bar{c} & \bar{d} & \bar{e} \\ 1 & 0 & 0 & 0 & 0 & 0 & 0 & 0 & 0 & 1 \end{array}$$

The literal superblocking matrix $\hat{B}$ can be constructed as before, by concatenation of the literal blocking matrices associated with individual cubes.

Theorem 16.12. *Let J be a minimum row cover of the literal superblocking matrix $\hat{B}$. The set $\{\ell_i | i \in J\} \cup \{\bar{\ell}_i | i + n \in J\}$ is a minimum set of literals that appear in any cover of $\mathcal{F}$ obtained by expanding F.*

Proof. Same reasoning as for minimum variable support. ■

For a (nontrivial) cube, minimum literal support is the same as minimum variable support.

Example 16.19. *Given the on-set cube $c^i = a\bar{b}d$ and off-set $r = \bar{a}\bar{b}\,\bar{d} \vee ab\bar{d} \vee ac\bar{d} \vee bcd \vee \bar{c}\bar{d}$, the literal superblocking matrix $\hat{B}^i$ is*

	a	b	c	d	$\bar{a}$	$\bar{b}$	$\bar{c}$	$\bar{d}$
$\bar{a}\bar{b}\,\bar{d}$	1	0	0	1	0	0	0	0
$ab\bar{d}$	0	0	0	1	0	1	0	0
$ac\bar{d}$	0	0	0	1	0	0	0	0
bcd	0	0	0	0	0	1	0	0
$\bar{c}\bar{d}$	0	0	0	1	0	0	0	0

The minimum column cover is $\{d, \bar{b}\}$, and thus $\bar{b}d$ is the maximum prime covering the cube $a\bar{b}d$.

To see the different operations in action, the following example shows how to perform Boolean division by applying the steps of the algorithm in Figure 16.6.

Example 16.20. *Given $F = a \vee bc$, by algebraic division we get $F/(a \vee b) = \emptyset$. By Lemma 16.2 and Lemma 16.3, $G = a \vee b$ is a Boolean factor and a Boolean divisor of the function represented by F. Moreover, it is easy to verify the Boolean division $F \div (a \vee b) = a \vee c$. Instead let us perform the Boolean division according to the algorithm in Figure 16.6.*

- *Set $x = a \vee b$.*
- *Generate the don't-care set $D_1 = \bar{x}(a \vee b) \vee \bar{a}\bar{b}x$.*
- *Generate the care on-set $F_1 = F\overline{D}_1 = (a \vee bc)(ax \vee bx \vee \bar{a}\bar{b}\bar{x}) = ax \vee bcx$. Let $C = \{c^1 = ax, c^2 = bcx\}$.*
- *Generate the care off-set $R_1 = \overline{F}\ \overline{D}_1 = (\bar{a}\bar{b} \vee \bar{a}\ \bar{c})(ax \vee bx \vee \bar{a}\bar{b}\bar{x}) = \bar{a}b\bar{c}x \vee \bar{a}\bar{b}\bar{x}$. Let $R = \{r^1 = \bar{a}b\bar{c}x, r^2 = \bar{a}\bar{b}\bar{x}\}$.*

$$
\begin{aligned}
(Q, R) &= \text{Boolean\_Division}(F, G) \\
&\Updownarrow \\
(Q_1, R_1) &= \text{Algebraic\_Division}(F, G) \\
D_1 &= (x \oplus G) \vee Q_1 x \\
R_2 &= \text{ESPRESSO}(R_1, D_1) \\
(Q_3, R_3) &= \text{Algebraic\_Division}(R_2, x) \\
(Q, R) &= (Q_3 \vee Q_1, R_3)
\end{aligned}
$$

Figure 16.9. A heuristic scheme for Boolean division that computes $(Q, R) =$ Boolean_Division(F, G).

- *Form the variable superblocking matrix using column order (a, b, c, x). Notice that c^1 and c^2 are positive unate cubes.*

$$
B = \begin{bmatrix} B^1 \\ B^2 \end{bmatrix} = \begin{bmatrix} a & b & c & x & \\ 1 & 0 & 0 & 0 & ax \\ 1 & 0 & 0 & 1 & \\ - & - & - & - & \\ 0 & 0 & 1 & 0 & bcx \\ 0 & 1 & 0 & 1 & \end{bmatrix} \tag{16.3}
$$

- *Find the minimum row cover $S = \{a, c, x\}$.*
- *Eliminate in F_1 all variables associated with b. So from $F_1 = ax \vee bcx$ we get $F_3 = ax \vee cx = x(a \vee c)$ (if we would notice $F_3 = x(a \vee c)$, then we could conclude $F \div (a \vee b) = a \vee c$ and we would be done).*
- *Simplifying F_3 (applying expand, irredundant cover), we get $F_4 = a \vee cx$.*
- *Thus the quotient is $H = F_1/x = c$, the remainder is $E = a$, and so we get the cover $xH \vee E = xc \vee a$.*
- *In summary, from $F = a \vee bc$ we got $a \vee cx = a \vee c(a \vee b)$.*

A question is, how do we force x in the final cover? By default, ESPRESSO does not guarantee that it will keep it; one must restrict/modify its operation to preserve x in the final cover. For instance, one could put MINVAR in the inner loop of ESPRESSO.

Another heuristic for Boolean division mentioned in [4, 5] interlaces algebraic division and simplification as shown in the scheme of Figure 16.9. For simplicity here, we omitted the minimum literal support step. How to perform algebraic division is the topic of the next section.

Example 16.21. *Performing Boolean division by two steps of algebraic division (with some help from* ESPRESSO *too !):*

$$
\begin{aligned}
F &= ab\overline{c}\overline{d} \vee ab\overline{e}\overline{f} \vee cd\overline{a}\overline{b} \vee cd\overline{e}\overline{f} \vee ef\overline{a}\overline{b} \vee ef\overline{c}\overline{d} \\
G &= ab \vee cd \vee ef \\
(Q_1, R_1) &= Algebraic\_Division(F, G) \\
(Q_1, R_1) &= (\emptyset, F) \\
D &= (ab \vee cd \vee ef) \oplus x \\
R_2 &= \text{ESPRESSO}(F, D) \\
R_2 &= \overline{a}\overline{b}x \vee \overline{c}\overline{d}x \vee \overline{e}\overline{f}x \\
(Q_3, R_3) &= Algebraic\_Division(R_2, x) \\
(Q_3, R_3) &= (\overline{a}\overline{b} \vee \overline{c}\overline{d} \vee \overline{e}\overline{f}, \emptyset) \\
F &= (ab \vee cd \vee ef)(\overline{a}\overline{b} \vee \overline{c}\overline{d} \vee \overline{e}\overline{f})
\end{aligned}
$$

16.3 Algebraic Division

The key motivations to introduce algebraic methods are:

(i) They treat logic functions like polynomials (often the irredundant prime representation is canonical, e.g., unate).
(ii) Fast methods for manipulation of polynomials are available (complexity from linear to quadratic).
(iii) There is a loss of optimality, but results are still quite good.
(iv) They can iterate and interleave with Boolean operations.

When introducing algebraic division, we have already noticed that it does not guarantee the uniqueness of quotient and reminder.

Example 16.22. *Given $F = (a \vee b \vee c)(d \vee e) \vee \overline{a}c$ and $G = d \vee e$, we can get either*

$$
\begin{aligned}
F &= (d \vee e)(a \vee b) \vee c(\overline{a} \vee d \vee e) = GH_1 \vee R_1, \text{ or} \\
F &= (d \vee e)(a \vee b \vee c) \vee c\overline{a} = GH_2 \vee R_2
\end{aligned}
$$

so H and R are not unique.

To achieve uniqueness, we introduce weak division, a restricted definition of algebraic division, where the quotient is the largest (maximum number of terms) expression H such that $F = GH \vee R$.

Definition 16.24. *Given two algebraic expressions F and G, a division of F by G is called* **weak division** *if:*

(i) *It is an algebraic division, that is, it generates expressions H and R such that HG is an algebraic product;*
(ii) *R has as few cubes as possible, or equivalently, H has as many cubes as possible;*

(iii) *The expressions $GH \vee R$ and F are identical (they are the same set of cubes when multiplied out).*

The quotient H resulting from weak division is denoted by F/G.

Theorem 16.13. *Given the algebraic expressions F and G, the quotient H and the remainder R generated by weak division are unique.*

Example 16.23. *Given $F = ac \vee ad \vee ae \vee bc \vee bd \vee be \vee \overline{a}b$, we have $F/a = c \vee d \vee e$, $F/b = c \vee d \vee e \vee \overline{a}$ and $F/(a \vee b) = c \vee d \vee e$. Notice that $F/(a \vee b) = F/a \cap F/b$. Notice that $F/(a \vee b) = (F/a)(F/b)$.*

To compute the quotient of an algebraic division F/G, where $F = \{c_1, \ldots, c_{|F|}\}$ and $G = \{a_1, \ldots, a_{|G|}\}$, we first define the quotient when the divisor is a cube as

$$h_i = F/a_i = \{b_j | a_i b_j = c_k \in F\}.$$

Example 16.24. *Let $F = abc \vee a\overline{bc} \vee abd\overline{e} \vee abg \vee b\overline{c}$; then $F/(ab) = c \vee d\overline{e} \vee g$.*

The following theorem argues that the quotient can be obtained by intersecting the quotients with respect to the single cubes of the divisor.

Theorem 16.14. *Given $F = \{c_1, \ldots, c_{|F|}\}$ and $G = \{a_1, \ldots, a_{|G|}\}$, then*

$$F/G = \{d_j \mid d_j \in F/a_i \text{ for all } a_i \in G\} = \bigcap_{i=1}^{|G|}(F/a_i).$$

Proof. Let $\{d_1, \ldots, d_s\} = \{d_j \mid d_j \in F/a_i$ for all $a_i \in G\}$; then

$$F = (d_1 \vee \cdots \vee d_s)(a_1 \vee \cdots \vee a_{|G|}) \vee R,$$

where R contains the remaining terms c_h: $R = F \setminus \{d_j a_i | j = 1, \ldots, s,\ i = 1, \ldots, |G|\}$.

We show by contradiction that $\{d_1, \ldots, d_s\}$ is the largest quotient; suppose not, then there exists $d \notin \{d_1, \ldots, d_s\}$, and $d(a_1 \vee \cdots \vee a_{|G|}) \in F$. Then $d \in F/a_i$ for all $a_i \in G$, which is a contradiction. ■

Theorem 16.14 suggests an algorithm to perform weak division, as outlined in Figure 16.10.

Example 16.25. *Given F and G*

$$F = ace \vee ade \vee bc \vee bd \vee be \vee \overline{a}b \vee ab$$

$$G = ae \vee b,$$

```
WEAK_DIVISION(F, G): {
    U = {u_j} - all cubes of F where only literals appearing in G have been
                kept
    V = {v_j} - all cubes of F with literals appearing in G removed
    /* note that u_j v_j is the j-th cube of F */
    V^i = { v_j ∈ V : u_j = a_i } /* one set for each cube of G */
    H = ∩_{i=1}^{|G|} V^i /* those cubes found in all V^i */
    R = F \ GH
    return (H, R)
}
```

Figure 16.10. An algorithm to perform weak division.

the algorithm for weak division in Figure 16.10 computes the following expressions:

$$\begin{aligned} U &= ae \vee ae \vee b \vee b \vee b \vee b \vee ab \\ V &= c \vee d \vee c \vee d \vee 1 \vee \overline{a} \vee 1 \\ V^{ae} &= c \vee d \\ V^{b} &= c \vee d \vee 1 \vee \overline{a} \\ H &= c \vee d = F/G \\ R &= be \vee \overline{a}b \vee ab. \end{aligned}$$

Finally, F can be factored as follows:

$$F = (ae \vee b)(c \vee d) \vee be \vee \overline{a}b \vee ab.$$

The time complexity of WEAK_DIVISION is $O(|F||G|)$.

McGeer and Brayton [32] investigated efficient implementations of the algebraic division algorithm, and they were able to show that its complexity can be reduced to $O((|F| + |G|)\log(|F| + |G|))$. Moreover, if F, G are already given as sorted cubes, that is, in the order of their binary encoding (e.g., $a\overline{b}de$ can be encoded by the binary number 0110110101), they reported a **linear** time algorithm such that F/G and R are produced in sorted order. In fact, the operations of algebraic division, multiplication, addition, subtraction, and equality test were all proved to be linear and stable.

Definition 16.25. *A* **stable** *algorithm produces its output in sorted order if it receives its input in sorted order.*

If all algorithms are stable, then we can start with a Boolean network, do an initial sort on each node, and then use only stable operations.

The steps of the $\mathcal{O}(n\ log\ n)$ algorithm are outlined here:

(i) Encode the cubes $a_i \in G$ as binary numbers $n_1, n_2, \ldots, n_{|G|}$.
(ii) Encode the cubes $c_i \in F$ restricted to the support of G, $c_{i|sup(G)} \equiv b_i$, as binary numbers $m_1, m_2, \ldots, m_{|F|}$.

(iii) Sort the cubes in the set $\{n_1, \ldots, n_{|G|}, m_1, \ldots, m_{|F|}\}$, in time $O((|F| + |G|)\log(|F| + |G|))$.
(iv) Define the set $I = \{i \mid \exists\, j$ such that $m_i = n_j\}$, denoting the cubes of F divided by a cube of G.
(v) Encode the set $\{d_i \equiv c_{i|sup(F)\setminus sup(G)} \mid i \in I\}$, as binary numbers $q_1, q_2, \ldots, q_{|I|}$.
(vi) Sort the cubes in the set $\{q_1, \ldots, q_{|I|}\}$, in time $O(|F| \log |F|)$.
(vii) Define the set $J = \{j \mid q_j$ appears $|G|$ times$\}$.
(viii) $F/G = \{c_{j|sup(F)\setminus sup(G)} \mid j \in J\}$.

Example 16.26. *Consider* $F = ac\overline{d} \vee a\overline{c}d \vee ae \vee bc\overline{d} \vee b\overline{c}d \vee be \vee ag \vee \overline{b}e$ *and* $G = a \vee b$. *The encoded cubes of* F *are (to ease readability, we report the binary encoding used in the algorithm with the common decoding convention* $01 \rightarrow 1$, $10 \rightarrow 0$, $11 \rightarrow 2$*):*

	ab cdeg
(1)	12 1022
(2)	12 0122
(3)	12 2212
(4)	21 1022
(5)	21 0122
(6)	21 2212
(7)	12 2221
(8)	20 2212

Notice that $sup(G) = \{a, b\}$, $sup(F) \setminus sup(G) = \{c, d, e, g\}$. *Then* $I = \{1, 2, 3, 4, 5, 6, 7\}$, *and the sorted* $\{q_i\}$ *are:*

	cdeg
(1 − 4)	1022
(2 − 5)	0122
(3 − 6)	2212
(7)	2221

Thus $J = \{1-4, 2-5, 3-6\}$ *and* $F/G = c\overline{d} \vee \overline{c}d \vee e$.

Algebraic division filters were devised to speed up algebraic division. The function f_j is not an algebraic divisor of f_i if any of the following cases is true:

(i) f_j contains a literal not in f_i.
(ii) f_j has more terms than f_i.
(iii) For any literal, its literal count in f_j exceeds that in f_i.
(iv) y_i is in the transitive fan-in of f_j.

16.4 Algebraic Divisors and Kernels

So far, we learned how to divide a given expression F by another expression G. But how do we find G? The problem is that there are too many Boolean divisors, so a practical strategy is to restrict the exploration to algebraic divisors, i.e., to restrict the problem to: given a set of functions $\{F_i\}$, find common weak (algebraic) divisors.

16.4.1 Kernels and Kernel Intersections

Definition 16.26. *An expression F is* **cube-free** *if no cube divides the expression evenly, that is, if there are no expressions G and cubes c such that $F = Gc$.*

Examples are: $ab \vee c$ is cube-free; $ab \vee ac$ and abc are not cube-free. Notice that a cube-free expression must have more than one cube.

Definition 16.27. *The* **primary divisors** *of an expression F are the elements of the set of expressions*

$$\mathcal{D}(F) = \{F/c \mid c \textit{ is a cube }\}.$$

Definition 16.28. *The* **kernels** *of an expression F are the elements of the set of expressions*

$$\mathcal{K}(F) = \{G \mid G \in \mathcal{D}(F) \textit{ and } G \textit{ is cube-free }\}.$$

In other words, the kernels of an expression F are the cube-free primary divisors of F.

Definition 16.29. *A cube c used to obtain the kernel $K = F/c$ is called a* **cokernel** *of K, and $\mathcal{C}(F)$ is used to denote the set of cokernels of F.*

Example 16.27. *The following table lists the kernels and cokernels of the following expression:*

$$\begin{aligned} Q &= adf \vee aef \vee bdf \vee bef \vee cdf \vee cef \vee g \\ &= (a \vee b \vee c)(d \vee e)f \vee g. \end{aligned}$$

Kernel	*Cokernel*
$a \vee b \vee c$	$df,\ ef$
$d \vee e$	$af,\ bf,\ cf$
$(a \vee b \vee c)(d \vee e)$	f
$(a \vee b \vee c)(d \vee e)f \vee g$	1

Brayton and McMullen established in [3] the fundamental Theorem 16.16 that motivates the role of kernels. It states that if two expressions F and G are such that $\mathcal{K}(F)$ and $\mathcal{K}(G)$ have at most one term in common, then F and G have **no common algebraic divisors with more than one term**.

Lemma 16.15. *Every divisor G of expression F is contained in a primary divisor, that is, if G divides F, then there exists $P \in \mathcal{D}(F)$ such that $G \subseteq P \in \mathcal{D}(F)$.*

Proof. (From [21]). Let c be a cube of F/G. Then $G \subseteq F/(F/G)$ and $F/(F/G) \subseteq F/c \in \mathcal{D}(F)$. ■

Example 16.28. *Let $F = ac \vee ad \vee bc \vee bd \vee ec \vee ed \vee cd$. Consider the algebraic divisor $G = a \vee b$ that is not a primary divisor, and let us build a primary divisor that contains G according to the lemma. >From $F/G = c \vee d$, take cube $c \in F/G = c \vee d$. Then $G = a \vee b \subseteq F/(F/G) \equiv F/(c \vee d) = a \vee b \vee e$ and $F/(F/G) = a \vee b \vee e \subseteq F/c = a \vee b \vee e \vee d \in \mathcal{D}(F)$. So given the algebraic divisor $G = a \vee b$, we found a primary divisor $P = a \vee b \vee e \vee d$ such that $G \subseteq P$, that is, all cubes in G are also cubes in P.*

Theorem 16.16. *Expressions F and G have a common divisor involving more than one cube if and only if there exists $K_F \in \mathcal{K}(F)$ and there exists $K_G \in \mathcal{K}(G)$ such that $|K_F \cap K_G| > 1$, i.e., $K_F \cap K_G$ is an expression with at least two terms (it is a not a single cube).*

Proof. (From [21].)

If part: If there are a kernel $K_F \in \mathcal{K}(F)$ and a kernel $K_G \in \mathcal{K}(G)$ whose algebraic intersection D is an algebraic expression with at least two cubes, then D is by definition a common divisor of F and G with at least two cubes.

Only if part: Let D be an algebraic divisor with at least two cubes. Then there is a cube-free expression E that divides D (either D is cube-free or we make it cube-free by dividing it by the largest cube divisor).

By Lemma 16.15, $\exists P_F \in \mathcal{D}(F)$ and $\exists P_G \in \mathcal{D}(G)$ such that $E \subseteq P_F \in \mathcal{D}(F)$ and $E \subseteq P_G \in \mathcal{D}(G)$.

Beacuse E is cube-free, P_F and P_G are also cube-free. This means that P_F and P_G are cube-free primary divisors, that is, they are kernels of their respective functions: $P_F \in \mathcal{K}(F)$ and $P_G \in \mathcal{K}(G)$.

From $E \subseteq P_F$ and $E \subseteq P_G$, it follows $E \subseteq P_F \cap P_G$ and, because E is nontrivial (it has at least two cubes), $P_F \cap P_G$ is also nontrivial. ■

In summary, if we compute the kernels of all functions and there are no nontrivial intersections, then the only common algebraic divisors left are single cube divisors (these are not the only common divisors of F and G, because there could be common Boolean divisors).

Example 16.29. *Let $F = a(bc \vee \overline{b}d) \vee e$, $G = a(bc \vee \overline{b}e) \vee d$. Then:*

$\mathcal{K}(F) = \{bc \vee \overline{b}d, a(bc \vee \overline{b}d) \vee e\}$, $\mathcal{K}(G) = \{bc \vee \overline{b}e, a(bc \vee \overline{b}e) \vee d\}$;

$\{K_F \cap K_G : K_F \in \mathcal{K}(F), \quad K_G \in \mathcal{K}(G), \quad K_F \cap K_G \neq \emptyset\} = \{bc, abc\}$ *are cubes, so F and G have no common nontrivial algebraic divisors.*

Now, let $F = abc \vee cd \vee e$, $G = ab \vee \overline{c}d \vee e$. Then:

$\mathcal{K}(F) = \{ab \vee cd \vee e\}$, $\mathcal{K}(G) = \{ab \vee \overline{c}d \vee e\}$;

$\{K_F \cap K_G : K_F \in \mathcal{K}(F),\ K_G \in \mathcal{K}(G),\ K_F \cap K_G \neq \emptyset\} = \{ab \vee e\}$ *is a common nontrivial algebraic divisor of F and G.*

Some subsets of kernels of interest in computations are defined by introducing the notion of level of a kernel.

Definition 16.30.

$$\mathcal{K}^n(F) = \begin{cases} \{k \in \mathcal{K}(F) \mid \mathcal{K}(k) = \{k\}\} & n = 0 \\ \{k \in \mathcal{K}(F) \mid \forall k_1 \in \mathcal{K}(k), k_1 \neq k \Rightarrow k_1 \in \mathcal{K}^{n-1}(F)\} & n > 0 \end{cases}$$

If $k \in \mathcal{K}^0(F)$, *then* k *is a* **level-0 kernel** *of* F.
If $k \in \mathcal{K}^n(F)$, *but* $k \notin \mathcal{K}^{n-1}(F)$, *then* k *is a* **level-*n* kernel** *of* F.

Intuitively, a kernel of level 0 has no kernels except itself. Similarly, a kernel of level n has at least one kernel of level $n-1$ but no kernels (except itself) of level n or greater. Thus,

$$\mathcal{K}^0(F) \subset \mathcal{K}^1(F) \subset \mathcal{K}^2(F) \subset \cdots \subset \mathcal{K}^n(F) \subset \mathcal{K}(F).$$

Example 16.30.

$$\begin{aligned} F &= (a \vee b(c \vee d))(e \vee g) \\ K_1 &= a \vee b(c \vee d) \in \mathcal{K}^1, \quad K_1 \notin \mathcal{K}^0 \\ K_2 &= c \vee d \in \mathcal{K}^0 \end{aligned}$$

$$\begin{aligned} F &= a(b\bar{d} \vee c(\bar{b} \vee d)) \vee b\bar{c}d \\ \mathcal{K}^0(F) &= \{\bar{b} \vee d, a\bar{d} \vee \bar{c}d, b\bar{c} \vee ac\} \\ \mathcal{K}^1(F) &= \{b\bar{d} \vee c(d \vee \bar{b})\} \cup \mathcal{K}^0(F) \\ \mathcal{K}^2(F) &= \{a(b\bar{d} \vee c(\bar{b} \vee d)) \vee b\bar{c}d\} \cup \mathcal{K}^1(F) \\ \mathcal{K}(F) &= \mathcal{K}^2(F) \end{aligned}$$

Figure 16.11 shows the pseudocode of KERNEL, an algorithm to compute all kernels, when invoked as KERNEL(0, F). The literals appearing in F are denoted by $\ell_i, i = 1, \ldots, n$.

Note the following facts:

(i) The test $(l_k \in c)$ is a **major** efficiency factor. It also guarantees that no cokernel is tried more than once.
(ii) This algorithm has stood up to all attempts to find faster ones.
(iii) The algorithm can be used to generate all cokernels.
(iv) A simple modification of the kerneling algorithm allows one to generate only the kernels of a certain level d. In case of level 0, one observes that if no kernels are found in the *for* loop, then the argument is a kernel of level 0.

```
KERNEL(j, G) {
    R = ∅
    if (G is cube-free) R = R ∪ {G}
    for (i = j + 1, ..., n) {
        if (ℓ_i appears only in one term) continue
        c = largest cube dividing G/ℓ_i evenly
        if (ℓ_k ∈ c, for some k ≤ i) continue
        else {
            R = R ∪ KERNEL(i, G/(ℓ_i c))
        }
    }
    return R
}
```

Figure 16.11. An algorithm to compute the kernels of F when invoked as KERNEL(0, F).

Example 16.31. *Figure 16.12 shows a fragment of the kerneling computation tree for* $F = a((bc \vee fg)(d \vee e) \vee de(b \vee cf))) \vee beg$. *Kernels and cokernels are listed next.*

co-kernel	**kernel**
a	$bcd \vee bce \vee bde \vee cdef \vee dfg \vee efg$
ab	$cd \vee ce \vee de$
abc	$d \vee e$
abd	$c \vee e$
abe	$c \vee d$
ac	$bd \vee be \vee def$
acd	$b \vee ef$
ace	$b \vee df$
ad	$bc \vee be \vee cef \vee fg$
ade	$b \vee cf$
adf	$ce \vee g$
ae	$bc \vee bd \vee cdf \vee fg$
aef	$cd \vee g$
af	$cde \vee dg \vee eg$
afg	$d \vee e$
b	$acd \vee ace \vee ade \vee eg$
be	$ac \vee ad \vee g$
e	$abc \vee abd \vee acdf \vee afg \vee bg$
eg	$af \vee b$
g	$adf \vee aef \vee be$
1	$abcd \vee abce \vee abde \vee acdef \vee adfg \vee aefg \vee beg$

Co-kernels abc *and* afg *generate both the same kernel* $d \vee e$.

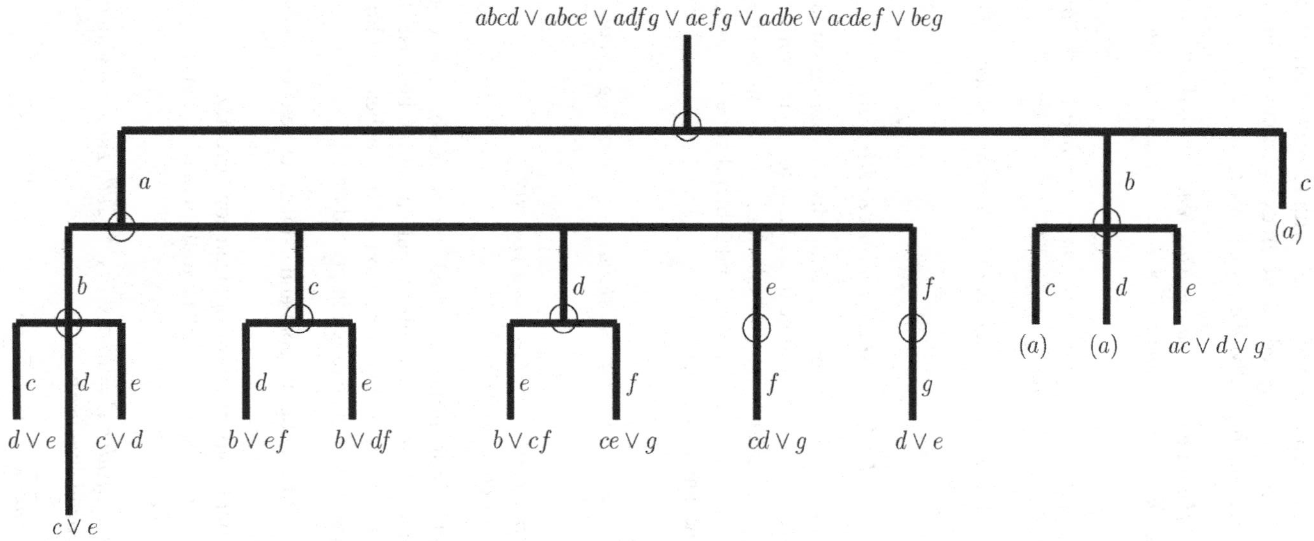

Figure 16.12. Kerneling illustrated for Example 16.31.

In [33], interesting properties of kernels with respect to prime factorization of logical expressions are investigated, where an expression is defined to be prime if it cannot be factored algebraically. We state some results.

Theorem 16.17. *(Unique Factorization Theorem for Weak Division.) An expression F has a unique prime factorization $F = \Pi_i F_i$, where each F_i is a prime expression.*

Theorem 16.18. *If P is a kernel of F and is a prime expression, then P is a kernel of exactly one of the prime factors of F.*

Theorem 16.19. *If K is a level-0 kernel of F, then K is a kernel of exactly one of the prime factors of F.*

16.4.2 Fast Extraction of Divisors

Too much time may be spent for extracting divisors based on kernels: indeed, some functions (e.g., symmetric functions) have too many kernels, and kernels need to be recomputed often (e.g., after substitution). This motivates the restriction to a subset of divisors: 2-cube divisors, and 2-literal cube divisors, introduced by Rajski and Vasudevamurthy [37, 43].

Example 16.32.
The expression $F = abd \vee \overline{a}\overline{b}d \vee \overline{a}cd$ has the following 2-cube and 2-literal divisors:

(i) $ab \vee \overline{a}\overline{b}$, $\overline{b} \vee c$, $ab \vee \overline{a}c$ *(2-cube divisors);*
(ii) $\overline{a}d$ *(2-literal cube divisor).*

An algorithm for finding such restricted divisors has been provided in [37]. The extraction of 2-cube divisors is a polynomial operation, because there are $\mathcal{O}(n^2)$ 2-cube divisors in an algebraic expression with n cubes. Additional attractive computational features of the extraction procedure are:

(i) Extraction of 2-cube divisors and 2-literal cube divisors is done concurrently.
(ii) Complement divisors are recognized concurrently.
(iii) The result can be expanded to multiple-cube divisors and single-cube divisors of any size.

Example 16.33. *Consider $F = abd \vee \overline{a}\overline{b}d \vee \overline{a}cd$. The following complement divisors are recognized concurrently:*

(i) $k = ab \vee \overline{a}\overline{b}$, $\overline{k} = a\overline{b} \vee \overline{a}b$ *(both 2-cube divisors);*
(ii) $k = ab \vee \overline{a}c$, $\overline{k} = \overline{a}\,\overline{c} \vee a\overline{b}$ *(both 2-cube divisors);*
(iii) $k = \overline{a}b \vee ac$, $\overline{k} = \overline{a}\overline{b} \vee a\overline{c}$ *(both 2-cube divisors);*
(iv) $c = ab$ *(2-literal cube)*, $\overline{c} = \overline{a} \vee \overline{b}$ *(2-cube divisor).*

The extraction procedure of 2-cube divisors has $\mathcal{O}(n^2)$ complexity and, given $F = \{c_i\}$, defines a set $\mathcal{D}_2(F) = \{d \mid d = \text{make\_cube\_free}(c_i \vee c_j), \forall c_i, c_j \in F\}$, which takes all pairs of cubes in F and makes them cube-free, by the operation *make_cube_free* that divides the terms of its argument by the largest cube common to them.

Example 16.34. *Given* $F = axe \vee ag \vee bcxe \vee bcg$, *then* $\mathcal{D}_2(F) = \{xe \vee g, a \vee bc, axe \vee bcg, ag \vee bcxe, xe \vee g\}$, *because* $\mathit{make\_cube\_free}(axe \vee ag) = xe \vee g$, $\mathit{make\_cube\_free}(axe \vee bcxe) = a \vee bc$, $\mathit{make\_cube\_free}(axe \vee bcg) = axe \vee bcg$, $\mathit{make\_cube\_free}(ag \vee bcxe) = ag \vee bcxe$, $\mathit{make\_cube\_free}(bcxe \vee bcg) = xe \vee g$.

Note that the functions are made algebraic expressions before generating double-cube divisors; also, not all 2-cube divisors are kernels.

Example 16.35. $\mathcal{K}(axe \vee ag \vee af) = \{xe \vee g \vee f\}$, *but* $\mathcal{D}_2(axe \vee ag \vee af) = \{xe \vee g, g \vee f, xe \vee f, \overline{g}\overline{f}\}$.

$\mathcal{K}(abd \vee \overline{a}\overline{b}d \vee \overline{a}cd) = \{ab \vee \overline{a}\overline{b} \vee \overline{a}c, \overline{b} \vee c\}$, *but* $\mathcal{D}_2(abd \vee \overline{a}\overline{b}d \vee \overline{a}cd) = \{ab \vee \overline{a}\overline{b}, \overline{b} \vee c, ab \vee \overline{a}c, \overline{a}d, \overline{a}b \vee a\overline{b}, b\overline{c}, \overline{a}\,\overline{c} \vee a\overline{b}, a \vee \overline{d}\}$.

Some lemmas proved in [43] allow us to establish what complements of elements in $\mathcal{D}_2(F)$ are also 2-cube divisors or 2-literal cube divisors to be added to $\mathcal{D}_2(F)$. Moreover, some lemmas established in [37] show the relations between single-cube divisors, double-cube divisors, and their complements, allowing the concurrent computation of all of them and their addition to $\mathcal{D}_2(F)$.

Given two algebraic expressions F and G, during decomposition we need to establish whether F has a complement cube divisor in G, and F has a common-cube divisor in G. A result from [37] states that given two algebraic expressions F and G, if certain structural relations are verified, then F has neither a complement double-cube divisor nor a complement single-cube divisor in G.

A fundamental task during decomposition is to establish whether two or more expressions have any common algebraic divisors other than single cubes. Theorem 16.16 stated that there are multiple-cube common divisors among different functions if and only if there are nontrivial intersections among their kernels, and justified the focus on kernels as a subset of algebraic divisors. The following theorem from [43] shows that the same argument also holds for 2-cube divisors, which are an even smaller subset of algebraic divisors.

Theorem 16.20. *Expressions F and G have a common multiple-cube divisor if and only if* $\mathcal{D}_2(F) \cap \mathcal{D}_2(G) \neq 0$.

Proof. **If part:** If $\mathcal{D}_2(F) \cap \mathcal{D}_2(G) \neq 0$, then there exists $D \in \mathcal{D}_2(F) \cap \mathcal{D}_2(G)$, which is a double-cube divisor of F and G. Then D is a multiple-cube divisor of F and G.

Only if part: Suppose $C = c_1 \vee c_2 \vee \ldots \vee c_m$ is a multiple-cube divisor of F and of G. Take any $E = c_i \vee c_j$, $c_i, c_j \in C$. If E is cube-free, then $E \in \mathcal{D}_2(F) \cap$

$\mathcal{D}_2(G)$. If E is not cube-free, then let $\widetilde{E}$ = make_cube_free($c_i \vee c_j$), given that $\widetilde{E}$ exists because F and G are algebraic expressions. Hence $\widetilde{E} \in \mathcal{D}_2(F) \cap \mathcal{D}_2(G)$. ■

As a result of Theorem 16.20, all multiple-cube divisor can be "discovered" from double-cube divisors.

Example 16.36. *Suppose that* $C = ab \vee ac \vee d$ *is a multiple divisor of* F *and* G. *If* $E = ac \vee d$, E *is cube-free and* $E \in \mathcal{D}_2(F) \cap \mathcal{D}_2(G)$. *If* $E = ab \vee ac$, *make_cube_free*$(E) = \widetilde{E} = b \vee c$ *and* $\widetilde{E} \in \mathcal{D}_2(F) \cap \mathcal{D}_2(G)$.

In summary, an algorithm for fast divisor extraction has the following steps:

(1) Generate and store all 2-cube divisors and 2-literal cube divisors and recognize complement divisors.
(2) Find the best (by value) 2-cube divisor or 2-literal cube divisor at each stage and extract it.
(3) Update the set of 2-cube divisors after extraction.
(4) Iteratively extract divisors until no more improvement occurs.

Experimentally, fast extraction of divisors is much faster and of comparable quality with respect to general kernel extraction.

16.5 Optimization of Boolean Networks by Division

We discuss how algebraic and Boolean division are used in modern multilevel logic synthesis systems such as sis to define restructuring operations on a Boolean network. These network operations are factoring, decomposition, substitution, extraction, and elimination, which are variously combined into sequences of optimization steps in synthesis tools.

16.5.1 Factoring and Decomposition

Factoring is the operation to convert a logical expression usually available in SOP form into a factored form with a minimum number of literals. From the state-of-art of computational tools, exact techniques are too expensive, and so one resorts to heuristic algorithms.

The abstract scheme of a factoring algorithm is shown in Figure 16.13. By choosing a divisor and a division operation, one can design a specific factoring algorithm. It must be said that this scheme is too simple-minded; a better generic factoring algorithm (still uncommitted as type of divisor and division) improving hastily chosen divisors is shown in Figure 16.14.

The next two examples (Examples 16.37 and 16.38) motivate the design of the generic algorithm in Figure 16.14. The following notation with reference to the algorithm in Figure 16.13 is used in Examples 16.37 and 16.38: F is the original function, D is the divisor, Q is the quotient, P is a partial factored form, and

```
FACTOR(F) {
    if (F has no factor) return F
    (e.g. if |F| = 1, or an OR of literals)
    D = CHOOSE_DIVISOR(F)
    (Q, R) = DIVIDE(F, D)
    return FACTOR(Q)FACTOR(D) ∨ FACTOR(R)
}
```

Figure 16.13. General structure of a factoring algorithm.

O is the final factored form by *FACTOR*. In the examples we assume algebraic operations only.

In the first example there is a problem due to the fact that the quotient is a single cube.

Example 16.37.

$$\begin{aligned}F &= abc \vee abd \vee ae \vee af \vee g\\ D &= c \vee d\\ Q &= ab\\ P &= ab(c \vee d) \vee ae \vee af \vee g\\ O &= ab(c \vee d) \vee a(e \vee f) \vee g\end{aligned}$$

O is not optimal because it is not maximally factored. It can be further factored to

$$a(b(c \vee d) \vee e \vee f) \vee g.$$

This problem occurs when the quotient Q is a single cube, and some of the literals of Q also appear in the remainder R.

A solution of the problem highlighted in Example 16.37 is the following:

(1) If the quotient Q is not a single cube, then done.
(2) If the quotient Q is a single cube, then pick a literal ℓ_1 in the cube that occurs in the greatest number of cubes of F.
(3) Divide F by ℓ_1 to obtain a new divisor D_1. Now, F has a new partial factored form

$$(\ell_1)(D_1) \vee (R_1),$$

and literal ℓ_1 does not appear in R_1.

Note that the new divisor D_1 contains the original D as a divisor because ℓ_1 is a literal of Q. When recursively factoring D_1, D can be discovered again.

In the second example there is a problem due to the fact that the quotient is not cube-free.

Example 16.38.

$$\begin{aligned} F &= ace \vee ade \vee bce \vee bde \vee cf \vee df \\ D &= a \vee b \\ Q &= ce \vee de \\ P &= (ce \vee de)(a \vee b) \vee (c \vee d)f \\ O &= e(c \vee d)(a \vee b) \vee (c \vee d)f \end{aligned}$$

Again, O is not maximally factored because $(c \vee d)$ is common to both products $e(c \vee d)(a \vee b)$ and $(c \vee d)f$. The final factored form should have been

$$(c \vee d)(e(a \vee b) \vee f).$$

A solution of the problem highlighted in Example 16.38 is the following:

(1) Make Q cube-free to get Q_1.
(2) Obtain a new divisor D_1 by dividing F by Q_1.
(3) If D_1 is cube-free, the partial factored form is $F = (Q_1)(D_1) \vee R_1$, and we can recursively factor Q_1, D_1, and R_1.
(4) If D_1 is not cube-free, let $D_1 = cD_2$ and $D_3 = Q_1 D_2$. We have the partial factoring $F = cD_3 \vee R_1$ (i.e., just start with c as the divisor). Now recursively factor D_3 and R_1.

Various kinds of factoring are obtained by choosing different forms of $DIVISOR$ and $DIVIDE$, as listed next.

- $CHOOSE\_DIVISOR$ can be any of the following:
  - $LITERAL$ – choose a literal, or the best literal.
  - $QUICK\_DIVISOR$ – choose a level-0 kernel.
  - $BEST\_DIVISOR$ – choose the best kernel.
- $DIVIDE$ can be any of the following:
  - Algebraic division.
  - Boolean division.

Three specialized factoring algorithms to mention are:

- $QUICK\_FACTOR$, whose divisor is a level-0 kernel (produced by $QUICK\_DIVISOR$) and whose division is weak division.
- $GOOD\_FACTOR$, whose divisor is a level-0 kernel (produced by $BEST\_KERNEL$, which looks at all kernels and picks a kernel k that, when substituted into the original form F, maximally reduces the total number of SOP literals of F and k), and whose division is weak division.
- $BOOLEAN\_FACTOR$, whose divisor is the same as for $GOOD\_FACTOR$ and whose division is Boolean division.

```
GFACTOR(F, DIVISOR, DIVIDE){
    D = DIVISOR(F)
    if (D = ∅) return F
    (Q, R) = DIVIDE(F, D)
    if |Q| = 1 return LF(F, Q, DIVISOR, DIVIDE)
    Q = make_cube_free(Q)
    (D, R) = DIVIDE(F, Q)
    second divide to improve divisor
    if (cube_free(D)) {
        Q = GFACTOR(Q, DIVISOR, DIVIDE)
        D = GFACTOR(D, DIVISOR, DIVIDE)
        R = GFACTOR(R, DIVISOR, DIVIDE)
        return (Q)(D) ∨ R
    } else {
        C = common_cube(D)
        return LF(F, C, DIVISOR, DIVIDE)
    }
}
```

```
LF(F, C, DIVISOR, DIVIDE){
    l = best_literal(F, C) /* most common literal */
    (Q, R) = DIVIDE(F, l)
    C = common_cube(Q) /* largest common cube */
    Q = cube_free(Q)
    Q = GFACTOR(Q, DIVISOR, DIVIDE)
    R = GFACTOR(R, DIVISOR, DIVIDE)
    return lC(Q) ∨ R
}
```

Figure 16.14. Improved general structure of a factoring algorithm.

Example 16.39. *Given the SOP form* $H = ac \vee ad \vee ae \vee ag \vee bc \vee bd \vee be \vee bf \vee ce \vee cf \vee df \vee dg$, *we show the results of some factoring algorithms previously mentioned:*

- *LITERAL_FACTOR:* $H = a(c \vee d \vee e \vee g) \vee b(c \vee d \vee e \vee f) \vee c(e \vee f) \vee d(f \vee g)$.
- *QUICK_FACTOR:* $H = g(a \vee d) \vee (a \vee b)(c \vee d \vee e) \vee c(e \vee f) \vee f(b \vee d)$.
- *GOOD_FACTOR:* $H = (c \vee d \vee e)(a \vee b) \vee f(b \vee c \vee d) \vee g(a \vee d) \vee ce$.

An excellent discussion of factoring can be found in [25, 44].

Decomposition is similar to factoring, from which it differs because each divisor is added as a new node in the network and the associated variable is substituted into the function being decomposed. Figure 16.15 shows the scheme

$DECOMPOSE(\mathcal{N}, f_i)$ {
  $k = CHOOSE\_KERNEL(f_i)$
  if ($k = \emptyset$) return
  $f_{m \vee \ell} = k$ /* create new node $m \vee \ell$ */
  $f_i = (f_i/k)y_{m\vee\ell} \vee (f_i/\overline{k})\overline{y}_{m\vee\ell} \vee r$
  /* $\mathcal{N}'$ is modified network */
  $\mathcal{N}'' = DECOMPOSE(\mathcal{N}', f_i)$
  $\mathcal{N} = DECOMPOSE(\mathcal{N}'', f_{m\vee\ell})$
}

Figure 16.15. General structure of a decomposition algorithm.

of a decomposition algorithm, based on choosing kernels as candidate factors. Similar to factoring, we can define

- *QUICK_DECOMP*: pick a level 0 kernel.
- *GOOD_DECOMP*: pick the best kernel.

In general, factorization yields a tree network, because it uses factors only once, whereas decomposition yields a generic Boolean network, because it uses factors more than once.

16.5.2 Substitution

Substitution (or **resubstitution**) checks whether an existing function f_i at node i in the network is the divisor of another function f_j or $\overline{f}_j$ at node j. If f_j is a divisor of f_i, then f_i is transformed into

$$f_i = qy_j \vee r.$$

and similarly for $\overline{f}_j$. Division may be either algebraic or Boolean, the former being more common. In practice, this is tried for each node pair of the network; if there are n nodes in the network, then one tries about $2n^2$ divisions (substitution is tried for positive and negative phase). Figure 16.16 shows f_j being substituted in node f_i, and Figure 16.17 provides the operational scheme.

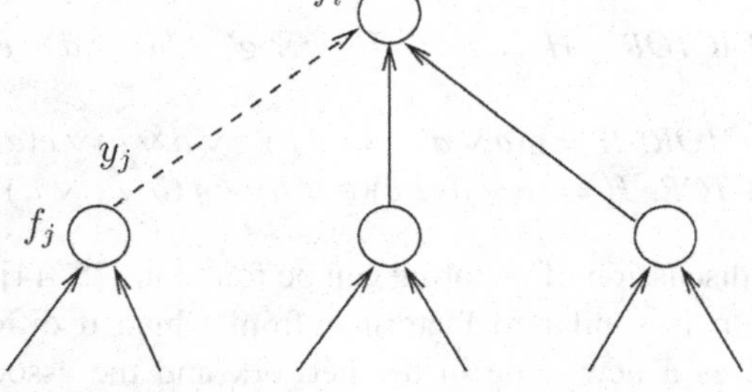

Figure 16.16. Illustration of substitution, where y_j (computed by node f_j) becomes a fan-in of node f_i.

```
SUBSTITUTE(N, f_i) {
    for each node f_j ≠ f_i {
        (h, r) = DIVIDE(f_i, f_j)
        if h ≠ 0 {
            f_i = y_i h ∨ r
            f_new = h /* create new node f_new */
        }
    }
}
```

Figure 16.17. General structure of a substitution algorithm.

Example 16.40. *Let us consider Boolean substitution of x into F.*

$$
\begin{aligned}
x &= ab \vee cd \vee e\\
F &= abf \vee \overline{a}cd \vee cdf \vee \overline{a}de \vee ef\\
D_1 &= \overline{x}(ab \vee cd \vee e) \vee x\overline{(ab \vee cd \vee e)}\\
F_1 &= xef \vee x\overline{a}cd \vee x\overline{a}de \vee xabf \vee xcdf\\
R_1 &= ab\overline{f}x \vee ae\overline{f}x \vee \overline{d}e\overline{f}x \vee \overline{a}\,\overline{c}\,\overline{e}\,\overline{x} \vee\\
&\quad \overline{b}\,\overline{c}\,\overline{e}\,\overline{x} \vee \overline{a}\,\overline{d}\,\overline{e}\,\overline{x} \vee \overline{b}\,\overline{d}\,\overline{e}\,\overline{x} \vee acd\overline{f}x\\
J &= \{a, d, f, x\} \text{ (minimum literal support)}\\
F_3 &= xf \vee x\overline{a}d \vee x\overline{a}d \vee xaf \vee xdf\\
F_4 &= xf \vee x\overline{a}d\\
H_1 &= f \vee \overline{a}d\\
F &= x(f \vee x\overline{a}d) = (ab \vee cd \vee e)(f \vee \overline{a}d)
\end{aligned}
$$

For efficiency, we use filters to prevent trying a division. The cover G is not an algebraic divisor of F if any of the following is true:

(1) G contains a literal not in F.
(2) G has more terms than F.
(3) For any literal, its count in G exceeds that in F.
(4) F is in the transitive fan-in of G.

16.5.3 Extraction

Extraction identifies common subexpressions that become new nodes in the Boolean network. Common subexpressions are multiple-cube common divisors obtained by computing the kernels of the functions in the network. The best kernel intersection may be chosen as the new factor; the cost function to evaluate the value of a kernel intersection may be the number of literals in the factored form for the network. Because exhaustive kerneling may be expensive, a good compromise is to look only for double-cube divisors. So the steps of extraction are as follows (see also Figure 16.18):

(1) Find **all** kernels of **all** functions.
(2) Choose one with best "value."

$EXTRACT(\mathcal{N})$ {
    repeat {
        for each node f_i {
            $\mathcal{N} = DECOMPOSE(\mathcal{N}, f_i)$
        }
        for each node f_i {
            $\mathcal{N} = SUBSTITUTE(\mathcal{N}, f_i)$
        }
        $\mathcal{N} = ELIMINATE(\mathcal{N})$ /* eliminate nodes with small value */
    } until (cost function does not improve)
}

Figure 16.18. General structure of an extraction algorithm.

(3) Create a new node with this as function.
(4) Algebraically substitute the new node everywhere.
(5) Repeat 1,2,3,4 until the best value is less or equal than a given threshold.

We can combine decomposition and substitution to provide an effective extraction algorithm. Figure 16.19 shows the extraction of a factor common out of three nodes, and the creation of a new node realizing the function of the selected factor.

Example 16.41. *An example of extraction follows:*

$$F_1 = ab(c(d \vee e) \vee f \vee g) \vee h$$
$$F_2 = ai(c(d \vee e) \vee f \vee j) \vee k.$$

Kernel extraction: $\mathcal{K}^0(F_1) = \mathcal{K}^0(F_2) = \{d \vee e\}$

$$l = d \vee e$$
$$F_1 = ab(cl \vee f \vee g) \vee h$$
$$F_2 = ai(cl \vee f \vee j) \vee k.$$

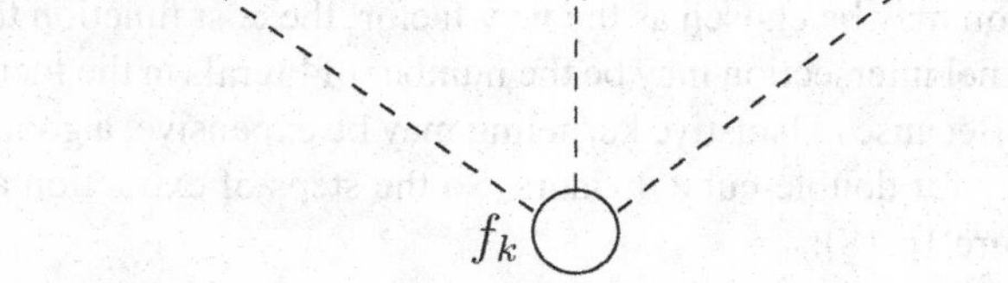

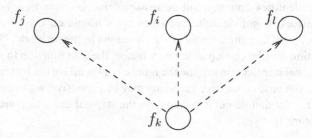

Figure 16.19. Illustration of extraction of node f_k out of nodes f_j, f_i, and f_l.

Kernel extraction: $\mathcal{K}^0(F_1) = \{cl \vee f \vee g\}$, $\mathcal{K}^0(F_2) = \{cl \vee f \vee j\}$, $\mathcal{K}^0(F_1) \cap \mathcal{K}^0(F_2) = cl \vee f$

$$m = cl \vee f$$
$$F_1 = ab(m \vee g) \vee h$$
$$F_2 = ai(m \vee j) \vee k.$$

Kernel extraction: *no kernel intersections at this point.*
Cube extraction: $\{am\}$

$$n = am$$
$$F_1 = b(n \vee ag) \vee h$$
$$F_2 = i(n \vee aj) \vee k.$$

16.6 Conclusions

In this chapter we covered the foundations of algebraic and Boolean division in multi-level logic synthesis. Variants of division have been proposed in the literature, e.g., algebraic division where the distributive law is augmented with annihilation ($a\overline{a} = 0$) and idempotency ($aa = a$), used in the system M32 [28], and an analogous proposal called coalgebraic division [26]; a form of constrained Boolean division, called extended algebraic division used in the system GLSS [12, 27], but many more could be listed. A way to perform Boolean division by redundancy addition and removal has been presented in [13].

To keep the chapter short we did not delve into the theory of optimal and near-optimal factored forms [9–11, 44].

Moreover, we altogether skipped rectangle covering, which refers to an elegant and useful reduction of common-cube extraction and kernel intersection extraction to the problem of covering a matrix with rectangles [6]. Rectangles in a matrix provide an alternate way of interpreting the kernels of a Boolean function, for example, given $F = ab\overline{d} \vee acd \vee bcd$, then its related cube-literal matrix

	a	b	c	d	$\overline{d}$
$ab\overline{d}$	1	1	0	0	1
acd	1	0	1	1	0
bcd	0	1	1	1	0

has the property that an arbitrary cube is a cokernel of F if and only if it is the cube corresponding to a prime rectangle involving at least two rows of the cube-literal matrix of F. The co-rectangle of the prime rectangle identifies the kernel. A prime rectangle is a rectangle of 1s not contained in any other rectangle. In this case, the prime rectangle $(\{1, 2\}, \{1\})$ corresponds to cokernel a and kernel $b\overline{d} \vee cd$, the prime rectangle $(\{1, 3\}, \{2\})$ corresponds to cokernel b and kernel $a\overline{d} \vee cd$, and the prime rectangle $(\{2,3\},\{3,4\})$ corresponds to cokernel cd and kernel $a \vee b$ (the trivial kernel F is not accounted by this interpretation).

Finally, it must be mentioned that a major effort in multilevel logic synthesis has been devoted to node simplification, by exploiting the don't-care set that expresses the restricted controllability and observability of a node in the network. Node minimization uses global network information, but the quality of the final result is affected by the starting point, obtained by preliminary network restructuring based on algebraic and Boolean division. To do justice to this subject would take another chapter; see [1, 4, 5].

Acknowledgments

The authors thank Prof. Yves Crama for very careful readings of and discussions on the drafts of this chapter, which helped a great deal in improving the text.

References

[1] D. Bostick, G. Hachtel, R. Jacoby, M. Lightner, P. Moceyunas, C. Morrison, and D. Ravenscroft. The Boulder optimal logic design system. In *Proceedings of the International Conference on Computer-Aided Design*, pp. 62–5, 1987.

[2] R. Brayton, G. Hachtel, C. McMullen, and A. Sangiovanni-Vincentelli. *Logic Minimization Algorithms for VLSI Synthesis*. Kluwer Academic, Dordrecht, 1984.

[3] R. Brayton and C. McMullen. The decomposition and factorization of Boolean expressions. In *Proceedings of the International Symposium on Circuits and Systems*, pp. 49–54, 1982.

[4] R. Brayton, R. Rudell, A. Sangiovanni-Vincentelli, and A. Wang. MIS: A multiple-level logic optimization system. *IEEE Transactions on Computer-Aided Design*, pp. 1062–81, 1987.

[5] R. Brayton, A. Sangiovanni-Vincentelli, and G. Hachtel. Multilevel logic synthesis. *Proceedings of the IEEE*, 78(2), pp. 264–300, 1990.

[6] R. K. Brayton, R. Rudell, A. Sangiovanni-Vincentelli, and A. Wang. Multi-level logic optimization and the rectangular covering problem. In *Proceedings of the International Conference on Computer-Aided Design*, pp. 66–9, 1987.

[7] R. K. Brayton. Factoring logic functions. *IBM Journal of Research and Development* 31(2), pp. 187–98, 1987.

[8] R. K. Brayton and C. McMullen. Synthesis and optimization of multi-stage logic. In *International Conference on Computer Design*, pp. 23–8, 1984.

[9] G. Caruso. Near optimal factorization of Boolean functions. *IEEE Transactions on Computer-Aided Design* 10(8), pp. 1072–8, 1991.

[10] G. Caruso. Boolean factoring of logic functions. *Proceedings of the 36th Midwest Symposium on Circuits and Systems* 2, pp. 1312–15, 1993.

[11] G. Caruso. An improved algorithm for Boolean factoring. *ISCAS '94, IEEE International Symposium on Circuits and Systems* 1, pp. 241–4, 1994.

[12] G. Caruso. An algorithm for exact extended algebraic division. *IEICE Transactions on Fundamentals of Electronics, Communications and Computer Sciences*, E86-A(2), pp. 462–71, 2003.

[13] S.-C. Chang and D. I. Cheng. Efficient Boolean division and substitution using redundancy addition and removing. *IEEE Transactions on Computer-Aided Design of Integrated Circuits and Systems* 18(8), pp. 1096–1106, 1999.

[14] O. Coudert. Two-level logic minimization: an overview. *Integration* 17(2), pp. 97–140, 1994.
[15] Y. Crama and P. L. Hammer. *Boolean Functions: Theory, Algorithms, and Applications*. Cambridge University Press, Cambridge, UK, 2010.
[16] J. A. Darringer, D. Brand, J. V. Gerbi, W. H. Joyner Jr., and L. Trevillyan. LSS: A system for production logic synthesis. *IBM Journal of Research and Development* 28(5), pp. 537–45, 1984.
[17] J. A. Darringer, D. Brand, J. V. Gerbi, W. H. Joyner, Jr., and L. Trevillyan. LSS: A system for production logic synthesis. *IBM Journal of Research and Development* 44(1), pp. 157–66, 2000.
[18] J. A. Darringer, W. H. Joyner, Jr., C. L. Berman, and L. Trevillyan. Logic synthesis through local transformations. *IBM Journal of Research and Development* 25(4), pp. 272–80, 1981.
[19] E. S. Davidson. An algorithm for NAND decomposition under network constraints. *IEEE Transactions on Computers*, C-18(12), pp. 1098–1109, 1969.
[20] R.C. de Vries. Comment on Lawler's multilevel Boolean minimization. *Communications of the ACM* 13(4), pp. 265–6, 1970.
[21] S. Devadas, A. Ghosh, and K. Keutzer. *Logic Synthesis*. McGraw-Hill, New York, 1994.
[22] S. Devadas, A. Wang, R. Newton, and A. Sangiovanni-Vincentelli. Boolean decomposition in multilevel logic optimization. *IEEE Journal of Solid-State Circuits*, pp. 399–408, 1989.
[23] D. Dietmeyer and Y. Su. Logic design automation of fan-in limited NAND networks. *IEEE Transactions on Computers* C-18(11), pp. 11–22, 1969.
[24] M. Fujita, Y. Matsunaga, and M. Ciesielski. Multi-level logic optimization. In R. Brayton, S. Hassoun, and T. Sasao, eds., pp. 29–63. Kluwer, Dordrecht, 2001.
[25] G. Hachtel and F. Somenzi. *Logic synthesis and verification algorithms*. Kluwer Academic, Dordrecht, 1996.
[26] W.-J. Hsu and W.-Z. Shen. Coalgebraic division for multilevel logic synthesis. In *DAC '92: Proceedings of the 29th ACM/IEEE Conference on Design Automation*, pp. 438–42, 1992.
[27] B.-G. Kim and D. L. Dietmeyer. Multilevel logic synthesis with extended arrays. *IEEE Transactions on Computer-Aided Design* 11(2), pp. 142–57, 1992.
[28] V. N. Kravets and K. A. Sakallah. M32: A constructive multilevel logic synthesis system. In *DAC '98: Proceedings of the 35th Conference on Design Automation*, pp. 336–341, 1998.
[29] A. Kuehlmann, V. Paruthi, F. Krohm, and M. Ganai. Robust Boolean reasoning for equivalence checking and functional property verification. *IEEE Transactions on Computer-Aided Design* 21(12), pp. 1377–94, 2002.
[30] E. L. Lawler. An approach to multilevel Boolean minimization. *Journal of the ACM* 11(3), pp. 283–95, 1964.
[31] P. McGeer, J. Sanghavi, R. Brayton, and A. Sangiovanni-Vincentelli. ESPRESSO-SIGNATURE: A new exact minimizer for logic functions. *IEEE Transactions on VLSI Systems* 1(4), pp. 432–40, 1993.
[32] P. C. McGeer and R. K. Brayton. Efficient, stable algebraic operations on logic expressions. In *Proceedings of International Conference on VLSI*, August 1987.
[33] P. C. McGeer and R. K. Brayton. Efficient prime factorization of logic expressions. In *Proceedings of the Design Automation Conference*, pp. 221–5, 1989.
[34] G. De Micheli. *Synthesis and Optimization of Digital Circuits*. McGraw-Hill, New York, 1994.
[35] A. Mishchenko, S. Chatterjee, and R. Brayton.DAG-aware AIG rewriting – A fresh look at combinational logic synthesis. In *Proceedings of the Design Automation Conference*, pp. 532–6, 2006.
[36] A. Mishchenko, S. Chatterjee, R. Jiang, and R. Brayton. FRAIGs: A unifying representation for logic synthesis and verification. ERL Tech. Rep., EECS Dept., UCB, Berkeley, CA, March 2005.

[37] J. Rajski and J. Vasudevamurthy. Testability preserving transformations in multi-level logic synthesis. In *Proceedings of the International Test Conference*, pp. 265–73, 1990.
[38] R. Rudell and A. Sangiovanni-Vincentelli. Multiple-valued minimization for PLA optimization. *IEEE Transactions on Computer-Aided Design* CAD-6, pp. 727–50, 1987.
[39] R. Rudell. *Logic synthesis for VLSI design*. PhD thesis, University of California, Berkeley, April 1989. Tech. Report No. UCB/ERL M89/49.
[40] E. Sentovich, K. Singh, L. Lavagno, C. Moon, R. Murgai, A. Saldanha, H. Savoj, P. Stephan, R. Brayton, and A. Sangiovanni-Vincentelli. SIS: A system for sequential circuit synthesis. Tech. Rep. No. UCB/ERL M92/41, Berkeley, CA, May 1992.
[41] E. Sentovich, K. Singh, C. Moon, H. Savoj, R. Brayton, and A. Sangiovanni-Vincentelli. Sequential circuit design using synthesis and optimization. In *Proceedings of the International Conference on Computer Design*, pp. 328–33, 1992.
[42] Berkeley Logic Synthesis and Verification Group. ABC: A system for sequential synthesis and verification. Release 61225, 2009. Available at *http://www-cad.eecs.berkeley.edu/~alanmi/abc*.
[43] J. Vasudevamurthy and J. Rajski. A method for concurrent decomposition and factorization of Boolean expressions. In *International Conference on Computer Aided Design*, pp. 510–13, 1990.
[44] A. Wang. *Algorithms for multi-level logic optimization*. PhD thesis, University of California, Berkeley, April 1989. Tech. Report No. UCB/ERL M89/50.
[45] G. Whitcomb. Exact factoring of incompletely specified functions. EE290LS Class Project Report, EECS Dept., UCB, May 1988.

17

Boolean Aspects of Network Reliability

Charles J. Colbourn

17.1 Introduction

We explore network reliability primarily from the viewpoint of how the combinatorial structure of operating or failed states enables us to compute or bound the reliability. Many other viewpoints are possible, and are outlined in [7, 45, 147]. Combinatorial structure is most often reflected in the simplicial complex (hereditary set system) that represents all operating states of a network; most of the previous research has been developed in this vernacular. However, the language and theory of Boolean functions has played an essential role in this development; indeed these two languages are complementary in their utility to understand the combinatorial structure. As we survey exact computation and enumerative bounds for network reliability in the following, the interplay of complexes and Boolean functions is examined.

In classical reliability analysis, failure mechanisms and the causes of failure are relatively well understood. Some failure mechanisms associated with network reliability applications share these characteristics, but many do not. Typically, component failure rates are estimated based on historical data. Hence, time-independent, discrete probability models are often employed in network reliability analysis. In the most common model, network components (nodes and edges, for example) can take on one of two states: operative or failed. The state of a component is a random event that is independent of the states of other components. Similarly, the network itself is in one of two states, operative or failed. The reliability analysis problem is to compute a measure of network reliability given the probabilities that each component is operative. The simple two-state model is sufficient for the consideration of connectivity measures. In the two-state model, the component's probability of operation or, simply, reliability could have one of several possible interpretations. The most common interpretations are the component's availability or the component's reliability. We use the term *reliability* to mean the probability that a component or system operates.

The input to network reliability analysis problems includes a network $G = (V, E)$, where V is a set of nodes and E is a set of undirected edges or a set of directed arcs. For connectivity measures, for each $e \in E$, p_e, the probability that e operates is input. In the general system reliability context, a system is made up of a set of components and a random variable, X_e, associated with each component, e. The value of X_e indicates the current performance level of e; the expected performance of the system is a function of the X_e values. In the network reliability context, the system is the network and the components are (typically) arcs or edges. A function Φ maps the states of the components into system states. Thus, $\Phi(\mathbf{x})$ is a random variable that provides information on the overall performance of the system. A variety of system reliability measures may be defined in terms of Φ.

We concentrate on two-state system reliability, in which the system is either operating or failed. In this case, suppose that T is the set of components. Then the reliability is completely defined by the family $\mathcal{P}$ of subsets of T, so that $P \in \mathcal{P}$ if the system operates when all components of P operate and all components of $T \setminus P$ fail. This *set system* representation has been widely examined.

A *stochastic binary system* (SBS) provides an equivalent general model in the reliability setting. In an SBS, each component in the component set, $T = \{1, 2, \ldots, m\}$, can take on either of two states: operative or failed. X_e has value 1 if e operates and 0 if e fails. Φ maps a binary component state vector $\mathbf{x} = (x_1, x_2, \ldots, x_m)$ into the system state by

$$\Phi(\mathbf{x}) = \begin{cases} 1 \text{ if } \mathbf{x} \text{ is an operating system state} \\ 0 \text{ if } \mathbf{x} \text{ is a failed system state.} \end{cases}$$

This leads to a third vernacular. Let ψ_i be a Boolean variable that represents the state of edge i. For each state $\mathbf{x} = (x_1, \ldots, x_m)$, define the conjunction $\Psi(\mathbf{x}) = \bigwedge_{x_i=1} \psi_i \bigwedge_{x_i=0} \overline{\psi}_i$. Consider the *Boolean function* $\Upsilon = \bigvee_{\Phi(\mathbf{x})=1} \Psi(\mathbf{x})$. The expression Υ is written here as a disjunction of conjunctive formulas, and hence is in *disjunctive normal form* (DNF). There are numerous sensible alternative ways to write an expression that is logically equivalent to Υ, and some offer either conceptually different or computationally more efficient formulations. One measure of the complexity is the size of the Boolean formula. The *size* is defined recursively as follows: $size(0) = size(1) = 0$; $size(\Upsilon) = 0$ if $\Upsilon = \gamma_i$ for some i with $1 \le i \le m$; $size(\overline{\Upsilon}) = size(\Upsilon) + 1$; $size(\bigvee_{i=1}^m \Upsilon_i) = size(\bigwedge_{i=1}^m \Upsilon_i) = m - 1 + \sum_{i=1}^m size(\Upsilon_i)$. When Υ is evaluated, its size indicates the number of Boolean operations of negation, conjunction, and disjunction required. As we shall see, however, the number of operations employed in evaluating Υ and the number employed in evaluating a reliability function that it supports can differ significantly.

An SBS is *coherent* if $\Phi(\mathbf{1}) = 1$, $\Phi(\mathbf{0}) = 0$ and $\mathbf{x}^1 \ge \mathbf{x}^2$ implies $\Phi(\mathbf{x}^1) \ge \Phi(\mathbf{x}^2)$. The third property implies that the failure of any component can only have a detrimental effect on the operation of the system. Thus, the set system $\mathcal{P}$ that is equivalent to Φ has the property that whenever $P \in \mathcal{P}$, every set P' satisfying $P \subseteq P' \subseteq T$ satisfies $P' \in \mathcal{P}$. In other words, $\mathcal{Q} = \{T \setminus P : P \in \mathcal{P}\}$ is closed under taking subsets – it is a *hereditary set system*, or equivalently a *simplicial complex*. Often this is abbreviated simply to *complex*.

A Boolean function with the property that the change of any set of Boolean variables from false to true does not change the truth value of Υ from true to false is *positive*. When $\mathbf{x}^1 \geq \mathbf{x}^2$, this implies that $\Psi(\mathbf{x}^1) \geq \Psi(\mathbf{x}^2)$. When Υ is a positive Boolean function, we can equally well write $\Upsilon = \bigvee_{\Phi(\mathbf{x})=1} \bigwedge_{x_i=1} \psi_i$ and omit the complemented variables; hence the term *positive*.

The fundamental computational problem for a set system $\mathcal{P}$ is to compute

$$Rel(\Phi, \mathbf{p}) = \Pr[\Phi(\mathbf{x}) = 1]$$

given some representation of $\Phi()$ of $\mathcal{P}$. For example, Φ could be represented as a Boolean formula Υ or as a graph. At times we consider reliability problems where $p_e = p$ for all e, in which case we replace $\mathbf{p}$ by p in the foregoing notation. For any complex, define a *pathset* as a set of components whose operation implies system operation, and a *minpath* as a minimal pathset. In the Boolean setting, when $\mathbf{x}^1 \geq \mathbf{x}^2$ and $\Psi(\mathbf{x}^2)$ is true, $\mathbf{x}^2$ is an *implicant* of $\mathbf{x}^1$. Thus, pathsets are implicants. An implicant $\mathbf{x}^1$ is *prime* if whenever $\mathbf{x}^1 \geq \mathbf{x}^2$ and $\Psi(\mathbf{x}^2)$ is true, $\mathbf{x}^1 = \mathbf{x}^2$. Thus, minpaths are *prime implicants* of the positive Boolean function Υ.

Dually, a *cutset* is a set of components whose failure implies system failure, and a *mincut* is a minimal cutset. Complements of cutsets form a complex exactly when complements of pathsets do; indeed, every cutset forms a transversal of all pathsets, and every pathset forms a transversal of all cutsets. In the Boolean setting, then, cutsets and mincuts correspond to implicants and prime implicants of $\overline{\Upsilon}$, the function that is true for all failed network states. The function $\overline{\Upsilon}$ is negative (i.e., $\mathbf{x}1 \geq \mathbf{x}2$ implies $\overline{\Upsilon}(\mathbf{x}1) \leq \overline{\Upsilon}(\mathbf{x}2)$) exactly when Υ is positive.

Network reliability measures are either the probability of certain random events or the expected value of certain random variables. The majority of the research in network reliability is devoted to connectivity measures, specifically to the *k-terminal measure*. A set of nodes K and a node $s \in K (k = |K|)$ are given. Given a network G and an edge reliability vector $\mathbf{p}$, the k-terminal reliability measure is defined as

$$Rel(G, s, K, \mathbf{p}) = \Pr[\text{there exist operating paths from } s \text{ to each node in } K].$$

Two important special cases of the measure are the *two-terminal measure* for which $|K| = 1$ and the *all-terminal measure* for which $K = V$. The two-terminal and all-terminal measures are denoted by $Rel_2(G, s, t, \mathbf{p})$ and $Rel_A(G, s, \mathbf{p})$, respectively. Node s is the *source* node, and the nodes in $K \setminus \{s\}$ are the *terminals*. When the appropriate values are obvious from the context, we may leave one or more of the arguments out of the $Rel()$ notation. Of course, other connectivity measures have been analyzed (see, for example, [6, 46]), but for concreteness we focus on the edge-based measures defined here.

In the Boolean perspective, two main issues arise. Among the multitude of equivalent Boolean expressions that all represent the network reliability, we prefer succinctness (small size of the formula) but also ease of evaluation. With the latter in mind, the evaluation of a DNF formula in which two or more of the conjuncts can hold simultaneously necessitates the determination of the event in which more than one hold, and thus the brevity of a formula need not result in faster evaluation. A natural goal, then, is to insist that the conjuncts themselves represent disjoint events. A DNF formula is *orthogonal* when at most one conjunct can be true for a given state. Crama and Hammer [52] describe orthogonalization of Boolean formulas; we review related work specific to reliability here.

We begin with a summary of the computational complexity of reliability problems, which is necessary to justify much of the research on exact methods and bounds. For exact methods, we focus on connections with Boolean methods (see [52] for an excellent introduction); for bounds we focus on structure associated with the simplicial complexes arising in reliability problems. These two topics are intimately related [26], but the connections deserve much more exploration; hence one of our motivations is presenting these two directions in a complementary manner. The structure of complexes arising in all-terminal reliability has been further elucidated by examining a conceptually simple game on graphs, the chip firing game. We outline recent developments in this area, demonstrating their application to bounding all-terminal reliability. We close with computational results to demonstrate the practical value of obtaining these improved characterizations of the complexes involved.

17.2 Computational Complexity

The computational problems most often studied are recognition problems and optimization problems. Reliability analysis problems are somewhat different in that they involve "counting." Consequently, the analysis of their complexity involves concepts extending the classes P, NP, and NP-complete used to analyze recognition and optimization problems.

We consider the special case of the reliability analysis problem that arises when all individual component reliabilities are equal, that is, $p_e = p$ for all components e. In this case, $Rel(\Phi, p)$ can written as a polynomial in p with the following form:

$$Rel(\Phi, p) = \sum_{i=0}^{m} F_i p^{m-i}(1-p)^i.$$

This polynomial is the *reliability polynomial*. The *functional reliability analysis problem* takes as input a representation of Φ and produces as output the vector $\{F_i\}$. The *rational reliability analysis problem* is to output the evaluation of $Rel(\Phi, p)$ when p is a given rational number.

The general term in the reliability polynomial, $F_i p^{m-i}(1-p)^i$, is the probability that exactly $m-i$ components operate and the system operates. Thus, we can interpret F_i as the number of operating system states having i failed components or, more precisely:

$$F_i = |\{\mathbf{x} : \textstyle\sum_k x_k = m - i \text{ and } \Phi(\mathbf{x}) = 1\}|.$$

Determining each of the coefficients F_i is a counting problem. Whereas the output of a recognition problem is "yes" or "no," the output of a counting problem is the number of distinct solutions. NP and NP-complete are classes of recognition problems. The corresponding classes of counting problems are #P and #P-complete. The counting problem is at least as hard as the corresponding recognition problem, and hence counting versions of NP-complete problems are trivially NP-hard. However, certain recognition problems are solvable in polynomial time despite their corresponding counting problems being #P-complete [162]. We simply classify problems as NP-hard or polynomial.

Given a complex, define:

$$\begin{aligned} m &= \text{number of components in the system,} \\ c &= \text{cardinality of a minimum cardinality cutset,} \\ C_c &= \text{number of minimum cardinality cutsets,} \\ \ell &= \text{cardinality of a minimum cardinality pathset,} \\ N_\ell &= \text{number of minimum cardinality pathsets.} \end{aligned}$$

In general, N_i is the number of pathsets of cardinailty i, whereas C_i is the number of cutsets of cardinality i. The coefficients of the reliability polynomial satisfy

$$\begin{aligned} F_i &= N_{m-i} \quad \text{for} \quad i = 0, 1, \cdots, m, \\ F_i + C_i &= \tbinom{m}{i} \quad \text{for} \quad i = 0, 1, \cdots, m, \\ F_i &= \tbinom{m}{i} \quad \text{for} \quad i < c, \\ F_c &= \tbinom{m}{c} - C_c \\ F_{m-\ell} &= N_\ell \\ F_i &= 0 \quad \text{for} \quad i > m - \ell. \end{aligned} \tag{17.1}$$

The reliability polynomial determines basic properties of the complex. For any complex, the functional and rational reliability analysis problems are NP-hard if one or more of the minimum cardinality pathset recognition problem, the minimum cardinality pathset counting problem, the minimum cardinality cutset recognition problem, the minimum cardinality cutset counting problem, or the problem of determining a general coefficient of the reliability polynomial are NP-hard.

We now consider the complexity of network reliability for the k-terminal, 2-terminal, and all-terminal operations. A minimum cardinality pathset for the k-terminal measure is a minimum cardinality Steiner tree. The associated recognition problem is NP-hard for both directed and undirected networks [83], so the associated functional and rational reliability analysis problems are NP-hard. Valiant [162] gives an alternate proof of this result by showing that computing $SN(K) = \sum F_i = |\{S : S \text{ is a subgraph that contains a path to each node in } K\}|$ is NP-hard. Here K is the set of terminals.

The minimum cardinality pathset and cutset recognition problems for 2-terminal are the shortest path and minimum cut problems; polynomial algorithms are known for both ([111] and [65]). Valiant [162] first showed that the 2-terminal reliability analysis problems are NP-hard. Provan and Ball [127] give

Table 17.1. Complexity results related to k-terminal, 2-terminal, and all-terminal reliability analysis (* $\Rightarrow$ polynomial, ! $\Rightarrow$ NP-hard).

	k-Term.	2-Term.	All-Term. Dir.	All-Term. Undir.
Min. card. pathset rec.	! [83]	* [111]	*	*
Min. card. cutset rec.	* [64, 65]	* [64, 65]	* [64, 65]	* [64, 65]
Min. card. pathset count.	! [83]	* [10]	* [91]	* [91]
Min. card. cutset count.	! [127]	! [127]	! [10]	* [10, 21]
Gen. term. count.	! [83]	! [162]	! [127]	! [127]

an alternate proof by showing the problem of determining the number of minimum cardinality s, t-cuts is NP-hard.

For the directed all-terminal measure (*reachability*), the minimum cardinal pathset and cutset problems are the minimum cardinality spanning arborescence and minimum cardinality s-directed cut problems; both are polynomially solvable [60]. Provan and Ball [127] showed that counting minimum cardinality s-directed cuts is NP-hard, so the associated reliability analysis problems are NP-hard. For the undirected case, the minimum cardinality pathset and cutset recognition and counting problems are all polynomially solvable. Nevertheless, Provan and Ball [127] showed that computing a general term in the reliability polynomial is NP-hard, so the undirected reliability analysis problems are NP-hard.

Table 17.1 summarizes the known complexity results for the five counting and recognition problems for the k-terminal, all-terminal, and 2-terminal problems.

In light of these negative results, much research has concerned the analysis of structured networks. The widest class of networks known to be solvable in polynomial time involve series-parallel graphs and certain generalizations. Recent research has addressed the complexity of reliability analysis over structured networks, specifically directed acyclic networks and planar networks. In [125], Provan shows that the undirected two-terminal reliability problem remains NP-hard over planar networks having node degrees bounded by 3, and the directed 2-terminal reliability analysis problems remain NP-hard over acyclic planar networks having node degrees bounded by 3. Vertigan [165] has shown that directed and undirected all-terminal reliability analysis problems are NP-hard when restricted to planar networks. There is a simple formula for the directed all-terminal reliability analysis problem of acyclic networks [10].

Polynomial time algorithms are only likely to exist for network reliability problems restricted to small classes of networks. Thus, much research has been devoted to network reliability bounds that can be calculated efficiently.

17.3 Exact Computation of Reliability

For general networks, all of the reliability measures considered here are #P-complete. This justifies the study of exponential time exact algorithms. There exist graph transformations that leave the values of various reliability measures unchanged, and these can often be used to simplify the network used in the exact computation of reliability.

We focus on the undirected problems here. An edge or arc e is *mandatory* if it appears in *every* minpath , and is *irrelevant* it it appears in no minpath. In the equivalent Boolean function Υ, e is irrelevant when its variable ψ_e is *inessential,* that is, appears in no prime implicant of Υ; it is mandatory when $\overline{\psi_e}$ is a prime implicant of $\overline{\Upsilon}$.

After irrelevant edges have been deleted, any bridge (edge cutset of size 1) that remains is mandatory. Let $G = (V, E)$ with terminal set $K \subseteq V$, and bridge $e \in E$ with operation probability p_e. The *contraction* $G \cdot e$ of an edge $e = \{x, y\}$ in G is obtained by removing e, identifying x and y, and making the resulting node a terminal whenever $K \cap \{x, y\} \neq \emptyset$. The *deletion* $G - e$ of an edge e is the graph obtained from G by simply removing the edge e. The reliability of G, $Rel(G)$, satisfies $Rel(G) = p_e Rel(G \cdot e)$ when e is a mandatory edge. Thus, the mapping from G to $G \cdot e$ is a reliability-preserving transformation *with multiplicative factor* p_e.

Two edges e, f having the same endpoints are in *parallel.* Because any minpath contains at most one of the two, and interchanging e and f is a natural bijection between the minimal pathsets containing one and the minpaths containing the other, the replacement of e and f by a single edge g having $p_g = 1 - (1 - p_e)(1 - p_f)$ is reliability-preserving. This is a *parallel reduction*. The notion of parallel reductions can be generalized when e and f are "substitutes"; see [7]. Indeed, for a positive Boolean function Υ, when for every Boolean formula f not involving two variables x and y, we find that $f \vee x$ is a prime implicant of Υ if and only if $f \vee y$ is, then x and y are substitutes.

Two edges $e = \{x, y\}$ and $f = \{y, z\}$ are in *series* when y is a node of degree 2. In this case, any mincut contains at most of e or f, and interchanging e and f is a natural bijection between the mincuts containing e and those containing f. Thus, a reliability-preserving transformation is obtained by removing the node y and the edges e, f and adding the edge $g = \{x, z\}$ with $p_g = p_e p_f$ *provided* that y is not a terminal node. This is a *series reduction*. More generally, when two edges are "complements," similar reductions can be applied; see [7]. Indeed, for a positive Boolean function Υ, when every prime implicant f of Υ has both or neither of x and y as implicants, then x and y are complements.

More sophisticated transformations have been studied extensively (see [45], for example).

17.3.1 State-Based Methods

When reliability-preserving transformations fail to reduce the network into a restricted class for which an efficient exact method is known, we are forced to resort to potentially exponential time methods. The first main class of these exact methods examines the possible states of the network.

A *state* of a network $G = (V, E)$ is a subset $S \subseteq E$ of operational edges. The conceptually simplest exact algorithm is *complete state enumeration*. Let $\mathcal{O}$ be the set of all operational states (pathsets). Then

$$Rel(G) = \sum_{S \in \mathcal{O}} \prod_{e \in S} p_e \prod_{e \notin S} (1 - p_e).$$

By generating all states and determining which is operational, the reliability is "easily" (but not efficiently) computed.

Of course, large groups of the states are easily seen to be operational without listing them out one by one. Hence, an immediate improvement is obtained by generating the states in a more clever way. A basic ingredient in this is the factoring theorem:

$$Rel(G) = p_e Rel(G \cdot e) + (1 - p_e) Rel(G - e)$$

for any edge e of G. *Factoring*, also called *pivotal decomposition*, was explicitly introduced in the reliability literature by Moskowitz [113] and Mine [108]. The corresponding operation on the Boolean function is the *Shannon expansion*, named after Shannon [145]. Factoring carried out until the networks produced have no edges is just complete state enumeration. However, some simple observations result in improvements. When $G - e$ is failed, any sequence of contractions and deletions results in a failed network, and hence there is no need to factor $G - e$. Moreover, although we may be unable to simplify G with a reliability-preserving transformation, we may well be able to simplify $G \cdot e$ or $G - e$.

Factoring with elimination of irrelevant edges, contraction of mandatory edges, and series, parallel, and degree-2 reductions forms the basis of many exact algorithms in the literature [137, 138, 141, 172, 174]. Satyanarayana and Chang [141] analyzed the number of times a factoring step is required in an optimal factoring strategy (see also [79, 173]). For complete graphs, complete state enumeration examines $2^{\binom{n}{2}}$ states, whereas a factoring algorithm using series and parallel reductions examines only $(n - 1)!$.

We state the factoring method more explicitly here:

procedure **factor** (**graph** G);
    apply reliability-preserving transformations to G to
        delete irrelevant edges, contract mandatory edges,
        apply series, parallel, and (perhaps) other reductions,
        maintaining G with the current probability on each edge and
        maintaining a multiplicative factor **mult** that results from the
        reductions;
    if G has only one terminal remaining **return(mult)**
    else
        select an edge e of G
        return(mult*(p_e***factor** $(G \cdot e) + (1 - p_e)$***factor**$(G - e)$))
end

Further improvements are possible by partitioning the graph into its biconnected or triconnected components at each step [174]. Despite these improvements, little of the network structure is employed. Whereas the number of states is exponential in the number of components (edges), the number of minpaths or mincuts may be much smaller and either serve to determine the structure completely. In analogy with Boolean functions, state-based methods examine all implicants when concentrating on the prime implicants suffices.

17.3.2 Path- and Cut-Based Methods

Once basic reductions are done, complete state enumeration could be applied to finish the computation. Unless the reductions have succeeded in dramatically reducing the size of the graph, however, this remains impractical. It may nevertheless be possible to generate all minpaths of the network.

17.3.2.1 Orthogonalization

Let $P_1, \ldots, P_h$ be an enumeration of the minpaths, and let E_i be the event that all edges/arcs in minpath P_i are operational. Unfortunately, the $\{E_i\}$ are not disjoint events, and hence we cannot simply sum their probabilities of occurrence. To be specific, $\Pr[E_1 \text{ or } E_2]$ is $\Pr[E_1] + \Pr[E_2] - \Pr[E_1 \text{ and } E_2]$. Equivalently, the corresponding Boolean function Υ is not orthogonal in general, because two of the conjuncts in $\bigvee_{i=1}^{h} \bigwedge_{e_i \in P_i} \psi_i$ may be true simultaneously. Every Boolean function can be written as an orthogonal formula in disjunctive normal form (in many ways, typically).

We examine the strategy of forming a set of disjoint events, termed *orthogonalization*. Let $\overline{E_i}$ denote the complement of event E_i. Now define the event $D_1 = E_1$, and in general, $D_i = \overline{E_1} \cap \overline{E_2} \cap \cdots \cap \overline{E_{i-1}} \cap E_i$. The events D_i are disjoint, and hence are often called "disjoint product" events. Moreover, $Rel(G) = \sum_{i=1}^{h} \Pr[D_i]$. In employing this approach, one must obtain a formula for $\Pr[D_i]$ in terms of the states of the edges/arcs. Each event E_i can be written as a Boolean expression that is the product of the states of the edges/arcs in minpath P_i. Hence, D_i can also be written as a Boolean expression. For this reason, algorithms using disjoint products are sometimes called "Boolean algebra" methods.

There is a wide variety of Boolean algebra methods based on orthogonalization; Crama and Hammer ([52], Chapter 7) give an excellent discussion. The pioneering paper here in the reliability literature is by Fratta and Montanari [68], although their procedure for orthogonalization is implicit in [94]. Subsequent improvements have employed two basic ideas (see [1–3, 8, 11, 35, 57, 75, 99–101, 120, 132, 133, 149, 160, 171, 175] for a sample). First, observe that the expression for event D_i is a complex Boolean expression, involving complements of events E_i and not just edge states and complements of edge states. Evaluation of D_i requires simplification of the Boolean expression to one that involves only edge states and their complements. Most methods are primarily concerned with making this simplification efficient and with producing resulting expressions that are as small as possible. Second, in order to make simplification easier, most methods employ some simple strategy for reordering the minpaths prior to defining the events $\{D_i\}$. The particular events defined depend heavily on the ordering of the minpaths chosen. A typical heuristic here is to sort the minpaths, placing minpaths with the fewest edges/arcs first; Balan and Traldi [4] show that this need not be optimal, however. Despite these heuristic improvements, there is no known polynomial bound in general for the length of the simplified Boolean expression for $Rel(G)$ in terms of the number of minpaths.

A polynomial bound for all-terminal reliability and reachability is achievable using an algorithm of Ball and Nemhauser [8] (see also [11, 13]) to produce a Boolean formula describing disjoint events based on minpaths in which the number of terms equals the number of minpaths.

Colbourn and Pulleyblank [49] give an algorithm for ordering the minpaths in k-terminal reliability problems so that the number of terms does not exceed the number of spanning trees. However, this may exceed the number of minpaths by an exponential factor; see also Chari [40]. Ball and Provan [11, 13] treat the optimal ordering of minpaths in a general setting.

An indirect way to compute the probability of obtaining a pathset is to compute the probability of obtaining a cutset. The reliability is then 1 minus the cutset probability. Let $C_1, \ldots, C_g$ be an enumeration of the mincuts, and let E_i be the event that all edges in mincut C_i fail. Again, inclusion-exclusion [82, 117] or orthogonalization to disjoint products [5, 14, 74, 131] can be applied. The potential advantage is that the number of mincuts is often much smaller than the number of minpaths. One can generate both minpaths and mincuts and proceed with the smaller collection of the two (see, for example, [57]). For mincuts, no current theory analogous to domination accounts for which states are relevant. This can be a serious drawback to approaches based on cutsets.

An algorithm due to Provan and Ball [128] determines two-terminal reliability in time that is polynomial in the number of mincuts. Their algorithm has a similar flavor to the dynamic programming strategy suggested by Buzacott [36–39]. Buzacott's algorithm applies more generally than the Provan-Ball strategy but does not perform nearly as well when the number of mincuts is small. Ball and Provan [12] develop a common extension of both methods.

Every algorithm mentioned here requires exponential time in the worst case, whether it enumerates states, minpaths, or mincuts. A complete graph on n nodes, for example, has $2^n - 1$ mincuts, n^{n-2} spanning trees, and $2^{\binom{n}{2}}$ states. If exponential algorithms are the best one can hope for, it is reasonable to consider algorithms that explore a *relatively* small number of states. Among the many algorithms mentioned here, three are noteworthy. Methods based on domination ensure that only relevant states are examined, and thereby improve on almost all other inclusion-exclusion methods. Methods based on factoring (for the undirected case) also generate only relevant states. The Satyanarayana-Chang approach obtains the best possible computation time using series and parallel reductions. In fact, in the all-terminal case, because the number of spanning trees exceeds the domination, the algorithm improves on all methods based on minpaths. Finally, methods based on disjoint products, although difficult to analyze in general, give two important algorithms: the Ball-Nemhauser algorithm to compute all-terminal reliability in time polynomial in the number of minpaths, and the Provan-Ball algorithm that computes two-terminal reliability in time polynomial in the number of mincuts.

This useful device of measuring complexity in terms of the number of minpaths, mincuts, or relevant states enables one to see that the methods singled out are indeed improvements on the vast remainder of reliability algorithms.

In the general context of Boolean functions, two challenging problems are to determine the fewest terms in an orthogonal DNF formula for a positive Boolean function, and to develop algorithms that orthogonalize to produce the fewest terms in time polynomial in the size of the output. The Ball-Provan and Ball-Nemhauser methods mentioned here for reliability address these questions for specific Boolean functions; for a more general treatment, see [26, 125] and Chapter 7 of [52].

17.3.2.2 Boolean Decomposition

Orthogonalization places the Boolean formula for reliability in a form that is "easily" evaluated, but the size of the formula is typically exponential in the number of components. The size of a smallest DNF formula is not indicative of the difficulty of computing the reliability, however. For this reason, Provan [126] develops a general theory relating the size of the Boolean expression to the complexity of reliability computations. Following Borodin [24], a Boolean function Υ has a *q-normal form decomposition* of *depth* $d \geq 0$ if $\Upsilon = \psi_j$ for some component x_j, or if $\Upsilon = \overline{\Upsilon_1}$, $\Upsilon = \Upsilon_1 \vee \Upsilon_2$, $\Upsilon = \Upsilon_1 \wedge \Upsilon_2$, or $\Upsilon = \Upsilon_1 \wedge \overline{\Upsilon_2}$, where Υ_1 and Υ_2 are Boolean functions having q-normal forms of depth $d - 1$. This definition is motivated by Boolean circuits in which d is the *depth* of the circuit, and the size of the formula indicates the number of logical gates employed by its implementation (cf. Chapter 11 in this volume). Provan [126] notes, however, that the complexity of evaluating Υ and the complexity of evaluating the associated reliability function can differ substantially. To remedy this, he restricts the q-normal form decomposition.

First, we say that Υ_1 and Υ_2 are *disjoint* if $\Upsilon_1 \wedge \Upsilon_2 = 0$; that Υ_1 *dominates* Υ_2 if $\Upsilon_1 \vee \Upsilon_2 = \Upsilon_1$; and that Υ_1 and Υ_2 are *independent* when they are defined on disjoint sets of variables. Then under the assumption of statistical independence of component failures,

(i) $Rel(\Upsilon, \mathbf{p}) = 1 - Rel(\overline{\Upsilon}, \mathbf{p})$;
(ii) $Rel(\Upsilon_1 \vee \Upsilon_2, \mathbf{p}) = Rel(\Upsilon_1, \mathbf{p}) + Rel(\Upsilon_2, \mathbf{p})$ when Υ_1 and Υ_2 are disjoint;
(iii) $Rel(\Upsilon_1 \wedge \Upsilon_2, \mathbf{p}) = Rel(\Upsilon_1, \mathbf{p}) \cdot Rel(\Upsilon_2, \mathbf{p})$ when Υ_1 and Υ_2 are independent; and
(iv) $Rel(\Upsilon_1 \wedge \overline{\Upsilon_2}, \mathbf{p}) = Rel(\Upsilon_1, \mathbf{p}) - Rel(\Upsilon_2, \mathbf{p})$ when Υ_1 dominates Υ_2.

With this in mind, a Boolean formula Υ is in *p-normal form* of *depth d* if $\Upsilon = \psi_j$ for some component x_j, or if $\Upsilon = \overline{\Upsilon_1}$, $\Upsilon = \Upsilon_1 \vee \Upsilon_2$ where Υ_1 and Υ_2 are disjoint, $\Upsilon = \Upsilon_1 \wedge \Upsilon_2$ where Υ_1 and Υ_2 are independent, or $\Upsilon = \Upsilon_1 \wedge \overline{\Upsilon_2}$ where Υ_1 dominates Υ_2, where Υ_1 and Υ_2 are Boolean functions having p-normal forms of depth $d - 1$.

Provan [126] establishes the crucial result: When Υ has a p-normal form decomposition, $Rel(\Upsilon, \mathbf{p})$ can be calculated in time that is linear in the size of the decomposition. This dramatically widens the class of Boolean functions beyond orthogonal DNF formulas for use in reliability computations.

17.3.2.3 Inclusion-Exclusion and Domination

Given a listing $P_1, \ldots, P_h$ of the minpaths of G, and again letting E_i be the event that all edges in minpath P_i are operational, the reliability is just the probability that one (or more) of the events $\{E_i\}$ occurs. Now $Rel(G) = \Pr[E_1 \text{ or } E_2 \text{ or } \ldots \text{ or } E_h]$, and hence

$$Rel(G) = \sum_{j=1}^{h} (-1)^{j+1} \sum_{I \subseteq \{1,\ldots,h\}, |I|=j} \Pr[E_I], \tag{17.2}$$

where E_I is the event that all paths P_i with $i \in I$ are operational. This is a standard inclusion-exclusion expansion. The algorithmic consequences of this formulation are immediate. Having a list of minpaths, one computes the probability of each subset of the minpaths occurring. To compute the reliability, one need only evaluate the foregoing sum. In doing so, observe that an odd number of minpaths contributes positively to the sum, whereas an even number contributes negatively. Arguably, this reduces our problem to the production of the set of all minpaths. This algorithm has been suggested by a number of authors; see, for example, [69, 90, 95, 97, 110]. A naive implementation of this approach is, in fact, much worse than complete state enumeration. The number of pathsets, h, may be exponential in n, and hence just the minpath generation requires exponential time. However, the more serious defect is that generating all subsets of minpaths in the naive manner takes 2^h time, which leaves us with a *doubly* exponential time algorithm for the reliability.

With a little care, this doubly exponential behavior can be avoided. Every subset of the minpaths corresponds to a subgraph whose edge set is the union of the edge sets of the minpaths. With this in mind, an *i-formation* of a subgraph is a set of i minpaths whose union is the subgraph. A formation is *odd* when i is odd, *even* when i is even. A graph having a formation is a *K-subgraph*. Every odd formation of the subgraph contributes positively to the reliability, and every even formation contributes negatively. The *signed domination* of G with terminal set K of nodes, $sdom(G, K)$, is the number of odd formations of G minus the number of even formations of G. The *domination* $dom(G, K)$ is the absolute value of the signed domination. We usually write $sdom(G)$ and $dom(G)$ with the terminal set K

understood. With these definitions, Satyanarayana and Prabhakar [144] simplified the expression for the reliability substantially:

$$Rel(G) = \sum_{H \subseteq G} sdom(H)\Pr[H],$$

where H varies over all states of G. This simplification is substantial because it entails only the generation of all states rather than the generation of all subsets of the minpaths. However, some effort is still required if we are to improve on complete state enumeration. In particular, we now require the signed domination of each state.

In each of the directed reliability problems, Satyanarayana and his colleagues [140, 142–144, 170] completely determined the signed domination of each state. We outline the derivation in the reachability case here [142, 143].

The first goal is to determine which states have signed domination zero, and can therefore be ignored in the reliability expression. With this in mind, a state (subgraph) is *relevant* whenever it contains no irrelevant arcs. A subgraph containing irrelevant arcs has no formations whatsoever, and hence has signed domination zero. Thus we restrict our attention to relevant subgraphs. Among the relevant subgraphs, many have signed domination zero: precisely the cyclic subgraphs (subgraphs with some directed cycle) [144]. Moreover, an acyclic relevant digraph with m arcs and n nodes has signed domination $sdom(G) = (-1)^{m-n+1}$.

This study of domination in directed reliability problems is a remarkably clever application of combinatorial arguments. A method that naively requires doubly exponential time has been reduced to requiring the generation of the acyclic relevant graphs, and a trivial calculation for each. In practice, this affords a substantial improvement on complete state enumeration. Nevertheless, the number of acyclic subdigraphs is potentially exponential in n. Hence, despite a very significant reduction in computational effort, a very large computational task remains.

The use of signed domination in undirected problems arises in quite a different way. In undirected problems, cyclic relevant graphs have nonzero domination. Thus, inclusion-exclusion algorithms using minpaths would require algorithms to compute the signed domination. However, the current algorithm to compute the signed domination of a single graph is the *same* as the algorithm that computes the reliability recursively using factoring. In fact, optimal factoring strategies using factoring with series and parallel reductions employ a number of factoring steps *equal* to the domination [141].

Classical bounds due to Bonferroni (see [123, 148]) can be obtained using the inclusion-exclusion formula (17.2). Truncating the sum after $\ell < h$ terms yields an upper bound when ℓ is odd and a lower bound when ℓ is even. The Bonferroni bounds require knowledge of all minpaths, an exponential quantity. Two-terminal bounds have been developed by Prékopa, Boros and Lih [123] that use "binomial moments" to improve upon the Bonferroni bounds.

From the Boolean perspective, the theory of domination does not yield an orthogonal DNF function as a consequence of the presence of numerical signed domination values in the evaluation. Nevertheless, inclusion-exclusion can lead to Boolean formulas whose p-normal form has size smaller than that obtained by orthogonalization [126].

17.4 Bounds on Network Reliability

Essentially all reliability problems of interest are #P-complete, and hence the fact that the exact algorithms described are quite inefficient comes as no surprise. Nevertheless, in assessing the reliability of a network, it is imperative that the assessment can be completed in a "reasonable" amount of time. The conflicting desires for fast computation and for great accuracy have led to a varied collection of methods for estimating reliability measures.

Boolean formulas and functions have arisen infrequently in these topics; the language of graphs, matroids, and complexes is much more prevalent. In part this arises because the bounding or approximation of a reliability function can be measured quantitatively, whereas bounding or approximation of a Boolean function is less natural. Nevertheless, we treat these topics in some detail.

Two main themes arise: the *estimation* of reliability by Monte Carlo sampling techniques, and the *bounding* of reliability. In the first, the goal is to obtain an accurate estimate of a reliability measure by examining a small fraction of the states, chosen randomly. This leads to a point estimate of the reliability measure, along with confidence intervals for the estimate. We do not discuss Monte Carlo methods here, but refer the reader to [7] for a somewhat dated survey. Bounding is different, both in technique and in result. Current techniques for bounding attempt to find combinatorial or algebraic structure in the reliability problem, permitting the deduction of structural information upon examination of a small fraction of the states. Unlike Monte Carlo methods, the states examined are not chosen randomly. The goal of bounding is to produce *absolute* upper and lower bounds on the reliability measure. It is perhaps misleading to draw a line between Monte Carlo methods and bounding techniques, because a number of the Monte Carlo methods employ bounding as a vehicle to limit the size of the sample space.

In this section, we treat bounds that are valid when every edge has the same operation probability p so that reliability can be expressed as a polynomial in p. A subgraph with operational edges $E' \subseteq E$ now arises with probability $p^{|E'|}(1-p)^{|E-E'|}$. Consequently, the probability of obtaining a subgraph depends only on the number of edges it contains. Then let N_i denote the number of operational subgraphs with i edges. The probability of network operation, denoted $Rel(G, p)$ or simply $Rel(p)$, is then

$$Rel(p) = \sum_{i=0}^{m} N_i p^i (1-p)^{m-i}.$$

Thus, the probability is a polynomial in p, which we saw before, called the *reliability polynomial*. This formulation is in terms of pathsets. Another formulation is obtained by examining cutsets. Letting C_i be the number of i-edge cutsets (leaving $m - i$ operational edges),

$$Rel(p) = 1 - \sum_{i=0}^{m} C_i (1-p)^i p^{m-i}.$$

Still another formulation, and probably the most common, is obtained by examining complements of pathsets. Let F_i denote the number of sets of i edges for which the $m - i$ remaining edges form a pathset. Then

$$Rel(p) = \sum_{i=0}^{m} F_i (1-p)^i p^{m-i}.$$

17.4.1 Basic Observations

The first goal in introducing the reliability polynomial is to obtain a compact encoding of reliability information to compare candidate topologies. Moore and Shannon [112] pioneered this approach in their study of electrical relays. Moskowitz [113] and Mine [108] employed a simple strategy for computing reliability polynomials in certain two-terminal problems on series-parallel networks. The application to computer networks, and the reliability polynomials introduced here, were studied by Kel'mans [87]. Kel'mans was apparently the first to make a fundamental observation about comparing two networks via their reliability polynomial. He proved that for two graphs G and H, their reliability polynomials may "cross"; that is, one may be more reliable for one value of p, while the other is more reliable for another value of p. Kel'mans [88, 89, 114] proved that for a given number of nodes and edges, in certain cases there is no graph that is most reliable for all edge operation probabilities. Thus, reliability is more than simple graph parameters; it is truly a function of link reliabilities.

17.4.2 Computing Some Coefficients Exactly

Using (17.1), the ability to compute the size ℓ of a minimum cardinality pathset and the size c of a minimum cardinality cutset enables us to determine a number of coefficients exactly. When ℓ is efficiently computable, further information can often be obtained by computing N_ℓ. In the k-terminal problem, this is truly hopeless; one would have to count minimal Steiner trees, a #P-complete problem. In the other two cases, however, efficient algorithms exist.

In the all-terminal problem, N_ℓ is the number of spanning trees. Kirchoff [91] developed an elegant method for counting spanning trees; for a computationally useful form, see [28]. In the 2-terminal problem, Ball and Provan [10] developed an efficient strategy for computing N_ℓ, the number of shortest s,t-paths. For any fixed $k \geq 0$, pathsets of size $\ell + k$ in the 2-terminal problem can be counted efficiently [27].

An efficient algorithm to compute c enables us to determine a number of additional coefficients exactly as well. This problem is tractable in each of the three cases of interest, using the same method in each case. Menger's theorem [105] guarantees that the minimum s, t-cut has size c exactly when the maximum number of edge-disjoint s, t-paths is c. This problem is easily recast as a network flow problem, with all edge capacities equal to 1.

Having computed c it would be useful to compute C_c, the number of minimum cardinality cutsets. In the 2-terminal case, computing just this coefficient is #P-complete [127]; because the k-terminal problem includes the 2-terminal problem, computing C_c in either of these problems is apparently intractable. The coefficient C_c can be calculated efficiently in the all-terminal case [9], using the fact that every n-node graph G has $C_c(G) \le \binom{n}{2}$ [21, 103]. In addition, for any i and any edge e of G, $C_i(G) = C_i(G \cdot e) + C_{i-1}(G - e)$. Consequently, there is an efficient method for counting cutsets of size $c + k$ for any fixed $k \ge 0$ [136].

17.4.3 Simple Bounds

The computation of many of the coefficients leaves us with many coefficients about which we have said nothing. Kel'mans [86, 87] observed that for all-terminal reliability, when p is close to 0,

$$Rel_A(p) \approx N_{n-1}p^{n-1}(1-p)^{m-n+1},$$

and when p is close to 1,

$$Rel_A(p) \approx 1 - C_c p^{m-c}(1-p)^c.$$

Phrasing the Kel'mans approximations as bounds, for all p,

$$N_{n-1}p^{n-1}(1-p)^{m-n+1} \le Rel_A(p) \le 1 - C_c p^{m-c}(1-p)^c.$$

For extreme values of p, either the term involving N_{n-1} or the term involving C_c dominates the remaining terms.

Now $N_i + C_{m-i} = \binom{m}{i}$, and hence $0 \le N_i, C_i \le \binom{m}{i}$. This leads us to a simple set of bounds first formulated by Jacobs [80] and improved to this current form by Van Slyke and Frank [163]:

$$Rel_A(p) \ge N_{n-1}p^{n-1}(1-p)^{m-n+1} + N_{m-c}p^{m-c}(1-p)^c + \sum_{i=m-c+1}^{m} \binom{m}{i} p^i (1-p)^{m-i},$$

$$Rel_A(p) \le N_{n-1}p^{n-1}(1-p)^{m-n+1} + \sum_{i=n}^{m} \binom{m}{i} p^i (1-p)^{m-i}.$$

In the lower bound, each "unknown" N_i is approximated by 0; the known coefficients are for $i < n-1$ (0), $i = n-1$ (the number of spanning trees), $i = m-c$ (the $(m-c)$-edge subgraphs whose complement is *not* a minimum cardinality cutset), and $i > (m-c)$ (all possible i-edge subgraphs). In the upper bound, the unknown N_i are approximated by $\binom{m}{i}$. The extension to 2-terminal reliability is straightforward: simply substitute ℓ for $n-1$ throughout. The extension to k-terminal reliability is complicated by the difficulty of computing ℓ. A lower bound is nevertheless obtained by underestimating N_ℓ as 0; an upper bound is obtained merely by using an underestimate for ℓ itself. These bounds are *extremely* weak and provide useful information only when p is very near 0 or 1.

17.4.4 Coherence

Ability to bound the unknown $\{N_i\}$ depends heavily on knowledge of the combinatorial structure of the collection of operational subgraphs. Most reliability problems of interest are coherent. Let $\mathcal{D} = \{D_1, D_2, D_3, \ldots, D_r\}$, where D_i is a set of edges for which $E - D_i$ is operational. The set $\mathcal{D}$ has a set for each operational subgraph, in which the edges *not* in the operational subgraph are listed. Then $\mathcal{D}$ is a hereditary family of sets (or *complex*) called the *$\mathcal{F}$-complex* (that is, if $S \in \mathcal{D}$ and $S' \subseteq S$, then $S' \in \mathcal{D}$). The fact that the family of sets produced is hereditary is precisely the property of coherence. (Recall that precisely in this case, the corresponding Boolean function is positive.) It is also no coincidence that F_i, defined earlier, is precisely the number of sets of cardinality i in $\mathcal{D}$. In fact, the reliability polynomial for a hereditary family $\mathcal{D}$ is completely prescribed by its *F-vector* $(F_0, F_1, \ldots, F_d)$ where $d = m - \ell$.

The property of coherence also suggests a particularly appropriate way of viewing the family $\mathcal{D}$, as a partial order whose relation is set inclusion. The key to using coherence in obtaining bounds is the following. Consider all of the i-sets in a hereditary family $\mathcal{D}$. Because the family is hereditary, there must be a number of $(i-1)$-sets contained in the collection of i-sets; the minimum number of such induced $(i-1)$-sets is a lower bound on F_{i-1}. The minimization of F_{i-1} as a function of F_i is a well-known problem in extremal set theory, apparently first studied by Sperner [150], who proved that $F_{i-1} \geq \frac{i}{m-i+1} F_i$. Birnbaum, Esary, and Saunders [20] first employed this in the reliability setting.

It has a very simple interpretation [15]: The fraction of operational subgraphs with i edges over all subgraphs with i edges is nondecreasing as i increases. Sperner's theorem can be used at little computational effort to improve the simple bounds when ℓ, c, $F_{m-\ell}$, and F_c are available. Then

$$Rel(p) \geq \sum_{i=0}^{c-1} \binom{m}{i} p^{m-i}(1-p)^i + F_c p^{m-c}(1-p)^c$$
$$+ \sum_{i=c+1}^{m-\ell} F_{m-\ell} \frac{\binom{m}{i}}{\binom{m}{m-\ell}} p^{m-i}(1-p)^i,$$

$$Rel(p) \leq \sum_{i=0}^{c-1} \binom{m}{i} p^{m-i}(1-p)^i + \sum_{i=c}^{m-\ell-1} F_c \frac{\binom{m}{i}}{\binom{m}{c}} p^{m-i}(1-p)^i$$
$$+ F_{m-\ell} p^\ell (1-p)^{m-\ell}.$$

This bounding technique applies to any coherent system and affords significant improvements on the simple bounds [45].

One method to improve on these bounds is to improve on Sperner's theorem. The best possible result in this direction was obtained by Kruskal [93] and independently by Katona [84]. See also [53, 66]. The Kruskal-Katona theorem places a lower bound $F_i^{i-1/i}$ on F_{i-1} given F_i; alternatively, it places an upper bound $F_{i-1}^{i/i-1}$ on F_i given F_{i-1}. The form of the bound is of little importance here, except to note that $x^{j/i}$ can be efficiently calculated, and that whenever $x \geq y$, $x^{j/i} \geq y^{j/i}$. Van Slyke and Frank [163] used the Kruskal-Katona theorem to bound individual coefficients in the reliability polynomial. Recall that F_c is $\binom{m}{c} - C_c$. For all–terminal reliability, we can therefore compute F_c exactly; in the remaining two cases, we cannot hope to compute F_c, but we can easily compute F_{c-1}. In general, let us suppose that we can compute a sequence of coefficients $F_0, F_1, \ldots, F_s$ efficiently. Then the Kruskal-Katona theorem gives us an upper bound on F_{s+1}. Then given an upper bound on F_{s+1}, we proceed in the same way to obtain upper bounds on F_{s+2}, F_{s+3} and so on.

Lower bounds are obtained symmetrically. We compute some sequence of coefficients $F_{m-\ell}, F_{m-\ell+1}, \ldots, F_m$ efficiently. For all-terminal and 2-terminal reliability, $F_{m-\ell}$ is the number of spanning trees and shortest paths, respectively. In the k-terminal problem, we can take $\ell = k - 1$ (for example) in order to compute this sequence. In any event let $d = m - \ell$. Knowing F_d, the Kruskal-Katona theorem gives a lower bound on F_{d-1}, namely $F_d^{d-1/d}$. This application of the Kruskal-Katona theorem, first done by Van Slyke and Frank [163], gives us the *Kruskal-Katona* bounds.

17.4.5 Shellability

The Kruskal-Katona theorem is best possible for hereditary families of sets. We therefore have no hope of improving on the Kruskal-Katona bounds without further restriction, which could arise in a number of ways. One would be to develop efficient algorithms for computing (or even bounding more tightly) one or more of the unknown F_i. Another would be to observe that the particular hereditary family that arises has some special combinatorial structure. This latter approach is promising, because although complements of pathsets in coherent systems produce a hereditary family, not all hereditary families arise in this way.

In fact, the $\mathcal{F}$-complex in an all-terminal reliability problem is a matroid, the *cographic matroid* of the graph. For now, we restrict our attention to the all-terminal problem. No progress appears to have been made on characterizing F-vectors of cographic matroids, and so one might ask what the F-vector of a matroid can be in general. Even on this problem, no progress has been made directly. However, we can identify a class of hereditary systems that are intermediate between matroids and hereditary systems in general, and results *are* available here.

Provan and Billera [130] prove a powerful result about the structure of matroids, which (together with later results) constrains their F-vectors; they observe that matroids are "shellable" complexes. The importance of the Provan-Billera result here is that it suggests the possibility of exploiting shellability to improve on the Kruskal-Katona bounds, which requires structure theorems for shellable systems. An *interval* $[L, U]$ is a family of subsets $\{S : L \subseteq S \subseteq U\}$. An interval partition of a complex is a collection of disjoint intervals for which every set in the complex belongs to precisely one interval. A complex is *partitionable* if it has an interval partition $[L_i, U_i]$, $1 \leq i \leq J$ with U_i a base for all i. Shellable complexes are all partitionable.

Ball and Nemhauser [8] developed the application of the partition property to reliability; see [26] for an interpretation in Boolean vernacular. Consider a shellable complex with b bases; let $\{[L_i, U_i]|1 \leq i \leq b\}$ be an interval partition for this complex. $[L_i,U_i]$ is a compact encoding of all sets in this interval; the probability that any one of these sets arises is then $p^{m-|U_i|}(1-p)^{|L_i|}$. In other words, $|L_i|$ edges must fail, and $m - |U_i|$ edges must operate; the state of the remaining edges is of no consequence. Every U_i is a base in the complex; hence the cardinality of each U_i is the same, the rank d of a base. Hence a Boolean formula for all-terminal reliability can be represented as a disjunction of conjuncts whose number is equal to the number of sets of maximum cardinality in the complex; a conjunct contains the Boolean variables corresponding to the negations of elements of L_i, and those corresponding to elements of U_i. This is an orthogonal DNF.

However, the ranks of the L_i are not all identical; therefore, define $H_i = |\{L_j : 1 \leq j \leq b, |L_j| = i\}|$. This gives rise to an *H-vector* $(H_0, \ldots, H_d)$. The coefficient H_i counts intervals in the partition whose lower set has rank i.

This gives yet another form of the reliability polynomial:

$$Rel(p) = p^{\ell} \sum_{i=0}^{d} H_i(1-p)^i.$$

Here, ℓ is the cardinality of a minimum cardinality pathset (spanning tree), and $d = m - \ell$ is then the rank of a base. More concretely, in an n-node m-edge graph, $\ell = n - 1$ and $d = m - n + 1$.

Naturally, any information about the H-vector also provides information about the reliability polynomial. However, to place the H-vector in appropriate context, it is worthwhile examining the relation between the H-vector and the F-vector for a shellable complex. The H-vector for any complex can be defined directly in terms of the F-vector (see, for example, [154]). In the partitionable case, however, the correspondence is easily seen combinatorially.

Consider the sets of rank k in the complex. These are counted by F_k. Now any interval $[L_i, U_i]$ accounts for certain of these sets. Let r be the rank of L_i. If $r > k$, the interval accounts for 0 of the sets in F_k; however, if $r \leq k$, it accounts for $\binom{d-r}{k-r}$ of the sets. Hence, we find that $F_k = \sum_{r=0}^{k} H_r \binom{d-r}{k-r}$. Equating the F-vector and

H-vector forms of the reliability polynomial gives an expression for H_i in terms of the F-vector, namely:

$$H_k = \sum_{r=0}^{k} F_r(-1)^{k-r}\binom{d-r}{k-r}.$$

This expression allows us to efficiently compute $H_0, \ldots, H_s$ from $F_0, \ldots, F_s$. Another obvious, but useful, fact is that $F_d = \sum_{i=0}^{d} H_i$.

Following pioneering research of Macaulay [104], Stanley [18, 152–154, 156] has studied H-vectors in an algebraic context, as "Hilbert functions of graded algebras." Stanley obtained a lower bound $H_i^{<i-1/i>}$ on H_{i-1} given H_i that is tight for shellable complexes in general; this in turn gives an upper bound $H_{i-1}^{<i/i-1>}$ on H_i given H_{i-1}. For our purposes, three things are important. First of all, for $k \geq j \geq i$, $x^{<k/i>} = x^{<j/i><k/j>}$. Second, given x, j and i we can compute $x^{<j/i>}$ efficiently. Third, whenever $x \geq y$, $x^{<j/i>} \geq y^{<j/i>}$.

Stanley's theorem can be used to obtain efficiently computable bounds on the reliability polynomial. Given a prefix $(F_0, \ldots, F_s)$ of the F-vector, we can efficiently compute a prefix $(H_0, \ldots, H_s)$ of the H-vector. Knowing this prefix, some straightforward bounds apply to shellable systems in general, given here in the all-terminal case:

$$Rel(p) \geq p^{n-1} \sum_{i=0}^{s} H_i(1-p)^i.$$

$$Rel(p) \leq p^{n-1}\left[\sum_{i=0}^{s} H_i(1-p)^i + \sum_{i=s+1}^{d} H_s^{<i/s>}(1-p)^i\right].$$

This exploits information about the size of the minimum cardinality cutset and, where available, the number of such cutsets. This simple formulation ignores a substantial piece of information, the number of spanning trees. This is introduced by recalling that $F_d = \sum_{i=0}^{d} H_i$. Ball and Provan [9, 10] develop bounds that incorporate this additional information; they suggest a very useful pictorial tool for thinking about the problem. Associate with each H_i a "bucket." Now suppose we have F_d "balls." Our task is to place all of the balls into buckets, so that the number of balls in the ith bucket, n_i, satisfies $n_i \leq n_{i-1}^{<i/i-1>}$.

How do we distribute the balls so as to maximize or minimize the reliability polynomial? These distributions, when found, give an upper and a lower bound on the reliability polynomial. Consider carefully the sum in the reliability polynomial: $\sum_{i=0}^{d} H_i(1-p)^i$. Because $0 < p < 1$, the sum is larger when the lower order coefficients are larger. In fact, for two H-vectors $(H_0, \ldots, H_d)$ and $(J_0, \ldots, J_d)$, whenever $\sum_{j=0}^{i} H_j \geq \sum_{j=0}^{i} J_j$ for all i, the reliability polynomial for the H_i dominates the reliability polynomial for the J_i.

This last simple observation suggests the technique for obtaining bounds. In the pictorial model, an upper bound is obtained by placing balls in the leftmost possible buckets (with buckets $0, \ldots, d$ from left to right); symmetrically, a lower bound is obtained by placing balls in the rightmost possible buckets. We are not totally without constraints in making these placements, as we know in advance the contents of buckets $0, \ldots, s$.

With this picture in mind, we give a more precise description. We produce coefficients $\overline{H}_i$ for an upper bound polynomial, and $\underline{H}_i$ for a lower bound polynomial, using the prefix $(H_0, \ldots, H_s)$ and F_d. The steps are:

(i) For $i = 0, \ldots, s$, set $\underline{H}_i = H_i = \overline{H}_i$.
(ii) For $i = s + 1, s + 2, \ldots, d$, set

$$\underline{H}_i = min\left[r : \sum_{j=0}^{i-1} \underline{H}_j + \sum_{j=i}^{d} r^{<j/i>} \geq F_d \right].$$

$$\overline{H}_i = max\left[r : r \leq \overline{H}_{i-1}^{<i/i-1>} \text{ and } \sum_{j=0}^{i-1} \overline{H}_j + r \leq F_d \right].$$

An explanation in plain text is in order. In each bound, we determine the number of balls in each bucket from 0 to d in turn; as we remarked, the contents of buckets $0, \ldots, s$ are known. For subsequent buckets, the upper bound is determined as follows. The number of balls that can go in the current bucket is bounded by Stanley's theorem and is also bounded by the fact that there is a fixed number of balls remaining to be distributed. If there are more balls remaining than we can place in the current bucket, we place as many as we can. If all can be placed in the current bucket, we do so; in this case, all balls have been distributed and the remaining buckets are empty. The lower bound is determined by placing as few balls as possible.

The method leads to a very powerful set of bounds, the *Ball-Provan* bounds:

$$Rel(p) \geq p^{n-1} \sum_{i=0}^{d} \underline{H}_i (1-p)^i,$$

$$Rel(p) \leq p^{n-1} \sum_{i=0}^{d} \overline{H}_i (1-p)^i.$$

Unlike the Kruskal-Katona bounds, in the case of the Ball-Provan bounds it is not generally the case that $\underline{H}_i \leq H_i \leq \overline{H}_i$. Brown, Colbourn, and Devitt [32] observe that a number of simple network transformations can be used to determine bounds $L_i \leq H_i \leq U_i$ efficiently. Incorporating these coefficient bounds on the H-vector into the Ball-Provan process can result in substantial improvements.

17.4.6 Polyhedral Complexes and Matroid Ports

The Ball-Provan bounds as developed here apply to all-terminal reliability and to reachability. For reachability, Provan [124] observes that the $\mathcal{F}$-complex is a "polyhedral complex" and uses a theorem of Billera and Lee [19] to obtain efficiently computable bounds on reliability that are tight for polyhedral complexes.

The matroid structure of the all-terminal problem and the polyhedral structure of reachability both lead to dramatic improvements over the Kruskal-Katona bounds for general coherent reliability problems. Building on a structure theorem of Colbourn and Pulleyblank [49], Chari [40] characterized 2-terminal complexes in terms of "matroid ports" and generalized these to develop some remarkable structure theorems about $\mathcal{F}$-complexes from k-terminal problems. For an n-node connected graph having k terminals and edge set E, $|E| = m$, let $\mathcal{F}$ be its $\mathcal{F}$-complex. The blocking complex $\mathcal{F}^*$ of $\mathcal{F}$ is $\{E \setminus S : S \in 2^E \setminus \mathcal{F}\}$. The F-vector $(F_0, \ldots, F_m)$ of $\mathcal{F}$ and the F-vector $(F_0^*, \ldots, F_m^*)$ of $\mathcal{F}^*$ satisfy $F_i + F_{m-i}^* = \binom{m}{i}$ for $0 \leq i \leq m$. Chari [40] shows that the subcomplex $\mathcal{F}^{(m-n+1)}$ obtained by removing all sets from $\mathcal{F}$ of cardinality exceeding $m - n + 1$ is a shellable complex. Hence, given bounds on the single coefficient F_{m-n+1}, the Ball-Provan strategy can be applied to this k-terminal problem in order to bound $(F_0, \ldots, F_{m-n+1})$. What about the remaining coefficients? Chari further proves that $\mathcal{F}^{*(n-2)}$, obtained from $\mathcal{F}^*$ by removing all sets of cardinality exceeding $n - 2$, is also shellable. Hence the Ball-Provan bounds can be applied again to bound $(F_0^*, \ldots, F_{n-2}^*)$, or equivalently to bound $(F_m, \ldots, F_{m-n+2})$.

All of the approaches developed for equal edge failure probabilities to date examine extremal results for complexes that are more general than those actually arising in reliability problems on graphs. It remains a very active area of research to determine least or most reliable graphs, rather than complexes, given the values of some specified graph parameters. Even the characterization of least reliable graphs for specified numbers of nodes and edges remains unresolved, however [23].

17.4.7 Chip Firing and Degree Bounds

Let us now return to undirected all-terminal reliability. Although the $\mathcal{F}$-complex is known to be shellable, not all shellable complexes arise from $\mathcal{F}$-complexes. Indeed, not all shellable complexes arise from matroid complexes. Most importantly, the complexes produced in the Ball-Provan bounding process do not arise, in general, from matroid complexes. This suggests the importance of obtaining a characterization of the complexes tighter than shellability. In the domain of Boolean functions, shellability has been restricted in a number of directions: to "aligned" functions [25], to "tree-shellable" functions [158, 159], and to functions with the lexico-exchange property [26]. The last class contains matroids ([52], Chapter 7) and hence applies directly to the all-terminal reliability problem [13, 129]. Nevertheless, we adopt the more traveled path using complexes.

Using techniques in commutative algebra, Stanley's characterization of H-vectors of shellable complexes employs a bijection between intervals in the shelling and a set of F_d monomials closed under divisibility (an *order ideal of monomials*). A further bijection that maps a monomial to a multiset that contains each variable a number of times equal to its exponent in the monomial maps intervals in a shelling to a collection of multisets, closed under taking submultisets (a *multicomplex*). We call this the $\mathcal{H}$*-multicomplex* of the graph. Shelling, in essence, tells us that this process yields a multicomplex, but tells us nothing about its structure. Stanley [155] conjectured that the $\mathcal{H}$-multicomplex is *pure*, in that every maximal multiset is of cardinality d (equivalently, every maximal monomial is of maximum degree).

Progress on characterizing $\mathcal{H}$-multicomplexes hinge on the development of a combinatorial mapping from shelling to the multicomplex. Brown et al. [29–31, 34] and Hibi [77, 78] establish relationships among the terms in the H-vector that are not implied by shellability alone. Chari [41, 42] developed connections with topological spaces to explain these relationships; his approach, although complementary to Stanley's, did not provide a simple mechanism for determining the $\mathcal{H}$-multicomplex.

Understanding more about the structure of the $\mathcal{H}$-multicomplex arises from a different research direction, which we explore next, following the development in [33].

Let $G = (V, E)$ be a connected multigraph without loops. Let $V = \{q\} \cup \{1, \ldots, n-1\}$. A *configuration* on G is a function $\theta : V \mapsto \mathbb{Z}$ for which $\theta(v) \geq 0$ for all $v \in V \setminus \{q\}$ and $\theta(q) = -\sum_{v \neq q} \theta(v)$. In configuration θ, vertex v is *ready to fire* if $\theta(v) \geq \deg(v)$; vertex q is ready to fire if and only if no other vertex is ready. *Firing* vertex v changes the configuration from θ to θ', where $\theta'(v) = \theta(v) - \deg(v)$ and, for $w \neq v$, $\theta'(w) = \theta(w) + \ell(v, w)$ where $\ell(v, w)$ is the number of edges connecting v and w. A configuration is *stable* when $\theta(v) < \deg(v)$ for all $v \neq q$; in such a configuration only q is ready to fire.

A *firing sequence* $\Theta = (\theta_0, \theta_1, \ldots, \theta_k)$ is a sequence of configurations in which, for $1 \leq i \leq k$, θ_i is obtained from θ_{i-1} by firing one vertex that is ready to fire. It is *nontrivial* when $k > 0$. We write $\theta_0 \rightarrow \theta_k$ when some nontrivial firing sequence starting with θ_0 and ending with θ_k exists. Configuration θ is *recurrent* if $\theta \rightarrow \theta$. Stable, recurrent configurations are called *critical*.

This is called a *chip firing game* on the graph G. Initial motivation for its study arises in so-called abelian sandpile models to study self-critical systems [16, 51]. Biggs [16, 17] (who called this the *dollar game*) and Merino [106] studied critical configurations and showed that the enumeration of critical configurations classified by the sum $\sum_{v \neq q} \theta(v)$ is an evaluation of the Tutte polynomial of the graph G. Indeed, Merino [106] establishes that the particular evaluation is precisely that corresponding to all-terminal reliability.

Merino [107] establishes much more. Consider the set $\mathcal{C}$ of all critical configurations of G. For each $v \in V \setminus \{q\}$, let x_v be an indeterminate. Now for each $\theta \in \mathcal{C}$, define a monomial m_θ as follows:

$$m_\theta = \prod_{v \in V \setminus \{q\}} x_v^{\deg(v) - 1 - \theta(v)}.$$

Merino proves the remarkable theorem that the set $\{m_\theta : \theta \in \mathcal{C}\}$ is the order ideal of monomials associated with the $\mathcal{H}$-multicomplex of G. Closure under division is easily seen by observing that if θ is a critical configuration with $\theta(v) < \deg(v) - 1$, the assignment to v can be increased by 1 to obtain another critical configuration.

That the number of critical configurations of degree δ is the same as H_Δ with $\Delta = (\sum_{v \neq q} \deg(v) - 1) - \delta$ follows from the fact that it is the same Tutte invariant.

Let us explore the first consequences of this. If we let $d_1, \ldots, d_n$ be the vertex degrees of the graph, the order ideal can be represented as monomials in $n-1$ variables $x_1, \ldots, x_{n-1}$ of degrees at most $d_1 - 1, \ldots, d_{n-1} - 1$. The selection of the nth vertex is precisely the choice of q, which is arbitrary. It seems natural to choose q to be either lowest degree or highest degree; although its choice does not impact the actual H-vector, it can affect the bound. We assume that d_n is maximum, and that $d_1 \geq \cdots \geq d_{n-1}$. In essence, then, a strategy like the Kruskal-Katona or Ball-Provan bounds results, if we can specify inequalities between numbers of multisets of cardinalities i and $i+1$ in a multicomplex. Kruskal and Katona determined these precisely when the multisets are sets; Macaulay [104, 151, 169] determined these precisely when there are no constraints on multiplicities in the multisets. When such restrictions on multiplicities are known, Clements and Lindström [44] proved a common generalization. We examine this next.

Consider multisets $\mathcal{M}$ of symbols $x_1, \ldots, x_{n-1}$ in which element x_i appears at most $e_i = d_i - 1$ times, and let $\mathcal{M}_r$ be the collection of multisets of cardinality r. Treating the order ideal as a multicomplex makes it a submulticomplex of $\mathcal{M}$. Clements and Lindström show that, assuming $d_1 \geq \cdots \geq d_{n-1}$, the subset of N_r sets in $\mathcal{M}_r$ that has the smallest shadow is obtained by taking the first N_r multisets in colexicographic order.

These have a simple description. Let $\binom{e_1,\ldots,e_k}{r}$ denote the number of ways to choose distinct multisets of size r from a multiset consisting of k types of elements, the ith type containing e_i indistinguishable elements. First find the value κ of k for which $\binom{e_1,\ldots,e_k}{r} \leq N_r < \binom{e_1,\ldots,e_{k+1}}{r}$. Next choose α to be the value of a for which $\sum_{j=0}^{a} \binom{e_1,\ldots,e_\kappa}{r-j} \leq N_r < \sum_{j=0}^{a+1} \binom{e_1,\ldots,e_\kappa}{r-j}$. It follows that all multisets of cardinality r containing x_i at most e_i times for $1 \leq i \leq \kappa$ and containing x_κ at most α times occur within the first N_r multisets of $\mathcal{M}_r$ in colexicographic order.

Now let $\widehat{N}_{r-\alpha-1} = N_r - \sum_{j=0}^{\alpha} \binom{e_1,\ldots,e_\kappa}{r-j}$. Choosing the first $\widehat{N}_{r-\alpha-1}$ multisets of $\mathcal{M}_{r-\alpha-1}$ inductively and adjoining $x_{\kappa+1}$ to each $\alpha+1$ times gives the next multisets in colexicographic order.

Computing the size of the shadow is straightforward. From the first $\sum_{j=0}^{\alpha} \binom{e_1,\ldots,e_\kappa}{r-j}$, multisets one obtains a shadow of $\binom{e_1,\ldots,e_\kappa,\alpha}{r-1} = \sum_{j=0}^{\alpha} \binom{e_1,\ldots,e_\kappa}{r-j-1}$ multisets. Proceed inductively for the remaining $\widehat{N}_{r-\alpha-1}$ sets, noting that the only new elements in the shadow are those that contain $x_{\kappa+1}$ exactly $\alpha+1$ times – all others appear already in the shadow.

This enables us to obtain an immediate improvement in the Ball-Provan process. Rather than computing upper pseudopowers, we compute the size permitted by the Clements-Lindström bound. This can have a dramatic effect [33].

17.4.8 Pure Multicomplexes

Stanley [155] conjectured that the $\mathcal{H}$-multicomplex is pure. The bijection with critical configurations enabled Merino [107] to prove this. The idea is straightforward. Consider a critical configuration, and a recurrent firing sequence for it. Imagine that each vertex has a number of "chips" equal to its current value in the configuration. Then firing involves the movement of some chips off the vertex to neighboring vertices. Observing that a recurrent firing sequence for a critical configuration fires each vertex exactly once, and keeping track of chips that are moved, it is easy to see that every unmoved chip could be deleted from the initial stable configuration. Now simple counting ensures that every monomial that is not of maximum degree must divide a monomial in the order ideal that has larger degree. This ensures purity (see [107] for details).

Again this has consequences for bounding. If we use only that the $\mathcal{H}$-multicomplex is pure, we can enforce the bound $H_{d-i} \geq H_i$ for $i \leq d/2$ [78]. This permits a modest improvement in the upper bound; unlike the Ball-Provan process, we are then constrained to place some balls in the rightmost buckets. However, purity is much more informative in conjunction with the degree bounds, as we see next.

17.4.9 M-Shellability

In the same manner that shellability (or at least partitionability) of the $\mathcal{F}$-complex permitted a representation as intervals, and ultimately as the $\mathcal{H}$-multicomplex, one can ask whether the $\mathcal{H}$-multicomplex admits a similar representation. Chari [42] first explored this. An *M-interval* in a pure multicomplex $\mathcal{M}$ is specified by a lower multiset L and an upper multiset U and contains all multisets containing L and contained in U; here U must be maximal. Then $\mathcal{M}$ is M-partitionable if it admits a partition into M-intervals. It is *M-shellable* if the M-intervals of an M-partitioning can be ordered so that the union of every prefix of M-intervals itself forms a multicomplex. Chari [42] conjectured and Merino [107] proved that the $\mathcal{H}$-multicomplex is M-shellable. Merino's proof employs a clever deletion/contraction argument.

Once again, more combinatorial structure is enforced. Unlike the Ball-Provan case, however, we are handicapped here in two ways. In the shelling of $\mathcal{F}$, one knows the number of sets of each cardinality in an interval simply by knowing the size of its lower set (and hence, the height of the interval). One also knows that the number of intervals to be produced equals F_d, a quantity that we can compute efficiently. In the M-shelling of $\mathcal{H}$, the height of an M-interval does not alone permit us to determine the number of multisets in the M-interval of each cardinality (except the smallest and largest, of course). Moreover, although we know that the number of M-intervals equals H_d, we have no general efficient method for its calculation (indeed, it is #P-complete [81]).

One strategy would be to bound H_d; another is to constrain the "shapes" of the M-intervals when we cannot determine them exactly. We make some preliminary observations. Knowing the cardinality of a lower multiset for an M-interval, we know its height. Without further information, the M-interval could be as thin as a chain, or arbitrarily thick. However, an M-interval this thin cannot be achieved in general as a consequence of the restriction on exponents of indeterminates in the corresponding monomials. Nor can it be arbirarily thick as a consequence of the limitation on the number of indeterminates. The thinnest admissible M-interval uses the fewest indeterminates allowed to the largest power allowed. The thickest uses instead as many indeterminates as possible, with exponents as equal as possible subject to the degree constraints. Using these observations results in the tightest bounds currently available for all-terminal reliability that can be computed efficiently [33].

A Boolean formulation of these approaches seems not to have been pursued but could prove quite interesting.

17.5 Bounds in General

In this section, we travel further afield from the well-studied connections with Boolean functions, but before so doing we lay some groundwork as motivation. Suppose that Υ is a positive Boolean function, for example one arising from a reliability problem. When edges fail with different probabilities, the $\mathcal{F}$-complex or the function Υ contains all of the information about reliability, given the operation probability of each edge. However, the F-vector and the reliability polynomial are no longer applicable; indeed, the aggregation of operating states that is essential to the methods using coefficient bounds does not work here.

Can one find Boolean functions Υ_L and Υ_U on the same set of variables as Υ so that every implicant of Υ is an implicant of Υ_U, and every implicant of Υ_L is an implicant of Υ? Of course the answer is affirmative, so we must ask more. We want Υ_L and Υ_U to be "close" to Υ, but our requirement for closeness is a specific one dictated by the reliability application. Indeed, our objective is to minimize the probability that $\Upsilon \wedge \overline{\Upsilon_L}$ holds in order to obtain a strong lower bound, and to minimize the probability that $\Upsilon_U \wedge \overline{\Upsilon}$ holds in order to obtain a strong upper bound. This appears to be a somewhat unusual question in the Boolean domain, but arises very naturally in the reliability setting. Moreover, for reliability the requirements on Υ_L and Υ_U are much weaker than those on implicants; we require only that the probabilities of $\Upsilon \wedge \overline{\Upsilon_L}$ and $\Upsilon_U \wedge \overline{\Upsilon}$ be nonnegative.

Of course one prefers choices of Υ_L and Υ_U that can be found efficiently, and also evaluated efficiently. Here the p-normal form from Section 17.3.2.2 comes into play.

This has led to a number of techniques for using the network structure to obtain bounds. We explore some major techniques here, referring the interested reader to [7, 45] for proofs and further discussion. We employ graph-theoretic language primarily, but encourage the reader to consider extensions to Boolean functions in general.

17.5.1 Edge-Packing

Let $G = (V, E)$ be a graph (or digraph or multigraph). An *edge-partition* of G into g graphs $G_1, \ldots, G_g$ with $G_i = (V, E_i)$, where the edge set E is partitioned into

g classes $E_1, \ldots, E_k$. An *edge-packing* of G by g graphs $G_1, \ldots, G_g$ is obtained by partitioning the edge-set E into $g + 1$ classes $E_1, \ldots, E_g, U$ and defining $G_i = (V, E_i)$. The main observation here is straightforward:

Lemma 17.1. *Let $G = (V, E)$ be a graph (or digraph or multigraph). Let Rel be a coherent reliability measure. Let $C_1, \ldots, C_s$ be an edge-packing of G by cutsets. Then*

$$Rel(G) \leq \prod_{i=1}^{s} \left(1 - \prod_{e \in C_i} (1 - p_e)\right)$$

where p_e is the operation probability of edge e.

Lemma 17.1 gives an upper bound, because the failure of any cut in the edge-packing causes G to fail, but the failure of G can occur even when no cutset in the packing is failed. Indeed there is a natural Boolean interpretation. Let $\Upsilon_U = \bigwedge_{i=1}^{s} \overline{\bigwedge_{e \in C_i} \overline{\psi_e}} = \bigwedge_{i=1}^{s} \bigvee_{e \in C_i} \psi_e$. Because every prime implicant of Υ contains, for every cutset of G, the positive of the variable of at least one edge in the cutset, it contains a transversal of $\{C_1, \ldots, C_s\}$ and hence is an implicant of Υ_U. The formula Υ_U is easily rewritten in p-normal form (from Section 17.3.2.2) of the same size, because failures of the chosen cutsets are disjoint.

Brecht and Colbourn [27] and Litvak and Ushakov [96] first studied edge-packing upper bounds for the 2-terminal reliability problem. Robacker [139] gives the necessary dual to Menger's theorem: the maximum number of edge-disjoint s, t-cuts is the length of a shortest s, t-path. Finding a maximum set of edge-disjoint mincuts is straightforward: simply label each node with its distance from s. If t gets label ℓ, form cutset C_i containing all edges between nodes labeled $i - 1$ and vertices labeled i, for $1 \leq i \leq \ell$. The result is ℓ edge-disjoint s, t-cuts. Finding a "good" set of mincuts for the edge-packing upper bound appears to be more difficult than for the lower bound [116].

For all-terminal reliability, finding a maximum packing by mincuts is NP-hard [47]. Thus it is particularly surprising that by directing the reliability problems, we *are* able to find a maximum arc-packing by cutsets for the reachability problem efficiently [60]. Thus an all-terminal reliability upper bound can be obtained by using the arc-packing bound for reachability.

For upper bounds, Lemma 17.1 has a "dual" form, interchanging the role of pathsets and cutsets:

Lemma 17.2. *If G has an edge-packing by g graphs $G_1, \ldots, G_g$, and Rel is any coherent reliability measure,*

$$Rel(G) \geq 1 - \prod_{i=1}^{g} (1 - Rel(G_i)).$$

An inequality results in Lemma 17.2 because there are operational states of G in which no G_i is operational. Now consider an edge-packing of G by $G_1, \ldots, G_g$.

If any G_i is itself failed, coherence ensures that $Rel(G_i) = 0$; in this event, the inclusion of G_i in the edge-packing does not affect the bound, and G_i can be omitted. Thus, we need only be concerned with edge-packings by pathsets.

In order to obtain efficiently computable bounds we must compute (or at least bound) $Rel(G_i)$ for each G_i. One solution to this, suggested by Polesskii [121], is to edge-pack G with minpaths. The reliability of a minpath is easily computed. This suggests edge-packing G with as many minpaths as possible, and then applying Lemma 17.2; this basic strategy has been explored extensively. Again there is a natural Boolean interpretation. Let $\Upsilon_L = \overline{\bigwedge_{i=1}^{g} \overline{\bigwedge_{e \in G_i} \psi_e}} = \bigvee_{i=1}^{g} \bigwedge_{e \in G_i} \psi_e$. Every prime implicant of Υ_L contains the conjunction of positive variables of all edges of at least one minpath and hence is an implicant of Υ. The formula Υ_U is easily rewritten in p-normal form (from Section 17.3.2.2) of the same size, because operations of the chosen pathsets are disjoint.

Polesskii [121] pioneered the use of edge-packing lower bounds, in the all-terminal reliability problem. Here an edge-packing by minpaths is a set of edge-disjoint spanning trees. Using a theorem of Tutte [161] and Nash-Williams [115], Polesskii observed that a c-edge-connected n-node graph has at least $\lfloor \frac{c}{2} \rfloor$ edge-disjoint spanning trees; hence when all edge operation probabilities are the same value p, the all-terminal reliability of the graph is at least $1 - (1 - p^{n-1})^{\lfloor c/2 \rfloor}$. When edge probabilities are not all the same, Polesskii's bound extends in a natural way. Using Edmonds's matroid partition algorithm [59, 61], a maximum cardinality set of edge-disjoint spanning trees, or its minimum cost analog [43], can be found in polynomial time. Applying Lemma 17.2 then yields a lower bound on all-terminal reliability. Naturally, to obtain the best possible bound from Lemma 17.2, one wants not only a large number of edge-disjoint minpaths, but also minpaths that are themselves reliable. Edmonds's algorithm need not yield a set of spanning trees giving the best edge-packing bound using minpaths.

Edge-packing in general was pursued much later. Brecht and Colbourn [27] and Litvak and Ushakov [85, 96] independently developed edge-packing lower bounds for two-terminal reliability. For 2-terminal reliability, minpaths are just s, t-paths. The maximum number of edge-disjoint s, t-paths is the cardinality of a minimum s, t-cut [56, 105]. Thus, using network flow techniques, a maximum edge-packing can be found efficiently [62, 65]. Even when all edge operation probabilities are equal, finding the best edge-packing is complicated by the fact that minpaths exhibit great variation in cardinality. Finding the best edge-packing by s, t-paths is NP-hard [134]. For this reason, heuristics have been examined to find "good" edge-packings [27].

For k-terminal reliability, a minpath is a subtree in which each leaf is a terminal, that is, a *Steiner tree*. Determining the maximum number of Steiner trees in an edge-packing is NP-hard [47]. For directed networks, edge-packing (more precisely, arc-packing) bounds can be obtained using directed s, t-paths found by network flow techniques for s, t-connectedness, and by using arc-disjoint rooted spanning arborescences found by Edmonds' branching algorithm for reachability [135].

17.5.2 Noncrossing and Consecutive Cuts

Lomonosov and Polesskii [102] devised a method that permits cutsets to share edges, while retaining an efficient method for computing the probability that one of the cutsets fails. For a graph $G = (V, E)$, a partition (A, B) of V forms a cutset, containing all edges having one end in A and the other in B. Two such cutsets (A, B) and $(\hat{A}, \hat{B})$ are *noncrossing* if at least one of $A \cap B$, $A \cap \hat{B}$, $\hat{A} \cap B$, and $\hat{A} \cap \hat{B}$ is empty. A collection of cuts is *noncrossing*, or *laminar*, if every two cutsets in the collection are noncrossing. In an n-node graph with k terminals, a set of noncrossing cutsets contains at most $n - 1 + k - 2 \leq 2n - 3$ noncrossing cuts [48].

A *cut basis* of an n-node graph is a set of $n - 1$ cuts $C_1, \ldots, C_{n-1}$ for which every cut can be written as the modulo 2 sum of these $n - 1$ cuts. A cut basis $C_1, \ldots, C_{n-1}$ in which $\sum_{i=1}^{n-1} |C_i|$ is minimum can be found efficiently that forms noncrossing cuts [70]. For any cut basis $C_1, \ldots, C_{n-1}$, the all-terminal reliability satisfies $Rel(G) \leq \prod_{i=1}^{n-1} \left(1 - \prod_{e \in C_i}(1 - p_e)\right)$ [102].

The use of cut bases for the k-terminal problem has been studied by Polesskii [122], generalizing the method outlined here. The restriction to a basis, however, limits the number of cuts that can be employed to one fewer than the number of terminals. A more general extension is obtained by permitting the use of sets of noncrossing cuts. Shanthikumar [146] used consecutive cuts in obtaining a two-terminal upper bound. See [63] for a further extension. The analog for k-terminal reliability (actually s, T-connectedness) is treated in [48]. The bound is obtained by establishing that the probability that none of the noncrossing cuts fails agrees with the k-terminal *nodal* reliability of a special type of graph, a *directed path graph*. A simple dynamic programming strategy then produces the bound in polynomial time.

Bounds using noncrossing cuts extend the edge-packing strategies essentially by considering a larger set of cuts, but still a polynomial number of them. In each case, a small Boolean formula in p-normal form can be determined using each of these techniques.

17.6 Concluding Remarks

Only a small part of the extensive literature on network reliability has been mentioned here; many more directions appear in [7, 45, 50, 52, 147]. Understanding combinatorial structure is essential in the development of both effective exact algorithms and efficiently computable bounds. More importantly, the logical structure of Boolean formulas to represent operating states and the structure of the related complexes provide complementary views of network reliability. Although our main goal has been to describe aspects of network reliability in which both viewpoints are profitable, we have included a number in which one or the other viewpoint appears to the exclusion of the other.

Orthogonal DNF formulas (more generally, formulas of small size in p-normal form) lead to exact methods and bounds, whereas partitionability of the corresponding complexes leads to more effective bounding techniques. The differences in their application to reliability mask an underlying similarity. Nevertheless, both viewpoints are fruitful. The Boolean approach suggests logical simplification of reliability formulas, whereas combinatorial properties of complexes lead to representations such as the $\mathcal{H}$-multicomplex that for the moment appear to have no obvious analog in the Boolean domain.

References

[1] J. A. Abraham. *An improved algorithm for network reliability*. IEEE Transactions on Reliability R–28, pp. 58–61, 1979.

[2] K. K. Aggarwal, K. B. Misra, and J. S. Gupta. *A fast algorithm for reliability evaluation*, IEEE Transactions on Reliability R–24, pp. 83–5, 1975.

[3] K. K. Aggarwal and S. Rai. *Reliability evaluation in computer-communication networks*. IEEE Transactions on Reliability R–30, pp. 32–5, 1981.

[4] A. O. Balan and L. Traldi. *Preprocessing minpaths for sum of disjoint products*. IEEE Transactions on Reliability 52, pp. 289–95, 2003.

[5] M. O. Ball. *Computing network reliability*. Operations Research 27, pp. 823–6, 1979.

[6] M. O. Ball. *Complexity of network reliability computations*. Networks 10, pp. 153–65, 1980.

[7] M. O. Ball, C. J. Colbourn, and J. S. Provan. *Network reliability*. In Handbook of Operations Research: Network Models, pp. 673–762. Elsevier North-Holland, Amsterdam, 1995.

[8] M. O. Ball and G. L. Nemhauser. *Matroids and a reliability analysis problem*. Mathematics of Operations Research 4, pp. 132–43, 1979.

[9] M. O. Ball and J. S. Provan. *Bounds on the reliability polynomial for shellable independence systems*. SIAM Journal on Algebraic and Discrete Methods 3, pp. 166–81, 1982.

[10] M. O. Ball and J. S. Provan. *Calculating bounds on reachability and connectedness in stochastic networks*. Networks 13, pp. 253–78, 1983.

[11] M. O. Ball and J. S. Provan. *Properties of systems which lead to efficient computation of reliability*. Proceedings of IEEE Global Telecommunications Conference, 1985, pp. 866–70.

[12] M. O. Ball and J. S. Provan. *Computing k-terminal reliability in time polynomial in the number of (s,K)-quasicuts*. Proceedings of Fourth Army Conference on Applied Mathematics and Computing, pp. 901–7, 1987.

[13] M. O. Ball and J. S. Provan. *Disjoint products and efficient computation of reliability*. Operations Research 36, pp. 703–16, 1988.

[14] M. O. Ball and R. M. Van Slyke. *Backtracking algorithms for network reliability analysis*, Annals of Discrete Mathematics 1, pp. 49–64, 1977.

[15] D. Bauer, F. Boesch, C. Suffel, and R. Tindell. *Combinatorial optimization problems in the analysis and design of probabilistic networks*. Stevens Research Report in Mathematics 8202, Stevens Institute of Technology, Hoboken NJ, 1982.

[16] N. L. Biggs. Chip firing and the critical group of a graph. Journal of Algebraic Combinatorics 9, pp. 25–45, 1999.

[17] N. L. Biggs. The Tutte polynomial as a growth function. Journal of Algebraic Combinatorics 10, pp. 115–33, 1999.

[18] L. J. Billera. *Polyhedral theory and commutative algebra*. In Mathematical Programming: The State of the Art, A. Bachem, M. Grötschel, and B. Korte, eds., pp. 57–77. Springer-Verlag, Berlin, 1977.

[19] L. J. Billera and C. W. Lee. *The number of faces of polytope pairs and unbounded polyhedra*. European Journal of Combinatorics 2, pp. 307–32, 1981.

[20] Z. W. Birnbaum, J. D. Esary, and S. C. Saunders. *Multi-component systems and structures and their reliability*. Technometrics 3, pp. 55–77, 1961.

[21] R. Bixby. *The minimum number of edges and vertices in a graph with edge-connectivity N and M N-bonds*. Networks 5, pp. 259–98, 1975.

[22] A. Björner. *Homology and shellability*. In Matroid Applications, N. White, ed. Cambridge University Press, Cambridge, UK, 1992.

[23] F. T. Boesch, A. Satyanarayana, and C. L. Suffel. *Least reliable networks and the reliability domination*. IEEE Transactions on Communications 38, pp. 2004–9, 1990.

[24] A. Borodin. *On relating time and space to size and depth*. SIAM Journal on Computing 6, pp. 733–44, 1977.

[25] E. Boros. *Dualization of aligned Boolean functions*. RUTCOR Research Report 9-94, Rutgers University, 1994.

[26] E. Boros, Y. Crama, O. Ekin, P. L. Hammer, T. Ibaraki, and A. Kogan. *Boolean normal forms, shellability, and reliability computations*. SIAM Journal on Discrete Mathematics 13, pp. 212–26, 2000.

[27] T. B. Brecht and C. J. Colbourn. *Lower bounds for two-terminal network reliability*. Discrete Applied Mathematics 21, pp. 185–98, 1988.

[28] R. L. Brooks, C. A. B. Smith, A. H. Stone, and W. T. Tutte. *The dissection of rectangles into squares*. Duke Mathematical Journal 7, pp. 312–40, 1940.

[29] J. I. Brown and C. J. Colbourn. *Roots of the reliability polynomial*. SIAM Journal on Discrete Mathematics 5, pp. 571–85, 1992.

[30] J. I. Brown and C. J. Colbourn. *On the log concavity of reliability and matroidal sequences*. Advances in Applied Mathematics 15, pp. 114–27, 1994.

[31] J. I. Brown and C. J. Colbourn. *Non-Stanley bounds for network reliability*. Journal of Algebraic Combinatorics 5, pp. 13–36, 1996.

[32] J. I. Brown, C. J. Colbourn, and J. S. Devitt. *Network transformations and bounding network reliability*. Networks 23, pp. 1–17, 1993.

[33] J. I. Brown, C. J. Colbourn, and R. J. Nowakowski. *Chip firing and all-terminal network reliability bounds*. Discrete Optimization 6, pp. 436–45, 2009.

[34] J. I. Brown, C. J. Colbourn, and D. G. Wagner. *Cohen-Macaulay rings in network reliability*. SIAM Journal on Discrete Mathematics 9, pp. 377–92, 1996.

[35] R. Bruni. *On the orthogonalization of arbitrary Boolean formulae*. Journal of Applied Mathematics and Decision Sciences 2, pp. 61–74, 2005.

[36] J. A. Buzacott. *A recursive algorithm for finding reliability measures related to the connection of nodes in a graph*. Networks 10, pp. 311–27, 1980.

[37] J. A. Buzacott. *The ordering of terms in cut-based recursive disjoint products*. IEEE Transactions on Reliability R–32, pp. 472–4, 1983.

[38] J. A. Buzacott. *Node partition formula for directed graph reliability*. Networks 17, pp. 227–40, 1987.

[39] J. A. Buzacott and J. S. K. Chang. *Cut-set intersections and node partitions*. IEEE Transactions on Reliability R–33, pp. 385–9, 1984.

[40] M. K. Chari. *Steiner complexes, matroid ports, and K-connectedness reliability*. Journal of Combinatorial Theory (B) 59, pp. 41–68, 1993.

[41] M. K. Chari. *Matroid inequalities*. Discrete Mathematics 147, pp. 283–6, 1995.

[42] M. K. Chari. *Two decompositions in topological combinatorics with applications to matroid complexes*. Transactions of the American Mathematical Society 349, pp. 3925–43, 1997.

[43] J. Clausen and L. A. Hansen. *Finding k edge-disjoint spanning trees of minimum total weight in a network: An application of matroid theory*. Mathematical Programming Studies 13, pp. 88–101, 1980.

[44] G. F. Clements and B. Lindström. *A generalization of a combinatorial theorem of Macaulay*. Journal of Combinatorial Theory 7, pp. 230–8, 1968.

[45] C. J. Colbourn. *The Combinatorics of Network Reliability*. Oxford University Press, New York, 1987.

[46] C. J. Colbourn. *Network resilience*. SIAM Journal of Algebraic and Discrete Methods 8, pp. 404–9, 1987.

[47] C. J. Colbourn. *Edge–packings of graphs and network reliability*. Discrete Mathematics 72, pp. 49–61, 1988.

[48] C. J. Colbourn, L. D. Nel, T. B. Boffey, and D. F. Yates. *Probabilistic estimation of damage from fire spread*. Annals of Operations Research 50, pp. 173–85, 1994.

[49] C. J. Colbourn and W. R. Pulleyblank. *Matroid Steiner problems, the Tutte polynomial and network reliability*. Journal of Combinatorial Theory B41, pp. 20–31, 1989.

[50] C. J. Colbourn and D. R. Shier. *Computational issues in network reliability*. In Encyclopedia of Statistics in Quality and Reliability, Wiley, New York, 2008.

[51] R. Cori and D. Rossin. On the sandpile group of dual graphs. European Journal of Combinatorics 21, pp. 447–459, 2000.

[52] Y. Crama and P. L. Hammer. *Boolean Functions: Theory. Algorithms, and Applications*. Cambridge University Press, Cambridge, UK, 2010.

[53] D. E. Daykin. *A simple proof of the Kruskal-Katona theorem*. Journal of Combinatorial Theory A 17, pp. 252–3, 1974.

[54] J. E. Dawson. *A collection of sets related to the Tutte polynomial of a matroid*. Lecture Notes in Mathematics 1073, pp. 193–204, 1984.

[55] J. deMercado, N. Spyratos, and B. A. Bowen. *A method for calculation of network reliability*. IEEE Transactions on Reliability R–25, pp. 71–6, 1976.

[56] G. A. Dirac. *Short proof of Menger's theorem*. Mathematika 13, pp. 42–4, 1966.

[57] W. P. Dotson and J. O. Gobien. *A new analysis technique for probabilistic graphs*. IEEE Transactions on Circuits and Systems CAS–26, pp. 855–65, 1979.

[58] T. A. Dowling. *On the independent set numbers of a finite matroid*. Annals of Discrete Mathematics 8, pp. 21–8, 1980.

[59] J. Edmonds. *Minimum partition of a matroid into independent subsets*. Journal of Research of the National Bureau of Standards 69B, pp. 67–72, 1965.

[60] J. Edmonds. *Optimum branchings*. Journal of Research of the National Bureau of Standards 71B, pp. 233–40, 1967.

[61] J. Edmonds. *Matroid partition*. In Mathematics of the Decision Sciences, G. B. Dantzig and A. F. Viennott, eds., pp. 335–45. American Mathematical Society, 1968.

[62] J. Edmonds and R. M. Karp. *Theoretical improvements in algorithmic efficiency for network flow problems*. Journal of the ACM 19, pp. 248–64, 1972.

[63] E. S. Elmallah and H. AboElFotoh. *Circular layout cutsets: An approach for improving the consecutive cutsets bound for network reliability*. IEEE Transactions on Reliability 55, pp. 602–12, 2006.

[64] S. Even and R. E. Tarjan. *Network flow testing and graph connectivity*. SIAM Journal on Computing 4, pp. 507–18, 1975.

[65] L. R. Ford and D. R. Fulkerson. Flows in Networks, Princeton University Press, Princeton, NJ, 1962.

[66] P. Frankl. *A new short proof for the Kruskal-Katona theorem*. Discrete Mathematics 48, pp. 327–9, 1984.

[67] L. Fratta and U. G. Montanari. *A recursive method based on case analysis for computing network terminal reliability*. IEEE Transactions on Communications COM–26, pp. 1166–77, 1978.

[68] L. Fratta and U. G. Montanari. *A Boolean algebra method for computing the terminal reliability in a communication network*. IEEE Transactions on Circuit Theory CT–20, pp. 203–11, 1973.

[69] Y. Fu and S. S. Yau. *A note on the reliability of communication networks*. Journal of SIAM 10, pp. 469–74, 1962.

[70] R. E. Gomory and T. C. Hu. *Multiterminal network flows*. SIAM Journal on Applied Mathematics 9, pp. 551–70, 1961.

[71] G. Gordon and L. Traldi. *Generalized activities and the Tutte polynomial*. Discrete Mathematics 85, pp. 167–76, 1990.

[72] C. Greene and D. J. Kleitman. *Proof techniques in the theory of finite sets*. In Studies in Combinatorics, G.-C. Rota, ed., pp. 22–79. MAA, 1978.

[73] E. Hänsler. *A fast recursive algorithm to calculate the reliability of a communication network*. IEEE Transactions on Communications COM–20, pp. 637–40, 1972.

[74] E. Hänsler, G. K. McAuliffe, and R. S. Wilkov. *Exact calculation of computer network reliability*. Networks 4, pp. 95–112, 1974.

[75] K. Heidtmann. *Statistical comparison of two sum-of-disjoint-product algorithms for reliability and safety evaluation*. Lecture Notes in Computer Science 2434, pp. 185–98, 2002.

[76] A. Heron. *Matroid polynomials*. In Combinatorics, D. J. A. Welsh and D. R. Woodall, eds., pp. 164–202. Institute for Mathematics and Its Applications, 1972.

[77] T. Hibi. *What can be said about pure O-sequences?* Journal of Combinatorial Theory A50, pp. 319–22, 1989.

[78] T. Hibi. *Face number inequalities for matroid complexes and Cohen-Macaulay types of Stanley-Reisner rings of distributive lattices*. Pacific Journal of Mathematics 154, pp. 253–64, 1992.

[79] A. B. Huseby. *On regularity, amenability and optimal factoring strategies for reliability computations*. Statistics Research Report 2001-4, University of Oslo, 2001.

[80] I. M. Jacobs. *Connectivity in probabilistic graphs*. Technical Report 356, Electronics Research Laboratory, MIT, 1959.

[81] F. Jaeger, D. Vertigan, and D. J. A. Welsh. *On the computational complexity of the Jones and Tutte polynomials*. Mathematical Proceedings of the Cambridge Philosophical Society 108, pp. 35–53, 1990.

[82] P. A. Jensen and M. Bellmore. *An algorithm to determine the reliability of a complex system*. IEEE Transactions on Reliability R–18, pp. 169–74, 1969.

[83] R. M. Karp. *Reducibility among combinatorial problems*. In Complexity of Computer Computations, R. E. Miller and J. W. Thatcher, eds., pp. 85–103. Plenum, New York, 1972.

[84] G. Katona. *A theorem of finite sets*. In Theory of Graphs, P. Erdös and G. Katona, eds., pp. 187–207, Akademia Kiadó, Budapest, 1966.

[85] V. A. Kaustov, Ye. I. Litvak, and I. A. Ushakov. *The computational effectiveness of reliability estimates by the method of nonedge-intersecting chains and cuts*. Soviet Journal of Computer Systems Sciences 24, pp. 70–3, 1986.

[86] A. K. Kel'mans. *Some problems of network reliability analysis*. Automation and Remote Control 26, pp. 564–73, 1965.

[87] A. K. Kel'mans. *Connectivity of probabilistic networks*. Automation and Remote Control 29, pp. 444–60, 1967.

[88] A. K. Kel'mans. *The graph with the maximum probability of remaining connected depends on the edge-removal probability*. Graph Theory Newsletter 9, pp. 2–3, 1979.

[89] A. K. Kel'mans. *On graphs with randomly deleted edges*. Acta Mathematica Acadiemae Scientiarum Hungaricae 37, pp. 77–88, 1981.

[90] Y. H. Kim, K. E. Case, and P. M. Ghare. *A method for computing complex system reliability*. IEEE Transactions on Reliability R–21, pp. 215–19, 1972.

[91] G. Kirchoff. *Über die Auflösung der Gleichungen auf welche man bei der Untersuchung der linearen Verteilung galvanischer Ströme geführt wird*. Poggendorffs Annalen der Physik und Chemie 72, pp. 497–508, 1847. (=*On the solution of equations obtained from the investigation of the linear distribution of galvanic currents*. IRE Transactions on Circuit Theory CT–5, pp. 4–7, 1958.)

[92] E. V. Krishnamurthy and G. Komissar. *Computer-aided reliability analysis of complicated networks*. IEEE Transactions on Reliability R–21, pp. 86–9, 1972.

[93] J. B. Kruskal. *The number of simplices in a complex*. In Mathematical Optimization Techniques, R. Bellman, ed., pp. 251–78. University of California Press, 1963.

[94] J. Kuntzmann. Algèbre de Boole. Dunod, Paris, 1965.

[95] C. Y. Lee. *Analysis of switching networks*. Bell System Technical Journal 34, pp. 1287–315, 1955.

[96] E. I. Litvak. *A generalized theorem on negative cycles and estimates of the quality of flow transport in a network*. Soviet Mathematics Doklady 27, pp. 369–71, 1983.

[97] P. M. Lin, B. J. Leon, and T. C. Huang. *A new algorithm for symbolic system reliability analysis*. IEEE Transactions on Reliability R–25, pp. 2–15, 1976.

[98] Ye. I. Litvak. *The probability of connectedness of a graph*. Engineering Cybernetics 13, pp. 121–5, 1975.

[99] M. O. Locks. Inverting and minimalizing path sets and cut sets, IEEE Transactions on Reliability R–27, pp. 107–9, 1978.

[100] M. O. Locks. *Recursive disjoint products, inclusion–exclusion, and min–cut approximations*. IEEE Transactions on Reliability R–29, pp. 368–71, 1980.

[101] M. O. Locks. *Recursive disjoint products: A review of three algorithms*. IEEE Transactions on Reliability R–31, pp. 33–5, 1982.

[102] M. V. Lomonosov and V. P. Polesskii. *An upper bound for the reliability of information networks*. Problems of Information Transmission 7, pp. 337–9, 1971.

[103] M. V. Lomonosov and V. P. Polesskii. *Lower bound of network reliability*. Problems of Information Transmission 8, pp. 118–23, 1972.

[104] F. S. Macaulay. *Some properties of enumeration in the theory of modular systems*. Proceedings of the London Mathematical Society 26, pp. 531–55, 1927.

[105] K. Menger. *Zur allgemeine Kurventheorie*. Fundamenta Mathematicae 10, pp. 96–115, 1927.

[106] C. Merino. Chip-firing and the Tutte polynomial. Annals of Combinatorics 1, pp. 253–9, 1997.

[107] C. Merino. The chip-firing game and matroid complexes. Discrete Mathematics and Theoretical Computer Science Proceedings AA, pp. 245–56, 2001.

[108] H. Mine. *Reliability of physical systems*. IRE Transactions on Circuit Theory CT–6, pp. 138–51, 1959.

[109] K. B. Misra. *An algorithm for the reliability of redundant networks*. IEEE Transactions on Reliability R–19, pp. 146–51, 1970.

[110] K. B. Misra and T. S. M. Rao. *Reliability analysis of redundant networks using flow graphs*. IEEE Transactions on Reliability R–19, pp. 19–24, 1970.

[111] E. F. Moore. *The shortest path through a maze*. Annals of the Computational Laboratory of Harvard University 30, pp. 285–92, 1959.

[112] E. F. Moore and C. E. Shannon. *Reliable circuits using less reliable relays*. Journal of the Franklin Institute 262, pp. 191–208, 263, 281–97, 1956.

[113] F. Moskowitz. *The analysis of redundancy networks*. AIEE Transactions on Communications and Electronics 39, pp. 627–32, 1958.

[114] W. J. Myrvold, K. H. Cheung, L. Page, and J. Perry. *Uniformly reliable graphs do not always exist*. Networks 21, pp. 417–19, 1991.

[115] C. St. J. A. Nash–Williams. *Edge–disjoint spanning trees of finite graphs*. Journal of London Mathematical Society 36, pp. 445–50, 1961.

[116] L. D. Nel and H. J. Strayer. *Two-terminal reliability bounds based on edge-packings by cutsets*. Journal of Combinatorial Mathematics and Combinatorial Computing 13, pp. 3–22, 1993.

[117] A. C. Nelson, J. R. Batts, and R. L. Beadles. *A computer program for approximating system reliability*. IEEE Transactions on Reliability R–19, pp. 61–5, 1970.

[118] J. von Neumann. *Probabilistic logics and the synthesis of reliable organisms from unreliable components*. In Automata Studies, C. Shannon and J. McCarthy, eds., pp. 43–98, Princeton University Press, 1956.

[119] L. B. Page and J. E. Perry. *A practical implementation of the factoring theorem for network reliability*. IEEE Transactions on Reliability R–37, pp. 259–67, 1988.

[120] K. P. Parker and E. J. McCluskey. *Probabilistic treatment of general combinational networks*. IEEE Transactions on Computers C–24, pp. 668–70, 1975.

[121] V. P. Polesskii. *A lower boundary for the reliability of information networks*. Problems of Information Transmission 7, pp. 165–71, 1971.

[122] V. P. Polesskii. *Bounds on probability of group connectedness of a random graph*. Problems of Information Transmission 26 (1990), 161–169.

[123] A. Prékopa, E. Boros, and K.-W. Lih. *The use of binomial moments for bounding network reliability*. In *Reliability of Computer and Communications Networks*, pp. 197–212. AMS/ACM, 1991.

[124] J. S. Provan. *Polyhedral combinatorics and network reliability*. Mathematics of Operations Research 11, pp. 36–61, 1986.

[125] J. S. Provan. *The complexity of reliability computations in planar and acyclic graphs*. SIAM Journal on Computing 15, pp. 694–702, 1986.

[126] J. S. Provan. *Boolean decomposition schemes and the complexity of reliability computations*. In *Reliability of Computer and Communications Networks*, pp. 213–228, AMS/ACM, 1991.

[127] J. S. Provan and M. O. Ball. *The complexity of counting cuts and of computing the probability that a graph is connected*. SIAM Journal on Computing 12, pp. 777–88, 1983.

[128] J. S. Provan and M. O. Ball. *Computing network reliability in time polynomial in the number of cuts*. Operations Research 32, pp. 516–26, 1984.

[129] J. S. Provan and M. O. Ball. *Efficient recognition of matroid and 2-monotonic systems*. In Applications of Discrete Mathematics, pp. 122–34, SIAM, Philadelphia, 1998.

[130] J. S. Provan and L. J. Billera. *Decompositions of simplicial complexes related to diameters of convex polyhedra*. Mathematics of Operations Research 5, pp. 579–94, 1980.

[131] S. Rai. *A cutset approach to reliability evaluation in communication networks*. IEEE Transactions on Reliability R–31, pp. 428–31, 1982.

[132] S. Rai and K. K. Aggarwal. *An efficient method for reliability evaluation of a general network*. IEEE Transactions on Reliability R–27, pp. 206–11, 1978.

[133] S. Rai, M. Veeraraghavan, and K. S. Trivedi. *A survey of efficient reliability computation using disjoint products approach*. Networks 25, pp. 147–63, 1995.

[134] V. Raman. *Finding the best edge-packing for two-terminal reliability is NP-hard*. Journal of Combinational Mathematics and Combinational Computing 9, pp. 91–6, 1991.

[135] A. Ramanathan and C. J. Colbourn. *Bounds on all–terminal reliability via arc–packing*. Ars Combinatoria 23A, pp. 91–4, 1987.

[136] A. Ramanathan and C. J. Colbourn. *Counting almost minimum cutsets and reliability applications*. Mathematical Programming 39, pp. 253–61, 1987.

[137] L. I. P. Resende. *Implementation of a factoring algorithm for reliability evaluation of undirected networks*. IEEE Transactions on Reliability R–37, pp. 462–8, 1988.

[138] M. G. C. Resende. *A program for reliability evaluation of undirected networks via polygon-to-chain reductions*. IEEE Transactions on Reliability R–35, pp. 24–9, 1986.

[139] J. T. Robacker. *Minmax theorems on shortest chains and disjoint cuts of a network*. Memo RM–1660–PR, The Rand Corporation, 1956.

[140] A. Satyanarayana. *A unified formula for analysis of some network reliability problems*. IEEE Transactions on Reliability R–31, pp. 23–32, 1982.

[141] A. Satyanarayana and M. K. Chang. *Network reliability and the factoring theorem*. Networks 13, pp. 107–20, 1983.

[142] A. Satyanarayana and J. N. Hagstrom. *A new algorithm for the reliability analysis of multi–terminal networks*. IEEE Transactions on Reliability R–30, pp. 325–34, 1981.

[143] A. Satyanarayana and J. N. Hagstrom. *Combinatorial properties of directed graphs useful in computing network reliability*. Networks 11, pp. 357–66, 1981.

[144] A. Satyanarayana and A. Prabhakar. *New topological formula and rapid algorithm for reliability analysis of complex networks*. IEEE Transactions on Reliability R–27, pp. 82–100, 1978.

[145] C. E. Shannon. *The synthesis of two-terminal switching circuits*. Bell System Technical Journal 28, pp. 59–98, 1949.

[146] J. G. Shanthikumar. *Bounding network reliability using consecutive minimal cutsets*. IEEE Transactions on Reliability R–37, pp. 45–9, 1988.

[147] D. R. Shier. *Network Reliability and Algebraic Structures*. Oxford University Press, New York, 1991.

[148] D. R. Shier, *Algebraic methods for bounding network reliability*. In *Reliability of Computer and Communications Networks*, pp. 245–59, AMS/ACM, 1991.

[149] S. Soh and S. Rai. *Experimental results on preprocessing of path/cut terms in sum of disjoint products technique*. IEEE Transactions on Reliability, 42, pp. 24–33, 1993.

[150] E. Sperner. *Ein Satz über Untermengen einer endlichen Menge*. Mathematische Zeitschrift 27, pp. 544–8, 1928.

[151] E. Sperner. *Über einen kombinatorischen Satz von Macaulay und seine Anwendung auf die Theorie der Polynomideale*. Abhandlungen aus dem Mathematischen Seminar der Universität Hamburg 7, pp. 149–63, 1930.

[152] R. P. Stanley. *The upper bound conjecture and Cohen–Macaulay rings*. Studies in Applied Mathematics 14, pp. 135–42, 1975.

[153] R. P. Stanley. *Cohen–Macaulay complexes*. In Higher Combinatorics, M. Aigner, ed., pp. 51–64. Reidel, Dordrecht, 1977.

[154] R. P. Stanley. *Hilbert functions of graded algebras*. Advances in Mathematics 28, pp. 57–83, 1978.

[155] R. P. Stanley. Combinatorics and Commutative Algebra, 2nd ed., Birkhäuser, 1996.

[156] R. P. Stanley. *An introduction to combinatorial commutative algebra*. In Enumeration and Design, D. M. Jackson and S. A. Vanstone, ed., pp. 3–18. Academic Press, New York, 1984.

[157] R. P. Stanley. A monotonicity property of h-vectors and $h^{\star}$-vectors. European Journal of Combinatorics 14, pp. 251–8, 1993.

[158] Y. Takenaga and N. Katougi. *Tree-shellability of restricted DNFs*. IEICE Transactions on Information and Systems E91, pp. 996–1002, 2008.

[159] Y. Takenaga, K. Nakajima, and S. Yajima. *Tree-shellability of Boolean functions*. Theoretical Computer Science 262, pp. 633–47, 2001.

[160] T. Tsuboi and K. Aruba. *A new approach to computing terminal reliability in large complex networks*. Electronics and Communications in Japan 58A, pp. 52–60, 1975.

[161] W. T. Tutte. *On the problem of decomposing a graph into n connected factors*. Journal of London Mathematical Society 36, pp. 221–30, 1961.

[162] L. G. Valiant. *The complexity of enumeration and reliability problems*. SIAM Journal on Computing 8, pp. 410–21, 1979.

[163] R. M. Van Slyke and H. Frank. *Network Reliability Analysis: Part I*. Networks 1, pp. 279–90, 1972.

[164] M. Veeraraghavan and K. S. Trivedi. *An improved algorithm for symbolic reliability analysis*. IEEE Transactions on Reliability R–40, pp. 347–58, 1991.

[165] D. Vertigan. *Bicycle dimension and special points of the Tutte polynomial*. Journal of Combinatorial Theory (B) 74, pp. 378–96, 1998.

[166] D. G. Wagner. *Zeros of rank-generating functions of Cohen-Macaulay complexes*. Discrete Mathematics 139, pp. 399–411, 1995.

[167] D. J. A. Welsh. Matroid Theory, Academic Press, New York, 1976.

[168] D. J. A. Welsh and C. Merino. *The Potts model and the Tutte polynomial*. Journal of Mathematical Physics 41, pp. 1127–52, 2000.

[169] F. Whipple. *On a theorem due to F. S. Macaulay*. Journal of London Mathematics Society 8, pp. 431–7, 1928.

[170] R. R. Willie. *A theorem concerning cyclic directed graphs with applications to network reliability*. Networks 10, pp. 71–8, 1980.

[171] J. M. Wilson. *An improved minimizing algorithm for sum of disjoint products*. IEEE Transactions on Reliability 39, pp. 42–5, 1990.

[172] R. K. Wood. *Polygon–to–chain reductions and extensions for reliability evaluation on undirected networks*. PhD thesis, University of California at Berkeley, 1982.

[173] R. K. Wood. *A factoring algorithm using polygon–to–chain reductions for computing k-terminal network reliability*. Networks 15, pp. 173–90, 1985.

[174] R. K. Wood. *Triconnected decomposition for computing K-terminal network reliability*. Networks 19, pp. 203–20, 1989.

[175] W.-C. Yeh. *An improved sum-of-disjoint-products technique for the symbolic network reliability analysis with known minimal paths*. Reliability Engineering and System Safety 92, pp. 260–8, 2007.

Printed in the United States
By Bookmasters